# FIRE SAFETY SCIENCE
## Proceedings of the First International Symposium

*Editors*

Cecile E. Grant
Civil Engineering Department
University of California, Berkeley, USA

Patrick J. Pagni
Mechanical Engineering Department
University of California, Berkeley, USA

**INTERNATIONAL ASSOCIATION
FOR FIRE SAFETY SCIENCE**

**HEMISPHERE PUBLISHING CORPORATION**
A subsidiary of Harper & Row, Publishers, Inc.

Washington     New York     London

**Distribution outside North America**
**SPRINGER–VERLAG**
Berlin     Heidelberg     New York     Tokyo

The cover flames were drawn by Dr. Y. Hasemi of the Building Research Institute, Tsukuba, Japan, based on photographs taken by Prof. E. E. Zukoski of California Institute of Technology, Pasadena, California, USA.

**FIRE SAFETY SCIENCE—Proceedings of the First International Symposium**

1 2 3 4 5 6 7 8 9 0    E B E B    8 9 8 7 6

The publisher, editors, and authors have maintained the highest possible level of scientific and technical scholarship and accuracy in this work, which is not intended to supplant professional engineering design or related technical services, and/or industrial or international codes and standards of any kind. The publisher, editors, and authors assume no liability for the application of data, specifications, standards, or codes published herein.

**Library of Congress Cataloging in Publication Data**

Main entry under title:

Fire safety science—Proceedings of the first international symposium

    Bibliography: p.
    Includes index.
    1. Fire prevention—Congresses. I. Grant, Cecile E.
II. Pagni, Patrick J. III. International Association
for Fire Safety Science.
TH9112.F5626  1986        628.9′22        85-30193
ISBN 0-89116-456-1   Hemisphere Publishing Corporation

**DISTRIBUTION OUTSIDE NORTH AMERICA:**
ISBN 3-540-16585-1   Springer-Verlag   Berlin

# Preface

This symposium evolved from discussions initiated by Dr. P. H. Thomas, Prof. K. Kawagoe, Prof. K. Akita, Dr. J. G. Quintiere, Dr. R. Friedman, and many others throughout the world. Symposium committees and technical areas were established at an informal meeting in May 1984 at Borehamwood, England. As stated in the Fall 1984 Call for Papers, the purpose of these symposia will be "to provide a forum dedicated to all aspects of fire research and their application to solving problems presented by destructive fire." Due to the enthusiastic response to that call, the original plan for a three-day meeting was expanded to a full week of simultaneous sessions, October 7–11, 1985.

The chapters that follow are arranged in ten technical sessions, as in the symposium program. The session chairs, listed on each session's title page, were responsible for coordinating a rigorous peer review that resulted in acceptance of approximately two-thirds of the submitted papers. We are particularly grateful to these ten scholars who gave so generously of their time and expertise, and who are primarily responsible for the quality of these proceedings.

Prof. T. Kubota presented the 1985 Howard W. Emmons Invited Lecture due to Prof. E. E. Zukoski's, hopefully short, illness. Invited papers were also given by Prof. H. W. Emmons, Prof. O. Pettersson, Dr. R. Friedman, Dr. E. Kendik, Dr. J. Unoki, Prof. W. Johnson, and Prof. D. J. Rasbash. The authors represented Australia, Austria, Belgium, Canada, Denmark, Federal Republic of Germany, France, Hong Kong, Japan, Luxembourg, Netherlands, Norway, Spain, Sweden, Union of Soviet Socialist Republics, United Kingdom, and United States of America.

The Arrangements Committee is to be congratulated for its faultless organization, which led to an enjoyable as well as a technically valuable symposium. We are particularly grateful to our host, the United States Department of Commerce–National Bureau of Standards. Dr. J. E. Snell, Director of the Center for Fire Research, and Dr. J. W. Lyons, Director of the National Engineering Laboratory, focused our efforts with their stirring remarks. Ms. S. Cherry, Ms. D. Cramer, and other members of the excellent staff at the Center for Fire Research provided invaluable support. The National Research Council of Canada kindly prepared a book of abstracts for each attendee. The Publications Committee furnished the editors with expert advice for over a year. Mr. W. Begell and Ms. F. Padgett at Hemisphere Publishing Corporation have been most cooperative. We especially appreciate assistance from Prof. R. B. Williamson and support from the College of Engineering at the University of California-Berkeley.

It was not possible at this first symposium to follow each chapter with printed comments. All the fire research related journals have invited submission of discussions related to these presentations. At the second symposium the session areas will be redefined. The chairs of the committees for the second symposium, planned for Tokyo, Japan in June 1988, have been appointed: Arrangements—Prof. T. Hirano; Publications—T. Wakamatsu; and Program—Dr. P. H. Thomas. It is intended that the proceedings of these symposia provide compendia of current progress in fire research.

*Berkeley, California*  *Cecile E. Grant*

*November 1985*  *Patrick J. Pagni*

# International Association for Fire Safety Science

At this symposium, a new international association was founded. The study of fire and the solutions to the problems it presents are multidisciplinary involving many professions and sciences. Physicists, chemists, statisticians, architects, actuaries, and many kinds of engineers and practitioners are all to be found working in the various fire research laboratories and organizations concerned with fire safety matters. Until this First International Symposium on Fire Safety Science, there did not exist any organization that provided a forum for them to regularly assemble on an international basis. Multinational work is a major feature of fire safety because similar problems arise internationally from comparisons between old and modern materials, configurations, and energy sources. In addition, the fire science personnel in most countries are few and, to achieve progress, we must collaborate with our peers in other countries.

Fire presents us with several unsolved problems. Some, like turbulence, are problems in basic physics; some, like the flammability of nonhomogeneous building materials, present engineering problems of considerable complexity. And most present questions of priority, societal responsibility, and cost. Fire problems are not solved only by applications of science. We are still relying heavily on law and regulations. Fire engineering is becoming established as a professional discipline; fire science is entering into higher education. The International Association for Fire Safety Science perceives its role to lie at the scientific bases for these developments. It will seek to cooperate with existing bodies, be they concerned with application or with the sciences that are fundamental to our interests in fire. It will seek to raise standards, to encourage and stimulate scientists to address fire problems, to provide the necessary scientific foundation, and to encourage applications aimed at reducing life and property loss.

The registrants at this first symposium are the charter members of the International Association for Fire Safety Science. An organizing committee was established at a business meeting on October 9, 1985. The members of this committee are: Dr. P. H. Thomas, UK, chair; Prof. R. W. Fitzgerald, USA; Dr. R. Friedman, USA; Dr. T. Z. Harmathy, Canada; Prof. T. Hirano, Japan; Prof. S. Horiuchi, Japan; Prof. K. Kawagoe, Japan; Dr. M. Kersken-Bradley, FRG; Mr. H. E. Nelson, USA; Prof. P. J. Pagni, USA; Prof. O. Pettersson, Sweden; Dr. J. G. Quintiere, USA; Prof. D. J. Rasbash, UK; Dr. P. G. Seeger, FRG; Dr. J. E. Snell, USA; Prof. Y. Uehara, Japan; Dr. J. Unoki, Japan; and Prof. R. B. Williamson, USA. The committee met on October 11, 1985 and elected the following association officers: Dr. P. H. Thomas, UK, chair; Dr. R. Friedman, USA, vice-chair; Prof. K. Kawagoe, Japan, vice-chair; Prof. O. Pettersson, Sweden, vice-chair; Prof. T. Hirano, Japan, secretary; and Dr. J. G. Quintiere, USA, treasurer. Finance and constitution subcommittees were also formed. The organizing committee will add to its membership as needed. The Japanese Association for Fire Science and Engineering will host the Second International Symposium on Fire Safety Science which will be held at Tokyo, Japan in June 1988. Please address inquiries about the association or the second symposium to the association officers: Dr. Philip H. Thomas, Fire Research Station, Borehamwood, Herts. WD6 2BL, UK; Prof. Toshisuke Hirano, University of Tokyo, 7-3-1 Hongo, Bunkyo-ku, Tokyo 113, Japan; Dr. James G. Quintiere, Center for Fire Research, National Bureau of Standards, Gaithersburg, MD 20899, USA.

# In Memoriam

**Professor Takashi Handa**
October 10, 1923–September 14, 1985

The fire research community is greatly saddened by the recent passing of Professor Takashi Handa of the Science University of Tokyo. He was a founder and dedicated supporter of fire research. He actively pursued international scientific exchange and collaboration. Many of us fondly recall his gracious hospitality to scientists visiting Japan.

He began his illustrious research career in the Applied Chemistry Department of Tokyo University in 1945. In 1962, he became Associate Professor of Physical Chemistry at Wayne State University in Detroit, Michigan. His long association with the Science University of Tokyo began in 1964. In 1966, he was appointed Professor there in the Department of Science. His many honors include the Mainichi Newspaper Company Award in 1974. He became the first Director of the Center for Fire Science and Technology of the Science University of Tokyo in 1981.

Professor Handa encouraged international progress in fire research by initiating a visiting lectureship in Fire Science at the University, by his leadership role in the United States–Japan Natural Resources Panel on Fire Research and Safety and by founding the new English-language Japanese journal, *Fire Science and Technology*. His innovative use of computer modeling for fire design and fire investigation is an example to us all on the benefits of these powerful new tools which he helped to develop.

We were looking forward to his chairing the Detection Session at this symposium. We deeply regret that he was not able to do so. He is missed.

# Symposium Committees

## ARRANGEMENTS COMMITTEE

Dr. James G. Quintiere (Chair)
Center for Fire Research
National Bureau of Standards

Dr. Tibor Z. Harmathy
Division of Building Research
National Research Council of Canada

Mr. Harold E. Nelson
Center for Fire Research
National Bureau of Standards

## PUBLICATIONS COMMITTEE

Prof. Patrick J. Pagni (Chair)
Mechanical Engineering Department
University of California, Berkeley

Prof. Toshisuke Hirano
Department of Reaction Chemistry
University of Tokyo

Dr. Paul G. Seeger
Forschungsstelle für Brandschutztechnik
Universität Karlsruhe (TH)

Prof. R. Brady Williamson
Civil Engineering Department
University of California, Berkeley

## PROGRAM COMMITTEE

Dr. Philip H. Thomas (Chair)
Fire Research Station
Building Research Establishment

Dr. Raymond Friedman
Factory Mutual Research Corporation

Prof. Kunio Kawagoe
Science University of Tokyo

Prof. David J. Rasbash
Unit of Fire Safety Engineering
University of Edinburgh

# Contents

## 1985 Howard W. Emmons Invited Lecture

## Fire Physics

## Structural Behavior

## Fire Chemistry

## People-Fire Interactions

## Translation of Research into Practice

## Detection

## Statistics, Risk, and System Analysis

*1985 Howard W. Emmons Invited Lecture*

# Fluid Dynamic Aspects of Room Fires

**E. E. ZUKOSKI**
Karman Laboratory of Fluid Mechanics and Jet Propulsion
California Institute of Technology
Pasadena, California 91125, USA

ABSTRACT

Several fluid dynamic processes which play important roles in the development of accidental fires in structures are discussed. They include a review of information concerning the characteristic flow regimes of fire plumes and the properties of the flow in these regimes, and a brief review of flow through openings and in ceiling jets. Factors which lead to the development of thermal stratification in ceiling layers are also discussed.

INTRODUCTION

A review is presented in this paper concerning several fluid dynamic processes which have a strong influence on the development of a fire within a compartment. We have selected only a few to discuss here in depth and more inclusive and detailed discussions of all or selected examples of these processes are given in Rocket (1975), Emmons (1978), Steckler et al (1982), Quintierre (1984) and Zukoski (1985).

To introduce the subject, consider the sketches of Figure 1 which outline the growth of a fire within a two room structure. In the initial stages of a fire within an enclosure, the plume of hot gas produced by the fire rises due to buoyancy forces and impinges on the ceiling to form a thin layer of hot gas. This layer spreads out across the ceiling and then grows in depth to form a more or less well defined hot gas layer, called here the ceiling layer. The plume entrains gas from the air in the room and acts as a primary source of material for the ceiling layer.

In the first moment of the fire, the ceiling layer will have negligible thickness and gas in the fire plume can only entrain air at the ambient temperature as it rises toward the ceiling. The average temperature of the gas at the level of the ceiling will be fixed by the rate of heat addition from the fire and the rate of entrainment of cool, unvitiated air from the room. When the fire is small, most of the material in the plume will be air and only a small fraction will be products of combustion.

As the thickness of the ceiling layer grows due to the addition of plume material, the plume still penetrates the layer and it

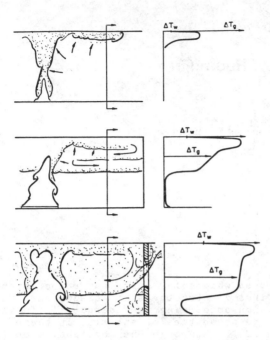

FIGURE 1.  Development of the ceiling layer.

impinges on the ceiling to form a relatively hot ceiling jet within
the ceiling layer. Consequently, it is useful to think of the gas
from the fire plume as being added at the top of the ceiling layer.
The gas in the plume will also begin to entrain fluid from the hot
gas in the ceiling layer.  Because the relative importance of the
entrainment from the hot layer grows as the layer thickness grows,
the temperature of the gas arriving at the ceiling will increase as
the layer thickness grows and this temperature increase will become
still larger if the heat release rate of the fire increases with
time.

     Thus, the temperature of the gas delivered to the ceiling will
be higher than that of the ceiling layer even when the layer itself
is well mixed and adiabatic.  In addition, convective and later
radiative heat transfer to the walls will reduce substantially the
temperature of the ceiling layer gas below the adiabatic value and
hence will further increase this difference.

     Because of these processes, we expect that there may be a
considerable temperature variation across the height of the layer.
The degree of stratification in the ceiling layer will depend in
detail on the flow within the layer, changes in the heat release
rate of fires, the enthalpy fluxes resulting from the flow of gas
into and out of the layer at openings, and heat loss to the walls.
Several of these processes will be discussed in a later part of this
paper.

     Later in the history of this fire, gas will flow out of the
doorway and will form a ceiling layer in the adjacent space.  The

2

plume formed at the doorway will have the same history of increasing temperature as that in the room of origin and many of the features described above will be repeated.

The model which is most widely used to describe the propagation of fire and the motion of smoke through buildings described above is the two-layer model. The basic assumption made in this model is that the gas in any space can be usefully separated into two horizontal and homogeneous layers: a hot ceiling layer and a cooler floor layer. In developing equations to describe the growth of these layers and changes in their properties with time, the assumption is made that only a single set of parameters such as temperature and species concentrations are required to describe each layer.

One of the great simplifications resulting from the use of this type of model is that the details of the flow within the layers can be ignored. However, it is evident from the above discussion that at least a rudimentary description of flow within the layer is required to allow a description of convective and radiative heat transfer from the gas to the surfaces of the room.

In addition, recent experiments suggest that large temperature gradients can exist in ceiling layers produced in large scale fires and laboratory test equipment. These experiments suggest that for certain conditions the formation of homogeneous layers in a room may be a very poor assumption, and they have led us to consider and to discuss here the factors which might cause the development of poorly mixed layers.

The purpose of this paper is to review briefly our understanding of several fluid dynamic processes which influence the progress of fires of the type described here. These are the fire plume, flows through openings and ceiling jets. We then examine some of the factors to determine the degree of stratification which can exist in the ceiling layer.

BUOYANT FIRE PLUMES

The diffusion flame and accompanying buoyant plume which are present in a room fire act as the engine which entrains fresh air from the lower layer and pumps products into the upper layer. The entrainment characteristics of the fire, the region of heat release made visible by combustion processes, and the adiabatic plume rising above it fix the temperature and mass transfer rate of hot gas into the upper layer. The flame height becomes an important scaling parameter because the entrainment characteristics of the flame and plume are different, and because the radiant flux to the fuel bed which controls the growth of the fire depends on the flame geometry. Finally, the height of the flame enters into the behavior of the adiabatic plume rising above it because the heat addition region or flame acts as the source for the plume.

The flow field produced by a large diffusion flame in the region in which rapid heat release occurs is not yet well understood. In contrast, the flow field of an isolated, adiabatic, and axisymmetric buoyant plume has been one of most frequently studied flows which occur in room fires and, over the past fifteen years, we have obtained the information required to produce good

3

predictions for the properties of this flow which are required in fire models.

However, our understanding of the entrainment process in this flow is no better now than that used in the integral models of Morton et al (1955). For example, the local entrainment rate of ambient fluid into a buoyant plume is almost twice that in a momentum dominated jet. Despite the fact that we can describe either of these flows in terms of simple algebraic formulae, we do not yet understand this difference in entrainment rates. Furthermore, these descriptions apply in the region far from the source of the plume, whereas in fire problems we are most often interested in the region near the source where the initial conditions are important.

Flow Regimes and Transition Criteria

Before proceeding, it is convenient to discuss the flow regimes of interest to us. In many flows arising in accidental fires, we must deal with diffusion flames which have source diameters of one to several meters. In these flames, the gaseous fuel is often generated by a pyrolysis process which produces maximum mass fluxes of 10 to 30 grams/s/m$^2$ with maximum velocities at the fuel surface of 1 to 2 cm/sec. In general, we characterize the flames and the plumes which rise above these fuel sources as being turbulent and buoyancy controlled flows.

Problems arise when we attempt to define criteria which will allow us to determine whether or not scaling laws derived from laboratory experiments will follow the scaling laws for full scale turbulent and buoyancy controlled flows. The transition between laminar and turbulent flows and the operational significance of the difference between unsteady laminar flows, which contain highly wrinkled laminar flames, and turbulent flows is often obscure and a body of data does not yet exist from which a convincing set of criteria can be developed to define the laminar to turbulent transition in these flames.

It is convenient to introduce two parameters which will be used here to describe the initial conditions in jets and plumes. The first is Q*, a dimensionless heat release parameter which has been found useful, Zukoski (1975), in describing large diffusion flames and plumes, and is discussed below in the Flame Height section. It is defined as,

$$Q^* = \dot{Q}/(\rho_\infty Cp_\infty T_\infty \sqrt{gD}\ D^2)$$
[1]

Here $\dot{Q}$ is the heat release rate of the fire and $\rho_\infty$, $Cp_\infty$, and $T_\infty$ are the density, specific heat at constant pressure and temperature of the ambient fluid, g is the acceleration of gravity and D is the burner diameter.

The second parameter is a Froude number given by Fro = $W_o^2$/gD when the source has a constant density $\rho_o$ and velocity Wo. The two parameters are related by

$$Q^* = [0.8(\ h_f\ /Cp_\infty T_\infty )(\rho_o /\rho_\infty\ )]\ \sqrt{Fro}$$
[2]

4

Here, hf is the heating value and the density of the fuel at the source. The product of density ratio and heating value differ by about a factor of ten, comparing hydrogen and propane, for typical fuels. For hydrocarbon fuels, the differences are much smaller.

The transition to turbulence in plumes has been described by Railston (1954), who proposed a transition criterion based on information obtained for transition in adiabatic plumes above a heat source, which can be put in the form,

$$Zt/D = 2000/[(Q^*)^{1/3}] [\sqrt{gD} D/\nu] \qquad [3]$$

Application of this result to fire problems indicates that most plumes of interest to us will be turbulent.

Becker and Liang (1978) give a review of ideas concerning transition in flames. They propose that a Reynolds number, based on their estimate of the properties of the flow at the top of the flame, be used to predict the transition to turbulence. For the low $Q^*$ range of primary interest here, this parameter is essentially the square root of a Grasshof number based on the flame length which can be put in the form,

$$Re = 0.14 (Zf \sqrt{[g \ Zf]} / \nu_\infty \qquad [4]$$

For the range, $1 < Zf/D < 5$ , they suggest 2500 as the transition Reynolds number when the kinematic viscosity, $\nu_\infty$ , is evaluated for air at standard conditions. This criteria reduces to the statement that the flame length should be greater than 0.2 m to insure that the flow will be turbulent. This choice appears to be entirely based on the data of Blinov and Khudiakov (1957) and we believe that it deserved further study. The applicability of this result to flows with larger or smaller ratios Zf/D is untested.

A second criterion concerns the relative importance of buoyancy forces with respect to the initial momentum flux in the flow. At the source of an adiabatic jet or flame plume which has some initial momentum, the momentum with which the fuel is injected may be so large that buoyancy forces are unimportant. In this region, the momentum flux will be close to the value at the source, and flow characteristics, such as entrainment rates, will be those of a momentum dominated jet with no influence of buoyancy. However, buoyancy forces act over the entire height of the flame and cause the momentum flux in the flow to increase with height. Therefore, at some point farther away from the origin, the momentum starts to increase rapidly above the initial value, where the flow becomes buoyant controlled and the flow characteristics will be those for a buoyant plume with no initial momentum flux but with some offset of origin.

For an adiabatic jet, this transition occurs at an elevation Zt which can be expressed as

$$Zt/D = (Constant) (Wo^2 /(g'D))^{1/2} \qquad [5]$$

Here, g' is the reduced acceleration of gravity, $g(\rho_\infty - \rho_0)/\rho_\infty$ , $\rho_\infty$ is the ambient density, and $\rho_0$ is the initial density in the jet.

Becker and Yamazaki (1980) and Becker and Liang (1980) have also observed this transition in their measurements of the momentum and mass fluxes in turbulent jet diffusion flames. They found that the region over which the initial momentum is important is related to a Froude number for the source Fro and a density ratio, $(\rho_o/\rho_\infty)$. Here $\rho_o$ is the density in the jet at the source and $\rho_\infty$ is the ambient gas density. A similar model, see Zukoski (1985), suggests that the density dependence should involve the mean gas density in the flame. Using either of these models, we expect that the transition will occur within a flame at an elevation Zt above the source given roughly by:

$$Zt/D = (\text{Constant}) (\rho_o/\rho_\infty)^{2/3} [Wo^2/(gD]^{1/3} \hspace{2cm} [6]$$

The transition is a broad one and a value for the constant lies between 1 and 5 depending on the experimental criteria used to define transition.

Transition of a different type arises when we examine the properties of buoyancy controlled flames. The dependence of the flame height on heat release rate, the initial momentum and source-diameter change markedly when the ratio of flame height to source-diameter is between 1 and 3. Thus, these data suggest that two regimes exist, one for tall flames and a second when the ratio of flame height to source-diameter is less than one.

Finally, based on measurements of entrainment rates and of temperature and velocity, we must distinguish at least two regions within a buoyant diffusion flame in which the the dependence of these properties on position and source-diameter are quite different. In flames which are more than three source-diameters high, this transition occurs at about 60% of the average flame height for the entrainment data of Cetegen et al (1984) and the velocity and temperature data of McCaffrey (1979). This transition is associated with the intermittency of the flame which begins to fall rapidly at these elevations.

The transitions and regimes described above are illustrated by the sketch of Figure 2 in which the dependence of flame height on the two flow parameters described above is shown for a wide range of values. To have a concrete example in mind, consider a flame produced by a constant fuel flow rate from a circular burner and examine the changes in the flame height as the diameter of the burner is increased from a small diameter to a very large one. When the diameter of the jet is very small, the Froude number for the source, Fro, is very much larger than one and the entire flame is momentum dominated. The flame length to diameter ratio Zf/D, see regime V in Figure 2, is roughly independent of the Froude number but has been shown to depend on the stoichiometric fuel-air ratio. For this reason, Q* is not an appropriate scaling parameter in this regime. The flame is a single well defined jet-like structure with considerable intermittency at the tip and a steady cone-like geometry at the base.

The broken line in Regime IV marks roughly the transition from momentum to buoyancy controlled flows suggested by the criteria given above. Presumably this transition line extends into the adiabatic plume region above the flame although no measurements are available to verify this conjecture.

6

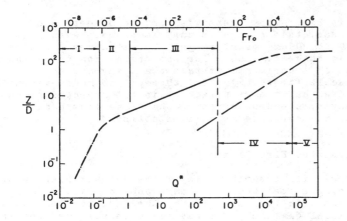

FIGURE 2.  Schematic diagram of flame length versus fuel flow rate parameters.

When the diameter of the jet is increased, the Froude number decreases and the flames cross this transition line.  Then, starting at the top of the flame, the influence of buoyancy will become more important and eventually the whole flow will be buoyancy controlled. In this regime, regime III, $Q^*$ appears to be an appropriate parameter to describe the flow.  Because the slope of the curve is about 2/5, the flame length is independent of the source diameter and depends primarily on the heat release rate.  The flame is still a well defined column but, as $Q^*$ decreases toward one, a regular pulsation at the base of the flame grows in amplitude and the length of the intermittent flame region at the tip of the flame grows as a fraction of flame length.  See, for example, Cetegen et al (1984).

For a still larger diameter source, a second transition, see regime II in Figure 2, is reached for values of $Q^*$ between 1.0 and 0.10.  In this transition region, the flame breaks up into a number of independent flamelets as $Q^*$ decreases and the flame height depends on the diameter and heat release rate.

Finally, there is evidence that a third regime exists for 1.0 > (Zf/D) , or $Q^* < 0.1$ .  In this regime, called regime I on Figure 2, the height of the flamelets appears to become roughly independent of the source-diameter and to depend only on the local heat release rate per unit area, i.e., on the fuel flow per unit area. This type of behavior seems reasonable in the limit of very large diameter sources if the transport of oxygen into gas above the center of the fuel bed is also independent of the diameter of the bed.  Experiments carried out with large laboratory fires, see the discussion in Wood et al (1971), and with very large scale fires, such as the the Flambeau Tests suggests that this is a reasonable assumption.  For further discussion, see Zsak and Zukoski (1985).

However, the transition between the last two regimes may not be so simply distinguished as that between the first two and the transition between turbulent and laminar flows may be important

here.  For example, in laboratory tests with source-diameters of the
order of a meter or less, this regime is only accessable when flame
heights are in the 10 to 20 cm height range.  These values lie close
to the 0.2 meter limit suggested above for the transition to laminar
flow and thus results of experiments at this scale may not be
applicable to large scale fires.  Finally, there is no good evidence
yet to support the idea that the Q# parameter is the appropriate one
in this regime.  However, the dependence of the flame length on the
heat release rate per unit area seems well established.


Characteristics of Isolated Fire Plumes

     In this section we will describe some of the characteristics of
flames formed above simple fuel beds such as a pool of liquid fuel
or a bed of glass beads with fuel supplied by the injection of
natural gas.  The initial diameter of these fires is well defined as
the diameter of the pool or bed and the flame lies above a well
defined level fixed at the top of the bed or pool.  Defining similar
geometric characteristics for flames formed above more complex fire
sources, such as a burning crib with a large height to width ratio,
is a more difficult task which has not yet been addressed
satisfactorily.  For example, when the crib height and the flame
height measured from the floor are comparable, then the definition
of the flame height becomes difficult and we should expect the
correlations given below to fail.

     In the following paragraphs of this section, we will restrict
our discussion to axisymmetric diffusion flames and plumes.
Information is available for adiabatic plumes rising above a line
source, but little is available for flames above a line source or
other geometries.

     Unsteady flow in flames.  One of the interesting
characteristics of many diffusion flames with a height to source-
diameter ratio greater than 1.0 is the unsteady flow they produce.
Immediately above the bed which fixes the origin of the fire, the
flame surface pinches in periodically toward the axis of symmetry
and produces a distinct structure in the flame which has the
appearance of a large irregular donut-shaped vortex ring.  This
structure rises slowly above the source and defines the top of the
flame when the fuel it contains burns out. In a flame with a large
height to diameter ratio, several of these structures are present at
any given time and for flames with a height to diameter ratio of two
to three diameters, only one structure is visible at a time.  For
flames with a height to diameter ratio less than one, the process
can not be seen in the visible flame, but does appear in shadowgraph
images of the rising plume of hot products.

     This process has been the subject of numerous investigations,
e.g. Thomas et al (1965) and recent reviews are given in Zukoski et
al (1981), Beyler (1984) and Cetegen et al (1984).  The dominant
frequency appears to scale as $\sqrt{g/D'}$ for a wide range of diameters
and fuels.  The fluctuations are most marked when $Zf/D$ is between
one and three, and the amplitude of the fluctuations decreases
markedly as $Zf$ increases from 3 to 20 source-diameters where the
initial momentum becomes important.  Due to the discrete nature of
these structures, the instantaneous height of the visible flame is
very unsteady and it fluctuates over a distance equal to 80% of the

8

average flame height when 1.0 < Zf/D < 3.0  and to 40%, when
6.0 < Zf/D < 20 .

In these ranges of Zf/D, photographic measurements of the
geometry of the flame taken by Cetegen et al (1984) show the flame
is always present at elevations in the lower part of the flame and
hence that the intermittency is one there.  However, above some
elevation, the intermittency decreases almost linearly to zero as
the elevation is increased.  The "average flame height" is defined
in this paper as the 50% intermittency point in contrast to "eye
averaged values" which typically record the top of the flame where
the intermittency approaches zero.  Because the fluctuations in
height of the top of the visible flame are so large, the averaging
process used in defining the average flame length Zf is important
and different techniques have lead to substantial variations in
values of Zf for the same fire.

The large size of these structures and the speed at which they
evolve suggests that they play an important part in fixing the
entrainment and heat release in the lower part of the flame.

Flame height measurements.  In contrast to momentum dominated
flows of regime V, for which the flame length is approximately
independent of the fuel flow rate but scales with the diameter of
the burner, the buoyancy controlled flame height depends on the heat
addition rate as well as the diameter.

Although the appropriate dimensionless parameters which should
be used in describing the flame height in regimes II and III have
not been certainly defined, Zukoski (1975) and Zukoski et al (1981),
have found that the parameter Q*, defined above in equation (1) is
satisfactory in regime II.  Note that in this parameter the
properties of the fire which appear are the total heat release rate
and the initial diameter.  The product $(\rho_\infty Cp_\infty T_\infty)$ is proportional to
the pressure, and hence could be evaluated within the flame or
outside in the ambient atmosphere without substantially affecting
the value of the parameter.  Although Q and D have been varied
greatly in buoyant flame experiments, the product $(\rho_\infty Cp_\infty T_\infty)$ has not
and thus there is no experimental evidence to support their
inclusion in this scaling parameter.  For this reason, a number of
investigators have used the dimensional parameter $Q/(D)^{5/2}$     to
correlate their data.

A selected set of data for flame heights obtained from various
investigators are shown in Figures 3 and 4.  In Figure 3, the data
are eye averaged values and the experimental data from a number of
sources, which are shown in this Figure, are in rough agreement for
the whole range of Q*.  However, the scatter is so large that they
may hide a more subtle dependence on parameters not used here.

The data for the lower values of Q* are shown in Figure 4.  The
data shown here are eye averaged values taken from Wood et al
(1971), Alvarez (1984), Kung and Stavrianidis (1985), and Blinov and
Khudiakov (1955), and average flame heights obtained by analysis of
video tapes by Cetegen et al (1984).  The error bars on the Cetegen
data show the 5% and 95% intermittency heights for these flames.

9

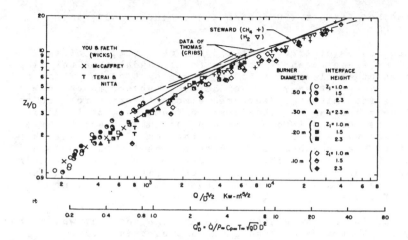

FIGURE 3.  Eye-averaged flame length.

The eye averaged data which appear in Figure 3 lie near the top of
these bars and hence are some 25% higher than the average values.
Burner diameters for the data reported in Figure 4 range from 0.5 to
23 m. and, despite the earlier correlations of Becker and Liang
(1978), our interpretation of the data for the very large gasoline
fires of Blinov and Khudiakov agree with other data for regimes II
and III in a reasonable manner.

     The curves fit to these data delineate regimes I, II, and III,
discussed above.  For values of Q* between one and several hundred,
a reasonable estimate of the average flame height for buoyant plumes
is given by

$$Zf/D = 3.3 (Q^*)^{2/5}$$  [7]

The value of the constant can easily be 25 % larger depending on the
method used to measure the flame height. Given the definition for
Q*, the dependence on burner diameter D drops out of the correlation
in this regime. This is satisfying, because we expect that the
diameter would be unimportant in a buoyant flow when the flame
height is large compared with the source diameter.

     When 0.1 < Q* < 1.0 , Cetegen et al (1984) find a correlation
of the form,

$$Zf/D = 3.3 (Q^*)^{2/3}$$  [8]

However, since this is a transition region, a continuously changing
slope between 0.4 and 2.0 is probably more reasonable.  There is
less agreement among various investigators concerning the proper
form of  this equation than that given for the high Q* regime.
However, the presence of a transition regime is unambiguous.

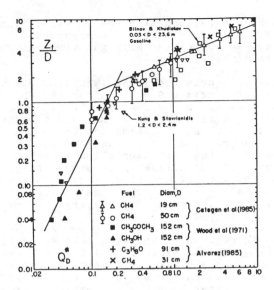

FIGURE 4.  Flame lengths for small Q*.


     For still smaller values of Q*, the data of Wood et al (1971)
and Cox and Chitty (1985) suggest that a correlation of the form,

$$Zf/D = (constant)(Q^*)^2 \qquad\qquad [9]$$

may be appropriate for regime I.  This correlation reduces to the
form

$$Zf \propto ( Q/D^2 )^2 \qquad\qquad [10]$$

     The constant in equation (9) is about 40 for the data of Wood
et al (1971) and Kung and Stavrianidis (1985), and about 15 for the
data of Cox and Chitty (1985).  We cannot explain the difference
between these two sets of results.  Several processes may be
responsible: the square and segmented burner of Cox and Chitty, Q*
may not be the appropriate variable in regime I, and finally
transition to turbulence may occur differently in the two sets of
experiments.

     The most well defined correlation presented here is for flames
which are more than three source-diameters high and the scatter here
is still at least 25%.  This scatter is so large that we cannot
definitely conclude that the Q* parameter is the appropriate one for
this regime.  Unfortunately, many accidental fires are in regime I
and II for which we have the least reliable data.

     Flow parameters.  The flame height is difficult to measure
because of the large fluctuations in the location of the top of the
luminous flame mentioned above.  These fluctuations in the visible
features of the flame are accompanied by similar fluctuations in
temperature and velocity of the gas and they make the determination

11

of local flow properties such as mass, momentum and enthalpy fluxes difficult to make accurately. This difficulty is compounded by experimental problems arising because of the high gas temperature and relatively low gas speeds involved.

Despite these difficulties, a number of experimentors, e.g., McCaffrey (1979), Kung and Stavrianidis (1985), and Cox and Chitty (1985), have made measurements using thermocouple probes to measure temperature and a variety of pitot-like probes to measure velocity. The influence of large fluctuations and radiation on the thermocouple data were often ignored and the averages of the fluctuating quantities produced by the instrumentation were used without correction. An example of centerline temperature and velocity profiles taken from the data of McCaffrey (1979) is given in Figure 5 for a 60 kW flame in regime II. The time averaged values for velocity are small, and those for the temperature are far below the adiabatic flame temperature.

These data cover flames in regimes II and show that velocity profiles change substantially as Q* decreases from 1.0 to 0.1 where as the temperature profiles for these regimes are very similar. Scaling laws for centerline temperature and velocity vary somewhat between these three papers. The general levels of temperatures and velocities are in reasonable agreement among them but there is considerable variation in the velocity data. Kung and Stavrianidis (1985) propose the use of an offset, of the type described below, for the origin for the vertical coordinate and Cox and Chitty (1985) use a slightly different scaling approach for small values of Q*.

Entrainment rates. The measurements of velocity and temperature described above are usually of insufficient extent or accuracy to allow estimates to be made of the heat release rates or the mass flux within the flame as a function of height above the burner for flames in the low Q* region.

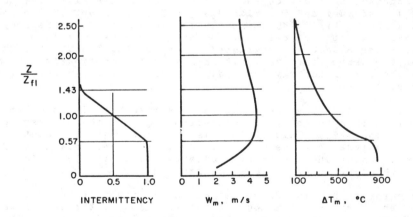

FIGURE 5.  Velocity on temperature profiles on centerline of flame.

Direct measurement of the plume mass flux have been made by several techniques. One, described in Zukoski et al (1980), uses a hood to create an interface similar to that expected in the two layer model to trap the plume flow, and a mass balance is performed on the ceiling layer, to measure the plume mass flux. The technique measured the flux into the layer, not the plume flux and there may be some difference between these two. Data obtained with this technique have been obtained for regime II and III flames and the plumes rising above them.

In the lowest part of the flame, the entrainment rate for regime III flames is independent of the fuel flow rate and is roughly proportional to the diameter of the burner. Data from Zukoski et al (1980) and Cetegen et al (1984), and that of Beyler (1984) are in agreement on these results but differ in magnitude by about a factor of two. Some of this spread is certainly due to differences in technique and some to differences in experimental constraints on the fire plumes.

Entrainment in the upper part of flames in regime III was more plume like and did depend on heat release rate and source-diameter. A typical set of data from Cetegen et al (1984) for a 19 cm burner is shown in Figure 6 to illustrate these two regions of entrainment. Note that for small elevations, the entrainment is independent of the heat addition rate but that for larger elevations, the data show a strong dependence on this parameter.

In regime II, entrainment for the entire flame is similar to that for the lower region of flames in regime III and is independent of the heat release rate.

The measurements discussed here are not definitive and more work is required. In particular, the role of the fluctuations in fixing the flux of mass, momentum and enthalpy in the flame should

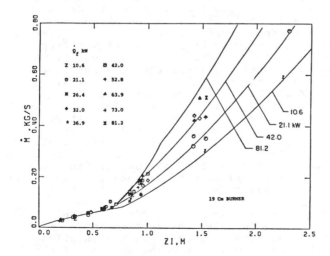

FIGURE 6. Plume mass flux versus height.

13

be clarified.  Attempts to predict these results from simple models have not been particularly convincing. For example, see Cetegen et al (1984).

Delichatsios and Orloff (1984) have made measurements of entrainment rates in jet diffusion flames using the method suggested by Ricou and Spalding (1961).  Values of Froude numbers used in these experiments ranged from 2 to 230 and values of $Q^*$, from 230 to 2600.  Given the large size of these parameters, these flows are clearly in the transition region between regimes III and V and hence we expect that they were strongly influenced by the initial momentum of the source.  Direct comparison of entrainment rates measured in these experiments with those measured with low $Q^*$ flames of interest to us here is inappropriate.  In addition, the absence in these experiments of the flow pulsations, observed at the base of low $Q^*$ flames, is most probably explained by the very high source Froude numbers used here.

Heat release distribution.  Tamanini (1983) has made direct measurements of the vertical distribution of heat release and species concentration in two flames with $Q^*$ values of 1.1 and 0.54. Because his technique involves quenching the flame at the measurement plane and the use of a coflowing air stream, we believe that it may introduce a significant distortion of the flow at that elevation and may also suppress the large structures which are characteristic of flames in this regime.  Data are presented for two flames which are in the upper part of regime II, a transition regime.  The data for the two flames differ substantially and, if accurate, suggest that substantial changes occur in the combustion processes in this regime.


Far Field Plume

The term far field is used to designate the adiabatic plume region which rises above the flame.  Extensive modeling work has been carried out since the 1950's following the paper by Morton et al (1950) and an algebraic model is available which allows a description of average values for parameters in the plume.  Given a point source of enthalpy in an infinite body of gas at a uniform temperature, it is possible to derive simple algebraic equations which define the entire flow.  When we assume that the profiles in temperature difference and velocity are Gaussian, the time averaged values of the Gaussian half width of the plume, and the time averaged values for the velocity and temperature difference on the centerline can be predicted with satisfactory accuracy.  These results can be put in a variety of forms, e.g. see Morton et al (1950) and Cetegen et al (1984) but the basic results are identical. The validity of the general form of this representation has been verified by a number of authors for the flow in the far field and, at most, there is still some argument over the values of some of the constants which appear in the equations.

However, an adiabatic plume rising above a fire has the top of the fire as its real source and, in order to use the simple formulation described above, we must be able to locate an effective source for the plume which clearly must lie below the top of the flame.

14

Given the fluxes of mass, momentum and enthalpy which characterize a distributed source, it is possible to find the elevation for the origin of a point source of enthalpy which produces a flow which is asymptotically identical to that produced by the distributed source. The two flows become identical in the limit of large distance from the source because, in the far field, the influence of the initial momentum and mass fluxes of the distributed source will become vanishingly small compared with the mass and momentum flux in the plume. See the discussions of Morton (1958) and Kubota (1982). This fictitious source allows us to use the simple algebraic representation for the point source to calculate plume properties in the region above the top of the flame.

To calculate the location of the effective source, we require estimates for the mass, momentum and enthalpy fluxes at the top of the flame. Kubota (1982) has carried out calculations of this type in the following manner. The mass flux was measured at the top of a flame with a known heat release rate, see Cetegen et al (1984). An estimate for the maximum velocity was obtained from a correlation of McCaffrey (1979) and was combined with the mass flux measurement and a Gaussian model for the flow to calculate the momentum flux. These data were then used with a numerical model of Kubota (1982) to calculate the elevation of the source for the plume.

Experimental schemes for locating the height of the effective source have been proposed, by Heskestad (1983), Cetegen et al (1984), and by Kung and Stavrianidis (1985), and the method suggested by Cetegen et al is illustrated in Figure 7. Here, experimental values for the offset $Z_o$ , divided by the source diameter D, are shown as a function of the ratio of flame height to burner diameter, $Z_f/D$. The offset is taken as the distance between the surface of the burner and the fictitious source, and is defined to be positive when the source lies beneath the burner. Note that the offset is small when $Z_f/D$ is small, and is negative and not more than 2 diameters even for $Z_f/D > 8$. Data from Heskestad (1983) and Kung (1984) obtained with large fires are shown here and agree satisfactorily with the data obtained with the smaller fires.

Values for the offset calculated by the method proposed by Kubota are shown here by the two points with error bars located close to the lower curve. They are in good agreement with the experimental values shown here and continue to agree up to values for $Z_f/D$ of 20. The error bars show the effects of changing the estimated velocity by a factor of two and they indicate that the results are not very sensitive to errors in velocity.

One of the interesting features of these plumes is that the entrainment rate of air measured at the top of the flame is that which would be predicted for a well developed buoyant plume. Thus, the transition between the flame and far field regions is very rapid. In addition, the measured entrainment rates are those predicted from a Boussinesq analysis even though the ratio of plume and air temperatures at this elevation is large.

Another of the interesting features of these fire plumes is that at the top of the visible flame the flow of air entrained into the fire plume is about 14 times more air than would be needed to react with the fuel flow from the source. This large excess is an important feature of the diffusion flames we are describing here

15

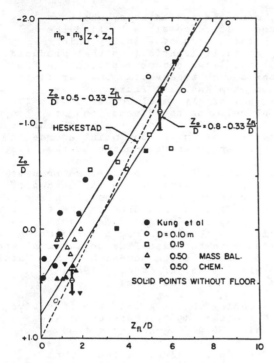

FIGURE 7. Offsets for field plume.

because it means that a flame which penetrates a large distance into a ceiling layer where the oxygen concentration is very low may still entrain sufficient oxygen from the lower and unvitiated layer to allow complete combustion to occur.

In summary, the height of the flame, $Z_f$, can be predicted from a knowledge of the heat release rate and burner diameter. Then this value can be used to predict the elevation of the effective source which in turn can be used to predict the entrainment rate and other properties of the adiabatic plume which rises above the fire. These predictions are applicable for the region above the visible flame and can be used at the top of the flame. Thus, many of the properties of the plume produced by a large diffusion flame, rising above the simple sources described here, can be predicted with confidence.

Data for entrainment into the flame itself are less reliable and more data of this type are needed. In the low $Q^*$ regimes, the data are least reliable in the range $Z_f/D < 3$ and this is just the region of most interest in accidental fires.

Finally it is clear from the discussion of the various flow regimes given here that the properties of the flow depend strongly

16

on the flow regime. Thus when we compare results obtained from
different experiments, and in particular when we compare small scale
laboratory experiments with large scale accidental fires, we must be
certain that the flames being considered are in the same flow regime
and that we are observing similar parts of the flame.

Plume Interactions

    Limitations of space make it impossible to deal with a number
of interactions between a buoyant fire plume and it's environment.
Several important interactions for which information is available
are briefly mentioned here. In each case, we consider the
interaction with the fire plume which includes both the flame region
and the adiabatic plume above it.

    Wall. When a fuel bed is in close contact with a wall or is
placed in a corner, the plume rising above the bed may attach to the
wall and the wall may then strongly influence the entrainment into
the plume. A simple reflection principal, see Zukoski et al,
(1981), suggests that the entrained air flow is reduced and the gas
temperature is increased. Hasemi and Tolunaga (1984) have presented
a review of the subject and new experimental work which shows that
this choice of the simple approach is not completely correct.

    Ceiling. The flow produced by a fire plume impinging on a
ceiling determines the convective heat transfer rates and the
initial properties of the ceiling jet formed by the plume flow.
This flow has been the subject of a number of papers such as Alpert
(1971), Zukoski and Kubota (1975), You and Faeth (1979), Cooper
(1982) and Sargent (1983) and the flow produced by this interaction
is discussed below in Section IV.

    Interface. The impingement of a buoyant plume on an interface
separating gas layers with large density differences has been
discussed by Baines (1971), Cooper (1982), and Zukoski and Mak
(1985). The problem here concerns whether or not the plume will
penetrate the interface and if it does penetrate, how much of the
plume material flows into the interface. Zukoski and Mak (1985)
suggest that when an average plume density, based on the mass-flux-
averaged plume temperature, is less than the density in the ceiling
layer, the plume will penetrate cleanly into the ceiling layer and
rise to the ceiling. No plume gas is left at the interface. When
this average plume density is less than that in the upper layer,
penetration will still occur but the mixed material produced by the
interaction will return to the original interface to produce a third
layer.

    Cross flow. Finally, the interaction of the fire plume with
cross currents and random disturbances is an important feature of
room fires which requires further clarification.


FLOW THROUGH OPENINGS

    The flow of gases through openings, such as windows and doors,
which are typical of those found in buildings, is discussed in this
section and we will not deal with flows through ducts such as those
used in air conditioning systems in which mixing between hot and

17

cold streams may be a dominant feature. Three examples of the flows which may be present during the development of a fire are illustrated in the sketches of Figure 8. They include flow through a doorway and window in a vertical wall, and flows though an orifice in a horizontal surface with a smoke layer below the opening.

The doorway and window flows shown here are the standard flows which are addressed successfully in most models. However, in some circumstances, the flow of fresh air from the second room into the lower layer of the fire room can cause entrainment of hot products into the lower layer in the fire room and thus can lead to substantial contamination of the air in this layer. This mixing process is illustrated here for the window flow, but can also occur in doorway flows and in general cannot yet be described adequately.

Finally consider the situation shown in Sketch (c) in which smoke generated on the first floor of a house flows through an opening in the ceiling. We assume that the house is well ventilated so that the pressure difference across the ceiling and at the opening is fixed by the depth of the hot smoke. As long as the depth of the hot gas layer is larger than the diameter of the hole, see the right hand side of Sketch c, the flow will be similar to that through an orifice. When the layer is thinner, see left hand side of Sketch c, the flow resembles that of a fluid across a broad crested weir and smoke may not fill the whole opening.

The plumes formed above or below the orifice for the configurations described above all resemble conventional turbulent, buoyant plumes. However, some procedure, such as that discussed above for the adiabatic plume rising above a flame, must be used to establish an effective elevation for the source of the plume.

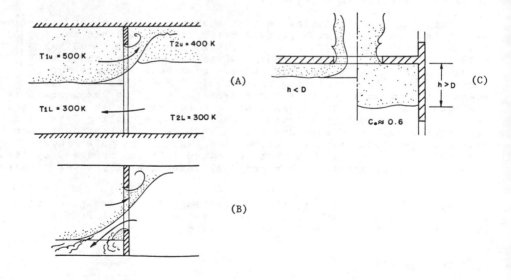

FIGURE 8. Flow through openings.

18

The calculation of the rate of flow of smoke through openings in vertical or horizontal surfaces is another area where models are available which are a step beyond simple dimensionless analysis. The information developed in studies of flow of gases or liquids through orifices, where buoyancy plays no part, and the flow of liquids across weirs, for which buoyancy is paramount, has been of great help here.

The first model for flows through a vertical surface such as a wall, which was brought to the attention of the fire research community, was developed by Kawagoe (1958) for the limiting case of a flow through an opening in a vertical surface which separated a hot gas inside of a room from ambient air on the outside of the opening. His model has been modified and extended over the past ten years by a number of authors, (e.g., Prahl and Emmons (1975), Zukoski (1975), Quintiere and DenBraven (1978) and Stechler et al (1980) to name a few) and present models are applicable, with accuracies at the 10 to 20% level, to the wide range of conditions which can exist when the two layer model is assumed to be applicable.

To illustrate the type of model used which has developed, consider that proposed by Zukoski (1975). The basic idea used in this model is that the flow through any opening can be modeled as an integral, over the area of the opening, of the flow through infinitesimal orifices each of which acts as an ideal isolated orifice in a wall. The velocity of the gas at the plane of the orifice, the direction of the flow, and the mass flux can be calculated by the conventional orifice equations given the local pressure difference across the orifice and properties of the gases on either side of the opening.

Even when density gradients in the vertical direction are allowed on either side of the surface, this model allows the flow to be calculated when the distribution of densities and a pressure difference at one elevation are specified. This model is best suited for the description of the flow through small openings in large walls and becomes less reasonable as the area of the opening approaches the area of the wall. However, even this type of flow can probably be treated by a modification of the analysis which has been developed to describe flow over a weir.

The models used to calculate flows through orifices typically contain a constant called the flow coefficient which is the ratio of the measured to theoretical flow rate and the coefficient is used to take into account a number of effects such as flow separation from sharp corners or viscous effects which are not covered by the theoretical model. The flow coefficients for flow through rectangular doorways and windows which are used in various fire models have values which range from 0.6 to 0.8 and a value of 0.7 seems a reasonable compromise. The use of a single constant ignores the effects on the flow of opening geometry, Reynolds number, radii of curvature of the lip of the opening, and approach stream velocity. These effects are usually in the 5 to 10% range. The large Reynolds number effects reported by Prahl and Emmons (1975) are incorrect and result from surface tension forces acting on the two immiscible fluids used in their tests.

For the flow through an opening in a vertical surface, the difference in densities on either side of the opening enters the calculation of the pressure difference across the opening but has no other effect even when pressure differences are small. However, if we consider the flow of gas through a horizontal surface, it is clear that some flow will occur even if the pressure difference across the wall is zero when the more dense fluid lies above the less dense. Thus, when the pressure drop across the surface becomes too small, the effects of buoyancy forces on the fluid near the opening will have an effect. We expect that the pertinent parameter here is the ratio of a gravitational force based on the opening diameter D and the pressure difference $\Delta P$:

$$\Delta \rho \; g \; D / \Delta P \qquad\qquad\qquad\qquad [12]$$

Here, $\Delta \rho$ is the density difference of the two fluids, and g the acceleration of gravity. Recent experiments suggest that when this ratio, becomes larger than 0.5, the flow coefficient will be influenced by buoyancy effects.

Observations of the flow, made in on-going salt-water/water experimental studies, for this zero pressure-difference case show that an unsteady flow develops in which puffs of fluid from either side of the ceiling flow into the other. Flow rates are not negligible, but a quantitative prediction for the magnitude of the flow in conditions similar to those encountered in building fires is not yet available.

Finally, when the depth of a layer of hot gas lying beneath an opening is smaller than the opening diameter, the flow through the opening will be similar to that shown on the left hand side of Figure 8c, with the additional assumption that the pressure difference across the ceiling of the first floor is imposed by the depth of the hot gas itself. For this condition, we expect that the flow coefficient will be as low as 0.4. However, when the thickness of the hot gas layer exceeds the diameter of the hole, see right hand side of Figure 8c, the flow will again be similar to that through an orifice and the coefficient will rise to values around 0.6.

Based on the information available for horizontal and vertical surfaces, it is reasonable to expect that the flow through surfaces with arbitrary orientations with respect to the gravitational vector can be described by the same orifice-like equations as were found to be useful for the two orientations described above. However, we would also expect that this approach will fail when the ratio of buoyancy force per unit area based on the opening scale approaches the imposed pressure difference across the opening.

FLOW IN CEILING LAYER

One of the great simplifications resulting from the use of the two layer model is that the details of the flow within the layers can be ignored. However, a rudimentary description of flow within the layer is required to allow the calculation of convective heat transfer from the gas to the walls of the room. In addition, strongly stratified layers have been observed recently in several full scale fire experiments and in similar smaller scale experiments

20

at Caltech.  The temperature gradients in the ceiling layer observed in these experiments were large and in some cases were almost constant across the height of the layer.  When this stratification is present, we clearly need a much better model to describe the flow within the upper layer.

These experimental results force us to consider the factors which lead to the development of the poorly mixed layers.  Our primary aim here is to make qualitative arguments concerning the development of stratification and to present some preliminary results of a model being developed to describe this process.  In the following paragraphs, we first consider the gravity current produced when the plume first reaches the ceiling and then the development of recirculating currents within the upper layer which have a strong influence on stratification in the layer.

Ceiling Jets

In the initial stages of a fire within an enclosure, the plume of hot gas produced by the fire rises due to buoyancy forces and impinges on the ceiling to form a layer of hot gas, called the ceiling jet.  This fluid spreads out from the impingement point to form a thin layer which spreads to reach the walls with a relatively constant thickness.  Later, this layer grows in depth to form the ceiling layer.  A similar process occurs in a room adjacent to the fire room where the source is a plume from a doorway.

This process is illustrated in Figure 9 for a two-dimensional flow into a hallway.  Near the plume impingement point, the flow entrains cooler fluid, but within a short distance buoyancy forces prevent further entrainment and the plume moves across the hall ceiling with constant mass flux and a constant thickness.  The velocity of the front and the temperature in the gas near the head of the current decreases with distance from the source due to heat loss

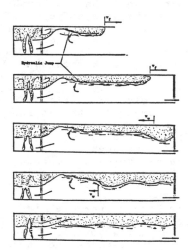

FIGURE 9.  Ceiling jet flow.

21

to the wall and wall shear.  The current reaches the far wall, is
reflected and returns to flood the entrainment region near the
plume.  When the hall length is more than three or four room-heights
long, the temperature in the reflected flow is typically much less
than that at the source because of convective heat transfer and this
difference remains large until the ceiling surface temperature rises
to a value comparable to the initial plume temperature. Flows of
this type have been described in a review by Zukoski and Kubota
(1984).

    We are developing a simple integral model to analyze the plume
and ceiling jet system which takes into account shear forces and
heat transfer at the wall, and entrainment from the gas beneath the
current.   An important parameter which appears in this analysis is
the Richardson number for the jet, which is defined as the ratio of
the gravitational forces acting over a height corresponding to the
thickness of the jet ,$\delta$ , to the  momentum flux in the layer or

$$Ri = ( \Delta\rho \ g \ \delta ) / (\rho_c \ Vc^2 )$$

    Here, $\Delta\rho$ is the difference in density between the fluid
adjacent to the jet and the jet fluid, g is the acceleration of
gravity, and $\rho_c$ and Vc are the density and characteristic velocity
in the ceiling jet.  When the Richardson number is large,
gravitational effects will be dominant and entrainment into the jet
will be supressed; when it is small, momentum effects will be
dominant and entrainment will be important.

    The integral model shows that the Richardson number of the flow
at the start of the ceiling jet is independent of both the height of
of the fire plume and the heat release rate of the fire, and that
the plume height is the proper scaling length for the initial
thickness of the ceiling jet.

    The analysis shows that near the plume impingement point where
the ceiling jet is formed, the Richardson number in the jet is less
than one.  However, as the jet entrains fluid, the Richardson number
grows with the square of the distance from the impingement point.
Thus, within a distance which is comparable to the plume height, it
typically reaches values larger than one.  In this high Richardson
number regime, no mixing or entrainment occurs between the fluid in
the jet and the adjacent fluid and the jet propagates across the
ceiling with constant mass flux.

    The propagation velocities for ideal inviscid, adiabatic jets
for which  the layer thickness,$\delta$ , is small compared with the room
height are proportional to $\sqrt{g'\delta}$ and the proportionality constant is
between 1 and 1.40.  Here, g' is $g[(\rho - \rho_c )/\rho_a ]$, the reduced
acceleration of gravity based on the difference between the
densities of the current, $\rho_c$ and that in the gas below the current,
$\rho_a$ .  Heat transfer effects enter primarily through a reduction in
this density difference ratio.

    Velocities for ceiling jets produced by large fires are
typically in the range of 0.2 to 2 m/s near the start of the jet.
Consequently in a room with a lateral scale of a few meters, fluid
in the jet will reach the side walls within a few seconds and the
complexities of this flow can be ignored in describing the
development of the layer.  However, when the room has a lateral

dimension which is very large compared with its height, the time scale for this process can become important.

Later on in the development of the fire, the effects of the heat transfer from the ceiling jet can also have a large effect on the flow within the upper layer. These will be discussed in the following section.

Stratification

The degree of stratification in the ceiling layer depends in detail on the flow within the layer as well as the action of the fire or doorway plumes, flows into and out of the layer at openings and heat losses to the ceiling and walls by convection and radiation. A number of processes can be identified as acting to increase or decrease the stratification of the gas in the layer and several of these are discussed in the following paragraphs with the aid of the sketches presented in Figures 1, 10 and 11.

Given the picture of the development of the fire in a compartment given in the introduction, it is clear that the processes of fire growth and heat transfer tend to produce a stratified layer and the real question is: Why do we get a well-mixed layer at all?

A number of processes act to reduce the degree of stratification. One of the most important of these is the entrainment, into the plume and the ceiling jet, of gas from the ceiling layer which produces a recirculation of gas within the ceiling layer. Since gas is entrained from all levels of the layer and mixed with the flow in the plume and ceiling jet, this entrainment process will act to reduce stratification and it is the primary process which may produce a well mixed ceiling layer.

A second factor which reduces stratification is the flow of gas out of the ceiling layer through a doorway, window, or other opening. For example, when this flow is withdrawn from the bottom of the layer, as it would be for a doorway with a soffit, high density gas near the interface will be the first to move down to reach the door soffit and hence will be the first to be removed.

A third factor, the flow path produced by the impinging plume and the resulting ceiling jet, is of major importance and can act either to increase or decrease stratification. The flow in the ceiling jet moves along the ceiling until it impinges on the side walls, and at that point it must turn downward. If the gas in the ceiling jet has sufficient momentum and a small enough buoyancy, with respect to the ceiling layer gas, it may flow down the side walls to reach the interface between the ceiling and floor layers before buoyancy forces overcome its downward momentum. This overshooting flow will cause strong mixing within the layer. However, if the momentum in the jet is small compared to gravitational forces, the ceiling jet will turn back over itself and flow back toward the plume. Less mixing will occur for this flow than for the previous one.

The parameter which controls the motion of the plume at the side wall is the Richardson number for the ceiling jet based on the

23

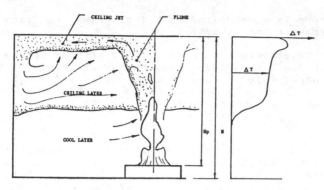

FIGURE 10.  Ceiling layer flow, well mixed.

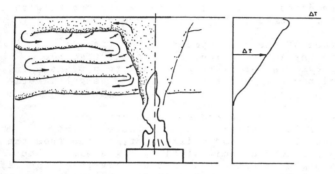

FIGURE 11.  Ceiling layer flow, poorly mixed.

density in the jet and in the adjacent ceiling layer.  When the Richardson number is still small, the jet will turn through 90 degrees at the corner and will have sufficient momentum to move down into the ceiling layer, a distance corresponding to many ceiling jet thicknesses.  This is the flow pattern shown in the sketch in Figure 10.  For low Richardson number flows, both the ceiling jet and the fire plume entrain fluid from the whole hot layer and a rapid recirculation results from the vortical motion.  This flow pattern and the entrainment into the plume and ceiling jet produces a well mixed ceiling layer with weak density gradients except near the interface with the cool layer. In this example, heat transfer can reduce the gas temperature, but the effects of this process are spread throughout the whole layer.

If the Richardson number of the ceiling jet is large when the gas impinges on the side wall, the fluid in the jet still must turn but it no longer has the momentum required to penetrate any appreciable distance downward.  Buoyancy forces act to force the flow to turn back toward the plume and move under the outflowing gas in the ceiling jet.  This type of flow pattern is shown in sketch of Figure 11.

Here, the flow forms a stream with a sinuous path and material from this stream is entrained in the plume, but little is entrained

24

into the ceiling jet because no entrainment is present in the high Richardson number part of that flow. Thus in this case, only a small part of the material in the recirculating flow is entrained into either flow as it approaches the interface to form the lower part of the ceiling layer. Because fluid is not entrained into the plume or ceiling jet, it undergoes little mixing and will retain its original temperature and composition. In a fire situation in which the material delivered to the ceiling by the fire plume experiences an increase in temperature, due to the growth of the fire and a decrease with time of convective heat transfer, this process will create a strongly stratified layer.

Thus, the pattern of the flow rather than the total heat transfer or the rate of recirculation is important here in fixing the stratification in the layer.

The two examples discussed above are limiting cases which are observed in room fires and the problem, not yet resolved to our satisfaction, is to predict at what radius the transition between them will occur in rooms with a geometry of interest to us. Estimates based on an idealized model for flow in the ceiling jet show that the Richardson number will be less than one at the origin of the ceiling jet and that it will grow roughly as the square of the distance from the plume. Thus we expect that the transition between the overturning flow of the well mixed case and the weakly mixed case will depend strongly on the geometry of the room and more weakly on other characteristics of the system. The idealized integral model indicates that the transition between low and high Richardson number behavior will occur at a radius, measured from the plume axis, of 1 or 2 plume heights. The transition point for a particular situation will depend on the values selected for the coefficients used to describe convective heat transfer, wall shear and entrainment rates, and hence is at least weakly dependent on the Reynolds number of the jet.

A number of small scale experiments suggest that this result is correct. For example, experiments carried out by Baines and Turner, (1969), and later by Tangren et al, (1978) with the salt-water/water modeling technique, suggest that the correct geometrical parameter to use in deciding which of these two flow patterns are to be expected is the aspect ratio for the room. Based on their experiments, Baines and Turner (1969), propose that the transition occurs when the aspect ratio, the ratio of lateral room dimension to plume height, is between one and two.

These experiments were carried out for a geometry in which the fire is located on the floor and at the center of a square room, and this configuration was also used in carrying out the integral model whose results are described above. If the plume is not located at floor level in the center of the room, the correct scale lengths are the mean distance from the location of the fire to the walls and the height of the plume.

This transition criteria is clearly a very crude one and a more refined definition would depend on other parameters of the system such as the Reynolds number of the flow in the ceiling jet and even the thermal properties of the ceiling material which can affect the influence of the heat transfer process. The influence of these parameters is now being investigated.

25

The flow through a doorway or other opening in a vertical wall which has a soffit substantially below the level of the ceiling of the adjacent room will produce a plume-like flow and a ceiling jet, see Figure 9. The major difference between this doorway plume and the fire plume is that the doorway plume will usually be much shorter than the fire plume and hence will produce substantially smaller rates of entrainment and recirculation. In addition, the plume height will usually be much smaller than the room height and consequently the transition between low and high Richardson flows will occur much closer to the plume impact point. Thus, for many reasonable room and soffit geometries, the flow in the adjacent space will be in the high Richardson number regime for which we expect strong stratification. Based on this discussion, we should expect that strong stratification will be more common in spaces adjacent to the room of fire origin than in the room containing the fire.

The material discussed above suggests that strong stratification may exist in many situations of interest to us. Stratification will affect energy transport processes by convection and radiation which depend on temperature differences, and mass transport through openings which depends on the density distribution in both layers.

The magnitude of these effects is time dependent: early in the history of the fire, stratification will often be important; later if mixing increases and outflow processes become important, the effects of stratification may be negligible.

However, when stratification is large, some modification must be developed for the two layer model to allow an accurate description of transport processes. As interest shifts to modeling flows in rooms adjacent to or far from the room in which the fire originates, the need for a more detailed description of flow within the ceiling layer will become more urgent.


CONCLUDING REMARKS

The discussion given above shows that the properties of the flows produced by large buoyancy controlled diffusion flames depend strongly on the flow regime for the flame. For example, data for flames in regimes II and III show that flame length dependence on $Q^*$, and entrainment rates and, to a lesser extent velocity and temperature on the flame centerline, change substantially between these regimes. Only flame length data are available for regime I flames, and they show again a large change in the dependence on $Q^*$.

The dimensionless parameter $Q^*$ , used here to correlate the flame height data and to define these regimes for the three buoyancy controlled flame regimes appears to be a satisfactory parameter for regimes II and III. However, the data are too scattered to allow a very precise test of its validity and its use in regime I is suspect although very little data are available for this regime.

For flows with larger Froude numbers, substantial changes occur as the transition from buoyancy controlled to momentum controlled flames occurs. The correct transition parameter here appears to be a Froude number, and not $Q^*$, and experimental results obtained with a wide range of fuel types would be useful here in defining the transition process more certainly.

The recognition that flow properties change substantially with flow regime is critical for those modeling diffusion flame phenomena.

A transition of a different type can occur in the ceiling layers of room fires - the transition between a well-mixed and a stratified ceiling layer. The present work suggests that this transition is most sensitive to room geometry and starts to occur when the ratio of the horizontal scale of the room to the plume height is greater than 2. Because plume heights in rooms adjacent to the fire room are often smaller than that for the fire plume or are entirely absent, we expect that stratification will be more prevalant in the adjoining rooms. Should this suggestion prove correct, some modification to the usual two-layer modeling approach must be developed to account for the influences of stratification on the transport of energy and mass.

REFERENCES

Albert, R. L.: "Fire Induced Turbulent Ceiling-Jet," FMRC Serial No. 19722-2, Factory Mutual Research Corp., Norwood, MA, 1971.

Alvarez, N. J.: Personal communication, 1985.

Baines, W. D. and Turner, J. J.: "Turbulent buoyant convection from a source in a confined region," J. Fluid Mech., 37, pp. 51-80. 1969.

Baines, W. D.: "Entrainment by a plume or jet at a density interface," J. Fluid Mech, vol. 68, part 2, pp. 309-320, printed in Great Britain, 1975.

Becker, H. A. and Liang, D.: "Visible Length of Vertical Free Turbulent Diffusion Flames," Combustion and Flame, 32, pp. 115-137, 1978.

Becker, H. A. and Yamazaki, S.: "Entrainment, Momentum Flux and Temperature in Vertical Free Turbulent Diffusion Flames," Combustion and Flame. 33, 1978.

Blinov, V. I. and Khudiakov, G. N.: Dokl. Akad. Nauk SSSR, 113, 1094. Reviewed by Hottell, H. C. (1959) Fire Res. Abstr Rev., 1, p. 41, 1957.

Cannon, J. B. and Zukoski, E. E.: "Turbulent Mixing in Vertical Shafts Under Conditions Applicable to Fires in High Rise Buildings," Technical Report No. 1, to the National Science Foundation, California Institute of Technology, Pasadena, California, 1975.

Cetegen, B. M., Zukoski, E. E., and Kubota, T.: "Entrainment in the Near and Far Field of Fire Plumes", Combustion Science and Technology, Vol. 39, pp. 305-331, 1984.

Cooper, Leonard Y.: "Convective Heat Transfer to Ceilings Above Enclosure Fires," 19th Symposium (Int.) on Comb./The Comb.

Institute, pp. 933-939, 1982.

Cox, G. and Chitty, R.: "Some Source-Dependent Effects of Unbounded Fires," Combustion and Flame, Vol. 60, pp. 219-232, 1985.

Delichatsios and Orloff: "Entrainment Measurements in Turbulent Buoyant Jet Flames and Implications for Modeling," Factory Mutual Report FMRC J.I. OKOJ2.BU, 1984.

Emmons, Howard: "Prediction of Fires in Buildings," 10th Symposium (Int.) on Comb./The Comb. Institute, pp. 1101-1111, 1978.

Hasegawa, H.K., et al: "Fire Protection Research for DOE Facilities: FY 83 Year-End Report," UCRL-53179-83, 1984.

Hasemi, Y., and Tokunaga, T.: "Some Experimental Aspects of Turbulent Diffusion Flames and Buoyant Plumes from Fire Sources Against a Wall and in a Corner of Walls," Combustion Science and Technology, 40, 0010-2202/84/4004-0001, pp. 1-17, 1984.

Heskestad, G.: Fire Safety Journal, 5, p. 109, 1983.

Kawagoe, K.: "Fire Behavior in Room Fires," Bldg. Research Institute, Tokyo, Japan, Report Number 27, 1958.

Kubota, T.: "Tubulent Buoyant Plume in a Stratified Media," Report prepared for The Center for Fire Research, National Bureau of Standards, Jet Propulsion Center Report, California Institute of Technology, 1977.

Kung, H., and Stavrianidis, P.: "Buoynat Plumes of Large-Scale Pool Fires," 19th Symposium (Int.) on Comb./The Comb. Institute, pp. 905-912, 1982.

McCaffrey, B. J.: "Purely Buoyant Diffusion Flames: Some Experimental Results," NBSIR 79-1910, Nat. Bur. Stand., Washington D.C., 1979.

Morton, B. R., Taylor, G. I., and Turner, J. S.: Proc. Royal Soc., A234, p. 1, 1950.

Morton, B. R.: "Forced Plumes," J. Fluid Mechanics, 5, pp. 151-163, 1958.

Prahl, J., and Emmons, H. W.: "Fire Induced Flow Through an Opening," Combustion and Flame, 25, pp. 369-385., 1975.

Quintiere, J. G., and DenBraven, K.: "Some Theoretical Aspects of Fire Induced Flows Through Doorways in a Room-Corridor Scale Model," NBSIR 78-1512, National Bureau of Standards, Washington, D.C., 1978.

Quintiere, J. G.: "A Perspective on Compartment Fire Growth," Combustion Science and Technology, 39, 0010-2202/84/3906-0011, pp. 11-54, 1984.

Railston, W.: "The Temperature Decay Law of a Naturally Convected Air Stream," Proc. Phys. Soc, 67B, pp. 42-51, 1954.

Ricou, F. P., and Spalding, B. P.: "Measurements of Entrainment by Axisymmetric Turbulent Jets," J. Fluid Mechanics, p. 11, 1961.

Rockett, John A.:, "Fire Induced Gas Flow in an Enclosure," Combustion Science and Technology, Vol. 12, pp. 165-175, 1975.

Sargent, W. S.: "Natural convection flows and associated heat transfer processes in room fires," Ph.D. Thesis, California Institute of Technology, 1983.

Steckler, K. D., Quintiere, J. G., and Rinkinen, W. J.: "Flow Induced by Fire in a Compartment," 19th Symp. (Int.) on Comb., The Comb. Inst., p. 913-920, 1982.

Steward, F. R.: Combust. Sci Technol., 2, p. 73, 1954.

Tamanini, F.: "Direct Measurements of Longitudinal Variation of Burning Rate and Product Yield in Turbulent Diffusion Flames," Combustion and Flame, v. 51, pp. 231-243.

Tangren, E. H., Sargent, W. S., and Zukoski, E. E.: "Hydraulic and numerical modeling of room fires," Jet Propulsion Center Report, California Institute of Technology, 1978.

Thomas, P. H., Baldwin, R., and Heselden, A.J.M.: "Buoyant Diffusion Flames: Some Measurements of Air Entrainment, Heat Transfer, and Flame Merging," Tenth Symposium (International) on Combustion, The Combustion Institute, pp. 983-996, 1965.

Wood, B. D., Blackshear, P. H., and Eckert, E. R. G.: Combustion Science and Technology, 4, p. 113, 1971.

You, H. Z. and Faeth, G. M.: "Ceiling Heat Transfer During Fire Plume and Fire Impingement," Fire and Materials, 3, p. 140, 1979.

Zsak, T. and Zukoski, E. E.: "Review of Flame Height Data," Report for the Center for Fire Research, National Bureau of Standards, Jet Propulsion Center Report, California Institute of Technology, 1985.

Zukoski, E. E.: "Convective Flows Associated with Room Fires," Semi-annual Progress Report for 1975 to the National Science Foundation, Grant No. GI 31892X1, California. Institute of Technology, Pasadena, California, 1975.

Zukoski, E. E., and Kubota, T.: "An Experimental Investigation of the Heat Transfer from a Buoyant Gas Plume to a Horizontal Ceiling, Part 2: Effects of a ceiling layer," NBS-GCR-77-98, Nat. Bur. Stand., Washington, D.C., 1975.

Zukoski, E. E.:"Development of a Stratified Ceiling Layer in the Early Stages of a Closed -room Fire," J. of Fire Materials, 2, p. 54, 1978.

Zukoski, E. E., Kubota, Toshi, and Cetegen, Baki,: "Entrainment in Fire Plumes," _Fire Safety Journal_, 3, pp. 107-121, 1980-81.

Zukoski, E. E., and Kubota, T.: "Two-Layer Modeling of Smoke Movement in Building Fires," _J. Fire and Materials_, 4, p. 17, 1980.

Zukoski, E. E. and Kubota, T.: Final Report to the Center for Fire Research, National Bureau of Standards, Jet Propulsion Center Report, California Institute of Technology, 1984.

Zukoski, E. E.,: "Algebraic Models for Jets and Plumes," Report for the Gas Research Institute, Jet Propulsion Center Report, California Institute of Technology, 1984.

Zukoski, E. E.,: "Smoke Movement and Mixing in Two Layer Fire Models," paper presentation at the 8th U.S. - Japan Panel on Fire Research and Safety, 1985.

SYMBOLS

| | |
|---|---|
| $C_p$ | Specific heat at constant pressure |
| D | Diameter of fire source |
| $Fr_o$ | Froude number, $W_o/gD$ |
| g | Acceleration of gravity |
| g' | Reduced acceleration of gravity, such as $g(o - o)/o$ |
| $h_f$ | Heating value of fuel, J/kG |
| $\dot{m}$ | Mass flux in plume |
| $\Delta p$ | Static pressure difference across an opening in a surface |
| $\dot{Q}$ | Heat release rate of fire |
| Q* | Dimensionless parameter, $Q/(o\ C_p\ T\ -/gD\ D)$ |
| Re | Reynolds number |
| T | Temperature |
| $T_p$ | Mass-flux averaged plume temperature, $(T + Q/m\ cp)$ |
| V | Horizontal velocity |
| $W_o$ | Initial source velocity |
| Z | Vertical coordinate |
| $Z_f$ | Height of top of flame |
| $Z_t$ | Elevation for transition |
| $\delta$ | Thickness of ceiling layer |
| $\rho$ | Density |
| $\nu$ | Kinematic viscosity |

Subscripts:

| | | | |
|---|---|---|---|
| o | Source value | c | Ceiling jet property |
| $\infty$ | Ambient gas property | a | Ceiling layer property |
| f | Fuel property | | |

# FIRE PHYSICS

Session Chair

**Prof. Toshisuke Hirano**
Department of Reaction Chemistry
Faculty of Engineering
The University of Tokyo
7-3-1 Hongo, Bunkyo-ku
Tokyo 113, Japan

# The Needed Fire Science

**H. W. EMMONS**
Division of Applied Sciences
Harvard University
Cambridge, Massachusetts 02138, USA

Since there have been many review articles on many aspects of Fire Science and its applications (1-10), this paper will stress the many aspects of Fire Science not yet complete or in some cases not even started. By now (1985), it has become broadly accepted that the way of the future in Fire Engineering is through various levels of modeling, aided by the modern computer. A number of models exist for various aspects of a fire in an enclosure (11-14) the most advanced of which are gradually becoming sufficiently general (15-21) to include a fire in any of man's structures. These include both zone and field models. At the same time the U.S. Forest Service has developed a fire model sufficiently accurate to have been incorporated into a special hand held calculator usable in the field (22).

Any fire model is, of course, based upon the available understanding of the phenomena of fire and the fire properties of the materials involved. The "understanding" may be based upon secure fire science, secure empirical correlations, or, if these are not available, then the best guess we can make. The ultimate purpose of fire science is to remove the guesswork. Similarly for data, we should know exactly what materials are present, in what amounts and what configurations, and we should have a complete handbook of the appropriate properties over the appropriate temperature range. Again it is frequently necessary to supplement available data by one's best guess. Anyone engaged in practical fire protection engineering will know that in design it is never known exactly what will be in a new building. And after a fire, it is very rare that the exact contents of the burned building is clear. In fact building changes over the years have generally not been documented, so the exact nature of the structure itself is often only illy known.

When we solve technical problems, it is only sensible to keep all phenomena (terms) down to some minimum size and to discard the rest. It takes a high level of scientific understanding of fire to know exactly what aspects of the fire environment are important and which are not.

This insight into the real needs of a practical problem is presented to the fire researcher in a difficult form. How deep do we have to carry our fire research? We all know that while turbulent flow occurs throughout fire dynamics, and a better understanding of fundamental turbulent flows would be useful, the basic turbulence problem is so complex and progress is so slow that little use of fire research funds for this purpose is justifiable. As fire researchers, we should follow such fundamental work (23) so as not to miss some important new discovery in the field. Various approximate methods of computing fire turbulent flows are being developed (24, 25, 26). These methods will play an increasingly important part in fire research. They will serve to check various approximate calculations used in zone model computer codes. They will become increasingly important as the computer continues to increase in capacity and speed. In the same way, chemical kinetics is at the core of all reactions (fire and otherwise) but again many very basic chemical kinetic problems should be followed but not made part of fire research. Those kinetic studies with a bearing on ignition, extinguishment, soot and toxic specie production are clearly important. It is part of the task of the fire researcher to choose his study on those parts that are important (and interesting) rather than merely interesting.

For example, it is well known that the pyrolysis of complex organic materials often produces hundreds of organic compounds (27, 28). Having learned this and measured them for a few cases, we can safely say that for now we have enough of such general chemical detail. We should always consider the answer to the question, "What would I do with the additional information if I had it?". Furthermore while the thousands of kinetic steps between these hundreds of compounds are of chemical interest, only the controlling steps or perhaps only overall kinetics in ignition, extinguishment, and special specie production are important for fire.


IGNITION

While the details of ignition can be traced for any specific case by balancing the kinetic heat production and dynamic heat loss, fire modeling at present does not use more than ignition temperatures. This approach will probably serve most practical purposes forever but special cases should be done better. The real world presents many peculiar cases of fire ignition (29-39) not now well understood so that a lot more ignition studies are needed to clear up these mysteries. These studies may be chemical kinetic, or dynamic but in most cases must eventually be both.


THE FIRE

Once a fire is started, we need sufficient information to *quantitatively* predict three things: the rate of fire spread over the given material in the given configuration, the rate of pyrolysis, and information about the pyrolysis products.

We know the general nature of the energy feedback system which heats and ignites material ahead of an advancing flame front and many special cases have been studied (40-51). The really important question before us is "Will it be possible to develop a sufficiently accurate general theory of flame spread that we can use it with confidence in place of correlated test results?". Flame spread is a near surface problem. The rate of pyrolysis often involves phenomena in depth. The simplest case involves evaporation of a liquid fuel or melt of the unzipping of a polymer. Knowing the feedback energy in excess of internal heat conduction together with a latent heat is sufficient to compute the rate of pyrolysis for simple cases. At present this same approach is often used generally even when not really appropriate.

Many solid fuels char and/or melt and this leads to great complications. Char generally adheres to the solid and protects the virgin material from direct receipt of the feedback energy. The chemical pyrolysis step now occurs below the char layer. Thermal transport processes now control the feedback; and mass transport processes control the removal of the gaseous pyrolysis products. Diffusion through the char and, when porous, deeper into the virgin fuel are initially important. As the char becomes thicker, its change of dimensions and resistance to flow cause it to crack. Finally the gaseous pyrolysis products

34

may further pyrolyze as they diffuse through the hot char or are overtaken as the pyrolysis region moves deeper into the solid fuel. Sometimes the resistance to pyrolysis gas flow is so high that pressure builds up below the char which is then expelled explosively. The analysis of the charring process has just begun (52-56). A proper treatment of cracking has not been attempted and nothing has been done with the theoretical and experimental study of, nor the statistics of the explosive removal of char. In fact, it is not yet clear whether or not the pyrolysis process can be adequately treated as an energy absorbing, infinite reaction rate front, or must be treated as a chemical rate controlled region of significant thickness. In either case, the thermal removal of absorbed water is an important first step for wood (53).

Some solids melt at temperatures near their ignition temperature. This may mean that a horizontal surface of the solid melts and pyrolyzes downward faster than the fire spreads horizontally over the surface (cellular polystyrene) (57). It also means that the fire behavior of melting solids is very sensitive to orientation. In fact thin sheets or cloth of such materials may fail to burn when hung vertically because they melt at a lower temperature than they ignite. Thus the flames coming up from the burning section melts the virgin material above the burning zone and the burning section falls away and burns out. Nothing has been done to study this process.

On other materials, wood, polyurethane foam, etc., the fire advances over the surface far faster than the advance in depth giving rise to a slightly dished burning area (53, 57).

The detailed chemical kinetics of pyrolysis is far too complex to expect successful prediction of the rate or composition of the pyrolysis products from empirical chemical kinetics or quantum mechanics in the near future. Realistically what do we really need to know? As the pyrolysis gases are released, they mix with air and burn (at least partially). We must know the amount of heat thereby released because this energy is what propagates the fire. We need to know the flame energy radiated. Since this depends primarily on the amount of soot present, it would be important to understand soot production. At present measured flame temperatures and emission coefficients are used to compute the energy radiated (58) or else we assume that some fraction of heat release (say 35%) (2, 59) is radiated away. Our fire predictions will be more accurate when we fully understand and can quantitatively predict soot production.

In a recent meeting on soot production nine excellent research papers were presented (1). The diverse results remind one of the nine blind men describing an elephant. This remark is not a reflection on the quality of soot research being done but is an observation on the complexity of the problem. Further work will eventually find the key which can put all of the present pieces together into an accurate soot algorithm.

Almost nothing is known about the lel and uel (flammability limits) of the pyrolysis products obtained from various fuels nor the heat released when they burn. The usual assumption that their heat of combustion equals that of the virgin fuel must, at least for charring fuels, be very bad.

The rate of pyrolysis is essential knowledge because the gases thus produced are the fuel in the flames. There are two basic approaches. First, there is the use of some simple theory. So long as the flow is laminar, the predictions using the Svab-Zeldovich transformation are in good agreement with experimental results (60-64). These theories assume a heat of pyrolysis using the feedback energy as the pyrolysing mechanism. This is adequate for evaporating liquid fuels and some non-charring solids but new theories with validating experiments

are needed for charring solids.  Second, there is direct use of experimental
data.  The Rate of Heat Release is directly measured at one or several radiant
levels and used to predict fire development (65, 66).  These are engineering
test methods and are in need of careful scientific study so their exact significance and use can be based on a thorough understanding of their limitations.
It is to be expected that the further development of both the theory and the
experiments will bring these two approaches together.

Various special problems have received some attention (67-70).  The problem
of smouldering is one devoid of all the complex dynamics of air supply and
flames.  However, only limited progress has been made (32, 71, 72) because the
complex pyrolysis chemistry plays a central role.

Once the flammable gases are released by pyrolysis they rise and burn in a
flame.  Although considerable effort has been expended (73, 74, 81, 82, 86, 122,
123, 131) we are a long way from understanding the full process of flame combustion.  All of the quantities required for fire modeling, soot production,
radiation, fraction of heat release, residual unburned fuel and toxic products
are now obtained from incomplete (for this purpose) experimental results (138).
All of this required information will be supplied by a complete theory of
diffusion flames although some of the information may forever best be obtained
empirically because of the joint complications of chemical kinetics and
turbulence.

Since many pyrolysis gases are not immediately burned, the content of
toxic substances needs to be known.  Carbon monoxide, hydrogen chloride, and
hydrogen cyanide are the toxic compounds sometimes considered.  While a few
other compounds acetaldehide, acrolein, are mentioned, they are almost never
measured.  In fact the compounds usually measured, $CO$, $CO_2$, $H_2O$, $O_2$, $(C_2H_2)_x$,
are selected as much by the existence of convenient instruments as by their
importance.

Such composition measurements are very important to be made after smouldering and flaming combustion as well as in pyrolysis alone.  In fact, such measurements in the flaming mode need to be known for fire vitiated atmospheres (not
nitrogen vitiated) since some burning always occurs in the hot layer of gas at
the ceiling.  There is only limited data of any of these types for the many fuel
types and fuel combinations encountered in practice.  At present all that can be
done in fire modeling is to use data from any source available and to use it as
though it were universal.


PLUMES

The simple point, line, and area source plume theory together with its
experimental validations is good enough for present fire modeling (75, 76).
This does not mean that there is no more to learn.  The conditions in the plume
close to the fuel surface are badly described by present theory (77).  The concentration of the pyrolysis products is high and probably absorbs a significant
fraction of the flame radiation on its  way back to the fuel surface.  This feedback radiant energy must be accurately known if we are to predict the pyrolysis
rate accurately.  There has been no careful detailed study of this radiation
absorption aspect of plume flow.

Recent measurements show that the ambient air entrainment rate immediately
above the fuel correlates with the vortex shedding frequency.  However, what it
is that determines this frequency is a mystery (77, 78).

A few studies of a plume against a wall, in a corner, or at a vent have been made (79, 80). None of these problems have been carried to the point that well validated algorithms exist for model fire predictions. This is especially true of the plume near a vent where no reliable data exists (80).

The theory of the burning by a flame in a plume is very unsatisfactory. The present theories which vary from simple entrainment ideas (81) to complex $\varepsilon$, k, g turbulence and probability controlled reactions (82) fall far short of the needed knowledge. Burning efficiency, residual products including unburnt fuels, soot content, temperature and radiation output are now all determined by a few measured results on a few fuels used as though they were universal. Many more measurements are needed but until a proper correlating theory is developed these measurements are not likely to result in a reasonably universal predictive system.

Many special cases have been considered (83-91). New 2- and 3-dimensional calculation methods used in research should greatly improve our understanding of special plume problems (90, 91).

## LAYERS

Fire models to date assume the formation of two homogeneous layers; a hot one above and a cold one below. The fire conditions for which this is adequate is not at present known. What is known is that the fire plume striking the ceiling flows outward in all directions as a ceiling jet. The initial plume has such small buoyancy that by the time it strikes the room walls the jet is cooled off and its residual momentum carries it downward setting up a general room circulation (92). This initial period is of little importance for further fire development but is important in fire detector operation. Fires often grow fast enough to almost immediately replace the general room circulation by hot and cold layers. The mixing between these layers appears to be unimportant except during hot layer burnout (30) in which much of the oxygen for burnout is buoyantly brought up from below. However mixing does occur where wall cooled hot layer gases flow downward into the cold layer and/or upward into the hot layer (93, 94) and where vent flows do some mixing by entrainment (86). The first of these effects need experimental validation while the second needs more studies under a wider range of circumstances.

The hot layer composition changes with time as products of partial combustion collect and the oxygen is more or less used up. At present only low accuracy is attainable not only because pyrolysis products are imperfectly known but also because the inefficiency of plume burning in fire vitiated air is not well understood.

A recent study (30) has shown that the LeChatelier (95) formula for the burning limits for the layer mixture of flammable gases holds for a thin (< 20 cm) hot layer independent of its oxygen content. Oxygen is drawn up from the cold layer by buoyancy as needed to burn the layer fuels. However the present limited results make it likely that for a much deeper layer the burning would be confined to a sublayer at the interface since only limited oxygen could be brought up from below. There is as yet no verification of this idea first because there has been no careful appropriate experimental work with a deep layer and second because no theory of any kind as yet exists for the layer interface combustion "plumes".

The cold layer while relatively insensitive to a fire cannot be ignored. It changes temperature by addition of small amounts of hot layer gases and is significantly heated by the floor (which in turn is radiatively heated from the flames and hot layer). Furthermore there is a slow accumulation of pyrolysis products by the mixed hot layer gases and most importantly by the gases from the pyrolyzing but not yet ignited hot floor. In the limit, one would expect the ventilation limited fire in a room to consist of a rich mixture hot layer extending to the floor with an air flame (air burning in a fuel atmosphere) issuing from the vent into the room at the floor while at the same time a fuel flame issues from the hot layer at the top of the vent onto the ceiling of the next room. There are no research results either theory or experiments on these problems known to this author.

## THE CEILING JET

The ceiling jet from a plume is of only small importance in a small room. In a very large room the assumption of two heterogeneous layers has a serious effect upon the fire growth prediction. Since a lot of feedback energy to ignite new fuels comes from the hot layer, an object in a large room far from the initial fire would ignite almost as soon as a nearby fuel if the hot layer were homogeneous.

This effect is especially bad when a fire plume issues from a fire room into a corridor. The time delay in reaching the far end of the corridor is often of major importance.

Most ceiling jet theories and measurements to date (96-99) assume a steady state. Only a few theoretical studies are nonsteady (100, 10, 101). A simple examination of a nonsteady ceiling jet in a corridor shows (102) that the jet front moves at the rate $u_f$

$$\frac{u_f}{\sqrt{gH}} = (1 - R)^{1/2} (1 - \Delta)$$

where

$$R = \frac{\rho}{\rho_a}$$

$$\Delta = \frac{\delta}{H}$$

relative to the fluid in front of it and has a depth related to the flow rate in the jet and below it given by

$$\Delta^2 - 2 \left\{ 1 - \frac{M_D + \frac{M_u}{R}}{(1-R)^{1/2}} \right\} \Delta + \frac{M_u}{R(1-R)^{1/2}} = 0 \quad .$$

But the flow and depth of the ceiling jet layer produced by the fire plume are not in general equal to these values (96). Thus there will generally have to be a jump in the initial shooting jet not to its tranquil flow regime but to the shooting flow appropriate to the advancing ceiling jet given by equations 1 and 2. New experimental studies in a long corridor with a smooth ceiling will be required to validate the above results. In addition, however, the nonsteady ceiling jet must eventually include friction, heat transfer, entrainment, and reaction before it is complete. Most of this experimental work has not been started.

It is, of course, possible to use various 3-dimensional nonsteady codes (20, 100) to analyze the ceiling jet growth. However, until computers with perhaps 1000 times present capacity become generally available (and cheap), these approaches are important research tools and for occasional special cases but are not practical for general fire engineering.

## VENTS

The buoyant flow from a fire room through a vent, in the form of an opening in a vertical wall has received considerable attention (103-106). Present knowledge is adequate in the sense that many other parts of fire science are in so much more urgent need of attention. However, a number of vent flow problems remain. For example, if there are two layers (hot and cold) on both sides of a vent the pressure differences can cause six different flow layers, two outflows and two inflows. This may or may not be of importance to the real fire, but for a fire model it can exist and the computer must be told what to do. Some available fire models already handle this case well. Furthermore, at present it is assumed that hot layer outflow goes into the next room hot layer while the cold layer inflow goes into the fire room cold layer. When there are fires and hot layers in both rooms some inflow may go up and some outflow may go down. In fact it may eventually prove to be more accurate to split each flow between the two layers in proportion to its temperature relative to that of the two layers. If this is required, it may be important also to split plume flows between the two layers. At present there is neither theory nor experiment available as a guide on what to do.

Although holes in a floor or ceiling are not very common, they present a special problem when they do occur. So long as there are multiple holes so that the flow through the horizontal opening is one way only (in or out), simple orifice formulas and correlated flow coefficients apply. This is the assumption always made so far, but it is not always correct. Suppose there is only one hole in a fire room and it is in the ceiling. An exact analysis will always permit an equilibrium one way flow all the way down to zero flow rate. Such solutions always exist, although it is clear that for small flows they are unstable. In particular zero flow through a horizontal vent with hot air below and cold air above is unstable and is replaced by two streams in and out. The criterion which indicates whether a one way flow or a two way in, out flow is the more stable is not known. Also incompletely known is the condition which leads to puffing--an instability in the time domain (107, 108). None of these problems are urgent but will all have to be clarified by proper scientific study eventually.

The flow through the fire room vent is at present treated in zone fire models as independent of dynamic effects inside the room. There are several cases when this is not true. As already mentioned the ceiling jet inside the fire room may arrive at the wall above a vent and dynamically turn downward. If the transome wall is short, the ceiling jet, sometimes including flames may considerably alter the vent flow. If there is a fire close to the vents its plume can greatly alter the vent flow (21, 80). No study, either theory or experiment, yet includes both buoyant vent flow and plume.

# EXTINGUISHMENT

Just as adding heat to a solid fuel in air can raise the temperature to ignition, so removing heat from a burning solid or liquid fuel can extinguish the fire. All the complications of phase change and pyrolysis, vaporization and charring, are present during extinguishment. Therefore as soon as we go beyond the fire triangle, remove heat, separate fuel and air (smother) alter the chemistry (fire retardant extinguishing agent), no general theory of extinction exists; there are many extinguishment theories for the many different cases. Almost none of these potential scientific studies have ever been classified.

There are many qualitative, semi-quantitative studies of the extinguishment of crib fires, pool fires, and building fires performed by pouring on water or other agents from a sprinkler, or hose (109-111). This is essential engineering but is short on science.

A pool fire of liquid fuel (say acetone) in a channel can be extinguished by cooling the channel. It is certainly possible to develop a criterion of how much cooling is necessary by combining present knowledge of diffusion flames and heat transfer. It is equally easy to run the experiment but neither has been done.

By adding a very fine spray to a stagnation point flow onto burning charcoal (112) it was found that .2% by mass of water can *increase* the burning rate by up to 30%. The cause appeared to be removal of a minute ash layer. This permits the speculation that a small hose stream on a large fire may be worse than useless. Where the hose stream hits an object directly the cooling and steam vitiation puts the adjacent fire out to be sure. However, the fine spray splash moving with the residual hose and fire pumped air into the next room may make that room burn faster. Furthermore the excess water in the first room may fall to a pile of burning charcoal on the floor and there--by the water gas reaction--produce a $CO-H_2$ mixture which will burn in the next room and enhance the fire. These notions have not yet been proven nor denied.

Only a few measurements of air pumping by hose streams have been reported (113) and nothing has been done on a theoretical prediction.

What happens as a drop of water approaches the surface of a burning object? It may evaporate before it gets there, it may splash, it may soak in and slowly evaporate as the burning is reestablished. This is an obvious study not done.

By putting water from a faucet into a frying pan containing very hot, smoking butter, the water flashes to steam, the butter is ejected as a fine spreay into the air and ignites as an explosive cloud. This actually happened and shows that some would be "extinguishment" can actually cause an ignition or increase an already bad fire.

A greatly enlarged science of extinguishment is sorely needed.

The scientific study of the performance of sprinkler systems has just begun (114-116). A sprinkler usually does not put out the fire directly because a strong buoyancy plume blows the drops away. The fire is controlled by wetting down all nearby fuels thus preventing fire spread. At the same time clouds of hot fire gases plus steam may open sprinklers at a considerable distance from the fire thus causing water damage and straining the sprinkler water supply system. Clearly a real understanding of these processes by the application of scientific methods is needed.

## HEAT TRANSFERS

All three modes, conduction, convection, and radiation are important in fire. Conduction of heat into the interior of a fuel controls the heating rate of its surface. Thus it controls ignition and extinguishment. Such conduction is at present computed by 1-dimensional linear heat diffusion. Only rarely is sufficient known of the temperature dependence of the thermal properties to take this nonlinearity into account. The more complete analysis including the evaporation of moisture and pyrolysis of the virgin fuel has only been attempted a few times (53-56, 117). No set of equations have yet been accepted as general. The widely varying analytical descriptions of diffusive (heat and mass) heating is probably partly responsible for the wide divergence of the reported heat of pyrolysis for wood $-80 < \Delta H < 444$ (118).

Conduction of heat is also the major effect in the weakening and collapse of steel structures in fires.

Convective heat transfer occurs wherever hot gases move over cold surfaces. The most important such heat transfer occurs where a plume hits a ceiling and produces a ceiling jet (119-121). Considerable progress has been made with this heat transfer problem.

A quite different situation applies at the base of a flame where the pyrolysis products diffuse out of the fuel surface into the gases (air, fuel, products of combustion) moving above it. The gases immediately above the surface because of its complex composition absorbs some of the feedback radiation on its way to the fuel. Thus the incident radiation at the fuel surface is reduced but because of the increased temperature of the gases there is an increase of convective heat transfer, thus partial compensation. This phenomena has not been studied in detail and the results are needed for more accurate pyrolysis rate predictions.

The floor of a fire room is heated by radiation from above. Near a vent the vent flow across the floor convectively cools it. However, away from the vent flow or other forced velocity a hot unstable layer of gas is formed which gives rise to vertical convection columns. There is no data on the distribution of the heat transfer coefficient over the floor and no adequate theory by which to calculate the cooling of the floor and heating of the cold layer.

Radiation heat transfer is by all odds the dominant mode of energy transfer in a fire. Because of this, there is intensive research on all aspects of radiation heat transfer. The physics of radiative emission and absorption by molecular and atomic bands and by soot particles (2) is well understood. Thus if an *assigned* volume of space is filled with a known temperature and composition distribution, the radiation from that volume to an external surface can be calculated with good precision. The theoretical formulas are well known but involve extensive calculations, too extensive for most practical purposes. The urgent need in this area of radiation from a known volume is a classification of the available approximate methods by time of calculation, precision of approximation, and spectral and other limitations. Perhaps still better methods may yet be found.

A far more serious problem, however, is to find out how to define the appropriate radiating volume and the temperature and composition within it.

Flame height correlation formulas exist (81, 122) and the base of the flame is set by the burning surface of a solid or liquid fuel. But here our knowledge of flame geometry stops. How good is the commonly used cone approximation (15)?

Is the cone hollow? Or should a cylinder approximation or some fancier approximation (123) be used? Since flames of importance in fires are very unstable and nonsteady, is there not some statistical analysis which could correlate the changes of flame shape with fuel, size, rate of energy feedback, vitiation and perhaps other variables?

A knowledge of internal flame temperature and composition distribution is slowly accumulating (2). We are nowhere near a predictive capability for these quantities in a form adequate for accurate radiative transfer prediction. It is for this reason that many studies are directed toward the measurement of overall flame radiation (123) with the ultimate hope that some correlating principle will be found which will skip all the internal physical and chemical complications.

And while moving in the direction of more gross overall approximations, perhaps we can go a step further to also skip all the complex view factor calculations by some such scheme as a radiation rate pool.

The question of how much spectral knowledge is required for fire purposes has never been settled. Yet much needed facts are yet to be learned. Figures 1 and 2 show the need for careful consideration of the radiation in fire testing. For both figures, opposite sides of the same block of maple were in succession covered with the same thin (2 mm) aluminum plate containing a 2 inch hole. In Fig. 1 the assembly was radiatively heated by a glowbar source at about 1000 K. In Fig. 2 a bank of tungsten wires in quartz tubes at about 2400 K were used. The differences in spectral response of aluminum and wood to the spectrally different heat sources is startling. We need to know more. But what and how much? Certainly some careful spectral measurements in real fires is a necessary start.

FIGURE 1. Maple block protected by an aluminum plate radiated by a glowbar unit.

FIGURE 2. Maple block protected by an aluminum plate radiated by tungsten filaments in quartz tubes.

## STRUCTURAL PROBLEMS

The heating and burning of a building structure can lead to its collapse either by decreased ultimate strength or increased plastic deformation (of steel structural elements) (beams, columns, reinforcing bar) or decrease of non-char cross section (wood or plastic). The science and engineering of structures has already reached an advanced stage of development. The new feature added by fire is the heating and burning of the structural elements. As soon as the fire itself can be predicted, the knowledge of heat transfer as discussed above can supply the external energy distribution to a steel member while the well developed theory of heat conduction can predict its internal temperature (124). Finally the internal temperature permits prediction of the member strength and yield (125-127). Steel failure and large deflection yield are active research areas in solid mechanics and progress in this field should be followed by fire scientists.

The failure of wooden structural members is closely tied in with the pyrolysis process. It is well known that a large wooden beam structure can withstand more hours of a fire than an inadequately insulated steel member structure. The reason is clear. Steel is very homogeneous so that steel structures are designed with a small safety factor while wood with its knots and other internal irregularities must have a much larger safety factor. Furthermore wood has a low thermal conductivity and is thermally protected by the adhering char formed by pyrolysis. Thus building collapse prediction is not far away for steel and reinforced concrete structures but for wood much additional fire science is required before fire engineering can be done properly.

Collapse is of little practical importance for fire development since a building is usually without value before collapse occurs (in fact it may have a negative value since the burned out remains have to be torn down and carried away). However, good fire collapse predictions are important for the safety of the fireman engaged in extinguishment.

There is one structural problem of importance to fire growth and this is the breaking of window glass. A fire in an enclosure should never be opened to a new supply of air until a fireman with a hose stream is ready to apply water through the new vent. However, glass does break as the window is heated with or without the application of cold water. Why? The only study of this problem was carried out by two Harvard seniors (128) in a small senior proejct. They subjected 6" × 7" window glass plates to a radiant heat source. As is well known, glass is relatively opaque to infrared radiation and thus becomes hot when radiated. Such a glass plate, fully exposed to uniform radiation (no shadow on the plate) gets hot all over but never breaks because the thermal expansion is uniform. (Actually the unshadowed glass sometimes did break because our radiation was not perfectly uniform). However window glass is as installed always held at its edges by putty or an opaque solid frame. A glass plate merely shadowed around the edge (no actual material contact to the glass) always breaks. Figure 3. It is easy to see why qualitatively. Glass is a very poor conductor and therefore if the edge is not heated but the center is, the edge is put into tension by thermal expansion. Thus a crack starts at some edge irregularity or small precrack and grows through the material in tension. It is interesting that the crack when it reaches the edge of the shaded area always bifurcates and travels through the glass along two internal stress paths. This glass breaking problem is one that has had no scientific study beyond that reported here.

## TOXICITY

This is a practical problem of major concern to fire protection engineers. However, it is often assumed to be a problem for the fire chemist and physiologist and of no concern to the fire dynamicist. This is way off the mark. It is true that the production of toxic specie is a chemist's problem and the effect on people of the toxic gases and their synergistic effects is a physiologist's problem. However, if fire engineering is ever to reach accurate design for fire safety, fire science must develop the knowledge required to predict the toxic properties in building escape routes. The present toxicity tests of materials in which a rat's (or other) nose is put into the rising plume of pyrolysis products from a heated sample of some solid gives important data.

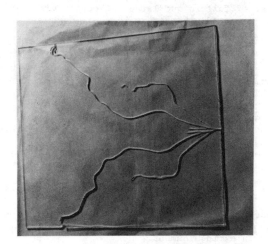

FIGURE 3. 4" square glass plate radiated by a Globar Unit with 1/4" of edge protected by a shadow.

44

However, the dynamicist, together with the chemist and physiologist must somehow follow the toxic property of the gases as they move about through the building. How much of the CO, $CH_3 = CH_2 - CHO$, and other toxic materials initially produced are burned in subsequent flames or are lost by absorption? How much of the HCl (if any) reacts with walls or dissolves in condensed moisture? How much of the HCN is absorbed on surfaces or on soot particles? How much of the soot originally formed is burned in subsequent flames or agglomerates into large particles and settles out on furnishings to decrease their value but are thereby removed from their otherwise toxic mission? Very few of the above questions have had any scientific study at all to date. Some fire studies have shown that HCl even when produced in abundance is rapidly removed to low levels (129). The mechanism of removal has not been identified. The agglomeration of soot particles has been studied in a static situation (130). However, nothing has been done with its transport throughout a building nor with its settling rate which has such a great effect on building furnishings.

Large numbers of studies of special cases have been carried out (129-138) but we do not yet have the capability of predicting the toxic hazard in an escape route during a fire.

## FIRE BRANDS

The spread of a fire by the movement of burning brands is usually thought of as a forest fire problem (139-141); which it is. However, it is not a problem without significance in structure fires. As a fire grows in a room, it moves quickly to the ceiling. It also heats ether flammable material throughout the room to their pilot ignition temperature. But where is the pilot? The original fire can so serve if it is close by. However, a more distant object may be pilot ignited by a burning brand from the original fire or from the ceiling area if there are flammable materials there. No studies of this problem currently exist. In fact there are no studies to answer the question of the available brands in a real fire. Since the formation of brands is an almost impossible problem deterministically, we need a suitable statistical study. Do new fuels generally ignite at their pilot ignition temperature or is it necessary for them to be heated to the higher spontaneous ignition temperature, or somewhere in between?

## REFERENCES

1.    Levine, R.S: Workshop on Flame Radiation and Soot, *Fire Technology*, 21, 14-58, 1985.

2.    DeRis, J.: "Fire Radiation--A Review," 17th Symposium (Int.) on Combustion, 1003-1016, 1978.

3.    Williams, F.A.: "A Review of Flame Extinction," *Fire Safety Journal*, 3, 3, 163-176, 1981.

4.    Emmons, H.W.: "Scientific Progress on Fire," *Ann. Rev. Fluid Mechanics*, 12, 223-236, 1980.

5.    Westbrook, C.K., and Dryer, F.L.: "Chemical Kinetics and Modeling of Combustion Processes," 18th Symp. (Int.) on Combustion, 749-767, 1981.

6.    Quintiere, J.: "The Spread of Fire from a Compartment--A Review," *ASTM Tech. Pub.*, 685, 139-168, 1980.

7. Jones, W.W.: "A Review of Compartment Fire Models," NBSIR 83-2684, 1-40, 1983.

8. Hall, A.R.: "Pool Burning--A Review," Rocket Propulsion Establishment TR#72/11, 0-82, 1972.

9. Birky, M.M.: "Hazard Characteristics of Combustion Products in Fires: The State of the Art Review," NBSIR 77-1234, p. 1-46, 1977.

10. *Flame Spread Volume*--Comb. and Flame, 32, 1-4, 1983.

11. Cox, G., and Kumar, S.: "Computer Modeling of Fire," BRI (England) IP2/83, 1-5, 1983.

12. Tanaka, T.: "A Model of Multiroom Fire Spread," *Fire Sci. and Tech.*, 3, 2, 105-122, 1983.

13. Visich, M.: "Consideration of Fire Development in an Enclosed Space," *J. Environ. Sys.*, 3, 3, 215-231, 1973.

14. Sekine, T.: "Room Temperature in Fire in a Fire-Resistive Room," BRI #29, 13-24, 1959.

15. Mitler, H.E., and Emmons, H.W.: "Documentation for Computer Fire Code V," The 5th Harvard Computer Fire Code, NBS-GCR-81-344, 1-183, 1981.

16. Gahm, J.: "Computer Fire Code VI," Home Fire Project TR#58, 1-110, 1983.

17. Chaix, J.M., and Galant, S.: "Modèle Numérique Instantaire de la Propagation d'un Feû en Compartment Ventile," Société Bertin et Cie, 1-87, 1979.

18. Reeves, J.B., and MacArthur, C.D.: "Dayton Aircraft Cabin Fire Model," vol. 1, Basic Mathematical Model, Report #FAA-RD-76-120, 1, 1-145, 1976.

19. Smith, E.E., and Clark, M.J.: "Model of the Developing Fire in a Compartment," ASHRAE Trans., 81, Part 1, 568-58, Part 2, Development of model, 11-48, 1975.

20. Pape, R., Waterman, T.E., and Eichler, T.V.: "Development of a Fire in a Room from Ignition to Full Room Involvement," RFIRES, IITRI Proj. J6485, 1-181, 1980.

21. Liu, V.K., and Yang, K.T.: "Undsafe II--A Computer Code for Buoyant Turbulent Flow in an Enclosure with Thermal Radiation," NBS-Grant G-7-9002 Report TR-7900, 2-78-3, 1-168, 1978.

22. Burgan, R.E.: "Fire Danger--Fire Behavior Computations with the Texas Inst. TI59 Calculator: Users Manual," USDA Forest Service, General Tech. Rep. INT-61, 0-25, 1979.

23. Kline, S.J., Cantwell, B.J., and Lilley, G.M.: "Complex Turbulent Flows," Vol. I, II, III, AFOSR-HTTM-Stanford Conference, 0-1550, 1980-81.

24. Patenkar, S.V., and Spalding, D.B.: "A Calculation Procedure for Heat, Mass, and Momentum Transfer in 3D Parabolic Flows," *Int. J. Heat and Mass Transfer*, 15, 1787-1802, 1972.

25. Launder, B.E., and Spalding, D.B.: "The Numerical Computation of Turbulent Flows, Computer Methods," in *Applied Mechanics*, 3, 269-289, 1974.

26. Rehm, R.G., and Baum, H.R.: "The Equations of Motion for Thermally Driven Buoyant Flows," *J. of Research NBS*, 38, 3, 297-307, 1978.

27. Banksten, C.P., Casanova, R.A., Powell, E.A., and Zinn, B.T.: "The Initial Data on the Physical Properties of Smoke Produced by Burning Materials with Different Conditions," *J. Fire and Flammability*, 7, 165-181, 1976.

28. Lipska, A.E., and Wodley, F.A.: "Isothermal Pyrolysis of Cellulose Kinetics and Gas Chromatographic/Mass Spectrometric Analysis of the Degradation Products," OCD Work Unit #2531C, NRDL-TR-68-89, 1-37, 1968.

29. Gross, D., and Robertson, A.F.: "Self-Ignition Temperatures of Materials from Kinetic Reaction Data," *J. Res. NBS*, 61, 5, 413-417, 1958.

30. Beyler, C.L.: "Ignition and Burning of a Layer of Incomplete Combustion Products," *Comb. Sci. and Tech.*, 39, 287-303, 1984.

31. Dubataki, P., Tingle, W.J., Ryszytiwskyj, W.P., and Tinelier, W.C.: "Self Ignition of Pyrolyzate-Air Mixtures," *Fire Research*, 1, 243-254, 1977.

32. Ohlemiller, T.J.: "Cellulosic Insulation Material, III. Effects of Heat Flow Geometry on Smoulder Initiation," *Comb. Sci. and Tech.*, 26, 3-4, 89-105, 1981.

33. Kashiwagi, T.: "Radiative Ignition Mechanisms of Solid Fuelds," *Fire Safety Journal*, 3, 3, 185-200, 1981.

34. Kashiwagi, To., and Kashiwagi, Ti.: "A Study of the Radiative Ignition Mechanism of a Liquid Fuel Using High Speed Holographic Interferometer," 19th Symp. (Int.) on Combustion, 1511-1521, 1982.

35. Kinbara, T., and Akita, K.: "On the Self-Ignition of Wood Materials," NAS-NRC Pub. #786, 256-270, 1961.

36. Thomas, P.H.: "On the Surface Ignition of Self-Heating Materials," *Comb. Sci. and Tech.*, 36, 5-6, 263-284, 1984.

37. Vega, J.M., and Liñan, A.: "Large Activation Energy Analysis of the Ignition of Self-Heating Porous Bodies," *Comb. and Flame*, 57, 3, 247-254, 1984.

38. Kashiwagi, T.: "Effects of Sample Orientation on Radiative Ignition," *Comb. and Flame*, 44, 1-3, 223-245, 1982.

39. Alvares, N., Blackshear, P., and Kanury, M.: "The Influence of Free Convection on the Ignition of Vertical Cellulosic Panels by Thermal Radiation," *Comb. Sci. and Tech.*, 1, 407-413, 1970.

40. Quintiere, J.: "A Simplified Theory for Generalizing Results from a Radiant Panel Rate of Flame Spread Apparatus," *Fire and Materials*, 5, 2, 52-60, 1981.

41. Frey, A.E., and Tien, J.S.: "A Theory of Flame Spread over a Solid Fuel Including Finite Rate Chemical Kinetics," *Comb. and Flame*, 36, 263-289, 1979.

42. Borgeson, R.A.: "Flame Spread and Spread Limits," NBS-GCR-82-396, 1-52, 1982.

43. Kashwagi, T., and Newman, D.L.: "Flame Spread Over an Inclined Thin Fuel Surface," *Comb. and Flame*, <u>26</u>, 163-177, 1976.

44. Schraufnagle, R.A., and Barlow, J.W.: "Flame Spread Rate Over Polyurethane Foams," Dept. Chem. Eng., U. Texas at Austin, 1977.

45. Sibulkin, M., Kulkarni, A.K., and Annamalai, K.: Burning on a Vertical Surface with Finite Chemical Reaction Rate," *Comb. and Flame*, <u>44</u>, 1-3, 187-199, 1982.

46. Fernandez-Pellow, A.C., Ray, R.S., and Glassman, I.: "Flame Spread in an Opposed Flow: The Effect of Ambient Oxygen Concentration," 18th Symp. (int.) on Comb., 579-589, 1981.

47. Kulkarni, A.K., and Sibulkin, M.: "Burning Rate Measurements on Vertical Fuel Surfaces," *Comb. and Flame*, <u>44</u>, 1-3, 185-186, 1982.

48. Fernandez-Pello, A. C.: "Flame Spread Modeling," *Comb. Sci. and Tech.*, <u>39</u>, 1-6, 119-134, 1984.

49. Fernandez-Pello, A. C., and Hirano, T.: "Controlling Mechanisms of Flame Spread," *Fire Sci. and Tech.*, <u>2</u>, 1, 17-54, 1982.

50. Carrier, A.F., Fendell, F.E., and Feldman, P.S.: "Wind Aided Flame Spread Along a Horizontal Fuel Slab," *Comb. Sci. and Tech.*, <u>23</u>, 41-78, 1980.

51. Atreya, A.: "Fire Growth on Horizontal Surfaces of Wood," *Comb. Sci. and Tech.*, <u>39</u>, 1-6, 40; 1-4, 163-194, 1984.

52. Carrier, A.F., Fendell, F.E., and Fink, S.: "Towards Wind Aided Flame Spread Along a Horizontal Charring Slab--The Steady Flow Problem," *Comb. Sci. and Tech.*, <u>99</u>, 999, 1983.

53. Atreya, A.: "Pyrolysis, Ignition, and Fire Spread on Horizontal Surfaces of Wood," NBS-GCR-83-449, 1-433, 1984.

54. Kung, H.C.: "A Mathematical Model of Wood Pyrolysis," *Comb. and Flame*, <u>18</u>, 185-195, 1972.

55. Lee, C.K., Chaiken, R.F., and Singer, J.M.: "Charring Pyrolysis of Wood in Fires by Laser Simulation," 16th Symp. (Int.) on Combustion, 1459-1470, 1976.

56. Delichatsios, M.A., and deRis, J.: "An Analytical Model for the Pyrolysis of Charring Materials," FMRC Report J.I0K0J1.Bu, 1-20, 1983.

57. Tan, S.C.: "A Study of Transient Horizontal Fire Spread over Cellular Plastics," Ph.D. Thesis, Harvard Univ., 1-209, 1983.

58. Markstein, G.H.: "Scanning-Radiometer Measurement of the Radiance Distribution in PMMA Pool Fires," 18th Symp. (Int.) on Comb., 537-547, 1981.

59. Souil, J.M., Joulain, P., and Gengambie, E.: "Experimental and Theoretical Study of Thermal Radiation from Turbulent Diffusion Flames to Vertical Target Surfaces," *Comb. Sci. and Tech.*, <u>41</u>, 1-2, 69-81.

60. Emmons, H.W.: "The Film Combustion of Liquid Fuel," *Z. Math. und Mech.*, <u>36</u>, 1/2, 60-71, 1956.

61. Pagni, P.J., and Shih, T.M.: "Excess Pyrolyzate," 16th Symp. (Int.) on Comb., 1329-1343, 1976.

62. Kim, J.S., deRis, J., and Kroesser, F.W.: "Laminar Free Convective Burning of Fuel Surfaces," 13th Symp. (Int.) on Comb., 949-961, 1971.

63. Adomeit, G., Hoeks, W., and Henriksen, K., "Combustion of a Carbon Surface in a Stagnation Point Flow Field," *Comb. and Flame*, 59, 3, 273-288, 1985.

64. Backovsky, J.: "The Theory of Boundary Layer Burning with Radiation," Ph.D. Thesis, Harvard Univ., 1-216, 1979.

65. Smith, E.E.: "Predicting Fire Performance Using Release Rate Information," Proc. Int. Conf. on Fire Safety, 229, 1979.

66. Bluhme, D., and Getka, R.: "Rate of Heat Release Test Calibration, Sensitivity and Time Constants of 1SO RHR Apparatus," Nordtest-Proj. 115-77, 1982.

67. Thomas, P.H.: "On the Rate of Burning of Cribs," BRI (England), Fire Res. Note 965, 1-17, 1973.

68. Orloff, L., and deRis, J.: "Modeling of Ceiling Fires," 13th Symp. (Int.) on Comb., 979-992, 1971.

69. Modak, A.T., "The Burning of Large Pool Fires," *Fire Safety J.*, 3, 3, 177-184, 1981.

70. Sibulkin, M., and Tewari, S.S., "Measurement of Flaming Combustion of Pine and Fire Retarded Cellulose," *Comb. and Flame*, 59, 1, 31-42, 1985.

71. Kimbara, T., Endo, H., and Sega, S.: "Combustion Propagation Through Solid Materials, I. Downward Propagation of Smouldering along a Thin Sheet of Paper, 11th Symp. (Int.) on Comb., 525-531, 1966.

72. Moussa, N.A., Toong, T.Y., and Garris, C.A.: "Mechanism of Smouldering of Cellulosic Materials," 16th Symp. (Int.) on Comb., 1976.

73. Bilger, R.W.: "Turbulent Jet Diffusion Flames," *Prog. in Energy and Comb. Sci.*, 1, 87-109, 1976.

74. Jeng, S.M., Chen, L.D., and Faeth, G.M.: "An Investigation of Axisymmetric Buoyant Turbulent Diffusion Flames," NBS-GCR-82-367, 1-83, 1982.

75. Morton, B.R., Taylor, G.I., and Turner, J.S.: "Turbulent Gravitational Convection from Maintained and Instantaneous Sources," *Proc. Roy. Soc. Ser.* A234, 1196, 1956.

76. Lee, S.L., and Emmons, H.W.: "A Study of Natural Convection above a Line Fire," *J. Fluid Mech.*, 11, 353-368, 1961.

77. Cetegen, B.M., Zukoski, E.E., and Kubota, T.: "Entrainment in the Near and Far Field of Fire Plumes," *Comb. and Flame*, 39, 1-6, 40; 1-4, 305-332, 1984.

78. Beyler, C.L.: "Development and Burning of a Layer of Products of Incomplete Combustion Generated by a Buoyant Diffusion Flame," Ph.D. Thesis, Harvard Univ., 1-169, 1983 (see p. 57).

**49**

79. Liburdy, J.A., Ahmad, T., and Faeth, G.M.: "An Investigation of the Overfire Region of Wall Fires," Eastern Section Conf., 1-15, 1976.

80. Emmons, H.W.: "The Ingestion of Flames and Fire Gases into a Hole in an Aircraft Cabin for Arbitrary Tilt Angles and Wind Speeds," Home Fire Project Report 52, 1-33, 1982.

81. Stewart, F.R.: "Prediction of the Height of Turbulent Diffusion Buoyant Flames," *Comb. Sci. and Tech.*, $\underline{2}$, 203-212, 1970.

82. Tamanini, F.: "An Integral Model of Turbulent Flame Plumes," 18th Symp. (Int.) on Combustion, 1081-1090, 1981.

83. Evans, D.D.: "Calculating Fire Plume Characteristics in a Two Layer Environment," *Fire Technology*, $\underline{20}$, 3, 39-63, 1984.

84. Heskestadt, G.: "Virtual Origins of Fire Plumes," *Fire Safety J.*, $\underline{5}$, 2, 109-114, 1983.

85. Quinture, J.R., Rinkinen, W.J., and Jones, W.W.: "The Effect of Room Openings on Fire Plume Entrainment," *Comb. Sci. and Tech.*, $\underline{26}$, 3-4, 193-201, 1981.

86. Zukoski, E.E., Kubota, T., and Cetegen, B.: "Entrainment in Fire Plumes," *Fire Safety J.*, $\underline{3}$, 107-121, 1980.

87. Rao, V.K.: "The Buoyant Plume above a Heat Source," *Atmo. Environ.*, $\underline{4}$, 557-575, 1970.

88. Varma, R.K., Murgai, M.P., and Ghildyal, C.D.: "Radiative Transfer Effects in Natural Convection above Fires--General Case," *Proc. Roy. Soc. London Ser.* A314, 195-215, 1970.

89. Tsang, G., and Wood, I.R.: "Motion of Two-Dimensional Starting Plume," *J. Eng. Mech. Div. Proc. ASCE*, 1547-1561, 1968.

90. Baum, H.R., Rehm, R.G., Barnett, P.D., and Carley, D.M.: "Finite Difference Calculations of Buoyant Convection in an Enclosure. I. The Basic Equations," *SIAM J. Sci. Stat. Comput.*, $\underline{4}$, 1, 117-135, 1983.

91. Rehm, R.G., Baum, H.R., and Barnett, P.D.: "Buoyant Convection Computed in a Vorticity, Stream Function Formulation," *J. of Research NBS* $\underline{87}$, 2, 165-185, 1982.

92. Torrance, K.E.: "Natural Convection in Thermally Stratified Enclosures with Localized Heating from Below," *J. Fluid Mech.*, $\underline{95}$, $\underline{3}$, 477-495, 1979.

93. Jaluria, Y.: "Buoyancy Induced Wall Flow Due to Fire in a Room," NBSIR 84-2841, 1-93, 1984.

94. Cooper, L.: "On the Significance of a Wall Effect in Enclosures with Growing Fires," *Comb. Sci. and Tech.*, $\underline{40}$, 19-36, 1984.

95. Coward, H.F., Carpenter, C.W., and Payman, W.: "The Dilution Limits of Inflammation of Gaseous Mixtures," *J. Chem. Soc.*, $\underline{115}$, 27, 1919.

96. Alpert, R.: "Turbulent Ceiling Jet Induced by Large Scale Fires," *Comb. Sci. and Tech.*, $\underline{11}$, 197-213, 1975.

97. Ellison, T.J., and Turner, J.S.: "Turbulent Entrainment in Stratified Flows," *J. Fluid Mech.*, 6, 423-448, 1959.

98. Atallah, S.: "Fire in a Model Corridor with a Simulated Combustible Ceiling, Parts 1, 2," BRI Note 620, 621, 1-31, 1-21, 1966.

99. Hwang, C.C., Chaiken, R.F., Singer J.K., and Chi, D.N.H.: "Reverse Stratified Flow in Duct Fires--A Two Dimensional Approach," 16th Symp. (Int.) on Combustion, 1385-1395, 1976.

100. Baum, H.R., Rehm, R.G., and Malholland, G.W.: "Prediction of Heat and Smoke Movement in Enclosure Fires," *Fire Safety J.*, 6, 3, 193-201, 1983.

101. Jones, W.W., and Quinture, J.: "Prediction of Corridor Smoke Filling by Zone Models," *Comb. Sci. and Tech.*, 35, 5-6, 239-253, 1984.

102. Emmons, H.W.: "A Perfect Fluid Non-Steady Ceiling Jet," to be published.

103. Prahl, J., and Emmons, H.W.: "Fire Induced Flows Thru an Opening," *Comb. and Flame*, 25, 3, 369, 1975.

104. Steckler, K.D., Baum, H.R., and Quintiere, J.: "Fire Induced Flows Thru Room Openings--Flow Coefficients," 20th Symp. (Int.) on Comb., 1984, to be published.

105. Quintiere, J., Den Braven, K.: "Some Theoretical Aspects of Fire Induced Flows through Doorways in a Room-Corridor Scale Model," NBSIR78-1512, 1-33, 1978.

106. Nakaya, I., Tanaka, T., and Yashida, M.: "A Measurement of Doorway Flow Induced by Propane Fire," 7th Meeting UJNR, Washington, 1983.

107. Satoh, K.: "Experiment and Finite Difference Study of Dynamic Fire Behavior in a Cubic Enclosure with a Doorway," BRI 55, 17-28, 1983.

108. Hasemi, Y.: "Thermal Instability in Transient Compartment Fire," *Fire Sci. and Tech.*, 2, 1, 1-16, 1982.

109. Rasbash, D.J.: "The Extinction of Fires by Water Sprays," *Fire Res. Abs. & Res.*, 4, 28, 1982.

110. Kung, H.C.: "Cooling of Room Fires by Sprinkler Spray," *J. Heat Transfer*, 99, 3, 355-359, 1977.

111. Takahashi, S.: "Extinction Mechanism and Efficiency by Sprays of Water and Chemically Improved Water," *Fire Res. Inst. (Japan) Ser.* 56, 7-11, 1983.

112. Bhagat, P.M.: "Wood Charcoal Combustion and the Effects of Water Application," *Comb. and Flame*, 37, 3, 275-291, 1980.

113. Heskestad, G., Kung, A.C., and Tottenkopf, N.F.: "Air Entrainment into Water Sprays and Spray Curtains," ASME 76-WA-FE40, 0-12, 1976.

114. Alpert, R.L.: "Calculated Interaction of Sprays with Large-Scale Buoyant Flows," *J. Heat Trans.*, 106, 2, 310-317, 1984.

115. Evans, D.D.: "Thermal Actuation of Extinguishing Systems," *Comb. Sci. and Tech.*, 40, 79-92, 1984.

116. Yuen, M.C., and Chen, L.W.: "On Drag of Evaporating Droplets," *Comb. Sci. and Tech.*, 14, 147-154, 1976.

117. Min, K., and Emmons, H.W.: "The Drying of Porous Media," Proc. 1972 Heat and Trans. and Fluid Mech. Inst., p. 1-18, 1972

118. Shivadev, U.K., and Emmons, H.W.: "Thermal Degradation and Spntaneous Ignition of Paper Sheets in Air by Irradiation," *Comb. and Flame*, 22, 223-236, 1974.

119. You, H.Z., and Faeth, G.M.: "Ceiling Heat Transfer during Fire Plume Impingement," *Fire and Materials*, 3, 3, 140-147, 1979.

120. Zukoski, E.E., and Kubata, T.: "An Experimental Investigation of the Heat Transfer from a Buoyant Plume to a Horizontal Ceiling," NBS-GCR-77-98, 1-73, 1975.

121. Cooper, L.Y.: "Heat Transfer from a Buoyant Plume to an Unconfined Ceiling," *J. Heat Trans.* 104, 3, 446-451, 1982.

122. Thomas, P.H.: "The Size of Flames from Natural Fires," 9th Symp. (Int.) on Comb., 844-859, 1963.

123. Orloff, L.: "Simplified Modeling of Pool Fires," 18th Symp. (Int.) on Comb., 549-561, 1981.

124. Lie, T.T.: "Temperature Distributions in Fire Exposed Building Columns," *J. Heat Trans.*, 99, 1, 113-119, 1977.

125. Becker, J.M., and Bresler, M.: "Reinforced Concrete Frames in Fire Environments," *J. Structural Div. ASCE*, 211-224, 1977.

126. Kawagoe, K.: "Damage of Structures in Full Size Fires," BRI (Japan) Ser. 29, 27-41, 1959.

127. Saito, H.: "Explosive Spalling of Prestressed Concrete in Fire," BRI (Japan) Ser. 22, 1-18, 1965.

128. Barth, P.K., and Sung, H.T.: "Glass Fracture under Intense Heating," Senior Proj. ES96r, Harvard Univ., 1977.

129. Stark, G.W., and Field, P.: "Toxic Gases and Smoke from Polyvinylchloride in Fires in the FRS Full Scale Test Rig," BRI (England), Fire Research Note 1030, 0-48, 1974.

130. Mulholland, G.W., and Baum, H.R.: "Effect of Initial Size Distribution on Aerosol Coagulation," *Phys. Rev. Letts.*, 45, 761, 1980.

131. Pagni, P.J., and Bard, S.: "Particulate Volume Fractions in Diffusion Flames," 17th Symp. (Int.) on Comb., 1017-1028, 1979.

132. Marikawa, T.: "Toxic Hazards of Acrolein and Carbon Monoxide Evolved under Various Combustion Conditions," BRI (Japan) Ser. 57, 21-27, 1984.

133. Tsuchiya, Y., and Boulanger, J.G.: "Carbonyl Sulfide in Fire Gases," *Fire and Materials*, 3, 3, 154-155, 1979.

134. Tsuchiya, Y.: "Significance of Hydrogen Cyanide Generation in Fire Gas Toxicity," *J. Comb. Toxicology*, 4, 271-282, 1977.

135.  Marikawa, T.:  "Effect of Water Vapor on Carbon Monoxide Evolution in Fire Conditions," BRI (Japan) Ser. 53, 1-8, 1983.

136.  Woolley, W.D., Fardell, P.J., and Buckland, I.G.:  "The Thermal Decomposition Products of Rigid Polyurethane Foams under Laboratory Conditions," BRI (England), Note 1039, 1974.

137.  Shafizadeh, F.:  "Thermal Behavior of Carbohydrates," *J. Polymer Sci.*, 36, 21-51, 1971.

138.  Tewarson, A., Lee, J.L., and Pion, R.F.:  "The Influence of Oxygen Concentration on Fuel Parameters for Fire Modeling," 18th Symp. (Int.) on Comb., 563-570, 1981.

139.  Tarifa, C.S.:  "On the Flight Paths and Lifetimes of Burning Particles of Wood," 10th Symp. (Int.) on Comb., 1021-1037, 1965.

140.  Tarifa, C.S.:  "Transport and Combustion of Fire Brands," Final Report Proj. FG-SP-114, Forest Service, U.S. Dept. of Agri., 1-90, 1967.

141.  Muraszew, A.:  "Firebrand Phenomena," Aerospace Corp. Report #ATR-77, (8165-01)-1.

# Flame Spread over Thin Layers of Crude Oil Sludge

**TAKUJI SUZUKI and NORIHITO KUDO**
Department of Mechanical Engineering
Ibaraki University
Ibaraki, Japan

**JUN'ICHI SATO and HIDEO OHTANI**
Research Institute
Ishikawajima-Harima Heavy Industries, Co., Ltd.
Tokyo, Japan

**TOSHISUKE HIRANO**
Department of Reaction Chemistry
University of Tokyo
Tokyo, Japan

## ABSTRACT

Flame spread over thin layers of crude oil sludge has been studied. Flame spread experiments were performed by using a tray of 12 cm wide, 60 cm long and variable depth, installed in a temperature control bath. The aspects of spreading flames, the movements of leading and trailing flame edges, and the temperature distributions near leading flame edges were examined for various sludge layer thicknesses. The mode of flame spread was found to change at a moment several minutes after the start of flame spread. At the first stage, the flame spread rate was observed to be almost constant, while at the second stage, it changed periodically. The former was about twice as large as the latter. Two different limiting thicknesses were revealed. One was the limiting thickness for the initiation of flame spread and the other was that for the continuance of flame spread. Based on these results, the mechanisms of flame spread were discussed. All the flame spread phenomena were shown to be consistently interpretable by assuming similar mechanisms of flame spread as indicated in our previous study.

## INTRODUCTION

Serious damage has been occasionally caused by a crude oil sludge fire in a cargo bay of a crude oil tanker during its repair. Since the growth rate of such a fire depends on the process of flame spread over it, knowledge of flame spread must be indispensable for developing a reliable fire protection method and/or establishing reasonable regulations or standards concerning the procedure for repair of the crude oil tanker.

Crude oil sludge is mainly the deposit of crude oil accumulated during voyages and remaining in the cargo bay after washing, so that it is a multicomponent combustible of non-definite composition. Further, a pasty crude oil sludge increases its fluidity as approaching the leading edge of a spreading flame[1,2]. Although the mechanisms of flame spread over single component liquid or solid combustibles have been examined in a number of previous studies[3-9], very little knowledge is available on flame spread over multicomponent combustibles like crude oil sludge.

Thus, in our previous studies, basic characteristics of flame spread over crude oil sludge were explored and analyzed[1,2]. The effects of sludge composition and/or temperature on flame spread phenomena were examined and the flame spread mechanisms were discussed[1]. Further, an analysis of flame spread over multicomponent combustibles was performed and the results were compared to those of the experiments on flame spread over crude oil sludge[2]. Knowledge obtained throughout these studies would be useful to understand the flame spread phenomena and to develop a test method for classifying the hazard of crude oil sludge.

Obviously, no fire occurs if crude oil sludge is completely removed from the floors, walls, and ribs in the cargo bay to be repaired. However, the complete removal of crude oil sludge is practically impossible and the grade of the procedure to remove crude oil sludge directly affects the cost. Therefore, suitable regulations or standards for it must be necessary.

To establish reasonable regulations or standards for the crude oil sludge removal, the information concerning the effect of sludge layer thickness on the flame spread phenomena must be needed. Therefore, in the present study, the flame spread phenomena over various thicknesses of crude oil sludge layers have been examined in detail.

EXPERIMENTAL APPARATUS AND PROCEDURE

A tray installed in a temperature control bath was used for the flame spread experiments(Fig. 1). The width and length of the tray were 12 cm and 60 cm, respectively, which were determined on the basis of the results of preliminary experiments. The width of 12 cm would be sufficient for neglecting the quenching effect of side walls of the tray on flame spread phenomena, and the length of 60 cm would be enough for observing typical variations of flame spread phenomena with time. The base plate of the tray was a brass plate and made to be movable so that the depth could be changed. The brim of the tray was set to be flush with a flat plate covering the temperature control bath.

The properties of crude oil sludge depend on the crude oil quality, cargo bay wall conditions, and washing procedure. It makes the properties to be non-definite. As discussed in our previous study[1], however, the flammability of sludge, of which hazard assessment is needed, must be in a certain range. The sludge used in the present study was similar to one of those examined in the previous study[1]. Its properties were examined by thermogravimetry(TG), differential thermal analysis(DTA), and derivative thermogravimetry(DTG) in the inert gas (Ar) atmosphere. The results are shown in Fig. 2. These properties are close to those of the sludge made by adding 10 % hexane to the standard sludge, which was used in our previous study[1]. The horizontally placed tray was filled with the sludge and its surface was finished to be flat by using a trowel. Then, a cotton string of 3 mm in diameter soaked with ethanol was laid on the sludge surface near the tray end brim to be parallel to it. The sludge was ignited by burning this cotton string. After igniting the sludge, the

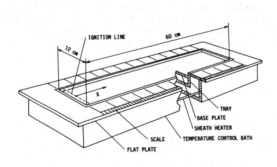

FIGURE 1. Tray and temperature control bath.

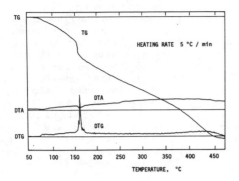

FIGURE 2. Properties of the sludge used in the present experiments, examined by thermogravimetry(TG), differential thermal analysis(DTA), and derivative thermogravimetry(DTG) in the inert gas(Ar) atmosphere.

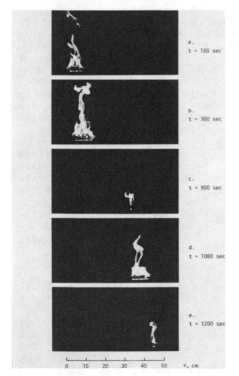

FIGURE 3. Processes of flame spread; $T_i$ = 20 °C, $\delta$ = 0.3 cm.

string was taken off to eliminate its effect on flame spread phenomena.

The aspects of spreading flames were examined by using a motor-driven 35 mm camera, and the movements of the leading and trailing flame edges, which are described in the following section, were measured by using a timer and the scale marked on the flat plate(Fig. 1).

The temperature distribution in the sludge during flame spread was measured by traversing a C-A thermocouple (wire diam.: 0.1 mm) connected to a high speed digital recorder.

RESULTS

Aspects of Flame Spread

A series of photographs representing flame spread over a thin layer of the sludge is presented in Fig. 3. In this case, the initial sludge temperature $T_i$ is 20°C and its thickness $\delta$ is 0.3 cm. t and x in the figure represent the time after ignition and the distance from the ignition line, respectively. For a few minutes after the start of flame spread, the flame size is observed to increase(a and b in Fig. 3). Then, the flame edge at the end brim departs from it and trails the leading flame edge. The approach of the trailing flame edge to the leading flame edge causes the reduction of flame size(c). After the flame size becomes to be several cm in the longitudinal direction, it starts to increase again(d). Thus, a flame is observed to spread over the sludge repeating the enlargement and reduction of the flame size.

Just after a flame starts to spread, the melted sludge surface beneath the flame is observed to be

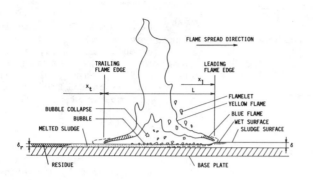

FIGURE 4. Schematic illustration of a spreading flame.

slightly higher than the unburned sludge surface. During flame spread, the sludge beneath the flame is consumed, so that the melted surface becomes lower than the unburned sludge surface, and after several minutes, the trailing flame edge starts approaching to the leading flame edge.

A typical aspect of a flame spreading over a thin layer of the sludge is depicted in Fig. 4. A blue leading flame edge followed by a luminous yellow zone is observed. The sludge melts just behind the leading flame edge and it is in the liquid state beneath the flame.

A blue flame oscillates back and forward over a few cm in front of the melting boundary of the sludge. The sludge surface, over which the blue flame propagates, becomes wet. In the region behind the leading flame edge, boiling of the melted sludge is observed. When the flame size is large, the boiling becomes intense and flamelets caused by the collapse of bubbles are observed over the main yellow flame zone.

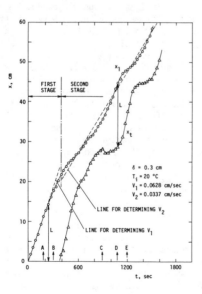

FIGURE 5. Position-time, x-t diagrams representing the movements of leading and trailing flame edges for a typical case.

Behind the trailing flame edge, an oscillatory propagating blue flame is also observed. The amplitude and the period of oscillation are larger than those for the preceding blue flame. The residue behind the trailing flame edge hardens to be non-fluid.

Movements of Leading and Trailing Flame Edges

The position-time, x-t, diagrams representing the movements of leading and trailing flame edges for a typical case are shown in Fig. 5. $x_l$ and $x_t$ are the locations of the leading and trailing flame

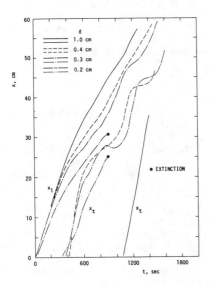

FIGURE 6. Position-time, x-t diagrams representing the movements of leading and trailing flame edges for various cases; $T_i$ = 20 °C.

The flame behavior for $\delta$ = 1 cm is not much larger value of $\delta$[1].

edges, respectively(Fig. 4). Points A-E represent the times when the photographs in Fig.2 were taken.

At an earlier stage (the first stage) of flame spread, the leading flame edge is found to spread at a constant rate $V_1$. In the following stage (the second stage), it is seen that the instantaneous spread rates of the leading and trailing flame edges fluctuate periodically and that the mean spread rate $V_2$ of the leading flame edge is smaller than $V_1$.

The movements of the leading and trailing flame edges for various sludge thicknesses $\delta$ were examined and the results represented by x-t diagrams are shown in Fig. 6. It can be seen that the characteristic of flame spread depends largely on $\delta$.

The period for the first stage decreases as $\delta$ decreases, and for $\delta$ = 0.2 cm, the flame stops spreading at about 15 minutes after the start of flame spread. so much different from that for a

The variation of the flame spread rate V with time can be determined from an x-t diagram. Figure 7 shows those obtained from the x-t diagrams in Fig. 6. The flame spread rate at the first stage is found to be constant and its period for $\delta$ = 1 cm is longer than that for $\delta$ = 0.2, 0.3, and 0.4 cm. No distinctive difference can be found between the variations of the flame spread rate at the second stage for values of $\delta$ from 0.3 to 1.0 cm.

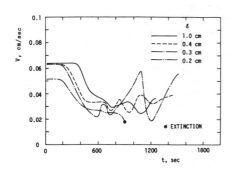

FIGURE 7. Variation of flame spread rate V with time; $T_i$ = 20 °C.

The values of $V_1$ and $V_2$ were examined for various cases. The results are shown in Fig. 8. $V_1$ for $\delta$ = 0.4 and 0.3 cm is found to be similar to that for $\delta$ = 1 cm, while $V_1$ for $\delta$ = 0.2 cm is a little smaller than those for a larger value of $\delta$. Further, $V_1$ for $\delta$ = 0.1 cm is zero, i.e., for $\delta$ = 0.1 cm no flame does spread at the first

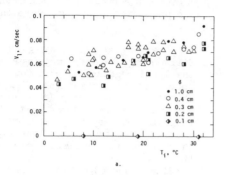

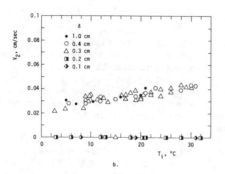

FIGURE 8. Flame spread rates $V_1$ at the first stage(a) and mean flame spread rates $V_2$ at the second stage(b).

stage. $V_2$ for $\delta$ = 0.4 and 0.3 cm is similar to that for $\delta$ = 1 cm. In this case, $V_2$ for $\delta$ = 0.2 cm and 0.1 cm is zero, i.e., for $\delta$ = 0.2 and 0.1 cm no flame does spread at the second stage.

Flame Size

In the first stage, the trailing flame edge is staying at the tray end brim, so that the longitudinal flame size L = $x_1 - x_t$ increases with time. In the second stage, the trailing flame edge moves oscillatory, and the variation of L is caused by this behavior of the trailing flame edge.

L at an arbitrary moment can be determined from an x-t diagram. The fluctuations of L were examined and are shown in Fig. 9. For $\delta$ = 1 cm, L continues to increase until t becomes more than 15 minutes. The period during which L continues to increase decreases as $\delta$ decreases. For first several minutes, L continues to increase and then, it decreases. After L decreases to several cm, it starts to increase again. In the succeeding period, L changes periodically.

Temperature Distribution in Sludge Layer

Figure 10 presents a typical temperature distribution in the sludge layer near the leading flame edge, which was measured at the moment when $x_1$ = 10 cm (in the first stage). The fluid-non-fluid boundary is assumed to be equal to the equi-temperature line for T = 50 °C. In the region within a few cm behind the leading edge of melted sludge, the melted sludge depth increases with the distance d from its leading edge and the temperature at a certain depth in the melted sludge increases with d. The temperature distribution in this region is very similar to that examined previously in the case of the flame spread over a thick layer of crude oil sludge[1].

DISCUSSION

Controlling Mechanisms of Flame Spread

The mode of flame spread is found to change within several minutes after the start of flame spread[Fig. 3, 5 and 6]. This transition has never

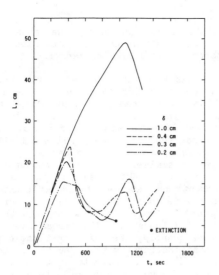

FIGURE 9. Variation of the longitudinal flame size with time; $T_i$ = 20 °C.

been observed in the studies on flame spread over single component liquid or solid combustibles in a quiescent atmosphere[3-9]. In the first stage, the flame spread rate is almost constant and the flame size continues to increase. On the other hand, in the second stage, the flame spread rate as well as the flame size becomes to fluctuate [Figs. 3, 7 and 9].

In our previous study, the mechanisms of flame spread over a thick sludge layer were discussed [1]. In the discussion, the flame spread rate was inferred to depend on the gas phase volatile concentration at the sludge surface near the leading flame edge. Since the gas phase volatile concentration depends on the sludge surface temperature, the flame spread rate must be controlled by heat transfer to the sludge surface. The experimental results of our previous study indicate that the dominant mode of heat transfer for flame spread over crude oil sludge is conduction through not-yet-liquefied sludge near the leading flame edge. This implies that the flame spread rate depends on the temperature distribution near the leading flame edge.

A marked difference was observed to exist between the levels of melted sludge surfaces behind the leading flame edges at the first and second stages. In the first stage, the surface of melted sludge is a little higher than that of not-yet-liquefied sludge, while in the second stage, it becomes a little lower. The location of the leading flame edge would depend on the aspect of the melting boundary, and the sludge melts due to heat transfer from the leading flame edge. Thus, temperature distribution in the not-yet-liquefied sludge in front of the melting boundary, must be closely related to the height of melted sludge surface.

It can be easily understood from the location of the leading flame edge that the rate of heat transfer to the surface of not-yet-liquefied sludge at the

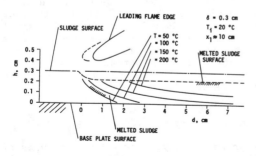

FIGURE 10. Temperature distribution in a sludge layer.

first stage is much larger than that at the second stage. Consequently, $V_1$ is larger than $V_2$[Fig. 9].

In the second stage, fluctuation of the flame spread rate was observed (Figs. 3, 5, 6 and 7). If the above discussion is reasonable, this fluctuation of the flame spread rate must be interpreted in a similar manner.

When the ejecting rate of flammable gas from the melted sludge surface becomes small and the flammable gas concentration over the surface decreases to be unable for sustaining a flame in the gas phase, no flame can appear. The trailing flame edge must be at the position where the rate of flammable gas ejection becomes insufficient for sustaining a flame in the gas phase. Thus, the trailing flame edge movement is closely related to the reduction of the height of melted sludge surface which would be caused by the evaporation of melted sludge. For a larger or smaller flame, the consumption of melted sludge occurs in a larger or smaller melted sludge pool area beneath the flame, respectively. At a moment when material vaporizable under the influence of quenching of the base plate is consumed up at the trailing flame edge, the trailing flame edge starts to move to forward. At this moment the height of melted sludge surface is lowest and the spread rate of the leading flame edge must be lowest. Consequently, the trailing flame edge approaches to the leading flame edge, so that the burning area decreases, and the pool area of melted sludge decreases. However, the rate of the melted sludge supply at the leading flame edge does not decrease instantly because it depends on the heat transfer from the leading flame edge. Thus, the height increases again and flame spread rate increases. This inferred behavior is consistent with the results shown in Figs. 5 and 6. Thus, the flame spread in this case is not inconsistent with that for a thick sludge layer.

Because the dominant mode of heat transfer for flame spread is conduction through the not-yet-liquefied sludge, the quenching effect of the base plate on the flame spread must exist. In a practical case, the structural material of the cargo bay is steel. Its thermal properties are not so much different from that of brass, which is used as the base plate material in the present experiment.

To reveal the effect of the base plate material on flame spread phenomena, the brass plate was replaced by a glass plate. The thermal conductivity of glass is about 1/60 of that of brass. The replacement of the base plate scarcely affected the values of $V_1$ and $V_2$. However, when the glass base plate was used, the flame behavior at the second stage changed a little, and the fluctuation of the flame spread rate vanished.

The thickness of residue behind the trailing flame edge was found to be less than 0.1 cm, which is small as compared to the value of more than 0.2 cm for the brass plate. This phenomenon is considered to be due to the small heat loss to the glass base plate. These results of flame spread over a crude oil sludge layer on a glass plate indicate that the flame spread mechanisms and dominant mode of heat transfer for flame spread are similar to those on a metal plate. Further, the limiting thicknesses on a glass plate are almost the same as those on a brass plate.

Possible Application of Results for Practical Purposes

As pointed out throughout the present study, there are two types of limiting thicknesses. One is the thickness below which the sludge cannot be ignited, and the other is that below which the spreading flame stopped to spread at a few minutes after starting to spread.

At both limiting conditions, the vapor concentration near the leading flame edge must be a limiting value for sustaining gas phase combustion under quenching conditions. At the thickness between these limits, 0.2 ∿ 0.3 cm, the flammable gas supplied to the gas phase must be sufficient for sustaining a flame just after starting flame spread, while that becomes insufficient after several minutes because of the consumption of the melted sludge.

Knowledge of limiting sludge thickness for flame spread is very important for practical purposes. Based on this knowledge, regulations or standards for washing cargo bays before repair works can be established.

Another phenomenon revealed in the present study and important for practical purposes is the finite flame size. When a flame starts from a point on a thin layer of sludge, the burning area is between the leading and trailing flame edges so that its shape must be annular. This is marked difference from the flame spread over a thick sludge layer. It makes the increasing rate of burning area for a thin sludge layer much smaller than that for a thick sludge layer.

These results can be applied not only for the sludge fire but also for the combustible material layer attached to non-flammable material, such as a carpet on a floor, a paint layer covering a wall, etc.


CONCLUSIONS

The aspects of flame spread over thin layers of crude oil sludge were examined in detail.

The mode of flame spread was found to change at a moment several minutes after the start of flame spread. At the first stage, the flame spread rate was almost constant and the flame size continued to increase. On the other hand, at the second stage, the flame spread rate as well as the flame size changed periodically. The amplitude of these periodic changes increased as the initial sludge thickness decreased. The flame spread rate at the first stage was about twice as large as the mean value of that at the second stage.

Two different types of the limiting thicknesses for the flame spread were revealed. One was the limiting thickness for the initiation of flame spread, and the other was that for the survival of the flame after the transition from the first stage to the second one.

The mechanisms of flame spread were discussed. It was shown that all these phenomena are interpretable by the mechanisms proposed in our previous study.

These results are inferred to be very useful for practical purposes.


ACKNOWLEDGEMENT

The authors would like to express their sincere thanks to Messrs. F. Ikeya and H. Terunuma for their help in conducting experiments.

REFERENCES

1.  Hirano, T., Suzuki, T., Sato, J., and Ohtani, H.:Twentieth Symposium(International) on Combustion, (in press).

2.  Ohtani, H., Sato, J., and Hirano, T.:AIAA 23rd Aerospace Sciences Meeting, AIAA-85-0396, 1985.

3.  Friedman, R.:Fire Research Abstracts and Reviews, 10, 1(1968).

4.  Magee, R. S. and McAlevy, III, R. F.:J. Fire and Flammability, 2, 271(1971).

5.  Sirignano, W. A.:Combustion Science and Technology, 6, 95(1972).

6.  Akita, K.:Fourteenth Symposium(International) on Combustion, p. 1075, The Combustion Institute, Pittsburgh, Pa, 1973.

7.  Williams, F. A.:Sixteenth Symposium(International) on Combustion, p. 1281, The Combustion Institute, Pittsburgh, Pa, 1976.

8.  Glassman, I. and Dryer, F.:Fire Safety J., 3, 123(1980).

9.  Fernandez-Pello, A. C. and Hirano, T.:Combustion Science and Technology, 32, 1(1983).

# Flow Assisted Flame Spread over Thermally Thin Fuels

**H. T. LOH and A. C. FERNANDEZ-PELLO**
Department of Mechanical Engineering
University of California
Berkeley, California 94720, USA

## Abstract

Small scale experiments have been conducted of the spread of flames over the surface of thin filter paper sheets in a mixed convective flow moving in the direction of flame spread. The rate of flame spread has been measured as a function of the flow velocity and oxygen concentration. It is found that after an initial accelerating stage, the flame spread rate becomes constant as the flame progresses over the fuel surface. The initial acceleration period becomes shorter as the oxygen concentration is increased. The spread rate is weakly dependent on the flow velocity for low gas velocities (mixed flow) but becomes independent of the flow velocity for forced flow conditions. The flame spread rate is practically linearly dependent on the oxygen concentration of the flow over the range of concentrations tested. The data for the rate of flame spread can be correlated in terms of an expression of the spread rate that is obtained from a simplified heat transfer model of the flame spread process.

## Introduction

Flame spread over solid combustibles is an important process during the growth of a fire. Although a considerable amount of research has been performed over the last few years on the flow assisted mode of flame spread, the complexity of the problem has so far prevented a complete understanding of the processes.

The concurrent mode of flame spread is characterized by a flame spreading over the surface of a solid combustible in a gaseous oxidizer

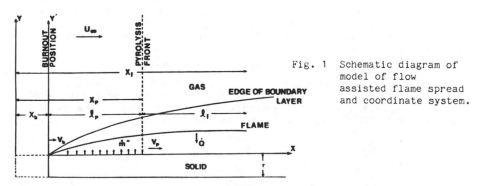

Fig. 1 Schematic diagram of model of flow assisted flame spread and coordinate system.

that flows in the same direction as that of flame propagation (Fig. 1). The fuel vapor that is not completely consumed by the upstream flame is induced by convection downstream ahead of the pyrolysis front. This fuel vapor keeps reacting with the oxidizer and thus extends the diffusion flame downstream from the pyrolysis front. The proximity of the flame and the post-combustion gases to the fuel surface favors the heat transfer to the unburnt combustible. As a consequence, the resulting flame spread process is generally rapid and hazardous.

Most experimental studies of the flow assisted mode of flame spread have been performed with thick fuels [1-6]. The only studies performed with thin fuels are the experiments of Markstein and de Ris [7] with cotton sheets, and Hirano and Sato [8] and Hirano et al [9] with paper sheets. In all cases the flames spread in an air flow. The reader is referred to the recent review of ref. [10] for an overview of these works. To the best knowledge of the authors, no studies have been performed on the spread of flames over thin fuel in a flow of varied oxygen concentration. Although the processes of flame spread over the surface of thick or thin combustible materials are basically the same, there are two major characteristics that differenciate one from another. One is that the fuel is consumed in the upstream region of the fuel sheet producing a propagating "burn out" front in addition to the propagating pyrolysis and flame sheet fronts common in both thin and thick fuels. Another difference is that in a thermally thin fuel the temperature is uniform accross its thickness, while in a thermally thick fuel its thickness does not affect the temperature distribution. These two basic differences may result in different controlling mechanisms of the flame spread processes for thin and thick combustibles.

In the present work an experimental study is carried out on the spread of flames over the surface of thin filter paper sheets in a forced flow with varied velocity and oxygen concentration moving in the direction of flame spread. The objective of the work is to provide basic information of the controlling mechanisms of flame spread. Of particular importance is the study of the influence of the burn-out process of the fuel on the flame spread rate.

Experiment

A schematic diagram of the experimental installation is shown in Fig. 2. The experiments are carried out in a small scale combustion tunnel with a test section that has a rectangular cross section of 0.127 m wide, 0.0762 m deep, and is 0.61 m long [6]. The walls of the test section are made of pyrex glass to allow optical access to the test area. The fuel specimens are 0.33 mm thick Whatman Chromatography paper, 0.076 m wide by 0.45 m long. The paper sheets are mounted in a metallic frame by inserting them in metallic spikes placed on the sides of the frame. The spikes are used to hold paper in slight tension to provide a flat surface [8]. In order to generate a flat plate flow over both surfaces of the fuel sheet, the paper is positioned in the middle of the test section 10 cm from the exit of the convergent nozzle. The paper sheets are dried in an oven and kept in a dessicator for at least forty-eight hours prior to performing the flame spread measurements.

The gas flow in the wind tunnel is supplied either from a centralized compressed air installation or from bottles of compressed oxygen and nitrogen. The gas flows of the individual gases are metered with calibrated critical nozzles and mixed at the settling chamber of the

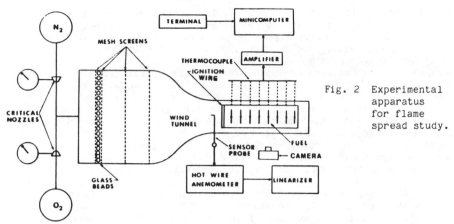

Fig. 2 Experimental
apparatus
for flame
spread study.

tunnel. Oxygen or nitrogen mixtures with concentrations accurate to within 1 percent are obtainable with the present installation. Maximum gas velocities obtainable in the present test section are of the order of 5 m/sec. The gas velocities are measured with a pitot tube instrument capable of measuring velocities down to 0.4 m/s, and with a hot wire anemometer. Prior to performing the experiments, extensive measurements were made of the velocity profiles along several planes of the test section to determine the characteristics of the flow. For the range of gas velocities used in these experiments, the flow showed a laminar character. The maximum flow Reynolds number at the downstream edge of the fuel specimen is of the order of $10^3$.

In the flow assisted mode of flame spread it is very important to have a uniform and well defined initiation of the flame spread process to assure two dimensionality [2,9,10]. In the present tests, the simultaneous ignition of the filter paper along its entire lower edge is achieved by means of an electrically heated nichrome wire. A thin layer of Duco-Cement is applied to the paper where it touches the nichrome wire to favor the uniform ignition of the fuel. The cement burns very quickly and does not affect the flame spreading process. To initiate the spread of the flame the following steps are taken. With the fuel sheet positioned in the test section, the gas flow is established at the predetermined velocity and oxygen concentration. During the process of fuel ignition, the gas flow is bypassed to have a quiescent gas region near the leading edge of the fuel specimen and thus facilitates its ignition. On ignition, the heated nichrome wire pyrolyzes and ignites a thin region of the paper, initiating the spread process. As soon as ignition is observed, the bypass is closed and the data acquisition started.

The behavior of the flame is recorded with both direct photographs and thermocouple probing. In the thermocouple probing method, the rate of flame spread is measured from the temperature histories of thermocouples placed at fixed distances along the fuel surface [3]. Eight chromel-alumel thermocouples 0.0762 mm in diameter are placed in grooves made on the filter paper sheets at distances 5.715 cm apart. The output from the thermocouples is amplified and processed in a real time data-acquisition system (PDP-11 mini-computer). The rate of spread of the pyrolysis front is obtained from the surface temperature histories by calculating the ratio of the distance between two consecutive thermocouples to the elapsed time of pyrolysis arrival to the thermocouples. The arrival of the pyrolysis front at the thermocouple position is characterized by the leveling of the temperature profile when the pyrolysis temperature of the fuel reaches an approximate constant value. The burnout location of the upstream fuel is recorded by observing the sudden decrease or increase in temperature due to the disappearance of the fuel or the contact of the trailing edge of the flame with the thermocouple, respectively. Motion pictures, taken at about 18 frames/second of the fuel surface, provide another means for quantitative measurement of the flame spreading process. The motion pictures are used to evaluate the locations of the pyrolysis and burnt-out fronts, and the flame tip. Illumination of the paper surface with tungsten lamps, combined with the proper choice of exposure, yield well defined pyrolysis front locations. Photographically, this corresponds to the onset of the blackening of the paper. The burn-out location is more difficult to determine because in the burning region the paper breaks and curls, making it difficult to establish the real location of the fuel's disappearance. To overcome this problem, the location of the fuel burn-out is first assumed to coincide with the position of the initiation of the paper break out. The tests are then repeated up to five times to obtain an average of this distance. The distance obtained with this method just after the initiation of the paper breaking is subsequently matched with the burn-out distance obtained prior to the onset of the paper break out. The resulting correction is then applied to the rest of the burn-out data. The accurate determination of the flame tip position is also difficult because of the fluctuations of the flame tip. The results presented here are average values of distance obtained from color photographs of repeated tests.

## Results and Discussion

The measurements of the distances of the pyrolysis front, $X_p$, the corresponding burn-out front, $X_b$, and flame tip, $X_f$, to the location of the flame spread initiation for flames spreading over the surface of filter paper sheets, are presented in Fig. 3 for several air flow velocities. The spread rates of each front can be deduced from these results by differentiating the corresponding distances with respect to time. The experiments are performed with the combustion tunnel in a vertical position to permit testing over the whole range of convective flow conditions (from free to forced). The pyrolysis front data is obtained using both the thermocouple and photography methods. The data for the burn-out front and flame tip location are obtained primarily from the photographs, because this method seems to provide more reliable results. As explained above, because the method used to determine the locations of the burn-out front and the flame tip are not very accurate, the burn-out and flame tip data presented here can only be considered as approximate. The velocity of 2 m/s is the maximum velocity for which the spread of the flame in air is observed. With thin fuels the length of the pyrolysis region remains relatively short due to the upstream consumption of the fuel, so the flame does not move into regions of larger boundary layer thickness. Consequently, extinction or non-flame propagation occurs at lower velocities than for the thick fuel sheets [6].

From the results presented in Fig. 3 it is seen that for air the flame spreading process is accelerative from ignition to approximately 15 cm downstream, becoming constant afterwards. This result follows the variation of the pyrolysis length $\ell_p = X_p - X_b$ and of the length of the flame over the not yet pyrolyzing fuel surface, thereafter called flame length $\ell_f = X_f - X_p$ with the distance from ignition. From Fig. 3 it is seen that

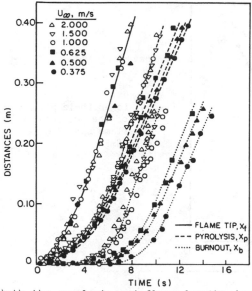

Fig 3 Measurement of the variation with time of the pyrolysis flame and burn-out distances for flames spreading over thin paper sheets in a concurrent air flow.

both the pyrolysis and flame lengths increase rapidly during the initial period of the flame spread process until burn-out of the fuel starts. After that, the rate of increase of these lengths decrease as the burn-out front progresses until finally they become practically constant at approximately 15 cm from ignition. The results of Fig. 3 also show that, for low flow velocity (mixed convection), the pyrolysis and flame lengths decrease as the flow velocity increases. Both lengths approximately become constant for forced flow conditions ($u_\infty > 1$ m/s). The flame spread rate follows the variations of these lengths, increasing with the flow velocity for mixed flow conditions and becoming pratically constant for forced flow.

In Fig. 4, a logarithmic plot of the spread rate data for the initial accelerative period is presented. It is seen that there is an approximate power law dependence beween the pyrolysis distance and time of the form $X_p \sim t^n$. The value of the exponent varies from 1.6 for natural convection to 2 for forced flow. The only theoretical models of the flow assisted mode of flame spread over thermally thin fuels that have been published to date are those of refs. [7 and 11] for natural convection and ref. [5] for forced convection. An overview of these models is given in ref. [12]. The analyses of refs. [7,6] are very similar and both predict a fourth power law dependence between the pyrolysis distance and time. This is in disagreement with the results of Fig. 4 for natural convection which gives a 1.6 power. On the otherhand, the analysis of ref. [5] for forced convection predicts a square power dependence for the pyrolysis distance with time which is in agreement with the results of Fig. 4. This last analysis, however, predicts a linear relationship between the pyrolysis distance and the free stream velocity. This prediction is in disagreement with the experimental results which show that the spread rate is practically independent of the flow velocity (Fig. 4). These comparisons indicate that the present theoretical models of the flow assisted spread of flames over thin fuels are not capable of predicting the process accurately and that some improved versions of these models are needed.

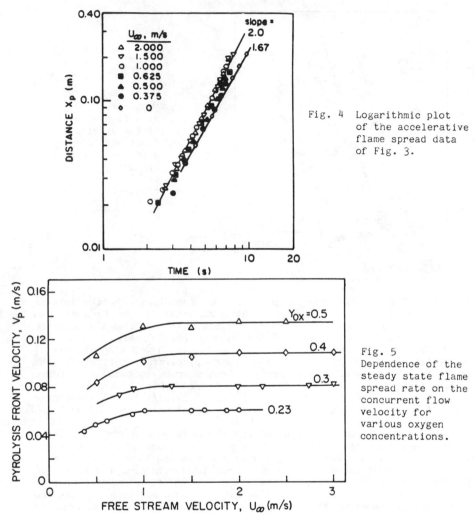

Fig. 4  Logarithmic plot of the accelerative flame spread data of Fig. 3.

Fig. 5 Dependence of the steady state flame spread rate on the concurrent flow velocity for various oxygen concentrations.

In Fig. 5 the rates of spread of the pyrolysis front, once the spread process has reached steady state, are presented as a function of the flow velocity for several oxygen concentrations. It is seen that for all oxygen concentrations the spread rate increases with the flow velocity for low flow velocities, and becomes practically independent of the gas velocity for forced flow conditions. Within the range of experimental conditions, the spread rate increases linearly with oxygen concentration.

In order to explain the nature of the above results, it is convenient to develop a simple model of the flame spread process over a thermally thin fuel. Assuming: that the primary controlling mechanism of flame spread is heat transfer from the flame to the non-burning material downstream from the pyrolysis front; that the heat flux from the flame is constant over the flame length and zero afterward; that the temperature of the fuel is uniform along its thickness (thermally thin); and that the combustible does

70

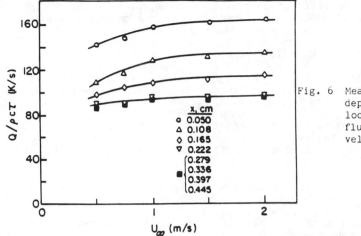

Fig. 6  Measurements of the dependence of the local surface heat flux on the air flow velocity.

occurance of flame extinction the range of gas velocities tested (1m/s to 2 m/s) is very small.  This, in conjunction with the fact that the experimental data has considerable scatter, and that the spread process may not behave as truly two-dimensional, suggests that the apparent independency of the heat flux with the flow velocity may not be totally true and that indeed the heat flux is dependent on the flow velocity as expected from the boundary layer analysis predictions.

The experimental results for the flame spread rate can be  explained phenomenologically with the help of Eq.(1) (or Eq.(2)) and the results of Figs. 3 and 6.  The initial accelerative period of the flame spread rate (Figs. 3 and 4) is due to the increase of the pyrolysis length which in turn results in the increase of the flame length (Fig. 3) and consquently of the total heat flux at the fuel surface.  The slight decrease of the local heat flux with the distance from ignition (Fig. 6) is counteracted by the increase of the flame length.  Once the fuel starts to be depleted in the upstream region, the rate of increase of the pyrolysis length decreases and so does the spread rate.  As the rate of spread of the burn-out front approaches that of the pyrolysis front, the pyrolysis length becomes constant.  This results in a steady state flame spread process.  Similarly, the increase of the spread rate with $u_\infty$ for low flow velocities is due to the increase of the surface heat flux (Fig. 6) which counteracts the slight decrease of the flame length (Fig.3).  For larger flow velocities [$u_\infty > 1$ m(s)], the pyrolysis and flame length and the surface heat flux become approximately constant, and consequently so does the rate of flame spread.

With regard to the independence of the flame spread rate on the gas velocity for forced flow conditions, it should be pointed out that the reasons given above for this result, i.e: the constancy of $\ell_p$, $\ell_f$ and $q''_f$, may not be totally accurate.  As explained before, the present measurements have some scatter due to experimental difficulties and the variation of the above parameters with $u_\infty$ is expected to be small due to the small range of velocities tested .  Therefore, there is the possibility that the above parameters are not truly constant but vary with length and velocity according to the boundary layer predictions.  Under these conditions the dependence of the flame spread rate on the flow velocity would be the result of the following mechanism.  As the flow velocity increases the not vaporize until its temperature reaches a given value; and that the spread process is two dimensional, an energy balance for a control volume in the solid downstream from the pyrolysis front (Fig. 1) gives

$$\rho \, c \, \tau \, V_p \, (T_v - T_\infty) = \dot{q}''_f \, \ell_f \qquad (1)$$

In the above equation, $\rho$, $c$, $\tau$, $T_v$ and $T_\infty$ are respectively the density, specific heat, thickness, pyrolysis and initial temperature of the fuel; $V_p$ is the spread rate of the pyrolysis front; $q''_f$ the heat transfer from the flame to the fuel surface by radiation and convection; $\ell_f$ is the flame length over the not yet pyrolizing surface. The spread rate can also be expressed in terms of the pyrolysis length by replacing $\ell_f$ in Eq. (1) by the relation $\ell_f = C \, \ell_p^n$ with $n \leq 1$, [2,11,12], ($n \cong 1.1$ from the results of Fig.3).

The heat flux at the surface, neglecting radiation, can be expressed in the form $\dot{q}''_f \sim \lambda_g (T_f - T_v)/\delta$, where $\lambda_g$ is the thermal conductivity, $T_f$ is the flame temperature and $\delta$ is the flame stand-off distance. Substituting this relation in Eq. (1), the following expressions is obtained for the rate of spread of the pyrolysis front,

$$V_p \sim \frac{\lambda_g(T_f - T_v)}{\rho c \tau (T_v - T\infty)} \, (\ell_f / \delta) \qquad (2)$$

The variation of $V_p$ with the pyrolysis front distance (or time) and with the flow velocity will depend on the respective variations of $\ell_f$ and $\delta$. The variation of the former parameter can be deduced from the results of Fig. 3 and of the latter from the results of Fig. 6.

In Fig. 6, the surface heat flux calculated from surface temperature histories at different locations along the fuel surface is plotted as a function of the free stream velocity. It is seen that the heat flux decreases initially with the distance from the fuel ignition location and becomes approximately constant after a distance of approximately 15 cm. This trend is in qualitative agreement with the predictions of boundary layer analyses of burning surfaces where a scaling law for $\delta$ of the form $\delta \sim \ell_p^a \, u_\infty^b$ can be deduced [12]. The initial increase of the pyrolysis length (Fig.3) results in an increase of the flame stand-off distance and, consequently, in a decrease of heat flux. Once $\ell_p$ becomes constant, so does $\delta$ and consequently $\dot{q}''_f$. From the dependence on $u_\infty$, it is seen that the heat flux increases with the velocity for low velocities, but becomes practically constant for $u_\infty > 1$ m/s. This last result shows agreement of the above dependence of $\delta$ on $\ell_p$ but seems to disagree with the predicted dependence on $u_\infty$. From the results of Fig. 3 it is seen that, for low velocities, as $u_\infty$ increases $\ell_p$ decreases, which results in a decrease of the flame stand-off distance and consequently in an increase of the heat flux. For large flow velocities, however, the pyrolysis length is practically independent of the flow velocity. Since the heat flux is also very weakly dependent on $u_\infty$, it appears that the heat flux is primarily dependent on $\ell_p$ and not on $u_\infty$. With regard to the apparent weak dependence of the heat flux on the flow velocity, it should be mentioned that the observed strong variation of the flame spread data with the flow velocity occurs during the transition from natural to forced convection, where markedly different flow patterns are expected. However, under forced flow conditions the variation of the flame spread parameters with the flow velocity is not very strong ($\sim 1/2$ power). Futhermore, because of the thickness of the boundary layer and consequently the flame stand-off distance decrease, which results in an increase of the heat transferred from the flame to the fuel. This increase in heat transfer has a dual effect. While it increases the heat flux at the non-burning fuel surface, it also increases the gasification rate of the burning surface. A larger mass burning rate causes an increase in the rate of spread of the burn-out front that tends to decrease the length of the pyrolysis region and conseqently of the flame length. As it is seen from Eq. (1) both effects -- the increase of the heat flux and the decrease of the pyrolysis length (or flame length) -- counteract each other, and, depending on their relative variation, the spread rate would either increase, remain constant, or even decrease. The results of Figs. 3 and 5 seem to indicate that at large velocities (forced flow) both effects would balance each other.

A correlation, using Eq. (2) of the flame spread rate data of Fig. 5 with $\ell_f/\delta$ assumed constant for forced convection, is presented in Fig. 7. The properties used in the computation of the correlation are $\lambda_g$ = 0.046 J/m-sec K, $c_p$ = 1.06 kJ/kg K, $\rho \tau$ = 18.5 x $10^{-3}$ g/cm$^2$. The flame temperature $T_f$ is calculated with the equation [15]

$$T_f = T_v + \frac{(T_\infty - T_v) + (Y_{O\infty}s/c_p)(\Delta H_c - L)}{1 + Y_{O\infty}s} \qquad (3)$$

this corresponds to the adiabatic flame temperature for constant specific heat of the products of combustion. The data used in calculating the flame temperature are: $s = 0.844$, $\Delta H_c = 1.674$ x $10^4$ KJ/kg fuel, $L = 753$ KJ/kg, $T_\infty = 298$ and $T_v = 618$ K. From the results of fig. 7 it is seen that except at low flow velocities ($u_\infty < 1$ m/s), where buoyancy effects become important, Eq. (2) correlates very closely the flame spread data of Fig. 5. Since Eq. (2) does not include radiation heat transfer from flame to fuel, the good correlation of the experimental data indicates that radiation is not an important mechanism of heat transfer for the present experiments. This is probably due to the small scale of the experimental set up, and to the fact that the fuel burn out limits the size of the flame. It is interesting to note that the analysis of ref. [14] for the spread of a flame over a thin fuel in an opposed flow predicts a spread rate that is

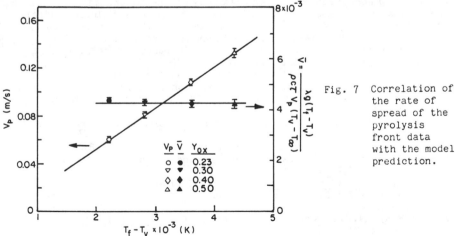

Fig. 7  Correlation of the rate of spread of the pyrolysis front data with the model prediction.

also given by an equation as in Eq. (3) and that is independent of the flow velocity. As it is explained in ref. [6] this is understandable. A lump energy balance of the gas phase for the opposed and concurrent modes of spread show that while the convective terms have opposite signs, the temperatures also have opposite signs, and both cancel each other. The result is a formula that is applicable to both models of heat transfer.

Conclusion

The measurement of the rates of flame spread over the surface of thin filter paper sheets and their comparison with currently available theoretical models show deficiencies in the predictive capabilities of the models. It appears that the major problem comes from the prediction of the variation with time and of the dependence on the flow velocities of the length of the pyrolysis region. This length determines in its turn the length of the flame and consequently the overall heat flux on the fuel surface. Thus an accurate prediction of the pyrolysis and flame lengths is imperative to predict accurately the rates of flame spread.

Acknowledgements

The authors gratefully acknowledge the comments and suggestions of Professors T. Hirano, K. Sato, and P.J. Pagni concerning the experimental procedure and results. The research was supported by the National Bureau of Standards under grant number: NB83NADA 4020.

References

1. Hansen, A. and Sibulkin, M.: Combustion Science and Technology, 9, p. 173 (1974).
2. Orloff, L., de Ris, J.N. and Markstein, G.H.: Fifteenth Symposium on Combustion, p. 183, The Combustion Institute (1975).
3. Fernandez-Pello, A.C.: Combustion Science and Technology, 17, p. 87 (1977).
4. Alpert, R.L.: ASME-AICHE National Heat Transfer Conference, ASME publication no. 79-MT-28 (1979).
5. Fernandez-Pello, A.C.: Combustion and Flame, 36, p. 63 (1979).
6. Loh, H.T. and Fernandez-Pello, A.C.:"A Study of the Controlling Mechanisms of Flow Assisted Flame Spread," The Twentieth (International) Symposium on Combustion, The Combustion Institute, (in press) (1984).
7. Markstein, G.H. and de Ris, J.H.: (a) Fourteenth Symposium (International) on Combustion, p. 1085, The Combustion Institute (1973); (b) also FMRC Report 20588 (1972).
8. Hirano, T. and Sato, K.: Fifteenth Symposium (International) on Combustion, p. 233, The Combustion Institution (1975).
9. Hirano, T., Noreikis, S.E. and Waterman, T.E.:Combustion and Flame, 22, p. 353 (1974).
10. Fernandez-Pello, A.C. and Hirano,T.:Combustion Science and Technology, 32, p. 1 (1983).
11. Fernandez-Pello, A.C.: Combustion and Flame, 31, p. 135 (1978).
12. Fernandez-Pello, A.C.: Combustion, Science and Technology, 39, p. 119 (1984).
13. Pagni, P.J.:Fire Safety Journal, 3, 2-4, 273 (1981).
14. de Ris, J.N.:Twelfth Symposium on Combustion, p. 241, The Combustion Institute (1969).
15. Altenkirch, R.A., Eichhorn, R. and Shang, P.C.: Combustion and Flame, 37, p. 71 (1980).

# Upward Turbulent Flame Spread

**K. SAITO**
Department of Mechanical and Aerospace Engineering
Princeton University
Princeton, New Jersey 08544, USA

**J. G. QUINTIERE**
Center for Fire Research
National Bureau of Standards
Gaithersburg, Maryland 20899, USA

**F. A. WILLIAMS**
Department of Mechanical and Aerospace Engineering
Princeton University
Princeton, New Jersey 08544, USA

ABSTRACT

Mechanisms and rates of upward spread of turbulent flames along thermally thick vertical sheets are considered for both noncharring and charring fuels. By addressing the time dependence of the rate of mass loss of the burning face of a charring fuel, a linear integral equation of the Volterra type is derived for the spread rate. Measurements of spread rates, of flame heights and of surface temperature histories are reported for polymethylmethacrylate and for Douglas-fir particle board for flames initiated and supported by a line-source gas burner, with various rates of heat release, located at the base of the fuel face. Sustained spread occurs for the synthetic polymer and not for the wood. Comparisons of measurements with theory aid in estimating characteristic parameters for the fuels.

## 1. INTRODUCTION

Upward flame spread on vertical surfaces is a critical aspect of accidental fires because of its inherent high speed and potential consequences of fire growth to surroundings. Many flammability test methods are configured to represent this hazard and attempt to assess the relative contribution for a material. Unfortunately no general test prescription exists to allow the prediction of a material's performance in upward flame spread. To that end, some research has been performed which can provide some guidance for achieving a generally applicable predictive model for upward spread.

Reviews on flame spread have included consideration of research on upward propagation.[1] There have been a number of theoretical and experimental investigations related to the subject of concurrent or upward spread.[2-13] Some of these concern spread along thermally thin materials, while others address thermally thick materials at scales (e.g., flame height) small enough for the spread processes to involve laminar flow. Here we are interested in thermally thick fuels that also are so thick that they are not completely consumed during the spread process, and we study planar two-dimensional upward spread at scales large enough for the flow to be turbulent. The systematic measurements of Kishitani et al.[13] were performed on polymethylmethacrylate and chipped-wood samples 25 cm high, a size corresponding approximately to conditions for onset of turbulence. The only earlier experiments on two-dimensional spread that lie well within the regime of turbulent flow are those of Orloff et al.[3], who employed polymethylmethacrylate slabs 356 cm high.

A motivation for the present study is to extend these turbulent-spread measurements to different materials and to different conditions of ignition. Transient pyrolysis of solids during upward spread contributes to making the process highly unsteady. The local mass-loss rates of thermoplastics will increase somewhat with time, while for charring materials they will eventually decrease with time. Indeed because of this, Delichatsios[14] suggests the existence of critical conditions for the occurrence of upward spread on vertical charring walls. In view of the possibility of the absence of spread, in the present experiments a diffusion-flame $(CH_4)$ line-burner was used at the base of the vertical sample to promote sustained spread. Results of measurements are reported herein for both polymethylmethacrylate and chipped wood. As a basis for interpretation of the experimental results, a theoretical development is first given that includes consideration of spread along charring fuels. The experiments and their results are then presented, and comparisons with theory are discussed.

## 2. THEORETICAL CONSIDERATIONS

### 2.1. Description of Normal Regression

Charring materials pose severe difficulties in attempts to describe their combustion. The complexity of normal burning of charring fuels is reflected in simplified models thereof.[14-18] Our current understanding of normal regression accompanied by char formation is incomplete, and more research on the subject is needed. The studies that have been performed demonstrate clearly that even under a constant rate of energy input the gasification rate remains time dependent. The rate of mass loss per unit area $\dot{m}''$ as a function to time $t$ (at a fixed location $x$) for a thick cellulosic material, whose face is exposed to a constant energy flux beginning at $t=0$, is shown schematically in Fig. 1 as curve 0. A rough approximation to experimental observations is $\dot{m}'' = \dot{m}''_o = $ constant during a fixed gasification time $t_b$ and zero otherwise, as illustrated in curve 1. Alternatively, various model simplifications[14,18] lead to the theoretically predicted inverse square-root dependence on $t-t_o$ indicated in the figure as curve 2. As a basis for describing upward spread we shall assume that after ignition each surface element of fuel exhibits the same gasification-rate history $\dot{m}''(t-t_o)$, independent of the location of the element. Although a general functional form of the relationship will be permitted in the analysis, the two simplified models illustrated in Fig. 1 as curves 1 and 2 will be kept in mind.

Responses of non-charring materials such as polymethylmethacrylate to constant rates of energy input are approximated well by curve 1 of Fig. 1 with $t_b = \infty$. Therefore the influence of charring can be viewed as providing a growing protective layer that introduces a finite characteristic gasification time $t_b$ (or $t_c \approx t_b/2$) for the material. If the rate of energy input were to vary with time then $\dot{m}''(t-t_o)$ would change accordingly, but a universal function still could exist. Therefore the hypothesis appears to have appreciable versatility. The additional complications associated with addressing questions of spread motivate the somewhat phenomenological level of description of the charring process adopted here.

### 2.2 Description of Spread Mechanisms

Spread occurs as a consequence of heating of the unignited portion of the fuel to a temperature at which vigorous pyrolysis begins. This heating is produced by convective and radiative heat transfer from the flames that bathe the fuel surface. Let x denote the vertical distance along the fuel surface, with x=0 at the base of the fuel, $x=x_p$ at the upper edge of the pyrolysis region and $x=x_f$ at the average height of the visible flame tip, as illustrated in Fig. 1. The heat transfer responsible for spread occurs in the region $x \geq x_p$. Over the size range of the

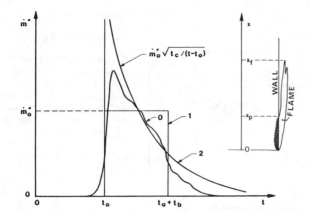

FIGURE 1.

Illustration of gasification response to a constant heat flux for a charring material and of the spread model.

experiments to be reported herein, for steady-state burning at the base of a vertical wall, the energy flux $\dot{q}''$ to the wall has been found experimentally[19] to correlate with $x/x_f$, and in a rough first approximation $\dot{q}'' = \dot{q}_0'' = $ constant $\approx 2.5$ W/cm$^2$ for $0 < x < x_f$ and $\dot{q}'' = 0$ otherwise, so that $x_f$ is a good measure of the distance over which the principal heat transfer occurs.

If this rough approximation is employed along with the further assumption that $x_f - x_p$ remains approximately constant during spread, then the upward spread velocity of the pyrolysis front is

$$v_p = 4(\dot{q}_0'')^2(x_f - x_p)/[\pi k \rho c (T_p - T_a)^2] \quad , \tag{1}$$

where $k$, $\rho$ and $c$ are the thermal conductivity, density and heat capacity, respectively, of the fuel, and $T_a$ and $T_p$ are the ambient and ignition (or pyrolysis) temperatures of the fuel. Equation (1) can be rewritten as

$$v_p = (x_f - x_p)/\tau \quad , \tag{2}$$

where the characteristic ignition time $\tau$ for spread depends only on fuel properties, the ambient temperature and the level of the heat flux to the fuel from the flame. As a simplification for describing time-dependent spread, we assume that Eq. (2) continues to apply with $x_f - x_p$ variable and that $\tau$ remains an approximately constant time characteristic of upward spread. Since Eq. (2) is not precisely derivable from the appropriate heat-conduction problem under the assumptions that have been introduced, its applicability will be tested herein from the spread data.

2.3. Flame-Height Correlations

Having hypothesized that the correlation of the heat-flux distribution with $x/x_f$ may lead to Eq. (2), we need an expression for $x_f - x_p$ to obtain $v_p$. By definition

$$x_p(t) = x_{po} + \int_0^t v_p(t_p)dt_p \quad , \tag{3}$$

where $x_{po}$ is the value of $x_p$ at an initial time $t=0$, and $t_p$ is the dummy variable of integration. Flame-height correlations are required for obtaining $x_f$. We allow for a pilot flame at the base of the wall releasing energy at the rate $\dot{Q}'$ per unit

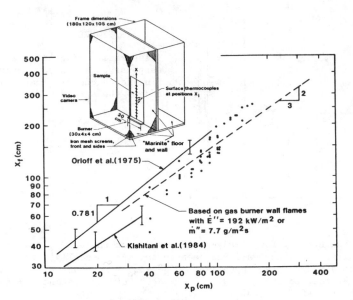

FIGURE 2.

Flame-height correlations with pyrolysis height for upward spread along thick sheets of polymethylmethacrylate, and schematic diagram of the experiment.

length. The total rate of energy release per unit length then is the sum $\dot{Q}'$ + $q\int_0^x$ m"dx, where m" is the previously introduced rate of mass loss per unit area of the fuel, and q is the heat released per unit mass of fuel consumed. Flame-height correlations are of the form

$$x_f = K[\dot{Q}' + q \int_0^{x_p} \dot{m}"dx]^n , \qquad (4)$$

where K and n are constants. Experiments on steady wall flames, over the size range of interest here, support Eq. (4) with n=2/3.[19] It is found that $K \approx 1$ cm/(W/cm for flames on open walls, while K is about 25% larger than this for flames on walls protected by side walls. Data for polymethylmethacrylate with $\dot{Q}'=0$ are useful for testing the applicability of Eq. (4) during upward spread because $\dot{m}"$ is approximately constant for the burning portion of this fuel, so that $x_f$ becomes proportional to $x_p^n$ according to Eq. (4). Results from Orloff et al.[3], from Kishita et al.[13] and from the present study are shown in Fig. 2. The methods employed for obtaining the present results will be described later.

It is seen from Fig. 2 that the slopes of the curves from the spread experimen tend to exceed those for steady wall burning. This may be caused by higher pyrolysis rates in the region of increased heat flux at the upper part of the pyrolyzing fuel during spread. The approach to a lower slope at large $x_p$ in the data of Kishitani et al. may be attributable to effects of the thickness of the fuel sheet (only 4 mm). That the data of Orloff et al. lie above ours is consistent with the fact that side walls were employed in their experiments but not in ours. From Fig. 2 it may be seen that within the accuracy of our data the exponent n in Eq. (4) may be taken to be unity if $\dot{Q}'=0$. Increasing $\dot{Q}'$ would be expected to decrease n toward the steady-state value of 2/3, shown by the dashed line in Fig. 2. Even for situations under which n approaches 2/3, use of n=1 in Eq. (4) may be thought to provid roughly correct theoretical predictions and sometimes will be introduced below for simplification.

For use in Eq. (2) a power-law dependence of $x_f - x_p$ would be simpler than tha of $x_f$ in Eq. (4). Unfortunately, the data show correlations for $x_f - x_p$ to be appreciably poorer.

## 2.4. Predictions of Spread Histories for Noncharring Fuels

For a noncharring fuel the application of Eqs. (2), (3) and (4) is relatively straightforward. Since $v_p = dx_p/dt$ and $\dot{m}''$ is approximately independent of x in the range $0 < x < x_p$, we obtain the first-order ordinary differential equation

$$dx_p/dt = [K(\dot{Q}' + \dot{E}''x_p)^n - x_p]/\tau , \qquad (5)$$

where $\tau$, n, K and $\dot{E}'' = q\dot{m}''$ are taken as known constants, and the pilot rate $\dot{Q}'$ is a known, experimentally adjustable function of time. For $n=1$ this equation is linear and easily solved; $x_p$ always increases linearly with t at early times (if $x_{fo} > x_{po}$ initially), and its long-time behavior depends on the sign of $K\dot{E}'' - 1$. If $K\dot{E}'' > 1$ (a condition applicable, for example, to polymethylmethacrylate which may be estimated from Fig. 2 to have $K\dot{E}'' \approx 2$) then $x_p(t)$ is acceleratory, and an exponential increase of $x_p$ with t is approached when $x_p \gg \dot{Q}'/\dot{E}''$ is reached. If $K\dot{E}'' < 1$ (a condition unlikely to be realistic for noncharring fuels since $x_f$ unaugmented by a pilot would be less than $x_p$) then $x_p(t)$ is deceleratory [unless forced by an acceleratory $\dot{Q}'(t)$], and eventually propagation of the pyrolysis front is predicted to cease at $x_p = K\dot{Q}'/(1-K\dot{E}'')$, corresponding to $x_f = x_p$. However, if Eq. (2) remains valid then the flame-height correlation of Eq. (4) with $n \leq 1$ certainly must fail for noncharring fuels before Eq. (5) gives $x_f = x_p$.

For $n < 1$ the implications of Eq. (5) are qualitatively similar in most respects to those just decribed for $n=1$; the only significant difference is that, unless $\dot{Q}'(t)$ is sufficiently strongly acceleratory, at sufficiently long times there is always a deceleration with an approach to a constant value of $x_p$. This may be seen directly, for example, from the explicit solution for $\dot{Q}'=0$ and $n=2/3$, viz.,

$$x_p = K^3(\dot{E}'')^2\{1 - [1 - x_{po}^{1/3}/(K\dot{E}''^{2/3})]e^{-t/3\tau}\}^3 , \qquad (6)$$

which exhibits a cubic acceleration, $x_p \sim t^3$, at small values of $t/\tau$ if $x_{po}$ is sufficiently small, but which always predicts that as $t/\tau \to \infty$, $x_p \to K^3(\dot{E}'')^2$, with $dx_p/dt \to 0$. More generally, Eq. (5) indicates that always $dx_p/dt = 0$ when

$$x_p/x_{p\infty} = (1 + A\, x_{p\infty}/x_p)^{n/(1-n)} , \qquad (7)$$

where $A = \dot{Q}'/(\dot{E}''x_{p\infty})$ is a nondimensional measure of $\dot{Q}'$, and where $x_{p\infty} = [K(\dot{E}'')^n]^{1/(1-n)}$ is the asymptotic value of $x_p$ as $t \to \infty$ if $A = 0$. Equation (7) possess a unique solution that approaches $x_p = x_{p\infty}[1+An/(1-n) + ...]$ for $A \ll 1$ and $x_p = x_{p\infty} \cdot A^n(1+n/A^{1-n} + ...)$ for $A \gg 1$. Whenever $dx_p/dt \to 0$ is encountered with this formulation the situation corresponds to $x_f \to x_p$, and either Eq. (2) or the flame-height correlation necessarily becomes invalid. Therefore use of the model for noncharring fuels should be restricted to conditions under which acceleratory propagation is predicted therefrom.

## 2.5. Predictions of Spread Histories for Charring Fuels

For a charring fuel the use of Eqs. (2), (3) and (4) is more complicated because of the variation of $\dot{m}''$ with x in Eq. (4). As an idealization of the present experiments assume that a constant value of $\dot{Q}'$, say $\dot{Q}'_o$, is employed to ignite the fuel and that the igniter provides a constant energy flux to the fuel over the flame height $K\dot{Q}_o'^n$ and zero flux elsewhere, leading to an initial pyrolysis height $x_{po} = K\dot{Q}_o'^n$ at the time of ignition, taken to be $t=0$. The initial condition for integration then will be $x_p = x_{po} \equiv K\dot{Q}_o'^n$, with $x_f = x_{fo} \equiv K[\dot{Q}'(0) + q\dot{m}''(0)\,x_{po}]^n$, where $\dot{m}''(0)$ is the value of the universal function $\dot{m}''(t-t_o)$ from Fig. 1 at $t=t_o$. Likely in correspondence with physical reality for most experiments, we assume that $K[q\dot{m}''(0)x_{po}]^n > x_{po}$, so that the initial pyrolysis-front velocity, $v_{po} = (x_{fo}-x_{po})/\tau$, is positive even if the pilot is turned off at the instant of ignition [i.e., even if $\dot{Q}'(0)=0$]. If this condition is not satisfied then unaugmented spread cannot occur, and the formulation is inappropriate, at least if $\dot{Q}'(0)$ is too small, as discussed earlier for noncharring fuels; note that if $n<1$ then in principle it is always possible to select $\dot{Q}_0'$ large enough to give a value of $x_{po}$ too large to satisfy this condition.

The integral in Eq. (4) may be written as

$$\int_0^{x_p} \dot{m}''dx = \int_0^{x_{po}} \dot{m}''dx + \int_{x_{po}}^{x_p} \dot{m}''dx = x_{po}\dot{m}''(t) + \int_0^t \dot{m}''(t-t_p)v_p(t_p)dt_p \quad , \tag{8}$$

where $v_p(t_p) = (dx/dt)_{t=t_p}$ is the velocity of the pyrolysis front at time $t_p$. Substitution of Eq. (8) into Eq. (4), followed by substitution of this result and of Eq. (3) into Eq. (2), gives a nonlinear integral equation for $v_p(t)$. Here we shall write the result only under the further approximation that $n=1$, in which case the integral equation becomes linear. Thus,

$$v_p(t) = v_{po} + \{\int_0^t [Kq\dot{m}''(t-t_p)-1]v_p(t_p)dt_p - Kq[\dot{m}''(0) - \dot{m}''(t)]x_{po} - K[\dot{Q}'(0)-\dot{Q}'(t)]\}/\tau . \tag{9}$$

With the nondimensionalizations $\xi = t/\tau$, $\xi' = t_p/\tau$ and $V(\xi)=v_p(t)/v_{po}$, Eq. (9) can be written as

$$V(\xi) = \int_0^\xi F(\xi-\xi')V(\xi')d\xi' + G(\xi) \quad , \tag{10}$$

where the kernel is

$$F(\xi) = Kq\dot{m}''(t) - 1 \quad , \tag{11}$$

and the forcing term is

$$G(\xi) = 1 - Kq[\dot{m}''(0) - \dot{m}''(t)]x_{po}/(x_{fo} - x_{po}) - K[\dot{Q}'(0) - \dot{Q}'(t)]/(x_{fo}-x_{po}) \quad , \tag{12}$$

in which $x_{po}/(x_{fo} - x_{po}) = [Kq\dot{m}''(0)]^{-1}$ if $\dot{Q}'(0)=\dot{Q}_0'$. Equation (10) is seen so be nonhomogeneous and of the Volterra type.

To gain an understanding of the character of the solution to Eq. (10) consider first the early-time behavior for $\xi<<1$. Since the assumption that $v_{po} > 0$ for $\dot{Q}'(0) = 0$ translates with $n=1$ to $Kq\dot{m}''(0)>1$, we see that $F(0)>0$, so that the short-time solution to Eq. (10) becomes $V(\xi) = 1+[F(0)+G'(0)]\xi + \ldots$, where $G'(0)=(K/v_{po})$ $[qx_{po}(d\dot{m}''/dt)_0 + (d\dot{Q}'/dt)_0]$, in which the subscripts o on the derivatives mean that they are to be evaluated at $t=0+$. If $\dot{Q}'(t)=\dot{Q}'(0)$ then $d\dot{Q}'/dt=0$, and from Fig. 1 it is then seen that $G'(0)$ will be zero or negative [since $(d\dot{m}''/dt)_0$ becomes zero or negative]. Thus, the initial motion of the pyrolysis front is acceleratory, i.e., $v_p(t)$ increases with $t$, if $F(0) + G'(0) > 0$, i.e., if [when $(d\dot{Q}'/dt)_0=0$] $Kq\dot{m}''(0)>$ $1 - Kq(d\dot{m}''/dt)_0(x_{po}/v_{po})$. This condition holds true for curve 1 of Fig. 1, which has $(d\dot{m}''/dt)_0 = 0$. It could be violated by a rapid initial rate of decrease of the gasification rate, like that of curve 2 of Fig. 1; curve 2 itself cannot be used here as $t \to t_0$ because it has $\dot{m}''(0) = \infty$ and therefore an infinite initial flame height. The observed experimental behavior of the gasification rate near $t=t_0$ (curve 0 of Fig.1) strongly suggests that $Kq\dot{m}''(0)>1-Kq(d\dot{m}''/dt)_0(x_{po}/v_{po})$ and that therefore the initial pyrolysis-front motion is acceleratory. In effect, at early times the extent of char-layer buildup is insufficient to decrease the gasification rate significantly, and the starting behavior is much like that of a noncharring fuel.

Although the initial development is expected to be acceleratory, the longtime behavior may differ appreciably from that of a noncharring fuel; a genuine tendency toward deceleration now may occur. To see this, first observe that in the absence of a strongly acceleratory driving pilot, $G(\xi)<1$ at sufficiently large values of $\xi$. Therefore if $V(\xi)$ has grown to a value sufficiently greater than unity, the nonhomogeneous term must become relatively unimportant in Eq. (10). Also, the early-time history becomes irrelevant, so that the lower limit of the integral effectively may be extended to $\xi' = -\infty$. The equation then admits a solution of the form $V(\xi)=V_0 e^{\alpha\xi}$, which by substitution requires

$$\int_0^\infty F(\xi) e^{-\alpha\xi} d\xi = 1 \quad . \tag{13}$$

This integral formula determines $\alpha$ in terms of nondimensional parameters, $a \equiv Kq\dot{m}_0''$ and $b \equiv t_b/\tau$, characteristic of the charring history.

For curve 1 of Fig. 1, use of Eq. (11) in Eq. (13) and evaluation of the integral gives

$$1 + \alpha = a(1-e^{-\alpha b}) \quad , \tag{14}$$

while curve 2, with the selection $t_c = t_b/2$, gives

$$1 + \alpha = \sqrt{\pi/2} \; a \sqrt{b} \sqrt{\alpha} \quad . \tag{15}$$

Necessary conditions for solutions $\alpha(>0)$ to exist are $ab > (1+b) + \ln(ab)$ for Eq. (14) and $a\sqrt{b} > 2\sqrt{2/\pi}$ for Eq. (15); both of these conditions require a sufficiently high level of the pyrolysis rate $(Kq\dot{m}_0'')$ and a sufficiently large ratio of the pyrolysis time to the spread time $(t_b/\tau)$ for accelerating spread to occur. A simple criterion for continued acceleration, derived from the model corresponding to curve 2 of Fig. 1, is

$$Kq\dot{m}_0''\sqrt{t_c/\tau} > 2/\sqrt{\pi} \quad , \tag{16}$$

which involves the pyrolysis rate more strongly than the burning time and is independent of the strength of the ignition source.

If the pyrolysis rate or duration is too small [e.g., if Eq. (16) is violated], then instead of the long-time exponential growth, Eq. (10) predicts deceleration toward $v_p=0$. The maximum value of $V(\xi)$, achieved prior to deceleration, is seen from Eq. (10) to be defined by $V(\xi) = [-G'(\xi)-\int_0^\xi F'(\xi-\xi')V(\xi')d\xi']/F(0)$, where the primes on F and G indicate differentiation. The character of the deceleration depends on the functional form of $\dot{m}''(t)$; spread may cease at a finite or infinite time. The kernel in Eq. (11) is not positive-definite but instead becomes negative at sufficiently large values of $\xi$, approaching $-1$ as $\xi \to \infty$. The nondimensional time of termination of spread, $\xi_t$, and the nondimensional extent of spread, $X_t \equiv \int_0^{\xi_t} V(\xi)d\xi$, may be estimated from the criterion $\int_0^{\xi_t} F(\xi_t-\xi)V(\xi)d\xi = 0$, obtained from Eq. (10). For a pyrolysis response like curve 2 of Fig. 1, it appears that when Eq. (16) is violated the deceleration is slow enough to give $\xi_t = \infty$, and for large values of $\xi$, $\int_0^{\xi_t} V(\xi')d\xi' \to X_t - \text{const.}/\sqrt{\xi}$, so $V \to \text{const.}/\xi^{3/2}$. Further study of Eq. (10) could aid in further clarification of both acceleratory and deceleratory upward spread of flames along charring fuels.


3. EXPERIMENTAL DESIGN

In the experiments the $CH_4$ flow rate to the pilot burner at the base of the vertically mounted fuel samples was set and held at a fixed value, while thermocouple, video and visual observations were made. In some of the tests the burner (whose energy release rate was adjustable between 8.6 and 52 kW) was turned off after ignition of the fuel was achieved.

Polymethylmethacrylate and (Douglas-fir particle-board) wood slabs 1.3 cm thick were the materials employed. In the apparatus, sketched in Figure 2, the wood or plastic sample was flush-mounted and fixed to a larger vertical Marinite (inert) wall. The sample and burner widths were 30 cm. Despite possible three-dimensiona edge effects, the flow and propagation were observed to have a two-dimensional cha acter. Screens provided shielding from laboratory drafts, and a video camera recorded the visible propagating flame. Surface thermocouples (0.1 mm diameter, Chromel-Alumel) were mounted along the sample vertical center-line such that their beads were just beneath the surface (i.e. within $\leq$ 1 mm of sample depth).

To define the pyrolysis front, auxiliary experiments of piloted ignition unde radiative heating for nominally 2 to 6 $W/cm^2$ were performed and showed this thermo couple/sample response at the onset of flaming to yield temperatures of approximat 315 ± 25°C for polymethylmethacrylate and 340 ± 50°C for the particle-board sample These temperature measurements also showed that the observed onset of bubbles in t polymethylmethacrylate samples occurred at a recorded temperature of about 320°C. Subsequently on the basis of more detailed measurements of ignition times and ignition temperatures as functions of externally applied radiant energy flux, it was found that in the range of the ignition times of the spread experiment, temperatur $T_p$ for onset of pyrolysis may be taken as approximately 320°C for both fuels, a value consistent with temperatures at ignition measured by the lower thermocouples mounted on the sample (those adjacent to the pilot flame).

Methods considered for defining the flame-tip heights $x_f$ included detecting the onset of thermocouple temperature rise (e.g., 10°C above ambient), short-time averaging of measurements made from video records, selection of individual video frames equally spaced in time and direct visual estimates. The methods based on temperature had uncertainties in detecting small temperature increases and in equating these to the presence of the flame tip rather than to influences of hot combustion products above the flame; the resulting values of $x_f$ exhibited the same trends as those of the other methods but were larger, and a temperature rise of 40°C provided best agreement of $x_f$ with results of other methods. The short-time averaging proved considerably more time-consuming but no more revealing than selection of individual frames, the method finally adopted. The results show oscillations in time about an evolving mean, with frequencies appreciably lower than woulc have been obtained had every frame been employed; i.e., the oscillations shown in $x_f$ are indicative only of their magnitudes.

4. EXPERIMENTAL RESULTS AND DISCUSSION

Sustained upward propagation occurred for the plastic, over the entire range of burner energy-supply rates, irrespective of whether the burner was turned off after ignition. However, even with the burner left on for ten minutes, no wood sample exhibited sustained propagation. The maximum height of the observed char (pyrolysis) front on the wood increased appreciably with increasing burner energy-supply rates (but not greatly with test duration if the test was sufficiently long) Thus Eq. (16) must be violated for the wood.

Representative experimental results are shown in Figs. 3 and 4 for the plastic and wood, respectively. For the plastic the first two thermocouple traces exhibit the temperature increase adjacent to the igniter flame; the remaining traces are similar to each other but displaced in time, showing upward pyrolysis propagation. For the wood the first two thermocouple traces are not greatly different from those for the plastic, although a continued gradual increase in temperature at long times is observed, possibly indicative of continuing char combustion. However, the remaining traces for the wood are quite different, showing an increase to a maximum temperature and a cooling thereafter, even though the burner remained on; the maximum temperature increase of the uppermost thermocouple was less than 100°C. Thus, by any criterion, the wood exhibited a maximum value of $x_p$, achieved at a finite ti

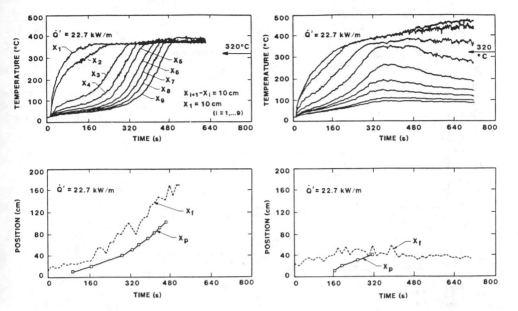

FIGURE 3.   Representative thermocouple traces and histories of $x_f$ and $x_p$ for polymethylmethacrylate.

FIGURE 4.   Representative thermocouple traces and histories of $x_f$ and $x_p$ for Douglas-fir particle board.

The histories of $x_p$ and $x_f$ shown in Fig. 3 for the plastic are both quite representative of acceleratory spread.  Spread velocities, obtained from curves like that of Fig. 3 for $x_p$, are shown in Fig. 5.  It is seen that the data on spread with the burner turned off agree well with the earlier results of Orloff et al.[3] Within the accuracy of the measurements, the slope of a line through our data points in Fig. 5 could be unity, corresponding to an exponential increase of $v_p$ with time, as first identified by Orloff et al.[3]  Thus, the data are consistent with n=1 in Eqs. (4) and (5).  However, the accuracy of the measurements is insufficient to conclude definitely that n=1; a line of lesser slope would also be consistent with the data.  In fact, within the accuracy of the measurements, Eq. (6) can correlate the data as well as any formula for n=1.  From the data we cannot distinguish with certainty between an exponential and power-law increase in $x_p$ with time.  Thus, we cannot draw conclusions about the applicability of steady-state flame-height correlations during spread.

From the results shown in Figs. 2 and 5 it is straightforward to calculate the $\tau$ of Eq. (2) for polymethylmethacrylate; $\tau$=170 s, with better than 5% accuracy. The observed spread histories are entirely consistent with Eq. (2) with a constant $\tau$, within the accuracy of the measurements.  Moreover, the value obtained for $\tau$ from the spread measurements is consistent with that which would be obtained from Eq. (1), with the independently inferred value $q''_o \approx 2.5$ w/cm$^2$, by independent estimates of properties of the polymer.  Therefore the general understanding of the spread mechanism for polymethylmethacrylate is on firm ground; the uncertainties involve only the accuracy with which Eq. (2) applies and the precise value of n in Eq. (4).

The burner-on data shown in Fig. 5 exhibit higher spread rates, as would be expected from Eq. (5), but are not accurate enough for conclusions to be drawn from numerical comparisons with theoretical predictions.

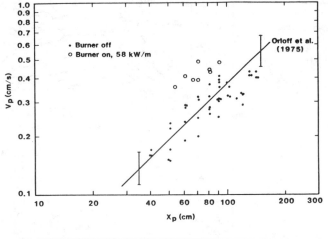

FIGURE 5.

Spread rate as a functi[on]
of pyrolysis position fo[r]
polymethylmethacrylate.

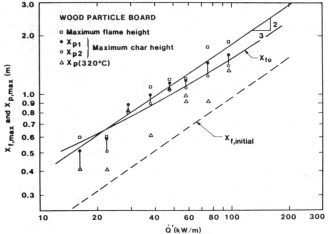

FIGURE 6.

Flame heights and pyro-
lysis heights for wood
as functions of the burn[er]
energy-supply rate.

The histories of $x_f$ and $x_p$ shown in Fig. 4 for wood clearly are not indicativ[e] of acceleratory spread. Although theoretically there should be an initial period of acceleration, the measurements cannot be made with sufficient refinement to ex-hibit it. The deceleration of spread and its eventual cessation are clear from Fig. 4. After a small increase to a maximum value the $x_f$ curve exhibits a gradual decrease toward the initial value associated with the burner flame alone. The $x_p$ curve terminates at a maximum value that depends on the selection of $T_p$; the first two points on this curve (and on the corresponding curve of Fig. 3 as well) should be ignored in making comparisons with the spread theories because they refer to positions below the initial burner flame-tip height. That the last six thermo-couple traces peak at the same time is a good indication that flame propagation should be considered to cease before $x_p$ reaches these thermocouples. Thus, only one of the thermocouple traces in Fig. 4 definitely corresponds to a position abov[e] the initial flame-tip height ($x_{po}$) and below the final maximum value of $x_p$. The data therefore clearly are insufficient for testing theoretical predictions of the manner in which $v_p$ decelerates to zero. An estimate of an average value for $v_p$ ca[n] be obtained and reasonably compared with $v_{po}$, defined above Eq. (8); uncertainties in $x_{fo}$ and in $\tau$ are great enough to make the agreement acceptable but the comparison nondefinitive.

Additional information obtainable from the results of the experiments with wood is shown in Fig. 6. Three different measures of the maximum value of $x_p$ are shown, one based on attaining a thermocouple temperature of 320°C, one $(x_{p1})$ based on observed darkening of the wood, and one $(x_{p2})$ based on a 2 mm char-layer thickness (after termination of the experiment). It is seen that the first of the three measures gives values appreciably lower than the last two. This is consistent with the first being more relevant to spread and the charring at higher positions being produced by continued heating of the sample by the burner in these relatively long experiments ($\sim$10 min.); charring occurs when wood is heated for long periods at energy fluxes too low to produce ignition.[20,21,22] With the 320°C points used to identify the maximum value of $x_p$, it is seen from Fig. 6 that if the initial value of $x_f$ is equated with $x_{po}$, as indicated above Eq. (8), then the total extent of spread is small (e.g., $x_p - x_{po} < x_{po}$ always). For the wood we may estimate 170 s $\lesssim \tau < 350$ s, 300 s $\lesssim t_b \lesssim 800$ s (so 150 s $\lesssim t_c < 400$ s), $K \approx 0.01$ m/(kJ/m s) (with the assumption that n=1),[19] $q \approx 10$ kJ/g, and $\dot{m}'' \approx 10$g/m$^2$s, so $Kq\dot{m}''_o \approx 1$, leading roughly to an equality in Eq. (16); thus the wood should be marginally capable of continued propagation, and sufficiently intense external radiant energy input should lead to continued spread.

The maximum flame height shown in Fig. 6 is roughly twice the initial flame height.[19] The initial flame height correlates well with Eq. (4) with n=2/3, as it must. Since the extent of spread is small we should expect the maximum $x_f$ to be close to the value $x_{fo}$ defined above Eq. (8); this is borne out, within the accuracies of the calculation and of the data, as seen in Fig. 6. Because of the small extent of spread, the power-law correlation for steady wall flames is much better for wood than for polymethylmethacrylate.

ACKNOWLEDGEMENTS

The authors wish to acknowledge the contributions of Ms. C. Carpentier and Ms. M. Harkleroad, who conducted the ignition experiments for identifying $T_i$; Mr. A. Flores, who developed the computer programs for reducing the flame spread data; and Mr. W. Rinkinen, who determined the flame heights from the video records and who assisted in the conduct of the flame-spread experiments. One author (FAW) wishes to acknowledge partial support from a National Sciences Foundation Grant (#INT-8403848).

REFERENCES

1. A. C. Fernandez-Pello and T. Hirano, Combustion Science and Technology 32, 1 (1983).

2. G. H. Markstein and J. de Ris, Fourteenth Symposium (International) on Combustion, The Combustion Institute, Pittsburgh (1973), pp. 1085-1097.

3. L. Orloff, J. de Ris and G. H. Markstein, Fifteenth Symposium (International) on Combustion, The Combustion Institute, Pittsburgh (1975), pp. 183-192.

4. M. Sibulkin and J. Kim, Combustion Science and Technology 17, 39 (1977).

5. A. C. Fernandez-Pello, Combustion Science and Technology 17, 87 (1977).

6. A. C. Fernandez-Pello, Combustion and Flame 31, 135 (1978).

7.   A. C. Fernandez-Pello, Combustion and Flame 36, 63 (1979).

8.   K. Annamalai and M. Sibulkin, Combustion Science and Technology 19, 185 (1979)

9.   R. L. Alpert, "Pressure Modeling of Vertically Burning Aircraft Materials," FAA-RD-78-139 (1979).

10.  A. C. Fernandez-Pello and C. P. Mao, Combustion Science and Technology 26, 147 (1981).

11.  L. Chu, C. H. Chen and J. S. T'ien, ASME Paper No. 81-WA/HT -42 (1981).

12.  G. Carrier, F. Fendell and S. Fink, Combustion Science and Technology 32, 161 (1983).

13.  K. Kishitani, S. Sugawara and K. Hamada, Saigai no Kenkyu 15, 133 (1984).

14.  M. A. Delichatsios, ASME Paper No. 83-WA/HT-64 (1983).

15.  D. B. Adarkar and L. B. Hartsook, AIAA Journal 4, 2246 (1966).

16.  H. C. Kung, "A Mathematical Model of Wood Pyrolysis," Factory Mutual Research Corporation, Report No. 19721-6 (1971).

17.  R. C. Corlett and F. A. Williams, Fire Research 1, 323 (1978).

18.  M. A. Delichatsios and J. de Ris, "An Analytical Model for the Pyrolysis of Charring Materials," Factory Mutual Research Corporation, Report No. J.I. OkOJ1.BU (1983).

19.  J. Quintiere, M. Harkleroad and Y. Hasemi, "Wall Flames and Implications for Upward Flame Spread," AIAA Paper No. 85-0456 (1985).

20.  F. J. Kilzer and A. Broido, Pyrodynamics 2, 151 (1965).

21.  F. A. Williams, Progress in Energy and Combustion Science 8, 317 (1982).

22.  M. J. Antal, Jr., "Biomass Pyrolysis: A Review of the Literature," Advances i Solar Energy p. 61 (1982).

# Thermal Modeling of Upward Wall Flame Spread

**YUJI HASEMI**
Building Research Institute
Ministry of Construction
Tatehara 1, Oho-machi, Tsukuba-gun
Ibaraki-ken, 305, Japan

ABSTRACT

This paper describes an engineering model of the upward turbulent flame spread along a vertical combustible surface based on a concept of ignition and flame spread as a result of inert heating of the solid to an ignition temperature. Experiments were made by using porous line burners to represent the wall flame heat transfer as a function of heat release rate and pyrolysis length. An exploratory analysis was made to correlate flame spread properties with thermometric material properties based on this model. This analysis seems to be consistent with current experimental work on turbulent wall flames.

INTRODUCTION

Control of the combustibility of wall is evidently essential to prevent a rapid fire growth in an enclosure, since fires tend to develop much faster along a wall than on a floor. Although most conventional theoretical work on flame spread has assumed an ideal laminar flame without radiation[1-3], a wall flame during a building fire is usually turbulent, and its spread is dominated by flame radiation[4,5]. While the turbulent flame spread may result from complicated interactions of chemical, kinetic and thermal processes[6], an engineering model of steady turbulent wall flame spread was formulated on the basis of a concept of ignition and flame spread as a result of inert heating of the solid to an ignition temperature[7,8]; Figure 1 shows its conception. This model assumes a one-dimensional flame spread in the x-direction and a one-dimensional thermal conduction in the solid normal to the surface. The location of the pyrolysis front is identified by $x_p$ where the surface temperature has reached an ignition temperature, $T_{ig}$. The aim of this paper is to formulate the flame spread velocity as a function of known material properties and correlate relevant thermal processes with material properties by experiments.

FORMULATION OF WALL FLAME SPREAD

If the formation of char at the fuel surface is negligible or ignored, the surface temperature at x, $T_w(x,0)$, can be represented by the convolution as

$$T_w(x,0) - T_o = \int_0^t \dot{q}_w''(x,t-\tau) \cdot \phi(\tau) d\tau \qquad (1)$$

where $\dot{q}_w''(x,t-\tau)$ is the heat flux applied to the surface at $(t-\tau)$ after the

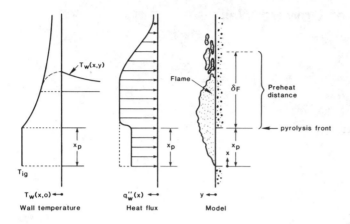

FIGURE 1. Schematic diagram of upward flame spread.

ignition. $\phi(\tau)$ is an impulse response of the surface temperature to heat application, and its functional form is dependent on wall conditions. The location of the pyrolysis front is given as a function of time by solving

$$T_{ig} - T_o = \int_0^t \dot{q}_w''(x, t-\tau) \cdot \phi(\tau) d\tau \tag{2}$$

for x. This equation can be solved for x only implicitly; however, it may be transformed into an explicit form for x by assuming some dependence of the location of pyrolysis front on time; most analysis on flame spread has dealt with a steady flame spread, i.e. $V_p = dx_p/dt = $ constant. Similar situation will be studied in this paper for the following two simple wall conditions. The limit of equation (2) as $t \to \infty$ should be considered for this analysis, since the steady flame spread velocity must not depend on time.

Semi-infinite Thick Combustible Wall

Ignoring the surface reradiation just for the simplicity, the impulse response function, $\phi(\tau)$, can be represented as $\phi(\tau) = (\pi k \rho c \tau)^{-1/2}$. Assuming $\dot{q}_w''$ is independent of flame spread velocity and replacing $\tau$ with $x_p$ by $x_p = -V_p \tau$, then equation (2) will become

$$T_{ig} - T_o = \int_{-\infty}^{x} \frac{\dot{q}_w''(x-x_p+L_p)}{\sqrt{\pi k \rho c V_p (x-x_p)}} dx_p \tag{3}$$

Equation (3) can be solved for $V_p$ as

$$V_p = \left\{ \int_0^\infty \dot{q}_w''(\xi+L_p)/\sqrt{\xi} d\xi \right\}^2 / \pi k \rho c (T_{ig} - T_o)^2 \tag{4}$$

where $\xi$ is the height above the pyrolysis front. This equation yields the Sibulkin-Kim relationship of wall flame spread[7], if the surface heat flux is represented as $\dot{q}_w'' = \dot{q}_o'' \cdot \exp(-\xi/\sigma)$ with constant $\dot{q}_o''$ and $\sigma$.

## Thermally Thin Wall With Newtonian Cooling

For a thin wall whose temperature is regarded as uniform in the normal direction to its surface(Figure 2), the impulse response function, $\phi(t)$, can be represented as $\phi(t) = \exp(-h_i t/\rho cd)$ where $h_i$ is the heat transfer coefficient. Substituting this into equation (2), the steady state flame spread velocity can be represented as

$$V_p = \frac{1}{\rho cd(T_{ig}-T_o)} \cdot \int_0^\infty \dot{q}_w''(\xi+L_p) \cdot \exp(-h_i\xi/\rho cdV_p)\,d\xi \tag{5}$$

Iteration is necessary to solve equation (5) for $V_p$. However, for an insulated wall,i.e. $h_i=0$, equation (5) yields

$$V_p = \frac{1}{\rho cd(T_{ig}-T_o)} \cdot \int_0^\infty \dot{q}_w''(\xi+L_p)\,d\xi \tag{6}$$

In equations (4) and (6), such parameters as $\rho,c,k$ and $T_{ig}$ are the material properties to be determined by thermometric measurements. The integrals appearing in these equations can be estimated by formulating the distribution of $\dot{q}_w''$ as a function of height above pyrolysis front.

The wall heat flux in the flame is believed to be governed by flame radiation(4);this suggests that flame height can be a scaling factor representing the distribution of wall flame heat transfer. On the analogy of unconfined fires, the flame height is expected to depend practically only on gross heat release rate per unit length of the pyrolysis zone, $Q_\ell$, and pyrolysis length, $L_p$ (e.g.10). Heat release rate is proportional to volatilization rate, $\dot{m}_v''$, which is to be determined by the heat balance around the fuel surface as

$$\dot{q}_w'' - \dot{q}_{rr}'' = \Delta H_G \cdot \dot{m}_v'' \tag{7}$$

The surface reradiation, $\dot{q}_{rr}''$, is related to the temperature of fuel surface, $T_{ig}$, and its emissivity. Although the material properties appearing in this model may not be obtained directly for some composite materials, the model will still serve as a paradigm to develop a testing method concerning the wall flame spread. Therefore, heat release rate can be again correlated with heat application to the fuel surface and such material properties as $T_{ig}$ and calorific potential.

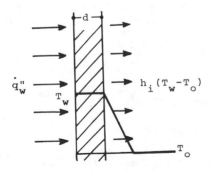

FIGURE 2.  Thermally thin wall model.

EXPERIMENTAL WALL FLAME HEAT TRANSFER CORRELATIONS

The central problem in estimating the flame spread properties from the present model is to formulate the distribution of wall flame heat transfer as a function of material properties. Since the intermittency of the flame is believed to dominate wall flame heat transfer due to radiation, flame height, either for flame tips and for solid flame, is expected to be a measure to represent the preheat distance of a burning wall. Thus, measurements of total heat flux to the wall surface and the visible flame height were made by using porous, methane, line burners against isothermal and thermally thin walls. Application of these burners were just for convenience; the behavior of flame near the burners may be different from that of vertical fuel. This difference will be discussed by comparing the present result with previous measurement on vertical wicks.

The isothermal wall had a smooth black-painted front surface, and it was water-cooled through coils on its back surface. Flat water-cooled side panels helped to maintain a two-dimensional flow pattern along the test wall. The surfaces of the test wall and the side walls were kept at a constant temperature, approximately 60°C. The widths of the methane burners were 0.0375m and 0.082m. The thin-wall experiments were made by using a porous methane burner 0.075m wide and 0.92m in length against one wall of a 2.80m x 2.80m x 2.18m enclosure covered with a ceramic fiber insulation board. A precise description of the enclosure is given elsewhere(11).

Total heat flux was monitored by Gardon-type flux meters. The reported values of total heat flux are the average of data recorded for 5 minutes at the intervals of 10 seconds. Heat release rate was estimated from the flow rate of the fuel gas, assuming complete combustion. Visible flame was monitored by a video camera during both experiments. The reported values of flame height are the average of the height of flame tips observed for 3 minutes at intervals of 0.5 seconds on the videotape by three observers. The height of solid flame seemed to be less consistent among the observers.


Flame Height Correlations

For unconfined turbulent diffusion flames, dimensional analysis based on the Froude number has established a dependence of the flame height on the size and intensity of fuel(10); $Q \equiv Q/\rho_\infty C_p T_0 g^{1/2} D^{5/2}$ may be a dominant dimensionless parameter concerning this problem(12). Similar analysis on a line fire results in a dependence of flame height on $Q_\ell^* \equiv Q_\ell/\rho_\infty C_p T_0 g^{1/2} D^{3/2}$, since the square root of the inertial force to buoyancy ratio for a constant maximum temperature at each height, $\theta_m$, can be represented as

$$(T_0/Dg\theta_m)^{1/2} \cdot u_c = (T_0/Dg\theta_m)^{1/2} \cdot Q_\ell/\rho_\infty C_p \theta_m D \propto Q_\ell/\rho_\infty C_p T_0 (gD^3)^{1/2} \qquad (8)$$

where the characteristic velocity, $u_c$, is defined as $u_c \equiv Q_\ell/\rho_\infty C_p \theta_m D$. Assuming the following functional expression for the flame height,

$$L_F = \gamma \cdot Q_\ell^{*n} \cdot D \qquad (9)$$

the power of $Q_\ell^*$ representing $L_F$, n, should approach 2/3 for a large value of $Q_\ell^*$, since the fuel size effect on flame height must vanish in the $Q_\ell^* \to \infty$ limit. The parameter, $\gamma$, will take a particular value according to the definition of flame height.

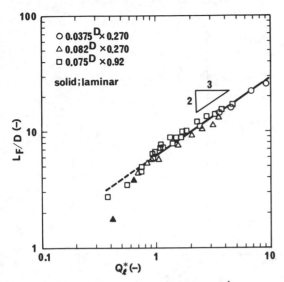

FIGURE 3. Height of flame tips vs. $Q_\ell^*$.

Figure 3 displays the dependence of $L_F/D$ on $Q_\ell^*$ based on the height of flame tips obtained from the three burners; the result shows that the height of wall flame is apparently proportional to 2/3 power of $Q_\ell^*$ for $Q_\ell^* \gtrsim 1$. The power of $Q_\ell^*$ representing the flame height appears to be slightly larger for $Q_\ell^* < 1$ than for $Q_\ell^* > 1$, i.e. $n \approx 0.8$. The parameter, $\gamma$, is approximately 6.0. This result seems to be very close to a recent analysis by Delichatsios(13). The height of solid flame was approximately $L_F \approx 2.8 Q_\ell^{*n} \cdot D$.

Heat Flux To The Wall Surface

A parameter characterizing the visible flame height for a wide range, $Q_\ell^{*2/3}$ $D = (Q_\ell / \rho_\infty C_p T_o g^{1/2})^{2/3}$, may be a characteristic scale length to classify the regions of an upward flow for the analysis of wall flame heat transfer. Figure 4 represents total heat flux to the isothermal wall against height normalized with $Q_\ell^{*2/3} D$. The data of total heat flux for a CO burner and for a saturated vertical wicks obtained by Faeth et al(9,14) are superimposed for reference. The arrows for the data of Liburdy and Faeth show the direction in which their data must move, because they used local heat flux estimated from temperature and velocity measurements instead of the heat release rate. The ratio of the local heat flux to the heat release rate, $Q_x/Q_\ell$, for the data of Liburdy and Faeth is estimated to be as low as 0.74; the length of each arrow shows the range of the correction.

The results of the five experiments are almost identical and they appear to cluster densely along one curve. For the analysis of the mechanism on wall flame heat transfer it may be convenient to classify the wall surface into the following three distinct regions.
I) $x/Q_\ell^{*2/3} D \lesssim 2.8$ (solid flame):
For $x/Q_\ell^{*2/3} \cdot D \lesssim 1$, $\dot{q}_w''$ increased with height for 0.0375m wide burner, while it was almost constant for 0.082m wide one. For $x/Q_\ell^{*2/3} \cdot D > 1$, $\dot{q}_w''$ is apparently constant with height, and it appears to be a weakly increasing function of $Q_\ell^*$ (Figure 5).

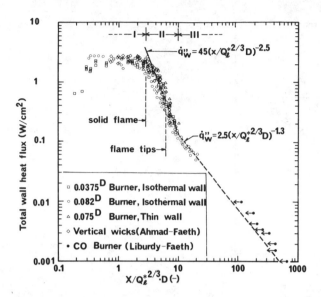

FIGURE 4. Total wall heat flux vs. normalized height.

II) $2.8 \leq x/Q_\ell^{*2/3}D \leq 10$(transition region):
    The slope is the steepest among the three regions; all data fall on the curve represented as

$$\dot{q}_w'' = 45(x/Q_\ell^{*2/3} \cdot D)^{-2.5} \tag{10}$$

This region may be characterized by the intermittency of flame. Recent wall flame radiation measurement by Kulkarni(14) using the same apparatus shows that the radiative heat flux to a wall surface decreases significantly in this region.

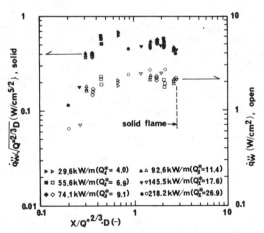

FIGURE 5. $\dot{q}_w''$ (open) and $\dot{q}_w'' / \sqrt{Q_\ell^{*2/3}D}$ (solid) vs. normalized height,(0.0375m$^D$ burner).

III) $x/Q_\ell^{*2/3} D \gtrsim 10$(plume region):

Although sufficient number of data have not been obtained for this region, the slope representing the dependence of heat flux on height is clearly less steep than in the transition region. Since the data of Liburdy and Faeth in Figure 4 must move slightly to the left in the figure, heat flux in this region will be represented approximately by

$$\dot{q}_w'' = 2.5 (x/Q_\ell^{*2/3} \cdot D)^{-1.3} \tag{11}$$

## DEPENDENCE OF FLAME HEIGHT AND FLAME SPREAD VELOCITY ON PYROLYSIS LENGTH

The present analysis has resulted in a good correlation between incident wall heat flux and the normalized height, $x/Q_\ell^{*2/3} \cdot D$; however, heat transfer within the pyrolysis zone should be still studied so that heat release rate can be represented as a function of material properties. For the heat transfer on a combusting surface, Orloff et al(5) measured radiative flux using a 3.56m high PMMA vertical slab and estimated convective heat flux; their measurement resulted in a proportionality of heat flux to $x^{0.2}$ for x<1.0m and to $x^{0.5}$ for x>1.0m. This proportionality is consistent with smooth wall heat transfer analysis for x<1m and with rough wall analysis for x>1m(16). This seems to be the only measurement of the distribution of incident flux within the pyrolysis zone of a wall fire. An exploratory analysis was made to correlate such flame spread properties as flame height and flame spreading velocity with material properties on the basis of the present model and Orloff et al's experiment.

Flame Height

By combining equations(7) and (9), flame height can be represented as a function of incident heat flux at vaporizing surface as

$$L_F = \gamma \left( \Delta H_c \int_0^{L_p} \dot{m}_v'' dx / \rho_\infty C_p T_o \sqrt{g L_p}^3 \right)^n \cdot L_p = \gamma \left\{ \Delta H_c \int_0^{L_p} (\dot{q}_w'' - \dot{q}_{rr}'') dx / \Delta H_G \rho_\infty C_p T_o \sqrt{g L_p}^3 \right\}^n \cdot L_p \tag{12}$$

where $\Delta H_c$ is the heat of combustion (J/g) and $\Delta H_G$ is the vaporization heat. Assuming $\dot{q}_w''$ is proportional to $x^{0.2}$ for x<1m, equation (12) implies that, if the surface reradiation is insignificant, flame height is proportional to $x^{0.76}$ for $Q_\ell^*$<1.0,i.e. n≈0.8, and to $x^{0.8}$ for $Q_\ell^*$>1.0,i.e. n≈2/3. This relationship seems to be consistent with previous measurements of the average height of flame tips of wall fires; Orloff et al found $L_F \propto x_p^{0.781}$ for $0.18<x_p<0.85$m of a vertical PMMA slab with side walls(4), and Kishitani et al obtained $L_F \propto x_p^{0.774}$ for $0.05<x_p<0.50$m of a PMMA slab without side walls(17). The asymptotes of this proportionality as x→∞ is $L_F \propto x_p$, since $\dot{q}_w''$ is approximately proportional to $x^{0.5}$ for x>1m. Using the following values of $\Delta H_c$, $\Delta H_G$, $\dot{q}_w''$ and $\dot{q}_{rr}''$ for PMMA,

$$\Delta H_c/\Delta H_G \approx 13, \quad \dot{q}_{rr}'' \approx 1 W/cm^2, \quad \dot{q}_w'' \approx 3x^j W/cm^2 (j \approx 0.2 \text{ for x<1m, } j \approx 0.5 \text{ for x>1m})$$

the flame height can be calculated from equation (12) as shown in Figure 8;the calculated values of $Q_\ell^*$ for $L_p$ < 1m were in the range of 0.17 ~0.21, and this implies that a solid flame would never cover the preheated region above a pyrolysis front for $L_p$ < 1m. Therefore, the incident wall heat flux is never constant, although it is assumed to be constant under the height of flame in some previous analysis (e.g.8). The higher flame height of Orloff et al than Kishitani et al's is probably due to the presence of side walls. The calculation resulted in flame heights somewhat close to Kishitani et al's.

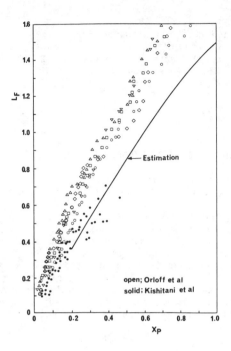

FIGURE 6. Experimental and calculated flame heights.

Flame Spread Velocity

Since the preheated region above a pyrolysis front is expected not to be covered with a solid flame for $L_p < 1m$ of a PMMA vertical slab, flame spread velocity, formulated as equation (4) or (6), can be calculated by using only equations (10) and (11). Figure 7 compares the flame spread velocity estimated from the material properties for a semi-infinite thick wall with the result of experimental work by Orloff et al and by Saito et al(18); this comparison demonstrates that the estimation tends to provide slightly larger velocity than the experiment. This seems to be reasonable, because a steady state flame spread velocity should give the upper limit of $V_p$ for growing wall flames.

CONCLUSIONS

Approximate relationships between the main wall flame properties and the thermometric fuel properties are derived from experiments using porous line burners against walls. The following conclusions can be drawn.
1) The flame height against a wall is proportional to $Q_{\ell}^{*n}D$, where n is approximately 0.8 for $Q_{\ell}^* < 1$ and 2/3 for $Q_{\ell}^* > 1$.
2) The incident heat flux to the wall surface is represented approximately as a function of $x/Q_{\ell}^{*2/3}D$.
3) Estimation of wall flame height and flame spreading velocity from thermometric material properties using the heat transfer correlations obtained through the present experiment was found to be consistent with previous experimental work.

94

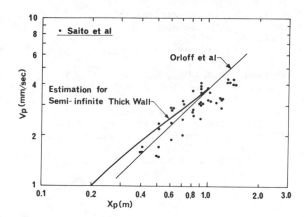

FIGURE 7. Experimental and calculated flame spread velocity.

ACKNOWLEDGEMENTS

The experiment introduced in this paper was conducted at the Center for Fire Research, NBS, Gaithersburg. The author would like to thank Dr.J.G.Quintiere and Mr.K.D. Steckler of NBS for the arrangement of the experiment. The author is also indepted to Mr.W.J.Rinkinen, Ms.M.Harkleroad and Mr.D.Wagger for the assistance in the experiment. Discussions of Dr.J.G.Quintiere and Dr.A.Robertson of NBS, and Dr.K.Saito of Princeton University are gratefully appreciated.

REFERENCES

1. de Ris,J.: Spread of a laminar diffusion flame. Twelfth Symposium (International) on Combustion,pp.241, 1968.
2. Fernandez-Pello,A.C,: A theoretical model for the upward laminar spread of flames over vertical surfaces. Combustion and Flame, 35, pp.135, 1978.
3. Annamalai,K., and Sibulkin,M.: Flame spread over combustible surfaces for laminar flow systems, Part I. Combustion Science and Technology, 19, pp.167, 1979.
4. Orloff,L., de Ris,J., and Markstein,G.H.: Upward turbulent fire spread and burning of fuel surface. Fifteenth Symposium(International) on Combustion, pp.183, 1974.
5. Orloff,L., Modak,A.T., and Alpert,R.L.: Burning of large-scale vertical surfaces. Sixteenth Symposium on Combustion, pp.1345, 1976.
6. Fernandez-Pello,A.C., and Hirano,T.: Controlling mechanism of flame spread. Combustion Science and Technology, 32, pp.1, 1983.
7. Sibulkin,M., and Kim,J.: The dependence of flame propagation on surface heat transfer II, Upward burning. Combustion Science and Technology, 17, pp.39, 1977.
8. Quintiere,J.G., and Fernandez-Pello,A.C.: A simplified model of radiating turbulent upward flame spread over the surface of a charring combustible. Eastern Section Meeting, Combustion Institute, 1982.
9. Liburdy,J.A., and Faeth,G.M.: Fire induced plume along a vertical wall, Part I. NBS Grant 5-9020, 1977.
10. Thomas,P.H., Webster,C.T., and Raftery,M,M.: Some experiments on buoyant diffusion flames. Combustion and Flame, 5, pp.359, 1961.
11. Steckler,K.D.: Fire induced flows through room opening-Flow coefficients.

Technical Research Report, Armstrong World Industries, 1981.

12. Zukoski,E.E., Kubota,T., and Cetegen,B.: Entrainment in fire plumes. Fire Safety Journal, 3, pp.107, 1980/81.

13. Delichatsios,M.A.: Modeling of aircraft cabin fires. Technical Report, Factory Mutual Research Corp., 1984.

14. Kulkarni,A.K.: Private communication. 1984.

15. Ahmad,T., and Faeth,G.M.: Fire induced plumes along a vertical wall, Part III. NBS Grant 5-9020, 1978.

16. Hasemi,Y.: Experimental wall flame heat transfer correlations for the analysis of upward flame spread. Fire Science and Technology, 4, 1984.

17. Kishitani,K., Sugawara,S., and Hamada,K.: Upward flame spread of building materials. "Saigai no Kenkyu", 15, pp.133, 1984( in Japanese).

18. Saito,K.: Private communication. 1985(to be presented at the First International Symposium on Fire Safety Science, NBS)

## TERMINOLOGY

$C_p$ : specific heat of air
$D$ : characteristic fuel size
$L_F$ : flame height
$L_p$ : pyrolysis length
$Q$ : heat release rate
$Q^*$ : dimensionless heat release rate$(Q/\rho_\infty C_p T_o \sqrt{gD^5})$
$Q_\ell$ : heat release rate per unit length
$Q_\ell^*$ : dimensionless heat release rate per unit length$(Q_\ell/\rho_\infty C_p T_o \sqrt{gD^3})$

$Q_x$ : local heat flux $\left( \int_0^\infty \rho \dot{C}_p u\theta dy \right)$

$T_{ig}$ : ignition temperature
$T_W$ : wall temperature
$V$ : flame spread velocity
$c_p$ : specific heat of wall material
$d$ : wall thickness
$g$ : gravitational acceleration
$h_i$ : heat transfer coefficient
$k$ : thermal conductivity
$\dot{m}''$ : fuel vaporization rate
$\dot{q}''_c$ : convective heat flux to wall surface
$\dot{q}''_{rr}$ : surface reradiation
$\dot{q}''_w$ : incident heat flux to wall surface
$t, \tau$ : time
$u$ : upward velocity
$x$ : height from the bottom of fuel
$x_p$ : location of pyrolysis front
$y$ : horizontal coordinate normal to wall surface
$\Delta H_c$ : heat of combustion
$\Delta H_G$ : gasification heat
$\phi(t)$ : impulse response of $T_W$ to $\dot{q}''_W$
$\rho$ : density
$\theta$ : excess temperature
$\xi$ : height from pyrolysis front

## Suffix

$c$ : characteristic value
$m$ : maximum
$o, \infty$ : ambience

# Effect of Sample Orientation on Piloted Ignition and Flame Spread

**A. ATREYA**
Department of Mechanical Engineering
202 Engineering Building
Michigan State University
East Lansing, Michigan 48824, USA

**C. CARPENTIER**
IRBAT, Paris, France
National Bureau of Standards
Gaithersburg, Maryland 20899, USA

**M. HARKLEROAD**
Center for Fire Research
National Bureau of Standards
Gaithersburg, Maryland 20899, USA

ABSTRACT

An experimental investigation is conducted to study the effect of sample orientation on piloted ignition and opposed-wind flame spread. Two types of wood (red oak and mahogany) were used for the purpose and two orientations (horizontal and vertical) were investigated. In the horizontal mode, axisymmetric fire spread over wood samples was studied and the corresponding piloted ignition tests were conducted on smaller samples of the same wood. In the vertical mode, lateral flame spread and piloted ignition tests were conducted in a radiant panel test apparatus.

The experimental data were reduced according to the thermal flame spread theory of deRis using the measured surface temperatures. It was found that as long as the temperatures are defined consistently with the thermal theory, the results are orientation independent within the measurement error. The reasons for this orientation independence are: (i) dominant re-radiative losses, and (ii) insensitivity of the flame spread rate to the induced air velocity at ambient $O_2$ concentrations.

Key words: Ignition, Flame spread, Sample orientation, Wood, Measurements.

INTRODUCTION

The ultimate goal of fire research is to provide a scientific and technical basis to minimize fire losses. Typically, in building fires, a wide variety of materials oriented at different angles relative to gravity burn under varying levels of externally supplied radiation. Hence, it is important to understand the effect of sample orientation and external radiation on fire initiation (i.e. ignition) and fire growth (i.e. flame spread).

The literature on both flame spread and ignition is abundant. Several reviews have also been published on these subjects. The pioneering work of deRis [1] on flame spread and the review by Fernandez-Pello and Hirano [2], the extensive work of Kashiwagi [3,4] on ignition and the review by Kanury [5] provide excellent sources of information.

Generally, two modes of flame spread (wind-aided and wind-opposed) and two modes of ignition (auto and piloted) have been recognized. A close relationship between piloted ignition and opposed-wind flame spread has also been established by Quintiere and coworkers [6,7]. However, in these previous studies, the effect due to changes in sample orientation on piloted ignition and opposed-wind flame spread has not received much attention. For instance, horizontal and vertical sample orientations provide very different buoyant flow configurations that would be expected to significantly affect both ignition and flame spread. Thus, the objective of this work is to experimentally investigate the effect of sample orientation on piloted ignition and opposed-wind flame spread mechanisms.

Horizontal and vertical sample orientations were examined for two kinds of wood (red oak and mahogany) under different levels of externally supplied radiation. In the horizontal mode, axi-symmetric fire spread over wood samples was studied and the corresponding piloted ignition tests were made on smaller samples of the same wood. In the vertical sample mode, lateral flame spread and piloted ignition tests were conducted in a radiant panel test apparatus. The results for the two types of wood were qualitatively similar, thus data for only mahogany are presented here.

EXPERIMENTAL PROCEDURE

The experimental arrangements used for the horizontal and vertical modes are schematically shown in Figures 1 and 2. Further details of the experimental setup are given in Reference 8 for the horizontal mode and Reference 9 for the vertical mode.

The Horizontal Mode

In the horizontal mode [Figure 1] the flame spread tests were conducted on samples two feet in diameter and 0.75 inches thick. The fire was started at the center of the sample and allowed to grow up to the edge under conditions of prescribed external radiation. Surface temperatures were measured by thermocouples installed perpendicular to the radial spread direction and the spread rate was determined by photographs taken during the tests. At least six tests were conducted on each type of wood. The data for the first six inch diameter and the last one inch were not used to avoid possible variations caused by ignition and edge effects.

The ignition tests in the horizontal mode were conducted on 3" x 3" x 0.75" samples exposed to a known external radiation flux. The sample edges

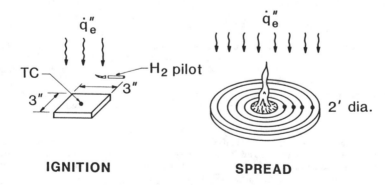

Figure 1  A schematic of the experiments in the horizontal mode. The black dots represent the thermocouples.

were shielded by aluminum foil to avoid any edge effects. To test for piloted ignition, a small hydrogen flame was lowered at regular intervals (one second) close to the pyrolyzing surface at a height no more than two millimeters. The pilot flame was off-center and the sample was very slowly rotated to avoid heating a particular spot. The surface temperature was continuously monitored and no measurable rise in surface temperature was observed because of the pilot flame. The exposure time required for sustained flaming was also recorded.

The Vertical Mode

In the vertical sample mode, lateral flame spread experiments were conducted on 6" x 31" x 0.75" samples [Figure 2] exposed to a known external irradiance. Ignition was instigated by an acetylene-air pilot positioned in the fuel plume above the sample. Flame position as a function of time was recorded by a video camera and the flame spread rate was determined by applying a running 3-point least square fit of the data. Thermocouples positioned at several locations along the sample monitored the surface temperature.

For piloted ignition tests in the vertical mode, 6" x 6" x 0.75" samples were used. These samples were exposed to different external irradiances and the time required for sustained flaming was recorded. As in the horizontal case, the surface temperature was also continuously monitored by thermocouples mounted on the sample surface.

SURFACE TEMPERATURE MEASUREMENTS AND THEIR INTERPRETATIONS

Measurement Method

For both piloted ignition and flame spread, surface temperature is a very important parameter. It is also very difficult to measure accurately. In the past, therefore, it has often been estimated by the use of a linear conduction theory (Simms [10]).

Primarily two methods of measuring surface temperature have been employed: (i) By mounting thermocouples on the sample surface (e.g. Gordon [11]; Kashiwagi [3]), and (ii) by using an infrared pyrometer (e.g. Smith et al [12]). The difficulty with using infrared pyrometers is that a knowledge

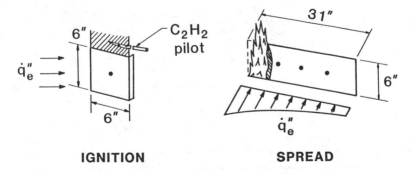

IGNITION                    SPREAD

Figure 2   A schematic of the experiments in the vertical mode. The black dots represent the thermocouples.

99

of surface emissivity, which changes as the thermal decomposition proceeds, is essential. Furthermore, the decomposition products and their exothermic reactions with oxygen interfere with the measurements. The use of surface thermocouples is also plagued with problems. This is because the measured values depend on the mounting method. Martin [13] estimated surface temperatures by extrapolating temperatures measured in depth. Gordon [11], on the other hand, had his thermocouples sprung lightly against the surface.

The method employed here is based on the observation that a thermocouple output is significantly reduced when not in contact with the surface. Thus the heat flows from the wood to the thermocouple junction when in contact. A method that produces the largest response to the same incident heat flux would then be the most correct (Beyler [14]). Due to large temperature gradients on both sides of the sample surface (in the gas and the solid phase), large errors are caused by either poor contact or by embedding the thermocouple in the solid.

Experimentally, the best compromise that was achieved is shown in Figure 3. Here the thermocouples used were made by electrically welding fine Chromel and Alumel wires 0.003" in diameter. The wires and the bead were then flattened to obtain a film thermocouple about 0.001" thick. A very fine incision was then made on the surface of the wood and the thermocouple was slid underneath this "skin" which was approximately 0.001" thick. The rest of the thermocouple was secured with as little wood glue as possible. The entire assembly was then pressed together and allowed to set. In the end, the thermocouple bead was visible through the "skin". This method was the most repeatable and gave the fastest and largest response to the same incident heat flux. The measured surface temperatures were repeatable to within ±5°C.

Piloted Ignition

The preceding technique was used to measure surface temperature for piloted ignition. A typical surface temperature-time curve obtained during the ignition experiments is shown in Figure 4. Also shown plotted on this

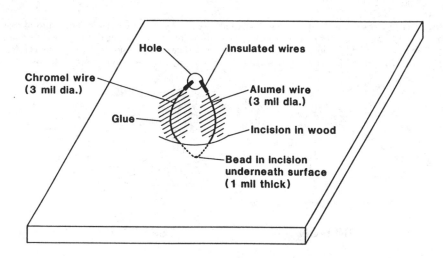

Figure 3 Method of surface temperature measurement.

curve is the time at which sustained flaming ignition was observed. This result is very similar to that obtained by Smith et al [12]. Note the sharp rise in the surface temperature due to flaming combustion and the fact that the time for the observed appearance of the flame plots on this sharp temperature rise. This, too, corroborates with Smith et al's [12] observation: "Sometimes the appearance of flame would produce a sudden large jump in millivolts, and other times the increase in millivolts would proceed the appearance of the flame." Since the time for visual flame observation typically has an error of plus or minus one second, large errors in piloted ignition temperatures are to be expected. Smith et al's measured ignition temperatures for pine blocks range from 343 to 571° C. This is clearly unacceptable for use in a thermal flame spread theory.

A closer look at the ignition process leads to a better understanding. An enlarged view of the temperature history (in Figure 4) during the last few instants before ignition is shown in Figure 5. A similar result was obtained for the horizontal case. This is shown in Figure 6. The flashes (unsustained momentary flaming) are more pronounced for the horizontal case than for the vertical case. This is because for the horizontal case the flashes occur in the middle of the sample, right where the thermocouple is, whereas in the vertical case they often did not cover the entire sample and remained close to the pilot flame far from the thermocouple. From Figure 6 it is also clear that there was enough time between the flashes for the surface to come to thermal equilibrium with the external radiation. Also note that the extrapolated surface temperature, caused by external radiation, at the time of sustained flaming is less than the momentary rise in temperature because of the flashes and yet sustained flaming was not achieved. In other words, for sustained flaming to occur, it is necessary for the surface temperature, caused by external radiation, to rise to some critical value. Any contributions due to gas phase exothermicity must not be included in determining this critical value. This is consistent with the concept of critical mass flux at ignition (Rashbash [15]) and implies that the required critical mass flux at ignition is produced by the solid indepth. The total heat contribution due to the flashes (proportional to the area underneath the peak) is small and limited to a thin surface layer. Also, this heat is quickly lost by reradiation. Furthermore, since the rise in surface temperature is faster for higher

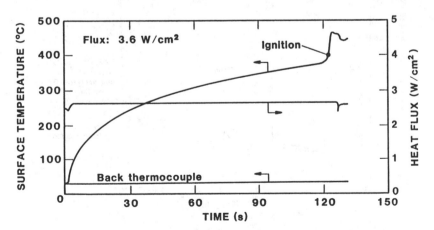

Figure 4 A typical surface temperature-time history for ignition. (Mahogany - Vertical Case)

101

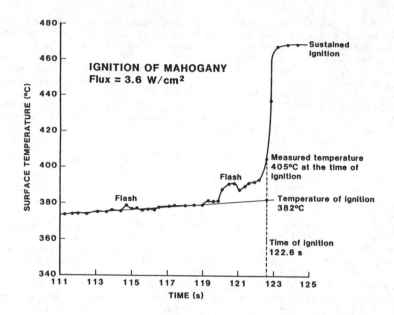

Figure 5 An enlarged view of the surface temperature-time history shown in Figure 4 at the time of sustained ignition.

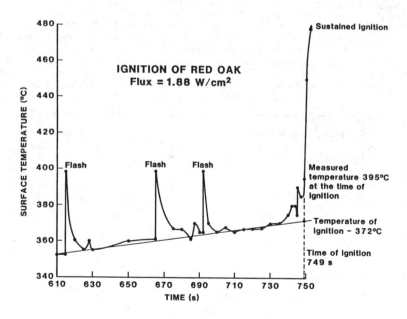

Figure 6 An enlarged view of the surface temperature-time history at the time of sustained ignition for the horizontal case.

heat fluxes, flashes and sustained ignition will be more closely spaced in time as is evidenced by a comparison of Figures 5 and 6. Hence, the surface temperature may not have time to come to equilibrium with the external radiation. This makes the flashes and ignition more difficult to distinguish.

Based on these observations, the critical surface temperature defined in this work is the temperature at the time of ignition achieved by external heating alone. Thus in Figure 6, 372°C rather than 395°C is the ignition temperature. This definition is consistent with the thermal flame spread theory.

Flame Spread

Surface temperatures measured during the flame spread experiments are shown in Figure 7 for the horizontal case and Figure 8 for the vertical case. The results for the two cases are qualitatively similar. Both figures show a sharp rise in surface temperature because of the arrival of the flame front. However, the source of "long-distance" heating is different. For the horizontal case (Figures 1 and 7), the rise in surface temperature prior to the arrival of the flame front takes place due to external radiation from the heaters and due to the radiation from the flame itself. Whereas, for the vertical case (Figures 2 and 8), the "long-distance" heating effects due to flame radiation are negligible because of the poor configuration factor, and the temperature rise is caused primarily by external radiation.

To determine the surface temperature consistent with the definition of ignition temperature and suitable for use in a thermal flame spread theory, consider the enlarged view shown inlaid in Figure 8. In Figure 8, $T_{ig}$ is the ignition temperature as defined in the previous section, $T_s(t_f)$ is the temperature when the flame visually arrives at the thermocouple location and

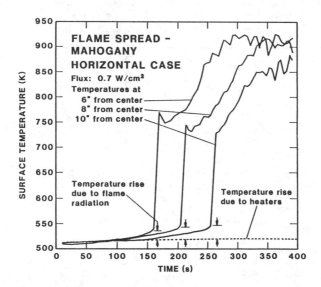

Figure 7 Measured surface temperatures during the flame spread experiments for the horizontal case at three locations.

$T_s(t_1)$ is the temperature rise due to external radiation alone. Thus, a suitable temperature for flame spread calculation is $T_s(t_1)$ extrapolated to time $t_f$, i.e. before the effect of gas-phase conduction is felt. A similar definition was used for the horizontal case (Figure 7); however, flame radiation was included in the "long-distance" heating effects. The rate of flame propagation will then depend upon how quickly the surface temperature ahead of the flame foot is brought up to the ignition temperature by the flame foot.


RESULTS AND DISCUSSION

Piloted Ignition

Figure 9 shows the measured surface temperatures for mahogany at the time of ignition as a function of external radiation for both the vertical and the horizontal samples. The plain bars represent the measured range (results of at least 6 experiments) of piloted ignition temperatures for the horizontal mode, whereas, the circles with bars are for the vertical mode. For the vertical mode, the error bars represent the net uncertainty in surface temperature for a single experiment caused by: (i) the error in measurement, and (ii) the ±1 sec uncertainty in observation of flaming combustion and the resulting uncertainty in temperature obtained from measured surface temperature profiles. These values are well within the range of surface temperatures reported in the literature (300-540°C). It also seems that the ignition temperature increases somewhat with decrease in external radiation. This is probably due to the depletion of surface reactants caused by charring.

It is important to note that surface temperature is an indirect measure of ignition. The actual ignition process is fairly complex. The solid must first chemically decompose to inject fuel gases into the boundary layer. These fuel gases must then mix with air and the local mixture ratio must be near or within the flammability limits. At this instant, a premixed flame, originating from the pilot flame, flashes across the surface of the solid through the fuel-air mixture formed in the boundary layer. To obtain ignition or

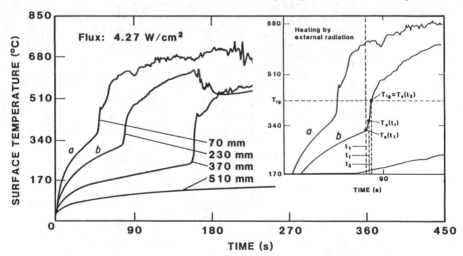

Figure 8   Measured surface temperatures at 4 locations during the flame spread experiments for the vertical case. INLAID: An enlarged view of curves 'a' and 'b'.

sustained flaming, which is marked by the establishment of a diffusion flame in the boundary layer, further heating of the solid is necessary. Evidence of this process can be seen in the measured surface temperature history depicted in Figure 6. Thus at the instant of ignition, pyrolysis gases must issue at a high enough rate to permit the establishment of a diffusion flame at a location far enough from the surface to avoid thermal quenching. Orientation relative to gravity was expected to significantly alter this process because of differences in the heat transfer to the surface and in the flow pattern of decomposition products and their mixing with entrained air. However, the results from both sets of experiments are within the error bars of only ± 30°C. Considering the error in the surface temperature measurements and the variation in the properties of wood from one sample to another, an average ignition temperature of 375°C seems to adequately represent both the horizontal and vertical modes. Experiments with red oak yield essentially the same conclusion, with the average ignition temperature being 365°C.

Since the ignition temperatures (as defined in this work) include only the effect of heating by external radiation, these results imply that at high temperatures (~650°K) necessary for ignition, convective losses (which depend on the sample orientation) are much less important than the re-radiative losses. Thus, the time required for the surface temperature to rise to the piloted ignition temperature is controlled primarily by re-radiation. Such measured times to ignition are shown in Figure 10 as a function of external radiation. Once again, the difference between the horizontal and the vertical modes is small. There is, however, a slight tendency for the ignition times to be shorter for the horizontal samples than for the vertical ones. This trend is consistent with Kashiwagi's [4] work on auto-ignition.

Figure 10 also shows that the time required for the surface temperature to reach the ignition temperature increases asymptotically to infinity as the

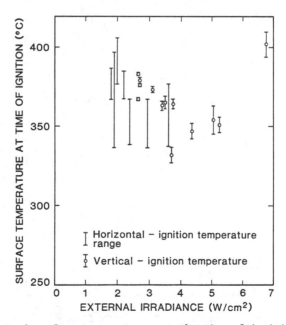

Figure 9 Measured surface temperatures at the time of ignition for different external radiation conditions.

external radiation is reduced. This asymptotic value can be found from the surface energy balance for an inert solid (assuming negligible thermal decomposition occurs prior to ignition) expressed as;

$$(-k \frac{\partial T}{\partial x}) = \dot{q}_e'' - h(T_s - T_\infty) - \varepsilon\sigma(T_s^4 - T_\infty^4). \tag{1}$$

Here the surface temperature '$T_s$' is replaced by the ignition temperature '$T_{ig}$' and as the time tends to infinity the left hand side of Equation (1) tends to zero. Hence, the minimum heat flux for piloted ignition is given by;

$$(\dot{q}_e'')_{min} = h(T_{ig} - T_\infty) + \varepsilon\sigma(T_{ig}^4 - T_\infty^4). \tag{2}$$

For wood, due to charring of the surface at low heat fluxes, '$\varepsilon$' is very nearly unity and 'h' calculated from convective heat transfer correlations is about 10 $W/m^2K$ for the horizontal case and 15 $W/m^2K$ (Quintiere [7]) for the vertical case. Using these values along with $T_{ig} = 375°C$ and $T_\infty = 20°C$, Equation (2) yields; $(\dot{q}_e'')_{min} = 1.32$ $W/cm^2$ for the horizontal case and $(\dot{q}'')_{min} = 1.50$ $W/cm^2$ for the vertical case -- a difference of only about 10%. Furthermore, this difference in the asymptotic values vanishes with only a ± 10°C change in the ignition temperatures which is well within the ± 30°C

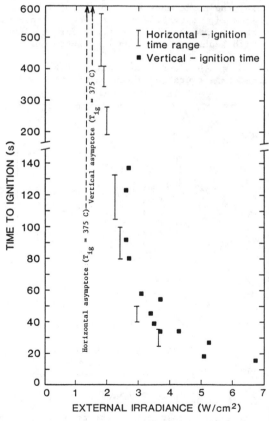

Figure 10   Measured time to ignition as a function of external radiation.

error in measurement.  It is, therefore, not surprising that the experimental results do not show any significant difference between the horizontal and vertical modes.


Flame Spread

The thermal theory (deRis [1]) for opposed-flow flame spread over a thermally thick solid gives the flame spread velocity 'V' by the formula:

$$V = \frac{V_a \, (\rho ck)_g \, (T_f - T_{ig})^2}{(\rho ck)_s \, (T_{ig} - T_s)^2} \tag{3}$$

where $(\rho ck)_g$ and $(\rho ck)_s$ respectively are the thermal responsivities of the gas and of the solid phase, '$T_f$' is the flame temperature and '$V_a$' is the opposed-flow gas velocity.  From the previous discussion, it is clear that '$T_{ig}$' is essentially the same for both the horizontal and the vertical cases.  However, it remains to be seen if the numerator of Equation 3 (represented by '$\phi$'), which explicitly contains the air velocity '$V_a$', is also invariant with changes in the sample orientation.

For appropriate definitions of '$T_s$' [see Figure 8, $T_s = T_s(t_1)$], Equation (3) can be used to correlate the data for both the horizontal and the vertical cases.  If $\phi/(\rho ck)_s$ is indeed a constant, then $V^{-1/2}$ must be linearly related to '$T_s$'.  Figure 11 shows the experimental results plotted in this manner for both the horizontal and the vertical cases.  Once again, within the experimental errors, the two cases are almost indistinguishable, although the flames look very different.

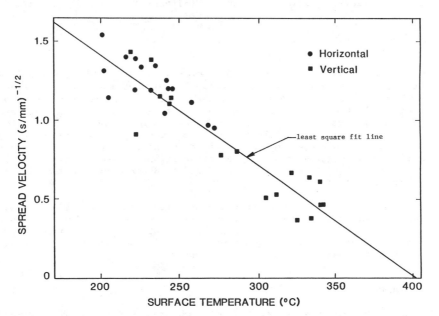

Figure 11  Flame spread rate correlation according to the thermal flame spread theory.

The fact that '$\phi$' for the two cases is the same (within experimental errors) was somewhat surprising because it is directly related to '$V_a$' (via Equ. (3)) which is different for the two cases. Similar results were also obtained by Fernandez-Pello et al. [16]. Their data shows that at ambient oxygen concentrations, the flame spread velocity is insensitive to the opposed-flow velocity, except near extinction. The nature of these results cannot be explained by a purely thermal theory used in the derivation of Equ. 3. As suggested by Fernandez-Pello, et al. [16], the increase in the flame spread rate at high opposed-flow velocities caused by the closer proximity of the flame is counteracted by the gas phase chemical kinetic effects. Thus, Equ. 3 and therefore '$\phi$' need to be modified by an appropriate correction factor which accounts for the gas phase chemical kinetic effects. This correction factor was found to be a function of the Damköhler number [16]. [Further improvements in Equ. 3 have also been suggested by Wichman, et al. [17] by incorporating a velocity gradient at the fuel surface and thus eliminating the uniform gas velocity assumption used in the derivation of Equ. 3.] Hence, these results are in agreement with the established mechanisms of opposed-flow flame spread (Ref. [2]). They confirm that for the present experimental conditions, the flame spread rate is controlled entirely by the processes taking place in the leading edge of the flame and that the flame geometry alters only the "long-distance" heating effects. Also, although the processes occurring at the flame foot are the result of a complex interaction between heat transfer and gas phase chemical kinetic effects, they can be lumped into the parameter '$\phi$' which is experimentally found to be approximately constant.

Finally, according to Equation (3), the intercept of the least square fit line with the x-axis [Figure 11], is the ignition temperature. This gives a value of 402°C for piloted ignition temperature. This number is within the range of the measured ignition temperatures, although on the high side of the scatter.


CONCLUSIONS

In this work the effect of sample orientation on piloted ignition and flame spread was experimentally investigated for two woods -- red oak and mahogany. Within the measurement error, the results appear to be orientation independent and seem to indicate that the relationship between ignition and flame spread, as assumed in the thermal flame spread theory, is valid. However, to be consistent with the thermal theory, both the ignition temperature '$T_{ig}$' and the surface temperature ahead of the flame foot '$T_s$' must not include gas-phase heating effects. For the horizontal case, flame radiation alters only the "long-distance" heating effects and hence must be included in the determination of '$T_s$'.

It was found that there are two reasons for orientation independence: (i) At high temperatures in question, heat loss by re-radiation dominates over the convective losses. (ii) At ambient $O_2$ concentrations, the flame spread rate is insensitive to small changes (~0.1 m/sec) in the induced air velocity that are caused by changes in the sample orientation.

ACKNOWLEDGEMENTS

The authors are indebted to Dr. J.G. Quintiere for his help and suggestions. The first author would also like to thankfully acknowledge the summer support provided by The Center for Fire Research, National Bureau of Standards. A part of this work was conducted at Harvard University with the direction and patient guidance of Professor H.W. Emmons who was the advisor of the first author.

REFERENCES

1.  deRis, J.N., Twelfth Symposium (International) on Combustion, The Combustion Institute, 241-252, 1969.

2.  Fernandez-Pello, A.C. and Hirano, T., Combustion Science Technology, Vol. 32, Nos. 1-4, 1-31, 1983.

3.  Kashiwagi, T., Fire Safety Journal, Vol. 3, 185-200, 1981.

4.  Kashiwagi, T., Combustion and Flame, Vol. 44, 223-245, 1982.

5.  Kanury, A. M., Fire Research Abstracts and Reviews, Vol. 14, No. 1, 24-72, 1972.

6.  Quintiere, J., Fire and Materials, Vol. 5, No. 2, 52-60, 1981.

7.  Quintiere, J., Harkleroad, M. and Walton, D., Combustion Science and Technology, Vol. 32, 67-89, 1983.

8.  Atreya, A., "Pyrolysis, Ignition and Fire Spread on Horizontal Surfaces of Wood," NBS-GCR-83-449 National Bureau of Standards, Washington, D.C., March 1984.

9.  Robertson, A.F., "A Flammability Test Based on Proposed ISO Spread of Flame Test," Third Progress Report, Intergovernmental Maritime Consultative Organization, IMCO FP/215, 1979.

10. Simms, D.L., Combustion and Flame, Vol. 7, 253-261, 1963.

11. Gardon, R., "Temperatures Attained in Wood Exposed to High Intensity Thermal Radiation," Fuels Res. Lab. Tech. Rep., No. 3, Massachusetts Institute of Technology, 1953.

12. Smith, W.K., and King, J.B., J. Fire and Flammability, Vol. 1, p. 272, 1970.

13. Martin, S., Tenth Symposium (International) on Combustion, The Combustion Institute, Pittsburg, PA, p. 877, 1965.

14. Beyler, C.L., Personal Communications, 1982.

15. Rasbash, D.J., Int. Symp. on Fire Safety of Combustible Materials, Edinburgh, p. 169, 1975.

16. Fernandez-Pello, A.C., Ray, S.R. and Glassman, I., Eighteenth Symposium (International) on Combustion, The Combustion Institute, Pittsburg, PA, p. 519, 1981.

17. Wichman, I.S., Williams, F.A., and Glassman, I., Nineteenth Symposium (International) on Combustion, The Combustion Institute, Pittsburg, PA, p. 835, 1983.

# Thermal Response of Compartment Boundaries to Fire

**J. R. MEHAFFEY and T. Z. HARMATHY**
Division of Building Research
National Research Council of Canada
Ottawa, Canada

ABSTRACT

A series of full-scale room burn experiments has been conducted on fully-developed fires to study their destructive potentials in terms of the thermal response of room boundaries. It is shown that normalized heat load is a convenient measure of the destructive potential (severity) of fire, irrespective of the nature of the boundary elements. A mathematical model for calculating normalized heat load yields satisfactory predictions, given the amount of combustible material in the compartment, its geometry, the thermal properties of the boundaries, and the size of ventilation openings.

INTRODUCTION

The conditions prevailing in a room during the fully-developed stage of a fire have been studied extensively over the last 25 years (1). Theoretical analyses (2,3) have shown that the normalized heat load is a measure of the destructive potential of fire (fire severity) and is capable of providing the link between the severity of real world fires and that of fire resistance tests. Although numerous room-burn experiments have been reported in the literature, however, there is a paucity of information that can be used to determine normalized heat load. To test the predictive capabilities of a model developed by the authors, it became necessary to conduct a series of full-scale room burn experiments. The normalized heat load values derived from such a test series show good agreement with model predictions and thereby prove the validity of the normalized heat load concept, irrespective of the thermal characteristics of the enclosure boundaries.

NORMALIZED HEAT LOAD

Normalized heat load is defined as

$$H = \frac{E}{\sqrt{k\rho c}} \tag{1}$$

where E is the heat absorbed per unit surface area of the enclosure during the fire, and $\sqrt{k\rho c}$ is the thermal absorptivity of the boundaries (k is thermal conductivity, $\rho$ is density, and c is specific heat). It has been shown (2) that the normalized heat load does not depend on the temperature history of the fire, and that it can be calculated from the maximum temperature rise within the enclosure boundary.

Usually the various boundaries of an enclosure are built of different materials. In such cases the thermal absorptivity of the enclosure is interpreted as the surface-weighted average of the thermal absorptivity for the individual boundaries.

$$\sqrt{k\rho c} = \frac{1}{A} \sum_i A_i \sqrt{k_i \rho_i c_i} \tag{2}$$

$$A = \sum_i A_i \tag{3}$$

Subscript i in this expression relates to information pertinent to the i-th boundary. The letter A stands for total boundary surface area. Numerical studies show that in a compartment lined with materials of differing thermal absorptivity the normalized heat loads imposed on various boundary elements are approximately identical (4). Loosely referred to as the "theorem of uniformity of normalized heat loads," this implies that the destructive potential of fire for all the boundaries is identical.

COMPARTMENT FIRE MODELS

A mathematical model of fully-developed compartment fires involving cellulosic fuels has been formulated (5). (Statistics indicate that combustible materials in buildings still consist predominantly of cellulosics.) According to the model, the nature of such fires depends primarily on five variables:

$G$  = total mass of fuel,
$h_c$ = height of compartment,
$A$  = total surface area of boundaries,
$\sqrt{k\rho c}$ = thermal absorptivity (thermal inertia) of boundaries,
$\Phi$ = ventilation factor.

The ventilation factor characterizes the rate of entry of air into a burning compartment. Under experimental conditions the rate of entry of air is not augmented by wind or drafts, and the ventilation factor can be expressed with the aid of the dimensions of the ventilation opening.

$$\Phi = \rho_a A_V \sqrt{g h_V} \tag{4}$$

where $\rho_a$ is the density of atmospheric air, $A_V$ is the area of ventilation opening, $h_V$ is the height of that opening, and g is the gravitational constant. It has been shown (6) that with cellulosic fuel a fire is ventilation-controlled if

$$\Phi < 0.263 \, \varrho G \tag{5}$$

Otherwise, it is fuel-surface-controlled. $\varrho$ is the specific surface of the fuel (for typically furnished rooms, $\varrho \simeq 0.13$ m$^{-2}$/kg).

In its most general form the compartment fire model requires the numerical solution (by iteration) of six coupled equations to determine the values of six process variables and the normalized heat load (5). If the normalized heat load is the only information required, however, one can bypass the iterative technique and employ the approximate formula (7)

$$H = 10^6 \frac{(11.0\delta + 1.6)G}{A\sqrt{k\rho c} + 935\sqrt{\Phi G}} \tag{6}$$

In this equation δ is, in a loose sense, the fraction of the fuel energy released inside the compartment. (This interpretation of δ is strictly appropriate only for ventilation-controlled fires. At very high air flow rates δ becomes a correction factor.) It is primarily a function of the height of the compartment ($h_c$) and the ventilation factor ($\Phi$) and is given by the expression

$$\delta = \begin{cases} 0.79 \ \sqrt{h_c^3/\Phi} \\ 1 \end{cases} , \text{ whichever is less} \qquad (7)$$

ROOM BURNS

Seventeen full-scale room burn experiments were conducted to test the theory and answer two questions:

1) Does the model of fully-developed compartment fires, in both iterative and analytic forms, predict accurately the normalized heat load?
2) Does the theorem of uniformity of normalized heat load hold?

A preliminary account of the first ten of these burns has been published (8). They were conducted in a concrete block room with one wall fitted with either a door or a window; wall, ceiling, and floor panels were attached directly to the concrete blocks. For this series of room burns the inner floor dimensions were in the range (2.39 to 2.47 m) × (3.59 to 3.66 m) and the inner height was in the range 2.54 to 2.56 m. The total surface area, A, of the ceiling, floor, and walls (excluding the opening) for each of the room burns is listed in Table I.

Eight wooden cribs supported 100 mm above the floor on bricks were distributed over the floor in each burn experiment. The cribs were constructed of pine sticks measuring 4 × 4 × 84 cm. To drive the fire to the fully-developed stage as quickly as possible the cribs were ignited simultaneously. The quantity of fuel per unit floor area was chosen to be representative of that found in practice. The total mass of wood for each burn, G, is listed in Table I. Fires involving approximately 130 kg of wood represented a fuel load per unit floor area of about 15 kg $m^{-2}$, typical of a hotel room; those involving approximately 240 kg of wood had a fuel load per unit floor area of about 27 kg $m^{-2}$, typical of an office room (9).

Five materials were chosen to line the room: brick, insulating fire brick, a fibrefrax board, gypsum wallboard and normal-weight concrete; thermal properties and thicknesses of linings are listed in Table II. As a set, they span the range of thermal properties likely to be encountered in buildings. The first three (brick, insulating fire brick, and fibrefrax board) were used most extensively since their properties remain relatively unchanged from test to test and walls did not have to be replaced after every experiment.

In each test at least two of the five materials were present as lining. In tests A-D the floor and ceiling were lined with fibrefrax and the walls with brick. In tests E-J, the floor and ceiling were lined with fibrefrax, the two long walls with brick, and the end walls with insulating firebrick. In tests K-M, the floor and ceiling were lined with fibrefrax and the walls with insulating firebrick. The last four tests, N-Q, were conducted by a colleague, K. Choi, as part of an independent research project (10). In these the ceiling was lined with fibrefrax, the floor with normal-weight concrete,

TABLE I

Summary - Room Burns

| Test | A<br>$m^2$ | G<br>kg | $\Phi$<br>kg s$^{-1}$ | $\sqrt{k\rho c}$<br>J m$^{-2}$s$^{-\frac{1}{2}}$K$^{-1}$ | $\psi$ +<br>$m^2$/kg | $h_C$<br>m | $\beta$ ++ | Burning*<br>mode |
|------|------|-------|------|------|------|------|------|------|
| A | 48.60 | 133.4 | 3.56 | 868 | 0.21 | 2.56 | 1.0 | VC |
| B | 48.60 | 236.6 | 3.56 | 868 | 0.21 | 2.56 | 1.0 | VC |
| C | 48.31 | 242.3 | 5.49 | 866 | 0.21 | 2.56 | 1.0 | VC |
| D | 48.31 | 133.1 | 5.49 | 866 | 0.21 | 2.56 | 1.0 | VC |
| E | 48.53 | 130.7 | 3.43 | 666 | 0.21 | 2.56 | 1.0 | VC |
| F | 48.26 | 135.7 | 5.25 | 667 | 0.21 | 2.56 | 1.0 | VC |
| G | 47.91 | 130.8 | 7.93 | 668 | 0.21 | 2.56 | 1.0 | FS |
| H | 48.53 | 240.3 | 3.43 | 666 | 0.21 | 2.56 | 1.0 | VC |
| I | 48.26 | 240.0 | 5.25 | 667 | 0.21 | 2.56 | 1.0 | VC |
| J | 47.91 | 240.0 | 7.93 | 668 | 0.21 | 2.56 | 1.0 | VC |
| K | 48.41 | 133.2 | 3.43 | 344 | 0.21 | 2.56 | 1.0 | VC |
| L | 48.14 | 242.1 | 5.25 | 344 | 0.21 | 2.56 | 1.0 | VC |
| M | 48.14 | 137.6 | 5.25 | 344 | 0.21 | 2.56 | 1.0 | VC |
| N | 46.74 | 240.0 | 3.31 | 725 | 0.21 | 2.54 | 1.0 | VC |
| O | 46.74 | 240.0 | 3.31 | 725 | 0.21 | 2.54 | 1.0 | VC |
| P | 46.74 | 240.0 | 3.31 | 725 | 0.21 | 2.54 | 1.0 | VC |
| Q | 46.74 | 240.0 | 3.31 | 725 | 0.21 | 2.54 | 1.0 | VC |

+ Exposed surface area of fuel is $\psi$G
* VC = Ventilation-controlled; FS = Fuel-surface controlled

and the walls with gypsum wallboard. Two of the walls had narrow strips (0.30 m wide × 2.54 m high) of insulating firebrick to facilitate comparison with tests E-M. The overall thermal absorptivity of the compartment for each burn is listed in Table I. This quantity was calculated employing Eq. (2), assuming that half the floor area was lined with wood. To calculate the normalized heat load, thermocouples were installed within the lining materials at depths indicated in Table II. Typically, nine thermocouples were embedded in each of the walls, in the ceiling and the floor.

The area of the ventilation opening was chosen to be approximately 9%, 12%, or 16% of the compartment floor area. In tests G and J this opening was a doorway 0.69 × 2.10 m. In all other cases the opening was a window whose approximate dimensions were either 0.7 × 1.2 m or 0.7 × 1.6 m. The ventilation factors, calculated from Eq. (4), are listed in Table I.

The quantity of fuel, the overall thermal absorptivity of the boundaries, and the size of the ventilation opening (and to a lesser degree the area of the boundaries) were varied from test to test, as shown in Table I. The normalized heat loads imposed on the boundaries by each fire have been calculated by both the iterative form of the model and its analytic form (Eq. (6)); the results are tabulated in Table III. The specific fuel area of the cribs employed in these burns was fairly large: $\psi \simeq 0.21$ m$^2$/kg. Consequently, only one fire, test G, was fuel-surface-controlled (cf. Eq. (5)); all others were ventilation-controlled.

TABLE II

Properties of Lining Materials
(Thermal properties averaged over temperature range 300 – 900 K)

| Material | $k$ (W m$^{-1}$K$^{-1}$) | $\rho$ (kg m$^{-3}$) | $c$ (J kg$^{-1}$K$^{-1}$) | $\sqrt{k\rho c}$ (J m$^{-2}$s$^{-\frac{1}{2}}$K$^{-1}$) | $k/\rho c$ (m$^2$ s$^{-1}$) | Thickness of lining material (m) | Thermocouple depth in lining x (m) |
|---|---|---|---|---|---|---|---|
| Normal-wt concrete | 1.68 | 2200 | 1300 | 2192 | $5.87\times10^{-7}$ | 0.100 | – |
| Brick | 0.80 | 1935 | 1025 | 1260 | $4.03\times10^{-7}$ | 0.057 | 0.027 |
| Gypsum wallboard | 0.27 | 680 | 3000 | 742 | $1.32\times10^{-7}$ | 0.0159 | 0.0159 |
| Insulating firebrick | 0.25 | 722 | 1000 | 425 | $3.46\times10^{-7}$ | 0.063 | 0.022 |
| Fibrefrax board | 0.072 | 318 | 750 | 131 | $3.02\times10^{-7}$ | 0.038 | 0.0127 |
| Wood | 0.15 | 550 | 2300 | 436 | $1.19\times10^{-7}$ | – | – |

## EXPERIMENTAL DETERMINATION OF NORMALIZED HEAT LOAD

The experimental values of the normalized heat load were determined from temperature measurements, using thermocouples installed in the lining materials. (As indicated earlier, the normalized heat load is related directly to the maximum temperature rise at critical depths within the boundary elements (2).) Both experimental results and model predictions are tabulated in Table III. It is clear that the iterative and analytic forms of the fire model provide satisfactory accuracy.

A comparison of the normalized heat load imposed on brick, insulating firebrick, and fibrefrax with the predictions of the iterative form of the fire model are provided in Figures 1, 2, and 3, respectively. Model predictions and experimental findings agree to within ±10% for both brick and insulating fire brick, despite the large differences in their thermal properties. Agreement for fibrefrax board is to within ±20% only.

To investigate the validity of the theorem of uniformity for normalized heat loads, those imposed on the different lining materials during the burns are compared in Figure 4. Fibrefrax is compared with insulating firebrick, fibrefrax with brick, and insulating firebrick with brick. The results for the material with the lower thermal absorptivity (in each pair) are plotted along the y-axis. Agreement between theory and experiment is apparently good.

## CONCLUSION

Experiments in which a series of full-scale room burns were carried out to study the thermal assault a fully-developed fire mounts against room boundaries have demonstrated the capability of a compartment fire model to assess the severity of fires in terms of the normalized heat load. The model predictions agree with experimental findings to within ±10%. It is believed that this is the first time the predictive capability of a model has been

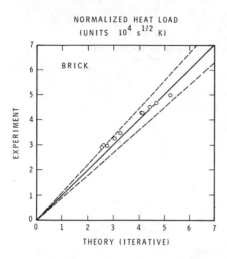

NORMALIZED HEAT LOAD
(UNITS $10^4$ $s^{1/2}$ K)

Figure 1. Comparison of normalized
heat loads, as derived from
experiments and from the model.
Lining material: brick. (Points
falling between dashed lines
represent agreement of better than
±10% between experiment and theory)

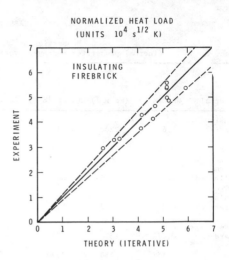

NORMALIZED HEAT LOAD
(UNITS $10^4$ $s^{1/2}$ K)

Figure 2. Comparison of normalized
heat loads, as derived from measurements and from the model. Lining
material: insulating firebrick.
(Points falling between dashed lines
represent agreement of better than
±10% between experiment and theory)

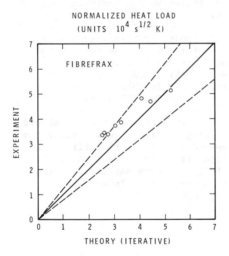

NORMALIZED HEAT LOAD
(UNITS $10^4$ $s^{1/2}$ K)

Figure 3. Comparison of normalized
heat loads, as derived from experiments and from the model. Lining
material: fibrefrax board. (Points
falling between the dashed lines
represent agreement of better than
±20% between experiment and theory)

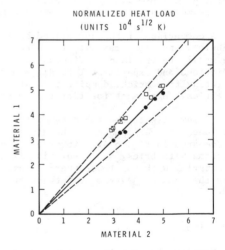

NORMALIZED HEAT LOAD
(UNITS $10^4$ $s^{1/2}$ K)

Figure 4. Comparison of measured
normalized heat loads imposed on
different lining materials. Fibrefrax
compared with insulating firebrick (Δ),
fibrefrax with brick (□), insulating
firebrick with brick (o)

116

TABLE III

Normalized Heat Load ($10^4$ $s^{\frac{1}{2}}K$)

| | Theory | | Experiment | | |
|---|---|---|---|---|---|
| Test | Analytic | Iterative | Brick | Firebrick | Fibrefrax |
| A | 2.69 | 2.76 | 2.97 | – | 3.41 |
| B | 4.30 | 4.44 | 4.52 | – | 4.71 |
| C | 4.02 | 4.09 | 4.31 | – | 4.83 |
| D | 2.50 | 2.53 | 2.91 | – | 3.39 |
| E | 3.16 | 3.27 | 3.48 | 3.34 | 3.87 |
| F | 2.99 | 3.05 | 3.27 | 3.30 | 3.75 |
| G | 2.65 | 2.61 | 3.00 | 2.97 | 3.47 |
| H | 5.12 | 5.26 | 4.99 | 4.88 | 5.17 |
| I | 4.63 | 4.69 | 4.68 | 4.66 | – |
| J | 4.15 | 4.15 | 4.29 | 4.30 | – |
| K | 4.58 | 4.60 | – | 4.14 | – |
| L | 6.11 | 5.92 | – | 5.38 | – |
| M | 4.16 | 4.12 | – | 3.77 | – |
| N | 5.02 | 5.17 | – | 5.13 | – |
| O | 5.02 | 5.17 | – | 5.17 | – |
| P | 5.02 | 5.17 | – | 5.34 | – |
| Q | 5.02 | 4.17 | – | 4.01 | – |

demonstrated. Further experiments are planned to study fuel-surface-controlled fires, the effect of compartment height, and the influence of non-charring fuels on the normalized heat load.

NOMENCLATURE

| | |
|---|---|
| A | total surface area of compartment, $m^2$ |
| $A_V$ | area of ventilation opening, $m^2$ |
| c | specific heat, $J\ kg^{-1}K^{-1}$ |
| E | heat absorbed per unit surface area, $J\ m^{-2}$ |
| g | gravitational constant, $\simeq 9.8\ m\ s^{-2}$ |
| G | total mass of fuel, kg |
| $h_C$ | compartment height, m |
| $h_V$ | ventilation opening height, m |
| H | normalized heat load, $s^{\frac{1}{2}}K$ |
| k | thermal conductivity, $W\ m^{-1}K^{-1}$ |
| $\sqrt{k\rho c}$ | thermal absorptivity, $J\ m^{-2}s^{-\frac{1}{2}}K^{-1}$ |

Greek Letters

| | |
|---|---|
| δ | fractional heat release from fuel inside the compartment |
| ρ | density, $kg\ m^{-3}$ |
| $\rho_a$ | density of atmospheric air, $kg\ m^{-3}$ |
| Φ | ventilation factor, $kg\ s^{-1}$ |
| $\psi$ | specific surface of the fuel, $m^{-2}\ kg^{-1}$ |

Subscript

| | |
|---|---|
| i | of or for the i-th compartment boundary |

REFERENCES

1. Harmathy, T.Z., and Mehaffey, J.R., "Post-flashover compartment fires," Fire and Materials, 7, 49 (1983).
2. Harmathy, T.Z., "The possibility of characterizing the severity of fires by a single parameter," Fire and Materials, 4, 71 (1980).
3. Harmathy, T.Z., "The fire resistance test and its relation to real-world fires," Fire and Materials, 5, 112 (1981).
4. Harmathy, T.Z., and Mehaffey, J.R., "Normalized heat load: A key parameter in fire safety design," Fire and Materials, 6, 27 (1982).
5. Harmathy, T.Z., "Fire Severity: Basis of fire safety design," in Fire Safety of Concrete Structures, American Concrete Institute, ACI Publication SP-80, p. 115 (1983).
6. Harmathy, T.Z., "A new look at compartment fires, Parts I and II," Fire Technology, 8, 196 and 326 (1972).
7. Mehaffey, J.R., and Harmathy, T.Z., "Assessment of fire resistance requirements," Fire Technology, 17, 221 (1981).
8. Mehaffey, J.R., and Harmathy, T.Z., "Fully developed fires: Experimental findings," Proceedings of the Third Symposium on Combustibility and Plastics, Society of the Plastics Industry of Canada, 24 and 25 Oct. 1983.
9. Pettersson, O., Magnusson, S.E., and Thor, J., "Fire engineering design of steel structures," Swedish Institute of Steel Construction, Stockholm, Bulletin 50 (1976).
10. Choi, K.K., "Effects of foamed plastic insulation on severity of room fires." To be published.

This paper is a contribution from the Division of Building Research, National Research Council of Canada.

# Fully Developed Compartment Fires: The Effect of Thermal Inertia of Bounding Walls on the Thermal Exposure

**B. BØHM**
Laboratory of Heating and Air Conditioning
Technical University of Denmark
Building 402A, DK-2800 Lyngby, Denmark

ABSTRACT

Based on an energy balance for the compartment gases the thermal exposure
in fully developed fires is determined as a function of opening factor,
fuel load and thermal inertia of the bounding walls. A conversion factor
is calculated which makes it possible to carry out all calculations in a
standard compartment by applying an equivalent fuel load and an equivalent
opening factor. The conversion factor is determined as the thermal inertia
in the standard compartment divided by the thermal inertia in the real compartment. The errors from this approximate method are estimated from the maximum
steel temperatures of an unprotected and a protected steel structure placed
inside the standard compartment and the real compartment. Nine different
compartments are used and apart from the compartment made of steel the greatest
error is smaller than 13%. The normalized heat load concept introduced by
Harmathy (7) is used as a principle of equivalent fire duration in the fire
test furnace, in the standard compartment and in a compartment with low
thermal inertia. The accuracy of the concept is estimated from the maximum
steel temperature of a protected structure.

Keywords

Fully developed compartment fires, thermal exposure, thermal inertia, normalized
heat load, equivalent fire duration, steel structures.

INTRODUCTION

For many years the fire resistance of building structures has been evaluated
in fire test furnaces. The temperature-time curve which is followed in these
tests is specified in standards, for instance ISO 834.

Ödeen (1) and Kawagoe (2) were among the first to calculate the thermal
response of structures. An energy balance for the compartment gases became
the basis for theoretical calculations of temperature-time curves in ventilation controlled fires. By computer simulations of a great number of fire
tests Magnusson and Thelandersson (3) were able to specify the net energy
release rate for the whole fire period and not only the fully developed
period. Their model which is based on wood fires solely became the basis
for differentiated design of steel structures, (4). For polyethylene fires
a model was proposed by Bøhm (5). When the energy release rate is prescribed

in the models (3,5) three main parameters determine the gas temperature-time curves: the opening factor $A\sqrt{H}/A_t$, the fuel load E and the thermal inertia $\sqrt{k\rho c}$ of the bounding walls.

The models assume a uniform compartment gas temperature although it is realized that temperature gradients will appear in real fires, (6). It should be remembered that the uncertainty associated with the thermal exposure in the compartment is superimposed on the uncertainty associated with the thermal and structural response of the structure itself.

The thermal inertia of the bounding walls influences the net heat flux to the walls and thereby the energy balance for the compartment gases. As a result gas temperature-time curves for design purposes will have to be calculated for a number of different bounding walls. To ease the application of these design curves for engineering purposes it would be helpful to restrict calculations to one type of compartment walls. An approximate method for "translation" of the thermal exposure in a specific compartment to that of a standard compartment is discussed in this paper.

The concept of normalized heat load introduced by Harmathy (7) is a measure of the destructive potential of a compartment fire. Compared with other principles of equivalent fire duration the importance of the normalized heat load concept is that it can be evaluated from the fire history and the fire compartment without knowledge of the actual structural element to be considered.

A comparison of the thermal response of structures exposed to either the standard fire curve ISO 834 or to the polyethylene design model will be reported here when the normalized heat load concept is applied as a principle of equivalent fire duration.


THE CONVERSION FACTOR $k_f$

Pettersson et al. (4) introduced a standard compartment made of brick and concrete with thermal inertia 1168 $W\sqrt{s}/m^2K$. The thermal exposure in other compartments is converted to the standard compartment by a conversion factor $k_f$. In the standard compartment (No. 1) an equivalent fuel load and an equivalent opening factor are applied:

equivalent fuel load      = $k_f$ x true fuel load E                          (1)

equivalent opening factor = $k_f$ x true opening factor $A\sqrt{H}/A_t$

The conversion factor is independent of the magnitude of the fuel load, and not very dependent of the opening factor, (4). The conversion factors were found by direct comparison of the gas temperature-time curves in the different compartments.

From the compartment energy balance and with the bounding walls treated as semi-infinite solids it appears (8) that the conversion factor can be evaluated as

$$k_f = \sqrt{k\rho c}_{No1}/\sqrt{k\rho c}_{true} \qquad\qquad (2)$$

where $\sqrt{k\rho c}$ is the thermal inertia of the bounding walls.

In Table 1 is shown the thermal inertia and the thermal conductivity of nine different, homogeneous compartments. Also shown in the Table are conversion factors either from reference (4) or calculated according to equation (2).

For long fire durations the walls can no longer be treated as semi-infinite solids. The thermal penetration time has been discussed by Harmathy (9) and McCaffrey et al. (10) among others. Here the critical time when the temperature on the unexposed side of the wall starts to rise is calculated as:

$$\tau_{crit} = 0.05d^2 \frac{\rho c}{k} \tag{3}$$

The critical time for 0.20 m wall thickness is shown in Table 1. The material properties are evaluated at $700^\circ C$.

It appears from Table 1 that the conversion factors given by Pettersson et al. are somewhat closer to unity than the values obtained from the equation (2).

For thermally light walls with low thermal inertia the conversion factor will be greater than unity, and consequently the thermal exposure is expected to be greater than in the standard compartment.

THERMAL EXPOSURE

To characterize the thermal exposure in a specific compartment the following parameters will be calculated:

- maximum gas temperature

- the integral of the incident flux and of the net heat flux from the start of the fire to the end of the fully developed period $\tau^*$

- maximum steel temperature of two steel structures with surface area/volume-ratio 100 m$^{-1}$. One structure is unprotected, the other is protected with 20 mm rockwool insulation (material No. 7).

TABLE 1. Material properties for compartments 1-9, conversion factors and critical time (0.20) m wall thickness)

| Compartment number and type | $\sqrt{k\rho c}_{700}$ W $\sqrt{s}/m^2$ K | $k_{700}$ W/m$^\circ$C | Conversion factor $k_f$ | | $\tau_{crit}$ minutes |
|---|---|---|---|---|---|
| | | | Ref(4) | $1168/\sqrt{k\rho c}$ | |
| 1 brick + concrete | 1168 | 0.814 | 1.00 | 1.00 | 68 |
| 2 concrete | 1537 | 0.809 | 0.85 | 0.76 | 119 |
| 3 autoclaved concrete | 295 | 0.207 | 3.00 | 3.96 | 68 |
| 4 wood-like | 436 | 0.150 | | 2.68 | 278 |
| 5 firebrick | 569 | 0.361 | | 2.05 | 83 |
| 6 rockwool A-batts | 134 | 0.404 | | 8.72 | 4 |
| 7 rockwool protection | 172 | 0.211 | | 6.79 | 22 |
| 8 steel | 13152 | 31.125 | | 0.09 | 6 |
| 9 non-existing | 436 | 1.500 | | 2.68 | 3 |

COMPUTER CALCULATIONS

Computer calculations were carried out with the polyethylene design model (5) to derive the errors when conversion factors are used. All material properties were evaluated as a function of temperature (8), and the thermal response of the steel structures was calculated according to Bøhm & Hadvig (11).

The greatest differences in maximum steel temperatures when either equation (1) is applied in the standard compartment or calculations are carried out in the real compartment are shown in Table 2. The table has been divided into two parts, eq. when the fire duration is shorter than the critical time (small fuel load), and when it is not (large fuel load). Apart from the compartment made of steel the greatest error is smaller than 13%. However, if a critical steel temperature is desired as $500^\circ C \pm 25^\circ C$ then the error should not exceed 5%.

For further details and in the cases with different homogeneous walls in the same compartment, non-homogeneous walls, and in the wood design case (3) the reader is referred to (8).

CONCEPT OF NORMALIZED HEAT LOAD

The normalized heat load (NHL) is calculated from

$$NHL = \int_o^{\tau^*} \frac{q''}{\sqrt{k\rho c}} \, d\tau \tag{4}$$

where $q''$ is the net heat flux to the bounding walls and $\tau^*$ is the duration of the fully developed fire.

Harmathy et al. (7) considered temperature-independent material properties which means that an effective value for the thermal inertia $\sqrt{k\rho c}$ must be evaluated, (12).

TABLE 2. Maximum errors in percentage of the value obtained from $k_f$-method, with $k_f = 1168/(\sqrt{k\rho c})_{700^\circ C}$ First figure for unprotected structure and second figure for protected structure. 3 different fuel loads have been applied for every opening factor.

| | $A\sqrt{H}/A_t$ $m^{\frac{1}{2}}$ | COMPARTMENT NUMBER | | | | | |
|---|---|---|---|---|---|---|---|
| | | 2 | 3 | 4 | 6 | 8 | 9 |
| $\tau^* < \tau_{crit}$ | 0.01 | 10/8 | 13/5 | | | | |
| | 0.04 | 4/2 | 4/7 | 6/6 | | | |
| | 0.12 | 3/4 | 3/5 | | | | |
| $\tau^* \geq \tau_{crit}$ | 0.01 | | 3/5 | | 11/11 | 132/185 | 9/6 |
| | 0.04 | | 4/6 | | | | |
| | 0.12 | | 3/5 | | | | |

122

In Figure 1 is shown the normalized heat load in the standard compartment as a function of opening factor and fuel load. The thermal exposure is calculated in case of a polyethylene fire (5). The duration of the fully developed fire is also indicated in the figure.

Harmathy (13) and Paulsen (14) investigated the heat transfer in fire test furnaces. The normalized heat load on the test specimen increases when the furnace walls are made of a material with low thermal inertia. However, this variation in normalized heat load is small compared with the variation in real compartments, due to different fuel loads and opening factors. On the left side of the vertical scale in Figure 1 is shown the equivalent fire duration according to ISO 834 in a fire test furnace built of fire brick (material No. 5). The furnace is a typical gasfired European wall furnace with mean beam length 1.2 m. The gas emissivity is calculated as a function of temperature and of the partial pressure of $CO_2$ and $H_2O$ (p · L = 0.3 m atm). The measurement error between true gas gemperature and thermo-couple signal is simulated, (11,14). The thermal exposure in this furnace is slightly less than the exposure from a black gas.

Equivalent Heat Load

As an example 60 minutes fire duration according to ISO 834 is considered. From calculation or Figure 1 the normalized heat load is $46{,}400\sqrt{s}$ K in the fire test furnace.

From the computer calculations with temperature dependent material properties it is possible (by interpolation) to find a fuel load so that the normalized heat load is equal to $46{,}400\sqrt{s}$ K. This is shown in Table 3 in the case of the standard compartment (No. 1) and in case of a compartment made of autoclaved concrete (No. 3)

The table also shows the maximum steel temperature for the two previously mentioned structures (calculated according to (5,11)) and also the charring of a wooden structure, calculated according to Hadvig (15). In compartment No. 1 the maximum temperatures for the protected structure are $-16/+43\,^{\circ}C$ compared with the steel temperature in the fire endurance test. The charring depths are 1-4 mm greater than in the fire endurance test.

In compartment No. 3 the maximum steel temperatures for the protected structure are $+2/+112\,^{\circ}C$ compared with the steel temperature in the test furnace.

By using the normalized heat load concept as a principle of equivalent fire duration the maximum steel temperature is thus determined with an accuracy of 30% (compartment No. 3 and opening factor $0.30 \ m^{1/2}$). In most cases the accuracy is much better.

Table 3 also shows that the normalized heat load concept cannot be used for unprotected structures. These structures are sensitive to the gas temperature-time history, eg. maximum gas temperature and the incident flux.

For a specific structure an equivalent fire duration can at least in principle be calculated so that the structure obtains exactly the same critical temperature in the fire test furnace as in the real fire. The normalized heat load on the other hand establishes a relationship between the thermal exposure in different compartments without any knowledge of the structures being required.

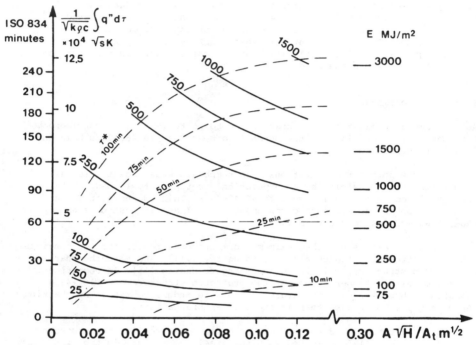

FIGURE 1. Normalized heat load in compartment No. 1 as a function of opening factor and fire load (——). Duration of fully developed fire indicated (---). Equivalent of fire duration according to ISO 834 shown on vertical scale. Normalized heat load in other compartments is estimated from:

$$E_{No.1} = k_f \times E_{true}$$

$$A\sqrt{H}/A_{t \, No.1} = k_f \times A\sqrt{H}/A_{t \, true}$$

TABLE 3. Thermal exposure in the fire test furnace and compartments Nos. 1 and 3 for a normalized heat load of 46,400 $\sqrt{s}$ K

| | $A\sqrt{H}/A_t$ $m^{\frac{1}{2}}$ | E [1] $MJ/m^2$ | $T_{g}$ max °C | $\int_o^{\tau*} H''d\tau$ $MJ/m^2$ | Max temp unprotected steel °C | Max temp protected steel °C | Charring of wood mm |
|---|---|---|---|---|---|---|---|
| furnace | ISO 834 [2] (60 min) | – | 974 | 266 | 924 | 371 | 32 |
| compartment no. 1 | 0.01 | 130 | 686 | 200 | 679 | 355 | – |
| | 0.02 | 152 | 861 | 235 | 851 | 365 | 36 |
| | 0.04 | 190 | 1047 | 285 | 1041 | 368 | 34 |
| | 0.06 | 228 | 1160 | 311 | 1152 | 380 | 35 |
| | 0.08 | 264 | 1237 | 342 | 1231 | 384 | 33 |
| | 0.12 | 335 | 1340 | 380 | 1334 | 395 | 34 |
| | 0.30 | 655 | 1532 | 455 | 1530 | 414 | 34 |
| compartment no. 3 | 0.01 | 49 | 1035 | 270 | 1030 | 373 | – |
| | 0.02 | 72 | 1213 | 349 | 1207 | 398 | – |
| | 0.04 | 116 | 1372 | 438 | 1369 | 425 | – |
| | 0.06 | 159 | 1457 | 497 | 1455 | 446 | – |
| | 0.08 | 204 | 1508 | 553 | 1507 | 462 | – |
| | 0.12 | 289 | 1566 | 597 | 1564 | 468 | – |
| | 0.30 | 650 | 1660 | 675 | 1660 | 483 | – |

[1] Interpolation

[2] Thermocouple measurement error included

CONCLUSIONS

The thermal exposure in fully developed compartment fires can be evaluated in a standard compartment with thermal inertia 1168 $W\sqrt{s}/m^2$K if an equivalent fire load and an equivalent opening factor are used. The equivalent values are obtained as the true values in the real compartment multiplied with the conversion factor $k_f$.

The conversion factor is calculated as $1168/\sqrt{k\rho c}_{700°C}$, where $\sqrt{k\rho c}$ is the thermal inertia in the real compartment, but the duration of the fire must be taken into consideration.

For homogeneous compartments the maximum error will be smaller than 13%, except for compartment No. 8 made of steel. Often the error will be smaller.

The normalized heat load concept, (7), can be used for well-protected structures as a measure of the destructive potential of a compartment fire. It can be used as a principle of equivalent fire duration for the transition from fire endurance tests to real fires with an estimated accuracy in maximum steel temperature of 30%.

NOMENCLATURE

| | |
|---|---|
| $A$ | opening area, $m^2$ |
| $A_t$ | total surface area of compartment, $m^2$ |
| $A\sqrt{H}/A_t$ | opening factor, $m^{1/2}$ |
| $c$ | thermal capacity, $J/kg\,^\circ C$ |
| $E$ | enthalpy of fuel per unit area of compartment (fuel load), $MJ/m^2$ |
| $H$ | opening height, m |
| $H''$ | incident radiation, $W/m^2$ |
| $k$ | thermal conductivity, $W/mK$ |
| $k_f$ | conversion factor |
| $L$ | mean beam length for furnace, m |
| $P$ | partial pressure, atm |
| $q''$ | net heat flux, $W/m^2$ |
| $T_g$ | gas temperature, $^\circ C$ |
| $\rho$ | density, $kg/m^3$ |
| $\tau$ | time, s |
| $\tau*$ | duration of fully developed fire, s |

REFERENCES

1. Ödeen, K.: "Theoretical Study of Fire Characteristics in Enclosed Spaces," Bulletin 10. Royal Institute of Technology, Stockholm, Sweden, 1963.

2. Kawagoe, K. and Sekine, T.: "Estimation for Fire Temperature-Time Curve in Rooms," BRI Occasional Report No. 11, Building Research Institute, Japan, 1963.

3. Magnusson, S.E. and Thelandersson, S.: "Temperature-Time Curves of Complete Process of Fire Development," Bulletin 16, Lund Institute of Technology, Sweden, 1970.

4. Pettersson, O., Magnusson, S.E., and Thor, J.: "Fire Engineering Design of Steel Structures," Publication 50, Swedish Institute of Steel Construction, SBI, 1976.

5. Bøhm, B.: "Fully-Developed Polyethylene and Wood Compartment Fires with Application to Structural Design," Laboratory of Heating and Air-Conditioning, Technical University of Denmark, 1977.

6. Bøhm, B.: "Non-Uniform Compartment Fires. Report No. 1, Preliminary Investigation of Fully Developed Fires," Laboratory of Heating and Air-Conditioning, Technical University of Denmark, 1983.

7.  Harmathy, T.Z. and Mehaffey, J.R.: "Normalized Heat Load: A Key Parameter in Fire Safety Design," _Fire and Materials_, 6, 1, 1982.

8.  Bøhm, B.: "On Thermal Inertia and Normalized Heat Load," Laboratory of Heating and Air Conditioning, Technical University of Denmark, 1984.

9.  Harmathy, T.Z.: "Design of Fire Test Furnaces," _Fire Technology_, 5, 2, May 1969.

10. McCaffrey, B.J., Quintiere, J.G., and Harkleroad, M.F.: "Estimating Room Temperatures and the Likelihood of Flashover using Fire Test Data Correlations," _Fire Technology_, 17, 2, May 1981.

11. Bøhm, B. and Hadvig, S.: _Fire Safety Journal_, 5, 281-286, 1983.

12. Williams-Leir, G.: _Fire and Materials_, 8, 2, 1984.

13. Harmathy, T.Z.: "The Fire Resistance Test and its Relation to Real-World Fires," _Fire and Materials_, 5, 3, 1981.

14. Paulsen, O.: "On Heat Transfer in Fire Test Furnaces," Laboratory of Heating and Air Conditioning, Technical University of Denmark, 1975.

15. Hadvig, S.: _Charring of Wood in Building Fires_, Technical University of Denmark, 1981.

# Full Scale Experiments for Determining the Burning Conditions to Be Applied to Toxicity Tests

**TAKEYOSHI TANAKA, ICHIRO NAKAYA, and MASASHI YOSHIDA**
Building Research Institute
Ministry of Construction
Japan

ABSTRACT

A series of full scale steady state fire experiments were conducted with a propane burner as the fire source in an attempt to determine the thermal and atmospheric conditions to which building materials are likely to be subjected in the event of a building fire. As a result of the experiments, a close coupling was revealed between the temperature elevation and the oxygen depletion in the room of origin. Based on this finding and the results of other experiments with a crib fire source, some thoughts are presented on the burning conditions to be applied to toxicity test apparatus.

INTRODUCTION

Since many kinds of artificial materials have recently entered our daily lives as building materials and furnishings etc., various test methods have been developed in many countries to discover and target for removal of those new materials that may cause extremely toxic hazard in the event of fire. Toxicity is a problem, however, so sensitive and complicated that, so far, little has been agreed among the people concerned with test methods how the evaluation of toxicity should be accomplished. Consequently, toxicity test methods which have been proposed differ in many ways. This inevitably causes discrepancies in evaluation even for the same material.

The U.S.A–Canada–Japan cooperative research on toxicity was, to a large extent, motivated by such conditions related to the current toxicity test methods. This study is, as a part of this toxicity project, intended to determine the conditions under which materials should be burned in a toxicity test. The most important factors that affect the burning behavior of materials and the production of toxicants are thought to be the thermal conditions imposed on the materials, and the atmospheric environment in which pyrolyzed volatiles from the materials undergo combustion. Therefore, the temperature, the heat transfer and the oxygen concentration in the room of origin were investigated in this study.

THE EXPERIMENT

At present, the most reliable way to approach the behavior of real fire is considered to be full scale fire experiments. However, a large number of full scale experiments can hardly be carried out in view of expense and labor. In order to get the best results from the limited number of full scale experiments available, it will be desirable to have insight on the basic relationship between fire behavior and the conditions that control the fire behavior before many tests are conducted. For this reason, a series of full scale tests were conducted with a propane gas burner as the fire source, since it is beneficial for various analyses to use a simple fuel whose properties are well known.

## Experimental Apparatus

The test apparatus employed for this series of experiments is, as illustrated in Fig.1, composed of a burn room, whose interior surfaces are lined with calcium silicate boards, and an adjacent room.

## Measurements

Room temperatures - As shown in Fig.1, three thermocouple trees, one in the room of origin and the other two in the adjacent room, were used for the temperature profiles in the rooms, and on each tree 20 suction thermocouples were located at every 10 cm spacing from the floor.

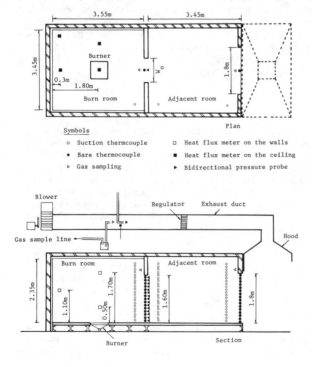

FIGURE 1. Experimental apparatus and instrumentation

<u>Opening temperatures</u> - Bare thermocouples, whose heights are the same as those of the suction thermocouples for room temperature, were used for the opening flow temperature measurement.
<u>Opening flow pressure</u> - Bidirectional pressure probes located at every 10cm of height were used to measure the doorway flow pressure and to produce the doorway flow rate together with the temperature measurement.
<u>Heat transfer</u> - Six total heat flow meters, Hycal model C-1300A, three of which were mounted on the ceiling and the other three on the walls, were used for heat transfer to the room surfaces.
<u>Gas concentrations</u> - The gas were sampled from a point right above the doorway, inside the burn room, and analyzed for $O_2$, $CO_2$ and $CO$ concentrations.


Test Conditions

The measurements were started after warming up the burn room for 20-40 minutes to attain a steady state condition of fire and continued for 2-4 hours. The tests were conducted for the following opening widths and fuel input rates:
<u>Opening width</u>       - 89, 59, 45, 29 (cm)
<u>Propane gas input</u> - 50, 100, 150, 200, 250, 300, 350 ($\ell$/min)


EXPERIMENTAL RESULTS


Temperature Profile in the Room of Origin

The temperature profiles in the burn room, which are exemplified in Fig.2 for 59cm and 29cm of doorway widths, demonstrate that the upper layer is fairly uniform in temperature and gets thicker as the doorway width gets narrower.

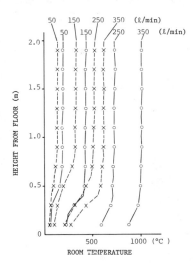

$-$X$---$ 59cm $-$o$-$29cm

FIGURE 2. Temperature profile
in the burn room

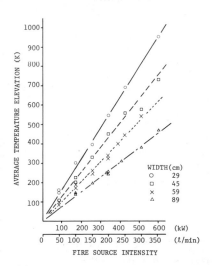

FIGURE 3. Burn room temperature rise and
fire source intensity

## Fire Source Intensity - Temperature Elevation

In this series of experiments, a proportional relationship was found between the fire source intensity(fuel input rate) and the average temperature elevation of the room of origin for each opening width as shown in Fig.3. It is suspected that longer warming up of the room could have brought more perfect linear relationship between the two.

## Doorway Flow Rate

The symbols in Fig.4 exhibit the flow rates through the opening, which were obtained by reducing the measured data in the same manner as Steckler's(2) and were plotted versus the average fire room temperature. It is observed in this figure that temperature dependence of the doorway flow rates is not significant for the region where temperature is above 200-300 °C. If we estimate doorway flow by a simple model based on uniform room temperature assumption, we obtain:

$$m_d = \frac{2}{3} \alpha \ BH^{3/2} \sqrt{2g} \rho_a \frac{(1-\frac{T_a}{T_G})^{1/2}}{\{1+(\frac{T_G}{T_a})^{1/3}\}^{3/2}} \tag{1}$$

The calculated results of Eq.(1) are shown also in Fig.4 by solid lines, by which the same tendency can be recognized in the dependence of the doorway flow rates on temperature. The reason why the discrepancy between the experimentally obtained flow rates and the theoretical estimation becomes larger as the doorway width gets wider may be attributed to the fact that in the case of the wide doorway, the burn room is stratified into two layers and so the single uniform temperature assumption no longer holds.

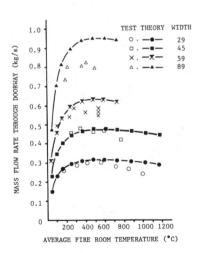

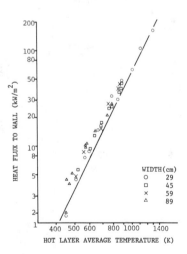

FIGURE 4. The doorway mass flow rate    FIGURE 5. The heat flux to wall surface

132

## Heat Transfer

Average heat fluxes to the walls are plotted in Fig.5 versus the temperatures averaged over what seemed to be the upper layers of the burn room There was no significant difference among the heat fluxes measured by the six heat flux meters on each location of the ceiling and the walls. Since the measurements were made using water cooled total heat flux meters, the heat fluxes shown here are considered to be almost the total incident heat fluxes to the walls. The solid line in this figure represents the theoretical radiant heat flux that the room gas would emit if it were black. The measured heat flux data almost always lie above this line. In most cases, the real fire gas will be close to black due to a significant amount of smoke particles involved but in this series of experiments, in which propane was the fuel used, it is not evident whether or not the fire room gas can be regarded as black, although a significant amount of soot production was observed. If the emissivity of the gas is small, the influence of wall surface temperature will appear in the incident heat flux to the wall. For instance, in the case of a room with uniform interior surface temperature filled with homogeneous gas, the incident heat flux $q''$ is:

$$q'' = \varepsilon_G \sigma T_G^4 + (1-\varepsilon_G)\varepsilon_W \sigma T_W^4 \tag{2}$$

where $\varepsilon_G$ and $\varepsilon_W$ are the emissivities of the room gas and the interior surface, respectively, and $T_G$ and $T_W$ are the gas and the interior surface temperatures. In the present experiments, the incident heat flux was almost as large as the theoretical black gas radiation either because the gas was sooty enough to be approximately black or because the wall surface temperature was as high as the room gas temperature due to its high insulation performance.

## Gas Concentration and Observation

In the experiments, the fire room gas was analyzed for $O_2$, $CO_2$ and $CO$, but the concentration of $CO$ was negligibly low in every case, probably because fire sizes were still too small to bring incomplete combustion even in the case of the narrowest opening.

Flames were not observed to come out of the burn room in any test although they seemed to extend and fill the room in the case of 29cm doorway width with 300 ℓ/min and 350 ℓ/min of fuel input.

## DISCUSSIONS

## Heat Transfer

If we look at the heat conservation in the room of origin in a steady state condition, we will have the following general equation:

$$Q_f = Q_d + Q_{dr} + Q_w \tag{3}$$

where $Q_f$ is the heat released in the room of origin, $Q_d$, $Q_{dr}$, and $Q_w$ are the heat losses due to opening flow, radiation through the opening and heat transfer to the wall, respectively. It will be interesting to investigate the contribution of each mode of heat loss in the total heat. The results shown in Fig.6 are the convective heat losses through the doorway in the tests estimated by the following equation:

$$Q_d = C_p \, m_d \, (T_G - T_a) \tag{4}$$

where $C_p$ is the specific heat of gas and $T_G$ is the average temperature of the room gas. It can be seen in Fig.6 that the heat loss due to the doorway convection is almost proportional to the fire source intensity, and is about 50% of the heat released in the room. The solid line drawn for reference in Fig.6 represents 50% of the heat input. The fraction of the heat loss due to convection tends to increase slightly as the doorway width becomes wider but the inclination is not so conspicuous. As was already shown in Figs.3 and 4, the room temperature elevation $T_G-T_a$ becomes smaller, while the doorway mass flow rate becomes larger, with the increase of the doorway width. This means that the former counterbalances the latter and thus convective heat loss through doorway turns out to be less dependent on doorway width. Once the convective heat loss due to doorway flow is known, the net heat transfer to the wall can be estimated.

Temperature Elevation and Oxygen Depletion in the Present Experiments

If combustion in the room of origin is vigorous, the room temperature inevitably gets high. At the same time, the oxygen concentration in the room must be low because vigorous combustion consumes a large amount of oxygen in the room. When a room has a large opening, in other words when the ventilation rate is large, the oxygen depletion in the room will be comparatively small and the temperature elevation must be small too because of the large amount of air that flows into the room and needs be heated up. Thus a strong relationship is suspected to exist between the temperature elevation and the oxygen depletion in the room of origin.

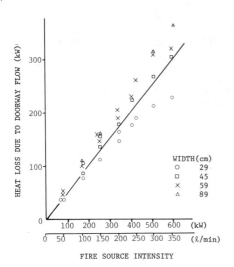

FIGURE 6. The fraction of the heat loss due to the doorway flow to the total heat release

134

In the above, it was seen that the fraction of convective heat loss associated with the opening flow to the total heat release($\equiv k$) was nearly constant in this series of experiments, that is:

$$Q_d = k \, Q_f \tag{5}$$

Noting that heat of combustion per unit of oxygen consumed is almost constant over a wide variety of fuels(3), $Q_f$ can be conveniently written as

$$Q_f = E \, \Delta X_{O_2} \, \frac{m_d}{W_{air}} \tag{6}$$

where $\Delta X_{O_2}$ is the difference of oxygen mole fraction between inflow air and outflow gas($=0.21 - X_{O_2}$), E is the heat of combustion of fuel per unit mole of oxygen consumed, and $W_{air}$ denotes the molecular weight of air. On the other hand $Q_d$ can be written as

$$Q_d = C_p \, m_d \, \Delta T \tag{7}$$

where $\Delta T$ is the room temperature elevation($=T_G - T_a$). From Eqs.(6) and (7),

$$\Delta T = \frac{k \cdot E}{C_p W_{air}} \, \Delta X_{O_2} \tag{8}$$

Equation (8) indicates that the temperature elevation should be proportional to the oxygen depletion and independent of the doorway flow rate. If 0.5 is taken for the value of k, Eq.(8) turns out to be as follows:

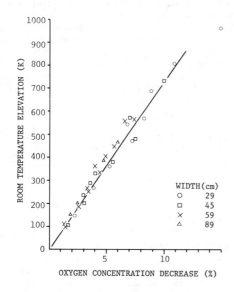

FIGURE 7. The temperature rise and the oxygen concentration decrease in the burn room

$$\Delta T = \frac{k \cdot E}{C_p W_{air}} \Delta X_{O_2} = \frac{0.5 \times 410}{1.0 \times 0.029} \Delta X_{O_2} = 7,069 \ \Delta X_{O_2} \tag{9}$$

This indicates that the temperature elevation rate will be about 70 K per 1% of the oxygen concentration decrease. With this in mind, the test data were plotted in Fig.7. It can be seen that this prediction nearly holds at least for small oxygen depletion. It is not obvious if this relationship remain valid for a larger decrease in the oxygen concentration. If this holds, then it follows that the temperature elevation cannot exceed about 1,500 K because oxygen concentration decrease must be less than 0.21.

Theoretically, the rate of temperature elevation per unit oxygen concentration decrease must depend on the level of thermal insulation performance of the room. For example, if the room is perfectly adiabatic, all the heat released by fire must be convected out of the room only by the doorway flow, which means k in Eq.(8) must be unity and the temperature elevation rate will be about 140 K. However, rooms are usually not insulated so perfectly. Therefore, it is likely that in a normal room fire, the temperature elevation rate is less than those of the present tests because the burn room employed in this series of experiments was not very large in size and was lined with calcium silicate boards, whose thermal insulation performance is among the highests of many construction materials.

Temperature Elevation – Oxygen Depletion in General Cases

As was demonstrated in Fig.7, an almost linear relationship was obtained between the temperature elevation and the oxygen concentration decrease for the steady state fire. However, it is not guaranteed that this relationship always holds since the results were only from limited test conditions. In different thermal or ventilation conditions or in nonsteady state conditions, the results may be different. In Fig.8, the data were taken from the crib fire experiments which were conducted while using the same apparatus and from other literatures(4,5) for the peak burning period and they were plotted together with the present test data. According to the data in Fig.8, the temperature elevation rates tend to scatter over a somewhat wider region than in the steady state case, but still appear to stay within a limited range covered by a narrow angle that has the origin of the coordinates as the apex.

If a room of origin filled with a single homogeneous gas layer is taken as a simple example, an approximate equation for the heat balance in the room can be given as follows:

$$E \frac{m_d}{W_{air}} \Delta X_{O_2} = C_p m_d (T_G - T_a) + A_W \{\sigma(T_G^4 - T_W^4) + h_c(T_G - T_W)\} \tag{10}$$

heat release     convective loss     heat loss to walls

where the radiation loss to the outside through the opening is neglected, the emissivities of the gas and the wall are assumed as unity, and inflow and outflow rates are assumed the same. When the wall is a good thermal insulator and small in area, the heat loss to the wall is small. In the case of a perfect thermal insulator, as was seen previously, it follows that $\Delta T = 14,000 \ \Delta X_{O_2}$. In an actual fire, however, $\Delta T = 7,000 \ \Delta X_{O_2}$ will virtually give the maximum temperature elevation rate attainable per unit oxygen concentration decrease by the reason given in the above.

On the other hand, it is much more difficult to determine the lowest limit of the temperature elevation rate. If we suppose a case in which the thermal conductivity of the room wall is extremely large, as large as metals for example, Eq.(10) can be reduced to

$$\frac{E(cA\sqrt{H})}{W_{air}} \Delta X_{O_2} = C_p(cA\sqrt{H})\Delta T + \frac{A_W}{2}\{\sigma(T_G^4 - T_a^4) + h_c\Delta T\} \tag{11}$$

where an attention was paid of the fact that doorway flow rate $m_d$ does not greatly vary with room temperature and hence can be determined almost only by $A\sqrt{H}$ of the doorway. Equation (11) shows that the temperature elevation rate still depends on wall area and opening dimensions even in the simplest case. It will be beneficial to use some mathematical modeling techniques for determining the lowest limit of temperature elevations, since so many experiments cannot be conducted using an apparatus with a large wall area.

From the empirical relationship between heat release rate and temperature rise that was discovered by McCaffrey and Quintiere(6), it is implied that temperature rise $\Delta T$ varies proportionally to $\Delta X_{O_2}^{2/3}$ rather than to $\Delta X_{O_2}$ (7). The data in Fig.8 do seem to support their prediction but we would like to refrain from any hasty conclusion and only value the empirical findings until we reach to a deeper understanding of the mechanism that bring this relationship between temperature elevation and oxygen concentration depletion.

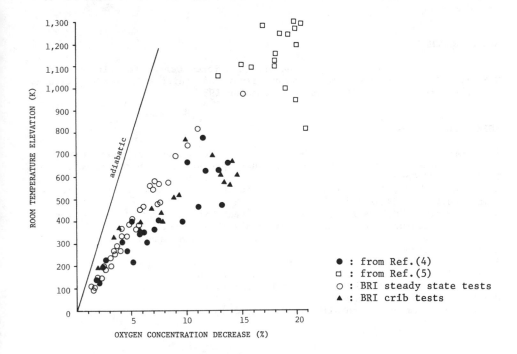

FIGURE 8. The relationship between the temperature rise and the oxygen concentration decrease in general case (peak values)

## CONCLUDING REMARKS

When we discuss the issues concerning a toxicity testing of building materials, etc. it will be the important to identify in what scenarios will the toxicity due to the burning of materials become a serious problem. And then the fire conditions that correspond to these scenarios need to be clarified. It will be reasonable to think that the conditions in developing or fully developed stage of fire which continue for a considerable time period are more important than a transient short time condition in view of the influence on people's evacuation.

In this series of steady state fire experiments, it was demonstrated that, in the room of origin, the heat transfer to the ceiling and the upper part of the walls maintain a considerable homogenuity, and a strong correlation exists between the temperature elevation and the oxygen concentration. The latter, of course, means a strong relation between the incident heat flux to the materials on the room interior surfaces and the oxygen concentration in the room, to which pyrolyzed volatiles come out, since a room temperature and the incident heat flux to the room boundaries are almost in one to one correspondence. This finding on the thermal and atmospheric conditions in the room of origin is not enough to pinpoint the conditions to be applied to a toxicity test apparatus. The conditions that are conceivable in fire at various stages still range over wide area. However, it can be said that this finding did narrow the conditions that must be considered in testing materials and will help remove the test conditions are unreasonable that in view of realistic fire situations.

## ACKNOWLEDGEMENT

The authors would like to thank Dr. Quintiere, Center for Fire Research, NBS, for his informative comments on our study and fruitful discussions with us during his technical visit to Japan in October 1984.

## REFERENCES

1. Bukowski,R.W., Jones,W.W. "Development of a Method for Assessing Toxic Hazard", 3rd Expert Meeting of Canada-Japan-U.S.A Cooperative Study on Toxicity, NRC Canada, October 1984.
2. Steckler,K.D., Quintiere,J.G. and Rinkienen,W.J. "Flow Induced by Fire in a Compartment", NBSIR 82-2520, 1982.
3. Hugget,C. "Estimation of Rate of Heat Release by Means of Oxygen Consumption Measurements", Fire and Materials. Vol 4. No2. 1980.
4. Quintiere,J.G.,McCaffrey,B.J. "The Burning of Wood and Plastic Crib in an Enclosure: Volume 1", NBSIR 80-2054, November 1980.
5. Bohm,B. "Fully Developed Polyethylene and Wood Compartment Fires with Application to Structual Design", Technical Univ. of Denmark, 1977.
6. Quintiere,J.Q. "A Simple Correlation for Predicting Temperature in a Room Fire", NBSIR 83-2712, June 1983.
7. Personal communication with Dr. Quintiere.

# Temperature Correlations for Forced-Ventilated Compartment Fires

**K. L. FOOTE**
Lawrence Livermore National Laboratory
L-442, Box 5505
Livermore, California 94550, USA

**P. J. PAGNI**
Mechanical Engineering Department
College of Engineering
Berkeley, California 94720, USA

**N. J. ALVARES**
Lawrence Livermore National Laboratory
L-442, Box 5505
Livermore, California 94550, USA

ABSTRACT

Force-ventilation compartments are a common environment for fire growth in sealed or high-rise structures. Currently, no method exists for reliably estimating the fire hazard in these enclosures. Using data from compartment fires in the forced ventilation facility at the Lawrence Livermore National Laboratory (LLNL), a simple correlation has been developed following the methods of McCaffrey, Quintiere and Harkleroad. The upper layer temperature rise above ambient, $\Delta T = T_u - T_\infty$, is given as a function of: the fire heat release rate, $\dot{Q}$, the compartment mass ventilation rate, $\dot{m}$, the gas specific heat capacity, $c_p$, the compartment surface area A and an effective heat transfer coefficient based on $\Delta T$, h. The nondimensional form of the best fit to the LLNL data is:

$$\Delta T/T_\infty = 0.63 \ (\dot{Q}/\dot{m}c_p T_\infty)^{0.72} \ (hA/\dot{m}c_p)^{-0.36}$$

This confirms the correlation suggested by McCaffrey et al. For their free-ventilation data the coefficient increases to 0.77. All the data are well fit by a coefficient of 0.7, while the powers remain unchanged. Alternatives for, limits on, and usefulness of such correlations are discussed.

INTRODUCTION

Several methods exist for evaluating the fire hazard in traditional enclosures [1-5]. Relatively little work, however, has been done for fires in modern forced-ventilated compartments [6]. One effort to optimally describe recent forced data is presented here. A temperature correlation similar to one developed by McCaffrey, et al. [1], for naturally ventilated fires is sought. Given the heat release rate of the fire, the properties of the walls, the ambient temperature, the ventilation rate and room geometry, the evolution of the upper layer temperature is estimated.

Supported at LLNL by USDOE Contract No. W-7405-ENG-48 and at Berkeley by USDOC-NBS CFR Grant No. 60NANB5D0552. Assistance from B. McCaffrey, J. Quintiere, E. Zukoski and J. Reed is gratefully acknowledged.

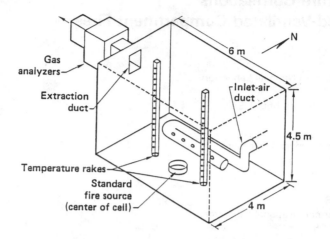

Fig. 1. Schematic of the LLNL test facility. The center of the exit opening
(0.65 m × 0.65 m) is 3.6 m above the floor. The inlet opening consisted of four
wall-facing horizontal rectangles (0.12 m high × 0.5 m) with centerlines 0.1 m
above the floor. All walls can be approximated as 50% $Al_2O_3$ - 50% $SiO_2$ refractory of thickness, $\delta$ = 0.10 m, with conductivity, $k_w$ = 0.46 W/m°K, specific
heat, $c_w$ = 1 J/g°K; and density, $\rho_w$ = 1607 kg/m$^3$. The wall thermal penetration
time is defined as $t_p = \delta^2/4\alpha_w$ where $\alpha$ = k/$\rho$c. Here $t_p \sim$ 2.5 hrs.


EXPERIMENT

The data used to develop the correlation come from a series of tests conducted during the summer of 1983 in the Lawrence Livermore National Laboratory
(LLNL) fire test cell shown in Fig. 1. A detailed description of the system and
its instrumentation has been published [6]. Fresh air was introduced at the
floor and pulled out near the ceiling by an axial fan. A constant exit gas flow
rate was controlled by a butterfly valve upstream of the fan and measured by a
sharp-edged orifice. The inlet flow rate was measured using a calibrated vane
anemometer. Methane gas was metered by a critical orifice into a 0.56 m
diameter rock-filled pan in the center of the test cell floor. This burner
allowed accurate control of the constant fuel flow rate and permitted a variety
of heat release rates for different tests. Additional instrumentation included
gas and surface temperature sensors, calorimeters, radiometers, combustion product and oxygen detectors, fuel and ventilation flow sensors, and a video camera
for recording the fire shape. All measurements were recorded and reduced using
an online data acquisition system.

The temperature correlation data were obtained from two thermocouple rakes
positioned at 1.5 m on either side of the fire source. Each rake supported 15
5-mil chromel alumel thermocouples spaced 0.3 m apart. Figure 2 shows the evolution of the gas temperature profile in the test cell at the East rake. The
upper layer is established quickly (within 150 s). The upper layer gas temperature was defined as the instantaneous average of the top four thermocouples
outside the ceiling jet. The quasi-steady approximation restricts $\dot{m}(t)$ and $\dot{Q}(t)$
to be slowly varying, i.e. no large changes (> 30%) may occur on time scales
fast compared with their respective characteristic times. For mass flow,

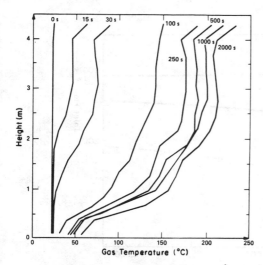

Fig. 2. Evolution of the compartment gas temperature profiles measured by the East rake shown in Fig. 1 for Test 6 (see Fig. 4). The upper layer height (~ 1.7m) and lower layer temperature may be defined as in Ref. 7.

$t_m = m/\dot{m}$ or $V/\dot{V}$. Here $V \sim 10^5 \ell$ and $\dot{V} \sim 250\ell/s$ give $t_m \sim 400s$. For energy, $t_Q = \dot{m}c_pT/\dot{Q}$. Using the ideal gas equation of state with $R = c_p-c_v$ and $\gamma = c_p/c_v = 1.4$ for air gives $t_Q = 3.5$ PV/$\dot{Q}$. Here $P \sim 1$ atm, $V \sim 10^5 \ell$ and $\dot{Q} \sim 250$ kW, with $10^{-2}$ $\ell$-atm/J, give $t_Q \sim 140$ s. Since temperature is correlated here, $t_Q$ dominates. Species correlations will be controlled by $t_m$.

Figure 3 shows the histories of the inlet flow, the upper layer gas temperature and the ceiling temperature above the fire from Test 6. Both temperatures rise quickly and then slow to their predicted approximately [1,6] $t^{1/6}$ and $t^{1/2}$ growth rates, respectively. The rapid temperature rise coincides with the air flow out the inlet duct. This flow reversal may prove useful in suppression strategies, as dampers could be closed after much of the oxygen initially in the compartment has flowed out, thus decreasing the time to suppression.

Upper layer temperature histories from the LLNL tests are shown in Fig. 4. The approximately constant heat release and ventilation rates were chosen to be representative of possible fires in ventilation-controlled rooms with ~ 7 room air changes per hour.

CORRELATION

Since the compartment is open to ambient at the inlet, its pressure, P, is fixed near one atmosphere. The compartment volume, V, is fixed. So from PV = mRT, and E ~ mRT the internal energy, E, must be constant. Since there is no energy storage in the gas in the compartment, the energy balance is quasi-steady,

$$\dot{Q} = \dot{m}c_p(T_u - T_\infty) + \dot{q}_{loss},\qquad(1)$$

assuming gas flows in at $T_\infty$ and out at $T_u$. All the time dependence of $T_u(t)$ comes from the time dependence of $\dot{Q}(t)$, $\dot{m}(t)$ and $\dot{q}(t)$. Following McCaffrey, et al. [1] the total wall heat loss is represented by a single thermal resistance,

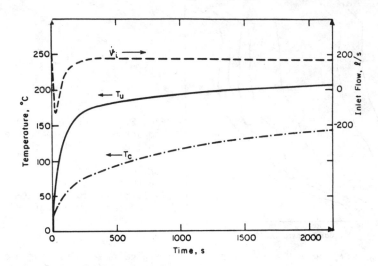

Fig. 3. Temperature and flow histories for Test 6 (see Fig. 4). $T_C$ is the ceiling temperature above the fire source, $\dot{V}_i$ is the inlet air volumetric flow rate showing a flow reversal just after ignition as the compartment dumps mass to accommodate its increased temperature.

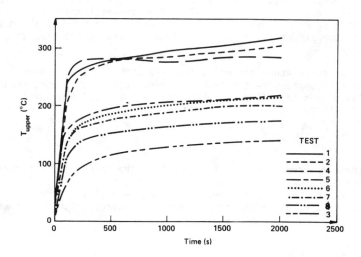

Fig. 4 Upper layer temperature histories for tests run at the following constant parameters:

| Test | 1 | 2 | 3 | 4 | 5 | 6 | 7 | 8 |
|---|---|---|---|---|---|---|---|---|
| $T_\infty$, °K | 307 | 303 | 282 | 298 | 295 | 298 | 300 | 284 |
| $\dot{Q}$, kW | 490 | 465 | 150 | 400 | 230 | 250 | 250 | 180 |
| $\dot{m}$, g/s | 300 | 160 | 190 | 220 | 110 | 175 | 325 | 180 |

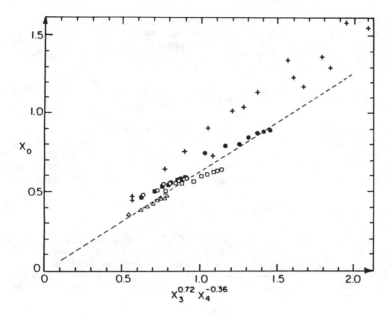

Fig. 5 Upper layer temperature data and correlation. The LLNL tests are: 1-●, 3-○, 6-■, 7-□, and 8-△. The NBS data (+) are from Ref. 8. The dashed line is Eq. (8).

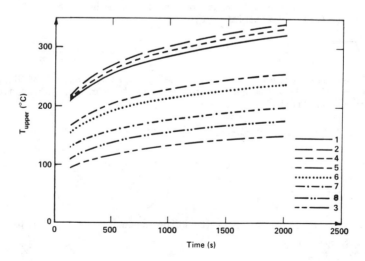

Fig. 6 Calculated upper layer temperatures using Eq. (9) with t ⩾ 120 s for comparison with Fig. 4.

$$\dot{q}_{loss} = hA(T_u - T_\infty) \ . \tag{2}$$

Many phenomena are lumped into h: the radiation and convection from the hot layer and the fire to the wall, the conduction through the wall and the convection outside the compartment. The approximation is made [1] that all thermal conductances are proportional to the wall conduction, which is smallest (by $\sim 1/3$) and therefore controlling, so that

$$h \sim h_k = (\rho_w c_w k_w/t)^{1/2} = 860 \ t^{-1/2} \ W/m^2 \mbox{}^\circ K \ , \tag{3}$$

where the last equality holds for the LLNL experiments for $t < t_p \sim 9000$ s. Even with the gross assumption that the overall heat transfer coefficient is proportional to $h_k$, there remain three time regimes with different $h_k(t)$. Before the thermal wave reaches the wall exterior, at $\sim t_p/4$, the wall can be treated as semi-infinite and the Rayleigh problem solution gives the local heat flux,

$$\dot{q}_w(x,t) = (\rho_w c_w k_w/\pi t)^{1/2} \ \Delta T \ exp(-x/(4\alpha_w t)^{1/2}) \ . \tag{4}$$

At the surface, $x = o$ and $\dot{q}_w \sim (\rho_w c_w k_w/t)^{1/2} \Delta T$, whence comes Eq. (3). After the finite wall thickness begins to play a role, series solutions for $\dot{q}$ are required [8]. Eventually, at $\sim 4t_p$, a linear temperature profile exists in the wall and $\dot{q} = \Delta T k_w/\delta$. Therefore, approximate h(t) as Eq. (3) for $t < t_p$ and as a constant, $k_w/\delta$, for $t > t_p$, and rely on the correlation to adjust for any inadequacies.

Substituting Eq. (2) into Eq. (1) and nondimensionalizing gives

$$\Delta T/T_\infty = (\dot{Q}/\dot{m}c_p T_\infty)(1 + hA/\dot{m}c_p)^{-1} \quad or \quad X_o = X_3(1 + X_4)^{-1} \ . \tag{5}$$

Equation (5) defines $X_0$ as the temperature rise in units of the absolute ambient temperature. $X_3$ is the fire strength in units of the flow sensible heat and $X_4$ is the ratio of the wall heat loss to the flow heat loss. $X_1$ and $X_2$ are reserved for different definitions [1],

$$X_1 = \dot{Q}/(A_0 \sqrt{g \ H_0} \ c_p \ \rho_\infty \ T_\infty) \quad and \quad X_2 = hA/(A_0 \ \sqrt{g \ H_0} \ c_p \ \rho_\infty) \ , \tag{6}$$

for free fires where only the opening geometry (area $A_0$ and height $H_0$), not $\dot{m}$, is known a priori. Only the dependence, not the form of Eq. (5), can be relied on, so a power law fit [1] is adopted,

$$X_o = C \ X_3^n \ X_4^m \ , \tag{7}$$

as a trial expression. The corresponding correlation for free flow [1] is,

$$X_o = 1.52 \ X_1^{0.65} \ X_2^{-0.39} \ . \tag{8}$$

Using this as an initial guess, an iteration process (see Fig. 5) led to,

$$X_o = 0.63 \ X_3^{0.72} \ X_4^{-0.36} \ . \tag{9}$$

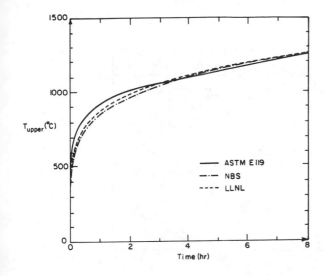

Fig. 7. Comparison of Eqs. (8 and 9) with the ASTM E119 Standard Time - Temperature Curve showing the similar time dependencies.

The data used to obtain Eq. (9) are the well-ventilated tests for which $\dot{Q}$ was known from both $\dot{m}_{fuel}$ and the $O_2$, CO and $CO_2$ exhaust mass flow rates. The dashed line in Fig. 5 is the fit of these points with a correlation coefficient of 0.95. A good overall fit to both the forced and free data is obtained with C in Eq. (7) as 0.70. For the NBS data (+) in Fig. 5, $\dot{m} \sim 0.1 \ (\rho_\infty \sqrt{gH_0} \ A_0)$. The average net exponent of $\dot{m}$ in Eq. (9) and $(\rho_\infty \sqrt{gH_0} \ A_0)$ in Eq. (8) is - 0.31. So if Eqs. (6 or 8) were expressed in terms of $X_1$ and $X_2$, the coefficient C would double. Comparing the 1.52 in Eq. (8) with $(0.7)(0.1)^{-0.31} \approx 1.5$, shows that these correlations verify previous free results [1].

Figure 6 shows predictions from Eq. (9) for comparison with the data in Fig. 4. Test 1 is a large, well-ventilated fire with the largest $\Delta T$ observed in the LLNL data; it is well described by Eq. (9), as is the smaller Test 3. Test 4 was initially a hot fire with little ventilation; it became underventilated and its heat release rate decreased accordingly. Tests 2 and 5, also underventilated, were not included in the fitting process, so their agreement is encouraging. Test 7 is a well-ventilated fire that quickly reached quasi-steady conditions. Test 8 was a small isopropanol pool fire which grew more slowly than the gas fires to a steady 150 kW.

Good agreement was also obtained for a separate series of tests with the fire source elevated 2.5 m above the floor. The correlation doesn't explicitly account for elevation. It can, however, be adapted to this situation by selecting A as the surface area of the hot layer, i.e. the floor and ceiling areas plus only the wall area in the hot layer. This allows some heat loss below the plane of the fire and still matches the observed higher temperatures.

Comparison of Eqs. (8 and 9) with the American Society for Testing and Materials' standard E119 time-temperature curve is shown in Fig. 7. Since the standard does not represent a specific compartment, the comparison had to be made by: assuming the maximum possible $X_{1or3} = 10$ (see Fig. 8), calculating $X_0$ = 4.2 from the ASTM curve at 8 hrs and solving for $X_{2or4}$. All earlier $X_{2or4}(t)$ = $(t/8)^{-1/2} X_{2or4}$ (8). Good agreement results.

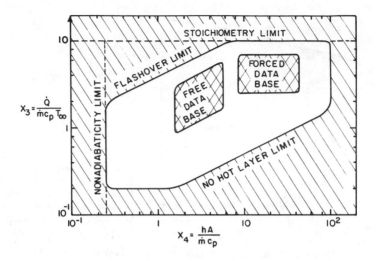

Fig. 8  Parameter space diagram showing data (XXX), possible extrapolation ( ), and regions where physical constraints prohibit application of the correlations (\\\\\\).

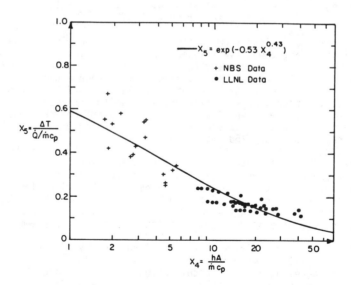

Fig. 9  The data trend in Figs. 4 and 8 and Eq. (1) suggest that the characteristic temperature is $\dot{Q}/\dot{m}c_p$, not $T_\infty$. An improved correlation is shown here along with data from tests 1,3,6,7 and 8 (●) and Ref. 9 (+).

Figure 8 shows a map of regions of special interest in $X_3$ - $X_4$ space. Test results define the inner regions on which the correlations is based. Surrounding this is the expected regime (unhatched) of valid extrapolation. Beyond this region is a space into which this correlation cannot go, for a variety of reasons, different at each boundary. $X_3$ cannot exceed 10 since $\dot{Q}_{max}$ = (13 kJ/gO$_2$)(0.23 gO$_2$/g air)$\dot{m}$, with $c_p \sim$ 1 J/gK and $T_\infty \sim$ 295K. At the other extreme an $X_3$ = 0.2 corresponds to a maximum $\Delta T$ of < 60°C since if $T_\infty$ = 295 it suggests $\dot{Q}/\dot{m}c_p$ = 59°C. This $\Delta T$ is not adequate to give a stratified upper layer in most compartments [10]. The same criterion gives the diagonal no-hot-layer limit along an isotherm at 50°C. The cutoff at $X_4 \sim 10^2$ is due to the short-time limit prior to layer formation, when only detail plume analyses are permitted [11]. The nonadiabaticity limit at $X_4 \sim$ 0.25 suggests that no wall will absorb less than 20% of O, since $X_4$ is physically the ratio of the wall heat loss, $hA\Delta T$, to the flow heat loss. The final limit, flashover at $\Delta T$ = 500°C, is based on the criterion suggested by McCaffrey et al. [1] and their data which show the correlation fails when $\Delta T$ > 500°C. Additional data are needed to sharpen the rather fuzzy boundaries portrayed in Fig. 8.

The general trend in Figs. 8 and 4 is that large $\dot{Q}$ leads to large $\Delta T$, i.e. large fires have hot layers and large losses. This suggests a characteristic temperature based on $\dot{Q}$ rather than $T_\infty$. The simple energy balance in Eq. (1) places all $T_\infty$ influence into $\Delta T$. Proper nondimensionalizing of Eqs. (1-3) gives the true characteristic temperature, $\dot{Q}/\dot{m}c_p$. Therefore let $X_5 \equiv X_0/X_3 \equiv \Delta T/(\dot{Q}/\dot{m} c_p)$ and seek $X_5(X_4)$. Several functional forms were considered. The data collapsed to nearly a single curve on a log-log plot of ln $X_5$ vs $X_4$, suggesting the form,

$$X_5 = \exp (-C' X_4^{n'}) , \qquad (10)$$

which gave the fit,

$$X_5 = \exp (-0.53 X_4^{0.43}) . \qquad (11)$$

As shown in Fig. 9, Eq. (11) fits both forced and free data with a correlation coefficient of 0.91. Also $X_5 \to 1$ as $X_4 \to 0$, i.e. $\Delta T = \dot{Q}/\dot{m}c_p$ as it should in the adiabatic limit, and $X_5 \to 0$ as $X_4 \to \infty$, i.e. $\Delta T \to 0$ as expected in the limit of large losses.

Figures 8 and 9 show that, if $\Delta T$ > 500°C is a valid flashover criterion [1], flashover can not occur for $X_4$ > 10. If $X_4$ < 10, flashover will occur only when

$$(\dot{Q}/\dot{m}c_p) \exp(-0.53 X_4^{0.43}) > 500°K . \qquad (12)$$

After flashover, the upper layer no longer has a uniform temperature due to local combustion.

CONCLUSIONS

1. The fact that one fit, $X_0 = 0.7 X_3^{0.72} X_4^{-0.36}$, describes both free and forced data to within ± 10% and that a single simpler fit, Eq. (11) incorporates both data sets, suggests that there is no essential difference between free and forced compartment fires.

2. The distinction is that $\dot{m}$ is known in the forced case and unknown in the free. This requires the use of $X_1$ and $X_2$, in free systems and $X_3$ and $X_4$, in forced cases.

3. This work then is a verification and extension of the temperature correlation of McCaffrey, Quintiere and Harkleroad [1]. The different range of the data in Fig. 8 is due, not to the forced flow, but rather to the 10 cm thick refractory walls and larger scale of the LLNL test cell.

4. Future work should examine extension of these fits to the full range suggested by Fig. 8 and should determine if the same flashover criterion, $\Delta T > 500°C$, applies uniformly [12]. An improved wall heat loss term should be pursued with a more accurate h(t) [13] and an area equal to the hot layer bounding surface area. The one parameter fit in Eq. (11) should be explored further.

REFERENCES

1. McCaffrey, R. J., Quintiere, J. G., and Harkleroad, M. F., "Estimating Room Temperatures and the Likelihood of Flashover Using Fire Data Correlations," Fire Technology, 17, 2, 1981.

2. Alpert, R. L., "Turbulence Ceiling-Jet Induced by Large-Scale Fires," Combustion Science and Technology, 11, 197, 1975.

3. Quintiere, J. G., "Growth of Fire in Building Compartments," in Fire Standards and Safety, ASTM STP 614, A. F. Robertson, Ed. American Society for Testing and Materials, 131, 1977.

4. Thomas, P. H., "Fire Modeling and Fire Behavior in Rooms," Eighteen Symposium (Int.) on Combustion, The Combustion Institute, p. 503, 1981.

5. Emmons, H. W., "The Prediction of Fires in Buildings," Seventeenth Symposium (Int.) on Combustion, The Combustion Institute, p. 1101, 1978.

6. Alvares, N. J., Foote, K. L., and Pagni, P. J., "Forced Ventilation Enclosure Fires," Combustion Science and Technology, 39, 55, 1984.

7. Quintiere, J. G., Steckler, K. and Corley, D., "An Assessment of Fire Induced Flows in Compartments," Fire Science and Technology, 4, 1, 1984.

8. Carslaw, H. S. and Jaeger, J. C., Conduction of Heat in Solids, p. 102, 2nd ed., Oxford University Press, London, 1959.

9. Quintiere, J. G. and McCaffrey, B. J., "The Burning of Wood and Plastic Cribs in an Enclosure," NBSIR 80-2054, Nov. 1980.

10. Zukoski, E. E. and Kubota, T., "Two-Layer Modeling of Smoke Movement in Building Fires," Fire and Materials, 4, 19, 1980.

11. Zukoski, E. E., "Fluid Dynamic Aspects of Room Fires," Plenary Lecture, This Symposium.

12. Thomas, P. H., Bullen, M. L., Quintiere, J. G. and McCaffrey, B. J., "Flashover and Instabilities in Fire Behavior," Combustion and Flame, 38, 159, 1980.

13. Quintiere, J. G., McCaffrey, B. J., and Kasiwagi, T., "A Scaling Study of a Corridor Subject to a Room Fire," Combustion Science and Technology, 18, 1, 1978.

# A Contribution for the Investigation
# of Natural Fires in Large Compartments

**E. HAGEN and A. HAKSEVER**
Institut für Baustoffe, Massivbau und Brandschutz
der Technischen Universität Braunschweig, FRG

ABSTRACT

To provide fundamental knowledge about fire development and fire spread in large
fire compartments, a project "Natural Fire" is established in "Sonderforschungsbereich 148" at the Technical University of Braunschweig. The full scale fire
experiments are carried out in a fire room with variable sizes between 500 m³
and 2000 m³, using the facilities of the Technical Research Centre of Finland.
The first test series, which should determine the behaviour of the burning rate
in dependence of the ventilation-conditions and the fire-load-configurations,
is presented and the experimental results are discussed in this contribution.

INTRODUCTION

The research project "Natural Fires"

The fire resistance of structural elements is usually determined by means of
fire-tests with respect to ISO-Fire Curve /1/. In this procedure, the structural
element which is being tested is subjected to a certain fire exposure under
defined structural boundary conditions. In practice, however, in a natural fire-
case, the structural elements are heated with respect to totally different non
steady temperature conditions. In order to determine the structural response of
bearing members in such a fire case, it is essential to predict at first the
development of natural fire taking into account the parameters which have an
influence on fire.

Being aware of the importance of the natural fires a new research project in
Sonderforschungsbereich 148 - special research group for fire problems - was
established at the Technical University of Braunschweig four years ago to investigate the phenomenon of natural fires both theoretically and experimentally.
The new project has the goal to research the development of fires not only in
the preflashover phase but also the fully developed and the cooling down phases
for significant boundary conditions. Besides that the investigations intend
to provide necessary information also for the non steady heat transfer conditions, thus a structural fire engineering design can be done as realistically
as possible.

Parameters influencing natural fires

Figure 1 shows the most important parameters which have a main influence on the
development of a natural fire. Especially the fire load, the ventilation conditions and the fire room itself are the essential parameters influencing the development of natural fires. The amount of fire loads as well as their configuration in a fire room, the size and the place of openings, the size of fire
room and the physical properties of walls, ceiling and floor are the main parameters which deserve a careful investigation. Bound to these parameters some
certain physical magnitudes are to determine especially concerning the burning
rate and the fire spread rate, the temperature fields, the components of the
hot gases and the heat and mass flow conditions in the fire room.

The research project analyses the interaction between the parameters given in
figure 1 mainly in an experimental way. The experimental program of the project
chiefly includes full-scale tests. The reason for this decision is that many
theoretical calculations use at present time the results of small-scale tests
on empirical basis, especially the burning rates of the combustibles used in
the models are obtained by means of tests carried out in a furnace with dimen-
dions less than $2.0^3$ m$^3$. Calculations based on such empirical data do not provide satisfactory results for natural fires in practice, which in fact has been
proved in Braunschweig with the analysis of small and full-scale fire tests /2/.

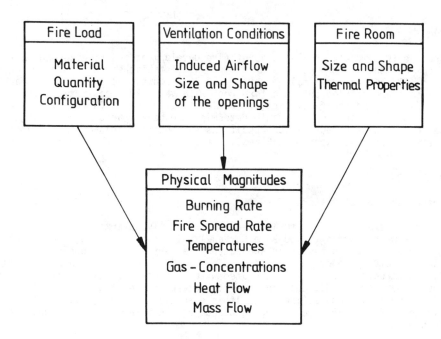

Fig. 1: Parameters influencing the development of natural fires and the physi-
        cal magnitudes to be measured

# EXPERIMENTAL PROGRAM

## General Scope

In order to carry out the experimental program on the basis of full-scale fire tests the facilities of the Technical Research Centre of Finland are being used where a fire room is available with a variable size between 500 m³ and 2000 m³. Within the scope of a carefully planned research program of the next three years, the development of the natural fires in large compartments will be investigated.

The fire loads mainly consist of pine wood in form of sticks with dimensions of 4x4x80 cm³. The sticks are nailed together to cribs, the ratio between wood and air inside the cribs amounts to 50%.

The first goal of the research program is to clear the influence of ventilation conditions and distribution of fire load on the fire development. Small-scale tests carried out in Braunschweig /2/ have shown that a very distinct interation exists between these two parameters during the fire: The ventilation conditions and the distribution of the fire load in connection with their density and physical properties determine the amount of oxygen which can participate in the burning process. For this reason the density, the distribution and the physical properties of the fire loads must be considered together with air flow conditions. Consequently experimental work planned has the intention to define a reasonable ventilation parameter taking into account these interaction processes in large fire rooms. Especially a relationship between natural ventilation and artificial ventilation conditions must be provided for large compartments.

## Determination of the burning rate

The burning rate is the most important physical parameter which should be determined carefully in the experimental tests. In order to predict the development of fire by means of a heat balance calculation as realistically as possible it is essential to know the burning rates in the ignition phase, during the fully developed fire and in the cooling down period of fire. For this reason, the prediction of the burning rate of the combustibles due to different fire influencing parameters gains a special attention for theoretical investigations.

Determination of the burning rate by means of the oxygen consumption. In the tests, two main possibilities are used to determine the burning rates of the wood cribs. First the unburned fire load was weighed by the help of a weighing platform during the fire. The derivation of the measured curve of the rest fire load against the time consequently provides the burning rates. However by means of this procedure information about the burning rates can only be obtained for a local area bound to size and shape of the weighing stage during a certain time of fire duration. Information about the total burning rate, especially in a large compartment can not be available.

In order to obtain a full information about the burning rates, the method of analyzing the exhaust gases /3,4/ is preferred in the tests. This method however implies to know as exactly as possible the amount of consumed oxygen per time unit. The procedure which is also used in the tests is explained in figure 2.

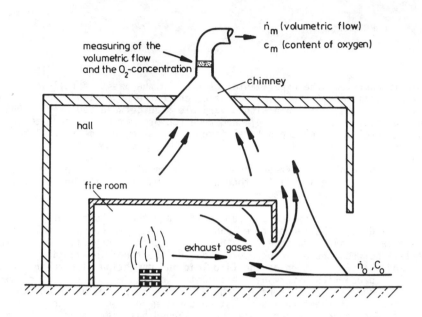

Fig. 2: Determination of the burning rate by means of hot gas analysis

Figure 2 illustrates the cross section of a fire room (the hatched painted wall) inside a hall building. In the chimney, on top of this building, a suction fan is installed which can produce under pressure inside the hall. Therefore all the gases going out of the hall pass through the chimney.

As measuring the volumetric flow $\dot{n}_m$ and the content of oxygen $c_m$ of the gases passing through the suction plant, it is possible to obtain the total energy released in the building. The total air flow in the plant, $\dot{n}_m$ consists of the incoming fresh air flow $\dot{n}_0$ with an oxygen concentration $c_0 \doteq 20.8\%$ and of the exhaust gases produced by the fire. Under the assumption that all the combustion gases are sucted into the duct, the knowledge of the measured values $\dot{n}_m$ and $c_m$ enables to determine the rate of heat release and consequently the total bur- ning rate of the fire loads in the fire room. The mathematical expression of this process is given by eq (1).

$$\dot{R} = \frac{1}{H_u} \frac{dQ}{dV_{O2}} \dot{n}_{O2}$$ 
Eq.(1)

$H_u$ is the lower calorific value of wood and the quotient $dQ/dV_{O2}$ represents the energy released with respect to volumetric unit of the oxygen consumed during the combustion. $\dot{n}_{O2}$ is the amount of oxygen consumed as given by Eq.(2)

$$\dot{n}_{O2} = \dot{n}_0 c_0 - \dot{n}_m c_m$$ 
Eq.(2)

However the amount of fresh air flow in the hall building cannot be measured. Therefore an additional relationship must be derived for the combustion reaction of cellulose as given in Eq.(3).

$$C_6H_{10}O_5 + 6O_2 \rightarrow 6CO_2 + 5H_2O$$ 
Eq.(3)

The eq.(3) shows that the total number of molecules produced during the combustion is 11 when 6 oxygen molecules are consumed. Consequently the alterna-
tion in the total gas flow inside the hall building can be written as in eq.(4)

$$\dot{n}_m - \dot{n}_0 = \frac{5}{6}\,\dot{n}_{02} \hspace{6cm} \text{Eg.(4)}$$

The eq.(4) combined with eq.(2) results into eq.(5) which provides the rate of
oxygen consumption.

$$\dot{n}_{02} = \dot{n}_m\,\frac{c_0 - c_m}{1 + 5/6\ c_0} \hspace{5cm} \text{Eq.(5)}$$

The eq.(5) indicates that the air flow $\dot{n}_m$ and the oxygen concentration $c_m$ are
the necessary parameters alone to be measured in order to determine the rate of
oxygen consumption during the fire and the eq.(1) can be used for the analysis
of burning rates.

Comparison  of the burning rates obtained by means of weighed rest fire load
and the hot gas analysis. Figure 3 shows the burning rates of a test obtained
by means of the hot gas analysis and also by means of weighing the rest fire
loads on a stage during the fire. In this test, all the wood cribs were loaded
up on the stage.

In figure 3, the full line belongs to the measurements of weight loss while the
dotted line presents the results from the hot gas analysis. The results from
the hot gas analysis show a certain delay with respect to the results of weight
loss measurements for the hot gases need a certain time to reach the suction
plant and to activate the measuring devices. However it is interesting to observe that both of the curves show a relatively good agreement and consequently
that the analysis of hot gases can provide reliable results.

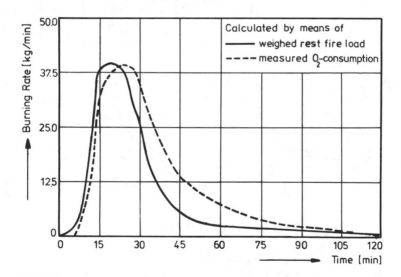

Fig. 3: Burning rates obtained by measuring the rest fire loads and the oxygen
consumption

Experiments carried out by variation of ventilation and fire load conditions

Presentation of experimental analysis. The research program has included 5
full-scale tests in 1983 in order to investigate the influence of the natural
ventilation conditions and of the distribution of the fire load on the development of real fires. The test fire room was made out of aerated concrete and has
the internal dimensions of 14.4 m x 7.2 m x 3.5 m. As ventilation conditions,
three different sizes and shapes of openings in the walls of the test room were
foreseen. Figure 4 gives necessary information about the tests with interesting
boundary conditions.

In the first test, a window opening with dimensions 5.25 m x 1.2 m in the front
wall was used, while in the second and the third tests a door opening was constructed at the same place. Another window opening with dimensions 10.5 m x
1.2 m and with a mean support was arranged at the long side of the fire room for
the 4. and 5. tests. Figure 4 gives additional informations about the distribution of fire loads and also about the ventilation factors which result from the
product of the opening area and the square root of the height of the opening.

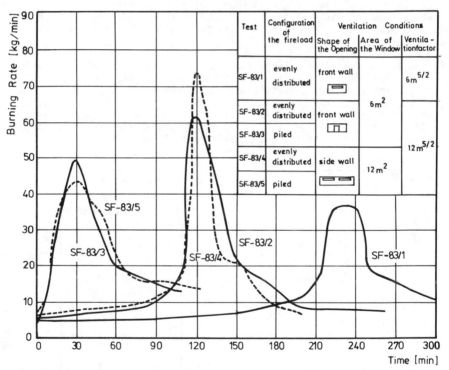

Fig. 4.: Burning rates of the full-scale tests and general information
concerning the tests

Small-scale tests carried out in Japan and Sweden have shown that the maximum
possible burning rates of wood cribs are linearly bound to the ventilation factor for ventilation controlled fires /5,6/. Theoretical investigations with
mass and heat balance calculations for this purpose also confirm the test results /7/. In order to find out whether such a relationship is also valid for
large compartments and in order to gain certainty if the ventilation factor

alone dominates the burning rates, tests have been carried out first with different ventilation factors but with the same size of openings. In the further step, the ventilation factor was kept fix by different opening areas. The fire loads consisted of 2000 kg of wood cribs in all tests have been either evenly distributed or concentrated on the middle area of the fire room.

Burning rates of the experiments. The burning rates of the mentioned full-scale tests are obtained by means of the hot gas analysis, the result is illustrated in figure 4. The tests represented by full lines had the openings of 6 m² at the front side, whereas the dotted lines represent the results of two tests with openings at the long side of the fire room.

The first two curves (SF-83/5, SF-83/3) with a maximum at the $30^{th}$ minute belong to the fire tests where the fire loads were concentrated in the middle of the floor. The curves SF-83/1 and SF-83/2, plotted with full lines show the calculation results from the tests which were carried out with the same size of the openings but different ventilation factors. On the contrary, the curves SF-83/2 and SF-83/4 belong to the tests with fix ventilation factor but the areas of the openings differ by factor two.

Conclusions drawn on burning rates of the tests. The following conclusions from fig. 5 can be drawn:

1. In case of evenly distributed fire loads the maximum values of the burning rates as well as the concerning times are influenced very distinctly by the ventilation factor as the comparison of the curves SF-83/1 and SF-83/2 proves this assertion. It can be stated that the maximum burning rates and the concerning times are proportionally influenced by the ventilation factor.

2. The maximum value of the burning rates by evenly distributed fire loads is dependent not only on the ventilation factor but also on the shape and size of the openings. This important result can clearly be seen by the pair of the curves SF-83/2 and SF-83/4. The curves show different maximums in the case of same ventilation factor with different sizes of openings. With increasing the size of the openings the burning rates have also an increasing tendency although the same ventilation factor is used. So it can be concluded that the maximum value of the burning rates of natural fires in large compartments is not only limited by the ventilation factor.

3. The maximum value of the burning rates and the concerning time are effectively influenced by the configurations of the fire loads as the comparison of the curves SF-83/2 and SF-83/3 shows clearly. In case of evenly distributed fire loads, the occurance of the maximum value of the burning rates is shifted to later fire durations due to lower fire spread rates especially at the beginning. The maximum burning rate is in this case obviously higher than in the case of concentrated fire loads on the middle part of the floor area. This can be explained by the fact that only a part of the fresh air coming through the opening penetrates the pile of wooden cribs and participates in the combustion, for the total surface of the fire load is relative small.

Therefore the burning rate of the concentrated fire load is limited due to two reasons: The first reason is the size of the opening of the enclosure and the second reason is the small surface of the fire load. Due to this limitation by the surface of the fire load the tests with concentrated fire loads do not show increasing burning rates with increasing sizes of the opening when the ventilation factor remains constant (see curves SF-83/3 and SF-83/5).

The burning rate of the wood cribs has proved in tests to be a function of the ventilation factor and, besides that of the size of the openings and of the effective surface of the fire loads. That means, that the ventilation factor and the surface of the fire loads influence both the maximum value of the burning rate and the fire spread rate in case of evenly distributed fire loads and as well as of concentrated fire loads. However in case of evenly distributed fire loads the size of the openings has an influence only in the maximum of the burning rate, whilst the fire spread rate remains uninfluenced (SF-83/2, SF-83/4).

## THEORETICAL INVESTIGATIONS

### Application of the heat balance calculations

Since 1976 the "Bundesministerium für Raumordnung, Bauwesen und Städtebau (Ministry of urban affairs)" has supported a research program concerning the development and spread of fires in small and large rooms. Within the activities of this research work calculation models have been developed to analyse the fires theoretically, however the reliability of the models has first been proved more effectively by means of the experimental investigation of the new project "Natural Fires" in Sonderforschungsbereich 148.

### Introduction of the heat balance program used

The heat balance program uses the following basic assumptions for the calculations /7/:

   - The fire room with only vertical openings has a homogeneous temperature distribution.

   - On the surfaces of the walls, the ceiling and the floor the heat transfer occurs with respect to one dimensional conditions and vertical to the heated surface.

   - There is a so-called neutral plane existing inside the fire room which provides a pressure balance on its level. Ebove this plane, overpressure and below it underpressure exists in the fire room.

With the help of these assumptions relationships for the energy- and mass balances were developed /7/. Results of the calculations are the temperature and mass-flow which can be obtained by the solution of nonlinear equations by means of an appropriate iteration procedure. Thus it is possible to calculate the temperature-time development in any large compartment on the basis of the results of experiments with the knowledge of the burning-rate time function derived from the tests.

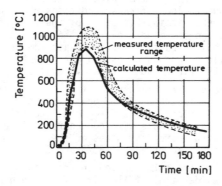

Fig. 5: Measured and calculated hot gas temperatured of test SF-83/3 (see fig.4)

Figure 5 shows the temperature-time curve measured by the test "SF-83/3" (see fig. 4) together with the results of the heat balance calculation. The dotted area indicates the temperature distribution measured at different levels inside the fire room. The calculation takes into account besides the special thermo-dynamical boundary conditions oxygen consumption in order to determine the burning-rate-time function.

Figure 5 illustrates that measured and calculated temperatures show a satisfactory agreement even during the cooling down period of the fire. However the calculated temperatures during the fully developed fire period are apparently low. The reason for this notice are some small inaccuracies in the measuring of oxygen consumption and as well as the assumption of evenly distributed temperatures in the heat balance model, which has generally not the validity for large fire rooms. In fact the inhomogenity in the temperature distribution is proved by fig. 6 for it illustrates different temperatures respectively in five different heights inside the fire room. Temperatures between the highest (25 cm below the ceiling) and lowest measuring points differ more than 200 K as it can be seen in figure 6.

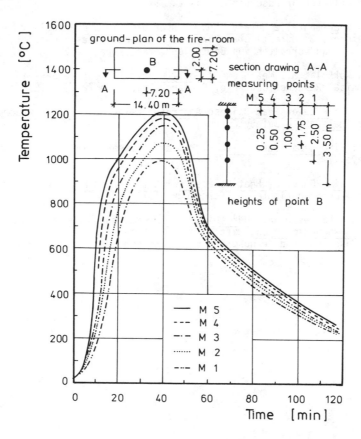

Fig. 6: Development of hot gas temperatures in different levels of test SF-83/3 (see fig. 4)

SUMMARY

Experimental investigations have been conducted to derive the following conclusions concerning the burning rates of the fire loads as wood cribs in large compartments:

The burning rate depends on the ventilation factor $A\sqrt{H}$, the window area and the fire load configuration or the surface area of the fire load (not speaking about the mass dependence and the dependence of the geometry of the enclosure). The ventilation factor has an effect on the maximum value of the burning rate and on the velocity of the extension of the flames. Comparing experiments with the same factor $A\sqrt{H}$ but different window areas one can see that the window area shows an additional effect on the maximum burning rate.

REFERENCES

1. DIN 4102, "Baulicher Brandschutz", Teil 2, 1977

2. Hagen, E.: Experimentelle Untersuchungen über die einen Brandablauf bestimmenden Parameter. 3. Öffentliches Forschungskolloquium des Sonderforschungsbereichs 148 "Brandverhalten von Bauteilen", Karlsruhe, 1982

3. Ahonen, A., Kokkola, M.: Measurements of the Rate of Heat Release in Room Fires. Technical REsearch Centre of Finland, Res. Note 72 1982 (Espoo, 1982)

4. Babrauskas, V.: Rate of Heat Release Apparatus using Polymethylmethacrylate and Gaseous Fuels. Fire Safety Journal 5, 1982

5. Kawagoe, K.: Fire Behaviour in Rooms. Building Research Institute, Report Nr. 27, Tokyo, 1958

6. Nilsson, L.: Time Curve of Heat Release for Compartment Fires with Fuel of Wooden Cribs. Lund Institute of TEchnology, 1974

7. Schneider, U., Haksever, A.: Wärmebilanzrechnungen für Brandräume mit unterschiedlichen Randbedingungen, Heft 46 der Schriftenreihe des Instituts für Baustoffe, Massivbau und Brandschutz der Technischen Universität Braunschweig.

# Some Field Model Validation Studies

**G. COX and S. KUMAR**
Fire Research Station
Borehamwood, Herts WD6 2BL, United Kingdom

**N. C. MARKATOS**
Thames Polytechnic
London SE18 6PF, United Kingdom

ABSTRACT

The three dimensional time-dependent field model known as JASMINE has been applied to several experimental fire configurations for validation purposes.
Comparisons of predictions with non-spreading experimental fires conducted in
a forced ventilated fire test cell (6 m x 4 m x 4.5 m), closed six-bed hospital
ward (7.3 m x 7.9 m x 2.7 m) and a railway tunnel, both forced and naturally
ventilated (390 m x 5 m x 4 m) are summarised in this paper.  The agreement is
shown to be quite satisfactory except in the immediate vicinity of the fire
source.  It is suggested that the model may now be used with some caution to
study smoke movement problems, however, improvements to the turbulence - chemistry
interaction at the source will be required before the spreading fire can be
reliably predicted.

Keywords:  Mathematical Model, Fire, Smoke, Validation, Tunnel, Hospital.

INTRODUCTION

Mathematical fire models will only enjoy widespread acceptance when and if
sufficient validation can be demonstrated for them.  The prospective user must
have enough confidence in a model to be able to apply it to situations, new
building types for example, in which there is no practical experience of fire
behaviour.  There is limited value in developing an accurate model that can only be
applied to one size of compartment for example.  It is the generality of the
field modelling approach to the problem that should offer significant advantages
in terms of the transportability of models to very different compartment types.
It is not the intention of this paper to describe in detail the mathematical
field model being employed in these studies.  The reader will find these elsewhere and in particular in refs 1 to 3.

Suffice it to say here that the so-called field modelling approach is essentially
a first-principles approach, solving the classical equations of motion for the
gas at discrete points in space and time, retaining as much rigour as is practically possible.  The need to treat turbulent mixing by a turbulence model,
and the use of numerical   methods to solve the equations, are the principle
points of departure from full rigour.  However these departures are not unique
to fire research problems and have evolved in many other areas of application
of computational fluid dynamics to engineering problems. Although not without
controversy, there is a growing body of evidence to support its successful
practical application.

The object of this contribution is to summarise some recent validation work with the Fire Research Station's field model known as JASMINE (for Analysis of Smoke Movement in Enclosures) in three distinctly different situations. Two of these concern rooms of similar size, in one case with no significant ventilation openings and in the other with forced ventilation. The third case concerns a railway tunnel both naturally and forced ventilated and with a slope from one end to the other.

These represent the most recent phase of a progressive validation and development study starting with its two-dimensional, non reacting steady-state predecessor through studies of the validity of traditional scaling relationships to the current three-dimensional, time-dependent version including chemical reaction. Some of these studies have been described in references 1 to 5.

II THEORETICAL FOUNDATION OF THE MODEL

Field models are based on the solution of the partial differential equation set describing the conservation of mass, momentum, heat, and species concentration etc., and therefore represent a first principles approach to the problem.

In fire problems the flow is generally dominated by buoyancy. This gives rise to large scale turbulent motion which controls the rate of diffusion of mass, momentum and the mixing of fuel volatiles with air. The rate of reaction of fuel and air is also controlled by this relatively slow turbulent mixing process rather than the faster chemical kinetics. The non-uniform buoyancy forces not only drive the flow but also increase turbulent mixing in the rising plume and inhibit it in hot stratified layers.

These characteristic features of fire have been incorporated in JASMINE. In summary, the model solves simultaneous non-linear partial differential equations for the nine dependent variables describing the system, ie, the three velocity components (u,v,w), the pressure (p), the enthalpy (h), the mixture and mass fractions of the fuel (f,$m_{fu}$), the turbulent kinetic energy (k) and its dissipation rate ($\epsilon$).

All these dependent variables, with the exception of pressure, appear as the subjects of differential equations of the general form:

$$\frac{\partial}{\partial t}(\rho\phi) + \text{div } (\rho\vec{u}\phi + \vec{J}_{\phi}) = S_{\phi} \tag{1}$$

where $\phi$ stands for a general fluid property and $\rho$, $\vec{u},\vec{J}$ ,$S_{\phi}$ are density, velocity vector, diffusive-flux vector and source rate per unit volume, respectively. The diffusive flux $\vec{J}_{\phi}$ is given by:

$$\vec{J}_{\phi} = -\Gamma_{\phi} \text{ grad}\phi \tag{2}$$

where $\Gamma_{\phi}$ denotes the 'effective exchange coefficient of $\phi$' determined from the turbulence parameters k and $\epsilon$. The pressure variable is associated with the continuity equation:

$$\frac{\partial\rho}{\partial t} + \text{div } (\rho\vec{u}) = 0 \tag{3}$$

The source term of the transport equation for $m_{fu}$ is calculated using Magnussen and Hjertager's extension of Spalding's eddy-break-up[8] concept. The values of $\Gamma_{\phi}$ and $S_{\phi}$ for each $\phi$ and details of the formulation of the finite difference equations from Eq (1) together with features of the numerical algorithm are given in refs 1 to 3. A six-flux radiation model is incorporated in JASMINE but in what follows, effects of radiant heat transfer have been ignored. Some calculations of the radiant contribution have been reported in ref 4.  The boundary conditions have been discussed in refs 1 - 3.

III  EXPERIMENTAL CONFIGURATIONS USED FOR VALIDATION STUDIES

(i)  Lawrence Livermore National Laboratory fire test cell.

The fire test cell at the Lawrence Livermore National Laboratory (LLNL), is 4 m x 6 m in plan and 4.5 m high, represented numerically by 12 x 13 x 14 grid nodes. A rectangular duct 0.65 m square centred 3.6 m above the floor provides forced extraction by control of an axial fan.  Air inlet is at low level through slots in a cylindrical duct close to one face of the compartment.  This has been approximated in the model by one slit 0.12 m high, 2 m long, 0.1 m above the floor. The particular fire examined here (designated Mod 8 by Alvares et al[9]) was produced by burning a spray of isopropyl alcohol formed from an opposed jet nozzle located at the centre of a steel pan of diameter 0.91 m.  The fuel is quickly evaporated and burnt before it contacts the pan surface.  The resulting fire is very similar to a natural pool fire.

The fuel injection rate was 13.1 gm/s adjusted to give a heat release rate of 400 kW, assuming efficient combustion.  The ventilation extraction rate, which was adjusted to 500 litres/s of air prior to ignition, had dropped to 400 litres/s just before test completion (20 minutes from ignition).  At this time, experimental conditions were assumed to have reached a quasi-steady state for which the predictions were made with a global heat transfer coefficient of 20 $W/m^2$ °K.  Complete oxidation of the fuel was assumed in a one-step combustion model for this and the following studies.

(ii)  Zwenberg tunnel

The experiments are those conducted in the disused railway tunnel at Zwenberg, Austria and described by Fiezlmayer[10].  The tunnel is 390 m long and 5 m across with a false ceiling 4 m above the floor.  There is a 2.18% gradient from the south to the north portal.  Since the fire was symmetrically situated in the experiment, the calculation domain occupied one half of the tunnel from the symmetry plane (through the axis of symmetry) down the centre of the tunnel across to the east wall, and contained a total of 1566 grid nodes distributed in a non-uniform manner with six nodes across the tunnel half-width, 9 nodes vertically and 29 along the length.  The tunnel was sealed at the south portal and an injection fan inserted in the end wall.  In the test cases considered here, the fan was either switched off (natural ventilation from the open north portal) or produced a forced air flow in the traffic space of 2 m/s and 4 m/s (cases of pure longitudinal ventilation).  All three ventilation cases involved fires of 200 litres of petroleum fuel burnt in a 2.6 m square tray.  The fire source in each case was situated 108 m from the south portal.  The fuel mass loss rates for the three ventilation conditions were calculated using the data of Yumoto[11] to allow for the increase of linear regression rate with increased air flow rate. The effect of the resolved components of the gravitational acceleration on the vertical and 'along the slope' directions were included in the formulation.

Steady state predictions were compared with the experimental data towards the end of the fire.  Hexane was considered as an idealisation for petrol.

The heat losses to the walls were calculated using a heat transfer coefficient varying linearly between 5 $W/m^2$ °K and 40 $W/m^2$ °K with gas temperature as suggested by the Harvard zone model[12].

(iii)  Six bed hospital ward

The experimental six bed hospital ward, in which bedding fires are being studied

studied by the Fire Research Station is 7.33 m x 7.85 m in plan and 2.7 m high, discretised into 14 x 13 x 11 grid nodes for its numerical representation. Six 1 kW radiators situated close to the north wall have been represented by one floorstanding 6 kW heater, 7.85 m in width and 0.5 m in height. The fire source was assumed to occupy a fixed area of 0.45 m x 0.5 m of polyurethane (PU) foam mattress, situated next to the pillow 0.25 m away from the east wall. The actual fuel was far more complex than just PU foam, comprising a small wood crib for ignition and including cotton, nylon and polyester bedding materials. However for modelling purposes, PU foam, being the major component, was considered to be a reasonable approximation for the fuel.

Prior to the start of the experimental fire, the steady-state natural convection conditions were established by leaving the heaters on for several hours. The bed was ignited by a small wood crib. Three distinct phases of mass loss were evident from load cell measurements. The first phase lasted for about 3 minutes where the wood crib was mainly involved and the heat output was estimated at approximately 5 kW, assuming a combustion efficiency of 0.65. In the second phase, up to 7 minutes from ignition, the PU foam mattress was more actively involved with the heat output rising to about 20 kW. In the third phase the pillow also became involved and the flames lengthened to around one metre and heat release rose to approximately 80 kW until about 12 minutes from ignition when the test was abandoned. Transient calculations were performed by assuming that these three phases could be represented by three stepped levels of mass release following a steady state calculation for the prefire conditions.

The heat losses to the walls were calculated as in the Zwenberg tunnel case.

IV  RESULTS

(i)  Lawrence Livermore National Laboratory fire test call.

One of the difficulties of validating this type of model is the considerable quantity of detailed information predicted and the very difficult task of obtaining enough experimental data to examine the details. Even in such a thoroughly instrumented experimental rig as the LLNL test cell it is only possible to compare predictions and experiments at comparatively few points.

Predictions of velocity vectors are shown on two orthogonal planes in a perspective view of the cell viewed from the east in Fig 1. The two planes are the horizontal one through the air inlet slot and the vertical one through the centre of the fire tray and extract duct (hereinafter known as the central plane). The processes of air entrainment into the fire, buoyant vertical acceleration of the plume, formation of a ceiling jet and the establishment of recirculation in the cell are clearly evident. Examination of vectors, on vertical planes parallel to that shown here, clearly demonstrates a leaning of the fire towards the south wall. A sample streamline is illustrated in Fig 1, starting at the inlet slot and rising three times before finally leaving at the extract duct.

The predictions of temperature, mass fractions of fuel, oxygen and product on the vertical plane through the centre of fire tray and exit duct are shown in Fig 2. The oxygen mass fraction above about 0.5 m from the floor is uniform at between 10% and 11% and product ($3 CO_2 + 4H_2 O$) mass fraction at between 16% and 17%. The unburnt fuel mass fraction is everywhere very low except at the source. Gas temperatures in the compartment show a steady increase from floor to ceiling in good agreement with the experiment. Over and within the fire source itself temperatures decay in the expected way but peak values are low compared with experimental determinations (around 1250 K). This may be the result of either the simplicity of the combustion model used or even the coarseness of the grid in the vicinity of the source (approximately 0.25 m cube). It is well known that

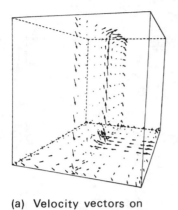

(a) Velocity vectors on
two orthogonal planes

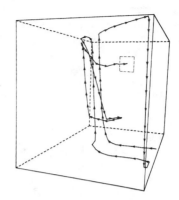

(b) Sample streamline
starting at inlet slot

FIG 1  Perspective view from the east

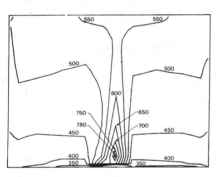

Absolute temperature contours (K)

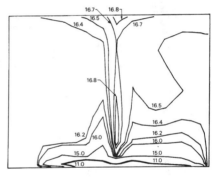

Percentage mass fraction of product

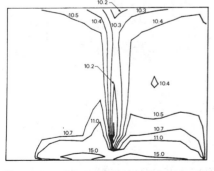

Percentage mass fraction of oxygen

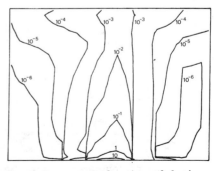

Percentage mass fraction of fuel

FIG 2  Gas property contours on central plane

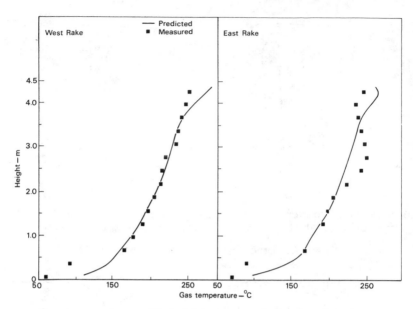

Gas temperature with height at two thermocouple rakes

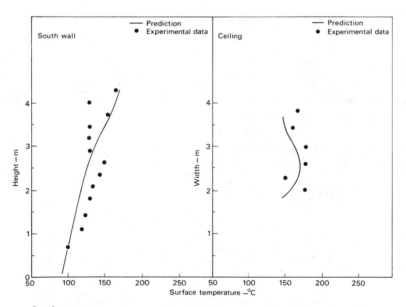

Surface temperature with height on south wall and ceiling

FIG 3  Temperature profiles in the LLNL test cell

164

fires exhibit strong combustion intermittency[13] and current research is directed towards more realistic modelling of the turbulence-chemistry interactions[14, 15]

Temperatures measured at two vertical thermocouple rakes situated 1.5 m either side of the fire tray on the central plane, and each comprising 15 thermocouples are compared with predictions in Fig 3. Wall surface temperature predictions are also compared in Fig 3 with measurements on the vertical centreline of the south wall and on the continuation of this line onto the ceiling to its centre (over the fire tray). Overall agreement can be seen to be reasonably good, although gas temperatures close to the floor and ceiling are predicted somewhat high.

Overall properties and some point determinations are compared in Table I. In general, agreement is seen to be excellent. The difference beween computed inlet and outlet mass flow rates corresponds to the 13 grams/s injection rate of fuel. The gas composition predictions are somewhat disappointing compared with the success of the thermal predictions. This again is likely to be due to the source prescription problem mentioned earlier.

The predicted pressure at the level of the extract duct differs markedly from the measurement. The reason for this is not clear but is likely to be caused by the contribution of dynamic pressure to the measurement. Lack of space here precludes further discussion but more details will be found in Ref 3.

TABLE I. Comparison of predictions and measurements of integral properties and some point determinations

| Property | Predicted | Measured |
|---|---|---|
| Mass outflow rate (kg/s) | 0.269 | 0.30 |
| Mass inflow rate (kg/s) | 0.257 | 0.24 |
| Exit gas temperature (°C) | 249 | 275 |
| Exit heat flow (kW) | 66 | 68 |
| Exit pressure above ambient (Pa) | 14.3 | -5 |
| Exit $O_2$ concentration (%) (dried gas) | 10.4 | 14 |
| Exit $O_2$ concentration (%) (dried gas) | 7.5 | 5.5 |
| Total heat loss to boundaries (kW) | 334 | 332 |

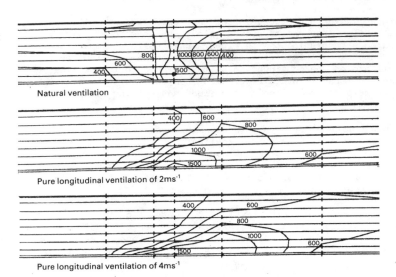

(a)  Temperature  contours (K)

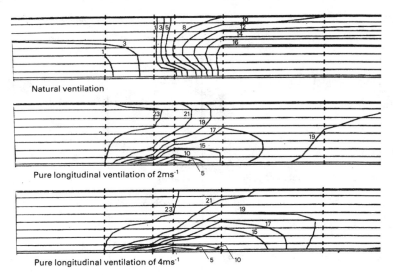

(b)  Percentage  oxygen  mass  fraction  contours

FIG 4  Predicted  gas  property  contours  on  the  central  plane
of  Zwenberg  Tunnel  (23m  long  magnified  section
around  fire)

TABLE II: Comparison of Measured and Predicted Gas Properties for the Case of Pure Longitudinal Ventilation of 4 m/s

| Measurement Station: Gas Property | | 1 | | 2 | | 3 | | 4 | | 5 | | 6 | | 7 | |
|---|---|---|---|---|---|---|---|---|---|---|---|---|---|---|---|
| | | Pred | Meas | Pred | Meas | Pred | Meas | Pred | Meas | Pred | Meas | Pred | Meas | Pred | Meas |
| Temperature (°C) | H | 131 | 88 | 198 | 220 | 224 | 227 | 394 | 312 | 12 | 176 | 10 | 12 | 10 | 12 |
| | M | 109 | 84 | 178 | 136 | 183 | 178 | 426 | 458 | 93 | – | 10 | 12 | 10 | 12 |
| | L | 86 | 80 | 140 | 120 | 152 | 120 | 341 | 312 | 638 | – | 10 | 12 | 10 | 12 |
| Volumetric Concentration (%) of $O_2$ (dried gas) | H | 19.3 | – | 19.1–19.2 | 19.3 | 19.0–19.1 | – | 19.7–17.7 | – | – | – | – | – | – | – |
| | M | 19.3 | 19.5 | 19.1–19.4 | 19.3–19.6 | 19.1–19.4 | 19.2–19.7 | 20.0–17.4 | – | – | – | – | – | – | – |
| | L | 19.3 | 19.5 | 19.2–19.5 | 19.6 | 19.3–19.6 | 19.7–20.1 | 20.3–18.2 | – | – | – | – | – | – | – |
| Volumetric Concentration (%) of $CO_2$ (dried gas) | H | 1.1 | – | 1.2 | 1.4 | 1.3– | – | 0.8–2.1 | 1.1 | – | – | – | – | – | – |
| | M | 1.1 | 1.0 | 1.2–1.0 | 1.1–0.9 | 1.3–1.0 | 1.3–0.8 | 0.6–2.4 | 2.0–5.8 | – | – | – | – | – | – |
| | L | 1.1 | 1.0 | 1.2–0.9 | 0.9 | 1.1–0.9 | 1.0–0.8 | 0.4–1.9 | 1.0–2.3 | – | – | – | – | – | – |
| Volumetric Concentration (ppm) of fuel | H | <1 | – | <1 | – | 3 | – | 1–60 | 250 | – | – | – | – | – | – |
| | M | <1 | – | <1 | – | 4 | – | 1–170 | 240–1750 | – | – | – | – | – | – |
| | L | <1 | – | <1 | – | 2 | – | 1–250 | 100–1100 | – | – | – | – | – | – |

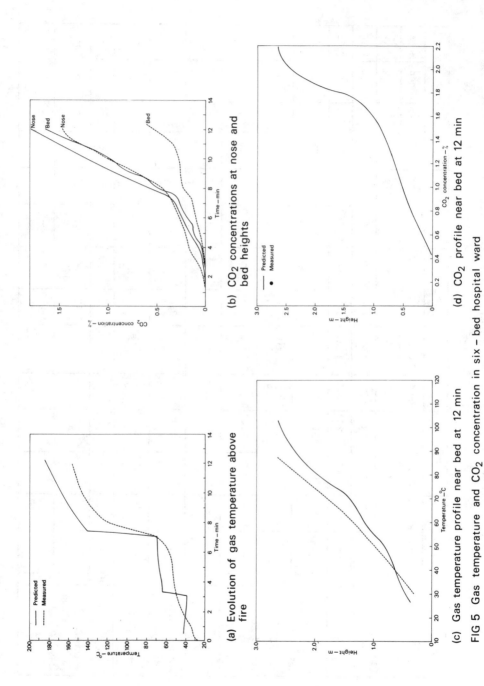

(a) Evolution of gas temperature above fire

(b) $CO_2$ concentrations at nose and bed heights

(c) Gas temperature profile near bed at 12 min

(d) $CO_2$ profile near bed at 12 min

FIG 5 Gas temperature and $CO_2$ concentration in six – bed hospital ward

168

(ii)  Zwenberg tunnel

Fig 4 shows the detailed temperature and oxygen mass fraction contours in an enlarged portion of  the tunnel around the source.

Detailed comparisons between the predictions and measurements for one test case (4 m/s) are shown in Table II.  It is clear from these tabulated results and from those for naturally ventilated and lower forced ventilated situations (not shown because of space limitations) that the agreement throughout is reasonably good except very close to the fire.  The discrepancy between the predictions and measurements immediately above the fire is due to radiant heat transfer (not included in the predictions) whereas that within the flames due to the coarse grid and the simple combustion model used, as mentioned earlier.

It is interesting to compare estimates of flame angles from the model with Thomas's empirical correlation[16].  Assuming 600° K or 21% oxygen mass fraction as the extent of flame front, the flame angles agree remarkably well.  At 2 m/s Thomas's correlation gives 58 degrees with respect to the vertical whilst the model gives 61 degrees and at 4 m/s both calculations give 68 degrees.  More details of this study will be found in Ref 6.

(iii)  Six bed hospital ward

Buoyancy driven flow in a sealed cavity is one of the most challenging problems in the field of computational fluid dynamics[17].  Therefore this particular test case is one of considerable interest.

Since the previous two test cases have already demonstrated a wide variety of predictive capabilities of this type of mathematical model, only direct comparison of the predicted transient results with the measurements is shown for this case.  In Fig 5a and b the evolution of predicted and measured gas temperatures and $CO_2$ concentration is shown at some selected locations.  The temperature 75mm below the ceiling, directly above the fire, has been chosen to demonstrate a worst case.  The three phases of fire development can be clearly seen particularly in the predictions.  At similar heights, further from the source, agreement is better as was experienced in the previous test cases.

The $CO_2$ concentrations shown are for 'nose' and 'bed' heights close to the bed and respectively 1.5 m and 0.9 m above the floor.  At nose height the predictions are in reasonable agreement but at bed height they are considerably in error.  Although detailed gas concentration profiles were not measured it appears that the model has overpredicted the extent of species diffusion in the vertical direction (Fig 5c), whilst the corresponding temperature profiles do agree reasonably well (Fig 5d).  With the complex solid fuel involved in the experiment, questions remain as to whether this difficulty occurs as a result of the over-simplified prescription of the source or again the problems of turbulence-chemistry interaction and coarseness of grid at the source.  These possibilities are currently under investigation.

V  CONCLUSIONS

It has not been possible, in the space available, to give full details of each validation study but more on the LLNL and Zwenberg tunnel comparisons will be found in Refs 3 and 6.  However it can be concluded that although not perfect, and by no means exhaustively validated, JASMINE does appear to offer a reliable tool for the prediction of detailed thermal properties of non-spreading fire problems where radiant heat transfer is relatively less important than convection.  Developments concentrate now more on improving details, at least for smoke movement problems, rather than the overall solution concept.

Several aspects still need improvement, particularly the treatment of the turbu-lence-chemistry interaction, if more accurate gas concentration predictions are to be made. Furthermore a more universal treatment for the calculation of heat losses to the boundaries is required to maintain the generality of solution technique aspired to by the field modelling approach. Clearly a more realistic treatment of the source will be necessary if the model is to be used to predict surface flame spread and mass burning rates.

ACKNOWLEDGEMENTS

The authors wish to acknowledge the contribution of Nicole Hoffmann to the hos-pital ward study. To conduct this work she was supported by a grant from the Department of Health and Social Security.

The work on tunnels was conducted under a contract to the Department of Trans-port. The paper forms part of the work of the Fire Research Station, Building Research Establishment, Department of the Environment, UK. It is contributed by permission of the Directors of the Building Research Establishment and the Transport and Road Research Laboratory.

REFERENCES

1. N C Markatos, M R Malin and G Cox : J Heat Mass Transfer 25, 63 (1982).

2. N C Markatos and G Cox : Physiochemical Hydrodynamics 5, 53 (1984).

3. G Cox and S Kumar : The Mathematical Modelling of Fire in Force Ventilated Enclosures, 18th DOE Nuclear Airborne Waste Management and Air Cleaning Conference, Baltimore, August 1984.

4. N C Markatos and K A Pericleous : An investigation of Three Dimensional Fires in Enclosures, in Fire Dynamics and Heat Transfer ed J Quintiere, ASME, 1983, p115.

5. S Kumar and G. Cox. 'The Application of a Numerical Field Model of Smoke Movement to the Physical Scaling of Compartment Fires' in Numerical Methods in Thermal Problems, ed R W Lewis, J A Johnson and W R Smith. Pineridge Press 1983, p837.

6. S Kumar and G Cox : Mathematical Modelling of Fire in Road Tunnels. Fifth International Conference on the Aerodynamics and Ventilation of Vehicle Tunnels, Lille, 1985.

7. B F Magnussen and B H Hjertager : Sixteenth Symposium (International) on Combustion. The Combustion Institute, 1976, p719.

8. D B Spalding : Thirteenth Symposium (International) on Combustion. The Com-bustion Institute, 1971, p649.

9. N J Alvares, K L Foote and P J Pagni : Comb Sci and Technol. 39, 55 (1984).

10. A H Fiezlmayer : Brandversuche in einem Tunnel, Bundesministerium fur Bauten und Technik, Heft 50, Vienna, 1976.

11. T Yumoto : J Japan Society for Fire Safety Eng. 10, No.3 (1971).

12. H E Mitler : 'The physical basis for the Harvard Computer Code'. Home Fire Project Tech Rept No.34, Harvard University (1978).

13. G Cox and R Chitty. Fire and Materials. 6, 127 (1982).

14. N L Crauford, S K Liew and J B Moss : Experimental and Numerical Simulation of a Buoyant Fire. Submitted to Combustion and Flame.

15. D B Spalding : Physiochemical Hydrodynamics. 4, 323 (1983).

16. P H Thomas : Ninth Symposium (International) on Combustion p844 (1963).

17. I P Jones and C P Thompson : Numerical Solutions for a Comparison Problem on Natural Convection in an Enclosed Cavity. UKAEA Harwell. Report No. AERE-R.9955 (1981).

# Turbulent Buoyant Flow and Pressure Variations around an Aircraft Fuselage in a Cross Wind near the Ground

**H. S. KOU and K. T. YANG**
Department of Aerospace and Mechanical Engineering
University of Notre Dame
Notre Dame, Indiana 46556, USA

**J. R. LLOYD**
Department of Mechanical Engineering
Michigan State University
East Lansing, Michigan 48824, USA

ABSTRACT

Two-dimensional numerical finite-difference calculations have been carried out to study the effects of cross wind speeds and the elevation of the fuselage on turbulent buoyant flow and pressure variations around an aircraft fuselage engulfed in a simulated fire in a uniform cross wind near the ground. Detailed velocity, temperature, smoke concentration, and pressure fields have been obtained and it is found that a major influence on the physical phenomena is the relative strength of the cross flow and the buoyant flow.

INTRODUCTION

In survivable aircraft accidents involving fires, there is a critical need to understand and predict flow of hot toxic gases inside the aircraft cabins. While the fire may be initiated inside the cabin, there is an equally important accident scenario in which a fire is initiated outside the fuselage due to fuel spill and spreads into the cabin through an opening (Emmons, 1982). In the latter scenario, the fire spread into the cabin is dictated by the flow and pressure fields surrounding the fuselage, which in turn depend on the location and strength of the fire source, the elevation of the fuselage, and the speed of the prevailing wind. The development of appropriate predictive schemes for such a scenario can ultimately provide needed inputs to the eventual development of fire-safety countermeasures. The purpose of this paper is to present some results of a recent numerical study addressing this very scenario to determine the physical effects of the various parameters on the flow and pressure fields surrounding the fuselage (Kou, 1984).

Similar to an earlier companion study dealing with fire and smoke spread inside aircraft cabins (Yang et al, 1984), the numerical computations are based on a two-dimensional, primitive-variable, differential field model which includes the effects of strong buoyancy, turbulence and compressibility. Flow, temperature, and smoke concentration fields are all calculated. An algebraic turbulence model which accommodates both local and stratification effects has been utilized. The fire located outside the fuselage, which is approximated by a circular cylinder, is simulated by a volumetric heat and smoke source with arbitrarily prescribed flame envelope and rates of local heat and smoke

---

H. S. Kou is currently at the Tatung Institute of Technology, Taipei, Taiwan, Republic of China.

generation. Calculations have been carried out for a fixed fire and smoke source which simulates the full-scale fire tests conducted at the Technical Center of the U. S. Federal Aviation Administration, but varying fuselage elevations and prevailing wind speeds.

## MATHEMATICAL FORMULATION AND NUMERICAL ANALYSIS

Figure 1 depicts the configuration for the solution field where a cylindrical coordinate system is employed in the vicinity of the simulated fuselage, and the far region away from the fuselage is represented by a Cartesian coordinate system. The prevailing wind is at a uniform speed $U_0$. For the speeds considered in this study, the turbulent wall layer next to the ground is very thin, and therefore is neglected. The flame envelope on the ground, which represents the fire and smoke source, is taken to simulate a real fire, and its dimension and strengths will be described later. The Cartesian coordinate system has its origin located at the left lower corner of the solution field. The origin of the cylindrical coordinate system is located at the center of the circular cylinder (simulated fuselage).

The conservation field equations for mass, momentum, energy and smoke concentration in terms of flux quantities are well known in both coordinate systems. For illustrative purposes, these equations in Cartesian coordinates can be non-dimensionalized by introducing the following definitions:

$$X = \frac{\bar{X}}{D}, \; Y = \frac{\bar{Y}}{D}, \; t = \frac{U_0 \bar{t}}{D}, \qquad U = \frac{\bar{U}}{U_0}, \; V = \frac{\bar{V}}{U_0}, \; T = \frac{\bar{T}}{T_0}$$

$$\rho = \frac{\bar{\rho}}{\rho_0}, \; \rho_\varepsilon = \frac{\bar{\rho}_\varepsilon}{\rho_0}, \; P = \frac{\bar{P} - \bar{P}_\varepsilon}{\rho_0 U_0^2}, \; S = \frac{\bar{S}}{S_0}, \; S^\circ = \frac{\dot{S} D}{\rho_0 U_0 S_0}, \; Q^\circ = \frac{\dot{Q} D}{\rho_0 U_0 T_0 C_{po}} \tag{1}$$

Here, the subscript o denotes reference quantities, and the subscript $\varepsilon$ represents the hydrostatic equilibrium conditions. It is noted here also that $S_0$ is simply a convenient reference concentration. The resulting non-dimensionalized equations, under the assumptions of no chemical reaction,

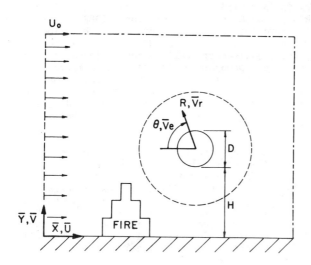

FIGURE 1. Coodinate System

174

negligible viscous dissipation and pressure work, and representations of turbulent flux quantities by means of effective (laminar plus turbulent) transport properties, can now be written as (Yang et al, 1984; Kou, 1984)

$$\frac{\partial \rho}{\partial t} + \frac{\partial (\rho U)}{\partial X} + \frac{\partial (\rho V)}{\partial Y} = 0 \tag{2}$$

$$\frac{\partial (\rho U)}{\partial t} + \frac{\partial (\rho U^2)}{\partial X} + \frac{\partial (\rho UV)}{\partial Y} = \frac{\partial}{\partial Y}\left[\frac{1}{Re_e}\left(\frac{\partial U}{\partial Y} + \frac{\partial V}{\partial X}\right)\right]$$

$$+ \frac{\partial}{\partial X}\left(\frac{4}{3\,Re_e}\frac{\partial U}{\partial X_e} - \frac{2}{3\,Re_e}\frac{\partial V}{\partial Y_e}\right) - \frac{\partial P}{\partial X} \tag{3}$$

$$\frac{\partial (\rho V)}{\partial t} + \frac{\partial (\rho UV)}{\partial X} + \frac{\partial (\rho V^2)}{\partial Y} = \frac{\partial}{\partial X}\left[\frac{1}{Re_e}\left(\frac{\partial U}{\partial Y} + \frac{\partial V}{\partial X}\right)\right]$$

$$+ \frac{\partial}{\partial Y}\left(\frac{4}{3\,Re_e}\frac{\partial V}{\partial Y} - \frac{2}{3\,Re_e}\frac{\partial U}{\partial X}\right) - \frac{\partial P}{\partial Y} - (\rho - \rho_\varepsilon)\frac{gD}{U_0^2} \tag{4}$$

$$\frac{\partial (\rho T)}{\partial t} + \frac{\partial (\rho UT)}{\partial X} + \frac{\partial (\rho VT)}{\partial Y} = \frac{\partial}{\partial X}\left(\frac{1}{Re_e\,Pr_e}\frac{\partial T}{\partial X}\right)$$

$$+ \frac{\partial}{\partial y}\left(\frac{1}{Re_e\,Pr_e}\frac{\partial T}{\partial Y}\right) + Q^\circ \tag{5}$$

$$\frac{\partial (\rho S)}{\partial t} + \frac{\partial (\rho US)}{\partial X} + \frac{\partial (\rho VS)}{\partial Y} = \frac{\partial}{\partial X}\left(\frac{1}{Re\,Sc_e}\frac{\partial S}{\partial X_e}\right)$$

$$+ \frac{\partial}{\partial Y}\left(\frac{1}{Re_e\,Pr_e}\frac{\partial S}{\partial Y}\right) + S^\circ \tag{6}$$

$$\rho T = \frac{U_0^2}{R_0 T_0} P + P_\varepsilon \tag{7}$$

$$\rho_\varepsilon = \exp\left[-\left(\frac{gD}{R_0 T_0}\right)Y\right] \tag{8}$$

where $Re_e$, $Pr_e$, and $Sc_e$ are the local effective Reynolds number, Prandtl number, and Schmidt number, respectively, defined by

$$Re_e = \frac{\rho_0 U_0 D}{\bar{\mu}_{eff}}, \quad Pr_e = \frac{\bar{\mu}_{eff} C_p}{\bar{k}_{eff}}, \quad Sc_e = \frac{\bar{\nu}_{eff}}{\bar{D}_{eff}} \tag{9}$$

where $\bar{\mu}_{eff}$, $\bar{k}_{eff}$ and $\bar{D}_{eff}$ are the effective (laminar plus turbulent) viscosity, conductivity and mass diffusion coefficient, respectively. It is understood that both $Q^\circ$ and $S^\circ$ are taken to be zero outside the flame envelope.

Before the above equations can be solved, closure models for the turbulent transport properties must be introduced. In the present study, an algebraic turbulence model for recirculating buoyant flows with wide variations in the turbulence level, which accounts for both local shear and stratification effects, is employed (Nee and Liu, 1978). It is given in a non-dimensional form by

$$\mu_{eff} = \frac{\bar{\mu}_{eff}}{\mu_0} = 1 + \frac{[(\frac{\partial U}{\partial Y})^2 + (\frac{\partial V}{\partial X})^2]^{1/2} (\frac{\ell}{D})^2}{2 + \frac{Ri}{Pr_t}} \tag{10}$$

where $\mu_0$ is the reference molecular viscosity and $\ell$ is a mixing length given by

$$\frac{\ell}{D} = K \left\{ \frac{(U^2 + V^2)^{1/2}}{[(\frac{\partial U}{\partial X})^2 + (\frac{\partial U}{\partial Y})^2 + (\frac{\partial V}{\partial X})^2 + (\frac{\partial V}{\partial Y})^2]^{1/2}} \right.$$

$$\left. + \frac{[(\frac{\partial U}{\partial X})^2 + (\frac{\partial U}{\partial Y})^2 + (\frac{\partial V}{\partial X})^2 + (\frac{\partial V}{\partial Y})^2]^{1/2}}{[(\frac{\partial^2 U}{\partial X^2})^2 + (\frac{\partial^2 U}{\partial Y^2})^2 + (\frac{\partial^2 V}{\partial X^2})^2 + (\frac{\partial^2 V}{\partial Y^2})^2]^{1/2}} \right\} \tag{11}$$

where K is an adjustable constant and Ri is the gradient Richardson number given by

$$Ri = \frac{gD(\frac{\partial T}{\partial Y})}{U_0^2 (\frac{\partial U}{\partial Y})^2} \tag{12}$$

Furthermore, the effective conductivity $\bar{k}_{eff}$ and effective mass diffusion coefficient $\bar{D}_{eff}$ are related to the effective viscosity $\bar{\mu}_{eff}$ by

$$\frac{1}{Pr_e} = \frac{\bar{k}_{eff}}{\bar{\mu}_0 C_p} = \frac{1}{Pr} + \frac{1}{Pr_t} (\mu_{eff} - 1) \tag{13}$$

$$\frac{1}{Sc_e} = \frac{\rho_0 \bar{D}_{eff}}{\bar{\mu}_0} = \frac{1}{Sc} + \frac{1}{Sc_t} (\mu_{eff} - 1) \tag{14}$$

where Pr and Sc are the molecular Prandtl and Schmidt numbers, respectively. In the present study, both the turbulent Prandtl number $Pr_t$ and the turbulent Schmidt number $Sc_t$ are taken to be unity.

The governing equations (2) through (8) and the turbulence model given in Eqs. (10) through (12) can be similarly written for the cylindrical coordinate system by noting that

$$U = - V_r \cos\theta + V_\theta \sin\theta \tag{15}$$

$$V = V_r \sin\theta + V_\theta \cos\theta \tag{16}$$

and the corresponding coordinate relations as represented in Figure 1, and they have been given by Kou (1984). For the sake of conserving space, they will not be repeated here. The geometrical linking of the two coordinate systems will be described later. The boundary conditions are easily prescribed. The velocity components on the ground and around the cylinder surface are zero, and the velocity U along X=0 is uniform at $U_0$. Theoretically, the solution regions are infinitely extended in both positive X and Y directions. The heat and smoke source located inside the flame envelope provides the driving force for the temperature and smoke concentration fields. The ground and the cylinder (fuselage) surface are taken to be thermally insulated, and at the same time there is no net smoke particle deposition there. Temperatures and smoke concentrations away from the heat source including those along X=0 are taken to be ambient at $T_0$ and S=0, respectively.

In the present study, the governing equations, subjected to the boundary conditions described above, have been solved numerically by finite differences based on a micro-control volume scheme with primitive variables originally introduced by Patankar and Spalding (1972) in a manner similar to that used in an earlier companion study dealing with aircraft cabin fires (Yang et al, 1984). However, the formulations of the finite-difference equations and the associated calculation algorithm are more complicated due to the hybrid (rectangular and cylindrical) grid system used. Details of the numerical scheme are given by Kou (1984) and also contained in a companion paper (Kou, Yang and Lloyd, 1985), together with the results of validation studies based on this numerical scheme. It suffices here to describe briefly several of the salient features of the numerical computations.

A typical calculation grid system for H/D = 1.0 is shown in Figure 2. It is seen that the circular cylinder is deliberately placed close to the left lower corner of the computational grid to accommodate the extent of the thermal plume in the region above the fire and shedding of the flow behind the cylinder. The boundary conditions at the top and right free boundaries are prescribed as follows: The gradients for both velocity components normal to the boundary are taken to be zero. Also at any point along either of the free boundaries, the gradients of temperature and smoke concentration are also taken to be zero if the flow is outward, and when the flow is inward, ambient values of T and S, i.e. T=1 and S=0 are specified at that point. These are sometimes known as the natural boundary conditions.

One unique feature of the grid system used here is the hybrid mesh depicted in Figure 2. While the computations in either the rectangular grid or the cylindrical grid can be carried out in a standard manner, overlapping interfacial cell regions are provided to allow continuation of calculations from one region into the other. Even though the present study is only concerned with long-time behaviors, calculations marching in time are carried out until the long-time solution is achieved. Within each time step, finite-difference solutions for the two coordinate regions are obtained alternately until

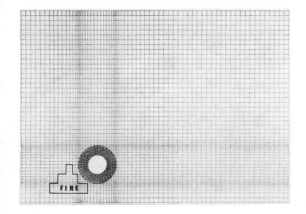

FIGURE 2. The Grid for the Case of Height Ratio H/D = 1.0.

convergence is reached. Calculations for each region are based on the conditions at the interface cells, and these values can be transferred from one region to the other by using standard bilinear interpolation formulae. Within each coordinate region, the computations proceed in a manner identical to that described by Yang et al (1984) and hence will not be repeated here. It is however important to note that the procedure does utilize upwind differencing for the convective terms and hence numerical errors due to false diffusion can be expected at large cell Peclet numbers (Patankar, 1980). However, due to relatively large effective viscosities encountered in the present problem, the cell Peclet number is found to be reasonably small, indicating that the effect of false diffusion should also be small.

The validity of the entire calculation procedure, including the part that deals with iterations involving both coordinate regions, has been established by a series of validation studies. Results of these studies have been given by Kou (1984), and are also included in a paper describing the details of the numerical procedure (Kou, Yang and Lloyd, 1985).

As will be described later, results for the buoyancy-dominated situations all show long-time oscillatory behaviors in the flow field. A question rises then as to whether such oscillatory behaviors are physically real. There are pervasive reasons that they are indeed real. The full-scale tests at the Technical Center of the U. S. Federal Aviation Administration also showed large-scale oscillations (Eklund, 1981). In preliminary calculations utilizing much coarser grid systems, such oscilltions have also been found (Kou, 1984). Even though no three-dimensional calculations have been made, there is evidence that the oscillatory flow field is predicted by both two-dimensional and three-dimensional calculations (Satoh et al, 1983). The resulting oscillatory frequencies are very similar, while the amplitudes for the three-dimensional cases are smaller. This is expected in view of the fact that two-dimensional flows are generally more stiff. Finally, it has also been found that the Richardson number in the turbulence model, equation (12), is important and affects both the frequency and amplitudes of the oscillations.


RESULTS OF SIMULATION STUDIES

The numerical procedure mentioned above is utilized in a series of simulation studies to determine the combined effects of the buoyant flow due to the fire and smoke source and the forced flow due to the prevailing wind speed on the velocity, temperature and smoke concentration fields in the neighborhood of the fuselage as a function of the wind speed and the elevation of the fuselage from the ground. Of particular interest is the resulting pressure variations, and smoke concentrations around the fuselage, since these are the physical parameters that will determine the fire and smoke hazards inside the aircraft cabin, if there is an opening present in the fuselage as a result of a survivable aircaft accident.

With reference to Figure 2, the diameter of the fuselage is taken to be 3.05 meters (10 feet), and the corresponding calculation domain is given by X=13 and Y=9. The center of the fuselage is located at X=3.5 from the left free boundary. The depth of the field is arbitrarily taken to be 6.10 meters (20 feet) and is only used to determine the strengths of the fire and smoke source in accordance with the full-scale tests performed at the Technical Center of the U.S. Federal Aviation Administration. As a result, the fire source is taken to be at a constant strength and has a total power output of 75,887 kw and a total rate of smoke generation of 0.1314 kg/sec, corresponding to that for a very

dirty fire. The location of the fire is taken to be fixed, and the fire
envelope shape is also fixed as shown in Figure 2 to approximate that for a real
fire. This approximation may indeed not be too realistic, especially in cases
where the forced flow is strong. However, since the effect of the cross flow on
the flame slape is not a priori known, it is difficult to model. This can only
be remedied by introducing a combustion model in the calculations. Simulation
studies have been carried out for fuselage elevations of H/D = 0.5, 1.0 and 1.5
and wind speeds of 3.05 m/sec (10 ft/sec), 6.10 m/sec (20 ft/sec) and 9.14 m/sec
(30 ft/sec). Calculations have alseo been carried out for the case of zero wind
speed corresponding to pure natural convection or fire plume situations and the
case of zero heat input which corresponds to pure forced flow over a circular
cylinder near a ground. Detailed results have been given by Kou (1984). Since
the limited space here does not permit their full presentation, only selected
results are given in the following:

Before these results are presented, however, it is expedient here to make
several general observations to facilitate the physical interpretation later on.
Firstly with the prescribed heat source, the range of wind speeds covered in
this study does span the complete region of interest ranging from forced-flow
dominated situations (9.14 m/sec) to natural-convection dominated flow (3.05
m/sec). Secondly, forced flow tends to stabilize the flow, while the buoyancy
effect due to the fire source destabilizes it. Thirdly, strong buoyancy effects
lead to long-time behaviors which are oscillatory, while time-independent
steady-state behaviors are obtained when forced-flow effects are dominant.
Fourthly, for elevation parameters H/D exceeding unity, the ground effect is not
significant. For H/D less than unity, the reduction in local flow area and the
ground friction become increasingly important and generally have the opposite
effects on the flow field. Lastly, in view of the similarity of the governing
equations and the boundary conditions, results for the temperature field are
very similar to that for the smoke concentration field. Consequently, separate
physical interpretations of these two fields are not necessary.

Figures 3(a) and (b) are typical smoke concentration contour plots for the case
of H/D = 0.5. Figure 3(a) is for a wind speed of 3.05 m/sec (10 ft/sec) and
Figure 3(b), for the speed of 6.10 m/sec (20 ft/sec). In both cases, buoyancy
effects are significant and long-time behaviors are oscillatory due essentially
to the instability of the fire plumes. The absolute time instants noted in
these figures have no special meaning, other than that reflecting the period
duing the oscillations. At the lower speed in Figure 3(a), the effect of the
wind on the fire plume is still minimal, and the fuselage is not much affected
by the fire. The temperatures on the surface are essentially ambient. When the
wind speed is doubled (Figure 3(b)), the fire plume is swept in the direction of
the wind and the fuselage is largely engulfed by the fire. This represents the
situation where both forced flow and natural convection play equally important
roles. As already noted previously, similar oscillatory behaviors of the smoke
field have been observed in the full-scale tests at the Technical Center of the
U. S. Federal Aviation Administration (Eklund, 1981). Unfortunately no direct
comparisons can be made because of the qualitative nature of the tests.

One important result in the present study, as pointed out previously, is the
pressure variation around the fuselage. Figure 4 describes such pressure
variations at $\theta = 0°$ and $180°$ for H/D = 0.5 and a wind speed of 9.14 m/sec (30
ft/sec). It is seen that the buoyancy effect induces a regular oscillation in
the long-time behavior and the period is about 4 seconds. Additionally, it has
been found that the pressure depressions at the bottom of the fuselage are
higher than that at the top. It is believed that the buoyancy in the fire

179

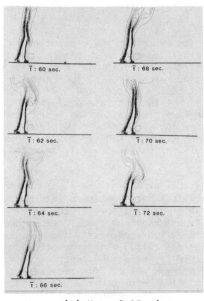

(a) $U_o$ = 3.05 m/sec          (b) $U_o$ = 6.10 m/sec

FIGURE 3. Oscillatory Smoke Concentration Fields for H/D = 0.5.

accelerates the flow more in the region below the fuselage, causing the pressure
to decrease there, despite the slight drag offered by the presence of the
ground. Figure 5 shows the effect of reducing the wind speed from 9.14 m/sec
(30 ft/sec) to 6.10 m/sec (20 ft/sec) at the same fuselage elevation H/D = 0.5.
The increased buoyancy effect in this case further destabilizes the flow and the
long-time oscillatory behavior is no longer very regular. The presence of the
ground seems to damp out some of the oscillations in the region below the

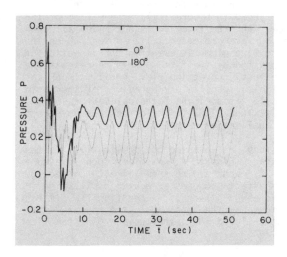

FIGURE 4. Pressure at $\theta$ = 0° and
180° for $U_o$ = 9.14 m/sec and
H/D = 0.5 (Heat On).

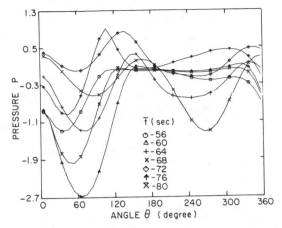

FIGURE 5. Pressure Variation Around Fuselage for $U_0$ = 6.10 m/sec and H/D = 0.5 (Heat Source On).

fuselage. From Figure 5 and Figure 3(b) , which depict the corresponding field behaviors, the period of oscillation can be approximately estimated at 24 seconds, and a more exact value can only be determined by performing a Fourier analysis of the irregular oscillations. Figure 6 shows the corresponding pressure variations for a case similar to that in Figure 8 except that the fuselage elevation is now increased to H/D = 1.5 Somewhat irregular oscillations are still present, but the behavior at the bottom and at the top of the fuselage are no longer that different, again attesting to the fact that for this value of H/D the ground loses much of its effect on the flow field. For this case, an approximate period is estimated at 28 seconds.

Figure 7 gives an illustration concerning the temperature variations around the fuselage, and it is understood that this same information can be interpreted as the smoke concentration variations. However, it should also be understood that T=1 would then correspond to S=0. Figure 10 is for the case of H/D = 0.5 and a wind speed of 6.10 m/sec (20 ft/sec). Here the wind speed sweeps the fire toward the fuselage, more under it than above. As a result, the temperatures are higher at the lower surface of the fuselage. It is also seen that during certain times in the cycle the fire plume actually stays away from the fuselage,

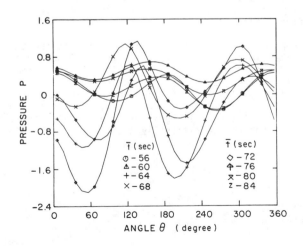

FIGURE 6. Pressure Variations Around Fuselage for $U_0$ = 6.10 m/sec and H/D = 1.5 (Heat Source On).

181

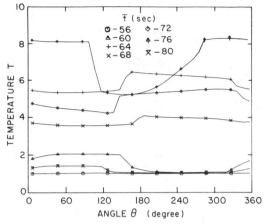

FIGURE 7. Temperature Variation Around Fuselage for $U_0 = 6.10$ m/sec and H/D = 0.5 (Heat Source On).

thus leaving the entire fuselage surface at essentially the ambient temperature. Also, during the other parts of the cycle, the surface temperature can become very high, especially at the front of the fuselage.

CONCLUDING REMARKS

A numerical study based on two-dimensional finite-difference calculations with a hybrid grid system has been carried out to study the effects of wind speed and elevation of an aircraft fuselage located downwind of a fire and smoke source on the pressure, temperature and smoke concentration variations at the surface of the fuselage. The following general conclusions can be drawn:

(a) The surface variations in pressure, temperature and smoke at the fuselage are very sensitive to the relative effects of the prevailing wind speed and the buoyant flow from the fire source. The buoyancy arises from the fire source destabilizes the flow, and causes the long-time behavior to become oscillatory. When the buoyancy effect is large, the oscillations can become rather irregular.

(b) For elevation parameters exceeding H/D = 1.0, the ground does not exert much effect on the physical characteristics in the immediate neighborhood of the fuselage. As H/D falls below unity, the ground effect becomes increasingly more important, especially when the fire is present.

The present study can only be considered as a first step in addressing the external fire problem in survivable aircraft accidents. In the real scenario, the flow field is necessarily three-dimensional, and there is always thermal radiation exchange between the fire, the smoke, the ground and the fuselage. The combustion process involving the spilled fuel is also present in the real phenomena. Systematic studies, both numerical and experimental, are needed to address these realistic effects so that the all important fire safety issues involved in survivable aircraft accident can be rationally dealt with.

NOMENCLATURE

| | |
|---|---|
| $C_p$ | isobaric heat capacity |
| D | cylinder diameter |
| D | mass diffusion coefficienbt |
| g | gravitational acceleration |

| | |
|---|---|
| H | fuselage elevation |
| K | constant |
| k | thermal conductivity |
| $\ell$ | mixing length |
| P | pressure |
| Pr | Prandtl number |
| $\dot{Q}$ | volumetric heat generation rate |
| $Q°$ | non-dimensional volumetric heat source strength |
| R | radius variable |
| Re | Reynolds number |
| Ri | Gradient Richardson number |
| $R_o$ | gas constant |
| S | smoke concentration |
| Sc | Schmidt number |
| $\dot{S}$ | volumetric smoke generation rate |
| $S°$ | non-dimensional volumetric smoke source strength |
| T | temperature |
| t | time variable |
| U,V | velocity components in X, Y directions, respectively |
| $U_o$ | reference velocity |
| $V_r,V_\theta$ | velocity components in R, $\theta$ directions, respectively |
| X,Y | Cartesian coordinates |

Greek Symbols

| | |
|---|---|
| $\theta$ | angle in cylindrical coordinates |
| $\mu$ | viscosity |
| $\rho$ | density |

Subscripts

| | |
|---|---|
| e | based on effective transport properties |
| eff | effective (molecular plus turbulent) |
| o | reference quantities |
| t | turbulent |
| $\varepsilon$ | hydrostatic equilibrium condition |

Superscript

| | |
|---|---|
| − | dimensional quantities |

ACKNOWLEDGEMENT

The authors gratefully acknowledge support of the research under U. S. National Bureau of Standards Grant NB81NADA 2000 through an interagency agreement with the Federal Aviation Administration.

REFERENCES

Eklund, T. (1981). personal communication.

Emmons, H. W. (1982). The ingestion of flames and fire gases into a hole in an aircraft cabin for arbitrary tilt angles and wind speed. Home Fire Project TR No. 52, Division of Applied Sciences, Harvard University, Cambridge, MA.

Kou, H. S. (1984). Turbulent buoyant flow and pressure variations around a circular cylinder in a cross uniform flow near the ground. Ph.D. dissertation, University of Notre Dame, USA, 233 pp.

Kou, H. S., Yang, K. T., and Lloyd, J. R. (1985). A numerical study of mixed convection for flow over a circular cylinder near a ground in the presence of a volumetric heat source. Paper under review for publication.

Nee, V. W. and Liu, V. K. (1978). An algebraic turbulence model for buoyant recirculating flow. Technical Report TR-79002-78-2, Dept. of Aerospace and Mechanical Engineering, University of Notre Dame, 107 pp.

Patankar, S. V. (1980). Numerical Heat Transfer and Fluid Flow, McGraw Hill Book Co.

Patankar, S. V. and Spalding, D. B. (1972). A calculation procedure for heat, mass and momentum transfer in three-dimensional parabolic flow, Int. J. of Heat and Mass Transfer, vol. 15, 1787.

Satoh, K., Yang, K. T., Lloyd, J. R. and Kanury, A. M. (1983). A Numerical Finite-Difference Study of the Oscillatory Behavior of Vertically Vented Compartments, Numerical Properties and Methodologies in Heat Transfer, Proc. 2nd Nat. Sym., ed. T. M. Shih, Hemisphere Publishing Corp., 517.

Yang, K. T., Lloyd, J. R., Kanury, A. M. and Satoh, K. (1984). Modeling of turbulent buoyant flows in aircraft cabins, Comb. Sci. and Tech., vol. 39, 107.

# Conditionally-Sampled Estimates of Turbulent Scalar Flux in a Simulated Fire

**N. L. CRAUFORD**
Schlumberger Cambridge Research Ltd.
Cambridge, England

**K. N. C. BRAY**
University Engineering Department
Cambridge, England

**J. B. MOSS**
Cranfield Institute of Technology
Bedford, England

ABSTRACT

A conditional sampling strategy is described in which temperature, measured by fine wire thermocouple, is used to trigger the acquisition of velocity data by LDA. The technique permits the determination of a range of scalar-velocity correlations, important to the development of flow field models, throughout a simulated buoyant fire. The experimental data is incorporated into flux estimates which permit the assessment of mass, energy and momentum balances in two distinctive, intermittently burning regimes in the buoyant flame.

INTRODUCTION

Natural fires are distinguished by the low initial momentum of the fuel source. Subsequent fuel-air mixing and combustion are then strongly influenced by buoyancy. The large-scale inhomogenities characteristic of buoyant mixing pose particular difficulties in respect of both mathematical models, seeking to predict rates of fire spread and flame shape, for example, and experimental techniques, providing detailed information to guide and evaluate such models. Several studies have been reported recently on turbulent flames subject to some degree of buoyant influence which attempt detailed diagnosis of flame structure and development (1-6). Probe techniques, notably fine wire thermocouples and ionisation probes, and laser Doppler anemometry have been variously employed in determinations of time-mean and fluctuating components of velocity and temperature. Such measurements are not free from ambiguity since flame features of differing scales and light scattering particles appear to propagate at different mean speeds (6). However, despite this ambiguity, the microscopic insights provided by LDA do discriminate between important alternative modelling prescriptions. The present paper seeks to expand this experimental base by reporting direct measurements of the turbulent heat flux, $\overline{u'T'}$, and related flow properties in a turbulent buoyant flame. The joint determination of fluctuating velocity and temperature is effected by conditional sampling of LDA velocity using temperature, measured by a fine wire thermocouple, to trigger data acquisition.

Some recent comparisons between numerical prediction and experiment have been encouraging (4, 7). Two-equation ($k$-$\varepsilon$) turbulence modelling, incorporating some buoyant influences and improved chemistry prescriptions, for example a conserved scalar formulation with flamelet chemistry, has permitted some macroscopic features, flame shape and mean properties, to be plausibly reproduced.

Considerable uncertainty however surrounds the most widely suitable representation of the large-scale, unsteady flame motion so characteristic of buoyant fires. Flame front flapping occurs, for example, at low frequency ($\sim 3$ Hz), on scales comparable with the dimensions of the fire and may be highly anisotropic. The turbulent transport which results from such behaviour and which in many respects more closely resembles a buoyant instability than fully-developed eddy transport is then poorly represented. Even the sense of the turbulent flux, positive or negative, is uncertain in different parts of the intermittently-burning turbulent flame. Preliminary investigations into the simultaneous determination of velocity and temperature were reported by Walker and Moss (6) and we report here the results of more extensive mapping of a simulated fire together with an assessment of their implications for control volume balances of mass, momentum and energy. The particular interest in fire studies is often to integrate the source into more general descriptions of air and combustion products movement in enclosures and this has focussed attention on overall property balances. The contribution of turbulent fluxes to such balances are frequently neglected. Joint measurements in buoyant flames have to date been restricted to non-combusting plume regions, for example Nakagome and Hirata (8).

Walker and Moss (6) discuss the advantages in buoyant flame applications of LDA signal processing and analysis by photon correlation in comparison with alternative techniques which rely on substantial artificial seeding. Although the correlation technique permits velocity determination at the low signal-to-noise levels characteristic of laser light scattered from naturally occurring seed particles, the velocity-time series is then made inaccessible. Direct cross-correlation between velocity and temperature series is thus not possible and a conditional sampling strategy at the time of data acquisition must be employed.

EXPERIMENTAL DETAIL

The simulated fire comprises a porous refractory burner, 25cm in diameter, fuelled by natural gas from the domestic supply. The exit fuel velocity is 0.017 ms$^{-1}$, giving an approximate heat output of 28kW and luminous flame height of roughly 0.9m. The burner is mounted on a traverse having two degrees of freedom, permitting movement of the flame in the vertical and horizontal directions. A minimum separation of 25cm is maintained between burner face and floor to avoid air movement restrictions which might disturb the flame. The burner is located in a test area 3m square by 4m high, enclosed by solid walls on three sides and protected from draughts by a fine nylon mesh, screening the fourth side.

Complete details of the LDA system and method of data reduction are described by Crauford (9). Briefly, the digital photon correlator constructs the auto-correlation function of light scattered from naturally occurring soot particles as they traverse the LDA fringe field. This function is an ensemble average of many scattering events and joint velocity – temperature measurements can only be obtained by direct conditioning of LDA velocity acquisition. The latter is triggered by an externally generated strobe signal and the auto-correlation function is constructed in a piecewise manner from the selected events. A fine wire thermocouple, positioned 1mm upstream of the LDA fringe volume, provides a continuous analogue signal which can be transformed into a conditioning function by two discriminating threshold levels. The transitions between levels lead to the sequential enabling and disabling of the correlator. The strategy is illustrated schematically in Fig.1.

The thermocouple employed was constructed from a platinum/platinum-10% rhodium wire combination, 50μm in diameter. On the scale of the flow disturbances studied, the necessary thermocouple positioning upstream of the LDA fringe volume in order to function as a trigger was not considered significant. By setting the temperature thresholds close together, with 50K separation, the digital correlator only accepts scattered light pulses from the LDA photomultiplier when the measured temperature lies within this narrowly defined range. Ensemble averaging of these conditioned events, namely local velocity $u$ given that temperature $T$ lies in the range $T_c - 25 \leqslant T_c \leqslant T_c + 25$ (K), permits a representation of the joint probability density function (pdf) for $u$ and $T$ to be constructed. This will be described more fully in the following section. Between eight and ten temperature gates giving distinct velocities could typically be established, spanning the complete range of observed temperature fluctuations. The frequency response of the thermocouple was boosted by electrical compensation for the effects of thermal inertia in order to resolve local flame structure as clearly as possible.

Joint determinations of velocity and temperature from which turbulent correlations can be estimated are reported here in the form of radial traverses at two axial stations, $z = 30$cm. and 77cm. These locations distinguish approximately the lower and upper boundaries of the intermittently burning regime in the flame studied. Below 30cm. the buoyantly-accelerated flame column is comparatively well-defined whilst beyond 77cm. continuous flame is not observed and significant dilution and partial mixing occurs.

Joint PDF Representation

The narrowly defined temperature windows permit the representation of the temperature pdf as a series of delta functions, each centred within particular temperature thresholds. By measuring the relative proportions of temperature window open to closed (cf. Fig.1(c)) and scaling to the value of the measured mean temperature we construct the pdf. The velocity, associated with each conditioning temperature window and measured by the LDA, is determined in the form of a mean value and accompanying rms fluctuation. Given the strengths of the delta functions in the representation of the pdf for temperature, we can use these to weight the contributions of the conditional velocities to the overall velocity statistics. Comparisons between such estimates and unconditional velocity measurements at the same location permit alternative prescriptions for the joint pdf $P(u,T)$ to be tested. Two prescriptions are described in this paper; that the conditional velocity pdf $P(u/T = T_c)$ is itself a delta function centred on the measured mean value or that $P(u/T = T_c)$ is Gaussian, having mean and variance as measured.

The joint pdf $P(u,T)$ is given by

$$P(u,T) \simeq P(u/T = T_c) \, P(T_c) \qquad\qquad ..(1)$$

where the discretised pdf for conditioning temperature is

$$P(T_c) = \sum_{j=1}^{n} a_j \, \delta \, (T_c - T_c^{(j)}) \qquad\qquad ..(2)$$

and where $n$ denotes the number of temperature windows and $a_j$ is the strength of the delta function corresponding to window-centre temperature $T_c^{(j)}$.

The conditional velocity distributions are then given either by

$$P(u|T = T_c) = \delta \, (u - \bar{u}^{(j)}) \quad \text{or} \qquad\qquad ..(3)$$

$$P(u|T = T_c) = \left\{ \sqrt{\pi} \, \sigma^{(j)} \right\}^{-1} exp \left\{ - \left[ u - \bar{u}^{(j)} \right]^2 / 2\sigma^{(j)2} \right\} \qquad\qquad ..(4)$$

where $\bar{u}^{(j)}$, $\sigma^{(j)2}$ denote the measured mean and variance associated with conditioning temperature $T_c^{(j)}$.

Not unexpectedly, the Gaussian form leads to a marginal distribution for velocity on integration over $T_c$ which predicts more closely the measured velocity statistics in the absence of conditioning. The joint pdf $P(u,T)$ corresponding to this case is illustrated in Fig.2. It must be emphasised that since we are here triggering velocity data acquisition and ensemble averaging over events distinguished by temperature and not merely partitioning a single velocity time series, these comparisons are not trivial. The plausible reproduction of unconditioned statistics is however crucial if we are to have confidence in the representation of the joint pdf and hence in estimates of velocity-temperature correlations.

Figure 2 reveals a significant spread of velocities associated with each individual conditioning temperature and whilst there is a discernible displacement of conditional mean velocity, $\overline{u}^{(j)}$, to higher values with increasing window temperature, there is also extensive overlapping. For conditioning temperatures in the range $298 \leqslant T_c \leqslant 1887$ (K), $1.5 \leqslant \overline{u}^{(j)} \leqslant 5.6$ (ms$^{-1}$), such spread suggests that temperature alone would prove an unsuitable diagnostic for local conditions – an important factor in relation to the 'two-velocities' observed in the flame.

The LDA system detects the motion of naturally occurring particles, largely of soot, as they pass through the fringe field. Small particles, produced in the vicinity of the reaction zone, exhibit comparatively small axial velocities ($\overline{u} < 2.6$ ms$^{-1}$) and are subject to the more direct influence of the flapping flame. More aged, larger particles, though fewer in number, are more strongly accelerated in the buoyant fuel/product regions ($\overline{u} < 4.6$ ms$^{-1}$). These two velocities can be distinguished at each location by varying the aperture of the collection optics – the larger the aperture the more significant the contribution from the large number of small particles accompanying the local flame zone and vice versa. Two distinct mean velocities are observed at most positions within the flame. The introduction of conditioning by local temperature as a further descriptor of local state, additional to particle number density and size, might be expected to offer some corroboration of this diagnosis. However, it proved impossible to overwhelm the seeding bias simply by displacing the temperature window and both velocities could be obtained over almost the complete range of conditioning temperature. Since temperature is not an unambiguous measure of local mixture state – the temperature mixture fraction relationship is double-valued, for example – this observation is not entirely unexpected, irrespective of any shortcomings in the measurement technique.

Table 1 presents comparisons at the $z = 30$cm. station between unconditional and conditional data using the alternative representations for the joint pdf described earlier. Whilst the temperature statistics are satisfactorily reproduced throughout, particularly in the centre of the flame, the velocities are less satisfactory. The marginal pdfs for velocity, cf. eqns.(3) and (4), constructed from conditional measurements, give mean velocities which compare reasonably well with unconditioned measurements – discrepancies $\lesssim 25\%$ – but more significant differences emerge in relation to the rms values. The delta function representation leads to substantial under-estimates whilst the Gaussian form in general leads to over estimation. Given the difficulties generally encountered in velocity measurement in buoyant fires – the large amplitude, low frequency flapping motion inevitably poses problems in relation to reproducibility even in the absence of conditioning – the Gaussian joint pdf representation was deemed sufficiently good for useful estimates of turbulent correlations to be made.

The estimates of $\overline{u'T'}$ are obtained by quadrature,

$$\overline{uT} = \iint\limits_{0}^{\infty} uT\ P(u|T = T_c)\ P(T_c) du\ d\ T_c \quad = \quad \overline{u'T'} + \overline{u}\,\overline{T} \qquad \qquad ..(5)$$

The turbulent flux is positive at all the stations investigated, peaking off-axis at $z$ = 30cm. but closer to the axis at $z$ = 77cm. The mean temperature gradient is small, but positive, at $z$ = 30cm. and then decreases such that $\frac{\partial \overline{T}}{\partial z} < 0$ at $z$ = 77cm. A simple gradient description

$$\overline{\rho\ u'T'} = -\frac{\mu_t}{\sigma_T}\frac{\partial \overline{T}}{\partial z}$$

would evidently prove unsatisfactory at the lower station. Since the occurrence of buoyantly accelerated fluid elements may quite simply reflect flame motion and local heat release, any relationship to the mean temperature gradient is of limited mechanistic significance.

Comparisons with other measurements are restricted to non-combusting heated plumes. The relative flux intensity,$\overline{u'T'}/(\overline{u'^2})^{\frac{1}{2}}(\overline{T'^2})^{\frac{1}{2}}$, is comparatively insensitive to the particular application as shown in Fig.3 and a broad measure of agreement is evident. At such levels the relative contribution of the turbulent heat flux, in comparison with the convective flux, might be expected to be small. The relative contributions are described in the following section.

## MASS AND HEAT FLUX BALANCES

In estimating fluxes the contribution of the density is complex. Whilst the flame is substantially isobaric from the standpoint of thermodynamic state, in using methane as fuel we admit density variations resulting from both molecular weight and temperature changes. Beyond $z$ = 30cm. the mean mixture fraction is everywhere small, $\overline{\xi} < 0.1$ (7), and therefore the influence of molecular weight may be assumed small. We shall suppose $\rho = \rho(T)$ such that $\rho T = \rho_0 T_0$, whence the mass flux crossing plane $z$ may be written

$$\overset{\bullet}{m}(z) = \int\limits_{0}^{\infty} \overline{\rho u}\ dA = \int\limits_{0}^{\infty}\left[\iint\limits_{0}^{\infty} \rho(T)u\ P(u,T)\ dT\ du\right] dA, \quad \text{and from eqns. (2) and (4)}$$

$$\overline{\rho u} = \sum_{j=1}^{n} a_j\ \rho(T_c^{(j)})\ u^{(j)} \quad = \sum_{j=1}^{n} a_j\ \rho_0 \left\{\frac{T_0}{T_c^{(j)}}\right\}\overline{u}^{(j)} \qquad \qquad ..(6)$$

Similar expressions apply to momentum and energy flux correlations

$$\overline{\rho u^2} = \sum_{j=1}^{n} \overline{u^{(j)2}}\ a_j\ \rho_0\left\{\frac{T_0}{T^{(j)}}\right\} \quad \text{and} \quad \overline{\rho u T} = \sum_{j=1}^{n} a_j\ \overline{u^{(j)}}\ \rho_0 T_0 = \overline{u}\rho_0 T_0 \qquad ..(7)$$

In the absence of detailed information of the type reported here it has been customary to neglect the turbulent contribution to such expressions. In such circumstances we would estimate the mass flux to be

$$\overset{\bullet}{m}_m(z) = \int\limits_{0}^{\infty} \overline{\rho}\ \overline{u}\ dA = \left\{\sum_{j=1}^{n} a_j\rho(T_c^{(j)})\right\}\sum_{j=1}^{n} a_j\ \overline{u}^{(j)} \qquad \qquad ..(8)$$

The turbulent contribution from eqns.(6) and (8) is then

$$\overline{\rho'u'} = \overline{\rho u} - \overline{\rho}\ \overline{u}\ = \rho_0 T_0 \sum_{j=1}^{n}\frac{a_j}{T_c^{(j)}}\left[\overline{u}^{(j)} - \sum_{j=1}^{n} a_j\ \overline{u}^{(j)}\right] \qquad ..(9)$$

189

The density-velocity correlation, estimated from eqn.(9), at two axial stations, $z = 30$ and 77cm, is illustrated in Fig.4. The correlation is negative at both stations across the complete flame and typically 10% of the mean flux, $\bar{\rho}\,\bar{u}$. The results quoted relate specifically to the upper mean particle velocity but relative flux levels, $\overline{\rho'u'}/\bar{\rho}\,\bar{u}$, are comparable for the lower velocity also. Despite the large-scale, buoyant flame motion the turbulent contribution is therefore indicated to be quite small. Whilst uncertainties surround the effect of molecular weight variations, not included in the density model, and the measurement of temperature fluctuations, these levels are believed to be broadly representative. The molecular weight effect might be incorporated using a flamelet profile and flapping flame model (cf.(11,12)) but in the present work we have elected to minimise the modelling element introduced directly into the data analysis.

Comprehensive mapping of the flame for velocity and temperature, including their joint determination described earlier, permits important features of zonal balances to be assessed. Fig.5 illustrates the data set in the form of radial profiles at axial stations $z = 0$, 30 and 77cm, the latter defining the lower and upper bounds of the intermittently burning regime. Additional data are reported elsewhere (7, 9).

Mass momentum and energy balances are investigated in terms of the properties summarised in Fig.6. The choice of a radius of 16cm. for the bounding cylinder is arbitrary. This particular value is broadly representative of the maximum flame width, attained in the neighbourhood of the $z = 77$cm. station, and permits the air drawn into the burning zone from beneath the burner to be featured in the 'source flow' at $z = 0$.

The mass flux crossing the $z = 0$ face, $\dot{m}(z = 0)$, comprises

$$\dot{m}(z = 0) = \dot{m}_f + \dot{m}_{a1} = \pi R_f^2 \,\rho_f\, u_f + 2\pi \int_{R=13}^{R=16} \rho_a\, u_{a1}\, R\,dR$$

since the turbulent contributions are negligible. Normalising with respect to the initial fuel flow, with $R^* = R/R_f$,

$$\dot{m}(z = 0)/\dot{m}_f = 1 + (\rho_a/\rho_f)\int_0^{1.5} (u_{a1}/u_f)\; dR^{*2} \qquad \qquad ..(10)$$

and substituting for the velocities (cf. Fig.5),
$$\dot{m}(z = 0)/\dot{m}_f = 11.5.$$

Given that stoichiometric methane combustion requires an air:fuel ratio of approximately 18:1 it might be argued that more than half the air required enters the flame through an annular region immediately surrounding the burner.However, the large amplitude flame flapping motion suggests that such literal interpretation of zonal balances of this type is potentially misleading. The mean radial inflow velocity, $\bar{v}$, in this region is shown in Fig.7. Whilst the fluctuating component is substantially larger than the mean, there are no significant temperature fluctuations and it appears plausible to neglect the velocity-density correlation. The mean flux into the reference cylinder over the height $0 \leqslant z \leqslant 30$ (cm), $\dot{m}_{a2}$, is estimated from

$$\dot{m}_{a2}/\dot{m}_f = (\rho_a/\rho_f)\; 2R^*_{16} \int_0^{30} \bar{v}\;(z)\; dz/u_f\, R_f \approx 36. \qquad \qquad ..(11)$$

The radial inflow induced by buoyancy is thus substantially greater than that drawn from below the burner and emphasises the importance of flame motion and fluctuating heat release on distributed burning.

The fluxes across the intermittently burning stations, $z$ = 30 and 77cm, are made rather ambiguous by the observation that the velocity field is characterised by two distinctive velocities. However the density and area weighting favours the wings of the flame where the differences between the two velocities is small. The discrepancies between flux estimates is accordingly rather smaller than might have been expected. Given the comparative sparseness of the conditioned data, these fluxes are estimated from

$$\dot{m}(z)/\dot{m}_f = \int_0^{1.5} \{\overline{\rho}\ \overline{u}\ +\ \overline{\rho'u'}\ \}\ dR^{*2}/\rho_f u_f \qquad ..(12)$$

with $\overline{\rho}\ \overline{u}$ determined from unconditioned measurements and $\overline{\rho'u'}$ alone by conditional analysis (cf. Fig.4). We find

$$\dot{m}(z = 30)/\dot{m}_f\ \simeq\ 48\ \text{(lower mean velocities) or}\ \simeq 50\ \text{(upper mean velocities)}$$

values which compare favourably with the sum of the fluxes into the cylindrical volume, bounded by the planes $z$ = 0 and 30cm, revealed in equations (10) and (11), namely 47.5.

Adopting the same approach to the volume bounded by planes $z$ = 30 and 77cms. we find

$$\dot{m}_{a3}/\dot{m}_f\ \simeq\ 73\ \text{and}\ \dot{m}(z = 77)/\dot{m}_f\ \simeq\ 114\ \text{(lower mean velocities) or 175 (upper mean velocities)}.$$

The difference between particle velocities is now significant across the complete radius and the flux estimates reflect this fact. Whilst the mean temperature is now falling (cf.Fig.5) and, increasingly, dilution mixing occurs, some burning is also observed. The flux across the plane $z$ = 77cm. is at least five times that required for complete combustion, a factor again emphasising the large-scale, intermittent nature of the flame.

Local apportioning of the scalar states giving rise to the two distinctive mean velocities has not proved possible. As indicated earlier the flux across $z$ = 30cm. is insensitive to the aperture setting although the lower mean velocity gives a slightly better mass balance. Scaling the upper/lower bound estimates crossing $z$ = 77cm. to the net inflow suggests the relative contributions from flame zone and burnt gas to be in proportions 8:1. This ratio seems unexpectedly high, suggesting that the flame zone, lower bound velocity is the more representative of mass average behaviour even close to the top of the visible flame. It should be noted however that upper and lower mean velocities converge as $z$ increases (6). Also, the large radial in-flow, $\dot{m}_{a3}$, has been estimated from $\overline{\rho}\ \overline{v}$ in the absence of the radial velocity-density correlation. At $z$ = 77cm. the turbulent contribution is less plausibly neglected than at $z$ = 30cm. in view of flame spread. This contribution is expected to be positive, $\overline{\rho'v'} > 0$, and so augment the radial inflow. Increased flux from this source will evidently shift the flame zone-to-burnt gas proportion towards the latter.

The momentum and energy balances are less complete. The mean sensible enthalpy flux, $\dot{q}(z)$, can evidently be calculated from conditioned measurements in a manner analogous to the mass fluxes reported earlier. For

$$\dot{q}(z)\ =\int_A Cp\ (\overline{\rho\ u\ T})\ dA,$$

assuming constant specific heat, $Cp$, and the density model introduced earlier $\rho T\ =\ constant\ =\ \rho_o T_o$ then

$$\dot{q}(z)/\dot{Q}_f\ =\ (\rho_o/\rho_f)(Q/CpT_o)\int_0^{1.5}(\overline{u}/u_f)\ dR^{*2} \qquad ..(13)$$

where $\dot{Q}_f$ is the fuel source energy flow rate and $Q$ is the heat of reaction.

The normalised enthalpy fluxes crossing planes $z$ = 30 and 77cm. are found

191

to be

$$\dot{q}(z = 30)/\dot{Q}_f \simeq 0.79 \text{ (upper mean)}, \simeq 0.59 \text{ (lower mean)}$$

$$\dot{q}(z = 77)/\dot{Q}_f \simeq 1.95 \text{ (upper mean)}, \simeq 0.96 \text{ (lower mean)}$$

Such estimates are dependent on the mean velocity profiles and the ambiguity introduced by upper and lower bounds is particularly marked at $z = 77$cm. Values substantially in excess of unity are clearly unphysical. We expect the zonal balances to be dominated by convective flux, chemical heat release rate ($\leqslant \dot{Q}_f$) and radiative loss. The lower mean velocity again suggests the more plausible behaviour. Modelling assumptions such as constant mixture specific heat and molecular weight introduce additional uncertainties into these estimates. The zonal balances suggest that these uncertainties may be broadly comparable with the radiative loss. Markstein (13) suggests from observation of propane flames over a range of fuel flow rates that the radiative power is approximately 0.21 – 0.25 times the total heat release. The radiative loss from methane flames, appropriate to the present experiment, is typically smaller since the contribution from luminous radiation is substantially reduced. Bearing this in mind, the energy releases in the two zones ($0 \leqslant z \leqslant 30$, $30 \leqslant z \leqslant 77$ (cm)) appear to be roughly comparable.

The momentum balance contains several terms which are inaccessible to measurement, in particular those relating to pressure gradients and buoyancy. The quadratic nature of the momentum flux, $\overline{\rho u^2}$, results in estimates which are strongly influenced by differences in particle velocities. Normalised flux estimates from the lower mean velocities are

$$\dot{P}(z)/\left\{ A_{ref} \ g \int_{z_0}^{z} \{\rho_0 - \overline{\rho}_{\mathscr{L}}\} \ dz' \right. \simeq 0.10 \ (0 \leqslant z \leqslant 30\text{cm}); \quad 0.26 \ (30 \leqslant z \leqslant 77\text{cm})$$

where $A_{ref}$ is the control volume cross sectional area and the centre-line mean density, $\overline{\rho}_{\mathscr{L}}$, is assumed to be given by

$$\overline{\rho}_{\mathscr{L}} = \rho_0 \ \{T_0/\overline{T}_{\mathscr{L} \ max} \ \} \quad (0 \leqslant z \leqslant 20 \ (\text{cm}))$$

$$= \rho_0 \ \{T_0/\overline{T}_{\mathscr{L}} \ \} \quad (z > 20 \ (\text{cm}))$$

and $\rho_0$ is the ambient air density.

The use of centre-line density to model the buoyant contribution will lead to an over-estimate of its effect since the mean density is a minimum there. Whilst the absolute values are evidently uncertain the relative contributions from the two zones are again broadly comparable. It was adjudged that a more detailed assessment of the buoyant contribution by integration of the radial profiles was unlikely to be more informative.

CONCLUSIONS

Gross inhomogenities in state and flow properties which are characteristic of buoyant flames make the turbulent flow field accessible to conditional sampling techniques. Temperatures measured by fine wire thermocouples are used to generate a strobe signal which triggers LDA data acquisition. Velocity statistics conditional on local temperature result. These have been interpreted as a joint probability density function for velocity and temperature. Results of an extensive mapping of the flow field are reported.

The joint pdf is used to estimate the axial turbulent fluxes arising in mass, energy and momentum balances at two axial stations. The turbulent velocity-temperature correlation is revealed to be positive at both stations although the gradient of mean temperature changes sign. The accompanying mass flux $\overline{\rho' u'}$ is negative but comparatively small, typically 10% of the flux estimated from mean properties, $\rho \, \bar{u}$.

Results of a simple control volume analysis for cylindrical volumes bounded by planes $z$ = 30 and 77cm. are presented. Some uncertainty is introduced into the analysis by the observation of two mean particle velocities. The smaller of these corresponds to the high number density of small particles in the vicinity of the burning zone whilst the higher velocity is associated with the lower density of larger, aged particles characteristic of fuel-product regions. Two zonal balances suggest that the lower of these velocities is the more representative in the region where the two velocities differ significantly.

The radial inflow, induced by large-scale flame motion and buoyancy, is shown to be the principal component in the mass balances. However, a significant air flow is induced from beneath the burner through an annular region immediately surrounding it. The mass flux through the top of the control volume at $z$ = 77cm. is at least five times that required for complete combustion, illustrating the large scale intermittent structure of the flame. The energy and momentum balances are incomplete but the flux estimates suggest that chemical heat release and buoyant acceleration are broadly comparable in the two zones. The radiative loss cannot be distinguished from the uncertainties introduced in modelling, notably the velocity ambiguity and molecular weight effects on the density field.

REFERENCES

1.  Becker, H.A. and Yamazaki, S. (1978). Entrainment, Momentum Flux and Temperature in Vertical Free Turbulent Diffusion Flames. Comb. and Flame 33, 123.

2.  McCaffrey, B.J. (1979). Purely Buoyant Diffusion Flames: Some Experimental Results. National Bureau of Standards, NBSIR 79-1910.

3.  Cox, G. and Chitty, R. (1980). A Study of Deterministic Properties of Unbounded Fire Plumes. Comb and Flame 39, 191.

4.  You and Faeth, G.M. (1982). Buoyant Axisymmetric Turbulent Diffusion Flames in Still Air. Comb.and Flame 44, 261.

5.  Gengembre, E., Cambray, P., Karmed, D. and Bellet, J-C. (1983). Turbulent Diffusion Flames with Large Buoyancy Effects. Paper presented at IXth ICODERS, Poitiers.

6.  Walker, N.L. and Moss, J.B. (1984). LDA Measurement in and around a Turbulent Buoyant Flame. Comb. Sci. and Tech. 17, 65.

7.  Crauford, N.L., Liew, S.K. and Moss, J.B. (1985). Simulated Buoyant Fire Modelling and Experiment. Comb. and Flame. (to appear).

8.  Nakagome, H. and Hirata, M. (1976). The Structure of Turbulent Diffusion in an Axi-symmetrical Thermal Plume. Heat Transfer and Turbulent Buoyant Convection (eds. Spalding, D.B. and Afgan, S.)

9.  Crauford, N.L. (1984). The Structure of an Unconfined Buoyant Turbulent Diffusion Flame. Ph.D. Thesis, Southampton University.

10. Ballantyne, A. and Moss, J.B. (1977). Fine Wire Thermocouple Measurements of Fluctuating Temperature. Comb. Sci. and Tech. 17, 65.

11. Roberts, P.T. and Moss, J.B. (1981). A Wrinkled Flame Interpretation of the Open Turbulent Diffusion Flame. Eighteenth Symposium (International) on Combustion, p.941, Combustion Institute.

12. Liew, S.K., Bray, K.N.C. and Moss, J.B. (1981). A Flamelet Model of Turbulent Non-premixed Combustion, Comb. Sci. and Tech. 27, 69.

13. Markstein, G.H. (1977). Scaling of Radiative Characteristics of Turbulent Diffusion Flames. Sixteenth Symposium (International) on Combustion,

p,1407, Combustion Institute.

14. George, W.K., Albert, R.L. and Tamanini, F. (1977). Turbulence Measurements in an Axisymmetric Buoyant Plume. Int.Journal Heat and Mass Transfer 20, 1145.

ACKNOWLEDGEMENTS

This paper forms part of the work of the Fire Research Station, Building Research Establishment, Department of the Environment, U.K. It is contributed by permission of the Director, Building Research Establishment. Helpful discussions with Mr.G.Cox, Fire Research Station are gratefully acknowledged. Crown Copyright 1985.

| Position, R(cm) | R = 0 Lower/upper velocities | | R = 4 Lower/upper | | R = 8 Lower/upper | | R = 12 Lower/upper | |
|---|---|---|---|---|---|---|---|---|
| $\bar{u}$ (ms$^{-1}$) | 2.27 | 3.94 | 1.73 | 2.75 | 0.71 | 0.74 | 0.32 | 0.33 |
| $\bar{u}_\delta (= \bar{u}_G)$ | 2.69 | 3.59 | 2.10 | 2.22 | 1.07 | 0.98 | 0.30 | 0.32 |
| $\overline{(u'^2)}^{\frac{1}{2}}$ | 0.75 | 0.93 | 0.55 | 0.96 | 0.35 | 0.37 | 0.24 | 0.24 |
| $\overline{(u'^2)}^{\frac{1}{2}}_\delta$ | 0.29 | 0.49 | 0.27 | 0.41 | 0.34 | 0.32 | 0.12 | 0.12 |
| $\overline{(u'^2)}^{\frac{1}{2}}_G$ | 1.11 | 1.11 | 0.80 | 0.80 | 0.60 | 0.60 | 0.26 | 0.26 |
| $\bar{T}$ (K) | 1225 | | 946 | | 757 | | 453 | |
| $\bar{T}_\delta$ | 1221 | | 943 | | 656 | | 401 | |
| $\overline{(T'^2)}^{\frac{1}{2}}$ | 304 | | 358 | | 311 | | 224 | |
| $\overline{(T'^2)}^{\frac{1}{2}}_\delta$ | 325 | | 353 | | 304 | | 158 | |
| $\overline{u'T'}$ | 93.51 | 114.61 | 88.70 | 114.54 | 92.68 | 84.77 | 20.90 | 20.18 |
| $\overline{u'T'}/\bar{u}_\delta \bar{T}_\delta$ | 0.028 | 0.026 | 0.045 | 0.055 | 0.128 | 0.128 | 0.174 | 0.157 |

* $\delta$, $G$ subscripts denote estimates from conditional data using delta function and Gaussian representations for velocity, cf. eqs.(3) and (4), respectively.

TABLE 1. Velocity and Temperature from Conditioned data at $z$ = 30cm.

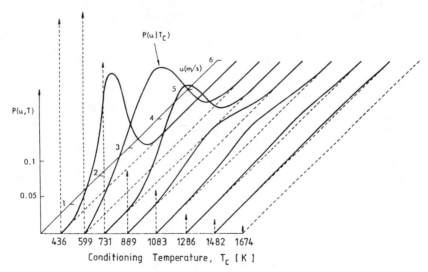

FIGURE 2. Joint Pdf for conditionally sampled velocity at $z$ = 77cm, $R$ = 0, upper mean velocity.

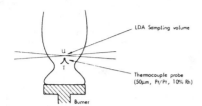

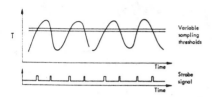

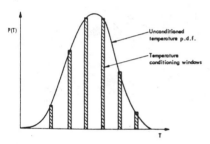

FIGURE 1. Schematic of conditional sampling strategy.

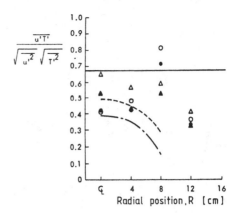

FIGURE 3. Heat flux intensity comparisons. Present experiments $z$=30cm, 0,● upper/lower mean velocities $z$=77cm, △,▲ —— ref.14, - - - - ref. 8

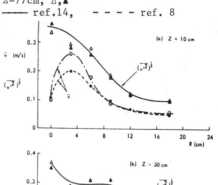

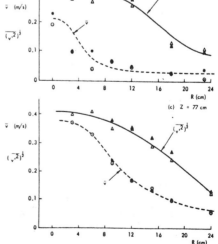

FIGURE 7. Radial component of mean and rms velocity.

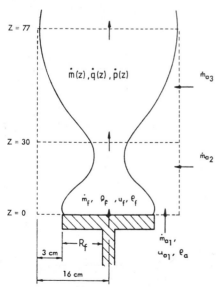

FIGURE 6. Nomenclature for control volume analysis.

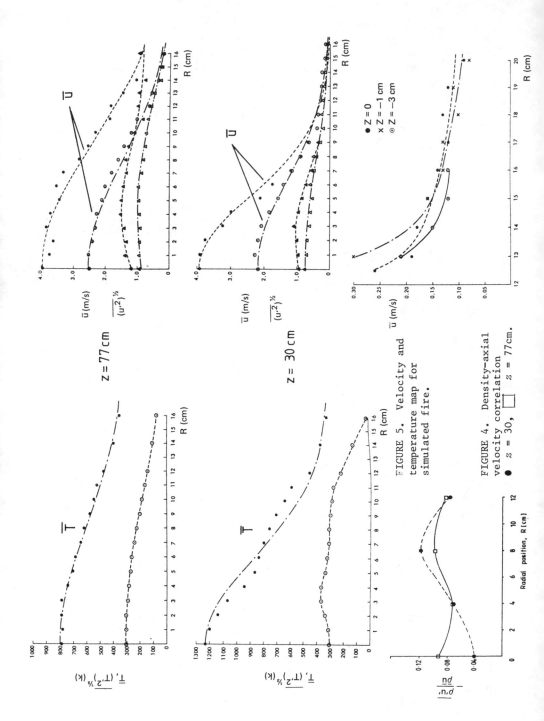

FIGURE 5. Velocity and temperature map for simulated fire.

FIGURE 4. Density–axial velocity correlation
● z = 30, ▢ z = 77cm.

# A Method for Calculating the Configuration Factor between a Flame and a Receiving Target for a Wide Range of Flame Geometries Relevant to Large Scale Fires

G. HANKINSON
British Gas Corporation Research and Development Division
Midlands Research Station
Wharf Lane, Solihull
West Midlands, England B91 2JW

ABSTRACT

To predict the thermal radiation field surrounding a fire, it is necessary to know the configuration factor between the flame and the receiving target. This requirement has previously imposed restrictions on the range of geometries that could be selected to represent the shape of flames. A method has therefore been developed which will enable the configuration factor to be calculated for any flame shape and has been applied to a particular geometrical configuration that can be used to represent a wide range of flame shapes associated with large fires. The method also allows the alignment of a target that is subjected to the maximum incident thermal radiation to be identified. This is of particular interest when studying radiant heat transfer.

## INTRODUCTION

Radiative heat transfer from the flames associated with large fires can endanger objects located near the flame. A quantitative determination of the thermal radiation field surrounding a fire is, therefore, necessary and this involves four major steps:

1. Specification of the size and shape of the flame.
2. Specification of the thermal radiation properties of the flame. In many cases flames are considered to emit uniformly over the whole of their surface.
3. Determination of the transmissivity of the intervening atmosphere. This will depend upon the atmospheric conditions, the emission characteristics of the flame surface and the mean path length from the object to the flame surface.
4. Calculation of the configuration factor. This enables the relative position and geometry of the flame and the receiving object to be taken into account.

The shapes of flames associated with large hydrocarbon pool fires have been studied by several groups of workers. These include American Gas Association[1] for spills into 1.83, 6.10 and 24.38m diameter earthen bunds, Raj et al.[2], U.S. Department of Transportation, for spills onto water, Mizner and Eyre[3], Shell, for spills into 20m diameter bunds and Moorhouse[4], British Gas Corporation for spills into rectangular bunds ranging in area from 37.2 to 185.8m$^2$ and length to width ratio from 1 to 2.5. The flames associated with these large pool fires were approximated using regular geometrical shapes. However, restrictions on the geometries adopted were imposed, in some cases, by the techniques available for calculating configuration factors between the flame and a target near the flame. The flame shapes defined in these studies range from an oblique cylinder of circular cross section, a tilted cylinder of circular cross section and an oblique cylinder of elliptical cross section.

Analytical expressions for calculating configuration factors have been derived using the geometrical determination technique[5] or the contour integral method[6]. Such equations are available for oblique cylinders of circular cross section and other simple shapes but are restricted to certain locations and orientations of the target. The contour integral method can also be used as a numerical technique, however, the identification of the contour that encloses that part of the flame surface contained within the field-of-view of the target can be extremely complex.

Another technique developed by Rein et al[7] for tilted cylinders and receiving targets located at ground level directly downwind of the flame is an area integral method. That part of the surface of the cylinder contained within the field-of-view of the target is divided into small parallelograms. Unfortunately the method makes no allowances for the differences in area of these small parallelograms as their position changes around the circumference of the cylinder. The expression used to calculate the area is accurate only for elements located normal to the wind direction on the upwind or downwind edges of the flame, or for the situation in which the angle of tilt of the flame from the vertical is zero. Inaccuracies using this method increase as the flame tilt increases and will further increase if the method is extended so that calculations can be performed for targets located at positions crosswind of the fire.

An area integral method has, therefore, been developed which overcomes the above difficulties and also extends the range of application to cover any solid geometrical shape. This is accomplished by dividing the whole of the surface of the solid shape used to represent the flame into small triangular elements. Conditions are then employed to select only those elements that can be seen from the position and orientation of the target. In some applications it is of great benefit to know the alignment of the target at a particular location that is subjeted to the maximum incident thermal radiation. When the emissive power of the flame is assumed to be uniform over the surface, this is identical to the alignment of the target that subtends the maximum configuration factor. A technique which enables this orientation of the target to be determined has been included in the method.


INCIDENT THERMAL RADIATION

The thermal radiation incident upon a target, 1, at any position and orientation relative to an emitting surface, 2, is dependent upon the surface emissive power, the transmissivity of the atmosphere between the surface and the target and the configuration factor as follows:

$$I = E \, F_{21} \, \tau \qquad\qquad (1)$$

where
$\quad$ I $\quad$ represents the incident thermal radiation $(kWm^{-2})$
$\quad$ E $\quad$ represents the surface emissive power $\quad (kWm^{-2})$
$\quad$ $F_{21}$ represents the configuration factor
$\quad$ $\tau$ $\quad$ represents the transmissivity of the atmosphere.

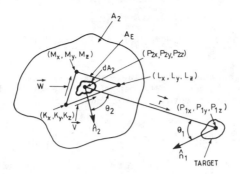

FIGURE 1.   Radiant heat transfer between a finite surface and an infinitesimal target.


CONFIGURATION FACTOR

The configuration factor for radiant heat transfer between an emitting surface, 2, of any shape and an infinitesimal target, 1, is given, with reference to Fig. 1, as follows:

$$F_{21} = \int_{A_2} \frac{\cos\theta_1 \cos\theta_2 \, dA_2}{\pi r^2} \tag{2}$$

where   $A_2$   represents the surface area of that part of the emitting surface that is both contained within the field-of-view of the target and contains the target within its own field-of-view.

To enable Eq. (2) to be solved by a numerical integration technique it is necessary to define the geometry of the surface.   This, in general, can be accomplished by dividing the whole of the surface area into small triangular elements and specifying the position in space of the nodal points relative to a set of Cartesian axes.

The configuration factor for radiant heat transfer between a small triangular element, E, and an infinitesimal target is given, with reference to Fig. 1, as follows:

$$F_{E1} = \frac{\cos\theta_1 \cos\theta_2 \, A_E}{\pi r^2} \tag{3}$$

where   $A_E$ represents the area of the triangular element.

$A_E$ is obtained from the vector product of the two vectors $\vec{V}$ and $\vec{W}$ which form two sides of the triangular element.   $\vec{V}$ and $\vec{W}$ are obtained from the positions of the nodal points as follows:

$$\vec{V} = (V_x, V_y, V_z) = \left[(L_x - K_x), (L_y - K_y), (L_z - K_z)\right] \tag{4}$$
$$\vec{W} = (W_x, W_y, W_z) = \left[(M_x - K_x), (M_y - K_y), (M_z - K_z)\right] \tag{5}$$

This gives:

$$A_E = \frac{|\vec{V} \times \vec{W}|}{2} = \frac{\left[(V_y W_z - V_z W_y)^2 + (V_z W_x - V_x W_z)^2 + (V_x W_y - V_y W_x)^2\right]^{\frac{1}{2}}}{2} \tag{6}$$

The length r of the connecting line between the centre-of-area of the small triangular element, E, and the target, 1, is the magnitude of the vector $\vec{r}$ which can be expressed in terms of the positions of the end points as follows:

$$\vec{r} = (r_x, r_y, r_z) = \left[(P_{1x} - P_{2x}), (P_{1y} - P_{2y}), (P_{1z} - P_{2z})\right] \tag{7}$$

where $(P_{2x}, P_{2y}, P_{2z})$ represents the centre-of-area of the small triangular element and is given by:

$$(P_{2x}, P_{2y}, P_{2z}) = \left[\frac{(K_x + L_x + M_x)}{3}, \frac{(K_y + L_y + M_y)}{3}, \frac{(K_z + L_z + M_z)}{3}\right] \tag{8}$$

and $(P_{1x}, P_{1y}, P_{1z})$ specifies the position of the target.

Therefore r is given by:

$$r = |\vec{r}| = \left[r_x^2 + r_y^2 + r_z^2\right]^{\frac{1}{2}} \tag{9}$$

To obtain the values of $\cos\theta_1$ and $\cos\theta_2$ it is necessary to define unit vectors, $\hat{r}_{12}$ and $\hat{r}_{21}$, that lie along the connecting line and are directed respectively from the triangular element to the target and vice versa. It is also necessary to define the unit normal, $\hat{n}_1$, to the target and to calculate the outward unit normal, $\hat{n}_2$ to the triangular element.

$$\hat{r}_{12} = \left[\frac{(P_{1x} - P_{2x})}{r}, \frac{(P_{1y} - P_{2y})}{r}, \frac{(P_{1z} - P_{2z})}{r}\right] \tag{10}$$

$$\hat{r}_{21} = \left[\frac{(P_{2x} - P_{1x})}{r}, \frac{(P_{2y} - P_{1y})}{r}, \frac{(P_{2z} - P_{1z})}{r}\right] \tag{11}$$

$\hat{n}_2$ is obtained in a similar manner to $A_E$ from the vector product of the two vectors $\vec{V}$ and $\vec{W}$ as follows:

$$\hat{n}_2 = \frac{\vec{V} \times \vec{W}}{|\vec{V} \times \vec{W}|} = \left[\frac{(V_y W_z - V_z W_y)}{2A_E}, \frac{(V_z W_x - V_x W_z)}{2A_E}, \frac{(V_x W_y - V_y W_x)}{2A_E}\right] \tag{12}$$

$\hat{n}_1$ specifies the orientation of the target and can be written as:

$$\hat{n}_1 = (n_{1x}, n_{1y}, n_{1z}) \tag{13}$$

$\cos\theta_1$ and $\cos\theta_2$ are determined from the scaler product of the unit vectors $\hat{r}_{21}$ and $\hat{n}_1$, and $\hat{r}_{12}$ and $\hat{n}_2$ respectively.

$$\cos\theta_1 = \hat{n}_1 \cdot \hat{r}_{21} = \left[\frac{n_{1x}(P_{2x} - P_{1x})}{r} + \frac{n_{1y}(P_{2y} - P_{1y})}{r} + \frac{n_{1z}(P_{2z} - P_{1z})}{r}\right] \tag{14}$$

200

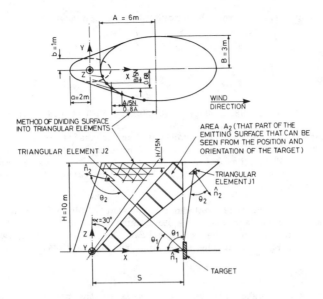

FIGURE 2. An oblique conical frustum of elliptical cross section showing the method used to define the area $A_2$.

$$\cos \theta_2 = \hat{n}_2 \cdot \hat{r}_{12} = \left[ \frac{(V_y W_z - V_z W_y)(P_{1x} - P_{2x})}{2A_E r} + \frac{(V_z W_x - V_x W_z)(P_{1y} - P_{2y})}{2A_E r} \right.$$
$$\left. + \frac{(V_x W_y - V_y W_x)(P_{1z} - P_{2z})}{2A_E r} \right] \tag{15}$$

The geometrical quantities that appear in the configuration factor relationship for a small triangular element, Eq. (3), are related to the positions in space of the nodal points of the element and the position in space and orientation of the target by Eqs. (4) to (15).

Solution of Eq. (2) is achieved by summing the contribution to the total configuration factor of each of the small triangular elemements that is both contained within the field-of-view of the target and contains the target within its own field-of-view. These elements can be identified by considering the values of $\cos \theta_1$ and $\cos \theta_2$ depicted on the oblique conical frustrum of elliptical cross section shown in Fig. 2.

Triangular element J1 is located outside the field-of-view of the target. This is defined by the condition that $\theta_1 > 90°$ or $\cos \theta_1 < 0$. Triangular element J2 is aligned such that the target cannot be seen from the position and orientation of the element. This is defined by the condition that $\theta_2 > 90°$ or $\cos \theta_2 < 0$. In both of these cases radiant heat transfer cannot take place between that particular section of the emitting surface and the target in its current position and orientation.

201

The solution of Eq. (2) can, therefore, be written as:

$$F_{21} = \sum_{\substack{\text{all elements} \\ \text{where } \cos \theta_1 > 0 \\ \text{and } \cos \theta_2 > 0}} F_{E1} \tag{16}$$

## ORIENTATION OF TARGET THAT SUBTENDS THE MAXIMUM CONFIGURATION FACTOR

Of particular interest when studying radiant heat transfer is the orientation of a target at a selected location that is subjected to the maximum incident thermal radiation. When the emission from the surface is assumed to be uniform, this is identical to the orientation of the target that subtends the maximum configuration factor.

Provided the plane surface containing the infinitesimal target does not intersect the emitting surface then the configuration factor for a particular location and orientation of the target can be related to the maximum configuration factor, $F_{21MAX}$, at the same location as follows:

$$F_{21} = F_{21MAX} \cos \phi \tag{17}$$

where $\phi$ represents the angle between the normals of the particular target and the target that subtends the maximum configuration factor.

If the plane surface containing the infinitesimal target does intersect the emitting surface then Eq. (17) does not hold, because $F_{21}$ is evaluated using only that part of the emitting surface contained within the field-of-view of the target. However, if the first condition imposed during the summation in Eq. (16), $\cos \theta_1 > 0$, is relaxed, then a quantity is calculated, $F'_{21}$, whatever orientation of the target is employed, such that:

$$F'_{21} = \sum_{\substack{\text{all elements} \\ \text{where } \cos \theta_2 > 0}} F_{E1} \tag{18}$$

and

$$F'_{21} = F_{21MAX} \cos \phi \tag{19}$$

$F'_{21}$, unlike $F_{21}$, can be considered as a vector component of $F_{21MAX}$ for all orientations of the target including those in which the plane containing the target intersects the emitting surface.

To determine the maximum configuration factor $F_{21MAX}$ and the orientation of the target that subtends the maximum configuration factor, $\hat{n}_{1MAX}$, for a particular location it is necessary to calculate $F'_{21}$ for three mutually perpendicular orientations of the target. The most convenient orientations to select are those such that the normals to the target lie parallel to the X, Y and Z axes of the Cartesian co-ordinate system employed. Hence $F'_{21x}$, $F'_{21y}$ and $F'_{21z}$ are obtained which can be considered as the X, Y and Z components of $F_{21MAX}$ the magnitude of which is given by:

$$F_{21MAX} = \left[ F_{21x}'^{\,2} + F_{21y}'^{\,2} + F_{21z}'^{\,2} \right]^{\frac{1}{2}} \qquad (20)$$

and

$$\hat{n}_{1MAX} = \left[ \frac{F_{21x}'}{F_{21MAX}}, \quad \frac{F_{21y}'}{F_{21MAX}}, \quad \frac{F_{21z}'}{F_{21MAX}} \right] \qquad (21)$$

APPLICATION TO FLAME SHAPES RELEVANT TO LARGE SCALE FIRES

The method has been employed to calculate the configuration factor for a solid geometry that can be used to represent a wide range of flame shapes relevant to large scale fires. This geometry is shown in Fig. 2 and in its most general form is an oblique conical frustrum of elliptical cross section. The equation describing the curved surface can be written as:

$$\frac{(x - z \tan\alpha )^2}{(z \tan\beta + a)^2} + \frac{y^2}{(z\, b \tan\beta /a + b)^2} = 1 \qquad (22)$$

where    a      represents half the dimension of the base of the flame in the X direction

         b      represents half the dimension of the base of the flame in the Y direction

         x, y and z      represent the X, Y and Z co-ordinates of a point on the curved surface

         $\alpha$      represents the angle of tilt of the axis from the vertical

         $\beta$      represents half the cone angle measured in the plane of the X and Z axes and is given by:

$$\beta = \tan^{-1} \frac{A - a}{H} \qquad (23)$$

where    A      represents half the dimension of the tip of the flame in the X direction

         H      represents the vertical height of the flame.

By selecting appropriate values for the dimensions contained within Eqs. (22) and (23) various shapes can be generated ranging from a right circular cylinder to an oblique conical frustrum of elliptical cross section as shown in Fig. 3.

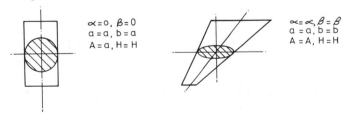

$\alpha = 0, \beta = 0$
$a = a, b = a$
$A = a, H = H$

$\alpha = \alpha, \beta = \beta$
$a = a, b = b$
$A = A, H = H$

a) Right circular cylinder        b) Oblique conical frustum of elliptical cross section

FIGURE 3. Wide range of flame shapes relevant to large scale fires.

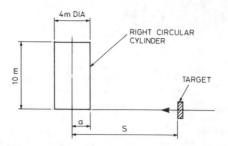

FIGURE 4. Geometrical arrangement used to examine the number of surface elements required to achieve a desired accuracy.


Tilted shapes can also be considered by appropriate positioning and orientation of the receiving target. More complex flame geometries can be formed using combinations from the range of basic shapes shown in Fig. 3.

Also shown in Fig. 2 is the technique employed to divide the surface into small triangular elements. The number of elements used within the numerical method depends upon the value selected for the integer N.


ACCURACY

The accuracy of the numerical technique was assessed by comparison with exact analytical solutions for certain simple geometries such as right and oblique circular cylinders[8] and circular and elliptical plane surfaces[5]. The accuracy of the computations is dependent upon the number of elements into which the surface is divided. Furthermore the number of elements required to achieve a desired accuracy increases as the distance of the target from the emitting surface decreases. This was examined by comparing the results from the numerical technique with the exact analytical solution for a right circular cylinder and a vertical target at ground level and aimed at the centre of the base of the cylinder as shown in Fig. 4.

The number of elements required for a range of different distances, S, from the centre line of the cylinder to the target to achieve an accuracy of ± 1% in the evaluation of the configuration factor is shown in Fig. 5.

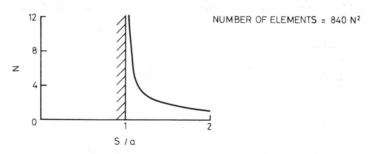

FIGURE 5. Variation of the number of surface elements required to achieve an accuracy of ± 1% in the evaluation of the configuration factor.

204

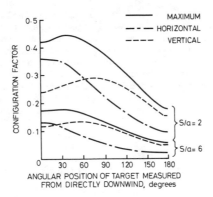

FIGURE 6. Variation of the configuration factor around the oblique conical frustum of elliptical cross section in Fig. 2.


The results shown in Fig. 5 confirm the validity of the computational method. They also show, however, that as the target approaches very close to the emitting surface the number of elements and consequently the computational time required to achieve a certain accuracy rapidly increases.

RESULTS FROM A TYPICAL APPLICATION OF THE METHOD

The configuration factor has been calculated for a flame shape defined by the oblique conical frustrum of elliptical cross section shown in Fig. 3. Targets were located at ground level at two radial distances from the centre of the elliptical base and at a range of angular positions. Three target orientations were considered. These were vertical such that the normal to the target was horizontal and aimed at the centre of the elliptical base, horizontal such that the normal to the target was vertical and aimed in an upward direction, and finally such that the target was orientated to subtend the maximum configuration factor.

The variation of the configuration factor as the target position changes from directly downwind through 180° to directly upwind is shown in Fig. 6.


DISCUSSION

The numerical technique developed is a general method and is capable of being applied to any solid geometrical shape. This need not necessarily be a regular shape but can comprise of combinations of regular shapes or even completely irregular shapes provided it can be accurately defined. In the case of geometries incorporating concave surfaces additional special conditions will be required to ensure that any particular element is not obscured from the target by other parts of the emitting surface.

The technique has been presented as a method for calculating the configuration factor between an emitting surface and an infinitesimal target. The thermal radiation incident upon a target is then given by Eq. (1) provided the surface emissive power can be considered to be constant and the transmisivity of the atmosphere can be adequately determined from a mean path length for the thermal radiation. However, the method is not restricted by these conditions. Large hydrocarbon fires involving fuels such as LPG and kerosene[3] are very sooty. This obscures large areas of the flame surface which effectively reduces the surface emissive power in those regions and the nett result can be a substantial reduction in the incident thermal radiation on objects located outside the flame. Such spatial variations in the flame surface emissive power can easily be incorporated by assigning a particular value to each element. The atmospheric transmisivity for individual elements can also be determined using the distance from the target to the centre-of-area of each element as the path length for the thermal radiation.

ACKNOWLEDGEMENT

The Author wishes to thank British Gas Corporation for permission to publish this paper.

REFERENCES

1. American Gas Association: "LNG Safety Program Phase II – Consequences of LNG Spills on Land," Project 1S-3-1, 1973.

2. Raj, P.K., Moussa, A.N. and Aravamudan, K.: "Experiments Involving Pool and Vapour Fires from Spills of Liquified Natural Gas on Water," Report. No. CG-D-55-79, U.S. Department of Transportation, 1979.

3. Mizner, G.A. and Eyre, J.A.: "Large-Scale LNG and LPG Pool Fires," I. Chem. E. Symposium Series No. 71, 1982.

4. Moorhouse, J.: "Scaling Criteria for Pool Fires derived from Large Scale Experiments," I. Chem. E. Symposium Series No. 71, 1982.

5. McGuire, J.H.: Heat Transfer by Radiation, Department of Scientific and Industrial Research and Fire Offices' Committee, Fire Research Special Report No. 2, London: Her Majesty's Stationary Office, 1953.

6. Sparrow, E.M. and Cess, R.D.: Radiation Heat Transfer, Brooks/Cole Publishing Company, 1966.

7. Rein, R.G, Jr., Sliepsevich, C.M. and Welker, J.R.: "Radiation View Factors for Tilted Cylinders," J. Fire and Flammability, Vol. 1, 1970.

8. Raj, P.K. and Kalelkar, A.S.: "Assessment Models in support of the Hazard Assessment Handbook (CG-446-3)," Technical Report Prepared for the U.S. Coast Guard, NTIS Publication # AD776617, 1974.

# Prediction of the Heat Release Rate of Wood

**WILLIAM J. PARKER**
Center for Fire Research
National Bureau of Standards, USA

ABSTRACT

A method for predicting the heat release rate of wood for different thicknesses, moisture contents, and exposure conditions is described. A model has been set up and calculations have been made on a microcomputer. Heat release rates and effective heats of combustion were measured as a function of time and external radiant flux on 12.5 mm thick dry vertical specimens of Douglas fir particle board. The calculated and measured curves are similar in shape and amplitude but differ significantly in time scale. The initial results with the model are promising.

INTRODUCTION

The model described in this paper embodies most of the features of the mass loss rate model originally developed by Kung [1] and modified by Taminini [2] and Atreya [3]. The chief distinguishing features of the present model over that of Atreya are: (1) it takes char shrinkage parallel and normal to the surface into account, (2) the thermal properties are a function of temperature and degree of char, (3) it allows for several components of the wood each with different decomposition rate constants, and (4) it takes the change in the heat of combustion of the volatiles for each of the components as a function of its individual degree of char into account.

This paper will describe the model and show some calculations and comparisons with the heat release rate and effective heat of combustion measured in the cone calorimeter [4]. For a more detailed discussion of the model and the experimental procedures see reference [5].

HEAT RELEASE RATE MODEL

The heat release rate model (1) breaks the specimen down into thin slices parallel to the surface as seen in figure 1, (2) calculates the mass loss rate for each slice, (3) multiplies this rate by the local heat of combustion of the volatiles generated, and (4) sums these contributions over the depth of the specimen to obtain the total heat release rate assuming complete combustion of the volatiles leaving the front surface.

The boundaries of each slice move as the specimen shrinks during the burning period so that no solid material crosses a boundary. The rear surface is impervious to mass flow. Flow into the cooler region of the specimen followed by condensation and subsequent evaporation is not treated here. All of the

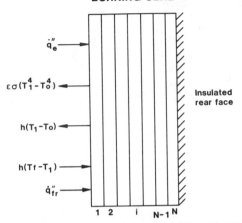

**BURNING SLAB**

$\dot{q}_e''$

$\varepsilon\sigma(T_1^4 - T_0^4)$

Insulated
rear face

$h(T_1 - T_0)$

$h(T_f - T_1)$

$\dot{q}_{fr}''$

1  2     i   N-1 N

- N slices of thickness $\Delta X_0$ of unit width and height
- Boundaries move as char shrinks

FIGURE 1.  Boundary conditions and subdivision of the burning vertical slab.

volatiles pass through the front surface.  The transit time of the volatiles
from their generation site to the surface is neglected.  However, the internal
convective heat transfer coefficient is assumed to be so large that the volatiles are maintained in thermal equilibrium with the char through which they
pass.  Secondary chemical reactions with the char layer are neglected.

The rate of mass generation is expressed as a function of the temperature and
the mass retention fraction for each component in each slice using an Arrhenius
type expression.  The temperature profile as a function of time is calculated
from the energy equation using the finite difference method.  The rate of change
of enthalpy, heat conduction, internal heat generation or absorption and convective cooling by the flow of volatiles are taken into account.  However, the
convective cooling by the volatiles is cut off in the fully developed char layer
since it is assumed that the gases will flow internally to the fissures and exit
the specimen from there.  The thermal properties are assumed to be a function of
the temperature and the total mass retention fraction of each slice.  The
instantaneous mass retention fraction profile is determined by accounting for
the cumulative loss of volatiles.

An adiabatic boundary condition is assumed at the rear surface while the front
face is exposed to a constant external radiant flux.  Both of these conditions
can be made more general later if desired.  The front surface also loses heat by
reradiation.  Up to the time of ignition it also loses heat by laminar free
convection.  After ignition the convective heat transfer from the flame to the
surface is taken into account along with flame radiation.  The convective heat
transfer coefficient both before and after ignition is multiplied by the blowing
factor due to the flow of volatiles through the surface.

Ignition is assumed to occur when the calculated total heat release rate reaches
30 kW/m$^2$ which is near the minimum heat release rate required to maintain a

flame on the surface. The flame is extinguished when the calculated heat release rate again drops below 30 kW/m² at the end of the flaming phase. Only the heat release rate during the flaming phase is treated by this model. During this period oxygen is assumed to be excluded from the solid. The oxidation of the surface prior to ignition and after flaming ceases is not considered.

The mass retention fraction $Z_i$ for each slice is defined by

$$Z_i = m_i/m_i^o = \frac{\rho}{\rho_o} \ell_x \ell_y \ell_z \tag{1}$$

where $m_i$ is the mass of each slice on an oven dry basis, $\rho$ is its density, and $\ell_x$, $\ell_y$, and $\ell_z$ are the contraction factors (i.e., the ratios of the thickness, width and height of the slice to their original values). The superscript and subscript o refer to the initial values.

The contributions of the individual wood components (cellulose, hemicellulose and lignin) are additive so that

$$Z_i = \sum_{k=1}^{n} Z_{i,k} \tag{2}$$

where $Z_{i,k}$ is the mass of the k th component of the i th slice divided by the original mass of the whole slice. It is necessary to know the reaction rate and heat of combustion of the volatiles for each component as a function of the temperature, $T_i$, and the component mass retention fraction, $Z_{i,k}/Z_k^o$. Here $Z_k^o$ is the original mass fraction of the k th component.

The moisture is taken into account by assigning a moisture retention fraction,

$$Z_{w_i} = m_{w_i}/m_i^o \tag{3}$$

where $m_{w_i}$ is the mass of adsorbed water located in slice i at any time.

The mass loss rate and heat release rate per unit area of the original specimen are calculated by the following formulas:

$$\dot{m}'' = \rho_o \frac{\Delta x_o}{\Delta t} \sum_{i=1}^{N} (Z_i^t - Z_i^{t + \Delta t}) \tag{4}$$

and

$$\dot{q}_{rel}'' = \rho_o \frac{\Delta x_o}{\Delta t} \sum_{i=1}^{N} F_i^t (Z_i^t - Z_i^{t + \Delta t}) \tag{5}$$

where the heat of combustion of the volatiles is given by

$$F_i^t = \sum_{k=1}^{n} F_{i,k}^t Z_{i,k}^t/Z_i^t . \tag{6}$$

The energy equation is first solved for the temperature. Then the rate equations are used to update the mass retention fraction for each component. The total mass retention fraction, $Z_i$, for each slice is substituted back into the energy equation which is then solved for the temperature at the next time step.

The increase in enthalpy of an interior slice is equal to the heat conducted in minus the heat conducted out minus the heat absorbed by the pyrolysis gases passing through on their way to the front surface minus or plus the heat absorbed or generated by pyrolysis and evaporation. The energy equation for an interior slice can be written,

$$
\rho_o \Delta x_o \left[ (z_i^{t+\Delta t} C_i + z_{w_i}^{t+\Delta t} C_{w_i}) \ T_i^{t+\Delta t} - (z_i^t C_i + z_{w_i}^t C_{w_i}) \ T_i^t \right]
$$

$$
= K_i \left( \frac{T_{i-1}^{t+\Delta t} - T_i^{t+\Delta t}}{\Delta X} - \frac{T_i^{t+\Delta t} - T_{i+1}^{t+\Delta t}}{\Delta X} \right) \Delta t \ \ell_y \ell_z
$$

$$
+ (C_{g_i} \dot{m}_{i+1}'' + C_{w_i} \dot{m}_{w_{i+1}}'' ) \left( \frac{T_{i+1}^{t+\Delta t} + T_i^{t+\Delta t}}{2} \right) \Delta t
$$

$$
- (C_{g_i} \dot{m}_i'' + C_{w_i} \dot{m}_{w_i}'') \ \frac{(T_i^{t+\Delta t} + T_{i-1}^{t+\Delta t}}{2}) \Delta t
$$

$$
- h_{p_i} \rho_o \Delta x_o \ (z_i^t - z_i^{t+\Delta t}) - L_v \rho_o \Delta x_o \ (z_{w_i}^t - z_{w_i}^{t+\Delta t}) \tag{7}
$$

$$
\text{where} \quad \dot{m}_i'' = \frac{\rho_o \Delta x_o}{\Delta t} \sum_{j=i}^{N} (z_j^t - z_j^{t+\Delta t}), \tag{8}
$$

$$
\dot{m}_{w_i}'' = \frac{\rho_o \Delta x_o}{\Delta t} \sum_{j=i}^{N} (z_{w_j}^t - z_{w_j}^{t+\Delta t}) \tag{9}
$$

$$
\text{and} \quad h_{p_i} = \sum_{k=1}^{n} \frac{z_{i,k}^t}{z_i^t} h_{p_k} . \tag{10}
$$

Here $\dot{m}''$ refers to the mass flow per unit original area of the specimen not the contracted area. The heat of pyrolysis of the slice, $h_{p_i}$, is equal to the weighted averages of the heat of pyrolysis of each component.

Radiative and convective heat transfer must be taken into account at the front surface (i=1). Radiation exchange with the external environment is assumed to take place at the surface. The exchange of radiation between interior slices is considered to be one component of the thermal conductivity. Prior to ignition the surface is cooled by laminar free convection. After ignition it is heated by the flame. The front surface temperature, $T_s$, for these calculations is obtained by linear extrapolation from the center points of the two slices nearest the surface

$$
T_s = \frac{3}{2} \ T_1 - \frac{1}{2} \ T_2 . \tag{11}
$$

The reradiation from the front surface is given by

$$\alpha\sigma \left(T_s^4 - T_o^4\right) = \alpha\sigma \left(T_s^2 + T_o^2\right) \left(T_s + T_o\right) \left(T_s - T_o\right). \tag{12}$$

The energy flow across the front surface is given by

$$\dot{q}''_{net} = \dot{q}''_e + \dot{q}''_{fR} - \omega \left(T_s - T_o\right) + h_f \left(T_f - T_s\right) B/(\exp(B) - 1) \tag{13}$$

where

$$\omega = \alpha\sigma \left(T_s^2 + T_o^2\right) \left(T_s + T_o\right) + h\,B/(\exp(B)-1) \ . \tag{14}$$

The other quantities are identified as follows: $h$ is the convective heat transfer coefficient from the surface to the ambient air which drops to zero after ignition, $h_f$ is the convective heat transfer coefficient from the flame to the surface which is equal to zero prior to ignition, $T_f$ is the flame temperature, and

$$B = \dot{m}''C_g/h \tag{15}$$

and thus accounts for blowing. Prior to ignition the radiant heat flux from the flame, $\dot{q}''_{fr}$, is zero. The energy equation for the first slice is given by

$$\rho_o \Delta x_o \left[\left(Z_1^{t+\Delta t}C_1 + Z_{w_1}^{t+\Delta t}C_{w_1}\right) T_1^{t+\Delta t} - \left(Z_1^t C_1 + Z_{w_1}^t C_{w_1}\right) T_1^t\right] =$$

$$\left[\dot{q}''_e + \dot{q}''_{fR} - \omega\left(\frac{3}{2} T_1^{t+\Delta t} - \frac{1}{2} T_2^{t+\Delta t} - T_o\right) + h_f\left(T_f - \frac{3}{2} T_1^{t+\Delta t} + \frac{1}{2} T_2^{t+\Delta t}\right)B/(\exp(B)-1)\right.$$

$$\left. -K_1\, \ell_z \ell_y\, \frac{T_1^{t+\Delta t} - T_2^{t+\Delta t}}{\Delta x}\right]\Delta t$$

$$+ \left(C_{g_1}\dot{m}''_2 + C_{w_1}\dot{m}''_{w_2}\right)\left[\frac{T_2^{t+\Delta t} + T_1^{t+\Delta t}}{2}\right]\Delta t$$

$$- \left(C_{g_1}\dot{m}''_1 + C_{w_1}\dot{m}''_{w_1}\right)\left[\frac{3}{2} T_1^{t+\Delta t} - \frac{1}{2} T_2^{t+\Delta t}\right]\Delta t$$

$$- h_{p_1}\rho_o\,\Delta x_o\,\left(Z_1^t - Z^{t+\Delta t}\right) - L_v\,\rho_o\,\Delta x_o\,\left(Z_{w_1}^t - Z_{w_1}^{t+\Delta t}\right) \tag{16}$$

The change in mass retention fraction for each component of slice i during the time step is given by

$$Z_{i,k}^{t+\Delta t} - Z_{i,k}^t = - \left(Z_{i,k}^t - Z_{f,k}\right) A_k \exp\left(- E_k/RT_i^{t+\Delta t}\right) \Delta t \tag{17}$$

where $Z_{f,k}$ is the residual mass fraction of component k when pyrolysis is completed in an inert atmosphere and $A_k$ and $E_k$ are the frequency factor and activation energy for component k. The total mass retention fraction for slice i is then given by

$$z_i^{t+\Delta t} = z_i^t + \sum_{k=1}^{n} (z_{i,k}^{t+\Delta t} - z_{i,k}^t). \qquad (18)$$

The change in the mass retention fraction for water during the time step is given by

$$z_{w_i}^{t+\Delta t} - z_{w_i}^t = - z_{w_i}^t A_w \exp\left(- E_w/RT_i^{t+\Delta t}\right)\Delta t \qquad (19)$$

These numerical equations were solved on a microcomputer. The parameters — $K$, $\rho$, $C$, $C_w$, $C_g$, $\alpha$, $A_k$, $A_w$, $E_k$, $E_w$, $F_k$, $h_{pk}$, $L_v$, $\ell_x$, $\ell_y$, and $\ell_z$ — are needed as input to the computer model. While some of these data can be obtained within the accuracy needed by consulting the literature the thermophysical properties ($K$, $\rho$, $\alpha$, $\ell_x$, $\ell_y$ and $\ell_z$) and the thermochemical properties ($A_k$, $E_k$, and $F_k$) were determined experimentally for the material of concern. Most of these parameters vary with temperature and mass retention fraction. For a discussion of the experimental methods see reference 5.

VERIFICATION WITH THE HEAT RELEASE RATE CALORIMETER

The input parameters along with their assigned values and sources for Douglas fir particle board are listed in Table 1. The front and rear surface temperatures, the mass loss rate, the heat release rate and the effective heat of combustion are all calculated as a function of time. The calculated and experimental curves for the heat release rate and the effective heat of combustion of a dry 12.7 mm specimen of Douglas fir particle board exposed at an external radiant flux of 50 kW/m$^2$ are compared in figures 2 and 3. The experimental curves were obtained with the cone calorimeter in the vertical orientation.

The calculated heat release rate curve jumps from zero to 30 kW/m$^2$ at ignition and continues to rise quickly to a sharp peak followed by a rapid decline to a minimum. A second large peak occurs when the thermal wave is reflected from the insulated rear face causing a more rapid rise in the average temperature of the slab. The second peak would be missing altogether if the rear face were maintained at some low temperature. The measured value of the first peak is about 25% lower than the calculated one. This ratio was also obtained for external radiant fluxes of 25, 75, and 100 kW/m$^2$. The second peaks are lower than the first peaks for the measured curves. This is partly due to the heat absorbed by the insulated backing board in the calorimeter. When the calculations were repeated using a heat of pyrolysis of 400 kJ/kg instead of zero the second peak was lower than the first peak and the first peak was closer to the measured value. The agreement is better than should be expected at this stage of the development of the model which needs better input data on the thermal properties at high temperature.

The calculated effective heat of combustion was approximately 12 MJ/kg which is the average value of the heat of combustion of the vapor over the full range of decomposition of the solid. It actually fell slowly with time and then rose rapidly as the final stage of char formation was approached. Except for the shift in time scale the agreement between calculated and measured values is within 25%.

DISCUSSION

For the limited comparisons made, the predicted and measured effective heats of combustion are very close. It is nearly constant at about 12 MJ/kg and independent of incident flux. The gross heat of combustion of the wood is 20 MJ/kg. The net heat of combustion after correcting for the evaporation of water is 18 MJ/kg. The much lower effective heat of combustion is due to dilution by the

TABLE 1. Input Data for Running the Model for Douglas Fir Particle Board

| Property | Value | Source |
|---|---|---|
| **Wood** | | |
| Activation Energy | 121 kJ/mol | 1 |
| Pre-Exponential Factor | $5.94 \times 10^7$ s$^{-1}$ | 1 |
| Heat of Pyrolysis | 0 | 2 |
| Original Density | 709 kg/m$^3$ | 1 |
| Heat Capacity | 1.11 (1 + 0.0067 T-273) x (0.54 + 0.46Z) kJ/kg.K | 3 |
| Thermal Conductivity | 1.24(0.35 + 0.65Z) x (1 + 6.8(T-To) x $10^{-4}$) x$10^{-4}$ kW/m.K | 1 |
| Absorptivity | 0.8  Z > 0.75;  1.0  Z < 0.75 | 1 |
| Contraction Factors* | $\ell_x = \ell_y = \ell_z = 1$ for $0.65 \leq Z \leq 1$ = $(0.65 - Z)^2$ for Z < 0.65 | 1a |
| Final Char Yield | 0.22 | 1 |
| **Volatile Pyrolysis Products** | | |
| Heat Capacity | $1.05 + 1.80 \times 10^{-4}$(T-273)  kJ/kg.K | 4 |
| Heat of Combustion | 20 [1.24 - Z] MJ/kg | 1 |
| **Water** | | |
| Activation Energy | 44 kJ/mole | 4 |
| Pre-Exponential Factor | $4.5 \times 10^3$ | 4 |
| Heat of Vaporization | 2400 kJ/kg | 4 |
| Heat Capacity | 4.19 kJ/kg.K, T < 373; | 5 |
| | $4.19 + 3.1 \times 10^{-5}$(T-373)$^2$kJ/kg.K, T > 373K | |

1  Experimentally determined in this project.

2  Assumed.

3  Temperature Dependence – Koch [6]; dependence on charring assumed based on published value for charcoal.

4  Atreya [3].

5  Based on data in reference 7.

\*  The contraction factors were found to be independent of direction along the surface. They were assumed to be the same through the depth. Departure from unity for contraction factors with Z > 0.65 were within the scatter of the data.

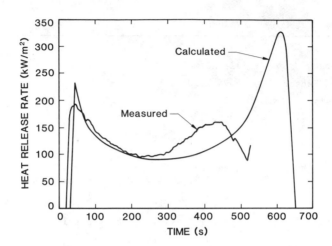

FIGURE 2.  Heat release rate of dry Douglas fir particle board at an external radiant flux of 50 kW/m$^2$.

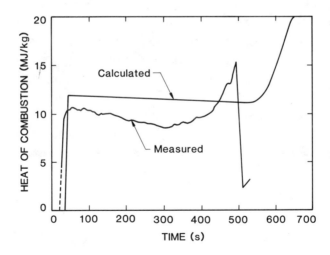

FIGURE 3.  Effective heat of combustion of dry Douglas fir particle board at an external radiant flux of 50 kW/m$^2$.

water released in the char forming process. The initial peak heat release rates are within 25%. The shape of the calculated and measured heat release rate curves are similar but the time scales are substantially different. This may be due to inadequate data on the thermal properties of the char at high temperature, the assumption of a single first order reaction for wood or any of the several assumptions used in the construction of the model.

This computer model is unique in that it (1) accounts for the change in the heat of combustion of the volatiles generated during the pyrolysis period, (2) accounts for char shrinkage, and (3) provides for different reaction rates for the different components of wood. However this last capability was not exercised during this first test of the model where a single first order reaction was assumed. This is the standard assumption used in the mass loss rate models.

As this work continues, improvements will be made in the model and in the experimental methods employed to obtain the input data. Furthermore a much broader data base will be obtained on the thermophysical and thermochemical properties of wood and wood char particularly at high temperature. In order to reliably describe the kinetic constants for the thermal decomposition of wood and the effective heat of combustion of its volatile pyrolysis products as it evolves with time it will be necessary to make these determinations on the individual wood components (cellulose, hemicellulose, and lignin).

ACKNOWLEDGMENT

This work was suported in part by the Federal Emergency Management Agency.

REFERENCES

1. Kung. H.C., "A Mathematical Model of Wood Pyrolysis," Combustion and Flame, Vol. 18, pp 185-195(1972).

2. Tamanini, F., "A Study of the Extinguishment of Wood Fires," Ph.D. Thesis, Harvard University, May 1974.

3. Atreya, A., "Pyrolysis, Ignition, and Fire Spread on Horizontal Surfaces of Wood, Ph.D. Thesis, Harvard University, May 1983.

4. Barauskas, V., "Development of the Cone Calorimeter - A Bench-Scale Heat Release Rate Aparatus Based on Oxygen Consumption," NBSIR 82-2611, Nat. Bur. Stand. (U.S.), Nov. 1982.

5. Parker, W. J., "Development of a Model for the Heat Release Rate of Wood A Status Report," NBSIR (to be published), Nat. Bur. Stand. (U.S.) 1985.

6. Koch, P., "Specific Heat of Oven-Dry Spruce Pine Wood and Bark," Wood Sci. 1:203-214, 1969.

7. Weast, R.C., Editor, "Handbook of Chemistry and Physics", 57th edition, 1976-1977, pp. D158-D159.

NOMENCLATURE

| | |
|---|---|
| A | frequency factor ($s^{-1}$) |
| B | mass transfer number |
| C | heat capacity (kJ/kg·K) |

| | |
|---|---|
| E | activation energy (kJ/mole) |
| F | heat combustion of volatiles (kJ/kg) (positive) |
| h | convective heat transfer coefficient (kW/m$^2$.K) |
| $h_p$ | heat of pyrolysis (kJ/kg) (positive when heat is absorbed) |
| K | thermal conductivity (kW/m.s) |
| $\ell$ | contraction factor |
| $L_v$ | latent heat of vaporization of water (kJ/kg) (positive) |
| m | mass (kg) |
| n | number of components |
| N | number of subdivisions of the specimen |
| $\dot{q}_e''$ | external radiant flux (kW/m$^2$) |
| $\dot{q}_{fR}''$ | flame radiation to the surface (kW/m$^2$) |
| $\dot{q}_{rel}''$ | rate of heat release (kW/m$^2$) |
| R | gas constant (8.314 kJ/mol/K) |
| t | time(s) |
| T | temperature (K) |
| Z | mass retention fraction |
| $\alpha$ | absorptivity and emissivity |
| $\Delta t$ | time step (s) |
| $\Delta x$ | thickness of each slice (m) |
| $\omega$ | defined by equation (14) |
| $\rho$ | density (kg/m$^3$) |
| $\sigma$ | Stefan Boltzmann constant $5.67 \times 10^{-11}$ (kW/m$^2$.K$^4$) |

# Some Critical Discussions on Flash and Fire Points of Liquid Fuels

HIROKI ISHIDA and AKIRA IWAMA
Institute of Space and Astronautical Science
4-6-1, Komaba, Meguro-ku, Tokyo, 153, Japan

Keywords: Pre-flash point, Flash point, Fire point, Vapor pressure, Lower
flammability limit, Raoult's law, Clausius-Clapeyron's law

ABSTRACT

The liquid fuel temperatures of the flash and fire points were discussed for their accurate measurement according to their definitions from the view point of spilled fuel fire hazard prevention. The flash and fire points for some hydrocarbon and alcohol fuels were measured at the open cup system with an electric spark ignition source, avoiding the external air blow. Measured results were compared with those by the usual methods of Tag Closed Cup and Pensky-Martens, and some influencing factors in the measurement were discussed. In the present study, pre-flash phenomenon, namely "Pre-flash point", was found at a little lower liquid temperature than the flash point. It suggests that the specification of the scale of flame appearance at flash phenomenon above the fuel pool surface should be introduced in the usual methods of flash point measurement. Presented consideration on flash point covers the theoretical prediction of flash point of the binary hydrocarbon fuel mixture.

## RECENT PHASES OF THE STUDIES

Many studies on the fire hazard prevention for liquid fuel have concentrated their serious interests to the flash point of the fuel for more than forty years. The flash point is defined as the fuel temperature at which the equilibrium fuel vapor concentration of its lower flammability limit is attained. It has been measured and investigated concerning such parameters as the latent heat of vaporization, the boiling temperature and the fuel vapor concentration of lower flammability limit for the purpose of theoretical prediction of the accurate fuel temperature when the flash phenomenon occurs (1-8). Many previous reports, however, have presented divergent measured values one another, which are due to complex parametric factors in their measuring apparatuses and methods for flash point such as the position and dimension of ignition source, mixing or no-mixing in vapor phase above pool, and fuel container condition (open or closed)(1-14). Then, not all investigations in this field are, of course, necessarily complete, as discussed in detail by Glassman et al.(16). It seems that some unsolved problems remain still in the usual methods of flash point measurement for the purpose of obtaining the accurate flash point (9,13-17). Although the fire point is defined as the lowest fuel temperature at which the diffusion flame is sustained above the fuel pool without any external heat supply (7,19,20), generally reliable data have not been published. We can see the considerable

The present address for Hiroki Ishida is Dept. of Mechanical Eng., Nagaoka Technical College. 888, Nishikatakai, Nagaoka, Niigata-pref., 940, Japan.

divergence among the fire point data in some previous papers, and the divergence seems due to more complex parametric factors in the measuring methods which have significant effects on the measured result than in flash point measurement(15-20).

## FAVORABLE MEASUREMENT OF FLASH AND FIRE POINTS

The purpose of the present study is to obtain the accurate flash and fire points for some hydrocarbon and alcohol fuels under the experimental condition according to their definitions, and to examine the measured results comparing with those in previous studies for clarifying the influencing factors in the measurement. Although different apparatuses and methods give different measured values, it is necessarily important to know the accurate fuel temperatures at flash and fire phenomena according to their definitions for the detailed discussions in the problems of the flame spread mechanism along liquid fuel surface (15-18) and of the ignition behavior of fuel droplet in the hot atmosphere (21-24) or on the hot solid surface (25,26). The accurate measurements of these temperatures are impotant also for the estimation and classification of the flammability hazard of liquid fuels. Figures 1 and 2 show the relations between the equilibrium vapor pressure and liquid temperature for alcohol and hydrocarbon fuels respectively, where $T_{fl}$ indicates the theoretical flash point. The essentials in all the methods for flash point measurement exist in the accurate measurement of $T_{fl}$.

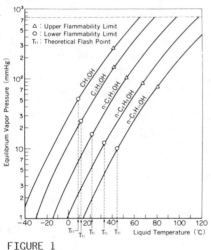 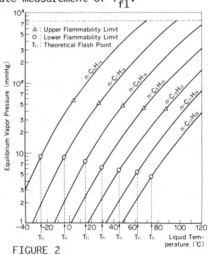

FIGURE 1          FIGURE 2

FIG.1 and 2: The Relation between Equilibrium Vapor Pressure and Liquid Fuel Temperature (from literature).

As predicted reasonably, there exists the boundary layer of fuel vapor concentration above liquid pool. Figure 3 shows the interferogram with He-Ne laser source (wave length; 6328 Å) and the concentration profile in the boundary layer of n-octane vapor above the liquid surface. Figure 3 shows that the equilibrium fuel vapor concentration adjacent to the pool surface agrees with that predicted from Fig.2. The result shown in Fig.3 suggests also that the disturbance of fuel vapor diffusion has a significant effect on the production of flammable concentration zone above the pool surface and on the measured results of flash point. We can expect that measured flash point approaches to the theoretical value $T_{fl}$ even by the open cup method, if the ignition source is placed closely to the pool surface and disturbing air blow is prevented. Also the flash point is to be measured as higher value with increase of the distance between the ignition source and the pool surface (6,16).

218

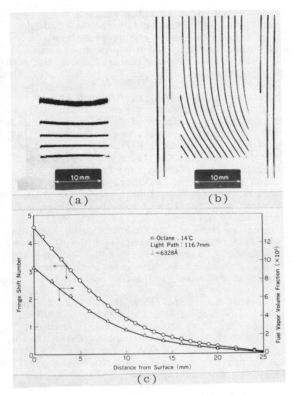

FIG.3: Interferograms (a) and (b) of Concentration Boundary Layer, and Calculated Concentration Profile (c) above n-octane pool at 14 °C.

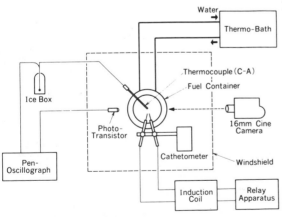

FIG.4: Schematic Drawing of The Measurement apparatus of Flash and Fire Points at Open Cup System with Electric Spark Ignition Source.

Figure 4 shows the measurement apparatus in the present study of flash and fire points with open cup and spark ignition source. The fuel was filled up to the lip level of the container cup (diameter 50 mm, depth 20 mm) and the fuel temperature was maintained constant by the temperature controlled cup. Phototransistor was used for the detection of flame when the flash phenomenon occurred , and thermocouple (C-A, 0.1 mm dia.) was taken for monitoring the fuel surface temperature. Spark electrode position above pool surface is accurately set at a height by a cathetometer, spark gap is 3 mm, and spark duration was set at 0.3 sec by relay circuit apparatus. Then, spark energy, approximately 100-300 mJ generated by the induction coil, is sufficient as the ignition source. For avoiding the disturbance of fuel vapor diffusion by external air blow, most of measurement equipments were enclosed in the windshield with PMMA plates. 16 mm movie was taken for observing the aspect of flash phenomenon in detail.

PRE-FLASH, FLASH AND FIRE PHENOMENA

In the present study, typical hydrocarbons and alcohols of which flash points are not so far from the room temperature were chosen from the view point of accurate measurement in the experiment. The "pre-flash" phenomenon can be observed for some fuels at a little lower liquid temperature than the flash point. It is instantaneous and faint but clear propagating appearence of a flamelet, and is due to local flame propagation through the stratified fuel vapor layer adjacent to the pool surface.

Figure 5 shows the pre-flash phenomenon on the pool surface of n-nonane, where a propagating stretch of flamelet appears in the picture No,2. Pre-flash is clearly different from the flame appearance such as small blue "halo"

219

of the burning fuel vapor which surrounds the pilot ignition flame as often observed in usual flash point measurements (e.g. Tag Closed Cup and Pensky-Martens methods), and such blue halo would not propagate across the pool surface. Figures 6 and 7 show the flash and fire phenomena on the n-nonane pool respectively, which follow after the pre-flash. Flame propagation area above the pool surface in the flash is clearly larger than that in the pre-flash. Although the full interpretation of "pre-flash" phenomenon, especially the reason why the flamelet propagates locally only above some part of pool surface and why it is extinguished on the way, can not be given in the present stage of this study, we can consider some important factors as that the position of ignition source (spark electrode) is the center of fuel pool surface of the tray, and there must exist the considerable escape of generated fuel vapor by outer diffusion near the periphery of tray. Fire phenomenon was determined as the flame sustaining of at least five seconds above the pool after the flash phenomenon. In these figures, a generated spark is imaged by halation shown as the small bright point. Flash phenomenon is in itself the ignition and flame propagation in gas phase where the fuel vapor concentration gradient exists, and the flame propagates through only the thin stratified gas layer close to the pool surface (27-30).

Then, the scale of flame appearance on the pool surface should be specified additionally in the usual methods of flash point measurement. On the other hand, for alcohol fuels (n-butanol and iso-butanol), such pre-flash and flash phenomena could not be observed. Glassman et al. also discussed the experimental result that the flash and fire phenomena occur at the same liquid temperature for alcohol fuels(16).

Table 1 shows the fuel temperatures at which pre-flash, flash and fire phenomena occurred, with the relation to the height of spark electrode above fuel surface. Table 2 shows the measured results of flash point by usual methods, Tag Closed Cup and Pensky-Martens methods, and theoretical flash point determined by the condition of lower flammability limit concentration on the curve of equilibrium vapor pressure. As shown in these tables, except alcohol fuel, the pre-flash point is 1-3 °C lower than the flash point, and the fire point is 3-7 °C higher than the flash point when the spark position is 3 mm above pool. For alcohol fuel, although it was very difficult to distinguish between flash and fire points in the present spark method, the fire point is, however, shown about 10 °C higher than the flash point by Tag Closed Cup method. For o-xylene, a bright flame and soot formation can be observed not only in the fire phenomenon but also even in the pre-flash and flash phenomena, which are usually observed for such fuels as aromatic compounds.

As evident from these tables, flash points by open cup method with spark ignition source are higher than those by Tag Closed Cup method. This is, as often discussed for closed cup method, due to the local heating of pool surface by small pilot ignition flame and the increase of air entrainment by the induced convection. On the other hand, fire points by the open cup method with spark ignition source are lower than those by Tag Closed Cup method, which is probably because the air entrainment is sufficient at open cup system but poor at closed cup system.

The heights of small pilot ignition flame (about 4 mm dia.) above pool surface are specified as 27-29 mm in Tag Closed Cup method (ASTM D-56), 19-21 mm in Pensky-Martens method (ASTM D-93) and 9-11 mm in Cleaveland Open Cup method ( ASTM D-92). They are very higher than the positions of spark ignition source in the present study, but the thermal energy of spark ignition source is very smaller than that of pilot flame. Then, the effect of local heating of pool surface by spark and/or thermocouple can be neglected in the present study.

Pre-flash, falsh and fire points rise 1-3, 2-5 and 2-10 °C respectively with increase of the height of spark ignition source. When the height was lower than 3 mm, it was often impossible to ignite the fuels in the present study. These results suggest lucidly that flash phenomenon occurs only when tne flammable mixture was produced around the ignition source, and the quenching layer above

220

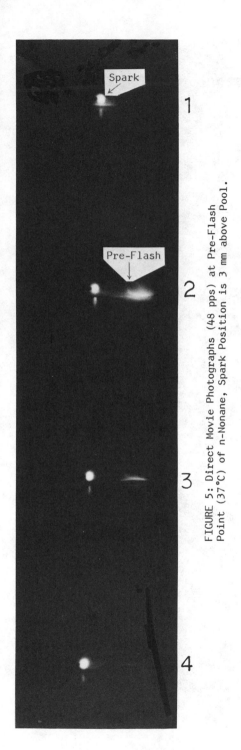

FIGURE 5: Direct Movie Photographs (48 pps) at Pre-Flash Point (37°C) of n-Nonane, Spark Position is 3 mm above Pool.

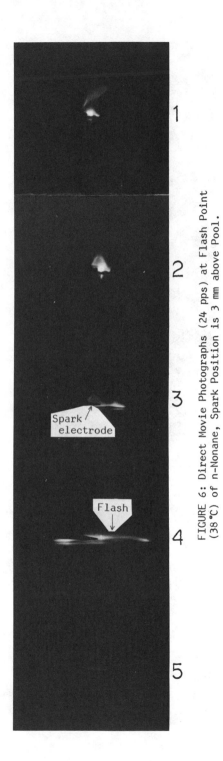

FIGURE 6: Direct Movie Photographs (24 pps) at Flash Point (38°C) of n-Nonane, Spark Position is 3 mm above Pool.

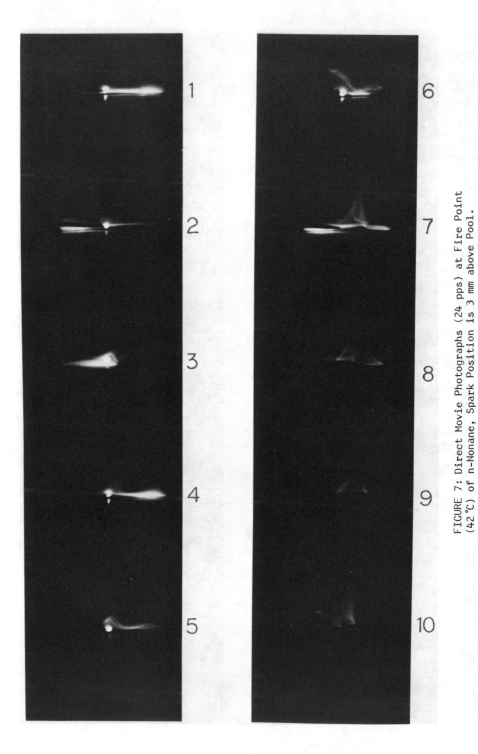

FIGURE 7: Direct Movie Photographs (24 pps) at Fire Point (42 °C) of n-Nonane, Spark Position is 3 mm above Pool.

TABLE 1: Measured Fuel Temperatures at Pre-flash, Flash and Fire Phenomena with The Relation to The Height of Spark Electrode above Pool Surface.

| Fuel | Pre-flash Point (°C) | | | | Flash Point (°C) | | | | | Fire Point (°C) | | | |
|---|---|---|---|---|---|---|---|---|---|---|---|---|---|
| | A | B | C | D | A | B | C | D | E | A | B | C | D |
| n-Octane | 16.6 | 14.8 | 17.8 | — | 17.4 | 17.0 | 20.1 | 20.4 | 12 | 20.8 | 17.6 | 23.0 | 22.8 |
| n-Nonane | 36.3 | 34.0 | 35.9 | 38.3 | 38.0 | 38.7 | 37.3 | 40.1 | 31 | 42.2 | 44.6 | 44.7 | 47.1 |
| n-Decane | 51.0 | 51.0 | 56.4 | 55.3 | 53.6 | 52.1 | 58.2 | 59.1 | 46 | 60.6 | 62.4 | 64.2 | 65.2 |
| n-C$_4$H$_9$OH | — | — | — | — | — | — | — | — | 29 | 38.1 | 42.9 | 44.7 | 48.7 |
| iso-C$_4$H$_9$OH | — | — | — | — | — | — | — | — | 28 | 28.8 | 32.3 | 35.7 | 38.5 |
| o-Xylene | 34.3 | 34.5 | 34.6 | 34.5 | 36.9 | 35.9 | 38.8 | 39.0 | 32 | 41.5 | 48.2 | 50.1 | 51.0 |

Spark electrode position above liquid surface: A= 3, B= 5, C= 7, D= 9 (mm)

E : Value from literature (Closed Cup method)

TABLE 2: Measured Flash and Fire Points by Tag Closed Cup and Pensky-Martens Methods.

| Fuel | Flash point (°C) | | Fire Point (°C) | | $T_{fl}$ (°C) |
|---|---|---|---|---|---|
| | C.C. | P.M. | C.C. | P.M. | |
| n-Octane | 17.1 | 28.0 | 24.5 | 37.0 | 13 |
| n-Nonane | 33.5 | 40.5 | 43.0 | 56.0 | 29 |
| n-Decane | 50.0 | 57.0 | 67.5 | 77.0 | 46 |
| n-C$_4$H$_9$OH | 37.0 | 46.0 | 48.0 | 56.0 | 32 |
| iso-C$_4$H$_9$OH | 30.3 | —— | 38.0 | —— | 26 |
| o-Xylene | 31.8 | —— | 43.3 | —— | 29 |

C.C. : Tag Closed Cup Method

P.M. : Pensky-Martens Method

$T_{fl}$ : Theoretical Flash point by Lower Flammability Limit on Vapor Pressure Curve

fuel pool where the flame can not propagate certainly exists. Glassman et al. reported that the quenching distance above liquid surface is very small (below 0.3 mm) for some alcohols and very large (about 3 mm) for some hydrocarbons[16]. Therefore, from the different measuring methods of flash point and consequent different measured results as mentioned above, the significance and/or application of flash and fire points data to the investigation of such problems as flame spread mechanism across the liquid surface, ignition of liquid droplet and ignition hazard classification of liquid fuels should be discussed separately.

PREDICTION OF FLASH POINT OF LIQUID FUEL MIXTURE

It is well known that the lower and upper flammability limit concentrations of multicomponent fuel vapor mixture can be estimated approximately by Le Chatelier's law. There are, however, hypothetical limiting conditions in Le Chatelier's law as that each component in the vapor mixture has the same activation energy, and each activation probability which is proportional to the concentration has the same proportional constant.

For the liquid mixture, it is very difficult to estimate the flash point because the composition of vapor mixture above liquid surface varies with the liquid temperature, although the flammable limit concentration of the equilibrium vapor mixture above liquid pool at its certain temperature can be calculated by Le Chatelier's law. On the other hand, for multicomponent liquid mixtures such as kerosene and JP-4, it has been suggested that the flash point is determined only by very small concentartion of dissolved highly volatile component (10,21). In the present study, theoretical prediction of the flash point for the liquid fuel mixture in which each component has largely different volatility was tried.

Table 3 shows the flash points, by Tag Closed Cup method, of the mixtures of n-butanol and n-decane, and of n-heptane ~ n-dodecane with various mixing ratios, where the predicted flash points were determined by the partial vapor pressure of the most volatile component corresponding to the lower flammability limit concentration. It was assumed that the flash phenomenon is ascribed only to the most volatile component. As shown in the results, the predicted flash points by such the assumption are not necessarily valid, except that the mixture dissolves highly volatile component compared with other ones.

The flash point of the binary mixture of liquid fuels of the same kind, of which vapor pressures are largely different, can be calculated theoretically as follows, on the assumption that the liquid mixture is an ideal solution and then Raoult's law can be applied.

$$p_1 = r_1 P_1 , \quad p_2 = r_2 P_2 \quad ( r_1 + r_2 = 1 ) \tag{1}$$

where, P is the inherent vapor pressure of the component, r is the mole fraction of the component in the liquid mixture, p is the partial vapor pressure of the component in the equilibrium vapor mixture above liquid, and subscripts 1 and 2 indicate the components 1 and 2 in the liquid mixture respectively.

Assuming that flash phenomenon is ascribed only to $p_1$, the partial vapor pressure of the more volatile and flammable component is estimated by Clausius-Clapeyron's law as follows, although the total equilibrium vapor pressure of the liquid mixture is $p_1 + p_2$.

$$P_1^* = C \exp \left( - \frac{\Delta H_1}{R T_1^*} \right) \tag{2}$$

For the temperature T of the liquid mixture,

$$p_1 = r_1 P_1^* \quad \text{at } T = T_1^* , \quad p_1 = P_1^* \quad \text{at } T = T_1^* + \Delta T$$

where, $P_1^*$ is the vapor pressure corresponding to the lower flammability limit concentration, T* is the flash point of component 1, $\Delta H_1$ is the latent heat of vaporization, C is the constant, R is the gas constant, $\Delta T$ is the elevation of flash point due to the solubilization of less volatile component 2.

Consequently,

$$C \exp \left( - \frac{\Delta H_1}{R T_1^*} \right) = r_1 C \exp \left( - \frac{\Delta H_1}{R (T_1^* + \Delta T)} \right) \quad \text{at } T = T_1^* + \Delta T. \tag{3}$$

Therefore,

$$T_1^* + \Delta T = \left( \frac{1}{T_1^*} + \frac{R}{\Delta H_1} \ln r_1 \right)^{-1} . \tag{4}$$

Fugure 8 shows the measured results of flash and fire points of the mixture of n-octane and n-decane, the calculated flash point is by the equation (4) where n-octane is the component 1, and the predicted flash point is by the partial vapor pressure of n-octane corresponding to the lower flammability limit (1.0 vol%) calculated by Raoult's law. As shown in these results, predicted flash point agrees well with the measured one if the volume fraction of n-decane is below about 0.7, but the calculated flash point is reliable only at high fraction of n-octane. If more accurate expression of vapor pressure equation is taken (31), and/or the difference between the theoretical flash point ($T_{fl}$) and the measured one by closed cup method can be reduced by the improved measurement, theoretical study on the flash point calculation will take long strides, which is the subject for future study.

TABLE 3: Measured Flash Points ($T_f$) and Predicted Ones of the Fuel Mixtures with Various Mixing Ratios.

| Volumetric mixing ratio | | Flash Point(°C) (Closed cup) | |
|---|---|---|---|
| n-C$_4$H$_9$OH | n-C$_{10}$H$_{22}$ | $T_f$ | Predicted |
| 1 | 1 | 34.0 | 35.5 |
| 3 | 1 | 34.5 | 34.5 |
| 1 | 3 | 34.5 | 40.7 |

| Volumetric mixing ratio | | | | | | Flash Point(°C) (Closed cup) | |
|---|---|---|---|---|---|---|---|
| n-C$_7$H$_{16}$ | n-C$_8$H$_{18}$ | n-C$_9$H$_{20}$ | n-C$_{10}$H$_{22}$ | n-C$_{11}$H$_{24}$ | n-C$_{12}$H$_{26}$ | $T_f$ | Predicted |
| | | | 1 | 2 | 3 | 28.5 | 44.6 |
| | | | 2 | 1 | 2 | 25.5 | 30.0 |
| | 1 | 2 | 3 | 2 | 1 | 37.5 | 50.0 |
| 1 | 1 | 2 | 3 | 2 | 1 | 24.3 | 34.0 |

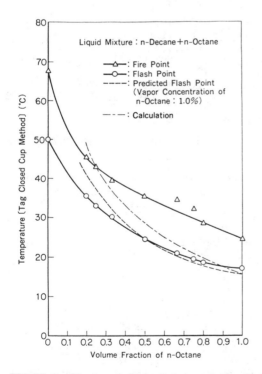

FIGURE 8: Flash and Fire Points of the Mixtures of n-Octane and n-Decane with Various Mixing Ratios by Tag Closed Cup method, Prediction and Theoretical Calculation.

CONCLUDING REMARKS

Above the pool surface of some liquid hydrocarbon fuels filled in the open cup, with spark ignition source in a quiescent atmosphere, "Pre-flash" phenomenon occurs at a little lower liquid temperature than the flash point. In pre-flash phenomenon, instantaneous faint but clear flamelet appears, and it propagates locally above pool surface,but is quenched on the way. Then, the scale of flame appearance above the liquid should be specified additionally in the usual methods of flash point measurement. The prediction of flash point of multicomponent liquid mixture by the partial vapor pressure of the most volatile and flammable component is not necessarily reliable, except the binary mixture. Comparing the calculated flash point of the binary hydrocarbon mixture by Clausius-Clapeyron's and Raoult's laws with the measured one suggested that the more accurate analysis of vapor pressure and the improvement of measuring method of the flash point are needed.

ACKNOWLEDGMENT

The authors express their sincere thanks to Mr. M. Kano for his great help in conducting the experiment. The authors wish also to acknowledge the help of drawing figures by Mr. S. Aoyagi.

REFERENCES

1) J.H. Burgoyne and J.F. Richardson; Fuel, 28, 150 (1949).
2) J.H. Burgoyne and G. Williams-Leir; Fuel, 28, 145 (1949).
3) J.B. Fenn; Ind. and Eng. Chemistry, 43, 2865 (1951).
4) R.M. Butler, et al.; Ind. and Eng. Chemistry, 48, 808 (1956).
5) W.A. Affens; J. of Chem. and Eng. Data, 11, 197 (1966).
6) J.H. Burgoyne, et al.; J. of the Inst. of Petroleum, 53, 338 (1967).
7) S.S.Penner and B.P. Mullins; Explosions,Detonations,Flammability and Ignition (AGARDograph NO.31, Pergamon Press, 1959), Part II.
8) M.G. Zabetakis; U.S. Bureau of Mines Blletine, NO.627 (1966).
9) J.T. Dehn; Comb. and Flame, 24, 231 (1975).
10) W.A. Affens and G.W. McLaren; J. of Chem. and Eng. Data, 17, 482 (1972).
11) W.A. Affens, et al.; J. of Fire and Flammability, 8, 141,152 (1977).
12) G.L. Nelson and J.L. Webb; ibid., 4, 210 (1973).
13) A.M. Kanury; Comb. Sci. & Tech., 31, 297 (1983). 14) J.E. Anderson and M.W. Magyagi; ibid., 37, 193 (1984). 15) K. Akita; Proc. of 14th Symp. on Comb.(Int'l) 1075 (1972). 16) I. Glassman and F.L. Dryer; Fire Safety J., 3,123 (1980/81).
17) H. Ishida and A. Iwama; Comb. Sci. & Tech., 36, 51 (1984).
18) J.H. Burgoyne, et al.; Proc.Roy.Soc., A308, 39, 55, 69 (1968).
19) A.F. Roberts and B.W. Quince; Comb. and Flame, 20, 245 (1973).
20) A.F. Roberts; Proc. of 15th Symp. on Comb.(Int'l) 305 (1974).
21) J.T. Bryant; Comb. Sci. & Tech., 10, 185 (1975).
22) M.M. EL-Wakil and M.I. Abdou; Fuel, 45, 177, 188, 199 (1966).
23) G.S. Scott, et al.; Analytical Chem., 20, 238 (1948).
24) L. Delfosse, et al.; Comb. and Flame, 54, 203 (1983).
25) Z. Tamura and Y. Tanazawa; Proc. of 7th Symp. on Comb. (Int'l), 509 (1958).
26) W-J Yang; Inst.of Space & Aeronaut.Sci. Univ.of Tokyo Report, 535 (1975).
27) I. Liebman, et al.; Comb.Sci.& Tech., 1, 257 (1970). 28) C.C. Feng, et al.; ibid., 10, 59 (1975). 29) M. Kaptain and C.E. Hermance; Proc.of 16th Symp. on Comb.(Int'l), 1295 (1976). 30) T. Hirano, et al.; ibid., 1307 (1976).
31) R.C. Reid, J.M. Prausnitz, T.K. Sherwood; The Properties of Gases and Liquids, 3rd ed., (McGraw-Hill, New York, 1977).

# STRUCTURAL BEHAVIOR

Session Chair

**Prof. Ove Pettersson**
Lund Institute of Technology
Division of Building Fire Safety and Technology
Box 118
S-221 00 Lund, Sweden

# Structural Fire Behaviour—
# Development Trends

O. PETTERSSON
Division of Building Fire Safety and Technology
Lund Institute of Science and Technology, Lund University
Box 118, S-221 00 Lund, Sweden

ABSTRACT

During the last two decades, a rapid progress has been made in the development
of analytical methods for a fire engineering design of load bearing and separating structures and structural members. Consequently, more and more countries
are now permitting a classification of structural members with respect to fire
to be formulated analytically as an alternative to the internationally prevalent
method of classification based on results of standard fire resistance tests.
In a long-term perspective, the development goes towards an analytical design,
directly based on a natural fire exposure, specified with regard to the combustion characteristics of the fire load and the geometrical, ventilation and thermal properties of the fire compartment.

Parallel to this progress, a further development is going on towards a reliability based structural fire engineering design. The development includes contributions related to a practical design format calculation, based on partial
safety factors, as well as to an evaluation, based on first order reliability
methods.

The paper describes and comments on these developments.

INTRODUCTION

During the last twenty years, important and rapid progress has been noted in
the development of analytical and computation methods for the determination of
the thermal and mechanical behaviour and the load bearing capacity of building
structures and structural members exposed to fire. Consequently, an analytical
design can be carried out today for most cases where steel structures are involved. Validated material models for the mechanical behaviour of concrete under
transient high-temperature conditions and thermal models for a calculation of
the charring rate in wood exposed to fire, derived during recent years, have
significantly increased the area of application of analytical design. To aid
this application, design diagrams and tables have been systematically computed
and published, giving directly, on the one hand, the temperature state of the
fire exposed structure, and on the other, a transfer of this information to the
corresponding load bearing capacity of the structure [1-27].

# METHODS OF STRUCTURAL FIRE DESIGN

The internationally applied methods for a fire design of load bearing structures and structural members may be described in outline with reference to the matrix set out in Figure 1 [25, 28]. The matrix is based on three models for thermal exposure ($H_1$, $H_2$ and $H_3$) in relation to three types of structural models ($S_1$, $S_2$ and $S_3$).

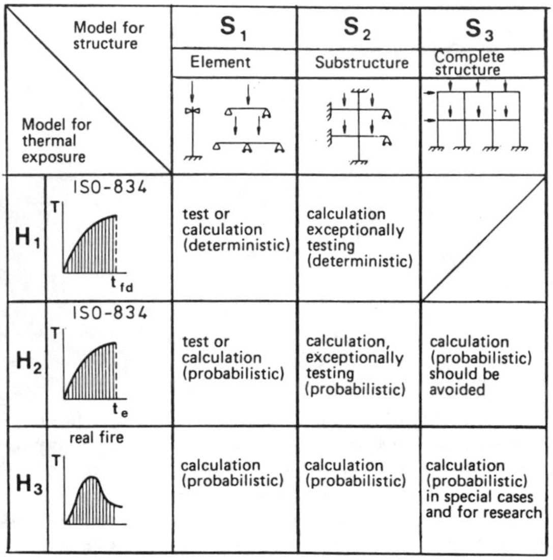

| Model for structure | $S_1$ Element | $S_2$ Substructure | $S_3$ Complete structure |
|---|---|---|---|
| **$H_1$** ISO-834 | test or calculation (deterministic) | calculation exceptionally testing (deterministic) | |
| **$H_2$** ISO-834 | test or calculation (probabilistic) | calculation, exceptionally testing (probabilistic) | calculation (probabilistic) should be avoided |
| **$H_3$** real fire | calculation (probabilistic) | calculation (probabilistic) | calculation (probabilistic) in special cases and for research |

FIGURE 1. Summary description of different methods for design of load bearing structures and structural elements under fire exposure conditions.

The thermal exposure conditions or models are defined as follows:

$H_1$ - A thermal exposure described by the standard temperature-time curve

$$T_t - T_o = 345 \log_{10}(8t + 1) \qquad (1)$$

as specified in ISO Standard 834, "Fire Resistance Tests - Elements of Building Construction". $T_t$ = furnace temperature at time t ($^\circ$C), $T_o$ = furnace temperature at time t = 0 ($^\circ$C) and t = time in minutes. The time of exposure $t_{fd}$ represents the time during which the structural element - or substructure - is required to fulfil its load bearing and/or separating function according to specifications in codes and regulations. The function may be verified either by test or by calculation.

$H_2$ - The same thermal exposure as for $H_1$ except that the duration of exposure $t_e$ is determined in each case for the characteristics of a particular compartment fire. Accordingly, $t_e$ represents an equivalent time of the standard fire exposure which produces the same effect upon the structural element or substructure with respect to the decisive limiting condition as the relevant natural fire. For protected and unprotected steel structures, the following approximate formula applies [29]:

$$t_e = 0.067 \frac{f}{\left(\dfrac{A\sqrt{h}}{A_{tot}}\right)^{\frac{1}{2}}} \qquad (min) \qquad\qquad (2)$$

where f = fire load density per unit area of the surfaces bounding the fire compartment $(MJ \cdot m^{-2})$, $A\sqrt{h}/A_{tot}$ = opening factor of the fire compartment $(m^{\frac{1}{2}})$, A = total area of the openings $(m^2)$, h = mean value of the heights of the openings, weighted with respect to each individual opening area (m) and $A_{tot}$ = total interior area of the structures enclosing the compartment, opening areas included $(m^2)$.

Eq. (2) is verified to be appropriate for use also for those reinforced concrete beams where the critical concern is yielding of the reinforcement under bending conditions. For other types of load bearing and separating structural elements, there are very few studies reported on the applicability of the formula.

$H_3$ – A thermal exposure determined by the conditions of a fully developed compartment fire with consideration given to: the combustion characteristics of the fire load, the ventilation of the fire compartment and the thermal properties of the structures enclosing the fire compartment. In the individual case, the exposure can either be calculated from the energy and mass balance equations for the compartment fire or be derived from curves or tables in manuals, giving the time variation of either the gas temperature within the compartment or the corresponding heat flux to the structure – cf., for instance, [2, 7, 8, 13, 17, 28] and further references given in these publications. Figure 2 gives an example of such data, taken from the Commentary 1976:1 to the Swedish Building Code.

The structural models are defined as follows:

$S_1$ – A simplification of a real structure by division into single elements such as beams and columns. The structural model may either be represented by a test specimen or dealt with analytically.

$S_2$ – A simplification of a real structure by division into substructures such as beam-column systems. The substructure thus derived is provided with well-defined, simplified conditions of support and/or restraint at its outer ends or edges. As with $S_1$, the structural model may either be dealt with analytically or - exceptionally - be represented by a test specimen.

$S_3$ – A complete real structure, e.g. a two or three dimensional frame, a beam-slab system or a column-beam-slab system. Such structural models are generally dealt with analytically, normally requiring the support of a computer.

The internationally most prevalent structural fire design is characterized by the combination $H_1-S_1$. The design is usually related to the results of the standard fire resistance test according to the ISO Standard 834 or some equivalent national standard. The fire resistance may also be derived analytically and this alternative is now officially being permitted in more and more countries. A few countries allow the application of the model combination $H_1-S_2$, normally by analytical methods. The combination $H_1-S_3$ involves a too great difference in the accuracy of simulation between the thermal exposure and structural models to be acceptable in practice.

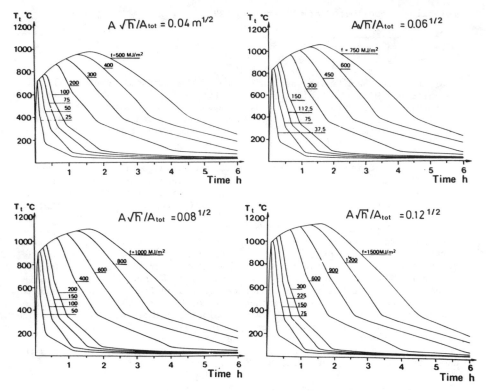

FIGURE 2. Examples of gas temperature-time curves for a natural fire as a function of fire load density f and opening factor $A\sqrt{h}/A_{tot}$ of the fire compartment. Enclosing structures, made of a material with a thermal conductivity $k = 0.81$ W·$\cdot m^{-1} \cdot {}^{\circ}C^{-1}$ and a heat capacity $\rho c_p = 1.67$ MJ·$m^{-3} \cdot {}^{\circ}C^{-1}$; fire compartment type A [7].

The substantial progress during the last twenty years in the development of analytical methods, referred to above, has considerably increased the possibility of performing a structural fire design, based upon the thermal exposure models $H_2$ and $H_3$ as an alternative to the conventional use, at present, of the thermal exposure model $H_1$.

A design directly based on a natural compartment fire exposure $H_3$ is generally characterized by an analytical treatment. For rapid practical application, it is necessary for systematized design data in the form of e.g. manuals to be available. Usually, then the model combination $H_3$-$S_2$ is used, and in certain cases the model combination $H_3$-$S_1$. Design according to the combination $H_3$-$S_3$ generally demands access to a computer. This combination is of central importance in the research context.

A structural design for thermal exposure of the $H_2$ type is based indirectly on a natural compartment fire, described by the temperature-time curve according to ISO 834, Eq. (1), with reference to the concept of equivalent time of fire exposure $t_e$. The structural behaviour, calculated for such an exposure, differs from the behaviour in a natural fire situation in cases where the heating histo-

232

ry is of significance. In the combination $H_2-S_1$, the design can be performed either analytically or on the basis of a furnace test according to ISO 834. In the combination $H_2-S_2$, analytical design is the normal procedure, and experimental verification is an exception. The combination $H_2-S_3$ may be questioned from a practical standpoint since it does not provide for the simplifications of a design developed using the model combination $H_3-S_3$.

## CHARACTERISTICS OF RELIABILITY BASED STRUCTURAL FIRE DESIGN

The most recent trend in the development of the structural fire design is to adopt modern loading and safety philosophy and include a probabilistic approach, based on either a system of partial safety factors (practical design format) or the safety index concept [25, 26, 28, 30-37]. For an everyday design, a direct application of the safety index concept then is too cumbersome and the more simplified practical design formats have to be used.

The fundamental components of such a reliability based structural fire design are

* the limit state conditions
* the physical model
* the practical design format
* deriving the safety elements.

Depending on the type of practical application, one, two or all of the following limit state conditions apply:

* Limit state with respect to load bearing capacity
* limit state with respect to insulation
* limit state with respect to integrity.

For a load bearing structure, the design criterion implies that the minimum value of the load bearing capacity R(t) during the fire exposure shall meet the load effect on the structure S, i.e.

$$\min\{R(t)\} - S \geq 0 \qquad (3)$$

The criterion must be fulfilled for all relevant types of failure. The requirements with respect to insulation and integrity apply to separating structures. The design criterion regarding insulation implies that the highest temperature on the unexposed side of the structure - $\max\{T_s(t)\}$ - shall meet the temperature $T_{cr}$, acceptable with regard to the requirement to prevent a fire spread from the fire compartment to an adjacent compartment, i.e.

$$T_{cr} - \max\{T_s(t)\} \geq 0 \qquad (4)$$

For the integrity requirement, there is no analytically expressed design criterion available at present. Consequently, this limit state condition has to be proved experimentally, when required, in either a fire resistance test or a simplified small scale test.

The physical model comprises the deterministic model, describing the relevant physical processes of the thermal and mechanical behaviour of the structure at specified fire and loading conditions. Supplemented with relevant partial safety factors, the physical model is transferred to the practical design format.

Related to an analytical fire design of load bearing structures, directly based on the natural compartment fire exposure – thermal exposure type $H_3$ – the practical design format can summarily be described according to the flow chart in Figure 3.

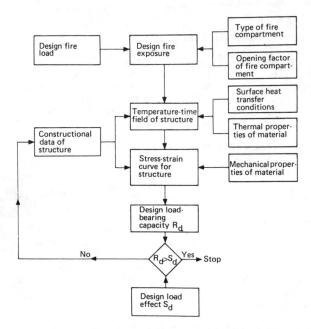

FIGURE 3. Flow chart for an analytical fire design of load bearing structures on the basis of a natural compartment fire exposure.

From the design fire load and the geometrical, ventilation and thermal characteristics of the fire compartment (opening factor and type of fire compartment), the design fire exposure is determined either by energy and mass balance calculations or from a systematized design basis. Together with design values for the constructional data of the structure and the thermal and mechanical properties of the structural materials, the design fire exposure provides the design temperature state and the related design load bearing capacity $R_d$ for the lowest value of the load bearing capacity during the relevant fire process.

The design format condition to be proved is

$$R_d - S_d \geq 0 \tag{5}$$

where $S_d$ is the design load effect at fire. Depending on the type of practical application, the condition has to be verified for either the complete fire process or a limited part of it, determined by, for instance, the design evacuation time for the building.

The probabilistic influences are considered by specifying characteristic values and related partial safety factors for the fire load, such structural design data as imperfections, the thermal properties, the mechanical strength and the loading. The partial safety factors then are to be derived by a probabilistic analysis, based on a first order reliability method.

The procedure of deriving the safety elements is further outlined in Figure 4, as exemplified for a timber structure [26].

Expressed in terms of a safety index $\beta$ - defined as the ratio of the mean value of the safety margin to its standard deviation - the design criterion then has the form

$$\beta_{fm} - \beta_r \geq 0 \qquad\qquad (6)$$

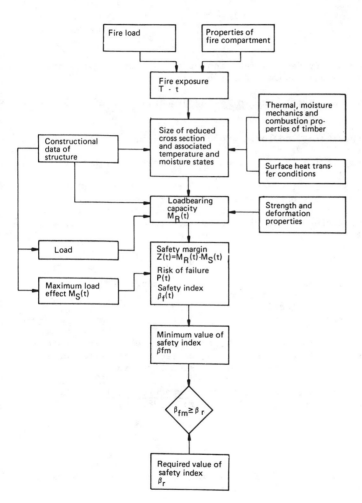

FIGURE 4. Derivation of partial safety factors for a fire exposed timber structure by a probabilistic analysis, based on a first order reliability method.

where $\beta_{fm}$ is the least value of the safety index for the structure during the relevant fire process and $\beta_r$ is the required value of the safety index.

The safety margin is defined by the formula

$$Z(t) = M_R(t) - M_S(t) \tag{7}$$

where $M_R(t)$ is the load bearing capacity at time t, expressed in terms of e.g. the bending moment at a critical section of the structure, and $M_S(t)$ is the corresponding bending moment related to the maximum load effect. The corresponding probability of failure P(t) and safety index $\beta_f(t)$ are given by the formulae

$$P(t) = \int_{-\infty}^{o} f_Z\{Z(t)\}dZ \tag{8}$$

$$\beta_f(t) = \phi^{-1}\{1 - P(t)\} \tag{9}$$

where $f_Z\{Z(t)\}$ = the probability density function of the safety margin Z and $\phi^{-1}$ = = the inverse of the standardized normal distribution.

In determining $Z(t)$, $P(t)$ and $\beta_f(t)$, the following probabilistic effects must be taken into account:

* The uncertainty in specifying the loads and of the model, describing the load effect on the structure
* the uncertainty in specifying the fire load and the characteristics of the fire compartment
* the uncertainty in specifying the design data of the structure and the thermal and mechanical properties of the structural materials
* the uncertainty of the analytical models for the calculation of the compartment fire, the heat transfer to and within the structure and its ultimate load bearing capacity.

The required value of the safety index $\beta_r$ depends on

* the probability of occurrence of a fully developed compartment fire
* the efficiency of the fire brigade actions
* the effect of an installed extinction system, if any
* the consequences of a structural failure.

In the design procedure according to Figure 3, the latter four influences can be accounted for by the partial safety factors allocated to either the design mechanical strength or the design fire load and design fire exposure.

For a structural fire design, based on the thermal exposure model $H_2$, the practical design format can be given in the following form [28, 36]:

$$\frac{t_f}{\gamma_f} \geq \gamma_n \gamma_e t_e \tag{10}$$

where $t_f$ is the fire resistance of the structural element, $t_e$ equivalent time of fire exposure – Eq. (2) – and $\gamma_f$, $\gamma_n$ and $\gamma_e$ partial safety factors, taking into account all uncertainties in the design system.

The partial safety factor $\gamma_e$ covers the uncertainties of the fire load and the fire compartment characteristics, including the uncertainties of the analytical model for a determination of the fire exposure. The partial safety factor $\gamma_f$

considers the uncertainties of the mechanical load and the thermal and mechanical properties of the structural element, including the uncertainties of the analytical models for a determination of the load effect, the transient temperature state and the load bearing capacity if the fire resistance is evaluated analytically. The additional partial safety factor (differentiation factor) $\gamma_n$ takes into consideration the effects related to the required safety index $\beta_r$, as listed above.

## THERMAL AND MECHANICAL BEHAVIOUR OF BUILDING STRUCTURES AT FIRE EXPOSURE

As stated in the introduction, the important progress in modelling the thermal and mechanical behaviour of fire exposed structures and structural elements during the last twenty years has considerably enlarged the area of application of an analytical fire design. Access to validated material behaviour models for transient high-temperature conditions then is a necessary prerequisite for a successful simulation of the real structural fire behaviour.

### Thermal Properties and Transient Temperature State

The transient heat flow within a fire exposed structure is governed by the heat balance equilibrium equation, based on the Fourier law

$$\underline{V}^T(\underline{\lambda}\underline{V}T) - \dot{e} + Q = 0 \tag{11}$$

where T = temperature, $\underline{\lambda}$ = symmetric positive definite thermal conductivity matrix, $\dot{e} = \partial e/\partial t$ = rate of specific volumetric enthalpy change, Q = rate of internally generated heat per volume, and t = time. The gradient operator $\underline{V}$ is defined as

$$\underline{V} = \begin{bmatrix} \dfrac{\partial}{\partial x} \\ \dfrac{\partial}{\partial y} \\ \dfrac{\partial}{\partial z} \end{bmatrix} \tag{12}$$

where x, y and z are Cartesian coordinates.

For isotropic materials

$$\underline{\lambda} = \lambda\underline{I} \tag{13}$$

where $\lambda$ = thermal conductivity, and $\underline{I}$ = identity matrix.

A solution of Eq. (11) requires the initial and boundary conditions to be specified. The initial condition is given by the distribution of temperature within the structure at a reference time zero. The boundary conditions are prescribed as temperature $T = T(x,y,z,t)$ or heat flow q on parts of the boundary $\partial V_T$ and $\partial V_q$, respectively. The total boundary is then

$$\partial V = \partial V_T + \partial V_q \tag{14}$$

The heat flow normal to the surface on the boundary $\partial V_q$ must satisfy the heat balance equation

237

$$q_n = -\underline{n}\underline{\underline{\lambda}}\nabla T \tag{15}$$

where $\underline{n}$ = outward normal to the surface.

At free surfaces, the heat flow $q_n$ is caused by convection $q_{nc}$ and radiation $q_{nr}$ and follows the formula

$$q_n = q_{nc} + q_{nr} = \alpha(T_s - T_t)^m + \varepsilon_r \sigma(\bar{T}_s^4 - \bar{T}_t^4) \tag{16}$$

where $\alpha$, m = convection factor and convection power, respectively - see, for instance, [38], $\varepsilon_r$ = resulting emissivity, varying with gas or flame emissivity, surface properties and geometric configuration, $\sigma$ = Stefan-Boltzmann constant, $T_s$ = surface temperature, $\bar{T}_s$ = absolute surface temperature, $T_t$ = surrounding gas temperature, and $\bar{T}_t$ = absolute surrounding gas temperature.

The solution of Eq. (11) is complicated by the fact that the thermal conductivity matrix $\underline{\underline{\lambda}}$ and the rate of specific volumetric enthalpy change $\dot{e}$ depend on the temperature T to an extent that cannot be diregarded. Further complications arise when the material undergoes phase changes during the heating and when the material has an initial moisture content.

Well-defined measurements of the thermal conductivity $\lambda$ for moist materials are difficult to undertake within the temperature range relevant at fire exposure, due to the complicated interaction between moisture and heat flow. As concerns the enthalpy e, the way evaporable water reacts to pressure has not been experimentally clarified and consequently, this influence has to be included in a simplified manner in calculating the transient temperature state of a fire exposed structure. Usually, all moisture is assumed to evaporate, without any moisture transfer, at the temperature 100°C or within a narrow temperature range - ref. 38 applies a range of 100 to 115°C - with the heat of evaporation giving a corresponding discontinuous step in the enthalpy curve. This simplification has proved to give acceptable results for most practical situations.

In reality, the evaporation of moisture in a fire exposed material is not comparable to that of a free water surface. Capillary forces, adhesive forces, and interior steam pressure will allow the temperature to increase during evaporation. During the heating of the structure, the moisture distribution changes continuously. Hence, it is not principally correct to include the effect of moisture content in the thermal properties. For a moist material, the heat transfer is combined with moisture transport and, from a strict thermodynamical point of view, these two transport mechanisms should be analysed simultaneously by a system of related partial differential equations. Consequently, Eq. (11) constitutes an approximation when applied to fire exposed structures made of materials that contain moisture.

For materials used, for instance, for fire protection of steel structures or in suspended ceilings, there are test methods developed for a determination of derived values, characterizing the fire behaviour of the product in an integrated way. Normally, the values are derived from test results by use of some analytical simulation model. As a consequence, the derived values do not represent any well-defined material or product properties but are influenced also by the characteristics of the analytical model, adopted for the evaluation. This leads to limitations with respect to a generalized application of the derived values.

Analytical solutions of the heat balance equilibrium equation (11) are feasible only for linear applications with simple geometries and boundary conditions. For a practical determination of the transient temperature state of fire exposed structures, numerical methods have been developed and arranged for computer calculations. The methods are based either on finite difference or finite element approximations. For the first group of numerical methods, reference can be made to [12, 39-44], and for the group employing finite element methods to [12, 38, 45-48].

The computer programmes can be used either directly as an advanced component in the fire design procedure or as a tool for calculations of diagrams and tables, facilitating a practical determination of the design temperature state for varying conditions of fire exposure and varying structural characteristics. For a thermal exposure according to the standard temperature-time curve, Eq. (1), such design aids are given in [5, 6, 9, 10, 13, 20, 23, 27] for steel structures and in [3-5, 11, 14-16, 19, 21] for concrete structures. A corresponding design aid is presented in [2, 7, 13, 17, 29, 49] for steel structures and in [7, 49] for concrete structures when exposed to a natural compartment fire with gas temperature-time curves according to Figure 2.

Fire exposed timber structures present special problems due to the continuous decrease of the effective cross section by combustion of the material. For a thermal exposure according to the standard fire resistance test, Eq. (1), a large number of tests, made in different fire engineering laboratories, verify an approximately constant rate of charring of $3.5 \text{ cm} \cdot \text{h}^{-1}$ for glued laminated timber beams and columns. The value is roughly applicable up to a charring depth equal to one quarter of the cross section dimension in the direction of charring. For a larger charring depth, the rate of charring increases.

Analytical models for a calculation of the charring rate and depth of wood at varying thermal exposure are presented in, for instance, [26, 50-54]. The refs. [51, 52, 54] also include a model for a determination of the temperature distribution within the uncharred part of the cross section. Ref. [53] includes diagrams giving the charring depth of a cross section at a natural compartment fire exposure, defined by the gas temperature-time curves according to Figure 2.


Mechanical Properties and Structural Behaviour

A reliable calculation of the mechanical behaviour and load bearing capacity of a fire exposed structure or structural element on the basis of the transient temperature state requires validated models for the mechanical behaviour of the materials involved within the temperature range associated with fires. It is important that the material behaviour models are phenomenologically correct with input information received from well-defined tests.

Available tests for a determination of the mechanical properties of materials at elevated temperatures can mainly be divided into two groups: steady state tests and transient state tests - Figure 5 [55]. Fundamental parameters are the heating process, application and control of load, and control of strain. These can have constant values or be varied during testing.

Figure 5 defines six practical regimes with mechanical properties as follows:

* steady state tests
  - stress-strain relationship (stress rate control, $\dot{\sigma} = \text{const}$)
  - stress-strain relationship (strain rate control, $\dot{\varepsilon} = \text{const}$)
  - creep (stress control, $\sigma = \text{const}$)

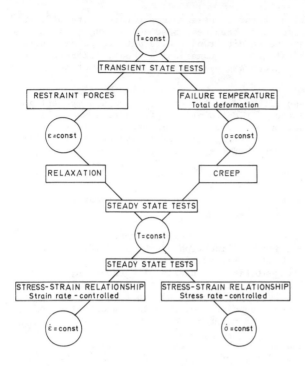

FIGURE 5. Different testing regimes for determining mechanical properties of materials at elevated temperatures [55].

    - relaxation (strain control, ε = const)

\* transient state tests
    - failure temperature, total deformation (stress control, σ = const)
    - restraint forces, total forces (strain control, ε = const).

The material properties measured are closely related to the test method used. Consequently, it is extremely important that reported test results always are accompanied by an accurate specification of the test conditions applied. For steels, there is analytical modelling technique available enabling a coupling of steady state and transient state tests [55].

For steel, validated mechanical behaviour models for transient, high-temperature conditions have been available for many years - cf., for instance, [55-59]. The models divide the total strain into thermal strain, instantaneous stress-related strain and creep strain or time dependent strain. Some of the models operate with temperature compensated time according to Dorn [56].

Analytical models for determination of the mechanical behaviour and load bearing capacity of steel beams, columns and frames exposed to fire are presented in, for instance, [12, 57-66]. The most general models are those put forward in [59, 63-66]. A simplified design aid, giving directly the load bearing capacity for a design temperature state or the critical temperature state for a design load effect, can be found in [2, 5-7, 9, 10, 13, 17, 20, 23, 27, 49, 57].

Simple formulae for the fire resistance of unprotected and protected steel columns, derived by Lie and Stanzak, are quoted in [8].

For concrete, the deformation behaviour at elevated temperatures is much more complicated than for steel. Stressed concrete involves special difficulties since considerable deformations develop during the first heating which do not occur when the temperature is stable. This effect has been confirmed by flexural, torsional and compressive tests and for moderate as well as high temperatures.

For practical applications, the total strain $\varepsilon$ can adequately be given as the sum of various strain components, phenomenologically defined with reference to specified tests and depending on the temperature T, the stress $\sigma$, the stress history $\tilde{\sigma}$, and the time t. For concrete, stressed in compression, then the following constitutive equation applies [67]

$$\varepsilon = \varepsilon_{th}(T) + \varepsilon_{\sigma}(\tilde{\sigma},\sigma,T) + \varepsilon_{cr}(\sigma,T,t) + \varepsilon_{tr}(\sigma,T) \qquad (17)$$

where $\varepsilon_{th}$ = thermal strain, including shrinkage, measured on unstressed specimens under variable temperature; $\varepsilon_{\sigma}$ = instantaneous, stress-related strain, based on stress-strain relations, obtained at a rapid rate of loading under constant, stabilized temperature; $\varepsilon_{cr}$ = creep strain or time-dependent strain, measured under a constant stress at constant, stabilized temperature; and $\varepsilon_{tr}$ = = transient strain, accounting for the effect of temperature increase under stress, derived from tests under constant stress and variable temperature.

For stressed concrete in a transient high-temperature state, the transient strain component $\varepsilon_{tr}$ usually plays a predominant role. Parameter formulations for each of the strain components and a practical guidance on the application of the material behaviour model at a time varying stress and temperature state are given in [67]. An alternative model formulation of the mechanical behaviour of concrete at transient elevated temperatures is given in [68].

In [69] an attempt is made to formulate a multiaxial constitutive model for concrete in the temperature range up to 800°C. The model can be characterized as isotropic, elastic-viscoplastic-plastic in the compression region. Brittle failure is assumed in the tensile region.

As stressed above, validated material behaviour models of the type described is a condition for getting reliable results from the calculation models and corresponding computer programmes for determination of the mechanical behaviour and load bearing capacity of fire exposed reinforced concrete structures. Comprehensive computer programmes for such a determination are presented in [12, 22, 65, 70-76]. The methods dealt with in [12, 22, 65, 72, 73, 75, 76] include secondary order effects. A computer programme for evaluating the fire response of reinforced concrete slabs is published in [77]. The programme is based on a non-linear finite element method coupled with a time-step integration and includes a combined bending and membrane action of the slab.

Simplified methods, facilitating the practical design of fire exposed, reinforced concrete beams and columns exposed to fire can be found in [3-5, 7, 8, 11, 14, 16, 19, 21, 39, 40, 78-80]. In [8], simple formulae are given for the fire resistance of concrete beams, columns, walls and slabs, based on an international survey - cf. also [14, 19]. The design aid has to be applied with due consideration of the fact that the analytical tool for a determination of the ultimate load bearing capacity of fire exposed concrete structures mainly covers the failure in bending. For other kinds of failure - shear, bond, anchorage and spalling – the present state of knowledge is still unsatisfactory. In a practical fire engineering design, it is therefore important to detail the structure in such a way that these types of failure will have a lower probability of occurrence than failure by bending.

For load bearing timber structures, the possibilities for an analytical modeling of the mechanical behaviour during fire exposure are essentially more limited than for steel and concrete structures. As mentioned earlier, validated analytical models are available for a calculation of the charring rate of wood under varying thermal exposure and approximate models also exist for an evaluation of the temperature distribution within the uncharred part of the cross section under the simplified assumption of no moisture content in the wood [51, 52, 54]. It is highly desirable that analytical models should be developed for the mass transfer of moisture and for the mechanical behaviour of wood under conditions of transient temperature and moisture content.

In [81], approximate formulae are derived for the fire resistance of laminated timber beams and columns, exposed to the standard temperature-time curve. A detailed structural design guide for the fire resistance of beams, columns, joints, floors, roofs and walls, based on classification and results of standard fire resistance tests is given in [24] which is a very comprehensive manual. A simplified design aid for laminated timber beams and columns, exposed to a natural compartment fire, is presented in [7, 26], and [82] supplements this design aid for beams with respect to the risk of lateral buckling during the fire exposure.


REFERENCES

1.  Lie, T.T.: Fire and Buildings, Applied Science Publishers, London, 1972.

2.  Pettersson, O., Magnusson, S.E. and Thor, J.: Fire Engineering Design of Steel Structures, Swedish Institute of Steel Construction, Stockholm, 1976 (Swedish edn., 1974).

3.  Gustaferro, A.H. and Martin, L.D.: PCI Design for Fire Resistance of Precast Prestressed Concrete, Prestressed Concrete Institute, Chicago, 1977.

4.  Institution of Structural Engineers: Design and Detailing of Concrete Structures for Fire Resistance, The Concrete Society, London, 1978.

5.  Bartélémy, B. and Kruppa, J.: Résistance au Feu des Structures – Béton, Acier, Bois, édn. Eyrolles, Paris, 1978.

6.  Fruitet, P.L.: Guide pour la Conception des Bâtiments à Structures en Acier, Office Téchnique pour l'Utilisation de l'Acier (OTUA), Paris, 1978.

7.  Pettersson, O. and Ödeen, K.: Brandteknisk dimensionering – principer, underlag, exempel, (Fire Engineering Design of Building Structures – Principles, Design Basis, Examples), Liber förlag, Stockholm, 1978.

8.  Harmathy, T.Z.: Design to Cope with Fully Developed Fires, from ASTM Symposium "Design of Buildings for Fire Safety", ASTM Special Technical Publication, 685, Washington, 1979, pp. 198-276.

9.  AISI: Fire Safe Structural Steel – A Design Guide, American Iron and Steel Institute, Washington, D.C., 1979.

242

10. Twilt, L. and Witteveen, J.: <u>Brandveiligheid Staalconstructies</u> (Fire Resistance of Steel Structures), Staalcentrum Nederland, 1980.

11. CSTB (Centre Scientifique et Téchnique du Bâtiment): <u>Méthode de Prévisions par le Calcul du Comportement au Feu des Structures en Béton</u>, Document Téchnique Unifié, Paris, 1980.

12. Dotreppe, J.C.: <u>Modèles Numériques pour la Simulation du Comportement au Feu des Structures en Acier et en Béton Armé</u>, Thèse d'Agrégation de l'Enseignement Supérieur, Université de Liège, 1980.

13. Thrane, E.J.: <u>Brandteknisk dimensjonering av bygningskonstruksjoner</u> (Fire Engineering Design of Building Structures), Tapir forlag, Oslo, 1981.

14. Kordina, K. and Meyer-Ottens, C.: <u>Beton-Brandschutz-Handbuch</u>, Beton-Verlag, Düsseldorf, 1981.

15. Reichel, V.: <u>Brandschutz-Anforderungen an Baukonstruktionen</u>, Staatsverlag DDR, Berlin, 1981.

16. American Concrete Institute: Guide for Determining the Fire Endurance of Concrete Elements, <u>Concrete International</u>, February 1981.

17. Pettersson, O., Magnusson, S.E. and Thor, J.: <u>Rational Approach to Fire Engineering Design of Steel Buildings</u>, Lund Institute of Technology, Division of Building Fire Safety and Technology, Report LUTVDG/(TVBB-3002), Lund, 1981.

18. Malhotra, H.L.: <u>Design of Fire Resisting Structures</u>, Surrey University Press, London, 1982.

19. CEB (Comité Euro-International du Béton): <u>Design of Concrete Structures for Fire Resistance</u>, Bulletin d'Information, N° 145, Paris, 1982.

20. CTICM (Centre Téchnique Industriel de la Construction Métallique): Méthode de Prévision par le Calcul du Comportement au Feu des Structures an Acier, Document Téchnique Unifié, <u>Revue Construction Métallique</u>, N° 3, Paris, 1982.

21. US Standard Building Code: <u>Calculated Fire Resistance</u>, Appendix P, 1982.

22. Forsén, N.E.: <u>A Theoretical Study on the Fire Resistance of Concrete Structures</u>, Doctor's Dissertation, Division of Concrete Structures, Norwegian Institute of Technology, University of Trondheim, 1982.

23. ECCS (European Convention for Constructional Steelwork): European Recommendations for the Fire Safety of Steel Structures: <u>Calculation of the Fire Resistance of Load Bearing Elements and Structural Assemblies Exposed to the Standard Fire</u>, Elsevier, Amsterdam-Oxford-New York, 1983.

24. Kordina, K. and Meyer-Ottens, C.: <u>Holz-Brandschutz-Handbuch</u>, Deutsche Gesellschaft für Holzforschung e.V., München, 1983.

25. Brozetti, J., Law, M., Pettersson, O. and Witteveen, J.: Safety Concept and Design for Fire Resistance of Steel Structures, <u>IABSE Surveys</u>, S-22/83, Zürich, 1983.

26. Jönsson, R. and Pettersson, O.: Timber Structures and Fire, Swedish Council for Building Research, Document D3:1985, Stockholm, 1985 (Swedish edn. 1983).

27. SIA (Schweizerischer Ingenieur- und Architekten-Verein): Feuerwiderstand von Bauteilen aus Stahl, Schweizerische Zentralstelle für Stahlbau, Zürich, 1985.

28. CIB W14: A Conceptual Approach Towards a Probability Based Design Guide on Structural Fire Safety, Report of CIB W14 Workshop "Structural Fire Safety", Fire Safety Journal, Vol. 6, No. 1, 1983.

29. Pettersson, O.: The Connection Between a Real Fire Exposure and the Heating Conditions According to Standard Fire Resistance Tests, European Convention for Structural Steelwork, Chapter II, CECM-III-74-2E.

30. Magnusson, S.E.: Probabilistic Analysis of Fire Exposed Steel Structures, Lund Institute of Technology, Division of Structural Mechanics and Concrete Construction, Bulletin 27, Lund, 1974.

31. Magnusson, S.E. and Pettersson, O.: Functional Approaches - An Outline, CIB Symposium "Fire Safety in Buildings: Needs and Criteria", held in Amsterdam 1977-06-02/03, CIB Proceedings, Prublication 48:120-145, 1978.

32. Schneider, U., Bub, H. and Kersken-Bradley, M.: Structural Fire Protection Levels for Industrial Buildings, presented at the 1980 Fall Convention, American Concrete Institute, San Juan, Puerto Rico, 1980.

33. Hosser, D. and Schneider, U.: Sicherheitskonzept für brandschutztechnische Nachweise von Stahlbetonbauteilen nach der Wärmebilanztheorie, Institut für Bautechnik, Berlin, 1980.

34. Magnusson, S.E. and Pettersson, O.: Rational Design Methodology for Fire Exposed Load Bearing Structures, Fire Safety Journal, Vol. 3, Nos. 2-4, 1981, pp. 227-241.

35. Pettersson, O.: Reliability Based Design of Fire Exposed Concrete Structures, Contemporary European Concrete Research, Stockholm, 1981.

36. DIN 18230, Baulicher Brandschutz im Industriebau, Vornorm, Teil 1 und 2, Berlin, 1982.

37. Gross, D.: Aspects of Stochastic Modeling for Structural Firesafety, Fire Technology, Vol. 19, 1983, pp. 103-114.

38. Wickström, U.: TASEF-2, A Computer Program for Temperature Analysis of Structures Exposed to Fire, Lund Institute of Technology, Department of Structural Mechanics, Report No. 79-2, Lund 1979.

39. Ehm, H.: Ein Beitrag zur rechnerischen Bemessung von brandbeanspruchten balkartigen Stahlbetonbauteilen, Dissertation, Technische Universität, Braunschweig, 1967.

40. Ödeen, K.: Fire Resistance of Concrete Double T Units, Acta Polytechnica Scandinavica, Ci 48, Stockholm, 1968.

41. Kordina, K., Schneider, U., Haksever, A. and Klingsch, W.: Zur Berechnung von Stahlbetonkonstruktionen im Brandfall, Institut für Baustoffe, Massiv-

bau und Brandschutz, Technische Universität, Braunschweig, 1975.

42. Coin, A.: Températures dans un Solide Hétérogène au Cours d'un Incendie, Serie: Information Appliquée, No. 28, Annales de l'Institut Téchnique du Bâtiment et des Travaux Publics, Paris, 1976.

43. Lie, T.T.: Temperature Distribution in Fire-Exposed Building Columns, Journal of Heat Transfer, Vol. 99, Ser. C, No. 1, 1977, pp. 113-119.

44. Rudolphi, R. and Müller, T.: ALGOL - Computerprogramm zur Berechnung zweidimensionaler instationärer Temperaturenverteilungen mit Anwendungen aus dem Brand- und Wärmeschutz, BAM-Forschungsbericht Nr. 74, Berlin, 1980.

45. Fisser, W.: A Finite Element Method for the Determination of Nonstationary Temperature Distribution and Thermal Deformation, Proceedings, Conference on Matrix Methods in Structural Mechanics, Air Force Institute of Technology, Wright Pattersson Air Force Base, Ohio, 1965.

46. Axelsson, K., Fröier, M. and Loyd, D.: FEMTEMP, datorprogram för analys av tvådimensionella värmeledningsproblem (FEMTEMP, Computer Programme for Analysis of Two-Dimensional Heat Conduction), Chalmers University of Technology, Division of Structural Mechanics, Publication 72:6, Gothenburg, 1972.

47. Becker, J., Bizri, H. and Bresler, B.: FIRES-T, A Computer Program for the Fire Response of Structures - Thermal, University of California, Fire Research Group, Report No. UCB FRG 74-1, Berkeley, 1974.

48. Iding, R., Bresler, B. and Nizamuddin, Z.: FIRES-T3, A Computer Program for the Fire Response of Structures - Thermal (Three Dimensional Version), University of California, Fire Research Group, Report No. UCB FRG 77-15, Berkeley, 1977.

49. Pettersson, O.: Theoretical Design of Fire Exposed Structures, Lund Institute of Technology, Division of Structural Mechanics and Concrete Construction, Bulletin 51, Lund, 1976.

50. Hadvig, S. and Paulsen, O.R.: One-dimensional Charring Rates in Wood, Journal of Fire and Flammability, Vol. 7, Oct., 1976.

51. Fredlund, B.: Modell för beräkning av temperatur och fuktfördelning samt reducerat tvärsnitt i brandpåverkade träkonstruktioner (Model for Calculation of Temperature, Moisture Distribution and Reduced Cross Section in Fire Exposed Wooden Structures), Lund Institute of Technology, Division of Structural Mechanics, Internal Report IR79-2, Lund 1979 (1977).

52. White, R.H. and Schaffer, E.L.: Application of CMA Program to Wood Charring, Fire Technology, Vol. 14, No. 4, 1978.

53. Hadvig, S.: Charring of Wood in Building Fires, Technical University of Denmark, Laboratory of Heating and Air-Conditioning, Lyngby, 1981.

54. Fredlund, B.: A Computer Program for the Analysis of Timber Structures Exposed to Fire, Lund Institute of Technology, Division of Building Fire Safety and Technology, Report LUTVDG/(TVBB-3020), Lund, 1985.

55. Anderberg, Y.: Predicted Fire Behaviour of Steels and Concrete Structures,

Lund Institute of Technology, Division of Building Fire Safety and Technology, Report LUTVDG/(TVBB-3011), Lund, 1983.

56. Dorn, J.E.: Some Fundamental Experiments on High Temperature Creep, Journal of Mechanics and Physics of Solids, Vol. 3, 1954.

57. Thor, J.: Deformations and Critical Loads of Steel Beams Under Fire Exposure Conditions, Lund Institute of Technology, Division of Structural Mechanics and Concrete Construction, Bulletin 35, Lund 1973.

58. Harmathy, T.Z.: Creep Deflection of Metal Beams in Transient Heating Processes, with Particular Reference to Fire, Canadian Journal of Civil Engineering, Vol. 3, No. 2, 1976.

59. Furumura, F. and Shinohara, Y.: Inelastic Behaviour of Protected Steel Beams and Frames in Fire, Research Laboratory of Engineering Materials, Report No. 3, Tokyo, 1978.

60. Witteveen, J., Twilt, L. and Bijlaard, F.S.K.: The Stability of Braced and Unbraced Frames at Elevated Temperatures, Symposium on Stability of Steel Structures, Liège, April 1977.

61. Beyer, R.: Der Feuerwiderstand von Tragwerken aus Baustahl. Berechnung mit Hilfe des Traglastverfahrens, Stahlbau, Vol. 46, No. 12, 1977.

62. Beyer, R. and Hartmann, B.: Eine Untersuchung des Lastfalles Brand bei Stahlrahmen, Kolloquium für Finite Elemente in der Baupraxis, Hannover, April 1978.

63. Iding, R.H. and Bresler, B.: Effect of Fire Exposure on Steel Frame Buildings (Computer Model FASBUS II), Final Report WJE 78124 Wiss, Janney, Elstner and Associates, Inc., September 1981.

64. Peterson, A.: Finite Element Analysis of Structures at High Temperatures, with Special Application to Plane Steel Beams and Frames, Lund Institute of Technology, Division of Structural Mechanics, Report TVSM-1001, Lund, 1984.

65. CEC Research 7210-SA/502-REFAO/CAFIR: Computer Assisted Analysis of the Fire Resistance of Steel and Composite Steel - Concrete Structures (Computer Program CEFICOSS), Technical Reports RT1-6, 1982/85

66. Forsén, N.E.: STEELFIRE - Finite Element Program for Non-linear Analysis of Steel Frames Exposed to Fire, Users Manual, Multiconsult A/S, Oslo, 1983.

67. Anderberg, Y. and Thelandersson, S.: Stress and Deformation Characteristics of Concrete at High Temperatures, Lund Institute of Technology, Division of Structural Mechanics and Concrete Construction, Bulletin 54, Lund, 1976.

68. Schneider, U.: Ein Beitrag zur Frage des Kriechens und der Relaxation von Beton unter hohen Temperaturen, Institut für Baustoffe, Massivbau und Brandschutz der Technischen Universität Braunschweig, Heft 42, Braunschweig, 1979.

69. Thelandersson, S.: On the Multiaxial Behaviour of Concrete Exposed to High Temperature, Nuclear Engineering and Design, Vol. 75, No. 2, 1983.

70. Becker, J.M. and Bresler, B.: FIRES-RC, A Computer Program for the Fire Response of Structures - Reinforced Concrete Frames, University of California, Fire Research Group, Report No. UCB FRG 74-3, Berkeley, 1974.

71. Bresler, B.: Response of Reinforced Concrete Frames to Fire. - Bresler, B. et al: Limit State Behaviour of Reinforced Concrete Frames in Fire Environments, University of California, Fire Research Group, Report No. UCB FRG 76-12, Berkeley, 1976.

72. Haksever, A.: Rechnerische Untersuchung des Tragverhaltens von einfach statisch unbestimmten Stahlbetonrahmen unter Brandbeanspruchung, Institut für Baustoffkunde und Stahlbetonbau der Technischen Universität Braunschweig, 1975.

73. Klingsch, W.: Traglastberechnung thermisch belasteter Stahlbetondruckglieder unter Anwendung einer zwei- und dreidimensionalen Diskretisierung, Institut für Baustoffkunde und Stahlbetonbau der Technischen Universität Braunschweig, 1975.

74. Anderberg, Y.: Fire Exposed Hyperstatic Concrete Structures - An Experimental and Theoretical Study, Lund Institute of Technology, Division of Structural Mechanics and Concrete Construction, Bulletin 55, Lund, 1976.

75. Haksever, A.: Zur Frage des Trag- und Verformungsverhaltens ebener Stahlbetonrahmen im Brandfall, Institut für Baustoffkunde und Stahlbetonbau der Technischen Universität Braunschweig, Heft 35, Braunschweig, 1977.

76. Quast, U., Hass, R. and Rudolph, K.: STABA-F; A Computer Program for the Determination of Load-Bearing and Deformation Behaviour of Uni-Axial Structural Elements under Fire Action, Institut für Baustoffe, Massivbau und Brandschutz, Technische Universität, Braunschweig, 1984.

77. Nizzamuddin, Z.: Thermal and Structural Analysis of Reinforced Concrete Slabs in Fire Environments, Thesis, University of California, Berkeley, 1976. - Nizzamuddin, Z. and Bresler, B.: Fire Response of Concrete Slabs, ASCE Proceedings, Journal of Structural Division, 105 (ST8), August 1979.

78. Haksever, A.: Stützenatlas im Brandfall - Ein praxisorientiertes Rechenverfahrens zur Bestimmung der Feuerwiderstandsdauer von Stahlbetonstützen im Brandfall, Sonderforschungsbereich "Brandverhalten von Bauteilen", Technische Universität, Braunschweig, 1978.

79. Klingsch, W. and Henke, V.: Feuerwiderstandsdauer von Stahlbetonstützen - baupraktische Bemessung, Sonderforschungsbereich "Brandverhalten von Bauteilen", Technische Universität, Braunschweig, 1978.

80. Hertz, K.: Design of Fire Exposed Concrete Structures, Report No. 160 - Stress Distribution Factors, Report No. 158, Technical University of Denmark, Institute of Building Design, Lyngby, 1981.

81. Lie,T.T.: A Method for Assessing the Fire Resistance of Laminated Timber Beams and Columns, Canadian Journal of Civil Engineering, Vol. 4, No. 2, 1977.

82. Fredlund, B.: Structural Design of Fire Exposed Rectangular Laminated Wood Beams with Respect to Lateral Buckling, Lund Institute of Technology, Department of Structural Mechanics, Report No. 79-5, Lund, 1979.

247

# Heat Conduction in Insulated Metal Roof Decks during Fire: A Computational Approach

**D. BREIN and P. G. SEEGER**
Forschungsstelle für Brandschutztechnik
an der Universität Karlsruhe (TH)
Hertzstr. 16, D-7500 Karlsruhe 21, FRG

ABSTRACT

The heat conduction through lightweight roofs is modeled with melting and decomposition processes involved. The derived set of equations is solved using
a numerical difference procedure, which has been enlarged widely for these
processes. The computer simulations for several roof assemblies are compared
with experimental results from large scale fire tests with flat lightweight
roofs exposed to point fire sources underneath the roof. The simulation
results show good agreement with data points as to the tendency of the curves.
The maximum temperatures reached show minor deviations from the test data and
give a good estimate for real situations. Differences among the curves are due
to scarcely known physical parameters with temperature and perhaps to some
simplifying assumptions whose validity will be studied in further work.

INTRODUCTION

The extent to which lightweight insulated steel deck roofs become involved in
a fire, esp. in large industrial premises, made it necessary to carry out
large scale fire tests with usual roof assemblies in addition to those tests
already mentioned in a literature review by Hofmann /1/. The test program was
planned with the intention to study the behaviour of a roof before flash-over
would occur in a fire test room. A series of six preliminary tests with relatively small point fire sources had been performed by Brein and Seeger /2/ in
1977, while a subsequent test series of another eleven large scale fires with
higher fire loads up to 400 kg wood cribs was completed at the Fire Research
Station in Karlsruhe recently. These fire loads had been concentrated to a
small part of the total area of the fire test room. The final test report by
Brein and Seeger /3/ and the conclusions drawn herein have led to proposals
for a better design of roof assemblies towards a higher fire safety.

The tests were accompanied by an ad hoc committee of experts with the intention to have a test program carried out which would satisfy practical needs.
A summary of the committee's work is given by Becker et.al. /4/ while an
interpretation of the design proposals with exemplary sketches is found in a
paper from Federolf /5/.

Descriptions of the weather-protected test building with a total area of 83 $m^2$
are found in /2,3/ . The roof assemblies included trapezoidally corrugated

steel decks of thicknesses .75 mm to 1.13 mm supported by steel beams 4 m above floor. These decks were covered by bituminous vapour barriers (which were sometimes omitted) and several types of combustible and non-combustible insulations of thicknesses ranging from 40 to 70 mm, depending on the individual thermal conductivity k of the tested specimen. The reason for a choice of several thicknesses was, that the product of the individual k times the thickness of the relevant insulation board should be kept constant and equal to that of expanded polystyrene foam with a bulk density of 20 kg/m$^3$ and 50 mm thickness. The insulation boards were then covered with two layers of bituminous roofing membranes according to technical rules. Further on in this paper the term "roof assembly" is used for the compound above the steel deck.

The tests have shown, that there are two main influences on the fire behaviour of lightweight roofs: the one is the construction itself, which comprises the type of fastening the decks at the side laps with each other and upon the supporting members or the design at the perimeter of the roof. As an example of the provisions made for a late collapse of the steel deck we have used rivets with a high melting point to fasten the steel decks to each other along the laps. The other main influence with respect to the time-dependent increase of surface temperature is the mechanical response of the insulation boards and of the roofing membranes to the developing fire underneath the roof since thin layers with relatively high thermal conductivity will lead to a faster increase in surface temperature than will thick layers with a relatively low thermal conductivity and so support ignition in an early stage of the fire.

Some insulations will melt at low temperatures of about 100 °C like expanded polystyrene foam, others will keep their shape up to very high temperatures of near 800 °C or more like mineral wool, others will show a behaviour like urethane rigid foam which pyrolyses and cracks reducing its thickness continuously while the fire is going on underneath.

The results of the fire test series mentioned in /2,3/ are very interesting but unsatisfactory as to the interpretation of the results for roofs with thicker insulations as prescribed by recent technical and administrative rules towards higher energy savings. Consequently, to diminish the costs for research on the fire behaviour of lightweight roofs, a research program has been initiated to calculate the time-dependent heat conduction through lightweight roofs with several types of insulation boards while exposed to a fire source of restricted area underneath. This work is being partially supported by the "Stiftung Volkswagenwerk", a foundation of a motor-car factory in the Federal Republic of Germany.

DEFINITION OF THE PROBLEM

The calculation of heat conduction through a lightweight roof exposed to fire from beneath firstly requires several simplifying assumptions concerning the governing mechanisms of heat transfer to and from the roof, i.e. the boundary conditions must be adequately set but such, that they can be handled by a computer program within a reasonable period of time.

In a real fire situation, the roof is being exposed to open flames or to the fire plume from combustion of e.g. stored goods. In the early stages of a fire, the main mode of heat transfer to the lower surface of the roof will be only by convection but in case, that the fire has grown, radiative transfer from open flames will perhaps become the dominant mode. The flames and hot gases impinging on the lower surface of the ceiling are being bent to the horizontal direction and flow out of the fire test room at a distance from the fire origin. It should be mentioned, that no ventilation openings had been installed in the tests referred to above and that the only opening was at the front side of the test room.

Heat passes through the steel deck by conduction. Depending on which cross-section of the roof is being considered, heat is then transferred to the lowest layer of the roof assembly by direct contact via the flanges or by radiation and by convection of horizontally flowing gases of decomposition inside the roof assembly via the ribs to that part of the lowest layer of the roof assembly, which is placed over the rib openings. The former heat conduction through the several layers of the roof assembly can be evaluated, if the melting mechanisms of the vapour barrier and the bituminous roofing membranes can be modeled. In addition the processes taking place as the insulations are being thermally cracked or molten have to be simulated. The processes considered lead to a reduction in shape and thickness and may also result in a partial outflow of molten products into the rib section of the steel deck if either no vapour barrier with a mechanically strong reinforcement is in use or if no vapour barrier is used at all. At the upper surface of the roof, radiation may be neglected as long as the temperatures are considerably lower than the ignition temperature of the roofing membranes. In this case, heat transfer by free convection has to be considered as the only heat transfer mechanism.

A lot of time-dependent temperature curves at several positions in the roofs has been measured and a lot of other data is available from the fire tests for comparison. But there is still a lack of knowledge about the dependence or physical parameters like the thermal conductivity k with temperature for most of the building materials dealt with in this paper. Therefore the values used here are approximations only on purpose to compare experimental with computed data. The reason for missing data on physical values at high temperatures is clear for materials commonly in use at room temperature.

THE MATHEMATICAL MODEL

Heat Conduction and Boundary Conditions

The mathematical model consists of the parabolic, partial-differential equation for the time-dependent heat conduction in solids, the relevant modeled equations for melting and thermal degradation processes and the boundary conditions at the lower and the upper surfaces of the roof assembly, respectively. The conventional "Fourier"- equation models the heat balance of a solid volume element. In the form used in this work, the equations that constitute the mathematical model are written as follows:

$$\frac{\partial}{\partial x}\left[k(T)\frac{\partial T}{\partial x}\right] + \frac{\partial}{\partial y}\left[k(T)\frac{\partial T}{\partial y}\right] + W = \rho c\frac{\partial T}{\partial t} \quad . \tag{1}$$

This is the two-dimensional heat flow equation, which has no closed-form solution because of the temperature dependence of the thermal properties of the materials used and also because of a nonlinearity of the boundary conditions. Among the schemes available to solve equation (1) on a computer numerically it was transformed into a set of difference equations. The well-known solution scheme will be described in short later in this paper.

The initial and boundary conditions are deduced from the practical problem. As an initial condition a unique temperature distribution was chosen in the roof assembly as a whole and equal to the constant ambient temperature in the test room before the ignition of the fire load.

The melting and degradation processes with variation of thickness of the insulation boards are modeled for one dimension in space. Consequently the description of boundary conditions is restricted to the (vertical) y-direction of the coordinate system. The boundary conditions at the lower surface of the roof depend on the roof section which is to be considered, i.e. either the region above the flanges or above the ribs of the steel deck. The first case comprises the derivation of a total heat exchange coefficient for both convective and radiative transfer to the steel deck and direct contact between the flange of the steel deck and the roof assembly. The second case comprises both modes below the steel deck and both modes between the ribs and the lower surface of the roof assembly as well.

For a comparison between the experiments and the computer results a thorough study of the heat transfer underneath the steel deck can be omitted because of test data available in the flange as well as in the rib region. The time-dependent temperatures measured are used as input values for the following boundary conditions at the lower surface of the roof assemblies, i.e. at the vapour barrier and at the insulation board, respectively:

Given the steel deck flange temperature, then

$$T_s = T_{fla} (t) \quad . \tag{2}$$

Given the temperature in the cavity of the ribs, then

$$h (T_g(t) - T_s) = -k \left[\frac{\partial T}{\partial y}\right]_s \quad . \tag{3}$$

The total heat transfer coefficient h in eq. (3) is the sum of the convective and the radiative heat transfer coefficients. The latter is evaluated using a zone method, see e.g. Hottel and Sarofim /6/.

At the interfaces between several layers of the roof assembly no special considerations are needed. This arises from the fact, that

$$- k_1 \left[\frac{\partial T}{\partial y}\right]_1 = - k_2 \left[\frac{\partial T}{\partial y}\right]_2 \quad . \tag{4}$$

At the upper side of the roof the boundary condition is given by an equation similar to eq. (3) with the restrictions, that only convection needs to be considered and that the ambient temperature is constant throughout the test simulation:

$$- h (T_s(t) - T_a) = k \left[\frac{\partial T}{\partial y}\right]_s \quad . \tag{3a}$$

The Melting Process

The melting process is modeled for the one dimensional problem. The consumption of heat of fusion is a characteristic of melting processes and if the materials are chemically pure ones, the other characteristic is a constant melting point at the solid-fluid boundary. Considering the materials used for lightweight roofs like bituminous roofing membranes or thermosetting expanded polystyrene foam (EPS), the assumption of a fixed melting point is a strongly simplifying assumption. Looking at the relevant roofing membranes a broad melting region will be found caused by the large range of hydrocarbons asphalt is consisting of. In the real

252

situation there arises an adequate difficulty with EPS which will firstly sinter forming cavities and then melt. Modeling of such processes is difficult and therefore as a first approximation the materials were assumed as pure with fixed melting points and a fixed heat of fusion. A reason for this simplification is the fact, that melting processes during real fires will often take place in a small time interval relative to the total duration of the test and so a too exact analysis would have been outside the scope of this work.

The heat input from one side will result in a continuously advancing phase boundary between the region still solid and the region already fluid of the specimen. Using a heat balance at the boundary, an expression for the phase boundary propagation velocity can be derived. As a simplifying assumption convection flows within the molten materials are assumed negligible because of their small thickness. The phase boundary itself is unable to store heat which leads to the following heat balance:

$$q_{in} = q_{out} + q_c \quad . \tag{5}$$

The terms in eq. (5) for input, output and consumption, respectively, are as defined in equations (6) to (8):

$$q_{in} = - k_{fl} \left[ \frac{\partial T}{\partial y} \right]_{ph,fl} \quad , \tag{6}$$

$$q_{out} = - k_{so} \left[ \frac{\partial T}{\partial y} \right]_{ph,so} \quad , \tag{7}$$

$$q_c = H_f \ \rho_{so} \left[ \frac{dy_{so}}{dt} \right] \quad . \tag{8}$$

Combining equations (6) to (8) with eq. (5) results in an expression for the phase boundary propagation velocity, which is a velocity term relative to the solid region of the material:

$$\frac{dy_{so}}{dt} = \frac{1}{H_f \ \rho_{so}} \left[ k_{so} \left[ \frac{\partial T}{\partial y} \right]_{ph,so} - k_{fl} \left[ \frac{\partial T}{\partial y} \right]_{ph,fl} \right] \quad . \tag{9}$$

In case of a material with a great difference in the densities of the solid and the fluid phases, respectively, the increase of the fluid phase is given by the following equation (10') :

$$\frac{dy_{fl}}{dt} = \frac{\rho_{so}}{\rho_{fl}} \frac{dy_{so}}{dt} \quad . \tag{10}$$

A typical example is EPS, where the density of the fluid phase is about 50 times the density of the solid phase.


Decomposition Kinetics

The pyrolysis and cracking of insulations of the urethane rigid foam type was modeled in a conventional manner:

$$- \frac{dm}{dt} = ( m - m_{fi}(T)) \ v(T) \quad . \tag{11}$$

253

## NUMERICAL SOLUTION

The numerical solution of eq.(1) with the boundary conditions and the expressions for the melting and degradation processes described above was performed with the Crank-Nicholson scheme for one dimensional appliances and with the Peaceford-Rachman scheme in case of two dimensions. The mathematical schemes were adopted from the literature, see e.g. Marsal /7/.

A short description of the solution procedure is concentrated upon the following remarks: using the difference scheme, the continuous governing differential eq.(1) is converted to a set of linear algebraic equations for each time step. This set of linear equations is formed by superposing a grid to the material or compound material in question and then by deriving the first and second order derivatives of the temperature values at the discrete set of grid points. The grid points need not to have a constant distance between each other. The distance is chosen variable according to the necessities of stability of the numerical solution. The set of equations thus derived forms a tridiagonal coefficient matrix which can be solved using a simple, well-known mathematical algorithm. The solution of the matrix leads to an array of temperature values and - in case of pyrolysis reactions - to an array of density values at each node of the grid. These are in turn used for the new difference quotients for a new time step and the resulting set of equations is solved as before.

In case of incorporated melting processes we have included an automatic forward time-step size such, that it can be diminished as soon as the melting point is reached. By this we avoid melting of the whole layer of material within one time step, which would bring down the accuracy of the computations. Details of the time-step size choice are beyond the scope of this paper. A full description is given by Brein /8/.

## RESULTS

The comparison of selected experimental results with the corresponding computer simulations is illustrated in Figs. 1-3 . In each of these figures the computer predictions are represented as broken lines, and the experimental results as continuous curves with symbols. The experimental curves are from the test report mentioned above /3/. The position 'F', where the temperatures had been measured using thermocouples, represents a region of the roof exactly above the fire source with the highest temperatures reached during the tests.

The computer simulations were carried out using the one dimensional approach for heat conduction through a roof assembly in the vertical direction. Horizontal gradients were neglected because of the ratio of vertical to horizontal spacing. Besides, the computations were restricted to one dimension, because melting processes with flow out of material into the rib region of the roof were modeled for this case only. The same is true for pyrolytic processes, where load bearing roof insulations will shrink in vertical direction mainly.

For the computer simulations the following assumptions were used:
- The boundary condition at the lower side of the roof assembly is the time-dependent heat flux at the position of thermocouple 'F2' in the cavity of the ribs. Above the upper side of the roof the ambient temperature of the relevant test date is chosen.
- Radiative transfer of heat from the gases within the rib region of the steel deck is neglected. The heat transfer from the ribs at temperature 'F2' to the

roof assembly due to radiation is derived using the zone method referred to above (see /5/).
- There is no heat loss of the gases within the rib section due to convective flows in horizontal direction.

Fig. 1 shows the comparison of a mineral-fibre board experiment with the simulation. In addition to the assumptions described above, the water vapour transfer through the insulation board was neglected. Flowing out of 50% of the molten part of the vapour barrier into the rib section was allowed. The trend of the computed curves is towards the real (measured) situation, but there are still differences among the curves, which can be explained with roughly approximated physical values, and perhaps with influences, which arise from the positioning of the termocouples during the tests or from water vapour in the mineral-fibre boards. The possible reasons are numerous and it seems difficult to weigh them adequately without thorough studies of each.

Fig. 2 shows the comparison of the EPS test with the computer simulation. With the real behaviour of an expanded polystyrene foam and the approximation of assuming a pure material in mind, the simulation result is rather good. The authors suggest, that the differences might mostly be due to a fixed thermocouple in the test and the formation of cavities within the insulation board, which might lead to questionable data. Another reason is supposed to be unreliable physical values at higher temperatures.

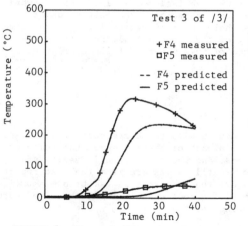

FIGURE 1. Mineral fibre insulation with vapour barrier.

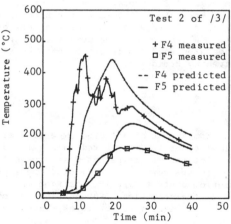

FIGURE 2. Expanded polystyrene foam with vapour barrier.

Fig. 3 shows the comparison of the urethane rigid foam test without vapour barrier with the simulation. Some assumptions concerning the kinetic parameters were incorporated in the simulations. These data were derived from small scale experiments carried out in a laboratory oven to determine the Arrhenius-rate constant, the activation energy and the final density. The heat of the decomposition reaction was evaluated using the formation enthalpies of plausible pyrolysis reactions. Fig. 3 shows a fairly good agreement between test and simulation, esp. with respect to the highest temperature value reached in the midst of the insulation board ( thermocouple 'F4'). But there are still differences as to the time scale.

Fig. 4 shows the variation of the specific mass loss with vertical distance with the time as parameter and thus demonstrates the degree of pyrolysis of the foam.

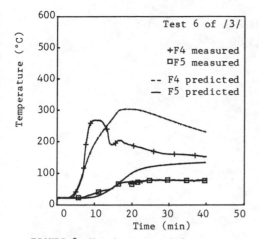

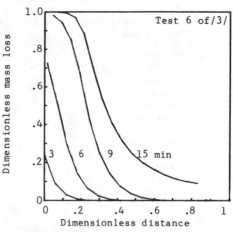

FIGURE 3. Urethane rigid foam,
no vapour barrier.

FIGURE 4. Variation of mass loss with
distance in the PU board.

## SUMMARY

The heat conduction through lightweight roofs is modeled with melting and decomposition processes involved. The derived set of equations is solved using a
numerical difference procedure, which has been enlarged widely for these processes.
The computer simulations for several roof assemblies are compared with experimental results from large scale fire tests with flat lightweight roofs exposed to
point fire sources underneath the roof. The simulation results show good agreement
with data points as to the tendency of the curves. The maximum temperatures reached
show minor deviations from the test data and give a good estimate for real situations. Differences among the curves are due to scarcely known physical parameters
with temperature and perhaps to some simplifying assumptions, whose validity will
be studied in further work.

## NOMENCLATURE

| | |
|---|---|
| E | activation energy (J/mole) |
| H | heat of ... (J/g) |
| R | universal gas constant (J/(mole K)) |
| T | temperature (K) |
| V | volume , volume element ($m^3$) |
| W | heat sink (J/($m^3$ s)) |
| c | heat capacity at constant pressure (J/(g K)) |
| f | frequency factor ($s^{-1}$) |
| h | heat transfer coefficient (J/($m^2$ s K)) |
| k | thermal conductivity (J/(m s K)) |
| m | mass (kg) |
| q | heat flux (J/($m^2$ s)) |
| t | time (s) |
| u | dimensionless mass loss ($-1$) |
| v | reaction rate constant ($s^{-1}$) |
| x | coordinate (m) |
| y | coordinate (m) |

z          coordinate (m)
μ          specific mass loss for a fully decomposed material ( - )
ρ          density ($kg/m^3$)

Subscripts

a          ambient
act        actual
c          consumption
f          ... fusion
fi         final
fl         fluid phase
fla        flange of steel deck
g          gas
in         input
max        maximum
min        minimum
o          initial
out        output
ph         at phase boundary
r          ... decomposition reaction
s          surface
so         solid phase

REFERENCES

1.  Hofmann,K., VFDB-Z 27, 4(1978).

2.  Brein,D. and Seeger,P.G., Fire and Materials, Vol. 3, No.3, 1979.

3.  Brein,D. and Seeger,P.G., Brandversuche an wärmegedämmten Stahltrapezprofil-
    dächern, Forschungsstelle für Brandschutztechnik, Karlsruhe (1982).

4.  Becker,W. et. al., VFDB - Z 33,2(1984).

5.  Federolf,S., VFDB - Z 33,2(1984).

6.  Hottel.H.C. and Sarofim,A.F., Radiative Transfer, McGraw-Hill (1967).

7.  Marsal,D., Die numerische Lösung partieller Differentialgleichungen,
    Bibliographisches Institut AG, Zürich (1976).

8.  Brein,D., Final report in preparation (1985).

# Measured and Predicted Behaviour of Steel Beams and Columns in Fire

**Y. ANDERBERG**
Lund Institute of Technology
Lund, Sweden

**N. E. FORSÉN**
Multiconsult A/S
Oslo, Norway

**B. AASEN**
NTH-Sintef
Trondheim, Norway

ABSTRACT

Analytical predictions of mechanical behaviour of fire-exposed steel structures are compared to experimental results obtained from three different research laboratories. Comparisons are made to axially free and restrained steel columns fire tested in Metz in France 1973-74, to simply supported steel beams fire tested in Germany and published in Stahlbau 1/1983 and to axially free and restrained steel columns fire tested in Trondheim in Norway 1984.

The measured time-temperature state of the steel structures is used as input information for the analytical prediction of the mechanical behaviour. For the analysis, the structural computer program Steelfire is used. Steelfire is a FEM-program and originates from NTH, Norway. The influence of the degree of axial restraint, load eccentricity and initial deformation in accordance to Dutheils formula are examined. Modelling of mechanical behaviour of steel is also presented.

Predictions and experimental results agree reasonably well, which illustrate the capability and reliability of the program Steelfire.

## 1. MECHANICAL BEHAVIOUR MODEL OF STEEL

It is generally proved that the deformation process of steel at transient high temperatures can be described by three strain components according to the constitutive equation

$$\varepsilon = \varepsilon_{th}(T) + \varepsilon_{\sigma}(\sigma,T) + \varepsilon_{cr}(\sigma,T,t) \tag{1}$$

where

$\varepsilon_{th}$ = thermal strain

$\varepsilon_\sigma$ = instantaneous, stress-related strain based on stress-strain relations, obtained under constant, stabilized temperature

$\varepsilon_{cr}$ = creep strain or time dependent strain, determined by ordinary creep tests at constant, stabilized stress and temperature

$\sigma$ = stress

T = temperature

t = time

A computer adapted mechanical behaviour model for steel, based on Eq. (1), is developed in Anderberg (1976) I1I and applied in Steelfire I2I.

The strain components are found separately in different steady state tests. It is shown that a behaviour model based on steady state data satisfactorily predicts behaviour in transient tests under any given fire process, load and strain history.

An analytical description of the $\sigma-\varepsilon$ curve as a function of temperature can be made in different ways as illustrated in Figs. 1 and 2. In the first case the curve is approximated by piecewise linear lines (used in Steelfire I2I) and in the second case by an elliptic branch placed between straight lines. In Magnusson (1974) I3I an analytical expression derived by Ramberg and Osgood (1943) I4I was used as follows

$$\varepsilon = \sigma/E_{0.T} + 3/7 * \sigma_{0.2}/E_{0.T} * (\sigma/\sigma_{0.2.T})^m \tag{2}$$

where

$E_{0.T}$ = modulus of elasticity at temperature $T°C$
$\sigma_{0.2}$ = yield strength or proof strength at room temperature
$\sigma_{0.2.T}$ = yield strength or proof strength at temperature $T°C$
m = temperature dependent factor.

For good convergence in computations a smooth curve is to prefer.

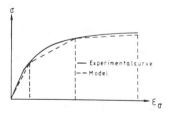

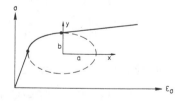

FIGURE 1.   Simplified model of the stress-strain curve for steel (used in Steelfire)

FIGURE 2.   Refined model of stress-strain curve for steel

Models of high temperature creep are in most cases based on the theory put forward by Dorn (1954) I5I, in which the effect of variable temperature is considered by the use of the concept temperature compensated time. The extension of the model to be applicable to variable stress can, for instance, be based on the strain hardening rule.

## 2. STRUCTURAL COMPUTER PROGRAM STEELFIRE

Steelfire is developed by Forsén (1983) I2I and the program analyses plane steel frames subjected to inplane loading and temperatures varying with time.

The analysis is based on a displacement formulation of the finite element method using straight beam elements. The nonlinear geometric effects (large displacements) are taken into account by updating the nodal coordinates of the structure during deformation. Nonlinear, temperature dependant material properties are considered and the current temperature distribution for each fire zone are recorded step by step from a temperature file.

## 3. EXPERIMENTAL INVESTIGATIONS, USED FOR ANALYTICAL SIMULATIONS

Three different experimental investigations are looked upon as follows.

a) At the Fire Research Station in Metz in France tests were carried out 1973-74 on steel column of box-girder profile (RHS) with different slendernesses $\lambda$ = 40, 80 and 120 with an effective length of 3.84 m at different degrees of axial restraint, load levels and rates of heating. These tests are analysed and reported by Magnusson (1973) I3I. In this paper, only 3 tests are analysed as shown in Table 1.

Table 1    METZ-COLUMNS

| Column test | Initial Load level kN | Degree of axial restraint $\gamma$ | Load eccentricity mm | Collapse temperature $^\circ$C |
|---|---|---|---|---|
| M1 - 3:4 | 460 | 1.0 | 0 | 376 |
| M2 - 5:4 | 460 | 1.0 | 7.2 | 319 |
| M3 - 6:3 | 230 | 0.8 | 0 | 480 |

The box-girder columns considered are characterized by steel 1411 with a yield stress = 343 MPa, a slenderness ratio = 80, with b = 127 mm, profile thickness t = 9.5 mm and by a slow rate of heating = 7°C/min. $\gamma$ represents the degree of axial restraint where $\gamma$ = 0 means full restraint and $\gamma$ = 1 no restraint at all.

Great difficulties to measure deformations were reported and the control of restraint was not perfect. Therefore the experimental results must be taken with care.

b) In Stahlbau 1/1983 I6I Reyer and Nölker present experimental results from beam and column tests carried out in FRG. In this paper a beam test is ana-

lysed. The beam consists of an IPE 80 profile (h = 80, b = 46 $t_{f}$ =
= 5.2, $t_w$ = 3.8 mm) of St 37 with a yield stress of 392 MPa. The beam has
a length l=1.14 m and is loaded by a concentrated load at the midsection.
The rate of heating is about 40 °C/min i.e. an uninsulated steel beam.
This rate of heating gives rise to thermal gradients over the section but
the value presented is an average value across the section as well as along
the beam. This simplification makes a comparison between test and prediction approximate.

c) A much more comprehensive and well documented experimental investigation is
carried out by Aasen (1985) I7I on steel columns. The main calculations are
focused on these tests.

The experimental program comprised 15 pinned and 5 axially restrained column
tests. All specimens were made of IPE 160 section and had a yield strength
$\sigma_{0.2}$ = 448 MPa. A complete description of the test series is given in
Aasen 1985 I7I and in this paper only 5 column tests are presented as illustrated in Table 2. The slenderness ratio is 92 and the length of the columns
is 1.7 m (b = 82, h = 160, $t_{f1}$ = 7.4, $t_w$ = 5 mm).


Table 2    AASEN-COLUMNS

| Column test | Initial load level kN | Degree of axial restraint $\gamma$ | Load eccentricity mm | Rate of heating T°C/min |
|---|---|---|---|---|
| A1 (16) | 98 | 1.0 | 0 | 7.7 |
| A2 (19) | 98 | 1.0 | 14 | 8.0 |
| A3 (20) | 97.9 | 1.0 | 20 | 8.7 |
| A4 (17) | 98.3 | 0 | 0 | 8.4 |
| A5 (18) | 196 | 0 | 0 | 8.4 |

The purpose of this study was to perform an experimentally well-defined simulation of the fire behaviour at an exposure according to a typical standard
fire test, Fig. 3. The tests were carried out by means of electrical heating
equipment using conventional laboratory facilities. As reference, a test specimen from I8I was chosen. The maximum applied load was adjusted according to
the recommendations of ECCS, I9I, assuming the original specimen with both
ends built in.


The experimental set-up is shown in Fig. 4. The test specimens were mounted in
a vertical position and bolted to end fixtures which acted as hinged bearings.
The column ends were braced against lateral displacement and torsion. The
loading was applied by a 400 kN Amsler hydraulic jack with a load cell located
at the top of the columns.

The heating was attained by 6 low voltage elements attached to the outside of
the flanges, 3 elements on each side. A typical heating element consisted of a
5.5 m stranded wire running in loops through ceramic beds. The test specimen
was finally insulated with blankets of ceramic fibres which is a material made
from synthetic mixes of aluminia and silica. The power unit provided a 60 V
supply for the heat input, totally 48 kW.

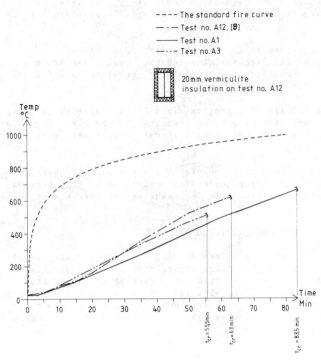

**FIGURE 3.** Standard fire curve and temperature curve for different tests

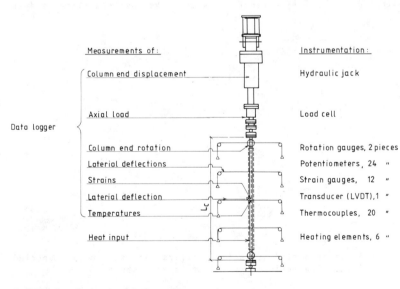

**FIGURE 4.** Experimental arrangements

The testing procedure comprised two phases. Firstly, the load was applied in increments at room temperature. Secondly, at a prescribed level the load was kept constant with increasing temperatures. Alternatively, the axial column end displacement was fixed in order to introduce an axial restraint, i.e. $\gamma = = 0$.

In general, the failure occurred by in-plane buckling about the weak axis. However, flexural-torsional buckling was observed when conducting tests with restraining beams. Due to a modification of the bracings at the end fixtures, the test rig became sensitive to torsion.

The temperature measurements showed approximately uniform temperatures along the central part of the specimens with steep gradients at the ends. Regarding columns Nos. 1-5, negligible cross-sectional temperature gradients were recorded.

In conclusion, the test procedure described in this paper allows representative steel columns to be tested under controlled conditions of high temperatures. Thus, it is possible to investigate the behaviour of fire exposed steel members in a simple way.

4. COMPARISON BETWEEN MEASURED AND PREDICTED BEHAVIOUR

The structural response is predicted by use of Steelfire for the structural members described in previous chapter. The temperature input for Steelfire is taken direct from measurements as an average value varying with time and representative for the whole structure. When the test is performed in a furnace with gas- or oilburners (tests in Metz and FRG), this simplification is very rough. When the heating is simulated by electrical elements this approximation is more adequate.

4.1 Beams

The structural response of a simply supported, fire-exposed steel beam (IPE 80) is illustrated in Fig 5 by deflection curves. The test conditions and the

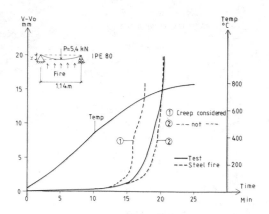

FIGURE 5. Measured and predicted deflection of fire-exposed simply supported beam tested in FRG I6I. The influence of creep is also shown

average temperature as function of time is also given in the figure. The predicted deflection process is shown for the cases with and without creep and
its importance is obvious. The predicted deflection curve is in satisfactorily
agreement with the measured curve where the failure time is 17.5 and 20.5 min
respectively. If creep is neglected, the agreement happens coincidentally to
be better than if creep is considered. Temperature approximations and the
incomplete documentation of the test make results somewhat uncertain.

## 4.2 Columns

### 4.2.1 Simply supported

The deflection behaviour of Metz-columns M1 and M2 is illustrated in Fig 6.
The load is 460 kN but only M2 is eccentrically loaded (e=7.2 mm) and the average temperature is shown. Even if measured deflection is somewhat uncertain
the predictions are quite close but failure time is attained about 8 min earlier in the tests. One important reason for that is the initial deformation of
the steel column which in accordance with Dutheils formula is $f=4.8\times10^{-5}$
$L,2/d = 11$ mm. Unintentional load eccentricity may also influence. From Fig
6 it can be seen that the test M1 coincides with the prediction of M2 representing a load eccentricity of 7.2 mm which is less than f!

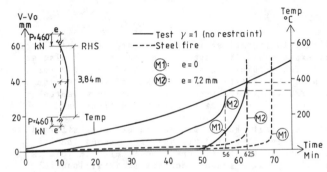

FIGURE 6. Measured and predicted deflection of from all sides fire-exposed
steel columns tested in Metz 131. The influence of eccentricity is also shown

The prediction of the Aasen tests A1, A2 and A3 with the load eccentricities 0,
14 and 20 mm respectively are shown in Fig 7. The load level is 98 kN and the
initial deformation as calculated above is only 2 mm. The temperature curve
for every individual test is given. The agreement between predicted and measured curves is very good and the initial deformations and unintensional load
eccentricity can easily explain the differencies.

The concordance is very much due to well performed tests under controlled conditions.

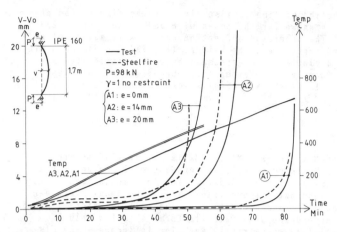

FIGURE 7. Measured and predicted deflection of from all sides fire-exposed steel columns tested in Norway I7I. The influence of eccentricity is also shown

## 4.2.2 Axially restrained

Full restraint is very difficult to accomplish in practise and it is also very hard to simulate and control experimentally a specific degree of restraint as experienced in the Metz-tests. Two predictions are illustrated in Fig 8 for the case full restraint $\gamma = 0$ at the initial load = 460 kN and for the case $\gamma = 0.8$ at the initial load = 230 kN. In the prediction of M3 the measured variation of axial force was followed in the calculation of deflection.

In the simulated case of full restraint the maximum restraint force attained about 900 kN after already 10 min and then it diminished and passed the initial force after 30 min which corresponds to failure time if minimum load is 460 kN. This rapid increase followed by a sudden decrease in axial load is very typical for axially restrained colums.

Deflection measurements are not made for the test M3 but failure temperature is 480°C compared to predicted 375°C. Experimental difficulties may have influenced the result. When the measured variation of the axial restraint force is followed in the calculation it is noticed that the prediction of the deflection process may differ very much from the measured curve. This is due to the very high sinsitivity to the restraint condition and the sudden decrease in restraint force but also due to the discrepancy between "real" and assumed mechanical and geometrical properties as well as the temperature distribution of the steel member.

The reliability of the column tests carried out by Aasen are comparatively good but one problem has arisen namely to obtain a complete restraint. The tests A4 and A5 with the load levels 98 and 196 kN at approximately full restraint have principally the same behaviour as the fully restrained Metz-column discussed above. The initial deformation is calculated to f = 2 mm.

The predicted behaviour of the test A4 (assumed eccentricity e = 1 mm) can be followed in Fig 9 where the axial force increases rapidly and attain a maximum value of 380 kN after 10 min. The test value is 345 kN attained after 22 min

266

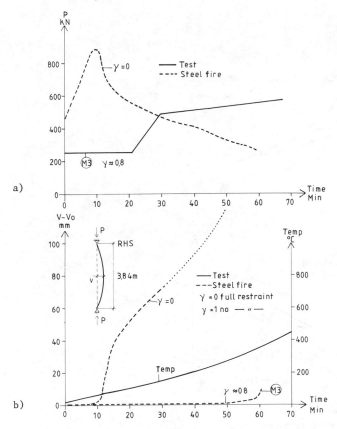

FIGURE 8. Measured and predicted behaviour of from all sides fire-exposed and axially restrained steel columns tested in Metz in France I3I
a) Axial restraint force    b) Deflection

but the discrepancy is due to inevitable movements in the loading arrangements at the supports. Therefore a prediction is made where the column was free to move axially 1 mm before full restraint was applied and the behaviour was very much influenced. The development of axial force started after 16 min and met the measured maximum value and followed after that the measured curve. Deflection curves came also very close to each other. If however a spring of appropriate stiffness (representative for load arrangement) was applied at the supports until $\Delta L$ attained 1 mm also the first part of the curve of axial force would be close. The observed phenomena illustrates the sensitivity of the degree of full restraint on the axial force.

The influence of increasing the eccentricity to 2 mm (compare to f = 2 mm) is also shown but will only influence the level of axial restraint force.

The result of the test A5 is given in Fig 9 but not predicted. In this test the higher initial load level (196 kN) results in a higher maximum load 420 kN occuring already after 18 min.

267

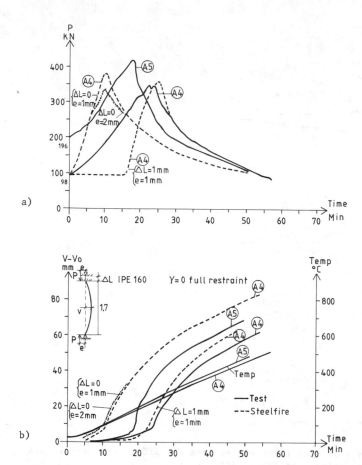

FIGURE 9. Measured and predicted behaviour of from all sides fire-exposed and axially restrained steel columns tested in Norway I7I
a) Axial restraint force        b) Deflection

## 5. HOW TO REDUCE THE DISCREPANCY BETWEEN MEASURED AND PREDICTED STRUCTURAL BEHAVIOUR

A detailed documentation of tests performed under thoroughly controlled conditions with an equipment of high reliability is a desirable situation for predicting structural response with good agreement to measurements.

In order to obtain this good concordance the theoretical modelling of the mechanical behaviour of the materials involved, the capability and reliability of the structural program is of decisive importance.

The equipment used for testing fire-exposed structures with complicated support conditions have been improved all the time and in this paper the importance of successful tests for good agreement is demonstrated.

It must be emphasized that a complete end restraint is almost impossible to obtain but this situation will never exist in practise either.

Electrical heating on steel structures gives the best control of temperature state with very small gradients along and across the member as a result [7].

Improvements in modelling can also be made. The $\sigma$-$\varepsilon$ curve ought to be a smooth curve instead of a piecewise linear relationship to improve convergence. Creep parameters are tabled for different kind of steels but there can be a considerable difference between tabled and 'real' values. There is a need for more data on creep properties.

## 6. CONCLUSIONS

* Steelfire has a good ability to describe real behaviour at fire.

* Test conditions are difficult to define beyond all doubts, especially at a high degree of restraint against axial deformation.

* Steelfire can be used to derive and describe real test conditions of experiments and thereby facilitate analysis and generalization of test data. However, the modelling is still a simplification or approximation of the real situation. It is always a coincidence if the predicted curve exactly follows the measured curve.

## REFERENCES

1. Anderberg, Y. : Fire-Exposed Hyperstatic Concrete Structures - An Experimental and Theoretical Study. Division of Structural Mechanics and Concrete Construction, Lund Institute of Technology. Bulletin 55, 1976. Lund.

2. Forsén, N.E. : Steelfire - Finite Element Program for Nonlinear Analysis of Steel Frames Exposed to Fire Users Manual. Multiconsult A/S, Oslo. 1983.

3. Magnusson, S.E. : Stålpelares Verkningssätt och Bärförmåga vid Brand. Säkerhetsproblemet vid Brandpåverkade Stålkonstruktioner. (Structural Behaviour and Loadbearing Capacity of Fire-exposed Steel Columns. Safety Problem of Fire-exposed Steel Structures). Division of Building Fire Safety and Technology. Lund Institute of Technology, Lund 1974. (In Swedish).

4. Ramberg, W. and Osgood, W. : Description of Stress-Strain Curves by Three Parameters. NACA Technical Note No 902, 1943.

5. Dorn J.E. : Some Fundamental Experiments on High Temperature Creep. Journal of Mechanics and Physics of Solids 3 (1954) 35,London.

6. Reyer, E & Nölker, A.: Zum Brandverhalten von Gesamtkonstruktionen des Stahl- und Stahlverbundbaues. 1 Teil: Verfahren, Eignungstests und Vergleichberechnungen zur experimenteller Untersuchung mit Grossmodellen (Fire behaviour of steel structures). Der Stahlbau 1/1983.

7. Aasen, B. : Buckling of Steel Columns at Elevated Temperatures. Dr.Ing Thesis, NTH, Trondheim 1985 (in preparation).

8. Knublauch et al : Berechnung der Stahltemperatur von Stahlstützen. (Calculation of Temperature for Steel Columns). Der Stahlbau 6 and 8, 1974.

9. ECCS : European Recommendations for the Fire Safety of Steel Structures. Elsevier SC. Pub. Co, Amsterdam - Oxford - New York, 1983.

# Structural Behaviour of Steel Frame in Building Fire

**K. NAKAMURA**
Building Research Institute
Ministry of Construction, Japan

**K. SHINODA**
Gifu Prefectural Government, Japan

**M. HIROTA**
Shimizu Construction Co. Ltd., Japan

**K. KAWAGOE**
Center for Fire Science and Technology
Science University of Tokyo, Japan

ABSTRACT

More than fifty experiments were made with model steel frames of two dimensional two story and of three dimensional three story in which girders or a column or both a girder and columns were heated by electric furnaces under a constant heating rate. These experiments made it possible to analyze thermal stress within the steel frame theoretically. Large thermal stress appeared in the frame corresponding to the "binding modulus" of the frame. The structure was more influenced by the thermal expansion generated by the heated girder than by that of the heated column in the elastic region. In some cases the buckling of heated girder or column occurred.

And also heated tests of columns in a six story full scale steel frame were made. The local buckling of column occurred from which the whole structure was influenced.

keywords: Thermal stress, Fire resistance, Building fire, Steel frame, Buckling

## 1. INTRODUCTION

As often pointed out, the accepted method of fire design based on standard fire resistance tests of structural elements has many problems. If fire design is going to keep pace with modern trends in other engineering design disciplines, we must appraise structural fire safety by an analytical structural fire engineering design method based on the structural behavior of the complete steel frames in fire. Recently, several such methods have been proposed.[1,2] These methods are important to the establishment of the analytical fire engineering design method. To investigate the applicability of these methods, a series of experimental studies based on heat tests of steel frames were started in 1981. These tests were:

1981: 22 Heat tests of a girder in a two dimensional two story model steel frame[3]

1982: 16 Heat tests of girders in a three dimensional three story
      steel frame[4]
1983: 13 Heat tests of girders or a column or both a girder  and
      columns in the same frame[5]
1984: 2  additional tests of 1983 and 4 heat tests of a column in
      a full scale six story steel frame.

This paper summarizes the experimental results of these heat
tests.


## 2. EXPERIMENTS OF MODEL STEEL FRAME

### 2.1 Experimental Set-up

(1) Model steel frame
    Fig.1 shows the two-dimensional model steel frame tested  in
1981. Many interesting results were obtained to use this frame[3],
but the test results showed us that the ratio of girder and column
length was not suitable for model tests and the effect of the
distortion of frame could not be eliminated. Then the new three-
dimensional model steel frame was designed as shown in Figs. 2-1
and 2-2. The columns and girders were connected by high-tension
bolts so that it is possible to change the element.
(2) Electric furnace
    Box shaped furnaces, composed of thermal insulation board in
which electric heat panels were set were used. Fig.3 shows
examples of the temperature distribution in the girder and column.
(3) Loading set-up
    The loading set-up was composed of hydraulic jacks and a
loading frame, as shown in Fig.2. The compressive loads were
applied on the top of columns by hydraulic jacks.

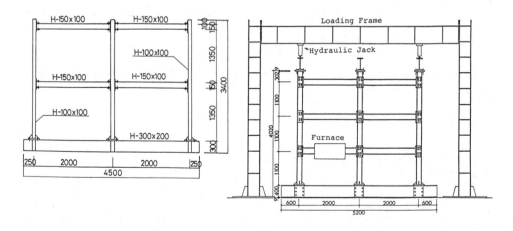

Fig.1 Two dimensional model     Fig.2-1 Three dimensional model

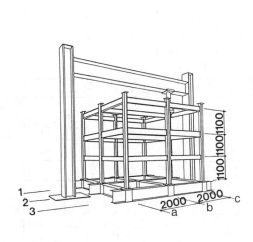

Fig.2-2 Three dimensional model

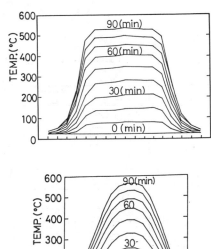

Fig.3 Temperature distribution
(upper:girder, lower:column)

## 2.2 Experimental Procedure

(1)Experimental model
　　Table 1 shows the types of frame and the positions of electric furnaces for each experiment in the series of 1982, and Table 2 shows one of in 1983.
(2)Experimental measurements
　　(a)Temperature: Room temperature, non-heated steel frame temperature, temperature of heated girder and column, and the ambient temperature in the furnace were measured by thermocouples(CA). (b)Strain: The compressive and tensile strains of the steel members were measured by strain gauges which were attached on the surface of steel members. They were located at the top and bottom of the columns and the near positions of both end of the girders. (c)Displacements: The horizontal and vertical displacements of the column-girder connection points were measured by the dial gauges.

## 2.3 Experimental Results

　　The distribution of bending moment inside frame were calculated from the measured values of strain gauges and the deformation of frame were obtained from horizontal and vertical displacements measured at the column-girder connection points. The examples of their diagrams are shown in Figs. 4 and 6.

273

Table 1 Experimental model of 1982

| A-1 | A-2 | B-1 | B-2 | B-3 | B-4 |
|---|---|---|---|---|---|
| 15.1 ton | 15.0 ton | | | | |
| B-5 | C-1 | C-2 | C-3 | C-4 | D-1 |
| D-2 | D-3 | E-1 | E-2 | F-1  F-2  F-3 | |
| | | 17  34  15 | 15  30  15 | P | F-1  0 t<br>F-2  15.0 t<br>F-3  23.7 t |

COLUMN
H-100×100 ——— H / ······· L
H-150×150 ——— H / ------- L

GIRDER
H-100×50
H-150×100
H-200×100

Table 2 Experimental model of 1983

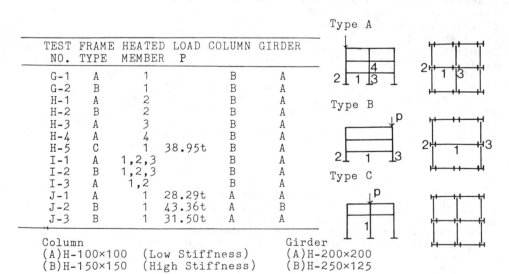

| TEST NO. | FRAME TYPE | HEATED MEMBER | LOAD P | COLUMN | GIRDER |
|---|---|---|---|---|---|
| G-1 | A | 1 | | B | A |
| G-2 | B | 1 | | B | A |
| H-1 | A | 2 | | B | A |
| H-2 | B | 2 | | B | A |
| H-3 | A | 3 | | B | A |
| H-4 | A | 4 | | B | A |
| H-5 | C | 1 | 38.95t | B | A |
| I-1 | A | 1,2,3 | | B | A |
| I-2 | B | 1,2,3 | | B | A |
| I-3 | A | 1,2 | | B | A |
| J-1 | A | 1 | 28.29t | A | A |
| J-2 | B | 1 | 43.36t | A | B |
| J-3 | B | 1 | 31.50t | A | A |

Type A
Type B
Type C

Column
(A)H-100×100  (Low Stiffness)
(B)H-150×150  (High Stiffness)

Girder
(A)H-200×200
(B)H-250×125

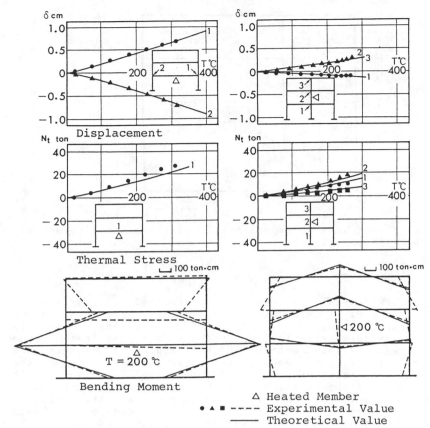

Displacement

Thermal Stress

Bending Moment

△ Heated Member
● ▲ ■ ---- Experimental Value
——— Theoretical Value

Fig.4 Theoretical and experimental Values of displacement, thermal
stress and bending moment

## 3. RESULTS AND DISCUSSION

### 3.1 Heat Tests of Column or Girder

Fig.4 shows some examples of the theoretical and experimental
values of displacement, thermal stress and bending moment when a
girder or a column was heated. The theoretical values were
obtained from the method of Saito[6-8] based on the idea of
coefficient of stiffness for expansion as shown in Appendix. T in
Fig.4 means the mean temperature rise of heated steel member. The
theoretical values agree with the experimental ones very well in
the elastic region.[4]

The buckling of heated girder occurred in the case of D-1,2
and 3 in Table 1. The comparison between experimental and
theoretical values of the buckling stress is shown in Fig.5. The
sum of the thermal stress increases as the temperature rise
increases and eventually reaches the theoretical values of the
buckling stress which decreases as the temperature rise.

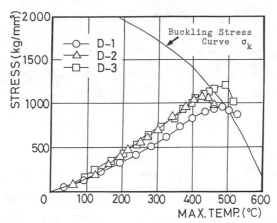

Fig.5 Comparison between experimental and theoretical values of buckling stress

## 3.2 Heat Tests of Both Columns and a Girder

The correlations between the theoretical values and the experimental ones of deformation and bending moment when columns and a girder were heated are shown in Fig.6. In this case, the theoretical values were obtained from a more sophisticated method named "direct stiffness method".[5] From Fig.6, it is seen that the theoretical values agree with the experimental values and that the thermal stresses resulted from the thermal expansion of the heated girder and the restriction of heated members imposed by the remainder of the frame. The results also show that the structure is more influenced by the thermal stress induced by a heated girder than that of heated columns in the elastic region.

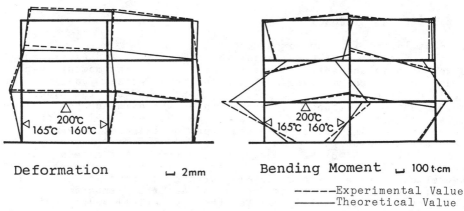

Fig.6 Correlation between theoretical and experimental value of deformation and bending moment

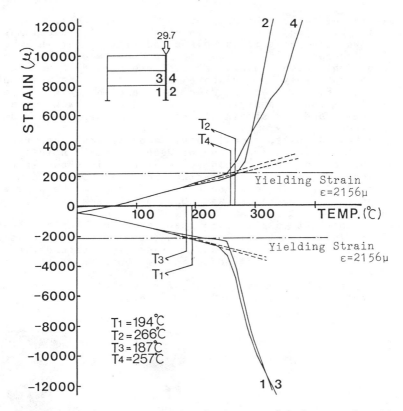

Fig.7 Relation between measured strain of column and estimated mean temperature of heated girder

## 3.3 Yield of Restraining Column

Damage to the restraining columns was different from that of the heated members. This was demonstrated in the experiment in 1984 as shown in Fig.7. On the basis of measurement of strain, it could be expected that the column would yield at their end. Fig.7 shows the relation between the measured strain of the column and the estimated mean temperature of heated girder. The following facts could be deduced; the bending strain in the column increases linearly as the increase of temperature of the heated girder, and the strain increases rapidly after the yielding occurs.

## 4. EXPERIMENTS OF FULL SCALE STEEL FRAME

Fig.8 shows the full scale six story steel frame of two spans used in heat tests. This structure was originally elected for the US-Japan cooperative study on the structural behavior under the earthquake force. After the completion of their tests, heat tests were made in this structure. The connected part of the columns and the girders was pin joint in one plane and rigid joint in another plane. The composite floor was composed of wide-flange steel girders and light weight reinforced concrete slabs. Four columns were heated separately by the above mentioned furnace which place are shown in Fig.8. No loading was applied. The vertical displacements of the column-girder connected points were measured by dial gauges, and the strain of the steel members were measured by strain gauges.

In all heated tests, local buckling of column occurred, as shown in Photo.1. The correlation between vertical displacements of top of the column and the mean temperature T is shown in Fig.9. The buckling of heated column started when the mean temperature T reached 350~430°C. This local buckling occurred at the part where the maximum steel temperature was about 550°C same as the model test of H-5.

It is important to attention that after the stop of heating the buckling displacement still continued under the cooling stage. For this shrinkage of column, whole braces from second floor to sixth floor in same plane of heated column buckled remarkably in case of K-2.

These phenomenon shows that the fire protection of column is very important for the structural fire safety.

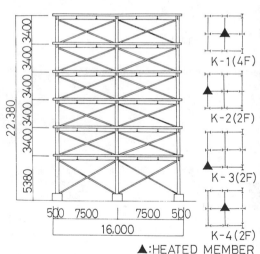

▲:HEATED MEMBER

Fig.8 Full scale six story steel frame of two spans

Photo.1 Local buckling of column ( K-1 )

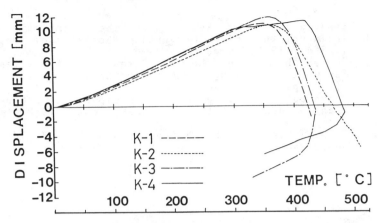

Fig.9 Correlation between vertical displacement of top of column
and mean temperature

## 5. CONCLUSIONS

The following conclusions may be drawn on the basis of
experimental results and analysis described above.
It is possible to predict the behavior of a structural steel
frame under a fire using the analytical theory on thermal stress
inside frame. Under some probable conditions, the structural damage
may actually occur in a fire due to large thermal stress.

## 6.ACKNOWLEDGMENTS

The authors wish to acknowledge their considerable debt to
Prof. Hikaru Saito, Chiba University, Hideki Uesugi, Assistant of
Chiba University, for their support and advice. The analytical
method in this paper is based on the theory proposed by Prof.
Saito. The authors are also grateful to undergraduate students of
Science University of Tokyo for fully support of the experiments.
The tests in 1982 was financially supported by SECOM Science and
Technology Foundation.

## REFERENCES

1. O.Pettersson, S. Magnusson and J. Thor: Fire Engineering Design
   of Steel Structures, Swedish Institute of steel Construction,
   1976.
2. "European Recommendations for the Calculations of the Fire
   Resistance of Load Bearing Steel Elements and Structural
   Assemblies Exposed to the Standard Fire", ECCS.
3. S.Koike et al., "Experimental Study on Thermal Stress within
   Structural Steel Work", Fire Science and Technology, Vol.2,
   No.2, 1982
4. N. Ooyanagi et al.,"Experimental Study on Thermal Stress within
   Steel-flames" Fire Science of Technology, Vol.3, No.1, 1983
5. M.Hirota et al., "Experimental Study on Structural Behavior of
   Steel Frames in Building Fire" Fire Science and Technology,

Vol.4, No.2, 1984

6. H. Saito, "Behaviour of End Restrained Steel Members under Fire", Bulletin of the Fire Prevention Society of Japan, Vol.15, NO.1, Jan., 1966.(in Japanese)
7. H. Saito, "Research on the Thermal Stress of Steel-frame", Prospectus for General Meeting of Architectural Institute of Japan, 1969.(in Japanese)
8. H. Saito, "Fire Safety of Steel-frame", Prospectus for General Meeting of Architectural Institute of Japan, Sept., 1962. (in Japanese)
9. H. Saito, "Thermal Stress of Steel Structure of Tall Buildings in Fire" Fire Science and Technology, Vol.3, No.2, 1983

APPENDIX

Thermal stress in steel-frame

Theoretical values of the thermal stresses in the frame can be obtained from the formula (1). Supposing that the columns are supported elastically, each coefficient of stiffness for expansion k in the formula (1) is calculated from the formula (2).

$$
\begin{bmatrix}
K_1 + \dfrac{E_{t_1}A_1}{L_1} & -\dfrac{E_{t_1}A_1}{L_1} & 0 \cdots 0 & 0 \\
K_1 & K_2 + \dfrac{E_{t_2}A_2}{L_2} & -\dfrac{E_{t_2}A_2}{L_2} \cdots 0 & 0 \\
\vdots & \vdots & \vdots & \vdots \\
K_1 & K_2 & \cdots \quad K_n + \dfrac{E_{tn}A_n}{L_n} & -\dfrac{E_{tn}A_n}{L_n} \\
K_1 & K_2 & \cdots \quad K_n & K_{n+1}
\end{bmatrix}
\begin{Bmatrix}
X_1 \\ X_2 \\ \vdots \\ X_n \\ X_{n+1}
\end{Bmatrix}
=
\begin{Bmatrix}
E_{t_1}A_1\left\{\alpha t_1 + \varepsilon_{11}\left(1 - \dfrac{E_1}{E_{t_1}}\right)\right\} \\
E_{t_2}A_2\left\{\alpha t_2 + \varepsilon_{12}\left(1 - \dfrac{E_2}{E_{t_2}}\right)\right\} \\
\vdots \\
E_{tn}A_n\left\{\alpha t_n + \varepsilon_{1n}\left(1 - \dfrac{E_n}{E_{tn}}\right)\right\} \\
0
\end{Bmatrix}
\quad (1)
$$

where, $E_i$; elastic modulus
$E_{ti}$; elastic modulus at high temperature
$L_i$; length of heated girder
$K_i$; coefficient of stiffness for expansion
$A_i$; area of cross section
$X_i$; horizontal displacements

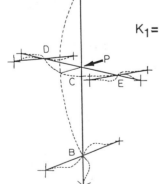

$$
K_1 = \left[ \frac{(a+b)^3}{3} - \frac{1}{8}\frac{2(a+b)Z_1Z_2 - bZ_1 - aZ_2 - a + b}{4Z_1Z_2 - 1} \right]^{-1} \times \frac{Z_{Ic}}{h_1 + h_2} \quad\text{---}(2)
$$

where, $Z_1 = \dfrac{\bar{k}_1 + \bar{k}_2 + \bar{k}_3 + \bar{k}_4}{\bar{k}_4}$ , $Z_2 = \dfrac{\bar{k}_4 + \bar{k}_5 + \bar{k}_6 + \bar{k}_7}{\bar{k}_4}$

$a = \dfrac{h_1 h_2^2}{(h_1 + h_2)}$ , $b = \dfrac{h_1^2 h_2}{(h_1 + h_2)}$

# Effects of Biaxial Loading on the High Temperature Behaviour of Concrete

**K. KORDINA and C. EHM**
Institut für Baustoffe, Massivbau und Brandschutz
Technische Universität Braunschweig
Beethovenstrasse 52, D-3300 Braunschweig, FRG

**U. SCHNEIDER**
Fachgebiet Baustoffkunde
Universität Gesamthochschule Kassel, FRG

ABSTRACT

Biaxial compression tests with ordinary concrete have been carried out under steady state and transient temperature conditions. The test results show that even small load levels in a second axis alter the mechanical properties of concrete significantly. The stress-strain relationships show a significant dependence on the temperature level and the stress ratio. The strength under biaxial compressive stress is higher than the strength under uniaxial compression. The volumetric strains increase with increasing stress ratios and increasing temperatures. The failure temperature of specimens being biaxially loaded is higher than the one of specimens that are uniaxially loaded. The modes of failure indicate that the tensile deformation is vital for the failure mechanism of concrete.

INTRODUCTION

During fires in buildings, concrete structures are exposed to mechanical and thermal stresses. With respect to mechanical stress states, there are uniaxial stresses, for example in centrically loaded columns, biaxial stresses, for example in beams, panels, slabs and shells and triaxial stresses. Regarding the temperature exposure, there are the ambient temperature, increasing and high temperatures when the fire is burning and decreasing temperatures, when the burning has gone out.
This paper deals with the behaviour of concrete under biaxial stress states and steady or transient high temperatures.

Up to now, only little is known about the strength and the deformation behaviour of concrete subjected to biaxial stresses at elevated temperatures. Our current knowledge about high temperature behaviour is mainly based on uniaxial experiments, see for example /1/. The knowledge about concrete behaviour under biaxial loading conditions is based on experiments being performed at ambient temperature as it is presented in, for instance, /2/.

In the following, at first the test equipment which was especially developed for these investigations will be presented and then some results concerning the strength and deformation behaviour of concrete under complex loading and temperature conditions will be discussed. The fracture behaviour under biaxial high temperature stress was already clearly presented in /3/.


EXPERIMENTAL EQUIPMENT

Loading Equipment

The loading frame of the test equipment (fig. 1) is designed as a welded steel construction of high stiffness. The mechanical loading of the specimens is achieved by four stress rate or strain rate controlled hydraulic cylinders each with a maximum force of 1000 kN. The heat resistant pressure pistons are cooled by water.

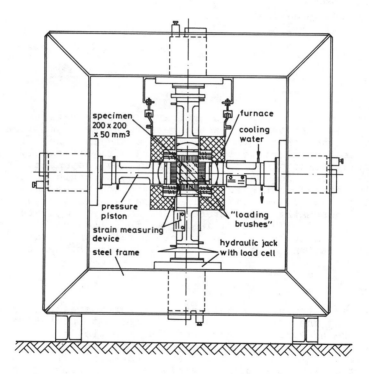

Fig. 1   Test rig for biaxial high temperature concrete tests


Shape of Specimens and Load Application

Investigations of the mechanical properties of concrete under bi-axial stress states at ambient temperature have shown that the uniform, two dimensional state of stress can best be achieved by the use of a disc-shaped specimen. This type of a specimen was chosen for the given investigations at high temperatures, too. The size of the specimens which are sawed out of cubes, results from concrete technological and experimental requirements. The optimum specimen size we obtained after different pre-investigations is 200 x 200 x 50 mm$^3$.

Particular problems in biaxial testing arise at high temperatures due to the load application onto the specimens. Generally a nearly unrestrained load application is required and the test equipment must provide a defined and homogeneous state of stress in the specimens. In order to minimize the restraining due to the friction between loading platens and the specimens' surfaces so-called "loading-brushes" were constructed as proposed by Hilsdorf /4/. Each load application steel platen is divided into 190 parallel rods having a distance of about 0,1 mm from each other. These rods follow the deformations of the loaded specimens' surfaces and therefore cause a considerable reduction of the transverse strain inhibition.

By comparing the uniaxial and biaxial high temperature compressive strength data determined with "loading-brushes" and with rigid steel platens, the reduction of the strain inhibition by using "loading-brushes" was proved. The strengths determined with brushes were more than 20 per cent lower than the strengths determined with rigid platens.

By additional experiments, it could clearly be proved that a defined uniaxial state of stress is achieved in the test specimens by using "loading-brushes". The measured compressive strength is the real uniaxial compressive strength.

Load Control

The control units of the test equipment fulfill the following important requirements:
- different types of loading and different stress paths can be realized;
- the central point of the specimen is in the centre of the loading during the whole test.

The last point especially apply to experimental conductions under biaxial high temperature conditions. For that the controls are put in the way that the deformations appearing during the heating and the loading of the specimen are synchronously accounted for by the loading units.

Load Measurements

In order to measure the forces in the specimens, four load cells are at disposal constructed between cylinders and the specimen. The application of two load cells per axis gives the possibility to control the correct installation of the specimens and to discover errors in the system.

Deformation Measurements

The deformation of the concrete specimens in the three axes are measured directly by means of a specially developed high temperature – deformation measuring device (fig. 2). With this device, the strains are transferred by transmitters out of the hot furnace regions. Inductive displacement transducers are installed in the region of ambient temperatures. The deformation measuring system for the determination of the deformations in the loaded axes is designed according to the principle of a dilatometer, that means thermal strains of the transmitters are compensated by a special geometric arrangement. With the measurements of the strains in the unstressed axis, the system is conducted according to a lateral extensometer. The thermal strains of the transmitters separately determined by calibration tests must be taken into consideration.

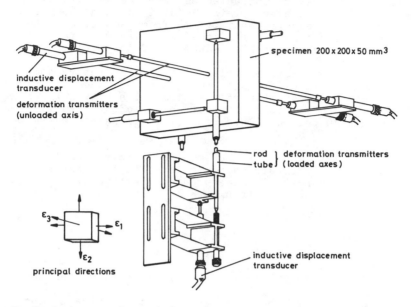

Fig. 2  High temperature deformation measuring system

Heating System and Temperature Measurements

The heating system consists of an electrical furnace and a 3- zone-PID programme controler regulating the heating and cooling of the specimens and the realization of prescribed temperature curves.

The temperature of the specimen is measured with NiCr-Ni – thermo-couples which are applied on the surface of the specimen by means of a heat-resistant glue. A sufficiently long hold time period of 2 hours and a moderate temperature increase of 2 K/min guarantee a uniform temperature distribution in the specimen.
A more detailed description of the experimental equipment is given in /3/ and /5/.

RESULTS

Compressive Strength

In the investigations to determine the biaxial compressive strengths at high temperatures all specimens were stored at least six months at 20°C/65% r.h.. The specimens were made of ordinary concrete with a cube strength of 41,0 N/mm$^2$ after 28 days. The biaxial compressive strengths are presented in figure 3. They are

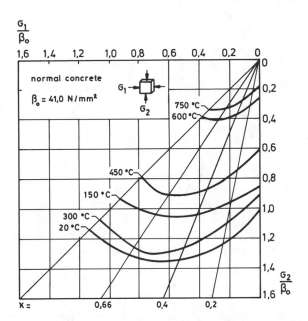

Fig. 3 Biaxial high temperature compressive strength of gravel
concrete related to the uniaxial strength at room temperature

related to the uniaxial compressive strength at ambient temperature
ß₀ at the time of testing.

For each particular temperature, the strengths were connected with
a failure envelope in the plane of principal stresses.

Concerning the biaxial high temperature compressive strength, the
following statements can be made:

1. The biaxial compressive strength of concrete is higher than the
uniaxial compressive strength. This comprises all temperatures and
all stress ratios κ . The increase of the strength compared to the
uniaxial strength is clear even with only small stress levels in
the second axis.

2. The biaxial compressive strength increases considerably with
increasing stresses in the second axis. The positive effect of the
stresses in the second axis obtains its greatest value with a
certain stress ratio which depends on the test temperature. A
further increase of the stress in the second axis beyond this point
consequently decreases the biaxial compressive strength.

3. The relativ increase of the strength under biaxial stress is
greater at higher temperatures than at ambient temperature. At
ambient temperature, the maximum increase of the strength comes to
20-40 per cent at a stress ratio of $\sigma_1 : \sigma_2 = 0,5$. At 750°C
this value is of the order of 200 per cent at $\sigma_1 : \sigma_2 = 1,0$.

This means that the maximum of the increase of the strength under
biaxial stress shifts with increasing test temperatures to higher
stress ratios. This is also valid for concretes with a higher cube

285

strength, as shown in /5/.


Mechanical strains

The mechanical strains $\varepsilon_1$, $\varepsilon_2$ and $\varepsilon_3$ at constant temperatures
are presented in the figures 4 and 5. The deformations depend on
the stress level, the stress ratio and the test temperature. The
deformations increase with increasing load in all three axes. At
high stress levels, they show a nonlinear relation in the direction
of the greatest principal stress.
With increasing temperatures, the development of $\sigma - \varepsilon -$ curves
indicates a decrease of the compressive strength. The decreasing
slope in the origin shows the decrease of the modulus of elasticity
at higher temperatures. The ultimate strains for all three axes are
shifted to greater values with increasing temperatures. A
particularly great increase of the deformations is observed from
600°C onward.
At higher temperatures, the $\sigma - \varepsilon -$ curves are highly non-linear: the
concrete changes its brittle behaviour into a soft and plastic
behaviour.

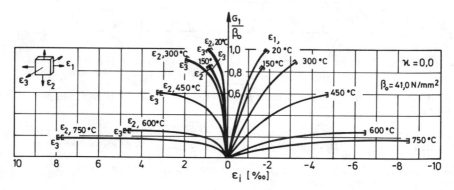

Fig. 4  Stress-strain relations of gravel concrete under uniaxial
compressive stresses

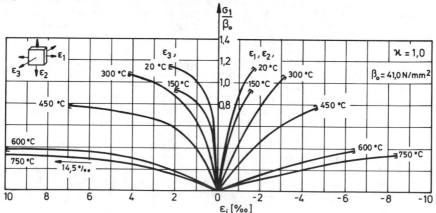

Fig. 5  Stress-strain relations of gravel concrete under
symmetrical biaxial compressive stresses

In the uniaxial case, figure 4, $\varepsilon_2$ and $\varepsilon_3$ are equal because of
equal Poisson's ratios.
In the biaxial case with $\sigma_1 : \sigma_2 = 0,4$ e.g., $\varepsilon_2$ has changed
and obtained positive values.
In the symmetrical biaxial case, $\sigma_1 : \sigma_2 = 1,0$ (figure 5)
$\varepsilon_1$ and $\varepsilon_2$ have the same value.
The strains in the unstressed axis, $\varepsilon_3$, reach extremely great
values at elevated temperatures and high stress ratios. The maximum
values of $\varepsilon_3$ appear with the stress ratio $\sigma_1 : \sigma_2 = 1,0$.


Total Deformations during Heating

With respect to the case of fire, it is important to investigate
the concrete behaviour at transient temperatures. Therefore high
temperature transient creep tests were performed.
In such tests, the total deformations of unsealed specimens are
measured under practical stress levels and defined non-steady temperature exposures.

Results of the deformation measurements in the uniaxial case for
different stress levels are presented in figure 6. The elastic
deformations at ambient temperatures are neglected in that figure.
As a comparison curve, the curve with a zero stress level, i.e.
pure thermal expansion is shown in figure 6, too.

Stress levels of about 70 per cent of the uniaxial strength
restrain the thermal extension in the direction of the loaded axis.
In this case, only slight extensions can be observed. With
continuously increasing temperatures comparatively great

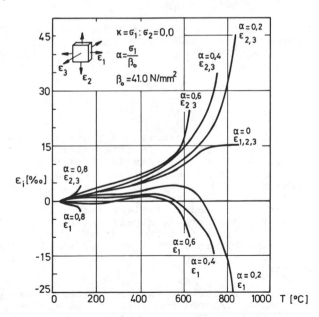

Fig. 6  Total deformations of gravel concrete during uniaxial
transient creep tests

287

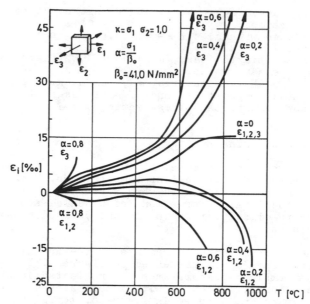

Fig. 7   Total deformations of gravel concrete during biaxial transient creep tests

compressive strains occur until the specimen fails. Stress levels above 70 per cent lead to a failure of the specimen at low temperatures of about 120°C to 130°C.

A fast increase of the lateral strains indicates the failure of the specimens. A symmetrical biaxial stress during heating (fig. 7) leads to extreme lateral strains in the free axis. The compressive strains in the stressed axes are not much greater than in the uniaxial case.


Failure Temperatures

In figure 8, the failure temperatures are presented. The individual stress ratios serve as a parameter in this graph. Three regions of failure can be distinguished: a stress level of 0 up to 70 per cent of the uniaxial compressive strength leads to failure temperatures above 600°C. Between 70 and 80 per cent, there is a very sensitive region with regard to the failure temperature. Only slight alterations in the stress level lead to failure temperatures that differ around several hundred centigrade. Stress ratios above 80 per cent consequently lead to failure at temperatures under 130°C. It has to be considered that in this region the time effect i.e. the duration of the stress exposure plays an important part in addition to the temperature effect.

With increasing stress ratios the failure temperatures increase. According to our tests, the optimum ratio is in the range of $\sigma_1$ : $\sigma_2 = 0,4$. The failure temperatures increase significantly in this case. In the region of high stress levels, this increase is of the order of several hundred centigrade.

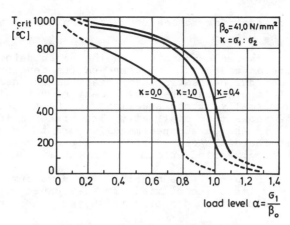

Fig. 8  Failure temperatures of gravel concrete

Restraining Stresses during Heating

Restraining forces arise in concrete specimens under transient
temperatures if mechanically or thermally caused extensions are
restrained. The time dependent restraining forces of different
concrete specimens are shown in figure 9. Predried and normal cured
specimens are compared.
Up to 120°C, the restraining forces rise considerably for a
normally stored concrete (20°C/65 % r.h.). After that a phase of
shrinking follows between 120°C and 200°C. During this phase the
restraining forces decrease. After passing a minimum at 200°C, the
restraining forces increase again with increasing temperature and
reach maximum values at 300°C to 350°C. These values are around 50
to 70 per cent of the failure load at ambient temperature. Finally
the restraining forces decrease continuously.
The behaviour of specimens that were predried at 105°C is something
different. At 200°C, a distinct stress maximum appears. The
restraining forces reach values between 70 and 90 per cent of the
failure load at 20°C.

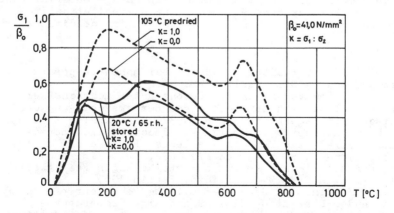

Fig. 9  Restraining stresses of gravel concrete during heating

Thereafter the restraining forces decrease rapidly. A second maximum appears at 650°C.

The values of the thermal expansion and the creep deformations of the concrete have an essential influence on the development of the restraining forces during heating. Vaporisation and dehydration processes are of importance. The development of restraining forces in predried and normally stored specimens in the temperature region of 20°C to 200°C shows this distinctly. The higher moisture content (20°C/65 % r.h.) in the cured specimens favours the creep of concrete so that there are considerably lower restraining forces in comparison with the predried specimens.

The stress ratio influences the quantitative development of the restraining forces, too. The biaxial stresses cause higher restraining forces of the order of 5 to 10 per cent compared to the uniaxial case.


CONCLUSIONS

Up to now, the experimental results can briefly be summarized as follows:
The biaxial high temperature tests have shown that even small load levels in the second axis alter the mechanical properties of concrete significantly. The strength under biaxial compression is higher than under uniaxial compression. This applies especially to higher temperatures. The strains increase with increasing stress ratios and with increasing temperatures.

The modes of failure indicate that tensile deformations are vital in the failure mechanism of concrete. The increase in strength and stiffness under biaxial compression is supposed to be due to the restraining of thermally induced and load induced microcracking. Studies on crack initiation and development will be performed to establish a theoretical background for the preliminary explanations.

Further experimental investigations are also necessary to proceed in the developing of analytical high temperature material models in order to describe the concrete behaviour as realistically as possible under extreme stresses and temperature conditions.

LITERATURE

/1/ Schneider,U.: Verhalten von Beton bei hohen Temperaturen. Heft 337 des DAfStB, Berlin 1982
/2/ Kupfer,H.: Das Verhalten des Betons unter mehrachsiger Kurzzeitbelastung unter besonderer Berücksichtigung der zweiachsigen Beanspruchung. Heft 229 des DAfStB, Berlin 1973
/3/ Ehm,C.; Schneider,U.; Kordina,K.: Fracture of Concrete under Biaxial High Temperature Tests. Proceedings of the 6th International Conference on Fracture, New Delhi 1984
/4/ Hilsdorf,H.: Die Bestimmung der zweiachsigen Festigkeit des Betons. Heft 173 des DAfStb, Berlin 1965
/5/ Ehm,C.; Kordina,K.; Schneider,U.: The Behaviour of Concrete under Biaxial Conditions and High Temperatures. RILEM International Conference on Concrete under Multiaxial Conditions, Toulouse 1984

# Influence of Restraint on Fire Performance of Reinforced Concrete Columns

**T. T. LIE**
Division of Building Research
National Research Council of Canada
Ottawa, Ontario, K1A 0R6, Canada

**T. D. LIN**
Portland Cement Association
Skokie, Illinois, USA

ABSTRACT

Experimental and theoretical studies were carried out on the effect of restraint on the fire resistance of reinforced concrete columns. Two tests were carried out on axially loaded columns fully restrained against thermal expansion. Both experimental and theoretical studies indicate that full restraint of axial thermal expansion has little influence on the fire performance of the columns. The maximum stresses in a fully restrained column at the time that the restraining load is maximum, are considerably lower than those at the time of failure of the column.

INTRODUCTION

In a fire the expansion of structural members due to heating is often restrained by the integrated surrounding structure. In the case of columns, restraint occurs if the expansion of the column is resisted by the floors above. As a consequence, additional load is imposed on the column; the greater the number of floors above the column, the greater the increase in load. The increase in load may have an adverse effect on the fire resistance of the column.

Theoretical studies [1] indicated that restraint against thermal expansion of a reinforced concrete column would not significantly affect its fire resistance. Facilities are now available for the testing of columns under restraint and for verification of theoretical results. Such tests have been conducted recently at the National Research Council of Canada as part of a study undertaken in cooperation with the Construction Technology Laboratories of the Portland Cement Association. The results of these tests, and the calculated results, are discussed in the present paper.

MAGNITUDE OF RESTRAINT

In a previous study [1], the magnitude of restraint was assessed by estimating the vertical stiffness of a floor slab as a function of its dimensions, and by assuming that the total vertical stiffness of the restraining structure is proportional to the number of floors above the column. Depending on the number of floors and their individual vertical stiffness, the magnitude of the restraining forces can vary from close to zero to values close to those present when column expansion is fully restrained.

In this study two tests were carried out in which the axial expansion of the columns was fully restrained. A full restraint condition was obtained by initially applying the maximum allowable load on the columns and preventing their expansion during the fire tests by controlling the load. The lengths of the columns were kept constant until the load, which increases initially but reduces later with reduction of column strength, had returned to its original value. Then the load was kept constant until the column failed. The maximum allowable load was determined according to ACI 318-83 [2], using a live-to-dead load ratio of 0.4 and the actual cylinder strength of the concrete on the test date.

TEST SPECIMENS[1]

The specimens were square, tied, reinforced concrete columns, made with siliceous aggregate. All were 3810 mm long and had a cross-section size of 305 × 305 mm. Twenty-five-mm diameter longitudinal reinforcing bars and 10-mm diameter ties were used. The location of the bars, which were welded to steel end plates, and the locations of the ties are shown in Figure 1.

The yield stress of the main reinforcing bars was 444 MPa and that of the ties was 427 MPa. The ultimate strength was 730 MPa for the main bars and 671 MPa for the ties.

The designed concrete mix had a strength of about 35 MPa. Its composition per cubic metre was as follows: cement , 325 kg; water, 140 kg; sand, 874 kg; coarse aggregate, 1058 kg.

The average compressive cylinder strength of the concrete of the two columns tested, measured on the test dates, was 42.6 MPa for column A and 36.7 MPa for column B. The moisture condition at the center of each column was approximately equivalent to that in equilibrium with air of 75% relative humidity at room temperature.

Chromel-alumel thermocouples, 0.91 mm thick, were installed at mid-height of the columns for measuring concrete temperatures at different locations in the cross-section.

CALCULATION PROCEDURE

The calculation of the fire performance of the columns involves the calculation of temperatures in the column, its deformations, and the stresses in it. The calculation procedure is described in detail in reference 3. Only a brief description of the method will be given here.

The column temperatures are determined by a finite difference method. The column cross section is divided into a large number of elements. For each element a heat and moisture balance is made. The effect of moisture on temperature is taken into account by assuming that in each element the moisture starts to evaporate when the element temperature reaches 100°C.

The load on the column during exposure to fire is calculated by a method based on a load-deflection analysis, which in turn is based on a stress-strain analysis of cross sections. In this method, the test columns, which are fixed at the ends during the tests, are idealized as pin-ended columns of reduced

---

[1]Detailed information on the test specimens is available.

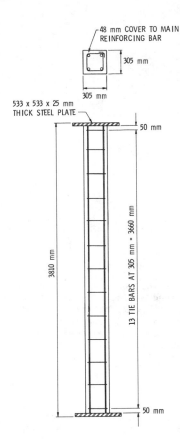

Figure 1.  Test column and location of reinforcing bars


effective length, KL (Figure 2), where K is the effective length factor and L
the unsupported column length.  The applied load on the test column is
intended to be concentric.  To represent imperfections in the column, an
initial deflection ($y_0$) of 2.5 mm is assumed.

The curvature of the column is assumed to vary from zero at pin-ends to a
maximum at mid-height according to a straight line relation, as illustrated in
Figure 2.  For any given curvature $\chi$, and thus for any given deformation, the
axial strain is varied until the axial force at mid-section times the
deflection is in equilibrium with the internal moment.  In this way a load vs
axial strain curve can be calculated for specific times during the fire
exposure.  From these curves the load needed to fully restrain the column can
be determined for each time.

In the calculations the material properties of the concrete and steel
given in Ref. 3 were used.  Stress-strain curves for the concrete and the
reinforcing steel used in the calculations are shown in Figures 3 and 4.

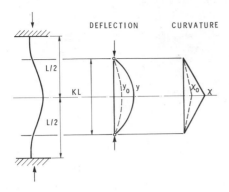

Figure 2.  Load-deflection analysis


TEST APPARATUS

     The tests were carried out by exposing the columns to heat in a furnace
specially built for the testing of loaded columns and walls.  The test furnace
was designed to produce the conditions to which a member might be subjected
during a fire, with respect to temperature, structural load, and heat
transfer.  It consists of a steel framework, supported by four steel columns,
and the furnace chamber inside the framework.  The characteristics and
instrumentation of the furnace are described in detail in Ref. 4.


TEST CONDITIONS AND PROCEDURE

     The columns were installed in the test furnace by bolting their steel
end-plates to a loading head at the top and a hydraulic jack at the bottom.
Concentric loads were applied to the columns about one hour before the fire
tests.  The load on column A was 1044 kN and that on column B, 916 kN.

     During the tests the heat input into the furnace was controlled so that
the average temperature followed as closely as possible the standard
temperature-time relation specified in ASTM-E119 [5] or CAN4-S101 [6].
Temperatures in the column were measured at various locations at mid-height.

     After application of the load and the start of the fire tests, the
lengths of the loaded columns were kept constant by controlling the load.  The
load increased initially, but decreased later with reduction of the strength
of the column.  After the load returned to its original value, it was kept
constant until the column failed.

     The length of each column was measured with differential transducers
attached to the furnace frame, one at the bottom level and one at the top
level of the column.  Inaccuracy in the length measurement due to deformations
of the column furnace structural frame, was eliminated by measuring those
deformations with strain gauges, and compensating column length accordingly.

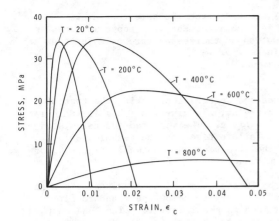

Figure 3.  Stress-strain curves for concrete at various temperatures
           ($f'_{co}$ = 35 MPa)

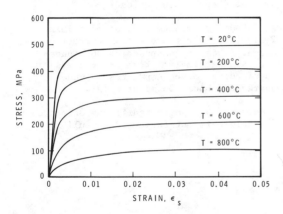

Figure 4.  Stress-strain curves for the reinforcing steel at various
           temperatures ($f_{yo}$ = 443 MPa)

The columns were considered to have failed and the tests were terminated
when the hydraulic jack, which had a maximum speed of 76 mm/min, could no
longer maintain the applied load.


RESULTS AND DISCUSSION

Column Temperatures

Previous studies [3] in which several columns were tested, showed that
the mathematical model used for the calculation of temperature in siliceous
concrete columns during fire exposure, gives reasonably accurate predictions.

The columns tested under restraint were made from the same concrete. Calculated temperatures were again in good agreement with those measured, as shown in Figure 5. Detailed information on the temperature as a function of depth in similar columns is given in Ref. 3.

Loads and Deformations

The measured axial applied loads and deformations during the test of columns A and B, under full axial restraint, are given in Table 1. Calculated loads to restrain the columns are also given. The measured and calculated loads to restrain the columns, their failure time and the measured axial deformations of the columns during the fire tests are shown in Figure 6 for column A, and in Figure 7 for column B.

During the test of column B, the furnace temperature dropped, after an exposure time of slightly more than 2½ hours, to below the standard fire temperature for about one half hour, due to power failure. The load on the column was kept above the calculated load by controlling the ram pressure. Because of the interruption, measured and calculated loads are not comparable for column B for the period after 2½ hour exposure time. For the period up to 2½ hours in the case of column B and for the entire test period in the case of column A, the calculated and measured loads are comparable. For these periods there is a good agreement between calculated and measured loads.

The failure time of column A was 3 hours 21 minutes. That of column B was 4 hours 2 minutes, or 3 hours 32 minutes, if a conservative correction is made for the power interruption of one-half hour by subtracting 30 minutes from the failure time. The calculated failure time for both columns was 3 hours 18 minutes. The failure time of companion columns, tested under comparable loads but unrestrained, averaged 3 hours 33 minutes [3].

The results indicate that full restraint has an insignificant effect on the fire performance of columns. If a column is not subjected to full

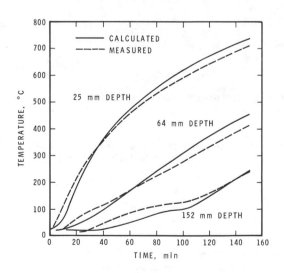

Figure 5.  Concrete temperatures in mid-height section along centerline at various depths

TABLE 1.  AXIAL LOADS AND DEFORMATIONS TO FAILURE
(COLUMNS A AND B, FULL RESTRAINT)

| Time (min) | Axial deformation (mm) | | Load (kN) | | | |
|---|---|---|---|---|---|---|
| | | | Column A | | Column B | |
| | Column A | Column B | Measured | Calculated | Measured | Calculated |
| 0 | 0 | 0 | 1044 | 1044 | 916 | 916 |
| 10 | 0 | 0 | 1513 | 1518 | 1410 | 1323 |
| 20 | 0 | 0 | 2049 | 1933 | 1892 | 1702 |
| 30 | 0 | 0 | 1985 | 2240 | 1889 | 1960 |
| 38 | | 0 | | | 1910 | 2067 |
| 40 | 0 | | 1967 | 2313 | | |
| 43 | | 0 | | | 1874 | 2060 |
| 45 | 0 | | 2010 | 2300 | | |
| 50 | 0 | 0 | 1967 | 2282 | 1874 | 2010 |
| 60 | 0 | 0 | 1935 | 2200 | 1878 | 1950 |
| 70 | 0 | 0 | 1892 | 2120 | 1892 | 1890 |
| 80 | 0 | 0 | 1846 | 2042 | 1860 | 1804 |
| 90 | 0 | 0 | 1853 | 1955 | 1832 | 1735 |
| 100 | 0 | 0 | 1767 | 1873 | 1814 | 1649 |
| 110 | 0 | 0 | 1724 | 1792 | 1696 | 1582 |
| 120 | 0 | 0 | 1635 | 1705 | 1649 | 1500 |
| 140 | 0 | 0 | 1413 | 1545 | 1413 | 1349 |
| 150 | 0 | 0 | 1303 | 1452 | 1324 | 1268 |
| 151 | | 0 | | | 1300 | 1260 |
| 160 | 0 | | 1167 | 1365 | | |
| 170 | 0 | | 1044 | 1272 | | |
| 172 | | -0.09 | | | 1120 | 1112 |
| 174 | -0.20 | | 1044 | 1240 | | |
| 178 | -0.57 | | 1044 | 1205 | | |
| 182 | -0.82 | | 1044 | 1170 | | |
| 186 | -1.27 | | 1044 | 1132 | | |
| 190 | -1.81 | | 1044 | 1100 | | |
| 194 | -2.46 | | 1044 | 1062 | | |
| 196 | -2.86 | | 1044 | 1044 | | |
| 197 | | +0.01 | | | 1103 | 916 |
| 198 | -3.27 | | 1044 | 1044 | | |
| 200 | -3.76 | | 1044 | 1044 | | |
| 210 | | 0 | | | 1103 | 916 |
| 220 | | 0 | | | 935 | 916 |
| 225 | | 0 | | | 916 | 916 |
| 230 | | -0.84 | | | 916 | 916 |
| 240 | | -1.80 | | | 916 | 916 |
| 242 | | -5.30 | | | 916 | 916 |

restraint and the surrounding structure carries some of the load when the column contracts, as is normally the case in practice, restraint is probably beneficial for the fire performance of the column.

Stresses in Concrete

In Figure 8 calculated stress and temperature distributions in the concrete section are shown for various times:  for 45 minutes, when the load on the restrained column reaches a maximum, for 120 minutes, and for 198 minutes, when the column would fail. The upper figures give the stress in an element in MPa; the lower figures (in brackets) give the temperature of the element in °C. Because the stresses and temperatures are symmetrical with respect to the centerline, only one half of this section is shown.

When the load reaches a maximum at 45 minutes, a maximum stress of about 22 MPa occurs in a region not far from the surface of the column. This stress is about 60% of the compressive strength of the concrete in this region. The concrete in this region is at temperatures in the range of 300 to 350°C and has hardly lost any strength.

As the exposure to fire proceeds, the region of high stresses moves towards the center of the column and the value of the stresses increases. The maximum stress after a two-hour exposure is about 24.5 MPa or 67% of the compressive strength of the concrete.

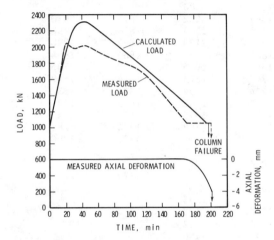

Figure 6.   Loads on column A and its axial deformation under full restraint during fire exposure

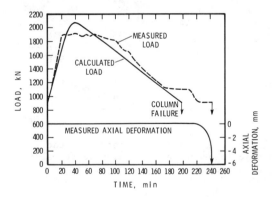

Figure 7.   Loads on column B and its axial deformation under full restraint during fire exposure

298

AT MAXIMUM LOAD: 45 min

| | | | | | |
|---|---|---|---|---|---|
| 4.1 | 9.7 | 12.4 | 13.6 | 14.0 | 14.1 |
| (787) | (672) | (620) | (598) | (590) | (588) |
| 9.7 | 20.5 | 21.8 | 21.9 | 21.8 | 21.8 |
| (672) | (443) | (340) | (296) | (281) | (277) |
| 12.3 | 21.6 | 21.2 | 19.9 | 19.1 | 18.7 |
| (620) | (340) | (205) | (146) | (123) | (115) |
| 13.4 | 21.4 | 19.5 | 16.8 | 15.5 | 15.0 |
| (598) | (296) | (146) | (87) | (69) | (63) |
| 13.7 | 21.1 | 18.1 | 14.8 | 13.0 | 12.3 |
| (590) | (281) | (123) | (69) | (46) | (41) |
| 13.7 | 20.8 | 17.1 | 13.6 | 11.4 | 10.5 |
| (588) | (277) | (115) | (63) | (41) | (33) |
| 13.6 | 20.5 | 16.6 | 12.8 | 10.6 | 9.6 |
| (588) | (277) | (115) | (63) | (41) | (33) |
| 13.5 | 20.3 | 16.5 | 12.7 | 10.6 | 9.7 |
| (590) | (281) | (123) | (69) | (46) | (41) |
| 13.0 | 20.1 | 17.1 | 13.6 | 11.9 | 11.4 |
| (598) | (296) | (146) | (87) | (69) | (63) |
| 11.9 | 20.2 | 18.7 | 16.6 | 15.4 | 14.9 |
| (620) | (340) | (205) | (146) | (123) | (115) |
| 9.3 | 19.3 | 20.0 | 19.7 | 19.5 | 19.5 |
| (672) | (443) | (340) | (296) | (281) | (277) |
| 4.0 | 9.3 | 11.8 | 12.8 | 13.2 | 13.3 |
| (787) | (672) | (620) | (598) | (590) | (588) |

152.5 mm

AT INTERMEDIATE TIME: 120 min

| | | | | | |
|---|---|---|---|---|---|
| 0.0 | 0.9 | 3.1 | 5.0 | 6.2 | 6.7 |
| (967) | (909) | (868) | (841) | (826) | (813) |
| 0.9 | 5.2 | 10.4 | 14.0 | 16.1 | 17.1 |
| (909) | (769) | (671) | (608) | (572) | (556) |
| 3.0 | 10.2 | 18.0 | 22.5 | 24.0 | 24.2 |
| (868) | (671) | (534) | (450) | (403) | (382) |
| 4.8 | 13.5 | 21.9 | 23.7 | 24.3 | 24.5 |
| (841) | (608) | (450) | (354) | (296) | (268) |
| 5.8 | 15.1 | 22.5 | 23.2 | 23.4 | 23.3 |
| (826) | (572) | (403) | (296) | (229) | (197) |
| 6.2 | 15.6 | 21.8 | 22.07 | 21.7 | 21.1 |
| (818) | (556) | (382) | (268) | (197) | (161) |
| 6.1 | 15.2 | 20.9 | 20.8 | 19.9 | 19.1 |
| (818) | (556) | (382) | (268) | (197) | (161) |
| 5.5 | 14.0 | 20.0 | 19.7 | 18.8 | 18.1 |
| (826) | (572) | (403) | (296) | (229) | (197) |
| 4.5 | 12.0 | 18.6 | 19.0 | 18.5 | 18.0 |
| (841) | (608) | (450) | (354) | (296) | (268) |
| 2.7 | 8.9 | 15.0 | 17.9 | 18.4 | 18.3 |
| (868) | (671) | (534) | (450) | (403) | (382) |
| 0.8 | 4.5 | 8.7 | 11.4 | 12.9 | 13.6 |
| (909) | (769) | (671) | (608) | (572) | (556) |
| 0.0 | 0.8 | 2.7 | 4.2 | 5.2 | 5.6 |
| (967) | (909) | (868) | (841) | (826) | (818) |

152.5 mm

AT FAILURE TIME: 198 min

| | | | | | |
|---|---|---|---|---|---|
| 0.0 | 0.0 | 0.0 | 0.0 | 0.0 | 0.0 |
| (1054) | (1016) | (987) | (966) | (952) | (946) |
| 0.0 | 0.0 | 2.2 | 5.4 | 7.8 | 9.0 |
| (1016) | (919) | (844) | (780) | (755) | (738) |
| 0.0 | 2.0 | 8.9 | 15.4 | 18.3 | 20.3 |
| (987) | (844) | (733) | (654) | (604) | (580) |
| 0.0 | 4.8 | 13.4 | 20.5 | 25.2 | 27.5 |
| (966) | (790) | (654) | (558) | (507) | (475) |
| 0.0 | 6.3 | 15.4 | 22.7 | 26.6 | 27.3 |
| (952) | (755) | (604) | (507) | (442) | (416) |
| 0.0 | 6.6 | 15.0 | 21.3 | 23.2 | 23.5 |
| (946) | (738) | (580) | (475) | (416) | (388) |
| 0.0 | 5.9 | 12.9 | 17.4 | 18.3 | 18.1 |
| (946) | (738) | (580) | (475) | (416) | (388) |
| 0.0 | 4.5 | 9.8 | 12.6 | 13.1 | 12.6 |
| (952) | (755) | (604) | (507) | (442) | (416) |
| 0.0 | 2.8 | 6.6 | 8.2 | 8.4 | 8.2 |
| (966) | (790) | (654) | (558) | (507) | (475) |
| 0.0 | 1.0 | 3.7 | 3.9 | 5.3 | 5.2 |
| (987) | (844) | (733) | (654) | (604) | (580) |
| 0.0 | 0.0 | 0.9 | 1.9 | 2.4 | 2.7 |
| (1016) | (919) | (844) | (780) | (755) | (738) |
| 0.0 | 0.0 | 0.0 | 0.0 | 0.0 | 0.0 |
| (1054) | (1016) | (987) | (966) | (952) | (946) |

152.5 mm

305 mm

Figure 8. Stresses (upper figure – MPa) and temperatures (lower figure – °C) in concrete section of restrained column, calculated for various times

The maximum stress continues to increase with the duration of fire exposure. Although the inward movement of the region of high stresses has a delaying effect, the temperature in that region also continues to increase with the duration of exposure. At the time of failure the temperature of the concrete in that region, which lies near the core of the column, will have reached values of 500–600°C; at this range the compressive strength of the concrete is reduced to about 70% of its initial strength. At the same time the stresses in that region will have reached the reduced compressive strength of the concrete.

Failure occurs after the load on the restrained column has returned to its initial value, as illustrated in Figure 9. The theoretical failure time and stress distribution in the column at that time are equal to those for an unrestrained column under the same load. In the tests only small differences in failure time were found between the restrained and unrestrained columns. Thus the additional load to fully restrain the columns did not cause damage or significant permanent deformations in the columns.

CONCLUSIONS

Experimental and theoretical studies indicate that full restraint of axial thermal expansion of reinforced concrete columns has little influence on the fire performance of the columns. The maximum stress in a fully restrained

299

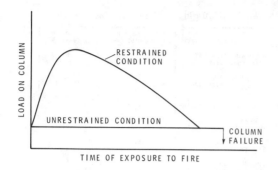

Figure 9. Loads on a restrained and an unrestrained column during exposure to fire

column when the restraining load is at the maximum is considerably lower than that near the time of failure of the column. The results suggest that restraint of thermal expansion of a reinforced concrete column is beneficial for its fire performance if the surrounding structure is capable of transferring part of the load to other supports, as is normally the case in practice.

REFERENCES

1. Allen, D.E. and Lie, T.T.: "Further Studies of the Fire Resistance of Reinforced Concrete Columns," National Research Council of Canada, Division of Building Research, DBR Technical Paper 416, NRCC 14047, Ottawa, 1974.
2. Building Code Requirements for Reinforced Concrete, ACI Standard 318-83, American Concrete Institute, Detroit, 1983.
3. Lie, T.T., Lin, T.D., Allen, D.E. and Abrams, M.S.: "Fire Resistance of Reinforced Concrete Columns," National Research Council of Canada, Division of Building Research, DBR Paper 1166, NRCC 23065, Ottawa, 1984.
4. Lie, T.T.: "New Facility to Determine Fire Resistance of Columns," Canadian Journal of Civil Engineering, 7: 3, 551-558, 1980.
5. Standard Methods of Fire Tests of Building Construction and Materials, ANSI/ASTM E119-83, American Society for Testing and Materials, Philadelphia, Pa, 1983.
6. Standard Methods of Fire Endurance Tests of Building Construction and Materials, CAN4-S101-M82, Underwriters' Laboratories of Canada, Scarborough, Ontario, 1982.

# Principles for Calculation of Load-Bearing and Deformation Behaviour of Composite Structural Elements under Fire Action

K. RUDOLPH, E. RICHTER, R. HASS, and U. QUAST
Institute for Building Materials, Concrete Structures and Fire Protection
and Special Research Department 148
Fire Behaviour of Structural Members
Technical University of Braunschweig, FRG

ABSTRACT

The fire behaviour of composite steel-concrete members was especially investigated by testing as well as by using calculation models for the temperature
rise and the load-bearing capacity including deformations and restraints. This
combined procedure resulted in well designed members within a comparatively
short period. The thermal and structural responses of composite structures are
studied theoretically-numerically. Analytical models for determining thermal
material properties for concrete and steel are described. Temperature-dependent
stress-strain relationships for concrete and structural steel are presented.
Finally the application of the analytical models in a computer program is shown.

KEYWORDS

Composite structures, computer design, computer programs, concrete, fire behaviour, fire protection, heat transfer, material behaviour, reinforcing steel,
structural analysis, structural steel, thermal analysis.

INTRODUCTION

Composite steel-concrete systems are structural elements, which are made of
concrete, structural steel and reinforcing steel. In recent years extensive
experimental and theoretical-numerical research has been carried out on the
load-bearing and deformation behaviour of composite uniaxial structures under
fire action by the "Institut für Baustoffe, Massivbau und Brandschutz der Technischen Universität Braunschweig". Fig. 1 shows some typical cross-sections of
investigated structures.

The computer Program "STABA-F" was developed to support the investigations
theoretically and numerically. Both material and geometric nonlinearities are
considered. In order to describe the behaviour of composite structures in fire
it is necessary to analyse the thermal and structural response under fire
action.

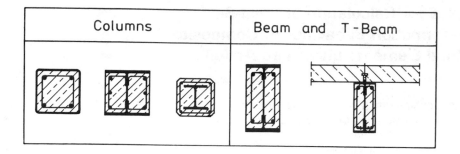

| Columns | Beam and T-Beam |

Fig. 1.  Typical cross-sections of composite structures

THERMAL ANALYSIS

The heat transfer from the fire to the structural element depends on the type of material and the surface of the member, the colour of the flames, the geometry and the material properties of the furnace walls and the ventilation conditions in the furnace.

Extensive investigations in the heating of structures in the furnaces for columns and beams of the "Institut für Baustoffe, Massivbau und Brandschutz der Technischen Universität Braunschweig" showed that there is sufficient correspondence between measured and calculated temperature distribution in a section, assuming for the convection coefficient of heat transfer $\alpha = 25$ W/m²K and for the resultant emissivity $\varepsilon = 0.3 - 0.7$ for concrete and $\varepsilon = 0.5 - 0.9$ for steel. Heat conduction is described by the well-known equation from Fourier, valid for homogeneous and isotropic materials. Applied to composite structures some simplifactions are necessary:

- water vaporizes as soon as reaching the boiling-point,
- movement of the steam is put together with other effects,
- consumption of energy for vaporizing the water and other peculiarities are taken into account in a simplified way by suitable design values for the specific heat capacity of concrete up to 200 °C,
- concrete is taken into account in a simplified way as a homogeneous material, the heterogeneous structure, as well as capillary pores and internal cracks, are lumped together.

A finite element method in connection with a time-step integration is used to calculate the temperature distribution in the cross-section /1/. The time steps have to be chosen quite small ($\Delta t = 2.5 - 5$ min), because the characteristic values of the thermal conductivity $\lambda$, specific density $\rho$ and specific heat capacity $c_p$ are very much dependent on temperature. Figs. 2 and 3 show the temperature-dependence of the thermal material properties for concrete with predominantly siliceous aggregates and for steel. To determine the temperature distribution a rectangular network is preferred with a maximum width of less than 20 mm. In the region of the structural steel it is advantageous to reduce the width of the network to the thickness of the structural steel profile. The elements of the cross-sectional discretization have corresponding thermal material properties of steel or concrete.

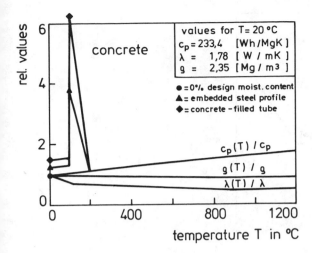

Fig. 2. Design values for thermal material properties for concrete with siliceous aggregates

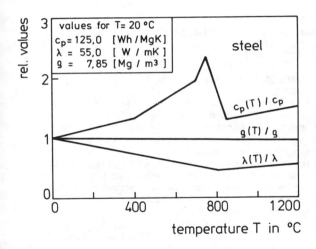

Fig. 3. Design values for thermal material properties for steel

Due to the element temperatures, the thermal strains for the cross-section elements are derived by using the temperature dependent thermal strain for concrete and steel as shown in Fig. 4.

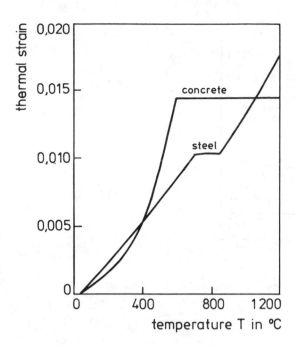

Fig. 4.  Design values for thermal strains of concrete with siliceous
         aggregates and steel


STRESS-STRAIN RELATIONSHIPS AT ELEVATED TEMPERATURES

At a fire situation the material is normally subjected to a transient process
with varying temperatures and stresses. To get material data of direct relevance for fire, transient creep tests are carried out. During the test the specimen is subjected to a certain constant load and a constant heating rate. From
these tests uniaxial stress-strain characteristics are obtained, which include
the temperature-dependent elastic strains and the comparatively large transient
creep strains. Fig. 5 shows the analytical model for determining the temperature-dependent stress-strain relationships for concrete and structural steel.
The model is also valid for other materials such as reinforcing steel and pre-
stressing steel. A complete stress-strain curve consists of different segments
described by equation (1) in Fig. 5. Material constants ($\varepsilon_o$, $\beta_o$, $\overline{E}_o$) are used
as starting-points for the parameters $\varepsilon_i$, $\sigma_i$ and $d\sigma_i/d\varepsilon_i$. The temperature dependence is introduced by varying these parameters with respect to temperature.
The coefficients $a_k$, $b_k$ and $c_k$ for calculating the material temperature dependence (see Fig. 5) are available at the "Institut für Baustoffe, Massivbau und
Brandschutz der Technischen Universität Braunschweig".

Theoretical stress-strain relationships for concrete and structural steel are
shown in Figs. 6 - 7. The input parameters for the calculation are listed in
Table 1.

| Material constants | | Concrete | Structural steel |
|---|---|---|---|
| $\varepsilon_o$ | $[-]$ | $-\ 10^{-3}$ | $10^{-3}$ |
| $\beta_o$ | $[N/mm^2]$ | $f'_c$ | 240 |
| $\bar{E}_o$ | $[N/mm^2]$ | $\beta_o \cdot 10^3$ | $2.1 \cdot 10^5$ |
| $f'_c$: specified compressive strength at T = 20 °C | | | |

Table 1.  Material constants applied to calculate the stress-strain
relationships at elevated temperatures for concrete and
structural steel (see Figs. 6 - 7)

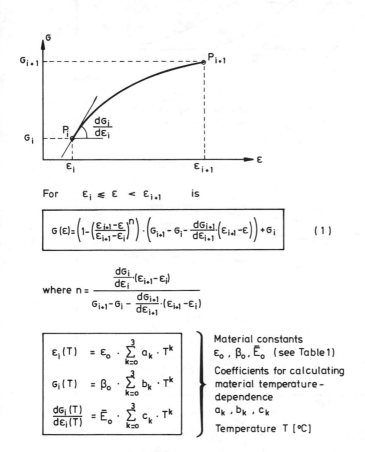

For $\quad \varepsilon_i \leqslant \varepsilon < \varepsilon_{i+1} \quad$ is

$$G(\varepsilon) = \left(1 - \left(\frac{\varepsilon_{i+1} - \varepsilon}{\varepsilon_{i+1} - \varepsilon_i}\right)^n\right) \cdot \left(G_{i+1} - G_i - \frac{dG_{i+1}}{d\varepsilon_{i+1}} \cdot (\varepsilon_{i+1} - \varepsilon)\right) + G_i \qquad (1)$$

where $n = \dfrac{\dfrac{dG_i}{d\varepsilon_i} \cdot (\varepsilon_{i+1} - \varepsilon_i)}{G_{i+1} - G_i - \dfrac{dG_{i+1}}{d\varepsilon_{i+1}} \cdot (\varepsilon_{i+1} - \varepsilon_i)}$

$$\varepsilon_i(T) = \varepsilon_o \cdot \sum_{k=0}^{3} a_k \cdot T^k$$

$$G_i(T) = \beta_o \cdot \sum_{k=0}^{3} b_k \cdot T^k$$

$$\frac{dG_i(T)}{d\varepsilon_i(T)} = \bar{E}_o \cdot \sum_{k=0}^{3} c_k \cdot T^k$$

Material constants
$\varepsilon_o$, $\beta_o$, $\bar{E}_o$ (see Table 1)

Coefficients for calculating
material temperature-
dependence
$a_k$, $b_k$, $c_k$

Temperature T [°C]

Fig. 5.  Principles for temperature dependent stress-strain relationships

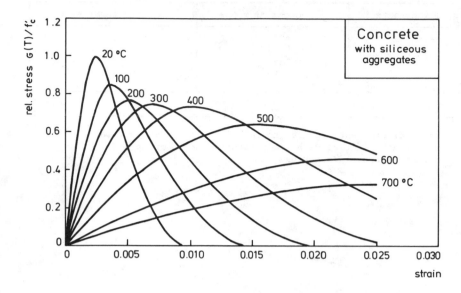

Fig. 6.  Stress-strain relationships at elevated temperatures for concrete
         with siliceous aggregates

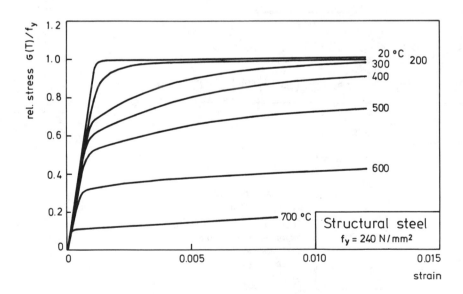

Fig. 7.  Stress-strain relationships at elevated temperatures for structural
         steel

STRUCTURAL ANALYSIS

The determination of load-bearing and deformation behaviour of composite structural elements under fire action can be described in four parts independent of
each other:

- geometric description of the composite structural cross-section,
- determination of the temperature distribution under fire action,
- determination of the nonlinear interaction between bending moment M and
  curvature 1/r dependent on normal force N and temperature distribution,
- determination of all forces and deformations in accordance with 2nd order
  theory and any boundary conditions and no matter how loaded.

The first step to determine a composite structural element should be the discretization of the cross-section. In order to ensure a detailed specification of
different parts of the cross-section (structural steel, concrete and reinforcement) rectangular and triangular elements are used /1/. The calculated temperature distribution of a composite cross-section, composed of a rolled shape
HE 240 B embedded in concrete and by 0.5 % reinforcement, after 30 minutes fire
exposure according to ISO 834 is shown in Fig. 8.

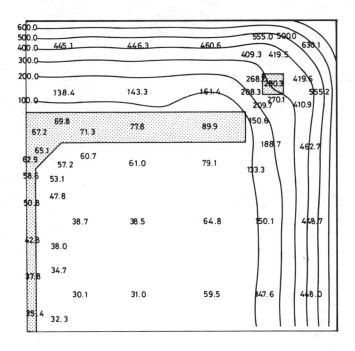

Fig. 8.  Calculated temperature distribution of a quarter of a cross-section,
         consisting of a rolled shape HE 240 B embedded in concrete and 0.5 %
         reinforcement, after 30 minutes fire exposure according to ISO 834

Knowing the temperature distribution and with assumption of the following simplifications

- the Bernoulli-Navier hypothesis,
- only uniaxial stresses are taken into account, shear stresses are neglected,
- there is no slip between concrete and steel,
- the stress-strain relationships are nonlinear elastic,

it is possible to determine the relations between loads and deflections of a bar. The same network as in the calculation of temperature is used. The stress causing strain $\varepsilon_i^\sigma$ at the location i follows from

$$\varepsilon_i^\sigma = \varepsilon^o + 1/r_y \cdot z_i + 1/r_z \cdot y_i - \varepsilon_i^{th} \tag{2}$$

With the actual temperature $T_i$ the stress $\sigma_i$ can be linked by the temperature dependent stress relationships (see Figs. 6 - 7). Fig. 9 shows the calculated temperature, the strain and the stress distribution in section A-A for a composite cross-section consisting of a concrete filled hollow-section with 3 % reinforcement after 30 minutes fire (ISO 834) exposure.

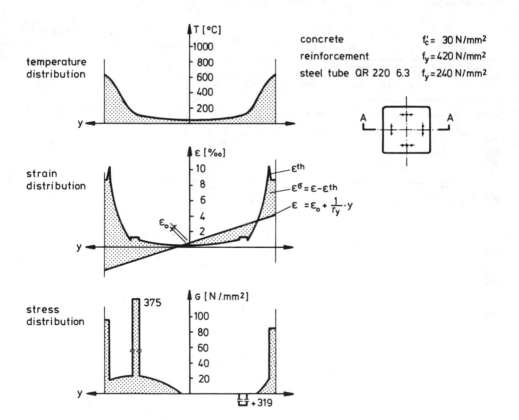

Fig. 9.   Temperature, strain and stress distribution in section A-A of a composite cross-section composed of a concrete filled hollow-section after 30 minutes fire exposure (N = 100 kN, $M_y$ = 25 kNm)

The integration of the stress distribution yields the following forces according to the temperature

$$N = \int_A \sigma \cdot dA \quad \cong \quad \sum_{i=1}^{n} \sigma_i \cdot \Delta A_i \qquad\qquad (3)$$

$$M_y = \int_A \sigma \cdot z \cdot dA \quad \cong \quad \sum_{i=1}^{n} \sigma_i \cdot z_i \cdot \Delta A_i \qquad\qquad (4)$$

$$M_z = \int_A \sigma \cdot y \cdot dA \quad \cong \quad \sum_{i=1}^{n} \sigma_i \cdot y_i \cdot \Delta A_i \qquad\qquad (5)$$

The Relations between the bending moment $M_y$ and the curvature $1/r_z$ of a composite cross-section is shown in Fig. 10. Every curve is dependent on the actual applied normal force N.

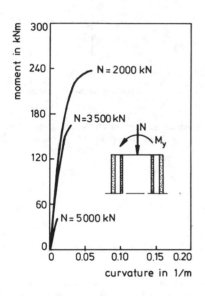

| | |
|---|---|
| concrete | $f'_c = 40\ \text{N/mm}^2$ |
| reinforcement 0.5 % | $f_y = 420\ \text{N/mm}^2$ |
| rolled shape HE 240 B | $f_y = 240\ \text{N/mm}^2$ |

Fig. 10.  Relationship between bending moment $M_y$ and curvature $1/r_z$
         dependent on the normal force N.
         Cross-section and temperature distribution shown in Fig. 8

An accurate evaluation of the load-bearing behaviour takes into account the influence of mechanical (nonlinear moment/curvature relationship) and geometrical nonlinear interaction (2nd order theory) between load and deformation. To determine bending Moment M, shear force Q, slope of the bar and deflection w the method of transferring these values from one division to the next is used. The initially unknown forces or deformations at the beginning of the structural element have to be determined by integration in that way that the compatibility condition at the end of the structural element is fulfilled. The definition of the stiffness as the gradient of the partially linear moment/curvature relationship results into a quick converging calculation algorithm. It is possible to determine the load-bearing and deformation behaviour with different moment/curvature relationships along the axis of the bar. Fig. 11 shows evaluated moments and deformations of a composite column under ultimate load after 30 minutes fire exposure. Recalculation of tests showed a good agreement of measurements and determined values /3/. In the research report /4/ the calculated ultimate loads are summed up in charts and diagramms.

concrete                     $f'_c$ =   40 N/mm²
reinforcement 0.5 %          $f_y$ =  420 N/mm²
rolled shape HE 240 B        $f_y$ =  240 N/mm²

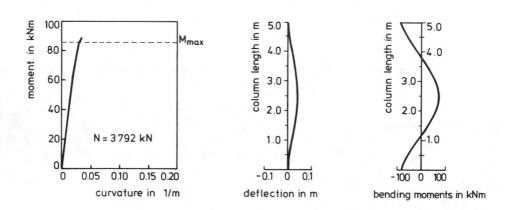

Fig. 11.   Ultimate limit state of a composite-column consisting of a rolled shape HE 240 B embedded in concrete and 0.5 % reinforcement after 30 minutes fire exposure

REFERENCES

/1/   Becker, J., Bizri, H., Bresler, B.: FIRES-T, A Computer Program for the Fire Response of Structures-Thermal, Berkeley, 1974.

/2/   Kordina, K., Klingsch, W.: Brandverhalten von Stahlstützen im Verbund mit Beton und von massiven Stahlstützen ohne Beton. Studiengesellschaft für Anwendungstechnik von Eisen und Stahl e.V., Forschungsbericht Projekt 35, 1984.

/3/   Haß, R., Quast, U.: Brandverhalten von Verbundstützen mit Berücksichtigung der unterschiedlichen Stützen/Riegel-Verbindungen. Studiengesellschaft für Anwendungstechnik von Eisen und Stahl e.V., Forschungsbericht Projekt 86.2.2, 1985.

/4/   Quast, U., Rudolph, K.: Bemessungshilfen für Stahlverbundstützen mit definierten Feuerwiderstandsklassen. Studiengesellschaft für Anwendungstechnik von Eisen und Stahl e.V., Forschungsbericht Projekt 86.2.3, 1985.

/5/   Dorn, T., Haß, R., Quast, U.: Brandverhalten von Riegelanschlüssen an Verbundstützen aus einbetonierten Walzprofilen und ausbetonierten Hohlprofilen. Studiengesellschaft für Anwendungstechnik von Eisen und Stahl e.V., Forschungsprojekt P 86.2.10, in Bearbeitung.

# Numerical Simulations of Fire Resistance Tests on Steel and Composite Structural Elements or Frames

**J. B. SCHLEICH**
ARBED-Research, Luxembourg

**J. C. DOTREPPE and J. M. FRANSSEN**
National Fund for Scientific Research
University of Liege, Belgium

ABSTRACT

A computer program for the analysis of steel and composite structures under fire conditions is presented. It is based on the finite element method using beam elements with subdivision of the cross section in a rectangular mesh. The structure submitted to increasing loads or temperatures is analyzed step-by-step using the Newton-Raphson procedure. The thermal problem is solved by a finite difference method based on the heat balance between adjacent elements. Comparisons are made between full scale tests on steel and composite structural elements and frames and the results given by the numerical simulations. The agreement appears to be quite good. Furthermore are discussed the new possibilities given by this numerical computer code.

INTRODUCTION

When a fire breaks out in a building, heavy damage to people and property will occur if the structural part of the building happens to collapse. A good fire resistance of the loaded structure is a non sufficient but necessary condition to preserve the integrity of a building, to allow the rescue of occupants and to give the fire-brigade the opportunity of an effective intervention.

The first way to determine the fire endurance of a structural element has been the full scale test in furnace. In Europe, the standard fire resistance test according to ISO-834 has been used quite intensively and, in many countries, it is still the only legal way to classify structural elements regarding their fire resistance. Nevertheless, a test procedure has several shortcomings concerning the maximum size of the element, the available loading capacities, the heating system and the restraint characteristics. A full scale test is long to prepare, expensive to perform and gives one result for particular values of the parameters. Furthermore, the behaviour of one single column or one single beam in a furnace will not necessarily give a good idea of the behaviour that the element would have if it was a part of the whole structure.

Therefore the need for analytical models of thermal and structural response has grown intensively. Considerable progress has been made in the development of simple analytical methods, particularly for steel and composite elements [1, 2, 3, 4] and in several countries the practical evaluation of the fire resistance can now be made through these simplified calculation models. Unfortunately, this type of method does not apply to all cases and it has severe limitations when more parameters have to be taken into account in order to simulate real situations.

During the last decade a considerable amount of work has been performed in the Department of Bridges and Structural Engineering of the University of Liège in order to develop models for the analysis of the structural behaviour under fire conditions. Most of the work has been conducted in the field of structural analysis for steel and mainly concrete structures. Many numerical examples have been performed using the model developed and practical conclusions have been drawn [5].

Presently new developments are realized at the University of Liège on steel and composite structures within a C.E.C. research [6] introduced by ARBED. The aim of this paper is to present some informations on the program itself, to show some results which have already been obtained and to discuss the limitations, the possibilities of the program and the applications that can be expected from now on. Since this work is still in progress, the present paper has to be considered as a status report on this research program [7].

THE PROGRAM CEFICOSS

CEFICOSS stands for " Computer Engineering of the FIre resistance for COmposite and Steel Structures". Indeed, though the program is suitable for reinforced concrete structures, it has essentially been developed for and applied to composite and steel structures.

Static Calculations

The program is a finite element program using a beam element with two nodes and six degrees of freedom, two translations and one rotation at each node. The shear displacements and the shear energy are not considered; the Navier-Bernoulli hypothesis is assumed. In one element, the stiffness properties and the internal forces are calculated in two cross-sections. Thus, the stiffness matrix of the element results from a Gaussian integration along the longitudinal axis. The cross section of this beam element is divided into subslices forming a rectangular mesh (Fig. 1). The kind of material, the temperature, the strain and the stress are different from one patch to another and the integrals on the cross section appearing in the stiffness matrix, the internal bending moments and centric forces are computed in a numerical way.

Stress-strain relations in the materials are non-linear and moreover are temperature dependent. In a structure submitted to fire loads, the materials are subjected to initial strains due to thermal effects ($\varepsilon_{th}$) and to creep effects ($\varepsilon_{cr}$). Thus the stresses are caused by the difference between the total strains ($\varepsilon_{to}$), derived from the nodal displacements, and the initial strains:

$$\sigma = \sigma \ (\varepsilon_{\sigma}) = \sigma(\varepsilon_{to} - \varepsilon_{th} - \varepsilon_{cr}) \qquad (1)$$

In this computer program, the creep strains are not explicitly taken into account, but have been considered indirectly in the stress-strain relationship. The strainhardening effect in the steel stress-strain relationship, which seems to be quite important, has been introduced in this computer code.

Due to the thermal effects, it may happen that the stress related strain ($\varepsilon_{\sigma}$) grows up in the first minutes of the simulation and then finally decreases. These return effects of the stress related strains must be taken into account. It has been assumed that the unloading branch of a stress strain curve is linear and that the plastic part of the strain ($\varepsilon_{pl}$) is not affected by a temperature variation (Fig. 2).

As the geometrical non-linearities have to be taken into account, the equilibrium conditions, based on the principle of virtual work and leading to the stiffness matrix of the structure, are written in an incremental form. Moreover, when estimating the displacements from one time step to the following one, each finite element is assumed to be straight in the starting step. Thus, the up-dated Lagrangian description is applied.

Due to the high non linear character of the analyzed structures, a single step process would lead to important discrepancies. Thus, within each time step, an iterative process has to be used in order to eliminate the out-of-balance forces and to restore equilibrium. Two different kinds of non-linearities have to be considered: continuous non-linearities like the variation of the material properties in function of temperatures, and discontinuities like the cracking of concrete. Due to these last discontinuities some sudden changes occur in the stiffness matrix which has to be reformulated at each step during the iterative process, consequently leading to a Newton-Raphson process.

Thermal Problem

A real fire in a building being a very complexe phenomenon, it is not easy to take into account every possible parameter able to influence the fire consequences. In the fire resistance tests, the gas temperature in the furnace is given as a function of time and, in fact, is the main parameter of the test.

The convection process between the air and the structure is obvious and leads to a boundary condition of the Newton type. Very important is the radiative mode of heat transfer which takes place between the walls of the furnace and the structural element. Thus, the temperature of the inside faces of the furnace and the temperature at the surface of the element should appear in the mathematical expression of that exchange. Yet, the temperature of the furnace walls is not directly known, but depends on the characteristics of every furnace. Therefore the gas temperature is considered instead of the temperature of the furnace walls. In fact, the approximation is not too bad because the walls of the furnace have a low thermal conductivity. The thermal gradient existing close to the inside face of the wall being very high, the temperature of the inside surface will be close to the gas temperature. Consequently, the boundary condition at the structural element surface is given by:

$$Q = \alpha \, (T_G - T_S) + \sigma^* \cdot \varepsilon^* \, (T_G^4 - T_S^4) \qquad (2)$$

$T_G$ = gas temperature (given as a function of time)
$T_S$ = temperature at the element surface (to be calculated)
$\alpha$ = coefficient of convection heat transfer (experimental value)
$\varepsilon^*$ = resultant emissivity (experimental value)
$\sigma^*$ = Stefan-Boltzmann constant
$Q$ = heat flow through boundary

The equations of the transient heat flow inside the structural element are well known. These equations are solved by a finite difference method

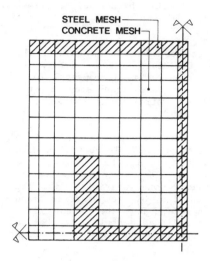

Fig. 1: Cross section of beam element with concrete and steel mesh.

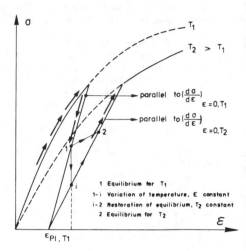

Fig. 2 : Return effect of stress related strain in function of temperature increase.

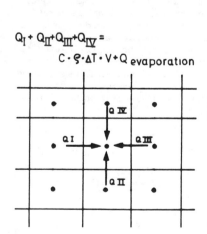

Fig. 3 : Heat balance between adjacent patches.

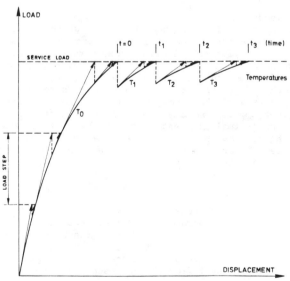

Fig. 4: Alternative thermal and static calculations at time t in order to restore equilibrium under simultaneous action of the service load and the temperature increase.

314

based on the heat balance (Fig. 3) between adjacent patches which moreover are identical to those of the structural analysis.

The main advantages of this method are:

- The equations have an immediate physical meaning so that f.i. the evaporation effect of moisture in concrete is easy to be simulated (Fig. 3).
- This method is explicit. The values of the temperature at a given time are obtained explicitly at the end of the previous time step. Furthermore, only n equations with one unknown per equation (n being the number of nodes of the mesh) have to be solved at each time step, whereas an implicit method would lead to a system of n equations with n unknowns per equation.

The main disadvantage of an explicit method lies in the fact that, for stability and precision reasons, a criterion relating the time step to the patch width must be satisfied. Due to the high thermal conductivity of steel, the thermal time step will be only about some seconds. It must be noticed that the arrangement of the program and the fact that the structural response of the structure must be calculated more or less every minute, do not allow at any rate to take profit of too long a time step which would have been obtained in case of an implicit analysis of the thermal problem.

Program Flow Chart

The solution principle is illustrated in figure 4 for a system with one degree of freedom. At ambient temperature, the load is applied step by step. After each load step, the equilibrium of the structure must be restored by the Newton-Raphson process. When the service load has been reached, it is kept constant all through the following fire simulation. In the cross section the temperatures of every patch are then calculated with a short time step derived from the stability condition mentioned previously. When the simulation of the fire test has reached a certain time of about one minute, the thermal analysis is stopped. Now the static part of the program calculates the displacements of the structure for the temperatures calculated at this time. Here again a Newton-Raphson process takes place in order to restore equilibrium.

This procedure composed of alternative thermal and static calculations goes on, up to the moment where equilibrium can no more be obtained. This moment is identical to the ultimate fire resistance time of the analysed structural element.

RESULTS OF FIRE TESTS

A lot of real tests have been performed in the last years, the results of which are available. Some of these fire resistance tests have been used during the development phase of this numerical software.

In that respect the structural behaviour of a reinforced concrete continuous T beam on three supports has been analysed and the theoretical results have been compared with test results obtained at the Technical University of Braunschweig [5, 8]. The loading and heating systems are presented in figure 5. The beam is loaded and heated unsymmetrically. The thermal program is applied according to the standard ISO temperature-time curve. The variation of the bending moment on the centric support of the T beam analysed here, is represented in figure 6. After approximately one hour the bending moment tends to become constant, which corresponds to the formation of a plastic hinge on the centric support. It can be observed that there is a good agreement between theoretical and experimental results. The corresponding

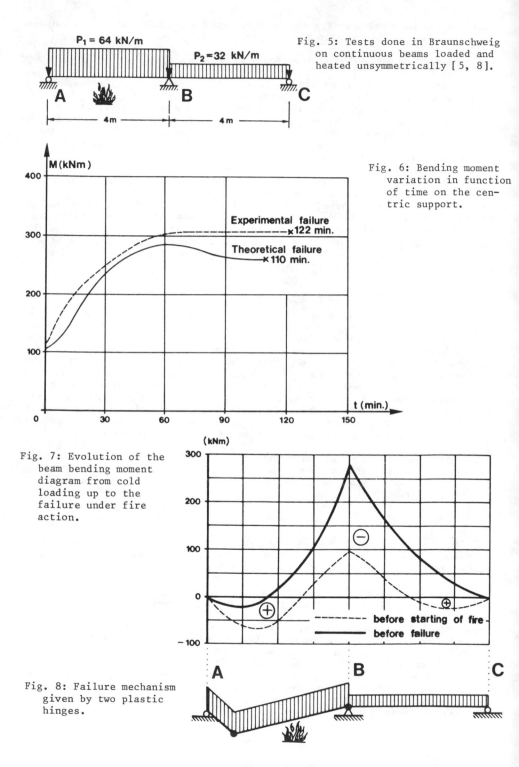

P₁ = 64 kN/m

P₂ = 32 kN/m

A    B    C

4m    4m

Fig. 5: Tests done in Braunschweig
on continuous beams loaded and
heated unsymmetrically [5, 8].

Fig. 6: Bending moment
variation in function
of time on the cen-
tric support.

M(kNm)

Experimental failure
×122 min.

Theoretical failure
×110 min.

t (min.)

Fig. 7: Evolution of the
beam bending moment
diagram from cold
loading up to the
failure under fire
action.

(kNm)

⊖

⊕    ⊕

before starting of fire
before failure

A    B    C

Fig. 8: Failure mechanism
given by two plastic
hinges.

316

evolution of the bending moment diagram is indicated in figure 7, showing the important redistribution of internal forces due to the thermal gradient. Before failure the diagram has only negative zones, except on the left hand side of the heated and most loaded span. This explains the failure mechanism represented in figure 8.

In order to verify the simulation results given by CEFICOSS and to estimate with greater accuracy the values of certain fundamental physical parameters, it was decided to perform new series of real fire tests based on the ISO-834 heating curve. Thus a better comparison was guaranteed between test and simulation results and most interesting informations got available on a new type of composite structure developed by ARBED [7].

Tests on Columns

At the University of Gent [9] two columns were tested under longitudinal load with an eccentricity of 180 mm around the weak axis. The steel profile in both tests was the heavy American wide flange shape W 14 x 16 x 500. Column 1.1. was not protected against direct fire action. This test made clear that a high massivity – the section factor F/V of this steel profile was 27 $m^{-1}$ – provides a good fire resistance even to bare steel profiles. Only numerical softwares, giving the temperature gradient through steel thickness, are able to predict correctly the behaviour of thick bare steel elements. The fire test gave a resistance time of 46 minutes, whereas the simulation by CEFICOSS gives 45 minutes.

Column 1.2. composed of the same profil as column 1.1., was loaded in exactly the same way. But column 1.2. was protected by an intumescent fire retardant coating [9] which contributed to give a resistance time of 145 minutes in the fire test. It has been shown that the program CEFICOSS is able to simulate correctly the behaviour of such a structure, provided that the thermal characteristics of the paint layer can be measured: the calculated fire resistance amounts to 134 minutes (92 %). The final thickness of the swollen intumescent coating has been measured after the test and introduced as a constant for the whole fire simulation. The value of the thermal conductivity has been given by the producer of the coating.

Two composite columns of the type AF30/120 were tested in Gent [9]. AF is a new type of composite cross section developed by ARBED [2, 3] and is composed of a rolled H-profile concreted between the flanges. This concrete contains longitudinal reinforcing bars which contribute to support loads. These columns were centrically loaded and the buckling occured around the weak axis of the steel profile. Column 1.3. which had no further protection on the exterior visible faces of the steel flanges, collapsed in the fire test only after 116 minutes, whereas the numerical simulation predicted a fire resistance time of 114 minutes.

Column 1.15. was identical to the previous one, with the exception of the steel flanges of the profile which were protected by a dry insulation layer of 25 mm thickness. The thermal properties of this insulation were not quite exactly known for higher temperatures. This could explain the slightly greater difference between the measured fire resistance time of 189 minutes and the value of 162 minutes (86 %) obtained by the CEFICOSS simulation.

The special cross-section column 1.4. (see fig. 9,a) behaved quite good in the fire test [9], as the resistance time attained 172 minutes, though four visible steel flanges were exposed directly to the fire action. This

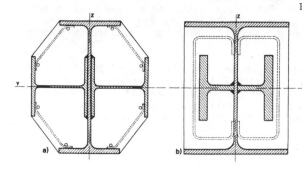

Fig. 9: Special composite cross sections, exclusively composed of rolled profiles, developed for the purpose of columns supporting centric loads and bending moments around Y or/and Z axis.

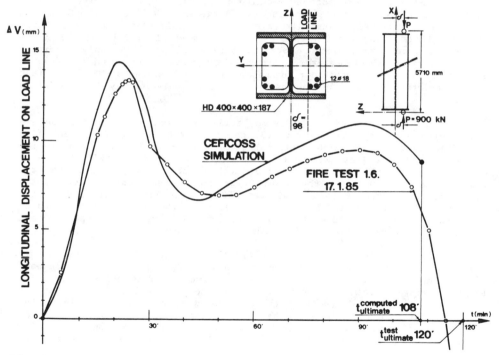

Fig. 10: Calculated and measured longitudinal displacement $\triangle V$ of a composite AF column, supporting the eccentric load P during the fire test [7, 10]

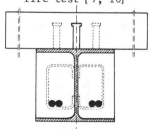

Fig. 11: Composite beam cross section.

column was composed of three rolled H-profiles, welded together and concreted between the flanges. The concrete of this octagonal cross section contained no reinforcing bars. CEFICOSS gave a fire resistance of 146 minutes (85 %).

Two columns of the type AF30/120 were tested at the University of Braunschweig [10], with a load eccentricity of 98 mm around the weak axis. Column 1.5. had a length of 3,74 m and collapsed in the fire test after 136 minutes, whereas the numerical simulation had given a fire resistance of 110 minutes. This difference is the worst numerical result we obtained.

Column 1.6. was identical to column 1.5. with the difference that it had a length of 5,71 m and that the load had been reduced. The fire test gave a resistance time of 120 minutes whereas the simulation by CEFICOSS predicted 108 minutes (90 %). Figure 10 shows the pretty good agreement between the calculated and measured longitudinal displacement of the column during fire action.

In order to allow higher bending moments around the weak axis, it is advantageous to replace reinforcing bars by T profiles welded on the web of the main H profile. Figure 9b presents such a cross section. Two columns of this type have been tested successfully in Braunschweig [10]. For column 1.7. f.i. the measured fire resistance time was 111 minutes, while the numerical simulation gave 114 minutes.

Tests on Beams

Four beams were tested in Gent [9]. They were composed of the AF composite profile supporting a concrete plate and fixed together by connectors welded on the upper flange of the steel profile (see fig. 11). In the test beam 2.11. the composite T beam was simply supported on both ends. When applying the deflection criterion f ≤ L/30 in order to define the fire resistance time, the comparison between the test result (171 minutes) and the simulation (149 minutes) appears to be quite good (87 %).

In the test beam 2.14., no connectors were placed between the AF profile and the covering plate. This was simply laid on the upper flange, had to be considered in the calculation of temperature distribution, but did not contribute to the static resistance of the lower AF cross section. Here again the fire resistance times measured in test (92 minutes) and computed by CEFICOSS (87 minutes) show a rather good agreement (95 %). Figure 12 gives the measured and simulated mid-span deflection of this composite beam.

The two composite beams 2.12 and 2.13 were tested with one end simply supported and the other fixed. In both cases a plastic hinge was formed close to the fixed end, which was confirmed by the numerical simulation.

Tests on Frames

One of the most interesting possibilities of CEFICOSS is the analysis of fire effect on frames. Of course, no furnace exists able to test a whole building under fire action. Yet, at the University of Braunschweig, the opportunity is given to test simple frames comprising one column and one beam. Thus, two frame-tests have been performed [11] confirming the numerical results given by the simulation program. The types of column and beam composing these two frames are shown on figures 9b and 11, whereas figure 13 shows the very good agreement between the measured and calculated horizontal displacement of the column pratically at mid-height.

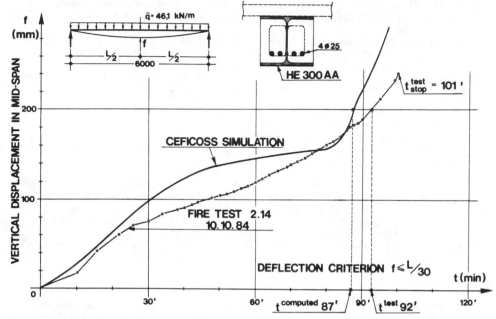

Fig. 12: Calculated and measured mid-span deflection f of a composite AF
beam during the fire test [7, 9].

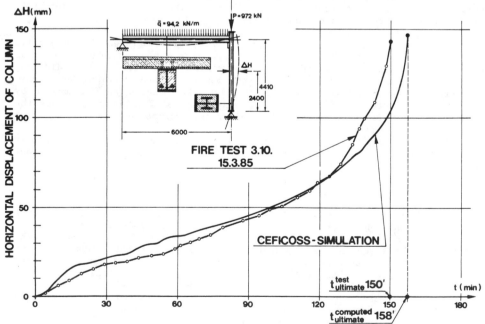

Fig. 13: Calculated and measured horizontal displacement ΔH of the frame
column 3.10 [7, 11].

320

Conclusion on Test and Simulation Results

Figures 10, 12 and 13 prove, and so do the similar curves drawn for the other tests, that a simulation by CEFICOSS is able to describe correctly the behaviour of a structure in a fire resistance test. Figure 14 is a graphic result presentation of all the tests that have been performed and simulated up to now. It can be noticed that the correspondence between theory and test results is quite acceptable [7].

APPLICATION FIELDS FOR CEFICOSS

Exploitation of Test Results

A lot of structural parameters have an influence on the final result of a fire resistance test. If one single test is performed, it provides one result for one value of all the parameters and nothing can be said with precision about the way the test result would change in function of one of these parameters. Even two or three real fire tests with different column lengths, will hardly give enough information on the behaviour of the column regarding the general slenderness effect.

However a computer program like CEFICOSS, eventually calibrated by a given number of real tests, will be of great help for the solution of such problems. It is intended to use this program in order to examine the problem of the simultaneous action of centric loads (N) and bending moments (M) in a fire environment. In that respect practical design tables and diagrams for the different fire classes F60, F90 and F120, will be established for composite columns under N/M interaction, and for composite beams under continuous load (see fig. 15).

Research

The most interesting application of this computer code may be the perspective offered for new research possibilities. It is intended not only to produce tables about the interaction between centric and bending forces, but it is also hoped to deduce some practical rules, allowing to take into account this phenomenon in a more simple manner. The correct behaviour of a whole structure in a fire can only be deduced from calculations based on a thermo-mechanical computer code like CEFICOSS. In this way, it will now be possible to simulate numerically the tests which cannot be executed in practice, because of the limitation in size of the fire test installations or because of prohibitive costs.

From a more practical point of view a lot of investigations could be done and errors avoided, when developing new kinds of structures. The optimum fire design of a structure becomes now feasible for a reasonable price. For instance, the localization of reinforcing bars or structural tees could be chosen in function of the highest possible efficiency.

Education

Results of fire simulations could be presented on graphic screens. Thus, the variations of bending moments in hyperstatic structures, or the changes of compressive and tensile zones in cross sections and due to thermal effects could appear graphically. This would help to make lectures in fire resistance more understandable, attractive and efficient.

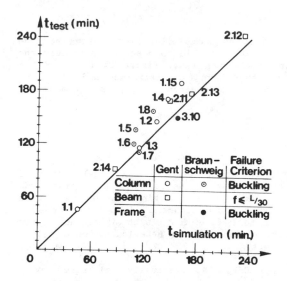

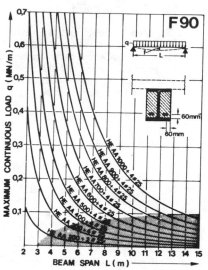

Fig. 14: Fire resistance times calculated by CEFICOSS simulation and measured on practical fire tests, of steel and composite structural elements [7, 9 10, 11].

Fig. 15: Maximum continuous total load q applicable to simply supported AF beams – having no connection to the concrete floor – in function of the beam span L and for the fire class F90 ($\sigma_y$ = 355 N/mm² ; $\beta_c$ = 35 N/mm²). Steel sections belong to the rolled H-profile series HEAA of ARBED.

CONCLUSIONS

It has been shown that the numerical software CEFICOSS is able to simulate in a correct way the structural behaviour in a fire resistance test and that it provides a pretty good estimation of the fire resistance time. This computer code will of course never be a substitute for real fire tests, because the real test is the only way allowing to detect local problems such as spalling of concrete, lack of adherence to reinforcing bars, bad behaviour of welded joints or local buckling. CEFICOSS is to be considered as a new tool, which at last makes feasible a lot of new investigations allowing to improve seriously our knowledge on the behaviour of structures under fire conditions [12].

Indeed CEFICOSS is a general, thermo-mechanical numerical computer code for the analysis of columns, beams or frames, composed of either bare steel profiles or steel sections protected by any type of insulation or even any types of composite steel-concrete cross-sections. Furthermore the ISO-834 standard fire curve, as well as any natural heating curve can be considered.

322

This new computer code, allowing to determine the structural fire safety, corroborates the idea expressed by P.J. DiNENNO [13] that "The next evolution of the application of computers in the delivery of a fire safety system appears to be in the area of predicting fire resistance." This numerical tool becomes available at suitable time as B. BRESLER [14] says "... that the acceptable level of fire safety should be determined by calculation, just as it is for other types of loading." Structural fire safety could be provided for a precisely imposed level without paying for excessive fire protection; this means that substantial cost savings can be foreseen by using this new fire safety approach.

REFERENCES

[1]  ECCS, TC3 - European Recommendations for the Fire Safety of Steel Structures - Elsevier; Amsterdam, Oxford, New York, 1983.
[2]  JUNGBLUTH O., FEYEREISEN H., OBEREGGE O., - Verbundprofil- konstruktionen mit erhöhter Feuerwiderstandsdauer - Bauingenieur 55, 1980.
[3]  SCHLEICH J.B., HUTMACHER H., LAHODA E., LICKES J.P., - A New Technology in Fireproof Steel Construction - Review Acier/Stahl/Steel Nr. 3, 1983.
[4]  SCHLEICH J.B. - Fire Safety, Design of Composite Columns / International Conference "Fire safe steel construction; practical design", Luxembourg, April 1984 - Revue Technique Luxembourgeoise Nr. 1, 1985.
[5]  DOTREPPE J.C. - Méthodes Numériques pour la Simulation du Comportement au Feu des Structures en Acier et en Béton armé - Thèse d'Agrégation de l'Enseignement Supérieur, Université de Liège, 1980.
[6]  DOTREPPE J.C., FRANSSEN J.M., SCHLEICH J.B. - Computer Aided Fire Resistance for Steel and Composite Structures - Review Acier/Stahl/Steel Nr. 3, 1984.
[7]  ARBED-Research, Luxembourg / Department of BRIDGES and STRUCTURAL ENGINEERING, University of Liège, Belgium - REFAO/CAFIR; Computer Assisted Analysis of the Fire Resistance of Steel and Composite Steel-Concrete Structures - C.E.C. Research 7210-SA/502, Technical reports 1 to 6, 1982/1985.
[8]  WESCHE J. - Stahlbetondurchlaufkonstruktionen unter Feuerangriff - Institut für Baustoffkunde und Stahlbetonbau, Technische Universität Braunschweig, 1974.
[9]  MINNE R., VANDEVELDE R., ODOU M. - Fire Test Reports Nr. 5091 to 5099 - Laboratorium voor Aanwending der Brandstoffen en Warmte-overdracht, University of Gent, April to June 1985.
[10] KORDINA K., HASS R. - Untersuchungsbericht Nr. 85636 - Amtliche Materialprüfanstalt für das Bauwesen, Technische Universität Braunschweig, April 1985.
[11] KORDINA K., WESCHE J., HOFFEND F. - Untersuchungsbericht Nr. 85833 - Amtliche Materialprüfanstalt für das Bauwesen, Technische Universität Braunschweig, Mai 1985.
[12] KLINGSCH W., SCHLEICH J.B. - Composite Steel-concrete Components, Present Time Acquirements and Future Possibilities - International Symposium "Steel in Buildings", Luxembourg, September 1985.
[13] DINENNO P.J. - Introduction/Guest Editor - Fire Safety Journal, Vol. 9 No. 1, May 1985.
[14] BRESLER B. - Analytical Prediction of Structural Response to Fire- Fire Safety Journal, Vol. 9 No.1, May 1985.

# Contribution to Fire Resistance from Building Panels

**B. J. NORÉN AND B. A.-L. ÖSTMAN**
Swedish Institute for Wood Technology Research
Box 5609, S-114 86 Stockholm, Sweden

ABSTRACT

The contribution from different types of building panels to the total fire resistance of a construction has been determined experimentally by testing in
furnaces of three different sizes. The agreement between a full size furnace
according to ISO 834 and a small size furnace is good, while a medium size furnace gives a somewhat higher fire resistance.

The panel thickness is the most important factor and has an almost linear relationship with the fire resistance, common for all wood-based boards and also
including gypsum boards. Other factors as panel density, moisture content, type
of adhesive and structural composition of the panel will also affect the contribution to the fire resistance. The behavior of various wood-based panels may
therefore be slightly different. The fire penetration rate as determined from
the measurements in a furnace exposure according to ISO 834 is slower than
0.9 mm/min for all boards with density over 400 kg/m$^3$.

The insulation criterion is usually determining the fire resistance, while the
panel integrity remains a little longer. This is especially true for inorganic
boards which may have a considerably longer time until the integrity criterion
fails, while the insulation criterion gives a fire resistance of the same order
of magnitude as wood-based boards.

The influence of some design factors as mineral insulation, studs and double
layers of panelling has also been studied, as well as varying thermal exposure.

INTRODUCTION

The fire resistance of elements of building constructions is usually determined
by the internationally accepted test method ISO 834, which tests in full scale,
e.g. wall elements of 3 x 3 m. The test elements form part of a furnace, in
which the temperature increase follows a standard time-temperature curve,
reaching e.g. 659 °C after 10 minutes, 821 °C after 30 minutes and 925 °C after
60 minutes. The elements may be load-bearing or not load-bearing during the
test depending on the application. The following three criteria are used:

- load-bearing capacity (withdrawn for non load-bearing elements),

- insulation, i.e. the temperature increase of the unexposed face shall be below an average of 140 °C and a maximum of 180 °C,

- integrity, i.e. the test element shall not crack or release flames.

The time in minutes until one of these criteria fails is by definition the fire resistance of the element.

Such a full scale testing is expensive. Smaller furnaces have been developed which might give useful information, especially for non load-bearing constructions, but they need verification with full scale in most cases. The design of these smaller furnaces may vary and they have not been standardized. Just a German proposal exists (DIN 4102).

The fire resistance can also be determined by theoretical calculations, but methods and basic data are still lacking for wooden constructions. A review (Pettersson and Jönsson) and a research program for developing analytical methods (Jönsson and Pettersson) have recently been presented.

A more simple approach for some types of constructions is to determine the contribution from different components to the total fire resistance of the construction. Such an approach has been proposed and plain applications have shown good agreement with full scale testing of the whole construction in spite of great simplifications (Hagstedt and Nyström; Dansk Brandværnskomité).

Building panels are components with great importance for the fire resistance. Relatively few studies have been published on the contribution from different types of panel. A recent review (Östman) shows that the contribution from panels is considered to be quite different in different studies. The main reason is differences in test methods and criteria. A suitable test for the contribution from panels shall be in accordance with the criteria for the fire resistance of the whole construction. The requirements for separating, non load-bearing construction are therefore applicable, i.e. insulation and integrity as mentioned above. This has been applied in a study carried out at the Swedish Institute for Wood Technology Research and summerized here. A full report is also available (Norén and Östman).

The aim is to determine the contribution from different building panels to the total fire resistance of a construction. The results might be useful for the fire classification of walls mainly. A large number of panels have been studied, most of them are wood-based, but different inorganic panels have also been included. Factors as panel thickness, density, type of adhesive have been studied, as well as the influence of design factors as mineral insulation, studs and double panel layers.


EXPERIMENTAL

Three furnaces of different sizes have been used, two in reduced scale and one in full scale. All furnaces have followed the standard time-temperature curve according to ISO 834. Some identical constructions have been tested in all furnaces to verify the agreement. Then most of the tests have been carried out in the small furnace.

The small furnace consists of a steel box with a side length of approximately 0.6 m. It has an inside, 70 mm thick ceramic insulation with density about 850 $kg/m^3$. It is heated by a gas burner which is operated manually and the standard fire curve is obtained within narrow limits. It may be operated with vertical or horizontal test elements. The size of these elements is 0.5 x 0.6 m.

The medium size furnace is walled with the same insulating brick as the small furnace, but heated with an oil burner. It has two openings, one vertical 1.24 x 1.23 m and also one horizontal, which was not used in these experiments. The size of vertical test elements is 1.7 x 1.6 m.

The large furnace is a conventional vertical furnace heated with oil burners. The size of the test elements is 2.4 x 3 m.

The test panels were different wood-based boards, i.e. particle boards, hardboards and plywood, with thicknesses between 3 and 30 mm. Some common inorganic boards were also evalutated, e.g. gypsum boards, cement bonded boards and boards made of calcium silicate. All boards were taken from commercial production, but chosen so that the influence of factors as thickness, density and adhesive could be studied systematically.

Composits were made with wood studs, 45 x 95 mm, and in some cases insulated with mineral wool with a density of 30 kg/m$^3$.

All test elements were stored to moisture equilibrium at 50 % r.h. and 23 °C before testing. Four elements were tested in the large furnace, five in the medium size furnace and about 50 different elements evaluated in duplicate tests in the small furnace. Most elements were oriented vertically, i.e. in wall position.

Thermocouples were applied on many places, mainly on the unexposed face of the panels, but also between panels, between panel and insulation and at studs. The signals from all thermocouples were stored on discs and then evaluated. Visual observations were also noted, especially cracking or other integrity failure.

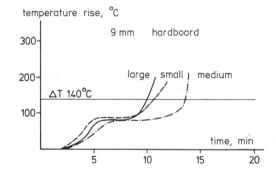

FIGURE 1.

Comparison of three different furnace sizes. Time-temperature curves for 9 mm hardboard of density 800 kg/m$^3$ with mineral wool insulation.

DIFFERENT FURNACES AND TEST CONDITIONS

Figure 1 compares the fire resistance determined in the three furnaces of different size. The test element is a wall consisting of a 9 mm hardboard on wood studs and insulated with mineral wool. The contribution from the hardboard panel as given in the figure is almost the same, within less than one minute, for the small and the large furnaces, while the medium size furnace gives a longer contribution. This is probably mainly due to differencies in time constants for the thermocouples controlling the furnace temperature. The same tendency was found also for other wall constructions.

The fire resistance was generally determined by the criterion of insulation with a few exceptions especially for very thin panels. The limit for the average temperature was usually reached at first, then the maximum temperature and somewhat later the panel cracked i.e. integrity failure. The repeatability when testing an identical element in the same furnace was good for all three furnaces and in the order of 0.6 minutes for the large furnace, 1.0 minutes for the medium size furnace and 0.5 minutes for the small furnace when testing a 10 mm thick panel.

The thermal exposure was varied during some experiments in the small furnace. Figure 2 shows that lower thermal exposure than the standard fire curve seems to give a better fire resistance, while a slightly higher thermal exposure seems to have little effect. However, higher thermal exposure than near the upper tolerance limit was hard to reach. But it still seems to be important not to have too low thermal exposure in order to get reproducable results. A slightly higher temperature than the standard fire curve seems to have less influence.

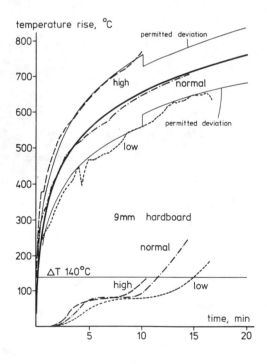

FIGURE 2.

Furnace temperatures and time-temperature curves for 9 mm hardboard of density 800 kg/m$^3$ at different thermal exposures.

INFLUENCE OF SOME COMMON PANEL PROPERTIES

The influence of some common panel properties as thickness, density and moisture content is summarized below. All data are for uninsulated panels. The influence of insulation and studs is given separately.


Thickness

The panel thickness is the most important factor for determining the contribution to the fire resistance. The relationship between fire resistance and
thickness is not exactly linear, but is better given by a parable through
origo:

$$b_m = 1.128t + 0.0088t^2$$

where $b_m$ is the contribution to fire resistance (in minutes).
      t  is the panel thickness (in mm).

The parable shape is probable because thicker boards get a thicker protective
char layer which means slower fire penetration and higher fire resistance.

Different types of wood-based panels might have slightly different fire resistance at equal thickness, see figure 3, which gives experimental data with
indications of panel types. The differences in fire resistance at equal thickness depend on different densities, structures and adhesives. Panels with high
density tend e.g. to have higher fire resistance.

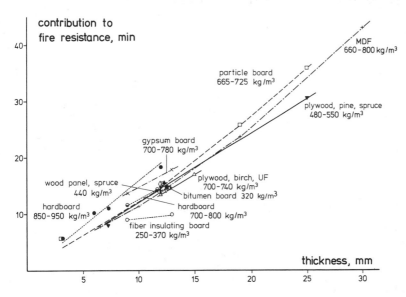

FIGURE 3.   Effect of panel thickness on the contribution to fire resistance
            with specification of different wood-based boards and including
            gypsum boards.

However, some panels with low density seem to have an unexpected high fire resistance, e.g. bitumen board which has higher fire resistance than ordinary insulating board at equal density and thickness. One reason is that insulating board cracks before the temperature criterion is reached, while the fire resistance for most other boards in figure 3, including bitumen board, is determined by the temperature criterion.

On the other hand, some panels with high density have an unexpected low fire resistance. One example is plywood made of birch and glued with urea resin which has low thermal stability and causes splitting of the plys.

Gypsum boards are also included in figure 3. They have only slightly higher fire reistance than most wood-based boards and about the same resistance as high density wood-based boards. The temperature criterion determines the fire resistance for gypsum boards as for most wood-based boards except insulating boards and very thin boards.


Density

The density has less influence than the thickness on the contribution to fire resistance. Figure 4 shows that three times higher density causes just twice the fire resistance or less. Fiber building boards which are manufactured at widely different densities were used for these experiments.

Higher density means higher thermal conductivity. For inorganic panels with high density this often also means that the temperature criterion determines its contribution to the fire resistance, sometimes much earlier than the integrity fails. But there are some exceptions e.g. some cement-bonded boards which crack early and before the temperature criterion is reached.

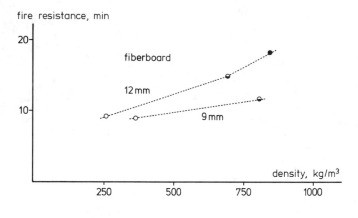

FIGURE 4.

Influence of panel density on contribution to fire resistance for wood-based fiberboards.

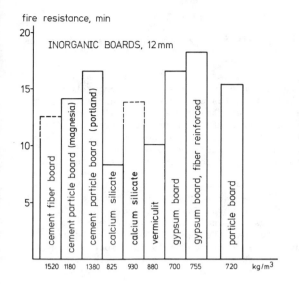

fire resistance, min

INORGANIC BOARDS, 12 mm

cement fiber board — 1520
cement particle board (magnesia) — 1180
cement particle board (portland) — 1380
calcium silicate — 825
calcium silicate — 930
vermiculit — 880
gypsum board — 700
gypsum board, fiber reinforced — 755
particle board — 720

kg/m³

FIGURE 5.

Contribution to fire resistance for some common inorganic boards. (The cement fiberboard and one of the calcium silicate boards were extrapolated from 10 to 12 mm.)

The fire resistance for some inorganic panels is given in figure 5. A fiber-reinforced gypsum board has the highest fire resistance followed by an ordinary gypsum board and a cement-bonded particle board, which has only 1 to 2 minutes longer time than ordinary particle board. Some inorganic panels have lower fire resistance, probably due to a low equilibrium moisture content.

Moisture content

All panels were in moisture equilibrium at 50 % r.h. and 23 °C before testing except in a few cases when the influence of moisture content was studied. Some panels were then either dried or given higher moisture content. The effect on fire resistance is given in figure 6, which shows a linear relationship with an increase of 0.4 min/%, i.e. 25 s/%.

fire resistance, min

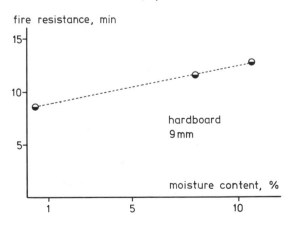

hardboard
9 mm

moisture content, %

FIGURE 6.

Relationship between contribution to fire resistance and moisture content for 9 mm hardboard of density 800 kg/m³.

Fire penetration rate

The rate of fire penetration through panels is given in <u>figure 7</u> as a function of panel density.

This rate, here called the fire penetration rate, is not identical with the charring rate, as used for solid wood, since it is based on a temperature criterion and not on charring. But these two rates are certainly related, and the charring rate is probably somewhat lower, as the panels are usually not charred on the unexposed face when the temperature criterion is reached. The fire penetration rate is, in spite of this, not much higher, 0.7 - 0.9 mm/min, than the charring rate usually applied for solid wood, 0.6 - 0.8 mm/min. The only panel with a fire penetration rate higher than 0.9 mm/min is insulating board with very low density.

The fire penetration rate may have a larger scatter with time than the charring rate usually determined for solid wood. There are two reasons for this. One reason is that the charring rate is constant only when a certain part of undestroyed wood remains, while the fire penetration rate is determined over the total panel thickness. The other reason is that most panels don't have an equal density over the thickness, but a gradient with higher density near the surfaces and lower density in the core.

But still the fire penetration rate seems to supply some useful information about the contribution to fire resistance from different panels.

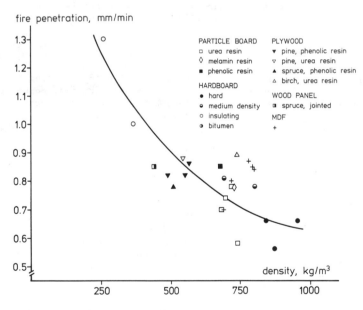

FIGURE 7.  Fire penetration rate for wood-based panels as a function of density.

## DESIGN FACTORS IN CONSTRUCTIONS

The contribution to fire resistance from building panels is besides different internal properties also depending on how they are used, i.e. on design factors in constructions as insulation, studs and double panel layers.

Insulation of a wall with mineral wool will slightly decrease the contribution of a panel to the fire resistance. Figure 8 gives some data for hardboard and gypsum board of equal thickness. This effect is easy to understand since the insulation will increase the temperature on the unexposed face of the panel. The panel will then dry quicker and its fire resistance will decrease. The difference in fire resistance between an insulated and an uninsulated panel seems to be slightly less for a hardboard than for a gypsum board.

Studs will also decrease the fire resistance of a panel, i.e. the contribution from a panel is less at studs than between studs, see figure 9. If the panel is jointed, the fire resistance is further decreased, because the panel shrinks when it is exposed to fire and the joint is enlarged. The fire can then penetrate much easier.

Double panel layers will decrease the fire reistance of the first exposed panel only slightly when compared to a single panel layer. The second panel then acts as insulation. This is true for both hardboard and gypsum board, but the effect is less than for insulation of mineral wool and for studs.

The total contribution of a double panel layer to the fire resistance is about twice the contribution of a single layer of hardboard and about three times the contribution of a single layer of gypsum board.

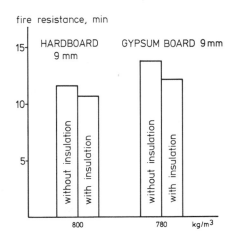

FIGURE 8.

Effect of insulation. Hardboard and gypsum board with 95 mm mineral wool insulation.

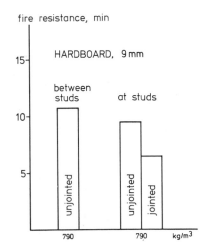

FIGURE 9.

Contribution to fire reistance at studs and between studs. The panel at studs is jointed or unjointed.

CONCLUSIONS

The contribution to the fire resistance from building panels can be evaluated
in a small scale furnace operating according to the standard time-temperature
curve in ISO 834. The contribution is equal to that evaluated in a full scale
furnace.

A thermal exposure slightly higher than the standard time-temperature curve
seems to have less effect on the fire resistance than a lower thermal exposure.

The panel thickness is the most important factor for determining the contribution to the fire resistance. Other factors are panel density, moisture content, type of adhesive and structural composition of the panel.

Inorganic panels as gypsum boards have only slightly larger contribution to
the fire resistance than wood-based boards.

The insulation criterion is determining the fire resistance in most cases
while the panel integrity remains a little longer, especially for some inorganic boards as high density mineral fiber boards.

Design factors in wall constructions as mineral wool insulation and studs will
slightly decrease the contribution to the fire resistance from building panels.

The data presented here are mainly applicable when evaluating the fire resistance of timber framed wall constructions. They may also form a basis for a
more theoretical approach.

REFERENCES

Dansk Brandværns-komité:
Beregningsmæssig bestemmelse av lette, ikke bærende pladebeklædte
træskeletvegges brandmotstandsevne.
Reports Dec. 1980 and Aug. 1982 (in Danish).

Deutsche Normen:
Brandverhalten von Baustoffen und Bauteilen. Kleingrüfstand.
DIN 4102, Teil 8 Entwurf, 1977.

Forest Products Laboratory:
Fire resistance tests of plywood-covered wall panels.
USDA For. Serv. Rep. No. 1257, 1956.

Hagstedt, J. and Nyström, P.:
Skiljande träkonstruktioner.
Statens råd för byggnadsforskning, Stockholm.
Report R 91:1977 (in Swedish).

Jönsson, R. and Pettersson, O.:
Träkonstruktioner och brand. Kunskapsöversikt och forskningsbehov.
Byggnadstekniskt brandskydd.
Tekniska högskolan i Lund. Report LUTVDG/(TVBB-3015), 1983.
(In Swedish.)

Kordina, K. and Meyer-Ottens, C.:
Holz-Brandschutz-Handbuch.
Deutsche Gesellschaft für Holzforschung e.V. München, 1983.

Meyer-Ottens, C.:
Behaviour of loadbearing and non-loadbearing internal and external walls of
wood-based materials under fire conditions.
Proc. Symposium No. 3 on Fire and structural use of timber in buildings. Joint
Fire Res. Org. HMSO London, 1967.

Norén, J. and Östman, B.A-L.:
Skivmaterials bidrag till brandmotståndet.
Swedish Institute for Wood Technology Research. Report No. 79, 1985.
(In Swedish.)

Pettersson, O. and Jönsson, R.:
Fire design of wooden structures.
Internatinal Seminar on three decades of structural fire safety, Fire Research
Station, UK Febr. 1983.

White, R.H.:
Wood-based paneling as thermal barriers.
USDA For. Serv. Res. Pap. FPL 408, 1982.

Östman, B.A-L.:
Fire penetration rate in different building boards. Critical literature
review.
Proceedings. New available techniques. The world pulp and paper week.
Stockholm, April 10-13, 1984.

# Reliability-Based Design of Structural Members for Nuclear Power Plants

**U. SCHNEIDER**
University of Kassel
Mönchebergstr. 7, D-3500 Kassel, Germany

**D. HOSSER**
König & Heunisch
Oskar-Sommer-Str. 15-17, D-6000 Frankfurt 70, Germany

ABSTRACT

The reliability of load bearing structural members in NPP has been proved by
1st order reliability theory. Studies on the temperature development in safety
related compartments and the failure probability of adjacent concrete members
were performed. Special effects e.g. due to forced ventilation and fire extinguishing systems are accounted for in the calculations with a post-flashover
heat balance model. Simplified failure models describing the load bearing capacity of concrete members were applied to check the reliability of different
types of building members. The results indicate that the member temperatures
and the life loads are the most important parameters. Therefore the practical
design may be performed by using partial factors for the temperatures and reduction factors for the life loads. It turned out that as a rule 90 minutes
fire duration acc. to DIN 4102 is a sufficient value for NPP fire compartments.

## 1. INTRODUCTION

Within a comprehensive research program SR 144 /1/, sponsored by the Federal
Minister of the Interior, the requirements for structural fire safety in NPP in
Germany were checked with respect to the effect of natural and standardized
fires. The report presents one part of the whole investigation program whereby
the following topics will be discussed:

- modelling and estimating of fire effects in safety related compartments of a
  NPP with a pressure water reactor (PWR),
- simplified structural models for the assessment of concrete building elements
  for NPP,
- probabilistic assessment of structural members under natural and standard
  fires.

## 2. HEAT MODELS FOR STRUCTURAL FIRE SAFETY

### 2.1 Fundamentals

The prediction of time-temperature curves for compartments in NPP is very complicated because there are many influencing parameters. Up to now, a complete
physical model for compartment fires is not at hand. The theoretical basis of
such calculation methods is given by the fundamental equations of heat and mass
transfer which are applied in the whole compartment or in certain parts of it.
The fire development is highly influenced by the following parameters:

- type and amount of combustible material,
- distribution, density and temperature of the fire load,
- oxygen supply to the burning surface of the material,
- ventilation of the compartment,
- thermal inertia of the building elements,
- influence of extinguishing systems.

For the sake of simplicity and with respect to the special question of fire design of structural members this report refers to a fire model for fully developed compartment fires /2/. The phase of ignition and fire spread in the model is accounted for by a simplified simulation procedure with variable spreading rates. The fire exposure of the structural members is related to the average
compartment temperatures. The energy balance in the model is described by six
energy terms comprising:

- combustion energy due to a ventilation controlled or fuelbed controlled fire,
- convective energy losses due to a natural or forced ventilation of the compartment,
- radiative energy losses through openings,
- energy losses due to the absorption of the building elements,
- intrinsic energy of the compartment gases and heat sinks,
- absorption energy losses due to the activation of extinguishing systems.

The developed  heat release model was calibrated by the data of fire tests derived from experiments with two types of fire loads. This limitation meets the
current practice as the combustible materials in German NPP are more or less
limited to oil in closed tanks and cable trays of different types and geometries.


2.2  Temperature-Time Curves

During the study 10 safety related areas in a NPP of a German PWR were investigated. The probable type of fire development was determined by a comprehensive
sensitivity study. Main results of the heat balance calculations are published
in /2/. In the following are discussed the results of calculations concerning a
cable fire in a cable room of 340 m² floor area and an oil fire in an oil tank
storage depot of 58 m², with an oil cup of 18 m².

For the cable room it was assumed that the initial fire area is 0.3 m². The
average spreading rate on the cable trays was fixed at a rate of 1.0 cm/s which
is an estimate on the safe side according to existing experiences. The whole
cable tray area comprises 56 m². Fig. 1 shows the temperature-time curves for
this room. Main parameter is the ventilation as the total fire load is fixed
with 9000 kg PVC. The results indicate significant differences with respect to
the different types of ventilation. The natural ventilation (2 doors open) leads
to unfavourable conditions. On the other hand the time delay for the activation
of the water spray system is comparatively large when the forced ventilation
system works.

The simulations of oil tank fires led to similar results. A natural ventilation
due to door openings leads to the worst cases (Fig. 2). In the calculations it
was assumed that the leakage oil burns in an oil cup whereby the burning rates
were controlled by the ventilation. The whole oil tank with 25.000 kg oil acted
as a heat sink in the program and the temperature increase of the oil was calculated by assuming a mixed temperature for the fluid. The boiling point of the
oil was not taken into account, but Fig. 2 indicates that boiling might occur
after 40 to 50 minutes if no active fire protection measures are undertaken.
The water spray system is early activated in this case and the suppression of

the fire is performed within 5 minutes if an automatic activation system is installed.

## 3. DESIGN MODELS FOR REINFORCED CONCRETE MEMBERS

### 3.1 General Assumptions

In the following, the reliability of load bearing reinforced concrete members e.g. a reinforced concrete ceiling, beam, wall and column (Fig. 3) under fire attack is investigated. Such elements are common for the switch-gear building, respectively. The following conservative assumptions are made:

- The simply supported members are designed for nominal loads (permanent and variable loads) according to the German standard DIN 1045 /3/.
- With respect to the fire safety a design according to the standard DIN 4102 /1/ is assumed.
- The temperature exposure is choosen according to i) the standard time-temperature curve and ii) a natural fire which might occur in a typical cable room of a switch-gear building under sufficient natural ventilation (Fig. 1).

In detail the calculations comprise the following tasks:

- Estimating the temperatures within the building elements with respect to the compartment temperatures and determining the related material properties.
- Calculating the load-bearing capacity of the building elements under external loads and temperature stresses.
- Performing a probabilistic reliability analysis with regard to the statistical values of the essential parameters.

### 3.2 Temperature Increase of the Structural Members

The temperature profiles of the members are determined by heat-balance calculations for each time step. From these profiles the mean value of the temperatures of the reinforcing steel bars in the tension zone and the mean value of the concrete element temperatures in the compression zone are computed. In the elements according to Fig. 3 mainly the mean temperature of the reinforcement in the tension zone is of interest. This does not hold for the temperatures in the column. Fig. 4 shows the average concrete $(T_b(t))$ and reinforcement $(T_s(t))$ temperatures of the four investigated members under standard (dott.lines) and natural (full lines) fire conditions. The temperature-dependent material properties were determined by the help of the design curves given in literature /4/.

### 3.3 Design Equations

The safety margin of reinforced concrete members like ceilings and beams may be determined as follows:

$$Z = M_R(t) - M_S \geq 0 \qquad (1)$$

$M_R(t)$ = bending moment capacity
$M_S$ = actual bending moment in the critical cross section

The bending moment capacity $M_R$ may be approximated by /5/:

$$M_R(t) = A_S \cdot \beta_S(T_s) \cdot z \qquad (2)$$

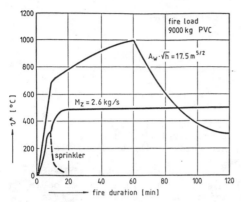

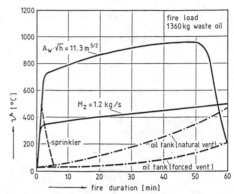

Fig. 1: Temperature-time curves of a cable room

Fig. 2: Temperature-time curves of an oil tank storage deposit

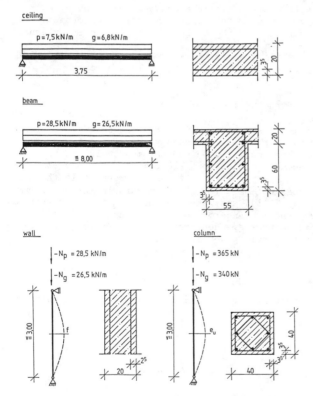

Fig. 3: Assumed building members and load actions for the probabilistic investigations

with area $A_S$ of tension reinforcement, temperature-dependent yield strength $\beta_S (T_s)$ and internal leverarm z. The actual bending moment is:

$$M_S = (g + p) \cdot 1^2/8 \tag{3}$$

with the distributed load g and p acc. to Fig. 3.

The bending moment capacity of summetrically reinforced walls and columns may be derived as follows:

$$M_R(t) = (1+1.5\cdot(0.1\frac{h'}{d}))\cdot(0.42\cdot A_S\cdot\beta_S(T_s)\cdot d+0.35\cdot N\cdot d) \tag{4}$$

for $N \leq N_1 = 0.4 \cdot \beta_R (T_b) \cdot b \cdot d$

and

$$M_R(t) = (1+1.5\cdot(0.1\frac{h'}{d}))\cdot(0.42\cdot A_S\cdot\beta_S(T_s)\cdot d+0.35\cdot N_1\cdot d)-0.26(N-N_1)\cdot d \tag{5}$$

for $N > N_1$ and $N = N_g + N_p$ according to Fig. 3.

Usually walls and columns are stressed by normal loads. In addition the columns are subjected to bending moments due to an unintended eccentricity $e_u$ after /3/. Walls are subjected to bending moments due to the one-dimensional fire attack. The action of external loads is determined by

$$M_S (t) = N \cdot e (1 + \frac{1}{8} \cdot \frac{N \cdot 1^2}{E\ I_w(t)}) \tag{6}$$

$e = e_u$ for columns and $e = f = \frac{1^2}{8} \cdot \alpha_T \cdot \frac{\Delta T}{d}$ for walls

$E\ I_w(t) =$ flexural stiffness $(kN/cm^2)$

4.   RELIABILITY ANALYSIS

4.1   Introduction to the Method

The reliability analysis of concrete members is performed for the ultimate limit states defined by the design models in chapter 3. The basic variables $X_i$ of the models are described by distribution functions. With the help of the error propagation law the mean value $m_Z$ and standard deviation $\sigma_Z$ of the safety margin Z according to equ.(1) is determined. The quotient

$$\beta = m_Z / \sigma_Z \tag{7}$$

is called safety index and is used as a measure of reliability. The scattering effects of the basic variables $X_i$ with respect to the safety index $\beta$ can be evaluated by the so-called weighting factors $\alpha_{X_i}$ :

$$\alpha_{X_i} = \frac{\partial Z}{\partial X_i} \cdot \sigma_{X_i} / \sqrt{\sum_{j=1}^{n} (\frac{\partial Z}{\partial X_i} \sigma_{X_i})^2} \tag{8}$$

with the partial derivatives $\partial Z/\partial X_i$ and the standard deviations $\sigma_{X_i}$. The safety index $\beta$ and the weighting factors $\alpha_{X_i}$ depend on the time t after fire occurance.

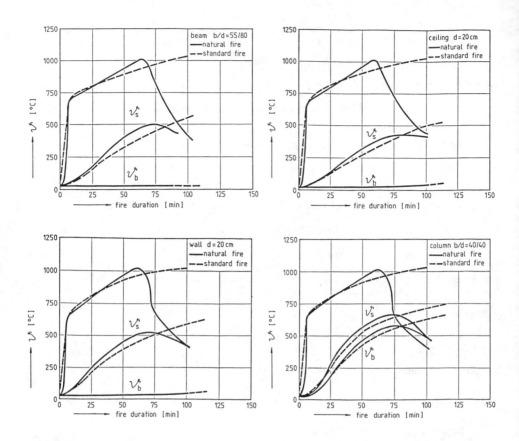

Fig. 4: Compartment and member temperatures of the investigated structural members

4.2 Random Variations and Uncertainties

Distribution functions for the most important parameters (basic variables) of the design models in chapt. 3 were chosen according to /5/. On Tab. 1 the nominal values of those parameters are defined as percentiles, in addition the types of distribution functions and the coefficients of variations are given. The numbers are in aggreement with those values which have been determined by performing several comparison calculations in the field of reinforced concrete design. The variation of the standard fire temperature-time curve are specified according to DIN 4102, part 2 /1/.

The variations of the temperature-time curve $T(t)$ in natural fires mainly depend on the

- specific rate of combustion $\dot{R}_{sp}$ in kg/m²h and the burning area,
- ventilation conditions, e.g. ventilation area $A_w$ or ventilation rate $M_z$,
- amount of combustible material, e.g. fire load $q$,

Tab. 1: Distribution assumptions for the basic variables of the design models

| Parameter | nominal value | distribution function | coefficient of variation | comment |
|---|---|---|---|---|
| T, stand. fire | mean | log. normal | 0.10 | DIN 4102 |
| T, nat. fire | mean | log. normal | 0.20 | Fig. 4 |
| reinf. temp. $T_s$ | mean | log. normal | 0.20 | Fig. 4 |
| concrete temp. $T_b$ | mean | log. normal | 0.20 | Fig. 4 |
| yield strength $\beta_s$ | 5 % | log. normal | 0.06 | acc. to /5/ |
| concrete strength $\beta_R$ | 5 % | log. normal | 0.15 | acc. to /5/ |
| steel mod. $E_s$ | mean | log. normal | 0.05 | acc. to /5/ |
| concrete mod. $E_b$ | mean | log. normal | 0.10 | acc. to /5/ |
| reinf. cover u | mean | log. normal | 0.25 | acc. to /5/ |
| perm. load g, Ng | mean | normal | 0.05 | acc. to /5/ |
| live load ⎫ | 98 % | extremal I | 0.40 | natural fire |
| load p, Np ⎭ | 57 % | one year max. | 0.05 | DIN 4102 |

in the respective compartment. Sensitivity studies in /5/ led to the conclusion that during a fully-developed fire the standard deviation $\sigma_T$ of the gas temperature T is nearly constant and may be evaluated as follows:

$$\sigma_{T(t)} = \sqrt{(\frac{\Delta T}{\Delta R_{sp}} \cdot \sigma_{R_{sp}})^2 + (\frac{\Delta T}{\Delta A_w} \cdot \sigma_{A_w})^2 + (\frac{\Delta T}{\Delta M_z} \cdot \sigma_{M_z})^2} \tag{9}$$

The standard deviations of the element temperatures $T_b(t)$ and $T_s(t)$ are lower than $\sigma_{T(t)}$. But the coefficients of variation approximately correspond to the coefficient of variation $V_{T(t)}$ of the gas temperature which depends on the fire duration. A maximum value $V_{T(t)} \approx 0.17$ has been estimated under different assumptions in /5/ and /2/.

In addition to the variations of the gas temperature one must consider the variations of the fire duration which is a - more or less linear - function of the fire load q. Under the assumption that in a NPP the determination of fire loads is individually performed a coefficient of variation $V_q = 0.1$ may be a sufficiently conservative estimate. As during the phase of a fully-developed fire the element temperatures increase nearly proportionally to the fire duration, the coefficient of variation $V_q$ results in an additional variation of the element temperatures $T_{b,s}$. Assuming independence of the random variations of T and q the overall coefficient of variation may be derived,

$$V_{T_{b,s}} = \sqrt{V_{T(t)}^2 + V_q^2} \tag{10}$$

yielding $V_{T_{b,s}} \approx 0.2$ (cf Tab. 1) as an average value.

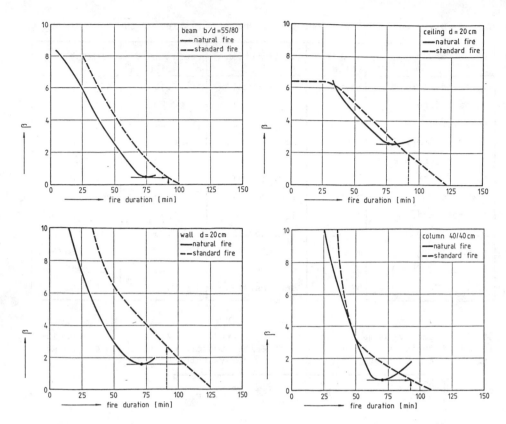

Fig. 5: Safety index ß of the investigated structural members

## 4.3 Results of the Reliability Analysis

The safety index ß of the investigated members are shown in Fig. 5 for standard and natural fire exposures. During standard fires the F 90 class elements fail in the range of 100 to 125 minutes (ß = 0), whereby beams and columns approach the lower and ceilings and walls the upper limit. The probabilities of failure after attaining the nominal fire duration of 90 minutes are of the order of 25 to 30 percent (beams, columns) and 0.5 to 2 % (ceilings, walls). In this case the failure probabilities under natural fire conditions are of the same order as under a standard fire after 85 to 90 minutes.

The overall variations ($\sigma_z$) during the natural fire are partly more than 100% larger than under standard fire conditions, due to i) the larger scattering of $T_{b,s}$ and ii) the additional influence of scattering loads. In both types of fire the variations of the member temperatures are dominating. Beside this, the yield strength of the steel, the concrete strength in connection with the analysis of columns and the live loads contribute to the overall variations; the remaining parameters are negligible.

## 4.4  Safety Elements for Fire Safety Design

From section 4.3 safety elements for practical fire safety design of structural elements may be derived. The required safety level is described by a target safety index ß depending on the occurence rate of fires in a compartment, the active fire protection measures taken to prevent a fire from spreading within the whole compartment, and perhaps the total damage due to a severe fire.

Informations on the influences of parameter uncertainties are taken from the weighting factors $\alpha_{X_i}$ of the basic variables $X_i$ in Tab. 2 (cf. section 4.2). The values $\alpha_{X_i}$ for the structural elements of Fig. 3 in the most critical phase of the natural fire as well as the respective values after the equivalent fire duration of the standard fire are summarized in Tab. 2.

Tab. 2:  Weighting factors $\alpha_{X_i}$ of the basic variables $X_i$ for the structural elements in the natural and the standard fire

| basic variable $X_i$ | $\alpha_{X_i}$ for structural elements | | | | | | | |
|---|---|---|---|---|---|---|---|---|
| | ceiling | | beam | | wall | | column | |
| | n | s | n | s | n | s | n | s |
| $\left.\begin{array}{c} T \\ T_s \\ T_b \end{array}\right\}$ | -0.990 | -0.837 | -0.971 | -0.881 | -0.971 | -0.911 | -0.853 | -0.617 |
| $\beta_R$ | 0.001 | 0.011 | 0.003 | 0.007 | 0.0 | 0.0 | 0.511 | 0.778 |
| $\beta_s$ | 0.126 | 0.244 | 0.076 | 0.151 | 0.040 | 0.098 | 0.004 | 0.01 |
| $E_s$ | 0.0 | 0.0 | 0.0 | 0.0 | 0.0 | 0.0 | 0.001 | 0.001 |
| $E_b$ | 0.0 | 0.0 | 0.0 | 0.0 | 0.0 | 0.0 | 0.001 | 0.001 |
| $h_s$ | 0.023 | -0.142 | 0.028 | 0.041 | 0.235 | 0.400 | 0.080 | 0.116 |
| $g, N_g$ | -0.012 | -0.100 | -0.053 | -0.098 | 0.001 | 0.013 | -0.017 | -0.026 |
| $p, N_p$ | -0.054 | -0.458 | -0.218 | -0.436 | 0.002 | 0.012 | -0.066 | -0.025 |

n = natural fire          s = standard fire

It is proposed to define design values of the basic variables with the target safety index ß and the following conservative weighting factors:

$\alpha_{T_b, T_s}$ = - 0.95  for natural fires and - 0.85 for standard fires,

$\alpha_{p, N_p}$ = - 0.25  for natural fires and - 0.45 for standard fires,

$\alpha_{\beta_R}$ = 0.5  for columns in natural fires and 0.75 in standard fires,

$\alpha_{\beta_s}$ = 0.25  for natural and standard fires,

$\alpha_{h_s}$ = 0.25  for compression members in natural fires and 0.45 in standard fires,

$\alpha$ = 0  for all other basic variables.

These design values may also be expressed with the help of the nominal values given in Tab. 1 and partial safety factors $\gamma$ as indicated in Tab. 3 for a range of the safety index $2.0 \leq \beta \leq 3.5$ (valid for nuclear power plant compartments). In practical applications the partial safety factors for $\beta_R$, $\beta_s$ and $h_s$ should be taken as ~ 1.0.

Tab. 3: Partial safety factors γ for the most important basic variables and design for natural and standard fires

| basic variable | ß | γ for structural element | | | | | | | |
|---|---|---|---|---|---|---|---|---|---|
| | | ceiling | | beam | | wall | | column | |
| | | n | s | n | s | n | s | n | s |
| $T, T_b, T_s$ | 2.0 | 1.43 | 1.18 | 1.43 | 1.18 | 1.43 | 1.18 | 1.43 | 1.18 |
| | 3.5 | 1.97 | 1.34 | 1.97 | 1.34 | 1.97 | 1.34 | 1.97 | 1.34 |
| $p, N_p$ | 2.0 | 0.46 | 0.63 | 0.46 | 0.63 | 0.46 | 0.46 | 0.46 | 0.46 |
| | 3.5 | 0.68 | 0.83 | 0.68 | 0.83 | 0.46 | 0.46 | 0.46 | 0.46 |
| $ß_R$ | 2.0 | 0.79 | 0.79 | 0.79 | 0.79 | 0.79 | 0.79 | 0.91 | 0.98 |
| | 3.5 | 0.79 | 0.79 | 0.79 | 0.79 | 0.79 | 0.79 | 1.02 | 1.16 |
| $ß_s$ | 2.0 | 0.93 | 0.93 | 0.93 | 0.93 | 0.93 | 0.93 | 0.93 | 0.93 |
| | 3.5 | 0.96 | 0.96 | 0.96 | 0.96 | 0.96 | 0.96 | 0.96 | 0.96 |
| $h_s$ | 2.0 | 1.0 | 1.0 | 1.0 | 1.0 | 0.88 | 0.78 | 0.88 | 0.78 |
| | 3.5 | 1.0 | 1.0 | 1.0 | 1.0 | 0.78 | 0.61 | 0.78 | 0.61 |

n = natural fire     s = standard fire

5.   CONCLUSION

The report describes the theoretical determination of failure probabilities for structural members in NPP. Models for the determination of fire effects are presented. The temperature-time curves are evaluated by post-flashover heat models, whereby the effect of forced ventilation in the safety related buildings of the power plants is accounted for. Based on the reliability analyses also safety elements for practical fire safety design of structural elements are derived depending on a target reliability. It seems sufficient to define partial safety factors for the member temperatures and reduction factors for the nominal live loads whilst all other parameters may be used with their nominal values.

6.   REFERENCES

/1/  DIN 4102: Brandverhalten von Baustoffen und Bauteilen, Beuth-Verlag, Berlin, Sept. 1977.
/2/  ABK/GRS: Optimization of fire protection measures and quality control in nuclear power plants. Final report of the research project SR 144/1 sponsored by the Federal Minister of the Interior. Draft December 1984 (in German).
/3/  DIN 1045: Beton und Stahlbeton, Bemessung und Ausführung, Beuth-Verlag, Berlin, Dez. 1978.
/4/  CEB (Comité Euro-International du Béton): Design of Concrete Structures for Fire Resistance, Bulletin d'Information No. 145, Paris, Ja. 1982.
/5/  Hosser, D. und U. Schneider: Sicherheitsanforderungen für brandschutztechnische Nachweise von Stahlbetonbauteilen nach der Wärmebilanztheorie, VfdB-Zeitschrift, Heft 1/82 und 2/82, 1982.
/6/  König,D.: D. Hosser; W. Schobbe: Sicherheitsanforderungen für die Bemessung von baulichen Anlagen nach den Empfehlungen des NABau, Bauingenieur Nr. 57, S. 69/78, 1982.
/7/  Hosser, D. und U. Schneider: Sicherheitskonzept für brandschutztechnische Nachweise nach der Wärmebilanztheorie, Forschungsbericht IfBt, Berlin, Dez. 1980.
/8/  KTA 2101.1: Brandschutz in Kernkraftwerken, Teil 1: Grundsätze des Brandschutzes, Regelentwurf, Köln, Juni 1984.

# FIRE CHEMISTRY

Session Chair

**Dr. Raymond Friedman**
Factory Mutual Research Corporation
1151 Boston-Providence Turnpike
Norwood, Massachusetts 02062, USA

# Some Unresolved Fire Chemistry Problems

RAYMOND FRIEDMAN
Factory Mutual Research Corporation
Norwood, Massachusetts 02062, USA

ABSTRACT

Six areas of fire science are selected for discussion which involve chemistry in an important way. The areas are: rate of pyrolysis of a solid combustible; generation of toxicants in a fire; fire luminosity (radiative output) and smoke generation; fire retardation of wood and synthetics; flammability of a hot gas layer in a compartment before flashover; and chemically active extinguishing agents. In each case, highlights of knowledge and remaining unsolved questions are touched upon, and 35 references to the recent literature are provided.

## INTRODUCTION

This paper will describe six important areas of fire science in which chemistry is heavily involved. In each case, highlights of the present state of knowledge will be indicated, references will be provided to current work, and unresolved problems will be mentioned.

Because of coverage elsewhere in this symposium, fire toxicology and fire detection will not be discussed here, although generation of toxicants and of smoke in flames will be discussed. In addition to the foregoing, the review will include rate of pyrolysis, flammability of the hot gas layer in a burning compartment, and mechanisms of retardation and extinguishment.

## RATE OF PYROLYSIS

For flaming fires involving solids the rate of burning is essentially the rate of pyrolytic gasification, referred to as the rate of pyrolysis. This rate is the net result of complex heterogeneous chemical kinetics, usually involving many reaction steps, some of which have high activation energies and are very temperature-dependent. These reaction steps have not yet been quantified completely for most combustible solids, and are not likely to be in the near future.

Fortunately, however, an approximate method may be used in many cases to describe the pyrolysis rate, which does not require knowledge of the chemical kinetics. It is based on four assumptions: (I) the overall pyrolytic gasification process is endothermic; (II) the overall heat of gasification is independent of the rate of gasification; (III) the overall activation energy of gasification is high enough so that the temperature at which gasification occurs is only weakly dependent on the rate of gasification; (IV) the burning is a "steady-state" process.

The approximate method is simply based on a heat balance. The rate of gasification is assumed to be the ratio of the net heat flux absorbed by the pyrolyzing region to the heat of gasification. This heat of gasification is considered to be a material property, and may be measured by applying a known heat flux to a weighed sample in an inert atmosphere. For a polymer, it would be equal to the sum of the heat needed to raise the polymer to the decomposition temperature, including the latent heat of any phase transitions, plus the heat of depolymerization to the monomer or oligomer vapors which form. For combustible substances, it is always substantially less than half the heat of combustion.

The validity of this concept for thermopolastic polymers is strongly supported by experiments of Tewarson and Pion (1) who find linear relations between burning rate and applied radiant heat flux for polystyrene, polyoxymethylene, polymethyl methacrylate, polyethylene, polypropylene, and flexible polyurethane foam, the slopes being consistent with independently measured heats of gasification. Vovelle et al. (2) have confirmed this result for polymethyl methacrylate. An extensive list of apparent heats of gasification for many substances has been published recently by Tewarson (3). The values are seen to range from one to four kilojoules per gram.

However, many combustible solids, including cellulosics and cross-linked synthetic polymers, pyrolyze with char formation. The char layer, which has low thermal conductivity, becomes progressively thicker as heating continues, so a "steady-state" approximation is an extremely crude way to describe the time-varying pyrolysis. A proper understanding of the burning of a charring substance will require substantial further research. Some complicating features will be discussed, in addition to the effect of the time-dependent thermal barrier just mentioned.

In general, many organic substances will release gases rapidly at 300-450°C, which would correspond to the temperature on the inner side of the char layer. The outer side of the char layer will be expected to be several hundred degrees hotter. This will have at least three likely consequences. Firstly, the pyrolysis gases passing through the pores of the char may undergo further chemical change, either depositing carbon on the char or removing carbon from the char. Secondly, the high surface temperature of the char and the corresponding high rate of radiant heat loss from the char to any cold ambient within the field of view will have a strong effect on the overall energy balance and may lead to extinguishment. Thirdly, when and if oxygen contacts the char surface, char oxidation (glowing combustion) will result, again with strong effect on the energy balance.

Another complication is structural, relating, in the case of wood, to the direction of the grain and consequent effect on the char. Even for homogeneous substrates, crack development in chars is common.

Yet other complications, chemical in nature, are the tendency of wood to pyrolyze by at least a two-stage mechanism (4) and the strong influence of even low concentrations of inorganic compounds to promote or inhibit char formation.

Various mathematical models of charring pyrolysis have been developed (5-7), but none attempts to treat all these factors.

GENERATION OF TOXICANTS

A computer model of a fire often requires an answer to the following question. Given that a specified material is burning at a known rate, what is the rate of generation of toxicants of interest (carbon monoxide, hydrogen cyanide, hydrogen chloride, acrolein, dioxin, etc.)? This is not only of interest from a toxicological viewpoint but also because the energy released per unit mass consumed in a fire is affected by incomplete combustion.

Of the various toxicants, carbon monoxide is the only one which has been proven to cause large numbers of fire fatalities (on the basis of carboxyhemoglobin data from autopsies) and deserves the major attention.

For a fire burning under conditions where excess air is present, thermodynamic considerations would predict essentially no carbon monoxide. Experimentally, it is found that the $CO/CO_2$ molar ratio in products of well-ventilated pool fires may be as low as 0.001 to 0.002 for oxygenated combustibles such as methanol, polyoxymethylene, and wood, or as high as 0.1 or 0.2 for aromatic or highly chlorinated combustibles, which tend to burn incompletely, producing black smoke as well as CO, even when ample air is available. Table I shows $CO/CO_2$ ratios for a variety of materials burning with adequate ventilation and shows a 200-fold variation of this $CO/CO_2$ ratio.

TABLE 1. Molar $CO/CO_2$ ratio found in various combustion products
(Pool configuration, well ventilated)*

| Substance | Molar $CO/CO_2$ Ratio |
|---|---|
| Methanol | 0.001 |
| Wood | 0.001-0.003 |
| Polyoxymethylene | 0.002-0.004 |
| N-heptane | 0.004-0.02 |
| Polymethyl methacrylate | 0.007-0.02 |
| Rigid polyurethane foam | 0.03 -0.04 |
| Coal | 0.05 |
| Styrene-butadiene rubber | 0.07 |
| Polystyrene | 0.08 |
| Polyvinyl chloride | 0.17 |
| Benzene | 0.18 |
| Polyethylene, 48% chlorinated | 0.27 |

*Unpublished data from Factory Mutual Research Corporation. Results such as these are sensitive to experimental conditions.

Three mechanisms have been proposed for incompleteness of combustion in a well-ventilated diffusion flame: (I) rapid heat loss by radiation from highly luminous flames; (II) partial quenching by the action of steep velocity gradients as in turbulent flames; and (III) partial quenching by heat conduction to adjacent cold surfaces. The ranking of combustibles in Table 1 strongly suggests a correlation between $CO/CO_2$ ratio and flame luminosity or soot-forming tendency (characteristic of aromatic or halogenated combustibles). Indeed, correlations between CO and smoke particle formation have often been reported. However, more detailed understanding is needed.

While the foregoing relates to well-ventilated fires, a fire burning in a compartment will generally fall outside this category. The upper portion of the compartment will become filled with an oxygen-deficient hot gas layer, and if the flames project into this layer the combustion chemistry is dramatically affected. Furthermore, some mixing occurs between the upper and the lower layer, which may affect combustion occurring entirely in the lower layer.

Experimental data on wood cribs and heptane pools burning with restricted ventilation give $CO/CO_2$ molar ratios of the order of unity in some cases, or two orders of magnitude higher than for full ventilation of the same combustibles. When $CO/CO_2$ ratios are high, high concentrations of a variety of organic molecules are also found (2), generally including toxicants.

The foregoing statements refer to _flaming_ combustion. In a fire, it is likely that radiant heat from a burning region will cause pyrolysis in a not-yet-burning region, with copious CO production. Lee et al. (8) have pyrolyzed wood by laser heating and found that the pyrolysis gases consisted of 75% CO, 15% $H_2$, and 10% $CH_4$. Yoshizawa and Kubota (9) reported that the pyrolysis gases from cardboard subjected to 30 W/cm contained 70% CO, 20% $CO_2$ and 10% light hydrocarbons; they found no $H_2$. Cullis et al. (10) report CO yields from cellulose pyrolysis under helium varying with flow rate and temperature. Furthermore, _smoldering combustion_ of porous cellulosic material or polyurethane foam will produce substantial CO; few quantitative data are available, presumably because smoldering is strongly affected by experimental conditions, especially air movement and impurities. However, it has been reported that cigarette smoke contains about 3 to 4 percent by weight of CO (11). Cullis et al. (12) have heated cellulose at heating rates of about $10°$/second, drawing air through, and found that the percentage of CO in the product gas increased and then decreased as the temperature increased with a maximum of about 20% CO at 450° C (the primary constituent being nitrogen). When 5% oxygen in nitrogen was used instead of air, the CO maximum was only 9% CO, under their conditions.

To summarize the experimental information on CO production in fires: radiative quenching appears responsible for CO yield from fully ventilated laminar flames; poorly ventilated flames produce as much CO as $CO_2$; pyrolysis and smoldering combustion of cellulosics produce CO as the dominant product.

Bilger and Starner (13) have tried a theoretical approach to the generation of CO in diffusion flames with a partial equilibrium model, which assumes equilibration of $CO + H_2O = CO_2 + H_2$ in the flame while burnout of CO is determined by three-body recombination of H, OH and O radicals in a partially equilibrated radical pool. Unfortunately, however, this model is in disagreement with predictions of complex numerical models by Miller and others (14, 15) which take into account the kinetics of more than 100 individual reaction steps as well as transport properties. These various models are tested by comparing with measured concentration profiles within a diffusion flame; they all predict that ultimately the CO is consumed. The next step would be to develop an even more complex model, perhaps including radiative quenching, which could predict the rate of release of unconsumed CO from the flame.

As for other toxicants, the chemical facts for HCℓ are somewhat simpler, as essentially all the chlorine in combustion products of chlorine-containing polymers is in the form of HCℓ. In the pyrolysis of polyvinyl chloride, the primary vapor products are HCℓ and benzene (16) with a carbon-rich char left behind.

Hydrogen cyanide, more than an order of magnitude more toxic than CO, would only be expected in combustion products of polymers containing CN groups, such as polyurethane and acrylonitrile polymers. Tewarson (17) has found $HCN/CO_2$ ratios of the order of 0.01 and $HCN/CO$ ratios of roughly 1/3 in combustion products of rigid polyurethane foam. Levin et al. [18] heated flexible polyurethane foam containing fire retardant to 800° C, with resulting flaming combustion; the HCN produced was 16% of the maximum possible yield, based on the 4.25% N content of the foam. On subsequent reheating of the resultant char, more HCN was produced. HCN yields would be expected to vary over the same wide range as CO yields, depending on ventilation conditions, flaming combustion vs. pyrolysis or smoldering, and flame luminosity (soot-forming tendency).

## FIRE LUMINOSITY AND SMOKE GENERATION

The luminous character of a diffusion flame of a given combustible may be quantified in terms of radiative emission, or absorptivity per unit depth, or soot volume fraction, or smoke point (ASTM D 1322) if the combustible is liquid or gaseous. Soot (carbonaceous particles) formed from pyrolysis of fuel vapors in the flame is the primary source of this luminosity. The radiative heat transfer rate from the flame to the pyrolyzing solid fuel and to the environment is strongly influenced by this luminosity. Of course, the radiative heat transfer rate from a flame to a target is also strongly affected by any intervening cold smoke.

The soot which forms may or may not subsequently be consumed by oxidation; the unconsumed soot becomes smoke in the fire products. However, the fire products may also contain aerosol particulates resulting from condensation of pyrolysis vapors, as well as ash particles in some cases. The smoke is important in that it will be a source or sink for infrared radiation in the fire compartment. Furthermore, it will obscure vision, hindering escape. Again, it provides an important means for detecting a fire. The smoke may be measured by optical transmission or scattering, by collection followed by mass measurement, or by ionization with measurement of a current.

Since a flame which is highly luminous will have a low smoke point and have a strong tendency to emit unburned carbonaceous material, there is a close connection between flame luminosity and smoke generation. The chemistry of the combustible material is of crucial importance. For example, on comparing two materials, polyoxymethylene and polystyrene, the former burns without soot formation while the latter burns with a yellow flame rich in soot, producing smoke-laden products. The latter flame produces more than 8 times as much thermal radiation as the former flame, at the same total heat release rate. Many materials, such as cellulosics, will burn with flames intermediate between these extremes.

Much more quantitative data are available for radiative and smoke-forming properties of gaseous fuels than for solid fuels. Markstein [19] has shown that the radiative fraction of the heat-release rate of a turbulent jet flame correlates very closely with the laminar smoke-point, for six hydrocarbon fuels. The radiative fraction of these six fuels ranges from 21% for ethane to 42% for 1, 3-butadiene. The corresponding smoke-point lengths are 245 mm and 20 mm. Methane would have an even greater smoke-point length (not measurable in Markstein's apparatus) and an even lower radiative fraction.

If these various fuels are burned in "air" of adjusted $O_2/N_2$ ratio so that the calculated adiabatic flame temperature of all fuels was the same, then Markstein finds that the radiative fraction of heat release for the laminar flame at the smoke point appears to be independent of the nature of the fuel ($\sim$24% for 2200 K and $\sim$37% for 2600 K). The significance of this is not yet completely clear, but there is hope that current research will yield some basic understanding of this soot-forming process in diffusion flames (20).

The foregoing refers to gases and vaporized liquids. De Ris (21) has suggested that an apparatus could be built to pyrolyze a small solid sample in an inert or reducing atmosphere and then to continuously burn the evolved vapors in a laminar diffusion flame so as to obtain a smoke point. This would lead to characterization of solid materials in regard to radiative output of their flames. A presently mising link would then be filled so that a fundamental relation between material properties and flammability could be developed.

A group of 20 recent papers on soot formation and flame luminosity were presented in a symposium in 1984 (22) and provide a picture of the current state of knowledge. A survey paper on fire radiation was presented by de Ris (23) in 1978.

As for the smoke produced in either flaming or smoldering combustion, an enormous amount of literature exists. Seader and Einhorn (24) have prepared a review article with 81 references. One generalization, for $\underline{\text{flaming}}$ combustion of nine different cellulosic and plastic materials, is that the optical density of the smoke ($= \log_{10} 1/T$, where T is the fraction transmitted) per meter of beam length is directly proportional to the particulate concentration $(g/m^3)$, the proportionality constant being 3.3 $m^2/g$. For $\underline{\text{nonflaming}}$ combustion/pyrolysis of 13 materials, the same relation is valid except that the proportionality constant is now 1.9 $m^2/g$ and the scatter is somewhat greater. On considering this proportionality constant from the viewpoint of the theory of absorption and scattering of light by particles, it is seen that the underlying factors are particle diameter and complex refractive index. Apparently, the variation of these properties from one smoke to another is not sufficiently great to cause large changes in the proportionality mentioned above. This needs confirmation.

For combustion of any given material, the percentage yield of particulates varies widely with conditions of heat flux, ventilation, ambient oxygen concentration, and extent of cooling of the smoke. For wood, the particulate yield may be as low as 0.2%, while it may be as high as 20% for polyisoprene or polystyrene. The factors influencing CO yield previously discussed will also influence smoke yield. However, the ratio of optical density to CO concentration is not absolutely constant from material to material, but varies by about a factor of three in the most extreme cases.

FIRE RETARDATION

Wood is the most extensively studied organic solid of interest to fire researchers. The literature is enormous. To summarize the highlights: on being heated in an inert atmosphere, wood pyrolyzes mainly in the range 250-400°C, leaving behind a char comprising 20 or 30% of the original weight. The volatile products include simple gases ($H_2O$, CO, $CO_2$) and a wide variety of organic molecules, including tars. The tar to char ratio is strongly affected by heating rate and by chemical additives. An additive which decreases the tar to char ratio is a fire retardant. It has been known for hundreds of years that inorganic salts are effective as fire retardants when impregnated into wood. The effective salts have ammonium, sodium, potassium, or zinc as cations and phosphate, borate, silicate, sulphate, or sulphamate as anions. The chemical mechanisms of retardation are not fully understood. Wood itself has two major chemical constituents, cellulose and lignin, which are quite different from one another in pyrolysis characteristics. The retardants are intended to: (1) reduce ignitability and flame spread; (2) reduce heat generation; (3) reduce smoke generation; (4) prevent afterglow. The practical problems in the use of the various retardants are: (1) leachability by water; (2) reduction of strength of structural wood; and (3) interference with gluing or painting.

Organic retardants have been developed more recently in the attempt to overcome the foregoing problems with inorganic salts. Organic compounds with retardancy power may contain phosphorous, boron, halogens, or nitrogen (generally as amines). Hundreds of such compounds have been reported in the literature (25-27).

Instead of impregnating wood, it may be protected with fire-retardant paints, including intumescent paints. Again, a large literature exists. The coating may be effective by virtue of providing a barrier, by releasing a suppressing gas, by affecting the char process, by providing a heat sink (such as aluminum hydroxide), or by any combination of these. These fire-retardant coatings often have special difficulty in surviving in wet or very humid environments.

Synthetic polymers as well as wood may be fire-retarded (27). One approach is to incorporate halogens, phosphorus, boron, silicon, or sulphur into the polymer structure. Alternately, additives may be introduced, of a great variety of types. One important class of retardants is a synergistic combination of an antimony compound and a chlorine compound. Apparently a volatile compound, either $SbCl_3$ or $SbOCl$, forms and inhibits the gaseous flame. An unfortunate by-product of halogenated retardant systems is the corrosive effect of the acid produced on any nearby sensitive equipment.

A major problem in evaluating fire retardants is the choice of the fire test method(s) employed. Small-scale tests are characterized by low levels of radiative heat flux impinging on the sample (unless a radiant heater is included in the test procedure). Retardants often appear effective in small-scale tests while they may be less effective in tests involving high radiative heat flux, such as would be present in a compartment approaching flashover.

FLAMMABILITY OF HOT GAS LAYER

In a compartment fire before flashover, there is a horizontal interface with air below and partly consumed combustion products above. A local fire plume is pumping combustion products, unburned combustibles and air into this oxygen-deficient upper region, but flames are not propagating along the horizontal interface.

In actual fires it is occasionally observed that, at a particular moment, a reddish flame initiated at the fire plume spreads across the compartment along the interface between the hot ceiling layer and the air below. The immediate consequences are a large increase in radiative heat flux to not-yet-ignited objects in the compartment, and a change in temperature and composition of gases leaving the compartment.

Presumably, the explanation of the phenomenon is that the hot layer underwent a transition from nonflammable to flammable, because of a change of temperature or composition or both. There was very little knowledge on how a relevant criterion for flammability could be specified, prior to a recent study by Beyler (28).

Beyler reviewed correlations for flammability limits of mixtures of both premixed and diffusion flames (Le Chatelier's rule, Burgess and Wheeler's constancy of adiabatic flame temperature at the limit, and Simmons and Wolfhard's study of diluents producing extinguishment of diffusion flames). Based on these, he proposed an "ignition index" requiring knowledge of the total hydrocarbons, the CO, and the hydrogen in the hot layer, and the hot-layer temperature. The value of the index increases with increasing concentrations of the combustible species. When the index reaches unity, combustion should be able to propagate along the interface, if an ignition source is available.

Beyler compared his predictions with experimental data obtained with a variety of liquid-fuel pool fires under a one-meter-diameter hood. He obtained very encouraging agreement. An extension of the model predicted that the critical condition would be reached when the fire size and location were such that the actual air entrained by the plume below the layer was less than a calculable fraction of the stoichiometric air required for the fuel supply.

Beyler did not treat the soot and aerosols present in the hot layer; furthermore the effect of reactant temperature was not explored. Follow-on studies of these effects as well as possible effects of volatile fire retardants should be investigated in future researches. The applicability of the correlation to solid combustibles should be investigated.

## CHEMICALLY ACTIVE EXTINGUISHING AGENTS

Many of the most common extinguishing agents, including water, aqueous foams, steam, carbon dioxide, nitrogen, powdered limestone, and sand, are well established to act by physical mechanisms, either serving as heat sinks or providing a barrier between combustible and air. However, two classes of extinguishing agents, namely completely halogenated carbon compounds, such as $CF_3Br$, $C_2F_4Br_2$, or $CF_2C\ell Br$, and chemical powders, such as ammonium or alkali metal salts or acid salts of carbonates, phosphates, or halides, are believed to act at least partially by chemical mechanisms. Research to delineate these mechanisms has been conducted sporadically over the past 40 years.

The action of compounds such as $CF_3Br$ is now fairly well understood (29-31). The agent decomposes in the flame, with some absorption of energy, but, more important, the species HBr is formed. HBr is believed to remove chain-propagating radicals H and OH by the reactions $H + HBr \rightarrow H_2 + Br$ and $OH + HBr \rightarrow H_2O + Br$. Thus, the rate of the key chain-branching reactions in the flame, $H + O_2 \rightarrow OH + O$ (followed by $OH + H_2 \rightarrow H_2O + H$), would be reduced, making extinguishment easier. The radical removal process may also be enhanced by the regeneration of the chain-breaking species HBr via Br recombination to $Br_2$ followed by $H + Br_2 \rightarrow HBr + Br$. In addition to the HBr effect, an agent such as $CF_3Br$ may also remove H atoms by formation of HF. Values for the rates of these various radical reactions are now known, to a reasonable accuracy, and the overall set of reactions may be modelled with a computer.

The increase in soot formation caused by the presence of halogens would increase the radiative heat loss from the flame and also encourage extinguishment. The actual extinguishment process depends in a complicated way on fluid-mechanical strain rates, often buoyancy-driven, and heat losses. However, the reduction in burning velocity of a fuel-air mixture, such as $CH_4$-air, when HBr or a compound decomposing to form HBr is added, may be measured as well as calculated from flame theory with good agreement. (It is known that addition of HF or $HC\ell$ is far less effective than addition of an equal volumetric percentage of HBr to a flame, thus providing further confirmation of a chemical mechanism.  Furthermore, $CF_4$ is much less effective than $CF_3Br$, suggesting that removal of H atoms by HF formation is not too important.)

The next step in more sophisticated understanding of extinguishment by bromine compounds might be to extend a computer-modeling approach, which Dixon-Lewis et al. (15) have applied to the strain-rate-induced extinction of a counterflow diffusion flame in the forward stagnation region of a porous cylinder, by including HBr in the kinetics.  Further progress could also be made by quantifying the radiative heat loss stimulated by halogen addition to a hydrocarbon flame.

The chemical mechanisms by which powders extinguish flames are much more obscure.  It is well established that effectiveness of a powder agent is correlated with specific surface area of the powder, but this leaves open the question of whether fine particles are more effective because they volatilize faster, or because there is more surface area for heterogeneous catalysis, or because of increased radiative heat loss, or simply because heat is absorbed more rapidly from the flame by a larger number of smaller particles passing through the flame.  Very possibly, all of these effects are important.

Tests with a variety of chemically different powders of the same particle size show differences of effectiveness; the most effective powders ($NH_4H_2PO_4$, $KHCO_3$, and $KHCO_3$ plus $CO(NH_2)_2$ ) all decompose readily to form gases at elevated temperature.  Clearly the endothermic decomposition will reduce the flame temperature and promote extinguishment, but the possibility that the volatile decomposition products enter into the flame chemistry cannot be excluded.  It has been suggested that potassium salts are particularly effective agents, possibly because KOH vapor destroys H or OH atoms by $KOH + H \rightarrow H_2O + K$, or $KOH + OH \rightarrow H_2O + KO$, but Friedman and Levy (32) added potassium vapor to methane (1:16 molar ratio) and found no effect on flame strength of a diffusion flame.

Some recent researches on extinguishment by powders are those of Mitani and Niioka (33), Kim and Reuther (34) and Hertzberg et al. (35).  The challenge for the future is to characterize the relative importance of thermal energy absorption, gas-phase or surface chemical kinetics, and radiative energy loss on the extinguishment process when powders are added to flames.


REFERENCES

1.  Tewarson, A. and Pion, R. F.:  Combustion and Flame, 26, 85-103, 1976.

2.  Vovelle, C., Akrich, R., and Delfau, J. L.: Combustion Sci. and Tech, 36, 1-18, 1984.

3.  Tewarson, A.:  in Flame Retardant Polymeric Materials, Vol. 3, 92-153, Plenum Press, New York, 1982.

4.  Vovelle, C., Mellottee, H., and Delbourgo, R.: Nineteenth Symposium (Int'l.) on Combustion, Pittsburgh, 1982, 797-805.

5.  Kung, H. C.:  Fifteenth Symposium (Int'l.) on Combustion, The Combustion Institute, Pittsburgh, 1975, 243-253.

6.  Kansa, E. J., Perlee, H. E., and Chaiken, R. F.:  Combustion and Flame, 29, 311-324, 1977.

7. Becker, H., Phillips, A. M., and Keller, J.: Combustion and Flame, 58, 163-189, 1984.

8. Lee, C. K., Chaiken, R. F., and Singer, J. M.: Sixteenth Symposium (Int'1.) on Combustion, The Combustion Institute, Pittsburgh, 1977, 1459-1470.

9. Yoshizawa, Y. and Kubota, H.: Nineteenth Symposium (Int'1.) on Combustion, The Combustion Institute, Pittsburgh, 1982, 787-793.

10. Cullis, C. F., Hirschler, M. M., Townsend, R. P., and Visanuvimol, V.: Combustion and Flame, 49, 235-248, 1983.

11. Summerfield, M., Ohlemiller, T. J., and Sandusky, H. W.: Combustion and Flame, 33, 268, 1978.

12. Cullis, C. F., Hirschler, M. M., Townsend, R. P., and Visanuvimol, V.: Combustion and Flame, 49, 249-254, 1983.

13. Bilger, R. W. and Starner, S. H.: Combustion and Flame, 51, 155-176, 1983.

14. Miller, J. A., Kee, R. J., Smooke, M. D., and Grcar, J. F., Sandia National Laboratories, Livermore, CA: Paper presented at Western States Section, The Combustion Institute, April 1984 (WSS/CI 84-20).

15. Dixon-Lewis, G., Fukutani, S., Miller, J. A., Peters, N., and Warnatz, J.: Twentieth Symposium (Int'1.) on Combustion, The Combustion Institute, Pittsburgh, 1984, 1893-1903.

16. Lum, R. M.: J. Appl. Polymer Sci, 20, 1635, 1976.

17. Tewarson, A. and Newman, J. S. This symposium.

18. Levin, B. C., Paabo, M., Fultz, M. L., Bailey, C., Yin, W., and Harris, S. E.: NBS IR 83-2791, November 1983.

19. Markstein, G. H.: Twentieth Symposium (Int'1.) on Combustion, The Combustion Institute, Pittsburgh, 1985, 1967-1973.

20. Kent, J. H. and Wagner, H. G.: Combustion Science and Technology, 41, 245-270, 1984.

21. De Ris, J.: Fire and Materials, in press, 1985.

22. Twentieth Symposium (Int'1.) on Combustion, The Combustion Institute, Pittsburgh, 1985.

23. De Ris, J.: Seventeenth Symposium (Int'1.) on Combustion, The Combustion Institute, Pittsburgh, 1979, 1003-1016.

24. Seader, J. D. and Einhorn, I. N: Sixteenth Symposium (Int'1.) on Combustion, The Combustion Institute, Pittsburgh, 1977, 1423-1445.

25. Lyons, J. W.: The Chemistry and Uses of Fire Retardants, Wiley-Interscience, New York, 1970.

26. Canadian Wood Council, Ottawa, Canada. Wood Fire Behaviour and Fire Retardant Treatment. Forest Products Laboratory, Ottawa, Canada, 1966.

27. Cf. Flame Retardant Polymeric Materials, Vol. 1, 1975, Vol. 2, 1978, Vol. 3, 1982, Plenum Press, New York.

28. Beyler, C. L.: Combustion Science and Technology, 39, 287-303, 1984.

29. Westbrook, C. K.: Nineteenth Symposium (Int'l.) on Combustion, The Combustion Institute, Pittsburgh, 1982, 127-141.

30. Safieh, H. Y., Vandooren, J., and Van Tiggelen, P. J.: Nineteenth Symposium (Int'l.) on Combustion, The Combustion Institute, Pittsburgh, 1982, 117-126.

31. Westbrook, C. K.: Combustion Science and Technology, 34, 201-225, 1983.

32. Friedman, R. and Levy, J. B: Combustion and Flame, 7, 195-201, 1963.

33. Mitani, T. and Niioka, T.: Combustion and Flame, 55, 13-21, 1984.

34. Kim, H. T. and Reuther, J. J.: Combustion and Flame, 57, 313-317, 1984.

35. Hertzberg, M., Cashdollar, K. L., Zlochower, L., and Ng, D. L.: Twentieth Symposium (Int'l.) on Combustion, The Combustion Institute, Pittsburgh, 1985, 1967-1973.

# Spatial Variation of Soot Volume Fractions in Pool Fire Diffusion Flames

**S. BARD and P. J. PAGNI**
Mechanical Engineering Department
University of California
Berkeley, California 94720, USA

ABSTRACT

Measurements are reported of the line of sight averaged soot volume fraction as a function of height within pool fire flames fueled by polystyrene (PS) and polymethylmethacrylate (PMMA). No-lip pools with diameters of 7.5 cm and 15 cm were examined and compared with Markstein's results at 31 cm and 73 cm. The multi-wavelength laser transmission technique used has been previously described. PS and PMMA represent two distinct cases. For the optically thick PS the soot volume fraction, $f_v$, appears to be independent of fuel scale at ~ 3.3 ppm. For the optically thin PMMA, $f_v$ increases substantially with fuel scale from ~ 0.2 to ~ 0.7 ppm. For both fuels, $f_v$ decreases only slightly with height and is well approximated as uniform throughout the flame. The correlation, $f_v/f_{v_{max}} = 1.5 \ (\kappa L)^{0.33}$ for $\kappa L \leqslant 0.3$ and $= 1$ for $\kappa L \geqslant 0.3$, where $\kappa L$ is the flame optical thickness, is inferred from this scant data base. It may provide $f_v$ for any fuel at any scale as well as estimates of the maximum possible conversion of fuel carbon to soot.

## INTRODUCTION

It is well established that flame radiation is a dominant mechanism for fire growth [1,2] and that carbon particles within the flame produce most of the radiation [3,4]. Since these particles are small compared to the infrared wavelengths emitted, the flame volume fraction occupied by soot, $f_v$, is the characteristic most important to flame radiation [5,6].

Neglecting blockage effects, the radiative energy flux from each element of flame surface area can be approximated as

$$\dot{q}'' = \varepsilon_f \ \sigma \ T_f^4 \ , \tag{1}$$

where $T_f$ is a mean flame temperature and the flame emissivity is

$$\varepsilon_f = \varepsilon + \varepsilon_g - \varepsilon \varepsilon_g \ . \tag{2}$$

Supported by the Center for Fire Research under USDOC-NBS Grant No. 60NAND5D0552. Valuable assistance from J. de Ris, M. Brosmer and J. Reed is gratefully acknowledged.

The gas emissivity, $\varepsilon_g$ is small, usually less than 20% of $\varepsilon_f$ [1]. Standard techniques [7,8] are available to evaluate $\varepsilon_g$ in terms of the $H_2O$ and $CO_2$ partial pressures and the flame mean beam length, L, [9,10]. The soot emissivity is given [5] by

$$\varepsilon = 1 - \exp(-\kappa L) , \tag{3}$$

where the soot absorption coefficient is defined as

$$\kappa = 36\pi \, F_a(\bar{\lambda}) \, f_v/\bar{\lambda} , \tag{4}$$

with the mean emitted wavelength,

$$\bar{\lambda} \, T_f = 0.40 \text{ cm } ^\circ K . \tag{5}$$

Orloff and de Ris [9] suggest $T_f \approx 1200^\circ K$ for most pool fires; Eq. (5) then gives $\bar{\lambda} = 3.3 \ \mu m$. The soot optical property effects are all in $F_a(3.3) \approx 0.044$ [11]. The only remaining unknown needed to quantify flame radiation is $f_v$. Since no fundamental theory of soot formation is yet available, experimental soot volume fractions are needed. Several techniques exist for measuring $f_v$ [3,6,12-16], one of which is described in the next section. Empirical answers to the following two questions are sought here: How does $f_v$ vary within the pool fire? and How does $f_v$ scale with pool fire size?

EXPERIMENT

The experimental procedure has been described [3,6]. Here 7.5 cm and 15 cm diameter pools formed of beads of polystrene, PS and polymethylmethacrylate, PMMA, with no lip [9] were scanned with height using the apparatus shown in Fig. 1. For each fuel 100 instantaneous intensity measurements at several wavelengths were stored and correlated with simultaneous laser pathlengths measured separately for each data point. The pathlength, $\ell$, was taken as the

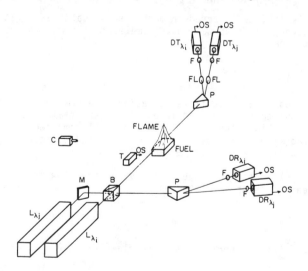

Fig. 1. Schematic of the multi-wavelength extinction apparatus. Shown are the lasers (L ), mirror (M), beamsplitter (B), prism (P), timer (T), camera (C), detectors (D), focusing lens (FL), filters (F) and output signals (OS).

width of continous luminosity in a timed videotape frame. When the laser beam passes through the flame, the transmitted intensity, $I$, is related to the initial intensity, $I_0$, by

$$I(\lambda)/I_0(\lambda) = \exp(-\tau(\lambda)\ell) \ . \tag{6}$$

The intensity and pathlength data $I$, $I_0$, and $\ell$ give experimental extinction coefficients, $\tau(\lambda)$. Results at each of two wavelengths give effectively two equations in two unknowns, the two parameters in the size distribution, $N(r)$, listed below. Since $\tau$ is not homogeneous along the pathlength, a line of sight average extinction coefficient and average soot volume fraction, are obtained here. If the radial variations in $f_v$ are important, local measurements are required [13]. From the viewpoint of calculating flame radiation, these averages contain the desired information.

The extinction coefficient is related to the Mie extinction efficiency, $Q(\lambda,m,r)$, of each particle of radius $r$, and to the particle concentration, $N(r)dr$, by

$$\tau(\lambda,m,r) = \int_0^\infty N(r) \, Q(\lambda,m,r) \, \pi r^2 dr \ , \tag{7}$$

where $m$ is the soot complex index of refraction [11]. Previous studies [6,17] suggest a Gamma size distribution with the constraint of a specified ratio of standard deviation to mean particle radius, $\sigma/r_m = 1/2$. In terms of the most probable radius, $r_{max}$, and the total particle concentration, $N_0$, the distribution is

$$N(r)/N_0 = (27r^3/2r_{max}^4) \, \exp(-3r/r_{max}) \ . \tag{8}$$

(See Fig. 2.) Once $N_0$ and $r_{max}$ are known from the extinction measurements, the soot volume fraction is

$$f_v \equiv \frac{4}{3}\pi \int_0^\infty N(r) \, r^3 dr = \frac{54}{38}\pi \, \Gamma(7) \, N_0 \, r_{max}^3 \approx 18.62 \, N_0 \, r_{max}^3 \ . \tag{9}$$

The results are summarized in Table I. The observed variation of $r_{max}$, $N_0$ and $f_v$ with height within the flame is as expected. The particle radii increase moderately while the concentrations decrease, due to coagulation and oxidation. The net effect of these complimentary variations is that the soot volume fraction decreases only slightly. Multi-wavelength results show $f_v$ is determined more accurately than either $r_{max}$ or $N_0$ [6,15]. The evolution of the particle size distribution is approximated by Fig. 2 for the 15 cm diameter PMMA pool fire. The change in $f_v$ is sufficiently small, that to a first approximation, $f_v$ is uniform throughout the flame. Markstein's results at larger scale [12,18] confirm this conclusion when corrected for optical properties and gas species contributions [19].

The soot absorption coefficient in column 5 of Table I comes from the $f_v$ of column 4 in Eqs. (4 and 5) with $T_f \sim 1200°K$ and $F_a \sim 0.044$. The soot emissivity in the last column is obtained from Eq. (3) with $\kappa$ and a mean beam length appropriate to the entire flame. The literature contains expressions for the flame height [12,20] and shape [9,10,21] which can be used to calculate L from

$$L = 3.6 \, V_f/A_f \ , \tag{10}$$

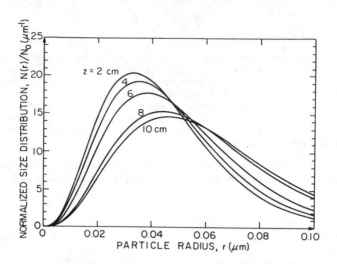

Fig. 2. Evolution of flame particulate size distribution with increasing height, z, for a 15 cm diameter PMMA pool fire.

TABLE I. Experimental Results

| Fuel/D,cm | $z$ cm | $r_{max}$ nm | $N_0 \times 10^{-9}$ cm$^{-3}$ | $f_v$ ppm | $\kappa$ m$^{-1}$ | $\varepsilon$ |
|---|---|---|---|---|---|---|
| PS/7.5 | 2 | 34 | 5.3 | 3.9 | 6.1 | 0.20 |
|  | 3 | 35 | 4.1 | 3.3 | 5.2 | 0.18 |
|  | 4 | 37 | 3.4 | 3.2 | 5.0 | 0.17 |
|  | 5 | 38 | 3.0 | 3.1 | 4.9 | 0.17 |
| PS/15 | 2 | 33 | 5.4 | 3.6 | 5.7 | 0.34 |
|  | 4 | 35 | 4.2 | 3.4 | 5.3 | 0.33 |
|  | 6 | 40 | 2.6 | 3.1 | 4.9 | 0.31 |
|  | 8 | 42 | 2.0 | 2.8 | 4.4 | 0.28 |
|  | 10 | 42 | 2.1 | 2.9 | 4.6 | 0.29 |
| PMMA/7.5 | 2 | 42 | 0.17 | 0.23 | 0.30 | 0.01 |
|  | 3 | 43 | 0.15 | 0.22 | 0.29 | 0.01 |
|  | 4 | 45 | 0.12 | 0.21 | 0.27 | 0.01 |
|  | 5 | 46 | 0.13 | 0.23 | 0.30 | 0.01 |
| PMMA/15 | 2 | 33 | 0.46 | 0.31 | 0.40 | 0.03 |
|  | 4 | 35 | 0.35 | 0.28 | 0.37 | 0.03 |
|  | 6 | 38 | 0.26 | 0.26 | 0.34 | 0.03 |
|  | 8 | 44 | 0.15 | 0.23 | 0.30 | 0.02 |
|  | 10 | 46 | 0.12 | 0.22 | 0.29 | 0.02 |

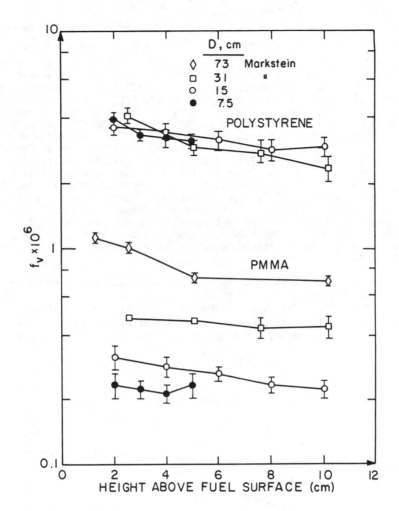

Fig. 3. Carbon particulate volume fraction, $f_v$, as a function of height above the fuel surface and flame scale for PS and PMMA. Markstein's values for larger scale flames are from Ref. 12. Contrast the constancy of the optically thick PS with the increase in $f_v$ for the optically thin PMMA.

where $V_f$ is the flame volume and $A_f$ is the flame surface area. However, to a good approximation, the results collapse to

$$L \approx D/2 \qquad\qquad\qquad (11)$$

where D is the pool diameter. The constancy of $\varepsilon$ for each pool fire in column 6 of Table I supports the uniform $f_v$ approximation. Figure 3 combines these data with those of Markstein [12] to display the difference between the optically thick PS and the optically thin PMMA. The PS $f_v$ is invariant with scale while the PMMA $f_v$ increases by over a factor of 3, roughly as $D^{1/2}$ [19] when the 31 cm and 75 cm data are corrected. The soot so dominates the gas in the polystrene flame that no correction of Markstein's Schmidt method data at 31 cm for $H_2O$ or $CO_2$ absorption is necessary [19]. Gas absorption or fluorescence has been shown [22] not to effect the multi-wavelength laser measurements described here.

Why does $f_v$ behave as shown in Fig. 3? Orloff and de Ris [9] have calculated radiative and convective energy fluxes to pool fire fuel surfaces for a wide variety of fuels and diameters. The ratio of radiation to convection is typically thirty. Therefore the fuel pyrolysis rate [21] is controlled by the flame radiation. Table I shows $\varepsilon$ increasing by 200% for PMMA as D increases from 7.5 to 15 cm. Similar increases occur at larger D. So the fuel pyrolysis rate is increasing as the flame radiation, given by Eq. (1), increases with D and $\varepsilon$. The increased pyrolysis rate causes the fuel mass fraction at the pool surface, $Y_{fw}$, to approach unity [23]. The literature [24,25] suggests that the soot formation rate is a nearly linear function of $Y_f$ and an exponential function of $T_f$. Assuming $T_f$ stays ~ 1200°K, the soot formation rate will increase as the fuel mass fraction profile, $Y_f(z)$, increases. Therefore, the $\varepsilon$ increase with optical path, given by Eq. (3), produces an increased surface $\dot{q}''$ which in turn increases $Y_f$. Through $Y_f$, the soot formation rate rises causing $f_v$ to increase, hence, the trend shown Fig. 3. Large $f_v$ increases occur where $\varepsilon$ is growing rapidly with D and $f_v$ remains constant for a given fuel after $Y_f$ and $\varepsilon$ become saturated.

This scenario can be quantified by a correlation such as the one shown in Fig. 4. Assume all effects of the fuel chemistry can be accounted for by a maximum soot volume fraction, $f_{v_{max}}$. Then seek a correlation of $f_v/f_{v_{max}}$ with $\varepsilon$, or equivalently by Eq. (3), with $\kappa L$, the flame optical thickness. The data suggest $f_{v_{max}}$ of 3.4 ppm for PS and 0.62 ppm for PMMA [19]. These values are

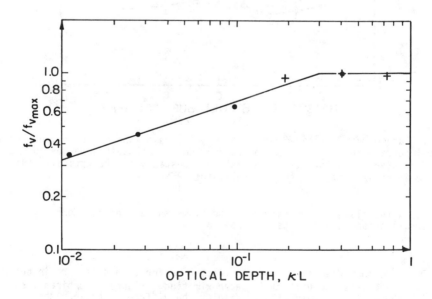

Fig. 4. Proposed correlation of $f_v$, normalized on a fuel characteristic, $f_{v_{max}}$, with optical depth. All pool fires may be described by this simple expression, although only data for PS (+) and PMMA (●) are available.

366

consistant with maxima found in free combusting boundary layer $f_v$ profiles [25]. The data in Fig. 4 were plotted using Eq. (11) for L and Eqs. (4 and 5) for $\kappa$. The resulting correlation is

$$f_v/f_{v_{max}} = 1.5\ (KL)^{1/3}\ ,\quad kL < 0.3\ ,$$

$$f_v/f_{v_{max}} = 1\ ,\quad kL > 0.3\ . \tag{12}$$

The one third power follows from an optically thin $f_v \sim D^{1/2}$ with $\kappa \sim f_v \sim D^{1/2}$ and $L \sim D$, so that $\kappa L$ goes as $f_v{}^3$.

Since the data are so sparse, Eq. (12) can only be regarded as speculation. However, it may permit one measurement of $f_v$ to provide the $f_v$ for all pool fires with that fuel. Six additional fuels $f_v$'s at small scale [6] are extrapolated in Table II to illustrate the use of Eq. (12). From Eqs. (4, 5 and 11) with $T_f \sim 1200°$ and $F_a \sim 0.044$, the small scale soot absorption coefficient and optical depth are calculated as in columns 3 and 4. Column 5 is $f_v/f_{v_{max}}$ from Eq. (12). Since $f_v$ is known, $f_{v_{max}}$ can be found as in column 6. With $f_{v_{max}}$ and D, Eqs. (4, 5, 11 and 12) give any other fires' $f_v$.

Using $f_{v_{max}}$ in Eqs. (4 and 5) gives the absorption coefficient in column 7. The pool fire diameter above which $f_v = f_{v_{max}}$, is found from $\kappa_{max}D_{max}/2 = 0.3$, as in column 8. For common fire fuels, other than PS, $f_v = f_{v_{max}}$ if $D > 1/2$ m. The flame emissivity approaches the black limit, $\varepsilon_f \to 1$, for $D \sim 5\ D_{max} \sim 2.5$ m since Eq. (3) gives $1 - \exp\ (-1.5) = 0.78$ and gas contributions make up the remainder.

The maximum fuel carbon convertible to soot is estimated in the last column of Table II [26]. Such numbers may be useful in the assessment of the nuclear winter problem. Let n reactions with $CO_2$ as a product occur for each reaction

TABLE II.   Uses of the correlation.  PS and PP had D = 7.5 cm.  All other fuels had D = 15 cm.  L = D/2.  $\kappa(m^{-1}) \approx 1.5\ f_v$ (ppm).  PU is the Products Research Committee's polyurethane.

| Fuel | $f_v$ ppm | $\kappa$ $m^{-1}$ | $\kappa L$ | $f_v/f_{v_{max}}$ | $f_{v_{max}}$ | $\kappa_{max}$ $m^{-1}$ | $D_{max}$ m | Max Soot % of Fuel C |
|------|------|------|------|------|------|------|------|------|
| PS | 3.3 | 4.8 | 0.181 | 0.85 | 3.9 | 5.6 | 0.1 | 20.8 |
| Octane | 0.46 | 0.67 | 0.050 | 0.52 | 0.83 | 1.3 | 0.5 | 6.3 |
| PU, GM-21 | 0.51 | 0.75 | 0.056 | 0.57 | 0.89 | 1.3 | 0.5 | 5.7 |
| PP | 0.27 | 0.39 | 0.015 | 0.30 | 0.73 | 1.3 | 0.5 | 5.4 |
| PMMA | 0.22 | 0.32 | 0.012 | 0.34 | 0.65 | 0.94 | 0.6 | 4.0 |
| Wood | 0.29 | 0.42 | 0.031 | 0.45 | 0.62 | 0.93 | 0.6 | 3.5 |
| Acetone | 0.11 | 0.16 | 0.012 | 0.34 | 0.32 | 0.47 | 1.3 | 2.2 |
| Ethanol | 0.07 | 0.10 | 0.008 | 0.30 | 0.24 | 0.33 | 1.8 | 1.9 |

with C (soot) as a product. Then $1/(n+1)$ is the fraction of the fuel carbon converted to soot. n is found by equating the measured soot mass per mass of gas product, $Y_s$, to a chemical $Y_s$. The solid soot density is assumed [25] to be $\rho \sim 1.2$ gm/cm$^3$ and the product gas density at 1 atm and 1200°K is $\rho_g \sim 0.3 \times 10^{-3}$ gm/cm$^3$. Therefore the measured $Y_s = f_v \rho/\rho_g \sim 4 \times 10^3 f_v$. In the $CO_2$ producing reaction let $M_g$ be the product mass. In the C producing reaction let $M_{gc}$ be the gas product mass and $M_s$ be the solid product mass. Then the chemical $Y_s = M_s/(nM_g+M_{gc})$. Equating $Y_s$'s and solving for n gives

$$\text{max soot/fuel carbon} = (n+1)^{-1} = M_g/(2.5 \times 10^{-4}M_s/f_{v_{max}} + M_g - M_{gc}) \tag{13}$$

For example, polypropylene, PP, is $C_3H_6$ which, with 4 $N_2$ per $O_2$ for air, gives $M_g = 690$, $M_{gc} = 222$ and $M_s = 36$. From Table II, $f_{v_{max}} = 0.73 \times 10^{-6}$, so n = 17.6 and Eq. (13) gives 5.4% of the fuel carbon as the maximum convertible to soot. These values are in good agreement with de Ris's estimates [1] of 18% for PS, 5.5% for PP and 1.9% for PMMA.


CONCLUSIONS

1. Within pool fires the vertical line of sight averaged soot volume fraction decreases sufficiently slowly with height to be approximated as constant throughout the flame.

2. Limited measurements suggest, that for a given fuel, $f_v$ is only a function of the optical depth of the pool fire flame, i.e.,

$f_v/f_{v_{max}} \approx 1.5 \ (\kappa L)^{0.33}$ for $\kappa L \leqslant 0.3$ and $f_v = f_{v_{max}}$ for $\kappa L \geqslant 0.3$ .


REFERENCES

1. de Ris, J., "Fire Radiation - A Review," Seventeenth Symposium (International) on Combustion, The Combustion Institute, 1003, 1979.

2. Emmons, H.W., "The Calculation of a Fire in a Large Building," 20th Joint ASME/AIChE National Heat Transfer Conference, Paper 81-HT-2, Aug. 1981.

3. Pagni, P.J., and Bard, S., "Particulate Volume Fractions in Diffusion Flames," Seventeenth Symposium (International) on Combustion, The Combustion Institute, 1017, 1979.

4. Tien, C.L., and Lee, S.C., "Flame Radiation," Prog. Energy Combust. Sci., 8, 41, 1982.

5. Yuen, W.W., and Tien, C.L., "A Simple Calculation Scheme for the Luminous-Flame Emissivity," Sixteenth Symposium (International) on Combustion, The Combustion Institute, 1481, 1977.

6. Bard, S., and Pagni, P.J., "Carbon Particulate in Small Pool Fire Flames," ASME J. Heat Transfer, 103, 2, 357, 1981.

7. Modak, A.T., "Thermal Radiation from Pool Fires," Combust. Flame, 29, 177, 1977.

8. Hubbard, G.L., and Tien, C.L., "Infrared Mean Absorption Coefficients for Luminous Flames and Smoke," ASME J. Heat Transfer, 100, 235, 1978.

9. Orloff, L., and de Ris, J., "Froude Modeling of Pool Fires," Nineteenth Symposium (International) on Combustion, The Combustion Institute, 885, 1983.

10. Orloff, L., "Simplified Radiation Modeling of Pool Fires," Eighteenth Symposium (International) on Combustion, The Combustion Institute, 549, 1981.

11. Lee, S.C., and Tien, C.L., "Optical Constants of Soot in Hydrocarbon Flames," Eighteenth Symposium (International) on Combustion, The Combustion Institute, 1159, 1981.

12. Markstein, G.H., "Radiative Properties of Plastic Fires," Seventeenth Symposium (International) on Combustion, The Combustion Institute, 1053, 1979.

13. Markstein, G.H., "Measurements on Gaseous - Fuel Pool Fires with a Fiber-Optic Absorption Probe," Combust. Sci. and Tech., 39, 215, 1984.

14. Santoro, R.J., Semerjian, H.G., and Dobbins, R.A., "Soot Particle Measurements in Diffusion Flames," Combust. Flame, 51, 2, 203, 1983.

15. Dobbins, R.A., and Mulholland, G.W., "Interpretation of Optical Measurements of Flame Generated Particles," Combust. Sci. Tech., 40, 175, 1984.

16. Shinotake, A., Koda, S., and Akita, K., "An Experimental Study of Radiative Properties of Pool Fires of an Intermediate Scale," Combust. Sci. Tech., 43, 85, 1985.

17. Lee, S.C., Yu, Q.Z., and Tien, C.L., "Radiative Properties of Soot from Diffusion Flames," J. Quant. Spect. Radiat. Transfer, 27, 387, 1982.

18. Markstein, G.H., "Scanning-Radiometer Measurements of the Radiance Distribution in PMMA Pool Fires," Eighteenth Symposium (International) on Combustion, The Combustion Institute, 537, 1981.

19. Brosmer, M., Private Communication.

20. Zukoski, E.E., "Fluid Dynamic Aspects of Room Fires," Plenary Lecture, this Symposium.

21. Modak, A.T., and Croce, P.A., "Plastic Pool Fires," Combust. Flame, 30, 251, 1977.

22. Bard, S., and Pagni, P.J., "Comparison of Laser-Induced Fluorescence and Scattering in Pool-Fire Diffusion Flames," J. Quant. Spect. Radiat. Transfer, 25, 453, 1981.

23. Pagni, P.J., "Diffusion Flame Analyses," Fire Safety J., 3, 273, 1981.

24. Glassman, I., and Yaccarino, P., "Temperature Effect in Sooting Diffusion Flames," Eighteenth Symposium (International) on Combustion, The Combustion Institute, 1175, 1981.

25. Pagni, P.J., and Okoh, C.I., "Soot Generation within Radiating Diffusion Flames," Twentieth Symposium (International) on Combustion, The Combustion Institute, 1045, 1985.

26. Rockett, J.A., private communication.

# The Involvement of Oxygen in the Primary Decomposition Stage of Polymer Combustion

C. F. CULLIS
Department of Chemistry
The City University, London, England

ABSTRACT

The effects of oxygen on the thermal decomposition of organic polymers are first described. Evidence for the involvement of oxygen in the primary decomposition stage of polymer combustion is then reviewed and factors which determine whether or not oxygen is present during this stage are considered. Some suggestions are also made as to ways in which it might be possible to exclude oxygen from the region where polymer breakdown first occurs and thus decrease the readiness with which polymeric materials are initially ignited. Finally the desirability is stressed of further studies being made of the oxidative pyrolysis of organic polymers.

INTRODUCTION

There are at least two essential requirements for the development of any fire; one is a source of ignition and the other is one or more flammable materials. Although the principal ignition sources responsible for starting fires are fairly well known (1,2), there is still relatively little statistical information about the precise types of material involved in fires (3). Nevertheless, most of the flammable components of residential, public and commercial buildings consist largely of either natural or synthetic organic polymers, and these materials are therefore nearly always implicated in, if not directly responsible for, fires in urban areas.

The likelihood of fire developing depends not only on the magnitude and effectiveness of the ignition source but also on the ease with which the adjacent materials can be ignited and the readiness with which they subsequently burn and spread flame to other parts of the structure under consideration. Although all organic polymers will burn under sufficiently vigorous experimental conditions, there are substantial variations in the combustion reactivity of different polymeric materials; and, in order to assess the risks of fire developing, it is important to be able to identify the principal factors determining the readiness or otherwise with which a particular polymer burns.

The combustion of organic polymers is a very complex process, which involves a number of interrelated, although conceptually distinct, stages taking place both in the condensed phase and in the adjacent gas phase (4). In flaming combustion, the polymers first suffers breakdown in the condensed phase to give combustible volatile products; these then enter the flame zone above the decomposing polymer where they burn in the gas phase yielding final combustion products and liberating heat; and finally at least some of this heat is transferred back to the polymer where it causes the evolution of a further supply of volatile breakdown products. Since in such a system it is in effect the gases formed from the polymeric material which burn rather than the polymer itself, factors which determine how readily a given polymer undergoes combustion will include its ease of breakdown and the extent to which its decomposition yields combustible gaseous products which can form flammable mixtures with the surrounding air or other gaseous oxidant.

Over the last 35 years or so, detailed studies have been made of both the kinetics and products of decomposition of numerous organic polymers, so that there is now a wealth of experimental evidence regarding the mechanisms of breakdown of polymeric materials and the nature and amounts of gaseous and other products formed from them (5-9). By far the greater part of the vast amount of work on thermal reactions of polymers in the condensed phase has been carried out either in a vacuum or in inert atmospheres. There is now however considerable evidence that, at least under certain combustion conditions and with some organic polymers, the initial breakdown of the polymer is a process which actively involves oxygen.

In this paper, therefore, an account is first included of the influence of oxygen on the thermal decomposition of some organic polymers. A review is then given of the evidence for the involvement of oxygen in the primary decomposition stage of the combustion of certain polymers and an attempt is made to identify at least some of the factors which determine whether or not oxygen is present during this stage. Suggestions are made too as to ways in which oxygen might be excluded from the region where breakdown of the polymer first occurs so as to decrease the readiness with which certain polymeric materials are initially ignited. Finally the need is emphasised for detailed studies of the kinetics and mechanisms of the oxygen-catalysed pyrolysis of many more organic polymers.

THE INFLUENCE OF OXYGEN ON THE THERMAL DECOMPOSITION OF ORGANIC POLYMERS

It has long been known that even stoichiometrically insignificant amounts of oxygen can considerably alter the rate and course of decomposition in the gas phase of hydrocarbons and other organic compounds. Probably one of the best-known examples of oxygen-catalysed pyrolysis is found with acetaldehyde (10-12), where as little as $10^{-3}$ vol% of oxygen can double the rate of decomposition, although methane and carbon monoxide remain the predominant breakdown products. Somewhat similar behaviour is observed with low molecular weight hydrocarbons (13-15), although the mechanism of action of the oxygen is complex and indeed the experimental results suggest that oxygen can act both homogeneously and heterogeneously and can terminate as well as initiate the free radical chains involved (16-18).

It is not perhaps surprising therefore that the kinetics and products of decomposition of many organic polymers are considerably affected by the presence of (often very small amounts of) oxygen. For example, with polypropylene, oxygen decreases the reaction temperature by some 200° and reduces the activation energy by about 150 kJ mol (19,20); and similar behaviour is found with poly-(vinylidene fluoride), where oxygen, in addition to causing a dramatic lowering of the activation energy and the pre-exponential factor, changes a normally second-order reaction to one of zero order (21). Among other common organic polymers whose decomposition is generally accelerated by oxygen are poly(vinyl chloride) (22), polyamides (23) (including wool (24)) and cellulose (25). Interesting behaviour is observed with styrene-based polymers, where the effect of oxygen appears to depend on the nature of the substituent groups attached to the aromatic ring (26); with polystyrene itself, oxygen considerably decreases the thermal stability but poly(p-hydroxystyrene) breaks down at virtually the same rate in the absence and presence of oxygen (27). Polybutadiene is another polymer whose decomposition is very little affected by oxygen (28). However the complexity of the action of oxygen is shown by the fact that the gas tends to retard the decomposition of other polymers, such as polyacrylonitrile (29,30) and certain polyurethane foams (31,32).

It is thus clear that, with many but not all organic polymers, oxygen affects (and usually promotes) the thermal decomposition process. However, in general, the effects of the gas are less marked with thermosetting polymers than with thermoplastic polymers (33).

## EVIDENCE FOR THE INVOLVEMENT OF OXYGEN IN THE PRIMARY DECOMPOSITION STAGE

Experimental evidence for the involvement of oxygen during the initial breakdown stage of polymer combustion is derived from several different types of measurement, which will now be considered in turn.

### Studies of Stationary Flames above Burning Polymers

Measurements have been made of the chemical composition of the gases at various positions in the flames above candles of different organic polymers with both air and oxygen-enriched air as the supporting gas (34,35). With polyethylene burning in air, it was found that close to (i.e. within 1 mm of) the surface of the molten polymer, at least 70% of the flame gases consist of nitrogen, with less than 2% of oxygen, appreciable quantities of oxides of carbon and water and small amounts of several low molecular weight hydrocarbons. In other words, nearly all the oxygen originally associated with the nitrogen in the air had undergone chemical reaction and disappeared within a very short distance of the polymer surface. The possibility was considered that the oxygen oxidised the gaseous decomposition products of the polyethylene to carbon oxides and water as soon as they entered the gas phase. However calculation showed that the residence time of the gases in the first millimetre above the surface was only 5-10 ms, and separate experiments revealed that, at the temperatures in this region (ca. 400°C), negligible oxidation of simple hydrocarbons would take place within so short a time (36). It was therefore concluded that the carbon oxides and water were formed by oxidation reactions within the surface of the polymer melt rather than in the gas phase. On the other hand, similar experiments with both poly(methyl methacrylate) and polyoxymethylene showed that there were still substantial amounts of unchanged oxygen in the flame gases immediately adjacent to the polymer and that little or no surface oxidation took place. In other words, these polymers appeared to undergo principally thermal decomposition to the corresponding monomers, which then reacted with oxygen within the flame.

More detailed analysis of the flame gases only 0.1 mm above the surface of burning polyethylene and polypropylene (37) also showed very little residual oxygen and substantial amounts of oxides of carbon and water, strongly suggesting that, with both these polyolefins, significant amounts of oxygen diffuse through the flame to the polymer surface and are then absorbed into a well-defined surface layer in which an appreciable proportion of the polymer undergoes oxidative decomposition. However, it may be possible to explain the presence of the final gaseous products of combustion of the polymer close to its surface in terms of the relatively rapid diffusion of these species from the flame zone to other parts of the combustion region.

Analysis of Oxygen in the Condensed Phase during the Combustion of Polymers

Perhaps the most direct evidence for the presence of oxygen in the surface layer
of burning polymers during combustion comes from experiments with both polyethylene and polypropylene (37). Burning candles of these polymers were extinguished
by an excess of nitrogen or argon and then cooled in a stream of the inert gas.
The oxygen content was then determined by neutron activation analysis of a
section sliced lengthwise through the samples. In the case of one polypropylene
sample, the oxygen content was about 0.3 wt%, 1.6 mm below the surface, but this
rose to 12 wt% at a depth of 0.1 mm, and extrapolation suggested an oxygen concentration of about 26 wt% at the surface.

Another technique which may well prove useful for studying condensed-phase processes in polymers undergoing combustion is in situ continuous $\gamma$-radiometry,
which can be used to monitor density changes in discrete isothermal layers of
relatively large blocks of materials during their decomposition (38). This
technique has recently been extended to the determination of the changing
elemental composition of burning polymers and, although it has been used so far
mainly to detect the chemical changes occurring in various types of wood, it can
also be applied to measuring the oxygen concentration during the combustion of
other polymers (39).


Measurement of Temperatures in the Condensed Phase of Burning Polymer Systems

More indirect evidence suggesting the intervention of oxygen in the decomposition stage of polymer combustion comes from measurement of the temperatures at
and near the surface of polymeric materials when they burn. With polyethylene
burning in air, for example, the temperature rose rapidly from about 200°C, 5 mm
below the polymer surface, to approximately 400°C at the surface (35). Separate
experiments have shown however that, in the absence of oxygen, polyethylene
suffers scarcely any decrease in weight at this latter temperature (40). On the
other hand, this polymer starts to break down at temperatures as low as 250°C
when oxygen is present (40), and indeed most high molecular weight hydrocarbons
undergo rapid oxidation in the liquid phase under these conditions (41,42). It
is thus reasonable to assume that oxidation of the molten polymer could take
place in the surface layer, whereas pure thermal decomposition could not.
Another interesting finding was that the temperature of the surface of burning
molten polyethylene increased rapidly when the oxygen content of the surrounding
atmosphere was increased, whereas the maximum flame temperature remained more or
less constant (35). This suggests that the additional oxygen becomes involved
primarily in a direct reaction with the polymer in the condensed phase rather
than in purely gaseous reactions. It is perhaps significant, however, that, in
contrast to the behaviour observed with polyethylene, the surface temperature of
burning poly(methyl methacrylate) decreases when the oxygen concentration is
increased (43).


Determination of the Influence of Oxygen on Rates of Gasification of Polymers

In experiments involving the thermal irradiation of polyethylene, an increase in
the oxygen concentration (from 0 to 40 vol%) in the surrounding atmosphere considerably increased the rate of gasification of the polymer (43). The results
of a theoretical study suggest too that, with poly(methyl methacrylate), both
oxidative and thermal degradation are responsible for the gasification of the
polymer (44).

On the other hand, the use of a radiative pyrolysis technique has shown that, when small-diameter rods of polystyrene decompose in the absence and presence of oxygen, the linear regression rates are the same (45). Interestingly, observation of the onset of ignition by means of high-speed cine-shadow photography, when a stream of hot oxygen was passed over a flat polymer plate, has indicated that ignition occurs mainly as a result of a gas-phase reaction in the boundary layer rather than at the surface of the polymeric material (46).

## Consideration of the Energy Balances involved in Polymer Combustion

A physical model has been proposed for the flaming combustion of organic polymers in which heat is transferred, mainly by conduction from the flame, to the polymer surface, where pyrolysis takes place to produce more gaseous fuel (47, 48). If all the oxygen in the surrounding atmosphere is consumed in the flame zone, so that decomposition of the polymer occurs in the absence of oxygen, the model predicts that the critical ratio of oxygen to inert gas needed to sustain combustion of the polymer should be a linear function of the heat capacity of the inert gas. Experiments with different inert gases showed that this was the case with poly(methyl methacrylate) and polyoxymethylene (49) but not with polytetrafluoroethylene (50). With the fluorinated polymer, it is therefore necessary to invoke exothermic oxidation reactions at the polymer surface in order to supply the missing energy. Consideration of the energy fluxes for both polyethylene and polypropylene burning in air (37) showed too that, on the assumption that a small amount of energy is provided by radiation from the flame envelope, it is possible to calculate the extents to which these polymers simultaneously undergo oxidative decomposition (which is exothermic (51)) and thermal decomposition (which is endothermic). Measurement of the composition profiles in an opposed-flow diffusion flame of polyethylene also indicate the appreciable contribution (i.e. up to 20%) which surface oxidation of the polymer makes to the total enthalpy required for production of the gaseous polymer breakdown products (52).

## FACTORS AFFECTING THE ACCESS OF OXYGEN TO THE SURFACE OF BURNING POLYMERS

The factors which determine whether or not oxygen is present at the interface between the condensed and gaseous phases during the flaming combustion of polymers are not at all well understood. There must however be a delicate balance between various physical and chemical processes, including the rate at which oxygen diffuses towards the burning polymer surface, the rate at which the gaseous polymer breakdown products diffuse away from the surface and the rates of the different chemical reactions in the condensed phase, in the gas phase and at the interface between them.

The general conclusion seems to be that, at least with some organic polymers, oxygen is involved in the decomposition stage of flaming combustion which generates combustible gaseous fuel. However the extent of oxygen involvement varies considerably with the nature of the polymer, the size, shape and orientation of the sample being burnt and the combustion conditions, including the flame geometry.

One important factor which clearly influences whether or not oxygen reaches the surface of the polymer undergoing combustion is the vigorousness of the emission of the gaseous decomposition products. Thus, with poly(methyl methacrylate), as the rate of gasification becomes larger and the resulting counterflow of gaseous products allows less and less oxygen to reach the condensed phase, the effect of oxygen on the gasification process becomes less marked (43). This competition between the diffusion of oxygen to the surface and the convective flow of

375

product gases away from the surface clearly influences the magnitude of the effect of gas-phase oxygen on the rate of polymer gasification. Such an effect, which depends on the distance over which oxygen has to diffuse to the surface, is thus not surprisingly affected too by the size and shape of the polymer sample undergoing combustion. Thus, in a fire situation, where large areas of polymeric materials are generally involved, the role of oxygen in the decomposition stage of polymer combustion may well be much less important than during the candle-like burning of relatively small polymer samples.

Another factor is no doubt the ease with which oxygen, after it has successfully diffused to the surface of the condensed phase, can dissolve in the polymer. The molten surface layer of certain polymers becomes less viscous in an oxygen-containing environment (43,53). In the case of poly(methyl methacrylate), it has been observed that the bubbles then burst more frequently, leaving large holes which oxygen can enter; in these circumstances, oxygen penetrates a deeper layer of near-surface material than it could reach by diffusion alone (43).

It has already been mentioned that oxygen has in general a smaller effect on the decomposition of thermosets than it has on thermoplastics (33). The probable reason for this is that the melting which generally occurs when thermoplastics are heated tends to increase the surface area exposed to oxygen; this means that oxygen and volatile decomposition products can diffuse more readily through the polymeric material. In contrast, the surface area of thermosets, which do not melt, remains unchanged, so that gaseous diffusion in and out of the polymer remains difficult. In this context, the decreased flammability of aged polyurethane foams compared with newer ones may be due to the higher crosslink density caused by storage which makes diffusion even more difficult (54). Indeed, in a fire environment, the high temperatures attained may lead to further extensive crosslinking and hence even less access for oxygen.

The greater solubility of oxygen in the liquid as compared with the solid phase of the polymer probably explains why the involvement of oxygen is greater when poly(methyl methacrylate) is heated slowly, so that it remains molten for a relatively long time, than when heating is more rapid (55).


CONCLUSIONS

For oxygen catalysis of the primary decomposition stage of polymer combustion, not only must oxygen be present in the immediate vicinity of the surface of the burning material but the gas must react chemically with the polymer in such a way as to accelerate its decomposition. Paradoxically, with some polymers, whose breakdown is quite strongly catalysed by oxygen, no oxygen is apparently present to take part in the decomposition stage of polymer combustion. Conversely, there are other polymeric materials, where the presence of oxygen has no effect on decomposition and yet appreciable quantities of the gas are nevertheless present at the condensed phase – gas phase interface of the burning polymer. It is only in systems where oxygen is present at the polymer surface and catalyses the primary decomposition stage that it may be possible to decrease the ease with which the polymer burns by excluding oxygen from the polymer-gas interface.

Exclusion of air or oxygen is of course a universal method of preventing altogether the combustion of any flammable material. However, although in most real fire situations removal of all oxygen is not practicable, it may nevertheless be possible in certain burning polymer systems to reduce or even eliminate completely diffusion of oxygen to the polymer surface and thus help to prevent ignition of the underlying material. Perhaps the most obvious way of achieving this is to incorporate into the polymer an additive which either decomposes (56) or volatilises (57) to yield a vigorous counterflow of non-flammable gaseous products. Another and perhaps simpler approach is to arrange that the surface of the polymer is or becomes completely covered with a non-flammable protective solid coating which effectively excludes oxygen (58).

On the basis of the simplest cycle for the flaming combustion of organic polymers, according to which the polymer first suffers simple thermal decomposition to yield combustible volatile products, good correlations might be expected between polymer flammability and (a) the thermal instability of the polymer, (b) the extent of formation of combustible gases therefrom and (c) the heat of combustion of these gaseous products (59). In practice such correlations do not exist (60). Since oxygen is often involved in the primary decomposition stage of polymer combustion, it is clearly necessary also to have quantitative information regarding the thermo-oxidative instability of polymeric materials as well as the nature, amounts and heats of combustion of the gaseous products of such initial oxygen-catalysed breakdown. It is thus very important, for a fuller understanding of the factors governing polymer flammability and of the fire risks associated with the use of polymeric materials, that at least some of the excellent detailed studies which have been made of the purely thermal decomposition of organic polymers should now be extended to conditions under which controlled quantities of oxygen are also present.

REFERENCES

1.  Eckert, E.R.G.: "The Fire Problem in the U.S." in Heat Transfer in Fires, Thermophysics, Social Aspects, Economic Impact, ed. P.L. Blackshear, 3-16, Wiley, New York, 1974.

2.  Birky, M.M., Halpin, B.M., Caplan, Y.M., Fisher, J.S., McAllister, J.M. and Dixon, A.M.: "Fire fatality studies" Fire Mater. 3, 211-7, 1979.

3.  Christian, S.D.: "Legislation in the U.K. relating to plastics, polymers and textiles" in Proc. Interflam '79, p.11, Heyden, London 1979.

4.  Cullis, C.F. and Hirschler, M.M.: The Combustion of Organic Polymers, Oxford Univ. Press, Oxford, 1981.

5.  Madorsky, S.L.: Thermal Degradation of Organic Polymers, Wiley, New York, 1964.

6.  Conley, R.T.: Thermal Stability of Polymers, Dekker, New York, 1970.

7.  Reich, L. and Stivala, S.S.: Elements of Polymer Degradation, McGraw Hill, New York, 1971.

8.  David, C.: "Thermal degradation of polymers" in Comprehensive Chemical Kinetics, ed. C.H. Bamford and C.F.H. Tipper, 14, 1-173, Elsevier, Amsterdam, 1975.

9.  Grassie, N. (ed.): Developments in Polymer Degradation, 1-3, Applied Science, London, 1977-81.

10. Letort, M.: "Effect of traces of oxygen on thermal decomposition of acetaldehyde" J. Chim. Phys., 34, 428-43, 1937.

11. Niclause, M. and Letort, M.: "Induced pyrolysis of gaseous acetaldehyde by traces of oxygen" Compt. Rend., 226, 77-8, 1948.

12. Letort, M. and Niclause, M.: "Oxidation and pyrolysis" Rev. Inst. franc. Petrole, 4, 319-25, 1949.

13. Martin, R., Dzierzynski, M. and Niclause, M.: "Pyrolysis of propane in presence of traces of oxygen", J. Chim. Phys., 61, 790-801, 1964.

14. Blakemore, J.E., Barker, J.R. and Corcoran, W.H.: "Pyrolysis of butane and effect of trace quantities of oxygen", Ind. Eng. Chem. Fundam., 12, 147-55, 1973.

15. Karnaukhova, L.I. and Stepukhovich, A.D.: "Kinetics of oxygen-initiated cracking of propane and butane", Zhur. Fiz. Khim., 49, 551-3, 1975.

16. Voevodsky, V.V.: "Thermal decomposition of paraffin hydrocarbons", Trans. Faraday Soc., 55, 65-71, 1959.

17. Poltorak, V.A. and Voevodsky, V.V.: "Kinetics of propane cracking in presence of oxygen", Zhur. Fiz. Khim., 35, 179-80, 1961.

18. Martin, R., Niclause, M. and Dzierzynski, M.: "Complex influence of traces of oxygen and surface effects in pyrolysis of propane", Compt. Rend. 254, 1786-8, 1962.

19. Chien, J.C.W. and Kiang, J.K.Y.: "Pyrolysis and oxidative pyrolysis of polypropylene" in Stabilisation and Degradation of Polymers, ed. D.L. Allara and W.L. Hawkins, Adv. Chem. Ser. No. 169, 175-97, Am. Chem. Soc., Washington D.C., 1978.

20. Chien, J.C.W. and Kiang, J.K.Y.: "Oxidative pyrolysis of polypropylene" Makromol. Chem., 181, 45-57, 1980.

21. Hirschler, M.M.: "Effect of oxygen on thermal decomposition of poly-(vinylidene fluoride)", Eur. Polym. J., 18, 463-7, 1982.

22. Gupta, V.P. and St. Pierre, L.E.: "Effect of oxygen on thermal decomposition of PVC", J. Polymer Sci., Polym. Chem. Edn., 17, 797-806, 1979.

23. Chatfield, D.A., Einhorn, I.N., Mickelson, R.W. and Futtrell, J.H.: "Analysis of products of thermal decomposition of an aromatic polyamide fabric", J. Polym. Sci., Polym. Chem. Edn., 17, 1367-81, 1979.

24. Ingham, P.E.: "Pyrolysis of wool and action of flame retardants", J. Appl. Polym. Sci., 15, 3025-41, 1971.

25. Cullis, C.F., Hirschler, M.M., Townsend, R.P. and Visanuvimol, V.: "Combustion of cellulose under conditions of rapid heating", Combust. Flame, 49, 249-54, 1983.

26. Still, R.H. and Whitehead, A.: "Thermal analysis studies on styrene polymers", J. Appl. Polym. Sci., 20, 661-79, 1976.

27. Still, R.H. and Whitehead, A.: "Thermal analysis studies on poly(p-methoxystyrene) and poly(p-hydroxystyrene)", J. Appl. Polym. Sci., 21, 1215-25, 1977.

28. Cullis, C.F. and Laver, H.S.: "Thermal degradation and oxidation of polybutadiene", Eur. Polym. J., 14, 571-3, 1978.

29. Grassie, N. and McGuchan, R.: "Thermal analysis of polyacrylonitrile", Eur. Polym. J., 6, 1277-91, 1970.

30. Grassie, N. and McGuchan, R.: "Effect of sample preparation on thermal behaviour of polyacrylonitrile", Eur. Polym. J., 7, 1091-104, 1971.

31. Koscielecka, A.: "Thermal degradation of polyurethanes", Polimery, 20, 537-9, 1975.

32. Benbow, A.W. and Cullis, C.F.: "The combustion of flexible polyurethane foams", Combust. Flame, 24, 217-30, 1975.

33. Nicholson, J.W. and Nolan, P.F.: Behaviour of thermoset polymers under fire conditions", Fire Mater., 7, 89-95, 1983.

34. Burge, S.J. and Tipper, C.F.H.: "Burning of polythene", Chem. Ind., 362-3, 1967.

35. Burge, S.J. and Tipper, C.F.H.: "Burning of polymers", Combust. Flame, 13, 495-505, 1969.

36. Sampson, R.J.: "Reaction between ethane and oxygen at 600-630°C", J. Chem. Soc., 5095-106, 1963.

37. Stuetz, D.E., Di Edwardo, A.H., Zitomer, F. and Barnes, B.P.: "Polymer combustion", J. Polym. Sci., Polym. Chem. Edn., 13, 585-621, 1975.

38. Nolan, P.F. and Brown, D.W.: "Use of continuous dynamic radiometry in study of thermal decomposition of wood" in Proc. 4th Int. Conf. on Thermal Analysis, 3, 251-63, 1975.

39. Nicholson, J.W. and Nolan, P.F.: "A $\gamma$-radiometric method for studying condensed-phase processes in polymers undergoing combustion", Fire Mater., in press.

40. Igarashi, S. and Kambe, W.: "Differential thermal analysis and thermo-gravimetric analysis of some polyethylenes", Bull. Chem. Soc. Japan, 37, 176-81, 1964.

41. Reich, L. and Stivala, S.S.: Autoxidation of hydrocarbons and polyolefins, McGraw-Hill, New York, 1969.

42. Rabek, J.F.: "Oxidative degradation of polymers" in Comprehensive Chemical Kinetics, ed. C.H. Bamford and C.F.H. Tipper, 14, 425-538, Elsevier, Amsterdam, 1975.

43. Kashiwagi, T. and Ohlemiller, T.J.: "A study of oxygen effects on non-flaming transient gasification of PMMA and PE during thermal irradiation" in Proc. 19th Int. Symp. on Combustion, 815-23, Combustion Institute, Pittsburgh, 1983.

44. Kumar, R.N. and Stickler, D.B.: "Polymer degradation theory of pressure-sensitive hybrid combustion", in Proc. 13th Int. Symp. on Combustion, 1059-72, Combustion Institute, Pittsburgh, 1971.

45. Brauman, S.K.: "Polymer degradation and combustion", J. Polym. Sci., Polym. Chem. Edn., 12, 125-35, 1968.

46. Pitz, W.J., Brown, N.J. and Sawyer, R.F.: "Structure of a polyethylene opposed flow diffusion flame", in Proc. 18th Int. Symp. on Combustion, 1871-9, Combustion Institute, Pittsburgh, 1981.

47. Martin, F.J.: "A model for candle-like burning of polymers", Combust. Flame, 12, 125-35, 1968.

48. Fenimore, C.P. and Martin, F.J.: "Burning of polymers" in The Mechanisms of Pyrolysis, Oxidation and Burning of Organic Materials, ed. L.A. Wall, N.B.S. Spec. Publ. 357, 159-70, 1972.

49. Fenimore, C.P. and Jones, G.W.: "Modes of inhibiting polymer flammability", Combust. Flame, 10, 295-301, 1966.

50. Fenimore, C.P. and Jones, G.W.: "Decomposition of burning polytetrafluoro-ethylene", J. Appl. Polym. Sci., 13, 285-94, 1969.

51. Edgerley, P.G.: "Oxidative pyrolysis of plastics", Fire Mater., 4, 77-82, 1980.

52. Kashiwagi, T., Macdonald, B.W., Isoda, H. and Summerfield, M.: "Ignition of a solid polymer fuel in a hot oxidising gas stream", in Proc. 13th Int. Symp. on Combustion, 1073-86, Combustion Institute, Pittsburgh, 1971.

53. Mikhailov, N.V., Tokareva, L.G., Buravchenko, K.K., Terekhova, G.M. and Kirpichnikov, P.A.: "Stabilisation of poly(ethylene terephthalate) melts", Vysokomol. Soedin, 4, 1186-92, 1962.

54. Eaves, D.E. and Keen, C.V.: "Thermally initiated oxidation and ignition of flexible polyurethane foams", Brit. Polym. J., 8, 41-3, 1976.

55. Kashiwagi, T.: unpublished results.

56. Chamberlain, D.L.: "Mechanisms of fire retardance in polymers", in Flame Retardancy of Polymeric Materials, ed. W.C. Kuryla and A.J. Papa, 4, 106-68, 1978.

57. Benbow, A.W. and Cullis, C.F.: "Halogen compounds and burning of polymers" in Proc. Combustion Institute European Symposium, Univ. Sheffield, 183-8, Academic Press, New York, 1973.

58. Rhys, J.A.: "Intumescent coatings and their uses", Fire Mater., 4, 154-6, 1980.

59. Cullis, C.F.: "Thermal stability and flammability of organic polymers", Brit. Polym. J., 16, 253-7, 1984.

60. Cullis, C.F. and Hirschler, M.M.: "Significance of thermoanalytical measurements in assessment of polymer flammability", Polymer, 24, 834-40, 1983.

# Oxidative Pyrolysis of Polymers before Flaming Ignition

**ALISON BAIGNÉE**
Department of Chemical Engineering
University of Ottawa, Canada

**FERRERS R. S. CLARK**
Division of Building Research
National Research Council Canada
Ottawa, Canada, K1A OR6

ABSTRACT

Thin films of polyethylene, polystyrene and poly(methyl methacrylate) were exposed to the wake of a lean hydrogen–oxygen flat flame. Exposure was terminated at times ranging up to the point when flaming combustion began. The films were then analyzed by infrared spectroscopy.

Oxidation of polyethylene and polystyrene occurred after a latency period but before ignition. The oxidation rate remained constant despite changes in flame–polymer separation and flame gas equivalence ratio. The latency period before oxidation observed for polyethylene depended on equivalence ratio and separation. No oxidation was observed with poly(methyl methacrylate).

A comparison of infrared spectra obtained by transmission through films with spectra obtained by internal reflection spectroscopy, demonstrated that oxidation of polyethylene and polystyrene before ignition is confined to within approximately 5 μm of the exposed surface.

KEYWORDS

Polyethylene, polystyrene, poly(methyl methacrylate), hydrogen–oxygen, flame, infrared spectroscopy, oxidative pyrolysis, ignition.

INTRODUCTION

Study of chemical changes that occur during polymer combustion has been primarily confined to indirect methods such as thermal analysis and to analysis of gases evolved during pyrolysis or formed in the gas phase during burning (1). Studies of the chemical changes in the condensed phase of polymer combustion systems have usually been made during steady-state candle-like combustion of polyolefins (2,3), by neutron activation analysis for oxygen. This work has resulted in a controversial theory that oxidative pyrolysis can be an important contributor to the energy balance at the surface of a burning polymer (4).

Transient phenomena that occur on burning polymer surfaces have not been extensively studied. Jakes and Drews (5) reported oxidation of polypropylene in samples removed from a slab along which a flame had been allowed to burn;

samples were analyzed by iodometry for peroxide and hydroperoxide and by hydrazine-hydrazone derivatization for carbonyl functions.

Nothing appears in the literature on condensed phase chemical changes during the ignition delay of polymers exposed to flames. In previous publications (6-8) Clark reported differences between the gas phase events surrounding such ignition of polystyrene, polyethylene and poly(methyl methacrylate). Spectrometric measurements of oxidation on the surface of those polymers, made during the ignition sequence using the same apparatus and exposure conditions, are reported in this paper.

## EXPERIMENTAL DESCRIPTION

### Materials

Low molecular weight poly(methyl methacrylate) powder (Aldrich 18, 223-0) was hot pressed onto aluminum foil (221°C, 8.6 MPa) to form films of 40-100 μm thickness. High density polyethylene pellets (Aldrich 181, 190-0) were hot pressed onto aluminum foil (154°C, 17 MPa) to form films of 60-70 μm thickness. Polystyrene pellets (Aldrich 18242-7) were hot pressed onto aluminum foil (204°C, 17 MPa) to form films of 50-80 μm thickness.

### Exposure

The burner used (Fig. 1) was after a design of Botha and Spalding (9), using principles described by Hunter and Hoshall (10). A lean hydrogen-oxygen flat flame was supported above the polymer film sample, which was sandwiched

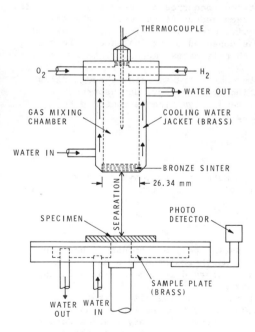

Figure 1. The burner

between a metal plate with a central circular orifice of 4.0 cm diameter and an asbestos mat 2 mm thick on the watercooled base.

The flame was in the form of an invisible flat circular disk 25 mm across and its wake extended to the polymer surface. Ignition of a polymer film was signalled by the first appearance of a very bright disk of light just above the polymer surface. Heat transfer was primarily by convection; at an equivalence ratio of 0.1 the total flux at the polymer surface had a maximum measured value of 19.7 kw·m$^{-2}$, of which no more than 3 kw·m$^{-2}$ was by radiation. (Equivalence ratio is defined as the volume of oxygen required for complete conversion of the hydrogen to water divided by the actual values of oxygen supplied.) The burner gases moved with a velocity that ensured laminar flow; the Reynolds number was about 558.

Infrared Spectroscopy

A Nicolet Model 6000 Fourier transform infrared spectrometer was employed, fitted with a broad bandwidth mercury-cadmium telluride (MCT) detector. Each spectrum was the average of 200 scans collected at 4.0 cm$^{-1}$ resolution with Happ-Genzel apodization and single order zero filling.

Spectral examination of the films was conducted in two ways. The object of the first, termed the reflection-transmission technique, was to direct the collimated beam of the spectrometer through the entire area of exposed film, to reflect off the aluminum foil, and through the film a second time. This was achieved by replacing an existing plane mirror near the detector with the film-foil laminate, complete with the metal plate with the 4 cm orifice used in the burner exposure. The collimated beam of the spectrometer is approximately 5 cm in diameter, and thus absorptions over the entire flame-exposed film were observed. Commercial accessories for the spectrometric examination of large areas provided coverage of less than 1/7 of the area covered by this technique.

The second method of spectroscopic inspection involved the use of a conventional internal reflection apparatus (a Wilks Model 9) fitted into the normal sample port of the Nicolet spectrometer. A nominal 30 reflection KRS-5 (45° bevel) element was employed throughout.

Spectral deconvolution

In an attempt to determine the mode of oxidation of polymer films, Fourier self-deconvolution techniques (11) were applied to bands observed in the carbonyl region of infrared spectra of sample films of each polymer after flame exposure. The intrinsic bandwidths of infrared absorptions of this class are often greater than the separation between bands. The deconvolution technique computes the bandshapes at reduced bandwidths, allowing resolution of otherwise hopelessly overlapping bands. The technique was found most valuable in this study in demonstrating that oxidation had not occurred; in this case deconvolution of spectra of films before and after flame exposure showed that no new bands had been formed. The technique also allowed the complexity of the oxidation process for polyethylene and polystyrene to be observed (Figs. 2, 3).

Data handling

Spectra obtained of exposed polymer films by the reflection transmission technique were stored for later processing. The transmission spectra were converted to absorbance spectra. Spurious bands due to adventitious water in

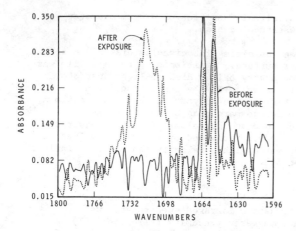

Figure 2. Deconvolved spectra of polyethylene before and after exposure

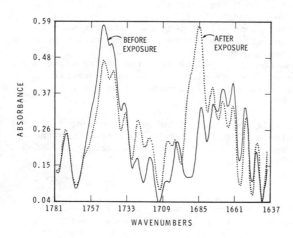

Figure 3. Deconvolved spectra of polystyrene before and after exposure

the system were removed by automated subtraction. Next a small positive y axis shift was added by coadding a straight line to each spectrum. A spectrum of unexposed polymer was then subtracted from that of the exposed polymer. At this stage the effects of oxidation were clearly apparent, especially in the carbonyl region of the spectrum. With suitable baseline correction, the area under the carbonyl peak was then calculated (1770–1635 cm$^{-1}$ for polystyrene, 1770–1683 cm$^{-1}$ for polyethylene). At very low levels of oxidation, noise in the spectrum occasionally led to negative areas being reported; this problem was avoided by the y axis displacement mentioned above.

RESULTS

Ignition delays

Previous work with this equipment (6-8) was conducted with thermally-thick slabs. It was of interest therefore to check that ignition delay times with polymer films were as reproducible as those recorded with polymer slabs. Table 1 shows the range of values obtained in runs of 10 exposures to ignition in various conditions; the reproducibility recorded gave confidence that different films exposed for varying times before spectroscopic analysis could be meaningfully compared.

TABLE 1.  Ignition delay times at equivalence ratio 0.1; figures are averages of 5 determinations

|  |  | Polyethylene | Polystyrene | Poly(methyl methacrylate) |
|---|---|---|---|---|
| Separation (mm) | 20 | 6.1 ± 0.2 | 13.0 ± 2.0 | 8.5 ± 0.5 |
|  | 15 | 5.3 ± 0.1 | 8.5 ± 0.5 |  |
|  | 10 | 4.4 ± 0.2 | 4.7 ± 0.2 |  |

Consistent with the increase in gas temperature (6) at the polymer surface as burner-polymer separation decreased, the ignition delays increased with separation.

Homogeneity

Oxidation before ignition was observed for polyethylene and polystyrene, but not for poly(methyl methacrylate) (Figs. 2, 3). For those films that oxidized, substantial variation in carbonyl concentration was observed across the exposed surface, ranging from no oxidation to a maximum, even at the end of the ignition delay. The reason for this is not clear but it could have been due to irregularities in the flat flame. However, spectroscopic examination by the reflection-transmission technique allowed measurement of the average degree of oxidation of the whole exposed film for a given exposure time.

Location of Oxidation

Films of polystyrene and polyethylene were exposed to the flame until oxidation was apparent in the reflection-transmission spectrum. The exposed and unexposed faces of each film were then examined by internal reflection spectroscopy. In this technique, the depth of penetration of the analyzing infrared beam is about 1 μm. The exposed faces of films of both polymers showed substantial oxidation products in both carbonyl (approximately 1700 cm$^{-1}$) and hydroxyl (approximately 3200 cm$^{-1}$) regions. By contrast, the unexposed faces of both polymers showed no sign of oxidation. These experiments demonstrate that oxidation upon exposure to flame did not penetrate right through the polymer films.

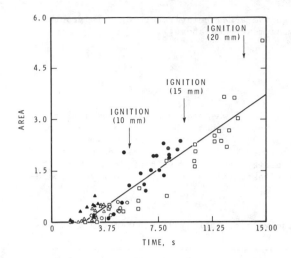

Figure 4.  Oxidation of polystyrene before ignition, relative area of carbonyl
band; 0.1 equivalence ratio:  o 10 mm separation,  • 15 mm
separation, ▫ 20 mm separation; ▵ 20 mm separation, 0.13
equivalence ratio, ▲ 20 mm separation, 0.15 equivalence ratio

The relative sizes of the O-H stretch and the adjacent C-H stretches were
approximately the same in the surface and by reflection-transmission,
indicating that the bulk of the oxidation was very close to the exposed
surface.

Variation of Flame-Polymer Separation and Flame Gas Equivalence Ratio

Figures 4 to 6 show the changes in carbonyl concentration in polystyrene
and polyethylene films for various exposure times at three different
burner-polymer separations and three different flame gas equivalence ratios.
Each point represents analysis of the carbonyl region of the spectrum of a
different, but very similar, film exposed to the flame and analyzed by the
reflection transmission technique.

Despite changes in flame-polymer separation and flame gas equivalence
ratio, the locus of all points on the oxidation-time plot for polystyrene was
constant to the limits of error (Figure 4).  Oxidation began after a very
similar period of exposure and oxidation to produce carbonyl functions
occurred at the same rate until ignition.  Time to ignition, as noted above,
was related to the flame-polymer separation.

The onset of oxidation of polyethylene was more sensitive to flame-
polymer separation, but once oxidation had begun, the rate of oxidation was
similar at the two separations at which the data scatter was small enough to
allow meaningful comparison (Figure 5).  Changing the equivalence ratio of the
flame also influenced the time of onset of oxidation; the richer the flame,
the shorter the initiation period.  After initiation the rate of oxidation was
high and similar for each fuel ratio (Figure 6).

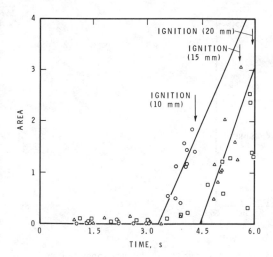

Figure 5. Oxidation of polyethylene before ignition at 0.1 equivalence ratio, relative area of carbonyl band: ○ 10 mm separation, △ 15 mm separation, ▫ 20 mm separation

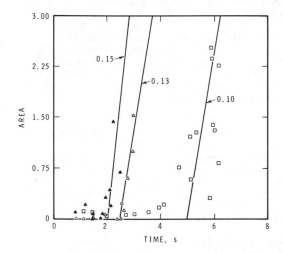

Figure 6. Oxidation of polyethylene before ignition at 20 mm separation, relative area of carbonyl band: ▫ 0.1 equivalence ratio, △ 0.13 equivalence ratio, ▲ 0.15 equivalence ratio

Poly(methyl methacrylate)

The analysis of this polymer was complicated by the presence of strong carbonyl absorptions in the infrared spectrum of the unexposed polymer. The

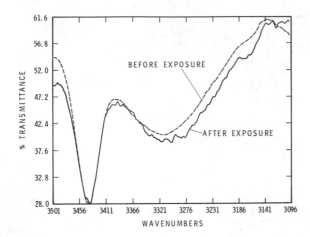

Figure 7. Poly(methyl methacrylate) just before ignition (9 s exposure) and before exposure

appearance of spectra of films before and after flame exposure was identical and this similarity persisted after implementation of the spectral deconvolution technique described above.

The spectrum of poly(methyl methacrylate) above 3000 cm$^{-1}$ is less cluttered. However, hydroxyl absorption is generally broader and less easily distinguishable from the baseline than are carbonyl absorptions. Nonetheless, unlike polyethylene and polystyrene, flame exposed poly(methyl methacrylate) showed no evidence of hydroxyl bands (Figure 7). Poly(methyl methacrylate) is not oxidized in the condensed phase under these conditions.

DISCUSSION

Hirschler recently queried the value of the study of chemical changes in polymers in fire conditions, arguing that much had already been determined about polymer pyrolysis (12). In reply, Kashiwagi and Ohlemiller pointed out that slow pyrolysis is not necessarily a good process to study in order to ascertain performance in fire conditions, since pyrolysis rates in fire conditions are at least two orders of magnitude faster.

The present study provides support to the latter view; the phenomena observed would have been difficult to predict from published reports on the (slow) pyrolysis of the polymers burned. Varying the conditions of fire exposure changed the mode of pyrolysis observed. Thus, in the radiant exposure conditions used by Kashiwagi and Ohlemiller (12), poly(methyl methacrylate) was responsive to oxygen concentration, while under the convective heat transfer conditions used in the present work, condensed phase oxidation was not observed.

Polystyrene and polyethylene show similarities and differences in their pre-ignition behaviour. Both exhibit rather constant rates of oxidation, over the range of conditions used. However, for polyethylene the oxidation process is initiated after a time which depends on equivalence ratio and flame proximity; these factors are less noticeable for polystyrene.

The two polymers undergo fundamentally different pyrolysis reactions; polystyrene decomposes by random- and end-chain scission into a mixture of monomer and oligomeric fragments, while polyethylene, in a random scission process (1), produces little monomer. The behaviour observed in this study can be accounted for by assuming that the surface of polyethylene is more sensitive to oxygen than that of polystyrene, because of a greater concentration of unsaturated sites.

One finding of possible industrial importance is that oxidation in these conditions was confined to very near the exposed surface. To the extent that pre-ignition oxidation should be restricted, the provision of fire retardant chemicals throughout a polymer may be an unnecessary and ineffective mode of protection. The results suggest surface treatment with fire retardants may be all that is required. If reactive fire retardants are required, surface grafting may be the method of choice.

However, the finding that oxidation occurs before ignition does not necessarily mean that control of this oxidation will result in increased resistance to ignition. Such an extrapolation must await the results of further study.

The authors acknowledge the assistance of Mr. Raymond Flaviani in conducting some of the experiments reported here. This article is a contribution from the Division of Building Research, National Research Council of Canada.

REFERENCES

1.  Cullis, C.F. and Hirschler, M.M.: The Combustion of Organic Polymers, Clarendon Press, Oxford, 1981.

2.  Stuetz, D.E., DiEdwardo, A.H., Zitomer, F. and Barnes, B.P.: "Polymer Combustion," J. Polymer Science: Polymer Chemistry Edition, 13: 585-621, 1975.

3.  Stuetz, D.E., DiEdwardo, A.H., Zitomer, F. and Barnes, B.P.: "Polymer Flammability I," J. Polymer Science: Polymer Chemistry Edition, 18: 967-985, 1980.

4.  Stuetz, D.E., DiEdwardo, A.H., Zitomer, F. and Barnes, B.P.: "Polymer Flammability II," J. Polymer Science: Polymer Chemistry Edition, 18: 987-1009, 1980.

5.  Jakes, K.A. and Drews, M.J., "The Role of Oxygen in the Flame Spread of Polypropylene," Preprints of Papers Presented by the Division of Organic Coatings and Plastics Chemistry at the American Chemical Society 180th National Meeting, San Francisco, California, August 1980, 286-290, 1980.

6.  Clark, F.R.S.: "Ignition of Poly(methyl methacrylate) Slabs using a Small Flame," J. Polymer Science: Polymer Chemistry Edition, 21: 2323-2334, 1983.

7. Clark, F.R.S.: "Ignition of Low-Density Polyethylene Slabs by a Small Flame," J. Polymer Science: Polymer Chemistry Edition, 21: 3225-3232, 1983.

8. Clark, F.R.S.: "The Role of Oxygen in the Ignition of Polystyrene by a Small Flame," J. Polymer Science: Polymer Chemistry Edition, 22: 263-268, 1984.

9. Botha, J.P. and Spalding, D.R.: "The Laminar Flame Speed of Propane/Air Mixtures with Heat Extraction from the Flame," Proc. Roy. Soc. Lond., Ser. A., A225 71-96, 1954.

10. Hunter, L.W. and Hoshall, C.H.: "An Ignition Test for Plastics," Fire and Materials, 4: 4, 201-202, 1980.

11. Kauppinen, J.K., Moffatt, D.J., Mantsch, H.H. and Cameron, D.G.: "Fourier Self-Deconvolution: A Method for Resolving Intrinsically Overlapped Bands," Applied Spectroscopy, 35: 3, 271-276, 1981.

12. Hirschler, M.M.: Comments to Kashiwagi, T. and Ohlemiller, T.J., "A Study of Oxygen Effects on Nonflaming Transient Gasification of PMMA and PE during Thermal Irradiation," Nineteenth Symposium (International) on Combustion, The Combustion Institute, Pittsburgh, p. 815-823, 1982.

# Heat of Gasification for Pyrolysis of Charring Materials

MERWIN SIBULKIN
Division of Engineering
Brown University
Providence, Rhode Island 02912, USA

ABSTRACT

The relationships between the heats of reaction, pyrolysis and gasification are examined for vaporizing and charring materials. During burning of thermally thick, vaporizing materials the heat of gasification $h_g \equiv \dot{q}''_{net}/\dot{m}''_{G,w}$ is equal to the heat of pyrolysis $h_p$. However, no simple relationship exists between $h_g$ and $h_p$ during the burning of charring materials. Solutions for pyrolysis of cellulose are obtained using a onedimensional, numerical model with material properties estimated from the literature. For this material it is found that after an initial transient $h_g$ is about twice $h_p$. For a semi-infinite slab the value of $h_g$ then remains nearly constant; for a finite slab the value of $h_g$ may decrease with time to values less than $h_p$. Changes in the assumed chemical reaction rate are found to have a minor effect on $h_g$. The heat of gasification is also relatively insensitive to changes in the char density and thermal conductivity. The variation of $h_g$ with time is nearly independent of the assumed heat of reaction, but the magnitude of $h_g$ varies almost directly with the value of the heat of reaction. Changes in the incident heat flux give corresponding changes in the gasification rate, but much smaller changes in the heat of gasification.

## INTRODUCTION

One of the modern developments in fire science has been the use of mathematical models. Such models are used to gain a more fundamental knowledge of fire phenomena, to interpret standard fire tests, and (in the future) to design buildings with a prescribed degree of fire safety. Diffusion flame models, e.g. Pagni (1980), Sibulkin et al. (1982), have been developed to predict burning rates and extinction limits. These models have also identified the material properties which most affect burning. Two of these parameters are the heat of combustion and the heat of gasification, and their ratio (as used in the B-number for example) is a key factor in burning rate calculations.

Most diffusion flame analyses to date have modeled simple fuels, such as polymethyl methacrylate (PMMA). For these fuels there is no accumulation of char at the fuel surface. Consequently, it can be shown that after an initial transient the surface temperature, heat of gasification and burning rate are independent of time (as long as the fuel remains thermally thick). On the other hand, materials which char (such as wood) show a dependence on time related to the growth of the char layer. We have begun to study the burning of a particularly simple charring material, i.e., cellulose (Sibulkin and Tewari, 1985) and are currently measuring its heat of gasification. To guide us in this study, we have evaluated a onedimensional model for pyrolysis of cellulose, and the results are presented in this paper. We first examine the relationships between the heats of reaction, pyrolysis and gasification. Numerical results are then presented for these parameters for cases of constant <u>incident</u> heat flux. The heat flux into the sample is not constant, however, because of radiative heat loss from the surface.[1]

The considerable effort which has been devoted to the kinetics of cellulose pyrolysis (e.g., Lewellen et al., 1977) is not discussed in this paper. A nominal reaction rate is taken from the literature, and the effect of variations in the assumed reaction rate are presented. A similar procedure is used for the still controversial heat of reaction for cellulose pyrolysis. The analysis assumes that the material has been dried (as was done in the burning experiments referenced above). It is also implicitly assumed that no oxygen is present at the surface to react with the char. At present this appears to be the best assumption for surfaces "shielded" by a diffusion flame. Possible surface reactions with carbon dioxide and water vapor are also excluded from consideration; these may become significant at higher temperatures.

## HEAT OF GASIFICATION ANALYSIS

The standard definition for the heat of gasification $h_g$ as used in diffusion flame theory is

$$h_g \equiv \dot{q}''_{net}/\dot{m}''_{G,w}. \tag{1}$$

Its relationship to the heat of pyrolysis is discussed below.

### Vaporizing Materials

We first consider the case of a thermally thick vaporizing fuel as shown in Fig. 1(a). For this case steady burning can occur at a mass loss rate $\dot{m}''_{G,w}$ and a surface regression rate $V_p = \dot{m}''_{G,w}/\rho_S$. By fixing the origin of the x coordinate in the moving surface and using a control volume analysis between $x = 0^-$ and $x = \infty$ (where $x = 0^-$ is in the gas phase just off the surface), one obtains the energy balance

$$\dot{q}''_{net} = \dot{m}''_{G,w}(h^*_{G,w} - h^*_{S,\infty}) \tag{2}$$

If one now <u>defines</u> the heat of pyrolysis $h_p$ as the difference in total enthalpy between the products of pyrolysis at the surface temperature $T_w$ and the virgin material at $T_\infty$ (per unit mass of volatile products), one has for a non-charring material

$$h_p \equiv h^*_{G,w} - h^*_{S,\infty}. \tag{3}$$

Combining Eqs. (1) and (2) shows that for the thermally thick case

$$h_g = h_p. \tag{4}$$

---

[1]Because of their low boiling points, surface radiation is normally unimportant for liquid fuels. It has been shown to be significant for fuels such as PMMA ($T_w \approx 650$ K), and is even more important for charring materials where $T_w \approx 900$ K.

For this reason the quanity defined by equation (1) has also been called "heat of pyrolysis" in the literature (which is a source of confusion).

An alternate definition of the heat of pyrolysis for vaporizing materials is obtained as follows. The relationship between the total and sensible enthalpies of species i at temperature T is given by

$$h^*_{i,T} \equiv h^*_{i,\infty} + h_{i,T} \equiv h^*_{i,\infty} + \int_{T_\infty}^{T} c_{p,i}\, dt$$

and the heat of reaction is $-h_{r,\infty} = h^*_{G,\infty} - h^*_{S,\infty}.$ Using these definitions, Eq. (3) becomes

$$h_p = (-h_{r,\infty}) + h_{G,w}. \tag{5}$$

(Here the convention has been used that $h_{r,\infty}$ is positive when exothermic while $h_p$ is positive when endothermic.) For a material such as PMMA, $h_p$ is slightly dependent on the burning rate through the dependence of $T_w$ on burning rate.

### Charring Materials

The comparable analysis for a thermally thick, charring material as shown in Fig. 1(b) is more complicated. The material is assumed to be dry and to maintain its orginal geometry. We begin by assuming a pyrolysis process of the form

Active Solid $(\sigma) \rightarrow$ Gaseous Volatiles (G) + Char $(\kappa)$. (6)

During pyrolysis the bulk density $\rho_\sigma$ of the active solid decreases from a value $\rho_S$ for the virgin solid to zero, while the char bulk density $\rho_\kappa$ increases from zero to its final value $\rho_C$. The density of gaseous products is neglected so that during pyrolysis

$$\rho = \rho_\sigma + \rho_k \tag{7}$$

and it is hypothesized that

$$\frac{d\rho_\kappa}{dt} = -\frac{\rho_C}{\rho_S}\frac{d\rho_\sigma}{dt} \tag{8}$$

Manipulation of Eqs. (7) and (8) gives the relations

$$\frac{d\rho_\sigma}{d\rho} = \frac{\rho_S}{\rho_S - \rho_C} \quad \text{and} \quad \frac{d\rho_\kappa}{d\rho} = -\frac{\rho_C}{\rho_S - \rho_C} \tag{9}$$

which are used subsequently.

Applying conservation of mass to a differential element of length dx and unit cross-sectional area gives

$$\frac{d\dot{m}''_G}{dx} = +\frac{\partial \rho}{\partial t} \tag{10}$$

393

(where $\dot{m}''_G$ is taken as positive in the negative x direction).

Applying conservation of energy to the same element gives

$$\frac{\partial}{\partial t}(\rho_\sigma h_\sigma + \rho_\kappa h_\kappa) = \left[-\frac{\partial \rho}{\partial t}\right]h_{r,\infty} + \frac{\partial}{\partial x}\left[k\frac{\partial T}{\partial x}\right] + \frac{\partial}{\partial x}(\dot{m}''_G h_G). \tag{11}$$

Intergration of Eq. (11) from x = 0 to ∞ gives

$$\int_0^\infty \frac{\partial}{\partial t}(\rho_\sigma h_\sigma + \rho_\kappa h_\kappa)\ dx = \dot{m}''_{G,w}h_{r,\infty} + \dot{q}''_w - \dot{m}''_{G,w}h_{G,w} \tag{12}$$

where $\dot{q}''_w \equiv \dot{q}''_{net}$. Expanding the integrand on the LHS of Eq. (12) and using Eq. (9) gives

$$\frac{\partial}{\partial t}(\rho_\sigma h_\sigma + \rho_\kappa h_\kappa) = (\rho_\sigma c_\sigma + \rho_\kappa c_\kappa)\frac{\partial T}{\partial t} + \left[\left[\frac{\rho_S}{\rho_S - \rho_C}\right]h_\sigma - \left[\frac{\rho_C}{\rho_S - \rho_C}\right]h_\kappa\right]\frac{\partial \rho}{\partial t}. \tag{13}$$

Substitution of Eq. (13) into Eq. (12) and application of Eq. (1) gives as the general result for the heat of gasification

$$h_g = (-h_{r,\infty}) + h_{G,w} + \frac{1}{\dot{m}''_{G,w}}\int_0^\infty \left[-\frac{\partial \rho}{\partial t}\right]\left[\left[\frac{\rho_C}{\rho_S - \rho_C}\right]h_\kappa - \left[\frac{\rho_S}{\rho_S - \rho_C}\right]h_\sigma\right]dx$$

$$+ \frac{1}{\dot{m}''_{G,w}}\int_0^\infty (\rho_\sigma c_\sigma + \rho_\kappa c_\kappa)\frac{\partial T}{\partial t}\ dx\ . \tag{14}$$

In Eq. (14) the first integral gives the change in sensible enthalpy between active solid and char while the second integral is the unsteady, energy storage term.

Again using the definition that $h_p$ is the difference in total enthalpy between the products of pyrolysis at $T_w$ and the virgin material at $T_\infty$ (per unit mass of volatiles), one obtains for a charring material

$$h_p = (-h_{r,\infty}) + h_{G,w} + \left[\frac{\rho_C}{\rho_S - \rho_C}\right]h_{C,w} \tag{15}$$

(which reduces to Eq. (5) for $\rho_C = 0$).

A comparison of Eqs. (14) and (15) shows that there is no simple relationship between the heat of gasification and the heat of pyrolysis for charring materials. Calculated values for $h_g$ and $h_p$ are presented in the next section.

394

## NUMERICAL RESULTS FOR CELLULOSE

Numerical solutions have been obtained for the propagation of a onedimensional pyrolysis wave into a charring solid. The method was developed by Kung (1972), and a modification of his computer program (Tamanini, 1976) was used. The finite-difference program solves Eqs. (10) and (11) for a pyrolysis reaction given by

$$- \frac{\partial \rho}{\partial t} = \rho_\sigma \ A \ \exp(-E/RT). \tag{20}$$

The material is assumed to be dry and to maintain its original geometry. Results are presented for a symmetrically heated slab having a half-width L which is subjected to a constant, external incident flux $\dot{q}''_{ex}$. Values of L = 1 cm and $\dot{q}''_{ex}$ = 50 kW/m$^2$, chosen to simulate our previous burning measurements (Sibulkin and Tewari, 1985), are used except as noted. The net heat flux is determined as part of the calculation from

$$\dot{q}''_{net} = \dot{q}''_{ex} - \epsilon\sigma(T_w^4 - T_\infty^4) . \tag{21}$$

The set of base values chosen for the properties of cellulose used in the calculations are given in Table 1; they are taken as best estimates from the literature. In particular, the kinetic parameters for cellulose are average values from Lewellen et al. (1977). The properties of cellulose char and both the rate and heat of reaction of cellulose pyrolysis are not well known; they are inter-related and depend on the external heating rate and the scale of the sample. Results are presented for the sensitivity of the heat of gasification to changes in the values of these properties.

The calculated values of surface temperature and net heat flux are shown in Fig. 2. The calculated values for $T_w$ in the vicinity of 900 K agree with measured surface temperatures made during flaming combustion. Of particular significance is the drop in $\dot{q}''_{net}$ to about 20 percent of the incident heat flux. The calculations were terminated when 1 percent of the material at the centerplane was pyrolyzed.

Profiles of density and temperature within the material are shown in Fig. 3. Pyrolysis occurs in a fairly narrow zone which increases from 1 mm to 2 mm as the wave moves inward. The temperature range for which 90 percent of pyrolysis takes place is shown by the solid segments of the temperature curves; the range of values is about 600 - 750 K. Note that the temperature at the centerplane has already begun to rise at t = 40 s, so that after this time the slab is not thermally thick.

Typical curves showing the time variation of pyrolysis rate $V_p \equiv \dot{m}''_{G,w}/(\rho_S - \rho_C)$ and heat of gasification $h_g$ given by the calculations are presented in Fig. 4. We have chosen to plot the pyrolysis rate in terms of $V_p$ since one can interpret the numerical values more easily, e.g., $V_p \times 10^5$(m/s) = 1 is a velocity of 1 mm in 100 s. Gasification begins after about 5 s, reaches a maximum at about 20 s and then begins to decrease. During this period the variation of $h_g$ is approximately the inverse of $V_p$ (see Eq. 1). Using a simplified, analytical model for pyrolysis, Delichatsios and deRis (1984) find that in the "final" period $V_p \propto t^{-1/2}$, and these numerical calculations give a similar result when the centerplane temperature remains at $T_\infty$ i.e. for L = ∞. For the case of L = 1 cm, the pyrolysis rate departs from the L = ∞ curve at about 70 s and then begins a rise which is a consequence of the increasing centerplane temperature. After the initial transient, $h_g$ is nearly constant for the L = ∞ case which would simplify the problem of modeling the complete (gas plus solid phase) diffusion flame. However, once the thermal wave reaches the centerplane of the material, the value of $h_g$ decreases as shown by the L = 1 cm case.

A comparison of the heats of reaction, pyrolysis and gasification is shown in Fig. 5 for the base case. The value of $(-h_{r,\infty})$ is constant at an assumed value of $0.5 \times 10^{+6}$ J/kg. The value of $h_p$ calculated using Eq. (15) is 3 to 4 times greater than $h_{r,\infty}$; it increases with time as $T_w$ increases. At $t \approx 50$ s, the value of $h_g$ is about twice that for $h_p$ but becomes less than $h_p$ after about 150 s. This difference in behavior of $h_p$ and $h_g$ emphasizes the importance of the unsteady aspects of pyrolysis of charring materials.

Considerable effort has gone into measuring the overall reaction rate for cellulose pyrolysis. A comparison of the kinetic parameters obtained by several investigators is given in Lewellen et al. (1977); over the temperature range 600 - 750 K where most of our pyrolysis occurs the variation in the reported reaction rates is less than a factor of 10. The sensitivity of the calculated values of $h_g$ to such a change in the pre-exponential factor A is shown in Fig. 6. It is found that factor of 10 changes in reaction rate have only a modest effect on $h_g$. Thus use of the current literature values should be satisfactory.

The density of cellulose char is known to depend on the rate of heating and on impurities in the material, which can be either natural minerals or artificial fire retardants. Values of the thermal conductivity of char are also uncertain but should tend to decrease as the char density decreases. The sensitivity of $h_g$ to a decrease in the product $\rho_C k_C$ by a factor of 4 is shown in Fig. 7. Fortunately, $h_g$ is relatively insensitive to changes in this parameter.

Widely varying values for the magnitude, and even the sign, of the heat of reaction for cellulose pyrolysis have been reported (Kanury and Blackshear, 1970). Some of these variations depend on whether the results are for an inert or oxidizing atmosphere, whether the samples were dried or at a specified relative humidity, and whether or not the results have been corrected to a standard temperature (Kung and Kalelkar, 1973). On Fig. 8 the effect on $h_g$ of varying $h_{r,\infty}$ from 0 to 2 times the base value is presented. The variation of $h_g$ with time is nearly independent of $h_{r,\infty}$, but the magnitudes of $h_g$ differ by about 3/4 of the change in $h_{r,\infty}$. Further experimental results are needed to resolve this uncertainty. (Direct measurements of $h_g$ for cellulose are in progress at our laboratory.)

The effects of changing the incident heat flux by up to a factor of 2 are examined in Figs. 9 and 10. Both the initial pryolysis delay and its peak value are seen in Fig. 9 to be strongly dependent on $q''_{ex}$. This is followed by a decreased sensitivity to $q''_{ex}$ after $t \approx 100$ s. The results for $h_g$ given in Fig. 10 show a similar dependence of the duration of the initial transient on $q''_{ex}$. However, after this delay the values of $h_g$ are remarkably independent of $q''_{ex}$. This suggests that in modeling the burning of charring materials it may still be possible to decouple the pyrolysis analysis from the gas phase analysis.

CONCLUSIONS

A theoretical analysis for onedimensional pyrolysis of thermally thick materials shows that for charring materials the heat of gasification $h_g$ is no longer simply related to the heat of pyrolysis $h_p$.

Numerical solutions of the pyrolysis problem using material properties for cellulose were made for a steady incident heat flux. It is found that:

(i) The net flux into the solid decreases to 20 percent of its initial value because of surface heat loss.

(ii) Pyrolysis occurs primarily in the temperature range $T = 600 - 750$ K; changes in reaction rate by a factor of 10 cause much smaller changes in $T_w$, $h_p$ and $h_g$.

(iii) Changes in the value assumed for the heat of reaction $h_{r,\infty}$ cause significant changes in $h_g$; additional research is needed in this area.

(iv) Although the rate of pyrolysis increases with increasing incident flux level $\dot{q}''_{ex}$, the values of $h_g$ after an initial transient are nearly independent of $\dot{q}''_{ex}$.

(v) After an initial transient, the value of $h_g$ for a semi-infinite slab remains nearly constant at about twice the value of $h_p$. Because the 1 cm half-thickness slab does not remain thermally thick, its value of $h_g$ decreases with time to values below $h_p$. (A similar decrease in $h_g$ occurs for vaporizing fuels which do not remain thermally thick.)

We believe these results show that the heat of gasification is a useful concept for modeling fires on charring fuels. However, unlike the heat of pyrolysis, its value cannot be determined from thermodynamic properties alone, but must either be calculated as in this paper or found from experiment. Further studies of both types are recommended.


## NOMENCLATURE

| | | | | |
|---|---|---|---|---|
| A | pre-exponential factor | | T | temperature |
| c | specific heat | | $V_p$ | velocity of pyrolysis wave |
| E | activation energy | | x | distance from surface |
| h | sensible enthalpy | | $\epsilon$ | emissivity |
| h* | total enthalpy | | $\rho$ | (bulk) density |
| $h_g$ | heat of gasification | | | |
| $h_p$ | heat of pyrolysis | | Subscripts | |
| $h_r$ | heat of reaction | | | |
| k | thermal conductivity | | C | final char value |
| L | slab half-width | | G | gas (volatile products of pyrolysis) |
| $\dot{m}''$ | mass flux per unit area | | S | initial solid value |
| $\dot{q}''_{ex}$ | external incident heat flux per unit area | | w | at surface |
| | | | $\kappa$ | char value during pyrolysis |
| $\dot{q}''_{net}$ | net heat flux per unit area | | $\sigma$ | active solid value during pyrolysis |
| | | | $\infty$ | ambient |
| R | gas constant | | | |
| t | time | | | |

## ACKNOWLEDGEMENT

It is a pleasure to acknowledge the assistance of Kenneth Siskind in carrying out the computations. This work was supported by the Center for Fire Research of the National Bureau of Standards under Grant NB83NADA4017.

## REFERENCES

1. Delichatsios, M. A. and deRis, J. (1984). An analytical model for the pyrolysis of charring materials. Factory Mutual Research, Technical Report, J.I.OKOJ1.BU.

2. Kanury, A. M. and Blackshear, P. L., Jr. (1970). Some considerations pertaining to the problem of wood burning. *Combustion Science and Technology* 1, 339.

3. Kung, H-C. (1972). A mathematical model of wood pyrolysis. *Combustion and Flame* <u>18</u>, 185.

4. Kung, H-C. and Kalelkar, A. S. (1973). On the heat of reaction in wood pyrolysis. *Combustion and Flame* <u>20</u>, 91.

5. Lewellen, P. C., Peters, W. A. and Howard, J. B. (1977). Cellulose pyrolysis kinetics and char formation mechanism. *Sixteenth Symposium (International) on Combustion*, The Combustion Institute, Pittsburgh, pp. 1471-1480.

6. Pagni, P. J. (1980). Diffusion Flame Analyses. *Fire Safety Journal* <u>3</u>, 273.

7. Sibulkin, M., Kulkarni, A. K. and Annamalai, K. (1982). Burning on a vertical fuel surface with finite chemical reaction rate. *Combustion and Flame* <u>44</u>, 187.

8. Sibulkin, M. and Tewari, S. S. (1985). Measurements of flaming combustion of pure and fire retarded cellulose. *Combustion and Flame* <u>59</u>. 31.

9. Tamanini, F. (1976). A numerical model for one-dimensional heat conduction with pyrolysis in a slab of finite thickness. (in) Factory Mutual Research, Serial No. 21011.7, Appendix A.

Table 1.   Base Values of Properties

$$
\begin{aligned}
c_S &= 1500 \text{ J/kgK} \\
c_C &= 1000 \text{ J/kgK} \\
c_{p,G} &= 1700 \text{ J/kgK} \\
k_S &= 0.2 \text{ W/mK} \\
k_C &= 0.08 \text{ W/mK} \\
\rho_S &= 500 \text{ kg/m}^3 \\
\rho_C &= 150 \text{ kg/m}^3 \\
E &= 1.5 \times 10^8 \text{ J/kg-mole} \\
A &= 1 \times 10^{10} \text{ s}^{-1} \\
h_{r,\infty} &= -5 \times 10^5 \text{ J/kg} \\
\epsilon &= 1.0
\end{aligned}
$$

$L = 1$ cm and $\dot{q}''_{ex} = 50$ kW/m$^2$ (except where noted).

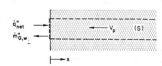

(a) Vaporizing fuel (e.g. PMMA)

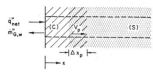

(b) Charring fuel (e.g., cellulose)

Fig. 1.   Control volumes used in analysis of pyrolyzing fuels.

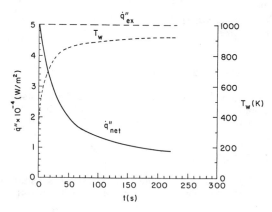

Fig. 2.   Surface temperature response and net heat flux. $L = 1$ cm.

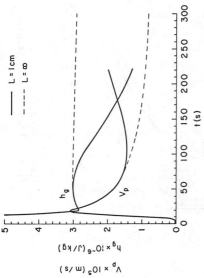

Fig. 3. Internal density and temperature profiles. L = 1 cm.

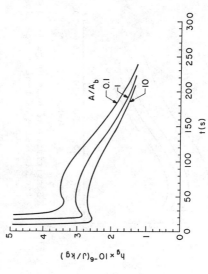

Fig. 4. Pyrolysis wave velocity and heat of gasification for base case (L=1cm) and for semi-infinite slab (L=∞).

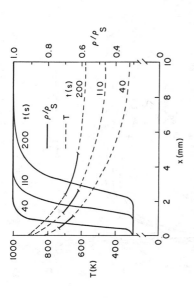

Fig. 5. Comparison of heat of reaction, pyrolysis and gasification. L = 1 cm.

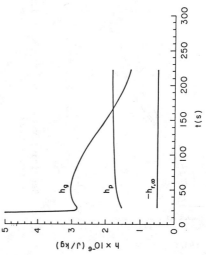

Fig. 6. Sensitivity of heat of gasification to rate of chemical reaction. L=1cm.

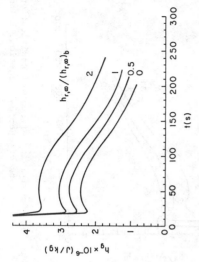

Fig. 7. Sensitivity of heat of gasification to properties of char. L = 1 cm.

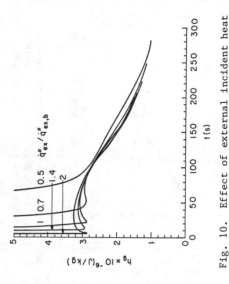

Fig. 8. Sensitivity of heat of gasification to value of heat of reaction. L = 1 cm.

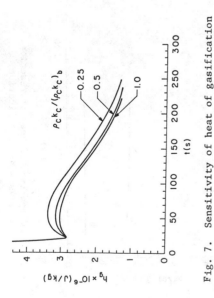

Fig. 9. Effect of external incident heat flux on pyrolysis wave velocity. L = 1 cm.

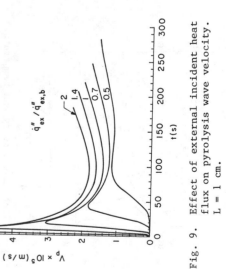

Fig. 10. Effect of external incident heat flux on heat of gasification. L = 1 cm.

# A Combustibility Study of Gaseous Pyrolysates Produced by Polyester/Cotton Blends

M. DAY, T. SUPRUNCHUK, and D. M. WILES
Textile Chemistry Section, Division of Chemistry
National Research Council of Canada
Ottawa, Ontario, Canada K1A OR6

ABSTRACT

The thermal decomposition and flammability limits of the gaseous pyroly-sates obtained from a series of polyester/cotton blends have been studied. Comparison of data obtained from intimately blended samples with control samples indicates that both physical and chemical interaction are taking place between the components during the condensed phase decomposition. Based upon the measured flammability limits and pressure rises on igni-tion, the importance of chemical interactions were demonstrated; however physical effects associated with heat transfer cannot be ignored. Greatest changes were observed with the 67/33 polyester/cotton blends, suggesting that this may be the most difficult to flame retard.

## INTRODUCTION

Clothing and textile related fires are responsible for the majority of repor-ted fire deaths in Canada and the United States. It is in an attempt to reduce the number of these fatalities that research into safer materials has been actively pursued over the last decade or so. Because polyester/cotton fibre blends combine many of the advantages associated with the individual fibres such as comfort, moisture absorption and durable press, they find widespread application in apparel and home furnishings. Unfortunately, whilst the polyester/cotton blends may have many physical and aesthetic attributes, the blends do cause problems from the flammability point of view in that it is not possible to predict the flammability behaviour of the blends from a knowledge of the burning behaviour of the individual components. One of the reasons for this lack of predictability is associated with large differences in the combustion behaviour of the components. For example, untreated cotton fabrics are relatively easy to ignite and burn, leaving a char. Polyester fabrics on the other hand are difficult to ignite because the thermoplastic material tends to shrink away from the ignition flame, and consequently, even when ignited, burn erratically with molten drips being capable of carrying the flame away from the fabric. When the two fibre types are blended together, however, the cotton is capable of forming a carbonaceous grid or "scaffold" which acts as a support for the melting

KEYWORDS
Polyester/Cotton, Flammability, Blends, Cotton/Polyester, Pyrolysates.

Issued As NRCC #24535

polyester enabling it to burn in much the same way as does the parafin wax at a candle's wick. In terms of the flammability behaviour of the polyester/cotton blends as measured by various test methods, although it has has been shown that blends may have oxygen-indices lower than those of the individual fibres[1], the results are dependant upon moisture content[2] and environmental test temperatures[3]. In terms of burn injury potential, however, several studies[4-6] have revealed that polyester/cotton blends represent less of a hazard than 100% cotton fabrics and indicate that increased polyester content causes reductions in the burn injury potential. These complexities arise because of the interactions between the two dissimilar components in the blend. Hendrix[2],[3] concluded that in the burning of polyester/cotton blends, cotton was the initial fuel which the polyester augmented as the burning proceeded causing a more vigorous burning of the blend. More recent work[7] however, indicates that the polyester controls the decomposition by melting and covering the cotton, thereby preventing its decomposition.

Obviously, some form of interactive behaviour is taking place during the burning of polyester/cotton blends. Miller[8] indicated that these interactions were chemical in nature and were occuring during the pyrolysis and combustion processes to produce more volatile fuels which were responsible for lower ignition times and faster burn rates than anticipated from the individual components. Meanwhile, another study[9] examining ignition characteristics has indicated that the interactions are more physical in nature than chemical and the observed differences in behaviour of blended fabrics may be attributable to heat transfer effects.

In an attempt to better understand the fundamental processes taking place in these polyester/cotton blends, we decided to investigate the decomposition and flammability limits of the gaseous pyrolysates produced using a technique developed in our laboratories[10]. In addition to measuring the decomposition and flammability limits of intimate polyester/ cotton blends, results have also been obtained when the solid components were separated.

EXPERIMENTAL

Materials

The cotton and polyester fabrics used in this study were obtained from Test Fabrics, Inc. The cotton was 100% bleached sheeting style 405 (176 $g/m^2$) while the polyester was a spun woven fabric style 767 (127 $g/m^2$). The samples were prepared by passing strips of fabric through a Wiley Mill (Model 4276) at 3,500 rpm to produce a fine fluff. The intimate blends were prepared by mixing the polyester and cotton fine fluffs together at predetermined weight ratios, blending the mixture on a ball mill followed by a pass through the Wiley Mill. Three intimate blends were produced containing 67%, 50% and 33% respectively of cotton.

Apparatus and Test Procedure

The experimental determinations were carried out according to the procedure used previously[10] employing the apparatus (see Fig. 1) described in detail in an earlier paper[11]. In the blended experiments, 200 mg of the combined fluff is weighed out accurately into one half of the divided crucible (e in Fig. 1) while in the separated experiments, the polyester is weighed out accurately into one half and cotton into the other half. The

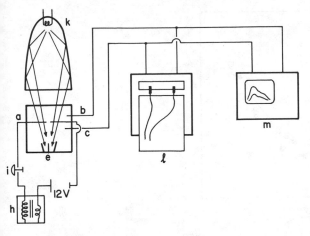

Figure 1 Schematic of the test equipment

crucible is then placed into the insulated brass combustion vessel which is 75 mm long and 50 mm in diameter. The sample is then heated in the sealed vessel by means of an external infrared heater (k) focused on the sample through a transparent mica window. After a predetermined heating time, the gaseous pyrolysate produced is subjected to a spark discharge from the electrodes (a) generated by a power supply and ignition coil (h). The strip chart recorder (l) and storage oscilloscope (m) monitor the pressure and temperature during pyrolysis and combustion. All studies employed a lamp voltage of 80 V to give an incident radiative pyrolysis heat flux of about 13 W/cm$^2$. It should be pointed out that each point represents one determination with each experiment being performed on a new sample. The weight of pyrolysate produced was determined from the loss in weight of the crucible contents after each experiment. In the case of the experiments performed on the separated samples, individual weight losses were determined by cleaning and dissolving the residue in each half separately.

RESULTS AND DISCUSSION

Gases or vapours which form flammable mixtures with air or oxygen do so within specific concentration ranges defined as the flammability limits. These limits known as the "lower and upper flammability limits" are a measure of boundary limits in which a mixture of the gas or vapour with air will, if ignited, just support flame propagation. Although widely used in hazard evaluation, these values may well be dependant upon: pressure, temperature, ignition source, equipment geometry etc. In our studies, we have taken the flammability limit concept and applied it to the gaseous pyrolysates produced from decomposing polymeric materials. However, in this latter application, the flammability limits determined are also dependant upon the thermal stability of the polymeric material and the heating rate employed to generate the pyrolysate. These factors become exceedingly important in the case of polymers whose decomposition products are dependant upon the thermal conditions employed, as is the case with polyester[12] and cellulose[13]. For these reasons, only one set of experimental and reproducible conditions have been employed throughout this study in order that useful comparisons and interpretations can be made.

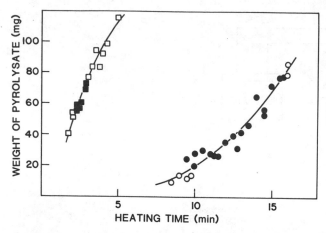

Figure 2 Weight of pyrolysate produced from cotton (□) and polyester (○) with combustible mixtures denoted ■ and ● .

The marked difference in the thermal stability of each component is immediately evident from Figure 2 in which the weight of pyrolysate produced (weight loss of sample) is plotted as a function of heating time. This figure clearly indicates that cotton undergoes a relative rapid weight loss to produce a gaseous pyrolysate which reaches its lower flammability limit in approximately 2 minutes. However, further heating produces more pyrolysate such that the gas-air mixture reaches the upper flammability limit after about 3 minutes and then enters the fuel rich zone. In contrast, the polyester has a much slower rate of weight loss than the cotton, such that heating for a period of 9.5 minutes is required before a sufficient concentration of pyrolysate in air is attained to obtain a combustible gas-air mixture. However, in the case of the polyester, the range of gaseous pyrolysate concentrations in air capable of supporting combustion is far greater than that observed with the cotton, and it is only after heating for a period of 15.5 minutes that the upper flammability limit is reached. Thus, based upon the observed behaviour of the individual components, it would be anticipated in terms of a polyester/cotton blend, the pyrolyste from the cotton will produce a combustible gas-air mixture, which will become too rich in cotton pyrolysate products before an appreciable pyrolysate has been generated from the polyester.

The actual behaviour of the blended polyester/cotton samples are presented in Figures 3-5 for the 33/67, 50/50 and 67/33 P/C systems respectively. These graphs provide not only the data for the intimately blended samples, but also results obtained in which the corresponding weights of the individual components are placed in each side of the divided crucible. In this way it was possible to compare the data obtained from the two systems under identical experimental conditions and elucidate any possible interactions taking place as a result of the intimate blending of the two components. Examination of the data presented in Figures 3-5 clearly indicates that changes are occuring because of the mixing of the two components together. When the components are blended together, the amount of gaseous pyrolysate produced is always greater than when the compo-

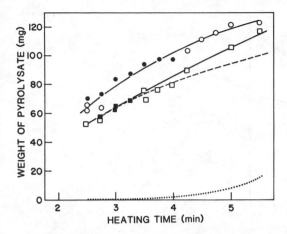

Figure 3 Total weight of pyrolysate produced from the 33/67 polyester/ cotton system for blended samples ( O ) and separated samples ( □ ) with combustible mixtures denotes ● and ■ . The pyrolysate contribution of cotton and polyester to the separated systems are shown ‑‑‑ and ... respectively.

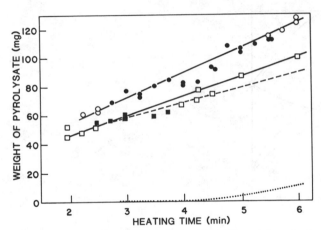

Figure 4 Total weight of pyrolysate produced from the 50/50 polyester/ cotton system for blended samples ( O ) and separated samples ( □ ) with combustible mixtures denotes ● and ■ . The pyrolysate contribution of cotton and polyester to the separated systems are shown ‑‑‑ and ... respectively.

nents are separated, with the difference appearing to increase as the amount of polyester in the system increases. This observation regarding decomposition and the production of pyrolysate is consistent with data reported in the literature[9] which was explained on the basis of physical interactions based upon a heat transfer mechanism. However, Figures 3-5 also indicate that in addition to changes in the amount of pyrolysates produced, blending of the two components also results in changes in the

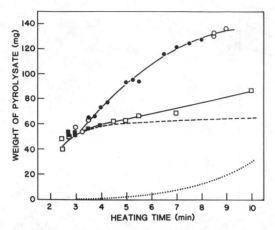

Figure 5 Total weight of pyrolysate produced from the 67/33 polyester/
cotton system for blended samples ( O ) and separated samples ( □ ) with
combustible mixtures denotes ● and ■ . The pyrolysate contribution of
cotton and polyester to the separated systems are shown --- and ...
respectively.

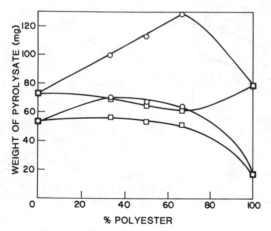

Figure 6 Flammability limits for the blended samples ( O ) and separated
samples ( □ ) as a function of polyester in the sample.

flammability limits different to those produced with the separated system.
In order to better appreciate these changes in the flammability limits as a
function of composition, the data have been replotted in Figure 6 as a
percentage of polyester in the systems.

    Figure 6 clearly indicates that the flammability limits for the separa-
ted system are distinctly different from those obtained with the blended
samples.    In the case of the separated samples, the flammability limits
correspond very closely to those of the pure cotton, as would be antici-

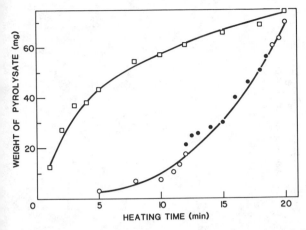

Figure 7 Total weight of pyrolysate produced from 100 mg sample of polyester ( O ) and a 200 mg of a 50/50 polyester/glass fibre blend ( □ ), with
combustible mixture denoted   •

pated based upon the decomposition behaviour noted in Figure 2.   Thus
the limits measured for the separated systems must be associated with the
pyrolysate generated from the cotton only, with no contribution from the
polyester.   Confirmation that the observed limits were due to the cotton
were made by performing experiments with cotton only, employing sample
weights of 100 mg and 67 mg when only slight changes in the limits were
obtained comparable to those measured for the separated systems.

The observation that the blended samples have flammability limits
vastly different from those obtained with the separated systems clearly
indicates that the components are not decomposing independantly.  In order
to further elucidate the nature of the interactions additional experiments
were performed with polyester blended with glass fibres to act as inert
non-combustible support.   The results obtained with 200 mg samples of a
50/50 polyester/glass fibre blend are given in Figure 7.   Also included in
this figure are the results obtained with 100 mg sample weights of polyester.   Immediately, the effect of the inert glass fibre on the decomposition
of the polyester becomes evident being responsible for a marked reduction
in heating time required to obtain gaseous pyrolysate (separate experiments with the glass fibres indicated no weight loss).   For example after
heating 100 mg of pure polyester alone for 7 minutes only 5 mg of pyrolysate has been produced, however, when supported by the glass fibre 50
mg has been produced in the same time period.   In the case of the pure
polyester, combustible pyrolysate gas/air mixtures are obtained with lower
and upper flammability limits consistent with expectation.   In the case of
the polyester/glass blend, however, no combustible gas/air mixtures were
obtained.   This result is rather suprising since although it would have
been anticipated that the rate of pyrolysate gas generating may increase
because of physical heat transfer effects of the inert support the change
in the composition of the gaseous pyrolysate such that no combustible gas
mixtures were obtained was unexpected.   However, in view of the large
differences in the flammability limits observed with the blended poly-
ester/cotton samples in comparison to the results with the separated
system significant changes in the chemical composition of the pyrolysate

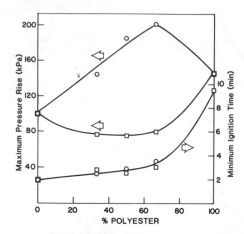

Figure 8 Minimum ignition time and maximum pressure rise on ignition for the blended samples (O) and separated samples (□) as a function of polyester in the sample.

gases must be occuring. Thus it appears that, in addition to physical interactions, chemical interactions must also be taking place during decomposition of these blends to produce a more combustible fuel than is generated in the separated system. This observation regarding the combustibility of the pyrolysate appears to confirm Miller's findings [8] which showed that the pyrolysates have a greater fuel value due to the production of larger amounts of ethylene and acetylene in polyester/cotton blends than would be predicted from the behaviour of the individual components.

The derivation of flammability limits for the gaseous pyrolysates along with the information concerning the thermal stability of the materials only represents part of the information obtainable from these combustibility studies. In terms of the decomposition rate, the time to obtain a combustible gas/air mixture is an important parameter in any flammability evaluation in that it reflects the dynamics of the burning process. In Figure 8 the heating time to obtain a combustible pyrolysate/air mixture has been plotted as a function of polyester content. Essentially the separated and blended systems appear to behave in a similar manner, without any obvious differences. If the interactions between the two components were purely physical in nature, and associated with a heat transfer process, it would be anticipated that shorter times would have been required for the blended systems. However, if chemical interaction in addition to the physical interaction were taking place, the different chemical composition of the gaseous pyrolysates could be sufficient to explain the observed results.

When combustible pyrolysate/air mixtures were obtained in this study, the pressure increase associated with the ignition of the pyrolysate are measured and can be plotted as a function of weight of pyrolysate. In all systems studied, these plots were similar to those obtained previously[10] and were characterised by Gausian shaped curves which increase sharply in regions close to the flammability limits to reach maximum values at some point close to the mid point between the limit values. For that reason, we have taken the liberty of reporting only the maximum pressure rises noted for each system studied. These results are presented in Figure 8.

Once again the behaviour of the polyester/cotton blended system are different to those obtained with the separated system. This observation is indicative that the chemical composition of the gaseous pyrolysate is different with the blended materials than found with the separated systems. The magnitude of the pressure rise on ignition of the pyrolysate/air mixtures may be taken as a useful indicator of the energy being released during the combustion process and is therefore capable of providing information on the potential heat feedback during the burning process. In the case of the separated systems, the energy released during ignition of the pyrolysate closely resembles that of the cotton alone, a fact to be expected based upon the data obtained which indicate that the cotton is the source of the pyrolysate. In the case of the polyester/cotton blends, the energy release as measured by the maximum pressure rise on ignition lie above the line connecting the values for the individual components with two of the blends having values greater than that for the polyester alone. These data while comparable to those obtained by Miller [8] appear to be in conflict with those reported by Yeh [14] who obtained a "minimum" heat release for a 45/55 polyester/cotton blend based upon calorimetric studies. It should be pointed out, however, that in our study, the ignition and combustion are occuring in a closed reaction vessel which by its nature limits the amount of air available for combustion. Although the pyrolysate/air mixtures are within the flammability limits in order to obtain ignition, the consumption of air during combustion undoubtedly will quickly result in oxygen starvation. In the previously reported studies [8, 14], the fabrics are being burnt in systems in which the oxygen starvation may not be a problem, although Miller's data do show some dependance upon air flow rate. However, irrespective of the cause of these discrepancies, the results are indicative of interactions between the two components.

CONCLUSIONS

The thermal decomposition of intimate blends of polyester/cotton is a complex process in which interactions take place in the condensed phase prior to the production of the gaseous pyrolysate which serves as the fuel for the combustion process. Because of the large changes in the flammability limits of the blends compared to the limits for the pure polyester and cotton, the chemical composition of the gaseous pyrolysate fuel must be significantly different. Based upon the results of this study, it appears that both physical and chemical interactions are occuring which can account for the changes in the chemical composition of the gaseous fuel being produced when these blends are heated during thermal decomposition leading to combustion. In view of this evidence of substantial chemical interactions in the blends, the difficulty in obtaining successful flame retardant systems for the polyester/cotton blends (based upon existing technology for the individual components) is understandable.

REFERENCES
1. Tesoro, G.C. and Meiser, C.H.: "Some Effects of Chemical Composition on the Flammability Behaviour of Textiles", Text. Res. J., 40 430-436, (1970).

2. Hendrix, J.E., Drake, G.L. and Reeves, W.A.: "Effects of Fabric Weight and Construction on Oxygen Index (OI) Values of Cotton Cellulose", J. Fire and Flammability, 3, 38-45 (1972).

3. Hendrix, J.E., Drake, G.L. and Reeves, W.A.: "Effects of Temperature on Oxygen Index Values", Text, Res. J., 41, 360 (1971).

4. Bercaw, J.R., Jordan, K.G. and Moss, A.Z.: "Estimating Injury from Burning Garments and Development of Concepts for Flammability Tests", Fire Standards and Safety, ASTM STP 614, Ed., Robertson, A.F.: American Society for Testing and Materials, p. 55-90 (1977).

5. Langstaff, W.I. and Trent, L.C.: "The Effect of Polyester Fibre Content on the Burn Injury Potential of Polyester/Cotton Blend Fabrics", J. Consum. Prod. Flam. 7 (1) 26-39 (1980).

6. Umbach, K.H.: "Comparative Studies of The Burning Behaviour of Textiles from Polyester/Cotton and Pure Cotton", Fire Mater., 5 (1) 24-32 (1981).

7. Barker, R.H. and Drews, M.J.: "Development of Flame Retardants for Polyester/Cotton Blends", NBS-GCR-ETIP 76-22, National Bureau of Standards, Washington, D.C. (Sept. 1976).

8. Miller, B., Martin, J.R., Meiser, C.H. and Gorgiallo, M.: "The Flammability of Polyester-Cotton Mixtures", Text. Res. J. 46 (7) 530-538 (1976).

9. Pintauro, E.M. and Buchanan, D.R.: "Ignition Process in Single and Multicomponent Polyester/Cotton Textile Structures", Text. Res. J. 49 (6) 326-334 (1979).

10. Day, M., Suprunchuk, T., and Wiles, D.M.:, "Flammability Limits of Some Polymer Pyrolysate-Air Mixtures", J. Appl. Polym. Sci., 28, 449-460 (1983).

11. Day, M., Suprunchuk, T. and Wiles, D.M.: "Combustion and Pyrolysis of Poly(ethylene terephthalate). I. The Role of Flame Retardants on Products of Pyrolysis", J. Appl. Polym. Sci., 26 3085-3098 (1981).

12. Day, M., Wiles, D.M.: "Influence of Temperature and Environment on the Thermal Decomposition of Poly(ethylene terepthalate) fibres with and without the Flame Retardant Tris (2,3-dibromopropyl) Phosphate", J. Anal. Appl. Pyrol. 7 65-82 (1984).

13. Shafizadeh, F, Farneaux, R.H., Cochran, T.G., Scholl, J.P. and Sakai, Y.: "Production of Levoglucosan and Glucose from Pyrolysis of Cellulosic Materials", J. Appl. Polym. Sci., 23 3525-3539 (1979).

14. Yeh, K., Valente, J.A.: "Calorimetric Study of Polyester/Cotton Blend Fabrics I. Combustion and Effects of Partial Retardants", J. Fire Ret. Chem. 6 92-124 (1979).

# TGA/APCI/MS/MS, A New Technique for the Study of Pyrolysis and Combustion Products

**YOSHIO TSUCHIYA**
Division of Building Research
National Research Council of Canada
Ottawa, Ontario, K1A OR6

ABSTRACT

The Fire Research Section, Division of Building Research, National Research Council of Canada has acquired a SCIEX TAGA 6000 APCI (atmospheric pressure chemical ionization)/MS/MS, an analytical instrument unique for its high sensitivity and high speed in analysis. The instrument is capable of monitoring simultaneously many types of gases generated in pyrolysis/combustion. Coupled with a Dupont 951 thermogravimetric analyser (TGA), it has been used for studying the pyrolysis products of polyacrylonitrile (PAN) at different stages of the pyrolysis process.

A 1 mg specimen was pyrolyzed in the TGA in a stream of nitrogen or air and the products were introduced to the APCI/MS/MS through a short glass capillary. The molecules of the product were ionized under atmospheric pressure, and analyzed in real time with three serial quadrupole mass filters. The main products were HCN, acetic acid, and a series of nitriles. The generation of each product is discussed in the light of the thermogravimetric analysis.

KEYWORDS

APCI/MS/MS, TGA/MS, polyacrylonitrile, combustion products, nitriles, HCN.

INTRODUCTION

A combination of thermogravimetric analysis (TGA) and mass spectrometry (MS) is a powerful tool for studying thermal degradation of materials. With a TGA, a specimen of the material can be thermally degraded at accurately controlled temperatures in a desired atmosphere. The effluent degradation products from the TGA can be analyzed by a highly sensitive MS. It is not surprising that many applications of TGA/MS, using various types of instrumentation, have been reported in the past 15 years, as shown in Table 1 (1-15), although the table is not intended to present a complete list.

The conventional MS, used in most of the earlier studies, had some limitations. The electron impact (EI) ionization extensively fragmented the molecules of the degradation products at the ion source, making the mass spectra complicated and restricting the use of the technique to the analysis of relatively simple degradation products. Chemical ionization (CI) has simplified the spectra, but the information on molecular weights supplied with

TABLE 1 TGA/MS Studies in Literature

| Author | Year | TGA | MS | EI/CI | mL/min | °C/min | Material |
|---|---|---|---|---|---|---|---|
| Zitomer | 1968 | Dupont 950 | Bendix T-O-F | EI | 100 | 15 | Polyethylsulfide |
| Chang | 1971 | Dupont 900/PE 881 GC | Dupont CEC 21-110B | EI | 60-80 | 10-15 | Ethyl vinylacetate |
| Gibson | 1972 | Mettler | Finnigan 1015 | EI | Vacuum | 2,4,6 | Geochemical samples |
| Mol | 1974 | Mettler I | UTI 100C | EI | Vacuum | 2,4,6 | PVC, ABS, PU, Polyester |
| Tsur | 1974 | PE TGS1 | Finnigan 1015 | EI | A, Air 18 | 16 | Polybenzimidazole |
| Kleinberg | 1974 | Cahn RH | Dupont 21-491 DF | EI | Oxid. | 5-25 | PVC, PMM, PU |
| Baumgartner | 1977 | Mettler TAC | Finnigan 3200 | EI-CI | 60-80 | 4 | Ca oxalate |
| Muller | 1977 | Mettler | Balzers Quad MS | EI | | 10 | Ca oxalate |
| Morisaki | 1978 | IR ray thermobalance | Quad MS Analyzer | EI | He, Air 150 | | PTFE |
| Shimizu | 1979 | Dir. insert. probe | Dupont 21-110B DF | CI | | | PVC, PMM |
| Chiu | 1980 | Dupont 990 | Dupont 21-104 | EI | He 60 | 5-10 | Ca acet. Polyacetal |
| Yuen | 1982 | Mettler TA1 | HP 5992 | EI | He, O$_2$/He | 15 | Ca oxalate, SB etc. |
| Chan | 1982 | Temp. prog. fraction. | Hitachi RMS-4 | EI | 50 | | PS, peanut oil |
| Dyszel | 1983 | PE TGS 2 | SCIEX TAGA 3000 | APCI | N$_2$ 80 | | Guar gums |
| Whiting | 1984 | Dupont 951/HP 5710 GC | LKB 9000 | EI | He 100 | 10 | Coal |
| NRC Fire | 1984 | Dupont 951 | SCIEX TAGA 6000 | APCI | N$_2$, Air 110 | 2-10 | PAN |

the use of CI is often not sufficient to identify the degradation products. In studying oxidative degradation, which is of particular interest to fire researchers, there is also a technical difficulty: an oxidative atmosphere causes drastic reduction in the life of the ion source.

In the present study an APCI/MS/MS (tandem) was used for the analysis of effluent degradation products of polyacrylonitrile (PAN) from the TGA. This MS is compatible with TGA, since the products can be analyzed in air under atmospheric pressure.

EXPERIMENTAL

Materials

An Orlon cloth that contained more than 85% polyacrylonitrile was used without further treatment. For comparison a 100% pure solid amorphous polyacrylonitrile (Cellomer Assoc. Inc. Cat# 134c) was also used.

Instrumentation

A Dupont 951 TGA and the SCIEX TAGA 6000 APCI/MS/MS were the two main instruments used in this study. The SCIEX TAGA 6000 APCI/MS/MS has been described elsewhere (16). Unlike other MS instruments, it employs a large volume of sample gas continuously introduced into the instrument (typically 2 L/min). Components of the gas are ionized by a corona discharge under atmospheric pressure using oxygen or water in the sample gas as the chemical reagent. A nitrogen flow forms a gaseous membrane between APCI source and the high vacuum analyzer section to prevent un-ionized molecules from going into the analyzer. The ionized molecules (parent ions) penetrate the membrane and are separated according to their mass/charge ratio (M/Z) at the first quadrupole mass filter. Argon gas flows perpendicular to the path of parent ions at the second mass filter. Parent ions colliding against argon atoms fragment with a pattern characteristic of the ion. The pattern is analyzed by the third mass filter.

The MS/MS and the TGA were coupled with a short glass capillary (0.5 mm dia., 20 mm long). The flow of effluent from the TGA to the MS/MS through the capillary was measured from the pressure difference across the capillary and controlled by a 'dump' valve fitted with a micrometer. The whole interface assembly was heated to prevent the condensation of degradation products. To the effluent flow, a make-up flow of 2 L/min of zero air (purified air with practically no organic gases) was added. The schematic diagram of the interface is shown in Fig. 1. A Hewlett Packard 5996 GC/MS/Data system with a CDS 100 pyrolysis unit was used separately to analyze the pyrolysis products and the results were compared with those from the APCI/MS/MS.

RESULTS AND DISCUSSION

The pyrolysis products of PAN analyzed by the pyrolysis GC/MS are shown in Fig. 2. The pyrolysis occurred in 3.6 atms of helium. The main products were HCN, a series of nitriles and acetic acid. The last was not generated when the experiment was repeated with 100% pure solid polyacrylonitrile. In Fig. 3, the results of TGA of PAN in atmospheric nitrogen are shown; they may be compared with the total ion vs time from MS/MS analysis of the effluent (Fig. 4). The total current varied in a manner similar to the first derivative of TGA curve. The total ion is the sum of each of the parent ions; the latter are shown in Fig. 5.

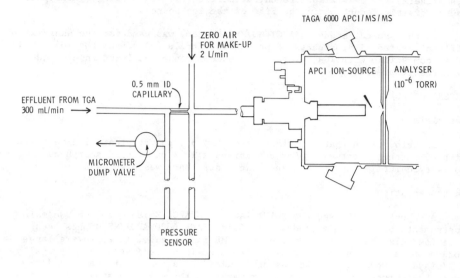

FIGURE 1. TGA/MS/MS interface.

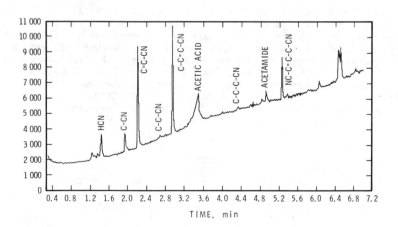

FIGURE 2. Pyrolysis products of PAN, pyrolysis/GC/MS.

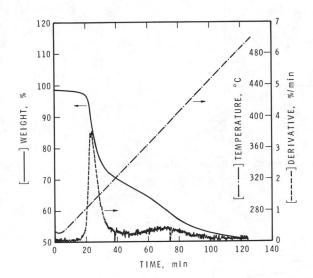

FIGURE 3. Thermogravimetric analysis of PAN in nitrogen.

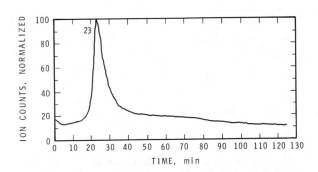

FIGURE 4. Pyrolysis of PAN in nitrogen, total ion in MS/MS analysis, positive mode.

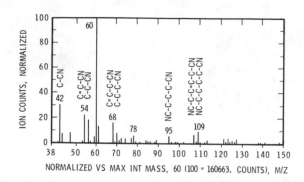

FIGURE 5. Pyrolysis of PAN in nitrogen, MS/MS analysis, parent scan, positive mode.

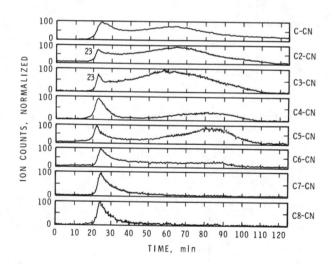

FIGURE 6. Pyrolysis of PAN in nitrogen, MS/MS analysis, single ion monitoring.

From both the GC/MS and the MS/MS experiments, it was clear that a series of nitriles were formed. M/Z = 60 was acetamide, which was very easily ionized and was dominating in the parent scan, but its actual concentration was smaller than that of the nitriles. In Fig. 6, generation of each nitrile is shown against time. Some nitriles were generated in two steps; the first step occurs at the same temperature (290°C) for all the nitriles; the second step at increasingly higher temperatures for higher nitriles. This phenomenon is not simply explained by the difference in their boiling points. For example, butane nitrile (shown as C3—CN in Fig. 6) and pentane nitrile (C4—CN) have boiling points of 117.6 and 140.7°C, respectively, while the recorded temperatures in the TGA at the peak of generation were 372 and 410°C, respectively. The elucidation of the mechanism is left for future studies.

The MS/MS can detect either positive ions or negative ions. In the present study, the positive ion mode was used for nitriles and the negative ion mode for acids. Fig. 7 shows MS/MS analysis in the negative mode. A parent scan at the peak is shown in Fig. 8; major components were HCN and acetic acid.

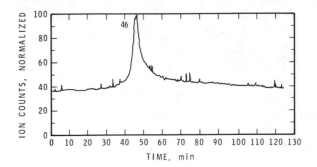

FIGURE 7. Pyrolysis of PAN in nitrogen, total ion in MS/MS analysis, negative mode.

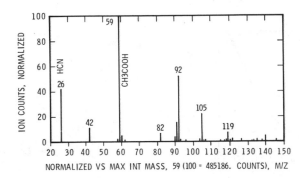

FIGURE 8. Pyrolysis of PAN in nitrogen, MS/MS analysis, parent scan, negative mode.

The TGA of PAN in air is shown in Fig. 9. PAN degraded in two steps, both in air and in nitrogen; the first step is at 290°C and the second at 400-500°C. In air, the weight loss in the first step was smaller and the second step was larger than the corresponding steps in nitrogen. Generation of HCN, as determined by the MS/MS, is shown in Fig. 10. The curve was similar to the first derivative of the TGA curve. The HCN was also generated in two steps.

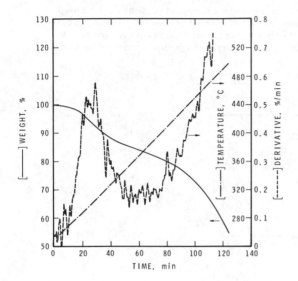

FIGURE 9. Thermogravimetric analysis of PAN in air.

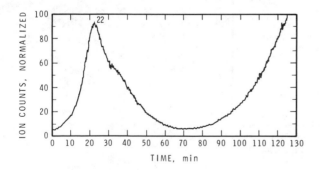

FIGURE 10. Pyrolysis of PAN in air, HCN generation, MS/MS analysis.

The first derivative of the TGA curve determined for the solid specimen, and the total ion current determined from the pyrolysis products in the gas phase, were nearly identical in shape in nitrogen and in air. Although this is expected, since the weight loss is a result of gasification (generation of gaseous products) of the solid specimen, the finding underlines the validity of the technique. The two-step generation of HCN may be explained by the unzipping of molecules in the first step and the decomposition of N-containing char in the second step. HCN is the major toxic component in the thermal decomposition products of PAN (17) and other N-containing organic materials (18).

CONCLUSION

The pyrolysis of PAN in nitrogen and in air was studied by TGA/APCI/MS/MS. HCN, a series of nitriles, and acetic acid were identified. The PAN degraded in two steps. Nitriles and HCN were generated in both steps in nitrogen and in air.

The TGA/APCI/MS/MS was found to be an effective tool for studying the pyrolysis of polymers. Pyrolysis products can be identified and their generation at different stages of pyrolysis can be closely observed. Further studies are planned.

ACKNOWLEDGEMENT

The author thanks J.B. Stewart for his assistance in running experiments and processing data. This paper is a contribution from the Division of Building Research, National Research Council of Canada.

REFERENCES

1.  Zitomer, F.: Thermogravimetric-mass spectrometric analysis, Anal. Chem., 40: 7, 1091, 1968.

2.  Chang, T.L., and Mead, T.E.: Tandem thermogravimetric analyzer-GC-high resolution MS system, Anal. Chem., 43: 534, 1971.

3.  Gibson, E.K. Jr., and Johnson, S.M.: Thermogravimetric-quadrupole mass spectrometric analysis of geochemical samples, Thermochim. Acta, 4: 49, 1972.

4.  Mol, G.J.: Simultaneous thermogravimetry and mass spectrometry in polymer characterization, Thermochim. Acta, 10: 259, 1974.

5.  Tsur, Y., Freilich, Y.L., and Levy, M.: TGA-MS degradation studies of some new aliphatic-aromatic polybenzimidazoles, J. Polym. Sci., Chem. ed., 12: 1531, 1974.

6.  Kleinberg, G.A., Geiger, D.L., and Gormley, W.T.: Rapid determination of kinetic parameters for the thermal degradation of high polymers utilizing a computerized thermogravimetric analyzer-mass spectrometer system, Makromol. Chem., 175: 483, 1974.

7.  Baumgartner, E., and Nachbaur, E.: Thermogravimetry combined with chemical ionization mass spectrometry: a new technique in thermal analysis, Thermochim. Acta, 19: 3, 1977.

8.  Muller-Vonmoos, M., Kahr, G., and Rub, A.: Quantitative determination of $H_2O$, CO and $CO_2$ by evolved gas anslysis with a MS, Thermochim. Acta, 20: 387, 1977.

9.  Morisaki, S.: Simultaneous thermogravimetry–mass spectrometry and pyrolysis-gas chromatography of fluorocarbon polymers, Thermochim. Acta, 25: 171, 1978.

10. Shimizu, Y., and Munson, B.: Pyrolysis/chemical ionization mass spectrometry of polymers, J. Polymer Sci., Chem., 17: 1991, 1979.

11. Chiu, J., and Beattie, A.J.: Techniques for coupling mass spectrometry to thermogravimetry, Thermochim. Acta, 40: 251, 1980.

12. Yuen, H.K., Mappes, G.W., and Grote, W.A.: An automated system for simultaneous thermal analysis and mass spectrometry, part I, Thermochim. Acta, 52: 143, 1982.

13. Chan, K.C., Tse, R.S., and Wong, S.C.: Temperature programmed fractionation inlet system for mass spectrometers, Anal. Chem., 54: 1238, 1982.

14. Dyszel, S.M.: Thermogravimetry coupled with atmospheric pressure ionization mass spectrometry, a new combined technique, Thermochim. Acta, 61: 169, 1983.

15. Whiting, L.F., and Langvardt, P.W.: On-column sampling device for thermogravimetry/capillary gas chromatography/mass spectrometry, Anal. Chem., 56: 1755, 1984.

16. French, J.B., Davidson, W.R., Reid, N.M., and Buckley, J.A.: "Trace monitoring by tandem mass spectrometry," in Tandem Mass Spectrometry, F.W. McLafferty, ed., John Wiley & Sons, New York, 1983.

17. Tsuchiya, Y., and Sumi, K.: Thermal decomposition products of polyacrylonitrile, J. Appl. Polym. Sci., 21: 975, 1977.

18. Tsuchiya, Y.: Significance of HCN generation in fire gas toxicity, J. Comb. Toxicol., 4: 271, 1977.

# Halogen-Free Flame-Retardant Thermoplastic Polyurethanes

**D. R. HALL, M. M. HIRSCHLER, and C. M. YAVORNITZKY**
BFGoodrich Chemical Company
Technical Center, P.O. Box 122
Avon Lake, Ohio 44012, USA

ABSTRACT

Studies have been made of the flammability of a thermoplastic polyurethane incorporating a halogen-free additive flame-retardant system. Clear-cut evidence of synergism has been shown for the interaction between the additives. Thermoanalytical studies have shown that the crucial reaction for synergism is the enhanced production of ammonia and that this only occurs when the two phosphorus- and nitrogen-containing additives are present. The overall pattern of thermal decomposition of the polymer-additives system shows slow but steady thermal activity over a wide range of temperatures; this is indicative of synergism in a condensed-phase mechanism. The most important flame-retardant effect is thus the formation of a charry intumescent foam, but there is also some physical and chemical gas-phase activity. The resulting thermoplastic polyurethanes can be produced with an LOI of over 40 and a UL94 ranking of VO, in the total absence of halogen.

INTRODUCTION

Thermoplastic polyurethanes are tough and versatile materials, with a good combination of processing and performance characteristics. Their levels of production have grown very rapidly in recent years, in particular in the U.S. They have risen from 13000 tons in 1975[1] to 38000 tons in 1984.[2] A typical application of these resins is in tubing for the enclosure of wires on board ships which are being reconditioned. It is essential for such a use that the polymers should have been flame-retarded, to help protect against ignition due to sparks and short-circuits. This flame retardance has traditionally been achieved by employing antimony oxide-halogen systems. It is of interest, however, to investigate other additives because some concern has been voiced about the emission of potentially corrosive hydrogen halides in some electrical applications.

The combustion of any organic polymer can be represented by a series of three interdependent and interrelated stages:[3,4] (a) thermal decomposition of the base polymer to yield flammable gaseous products; (b) flaming combustion of these products leading to the generation of heat and (c) transport of this heat back to the polymer surface to restart the cycle. Flame retardance can thus be achieved by interacting with one or more of these stages. The mechanism of action of antimony/halogen flame retardants is a mixture of effects on the first two stages, viz. the additives change the decomposition pattern of the resin (inter alia by slowing down the rate of breakdown),[3,7] and they inhibit the flame propagation reactions by the scavenging action of antimony and halogen species in the gas phase.[3,4,8,9] Such systems are, however, sometimes associated with dripping products and with large emissions of smoke particles.

Intumescent systems act by creating a charry foam, blown up by the action of evolved non-flammable gaseous products, which insulates the bulk of the polymer from the effect of further heat.[3,10-12] Such an additive system is thus potentially more complete than the one described previously because it can act by interfering with all three stages of the combustion process. A typical intumescent system contains a "carbonific" (which is the source of the carbon producing the charry residue), a "spumific" (which generates the non-flammable gases) and a "catalyst" (which promotes decomposition of the carbonific compound, typically by forming an inorganic compound at temperatures between 475 and 525 K). In many cases, more than one of these functions is performed by the same compound.

The present paper presents an investigation to search for a flame-retarded thermoplastic polyurethane (TPU) which is halogen-free. The additive systems used are phosphorus-based and contain up to three individual compounds in combinations chosen so as to achieve synergistic interactions leading to better overall performance than would have been possible with any individual additive. Ammonium polyphosphate is known to be an additive which can induce the formation of an intumescent char in thermoplastic materials;[13-15] it was therefore thought possible that it would be a much more effective flame retardant in the presence of a synergistic co-additive.

It has recently been shown that thermoanalytical measurements are a very useful source of data to allow interpretations to be made of the mechanism of flame-retardant action.[6,7] Thermal analysis was thus the main tool used to obtain non-flame data for analysis of the results of the present work.

EXPERIMENTAL

Materials

The thermoplastic polyurethane (TPU) used is a Shore 85A hardness, polyether urethane from 1000 Mw polytetramethylene ether glycol, 4-4 diphenylmethane diisocyanate and 1,4-butane diol (an ESTANE® resin); all three flame-retardant additives used, viz. bis melaminium pentate (Borg-Warner)[16] (MAP), ammonium polyphosphate (APP) and isopropylphenyl diphenyl phosphate (Kronitex 100, FMC Corp.) (KX) are phosphorus-based.

Procedure

Flame-retardant additive densities were determined by air displacement pycnometry for volumetric evaluations. Liquid additive KX was presorbed into the TPU at 293 K. The remaining additives, including lubricant and pigment, were incorporated by melt-mixing at 448 K in a 54 $cm^3$ batch-size Brabender internal mixer.

Flammability

The flammability of the system was determined by measuring the limiting oxygen index (LOI) on equipment designed to meet ASTM D2863 flammability test requirements and the UL94 vertical flammability ranking on similarly appropriate equipment. The standard UL94 test classifies samples into VO, V1 or V2 categories and it was used in order to investigate the flame spread time and the dripping characteristics of the sample (at 3 mm thickness) following piloted ignition.

Thermal Analysis

A DuPont 1090 thermogravimetric analyzer was used for all experiments; a heating rate of 10 deg min$^{-1}$ and an atmosphere of nitrogen (at a flow rate of 180 cm$^3$ min$^{-1}$) were used throughout. Char yields were compared with those obtained from a BFGoodrich proprietary smoke/char test,[17] which measures smoke production (photometrically) and char residue following the rapid (30 s) combustion of small samples ignited by a pencil tip propane flame.

RESULTS AND DISCUSSION

The pure thermoplastic polyurethane decomposes thermally by a mechanism involving four stages, the temperature ranges of which were 543-635, 635-656, 656-756 and 756-832 K (Table 1). The most important aspects of this decomposition are: a) the fact that the first two stages are very fast, and account for over 60% of the total weight and b) the fact that the final weight loss stage, which corresponds to char pyrolysis is, on the other hand, quite slow and accounts for ca. 15 wt%. The TPU starts decomposing substantially only at a very high temperature since the temperature at which 1.0 wt% has been lost[6,7] ($T_{1\%}$) is 543 K, which suggests that depolycondensation has already occurred before decomposition started, but that it has led to involatile fragments (by comparison with other polyurethanes[18,19]).

The thermal decomposition of MAP is very complex: it consists of seven stages (Table 1). The two most important stages are the production of ammonia (corresponding to ca. 23 wt%) and the char pyrolysis (corresponding to some 40% of the weight of the additive). The ammonia stage (maximum rate at 590 K) is the fastest breakdown in an otherwise rather slow decomposition covering a wide temperature range (335-1115 K) (Figure 1). APP breaks down in three stages (Table 1) corresponding, successively, to the slow elimination of ammonia (at virtually the same temperature as MAP eliminates ammonia; maximum rate at 581 K), to the slow elimination of water (these two stages accounting together for ca. 25% of the initial weight) and to a very fast breakdown of the remaining cross-linked phosphorus-containing residue (ultraphosphate), occurring much later. This is similar to what has been shown by other authors.[14] The thermal decomposition of KX is very simple: it volatilizes in a single stage (and probably unchanged) at 543 K.

Figure 2 shows the thermal decomposition pattern of the pure TPU and of systems containing the resin with 20 vol% of each additive, individually. The object of this is to try to understand the individual effects of the additives so as to be able to interpret the effects of their combinations. It has been shown that synergism in flame retardance is almost invariably accompanied by clear changes in the thermogravimetric pattern.[6,7,20-24] It is perhaps somewhat surprising that the effects of MAP and APP on the resin are very similar: both decrease the thermal stability of the system (although APP is significantly more effective in doing this, as shown by $T_{1\%}$, in Table 1) while they promote the first decomposition stage of the resin at the expense of the second one and, particularly, of the third one, which virtually disappears. Furthermore, Table 1 shows that char pyrolysis occurs at a much higher temperature in the presence of either of these additives (temperature range 880-1100 K, rather than 750-930 K), while there does not seem to be very much additional char formed. The N-free additive, KX, has very little effect: it simply volatilizes just before polymer breakdown (maximum rate of this stage at ca. 550 K). Furthermore, the char pyrolysis stage occurs at the same temperature range than in the pure TPU as distinct from the cases of the other two additives. One effect of KX is that it brings together the first two weight loss stages of the TPU.

Figure 3 shows the comparison, for a large number of systems, between the char (+ residue) remaining after the smoke/char test[17] (in the ordinate) and the char (+ residue) formed under thermogravimetric (TGA) conditions (in the abscissa); the base resin is the same TPU in all cases. It is clear that additives can

TABLE 1. Thermal analyses of polymer, additives and polymer-single additive systems

| | TPU ESTANE | TPU + 20% MAP | TPU + 20% APP | TPU + 20% KX | MAP | APP | KX |
|---|---|---|---|---|---|---|---|
| Initial Weight (mg) | 18.28 | 20.21 | 24.86 | 15.88 | 9.80 | 9.55 | 13.86 |
| Total Weight Lost (%) | 99.1 | 98.9 | 99.3 | 99.4 | 97.5 | 95.1 | 99.4 |
| $T_D$ (K) | 441 | 378 | 459 | 456 | 336 | 341 | 441 |
| $T_{1\%}$ (K) | 543 | 445 | 508 | 488 | 353 | 496 | 465 |
| **First Loss** | | | | | | | |
| Wt Loss (%) | 1.5 | 1.5 | 1.5 | 5.0 | 4.0 | 9.5 | 99.4 |
| $DTG_{max}$ (% min$^{-1}$) | 0.55 | 0.30 | 0.50 | 1.30 | 1.30 | 1.60 | 27.30 |
| T ($DTG_{max}$) (K) | 543 | 436 | 507 | 413 | 366 | 581 | 546 |
| T range (K) | 441-543 | 378-486 | 459-517 | 456-575 | 336-382 | 341-603 | 441-565 |
| **Second Loss** | | | | | | | |
| Wt Loss (%) | 47.0 | 37.5 | 30.5 | 17.5 | 2.5 | 8.5 | |
| $DTG_{max}$ (% min$^{-1}$) | 10.50 | 11.80 | 9.90 | 4.00 | 0.90 | 0.90 | |
| T ($DTG_{max}$) (K) | 612 | 578 | 555 | 554 | 408 | 617 | |
| T range (K) | 543-635 | 486-597 | 517-577 | 525-576 | 382-439 | 603-693 | |
| **Third Loss** | | | | | | | |
| Wt Loss (%) | 13.5 | 21.4 | 28.0 | 41.0 | 6.0 | 77.1 | |
| $DTG_{max}$ (% min$^{-1}$) | 9.20 | 6.70 | 5.40 | 8.80 | 1.50 | 13.2 | |
| T ($DTG_{max}$) (K) | 645 | 612 | 600 | 612 | 509 | 836 | |
| T range (K) | 635-656 | 597-637 | 577-650 | 576-651 | 439-533 | 693-973 | |
| **Fourth Loss** | | | | | | | |
| Wt Loss (%) | 21.5 | 15.5 | 6.0 | 21.0 | 23.0 | | |
| $DTG_{max}$ (% min$^{-1}$) | 5.00 | 3.40 | 0.80 | 4.20 | 4.90 | | |
| T ($DTG_{max}$) (K) | 673 | 644 | 674 | 681 | 590 | | |
| T range (K) | 656-756 | 637-766 | 650-761 | 651-761 | 533-637 | | |
| **Fifth Loss** | | | | | | | |
| Wt Loss (%) | 15.6 | 6.0 | 9.8 | 14.9 | 14.5 | | |
| $DTG_{max}$ (% min$^{-1}$) | 1.50 | 0.80 | 1.30 | 1.70 | 1.90 | | |
| T ($DTG_{max}$) (K) | 834 | 781 | 852 | 830 | 693 | | |
| T range (K) | 756-932 | 766-880 | 761-885 | 761-970 | 637-728 | | |
| **Sixth Loss** | | | | | | | |
| Wt Loss (%) | | 17.0 | 23.5 | | 8.5 | | |
| $DTG_{max}$ (% min$^{-1}$) | | 1.30 | 2.20 | | 2.10 | | |
| T ($DTG_{max}$) (K) | | 1005 | 956 | | 756 | | |
| T range (K) | | 880-1100 | 885-1115 | | 728-787 | | |
| **Seventh Loss** | | | | | | | |
| Wt Loss (%) | | | | | 39.0 | | |
| $DTG_{max}$ (% min$^{-1}$) | | | | | 1.90 | | |
| T ($DTG_{max}$) (K) | | | | | 1014 | | |
| T range (K) | | | | | 787-1135 | | |

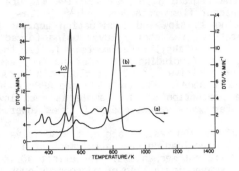

FIGURE 1. Differential thermogravimetric analysis of the additives. (a) MAP; (b) APP; (c) KX. A zero value for the ordinate of curve (a) is at a DTG of 1 % min⁻¹.

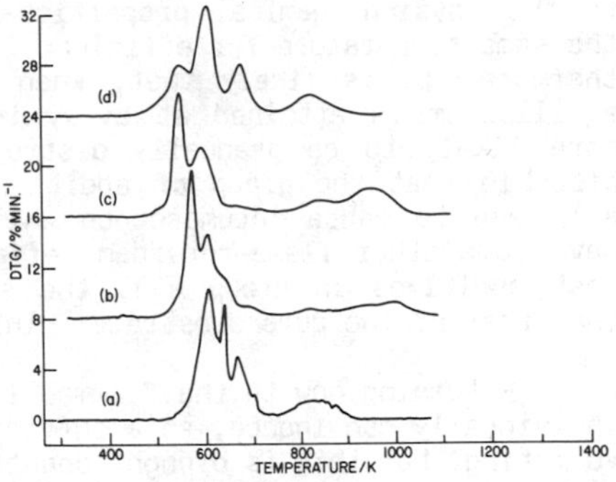

FIGURE 2. Differential thermogravimetric analysis of the polymeric substrate in the absence and presence of 20 vol% of each additive individually. (a) TPU; (b) TPU + 20 vol% MAP; (c) TPU + 20 vol % APP; (d) TPU + 20 vol% KX. Zero values for the ordinates of successive curves are at 6 % min⁻¹ intervals.

fall into two major categories in this respect: (a) those which form much more char on burning than on being heated slowly (roughly twice as much, on average) and (b) those which produce at least as much (or more) char under thermoanalytical conditions than on burning. The dashed line (of slope 1.0) separates the two groups of additives. This result is relevant to the phenomenon of intumescence. It has been shown that it is important to match the temperatures of gas evolution

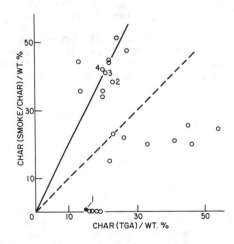

FIGURE 3. Relationship between thermoanalytical and combustion char. ○ : pure polymer; ○ : polymer containing additives; ── : slope for intumescent additives; --: slope of 1.0. (1) TU + 20 vol % KX; (2) TPU + 20 vol% APP; (3) TPU + 30 vol% APP; (4) TPU + 20 vol% MAP.

to the physico-chemical properties of the degrading polymer-additive mixture at the same temperature for efficient intumescent flame retardance.[10,14,25] Furthermore, it is likely that, when heating is slow and a situation approaching equilibrium is attained at every temperature, the charry layer being formed is more likely to be gradually destroyed than in short experiments. It is, thus, probable that the group of additives in a) (including both MAP and APP) would all tend to cause intumescence while the additives in b) (including KX) would have some other flame-retardant effects (or none). For both MAP and APP (as for most additives in group a)), the smoke production of the sample is much lower than that of the pure substrate: integrated smoke values are 5-10 times lower.

Returning now to the flammability of the TPU resin, each of the additives individually can induce, at a sufficiently high loading, the attainment of a UL94 VO rating, but this is brought about by different mechanisms. The effect of KX is probably concentrated in the free-radical scavenging action of phosphorus-containing species, which have been volatilized. This gas-phase mechanism is a well established one for thermoplastic polymers in the presence of organic phosphorus-containing flame retardants.[3] Thermal analysis results and the lack of any major effect on char formation or smoke production (by the smoke/char test) suggest thus that KX does not have a condensed-phase mechanism but may be an aid in a multi-component additive system, by its gas-phase activity.

APP, a potential spumific agent for intumescence, increases char production and decreases smoke generation, but has a very low efficiency as a flame retardant Its monomer, ammonium phosphate, is a typical intumescent catalyst.[3] Its efficient interaction (in the thermoanalyzer) with the substrate suggests that the polymeric form might also be a potential catalyst for intumescence. Furthermore, at 30 vol% additive loading, the TPU + APP system shows a breakdown stage (between the two first polymer stages) where volatilization of ammonia occurs and which accounts for up to 10 wt% of the entire system. This same system also shows further increases in char production from the resin (from 15 wt% to 35 wt% of resin), presumably because of the formation of the intumescent charry layer across the surface of the condensed phase. This is of low efficiency, unfortunately, because it follows a rather fast first weight loss stage for the resin.

MAP is a potential carbonific; it also decreases both the flaming characteristics of TPU and the smoke generated from it, while increasing char production. Furthermore, it is capable of releasing inert gases (ammonia in this case) and it contains a structural part (melamine phosphate) which is a potential intumescent catalyst.[3]

Consequently, over 50 systems were tried in which various combinations of TPU with MAP and APP were investigated for flammability, both in the absence and in the presence of KX. The LOI values determined for the systems were fitted by a polynomial function of a linear transformation of the composition coordinates and represented in two triangular diagrams (Figure 4, in the absence of KX, and Figure 5, in the presence of KX).[9,26] Coefficients of the polynomials were calculated by least-squares regression analysis and the quality of the fit assessed by the value of the root mean square differences between the calculated and observed LOI values ($\sigma$), by the multiple regression correlation coefficients (RCC), by examination of the individual differences between calculated and observed LOI values and by comparison with experimental trends. The vertices of the triangles shown correspond to (A) 100% TPU; (B) 62.92 vol% TPU; 37.08 vol% APP; (C) 62.92 vol% TPU; 37.08 vol% MAP. In both cases there is very clear evidence of synergism with LOI values of well over 40 (as compared with values of 22.1 for the pure resin and of well under 30 for maximum loading of each individual additive). The mathematical fit of the data is more than adequate so that the triangular diagrams shown (cubic models) are good representations of the trends. It

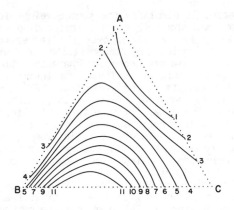
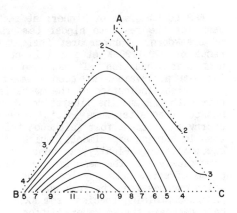

FIGURE 4.   The flammability (by LOI) of   FIGURE 5.   The flammability (by LOI) of
the systems containing TPU, MAP and APP.   the systems containing TPU, MAP, APP and
(RCC = 0.867).                              KX.   (RCC = 0.936).
A:   (1.00;  0.00;  0.00);  B:   (0.63;  0.37;  0.0);  C:   (0.63;  0.00;  0.37).   LOI
Values:  1,  23.0;  2,  25.0;  3,  27.0;  4,  29.0;  5,  31.0;  6,  33.0;  7,  35.0;  8,  37.0;
9,  39.0;  10,  41.0;  11,  43.0.

can be calculated from these diagrams that optimum volumetric ratios of APP to MAP
(i.e. those which will yield highest LOI values) lie between 1.5 and 3.0; furthermore the diagrams show that LOI values of well over 30 can be obtained without
very high levels of additives.   The third additive (KX) can be effective in
decreasing the flammability of the resin at relatively low total loadings, but
becomes quite unnecessary, for flammability purposes, at very high loadings.

        Table 2 shows that UL94 VO rankings can be obtained for loadings as low as
19 vol% of additives but that, at higher loadings (in the presence of some KX),
VO spectra can be obtained for wider ranges of APP to MAP ratios.

TABLE 2.   UL94 vertical flammability rankings

|              | PNB/PNA | PNB/PNA | PNB/PNA | PNB/PNA | PNB/PNA |
|--------------|---------|---------|---------|---------|---------|
|              | 1/0[a]  | 3/1[a]  | 1/1[a]  | 1/3[a]  | 0/1[a]  |
| 19 vol %[b]  | V2      | VO      | V2      | V2      | V2      |
| 23 vol %[b]  | V2      | VO      | VO      | V1      | V2      |
| 26 Vol %[b]  | V2      | VO      | VO      | VO      | V2      |
| 30 vol %[b]  | V2      | VO      | VO      | VO      | V2      |
| 33 Vol %[b]  | VO      | VO      | VO      | VO      | V2      |
| 37 vol %[b]  | VO      | VO      | VO      | VO      | V1      |

[a] Volumetric ratio of additives; [b] Total additive level.

        Figures 6 and 7 show the differential thermogravimetric pattern of a series
of systems at constant total loading where the ratios of MAP to APP are gradually
varied.   It can be seen that there is, in both diagrams, a noticeable change: the
stage corresponding to ammonia liberation (at ca. 600 K) is almost negligible when
either MAP or APP are missing.   However, it is the dominant stage when the ratio

427

of APP to MAP is 1 or higher; at such ratios it displaces the second weight loss peak for the resin to higher temperatures and slows down the corresponding rate of breakdown. Furthermore, thermal activity is continuous over the temperature range of 520-1190 K at a fairly constant and quite low decomposition rate with high char production. It has been shown that these are necessary characteristics for the achievement of good flame-retardant synergism.[6] Thus, the thermograms are a clear indication of the mechanism that results in the synergism: MAP interacts with APP and the liberation of ammonia is enhanced. This, in turn, since it has occurred after some initial breakdown of the polymeric structure, will form a charry intumescent foam (although this char is less noticeable in thermoanalytical than in burning experiments) which will protect the bulk of the resin from further decomposition. Finally, if heat continues to be applied, the charry layer breaks down slowly. In the meantime, the flame reactions are also slowed down by gas-phase effects, both physical (due to the cooling effects of some of the ammonia which escapes from the foam and is non-flammable) and chemical (due to the phosphorus-containing species from KX). Such an increased char production will also help decrease the smoke production tendency,[27] since it is clear that the carbon needed for smoke is partly being retained in the condensed phase.

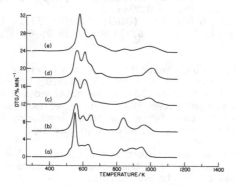 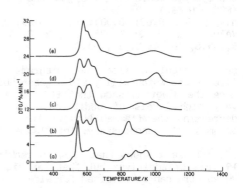

FIGURE 6. Differential thermogravimetry of the TPU in the presence of 30 vol % of varying proportions of MAP and APP.

FIGURE 7. Differential thermogravimetry of the TPU in the presence of 30 vol % (with respect to TPU) of varying proportions of MAP and APP, and in the presence, additionally, of KX.

(a) 30.0 vol% APP; (b) 22.5 vol% APP + 7.5 vol% MAP; (c) 15.0 vol% APP + 15.0 vol% MAP; (d) 7.5 vol% APP + 22.5 vol% MAP; (e) 30.0 vol% MAP. Zero values for the ordinates of successive curves are at 6 % min$^{-1}$ intervals.

At high loadings of MAP and APP (e.g. 30 vol% or higher) the gas phase activity of KX becomes mostly superseded and is thus scarcely noticeable. This transpires both from the thermoanalytical pattern (Figures 6 and 7) and from the flammability pattern (Figures 4 and 5).

An excellent correspondence between the findings of the thermoanalytical and flammability measurements is another example of the good extrapolatability of thermal analysis as a first-order approximation model for some of the phenomena occurring in burning polymers. This has helped to establish a model for the synergistic flame-retardant mechanism of an additive system, for a thermoplastic polyurethane, which is halogen-free and exhibits extremely low flammability and smoke under the conditions investigated.

Caveat:  The flammability test data reported here represent laboratory small-scale measurements and are not necessarily indicative of the response of the materials to a full-scale fire scenario.

CONCLUSIONS

Thermoanalytical measurements and flammability determinations have shown the synergistic flame retardant activity of an additive system for a thermoplastic polyurethane containing phosphorus and nitrogen but no halogen.  The flame-retardant acts on all three stages of the combustion cycle, but its most important effect is the formation of an intumescent charry foam on the surface of the polymer.  Very low flammability is achieved with LOI values of over 40 and a UL94 ranking of VO accompanied by high char and low smoke.

REFERENCES

1.  Modern Plastics, 54(1), 49 (1977).

2.  Modern Plastics, 62(1), 61 (1985).

3.  Cullis, C. F. and Hirschler, M. M., "The Combustion of Organic Polymers," Oxford University Press, Oxford, 1981.

4.  Hirschler, M. M., in "Developments in Polymer Stabilization-5" (Ed. G. Scott), p. 107, Applied Science Publ., London, 1982.

5.  Learmonth, G. S. and Thwaite, D. G., Brit. Polymer J. 2, 104 (1970).

6.  Hirschler, M. M., Europ. Polymer 19, 121 (1983).

7.  Cullis, C. F. and Hirschler, M. M., Polymer 24, 834 (1983).

8.  Hastie, J. W., J. Res. Natl Bur. Stand. 77A, 733 (1973).

9.  Antia, F. K., Baldry, P. J. and Hirschler, M. M., Europ. Polymer J. 18, 167 (1982).

10.  Vandersall, H. L., J. Fire Flammability 2, 97 (1971).

11.  Rheineck, A. E., J. Paint. Technol. 44 (567), 35 (1972).

12.  Hindersinn, R. R. and Witschard, G., in "Flame Retardancy of Polymeric Materials - Vol. 4" (Ed. W. C. Kuryla and A. J. Papa), p. 1, Marcel Dekker Publ., New York, 1978.

13.  Camino, G., Grassie, N. and McNeill, I. C., J. Polymer Sci., Polym. Chem. 16, 95 (1978).

14.  Camino, G., Costa, L. and Trossarelli, L., Polymer Degrad. and Stabil. 6, 243 (1984).

15.  Camino, G. Costa, L. and Trossarelli, L. Polymer Degrad. and Stabil. 7, 25 (1984).

16.  Halpern, Y., U.S. Pat. No. 4154930, May 15, 1979 (to Borg-Warner Corp.).

17.  W. J. Kroenke, J. Appl. Polymer Sci. 26, 1167 (1981).

18. Ballistreri, A., Foti, S., Maravigna, P., Montaudo, G. and Scamporrino, E., Makromol. Chem. 181, 2161 (1980).

19. Montaudo, G., Puglisi, C., Scamporrino, F. and Vitalini, D. Macromolecules 17, 1605 (1984).

20. Donaldson, J. D., Donbavand, J. and Hirschler, M. M., Europ. Polymer J. 19, 33 (1983).

21. Hirschler, M. M. and Tsika, O., Europ. Polymer J. 19, 375 (1983).

22. Cullis, C. F., Hirschler, M. M. and Khattab, M. A. A. M., Europ. Polymer J. 20, 559 (1984).

23. Broadbent, J. R. A. and Hirschler, M. M., Europ. Polymer J. 20, 1087 (1984).

24. Hirschler, M. M. and Thevaranjan, T. R., Europ. Polymer J. 21, 371 (1985).

25. Chang, W. H., Scriven, R. L. and Ross, R. B., in "Flame-retardant polymeric materials, Vol. 1" (Ed. M. Lewin, S. M. Atlas and E. M. Pearce), p. 399, Plenum Press, New York, 1975.

26. Antia, F. K., Cullis, C. F. and Hirschler, M. M., Europ. Polymer J. 18, 95 (1982).

27. Cullis, C. F. and Hirschler, M. M., Europ. Polymer J. 20, 53 (1984).

# Major Species Production by Solid Fuels in a Two Layer Compartment Fire Environment

C. L. BEYLER
Division of Applied Science
Harvard University
Cambridge, Massachusetts 02138, USA

## ABSTRACT

Major species production rates from burning polyethylene, poly(methyl methacrylate), and ponderosa pine were measured in a two layer compartment fire environment. Production rates were found to be correlated by the fuel to oxygen ratio, where the fuel supply rate is the fuel volatilization rate and the oxygen supply rate is the entrainment rate of oxygen between the fuel surface and the hot/cold layer interface. The results are similar to previous results with simple gaseous and evaporating liquids and support the observation that carbon monoxide production under fuel rich conditions is greater for oxygenated hydrocarbons than hydrocarbons.

Carbon monoxide yields for wood four times as large as the present results have been reported in the literature. These very large literature values were the result of sampling within the reaction zone. The present results and reanalysis of literature results obviate the need to postulate three different thermal decomposition mechanisms for wood at different fuel to oxygen ratios as has been proposed in the literature.

An analysis of literature results indicates that under fuel rich conditions compartment residence times of 10-15 seconds are required for combustion to final products. Under fuel lean conditions far lower residence times are required, probably due to the enhanced mixing under these conditions.

## INTRODUCTION

Investigations over the past twenty years have demonstrated that restricted ventilation can increase the production of carbon monoxide by an order of magnitude or more[1-6]. This can have significant effects on the extent to which toxic products will cause other areas of the building to become untenable. A recent investigation[7] showed that the production rate of major chemical species per unit mass of fuel in a two layer environment can be expressed as a function of the fuel to oxygen ratio, where the fuel supply rate is the generation rate of fuel volatiles and the oxygen supply rate is the oxygen entrained into the flame between the fuel surface and the hot/cold layer interface. Fuels used in the investigation were gaseous and liquid fuels, and the correlations appeared insensitive to the thermo-fluidic details of the flame. The production of carbon monoxide under fuel rich conditions was found to be a strong function of the the fuel chemical structure with the results following the general ranking: oxygenated hydrocarbons > hydrocarbons > aromatic hydrocarbons.

In this work the methods developed for determining the major chemical species production rates under restricted ventilation conditions in a two layer environment using gaseous and liquid fuels were used to study solid fuels of direct interest in fires in buildings. Experiments were performed using polyethylene, poly(methyl methacrylate), and ponderosa pine. Where available, the present results are compared with existing data.

## EXPERIMENTAL APPARATUS and PROCEDURES

The apparatus used in these experiments included the hood system in which the hot gas layer was formed, the load cell system, and the gas sampling and analysis system. The hood system included the hood in which the

The present address for C. L. Beyler is Center for Firesafety Studies, Worcester Polytechnic Institute, Worcester MA 01609.

hot gas layer was confined, the exhaust system, and instrumentation to control and measure the exhaust rate. The one meter diameter hood was constructed of sheet metal and the ceiling was made up of 0.012m thick ceramic fiberboard backed by 0.025m of ceramic fiberblanket. Fiberblanket(0.025m) was also used to insulate the outer vertical surfaces of the hood. Gases were exhausted around the full periphery of the hood 0.3m above the base of the 0.48m deep hood through a vertical plenum system. The exhaust was measured using a 0.044m diameter orifice meter with flange taps in the 0.15m exhaust duct. Temperature at the orifice meter was measured by thermocouple and the pressure drop was measured using an electronic manometer.

Specimen mass was measured using a water cooled LVDT load cell separated from the specimen by several layers of low density insulation. Mass loss rate was determined numerically and the average mass loss rate over the steady burning period was used in the analysis. The load cell and specimen were positioned on the centerline of the hood and were mounted on a piston assembly to allow vertical positioning relative to the hood.

Hood exhaust gases were sampled at the orifice meter to insure a well mixed sample. Measurements of chemical composition at the hood exit indicated that no measurable chemical reactions occurred in the exhaust system. Particulates were trapped immediately upon sampling and the sample gas was transported to the gas analyzers via 0.6cm O.D. impolene tubing maintained at or above 70°C.

The gas sample stream was analyzed using continuous analyzers for oxygen (polaragraphic,Beckman OM-11), carbon monoxide and dioxide (nondispersive IR absorption, Infrared Industries 702), total hydrocarbons(THC) (flame ionization detector(FID), $H_2$/He carrier gas, Shimazdu), and water (dew point hygrometer, General Eastern 1200APS). Hydrogen was analyzed by gas chromatography. Flows to the oxygen, carbon monoxide/dioxide, and hydrogen analyzers were passed through a -5°C cold trap and the resulting measurements were corrected for the volume of water removed(no correction for trapped hydrocarbons was made). Flows to the water and total hydrocarbon analyzers were maintained at or above 70°C and no corrections were required. The total hydrocarbon data was interpreted as $CH_2$. An H/C ratio of two was assumed in order to represent general hydrocarbons.

The operation of the apparatus involved three basic parameters which were controlled by the experimenter: the vertical position of the specimen, the specimen surface area, and the hood exhaust rate. Generally, the last two were chosen and the vertical position of the specimen was set so that the layer/air interface was below the exhaust lip and above the base of the hood. The layer interface was maintained 10-15cm above the base of the hood. In this way all gases leaving the layer were contained in the exhaust flow and no additional ambient gases were contained in the exhaust flow.

Experiments were performed using 0.025m thick low density polyethylene(LDPE) in 0.2, 0.23, 0.255m diameter pans. After ignition the polyethylene burned slowly until the whole specimen had melted. The burning rate then increased to a steady rate. Poly(methyl methacrylate) (PMMA) was obtained as 0.025m thick slab stock and was burned with fuel surface areas of 0.023-0.061m$^2$. Ponderosa pine was burned in cribs designed to be fuel surface controlled in open burning. Each of the three to five layers was made up of three, 0.038m thick, 0.2m long sticks. All fuels were ignited using a propane torch.


CORRELATION CONCEPTS

The mass production rate of a species i per unit mass of fuel volatilized is known as the mass yield, $Y_i$. If the chemical composition of the fuel volatiles is known, the yield can be normalized by the maximum possible mass production rate of species i per unit mass of fuel volatiles,$k_i$. This is known as the normalized yield of species i, $f_i$. As the chemical composition of the fuel volatiles is not generally known for solid fuels, especially those that


TABLE 1

| Fuel | Empirical Formula | $(F/O)_{stoich}$ (mass units) | $k_{O2}$ | $k_{CO2}$ (mass units) | $k_{H2O}$ |
|------|-------------------|------------------------------|----------|------------------------|-----------|
| PE | $CH_2$ | 0.29 | 3.43 | 3.14 | 1.28 |
| PMMA | $C_5H_8O_2$ | 0.52 | 1.92 | 2.20 | 0.72 |
| Pine | $C_{0.95}H_{2.4}O$ (a) | 0.83 | 1.13 | 1.40 | 0.72 |

(a) estimated from low (F/O) yield data


432

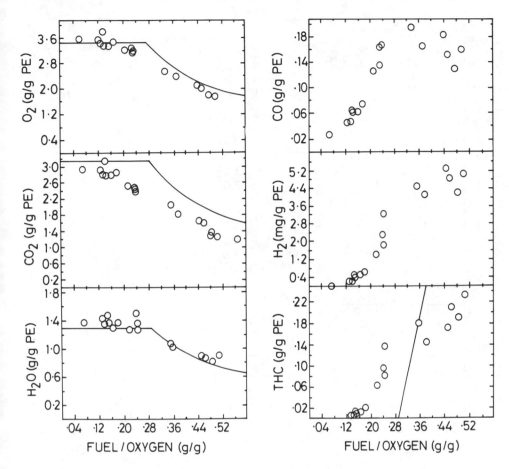

Figure 1. Mass yield of major species for polyethylene as a function of the mass fuel to oxygen ratio.
Solid lines are the prediction of Equation 1 using constants in Table 1.

char, the present results will be presented in terms of mass yields. The yield of oxygen refers to the consumption rather than the production of oxygen.

In two layer compartment fire environments the chemical species yields will be a function of the mass fuel to oxygen ratio, (F/O), where the fuel supply rate is the fuel volatilization rate and the oxygen supply rate is the rate of oxygen entrainment between the fuel surface and the hot/cold layer interface. This can be normalized by the stoichiometric fuel to oxyegn ratio, (F/O) $_{stoich}$, if the chemical composition of the fuel volatiles is known. This normalized fuel to air ratio is known as the equivalence ratio, Ø. The equivalence ratio is the reciprocal of the stoichiometric fraction used by Tewarson[5,9].

The simplest possible model for combustion of a C,H,O containing fuel is given by

F+ $O_2$ -> products + excess oxygen      for Ø<1
F+ $O_2$ -> products + excess fuel      for Ø>1

where the products are carbon dioxide and water. Taking excess fuel as total hydrocarbon, the following

433

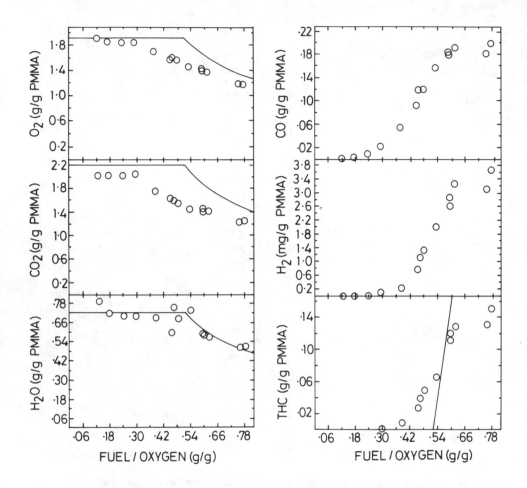

Figure 2. Mass yield of major species for poly(methyl methacrylate) as a function of the mass fuel to oxygen ratio. Solid lines are the prediction of Equation 1 using constants in Table 1.

normalized yield expressions result

$$f_{CO2} = f_{H2O} = f_{O2} = 1 \, , \, f_{THC} = 0 \qquad \text{for } \emptyset < 1$$
$$f_{CO2} = f_{H2O} = f_{O2} = 1/\emptyset \, , \, f_{THC} = 1 - 1/\emptyset \qquad \text{for } \emptyset > 1 \qquad (1)$$

These relationships provide a good first estimate of the production rate of carbon dioxide, water , total hydrocarbons, and consumption of oxygen. To within the approximation that the heat of reaction of oxygen is a constant, the oxygen normalized yield is also the combustion efficiency.

RESULTS AND DISCUSSION

Major species results for low density polyethylene, poly(methyl methacrylate), and ponderosa pine are shown in Figures 1-3. The yield relations of Equation 1 are also plotted using constants given in Table 1. The constants for polyethylene and poly(methyl methacrylate) are those for pure polymer. The constants chosen for 1

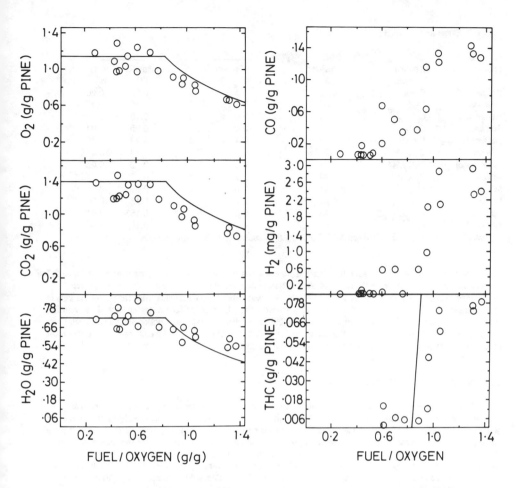

Figure 3. Mass yield of major species for ponderosa pine as a function of the mass fuel to oxygen ratio. Solid lines are the prediction of Equation 1 using constants in Table 1.

pine volatiles were determined from the low fuel to oxygen ratio yields of carbon dioxide, water, and oxygen. The empirical chemical formula chosen was that formula which reasonably represented all three yields simultaneously, but no formal fitting procedure was employed.

The results for all three solid fuels exhibit the same qualitative features previously observed with gaseous and liquid fuels. In particular the carbon dioxide, water, and oxygen generally follow the simple relations of Equation 1 with carbon dioxide deviating to an extent consistent with the production of carbon monoxide and soot. As previously observed for other fuels, the water yield remains at the low equivalence ratio yield quite near the stoichiometric fuel to oxygen ratio. In general the total hydrocarbon yields at high fuel to oxygen ratios are less than expected on the basis of Equation 1. For oxygenated fuels this has previously been observed and is due to the reduced response of flame ionization detectors to oxygenated hydrocabons[7]. Further, using $CH_2$ as a formula for oxygenated hydrocarbons also underestimates the mass production rate. However, the total hydrocarbon yields at high fuel to oxygen ratios for polyethylene are lower than expected on the basis of previous work. This may result from condensation of high molecular weight oligomeric products of thermal decomposition of polyethylene.

TABLE 2. Summary of results

| Fuel | Empirical Formula | Ø<0.7 Normalized $CO_2$ Yield | %$O_2$ | %CO | CO | $H_2$ |
|---|---|---|---|---|---|---|
| | | | Vol. Pct. | | Normalized Yields | |
| Propane | $C_3H_8$ | 0.95 | 0.5 | 1.8 | 0.12 | 0.06 |
| Propene | $C_3H_6$ | 0.88 | 2.0 | 1.6 | 0.10 | 0.03 |
| Hexanes | $C_6H_{14}$ | 0.93 | 3.0 | 1.6 | 0.10 | 0.03 |
| Toluene | $C_7H_8$ | 0.78 | 8.0 | 0.7 | 0.05 | 0.01 |
| Methanol | $CH_3OH$ | 0.96 | 0.1 | 4.8 | 0.27 | 0.10 |
| Ethanol | $C_2H_5OH$ | 0.97 | 0.1 | 3.6 | 0.18 | 0.075 |
| Isopropanol | $C_3H_7OH$ | 0.96 | 2.0 | 2.4 | 0.12 | 0.05 |
| Acetone | $C_3H_6O$ | 0.94 | 0.7 | 4.4 | 0.21 | 0.045 |
| PE | $CH_2$ | 0.91 | 2.0 | 1.7 | 0.09 | 0.035 |
| PMMA | $C_5H_8O_2$ | 0.91 | 2.0 | 3.0 | 0.135 | 0.04 |
| Pine | $C_{0.95}H_{2.4}O$ | 0.93 | 2.0 | 3.2 | 0.155 | 0.03 |
| Methane(a) | $CH_4$ | -.-- | -.-- | 1.6 | 0.10 | -.-- |

(a) calculated fron the data of Cetegen,B.M., Zukoski,E.E., Kubota,T., NBS-GCR-82-402.

The results of the present solid fuel data and previous gas and liquid fuel data are summarized in Table 2.The present data are consistent with the previous observation of higher normalized carbon monoxide yields for oxygenated hydrocarbon fuels than hydrocarbon fuels. However, the carbon monoxide yield for ponderosa pine is less than expected based on its high O/C ratio. This probably results from the high water content of the volatiles which contributes to the O/C ratio, but does not contribute to carbon monoxide production. While some relation exists between O/C and H/C ratios and carbon monoxide yields, the chemical structure of the fuel volatiles must be considered in order to explain the carbon monoxide yields observed.

## COMPARISON WITH COMPARTMENT FIRE DATA

Of the fuels used in the present investigation, compartment fire data including fuel mass loss rate and species concentration measurements appear to exist only for wood. Tewarson[9] has recently reviewed and analyzed the existing wood data. His analysis led him to the conclusion that at least three mechanisms of thermal decomposition were required to explain the mass yield of carbon monoxide as a function of the equivalence ratio. His analysis of the data led to calculated carbon monoxide yields of up to 0.6gms/gm of fuel volatilized, over four times the maximum observed in this study. Yields of this magnitude were principally deduced from the small scale compartment fire data of Gross and Robertson[1], Tewarson[3], and Croce[6]. In these investigations the species measurements were made inside the compartment with uncooled probes.

The analysis of Tewarson[9] was based on several assumptions which may not have been satisfied. The most significant in terms of the present discussion is the assumption that the composition measurements made in the compartment were representative of the products of combustion. In their paper Gross and Robertson[1] indicated that errors in species concentration measurements resulted from the use of an uncooled probe. The magnitude of the errors incurred is best illustrated by an example presented by Gross and Robertson themselves[10]. In the same experiment both a water cooled and an uncooled probe were used for sampling at approximately the same location in the compartment. The uncooled probe measured 4.1% CO, 7.4% $CO_2$, and 12.8% $O_2$, while the cooled probe measured 1.2% CO, 1.5%$CO_2$, and 20.4% $O_2$(all measured on a dry basis). Clearly, the sample was drawn from a region of chemical reaction and the reaction continued within the uncooled probe.

If we assume that the sample drawn from the compartment is likely to be drawn from the reaction zone if the flame volume is large, a plot of the mass carbon monoxide yield as a function of the ratio of the flame volume to the enclosure volume may indicate in which experiments the chemical sampling was done incorrectly for the present purposes. While no measurements of flame volume were made, an estimate can be made by assuming the heat release per unit volume is the same as Orloff and deRis [11] found in the open, 1200kW/m$^3$. Using the measured mass loss rate and an estimated heat of combustion for wood volatiles of 15kJ/g[12], such a plot for Gross and Robertson's small enclosure data is shown in Figure 4. All the data with mass carbon

436

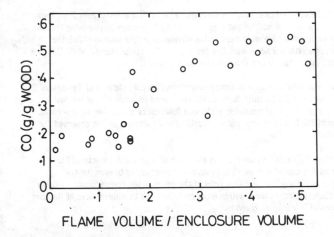

Figure 4. Mass carbon monoxide yields calculated by Tewarson[9] from the small scale enclosure data of Gross and Robertson[1] plotted as a function of the ratio of the estimated flame volume to the enclosure volume.

monoxide yields greater than 0.25 result from experiments in which the flame volume estimate is greater than 16.5% of the enclosure volume. Figure 5 is a plot of the existing wood compartment fire data as analyzed by Tewarson with the data with estimated flame volumes greater than 16.5% of the compartment volume indicated as solid symbols. In analyzing the data Tewarson used a $k_{O2}$ for the original wood and applied it to the wood volatiles. As a result stoichiometric conditions correspond to a stoichiometric fraction of 0.7 rather than the expected 1.0. Figures 4 and 5 show that the very large carbon monoxide yields deduced resulted from incorrect gas sampling.

The present attempt to identify inappropriate data is an after the fact analysis based on incomplete information and may not be sufficiently severe. Even where the chemical composition was not measured in the reaction zone, it may not be a representative sample of the exhaust gases due to unmixedness. It is also of note that the carbon monoxide yields from the compartment data are approximately 20% overestimated due to the lack

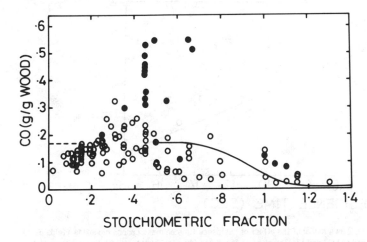

Figure 5. Mass carbon monoxide yields calculated by Tewarson[9] from data in References 1,3 and 6. Filled symbols indicate data which were sampled from the reaction zone based on the criterion that the estimated flame volume was greater than 16.5% of the enclosure volume. The solid line is the present results with the water correction removed and the dashed line is the oxidative pyrolysis data of Tewarson[9].

of a correction for water removed from the sampled gases before analysis. Finally, the exhaust rate from the compartment was not measured in any of the original studies but was inferred by Tewarson by requiring that the carbon dioxide and oxygen yields as a function of equivalence ratio in the compatment fire studies match that determined in Tewarson's flammability apparatus and some duct fire tests at the Bureau of Mines. While this is a plausible assumption, it clearly introduces uncertainty into the correlation of Figure 5.

In the light of the the present data and the limitations of the compartment fire correlations of Tewarson, it is clear that Tewarson's postulate of changes in the thermal decomposition of wood as a function of the fuel to oxygen ratio is not required to explain the mass carbon monoxide yield as a function of the fuel to oxygen ratio. The carbon monoxide yield as a function of the fuel to oxygen ratio is qualitiatively similar to that observed for simple evaporating liquids and gaseous fuels.

While the measurements from within the reaction zone made in the work of Gross and Robertson[1], Tewarson[3], and Croce[6] were made with uncooled probes, it is nonetheless interesting to examine the relationship between the carbon monoxide yield and the residence time of the gases in the compartment. Takeda and Akita[13] have proposed that reductions in heat release within a compartment due to unmixedness of fuel and oxygen may be described by a combustion efficiency, $\mu$, given by

$$\mu = 1 - \exp(-t_{res}/t_{mix})$$

where $t_{res}$ is a residence time, and $t_{mix}$ is a required mixing time for combustion. Motivated by this stirred reactor type expression, it is proposed that the effect of limited residence time in the compartment on carbon monoxide yield can be expressed as

$$Y_{CO}(t_{res}) - Y_{CO}(\infty) = [Y_{CO}(0) - Y_{CO}(\infty)] \exp(-t_{res}/t_{mix}) \qquad (2)$$

where $Y_{CO}(\infty)$ is the carbon monoxide yield uneffected by limited residence time, $t_{res}$ is the residence time given by the volume of the enclosure divided by the volume exhasut rate, and $t_{mix}$ is a yet to be defined mixing time. As all of Gross and Robertson's data[1] in their small compartment are in the fuel rich regime where a constant $Y_{CO}(\infty)$ is anticipated on the basis of all fuels examined to date, these data present a particularly simple system with which to test Equation 2. The pine results of the present investigation indicate that $Y_{CO}(\infty)=0.14$. If we remove the water correction from the present results to be consistent with Gross and Robertson's data, $Y_{CO}(\infty)$ is increased to 0.17. Figure 6 shows a plot of $\ln(Y_{CO}(tres)-0.17)$ as a function of the residence time for Gross and

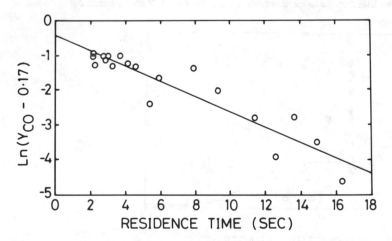

Figure 6. Evaluation of Equation 2 as a model of the effect of residence time on mass carbon monoxide yields using the small scale data of Gross and Robertson[1]. The slope of the plotted line is consistent with $t_{mix}$=4.5 seconds. Robertson's small enclosure data with residence times less than 18 seconds. Data with larger residence times were excluded as the choice of $Y_{CO}(\infty)$ dominates the behavior of the plot for large residence times. The slope of this plot is consistent with $t_{mix}$=4.5 seconds.

While the definition of the residence time used here is different than that used by Takeda and Akita, Figure 6 is direct chemical evidence of the effect of residence time on chemical species yields. The determined mixing time of 4.5 seconds should be considered as indicative of the order of magnitude of the required mixing time, rather than a quantitative determination, given the assumptions required for its determination and the quality of the original data.

It is of interest to note that all the fuel rich carbon monoxide yields determined from the data of References 1,3, and 6 are correlated by a plot of carbon monoxide yield as a function of the residence time. However, under fuel lean conditions the carbon monoxide yields show no correlation with residence time. Further, for fuel rich conditions the requirement that the flame volume be less than 16.5% of the enclosure volume is essentially equivalent to requiring the residence time to be greater than 7.5 seconds. For fuel lean conditions, requiring a residence time greater than 7.5 seconds excludes data which is not rejected by the volume ratio requirement of 16.5%. This indicates that the effect of residence time on product yields and combustion efficiency is different under fuel lean and fuel rich conditions. This result is not in accordance with the model of Takeda and Akita, but must be regarded as somewhat speculative given the quality of the data in References 1,3, and 6 for present purposes. It is significant to note that Takeda and Akita presented no chemical evidence for their model. Rather, they depended upon the results of calculations of the volatilization rate using a model incorporating their combustion efficiency expression to validate their model.

The force responsible for mixing in the hot layer is characterized by the momentum of the plume flow as is enters the hot layer, if all the momentum of the plume flow is dissipated within the compartment. The intensity of mixing processes in the layer under geometrically similar conditions is expected to be proportional to the ratio of the momentum flux to the mass flux as the flame enters the layer(the ratio of the force dissipated to the mass of gases involved). This ratio is simply the mass averaged flame gas velocity. According to the data and analysis of Cox and Chitty[14], the mass averaged veolcity increases as $Z^{1/2}$ in the continuous flaming region of open flames and becomes constant in the intermittent region. Under fuel rich conditions the layer interface is always in the continuously flaming region of the flame. As the required mixing time is expected to vary inversely with the mixing intensity, this indicates that the required mixing time will be greatest under fuel rich conditions. Both the required mixing time and the fraction of fuel not burned in the plume will decrease as the fuel to oxygen ratio decreases.

The $Z^{1/2}$ dependence of the mass averaged velocity also indicates that the required mixing time may decrease with increasing compartment scale under fuel rich conditions. However, fuel source size effects were not included in the analysis[14] and such effects may have equally strong scale effects. Scale effects cannot be examined using existing data as there are virtually no fuel rich, low residence time data available at large scale. This strong bias of low residence times for small compartments results from the use of very large cribs in small compartments to increase the steady burning period. The small scale data of Croce[6] suffers least from this tendency due to the use of Froude modeling concepts in experiment scaling.

For the purposes of toxic products production and transport, restricted compartment residence time results are not of direct relevance, as reaction will continue outside the compartment in regions which will be untenable on a thermal basis alone. That is to say a flame will extend from the the compartment vent under these conditions. The efficiency of the external flame in destroying toxic products then becomes a relevant consideration. If the air used for this combustion is unvitiated, it may be reasonable to expect the combustion to be essentially as efficient as burning in the open. However, this remains to be shown.

CONCLUSIONS

Major species production rates measured for polyethylene, poly(methyl methacrylate), and ponderosa pine as a function of the fuel to oxygen ratio are qualitiatively similar to that previously measured for simple gaseous and evaporating liquid fuels. The normalized carbon monoxide yield results support the observation that oxygenated hydrocarbons produce carbon monoxide more efficiently than hydrocarbons under fuel rich conditions.

It is not necessary to postulate multiple thermal decomposition mechanisms to explain the carbon monoxide yield of wood as suggested by Tewarson[9]. The carbon monoxide yield is qualitatively similar to simple gaseous and evaporating liquid fuels. Previous measurements indicating carbon monoxide yields as much as four times greater than the present data were the result of sampling within the reaction zone. Such measurements can not be used directly in the prediction of the production and spread of toxic products within a building.

Analysis of existing wood crib compartment fire data indicates that compartment residence times of 10-15 seconds are required for reaction to final products under fuel rich conditions. Under fuel lean conditions far shorter residence times appear to be required, indicating more intense mixing under these conditions.

REFERENCES

1.  Gross,D., Robertson, A.F., *Tenth Symposium(International) on Combustion* , The Combustion Institute, Pittsburgh, Pa.(1965), 931.

2.  Rasbash,D., Stark,W., The Generation of Carbon Monoxide by Fires in Compartments, Fire Research Note 614, Fire Research Station, Borehamwood, England, 1966.

3.  Tewarson,A., *Combustion and Flame*, 19, (1972) 101.

4.  Tewarson,A., *Combustion and Flame*, 19, (1972) 363.

5.  Tewarson,A., Steciak,J., *Comb. and Flame*, 53, (1983) 123.

6.  Croce, P., Modeling of Vented Enclosure Fires Part 1. Quasi-steady Wood Crib Source Fire, Factory Mutual Research Corp., Technical Report FMRC J.I. 7AOR5.GU, July 1978.

7.  Beyler,C.L., accepted for publication, *Fire Safety Journal* , 1985.

8.  Beyler, C.L., *Development and Burning of a Layer of Products of Incomplete Combustion Generated by a Buoyant Diffusion Flame*, Ph.D. Thesis, Harvard University, 1983.

9.  Tewarson, *Twentieth Symposium(International) on Combustion* , The Combustion Institute, Pittsburgh, Pa. (1985).

10. Gross,D., Robertson,A.F., NBS Report 8147, National Bureau of Standards, Washington D.C., 1964.

11. Orloff, L., deRis, J., *Nineteenth Symposium(International) on Combustion* , The Combustion Institute, Pittsburgh, Pa. (1982), 885.

12. Atreya,A., *Pyrolysis, Ignition, and Fire Spread on Horizontal Surfaces of Wood*, Ph.D. Thesis, Harvard University, 1983.

13. Takeda,H., Akita,K., *Nineteenth Symposium(International) on Combustion* , The Combustion Institute, Pittsburgh, Pa. (1982), 897.

14. Cox,G., Chitty,R., *Combustion and Flame*, 39, (1980) 191.

ACKNOWLEDGEMENTS

The author would like to thank Prof. Howard Emmons for his guidance through the course of this work. This research was supported by the United States Department of Commerce, National Bureau of Standards, Grant No. 60NANB4D0010.

# The Formation of Carbon Monoxide from Diffusion Flames

**S. LOMAX and R. F. SIMMONS**
Department of Chemistry
The University of Manchester Institute of Science and Technology
Manchester M6O 1QD, England

ABSTRACT

The conditions have been determined under which propane diffusion flames burning around sintered metal hemispheres in an atmosphere of air and nitrogen give a high concentration of carbon monoxide in the exhaust gases. Concentration profiles have also been obtained through representative flames and across the burner housing above the burner. The results show that when the flame burns with a deficiency of oxygen, the central core of the products contains a high concentration of propane as well as propylene, ethylene, acetylene and methane. In addition, just above the top of the burner, the oxygen concentration has fallen to almost zero, and even well away from the burner it is only a tenth of its original level. It is clear that pyrolysis reactions and secondary oxidation processes occur in the region above the burner and that it is these processes which are responsible for the increased formation of carbon monoxide which occurs with flames burning in a deficiency of oxygen.

INTRODUCTION

There has been much concern about the level of toxic gases in the vicinity of a fire, and a particular example is the formation of carbon monoxide (CO). The relative levels of CO and carbon dioxide ($CO_2$) have been determined in the combustion products from a model fire burning in a compartment [1], and the experimental conditions simulated the typical dimensions of a living room, its contents and the degree of ventilation likely to be encountered in a real fire. The formation of CO was favoured by vitiated conditions in the compartment, and up to 15% of the total oxides of carbon was CO.

This type of fire involves a diffusion flame and there are two basic mechanisms which can be envisaged by which the formation of CO can be enhanced in such flames. When a diffusion flame burns in a vitiated atmosphere the maximum temperature in the reaction zone is relatively low and the flame consists of a thin blue luminous region [2]. As a result, there is the possibility that CO can escape through the reaction zone into the surrounding atmosphere without being converted to $CO_2$. The other possibility is that the hot combustion products from the flame still contain unburnt fuel, or more likely the products of its pyrolysis, and that these react subsequently with residual oxygen when mixing with the surrounding atmosphere occurs.

The present work was undertaken to examine the relative importance of these two mechanisms. Sintered hemispherical burners were used to determine the experimental conditions under which a laboratory scale diffusion flame gives a high overall [CO] in the combustion products when propane is the fuel, and concentration profiles have been determined through diffusion flames burning on cylindrical burners and through the combustion products at two heights above the burner.

EXPERIMENTAL

Measurement of Overall Concentrations

The hemispherical burners (see Fig. 1) were made by cold pressing stainless steel powder directly onto a threaded backplate [3]. The required flow of propane and nitrogen was fed through the backplate by a 6 mm o.d. stainless steel tube which was water-cooled and a small cup was situated 3.5 cm above the burner to collect any water that condensed on the cooling jacket. The top of this cooling jacket was sealed into a water-cooled cone which was slightly smaller than a B45 glass joint. Two grooves were cut into the wall of this cone and Viton O-rings located in these grooves ensured a gas-tight seal when the burner was positioned in the burner housing [4].

The sintered hemisphere was situated about 5 cm below the rim of a 7.5 cm diameter Pyrex tube (94 cm long) which, in turn, was sealed axially inside the burner housing (10 cm diameter x 110 cm high) to give a chamber 30 cm high above the rim of the inner tube. The required flow of oxygen and nitrogen was fed up this inner tube, and a flow of nitrogen was fed up the annular jacket to minimise any recirculation of combustion products in the upper part of the burner housing. This arrangement gave a very stable counterflow diffusion flame and the dimensions of the apparatus ensured that all gas flows were laminar.

The combined gas flows passed out of the housing through a 26 mm i.d. tube, 4 cm from the top of the housing, and the major part of the water was removed by passage through two water-cooled condensers. Part of the flow was also passed through two traps cooled to -80 $^{\circ}$C, to remove the final traces of water vapour, and thence to a Pye 104 gas chromatograph, where CO was determined using a 2.5 m column of Molecular Sieve 5A and $CO_2$ using a 1.5 m column of Poropak Q. All quantitative measurements were made using peak areas from a Hewlett-Packard 3352 integrator and data system in conjunction with calibrations obtained by injecting samples of known composition. In a typical

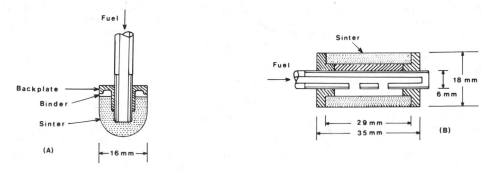

FIGURE 1. Details of burner construction.
(A) hemispherical burner, (B) cylindrical burner.

set of measurements the interval between determinations was about 30 minutes, which corresponded to at least 10 volume changes for the whole system, and thus a steady state always existed throughout the system when measurements were made.

The equivalence ratio through a diffusion flame varies from fuel rich to lean, but the overall experimental conditions for a given flame can be defined in terms of the overall stoichiometry ($\phi$), which is given by the oxygen flow required for stoichiometric combustion of the fuel divided by the actual oxygen flow. Thus a value of $\phi = 2.0$ implies that the oxygen flow was only half that required for stoichiometric combustion.

Concentration Profiles

In this part of the work, the burner was a hollow sintered cylinder [5], whose ends were sealed with stainless steel plates which were locked into position with a threaded pipe which passed through the centre of the cylinder (see Fig. 1). This pipe also served as the fuel inlet pipe and the upper part of the annular space was blocked off to restrict fuel flow to the bottom half of the cylindrical surface. This arrangement gave a flame which was parallel to and at a uniform distance from the burner surface, so that the lateral positioning of the probe was less critical than with a hemispherical burner. The probe was made from 6 mm o.d. quartz tubing ; the orifice was $\sim 70\,\mu$m, with an external diameter at the tip of $\sim 1$ mm, and the angle subtended at the tip was $\sim 15°$. A VG Q8 quadrupole mass spectrometer was used for the analyses; a background pressure of $10^{-9}$ Torr was routinely obtained in the mass spectrometer and the sample pressure was normally about $10^{-8}$ Torr.

The cracking patterns were determined for each of the species Ar, $O_2$, CO, $CO_2$, $C_3H_8$, $C_3H_6$, $CH_4$, $C_2H_4$, $C_2H_2$, and $H_2O$, together with their sensitivities (relative to argon). These enabled the mass spectral data to be converted to

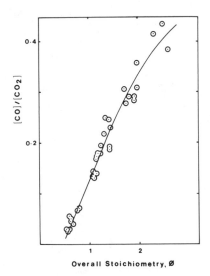

FIGURE 2. Variation of $[CO]/[CO_2]$ in the exhaust stream with overall stoichiometry.

composition profiles [5]; the peaks at m/e = 20(Ar), 32($O_2$), 29($C_3H_8$), 44($CO_2$), 39($C_3H_6$), 25($C_2H_2$), 27($C_2H_4$), 28(CO), 15($CH_4$) and 18($H_2O$) were used for the individual species after subtracting contributions from any other species to that peak. Argon was used as diluent to avoid the problem of identifying CO in the presence of $N_2$.


RESULTS

Overall Concentrations

For the initial set of determinations, the air flow was kept constant and $\phi$ varied by altering the flow of propane to the burner. In addition, to avoid any possible complication from carbon formation in the flame, nitrogen was added to the air to eliminate the carbon zone, and the proportion of this nitrogen was kept constant. This gave a flame which consisted of a single blue reaction zone, and flames were burned for $0.55 < \phi < 2.55$. The results in Fig. 2 show that as $\phi$ increased from 0.55 – 1.5 the ratio [CO]/[$CO_2$] increased linearly. At this point, the CO accounted for about 17% of the total oxides of carbon, but increasing $\phi$ above 1.5 only produced a modest increase in CO formation. For a given value of $\phi$ the actual gas velocities of the oxidant and fuel streams had a negligible effect on CO formation, but increasing the oxygen content of the oxidant flow from 16 – 21% for $\phi = 2.0$ produced a 30% increase in the [CO]/[$CO_2$] ratio. In contrast, when $\phi < 1.0$ the formation of of CO was not affected by the oxygen content of the oxidant stream.

The gas chromatographic results also enabled the residual oxygen in the exhaust stream to be determined and some typical results are shown in fig. 3. With $\phi < 0.65$, the residual oxygen was that expected on the basis of stoichiometric combustion of the fuel, but even with $\phi = 0.8$, i.e an overall 25% excess of oxygen, the experimental curve had begun to deviate from the theoretical line. It is particularly striking that for $\phi > 1.5$ about 10% of the

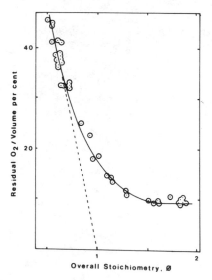

FIGURE 3. Variation of the residual oxygen in the exhaust stream with overall stoichiometry.

original oxygen remained.

To obtain further information about the overall conversion of the propane to CO and $CO_2$, the carbon balance was determined for a series of flames. This was done by adding a trace of neon to either the nitrogen flowing up the outer jacket of the burner housing or to the fuel itself, so that the ratio of the inlet flows of propane and neon gave the expected total flow of CO and $CO_2$ relative to neon in the exhaust gas if all the carbon was converted to oxides of carbon. Comparison with the experimental results (including unburnt propane) then gave the carbon balance. In practice, identical results were obtained from the two ways and when there was an excess of oxygen the carbon balance was close to unity. In contrast, for $\phi > 1.5$, about 35% of the initial propane was not accounted for. It is also striking that for these overall rich conditions about 10% of the carbon was still present as propane. There were also a few small peaks in the gas chromatograph which could not be identified with certainty, but these could not have accounted for the discrepancy. The most likely explanation for the missing carbon is the formation of formaldehyde which was present in the water condensed out from the exhaust stream, although this was not determined quantitatively.

Concentration Profiles

Figures 4 and 5 show the concentration profiles along a line at right angles to the midpoint of the burner surface for a flame with $\phi = 2.0$. The start of the profile corresponds to the position at which the tip of the probe was just touching the burner surface, namely 0.5 mm from the surface. It will be seen that the $[C_3H_8]$ falls rapidly to zero at about 4 mm, at which point the $[O_2]$ begins to rise, reaching a steady value at d = 15 mm. The $[O_2]$ found experimentally at this point was that expected from the known input flows of

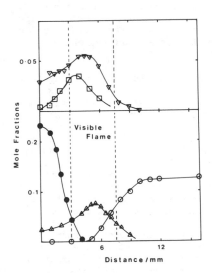

FIGURE 4. Concentration profiles for the major species through a propane diffusion flame ($\phi = 2.0$). ● $C_3H_8$; ⊙ $O_2$; △ $H_2O$; ⊟ CO; ▽ $CO_2$.
Input flows to:
     Burner: 400 cm$^3$ min$^{-1}$ $C_3H_8$ + 1020 cm$^3$ min$^{-1}$ Ar.
     Jacket: 1000 cm$^3$ min$^{-1}$ $O_2$ + 7480 cm$^3$ min$^{-1}$ Ar.

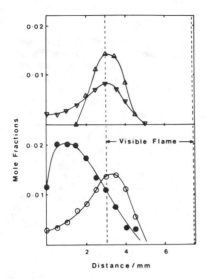

FIGURE 5. Composition profiles for the minor species through a propane diffusion flame ($\phi = 2.0$). Input flows as for Fig. 4.
● $C_3H_6$; △ $C_2H_4$; ◎ $C_2H_2$; ▽ $CH_4$.

oxygen and argon to the jacket. In contrast, the $[C_3H_8]$ found at the burner surface was nearly 20% lower than expected from the input flows and this must arise from the forward diffusion of $C_3H_8$ into the flame as a result of the very steep concentration gradient for $C_3H_8$ between the burner surface and d = 4 mm. The profiles for CO, $CO_2$ and $H_2O$ have the expected shapes, reaching their maxima at 3.8, 4.8 and 5.4 mm respectively. Figure 5 shows the profiles for the minor species; $C_3H_6$ is formed very close to the burner surface, while $C_2H_4$, $C_2H_2$ and $CH_4$ all reach their maximum at the inner edge of the flame.

The profiles through the other flames, which had values of $\phi$ ranging from 0.7 to 2.0, showed essentially the same spatial separation and maximum concentrations for $CO_2$ and $H_2O$. As $\phi$ decreased, however, the maximum concentrations for CO and the intermediate hydrocarbons decreased. For example, with $\phi = 1.0$ the maximum concentrations of $C_2H_4$, $C_2H_2$, and $CH_4$ were about a factor of two lower, while the maximum for $C_3H_6$ was only about a quarter of its value in the flame with $\phi = 2.0$. Comparison of the [CO] immediately outside the visible region of the flame showed that CO persisted further out when there was a deficiency of oxygen, but the actual [CO] in this region was not markedly affected by the value of $\phi$. As a result, it seems most improbable that the high CO levels obtained when the flame burned in a deficiency of oxygen could arise through the escape of CO through the flame into the surrounding atmosphere.

There was one major differnce in the appearance of the flames burning in a deficiency of oxygen from those with $\phi < 1$. With the latter, there was a complete envelope of flame surrounding the cylinder, but the flames burning in a deficiency of oxygen had an open top. This difference is reflected in the concentration profiles higher up the burner housing. Figures 6 and 7 show the profiles for the flame with $\phi = 2.0$ for a height of 4.8 mm above the top of the burner. In these profiles, the same zero position has been used as for the profiles through the flame, so that the position vertically above the midpoint of the top of the burner has a distance co-ordinate of -9.5 mm. Figure 6 shows

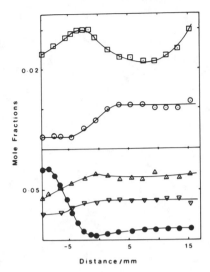

FIGURE 6. Composition profiles for the major species across the burner housing 4.8 mm above the burner ( $\phi = 2.0$ ). Input flows as for Fig. 4 and 5.
● $C_3H_8$; ○ $O_2$; △ $H_2O$; ▢ $CO$; ▽ $CO_2$.

that the steady oxygen level towards the housing wall has fallen to a tenth of its initial concentration, and it is even lower immediately above the burner. It is striking that the $[C_3H_8]$ immediately above the burner is still a third of its initial level (see Fig. 4). The profiles for $CO_2$ and $H_2O$ are now much flatter and their average concentrations are only slightly lower than the maximum concentrations found in the flame. The $[CO]$ passes through a shallow maximum at $-3$ mm, and the average $[CO]$ is significantly higher (relative to $[CO_2]$) than in the flame. The most probable explanation of the increase in $[CO]$ as the wall of the housing is approached is that some recirculation of combustion products occurs. Qualitative support for this view comes from the fact that although the flame itself had only a very weak carbon zone, a brown film of carbon was deposited on the housing wall and the probe tip became restricted by carbon after a period of use well outside the flame at this level.

Figure 7 shows that the profiles for $C_3H_8$, $C_2H_4$, $C_2H_2$ and $CH_4$ 4.8 mm above the burner are also much broader than lower down in the flame and, apart from $C_3H_6$ the concentrations are very similar to those in the flame. Thus the central core of combustion products contains a high concentration of these pyrolysis products. The concentrations of these species also have a tendency to rise as the housing wall is approached and this is a further indication that recirculation of combustion products is occurring.

The profiles 1 cm above the top of the burner for the same flame show an intermediate situation. The $[C_3H_8]$ above the burner is about half the value shown in Fig. 4 for near the burner surface, while the $[O_2]$ well away from the flame has already fallen to the value found at 4.8 cm above the burner. The profiles for $CO$, $CO_2$ and $H_2O$ are already very much broader and significant levels of $C_2H_4$, $C_2H_2$ and $CH_4$ are also present well away from the flame.

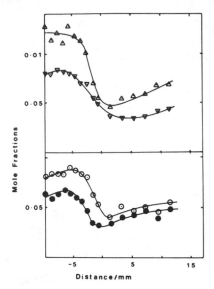

FIGURE 7. Composition profiles for the minor species across the burner housing 4.8 mm above the top of the burner ($\phi = 2.0$). Input flows as for Fig. 4 and 5. ● $C_3H_6$; △ $C_2H_4$; ☉ $C_2H_2$; ▽ $CH_4$.

DISCUSSION

The present results show that high concentrations of CO can be present in the exhaust stream from a diffusion flame burning in a vitiated atmosphere, and the levels observed are comparable to those obtained from model fires burning under similar conditions. It is clear that the ratio $[CO]/[CO_2]$ in the exhaust stream is primarily controlled by the relative flows of fuel and oxygen, but the actual velocity of the oxidant stream or of the fuel from the burner surface has a negligible effect on the formation of CO. In contrast, increasing the oxygen content of the oxidant flow actually increases CO formation when there is an overall deficiency of oxygen. The primary reason for this behaviour can be seen from the composition profiles, which clearly show that when $\phi > 1$ part of the fuel escapes upwards through the open top of the flame, and its subsequent reaction in the region above the burner results in the formation of more CO.

This possibility has been confirmed experimentally by adding trace amounts of $C_3H_8$ to the oxidant flow of a flame with $\phi = 0.7$; Fig. 8 shows that this resulted in a marked increase in CO formation. Any $C_3H_8$ carried into the flame would have been converted to $CO_2$, since there was an excess of $O_2$ present, and thus the CO must have been formed when the hot combustion products from the flame mixed with the excess oxidant in the upper part of the burner housing. This would also explain the formation of formaldehyde; this is unlikely to survive the high temperature occurring in the flame, but it is a probable product of combustion at the lower temperature existing in the upper part of the burner housing.

The present paper only reports the results that are pertinent to the formation of CO and a more detailed report of the concentration measurements

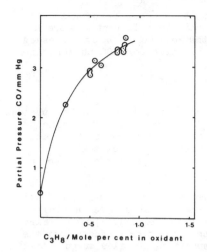

FIGURE 8. Effect on CO formation of adding $C_3H_8$ to the oxidant flow ($\phi$ = 0.7).

and their mechanistic interpretation will be made elsewhere. Nevertheless, a brief discussion of the important reactions is appropriate. In the pyrolysis region of the flame (i.e. between the burner surface and the inner luminous edge of the flame), $C_3H_6$, $C_2H_4$, and $CH_4$ were all detected and the following reactions represent the most plausible mechanism for their formation.

$$X + C_3H_8 = HX + C_3H_7 \qquad (1)$$

$$i\text{-}C_3H_7 = H + C_3H_6 \qquad (2)$$

$$n\text{-}C_3H_7 = CH_3 + C_2H_4 \qquad (3)$$

$$CH_3 + C_3H_8 = CH_4 + C_3H_7 \qquad (4)$$

In this scheme X is any radical species which can abstract a hydrogen atom from $C_3H_8$. The most probable route for the formation of $C_2H_2$ is hydrogen abstraction from $C_2H_4$ and the subsequent pyrolysis of $C_2H_3$. The same basic mechanism can also account for the consumption of $C_3H_8$ which occurs above the burner.

The maximum in the profile for CO at –3 mm and a height of 4.8 mm above the burner (see Fig. 6) clearly indicates that CO is still being formed in this region. In addition, although the profiles for $CO_2$, $H_2O$ and $O_2$ are quite flat moving out into the burner housing, their concentrations fall in the central core of the combustion products immediately above the burner. As the $O_2$ level is already very low, it is unlikley that any direct reaction involving oxygen is a major route for the formation of CO, but both $CO_2$ and $H_2O$ could be involved in its formation. $CO_2$ is diffusing back into a region which is fuel rich and thus there is the possibility that some CO is formed by the water gas equilibrium which will be maintained by reaction (5).

$$CO_2 + H = CO + OH \qquad (5)$$

Reaction (5) has an activation energy of 26.5 kcal $mol^{-1}$[6] and thus it will not be a very fast process, but part of the CO formation almost certainly arises from this route. The fall in the [$H_2O$] in the central core of combustion products above the burner also indicates that it is being removed in this region. It is probable that the combustion products in this central core are in thermal and chemical equilbrium and hence the removal of OH by processes such as reaction (6) will not only remove $H_2O$ but also form CO through the intermediate formation of formaldehyde [7].

$$OH + C_2H_4 = CH_3 + CH_2O \quad (6)$$

There was no clear evidence, however, from the mass spectra that formaldehyde was present, but its major peaks (at $m/e = 29$ and 28) coincide with those of other species. Thus a high concentration would need to be present before there was any significant effect on the ratio of the peaks at $m/e = 29$ and 43; within experimental error, this ratio was always that expected for $C_3H_8$.

The situation which exists when a real fire burns in a compartment will be different to that existing in the present work, but there are sufficient similarities to enable the present results to be applied to the real fire situation with confidence. In the present work the combustion products left the burner housing at the top, but even so, the composition profiles suggest that some recirculation of combustion products occurred near the wall of the housing. It is also clear from the results, that secondary reactions in the region above the burner is important, both from the point of view of pyrolysis of excess fuel and also for the formation of CO. In the initial stages of a real fire burning in a room there will be a closed envelope of flame, but as the fire develops and grows, the plume of hot gases from the fire will reach the ceiling of the compartment and spread out sideways. At this stage, the fire is essentially an open topped diffusion flame, so that the central core of the plume of hot gaes from the fire will be comprised mainly of pyrolysis products. These can only leave the compartment through any openings that may be present, and these will also admit the air which sustains the fire. In a closed compartment, therefore, these hot products must mix with the atmosphere in the compartment by recirculation, and some recirculation will also take place even when there is a ventilation opening present. The same type of situation will then exist as in the present work from the point of view of the occurrence of secondary reactions. The details of the mixing processes will be different in the two cases, however, and in the compartment fire they are likely to be less reproducible, unless the experimental conditions are very carefully controlled. Nevertheless, the present work clearly shows that the high [CO] arises from secondary reactions which occur in the housing above the flame and essentially the same situation can occur when a fire burns in a confined space.

This paper forms part of the work of the Fire Research Station, Building Research Establishment, Department of the Environment. It is contributed by permission of the Director, Building Research Establishment.

REFERENCES

1. Raferty, M.M.: Building Research Establishment, Note N53/81.
2. Simmons, R.F. and Wolfhard, H.G.: Combustion & Flame, 1, 155 (1957).
3. Lomax, S.: Ph.D. Thesis, University of Manchester, 1985.
4. Crowhurst, D. and Simmons, R.F.: Fire & Materials, 7, 62 (1983).
5. Crowhurst, D. and Simmons, R.F.: Combustion & Flame, 59, 167 (1985).
6. Westley,F.: "Table of Recommended Rate Constants for Chemical Reactions Occurring in Combustion," NSRDS-NBS 67 (1980).
7. Baldwin,R.R.and Walker R.W.: Eighteenth Symposium (International) on Combustion," p. 819, The Combustion Institute, Pittsburgh,1981.

# Scale Effects on Fire Properties of Materials

**A. TEWARSON and J. S. NEWMAN**
Factory Mutual Research Corporation
1151 Boston-Providence Turnpike
Norwood, Massachusetts 02062, USA

ABSTRACT

The scale effects on fire properties have been examined for materials in pool-like, box-like, and crib-like configurations.

For turbulent fires of a material, with varying sizes and geometrical arrangements, a chemical similarity was found for each specified value of ventilation and decomposition mode. The decomposition mode in the combustion of the material was found to be important for the production of CO and particulates.

## I   INTRODUCTION

Mathematical fire models are now available to evaluate the fire performance of materials in buildings and to estimate the hazards presented by such fires and required protection. Fire models require numerous input parameters, some related to materials and others related to environment. The input parameters are generally quantified in small-scale experiments; it is thus necessary that fire scale effects (if any) on the parameters be known and corrected accordingly.

Input parameters related to materials are needed by the models to describe: 1) fire initiation and growth; 2) mass flow of fuel vapors from the surface of the material; 3) generation of heat and chemical compounds; 4) light attenuation by particulates; 5) biological and corrosive effects of chemical compounds; and 6) efficiency of fire detection, suppression, and extinguishment.

In this paper, parameters describing the generation rates of heat and chemical compounds, and light attenuation by particulates have been considered.

## II   CONCEPTS

The concepts have been described in detail elsewhere[1-7]. As the fire scale changes, the mass flow rate of fuel vapors from the surface of the material and generation rates of heat and chemical compounds change. The parameters describing these rates, however, should be independent of the fire scale, if fires are turbulent and the decomposition chemistry of the material remains invariant.

The total mass flow rate of the mixture of fire products and air, $\dot{m}_t$, can be expressed as:

$$\dot{m}_t = \dot{m}_a + \dot{m}_f \quad , \tag{1}$$

where $\dot{m}_a$ = mass flow rate of air (g/s); and $\dot{m}_f$ = mass flow rate of fuel vapors (g/s). If we express the ratio between $\dot{m}_f$ and $\dot{m}_t$ as $X_f$, then the following relationships can be derived for the generation rates of heat and chemical compounds:

$$\dot{G}_j = X_j \dot{m}_t = Y_j X_f \dot{m}_t \quad , \tag{2}$$

$$\dot{Q}_A = (H_T/k_j) X_j \dot{m}_t = (H_T/k_j) Y_j X_f \dot{m}_t \quad , \tag{3}$$

and, 

$$\dot{Q}_C = \chi_C H_T X_f \dot{m}_t = \Delta T c_p \dot{m}_t \quad , \tag{4}$$

where, $\dot{G}_j$ = mass generation of a chemical compound (g/s); $X_j$ = mass fraction of the compound (g/g); $Y_j$ = mass of the compound generated per unit mass of the fuel vapors defined as the yield of the chemical compound (g/g); $\dot{Q}_A$ = generation rate of heat associated with chemical reactions in the fire, defined as the actual heat release rate (kW); in Eq (3), compound j is associated with complete combustion; $H_T$ = net theoretical heat of complete combustion per unit mass of the fuel vapors (kJ/g); $k_j$ = maximum possible theoretical yield of the compound (g/g); $\dot{Q}_C$ = convective heat release rate (kW); $\chi_C$ = convective component of combustion efficiency; and $c_p$ = average specific heat of the mixture of fire products and air at the gas temperature (kJ/g K).

The theoretical heat release rate, $\dot{Q}_T$, and the theoretical generation rate of a chemical, $\dot{G}_{Tj}$, can be expressed as

$$\dot{Q}_T = H_T X_f \dot{m}_t \quad . \tag{5}$$

$$\dot{G}_{Tj} = k_j X_f \dot{m}_t \quad . \tag{6}$$

The combustion efficiency, $\chi_A$, is defined as the ratio between $\dot{Q}_A/\dot{Q}_T$, and the generation efficiency of a chemical compound, $f_j$, is defined as the ratio between $\dot{G}_j$ and $\dot{G}_{Tj}$. Thus

$$\chi_A = f_j = Y_j/k_j \quad . \tag{7}$$

Equation (7) implies that the ratios between various molecules of chemical compounds generated in the fire are conserved for various degrees of completeness of combustion. From Eqs (3) and (7):

$$\dot{Q}_A = H_T \, \chi_A \, X_f \, \dot{m}_t = H_T \, f_j \, X_f \, \dot{m}_t \quad . \tag{8}$$

In Eqs (4) and (8) $\chi_C \, H_T$ and $\chi_A \, H_T$ can be defined as convective and actual heat of combustion ($H_C$ and $H_A$ respectively). In a similar fashion, $\chi_R \, H_T$ can be defined as the radiative heat of combustion, $H_R$, where $\chi_R$ is the radiative component of the combustion efficiency.

The fraction of light attenuated by particulate, $I/I_0$, can be expressed as

$$D = \sigma_\lambda = \xi_\lambda \, X_S \, \rho_T = \xi_\lambda \, Y_S \, X_f \, \rho_T \quad , \tag{9}$$

where $D = (1/\ell) \ln (I_0/I)$, defined as the optical density per unit path length $\ell$; $\sigma_\lambda$ = extinction coefficient of particulates $(m^{-1})$; $\xi_\lambda$ = specific extinction coefficient of particulates $(m^2/g)$; $\rho_T$ = density of fire products and air $(g/m^3)$; and $\lambda$ = wave length of light $(\mu)$.

In order to examine the scale effects on the parameters, experiments can be performed at various fire scales and measurements made for $\dot{m}_f$, $\dot{G}_j$, $\dot{Q}_C$, and $D$. The accuracy of such an examination can be enhanced if the fire products and air, downstream of the reaction zone of the fire, are captured and are well mixed before the analytical measurements.

If the fire products and air are distributed nonuniformly and if heat losses and $\dot{m}_t$ values are unknown, such as in enclosure fires, then inter-relationships between $\dot{m}_f$, $\dot{G}_j$, $\dot{Q}_C$, and $D$ are needed for examining the scale effects.

2.1 Interrelationships Between Mass Flow Rate of Fuel Vapors, Generation Rates of Heat and Chemical Compounds, and Light Attenuation by Particulates

From Eqs (2), (4), and (7)

$$\Delta T/X_j = (1/c_p) \, (H_T/k_j) \, (\chi_C/f_j)$$

$$= (1/c_p) \, (H_T/k_j) \, (\chi_C/\chi_A) \quad , \tag{10}$$

where $j$ is a compound produced (or consumed) in those reactions where heat is also produced. For example, for oxygen $f_0$ is defined as the depletion efficiency of oxygen.

From Eqs (2) and (9):

$$D/X_j = \xi_\lambda \, \rho_T \, (Y_S/Y_j) \quad . \tag{11}$$

From Eqs (4) and (9):

$$\Delta T/D = (1/\xi_\lambda \, c_p \, \rho_T) \, (\chi_C/f_s) \, (H_T/k_s) \quad . \tag{12}$$

453

In turbulent fires, $\Delta T/X_j$, $D/X_j$, and $\Delta T/D$ would be expected to be conserved from one location to another, because various ratios between $H_T$, $\chi_i$, $k_j$, $f_j$, $Y_j$, $Y_s$, and $\xi_\lambda$ are conserved. For turbulent fires, the radiative component of the combustion efficiency, $\chi_R$, is also expected to be conserved.

In enclosure fires, $\Delta T$, $X_j$, and $D$ can be measured at various locations and the scale effects can be examined using Eqs (10) to (12), irrespective of mixing and dilution of the chemical compounds. However, in the analyses, it is necessary to differentiate between effects related to fire scale, to fire chemistry, and to heat losses.

## 2.2 Factors Affecting the Mass Flow Rate of Fuel Vapors, Generation of Heat and Chemical Compounds, and Light Attenuation by Particulates

2.2.1 Fire Scale. Fire scale affects the mass flow rate of fuel vapors from the surface of the material, due to variations in the flame radiative heat flux, $\dot{q}_{fr}''$. As fire scale increases, $\dot{q}_{fr}''$ and the corresponding value of the mass flow rate of fuel vapors per unit surface area of the material, $\dot{m}_f''$, increase and finally approach their asymptotic values; $\dot{m}_f''$ also increases with the external heat flux, $\dot{q}_e''$. Variations in the oxygen concentration of the environment also affect $\dot{q}_{fr}''$ and $\dot{m}_f''$. When the oxygen concentration in the environment is decreased below the ambient value, $\dot{m}_f''$ and $\dot{q}_{fr}''$ decrease until the flame extinction limit is reached. When the oxygen concentration in the environment is increased above the ambient value, $\dot{m}_f''$ and $\dot{q}_{fr}''$ increase until they reach their respective asymptotic values (oxygen concentration greater than about 30 percent with a material surface area of about 0.008 $m^2$).[3]

2.2.2 Fire chemistry. Fire chemistry is affected by fire ventilation. We define a ventilation parameter, $\phi$, as

$$\phi = \dot{m}_a/\dot{m}_f k_a , \qquad (13)$$

where $k_a$ = theoretical mass of air consumed per unit mass of fuel vapors, also defined as the stoichiometric mass air-to-fuel ratio (g/g).

When $\phi > 1$, fires are defined as well ventilated and when $\phi < 1$, fires are defined as underventilated. From Eqs (1) and (13):

$$X_f = 1/(1 + k_a \phi) . \qquad (14)$$

The parameters, $H_i$, $\chi_i$, $Y_j$, $f_j$, and $\xi_\lambda$, are strong functions of the chemical structure of the material, as well as the decomposition processes followed by the material at various fire stages. In turbulent fires, these parameters are expected to be conserved for noncharring materials but vary with the extent of char formation, at various fire stages, for the char-forming materials.

2.2.3 Heat Losses. Values of $\Delta T$ are affected by heat losses. Laboratory-scale experiments can be designed such that heat losses are negligibly small. In enclosure fires, the heat losses could be quite significant, and, if the losses are not accounted for, then $\Delta T$ values would be in error.

Furthermore, in laboratory-scale experiments the distribution of $\dot{Q}_A$ into $\dot{Q}_C$ and $\dot{Q}_R$ can be quantified quite accurately, whereas in enclosure fires this distribution is not well defined. The distribution, however, can be defined in terms of the heat flow out of the enclosure, $\dot{Q}_g$, and heat lost within the enclosure, $\dot{Q}_\ell$. The relationship developed for $\dot{Q}_C$ and $\chi_C$ will still apply to $\dot{Q}_g$ and $\chi_g$ (combustion efficiency component associated with the temperature of the gas flowing out of the enclosure).

## III  EXPERIMENTS

Experiments were performed in three combustibility apparatuses: a 10-kW[1-7] scale, a 500-kW scale[8], and a 5000-kW scale apparatus[9].

In the apparatuses, the mass flow rate of the fuel vapors is monitored. The fire products are diluted and well mixed with ambient air as they are captured in the sampling ducts of each apparatus. The maximum flow rates of the mixture of fire products and air used in the 10-, 500-, and 5000-kW scale apparatuses are about 0.03, 2.0, and 28 $m^3/s$ respectively. The measurements made in the sampling ducts include: total mass and volumetric flow rate of the mixture of fire products and air; gas temperature; light attenuation by particulates*; and concentration of $CO_2$, CO, hydrocarbons, and water*. The output from all the instruments is stored and analyzed by a computer.

## IV  RESULTS

### 4.1  Physical Similarity in Terms of Mass Flow Rate of Fuel Vapors from the Surface of the Material

The average steady state values of $\dot{m}_f''$, including the asymptotic values obtained by increasing either the surface area of the material or by increasing the oxygen concentration in the environment are listed in Table 1. A reasonable agreement can be noted between the asymptotic values of $\dot{m}_f''$ obtained by two different types of experiments. The data in Table 1 suggest that the fire scale effects on $\dot{m}_f''$ due to flame radiative heat flux can be compensated in the laboratory-scale fires by using either oxygen concentration in the environment greater than the ambient value or by using external heat flux.

### 4.2  Chemical Similarity in Terms of $H_i$, $Y_j$, and $\xi_\lambda Y_s$

The data for $H_i$, $Y_j$, and $\xi_\lambda Y_s$ are listed in Tables 2 and 3. An examination of the data indicates that, within the experimental variation of the data, there is a chemical similarity between various scale fires of each material under a specified mode of decomposition for well ventilated fires. For cellulosic material, flaming mode is defined as the initial fire stage, where flames are relatively taller; flaming/smoldering mode is defined as a mode which follows the initial flaming mode. In the flaming/smoldering mode, flames are relatively shorter accompanied by surface glowing; flames are intermittent and do not cover the entire surface of the material.

---

* not measured in the 5000-kW scale apparatus

TABLE 1. Average Steady State Values of the Mass Flow Rate of Fuel Vapors per Unit Surface Area of the Material[a]

| Material | Surface Area ($m^2$) | $\dot{m}_f''$ ($g/m^2 s$) | |
|---|---|---|---|
| | | $X_0 = 0.233$ | $X_0 > 0.233$[b] |
| **Pool-like Configuration** | | | |
| Polyethylene/42% Chlorine | 0.008 | – | 7 |
| Polyvinyl Chloride | 0.008 | 15 | 16 |
| Polyoxymethylene | 0.008 | 6 | 16 |
| Methanol | 0.008 | 16 | 20 |
| | 2.37 | 20[b] | – |
| Flexible polyurethane foams | 0.008 | – | 21 to 27 |
| Rigid polyurethane foams | 0.008 | – | 22 to 25 |
| Polypropylene | 0.008 | 7 | 24 |
| Polyethylene | 0.008 | 8 | 26 |
| High temp. hydrocarbon fluids | 0.008 | – | 27 to 30 |
| | 2.37 | 25 to 29[b] | – |
| Polymethylmethacrylate | 0.008 | 12 | 28 |
| | 2.37 | 30[b] | – |
| Polystyrene | 0.008 | 14 | 38 |
| | 0.93 | 34[b] | – |
| Heptane | 0.008 | 36 | 63 |
| | 1.17 | 66[b] | – |
| **Three-Dimensional Arrangement** | | | |
| Corrugated paper boxes | 0.06[c] | 9;14[e] | – |
| with shredded paper | 53[d] | 14 | |
| Corrugated paper boxes | 0.06[c] | 14[e] | – |
| with a foam | 53[d] | 10 | – |
| Pine wood cribs | 0.26 to 11 | 8 to 11 | – |

[a] no external heat flux unless specified;

[b] asymptotic value;

[c] total exposed surface area of single box (0.1 x 0.1 x 0.1 m);

[d] total exposed surface area of sixteen boxes (four boxes on one pallet load, two pallet load wide x two pallet load deep x two tiers high, each box dimensions 0.53 x 0.53 x 0.53 m);

[e] 25 $kW/m^2$ of external heat flux on four sides.

For cellulosic materials, chemical similarity in terms of $H_i$ and $Y_{CO_2}$ is maintained for both flaming and flaming/smoldering modes. Chemical similarity in terms of $Y_{CO}$ and $\xi_\lambda Y_S$, however, is maintained only if the decomposition mode does not change. For example, the average value of $Y_{CO} = 0.0059$ g/g in the flaming mode (except for cellulose and densely packed paper box) and 0.097 g/g for the flaming/smoldering mode. The average value of $\xi_\lambda Y_S$ in the flaming mode is 0.0075 $m^2$/g for cellulose and paper box and 0.078 $m^2$/g for oak, fir, and pine. In the flaming/smoldering mode, the average value of $\xi_\lambda Y_S = 0.67$ $m^2$/g for paper box, 0.013 $m^2$/g for cellulose, and 0.15 $m^2$/g for pine-wood crib. In oxidative-pyrolysis, the average value of $\xi_\lambda Y_S = 0.80$ $m^2$/g for oak and pine.

TABLE 2.  Average Yields of $CO_2$, CO, and HCN and Heat of Combustion for
Well Ventilated Fires of Materials in Pool-Like Configuration

| Material | Surface Area $(m^2)$ | $Y_j$ (g/g) | | | $H_1$ (kJ/g) | | |
|---|---|---|---|---|---|---|---|
| | | $CO_2$ | CO | HCN | Actual | Conv. | Rad. |
| Methanol | 4.68 | 1.29 | < 0.001 | – | 18.7 | 15.6 | 3.1 |
| | 2.32 | 1.30 | < 0.001 | – | 18.8 | 15.7 | 3.1 |
| | 0.008 | 1.32 | < 0.001 | – | 19.4 | 17.1 | 2.3 |
| Rigid Polyurethane Foam | 7 | 1.50 | 0.027 | 0.010 | 16.4 | 10.8 | 5.6 |
| | 0.008 | 1.51 | 0.036 | 0.012 | 15.8 | 6.5 | 9.3 |
| Polymethylmethacrylate | 2.37 | 2.11 | 0.008 | – | 24.2 | 15.8 | 8.4 |
| | 0.073 | 2.10 | 0.010 | – | 23.8 | 14.9 | 8.9 |
| | 0.008 | 2.15 | 0.011 | – | 24.4 | 17.9 | 6.5 |
| High Temperature Hydrocarbon Fluids | 2.37 | – | – | – | 35.6 | 23.8 | 11.8 |
| | 0.008 | – | – | – | 38.2 | 25.1 | 13.1 |
| Heptane | 0.92 | 2.83 | 0.015 | – | 41.2 | 26.8 | 14.4 |
| | 0.059 | 2.92 | 0.0090 | – | 42.5 | 26.2 | 16.3 |
| | 0.041 | 2.82 | 0.0081 | – | 40.9 | 26.9 | 14.0 |
| | 0.008 | 2.84 | 0.0091 | – | 40.8 | 24.8 | 16.0 |

4.3  Chemical Similarity in Terms of the Ratios of Generation Rates of Heat
and Chemical Compounds, Light Attenuation by Particulates, and Depletion
Rate of Oxygen

The generation of $CO_2$ and depletion of $O_2$ are associated with the generation of heat; an interrelationship between $\Delta T$, $X_{CO_2}$, and $X_O$ is thus expected:

$$\frac{(\Delta T/T_a)/(X_{CO_2} - X_{CO_2,a})}{(\Delta T/T_a)/(X_{Oa} - X_O)} = (k_O/k_{CO_2}) \, (f_O/f_{CO_2}) \tag{15}$$

where $X_{CO_2}$ and $X_O$ = measured mass fractions of $CO_2$ and $O_2$ respectively; $X_{CO_2,a}$
and $X_{Oa}$ = measured ambient mass fractions of $CO_2$ and $O_2$ respectively; $k_O$ and
$k_{CO_2}$ can be calculated from the measured elemental compositions of the materials.  Chemical similarity in terms of $f_O/f_{CO_2}$ thus can be established for
various fire scales, irrespective of mixing and dilution of the compounds.

Our laboratory-scale data indicate that $f_O/f_{CO_2}$ is conserved for numerous
materials that we have examined, under a variety of experimental conditions,
where $f_O$ = 0.98 $f_{CO_2}$.  The data shown in Figure 1, for the enclosure fires of
wood cribs under a variety of ventilation conditions, enclosure sizes, and
crib sizes support our conclusion.

The data for $f_O/f_{CO}$ indicate a strong dependency on the chemical structure of the material, decomposition chemistry, and fire ventilation; the
effect of ventilation on this ratio is shown in Figure 2 for pine wood crib
fires in enclosures and in our 10-kW scale apparatus.  Data for cellulose and
red oak have also been included in the figure.  The data in Figure 2 indicate
that $f_O/f_{CO}$ is conserved for each value of $X_f$ for various fire scales.  With

457

TABLE 3. Average Yields of $CO_2$, CO, and Gaseous Hydrocarbons, Heat of Combustion, and Specific Extinction Coefficient for Well Ventilated Fires of Cellulosic Materials

| Material | Configuration | Total Exposed Surface Area (m²) | $\dot{q}_e''$ (kW/m²) | Combustion Mode | $H_1$ (kJ/g) Actual | $H_1$ (kJ/g) Conv. | $H_1$ (kJ/g) Rad. | $Y_j$ (kJ/g) CO | $Y_j$ (kJ/g) CO₂ | $Y_j$ (kJ/g) HC[a] | $\xi_\lambda \, y_s$ [b] (m²/g) |
|---|---|---|---|---|---|---|---|---|---|---|---|
| Corrugated Paper | Box with sheets of paper (25% by weight) | 0.06 | 0 | Flaming | 11.1 | 9.4 | 1.7 | 0.0082 | 1.12 | c | 0.0053 |
| Corrugated Paper | Box with shredded paper (62% by weight) | 0.06 | 0 | Flaming | 12.4 | 9.0 | 3.4 | 0.019 | 1.25 | 0.001 | 0.0067 |
| | | | | Flaming/Smold. | 10.8 | 6.8 | 4.0 | 0.12 | 1.11 | 0.002 | 0.062 |
| | | | 10 | Flaming | 11.6 | 8.7 | 2.9 | 0.021 | 1.17 | 0.001 | 0.0067 |
| | | | | Flaming/Smold. | 11.2 | 6.8 | 4.4 | 0.091 | 1.16 | 0.002 | 0.065 |
| | | | 21 | Flaming | 9.9 | 7.9 | 2.0 | 0.030 | 1.00 | c | 0.0062 |
| | | | | Flaming/Smold. | 11.4 | 7.2 | 4.2 | 0.091 | 1.18 | 0.004 | 0.058 |
| | | | 31 | Flaming | 11.4 | 8.0 | 3.4 | 0.018 | 1.16 | 0.001 | 0.012 |
| | | | | Flaming/Smold. | 11.8 | 7.8 | 4.0 | 0.110 | 1.27 | 0.021 | 0.082 |
| Corrugated Paper | Boxes with shredded paper (19% by weight) on wood pallets (38% by weight)[d] | 53 | 0 | Flaming | 12.3 | 9.2 | 3.1 | 0.0085 | 1.25 | c | ND |
| Cellulose Powder | Pool | 0.007 | 50 | Flaming | 12.8 | 9.1 | 3.7 | 0.0021 | 1.29 | c | 0.0057 |
| Cellulose Filter Papers | Pool | 0.007 | 50 | Flaming/Smold. | 10.2 | 5.8 | 4.3 | 0.095 | 1.05 | 0.001 | 0.013 |
| Red Oak | Pool | 0.007 | 31 to 71 | Flaming | 13.3 | 7.8 | 5.5 | 0.0039 | 1.34 | c | 0.086 |
| | | | | Oxid. Pyrolysis | NA | NA | NA | 0.096 | 0.14 | 0.041 | 0.85 |
| | | | | Pyrolysis | NA | NA | NA | 0.054 | 0.12 | 0.029 | 0.91 |
| Douglas Fir | Pool | 0.007 | 31 to 60 | Flaming | 13.0 | 8.1 | 4.9 | 0.0036 | 1.31 | c | 0.084 |
| Pine | Wood-crib | 0.062 | 0 to 60 | Flaming | 12.2 | 8.7 | 3.5 | 0.0054 | 1.31 | c | 0.064 |
| | | | | Flaming/Smold. | 10.7 | 8.0 | 2.7 | 0.092 | 1.18 | 0.019 | 0.15 |
| | | | | Oxid. Pyrolysis | NA | NA | NA | 0.18 | 0.23 | 0.060 | 0.68 |

a   Total gaseous hydrocarbons

b   $\lambda = 0.63 \, \mu$

c   $< 0.001$

d   16 boxes, four boxes on one pallet load; 2 pallets wide x 2 pallet loads deep x 2 tiers high        ND   not determined

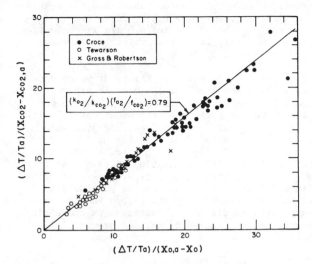

FIGURE 1. Chemical similarity in terms of the ratio of generation efficiency of $CO_2$ and depletion efficiency of $O_2$. Solid line is theoretical relationship. Experimental data taken from Refs. 10-12. Enclosure sizes from 0.21 to 22 $m^3$; exposed surface areas of cribs from 0.062 to 11 $m^2$.

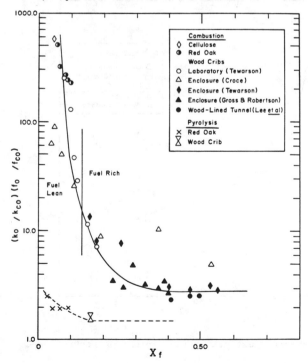

FIGURE 2. Dependency of the ratio of depletion efficiency of $O_2$ and generation efficiency of CO on fuel richness of the fire (Pyrolysis $X_0 = 0.11$).

increasing values of $X_f$ (fire conditions changing from well ventilated to underventilated), $f_O/f_{CO}$ approaches an asymptotic value (about 2.3 for combustion and about 1.2 for oxidative pyrolysis, for $X_f > 0.3$). If we assume $X_O = 0$, then the values of $f_{CO}$ and $Y_{CO}$ would be about 0.4 and 0.5 g/g in combustion and about 0.8 and 1.0 g/g in oxidative pyrolysis respectively. The maximum possible theoretical yield of CO for wood is 1.23 g/g.

The relationship between $\xi_\lambda$ $Y_S$, and $X_f$ is shown in Figure 3, where data for pine wood cribs from enclosure fires and from our 10-kW scale apparatus, together with cellulose and paper, have been included. The $\xi_\lambda$ $Y_S$ values for paper products are lower than the value for pine wood crib under natural air flow. With increasing value of $X_f$, $\xi_\lambda$ $Y_S$ values for red oak and pine wood crib increase, reaching a maximum value of about 0.12 m$^2$/g for 0.13 g/g $< X_f >$ 0.20 g/g, indicating a maximum generation efficiency of particulates at this slightly fuel rich condition (for stoichiometric combustion of pine wood crib, $X_f = 0.13$ g/g.) For $X_f > 0.20$ g/g. in the enclosure fires of pine wood cribs, $\xi_\lambda$ $Y_S$ decreases with increase in $X_f$, indicating a decrease in particulate formation. This suggests that the major fraction of the carbon in the fuel vapors is converted into the oxygenated species as $X_f$ increases in the enclosure fires, which is in agreement with the conclusion derived previously[6] on the basis of the carbon atom balance.

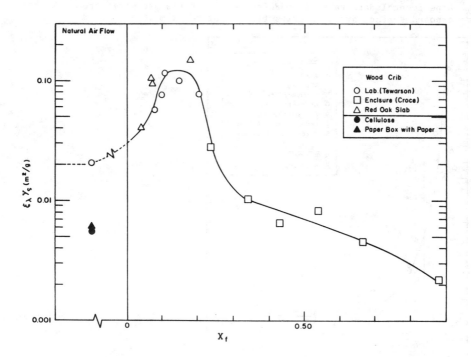

FIGURE 3. Dependency of the specific extinction coefficient of particulates on fuel richness of the fire.

## V CONCLUSIONS

1. Experimental results obtained from the 10-, 500-, and 5000-kW scale apparatuses and enclosure fires of various sizes, for materials in pool-like, box-like and crib-like configurations, suggest that in turbulent fires of each material, under a specified value of the ventilation parameter and mode of decomposition, chemical similarity is maintained in terms of combustion efficiency and generation efficiency of chemical compounds and specific extinction coefficient of particulates.

2. The generation efficiency of CO and extinction coefficient of particulates was found to be very sensitive to the decomposition mode of cellulosic materials. The combustion efficiency and the generation efficiency of $CO_2$ was found to be less sensitive to the decomposition mode.

3. In the laboratory-scale experiments, it was possible to compensate the fire scale effects due to flame radiation by increasing the oxygen concentration in the environment above the ambient value.

4. The specific extinction coefficient of particulates for wood was found to reach a maximum value of about 0.12 $m^2$/g under slightly fuel-rich conditions.

5. The maximum possible yield of CO from wood fires was estimated to be about 0.5 g/g in combustion and about 1.0 g/g in oxidative pyrolysis.

## ACKNOWLEDGEMENT

The financial support under Grant No. NB83NADA4021 from the U.S. National Bureau of Standards, Center for Fire Research, Washington, D.C. is deeply appreciated.

## REFERENCES

1 Tewarson, A., "Physico-Chemical and Combustion Pyrolysis Properties of Polymeric Materials," National Bureau of Standards, Washington, D.C. NBS-GCR-80-295 (1980).

2 Tewarson, A., Flame Retardant Polymeric Materials, Volume 3, Lewin, Atlas and Pearce (editors). Plenum Press, New York, (1982).

3 Tewarson, A., Lee, J. L., and Pion, R. F., "The Influence of Oxygen Concentration on Fuel Parameters for Fire Modeling," Eighteenth Symposium (International) on Combustion, p. 563, The Combustion Institute (1980).

4 Tewarson, A.,"Quantification of Fire Properties of Fuels and Interaction with Fire Environment," National Bureau of Standards, Washington, D.C., FMRC J.I. OEON6.RC (1982).

5 Tewarson, A. and Steciak, J., "Fire Ventilation," Combustion and Flame, _53_, (1983).

6 Tewarson, A., "Fully Developed Enclosure Fires of Wood Cribs," Twentieth Symposium (International) on Combustion. The Combustion Institute (in press).

[7] Tewarson, A., "Scale Effects on Fire Properties of Materials," National Bureau of Standards, Washington, D.C., FMRC J.I. 0J4N2.RC (1984).

[8] Newman, J. S., "Standard Test Criteria for Evaluation of Underground Fire Detection System," U.S. Bureau of Mines, Pittsburgh, PA, FMRC J.I. 0G2N4.RC (1984).

[9] Heskestad, G., "A Fire Products Collector for Calorimetry into the MW Range," Factory Mutual Research Corporation, Norwood, MA, FMRC J.I. 0C2E1.RA, June (1981).

[10] Gross, D. and Robertson, A. F., Experimental Fires in Enclosures," Tenth Symposium (International) on Combustion, p. 931, The Combustion Institute (1965).

[11] Tewarson, A., "Some Observations on Experimental Fires in Enclosures - Part I. Cellulosic Materials," Combustion and Flame, 19, 101 (1972).

[12] Croce, P. A., "Modeling of Vented Enclosures Fires, Part 1 - Quasi-Steady State Wood Crib Fire," Factory Mutual Research Corporation, Norwood, MA, FMRC J.I. 7AOR5.GU (1978).

[13] Lee, C. K., Chaiken, R. F., Singer, J. M. and Harris, M. E., "Behavior of Wood Fires in Model Tunnels under Forced Ventilation Flow," U. S. Bureau of Mines, Pittsburgh, PA, Report of Investigations 8450, 1980, available from Supt. of Docs. No. I28.23:8400.

# Critical Ignition Temperatures of Wood Sawdusts

TAKASHI KOTOYORI
Research Institute of Industrial Safety, Ministry of Labor
5-31-1 Shiba, Minatoku, Tokyo, 108 Japan

ABSTRACT

Critical ignition temperatures were estimated for fifteen species of wood sawdust slabs, being stacked in air at normal pressure, with an adiabatic self-ignition testing apparatus following a procedure. Assuming each stack to be in the form of an infinite slab of a thickness of 60.96 cm (2 feet), the temperature is estimated to range from 118 for Zelkova to 142 °C for Sitka spruce. These values are in reasonable agreement with a few real data observed on wood sawdust stacks of similar sizes. Zelkova and Western red cedar are relatively easy to ignite, meanwhile Sitka spruce and Western hemlock are relatively hard to ignite. Douglas fir and Port Orford cedar are medium in ignitability.

KEY WORDS: Critical ignition temperature; wood sawdust; self-ignition; adiabatic oxidation reaction; thermal diffusivity; relative ignitability.

INTRODUCTION

Many research reports have appeared on various aspects of pyrolysis and combustion phenomena of wood or woody materials. Instances, however, have been only a few so far where the critical ignition temperatures of bulky stacks of wood sawdusts were actually measured. As for that point, the works by Gross et al.[1], by Akita[2], and lately by Anthony et al.[3], by John[4] and by Schliemann[5] are quite precious.

In the present work, critical ignition temperatures of fifteen species of wood sawdust slabs, assuming each to be stacked in hot air surroundings at normal pressure, were estimated with an adiabatic self-ignition testing apparatus following a procedure.

A detailed description of the apparatus was given elsewhere[6]. So the procedure to estimate the critical ignition temperature of any stack of self-heating substance and results when the procedure was applied to wood sawdust slabs are mainly described in this report.

Wood species tested are as follows: Telaling (a Southeast Asian lumber), Meranti (ditto), Sawara cedar, Japanese cedar, Japanese red pine, Japanese cypress, Zelkova, Paulownia, Western red cedar, Douglas fir, Port Orford cedar, Alaska yellow cedar, Western hemlock, Sitka spruce and Eli ayanskya (a Siberian lumber).

AN EQUATION HOLDING FOR THE SELF-HEATING PROCESS UNDER AN ADIABATIC ZEROTH ORDER

## ASSUMPTION

When a sample heats spontaneously at a temperature T, eqn. (1) holds generally under an adiabatic zeroth order assumption, based on the principle of energy conservation.

$$c\rho\frac{dT}{dt} = \Delta H \cdot A \, \exp[-\frac{E}{RT}] \tag{1}$$

If we assume that the temperature rise rate, dT/dt, remains effectively constant within small temperature ranges near T, we can easily integrate eqn. (1) and get the following equation after taking natural logarithms on both sides of an equation obtained by the definite integral of eqn. (1).

$$\ln \Delta t = \frac{E}{RT} + \ln[\frac{\Delta T \cdot c\rho}{\Delta H \cdot A}] \tag{2}$$

where $\Delta t$ is the time required for the sample temperature to rise from T to $(T + \Delta T)$. Equation (2) can be applied to the adiabatic self-heating curve recorded precisely, by an adiabatic self-heating process recorder such as the apparatus used here, in temperature ranges of a few degrees above T. Then the gradient, a, and intercept, b, of an empirical formula, $\ln \Delta t = (a/T) + b$, can be expressed as,

$$a = \frac{E}{R} \tag{3}$$

$$b = \ln[\frac{\Delta T \cdot c\rho}{\Delta H \cdot A}] \tag{4}$$

respectively.


## A RELATION DERIVED FROM FRANK-KAMENETSKII'S CRITICAL CONDITION FOR THERMAL IGNITION AND EQUATION (2)

Frank-Kamenetskii's critical parameter $\delta_c$ for thermal ignition is expressed as,

$$\delta_c = \frac{\Delta H \cdot E r^2 A}{\lambda R T_c^2} \exp[-\frac{E}{RT}] \tag{5}$$

Frank-Kamenetskii's critical condition for thermal ignition holds on a balance between the heating process, again based on a zeroth order reaction, and the heat loss process under non-adiabatic conditions. Incidentally eqn. (2) is concerned with the heating process alone, since it stands on the adiabatic conditions. So far as the heating process is concerned, however, both equations deal with the identical slow heating process in the subcritical state prior to ignition. Therefore it can be thought that factors such as E, A and $\Delta H$ appearing in eqn. (2) and corresponding those in eqn. (5) represent the same physical quantities, respectively.

Taking natural logarithms on both sides of eqn. (5) and rearranging them gives,

$$\ln T_c + \frac{(E/R)}{2T_c} = \ln r + \frac{1}{2}\ln[\frac{(\Delta H \cdot A)(E/R)}{\delta_c \cdot \lambda}] \tag{6}$$

substituting eqns. (3) and (4) in eqn. (6) for (E/R) and $\ln(\Delta H \cdot A)$ gives,

$$\ln T_c + \frac{a}{2T_c} = \ln r + \frac{1}{2}[\ln(\frac{a \cdot \Delta T \cdot c\rho}{\delta_c \cdot \lambda}) - b] \tag{7}$$

464

since thermal conductivity is expressed in terms of thermal diffusivity and heat capacity per unit volume, as in eqn. (8),

$$\lambda = \alpha \cdot c\rho \tag{8}$$

eventually we obtain,

$$\ln T_c + \frac{a}{2T_c} = \ln r + \frac{1}{2}[\ln(\frac{a \cdot \Delta T}{\delta_c \cdot \alpha}) - b] \tag{9}$$

We see from eqn. (9) that we may dispense with direct knowledge of the individual values of $\Delta H$, E, A, $\lambda$ and c of the substance, so far as the estimation of $T_c$ is in question. The only data required to estimate $T_c$ value is the thermal diffusivity associated with the material, besides data, obtained with the apparatus, such as a, b and $\Delta T$. values of r and $\delta_c$ can be assigned as desired. Methods of determining thermal diffusivity for materials are well established[7]. As a matter of course the packing density $\rho$ of the substance in the sample cell of the self-ignition testing apparatus and of the thermal diffusivity measuring apparatus should be as equal as possible to that in the practical conditions.


STARTING PROCEDURE

A block diagram of the apparatus used here is shown in Fig. 1.  Measuring procedure for oxidatively heating substances has been described elsewhere[6], but a part of it was modified in this work because of the nature of wood sawdust.

300 mg of wood sawdust sample is packed in the cell that has a hole at the bottom.  Next the thermocouple for measuring sample temperature is inserted into

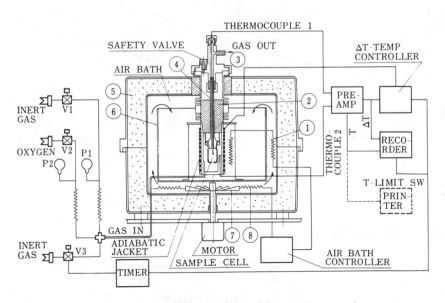

FIGURE 1.  Block diagram of the adiabatic self-ignition testing apparatus.  1, Thermometer; 2, glass wool; 3, sample cell assembly; 4, silica tube; 5, glass wool; 6, wind guide plate; 7, fan; 8, heater.

the sample. Then the sample cell assembly is set in the adiabatic jacket, and nitrogen gas is supplied at a rate of 2[ml/min]. The initial starting temperature is selected as desired by setting a temperature dial on the air bath. Then the power is switched on, that is, bath heating is started. After that the apparatus is left for 270 minutes at the starting temperature in nitrogen atmosphere, to remove all of the moisture from wood sawdust and to establish thermal equilibrium. While moisture-removal process lasts, the measurement can not be started, since the endothermic effect due to vaporization of moisture masks the exothermic phenomenon due to oxidation reaction of the substrate. Then, zero-suppression procedure is carried out, and immediately after that, the supply of air at a rate of 2[ml/min], adiabatic control and recording of the heating process are commenced.

ESTIMATION PROCEDURE OF CRITICAL IGNITION TEMPERATURE OF WOOD SAWDUST

Wood strips are first prepared by cutting some 2.5 mm thick slabs from visually uniform pieces of wood block of each species using a band saw. The strips are next cut into some 2.5 mm by 2.5 mm by 2.5 mm chips with a nipper. Chips are then pulverized using an ultra centrifugal mill and size-graded to 35 to 60 mesh by sieving.

Adiabatic oxidation heating curves of each species of wood sawdust are recorded at starting temperatures ranging from 150 to 178 °C, following the starting procedure stated above. The curves for Port Orford cedar sawdust are shown in Fig. 2 as an example. In these cases it is seen that the rate of adiabatic oxidation reaction is gradually accelerated, as nitrogen atmosphere

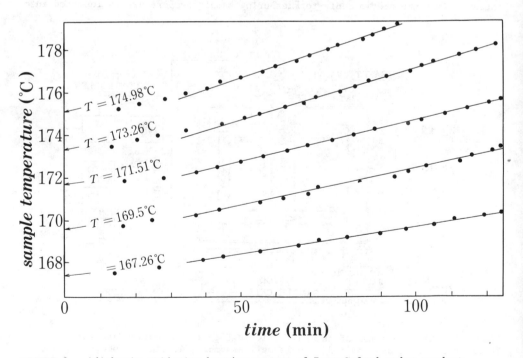

FIGURE 2. Adiabatic oxidation heating curves of Port Orford cedar sawdust.

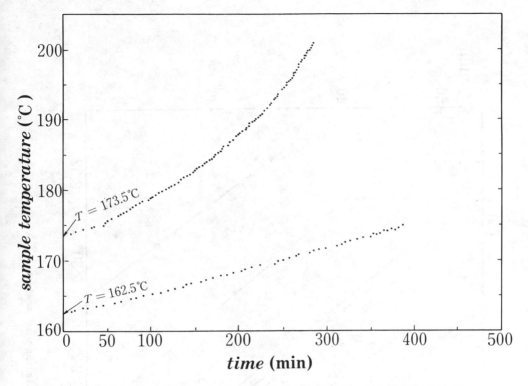

FIGURE 3.  Adiabatic oxidation heating curves of Western red cedar sawdust
with wider temperature increments.

around and in the sample cell is replaced by air for about 30 minutes after the
start.

Adiabatic oxidation heating curves of wood sawdusts recorded with this
apparatus appear in most cases approximately linear.  For instance, a curve is
almost linear at temperature levels of 160 to 175 °C in the case of Western red
cedar sawdust, though the temperature rise rate begins to accelerate gradually at
levels in excess of 180 °C, as shown in Fig. 3.  The same thing has been reported
also by Akita[2], by Anthony et al.[3] and by Schliemann[5].  Therefore a decision
was made in this work that the gradient, $\Delta T/\Delta t$, of the linearly rising part of a
curve was used as a physical quantity corresponding to the rate of the oxidatively
heating reaction of wood sawdust.

The time, $\Delta t$ min, required for the sample temperature to rise by a $\Delta T$ value
of 1.25 K is determined on each heating curve.  Then $\ln \Delta t$ vs. $1/T$ plot is made
for each species.  The entire plots for fifteen species of wood sawdust are given
in Fig. 4.  We can see roughly from Fig. 4 the relative liability of these wood
sawdusts to heat oxidatively, or to ignite ultimately.

Values of coefficients, a and b, of eqn. (2) are determined from these plots
by the least-squares method with a computer and are listed in TABLE 1.

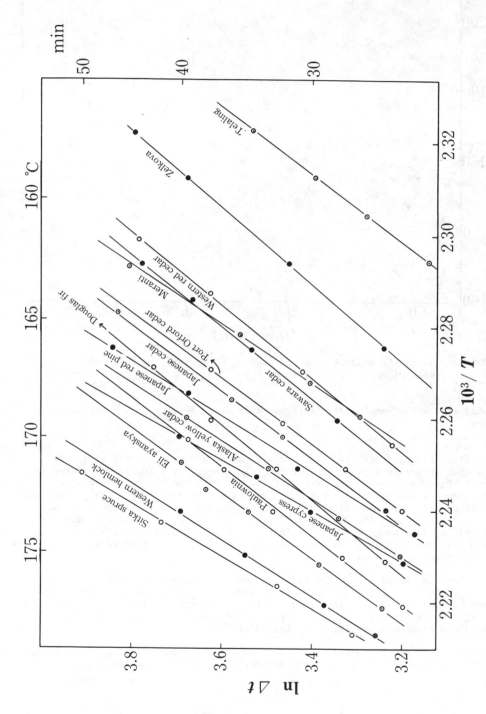

FIGURE 4. The entire ln Δt vs. 1/T plots for fifteen species of wood sawdust.

TABLE 1 Adiabatic experimental parameters of slow oxidation reaction and critical ignition temperatures of 15 species of wood sawdust, when each stacked in a slab, 60.96 cm thick.

| Wood species | a | b | $\alpha$ | E | $\rho$ | $T_c$ |
|---|---|---|---|---|---|---|
| | | | | | 0.279g/cm$^3$ | 118°C |
| Zelkova | 11,899 | -23.838 | 0.14 | 99 | 0.279 | 118 |
| Telaling | 13,736 | -28.379 | 0.15 | 114 | 0.214 | 119 |
| Western red cedar | 12,487 | -24.936 | 0.16 | 104 | 0.153 | 124 |
| Sawara cedar | 12,536 | -24.994 | 0.16 | 104 | 0.173 | 125 |
| Douglas fir | 12,430 | -24.474 | 0.14 | 103 | 0.197 | 127 |
| Meranti | 15,998 | -32.881 | 0.15 | 133 | 0.185 | 130 |
| Port Orford cedar | 13,906 | -27.950 | 0.17 | 116 | 0.138 | 131 |
| Japanese cedar | 13,944 | -27.994 | 0.18 | 116 | 0.145 | 132 |
| Paulownia | 13,110 | -25.887 | 0.17 | 109 | 0.173 | 133 |
| Japanese red pine | 16,483 | -33.656 | 0.13 | 137 | 0.253 | 133 |
| Eli ayanskya | 14,478 | -28.882 | 0.16 | 120 | 0.138 | 135 |
| Alaska yellow cedar | 16,124 | -32.765 | 0.17 | 134 | 0.138 | 136 |
| Japanese cypress | 17,644 | -36.123 | 0.16 | 147 | 0.131 | 138 |
| Western hemlock | 16,505 | -33.267 | 0.17 | 137 | 0.131 | 141 |
| Sitka spruce | 17,092 | -34.517 | 0.17 | 142 | 0.131 | 142 |

Reference data

Mixed hardwood sawdust[3], when stacked in a 61 cm cube, but in cool surroundings, 0.178  135 (obs)

Kiefernholz (German pine) sawdust[4, 5], when stacked in a sphere, 1 m diameter, 0.125  130 (obs)

The $\alpha$ value of wood sawdust was measured at room temperature and at 100 °C with the stepwise heating method developed by Araki et al.[7]. Values at both temperatures were much the same for each species, so the value is given in TABLE 1 as an arithmetical mean, respectively.

The form of the stack was assumed to be an infinite slab, which has a $\delta_c$ value of 0.88, of a thickness of 60.96 cm (2 feet), which is equal to 2r.

Substituting these data in eqn. (9), $T_c$ values are obtained and are presented in TABLE 1. Wood species are arranged in order, wood sawdust of high critical ignition temperature being listed near the bottom.

RESULTS AND DISCUSSION

A possibility of heat loss due to air flow in the sample cell could not be ignored, though it would be almost perfectly covered by the adiabatic control. If that effect were serious, the true $T_c$ value might be several degrees lower than the one determined by the procedure. Furthermore there might be some uncertainty in ln $\Delta t$ vs. 1/T plots, though it would be considerably compensated by the least-squares method. As for that problem, to increase the number of the point on each plot from 4 or 5 to 10, e.g., may be helpful to increase the certainty of the plot. It is also not very pleasant that we can not make the packing density constant for each species because of the nature of wood sawdust.

For all those problems, $T_c$ values of wood sawdust slabs estimated by calculation, following the procedure introduced above, agreed in the result reasonably with a few real data observed on stacks of similar sizes, though the species are different from those used in this work[3, 4, 5]. It thus appears that those problems anticipated above are practically negligible with wood sawdusts, at least. Then we can say that the procedure is applicable to estimate relatively and quantitatively the ignitability of any stacks of wood sawdusts.

Assuming wood sawdust to be stacked in a form of infinite slab at a thickness of 60.96 cm (2 feet), the $T_c$ value is estimated to range from 118 for Zelkova to 142 °C for Sitka spruce. If the slab is 20.32 cm (8 inch) thick, the value is estimated to range between 147 for Telaling to 167 °C for Sitka spruce. Zelkova and Western red cedar are relatively easy to ignite, meanwhile Sitka spruce and Western hemlock are relatively hard to ignite. Paulownia may also belong under a category that is hard to ignite. Douglas fir and Port Orford cedar are medium in ignitability.

The oxidatively heating process of wood sawdusts proceeds showing an almost linear temperature rise rate in temperature ranges up to about 180 °C under adiabatic conditions.

The relative liability of wood species to heat oxidatively is scarcely recognized by thermal analysis such as TG-DTA, meanwhile it can be clearly distinguished from one species to another under adiabatic conditions.

Wood is a natural product and its composition is not homogeneous, so $T_c$ values listed in TABLE 1 are by no means of absolute character. Nevertheless the author thinks that, although there may exist still some uncertain factors such as those stated at the beginning of this chapter, a procedure to estimate $T_c$ values of wood sawdust stacks has been fundamentally established. The ignitability of any self-heating substances can be relatively and quantitatively determined by applying the critical ignition temperature as the criterion. So it would be interesting to see if the procedure is applicable also to other self-heating substances.

SYMBOLS USED

A       frequency factor $[mol/cm^3\ min]$

a       coefficient expressed by eqn. (3)

b       coefficient expressed by eqn. (4)

c       molar heat capacity [J/mol K]

$\rho$       molar density $[mol/cm^3]$

$c\rho$      heat capacity per unit volume $[J/cm^3\ K]$

E       apparent activation energy [J/mol]

$\Delta H$      molar heat of reaction [J/mol]

R       universal gas constant [J/mol K]

r       size factor [cm]

T       initial starting temperature [K]

$T_c$      critical ignition temperature [K]

t       time [min]

$\delta_c$      Frank-Kamenetskii's critical parameter for thermal ignition

$\alpha$       thermal diffusivity $[cm^2/min]$

$\lambda$       thermal conductivity [J/cm min K]

REFERENCES

1.  Gross, D. and Robertson, A.F.: "Self-Ignition Temperatures of Materials from Kinetic-Reaction Data," J. Research Natl. Bur. Standards, 61: 413-417, 1958.

2.  Akita, K.: "Studies on the Mechanism of Ignition of Wood," Rep. Fire Research Inst. Japan, 9: 1-106, 1959.

3.  Anthony, E. J. and Greaney, D.: "The Safety of Hot Self-Heating Materials," Combustion Science and Technology, 21: 79-85, 1979.

4.  John, R.: "Zur Selbstentzündung von Holz. Aufheizphase und freiwerdender Wärmestrom," in Origin and Spread of Fire (latest state of research) - No. 1, 4th Internatl. Fire Protection Seminar - Zürich, 59-84, Brand-Verhütungs-Dienst für Industrie und Gewerbe, Zürich, 1973.

5.  Schliemann, H.: "Die Anwendung der Differential-Thermoanalyse (DTA) zur Einschätzung des Selbstentzündungsverhaltens von Holz," Holzindustrie, 30: 5, 140-142, 1977.

6.  Kotoyori, T. and Maruta, M.: "An Adiabatic Self-Ignition Testing Apparatus," Thermochimica Acta, 67: 35-44, 1983.

7.  Araki, N.: "Measurements of Thermophysical Properties by a Stepwise Heating Method," Internatl. J. of Thermophysics, 5: 1, 53-71, 1984.

# Influence of the Thickness on the Thermal Degradation of PMMA

CHRISTIAN VOVELLE, ROBERT AKRICH, JEAN–LOUIS DELFAU,
and SILVIE GRESILLAUD
Centre de Recherches sur la Chimie de la Combustion
et des Hautes Températures. C.N.R.S.
1c Avenue de la recherche scientifique
45045 Orleans-Cedex, France

ABSTRACT

The influence exerted by the initial thickness on the thermal degradation of PMMA has been studied experimentally. The mass loss rate and the temperature profile in the solid have been measured continuously for .6, 1.0, 1.5, 3.0 and 5.0 cm thick specimens. All the experiments have been performed under nitrogen. The evolution of the thermal degradation has been interpreted in term of the two variables $R = \dot{m}''/(\dot{Q}'' - \dot{Q}''_1)$ mass loss rate per unit heat flux absorbed by the material and $q'' = \int_0^t (\dot{Q}'' - \dot{Q}''_1) \times dt$ amount of heat accumulated into the solid.

Two conclusions have been drawn from the experimental results :
  -there exists a critical thickness below which the maximum mass loss rate depends on the initial thickness of the specimen.
  -since this variation results from a change in the heat loss from the backside, it does not affect the curve $R = f(q'')$ except for very thin materials.

INTRODUCTION

This paper is concerned with the prediction of the mass loss rate of solid materials exposed to a radiant heat flux. This problem is of primary importance in fire development since it governs the height of the flame of the first burning object and therefore the secondary effects associated with this flame (air entrainment, flow rate of hot gases in the higher part of the room, radiation...). The thermal degradation of solid materials also plays a role in the prediction of the ignition delay of other objects (targets) or in the calculation of the rate of propagation of flames along surfaces.

Heat transfert in solid materials associated with pyrolysis and liberation of gaseous products has been extensively studied either theoretically (1,2,3) or experimentally (4,5,6). These studies lead to a good knowledge of the physical phenomena involved in thermal degradation. On the other hand, an examination of these investigations clearly shows that the kinetic parameters which govern the calculation of the rate of pyrolysis of solid materials are far from being well known (7). As a result, the use of a detailed model to calculate the mass loss rate of solid materials in a fire situation presents two inconveniences :
  (i)  the resolution of these models is time consuming, and
  (ii) the results obtained are not specific enough.

Presently, a simpler procedure derived from Tewarson's work is used in the computer codes for fire development (9,10). Tewarson and Pion (8) have shown that for various materials, a linear relationship is observed between the mass burning rate and the heat flux absorbed by the material :

$$\dot{m}'' = (\dot{Q}'' - \dot{Q}''_{1s})/L_G \qquad (1)$$

with :

$\dot{m}''$ : mass burning rate per unit surface ($g.cm^{-2}.s^{-1}$)
$\dot{Q}''$ : incident heat flux ($W.cm^{-2}$)
$\dot{Q}''_{1s}$ : heat loss from the surface ($W.cm^{-2}$)
$L_G$ : heat of gasification of the material ($J.g^{-1}$)

The main advantage of this relationship is simplicity since only one quantity ($L_G$) is required for a given material. However it must be noticed that in the study of Tewarson, only the stationary state, characterized by a constant value of the mass burning rate has been considered and since the development of a fire in a room involves mainly transient phenomena, it was worth studying how the Tewarson relationship could be extended to non-stationary conditions. Therefore, a few years ago, an experimental study of the thermal degradation of two materials : PMMA and particle board, exposed to a radiant heat flux has been undertaken in our laboratory.

In a first step, the mass loss rate for each material was continuously measured during constant heat flux experiments in the range $0-4$ $W.cm^{-2}$. This study showed that when the variables $R=\dot{m}''/(\dot{Q}'' - \dot{Q}''_{1s})$ and $q''=\int_0^t(\dot{Q}'' - \dot{Q}''_{1s})xdt$ (with t= duration of the exposition to the radiant flux) were used to describe respectively the "response" of the material and the progress of the overall thermal degradation process a unique curve was obtained for different values of the incident radiant flux ($\dot{Q}''$) (11).

This curve increases continuously towards a maximum value corresponding to the stationary state. The continuous increase can be easily explained since during the transient heating stage, the heat absorbed by the material is partly transferred by conduction into the solid, and if the temperature is high enough, partly used to the pyrolysis of the material. Of course, a progressive increase of the second phenomenon and a progressive decrease of the first one is observed when the material is exposed to a constant heat flux.

From the unique curve obtained for experiments performed with different values of the incident radiant flux, it was concluded that the evolution of the ratio : heat used for the pyrolysis of the material/heat transferred by conduction is reproduced with similitude.

This similitude is interesting, since it shows that even during the transient stages preceding the stationary state, the mass loss rate of a given solid material can be predicted.

In a second step, the behavior of the same materials : PMMA and particle board, exposed to a variable radiant heat flux was considered (12). A linear variation with time of the radiant heat flux was produced. In term of the two variables :

$$R = \dot{m}''/(\dot{Q}''(t)-\dot{Q}''_1(t)) \text{ and } q'' = \int_0^t (\dot{Q}''(t)-\dot{Q}''_1(t))xdt$$

the progress of the overall thermal degradation process was represented by the same curve as those obtained during the constant heat flux experiments.

Therefore, even when more realistic conditions, characterized by a conti-nuous increase of the incident radiant flux are considered the diagram R=f(q") can be used to predict the mass loss rate of the solid material, at least for the two materials studied.

Of course, some additional experiments performed with other materials are needed if one want to generalize the use of the diagram R = f(q") to predict the "response" of a solid material to thermal irradiation.

Another point had to be clarified. The previous experimental studies were performed with only one thickness of the specimens : 1.5 cm for PMMA and 1.2 cm for particle board and the observation of a stationary state with this range of thickness could be questioned. The work presented in this paper aimed at an answer to this point. The evolution of the rate of thermal degradation of PMMA has been studied for various thickness of the specimens in the range .6 - 5 cm.

EXPERIMENTAL

The experiments have been carried out in a stainless steel chamber enclo-sing the radiant heat source, the specimen and the electronic balance (fig.1).

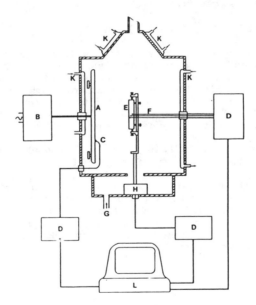

Figure 1 : Experimental device.
A : Radiant panel. B : Voltage regulator. C : Thermocouple. D : Analog-digital converter. E : Specimen. F : Thermocouples used for the measurement of the temperature profiles. G : Gas inlet. H : Electronic balance. K : Water cooling. L : Micro computer.

An electrical panel of dimensions close to .4 x .4 m was used as a radiant source. The size of the specimens were small compared to the dimensions of the panel (.1 x .1 or .1 x .06 m) in order to ensure an homogeneous heat flux on the surface.

In this study, all the experiments were conducted with an incident radiant flux close to 3 W.cm$^{-2}$. This heat flux was controlled by a water cooled fluxmeter.

Kashiwagi pointed out that it is only when the flow rate of the pyrolysis products is low that the rate of thermal degradation of PMMA is affected by the oxygen content of the atmosphere surrounding the material (13). The same result was observed in our previous study (11). With 3 W.cm$^{-2}$, the effect of oxygen would only be observed during a short period corresponding to the beginning of the liberation of gaseous products, the subsequent pyrolysis steps being unaffected. Therefore, in this work the thermal degradation of PMMA was only studied in a nitrogen atmosphere. The nitrogen flow rate in the stainless steel chamber was equal to 3 m$^3$h$^{-1}$.

The main part of the experiment consisted in the continuous measurement of the mass loss rate of the specimen. The calculation of the variables R and q" requires also a knowledge of the heat loss from each face of the material. Radiation as well as convection were taken into account in the calculation of these quantities :

$$Q"_{1S}(t) = \varepsilon \sigma (T_S{}^4(t) - T_{\infty}{}^4) + h\ (T_S(t) - T_G(t)) \quad (2)$$

$$Q"_{1B}(t) = \varepsilon \sigma (T_B{}^4(t) - T_{\infty}{}^4) + h\ (T_B(t) - T_G(t)) \quad (3)$$

$T_S(t)$ and $T_B(t)$ represent the variation with time of the temperature of each face of the specimen. These quantities were not obtained from direct measurements. They were evaluated by extrapolation of the temperature profile measured with thermocouples located in small holes (.5 mm diameter) drilled parallel to the face exposed to the radiant flux, at increasing distances from this face. Twelve "thermocoax" thermocouples (external diameter = .34 mm) were used for the measurement of the temperature profiles.

The value of $\varepsilon$ (.8) and h (5x10$^{-4}$ W.cm$^{-2}$.K$^{-1}$) used previously (12) have been used again for the calculation of $\dot{Q}"_{1S}$ and $\dot{Q}"_{1B}$.

The first objective of this work was to verify that the PMMA specimens used previously were thick enough for a stationary state to be observed (these specimens were 1.5 cm thick). Thicker specimens (3.0 and 5.0 cm) were studied to clarify this point. To complement this work, it was interesting to consider the behavior of thinner specimens also, so that additional experiments were conducted with specimens of thickness equal to 1.0 and .6 cm.


RESULTS AND DISCUSSION

The variation with time of the mass loss rate for samples with five thickness has been plotted on figure 2. These curves seem to indicate that the thickness used previously could lead to a stationary state since a maximum value close to 1.1x10$^{-3}$ g.cm$^{-2}$.s$^{-1}$ is obtained for 1.5, 3.0 and 5.0 cm. The plateau observed on the 5.0 curve confirms that this maximum value corresponds effectively to a stationary state.

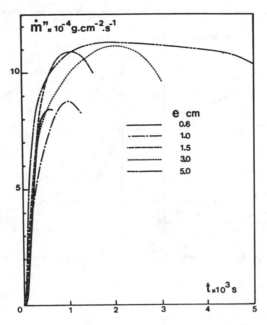

Figure 2 : Influence of the initial thickness on the variation with time of the mass loss rate. PMMA under nitrogen.

On the other hand, a difference is observed with the thinner specimens : the maximum mass loss rate is only $8.5 \times 10^{-4}$ $g.cm^{-2}.s^{-1}$ for 1.0 cm and slightly lower ($8 \times 10^{-4}$ $g.cm^{-2}.s^{-1}$) for .6 cm. It must be noticed that the thinner specimens (.6 and 1.0 cm) have been exposed to a radiant flux of 2.8 $W.cm^{-2}$ instead of 3.0 $W.cm^{-2}$. This difference in the incident radiant flux is partly responsible for the lower value measured for the maximum mass loss rate. However, from the linear dependence observed previously for PMMA between the maximum mass loss rate and the incident radiant flux, the decrease of the flux from 3.0 to 2.8 $W.cm^{-2}$ would lead to a corresponding decrease of the maximum mass loss rate from 1.1 $g.cm^{-2}.s^{-1}$ to 1 $g.cm^{-2}.s^{-1}$. Hence, the difference observed with the thicker specimens can not be totally explained by a difference in the radiant flux but is also an effect of the thickness.

It was interesting to consider whether the decrease in the maximum mass loss rate for thickness lower than a critical value was also observed when the variables R and q" are used to describe the evolution of the thermal degradation process. Prior to the calculation of these variables, the heat loss from each face had to be evaluated from the variation with time of the temperature profiles in the material. These profiles have been plotted on figures (3-7). To illustrate the degree of accuracy of the temperature measurements, the points resulting from four distinct experiments have been plotted for the 5.0 cm thick specimens (fig.7). The variation with time of the temperature of each face was obtained by extrapolation of these profiles. The temperatures of the exposed surface for the five thickness studied have been plotted on figure 8, while the corresponding curves for the rear face of the specimens have been plotted on figure 9.

The profiles measured in the thicker specimens (3.0 and 5.0 cm) show very

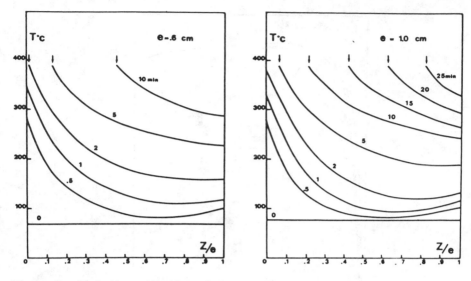

Figure 3 : Evolution with time of the temperature profile. $\dot{Q}''$=2.8 W.cm$^{-2}$

Figure 4 : Evolution with time of the temperature profile. $\dot{Q}''$=2.8 W.cm$^{-2}$

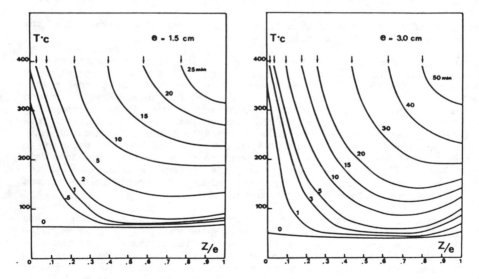

Figure 5 : Evolution with time of the temperature profile. $\dot{Q}''$= 3.0 W.cm$^{-2}$

Figure 6 : Evolution with time of the temperature profile. $\dot{Q}''$= 3.0 W.cm$^{-2}$

clearly that despite the water cooling of the walls of the chamber, the temperature of the gas behind the specimen increases up to a value close to 200°C.
As a result of this temperature increase, a heating of the rear face of the
specimen is observed. Since the temperature of the gas phase was not measured
during these experiments, we have considered that the term $T_G(t)$ in the equa-

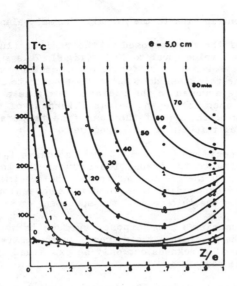

Figure 7 : Evolution with time of the temperature profile. $\dot{Q}''= 3.0$ W.cm$^{-2}$

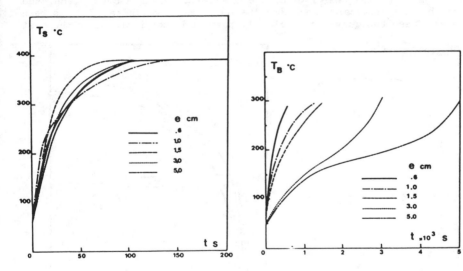

Figure 8 : Variation with time of the temperature of the exposed surface.

Figure 9 : Variation with time of the backside temperature.

tions 2 and 3 is equal to $T_B(t)$ up to 200°C and remains equal to this value when $T_B(t)$ becomes higher.

This assumption leads to keep the convective transfer between the gas and the material to a null value until $T_B(t)$ becomes higher than 200°C. A more rigorous procedure would require a calculation of the heat transferred into the solid by convective heating of the surface. However a rough estimation of this term shows that its contribution to the overall thermal balance is negligible.

The temperature of the face exposed to the radiant flux increased very rapidly up to a constant value equal to 390°C (fig.8). This surface temperature is close to the value measured in different experimental conditions (13, 14). It can be seen that the thickness does not affect the rate of increase of the temperature of the face exposed to the radiant flux . As a result, the heat balance for this face does not depend on the thickness of the material. Obviously, the variation with time of the rear face temperature is strongly dependent on the initial thickness of the specimens.

For the thicker specimens (3.0 and 5.0 cm) the curve $T_B = f(t)$ exhibits an inflection corresponding to the transition between the initial convective heating of the material by the gas located behind the specimen and the heating by conduction from the exposed surface when the regression of this surface decreases to a low value the thickness of the specimen.

Since the heat transfer at the exposed surface is not affected by the initial thickness of the specimens, the decrease in the maximum mass loss rate observed under a critical thickness can only be explained by a difference in the heat balance at the rear face.

This is illustrated in Figure 10 where the variation with the initial thickness of the maximum mass loss rate has been compared to the variation of the heat loss from the rear face of the specimen. For this comparison, the value of the heat loss calculated when the mass loss rate is maximum has been considered. These two curves show clearly that it is the increase in the heat

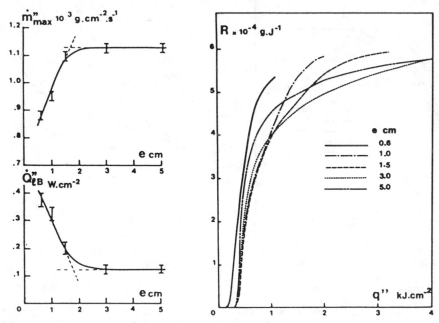

Figure 10 : Variation with e of $\dot{m}''_{max}$ and $\dot{Q}''_{1B}$.

Figure 11 : Variation of the ratio $R = \dot{m}''/(\dot{Q}'' - \dot{Q}''_1)$ vs $q'' = \int_0^t (\dot{Q}'' - \dot{Q}''_1) \times dt$.

loss from the backside of the specimens when the initial thickness is lower than a critical value, which is responsible for the decrease in the maximum mass loss rate. From these diagrams, a value close to 1.7-1.8 cm can be estimated for the critical thickness.

The dependence of the maximum mass loss rate on the heat loss from the backside can be also evidenced by considering the variables R and q". The curves R=f(q") have been plotted on figure 11 for the five thickness studied. It can be seen that the ratio R tends towards an unique value comprised between 5.5 and 6.0 g.J$^{-1}$. This maximum value corresponds to the stationary state and, according to the Tewarson relationship, is equal to the inverse of the heat of gasification of the material. This leads to a value in the range 1666-1818 J.g$^{-1}$, in close agreement with the value measured by Tewarson for PMMA.

For the 1.0 to 5.0 thick specimens, the differences observed in the evolution of the curves R=f(q") must be attributed to the accuracy of the measurements, especially to the determination of $T_s(t)$. On the other hand, for the thinnest specimens, the ratio R starts to increase for lower values of q". In that case, the increase in the heat loss from the backside is fast enough to change the overall heat balance the moment the exposure to the radiant flux starts.

CONCLUSION

In a previous study, we showed that it was very convenient to use the variables $R=\dot{m}''/(\dot{Q}'' - \dot{Q}''_1)$ and $q''=\int_o^t(\dot{Q}'' - \dot{Q}''_1)x dt$ to describe the thermal degradation of PMMA and particle board. The main advantage of these two variables is that a unique curve is obtained for experiments performed with different values of a constant incident radiant flux or with a variable radiant flux. This curve exhibits a maximum, and the main objective of this work was to check whether this maximum can be attributed or not to a stationary state. To give an answer to this question, some new experiments have been performed with PMMA specimens of thickness ranging between .6 and 5.0 cm.

The simultaneous mass loss rate and temperature profiles measurements have led us to the conclusion that under a critical thickness, an increase of the heat loss from the backside of the specimens and a corresponding decrease in the maximum mass loss rate is observed. With the experimental conditions used in this study (thermal degradation of PMMA under nitrogen, incident radiant flux close to 3.0 W.cm$^{-2}$) this critical thickness is equal to 1.7-1.8 cm. It can be considered that the conclusions drawn from the previous experiments remain valid, despite of a thickness of the specimens (1.5 cm) slightly too low.

The main interest of this work is that it confirms that the use of the variables R and q" to represent the thermal degradation of a solid material (at least for PMMA) leads to a simplification since except for very thin materials, the evolution of these quantities is not affected by the initial thickness of the material.

NOMENCLATURE

e   : specimen thickness (cm).
$L_G$ : Heat of gasification of the solid (J.g$^{-1}$).
$\dot{m}''$ : Mass loss rate per unit surface (g.cm$^{-2}$.s$^{-1}$).

$\dot{Q}''$   : Incident radiant flux $(W.cm^{-2})$.
$\dot{Q}''_l$  : Total heat loss $(W.cm^{-2})$.
$\dot{Q}''_{ls}$ : Heat loss from the exposed surface $(W.cm^{-2})$.
$\dot{Q}''_{lB}$ : Heat loss from the backside $(W.cm^{-2})$.
$q''$   : Heat accumulated into the solid $(J.cm^{-2})$.
$t$    : duration of the exposition to the radiant flux (s).
$T_s$  : Surface temperature (K).
$T_\infty$  : Ambient temperature (K).
$T_B$  : Backside temperature (K).
$T_G$  : Temperature of the gas phase surrounding the material (K).
     : emissivity of the surface.
$h$    : Convective heat transfer coefficient $(W.cm^{-2}.K^{-1})$.
$z$    : Distance from the exposed surface (cm).


REFERENCES

1. Kansa, E.J., Perlee, H.E. and Chaiken, R.F.: "Mathematical model of wood pyrolysis including internal forced convection," Combustion and Flame, 29, 311-324, (1977).
2. Kung, H.C.: "The combustion of vertical wooden slab," 15th Symposium (international) on Combustion, The Combustion Institute, Pittsburgh, 243-253, (1974).
3. Kashiwagi T.: "A radiative ignition model of a solid fuel," Combust. Sci. and Technol., 8, 225-236, (1974).
4. Kashiwagi T.: "Experimental observation of radiative ignition mechanisms" Combust. and Flame, 34, 231-244, (1979).
5. Becker, H.A., Phillips, A.M., and Keller, J.: "Pyrolysis of white pine," Combust. and Flame, 58, 163-189, (1984).
6. Petrella, R.V.: "The mass burning rate of polymers, wood and organic liquids," J. of Fire and Flammability, 11, 3, (1980).
7. Vovelle, C., Mellottee, H., and Delbourgo, R.: "Kinetics of the thermal degradation of cellulose and wood in inert and oxidative atmosphere" 19th Symposium (international) on Combustion, The Combustion Institute, Pittsburgh, 797-805, (1983).
8. Tewarson, A., and Pion, R.F.: "Flammability of plastics. I Burning intensity," Combust and Flame", 26, 85-103, (1976).
9. Mitler, H.E., and Emmons, H.W.,:Documentation for CFC V, the Fifth Harvard Computer Code, NBS-GCR-81-344, (1981).
10. Curtat, M., and Bodart, X.: Point sur le Modèle C.S.T.B. de Développement du Feu dans une Pièce unique. C.S.T.B., Centre de Recherches de Marne la Vallée, (1983).
11. Vovelle, C., Akrich, R., and Delfau, J.L.: "Mass loss rate measurements on solid materials under radiative heating," Combust. Sci. and Technol. 36, 1-18, (1984).
12. Vovelle, C., Akrich, R., and Delfau, J.L.: "Thermal degradation of solid materials under a variable radiant heat flux," 20th Symposium (international) on Combustion, The Combustion Institute, Pittsburgh, In Press.
13. Kashiwagi, T., and Ohlemiller, T.J.: "Study of oxygen effects on nonflaming transient gasification of PMMA and PE during thermal irradiation," 19th Symposium (international) on Combustion, The Combustion Institute, Pittsburgh, 815-823, (1982).
14. Fernandez-Pello, A., and Williams, F.A.: "Laminar flame spread over PMMA surfaces," 15th Symposium (international) on Combustion, The Combustion Institute, Pittsburgh, 217-231, (1974).

# Differences in PMMA Degradation Characteristics and Their Effects on Its Fire Properties

T. KASHIWAGI, A. INABA, and J. E. BROWN
Center for Fire Research
National Bureau of Standards

ABSTRACT
Thermal degradation and thermal oxidative degradation characteristics of Plexiglas G and Lucite were determined using thermogravimetry. The results show that degradation rate of Plexiglas G is sensitive to gas phase oxygen but that of Lucite is much less so. Comparison of derivative thermogravimetry curves between the two samples indicates that at low temperatures Plexiglas G is more stable with respect to degradation in nitrogen. Lucite is initially more stable with respect to degradation in air than is Plexiglas G. A similar trend was observed in a nonflaming gasification study using external radiative heating. It appears that the chemical nature of the degradation processes of the two samples is the same for slow heating thermogravimetry and for more rapid heating (gasification study) simulating a fire environment. In piloted radiative ignition at 1.8 W/cm$^2$, the ignition delay time of Plexiglas G is about 15% less than that of Lucite. Increasing the radiant flux reduces the difference in ignition delay time between the two samples. The downward flame spread velocity of Lucite is about 20% faster than that of Plexiglas G, but the difference in burning rate between the two samples is very small.

INTRODUCTION
Materials involved in fire are generally categorized only by their general chemical structure, for example, polystyrene, polyurethane, polyethylene, etc. However, within any one such general category there are significant property differences. These include molecular weight, impurities, plasticizer, additives, copolymer and so on. At present, it is not clear whether these differences affect such fire properties as ignition, flame spread and burning rates. This study examines whether such differences can have significant effects on fire properties for one type of material, poly(methylmethacrylate) (PMMA).

PMMA is studied because of its high purity polymer composition (it contains no plasticizers, it is not a copolymer and it generally contains few additives) and also because it has a relatively simple and well understood degradation mechanism (1-3). Therefore, this examination of the behavior of PMMA from two different manufacturers is only a first look at the impact of differing material source; more complex commercial polymers have greater potential to behave differently. Another advantage of using this material is that the principal degradation product is the monomer even when PMMA is manufactured by different companies. Therefore, it seems reasonable to assume that the gas phase behavior should be the same for different PMMA samples and that any observed differences are probably due to their condensed phase degradation characteristics.

---

Contribution from the National Bureau of Standards, not subject to copyright.

A. Inaba is a guest worker from National Research Institute for Pollution and Resources, Tsukuba, Japan.

This study consists of three parts. First, thermogravimetry (TG) is used to study a small PMMA sample at low heating rates: differences in global degradation chemistry between selected samples are examined. Second, nonflaming gasification of the samples under well-controlled thermal radiation conditions simulating material behavior in fire environments is examined. In this part of the study, the gasification process includes not only the degradation chemistry but also effects due to heat and mass transport processes. Third, some fire properties of selected samples are measured to determine whether they differ among samples. The measured fire properties are piloted radiative ignition delay time, downward flame spread velocity, and mass burning rate.

EXPERIMENTAL SECTION

Materials. The PMMA materials used in this study were commercial Plexiglas G (Rohm and Haas, Inc.)* and Lucite (E.I. Dupont de Nemours & Co.) in sheet form. The specimens for the TG study were disk-shaped, about 5.5 mm in diameter and about 200 μm thick; they were milled from commercially available sheet stock. The specimens for nonflaming gasification and also for piloted ignition under external radiative heating were 4 cm x 4 cm x 1.2 cm thick; they were 10 cm x 10 cm x 1.2 cm thick for the study of burning rates and 10 cm width x 30 cm length x 1.2 cm thick for the study of flame spread. All samples with 1.2 cm thickness had the original unmachined surface.

Thermogravimetry. Weight loss from the sample was measured using a Mettler Thermoanalyzer TA 2000. Heating rates of 0.5, 0.7, 1, 2, 3 and 5°C/min were used to obtain the overall kinetic constants for weight loss. Sample weight, temperature and time were simultaneously recorded with a computer. The reproducibility of TG and DTG was generally excellent and temperature at a peak weight loss rate could be reproduced within 2°C.

Apparatus for Nonflaming Gasification Under Radiative Heating. A detailed description of the experimental apparatus and procedure for the radiative heating pyrolysis study has previously been reported (4). Briefly, uniform thermal radiation from an electrically heated graphite plate was used to irradiate a vertically mounted sample at radiant flux of 2.2 W/cm² in a specified gas environment. No flaming occurred during the irradiation period. The sample was mounted on an electromechanical balance which could sense a 1 mg change in a total weight of up to 50 g. A 25 μm wire diameter chromel-alumel thermocouple was laid across the front surface of the sample with the junction near the center of the sample. To assure good contact between the thermocouple and the sample surface, the thermocouple was heated electrically and simultaneously pressed into the surface prior to a test. Any increase in temperature of the thermocouple by direct absorption of the external radiation was at most 5°C for the radiant fluxes used in this study (5). This magnitude of temperature increase is comparable to the reproducibility of the measured surface temperature. The reproducibility of the data is within 5% for mass flux and within 3% for temperature.

Apparatus for Piloted Radiative Ignition. The same apparatus as described above was used. Again the sample was mounted vertically. The only addition was a pilot electrically heated platinum wire (0.375 mm wire diameter) over and across the top edge of the sample. The distance between the bottom of the spiral wire (about a 6 mm diameter spiral) and the top edge of the sample was about 1.5 cm. The reproducibility of ignition delay time is within 5%.

Downward Flame Spread Study. A vertically mounted sample was supported between two marinite plates along its vertical edges. Flame spread was

---

*In order to adequately describe materials it is occasionally necessary to identify commercial products by manufacturer's name. In no instance does such identification imply endorsement by the National Bureau of Standards nor does it imply that the particular product is necessarily the best available for that purpose.

initiated by igniting an adhesive cement on the top edge of the sample with a match. Flame spread down over both sides of the sample. A 25 μm wire diameter chromel-alumel thermocouple spread across the surface of the sample parallel to the flame front, with the junction near the center of the sample, was used to measure the local surface temperature history. The reproducibility of flame spread rate is within 5%.

Mass Burning Rate Study. A sample was mounted horizontally on noncombustible fiber insulation. Flame spread along side edges of the sample was inhibited by cementing 1 mm thick pieces of cardboard on them so as to promote one-dimensional burning as much as possible. A 25 μm wire diameter chromel-alumel thermocouple with the junction near the center of the sample was used to measure surface temperature, and an electromechanical balance was used to measure the change in weight of the sample. The reproducibility of mass buring rate is about 5%.

RESULTS AND DISCUSSION

1. Thermogravimetry

Weight loss. Derivative thermogravimetry (DTG), i.e., weight loss rate vs temperature, of Lucite and Plexiglas G was measured at various heating rates. The DTG results were obtained by taking the time derivative, $(d(W/Wo)/dt)$, of the ratio of the sample weight, W, to the initial sample weight, Wo. Typical DTG results for Lucite and Plexiglas G in nitrogen and in air are shown in Fig. 1. The results for Lucite show that rapid weight loss starts slightly below 260°C in nitrogen and also in air at a heating rate of 5°C/m indicating small dependency on oxygen.

The results for Plexiglas G degrading in nitrogen show that a small weight loss appears around 160°C and rapid weight loss starts at about 300°C at a heating rate of 5°C/m. The early small weight loss may be caused by volatilization of unreacted monomer in the sample. Weight loss of Plexiglas G degrading in air starts rapidly at about 230°C and this is followed by a complex pattern of weight loss rate changes with increases in temperature, instead of by one major peak, as observed in nitrogen. Four DTG peaks are observed at all different heating rates. The difference in the temperatures at which rapid weight loss begins between nitrogen and air is about 70°C at a heating rate of 5°C/m. The thermal degradation of Plexiglas G is very sensitive to gas phase oxygen.

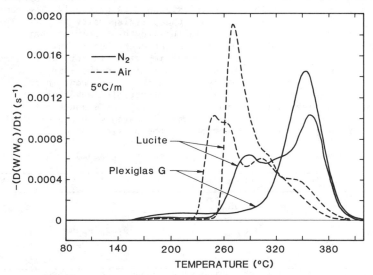

FIGURE 1. DTG curves in nitrogen and in air.

Neglecting the small early weight loss which starts at about 160°C for both samples, one sees that Lucite starts to lose weight in nitrogen about 50°C lower than does Plexiglas G (i.e., 250°C vs 300°C) at a heating rate of 5°C/m. The second peak of weight loss rate for Lucite nearly overlaps the major peak for Plexiglas G. This behavior indicates that Lucite degrades easier in nitrogen than does Plexiglas G at low temperatures. The comparison of results in air indicates that Plexiglas G starts to lose weight rapidly at about 230°C compared to about 260°C for Lucite. The rate of weight loss for Lucite is very large from 260°C to 300°C at a heating rate of 5°C/m. The pattern of weight loss rate above about 320°C is roughly the same for both samples. This behavior indicates that Plexiglas G degrades in air at a lower temperature than does Lucite, but the difference between them is relatively small. These qualitative trends are also observed with TG and DTG at the low heating rates.

Global kinetic rate constants. Global kinetic rate constants based on the rate of sample weight loss were determined from the DTG results at the various heating rates for Plexiglas G and Lucite. The relatively simple Kissinger's method (6) was used to determine the kinetic constants. The relation derived by Kissinger is as follows:

$$\ell n \left( \phi / T_m^2 \right) = \ell n \left( nRAW_m^{n-1} / E \right) - E/RT_m$$

$\phi$ is the heating rate in the TG experiment, $T_m$ is the temperature at the maximum rate of weight loss, R is the universal gas constant, E is the activation energy, A is the pre-exponential factor, $W_m$ is the fraction of the sample weight at the maximum rate of weight loss, and n is the apparent order of the reaction with respect to the sample weight. Activation energies were determined from the slopes of the straight lines. The results for both samples are listed in Table 1 and they confirm the above indication that the degradation of Plexiglas G is more sensitive to gas phase oxygen than is Lucite.

Molecular weight of the two samples. The molecular weight distributions of both samples were measured using gel permeation chromatography. The molecular weight distribution of Lucite is broad with a shoulder in a low molecular side. The number average molecular weight is 179,000 and the polydispersity (ratio of weight average molecular weight against number average molecular weight, a measure of the width of the molecular weight distribution) is 4.2. The molecular weight distribution of Plexiglas G is unimodal. The number average molecular weight is 402,000 and the polydispersity is 2.2. This indicates that there are differences in the length of polymer chains between the two samples and some complex molecular weight distribution for Lucite.

Sample purification effects. Another important factor, which strongly affects the degradation of polymers, is impurities in the sample. It is expected that the commercial samples used in this study contain some impurities: unreacted initiator, monomer, ultraviolet absorber, etc. The effects of impurities on the degradation of Plexiglas G and Lucite were examined by TG of purified samples; the results were compared with the original samples. The purification procedure is described in the previous study (8).

TABLE 1. Activation energy of TG weight loss (first peak)

| | Nitrogen | | Air |
|---|---|---|---|
| Lucite | 84 kJ/mol | | 95 kJ/mol |
| Plexiglas G* | 210 kJ/mol | First Peak | 174 kJ/mol |
| | | Second Peak | 156 kJ/mol |
| | | Third Peak | 114 kJ/mol |

* Values were determined in the authors' previous study (7).

The weight loss and DTG of the purified Plexiglas G and Lucite degraded in nitrogen and in air were compared with those of the original samples degraded in nitrogen and in air, respectively. The effects of sample purification on the DTG of both samples degrading in nitrogen is very small, except for an increase in stability at low temperatures probably due to removal of unreacted monomer from the sample. However, the effect of the purification of Plexiglas G on weight loss in air is significant. As shown in Fig. 2, there is only one peak in the DTG for purified Plexiglas G compared with the four peaks observed for the original sample. Furthermore, the stability of the sample increases significantly with an increase in the threshold of weight loss from about 210°C to 260°C at a heating rate of 2°C/m. There is little effect of the purification of Lucite on weight loss in air.

## 2. Nonflaming Gasification Due to Radiative Heating

The above TG study was based on small samples heated at rates much slower than those applicable to fire. The objective was to determine chemical behavior of the sample under conditions in which effects of mass and heat transport processes on weight loss are minimized. Further study is needed to demonstrate the impact of differing chemical behavior under conditions similar to those in fire, i.e., higher heating rates and substantial transport processes. For this reason, both samples were heated by a well-defined thermal radiation flux.

<u>Surface temperature.</u> Sample surface temperatures were measured when both samples were heated at a thermal radiant flux of 2.2 $W/cm^2$ in air and in nitrogen. The results are shown in Fig. 3. The range of measured surface temperatures is about the same as that used in the above TG study, although the heating rates were much higher. The results indicate that the surface temperatures of both samples heated in nitrogen are slightly higher than those heated in air. This trend is consistent with a previous study (4), indicating that the overall gasification process of Plexiglas G and of Lucite is endothermic even in air. Surface temperature curves of both samples heated in nitrogen show more roughness compared with those heated in air. This was caused by the more violent rupture of larger bubbles in nitrogen than in air due to the more viscous (higher molecular weight) molten polymer layer near the surface in

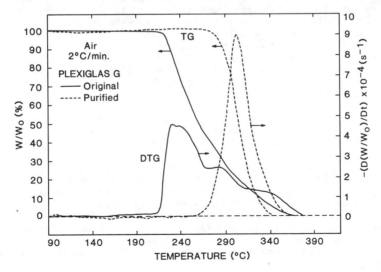

FIGURE 2. Comparison of TG and DTG curves or original Plexiglas G against purified Plexiglas G degrading in air.

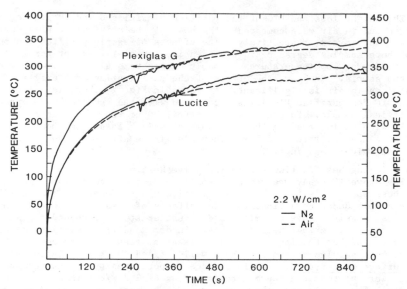

FIGURE 3. Comparison of surface temperature history with time heated at radiant flux of 2.2 W/cm² in nitrogen and in air.

nitrogen than in air (9). Overall, Fig. 3 shows that there is little difference in surface temperature history between Plexiglas G and Lucite heated in air or in nitrogen at the same radiant flux. This indicates that thermal properties of the two samples should be the same. This is confirmed: the specific gravity is 1.19*, the specific heat is 1.5 J/g°C* and the thermal conductivity is 0.13 J/ms°C** for both samples.

Gasification rates. The dependence of mass flux on time is shown in Fig. 4(a) at a radiant flux of 2.2 W/cm² in nitrogen and in air. These results are calculated by dividing the time derivative of the measured transient weight by the front surface area of the sample. The results indicate that the mass flux of Lucite increases slightly when it is degrading in air compared to degradation in nitrogen. The mass flux increases significantly before 500 seconds when Plexiglas G is degrading in air compared to degradation in nitrogen. These trends, weak effect of gas phase oxygen on weight loss for Lucite and a strong effect for Plexiglas G, are consistent with those obtained from the above TG study.

In Fig. 4(b), the mass flux of Lucite degrading in nitrogen from 240 to 720 s, is significantly larger than that of Plexiglas G and beyond 720 s the mass flux is about the same for both samples. In this time range, the surface temperature is in the range between 275 and 350°C as shown in Fig. 3. The TG study shown in Fig. 1, also shows that at low temperatures (250 ~ 300°C), at a heating rate of 5°C/m, Lucite loses weight more rapidly than does Plexiglas G. Above 320°C the weight loss rates for both samples are about the same. This observation is qualitatively consistent with nearly equal mass fluxes measured beyond 720 s.

Comparison of the change in mass flux with time between Lucite and Plexiglas G degrading in air is complex. Although the difference in mass flux between the two samples is small, there are three regimes similar to those

---

*Manufactor's values.
**Our measured values.

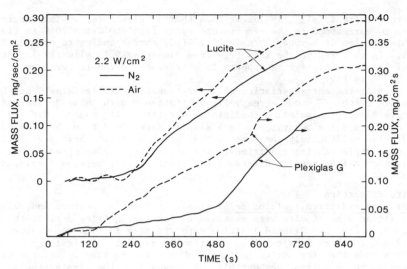

FIGURE 4(a). Comparison of history of mass flux at radiant flux of 2.2 W/cm$^2$ between nitrogen and in air.

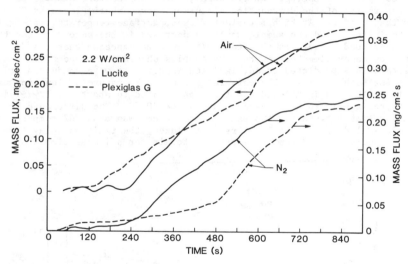

FIGURE 4(b). Comparison of history of mass flux at radiant flux of 2.2 W/cm$^2$ between Lucite and Plexiglas G.

observed in the TG comparison shown in Fig. 1. The first regime is between 120 and about 380 s corresponding to 225°C to 300°C determined from Fig. 3. In this regime the mass flux of Plexiglas G is larger than Lucite. This regime appears to correspond to the temperature range from 230°C to 260°C in Fig. 1. In making such comparisons, the heating rate is generally higher in the radiative heating experiment; this shifts the reaction to higher temperatures. In addition, the mass flux from the radiative heating experiment is integrated over the same in-depth temperature distribution. The second regime is between 380 and 720 s in Fig. 4(b), corresponding to 300°C to 350°C in Fig. 3. In this regime the mass flux from Lucite is slightly larger than that of Plexiglas G. This regime appears to correspond to the temperature range from 260°C to 320°C in Fig. 1. The third regime is beyond 720 s in Fig. 4(b), corresponding to about 350°C in Fig. 3. In this regime the mass flux from Plexiglas G is slightly larger than that of Lucite. This regime appears to correspond to the temperature range above 320°C in Fig. 1.

Overall, the characteristics of the degradation of Plexiglas G and Lucite determined by the TG study agree well qualitatively with those determined by radiative heating, simulating conditions in a fire. It appears that the chemical degradation behavior of both samples determined from the slow heating TG study is qualitatively the same as that which controls their degradation under more rapid heating conditions. The above study shows clearly that there are distinct differences in degradation characteristics between Plexiglas G and Lucite.

## 3. Fire Properties

Piloted radiative ignition delay time. Surface temperature and weight loss for Plexiglas G and Lucite were measured during the ignition period; the results are shown in Fig. 5. Piloted ignition occurred at a surface temperature of about 275°C for both samples. The ignition delay time for Plexiglas G is about 45 s less than that for Lucite. Since the controlling step in piloted radiative ignition is the supply of combustible fuel gases from the sample, the larger mass flux from Plexiglas G compared to that for Lucite causes Plexiglas G to ignite earlier than Lucite. At 1.8 W/cm$^2$, both samples were exposed mainly to the first regime of the degradation discussed above, where Plexiglas G degrades faster than Lucite. At higher radiant fluxes, surface temperature at ignition tends to increase and the sample would be degraded in the second or the third regime described above. In this case, it would be expected that the difference in ignition delay time between the two samples should then become less. To demonstrate this predicted trend, ignition experiments were conducted at 4 W/cm$^2$ for both samples. The measured surface temperature at ignition was close to 300°C, and ignition delay time was reduced to about 60 s, yielding less time for the sample to be exposed to air above 200°C. The difference in ignition delay time between the two samples at 4 W/cm$^2$ was about 6% compared to about 15% at 1.8 W/cm$^2$. At lower radiant fluxes, the ignition delay time becomes longer, and the sample is exposed to air for a longer time above 200°C.

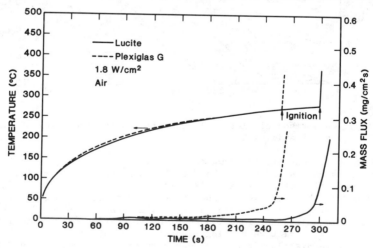

FIGURE 5. Comparison of surface temperature and mass flux histories between Lucite and Plexiglas G under piloted radiative ignition.

In this case, the differences in oxidative degradation characteristics between the two samples noticeably affect ignition delay times for the samples. For this reason, it is expected that the minimum radiant flux to cause pilot ignition would be less for Plexiglas G than Lucite. However, at higher radiant fluxes, the differences in oxidative degradation characteristics between the two samples become unimportant.

Downward flame spread velocity. Some phenomenological differences in flame front behavior between the two polymer samples were noticed. The flame front shape over the Lucite surface was uniform and the flame spread uniformly over the surface. However, the flame front shape over the Plexiglas G surface was sometimes disturbed by an accumulation of a black tar-like material. The black tar-like material was apparently formed on the violently bubbling surface behind the flame front possibly in conjunction with the deposition of soot-like particles from the flame. Since the surface temperature of the burning Plexiglas G is high, as shown in Fig. 6, the melt viscosity of the molten layer is quite low and the molten layer ends up slowly flowing down toward the flame front. This slow flow of the molten layer accumulates black tar-like material near the flame front. The black tar-like material appeared not to gasify or burn under this condition; occasionally it glowed. Near the flame-front, the black tar-like material formed several small spheres (diameter up to 3-4 mm). Sometimes these small black tar-like spheres obstructed the spread of the flame front; at other times they enhanced the spread of the flame front as they slid down and pulled the molten polymer locally ahead of the rest of the flame front. These effects caused a ragged flame front. It is not clear at present what net effect these black tar-like spheres have on flame spread velocity. Since a plot of flame front location with time shows a reasonably straight line, overall flame spread occurs with a constant speed. The black tar-like spheres were never observed in flame spread over Lucite.

The measured flame spread velocity over the Plexiglas G surface was $4.2 \times 10^{-3}$ cm/s; it was $5.0 \times 10^{-3}$ cm/s over the Lucite surface. Therefore, downward flame spread over the Lucite surface is about 20% faster than over Plexiglas G. Assuming that the black tar-like material does not affect the average flame spread velocity, one must look elsewhere to see what causes this difference in flame spread velocity. Since there are no differences in values

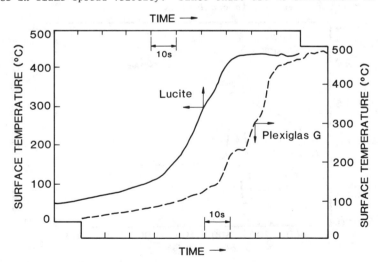

FIGURE 6. Surface temperature history across approaching downward traveling flame front.

491

of thermal properties between the two samples, the amounts of heat transferred through the sample from the flame front to the surface ahead of the flame should be the same for both samples. The surface temperature rise across the approaching flame front as shown in Fig. 6 indicates that the heating rate is about 10°C/s. With this heating rate, it takes about 20 s for the sample surface to heat from 100°C to 300°C. It appears that the difference described in the above first stage degradation in air is not detected during this rapid heating process. Although the surface temperature at the flame front was not precisely determined, it appears to be above 300°C from visual observation of the flame front location relative to the thermocouple bead. This puts the degradation in the second regime in air in which Lucite degrades faster than Plexiglas G or in nitrogen without any oxidative degradation where Lucite also degrades faster than Plexiglas G (such as shown in Figs. 1(a) and 4(b)).

    Burning rate. Material burning was initiated by piloted radiative ignition using an electrically powered cone shape heater (10). This caused sample burning over almost the entire top surface shortly after ignition. This ignition period corresponds to the period up to about 220 s in Fig. 7. Once ignition occurred, electric power to the heater was turned off and the heater was moved away so as not to interfere with the experiment. This period corresponds to about 220 s to 300 s in Fig. 7. Burning conditions without the heater are defined as those after about 300 s. The difference in measured mass flux during the burning period between Plexiglas G and Lucite is very small, although repeated experiments always indicate that the mass flux from Lucite is slightly larger than that from Plexiglas G. This is because the surface temperatures of both samples were in the range of 350°C to 400°C. These temperatures are so high that the difference in degradation behavior between the two samples at low temperatures (as shown in Figs. 1 and 4) is washed out.

    All results and discussion described in this paper apply only to the comparison between Plexiglas G and Lucite. Although the differences in measured fire properties between these two samples are small, it does not necessarily mean that there are small differences in fire properties among other polymers even including PMMA from another manufacturer. The intention of this work is to raise awareness of differences among the same generically classified polymeric materials and the fact that these differences may cause significant differences in fire properties under certain conditions.

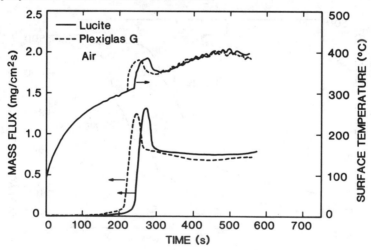

FIGURE 7. Comparison of surface temperature and mass flux histories between burning Lucite and Plexiglas G including initial piloted ignition.

## CONCLUSIONS

(1) The TG study shows that the degradation rate of Plexiglas G is sensitive to gas phase oxygen, but that of Lucite is less so. Comparison of DTG curves between Plexiglas G and Lucite indicates that Plexiglas G is more stable with respect to degradation in nitrogen at low temperatures than Lucite, but Lucite is more stable initially with respect to degradation in air than is Plexiglas G.

(2) The nonflaming gasification study under external radiative heating indicates that the comparative mass flux behavior for the two polymers agrees qualitatively with that of the TG study although the heating rates together with the heat and mass transport processes are different from the TG study. Therefore, the chemical nature of the degradation process of each of the two samples is the same for slow heating TG and the more rapid heating gasification study simulating a fire environment.

(3) In piloted radiative ignition at 1.8 $W/cm^2$, the ignition delay time of Plexiglas G is about 15% less than that of Lucite. Increased radiant flux reduces the difference in ignition delay time between the two samples because shorter ignition delay times reduce the time for the sample to be exposed to air above 200°C.

(4) Downward flame spread velocity over Plexiglas G is about 20% slower than over Lucite. However, the difference in burning rate between the two samples is negligible. This indicates that, if these two samples are heated rapidly to high degradation temperatures, differences in the chemical nature of their degradation would not be detected. The difference may become important when the two samples are heated at low temperatures for a sufficiently long period.

## ACKNOWLEDGMENTS

The authors would like to thank Messrs. William Wooden, William Twilley, Gerald King and Emil Braun for their assistance in conducting the experiments and Mr. R. Lawson for the measurement of thermal conductivities.

## REFERENCES

1. N. Grassie and H.W. Melville, Proc. Roy. Soc. (London) A199, 14 (1949).
2. H.H.G. Jellinek, "Degradation of Vinyl Polymers", Academic Press, New York, 74 (1955).
3. J.R. MacCallum, Makromol. Chem. 83, 137 (1965).
4. T. Kashiwagi and T.J. Ohlemiller, Nineteenth Symposium (International) on Combustion, The Combustion Institute, 815 (1982).
5. T. Kashiwagi, Combust. Flame 44, 223 (1982).
6. H.E. Kissinger, Anal. Chem., 29, 1702 (1957).
7. T. Hirata, T. Kashiwagi and J.E. Brown, "Thermal and Oxidative Degradation of Poly(methyl methacrylate) – Weight Loss", Macromolecules, 18, 131 (1985).
8. T. Kashiwagi, T. Hirata and J.E. Brown, "Thermal and Oxidative Degradation of Poly(methyl methacrylate) – Molecular Weight", to appear in Macromolecules.
9. J.E. Brown, T.J. Ohlemiller and T. Kashiwagi, "Polymer Gasification and Ignition", presented at Symposium on the Chemistry of Combustion Processes, 185th ACS National Meeting, Seattle, Washington, May 1983.
10. V. Babrauskas, "Development of the Cone Calorimeter, A Bench-Scale Heat Release Rate Apparatus Based on Oxygen Consumption", NBSIR 82-2611, National Bureau of Standards, 1982.

# PEOPLE–FIRE INTERACTIONS

Session Chair

**Dr. Paul G. Seeger**
Forschungsstelle für Brandschutztechnik
Universität Karlsruhe
Hertzstrasse 16
D 75000 Karlsruhe 21, Federal Republic of Germany

# Methods of Design for Means of Egress: Towards a Quantitative Comparison of National Code Requirements

**EZEL KENDIK**
COBAU Ltd.
Vienna, Austria

ABSTRACT

This paper provides a brief review of the modelling of people movement during the egress from buildings and discusses some of the questions raised by each type of modelling. Furthermore, it compares the predictions of a selected calculation method with regulatory requirements on means of escape in various countries.

INTRODUCTION

The increasing complexity of buildings concerning functions, size and configurations require a broader attention to the problems related to egress. Over the last two decades there has been considerable activity in modelling egress from buildings. According to the overall tendency in the technical literature the available models can be divided into two categories, viz. movement models and behaviour models. Although the former studies are generally concerned with the exiting flow of a buildings' occupants the design concepts show a somewhat dispersed variation.

The behavioural models as they have been developed, are essentially of two types, conceptual models which have attempted to include the observed, empirical and reported actions from collective interview or questionnaire studies, by Canter (1) and by Wood (2), and computer models for the simulation of the behaviour of the human individual in the fire incident. The conceptual models have attempted to include a theoretical desgin in the model which attempts to provide some understanding of decision making, and alternative choice processes of the individual involved with a fire incident situation. Most of the current models that have been developed of this type would probably be identified as describing the process of the participant in the fire incident as an information seeking and processing model. (after J. Bryan, ref.3)

The current models evolving from people movement may be classified as follows:

1. Flow models based on the carrying capacity of independent egress way components;
2. Flow models based on empirical studies of crowd movement;

3.  Computer simulation models; and
4.  Network optimization models.

This presentation will primarily be concerned with the first and second items, since the former is still world-wide governing the regulatory approaches covering the exit geometry whilst the latter studies are supported by extensive research work conducted in real-world settings. But, before we turn to our primary concern two other models supported by U.S. National Bureau of Standards should be acknowledged.

## BFIRES II: A BEHAVIOUR BASED SIMULATION OF EMERGENCY EGRESS DURING FIRES

This model by F.Stahl (4),(5) is a dynamic stochastic computer simulation of emergency egress behaviour of building occupants during fires. It is a modified and expanded version of BFIRES I (6), which was originally developed for the application to the health care occupancy. The model is not calibrated against real-world events, but a sensitivity analysis of the model proved that BFIRES outcomes are sensitive to (a) floor plan confuguration, (b) occupants' spatial locations at the onset of the emergency event, (c) the existence of any impairments to occupants' mobility, (d) occupants' familiarity with the building layout, and (e) permissible levels of occupant density.

The most interesting finding of this sensitivity analysis is that, when the individuals vary on the basis of occupant parameters (mobility impairment and and knowledge of safe exit location) the effects of variation in enviromental parameters (occupant density and spatial subdivision) disappear. As a result of this Stahl suggests that occupants unfamiliar with the building's physical layout will not be helped by designs providing shorter and more direct egress routes. This challenges the traditional design conventions.

The concept and structure of the model is described by Stahl as follows:

BFIRES conceptualizes a building fire event as a chain of discrete "time frames" and for each such frame , it generates a behavioural response for every occupant in accordance with their perceptions of a constantly changing environment. When preparing a behavioural response at Ti, a simulated occupant gathers information-which describes the state of the environment at this point in time. Next, the occupant interprets this information by comparing current with previous distances between the occupant, the fire threat, and the exit goal and by comparing "knowledge" about threat and goal locations possessed by the occupant, with amounts possessed by other nearby simulated persons. Current locations of physical barriers and of other occupants are also taken into account...The selection of a behavioural response (i.e. the decision to move in a particular direction) results from the comparison of available move alternatives with the occupant's current move criteria.

Here, the choise of exits and the selection of alternative moves appear to be critical . In the first report of BFIRES (6) it is suggested that, as the literature in human behaviour in fires (or fire drills) provide no guidance, that , if 60% (or more) of the occupants inhabiting a space favor a particular exit from the space, they will "convince" the remaining occupants of the quality of their opinion, and all the occupants will seek the exit. This option is not necessarily consistent with the human nature. The opposite choice might be that the majority follows one person.

About the criteria of selecting alternative moves, Stahl writes as follows:

To date, it has not been possible to calibrate computed values of the probability that an occupant will, during a given time frame select some move alternative, against data from actual fire situations. This is because no data on human behaviour during fires exist to describe emergency decision making processes at so fine a level of detail. Considerable research will be necessary to understand the mechanism by which people under emergency conditions perceive alternative courses of action, relate such alternatives to broader egress strategies and then select appropriate actions.

In spite of the limitation, that the model deals with maximum 20 persons in a simulation, it appears to be the only computer program attempting to simulate the individuals' information processing, decision making and responses to a migrating fire threat, like smoke and toxic agents.

EVACNET: A COMPUTERIZED NETWORK FLOW OPTIMIZATION MODEL  (8),(9)

This model, developed by R.L.Francis et al. determines an evacuation routing of the people so as to minimize the time to evacuate the building. Network models are not behavioural in nature. Rather they demonstrate a course of action which, if taken could lead to an evacuation of a building in an "appropriate" manner. The model represents the building's evacuation pattern as it changes over time, in discrete time periods. The model is able to answer several "what if" questions like "how should the building be evacuated if the fire breaks out on the tenth floor or what if more stairwells are added. (9)

The static network model is basically a transshipment model, where origins represent work centers, transshipment nodes represent portions of the building and destinations represent the building exits. The static capacity of the node gives the maximum number of persons simultaneously allowed to stay in this space.  The nodes are connected by arcs, of which the dynamic capacities are upper bounds on flow rates. Based on J. Pauls'"effective width" model, the model assumes constant flow rates in stairwells for a given number of occupants in the building. This assumption that the stairwell flow rates are independent of stairwell usage, appears to be a limitation of the network flow model, since its approach is somewhat contradictory to the effective width model.  Pauls' equation predicts the mean flow rate for the assessment of the overall evacuation performance, while the network model looks at the evacuation pattern every ten seconds.

The network flow optimization model is able to deal with large number of people as well as with complex buildings.

# FLOW MODELS BASED ON THE CARRYING CAPACITY OF INDEPENDENT EGRESS WAY COMPONENTS

The historical development of carrying capacity investigations has been already broadly reviewed by F.Stahl and J.Archea (10), (11) and J.Pauls (12),(13), in several publications. Hence, this presentation will be confined to the discussion of the calculation methods based on these investigations.

An early NFPA document recommended as a guideline for stair design an average flow rate of 45 persons/minute/22" width unit. (after ref.10) In 1935, in a publication of the U.S. National Bureau of Standards, test results about measurements of flow rates through doors corridors and on stairs under non-emergency conditions were presented. There,for different types of occupancy the measured maximum flow rates varied between 23 and 60 persons/min/ unit stair width, and 21 and 58 persons/min/unit door or ramp width. (14) Up to date, the NFPA Life Safety Code 101 (15) maintained the unit exit width concept together with the travel distances and the occupant load criteria. But, for some reason the time component is left out in the present code.

In the U.K. the first national guidance for places of public entertainment was produced in 1934 (16) ; the recommendations in which had been "based not only on experience gained in the U.K., but on a study of disasters which have happened abroad and of the steps taken by the authorities of forein countries". (17) In this document the following formulae for the determination of total width of exits required from each portion of a building were provided reflecting the concept of the unit exit width:

$$A= Z \text{ (Floor area in sq f)} / E \quad B \quad C \quad D \qquad (1)$$

A is the number of the units of exit width required;
B is a constant as to the construction type of building;
C is a constant for the arrangement and protection of the stairs;
D is a constant for the exposure hazard;
E is a factor dependent apon height of floor above or below ground level;
Z is the class of user of the building (closely seated audience etc.).

$$N = A/4 + 1 \qquad (2)$$

N is the number of exits required. In this document it was also stated, that about 40 persons per minute per unit exit width downstairs or through exits is an appropriate figure in connection with these formulae.

In fact the width of exits had been discussed ten years previously in a document for the fire protection in factories, (7), where it was reported that tests in the U.K. and in America had found that on average 40 persons per foot of width per minute was possible for "young and active lads" moving "through door-ways with which they were aquainted", but that figure would have to be reduced very conciderably for theatre audiences, it was considered that in factories a figure of 20 persons per foot of width per minute was quite safe under conditions ruling in a factory.(after ref.17)

40 persons/min/unit of exit width is also recommended in the Post-War Building Studies No.29. (18). In this report another calculation method is suggested. (Appendix II) The width of staircases in the current GLC Code of Practice (19), as well as in the BS 5588 Part 3 (20) are computed by this method (21), which calculates the total population a staircase can accomodate based on the following assumptions:

1. Rate of flow through an exit is 40 persons per unit width per minute;
2. Each storey of the building is evacuated on to the stairs in not more than 2.5 min. (This average clearance time was proposed after an evacuation experience during a fire in the Empire Palace Theatre in Edinburgh in 1911; (18)
3. There is the same number of people on each storey;
4. Evacuation occurs simultaneously and uniformly from each floor;
5. In moving at a rate of 40 persons/unit width/min, a staircase can accomodate one person per unit width on alternate stair treads and 1 person per each 3 sq. ft. of landing space;
6. The storey height is 10 ft;
7. The exits from the floors on the stairs are the same width as the stairs; and
8. People leaving the upper floors are not obstructed at the ground floor exit by persons leaving the ground floor.

$$P = (\text{staircase capacity})(\text{nu.of upper storeys}) + (t_e - t_s) \, r \, w \qquad (3)$$

$t_e$ is the maximum permissible exit time from any one floor onto the staircase (taken as 2.5 min.);
$t_s$ is the time taken for a person to traverse a storey height of stairs at the standard rate of flow (predicted as 0.4 min);
$r$ is the standard rate of flow (taken as 40 persons/unit/min); and
$w$ is the width of staircase in units.

The staircase capacity is predicted after point 5 of the above assumptions.

This method of calculation predicts with increasing number of storeys fewer persons per floor.

K.Togawa in Japan (1955), whose studies are hardly accessible, was apparently the first researcher who attempted to model mathematically the people movement through doorways, on passageways, ramps and stairs. (after Pauls,(13), Stahl and Archea,(10), and Kobayashi,(22) ) He provided the following equation:

$$v = V_o \, D^{-0.8} \qquad (4)$$

$v$ is the flow velocity;
$V_o$ is a constant velocity (1.3 m/sec, which is apparently the velocity under free flow conditions); and
$D$ is the density in persons per sq m.

Hence, the flow rate N is given by

$$N = V_o \, D^{0.2} \qquad (5)$$

This N is the same as the specific flow "q" referred to later.

Based on the data from the investigations by Togawa and the London Transport Board (23) S.J. Melinek and S.Booth (24) analysed the flow movement in buildings and provided the following formulae:

1. The maximum population M which can be evacuated to a staircase, assuming a permitted evacuation time of 2.5 min, is given by

$$M = 200 \ b + (18 \ b + 14 \ b^2) \ (n-1) \qquad (6)$$

b is the staircase width in m; and
n is the number of storeys served by the staircase.

This equation predicts higher number of persons than the method presented in the Post-War Building Studies No.29.

If the population Q and the staircase width b are the same for each floor then the minimum evacuation time is the larger of $T_1$ and $T_n$ where

$$T_1 = n \ Q/(N' \ b) + t_s \qquad (7)$$

$$T_n = Q/(N' \ b) + n \ t_s \qquad (8)$$

$T_1$ corresponds to congestion on all floors and $T_n$ to no congestion. Melinek and Booth suggested as typical values of N' and $t_s$ 1.1 persons/sec/min and 16 sec. Compared with evacuation tests in multi-storey buildings the method predicted in most cases evacuation times which are too low.

A further application of the unit width concept has been the mathematical model of W.Müller in East Germany. (25),(26), (27). Assuming a flow rate of 30 persons/min/0.6 m stair width Müller provided the following equation for the assessment of the total evacuation time in multi-storey buildings:

$$t = (3 \ h_G \ / \ v) + (P \ / \ (b \ f_o \ / \ 0.6)) \qquad (9)$$

$h_G$ is the floor height;
P is the number of persons in the building;
b is the stair width in m;
v is the flow velocity down stairs of 0.3 m/sec; and
$f_o$ is the flow rate/unit stair width of 0.6 m.

The minimum evacuation time via the staircase is

$$t = 10 \ h_G + 15 \ h_G \ n \qquad (10)$$

Müller suggested the limitation of building height rather than to widen the staircases.

FLOW MODELS BASED ON EMPIRICAL STUDIES OF CROWD MOVEMENT

During the last decade Jake Pauls (Canada) developed the "effective width" model. This model is based upon his extensive empirical studies of crowd movement on stairs as well as the data about the mean egress flow as a function of stair width. In this context he conducted several evacuation drills in high-rise office buildings and observed normal crowd movement in large public-assembly buildings. The model describes the following phenomena (13), (29), (30):

1.  The usable portion of a stair width , i.e. the effective width of a stair begins approximately 150 mm distance from a boundary wall or 88 mm distance from the centerline of a graspable handrail. (edge effect)
2.  The relation between mean evacuation flow and stair width is a linear function and not a step function as assumed in traditional models based on lanes of movement and units of exit width. The evacuation flow is directly proportional to the effective width of a stair.
3.  Mean evacuation flow is influenced in a nonlinear fashion by the total population per effective width of a stair.

Pauls provides the following equation for the evacuation flow in persons per metre of effective stair width:

$$f = 0.206 \ p^{0.27} \tag{11}$$

p is the evacuation population per metre of effective stair width. The total evacuation time is given by

$$t = 0.68 + 0.081 \ p^{0.73} \tag{12}$$

This calculation method has been recently accepted for an appendix to the NFPA Life Safety Code, 1985 edition.

Now, we turn to another flow model developed by Predtechenskii and Milinski in the Soviet Union. (31) This method is a deterministic flow model, which predicts the movement of an egressing population on a horizontal or a sloping escape route instantaneously in terms of its density and velocity.

Predtechenskii and Milinski measured the flow density and velocity in different types of buildings nearly 3600 times under normal environmental conditions. Their observations indicated, that the flow velocity shows a wide variation, especially in the range of lower densities. The following equation relating the ratio between the sum of the persons' perpendicular projected areas (P f) and the available floor area for the flow, estimates the flow density homogeneously over the area of an escape route:

$$D = P \ f \ / \ b \ l \tag{13}$$

P  is the number of persons in the flow;
f  is the perpendicular projected area of a person;
b  is the flow width, which is identical with the width of the escape route; and
l  is the flow length.

Note D has no dimensions.

The egress populatian passing a definite cross section on an escape route of the width of b, is referred to as flow capacity.

$$Q = D \ v \ b \qquad\qquad m^2 \ min^{-1} \tag{14}$$

Here, v is the flow velocity. Another flow parameter is the flow capacity per metre of the escape route width, which is defined as the specific flow:

$$q = D\ v \qquad\qquad m\ min^{-1} \qquad\qquad\qquad (15)$$

The efficiency of an evacuation depends on the continuity of the flow between three restrictions, viz. the horizontal passages, doors and stairs. Hence, the main condition for the free flow is the equivalence of flow capacities on the successive parts of the escape route:

$$Q_i = Q_{i+1} \qquad\qquad\qquad (16)$$

If the value of the specific flow q exceeds the maximum, the flow density increases according Predtechenskii and Milinski to a maximum value, which in effect leads to queuing at the boundary to the route i+1. At this stage, the flow consists of two parts, viz. of a group of persons with the maximum flow concentration who has already arrived at the critical section of the escape route, and the rest of the evacuees approaching by a higher velocity and a density less than $D_{max}$. In this case the rate of congestion is given by the following equation:

$$v''_{STAU} = (q_{Dmax}\ b_{i+1}\ /\ b_i - q_i)/(D_{max} - D_i) \qquad\qquad (17)$$

$q_{Dmax}$    is the specific flow at the maximum density;
$b_{i+1}$    is the width of the congested flow;
$b_i$     is the initial width of the flow;
$q_i$     is the initial value of the specific flow; and
$D_i$     is the initial flow density.

After the last person moving at the higher velocity reaches the end of the queue, the congestion diminishes at

$$v_{STAU} = v_{Dmax}\ b_{i+1}\ /\ b_i \qquad\qquad\qquad (18)$$

where $v_{Dmax}$ is the flow velocity at the maximum density.

This calculation method has been mainly applied by Predtechenskii and Milinski to the evacuation of auditoriums and halls.


A MODEL FOR THE EVACUATION OF MULTI-STOREY BUILDINGS VIA STAIRCASES

Kendik (32)-(35) developed an egress model based on the above work. This has been calibrated against the data from the evacuation tests carried out by the Forschungsstelle für Brandschutztechnik at the University of Karlsruhe. (36) If the following simplifications

1.  The length l of the partial flow built up by the occupants of each floor (defined between the first and the last persons of the flow) is assumed to be equivalent to the greatest travel distance along the corridor;
2.  The number of persons as well as the escape route configurations are identical on each storey; and
3.  Each partial flow attempts to evacuate simultaneously, and enters the staircase at the same instant.

are introduced into the general mode the flow movement via staircases shows some regularities:

1. If the evacuation time on the corridor of each floor, $t_F$, is less than the evacuation time on the stairs per floor, $t_{TR}$, then the partial flows from each floor can leave the building without interaction. In this case, the total evacuation time  is given by the following equation:

$$t_{Ges} = t_F + n\ t_{TR} \qquad (19)$$

$t_F$      is the evacuation time on the corridor of each floor;
$n$      is the number of the upper floors; and
$t_{TR}$      is the evacuation time on the stairs per floor.

2. If the evacuation time on the corridor of each floor, $t_F$ exceeds the evacuation time on the stairs per floor, $t_{TR}$, then the partial flows from each floor encounter the rest of the evacuees entering the staircase on the landing of the storey below. Even though this event causes the increase of density on the stairs, the capacity of the main flow remains under the maximum value, $Q_{max}$, which indicates, that the stair width is still appropriate to take up the merged flow, i.e. if

$$t_F > t_{TR}\ , \text{ and}$$

$$q_{TR;n-1} = (\ Q_{T;n-1} + Q_{TR}\ )\ /\ b_{TR}\ < q_{TR;max} \qquad (20)$$

where
$q_{TR;n-1}$      is the value of the specific flow on the stairs after the merging process,
$Q_{T;n-1}$      is the flow capacity through the door to the staircase on each floor,
$Q_{TR}$      is the initial flow capacity on the stairs, and
$q_{TR;max}$      is the maximum flow capacity on the stairs,

then the total evacuation time is given by

$$t_{Ges} = t_F + n\ t_{TR} + m\ dt \qquad (21)$$

where the last term of the equation relates the delay time of the last person from the top floor. The factor m is the number of patterns of higher density, which reduces during the course of the evacuation process. m can be assessed by an iteration.

3. If the value of the specific flow on the stairs exceeds the maximum during the merging of the partial flows at the storey (n-1) congestion occurs on stairs as well as at the entry to the staircase. In this case, the total evacuation time of a multi-storey building is determined by the following equation:

$$t_{Ges} = t_{TR;STAU} + (n-1)(l_{TR}/v_{TR;n-1}) + (n-2)\ dt \qquad (22)$$

$t_{TR;STAU}$      is the length of time required for the flow to leave the floor level (n-1);
$l_{TR}$      is the travel distance on the stairs between adjoining storeys;
$v_{TR;n-1}$      is the velocity of the flow emanating from the con-

```
               gested area at the floor level (n-1);
dt             is the delay time due to congestion; and
n              is the number of the upper floors in the building.
```

The total evacuation time $t_{Ges}$ is influenced in a non-linear fashion by the projected area factor (or the density increase).

The above results follow from the three simple situations described earlier. Recently, Kendik prepared a computer program in Basic language for a HP 150 personel computer based on the described egress model. The program enables the user to change the dimensions of the building's means of egress and the occupant load easily and work out the influence of the variation on the complete circulation system.

Kendik's egress model addresses the time sequence from when people start to evacuate the floors until they finally reach the outside or an approved refuge area in the building within the available safe egress time. Hence, it doesn't consider the time prior to their becoming aware of the fire nor their decision-making processes. But, it can cope with the problem of the potential congestion on stairs and through exits including the interdependencies between adjacent egress way elements, which appear to be a major problem, especially in case of high population densities.

The method differs from other egress models mainly in its flexibility in predicting the variation of the physical flow parameters during the course of the movement. In this it does not assign fixed values to the flow density or velocity for each individual or seperate groups but considers them to be a single group of a certain mean density on each section of the escape route.

A QUANTITATIVE COMPARISON OF NATIONAL CODE REQUIREMENTS ON MEANS OF EGRESS

As already mentioned elsewhere in this paper the regulatory requirements covering the exit geometry in several countries involve explicitly or implicitly the unit exit width concept accompanied by other criteria such as travel distances, occupant load, total number of occupancy, dead ends or maximum floor area. At the present moment only the building codes in Soviet Union (37) require a mathematical proof for the width of escape routes in buildings where the travel distance to one exit is more than 25 metres and the occupancy per floor using an exit exceeds 50 persons. The building codes in the Soviet Union as well as the new building codes in East Germany (38) use the flow model of Predtechenskii and Milinski under free flow conditions.

The building codes selected for inclusion in this study have been the Greater London Council Code of Practice (19), NFPA 101 New Business Occupancies (15), the German Building Codes for High-rise Buildings and Assembly Occupancies, the Japanese Design Guideline for Building Fire Safety (from ref.22), the Russian Building Codes (37) and the Building Codes for Vienna (41). The requirements in these codes have been compered with the predictions of the egress model developed by Kendik based on the data after Predtechenskii.

An example from an earlier paper (42) illustrates how the calculation method have been employed for this purpose:

According to the National Fire Codes (101-316, Chapter 26) the capacity of stairs, outside stairs and smokeproof towers for new business occupancies has to be one unit for 60 persons. (120 persons per 1.12 m). Furthermore, it is written that "for purposes of determining required exits, the occupant load of business buildings or parts of buildings used for business purposes shall be no less than one person per 100 sq ft (9.29 sqm (sic)) of gross floor area and the travel distance to exits, measured in accordance with Section 5-6, shall be no more than 200 ft (60.96 m (sic)). Not less than two exits shall be accessible from every part of every floor". After Section 5-6.1 the maximum travel distance in any occupied space to at least one exit, shall not exceed the limits specified for individual occupancies, in this case 200 ft.

These provisions might permit one to design a multi-storey office building of roughly 2400 sq m per floor with two remote exits each with a width of two units and circa 240 occupants per floor.

Assuming the stairs to be used at capacity levels and the widths of all exits (doors and stairs) as well as the escape routes leading to the staircases to be identical, the described flow model predicts for new business occupancies, that the last person from a floor enters the staircase after 2.15 min under congested flow conditions. The number of persons moving in the overcrowded flow would be 37. This means that a protected lobby of at least 10.5 m2 (37 x 0.28 m2) or two staircases with a width of 1.20 m were necessary to accomodate 120 persons per floor. In the latter case the exiting time of the last person from a storey would be 1 min. Without interaction of flows a staircase with a width of 1.12 m (2 units) would be able to accomodate 35 persons per floor. In this case the egress time from a floor would be about 0.4 min.

Time is an important criterion for the flexible and cost effective design of escape routes. Figure 1 illustrates the comparison of the calculated stair capacities with the requirements of various building codes on means of escape. Here, the calculated number of persons per floor are predicted under the assumption that the egress time from a floor will be 1 min. The horizontal axis gives the number of persons a staircase with a certain width would accommodate required in various codes, while the vertical axis are the predicted figures. It is interesting to notice, that most of the investigated code provisions relating stair capacity lie under the reference line. This might indicate, that the requirements of the existing codes imply floor evacuation times greater than 1 min. (In one case up to 5 min, ref.19).

The correlation between the reference line and required number of persons in regulations would change in accordance with the egress time from a floor. Namely, if the available time for all occupants to evacuate one floor is expected to be about 2 min for the above example the required stair capacity would suffice to accomodate the given occupancy. If the available evacuation time is expected to be 3 min the required stairs widths are likely overestimated for the given occupancy.

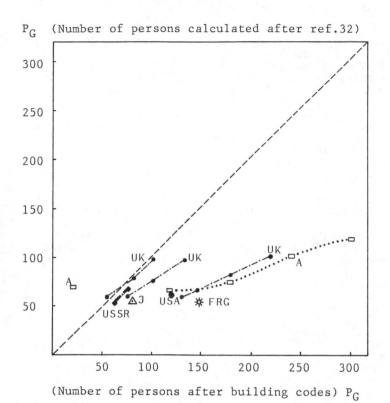

P_G (Number of persons calculated after ref.32)

(Number of persons after building codes) P_G

FIGURE 1.  Graphical representation of the calculated stair capacity against requirements of building codes on means of escape.

CONCLUSIONS

Recently, there has been considerable activity in modelling egress from buildings. The numerous methods available are basicly either behaviour or movement models. All of them appear to make several assumptions, partly to overcome the gaps in the technical literature, which makes their validation against real-world events or fire drills necessary. In fact, only a few of these models are calibrated in this manner and able to provide quantitative results.

The physical structure of a building is apparently an elementary determinant of its occupants' behavioural responses and actions to the changing environmental conditions in terms of time. The time needed to reach a place of safety inside or outside the building might strech from the time people need to escape by their own unaided efforts, as very often stated or implied in most of the national fire codes, until the time handicapped as well as non-handicapped persons need to be rescued. Hence, the critical nature of time requires an analysis that enables the designers to select an appropriate egress system and to estimate the escape facilities by exploiting performance-oriented calculation methods.

This paper also provided a quantitative comparison of the predictions of a selected flow model with the requirements of various codes that do not employ such methods but appear to be based on experience and judgment. In this way time should be regarded as a design component for means of escape in order to improve cost effectiveness and design flexibility.

REFERENCES

1. Canter, D.: Human Behaviour in Fires. Guilford, University of Surrey, Department of Psychology, U.K. 1978.

2. Wood, P.G.: The Behaviour of People in Fires. BRE Fire Research Station, Fire Research Note 953, UK, 1972.

3. Bryan J.L.: Implications for Codes and Behaviour Models from the Analysis of Behaviour Response Patterns in Fire Situations as Selected from the Project People and Project People II Study Programs. NBS-GCR-83-425, sponsered by National Bureau of Standards, Department of Commerce, Washington, March 1983.

4. Stahl, F.: BFIRES/Version 2: Documentation of Program Modifications. NBSIR 80-1982, U.S.Department of Commerce NBS Center for Building Technology, Washinton D.C., March 1980.

5. Stahl, F.: BFIRES II: A Behaviour Based Computer Simulation of Emergency Egress During Fires. Fire Technology, Feb.1982, p.49.

6. Stahl, F.: Final Report on the "BFIRES/Version 1" Computer Simulation of Emergency Egress Behaviour During Fires: Calibration and Analysis. NBSIR 79-1713, U.S.Department of Commerce, NBS, Center for Building Technology, Washington D.C. , March 1979.

7. Home Office: Fire Protection in Factories. Safety Pamplet No. 13. London. HMSO. 1928.

8. Francis, R.L. and Saunders P.B.: EVACNET: Prototype Network Optimization Models for Building Evacuation. NBSIR 79-1738. Operations Research Division, Center for Applied Mathematics, U.S.Department of Commerce, NBS, Washington D.C. 1979.

9. Chalmet, L.G., Francis, R.L., Saunders, P.B.: Network Models for Building Evacuation. Fire Technology, Feb. 1982, p.90.

10. Stahl, F.I. Archea, J. : An Assessment of the Technical Literature on Emergency Egress from Buildings. NBSIR 77-1313, Center for Building Technology, National Bureau of Standards, U.S. Dept. of Commerce, Washington, D.C. 1977.

11. Stahl, F.: Time Based Capabilities of Occupants to Escape Fires in Public Buildings: A Review of Code Provisions and Technical Literature. NBSIR 82-2480. U.S.Dep. of Commerce, NBS Center for Building Technology, Washington D.C., 1982.

12. Pauls, J.L. : "Building Evacuation: Research Findings and Recommandations. In Fires and Human Behaviour," ed. D. Canter, New York, John Wiley and Sons (1980), p. 251.

509

13. Pauls, J.: Development of Knowledge about Means of Egress. Fire Technology, Volume 20, Number 2, May 1984.

14. National Bureau of Standards: Design and Construction of Building Exits. US Dep. of Commerce, NBS M151, Washington D.C.,1935.

15. National Fire Codes, Volume 9, Code for Safety to Life from Fire in Buildings and Structures, NFPA 101. 1982.

16. Home Office: Manual of Safety Requirements in Theatres and Other Places of Public Entertainment. London, HMSO, 1935.

17. Read, R.: Means of Escape in Case of Fire: The Development of Legislation and Standards in Great Britain. To be published.

18. Post-War Building Studies No.29, Fire Grading of Buildings, Part III: Personal Safety. London, HMSO 1952.

19. GLC Code of Practice, Means of Escape in Case of Fire. UK,1974.

20. British Standard BS 5588: Fire Precautions in the Design and Construction of Buildings, Part 3. Code of Practice for Office Buildings. British Standards Institution, 1983, p.20.

21. Tidey, J.: Greater London Council; private communication. 1983.

22. Kobayashi, M. : Design Standards of Means of Egress in Japan. International Seminar on Life Safety and Egress at the University of Maryland, MD, U.S.A. November, 1981.

23. London Transport Board: II. Report of the Operational Research Team on the Capacity of Footways. London Transport Board Research Report, UK, 1958.

24. Melinek, S.J., Booth, S.: An Analysis of Evacuation Times and the Movement of Crowds in Buildings. BRE, Fire Research Station CP 96/75, 1975.

25. Müller, W.L. : Die Beurteilung von Treppen als Rückzugsweg in mehrgeschossigen Gebäuden. Unser Brandschutz 16, (1966), Nr.8, wissenschaftlich-techn. Beilage Nr.3 p.65; Unser Brandschutz 16,(1966), Nr.11, wissenschaftlich-techn. Beilage Nr.4, p.93.

26. Müller, W.L. : Die Überschneidung der Verkehrsströme bei dem Berechnen der Räumungszeit von Gebäuden. Unser Brandschutz 18 (1968) Nr.11, wissenschaftlich-technische Beilage Nr.4, p.87.

27. Müller, W.L.: Die Darstellung des zeitlichen Ablaufs bei dem Räumen eines Gebäudes. Unser Brandschutz 19 (1969) Nr.1, wissenschaftlich-technische Beilage Nr.4, p.6 .

28. Pauls, J. and Jonas, B.  : Building Evacuation: Research Methods and Case Studies. In Canter (ed.) Fires and Human Behaviour. Wiley a. Sons Ltd., Chapter 13, 1980.

29. Pauls, J.: Effective-width Model for Crowd Evacuation. VFDB,6th International Fire Protection Seminar, Karlsruhe, FRG, 1982.

30. Pauls, J.: The Movement of People in Buildings and Design Solutions for Means of Egress.Fire Technology,Vol.20,Nr.1,1984,p.27.

31. Predtechenskii, W.M., Milinski, A.I.: Planning of Foot Traffic Flow in Buildings. Published for National Bureau of Standards by Amerind Publishing Co. Pvt. Ltd., New Delhi, 1978.

32. Kendik, E.: Die Berechnung der Räumungszeit in Abhängigkeit der Projektionsfläche bei der Evakuierung der Verwaltungshochhäuser über Treppenräume. VFDB, 6th International Fire Protection Seminar, Karlsruhe, FRG; 1982.

33. Kendik, E.: Determination of the Evacuation Time Pertinent to the Projected Area Factor in the Event of Total Evacuation of High-Rise Office Buildings via Staircases. Fire Safety Journal, 5 (1983) p.223.

34. Kendik, E.: Die Berechnung der Personenströme als Grundlage für die Bemessung von Gehwegen in Gebäuden und um Gebäude. Technical University of Vienna, Ph.D., 1984. (unpublished)

35. Kendik, E.: Assessment of Escape Routes in Buildings- Discussion of a Design Method for Calculating Pedestrian Movement. Paper presented at the 1984 Annual Conference on Fire Research held at the Center for Fire Research , National Bureau of Standards, Gaithersburg, Maryland, Oct.1984, unpublished.

36. Seeger, P. John, R. : Untersuchung der Räumungsabläufe in Gebäuden als Grundlage für die Ausbildung von Rettungswegen, Teil III: Reale Räumungsversuche. Stuttgart, Informationszentrum für Raum und Bau der FgG,1978, p.395

37. SNIP II-82-80, Baunormen und Bauvorschriften Teil II: Projektierungsnormen Blatt 2: Brandschutznormen für die Projektierung von Gebäuden und baulichen Anlagen.Staatliches Komitee für Bauwesen (GOSSTROI) der UdSSR. Verlag für Bauwesen, Moskau, 1980.

38. Technischer Materialien zur Standardisierung (TMS)-Bautechnischer Brandschutz, Evakuierung von Personen aus Bauwerken. Brandschutz- Explosionsschutz, Aus Forschung und Praxis 3. Staatsverlag der DDR, Berlin, 1980. p.144.

39. Muster für Richtlinien über die bauaufsichtliche Behandlung von Hochhäusern. Fachkommission Bauaufsicht der ARGEBAU, FRG; 1979.

40. Versammlungsstättenverordnung-VStättVO Baden-Württemberg vom 12.Februar 1982. Vorbeugender Brandschutz, Hg.VFDB, 5.3 Bauten besonderer Art- Teil II: Versammlungstätten, Wiesbaden, 1982.

41. Landesgesetzblatt für Wien, Jahrgang 1976, 18.Gesetz: Bauordnung für Wien, § 106: Stiegen, Gänge und sonstige Verbindungswege. Vienna, 1976.

42. Kendik, E.: Assessment of Escape Routes in Buildings and a De-Design Method for Calculating Pedestrian Movement. Presented at SFPE'S 35th Anniversary Engineering Seminar, Chicago, May 1985.

511

# Leadership and Group Formation in High-Rise Building Evacuations

**B. K. JONES**
Division of Building Research
National Research Council Canada

**J. ANN HEWITT**
dh Access Research Associates, Ottawa, Canada

ABSTRACT

This study addresses group formation and leadership during the evacuation of a high-rise office building due to fire. Rather than focussing on the psychological parameters of individual evacuee behaviour, the authors concentrate on the social context and organizational characteristics of the occupancy within which decisions about evacuation strategy, group formation and questions of leadership are made. A distinction is drawn between "emergent" (situational) and "imposed" (authoritative) leaders and between the processes of status emergence (achievement of influence) and status maintenance (retention of influence). Both leadership and group formation can be viewed not only in terms of psychological processes but also as the interaction between the normal organizational structure and the roles people assume and play within their group.

INTRODUCTION

This paper represents an analysis of the impressions, interpretations and actions of a number of people who were required to evacuate a twenty-seven-storey office building because of fire. Two researchers from the Division of Building Research, National Research Council of Canada, conducted in-depth interviews with forty of the participants in the evacuation as well as with ten firefighters who came into contact with evacuees during the course of their duties. These interviews, including the respondents' descriptions of the event and questions posed by the interviewers for clarification, were tape recorded and subsequently transcribed.

The transcripts were analyzed on the basis of a number of categories including communication (formal, intragroup and intergroup), fire experience and training, movement through smoke and wayfinding, group formation, and leadership. The method of analysis used in this study draws on a number of sources including Breaux (1976), Canter (1980) and Sime (1980). This paper reports on a single issue or category of analysis: leadership and group formation. A more detailed account of the incident is currently in preparation.

Sime (1983) takes the position that in order to maximize an occupant's ability to escape, the building designer must consider the important psychological aspects of people's response to a fire. The authors would go a step further by suggesting that, in addition to the psychological aspects, one must take into account the social and organizational characteristics of the occupancy, including what a person knows (or believes) of the situation, whether the person is alone or part of a group, the normal roles that people hold within the occupancy, and the organizational structure or framework. One factor that appears to be related to the chosen evacuation strategy of an occupant is the presence of leadership and the form which that leadership takes.

A great deal of the social psychological research on the question of leadership and its effect on group decision-making has been carried out in laboratory settings. These experiments have tended to focus upon the psychological traits or personality characteristics of individual group members and the context was considered, for the most part, only as it provided a framework for addressing these questions (Shaw, 1973). Current approaches, however, place more emphasis on situational variables such as the internal structure of group organization, the established role relationships within the group, and the broader organizational framework within which group formation and leadership occur. These issues are considered crucial to any attempt at understanding how groups cope with an external problem or threat. Hollander (1971, p. 496), for example, holds that "leaders are made by circumstance" and draws a distinction between "emergent" (situational) and "imposed" (authoritative) leaders. Imposed leadership is determined by authority or by virtue of a person's position in the organizational hierarchy, whereas the situational approach conceives of leadership in terms of the function to be performed rather than in terms of the persisting traits of the leader. Hollander also hypothesized that the relationships by which leadership is produced or retained within groups may be further distinguished by:

> "... studying the interrelated processes of status emergence, concerning factors at work in the achievement of influence, and status maintenance, covering those which allow the retention of influence [author's emphasis] .... The retention of leadership necessarily depends somewhat upon others' perceptions of competence and effectiveness" (Hollander, 1971, p. 498).

Leadership in most organizations is of the imposed type, where group membership and structure, as well as the task requirements of leaders, are set. One of the aims of this study is to look at how this system operates under conditions of threat, when the structure of the established emergency plan is absent. This would entail examining not only the performance of groups and their leaders in the incident but also the perceived competence of the leaders in terms of the specific tasks of the group at the time and the adherence of the individual group members to the pre-emergency organizational procedures.

Although the following case studies do not exhaust the list of evacuees from the building, they do represent distinct leadership situations. The case studies include a situation where leadership was imposed and status maintained; one in which leadership was not retained and new leadership and status emerged; and finally one in which two imposed leaders directed a group's actions until a delay in egress led to a split in the group and in the leadership. This analysis is based upon the accounts of participants. It is not suggested that any actors had or did not have "leadership potential", an approach which would necessitate a discussion of personality traits. Instead the approach taken in this paper is centred on the fire situation itself and the strategy for evacuation selected by leaders and groups.

## BUILDING AND FIRE INCIDENT

The building in which the fire took place is located in Ottawa, Canada. It consists of a twenty-seven-storey central core office tower, a nine-storey above-ground parking garage, and a ground level shopping mall. At the time of the fire, the building was approximately ten years old, with a rentable floor area of approximately 130,000 m². The building has two separate and separated banks of elevators, one bank serving floors 1 to 14, the other serving floors 15 to 27, and four exit stairwells located at each end of the core. The office space is a typical open plan design with acoustical screens dispersed throughout the floor areas.

The on-site fire protection facilities consist of a fire alarm system connected to the central station of an off-premises alarm company, a klaxon alarm horn system and a 'live' voice communication system with speakers on all floors and in the stairwells. Manual pull stations and two emergency telephones on each floor are connected to the fire control room on the building's main level. The building is equipped with a sprinkler system, covering the entire basement and main plaza levels as well as the waste storage rooms and garbage chute on each floor, heat detectors in the service rooms, smoke detectors in the return air ducts, a standpipe and hose system, portable fire extinguishers, and a diesel generator to provide emergency power and lighting throughout the building.

## FIRE INCIDENT

At approximately 21:00 hours on a Wednesday in late November a smoke alarm on the third floor sounded a general alarm throughout the building and alerted the Fire Department. The fire alarm sounded for one full minute, although a number of occupants raised questions as to how long the alarm actually did ring. At the end of this period, the alarm was automatically silenced and a building security officer, using the voice communication system, notified the occupants that the cause of the alarm was being investigated and that they should remain at their work stations and await further instructions. At almost the same time as the voice announcement was made, a number of occupants on the fire floor (third) and on the upper floors (fifteenth to twenty-third) reported either seeing or smelling smoke, thus becoming aware that a fire actually did exist. A number of these people who saw or smelled the smoke, particularly those on the upper floors, decided to ignore the announcement to await further instructions and proceeded to evacuate the building via the stairwells and elevators. Other occupants followed the instructions given over the voice communication system and waited for further instructions.

Approximately five to ten minutes after the original standby message, another announcement was made over the communication system instructing all occupants to evacuate the building. A number of the evacuees who were interviewed were vague concerning the length of time between the two announcements, with some insisting that there was an interim announcement telling the third floor occupants to evacuate. With this general notice to evacuate, those who had waited proceeded to leave the building via the stairs and elevators. By this time, however, many of the people from the upper floors who had evacuated at the first sign of smoke, were already coming back up the stairs, having found the smoke in the stairwells too dense or too irritating (in their opinion) to continue. The evacuees who used the elevators encountered no difficulty in reaching the main plaza level. Many of those using the stairwells had to make several attempts, due to smoke

515

conditions, before they reached the ground floor. One group of government employees from floors 20 to 23 took forty to fifty minutes to reach the main plaza level, utilizing two different stairwells and ascending and descending some sixty flights of stairs in the process.

The number of people in the building at the time of the fire is open to question, although the Office of the Dominion Fire Commissioner (now the Fire Commissioner of Canada) estimated it to be 128. Of this number, thirty were federal employees, thirty were employees of a large private company, sixty-four were employees of the building's property management firm (mostly cleaning staff under contract), and four were employees of a law firm. The Ottawa Fire Department received the alarm at 21:00 and arrived at the building approximately two minutes later. The fire was brought under control by about 23:42, with the all clear given at 03:11. There were no fatalities but twenty-one of the evacuees and six firefighters were taken to hospital and treated for smoke inhalation. All but one were released the same evening; the remaining person was admitted overnight but was released from hospital the following morning.

Case No. 1:  Printing Unit, 23rd floor

Description.  A printing unit on the twenty-third floor constituted a separate and well defined group. It consisted of six people: two men, one of whom was a supervisor, and four women. Most of the group members, except the supervisor, reported hearing the alarm. He was working in an enclosed office on the shop floor but due to the noise of the machinery he remained unaware that the alarm was ringing. The other man, who served as assistant foreman, also had difficulty hearing the alarm. All of the women except one interpreted the alarm as false and continued to work. One woman stopped what she was doing and waited for the voice announcement which she believed was coming. No such announcement was made until the group had already reached the main lobby.

The group continued working until one of the women mentioned to her co-workers that she thought she could smell smoke. Upon looking into the corridor the three women could see smoke pouring from the freight elevator shaft and one of them informed the supervisor. When he saw the smoke he told the women to gather their personal belongings and to wait for him in the elevator lobby. While he shut down the machinery and locked the shop area, the assistant foreman went to inform some other people on a different part of the floor that the alarm signalled a real fire. Both men rejoined the waiting group and took the same elevator, reaching the main lobby on the ground floor without mishap.

Discussion.  The leadership pattern in this group represents an explicit example of the 'imposed' type, that is, leadership by virtue of position within the organizational framework. The members of the group abided by the supervisor's decisions, although a majority of them later expressed reservations about the course of action he selected. Several women reported thinking that they should not use an elevator as a means of evacuation, but no one voiced this concern during the egress. During the interview, the supervisor stated that his decision to take the elevator was a "quick one" and that he "didn't think too much about it". Most members of the group had participated in at least one fire drill in this or some other building. They characterized the training they received, however, as consisting of gathering in the corridor and following the instructions of the officer in charge. The supervisor, on the other hand, had never particpated in a drill nor had he ever read anything distributed by the building's fire safety committee about what to do in the event of a fire. Moreover, he had never been told what his duties as a supervisor were in an emergency which occurred after normal working hours when the structure imposed by the emergency plan was absent, even though the printing unit often worked late. This lack of training is evidenced by the fact that using the stairs or the emergency telephones did not occur to him.

However, the supervisor's leadership of the group remained intact and his instructions were followed although the proper procedures were known by some group members. For example, while waiting for the assistant foreman to return from warning other floor occupants, several women became nervous and began shouting for him to hurry. One woman was especially angry at the delay ("I could have killed the guy") and urged the group to leave him behind. This suggestion was ruled out by the supervisor and they continued to wait for his return. In general, the leadership pattern apparent in this group was of the imposed type, that is, imposed by the organizational structure. A new leader did not emerge from the membership as a result of either perceived superior knowledge or force of personality. In the interviews, several people said that during the fire they considered some of his instructions with trepidation. He maintained his status as leader nevertheless and they accepted his directions, or as one woman stated: "I just did what the boss told me".

Case No. 2:  Purchasing/Stores; Mail Room, 15th floor

Description. Five people were working in two separate rooms on the fifteenth floor at the time of the fire: two men in stores and two men and a woman in the mailroom. Everyone reported hearing the alarm when it first sounded. The male supervisor from the mailroom went into the corridor to listen for the announcement that would come over the public address system. When he saw smoke escaping from the freight elevator shaft, he shouted a warning to the two men in the other room and returned to his office and told his staff to get ready to leave. One of the men from the purchasing/stores unit, however, had already been in the corridor and smelled smoke. He returned to tell his co-worker about it and said he was leaving the building immediately. As he was gathering his belongings, the other man went to check the corridor for himself and had just noticed the smoke when he heard the supervisor's warning. The five people joined together and decided to evacuate via the nearest exit stairs.

They started their descent, but upon reaching the twelfth floor they could see smoke coming up the stairwell towards them. By the time they reached the fifth floor the smoke was quite thick and irritating to the eyes and throat. At the fourth floor they found the smoke too dense to continue and decided to change to another stairwell. The group began ascending the stairs, trying to find an unlocked door from the stairwell to the office space. They found a door open on the ninth floor, crossed the office area, and entered the stairwell on the opposite side of the building. Here they met a group of women from the cleaning staff and together they began to walk down the stairs. Within a floor or two they encountered smoke as thick as in the other stairwell. Returning to the ninth floor, they entered the office area and argued about what to do next.

The supervisor, supported by the man from the mailroom and one of the men from purchasing/stores, wanted to go up the stairs to the roof and find fresh air. The other man from stores tried to convince them to try to make it down the stairs. He argued that if they went back down they would have only nine floors to go and that, as the smoke was rising, the higher they climbed the more smoke they would eventually have to go through. The three men were adamant about trying for the roof; the female members of the group did not appear to take part in the debate. Here the group split apart. One group, consisting of three men, left and continued up the stairs toward the roof. In their ascent they encountered another group and their subsequent actions are described in case number three. The other group consisted of one man from the purchasing/store room, the woman from the mailroom and the female cleaning staff. He led them to a washroom where they wet their sweaters, and covering their faces with them, proceeded down the stairs to the main lobby without much difficulty.

Discussion. Both imposed and emergent leadership were evident in this group as well as status maintenance and status emergence. From the beginning, the supervisor displayed characteristics of imposed leadership. It was he who checked the corridor, alerted the two men in the purchasing/store room, told the two people in his office that it was time to leave and locked the office after them. He led the group to the stairwell and although the two men from the other office did not come under him in the organizational structure they seemed willing to follow his lead. When the group reached the fourth floor it was apparently the supervisor who felt most strongly that the smoke was impassable and that they should change stairwells. The rest followed him as he went back up the stairs trying to find an unlocked door. The man from the purchasing/store room, who later left the group, reported thinking that if the stairwell was impassable the person on the public address system would have mentioned it. Nevertheless, he went along with the supervisor, who maintained his leadership status. After they had come back up the second stairwell, however, the leadership pattern underwent some change. The supervisor, supported by two of the men, stated that the group should try for the roof. The other man from purchasing/stores argued that this course of action was foolish and tried to convince the group to make for the ground floor. Although this man displayed a fair degree of knowledge about the properties of fire and smoke movement, the supervisor and the two men disregarded his advice. He, in turn, rejected the decision by the supervisor. In the interview he stated that he was willing to go along with the group and follow the lead of the supervisor as long as their actions corresponded with what he thought they should do. When the supervisor advocated a course of action which he believed to be wrong, he withdrew his support. When the three men left for the roof the women members remained with this man. One of the women said he seemed to make sense and she thought he knew what he was doing. The man later stated that he was nervous but that: "as long as I'm involved with someone else, I don't think about myself". He also said that he would not accept a post as a fire warden because he doesn't want to feel obligated to help other people. Moreover, people lack respect for him because of the low status position he holds and consequently they wouldn't follow his instructions.

Case No. 3:  Communications Unit - 23rd floor

Description. At the time of the fire alarm, fourteen people were working on the twenty-third floor. This case study pertains to seven of those people (two men and five women); the remainder did not form part of this particular group but left the building on their own as part of other groups. Of these seven people, five had been working together in one large office and two had been working in a separate area. In this group, the two men were senior to the five women in the organizational hierarchy and were supervising the evening work. After hearing the alarm, one of the men tried unsuccessfully to reach the fire control centre via the emergency telephone. Both men instructed the women to lock up and prepare to evacuate, while they checked

518

the adjacent floors to ensure that no one remained. The group then proceeded down the stairwell. At the fifteenth floor, they encountered three men coming up the stairs, who informed them that the stairwell was impassable at the lower levels due to smoke. The group, now consisting of ten members, took refuge on the fifteenth floor until increasing smoke levels led to a decision to proceed to the roof. On their ascent, they met another male evacuee, who joined the group in its climb. When this group of eleven people reached the twenty-seventh floor, one of the senior male employees and another male went up the short flight of stairs to the roof door. They found this door locked and shouted the information to those waiting below.

At this point, the group divided, and three of the men, who had not been part of the original group, descended on their own. The remaining eight people (three men and five women) returned to the twenty-seventh floor. One of the senior male employees provided the group with wet paper towels from the washroom. The group then split for a final time; one male accompanied by three females started down the stairs. The others remained on the twenty-seventh floor for a short period of time and then proceeded down the stairs themselves. They became the last occupants to leave the building, having taken some fifty minutes to do so.

Discussion. This situation presents a good example of imposed leadership. Both senior male employees influenced the egress strategy adopted by the group. Both were involved in informing the women working under their supervision to leave the building and at various times the two men attempted to contact the security personnel on the ground floor via the emergency telephones.

Quite clearly, the men perceived themselves to be the decision-makers, and this self-percepton was reinforced by the women members of the group. With few exceptions, group members reported that decisions on egress strategy were made by the male supevisors. The two leaders described not only their choices but also their decision-making role as "natural"; stressing their sense of responsibility for those working under them. Until the group reached the top floor, decisions appear to have been made jointly between these two men. The group followed their directions or, as one leader stated: "They did what we told them". Group members described the two males as knowledgeable, 'level-headed' and calm and stated that they simply followed instructions.

The first key decision vis-à-vis the egress strategy was to accept the information given to them by the three men they encountered on the fifteenth floor. As the group was already encountering smoke in the stairwell, both men considered it reasonable that descent would not be possible. In choosing to stop descent, they retained their status.

The second key decision was to proceed to the roof. Again, this was seen as an effective solution by group members in general. When this solution proved untenable, the leadership split. If leadership is viewed as that behaviour which best helps a group attain its goals, then in this case the leadership had failed, and the solidarity of the group dissipated. Ultimately, supervisors and their immediate staff exited the building together. Until the roof door was found to be locked, the women had been willing to follow either or both of the men. Following the failed attempt to open the roof door, these women reformed according to their normal organizational relationships. In their accounts of the group breakup, both the women and the male leaders expressed some confusion as to exactly why the group had split. Some suggested that it seemed more reasonable to descend in smaller groups; others felt that there wasn't a reason to wait any longer. At this point, leadership with respect to evacuation strategy had become irrelevant. There was no decision to be made, for the only remaining option or solution was descent. Not surprisingly, people descended with those with whom they were most familiar.

Case No. 4: Private Corporation, floors 2 to 12

Description. There were approximately twenty-four employees (ten female and fourteen male) of this company distributed over eleven floors at the time of the fire. The majority of these individuals were working alone and, apart from those nearest the fire floor, most people did not begin to evacuate until the general order was broadcast over the public address system. The stairs were the chosen means of egress for all of the occupants and although individuals may have encountered other people in the stairwells, it seems, from the interviews, that they considered this to be due to circumstance rather than choice. In other words, the occupants did not seek to form groups even when a number of individuals were working within sight and hearing of one another. For the most part, each person decided on which evacuation strategy to follow and acted upon this decision without consultation. Consequently, neither imposed or emergent leadership nor issues of status maintenance or emergence appeared to play a significant role in the evacuation of these occupants.

Discussion. A number of reasons may account for the lack of group formation and the non-emergence of leadership. For example, the majority of individuals who were working after regular office hours occupied positions at the middle management level in the organization's hierarchy. Most of their work was normally carried out on an individual basis, such as computer programming and analysis, systems or engineering design. In other words, they were used to making decisions on their own without consultation and many held supervisory posts in their own right. A second reason could relate to the fire safety training they had received. In addition to participating in the drills and exercises given by the building management company and the Office of the Fire Commissioner of Canada, the corporation provided extra training in the form of audio-visual presentation and question and answer sessions. Although the employees of the private firm, like the federal government employees on the upper floors, stated that the training they had received did not prepare them for the "real thing", the former individuals, on the whole, took the more effective action. In the interviews, the corporation's employees consistently displayed a better grasp of the physical properties of fire and sound evacuation and safety practices. For instance, unlike a number of people from the upper floors, they did not assume that the fire was in the stairwell when they encountered smoke during their descent of the stairs, nor did they confuse the concepts of 'fire proof' and 'fire resistant'. In addition, they knew why the elevator was not an advisable method for evacuation during a fire. These people had no reason to form groups or to seek leaders in order to decide on the best course of action. They knew what to do and it was simply a matter of getting on with it.

CONCLUSION

In recent years, research carried out in the field of human behaviour and fire has indicated that numerous fire safety plans and strategies are often based on unsubstantiated assumptions about human and organizational response to emergencies or are predicated upon the belief that evacuation is a relatively simple process largely controlled by alarms, exit facilities and training programmes. In point of fact, research has demonstrated that

520

behaviour in fire is far more complex than many such assumptions would suggest (see Stahl and Archea, 1977; Canter, 1980; Pauls and Jones, 1980). Sufficient evidence exists to suggest that not only are the psychological parameters important but also the social and organizational characteristics of the occupancy, including its organizational structure or framework and the normal roles that people hold within the organization.

The present report focuses on two of these organizational issues - group formation and leadership - and seeks to examine these phenomena in relation to the context or situation within which they emerge. Specifically, it looked at the notions of emergent (situational) and imposed (authoritative) leadership as well as the processes of status emergence (achievement of influence) and status maintenance (retention of influence). The data seem to indicate that the presence of leadership and the form that it takes do affect the evacuation strategy adopted by a particular group. For the most part, the leadership of the groups studied corresponded to the roles assigned by the organization. In other words, the influence on group actions exercised by certain individuals was strongly related to the position these people occuped in the organizational hierarchy. Evidence of this is provided by the behaviour of group members who accepted the legitimacy of this arrangement, relinquished decision- making to these individuals and generally followed their directions despite later voicing opinions that some of the decisions made were not the most appropriate ones.

There were exceptions however. In one instance, an imposed leader failed to retain his influence over part of the group he was directing and a new leader emerged. The split that developed was apparently influenced by the group's perception of their respective competence and effectiveness. It seems that people are willing to go along with imposed or authoritative leadership only so far. If the proposed action is contrary to some strongly held opinions or contradicts some 'factual' information held by group members, or if the course of action chosen by the leader results in a situation where people perceive themselves to be more at risk than before, then the tenure of the existing leader may be challenged.

The data also indicate that both leadership and group formation are related to the fire training people receive and the normal roles they occupy in the organizational structure. The amount of knowledge occupants believed they possessed about what to do in a fire appeared to be an important element. Those who characterized their training as inadequate or their knowledge of fire as meagre were usually the people who sought consultation with others as to how to interpret the situation and the proper course of acton to follow. They also tended to rely on leaders, either imposed or emergent, for guidance. Unfortunately, many of the leaders in this incident did not possess any better knowledge of the proper course of action than those who sought their advice and consequently led the groups into situations that increased their risk.

One group of the study population, however, did not seek consultation or attempt collective action but acted, more or less, on an individual basis. In addition to occupying relatively higher status positions and having no one under their charge at the time of the incident, these individuals had the benefit of the most comprehensive fire training programme given to the building's occupants. The fact that the fire occurred when the emergency plan was not in operation made no difference to these occupants, for they knew what to expect and were prepared for any eventuality. These findings are in accord with what Swartz (1979) discovered in his study of the Beverly Hills Supper Club fire, where people retained the same general roles they had before the fire. In that incident, staff members took care of patrons at the tables, rooms and stations to which they were assigned; patrons looked to the staff for guidance.  This led Swartz (1979, p. 73) to recommend that "fire safety plans... should examine the roles that people normally play and not seek to prescribe emergency actions that are contrary to these roles".  Given the tendency of occupants to follow the directions of their immediate supervisor, it would appear crucial to ensure that these people received sufficient training to adequately fulfil this leadership role, especially when the structure of the emergency plan is absent.  Moreover, even when the plan is in operation, assigning the task of floor fire warden to employees who occupy low status positions in the organization's hierarchy, a common enough practice, may create conflict or uncertainty among the occupants by pitting two imposed leaders against each another.

Normal organizational and occupancy structures are important in the behaviour of building occupants in a fire situation in general, and specifically leadership and group formation are strongly influenced by the organizational and situational contexts.  The data upon which these conclusions are based, however, were derived from a single incident and further investigation of these phenomena would be well justified.

This paper is a contribution of the Division of Building Research, National Research Council Canada.


REFERENCES

Breaux, J., Canter, D. and Sime, J.:  "Psychological Aspects of Behaviour in Fire Situations," International Fire Protection Seminar (5th), Vol. 2, Karlsruhe, West Germany.  Vereinigungzur Forderung des Deutschen Brandschutzes e.v., 1976.

Canter, D. (ed.):  Fires and Human Behaviour, John Wiley and Sons, London/New York, 1980.

Hollander, E.P. (ed.):  Current Perspectives in Social Psychology, 3rd. ed., Oxford University Press, London/New York, 1971.

Pauls, J.L. and Jones, B.K.:  "Research in Human Behaviour," Fire Journal, 73:3, 35-41, 1980.

Shaw, M.:  Group Dynamics: The Psychology of Small Group Behaviour, 2nd ed., McGraw-Hill, New York/Toronto, 1976.

Sime, J.D.:  "The Concept of Panic," in Fires and Human Behaviour, ed. D. Canter, John Wiley and Sons, London/New York, 1980.

Sime, J.D.:  "Affiliative Behaviour During Escape to Building Exits," Journal of Environmental Psychology, 3:1, 21-41, 1983.

Stahl, F. and Archae, J.:  "An Assessment of the Technical Literature on Emergency Egress for Buildings," NBS Report NBSIR77-1313, Center for Building Technology, National Bureau of Standards, U.S. Dept. of Commerce, Washington, D.C., October, 1977.

Swartz, J.A.:  "Human Behaviour in the Beverly Hills Fire," Fire Journal, 73:3, 1979.

# A Case Study of Fire and Evacuation in a Multi-Purpose Office Building, Osaka, Japan

**S. HORIUCHI**
Department of Architecture
Kansai University, Suita, Japan

**Y. MUROZAKI**
Department of Environmental Planning
Kobe University, Kobe, Japan

**A. HOKUGO**
Fire Information and Research Center
Mitaka, Japan

ABSTRACT

On April 4, 1984 a fire occurred at the Science and Technology Center of Osaka. The building was a typical multi-purpose office building which contained the office of various learned societies whose occupants were regular users of the building, and assembly halls used by people less familiar with the premises. The purpose of this study is to form a basis for future guidlines for the evacuation of multi-purpose buildings, a building type which has become increasingly common in recent years. Our research group conducted a survey of people who were in the Center at the time of the fire using a questionnaire, and obtained detailed information about the fire and the various actions the evacuees took. On the basis of our survey we sought to analyse the characteristics of the evacuees, their reactions to spatial conditions during the evacuation, and how they experienced the sequence of events throughout the emergency. An important result of our analysis that emerged very dramatically was the difference between regular users of the building and those less familiar with it. The differences were; action upon becoming aware of the fire, criterion for selecting escape routes, and ability to effectively reach an exit.

TABLE 1. Floor by floor breakdown of the Building and the occupants.

| floor | floor area (m²) | area burned (m²) | main use | number of occupants | number of evacuees | number of rescued by window | method of rescue | number of questionnaires collected(%) |
|---|---|---|---|---|---|---|---|---|
| 8 | 1,177 | – | assembly halls | 70 | 70 | – | – | 61(87.1) |
| 7 | 1,177 | – | restaurant,offices | 94 | 94 | – | – | 67(71.3) |
| 6 | 1,177 | – | offices,a.halls | 85 | 83 | 2 | ladder truck | 51(60.0) |
| 5 | 1,177 | – | offices | 63 | *61 | 2 | ladder truck | 39(61.9) |
| 4 | 1,177 | – | assembly halls | 254 | 158 | 96 | ladder truck | 170(66.9) |
| 3 | 1,177 | 473 | offices | 50 | 48 | 2 | portable ladder | 38(76.0) |
| 2 | 1,176 | – | exhibition halls | 11 | 11 | – | – | – |
| 1 | 1,124 | – | exhibition halls | 14 | 14 | – | – | – |
| B1 | 1,161 | – | restaurant | 29 | 29 | – | – | 23(79.3) |
| B2 | 1,150 | – | mechanical rooms | 9 | 9 | – | – | 7(77.8) |
| EV | – | – | – | – | – | – | – | 2 |
| | 12,485 | 473 | (total) | 679 | 577 | 102 | – | 458(67.5) |

*seven of these evacuated via life chute

## METHOD

Our research group began our investigation two weeks after the fire. We wrote a 23 item questionnaire and the Fire Defense Board of Osaka distributed it to the 679 individuals who were in the Science and Technology Center at the time of the fire. Of the questionnaires distributed, 458 were returned to the Fire Defense Boad (see TABLE 1.). Each step of the evacuation was analyzed and delineated.

## ANALYSIS

The Fire, the Building, and the Occupants

Time. April 4, 1984, appoximately 11:25 am.

Place. Osaka, West District, Science and Technology Center of Osaka, 3rd floor, a hallway near the west stairs.

The burned area. 473 m² of the floor and 62 m² of the cieling surface on the 3rd floor, 88 m² of the exterior wall surface on the 4th floor.

Injured. 8 persons. All sustained carbon monoxide poisoning, all alive.

Cause. Arson suspected.

The building. A reinforced concrete structure 8 stories high. Essential facilities (elevators, main stairs, lavatories etc.) located in the center core. In addition to central stairs, enclosed stairs on the east and west ends of the building. These enclosed stairs are protected by fire doors which are kept closed.

Occupants. The regular users of the building occupied offices on the 3rd and 5th floors. The assembly halls on the 4th and 6th floors were being used for new employee training sessions, and the occupants were not familiar with the building. The occupants of the 7th and 8th floor were attending conferences and were not regular users of the building although most had been in the building before (see FIGURE 1.).

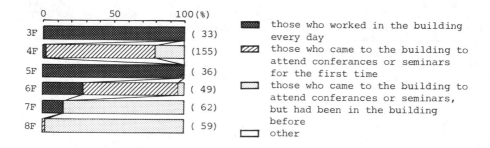

FIGURE 1.  Characteristics of the occupants

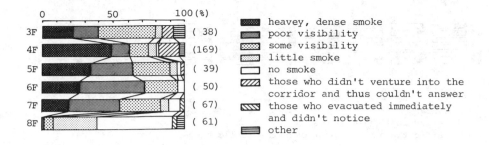

FIGURE 2. The amount of smoke in the corridor when evacuees became
aware of the fire

## The Fire and the Spread of Smoke

Discovery of fire. An employee of the Center working on the 3rd floor heard
an explosive sound and rushed to investigate. The employee found the floor of
hallway near the west stairway in flames, and the ceiling engulfed in smoke.
An employee in charge of fire prevention who was working on the same floor
told another employee to get a fire extinguisher and try to use it. He then
ran to the 1st floor security office and told the security officer to inform
the fire department. At the same time he alerted all of the floors over the
public address system. The contents of the announcement he repeated was, "Do
not use the Central stairway. Plese evacuate via the east or west emergency
stairways." The fire department records his call at approximately 11:32 in
the morning.

The arrival of the fire department. At 11:39 the first group of fire fighters
arrived and black smoke was pouring from a window on the south side of the 3rd
floor. There were several hundred evacuees already in the streets surrounding
the building. Many occupants unable to evacuate could be seen waving from
windows between the 3rd and 6th floors.

The spread of smoke. Because the fire shutters for the central stairway were
not closed, smoke spread to the floors above. FIGURE 2. shows the spread of
smoke to the various floors at the time when the occupants became aware of the
fire. The results of our questionnaire show that occupants of the 3rd floor
did not see heavy smoke. This can probably attibuted to the relatively early
discovery of fire by the occupants of the 3rd floor.

## Awareness of the Fire

In our questionnaire we asked, "How did you become aware of the fire ?", and
gave a series of multiple choice answers, letting the participants choose as
many answers as they felt were appropriate. We also asked, "What made you
believe it was a real fire ?", but asked them to chose only one answer from
the list available. The answers to these questions are compiled on a floor by
floor basis in FIGURE 3.

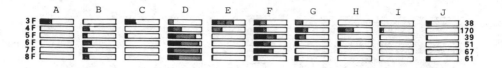

A:seeing the fire
B:smelling
C:hearing explosive sounds
D:hearing the announcement over the public adress system
E:hearing others yelling
F:opening their door and finding smoke in the corridors
G:smoke entering the room
H:being alerted by seminar leaders
I:being alerted by others (other than seminar leaders)
J:other

■ those realizing the fire to be real in the above way
▨ those who answered that the above was a secondary reason for
  believeing the fire to be real
▢ did not find the above to be the reason for believeing the
  fire to be real

FIGURE 3.  Analysys of how occupants became aware of the fire


31.6% of the occupants of the 3rd floor, where the fire broke  out,  reported
hearing an explosive sound and wondering what it was.   52.6% of the occupants
reported  hearing others trying to extinguish the fire and knew something  was
wrong.   13.2%  of  the occupants heard noises and knew immediately it  was  a
fire.   31.6%  acutually  saw  the fire and thus grasped the  reality  of  the
emergency.
Occupants on the upper floors(4-8) heard the announcement of the fire,  opened
their  doors,  and  found the hallways filled with smoke.  Most  grasped  the
reality  of  the emergency in this way.   Bcause of proximity to the source of
the  fire,  smoke  spread to the 4th floor faster than to  the  other  floors.
Training  sessions  were in progress and the doors to most of the  rooms  were
closed, so most trainees did not become aware of the fire untill their leaders
alerted them.   By this time,  smoke was thick in the hallways.   Occupants of
room  404,  which exited into an area adjacent to the central stairway,  found
smoke so thick that their leaders told them not to exit.
When  occupants  of the 8th floor became aware of the fire  relatively  little
smoke had spread into the corridors.   Most of the occupants on the 8th  floor
became aware of the fire by the public address.


Actions Taken

FIGURE 4.  shows the various courses of action taken by the occupants.    Five
patterns emerged:
a) those who thought to extinguish the fire before evacuating
b) those  who  thought to set the fire alarm and/or  evacuate  others  before


**526**

| | |
|---|---|
| occupants who thought only to evacuated | evacuated(18 persons)<br>evacuated---------- returned------┌---- rescued(9 persons)<br>　　　　　　　　　　　　　　　　　└---- evacuated(6 persons)<br>helped by others--┌-- evacuated(27 persons)<br>　　　　　　　　　　└- collected<br>　　　　　　　　　　　personal affects--- evacuated(44 persons)<br>collected　　　　　┌- evacuated(59 persons)<br>personal affects--├-- helped by othere--- evacuated(67 persons)<br>　　　　　　　　　　└- evacuated----------returned--------- rescued(7 persons)<br>　　　　　　　　　　┌- evacuated(7 persons)<br>confirmed fire----├-- helped by others--- evacuated(6 persons)<br>　　　　　　　　　　└- collected<br>　　　　　　　　　　　personal affects--- evacuated(8 persons) |
| occupants who experienced confusion and disorientation and then thought to evacuate | 　　　　　　　　　　┌- evacuated(29 persons)<br>confused---------├- helped by others--- evacuated(5 persons)<br>　　　　　　　　　　└- collected<br>　　　　　　　　　　　personal affects-┌- evacuated(8 persons)<br>　　　　　　　　　　　　　　　　　　　└- helped by others-- evacuated(8 persons) |
| occupants who thought to set the fire alarm and/or help others and then evacuated | confirmed fire------ alerted------┌---- collected<br>　　　　　　　　　　　　　　　　　│　　personal affects-- evacuated(5 persons)<br>collected　　　　　　　　　　　　　└---- led others-------- evacuated(12 persons)<br>personal affect ---- led others--------- evacuated(9 persons) |
| occupants who thought to extinguish the fire, and then evacuate | 　　　　　　　　　　　　　　　　┌-- led others(5 persons)<br>confirmed fire------ tried to-------[<br>　　　　　　　　　　 extinguish　　└-- evacuated(7 persons) |
| occupants who chose to wait for fire fighters to rescue them | confused---------┌-- rescued(5 persons)<br>　　　　　　　　　└-- collected　　　　　┌-- life chute(2 persons)<br>　　　　　　　　　　personal affect-└--- helped by others-- rescued(4 persons)<br>collected　　　　　┌-- rescued(13 persons)<br>personal affect--└--- helped by others--- rescued(9 persons) |

Explanatory notes
　confused=An unstable psychological condition. Experienced confusion and disorientation
　confirmed fire=Tried to find the source of the fire or went into the hallway to confirm the fire
　tried to extinguish=tried to extinguish the fire with a fire extinguisher
　alerted=tried to set an alarm or alert fire officials
　collected personal affects=spent some time arranging personal work area or collecting personal affects
　led others=went into the corridor and led others to the stairs
　evacuated=went into the hall and attempt to evacuate
　returned=went into the corridor but returned to the room because of heavy smoke
　rescued=rescued by hook and rudder truck
　life chute=evacuated via life chute
　helped by others=those who recieved aid from others

FIGURE.4. Major patterns of evacuation

evacuating.
c) those who thought only to evacuate
d) those who experiensed confusion and disorientation and then thought to evacuate
e) those who chose to wait for fire fighters to rescue them.

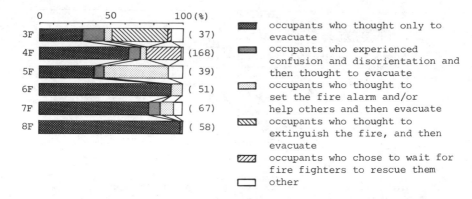

FIGURE 5.   Floor by floor breakdown of evacuation patterns

The distribution of occupants from these five groups and the floors they occupied is shown in FIGURE 5.   Those who thought to extinguish the fire(a) were all occupants of the 3rd floor.   Those who sought to alert or help others(b)   were   largely from the 4th floor.    Those who thought only to evacuate(c)   were from floors 4,6,7 and 8.   That occupants of the 3rd and 5th floors   were   regular of the building and had their own offices probably accounts for the large proportion of them who thought extinguish the fire or alert and help others (see FIGURE 1.).
On other floors where attendants of training sessions were largely unfamiliar with the building, the thought of direct evacuation was predominant.


Selection of Evacuation Routes

FIGURE 6.   shows the responses to the following question : "How did you decide whether to evacuate through the hallway or whether to stay where you were ?" Occupants   on   the highest floors were found to have evacuated quite  quickly. It  can be assumed that the absense of dense smoke on the highest floors  (see FIGURE 2.)    made it possible for occupants to quickly make the decision to

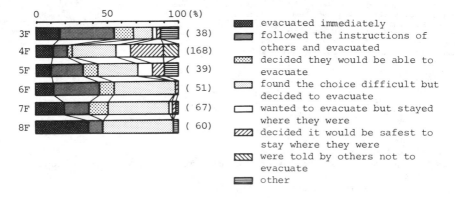

FIGURE 6.   Selection of evacuation routes

528

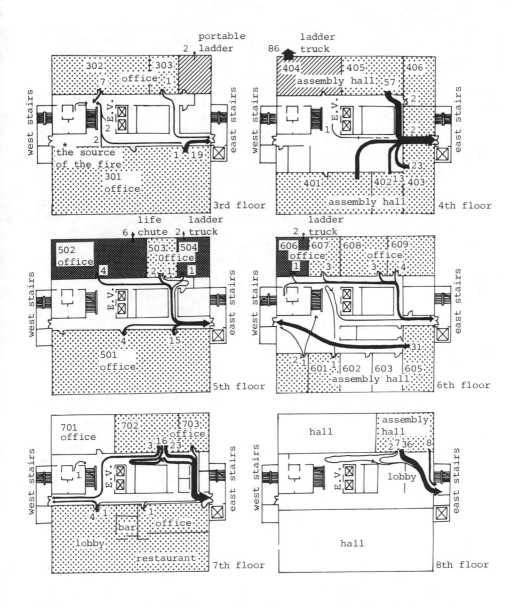

occupants of same room divided, some evacuating via the corridor, others via the window

all occupants evacuated via the corridor

all occupants evacuated via the window

FIGURE 7. Choice of evacuation route

TABLE 2.  Case when occupants of the same room devided,some
         evacuating via the window,others via the corridor

| room | Those who evacuated via corridor | | Those who evacuated(rescued) via window | |
|---|---|---|---|---|
| | sex age working position | reason for choosing evacuation route | sex age working position | method of evacuation (rescue) |
| 606 | F  24 clark, regular user | knew stairway | M  39 office manager, not regular user | ladder truck |
| 504 | F  44 clark, regular user | announcement | M  80 managing director regular user | ladder truck |
| 502 | F  30 clark, regular user | always used stairs | M  37 office manager, regular user | life cute |
| | F  36 clark, regular user | knew stairway always used stairs | M  41 office manager, regular user | " |
| | F  39 clark regular user | knew stairway | M  34 manager regular user | " |
| | M  38 office manager, regular user | always used stairs | M  56 fast-time employer regular user | " |
| | | | M  35 manager, regular user | " |
| | | | M  48 driver, regular user | " |
| | | | M  43 acting branch manager regular user | " |

exit,  and  then to traverse the corridors into the emergency stairwells  with
little  trouble.    The upper floors also had a higher proportion of occupants
to be instructed by others to evacuate.   On floors 3, 5 and 6 where occupants
were  largely regular users of the building,  those who chose to exit via  the
hallway were the majority instead of dense smoke.

In contrast, many occupants on the 4th and 5th floors decided not venture into
the  hallways or decided it was safe to stay where they were.   The smoke  was
very dense on these floors by the time occupants became aware of the fire, and
it  can  be concluded that the amount of smoke directly contributed  to  their
choice to stay where they were.

FIGURE 7. shows the results to the question, "Did you evacuate via the hallway
or  did you evacuate via the window,  using the hook and ladder truck  or  the
life  chute ?"  Occupants of the same rooms on the 5th and  6th  floors  were
divided  into both groups.   Besides the amount of smoke,  other factors  were
found  to play a role in how the occupants chose to evacuate.  TABLE 2.  shows
details of these occupants.   Most of those who evacuated via the hallway were
women employees who were familiar with the building.

That they knew the location of the emergency stairs or often used them can  be
cited  as  a significant reason for their choice.   Those who elected  not  to
evacuate were all men employed in managerial positions.  Occupants of room 606
were  largely unfamiliar with the building and those in room 504 were  elderly
men.

It  can be said that sex,  job and familiarity with the building are among the
factors that contribute to the choice of evacuation route.

those who work in the building (mainly 3rd and 5th floor, 102 persons)

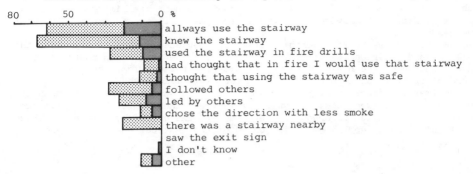

80    50    0 %
allways use the stairway
knew the stairway
used the stairway in fire drills
had thought that in fire I would use that stairway
thought that using the stairway was safe
followed others
led by others
chose the direction with less smoke
there was a stairway nearby
saw the exit sign
I don't know
other

those who visited the building to attend a conference or employee's trainning for the first time (mainly 4th and 6th floor, 84 persons)

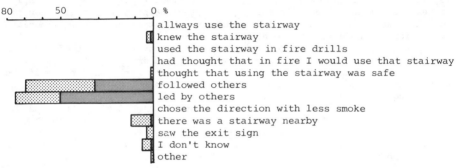

80    50    0 %
allways use the stairway
knew the stairway
used the stairway in fire drills
had thought that in fire I would use that stairway
thought that using the stairway was safe
followed others
led by others
chose the direction with less smoke
there was a stairway nearby
saw the exit sign
I don't know
other

those who visit the building to attend conferences or employee's trainning, have used the building (140 persons)

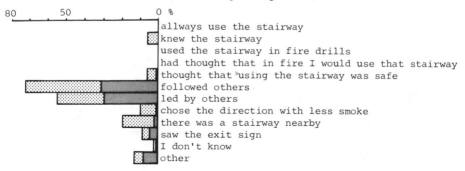

80    50    0 %
allways use the stairway
knew the stairway
used the stairway in fire drills
had thought that in fire I would use that stairway
thought that using the stairway was safe
followed others
led by others
chose the direction with less smoke
there was a stairway nearby
saw the exit sign
I don't know
other

FIGURE 8.  Reasons for choice of the evacuation route

Reasons for Choice of Evacuation Route

FIGURE 7. shows the plans of floors above the 3rd floor. Generally, the occupants on these floors evacuated via the east emergency stairway. On the 6th floor, occupants of room 605 found that the poor visibility due to smoke made it difficult to navigate any but a straight corridor, and chose the west emergency stairway.

FIGURE 8. shows the relationship between the evacuee's familiarity with the building and their choice of evacuation route. The evacuation route is most likely chosen on the basis of amount of smoke in the corridor, but as on floor 3 and 5, regular users used staires theywer familliar with.

In contrast, occupants of floors 4, 6, 7 and 8 were less familiar with the building and allowed others to guide or instruct them as to their evacuation route. Many simple followed other evacuees to the emergency stairs. These cases illustrate decision making based on the instructions or guidance of others.


CONCLUSION

The results of this study are summarized below:

a) Regular users of the building will act in various ways that include trying to extinguish the fire, alterting others, or helping others to evacuate. In the case of those who are less familiar with the building, immediate evacuation is the normal pattern.

b) The choice of evacuation route depends mostly on the amount of smoke, but sex, job and familiarity with the building are important factores.

c) The choice of evacuation route will often be a regularly used route if the evacuee is familiar with the building. For those not familiar, following or relying on others is the norm.

d) If familiar with the building, occupants have little difficulty finding exits even in heavy smoke. If the location of the stairs is not known, finding an exit can be of great difficulty.

e) In all phases of the evacuation process, familiarity with the building was found to be the primary determinant of speed and ease of evacuation.

# Movement of People on Stairs during Fire Evacuation Drill—Japanese Experience in a Highrise Office Building

**MASAHIRO KAGAWA, SATOSHI KOSE, and YASABURO MORISHITA**
Building Research Institute
Oho, Tsukuba, Ibaraki 305 Japan

ABSTRACT

Behaviour of people during a simulated fire evacuation practice in a high-rise office building in Tokyo was observed through video recording: subjective evaluation by the evacuees was also collected. About a fifth of the residents present in the building evacuated through two emergency stairs to the ground. The result suggests that stagnation of flow within the stairwell was noted in several spots even at this small number of evacuees. It was pointed out that doors opening into the stairway obstructed the smooth flow of people in the stairwell. Simultaneous total evacuation seems impractical in highrise buildings; reasonable procedures for selective evacuation need to be developed. It is also concluded that more frequent education on fire safety is necessary for the building users to get accustomed to the evacuation procedures and facilities.

KEYWORDS: Emergency stairs, evacuation, fire safety, highrise building, human behaviour, office building, questionnaire, video recording.

INTRODUCTION

Modern buildings are not completely safe from risks of fire or other disasters in spite of multiple technical measures provided. It is supposed to be the best way of assuring safety for the residents to evacuate from the building to the ground. It is not a common practice in Japan, however, to conduct such a fire drill without prior notice. Yet, not a small number of office buildings practice a scheduled evacuation on the occasion of "Disaster Prevention Day: September the 1st". The authors had the chance of making video recording of an evacuation practice in a highrise office building. A questionnaire survey was also done to examine subjective evaluation by the evacuees to the practice and their attitude to the fire risk in the building. The result of the study and the implication arising from this is discussed.

DETAIL OF THE EVACUATION PRACTICE

The highrise office building where this observation was conducted is in the Shinjuku Metropolitan Area in Tokyo. It has 53 stories and 5 basement stories: Floors between the 6th and the 28th are used by the owner, and the remaining

---

Masahiro Kagawa was a visiting researcher from April 1984 for one year. Address all correspondence to Taisei Corporation, Shinjuku, Tokyo 160-91, Japan.

floors are let to a lot of smaller tenants. Shops and restaurants are in two basement stories and also on the 53rd story. The typical floor plan of the building is in FIG.1.

The evacuation practice was conducted on Tuesday the 4th September 1984 (because the 1st was Saturday), and it was estimated that about 1,500 persons, a fifth of the residents present in the building, participated in the practice. It has been notified in advance that an evacuation practice is to be started at about 14:30. It has been supposed that following to a large earthquake, fires will break out at several floors in the building, and an order of total evacuation will be announced: Then, evacuees from each floor are guided by their group leaders to the emergency stairs, and they are to climb down the stairs.

The lobbies for access to the two emergency stairs are designed large enough to temporarily hold all residents of that floor protected from smoke and fire. The purpose of the observation is, therefore, to reveal the flow condition of the evacuees in the stairwell, and especially to reveal the mixing condition of the evacuees from different floors at the entrance door to the stairwell. To fully accomplish the aim of the study, four paired video cameras were set up in and out of entrance doors to the stairwell. Two sets were equipped with two cameras and a camera wiper, and the visual data were conbined directly into one video frame. Several other cameras were used to monitor the general flow of people during evacuation. One of them was placed on the first floor to record the final outflow of the evacuees from the stairwell. The positions of the cameras are identified in FIGS.2 and 3. Although the building had two main emergency stairwells, both of which were used this time, only the east one was selected as the object of observation. It was unavoidable due to shortage of equipments and staff. To compensate for this drawback, and also to try to simulate condition of congestion during total evacuation, the evacuees of successive floors between the 25th and the 28th where the cameras were mainly situated were to use the east stairwell as the emergency stair. Originally, the evacuees from two consecutive floors were supposed to use east and west stairwells alternately during the practice.

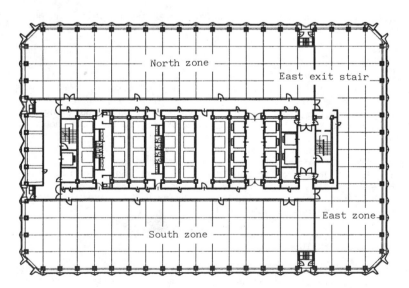

FIGURE 1. Typical floor plan of the building.

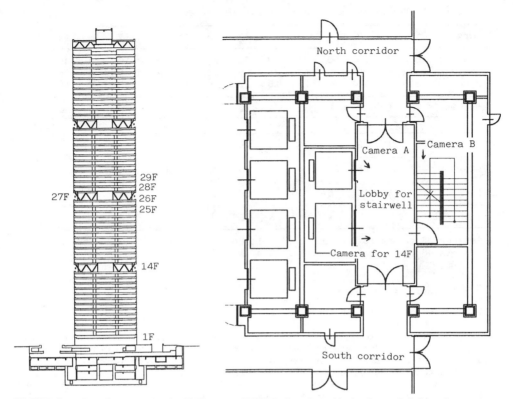

FIGURE 2.  Section of the building.     FIGURE 3.  Detailed plan of lobby for
east emergency stairwell.

In addition to these cameras, two observation staff walked down the stair both with a video camera, thus directly recording the conditions in the stairwell, and four other staff walked down the stair with evacuees, all with a portable tape recorder, just the same way as Pauls and Jones(1) did in 1972. They were all instructed to start climbing down with the last evacuee of the floor, and were to report noteworthy findings during their traverse.

The questionnaire survey was aimed at obtaining the attitude of the evacuees to the evacuation practice: Evaluation of the procedure of practice, subjective response to the evacuation condition, and their characteristic data were collected. The usual practice of stair use was also asked. To facilitate the comparison of data with those of Pauls (private communication), the questionnaire form was kept identical where possible. Taking into consideration of the possibility of good response rate to the questionnaire, the questionnaire sheets were distributed only to those who participated in the practice and who belonged to the owner-tenant company (T Corporation) and to a medium sized corporation (YEW Company) that rented three stories between the 48th and the 50th. These two organizations are supposed to be the best kind of tenants/residents in the building; other tenants are suspected to have less keen interest to the life safety in case of emergency. The number of expected participants and the stairwell selection for T Corporation and YEW Company are listed in TABLE 1. For other participants, no detailed information was available.

TABLE 1. Number of expected participants in the evacuation.

| Story | 6 | 7 | 8 | 9 | 10 | 11 | 12 | 13 | 15 | 16 | 17 | 18 | 19 |
|---|---|---|---|---|---|---|---|---|---|---|---|---|---|
| Total | 10 | 16 | 50 | 50 | 50 | 50 | 30 | 50 | 50 | 50 | 10 | 10 | 45 |
| Male | 5 | 9 | 30 | 30 | 40 | 35 | 20 | 35 | 35 | 45 | 5 | 5 | 35 |
| Female | 5 | 7 | 20 | 20 | 10 | 15 | 10 | 15 | 15 | 5 | 5 | 5 | 10 |
| Stairwell | W | E | W | E | W | E | W | E | E | W | W | W | W |

| Story | 20 | 21 | 22 | 23 | 24 | 25 | 26 | 28 | Subtotal | 48/50 | Total |
|---|---|---|---|---|---|---|---|---|---|---|---|
| Total | 45 | 65 | 45 | 25 | 4 | 70 | 40 | 70 | 835 | 70 | 905 |
| Male | 35 | 50 | 35 | 20 | 2 | 55 | 25 | 55 | 606 | NA | |
| Female | 10 | 15 | 10 | 5 | 2 | 15 | 15 | 15 | 229 | NA | |
| Stairwell | E | W | E | W | W | E | E | E | ** | NA | |

**Number of people to use east stairwell between the 6th floor and the 28th floor is 486, and those to use west one is 349.

RESULTS

Observation of People's Movement

The first public announcement stating that a large earthquake occurred was released at 14h24m35s, and the second one reporting the simulated outbreak of fires at the 3rd, the 29th, and the 43rd stories was done at 14h29m02s. This second announcement urged all the residents to evacuate through emergency stairs without delay. The first evacuees (perhaps from the 3rd floor) came out of the exit door of the stairwell to the first floor lobby at 14h29m44s, just 42 seconds after the announcement started, and others followed. It took about 16 minutes for almost all the evacuees to come down to the ground. The flow of evacuating people came out in groups headed by their leaders. The positions of the observer-staff with time are plotted in FIG.4. The total number of evacuees that came out of the ground floor stair exit is given in FIG.5. Flow rate out of the stair is also plotted.

Result of Questionnaire Survey

Seven hundred sixty-seven questionnaire sheets were collected out of 905 sheets that were delivered through group leaders. Fifty-five persons that responded to the questionnaire work for YEW Company, and the rest work for T Corporation. The age/sex distribution in TABLE 2 suggests that females over the age of 35 are comparatively rare among the samples, while females dominate under 25 years of age. It seems that younger people are a little over-represented in the sample. As to their subjective evaluation of the ability to descend long stairs, there is a tendency that older people report smaller number of stories for their maximum ability. Nearly sixty percent of subjects estimate that they can descend over 20 stories, but only just 30 percent of people actually have the experience of climbing down such a long distance. Their most popular use of stairs is going a couple of stories up and down. They use stairs instead of elevators mostly because elevators are crowded or going by stairs are faster. But about 14% of people rarely or never use stairs during usual time.

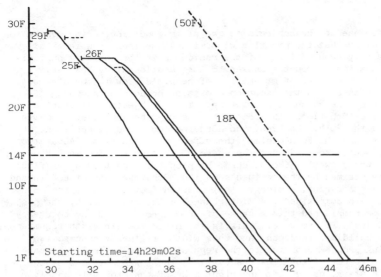

FIGURE 4. Movement of observers with evacuees in the east emergency stair.

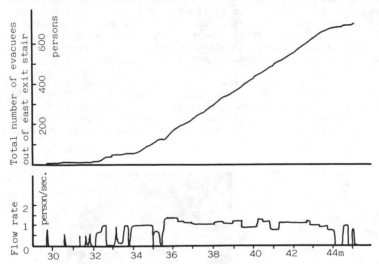

FIGURE 5. Number of evacuees out of east emergency stair, and flow rate of people out of stairwell at the first floor.

TABLE 2. Age-sex distribution of evacuees that answered questionnaire.

| Age<br>Sex | 15–24 | 25–29 | 30–34 | 35–54 | 55– | Total |
|---|---|---|---|---|---|---|
| Male | 65 | 86 | 99 | 259 | 17 | 526 |
| Female | 126 | 62 | 16 | 17 | 1 | 222 |
| Total | 191 | 148 | 115 | 276 | 18 | 748* |

*In addition to this, there are 19 subjects with age or sex unknown.

DISCUSSION

From the response of the subjects to the stair condition during evacuation practice, it is clear that the practice went without serious trouble. This was expected because only about a fifth of the residents in the building partici- pated in the evacuation. From the answer to the questionnaire, however, it is evident that there were cases of stagnation of people's flow in the stairwell. About 30% of subjects stated the occurrence of slowing down or stopping of people's flow during the evacuation practice. Those subjects that started descending from the 15th floor, the 18th (not observed because they used west emergency stair), the 20th, the 22nd, the 23rd (not observed either), the 25th, the 26th, the 28th and the 50th claimed the effect of stagnation. About two thirds of the subjects reported that the speed of descent was somewhat slower for them when it was slowest. The blockade by people coming from upper floors was significantly claimed by the residents of the 13th, the 16th, the 25th and the 26th. On the other hand, residents that started from the 15th, the 18th, the 22nd and the 49th complained the disturbance in the stairwell by the people coming into. Some complained this because of the narrowness of the emergency stairs, which is just 1200 mm wide. This is the minimum stair width requirement for large office buildings in Japan, and this width originally assumed two files flow of evacuating people according to the regulations, as Pauls(3) discussed. However, from the result of the flow observation, there was only one evacuee on every step of the stair, using left and right side of the step alternately (staggered files), or two persons on the same step and the succeeding persons on the next but one step just as shown in FIG.6, thus resulting density on the stair was 1 person per step in either case (this density is around 3 persons per square metre). Higher density was rarely observed even during stagnation of people's flow. It was natural because this evacuation was an exercise notified in advance, and practically no psychological stress was felt by the evacuees.

Judging from the traverse of observer-evacuees in FIG. 3, the average speed of descent was found to be around 16 seconds for a standard floor height of 3650 mm, without noticeable disturbance in the stairwell. This is close to the speed of climbing down during unobstructed stair use as Kose et al.(4) reported. For some cases, however, it took 20 seconds or more to descend one story, indicating

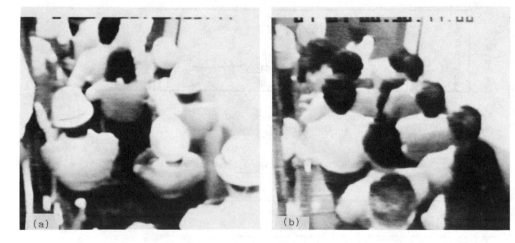

FIGURE 6. Arrangement of people on the flight during maximum flow on the 25th floor. Left figure (a) at time 14h33m16s, and right one (b) at 14h34m00s.

the effect of congestion in the stairwell. It occurred at several stories at
different times. On the 25th floor, the front of people coming into met with
the front of people descending just on the landing, and people coming down from
upstairs waited on the upper flights to let evacuees of the 25th floor to come
in, for a short period. In other floors where the door was initially closed,
however, it was difficult for the evacuees to come into the stairwell because
the door had to be opened against the flow of people on the flights and landings. The door was opened when there was a gap between flow of people in the
stairwell.

Although none of the observers who descended with the evacuees directly
reported the condition of congestion, it was clearly witnessed in the recorded
picture at the 14th, the 25th and the 26th floor where evacuating people were
stopped several times for 10 or 15 seconds due to congestion in the lower
floors. It is regrettable that observer-evacuees started descending too late to
encounter such cases of flow stagnation in the stairwell.

If the observed speed is used, it becomes clear that it will take at least
around 13 minutes for the residents to evacuate to the ground from the 53rd
story without slowing down of the people's flow. During the actual evacuation,
however, flow stagnation or at least slowing down of the flow will inevitably
occur. Two factors will contribute to this. First, each stairwell could hold
only 40 persons per floor according to the observed density, whereas at least
200 persons are expected to evacuate from each floor through two emergency
stairs. Supposing that just a half of the evacuees come to the east stairwell,
this amounts to 100 persons. Pauls(5) discussed that flow capacity of stairwell
decreases if flow density is exceeded 2 persons per square metre, and there is a
good reason to assume this condition to happen. Secondly, during the mixing of
people's flow on the landings, there may arise a standstill on either side to
let the others to go forward, as happened in the observation. This also leads
to a flow stagnation in the stairwell. Therefore, it does not seem to make
sense to assume total evacuation in case of emergency. It is highly probable
from the present result that a standstill in the stair may appear if 3 or 4
neigbouring floors were simultaneously evacuated. From this point of view, the
stair width is clearly insufficient.

There is no explicit statement in the Japanese Building Codes as to whether
total evacuation is needed for highrise buildings or not. In practice, the
underlying assumption seems: Fires will not occur which require total evacuation
because various countermeasures are provided; secondly, all the residents of one
floor could take a temporary refuge in the adjoining lobby spaces of emergency
stairs free of smoke and fire, before they are totally evacuated from the
building; and thirdly, such a total evacuation that requires simultaneous use of
the stairs is unlikely to happen. It is still debatable, however, if there
might be cases when immediate evacuation is desirable, such as a bomb threat or
post-earthquake fires.

Personal Factors Affecting Emergency Preparedness

The general knowledge of evacuation procedures or various escape facilities
provided in the building deeply depends upon the length of time after their
initiation of work in the building and also upon the role the subjects are
supposed to play during emergency. If they were designated as members of the
private fire brigade, they tended to study beforehand the procedures of the
evacuation, and thus they understood the importance of repeated practice of fire
evacuation. In contrast, previous experience of fire or bomb threat has had no
effect on the attitude of the residents toward the necessity of evacuation
practice. Although there exists a brief manual explaining the details of the

evacuation procedures and facilities, it was distributed only once when the building was completed. Thus, the younger subjects have had no chance of reading the manual or listening to the explanation.

CONCLUSION

Result of the present observation suggests that effective procedures for selective evacuation that will cause little congestion in the emergency stair-well should be developed. Simultaneous total evacuation as it is now considered might cause more trouble than selective evacuation for the residents that need most urgent evacuating action. This is vital for Japanese highrise buildings to cope with earthquake induced multiple fires. Evacuation procedures that positively utilize such spaces as stories for machinery that are designed as areas of temporary refuge need development.

It is advisable, on the other hand, that a brief manual explaining procedures in case of emergency is repeatedly printed and distributed to all workers in the building. This will certainly assist the residents of the building to realize the importance of fire and other safety measures.

ACKNOWLEDGEMENTS

The authors would like to express their sincere thanks to the management staff of the building who allowed the authors to conduct observation during the evacuation practice. Thanks also go to a lot of colleagues in the Building Research Institute for their assistance to set up the equipments on the site. The work was conducted as a part of the Research Project on the Development of Design System for Building Fire Safety, financed by the Ministry of Construction.

REFERENCES

1. Jake L. Pauls and Brian K. Jones, "Building evacuation: Research methods and studies", in Fires and Human Behaviour, edited by David Canter (John Wiley and Sons, Chichester, 1980), pp.227-249.

2. Jake L. Pauls (private communication).

3. Jake L. Pauls, "Building evacuation: Research findings and recommendations", in Fires and Human Behaviour, edited by David Canter (John Wiley and Sons, Chichester, 1980), pp.251-275.

4. Satoshi Kose, Yoshihiro Endo and Hidetaka Uno, "Experimental determination of stair dimensions required for safety", in Proceedings of the International Conference on Building Use and Safety Technology, (National Institute of Building Sciences, Washington,D.C., 1985), pp.134-138.

5. Jake L Pauls, ibid., pp.272-274.

# Computer Simulations for Total Firesafety Design of the New Japanese Sumo Wrestling Headquarters and Stadium (Kokugikan)

**HIROOMI SATO and TOMIO OUCHI**
Kajima Institute of Construction Technology, Japan

ABSTRACT

This paper reports the application of total firesafety system consist of simulation programs for fire behaviour, smoke propagation and evacuation behaviour, to the Kokugikan´s firesafety design. This building is composed of a large central space with seats for 11,500 spectators.

Assuming that fire occured in the office room and the generated smoke flowed into the large central space, various simulation were conducted.

As a result, it is predicted that the time when smoke will reach to the spectator seats is about 9.5 minutes for 2nd to 3rd floors and over 15 minutes for 1st floor even if the top vent is shut. On the other hand, from evacuation simulation results, 5,000 spectators on the 2nd to 3rd floors can arrive at the roof evacuation place about 3-4 minutes after the start of evacuation, and 6,500 spectators on the 1st floor can arrive at outdoors about 10 minutes.

Through the comparison of the smoke accumulation speed and evacuation behaviour, it is confirmed that 11,500 spectators can arrive at outdoors without danger from smoke and that there is no necessity to change the original plan.

This building was completed in January, 1985.

KEY WORDS

Computer simulation, Firesafety design, Fire, Smoke propagation, Evacuation

## INTRODUCTION

Factors concerning building firesafety are multifarious and include the problem of human safety, so that it is difficult to verify and find "absolute" firesafety by full scale and realistic experiments. Therefore, it is desirable to establish a systematic method for the assessment of "relatively" risk free and rational fire protection measures taking the properties of objective buildings and their occupants into consideration.

So the authors have proposed a total firesafety system by means of computer simulations based on engineering knowledge.[1] We considered this system as a time-sequence interaction system composed of six submodels for building, fire, smoke propagation, evacuation, information and equipment. We developed several computer simulation programs to calculate the behaviour of fire, smoke, and occupants during every minute of the fire taking account of the other submodels.

This paper reports the application of these simulation programs to the Kokugikan's firesafety design.

## OBJECTIVE BUILDING

The "Kokugikan" is the new headquarters of the Japanese national sport "Sumo", which was completed in January, 1985 in Tokyo, Japan. The basic plan and section are shown in FIGURES 1-2. This building as shown in FIGURE 1 is composed of a large central space with seats for 11,500 spectators and an entrance hall, lobby, office rooms, as well as Sumo-chayas (which are shops supplying drinks and lunch snacks), Sumo-training rooms, and some restaurants. The height of the large central space for spectators is about 40 meters. The area containing spectators' seats is about 4,250 square meters.

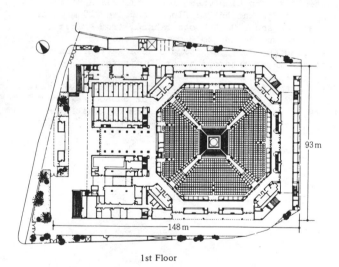

1st Floor

FIGURE 1.  Plan

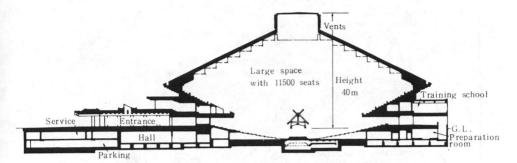

FIGURE 2.  Section


POST-FLASHOVER FIRE IN "NIPPON SUMOKYOKAI" OFFICE ROOM

In the main central space, as the spectators act in a sense as a warning system and also as a fire fighting system so that, even if a fire occurs they can easily detect and extinguish a fire in its early stages. Therefore, the authors supposed it is not necessary to discuss the behaviour of fire in this area. But in rooms such as the office room for Nippon Sumokyokai and the Sumo-chayas, there is some risk of a fire break out when the staff are out just as there is in the case of any ordinary building. So, in this paragraph, we will discuss the case of a post-flashover fire in the office room.


Simulation Method

We can calculate post-flashover fire temperature in the compartment by means of the well-known heat balance equation (1) proposed by Dr. Kawagoe et al whereby for any given instant "t" during the course of the fire after flashover.

$$I_C = I_L + I_W + I_R + I_B \tag{1}$$

The treatment on the right side of equation (1) are nearly equal to ones in Kawagoe model.

For the treatment on the left side, the term $I_C$ is given as the product of the burning rate R (kg/hr) and the heat of combustion of fuel $Q_1$ (J/kg), and we assumed that R was based on the type of combustion, and $Q_1$ was changed by the incomplete combustion factor.

Here, for R, the assumption was introduced that the fire behaviour in compartment develops either ventilation controlled or fuel controlled combustion, which is determined by the relation between the air supply through the opening of compartment and the necessary air based on the fuel surface area at any fire situation.

Namely, the burning rate at any fire situation R, as shown in dash line in FIGURE 3, is assumed to be chosen the less of the burning rate at the stage of fuel controlled combustion $R_{fuel}$ and the burning rate at the stage of ventilation controlled combustion $R_{air}$.

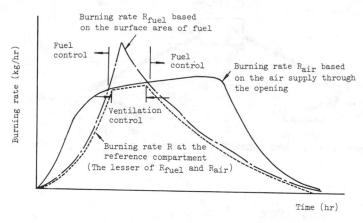

FIGURE 3. Transition model of combustion type

Calculation Results

From the many survey results which were sponsored or conducted by the Building Research Institute Japan (in 1983), the Society of steel construction of Japan (in 1973), etc, we know that the magnitude of fire loads in offce rooms is in the range of 7 to 53 kg/m². So, in this case the magnitude of the fire load of the objective office room was tentatively, considered to be 30 kg/m² and though there are various types of combustible material in the office room, the materials are regarded as being wooden.

The simulation result is shown in FIGURE 4. It is seen that the maximum temperature in this room is 810°C after 35 minutes.

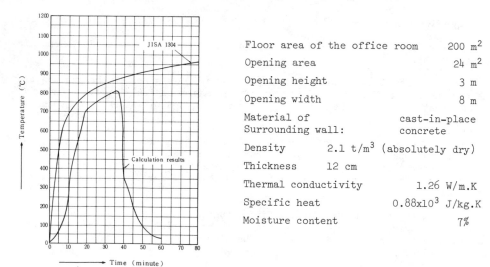

| | |
|---|---|
| Floor area of the office room | 200 m² |
| Opening area | 24 m² |
| Opening height | 3 m |
| Opening width | 8 m |
| Material of Surrounding wall: | cast-in-place concrete |
| Density | 2.1 t/m³ (absolutely dry) |
| Thickness | 12 cm |
| Thermal conductivity | 1.26 W/m.K |
| Specific heat | $0.88 \times 10^3$ J/kg.K |
| Moisture content | 7% |

FIGURE 4. Post-flashover fire temperature in the objective office room

SMOKE ACCUMULATION SPEED IN A LARGE SPACE

Though the smoke generated by combustion in the above mentioned office room fire may be released to all places around the room from its openings it is assumed that all of smoke released flow into the neighboring large central space. In order to estimate smoke accumulation speed in the top of the space, we enlarged and applied Dr. Tanaka´s theory[3] as follows.

Simulation Method

Simulation model takes into consideration the concept of the roof inclination and heat loss throuth the roof.
The temperature of layer gases Ts is give by

$$\frac{dTs}{dt} = \frac{T_S}{C_p \gamma_0 T_0 V_2} [Q - C_p M_2 (T_S - T_0)] \tag{2}$$

Where $Q = Q_f - Q_L$

The depth of a layer Zs is given by substituting $V_2$ obtained from equation (3) into the ralation between height $Z(= Z_1 + Zs)$ and volume $V(= V_1 + V_2)$. Here the actual roof inclination is considered.

$$\frac{dV_2}{dt} = \frac{Q}{C_p \gamma_0 T_0} + \frac{M_2 T_0 - G_S T_S}{\gamma_0 T_0} \tag{3}$$

While the mass balance in a lower region is,

$$\frac{Q}{C_p T_0} - \frac{T_S}{T_0} G_S + (G_0 + M_1) = 0 \tag{4}$$

Where $G_S$ and $G_0$ given by pressure P of a room at its floor must satisfy the requirement of the equation (4).

With regard to the plume mass flow, Tsujimoto´s equation based on Yih´s was applied. Namely, the plume mass flow rate $M_1$ and mass flow rate at a hot layer $M_2$ are given as follows.

$$M_1 = 0.153 \left(\frac{\gamma_0^2 gQ}{C_p T_0}\right)^{1/3} Z_0^{5/3} \left(1 + m \frac{\Delta T}{T_0}\right)^{-2/3} \tag{5}$$

$$M_2 = 0.153 \left(\frac{\gamma_0^2 gQ}{C_p T_0}\right)^{1/3} (Z_0 + Z_1)^{5/3} \left(1 + m \frac{\Delta T}{T_0}\right)^{-2/3} \tag{6}$$

While $\Delta T/T_0$ in the height $Z´$ is,

$$\Delta T/T_0 = 3(1 - 3F) + \sqrt{(1 - 3F)^2 + 1} \tag{7}$$

$$1/F = 11.0 \times \left(\frac{Q}{\gamma_0 C_p T_0}\right)^{1/3} \cdot g^{-1/3} Z´^{-5/3} \tag{8}$$

Further $G_0$ and $G_S$ are calculated by general ventilation calculation method.

Calculation Results

By means of solving above-mentioned simultaneous equations (2) to (8) etc. in any given instant, "t", we got the smoke accumulation speed in the large space.

The conditions in this calculation are as follows.

(a) As mentioned above, it is assumed that all of smoke released flows into the large central space. So, from the result of fire simulation, $Q_f$ is given as time-function. The area of fire source at the ground level is regarded as 3 x 3 m taking account of the area of objective door opening in this large central space. Therefore, the height of the virtual point source is regarded as 4.5 m beneath the standard floor level + 6.0 meter.

(b) Temperature of ambient air is 22°C with no wind.

(c) Except doors facing to a fire room, it is supposed that all doors in this space are kept open for the purpose of evacuation. So, neutral height of openings is 3.5 m, and effective area of inlet air $\alpha_i A_i$ is 140 m² . ($\alpha_i$ = 0.65)

(d) Actual top vent area $A_s$ is 177 m² and $\alpha_s$ is regarded as 0.4 taking account of the efficiency of louver, therefore $\alpha_s A_s$ = 70.8 m². Distance from the standard floor level (G.L.-4.0 meter) to the vent is 36.8 m.

For comparison with the above conditions, the cases of $\alpha_s A_s$ = 0 and $\alpha_s A_s$ = 35.4 m² were calculated. Calculated results are shown in FIGURE 5. It is seen that the depth of a layer $Z_s$ is about 13 m at 10 minutes after ignition if the top vent is kept open.


EVACUATION BEHAVIOUR FROM THE LARGE CENTRAL SPACE

When the staff failed to extinguish the above-mentioned fire and the smoke began to flow into the central space, the 11,500 spectators need to move to protect their lives to the safest area, i.e. out-of-doors. In this paragraph, we discuss how many minutes it takes for 11,500 spectators to arrive out of doors by way of the prescribed evacuation route and whether there is any interference on evacuation behaviour or not.

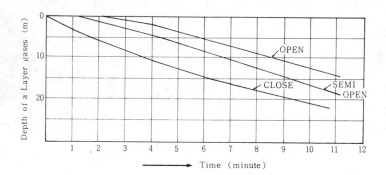

FIGURE 5. Smoke accumulation speed in the large space (calculation results)

## Simulation Method

Simulation model is constructed using the investigation results and calculation results for 9,000 spectators behaviour watching sumo wrestling matches at the Osaka-prefectural gymnasium as follows.

Evacuation route and its zoning. We divided this building into small blocks in accordance with evacuation route, such as chairs, box seats (usually 4 persons per one box seat), passage, lobby, corridor, staircase and out-door safety area. We considered each block to be a branch, and supposed a network structure, as shown in FIGURE 6 to be composed of branches. It is supposed that 2 to 5 boxseats and 3 to 28 chairs are respectively regarded as the unit evacuation zone, and use the same passage. The zonings of passage blocks were carefully zoned taking account of the existence of junctions and the divergence.

As concerns junctions, if the lower part is not wide enough for the evacuation behaviour, it is assumed that the ratio of junction number is proportional to the width of each part of the upper stream. The ratio of the divergence numberis also proportional to the width of the lower part.

Spectators' walking speed and capacity of each branch. The relations between the walking speed and crowd density using in this model are shown in FIGURE 7, which are based on the past imvestigations.[1,4] And the distribution of evacuation start time is shown in FIGURE 8. Here, as for box seats, this distribution was obtained from the investigation results at the Osaka-prefectural gymnasium. For chairs, the distribution was asumed as FIGURE 8 (b) in consideration of investigation results.

Passing time of each branch is based on the length of each branch and evacuee's walking speed (based on FIUGRE 7) in each brance. As concerns the capacity of each branch, it is supposed that in the lobby it is 3.0 persons/$m^2$ and in passages 3.5 persons/$m^2$. (In narrow lateral passages, 1 person/50 cm)

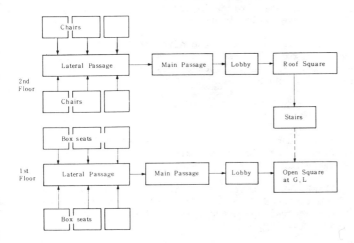

FIGURE 6. Block diagram of evacuation route

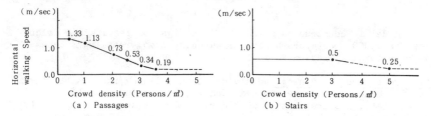

FIGURE 7. Relation between the horizontal walking speed and crowd density

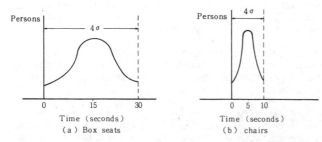

FIGURE 8. Distribution on time of evacuation start

Calculation Reselts

We calculated the case where it is impossible to use one of the main passages on the 1st floor for evacuation as a result of the office room fire.

Examples of the calculation results are shown in FIUGRE 9. Namely, it is seen that 6,500 spectators on the 1st floor could arrive out-of-doors 593 seconds after the start of the evacuation, and 5,000 spectators on the 2nd to 3rd floors could arrive at the roof evacuation place in 208 seconds. We made sure that was no interference on their evacuation route.

Therefore, from FIGURE 5 and FIGURE 9, we can confirm that even if all of smoke generated by the neighbouring office room fire flows into the large central space and even if the top vent is shut, 11,500 spectators can arrive safety out of door without danger from the smoke.

SUMMARY

In this paper, we have reported on the application of a total firesafety system composed of simulation system of fire behaviour, smoke propagation, and evacuation behaviour to the Kokugikan´s firesafety design.

We calculated the post-flashover temperature in the office room which is the most risky in case of an outbreak of fire in this building, and the smoke accumulation speed in the large central space according to the heat energy withdrawn from the room during the fire.

Finally, we discussed how fast 11,500 spectators can arrive out-of-doors by way of the prescribed evacuation routes. Through comparison of the smoke accumulation speed and evacuation behaviour during every minute, we confirmed that there was no necessity to change the original plan.

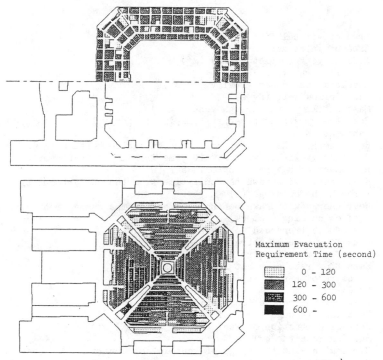

FIGURE 9. Calculation result (maximum evacuation requirement time)

ACKNOWLEDGEMENTS

This research was conducted by the Firesafety Systems Development Group of Kajima Corporation. The members of group except authors are M.Nagatomo Dr.Eng., K.Muta Dr.Eng., S.Hayakawa, S.Togari, K.Kageyama, K.Nakazawa, Y.Hara, H.Kurioka, Y.Sassa et al. The assistance of "Nippon Sumo Kyokai (Japan Sumo Association)" that permitted the presentation of this paper is gratefully acknowledged.

REFERENCES

1. K. Muta. H. Sato et al; "Study on a Total Fire Safety System", CIB Symposium, System Approach to Fire Safety in Buildings, Tsukuba, Japan, 29 and 30 August, 1979.

2 H. Sato et al; "Surrounding Materials and Fire Behavior (No.1 to 4)", Architectural Institute of Japan Annual Meeting Abstract Report, Fire Division, 1976, 1977, 1979, 1980. (In Japanese).

3. T. Tanaka; "Discussion on Modeling of Pre-flashover Behavior" Japanese Association of Fire Science and Engineering Annual Meeting Reports, 1976 etc. (In Japanes).

4. J.J. Fruin; Hokosha no Kukan, Translated into Japanese by M. Nagashima, Kajima Inst. Pub. Co., 1974, Original Title; PEDSTRIAL Planning and Design, 1971.

NOMENCLATURE

$A_f$       Area of virtual fire source (m$^2$)
$A_i$       Area of inlet air (m$^2$)
$A_s$       Area of vent (outlet gases) (m$^2$)
$C_p$       Specific heat (J/Kg·K)
$g$       Acceleration due to gravity (cm/s$^2$)
$G_o$       Mass inflow rate (Kg/s)
$G_s$       Mass outflow rate through vent (Kg/s)
$I_B$       Heat energy stored per unit time in the gas volume contained in the compartment (W)
$I_C$       Heat energy released per unit time during combustion (W)
$I_L$       Heat energy withdrawn per unit time from the compartment owing to the replacement of hot gases by cold air (W)
$I_R$       Heat energy withdrawn per unit time from the compartment by radiation through the openings (W)
$I_W$       Heat energy withdrawn per unit time from the compartment through wall, roof or ceiling etc. (W)
$m$       $\Delta Tm/\Delta T=0.5$ (regarded to Gauss distribution)
$M_e$       Mass flow rate of entrained gas (Kg/s)
$M_1$       Mass flow rate in fire plume (Kg/s)
$M_2$       Mass flow rate at a hot layer (Kg/s)
$P$       Pressure of a room at its floor (Pa)
$Q$       $Q_f-Q_L$ (W)
$Q_f$       Heat release rate of combustion of fuel (W)
$Q_L$       Heat loss through the roof (W)
$Q_1$       Heat of combustion of fuel (J/Kg)
$R$       Burning rate (Kg/hr)
$R_{air}$       Burning rate at the stage of ventilation controlled combustion (Kg/hr)
$R_{fuel}$       Burning rate at the stage of fuel controlled combustion (Kg/hr)
$S_A$       Surface area of wood crib (m$^2$)
$S_{AS}$       Opening area at bottom of wood crib (m$^2$)
$T_o$       Temperature of ambient air (K)
$T_s$       Temperature of layer gases (K)
$\Delta T$       Temperature rise at the center of any plume section (°C)
$\Delta T_m$       Mean temperature rise at above section (°C)
$T_M$       Time of maximum rate of combustion (hr)
$V_1$       Volume of ambient air (m$^3$)
$V_2$       Volume of layer gases (m$^3$)
$Z$       Distance from floor to vent (m)
$Z'$       $Z_o + Z_1$ (m)
$Z_o$       Distance from virtual point source to floor (m)
$Z_1$       Distance from floor to layer (m)
$Z_s$       Depth of layer (m)
$\alpha i$       Effective coefficient for area of inlet air
$\alpha s$       Effective coefficient for area of outlet air
$\gamma_o$       Density of ambient air (Kg/m$^3$)
$\gamma_s$       Density of layer gases (Kg/m$^3$)

# Evacuating Schools on Fire

A. F. VAN BOGAERT
State School Building Fund
Ministry of Education
B-1040 Brussels, Belgium

ABSTRACT

Starting from the statement that time is the overruling factor in fire circumstances, this study dissects the egress operation in the
- physical domain : threats of untenable conditions and cut off exit ways.
- mathematical domain : users' velocity and egress flow rates as factors of the
  evacuation time; pre-calculation of the evacuation time in the building design
  stage; prediction of place and duration of traffic congestions : potential
  sources of panic; remedial measures in building outlay.
- psychological domain : importance of self-control and adaptation ability in the
  phases of discovering the fire, the shockmoment and decision-making reaction.
  There is a complaint about the nowadays poor character-building school education.
- practical domain : special school risks versus specific advantages expected
  to further smooth egress operations; evacuation strategy and tactics; the importance to inform and direct the evacuees (e.g. by megaphone) and to organize the
  evacuation in priority order according to local threats.

The paper considers the problems in special institutions attented by ambulatory and non-ambulatory impaired pupils. It motivates a drastic raise of night
staff to pupils ratio and proposes a specific evacuation scheme, concentrated on
the rescue of non-ambulatory users by accurate organization of thoroughly trained
staff teams.

Key Words : ambulatory user; boarding school; congestion; design; discovery,
escape routes, evacuation; fire drill; flow rate; human behaviour; layout; nervous
stress; passage unit; personnel-pupils ratio; reaction; rescue; school; shock
moment; special school; time; velocity.

## 1. INTRODUCTION

For a fire to break out and to extend, four accomplices need to agree :
1 - combustible matter; 2 - oxygen; 3 - an ignition source; 4 - time. But as soon
as the fire has started, we have but a handful of seconds to act on the oxygen
factor; we can try during a couple of minutes to quench the flames, and after the
third minute we can only attempt to cope with the fourth factor : time. The over-
whelming role of the time factor in a fire disaster appears from this formula,
developed in Canada (ref. 1) :

$\dfrac{Tr + Ts}{Tc} \leq 1$, in which Tr = the time needed to discover a fire and to react; Ts =
the time the occupants need to secure themselves; Tc = the time the fire needs to
create untenable living conditions or to cut evacuation routes. If the result of

this division is less than or equal to 1, safety conditions are fulfilled. Therefore the numerator value Tr + Ts should be kept as small as possible. There are no other controls over Tr than human advertence and detectors. Ts largely depends on the built environment, the characteristics of the escape routes, the users' presence of mind and capabilities. Tc is strongly influenced by the layout and construction of the building and by the oxygen supply.

In a school, apart from giving the alarm and calling the fire brigade, there is little that users, surprised by a fire, can do. But this little is of life importance. Hence, this trial to find the most appropriate egress conditions by identifying the factors that govern the extent of Ts and by analysing the evacuation episodes, with occasional hints to remedial design aspects.


## 2. THE FLIGHT TO SAFETY

The Ts action : flight to security, is a complicated mechanism ruled by the number of users, their physical conditions and moving velocity, the flow rate of exit ways, the horizontal and vertical distances to be covered.

2.1. Number of users : the connection is evident. But existing regulations, because their aim is to determine the stair width, neglect the number of ground floor users. Yet it is important that these should not obstruct the ground floor exits when the upper floors users already arrive. Direct ways out must be provided for the former, apart from a direct exit, at the bottom of each staircase, for the latter.

2.2. The physical condition, as a rule, does not cause any problem in schools, except for the kindergarten (which however is always ground floor located) and for the 0,3 % mentally normal but physically impaired, who should be helped by their fellow-pupils.

2.3. Velocity (V) varies according to horizontal (h) or vertical (v) traffic and, in the latter case, according to an upward (s) or downward (d) sense. The following normal average values resulted from measurings and checkings in schools (ref. 2) : - horizontal velocity Vh : 90 m/min or 1,5 m/s
                              - vertical downward Vd : 132 stairs/min or 2,2 stairs/s
                              - vertical upward Vs : 108 stairs/min or 1,8 stairs/s.

2.4. Consequently, flow rates (D), expressed in number of persons (p) per passage unit (u) per second (s) also vary according to the same conditions.
- Horizontal rate Dh : Admitting the distance between two succeeding evacuees be 1 m, the flow rate is 1,5 p/u/s.
- Vertical downward rate Dd : The velocity is 2,2 stairs/s. Each moving evacuee occupies 2 stairs. Flow rate Dd attains 1,1 p/u/s.
- Vertical upward flow rate Ds : Likewise velocity Vs 1,8 stairs/s results in a flow rate Ds = 0,9 p/u/s.

2.5. Diagram 1 (ref. 3) compares the average horizontal and vertical flow rates D per u per s, visualized in points a, b, c, d (respectively 1 - 2 - 3 - 4 passage units). These differences in flow rates may cause congestions in the escape routes at each transition from a higher to a lower rate and this occurs to an extent of 1.5 - 1.1 = 0.4 p/u/s at transitions from Dh to Dd, and 1.5 - 0.9 = 0.6 p/u/s at transitions from Dh to Ds.

If we compare, in a horizontal vision, the positions of points Hc (horizontal flow rate in a 3-passage corridor) with Vdd (downward flow rate in a 4-passage staircase) and with Vse (upward flow rate in a 5-passage staircase), we notice

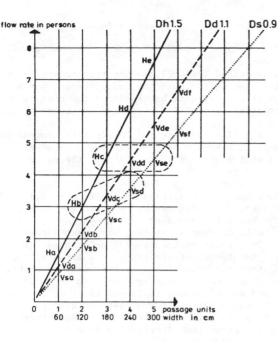

Diagram 1 : Normal evacuation flow
rate D per passage unit per s

that these three segments attain an almost equal flow rate of 4.5 p/s. So, logically, to avoid flow rate congestions, we should build school staircases wider than corridors. This might shorten but not prevent another kind of traffic jam : the cumulation congestion, which occurs at the knot of a corridor with a staircase, when upstream users and those of the corridor level meet at the top of a downward stair-flight (fig. 2).

Obviously, cumulation congestions can only be neutralized by building stair-cases wider and wider as they draw near the exit level, imitating the river that extends as it absorbs its affluents. Starting from a different viewpoint, H.H. Kiehne (ref. 4) comes to a similar suggestion, logically unattackable, but, for economical reasons, not to be practised in schools. Still, it is during these inactive halts that panic is sneaking around among a crowd not familiar with the building (warehouses, hotels, restaurants..). But even in a school environment,

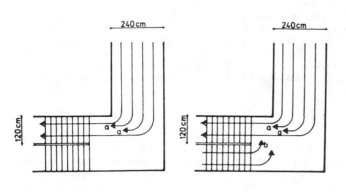

Fig. 2.1. : Flow rate
congestion (a)
Fig. 2.2. : Cumulation
congestion (b)

well known by its users, these stagnations might soon entail fear, degenerating into senseless behaviour.

Two positive actions can be taken :
- Since a relatively simple computing method (ref. 3) enables the architect not only to pre-calculate the evacuation time of the building under design, but also to predict place, time and duration of congestions, he can judge the need to add one, maybe two staircases to his project : precaution that will considerably shorten the traffic halts.
- During periodic fire-drills, pupils will experience and retain that congestions will occur where, when, for how long. The knowledge that these are inherent aspects of the lifesaving operation will lessen or even eliminate the psycho-physiological effects of these congestions.

2.6. <u>Design interferences</u> : Some hints to designers emerge from §§ 2.1 to 2.5:
- The every-day traffic routes are also the escape routes. - Avoid the need to build emergency stairs. - Divide the building, and even each compartment, into evacuation sectors, i.e. the areas to be discharged by each staircase and each outdoor exit. - Throughout the building, these sectors should be balanced in importance according to the prognosticated number of their users. - Avoid mixing stairway exits with those that discharge the groundfloor. This means that wherever possible, the staircase-bottom should have a doorway direct to open air.
- All stairways are completely encaged and closed by walls and doors of 1 or 2 h fire resistance, according to the height. - The wider you plan the stairs, the fewer you have to provide, but the more you further cumbersome concentrations of evacuees, the fewer are the chances to spread the alternative exit ways. - So, whenever possible, use a stairwidth of 2 passage units with railings on both sides. Stairs of 3 passage units are dangerous, the central user having no rail protection. In case of 4 passage units, the stairs have two side and one central railings.

2.7. <u>Evacuation Time</u> : Figure 3 shows the evacuation scheme of a 4 levels school, divided into 2 fire compartments and occupied by 590 users (90 + 60; 66 + 62; 80 + 73; 82 + 77). There is a 1.20 m wide staircase (2 u) at each extremity. The ground floor has 4 issues, 1.80 m wide (3 x 4 = 12 u). One staircase is under untenable conditions. Basing on distances, on the respective V and D values, and proceeding by phases, the prediction computings result in the partial and complete evacuation times of
- 30 s for the ground floor - 88 s for the 1st floor - 148 s for the 2nd floor - 210 s for the 3rd floor - 246 s for the whole building. From the 40th second, the 2nd and 3rd floor students will be caught in a congestion that will last 48 s for the former and 108 s for the latter (respectively 40th s = $C_i$ to 88th s = $C_f$; and 40th s to 148th s = $C_f$).

It should be noted that the same number of users, otherwise spread over the three upper floors, will experience congestion times that differ from the above scheme, for the duration always depends on the number of predecessors. But in spite of different congestion delays (and of their different total) the entire evacuation time being the sum of moves and stops, will be constant. The same phenomenon evidently occurs when the upper floors evacuation is operated in a different order. Anyway, a school should be evacuated within 5, a boarding school within 7 minutes.


3. HUMAN FACTOR

Yet, a school is not an hour glass, the grains of which obey to gravity. The matter that flies from a fire is made of muscles and nerves. The precedent schemes, impersonal and mathematic, should be confronted with a graph showing

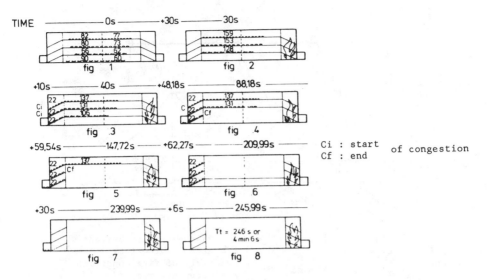

TIME

| | | | | |
|---|---|---|---|---|
| 0s | +30s | 30s | | |

fig 1    fig 2

+10s — 40s — +48,18s — 88,18s

fig 3    fig 4

+59,54s — 147,72s — +62,27s — 209,99s

Ci : start    of congestion
Cf : end

fig 5    fig 6

+30s — 239,99s — +6s — 245,99s

Tt = 246 s or 4 min 6 s

fig 7    fig 8

Figure 3 : Chrono-analytic evacuation scheme

the nervous tension evolution of people caught by a fire (ref. 5). This evolution covers 3 phases : a. discovery of the fire - b. shock moment - c. reaction.

3.1. Discovery conditions (a) may widely vary, from effective presence on the spot, and consciousness, to sleep and being at some distance, recognizing physical signals or perceiving the alarm. But the common denominator of these conditions is : surprise. Quick adaptation abilities will be of utmost importance in the next moments.

3.2. Discovering a fire causes a psychological shock (b), rushing up a sudden stress to suspense that may preclude people from thinking logically. Physical perturbations and psychological troubles may occur and produce a gap of impotence wide enough to lose precious time.

3.3. Reaction conditions (c) are in reliance to one's preparedness for fire circumstances, to the degree of familiarity with the premises, to other people's presence, to physical smoke and heat conditions. Life danger may lame any initiative, but can also push people to extraordinary actions to save either themselves or others and to take outstanding risks.

3.4. Conditions may change as time proceeds or as occupants travel to the exit : they may be caught in poisonous gases (escape trial 1 in diagram 4) or stopped by flames or heat (trial 2) or be upheld in a congestion (trial 3). Such incidents entail new developments in the evolution, re-passing through stages a - b - c. If there seems to be no outcome any more, nervous stress rises to untenable suspense, capable of pushing individuals or groups to desperate acts. Selfpossessed characters will decide to wait for fire brigade rescue.

3.5. This analysis implies that behaviour in fire is determined by a large variety of diverging factors. As one fire differs from another, as different persons have opposite reactions to similar stimuli or similar reactions to different circumstances, the margin of overall correct action principles is narrow. No doubt correct reactions can only spring from a constructive attitude gained and trained by education. What really matters, is to persevere in such

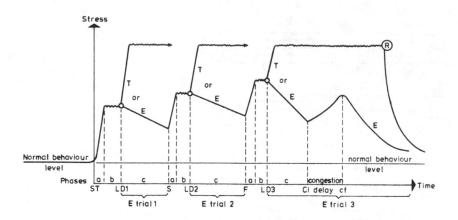

| | |
|---|---|
| Ci | Congestion delay (start) |
| Cf | Congestion delay (end) |
| E | Escape |
| F | Escape cut off by flames |
| LD | Life danger |

| | |
|---|---|
| R | Rescue by fire brigade |
| S | Escape cut off by smoke |
| St | Start of fire conditions |
| T | Occupants trapped |

Note : Much more complex situations may occur, but they will nevertheless be composed by the essential elements shown above.

Diagram 4 : Nervous stress on human behaviour in fires.

attitude throughout the three phases, keeping up self-control and command of mental forces in all circumstances : this is part of the character-building program of school education that should start at an early age but which nowadays seems to suffer from neglection.


4. PRACTICE

　　4.1. The above wanderings through physical, mathematic and psychological domains indicate that life in schools is not more dangerous than elsewhere. Still, some specific risks are permanent defiances, such as
- the dense occupancy (the densiest after theatres, movies and churches);
- the duration of occupancy (possibly up to 60 h/week, owing to evening use);
- the presence of a boiler house, collective kitchen, laboratories, workshops, extended energy-nets;
- the age and the behaviour of the young users.

　　However, the school population disposes of some trumps that could not concur elsewhere : the users know the premises; they are attended by educators who in case of need take initiatives to react (cf. phase c); knowing each other, the users shape a solid community; they are young and alert.

　　4.2. This offers a reliable safety base, on condition that positive attitude and efficient actions in case of calamity be carefully prepared according to the 3 phases : discovery, shock moment, reaction.
Phase a : discovery
　　1. Fire is signalled to the schoolhead. He calls the fire brigade and gives the alarm.
Phase b : shock moment

2. From the alarm signal onwards, the teacher's attitude is of utmost
importance to the pupils' composure and reassurance.
Phase c : <u>reaction</u>
3. Stop all avocations.  4. Shut windows and doors.  5. Check corridor conditions on opaque smoke and excessive heat; if so, remain in room, keep door
shut and fill gaps with moist paper or cloth.  Wait for rescue by fire
brigade.  6. If corridor conditions are tenable, leave room, shut doors,
7. Teacher and/or fellow pupils assist impaired pupil.  8. Each teacher
takes the lead of his group but stays at the rear to command and control.
9. Get to the usual staircase.  10. If not practicable, turn to an other
staircase.  11. Proceed crawling in smoky atmosphere.  12. Don't use lifts.
13. Do not overtake other fugitives.  The group should stay homogeneous, as
it left the room.  14. In case of congestion, close up as to shorten queues.
15. If no staircase is practicable, take refuge in the next fire compartment.
16. Teacher checks completeness of group.

4.3. These principles shape the essence of evacuation <u>strategy.</u> But in action,
<u>tactics</u> must constantly adapt to circumstances in rapid evolution, near by or at a
distance.  Ignorance of somewhat remote developments may put people on the wrong
way.  Therefore, whenever possible, the evacuating users should receive <u>information.</u>
Concentrated terms, in a calm voice (from a megaphone or a telecom system fed by an
autonomous energy source) can save lives.  Moreover the same procedure allows, from
the start, to organize the <u>evacuation in a priority order</u> according to the urgence
of threat :  1. the compartment and the floor where the fire started; 2. the floors
above and 3. the floors beneath the fire.


5. SPECIAL SCHOOLS AND SPECIAL BOARDING SCHOOLS

   Here, the problem must be reconsidered entirely.  The deficiencies of the
handicapped fall into the areas of perception, response and mobility and call for
special precautions as to the design and construction of the building, the number
and the qualification of educators, the rescue in case of fire.

These deficiencies take different forms which moreover may combine into complexes
of mental or physical or both mental and physical impairment.  A fire in such
institutions threatens heterogeneous groups of pupils - with slow and confused
perception, - some of whom do not understand what is going on, - who react in
unpredictable ways, - who are subject to strong emotions, - and a number of whom
are immobilized by handicaps.

   This latter aspect takes a primordial importance, as in all deficiency
categories (mental, emotional, functional, physical, sensorial) occur a number of
children with mobility problems that prevent them from leaving, without help, a
building on fire.  Investigations revealed that this percentage may vary from 10
to 80 % (ref. 6) : the regular school evacuation procedure is out of the question.

   5.1. Considering the variety of individual capacities, <u>the specific egress
pattern</u> covers three phases (figure 5) :
- The ambulatory pupils (A) can be evacuated (E1) in a group by the educator
  to a safe place where they are guarded (G1).
- To comfort, to guard (G2) and to prepare the waiting non-ambulatory (N-A)
  pupils are combined actions that require one or several more staff members.
- Evacuation (E2) of the non-ambulatory (N-A) pupils must be done individually
  and successively.  Once brought outdoors, these pupils must also be guarded
  (G3) and comforted.  This may be combined with G1.  Figure 5 represents the
  egress scheme for a group of 20 pupils (8A + 12 N-A); in phase E2 two staff
  members are operating : they each evacuate one N-A, in six repetitions.

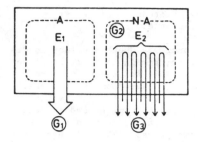

.Figure 5 : Evacuation pattern for impaired pupils.

If only one educator were available, he should travel twelve times, during twice as much time, - assuming that there are no corpulent adolescents who need double help.

    5.2. It is obvious that the egress time of N-A pupils is inversely proportional to the number of operating staff. In daytime, teachers, tutors, administrative, medical and social and service personnel (all having to take part in egress operations) compose an average personnel/pupils ratio of 1 to 4, which offers reasonable chances to perform an evacuation in due time. At night, conditions are substantially worse, as only the administrator and the night tutors are present, the latter in an average ratio of 1 to 17 pupils. In case of calamity, mutual assistance among tutors (cf. figure 5) is virtually excluded, since they should not leave the pupils entrusted to them. If outside assistance does not show up almost immediately, there will be victims. The tutors/pupils ratio should be raised to 1 to 10, plus an infant nurse (not group bound as the tutors are) per 30 pupils. This arrangement would result in a 1 to 7.5 ratio.

    5.3. The correct evacuation proceeding is founded on four basic actions : 1) Next to the alarm issue and the fire-brigade warning, the maximum of available personnel is directed toward the rooms that hold the pupils most threatened by smoke or fire. 2) One or more teams of staff members bring groups of ambulatory users to a safe place, preferably outside, whilst 3) other teams prepare the non-ambulatory to be evacuated individually. 4) The actual N-A egress phase is operated by a third range of teams, either chainwise or in shuttle travels (users in wheelchairs, borne, transported on stretchers, dragged along on blankets, even rolled off in their mobile beds).

    This model for handicapped children's evacuation corresponds, mutatis mutandis, to the egress scheme advised for hospitals and homes in a study done at the request of the LEUVEN University (ref. 7). Diagram 6 clearly shows that all action weight is concentrated on the supply of manpower capable of participating in the N-A pupils'rescue (Actions 8 - 8.1 - 8.2 - II - III - IV).

    5.4. Where to start and which impaired pupils should be helped first are crucial questions. No doubt the immediate concern goes to the most threatened user, i.e. the one being in the most critical position owing to the proximity of fire, smoke, toxic gases. Once this pupil is ready for evacuation, there is a second most threatened, a third..., - priority being always determined by the most endangered spot (ref. 8).

    5.5. Where to evacuate impaired users ? In any case : to a place of safety. The radical answer is : the outdoors. A. Tait's relevant statement on this matter is : "The ideal arrangement is the provision of escape routes so arranged that

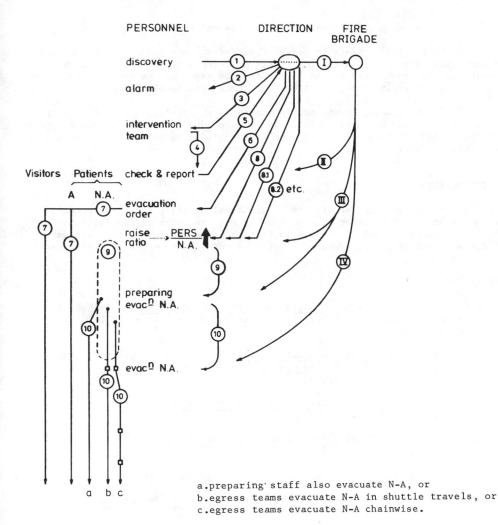

| PERSONNEL | DIRECTION | FIRE BRIGADE |

discovery

alarm

intervention team

Visitors   Patients   check & report

A   N.A.

evacuation order

raise ratio → PERS / N.A.

preparing evac⁰ N.A.

evac⁰ N.A.

a   b   c

a. preparing staff also evacuate N-A, or
b. egress teams evacuate N-A in shuttle travels, or
c. egress teams evacuate N-A chainwise.

Diagram 6 : Hospitals and Homes evacuation scheme (ref. 7)

any occupant in a building can turn his of her back on a fire and proceed to make their way without hindrance, in the direction of open air until safety is reached" (ref. 9). This radical evacuation model can only be performed by N-A occupants if they are located on the ground- or an other egress level and do not have to tackle steps and stair problems on their way out.

In the limited evacuation scheme, egress from endangered rooms provisionally does not go farther than the next compartment which is not (yet) threatened by flames of smoke, owing to the fire-resistive capacities of its walls, ceilings, floors and doors. Thus such a compartment is a provisional shelter for the occupants of a nearby endangered compartment.

One cannot make a preliminary choice between radical and limited egress.

Fire conditions, even in the same building, may well impose either of the schemes.

5.6. Limitation of risks in cases of extended smoke and flame spread, which require total evacuation, undoubtedly call for arrangements in time and space whereby the NA handicapped should be located, by day : on the egress level or on the floor immediately above; by night : on the egress level only (ref. 6 and 10).

5.7. Drills : Owing to the pupils' manifold shortcomings, fire-drills in special schools and boarding schools should more than anything else aim at preparing the personnel members to coordinate their exiting tasks and behaviour and to efficiently manage the egress.


6. CONCLUSION

As a conclusion, let architects, educational building officers, school directors and staff remember that in regular and even more in special schools, the succes of evacuating a building on fire is never measured after the number of rescued, but after the number of victims. And each time one is one too many.

REFERENCES

1.  R.S. Ferguson : User-need Studies to improve Building Codes. Technical Paper n° 368, Division of Building Research, National Research Council of Canada, 1972.

2.  A.F. Van Bogaert : Logica en Actie in de Scholenbouw, Ed. Simon Stevin, Brussel, 1972, and Prospective dans la construction scolaire. Ed. Vander, Leuven & Cesson (France) 1974.

3.  id. : Fire and Evacuation Times - Evacuatietijden bij Brand. Ed. Story Scientia Gent, 1978.

4.  H.H. Kiehne : Zum Problem der Treppenbreiten in Hochhäusern, in : Schadenprisma nr. 2/75, Berlin 1975.

5.  A.F. Van Bogaert : Man Facing Fire. School Building Fund (SBF), Brussel 1977.

6.  id. : Fire Prevention in Schools and Boarding Schools for Handicapped. SBF Brussel 1978 and in NBSIR n° 2070, Washington DC, 1980.

7.  id. : Menselijk gedrag bij brand in Gezondheidsinstellingen. Katholieke Universiteit Leuven and SBF Brussel, 1981.

8.  J. Archea : The Evacuation of non-ambulatory Patients from Hospital and Nursing Home Fires : A Framework for a Model. Ed. NBS, Washington, 1979.

9.  A. Tait : Fire safety Legislation and Management. Univ. of Edinburgh, 1975.

10. BIN (Belgian Institute of Normalization) : Norm NBN S 21 - 204 : Fire Protection in Buildings - School Buildings - General Requirements and Fire Reaction, Brussel, 1982.

# Perceived Time Available: The Margin of Safety in Fires

**JONATHAN D. SIME**
School of Architecture
Portsmouth Polytechnic, United Kingdom

## ABSTRACT

The knowledge people have of the degree of fire threat is considered to be important to an understanding of the time people need for response. 'Time needed' is a psychological concept which varies in terms of the 'perceived' time available (PTA) to carry out various actions. A distinction is made between the objective and subjective (perceived) availability of escape (or refuge) options: termed ODF and SDF 'Degrees of Freedom'. Analyses are presented of the time lag between the fire growth (ODF) and a person's knowledge of the degree of threat (SDF), also the diminishing availability of escape routes. The time components of different 'Required Safe Egress Time' equations are discussed. An alternative equation is discussed. Two integrated levels of analysis are necessary in monitoring the timing of people's actions in fires (A) the psychological/ social goal of the action: (B) the location of the action as a person moves around a building. Attempts to measure the duration of behaviour are discussed together with recommendations for future research. It is suggested that PTA should feature in any calculations of margins of safety in equations contrasting time needed and time available.

## INTRODUCTION

Fire tragedies involving a large number of fatalities usually have one simple feature in common. A serious delay occurred in the occupants of the building becoming or being made aware of the encroaching danger. In the Summerland Fire, Isle of Man, U.K. 1973 (in which 50 people died) and the Beverly Hills Supper Club Fire, U.S.A. 1977 (in which 164 died), there was some 20 minutes from the time of the initial discovery of the fire by a staff member to the point when most people became fully aware of a direct danger to their lives. What had been sufficient time available for egress turned into a situation in which the time needed for everyone to leave was too great to guarantee their safety. In this respect the concepts of time 'needed' (or 'required') and time available for egress, juxtaposed in safety planning and design formulae, could be usefully complemented by attention to the time available as perceived by people in fires. This concept of perceived time available (PTA) is a central focus of this paper and helps in understanding patterns of movement preceding any attempt to escape.

---

Acknowledgement: Several analyses presented in this paper were carried out by my former colleague in the Fire Research Unit, University of Surrey: Dr. John Breaux. I would like to acknowledge his contribution.

The aim of the paper is to shift the focus away from measures of time needed which concentrate almost exclusively on egress. For important time can be lost in the period preceding egress. Thinking of the perceived time available (PTA) highlights the incomplete knowledge about the fire threat people in different parts of a building are likely to have at different stages. PTA refers to the time an individual feels is available, for example, to reach an exit and knowledge of the time available to carry out the action. PTA can also refer to a person's view of the likelihood that other people will be able to carry out an action in time to avoid injury. People in fires often pursue investigative and affiliative actions (moving towards other people) in an attempt to reduce uncertainty. The amount of information about a fire influences the timing of people's escape and effectiveness of their actions.

OBJECTIVE AND SUBJECTIVE DEGREES OF FREEDOM

Figure 1 is a representation of the rate of fire growth in relation to the diminishing availability of escape or refuge options in a hotel fire. (1) The options, termed objective and objective (perceived) 'degrees of freedom' by Breaux (1), diminished rapidly once a fire took hold. As he points out "The unfortunate aspect is that once a fire reaches the stage of giving off noises, it is approaching or has reached the stage when one can expect non-linear growth".

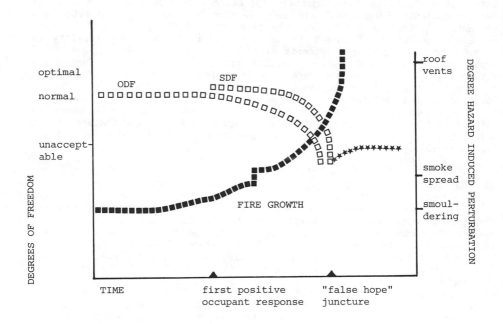

Figure 1: from Breaux (1) Dependency of ODF (Objective Degree of Freedom) and SDF (Perceived Degree of Freedom) on growth of fire for "Unsuccessful" occupants.

Although we were not able to measure the durations of actions in this fire our attempts in general to carry out statistical analyses on act sequences in fires (2, 3, 4) have explored the temporal sequence of acts. In another study of the hotel fire represented in Figure 1, I traced (mapped out) the

path and distance of movement on the architectural plan of the building (3). The sources of data were interview transcripts from the 33 fire survivors. The fire was chararacterised not only by guests in the hotel not being aware of the fire spreading up the central staircase until it was too late to leave by a 'normal route', but the misinterpretation of early cues to the fire's existence.

Movement by the hotel occupants generally involved individuals opening their hotel room door to investigate noises, only to find the fire had already cut off the main stairway route to safety. The fire began sometime before 2.00 am. It is estimated by the fire brigade, who arrived at 2.11 am, that within six minutes of the fire being first discovered by one of the residents it had involved the central stairway enclosure.

The 'time lag' between the objective growth of a fire and people's know-ledge of this and available escape routes from different parts of a building is characteristic of many serious fires. In fires we studied ranging from domestic fires in buildings of one or two storeys, to large-scale multiple-occupancy fires, the early stage of recognition was often characterised by ambiguous information cues. In a number of cases there was a serious delay in people taking these cues seriously before they realised there was a fire. Unknown to the occupants the fire was already reducing the time available to reach safety.

A further analysis (2) showed that the likelihood of reaching safety without assistance in the hotel fire example depended on accurately interpre-ting the early ambiguous cues. Breaux describes the 'false hope juncture' along the abscissa of Figure 1 as, "a situation for several individuals in which the ability to evade the fire was so curtailed that they were conside-ring jumping from a fourth floor window". In this case, the respondents subjective degrees of freedom (i.e. alternatives considered viable) appear to have changed to admit an option previously not entertained.

Figure 2 represents an analysis by Breaux (2) of escape reduction as a function of time for the hotel fire together with data recorded for the Beverly Hills Supper Club Fire (5, 6). Both events share a similar time scale. Point A represents the 'design potential' for the English structure and those rooms in the American nightclub. As Breaux points out "ironically, by mere virtue that people are using the buildings there is already a reduction in possible escape routes. This is due to such measures as locking doors to keep out pilferers, gatecrashers and others for whom entry is unauthorised. This in itself is interesting for it means that certain structures are already affected before the fire".

In both cases the fire begins somewhat after B in Figure 2. As Breaux points out "In total, the curve which describes progressive route reduction is of a history or sigmoid type. Of primary interest concerning the fire is that section defined by the function $Y = ax^b$ which extends from 'B' onwards. The meaning of this environmental relationship is, simply, ways to escape disappear at an ever faster rate".

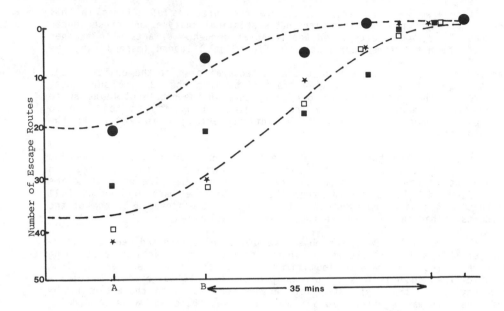

Figure 2: from Breaux (2) Escape route reduction as a function of time. Areas indicated are Cabaret Room ✳ , Empire Room ■, and Kitchen □ (data from Best, (5). English hotel fire data ● . Design potential and pre-fire status indicated by A and B respectively.

The behaviour of people, the escape route capacity, the fire growth are interrelated. This is best illustrated by the fact that the main reason for a rapid change in the escape possibilities of people is often the point when a person investigating a source of ambiguous information opens a door to the room of fire origin. Either the person forgets to close the door or is often unable to as the fire, which has been developing for some time in an enclosed space, rapidly escapes into the building from where the person has come. Time needed expands or reduces by virtue of a person's knowledge of the threat (i.e. perceived time available: PTA).

STAGES IN REQUIRED SAFE EGRESS TIME

The terms 'time needed' and 'time available' derive directly from recent attempts to 'model' what is going on in fires. A number of people have elaborated an equation for estimating life safety, in which the basic principle adhered to is that the time it takes for an individual or building population to reach safety, should not exceed the time it takes for fire conditions to make it possible to reach safety. The simplest form of this time needed versus

time available equation (7) is:

$$\frac{Te}{Tc} < 1 \qquad\qquad\qquad \text{Equation (a)}$$

where  Te = Time required for escape
       Tc = Time required for the toxic environment to reach a critical or
            untenable state

Under the heading of 'escape route analysis' Marchant (8) presented an equation which breaks down the 'time required' into 3 components:

$$\frac{Tp + Ta + Trs}{Tf} \leqslant 1 \qquad\qquad \text{Equation (b)}$$

where Tp = Elapsed time from ignition to perceive that a fire exists
      Ta = Elapsed time from perception to the beginning of safety action
      Trs = Elapsed time from initiation of safety action to reach a place of safety
      Tf = Elapsed time from ignition for the fire to develop untenable environmental conditions

Stahl, Crosson and Margulis (9) have broken down the 'time required' component even further, concentrating on the sensation and perception of a fire and evaluation of the degree of life threat. As Stahl et al acknowledge, their expanded form of equation (1) may be difficult to apply and estimate accurately in fires.

$$\frac{Ta + Ts + Tt + Te + Ti + Tx}{Tc} < 1 \qquad\qquad \text{Equation (c)}$$

where Ta = Time required for sensation of a stimulus from the fire environment
      Ts = Time required to become aware of this sensation
      Tt = Time required to become aware of the sensation as a potential life threat
      Te = Time required to evaluate the quality and extent of the life threat
      Ti = Time required to initiate effective actions
      Tx = Time required to follow-through and complete actions leading to safety

This equation (c) is clearly closer than equation (a) to the concept of perceived time available (PTA). Some confusion is possible in equations (b) and (c) in the definition of what constitutes a 'safety' or 'effective' action. As they stand, these terms could encompass certain actions prior to escape (e.g. contact fire brigade, warn others) as well as escape behaviour itself. One might assume that Ta (equation b) = Ti (equation c) and Trs (b) = Tx (c). Without clearer definitions it is possible to make assumptions with which the authors of these equations might disagree or at least would acknowledge are open to discussion.

Cooper (10) has been working on the 'time available' aspects of a model of fire growth. The aim has been to develop a quantitative criterion for safe building design, which he terms the 'designed safety egress criterion', viz: relative to a potentially hazardous fire, a building is of safe design if

$$ASET = Thaz - Tdet > RSET \qquad (4) \qquad\qquad \text{Equation (d)}$$

This principle expressed in the same terms as equations (1) to (3) gives:

$$\frac{RSET}{ASET} < 1 \qquad\qquad \text{Equation (d)}$$

where RSET = <u>Required Safe Egress Time</u>: the length of time subsequent to alarm which is actually required for safe occupant egress from threatened spaces

ASET = <u>Available Safe Egress Time</u>: the length of time interval between fire detection/alarm, Tdet and the time of onset of hazardous conditions, Thaz

Cooper concentrates on possible ways in which to estimate ASET. He writes "Assuming a capability for estimating RSET, the results of this study, through application of the designed safety egress criterion, would lead to rational evaluations of building safety." Where there is an immediate problem in this model is that it refers to egress movement following an 'alarm'.

To take account of the pattern of psychological and social responses in a variety of fires, any principle of safety design has to accommodate rather than discount the time it takes to respond to an 'alarm' or 'cue'. Unless there is a highly efficient communication system (public announcement system) people in different parts of a building are not, or cannot by definition be, simultaneously aware of the degree of a threat. The ASET model is currently addressed primarily to the issue of egress systems and forms of fire resistance designed into the building. Yet what is crucial is not only the time it takes to move to safety but when this movement begins.

A REVISED RSET EQUATION

The following equation, which concentrates on RSET, is intended to accommodate both the perceived time component <u>and</u> the pattern of movement prior to and during movement to safety. It is being used as the basic framework by a research team consisting of Maclennan, Pauls and Sime to develop a model of human behaviour applicable to assessing the adequacy of part 24 of the Australian Model Uniform Building Code, (a research project commissioned by the Australian Uniform Building Regulations Co-ordinating Council (12). The research involves monitoring evacuation times from office blocks (above 6 storeys in height). A definition of this model is as follows:

$$\frac{Tr + Tc + Te}{Tf} < 1 \qquad \qquad \text{Equation (e)}$$

where Tr = Recognition Phase: time from being alerted by a cue to knowing there is an emergency fire (includes acts such as investigate)

Tc = Coping Phase: time from knowing there is an emergency to beginning of escape (includes acts such as fight fire)

Te = Escape/Evacuation phase: time from the end of the coping phase i.e. beginning escape (evacuation) to leaving the building

RSET = Tr + Tc + Te
ASET = Tf (as in equation b)

Two levels of definition are necessary if one is to 'model' the interrelationship between behaviour in an emergency and a building's design.

Level (A) = Psychological/Social goal
(B) = Location of movement through building

Mroom  = Movement within room (before leaving through exit)
Mcorr  = Movement within corridor (before leaving through exit onto stairway)
Mstair = Movement down staircase
Mgrnd  = Movement on ground floor (before exiting from building)

Any reference to (1) Goal is qualified by referring to (2) Movement. Both are essential for an understanding of the temporal and spatial aspects of sequences of actions in an emergency. Unfortunately, research to date has concentrated almost exclusively on (1) or (2).

In terms of the Australian research on high-rise office buildings, the primary distinction will be between movement which takes place on the upper floor of a building and that on a protected stairway. These phases are termed:

MH = Horizontal component of evacuation: movement on upper floor prior to leaving through an exit from the floor (Mroom + Mcorr)

MV = Vertical component of evacuation: from entry into the protected exit system to the time of entry into open space or an approved refuge area (normally involves movement down stairway and on the floor of final access to the outside). (Mstair + Mgrnd)

The task will be to try to establish broad time estimates for these movement phases. In equation (e) RSET includes a recognition phase (Tr), and ASET a detection phase (Tdet).

McClennan (11) uses the term 'Exit Access' to refer both to the psychological/social factors involved in people moving towards an exit from a space and the physical design of access to an area of 'relative' or 'complete' safety. The aim here is to develop a design model which integrates research on psychological/social responses and the timing of physical movement to and through exits.

MEASURING THE TIME IT TAKES TO RESPOND IN FIRES

A recent overview of egress research in relation to the N.F.P.A.'s 1976 Life Safety Code (9) reflects limited knowledge of the time people expend in the early stages of a fire. Research on building evacuations (12) has been able to use 'time' directly as a measure of movement. Yet the time to act in the Tr and Tc stages has been ignored or inferred to date due to the problems in (a) measuring time directly in fires (b) reliability of time estimates given by fire survivors.

In one questionnaire survey (13) people were asked to estimate the time it took for them 'in leaving' buildings on fire. The average estimate of time to leave of 1.92 minutes was inevitably nearer to 'perceived' than actual time. Moreover the question of what constitutes 'leaving' is open to different interpretations since in low-rise domestic fires (a high percentage of the sample) leaving a house can take place in the Tr, Tc or Te stages and can be followed by re-entry more than once. The few time experiments which have been conducted (14)(15)(16) generally suggest it would be difficult to obtain reliable estimates from individuals of the actual distances they travel in fires. This difficulty would be reduced in research of behaviour in fires if the 'estimates' were compared with objective physical measures of the distances along routes through the building which people indicated they followed.

Experiments on potential entrapment simulate the scramble for an exit assumed to occur when people are faced by the diminishing possibility of everyone reaching safety in a fire (17) but bear little similarity to the time it takes to reach an exit in a fire. This experimental research of RSET/ASET ignores the ambiguity of fires and delays in warning which make such behaviour so unavoidable (18).

It may well be that the greatest potential for incorporating valid time estimates into an RSET/ASET model is in the area of simulation research in which time can be measured directly. Some research has measured the time it takes to evacuate a ward (19) and to enact different sequences of actions (20) in fire role playing exercises. Recording times for the early phases of monitored evacuations (21) will yield useful information. If the behaviour is also found to occur in researched fires, it could provide a useful estimate of the likely duration of acts which otherwise could not be recorded. The validity of role playing exercises and advice, on how to decrease needed time and increase available time in emergency planning (22), depends very much on the adequacy of research on actual behaviour in emergencies. By 'mapping' distances moved by people in fires (using architectural plans) (3) it may be possible to assign estimated travel times to these distances, based on knowledge of the fires and knowledge of characteristic speeds of 'pedestrian movement'. Video disc simulations of escape routes, as potentially viewed by escapees in fires, could also incorporate a time component (23).

IMPROVING THE MARGIN OF SAFETY

What one wants to move away from is the tenous basis for time estimates enshrined in design yardsticks. For example, an assumption in the British codes is that a fire compartment should be evacuated in 2½ minutes if people's safety is to be assured (24). The recent fire tragedy at Bradford City Football Club (May 1985) highlights the problems in concentrating on the 2½ minutes escape criterion without taking into account the time to begin movement (25). This period of time may exceed 2½ minutes and suggests the importance of a prompt warning to evacuate being put over a public announcement system.

The reason for emphasizing PTA, as an essential component of a RSET/ASET model, was my concern to point out the danger in considering that a direct mathematical or predictive relationship can be found between time needed and time available, independantly of the knowledge building users have at different stages about the fire's growth. Essentially, the current problem in comparing RSET and ASET lies in the assumption that, once an automatic smoke detector or alarm is set out off, the occupants of a building recognise there is a fire and will start their escape immediately. This is rather like equating people with inanimate objects or 'snooker balls', which upon external impact from an outside physical force (the 'cue'), will be propelled towards a an exit (the 'pocket') (26). This physical-science model of people's movement assumes that $T_{det} = T_r$. In contrast, there is likely to be a time lag between $T_{det}$ and $T_r$; thus $T_{det} \neq T_r$. The potential mismatch between $T_{det}$ and $T_r$, the time it takes people to interpret and respond to initial changes in the social and physical environment of a fire, has to be reduced. The margin of safety, referred to in the title of this paper, is the time lag between the 'objective' and 'subjective' degrees of freedom (i.e. the actual and perceived: state of the fire and availability of exits).

In conclusion, time required is an 'elastic' psychological concept in the sense that for the people in a fire it expands or shrinks according to perceived time available (PTA). The characterisation of time required as a subjective concept, as well as an objective one (referring to the time it takes to move around and escape from a building), is not one with which those working on the physical fire growth and design parameters of the ASET model are likely to feel comfortable. It is, however, a more accurate reflection of the limits of knowledge of a fire threat, which explain much of people's behaviour and timing of escape in fires. The key to improving the margin of safety undoubtedly lies as much in efficient forms of information and communication about the state of the fire at different stages, as in the building's design. Time available to escape can be increased by helping people to respond promptly in the Tp (perception) phase and use the time economically in the Tc (coping) phase of response. Paradoxically, one has to try to reduce rather than increase the perceived time available, if one wants to prompt people to begin their escape in time to reach safety.

REFERENCES

1.  Breaux, J.J. Analysing Complex Data: the Description and Analysis of Dynamic Behaviour in Fire Situations. Annual Conference of British Psychological Society. Exeter 31 March - 4 April, 1977.

2.  Breaux, J.J. On Analysing and Interpreting Behaviour in Fires. Fire Research Unit, Psychology Department, University of Surrey. Mimeo, 1979.

3.  Sime, J.D. Escape Behaviour in Fires: 'Panic' or Affiliation ? Doctoral thesis, Psychology Department, University of Surrey (unpublished), 1984.

4.  Canter, D., Breaux, J.J. and Sime, J.D. Domestic, Multiple Occupancy and Hospital Fires. In D. Canter (Ed.) Fires and Human Behaviour. Ch.8, 117-136, Chichester/New York: Wiley, 1980.

5.  Best, R.L. Investigation Report: The Beverly Hills Supper Club Fire, Southgate, Kentucky, May 28th 1977, NFPA (National Fire Protection Association) Fire Investigations Department (in co-operation with National Fire Prevention and Control Admin. and NBS (National Bureau of Standards) Draft Report, 1977.

6.  Berlin, A.N. A Modelling Procedure for Analysing the Effect of Design on Emergency Escape Potential. In B.M. Levin and R.L. Paulsen (Eds.) Second International Conference on Human Behaviour in Fire Emergencies: October 29 - November 1, 1978. Proceedings of Seminar, NBS. Report NBSIR 80-2070, 13-41, Washington D.C., National Bureau of Standards, 1980.

7.  Caravaty, R.D. and Haviland, D.S. Life Safety from Fire: A Guide for Housing the Elderly. Troy, N.Y.: Rensselaer Polytechnic Institute, Center for Architectural Research, 1967.

8.  Marchant, E.W. Modelling Fire Safety and Risk. In D. Canter (Ed.) Fires and Human Behaviour. Chichester/New York: Wiley. Ch.16, 293-314,1982.

9.  Stahl, F.I., Crosson, J.J. and Margulis, S.T. Time-based Capabilities of Occupants to Escape Fires in Public Buildings: A Review of Code Provisions and Technical Literature, National Bureau of Standards, Report NBSIR 82-2480, 1982.

10. Cooper, L.Y. A Concept for Estimating Available Safe Egress Time in Fires. Fire Safety Journal, 5. 135-144.

11. Maclennan, H.A. Current Research Relating Time Required and Time Available for Egress. N.F.P.A. National Fire Protection Association's Fall Meeting. San Diego, U.S.A. 12-15 November, 1984.

12. Pauls, J.L. Building Evacuation: Research Findings and Recommendations. In D. Canter (Ed.) Fires and Human Behaviour. Chichester/New York: Wiley. Ch.14, 251-276, 1980.

13. Bryan, J.L. Smoke as a Determinant of Human Behaviour in Fire Situations (Project People). Center for Fire Research, National Bureau of Standards Program for Design Concepts, Grant No.4 - 9027, 1977.

14. Langer, J. Wapner, S. and Werner, H. The Effect of Danger upon the Experience of Time. American Journal Psychology. March 74 (1), 94-97, 1961.

15. Werner, H. and Wapner, S. Changes in Psychological Distance under Conditions of Danger. Journal Personality, 24, 153-167, 1955.

16. Sadalla, E.K. and Magel, S.G. The Perception of Traversed Distance. Environment and Behaviour, 12 (1), 65-79, 1980.

17. Guten, S. and Vernon, L.A. Likelihood of Escape, Likelihood of Danger and Panic Behaviour. Journal of Social Psychology 87, 29-36, 1972.

18. Sime, J.D. The Outcome of Escape Behaviour in the Summerland Fire: Panic or Affiliation ? International Conference on Building Use and Safety Technology. Conference proceedings, Los Angeles 12-14 March 1985. Washington D.C: National Institute of Building Sciences (NIBS), 1985.

19. Hall, J. Patient Evacuation in Hospitals in D. Canter (Ed.) Fires and Human Behaviour. Ch.12. 205-226, Chichester/New York: Wiley, 1980.

20. Pearson, R.G. and Joost, M.G. Egress Behaviour Response Times of Handicapped and Elderly Subjects to Simulated Residential Fire Situations. National Bureau of Standards. Report NBS-GCR-83-429, 1983.

21. Sime, J.D. Post-evacuation Walkthrough (P.E.W.) Instruction Manual. New South Wales Institute of Technology, Sydney, Australia, 1985.

22. Groner, N.E. A Matter of Time - A Comprehensive Guide to Fire Emergency Planning for Board and Care Homes. National Bureau of Standards. Report NBS -GCR-82-408, 1982.

23. Sime, J.D. The Fire Game: Future Directions for Research and development. Fire Protection. February, 1985.

24. Ministry of Works Fire Grading of Buildings: Means of Escape, Part 3, Personal Safety, Post War Building Studies, 29, London, HMSO, 1952.

25. Home Office/Scottish Home and Health Department Guide to Safety at Sports Grounds (Football). London: HMSO (Her Majesty's Stationery Office), 1976.

26. Sime, J.D. Designing for People or Ball-bearings. Design Studies(6)3, 1985.

# Initial Reactions to a Fire from a Simple Robotic Device

**J. J. BREAUX**
Advisor in Computation and Planning
Asturiana de Zinc, Spain

## 1. INTRODUCTION

It can be argued that an understanding of human response when confronted with catastrophic events should provide more than post hoc rationales and explanations of people's actions. Although an important facet in the interpretation of a given event, emphasis on clarifying the past does not in itself result in methods aimed at, or capable of, resolving potential behaviour in future or hypothetical events of a similar nature. Such future-oriented systems should encompass actual human behaviour to a large degree and as such should not be confused with global simulations in which people are cast as uniform objects in the evaluation of person-hazard outcome (e.g., Stahl, 1975; Melinek & Booth, 1975; Berlin, 1981). Clearly, simulations of this type seek to answer specific questions as, for example, whether a given structure can cope with a given flow of people. Accordingly, the assumption that the target population can be regarded as homogeneous performs the service of allowing one's equations to fully assess worst-case events in which the structure's integrity is being "pushed to the limit" under generalised gasflow hypotheses. However, should one wish to project or understand the behaviour of an individual or isolated group under these models, forthcoming extrapolations or accounts are bound to the collective flow and movement premises inherent in the method. Although these systems might mimic global patterns of action when the target population is sufficiently large and constrained, as in traffic problems (Baerwald, 1965) or large-scale disasters (Fritz & Mathewson, 1957), they can tell the investigador little or nothing at all about motives and actual behaviour. That this latter information is critical to an understanding of how events transpire and, further, how individuals in fact influence the development of the hazard has been noted elsewhere for the case of fires (Breaux et.al., 1977; Canter, 1980; Keating & Loftus, 1984). In part, the purpose of the present contribution to the symposium is to emphasise the importance, utility and extended scope of predictive models designed to mimic or otherwise emulate the inferred reasoning and ensuing behaviour of people subjected to stressful events.
    The present paper is the result of a project to develop a fully automated system capable of generating both "intelligent" analysis of an hazard and coherent activity in response to it. Of further importance at the design stage was the ability to tune or configure the model to reflect the circumstances and behavioural tendencies of hypothetical individuals with the aim of assessing the impact of such factors on the reasoning and activity generated. A specification of the entire model and associated algorithms is due for publication in late 1985 (Breaux, in press).

In what follows a specific component of the system is discussed. This unit, referred to as the "Priority controller", has been selected for two reasons. First, it illustrates the design feature of adjusting the logic to simulate a given type of person or persons. Second, the implications for behaviour embodied in this component relate well to the observation that in fires a person's reactions and ultimate fate are often a function of more than just the hazard and the individual's desire to evade it. Often, other objects, animate or otherwise, can be shown to have "driven" the behaviour under observation.

The style of exposition to be followed is intentionally "wordy" with the relevant mathematics inserted where required by the narrative. This follows from the fact that what often appears "obvious" is, upon closer scrutiny, highly complex in both derivation and execution. The reader will also note a tendency to antoromorphise and explain "from the ground up". The author regrets these conventions but considers their use beneficial in conveying a "feeling" for the subject.

## 2. PRELIMINARY CONSIDERATIONS

The global model initially assumes that an individual's response to a known fire (or similar hazard) is a function of one or several objects perceived as threatened. These objects can vary in number and include the self, other people, pets and material possessions. It is assumed that for any given context such an implicit list of objects exists and is brought into awareness by the appreciation of a generalised threat. The relative importance of each object, the degree to which it is perceived as threatened and the effort associated with the reduction of that threat contribute to a decision in which one of the objects (or a number in succession) become the locus of primary concern in terms of which activity is subsequently planned. It is obvious that such objects can be defined as important in different ways across a variety of situations. The model requires access to a ranked list. This list is passed to the decision making Priority controller from a subcomponent called the Major priorities unit.

That people can rank a set of objects in this manner seems reasonable although one might question whether an implicit list has any basis in reality. Many young fathers, for example, indicate little difficulty in ranking objects of this nature when the criterion is "general importance assuming a risk to all objects". Typically, children come first followed by or tied with wife. Lower in the hierarchy one finds car, possessions and house, among others. The model assumes the existence of such an ordered list prior to the fire which is "activated" once a threat is perceived. Its importance is sufficient to merit detailed consideration.

There are numerous ways of deriving and representing ranked priority lists ranging from unweighted rank orders (Diamond, 1959), to adjusted ratio scales (Phillips, 1977) and magnitude models (Curry, 1977). For reasons to become apparent below the modelling process is best served by ratio scale vectors yielding a fair representation of reality given minor inconsistencies in derivation. With few exceptions the first stage in establishing a ratio scale for a set of n objects or elements consists of obtaining a fundamental number of pairwise comparisons (1) which in turn are used to construct a two-dimensional matrix. The cells of this matrix will have values which reflect the relationship between any two objects with respect to the scale or dimension used to compare them. Certain ground rules are imposed on this matrix. Thus, the comparison of an object with itself should constitute an identity relation (not always the case with correlation matrices). Further, comparisons should be consistent. That is, if an object alpha is regarded as more important than an object beta then beta, on its own, must be regarded as less important than alpha. This also implies that for more than two objects the matrix should exhibit transitivity. Accordingly, if A is regarded as more important than B and, further, B is regarded as more important than C, then the matrix must indicate that the cell relating A and C shows the former as being more important. This is the condition of qualitative transitivity. One can also posit quantitative transitivity. If strictly adhered to and if A were regarded as twice as important as B then, logically, B should be regarded as half as important as A. This condition can and should be relaxed insofar as one cannot expect repondents to be absolutely consistent in the quantitative sense. Fortunately, it has been demonstrated that the eigenvector associated with a comparison matrix of moderate size is relatively insensitive to minor consistency violations (Wilkinson, 1965). This has been put to advantage in a subsequent demonstration indicating that the eigenvector corresponding to the maximum eigenvalue of such a matrix yields a cardinal ratio scale for the elements compared (Saaty, 1976; see also Keyfitz, 1968). This type of scale has several advantages not the least of which is that it can be readily concatenated with other indices thereby allowing for the adjustment of scales and set functions, a capability required by the model. Indeed, it is this type of vector which characterises output from the Major priorities unit for eventual use by the Priority controller.

       A specification of such a vector can be illustrated using data obtained from a 48 year old widowed woman with a pet dog and a number of valued possessions. Following a short discussion it was found that three objects were a cause of primary concern to her; the dog, a number of documents including photos (kept in a box in the kitchen) and several pieces of jewelry given to her by her parents and late husband (kept in the main bedroom). Empoying a ten-point scale for gauging relative importance (Saaty & Khouja, 1976) she was asked to make a series of comparisons involving the three objects assuming an equivalent threat to each. This produced the following:

|  | Dog | Documents | Jewelry |
|---|---|---|---|
| Dog | 1 | 4 | 6 |
| Documents |  | 1 | 3 |
| Jewelry |  |  | 1 |

The entries reflect the fact that the dog is regarded as more important to her than documents and demonstrably more important than jewelry. Similarly, the documents are judged as weakly more important than jewelry. The "1's" in the main diagonal indicate identity relations, that is, the assumption that an object when compared with itself, given the same criterion, is neither more or less than itself. Clearly, such a matrix is not symmetrical and the missing values, intentionally excluded from the comparison task, are assigned the corresponding reciprocals of the principal values. This gives:

|  | Dog | Documents | Jewelry |
|---|---|---|---|
| Dog | 1 | 4 | 6 |
| Documents | 1/4 | 1 | 3 |
| Jewelry | 1/6 | 1/3 | 1 |

One is now in a position to solve the eigenvector problem which requires a solution to the following equation:

$$Ax = \lambda_{max} x ,$$

where A is the square matrix, x a column vector and $\lambda_{max}$ the maximum eigenvalue. In fact one is primarily interested in x. Some readers will be more familiar with the form,

$$(A - \lambda_{max} I)x = 0 ,$$

where I is the identity matrix, that is, $AA^{-1} = I$. For the above,

$$x = \begin{pmatrix} .690 \\ .218 \\ .092 \end{pmatrix} .$$

This provides a robust, one-dimensional indication of relative importance. It is this relative assessment of objects which shall be assumed to reside in a potential victim's head, being cued upon the perception of threat. Of course, such a precise specification will not characterise a real person but should nonetheless adequately reflect subjective relative importance. In any event, the vector per se is intended for the modelling process.

## 3. PRIORITY CONTROLLER - SPECIFICATION

The main task of the controller is to provide a revised ranking of relevant objects given new information which is a function of the hazard. That object with the highest adjusted rank will subsequently become the focus of short or long-term planning. In order to effect this adjustment of the vector derived in the preceding section, the controller processes data through a two-phased procedure. First an integration of hazard-related information is accomplished by evaluating the objects on a number of criteria specific to the situation. The second phase consists of postulating a decision function which places the criteria in perspective and merges these to obtain a final revised ranking. Crucial to this undertaking is the representation of objects in a given criterion. In the model this involves the use of inclusion functions.

An inclusion function allows one to specify, for a given object, the degree to which it "participates" or shares in a criterion. The notion of degree is important in this context and requires a specialised approach.

Traditionally, a criterion such as threat might have been applied to an object according to classical set theory. This would result in discrete appraisals, for example, "the object is threatened" or "the object is not threatened". The all-or-none status of these statements often poorly describes how people view a goal or object. Items may be perceived as only partly represented in a set (the set of threatened objects, say). This is often apparent in the study of people's behaviour in fires. They frequently indicate differential concern or worry for a variety of objects important to them and, further, perceive the hazard as posing a differential threat to these objects. It can be argued that only in special cases does one encounter totally dichotomised threat perception, that is, a mental allocation of objects to either the threatened or non-threatened category. The importance of obtaining an adequate and reasonable representation of this process for the model has resulted in the inclusion of techniques normally associated with soft or "fuzzy" set theory (Zadeh, 1971). This allows for a graded specification of how much an object belongs to a set which itself may be vague. By stressing the degree of inclusion one is more likely to avoid over-aggregated models and solutions based on functions with coarse discrimination.

To make the use of these functions somewhat more apparent, the above mentioned woman can be placed in an hypothetical event. Assumed is a fire in the main bedroom which is next door to her living room (sitting room). Her dog is in the guest bedroom. The kitchen is at some remove from this latter location. (This example, including plans of the apartment, is discussed at length elsewhere, Breaux, in press).

The model assumes that once a threat is perceived to exist two functions are subjectively evaluated. The first of these assesses the degree to which each of the objects in the Major priorities vector is threatened. This vector may include the person making the assessment as well, but for simplicity this will be treated somewhat differently in the present paper. Accordingly, an inclusion function for "threat to object" can be advanced. Continuing with the above example one might have:

$$F_{\substack{\text{threat to}\\\text{object}}} = \left\{ \frac{.6}{\text{Dog}} , \frac{.2}{\text{Documents}} , \frac{.9}{\text{Jewelry}} \right\} .$$

The metric for evaluation is a scale ranging from 0 (no threat) to 1 (absolute threat). Given this event and its corresponding threat function it can be seen that since the fire is in the main bedroom, jewelry is most threatened. The close physical proximity of the guest bedroom to the hazard source places the dog next, with documents (in the kitchen) least threatened. The ability to postulate this function presupposes outside information which in the model is provided by the External features component. For the moment this information is assumed as given. Further, when automated, the system assigns threat values which are inversely proportional to the object's distance from the source of the hazard.

The second function to be advanced assumes that the individual assesses the difficulty of reducing threat to each of the objects. In order to render this function compatible with the decision process to be discussed below, it is stated in terms of the 'ease' with which threat reduction can be effected. Given the context and perceived degree of threat, our respondent's "ease of threat reduction" function might be:

$$F_{\substack{\text{ease threat} \\ \text{reduction}}} = \left\{ \frac{.5}{\text{Dog}} , \frac{.9}{\text{Documents}} , \frac{.1}{\text{Jewelry}} \right\} .$$

The metric ranges from 0 (least ease, that is, greatest difficulty) to 1 (least possible difficulty). Thus, the threat associated with documents is seen as the easiest to relieve (they are far from the fire, the kitchen near an exit), followed by dog (nearer the fire). The threatened status of jewelry is most difficult to alter since this object is in the same room as the fire and would incur considerable risk. In the model this function is not simply correlated with distance from the fire. A Resource component attenuates the function given such factors as availability of fire-fighting materials as well as proximity to the hazard.

At this point it should be stated that there are a number of ways whereby one could attempt to alleviate threat to objects whether these include the self or not. This is the subject of the Plan generator. Presently the primary concern is with the assumption that if an individual values a number of objects and these are perceived as threatened then an improvement in their status is contemplated.

Both functions must now be adjusted for object importance. That is, degree of inclusion in the threat and ease functions are weighted for relative importance insofar as the objects are of unequal salience to the victim. How this might occur in reality is questionable. For the purpose of the model a multiplicative process is assumed and, further, is considered a reasonable approximation. This is accomplished by post-multiplying the inclusion functions by the Major priorities vector thereby adjusting the value of the former, that is,

$Fx = F'$ .
Continuing with the above example:

$$F_{\text{threat}} = \left\{ \frac{.6}{\text{Dog}} , \frac{.2}{\text{Documents}} , \frac{.9}{\text{Jewelry}} \right\} . \begin{pmatrix} .690 \\ .218 \\ .092 \end{pmatrix} =$$

$$F'_{\text{threat}} = \left\{ \frac{.414}{\text{Dog}} , \frac{.044}{\text{Documents}} , \frac{.083}{\text{Jewelry}} \right\} .$$

And,

$$F_{ease} = \left\{ \frac{.5}{Dog} , \frac{.9}{Documents} , \frac{.1}{Jewelry} \right\} \cdot \begin{pmatrix} .690 \\ .218 \\ .092 \end{pmatrix} =$$

$$F'_{ease} = \left\{ \frac{.345}{Dog} , \frac{.196}{Documents} , \frac{.009}{Jewelry} \right\} \cdot$$

The new functions (F') can be regarded as placing in perspective the concern associated with objects given perceived hazard impact. This method of adjusting the "raw" inclusion functions realises or mimics two phenomena often detected in the accounts of those who have been through such an ordeal. First, the degree to which a valued object is regarded as threatened cannot truly be understood as entailing an objective process. The role or value of that object in one's life will colour or distort its status in more circumscribed events. This can be expected to be especially so relative to children and other loved ones. In the model the relative importance of objects, in general, is given by x. The impact of the event is given by the inclusion functions which are "personalised" by x. In deriving the $F'_{threat}$ function there is the implicit assumption that perceived threat depends on sensory data (including extrapolations) interacting with an object's meaning to the perceiver. In the example, this has resulted in dog being relatively most threatened.

The second phenomenon brought out by the above concerns the "ease of threat reduction" function. In the example the unadjusted function for ease of threat reduction places documents in the best position. Were one's actions relative to an object based solely on this dimension it might be expected that documents would be an initial target. By adjusting this function for object importance (x) a distortion is postulated which improves the position of dog ($F'_{ease}$). In this case the highly disproportionate importance of an object has attenuated the level of difficulty associated with improving its threatened status.

In real life there is usually a rapid appreciation of the true difficulty in reducing threat to an object. This occurs once action is taken relative to that object. Such statements as, "I didn't expect it to be so hot" or " the smoke didn't seem so bad until you got into it" reflect the subsequent reappraisal. In the model the ease function is highly sensitive to elapsed time which in turn is correlated with the fire growth functions employed to drive the event.

## 4. PRIORITY CONTROLLER - DECISION STAGE

As noted previously, the role of the adjusted F functions (the criteria) is to provide a basis for a subsequent decision. This will result in one of the objects being nominated as an immediate focus of primary concern. Other objects may follow or take its place but at a given instant the recognises but one. (The Plan generator can specify a temporal series of such objects.) Because the model employs multiple criteria, their relative importance prior to the decision must be clarified. This can be hypothesised or based on interviews. For example, the respondent asserted that she would take "great risks" to save the dog if it were threatened. This stated predisposition could be taken to imply that threat was considerably more important than ease as a predecision variable. This might be represented as a matrix in which criteria are scaled for relative influence, for example:

|        | Threat | Ease |
|--------|--------|------|
| Threat | 1      | 7    |
| Ease   | .143   | 1    |

for which the primary eigenvector is:

$$\begin{pmatrix} .88 \\ .12 \end{pmatrix} \, .$$

This vector becomes the model's representation of relative importance. It now remains to adjust the F' functions for differential influence. By effecting this adjustment now the difficulty in projecting the subsequent decision is reduced. This is because irrespective of the decision function used to combine the criteria, these (the criteria) will already have been equated or made comparable.

One method of placing the criteria in perspective derives from the technique of exponential quantification in fuzzy set theory (cf. Zadeh, 1976). Accordingly, concepts are regarded as definable in terms of exponential relations. Applying an extension of this approach (cf. Yager, 1977) the preceding vector ($\begin{smallmatrix} .88 \\ .12 \end{smallmatrix}$) can be used to define a scalar, $\alpha \geq 0$, such that,

$$F'^{\alpha} = \left\{ \frac{x^{\alpha}}{\text{object a}} \, , \, \frac{y^{\alpha}}{\text{object b}} \, , \, \frac{z^{\alpha}}{\text{object c}} \right\} \, .$$

Where x, y, and z are the values for objects a, b, and c in criterion F'. The values of $\alpha$ are given by:

$$nx = \begin{pmatrix} nx_1 \\ nx_2 \\ \vdots \\ nx_n \end{pmatrix} \, ,$$

where n is the number of criteria (in this case two, threat and ease) and $x_i$ the vector value corresponding to the $i^{th}$ criterion. For the present example this yields:

$$\begin{pmatrix} 2^{\cdot} & .88 \\ 2^{\cdot} & .12 \end{pmatrix} = \begin{pmatrix} 1.76 \\ .24 \end{pmatrix} \, .$$

The resulting $\alpha$'s, one corresponding to $F'_{threat}$ (1.76) and the other to $F'_{ease}$ (.24), are the operative exponents used to adjust the criteria for decision influence. The role of this manipulation with respect to exponential quantification is best understood by regarding the criterion importance vector ($\begin{smallmatrix} .88 \\ .12 \end{smallmatrix}$) as reflecting a superordinate criterion (impact on decision) in terms of which the $F'$ are expressed. The effect of $\alpha > 1$ is to further accentuate an important object as one might expect if the criterion is relatively influential in reaching a decision. Conversely, for $\alpha < 1$ the effect is to level or deaccentuate object values for that criterion. For $\alpha = 1$, the inclusion function is unaltered.

Continuing, for $F'_{threat}$ and $F'_{ease}$ one has:

$$F'^{(1.76)}_{threat} = \left\{ \frac{.218}{\text{Dog}} \, , \, \frac{.004}{\text{Documents}} \, , \, \frac{.013}{\text{Jewelry}} \right\} \, .$$

$$F'^{(.24)}_{ease} = \left\{ \frac{.775}{\text{Dog}} \, , \, \frac{.676}{\text{Documents}} \, , \, \frac{.323}{\text{Jewelry}} \right\} \, .$$

Since the criteria are now fully adjusted their contribution to the decision can be based on a simple conjunctive algorithm. The final selection of an object is based on a decision function, D, in which the highest ranked element is chosen. The algorithm most suitable to produce D given the criteria is a matter of debate. initially appealing would be D based on additive worth. This has been avoided and instead a maximin strategy selected. For the above this implies:

$$D = Min(F'_{threat} \cap F'_{ease}) \text{ for all objects, viz.}$$

| Threat | (.218 | , .004 | , .013) |
|--------|-------|--------|---------|
| Ease   | (.775 | , .676 | , .323) |
| D =    | (.218 | , .004 | , .013) . |

It is apparent that the dog receives the highest rank. In fact, both the additive worth and maximin approaches select the dog given these inclusion functions. However, under more complex conditions the maximin strategy can be used to introduce an element of "hedging" when acting on minimal information, as when one is a victim in a strange environment. Given this approach one is essentially basing a decision on a set (D) of least attractive values. In the present example this has resulted in th highest ranking for the dog and, given this, the model would now start to generate activity in the service of that object.

Over the course of the event we can expect D to change given a variety of circumstances. The present exercise constitutes a single pass through the logic. For a given fire numerous cycles can be expected and an example indicating this is presented elsewhere (Breaux, in press).

## 5. CLOSING COMMENTS

The design of robotic devices capable of exhibiting coherent behavioural structure, if only at the level of action strings on a computer printout, requires a departure from classical statephase transition schemes based on inherent Markovian principles. Even where semi-Markov assumptions are employed, thereby incorporating time as a factor, it can be argued that such methods are better suited to the retrospective analysis of an event as opposed to its projection in future time.

It is likely that the manner in wich human beings make decisions, as well as the implications of these for related activity, bear little resemblance to systems based on a finite (even if long) history. It is here where the present paper has attempted to indicate, in highly simplified terms, a way out. This need not imply that the use of transition spaces be disregarded. However, at the very least, it indicates that the successive selection of such spaces is arrived at in a manner more consistent with what is known about human potential. Although not treated herein, the present model does make use of low-level state transition matrices where applicable (associated with a Plan generator and Plan executor). These inclusions can be regarded as locally valid manifolds whose necessity is irreducible given spatial and 'logical' constraints on the individual(s) modeled. that is, in certain cases they can be regarded as primitive sets whose coherence does not require derivation. (For example, to get from one point to another might require a series of directed steps of an invariant nature.)

In terms of computation the model discriminates two processes which can be run in parallel. The target' logic and behaviour, and the nature of the hazard. For fires the latter is given by inputted parameters of a fire growth function whose derivative is used to progressively invalidate use of the building or area. Because the target and hazard processes are reciprocally contrient, people (that is, hypothetical entities) can influence the progression of the hazard, for better or for worse. Similarly, depending on circumstances, the hazard parameters will account for target behaviour.

Parts of this system have been executed on an Hewlett-Packard 3040 and an IBM-XT. A real-time robotic device would require somewhat greater sophistication. However, this would not necessarily imply a LISP or PROLOG machine insofar as the relevant algorithms indicate structural integrity. This is considered to be an important point. Even as concerns string processing it can be argued that matrix localisation of string components is subsidiary to the manner in which these are pointed to (in computer memory). Contrary to popular belief, "If-Then" rule systems might not require list or logical processing systems for their realisation. The "bottom line" is who or what can do it in time.

### Footnote

(1) Experience with the present and alternative comparison techniques has indicated to the author that people find it easier (and are more consistent) when making "greater than" comparisons than "less than" judgements where these additionally require some numerical or verbal quantification. For this reason comparisons characterised by the former distinction are taken as the base data with converse instances being assigned the reciprocal of the corresponding value. This has the further advantage of limiting the number of comparisons to be made which in certain circumstances can be more than the respondent is willing to tolerate. In the example this number, C, is limited to:

$$C = \binom{n}{2},$$

where n is the number of objects. This is equivalent to saying that the number of essential comparisons is given by:

$$\frac{n!}{[(2!)][(n-2)!]} = \frac{6}{2} = 3$$

## REFERENCES

Baerwald, J. Traffic Engineering Handbook. Washington, D.C.:
Institute of Traffic Engineers, 1965.

Berlin, G. A Modeling Procedure For Analyzing The Effect Of Design On
Emergency Escape Potential. N.F.P.A., Boston, 1981.

Breaux, J., Canter, D. & Sime, J. Human Behaviour in Domestic Fires.
In S. Burman & H. Genn (Eds.) Accidents In The Home.
London: Croom-Helm, 1977.

Breaux, J. Robotica Cognoscitiva (Cognitive Robotics). In Press, Leon:
I.E.B. Spain.

Canter, D. (Ed.) Fires And Human Behaviour. New York: Wiley, 1980.

Curry, R.   Worth Assessments of Approach to Landing. IEEE Transactions
on Systems, Man, and Cybernetics, 1977, 7, 395-398.

Diamond, S. Information And Error. New york: Basic, 1959.

Fritz, c. & Mathewson, J.  Convergence Behavior in Disasters.
Washington, D.C.: National Academy of Sciences, 1957.

Keating, J. & Loftus, E. Post Fire Interviews: Development and Field
Validation of the Behavioral Sequence Interview Technique.
Gaithersburg, MD.: National Bureau of Standards, 1984.

Keyfitz, N. Introduction to the Mathematics of Population. Reading, Mass.:
Addison-Wesley, 1968.

Melinek, S. & Booth, S.   An Analysis of Evacuation Times and the Movement
of Crowds in Building Research Establishment,  Borehamwood, 1975.

Phillips, J.   A Simple Method of Obtaining A Ratio Scaling of a Hierarchy.
Behav. Res. & Therapy, 1977, 15, 285-295.

Saaty, T. A Scaling Method for Priorities In Hierarchical Structures.
Manuscript, 1977.

Saaty, T. & Khouja, M.  A Measure of World Influence. J. peace Sci.,
1976, 2, 31-47.

Stahl, F. Simulating Behavior in High-Rise Building Fires.
Gaithersburg, MD.: National Bureau of standards, 1975.

Wilkinson, J.  The Algebraic Eigenvalue Problem. Oxford: Clarendon
Press, 1965.

Yager, R.  Multiple Objective Decision-Making Using Fuzzy Sets.
Int. J. Man-Machine Studies, 1977, 9, 375-382.

Zadeh, L.  Quantitative Fuzzy Semantics. Inf. Sci., 1971, 3, 159-176.

# Towards an Integrated Egress/Evacuation Model Using an Open Systems Approach

HAMISH A. MACLENNAN
School of Building Studies
The N.S.W. Institute of Technology, Australia

ABSTRACT

Egress Research is fragmented through being too narrowly focussed. This research can be categorised into Behaviour and Movement. The Paper discusses the various evacuation models available and how they have not made the necessary allowances for all the aspects of egress. A new model in the process of development is briefly discussed that will use an open systems approach to egress from buildings.

INTRODUCTION

Egress research appears to have been focused into narrow compartments (Pauls 1984). Stahl and Archea (1977) identified these areas prior to this as being;

1.  Field studies of the use of circulation facilities in non emergency conditions;
2.  Laboratory studies (eg. sign visibility in smoke);
3.  Post incident surveys of human behaviour in emergencies.

The results of the studies are well recorded in the literature. They are still somewhat fragmented, but have been reviewed and compared with codes by such writers and researchers such as Stahl and Archea (1977), Pauls (1984), and Stahl, Crosson and Margulis (1982). Fragmentation can be overcome via integration. Certain models have already been developed, but the writer would maintain that the only valid ones will be those which view egress and/or evacuation as an open system so that all aspects of egress in an emergency environment can be assessed. The Search and Rescue Model (Alvord-1983) developed by the National Bureau of Standards seems to have adopted this approach in part. Stahl's model B Fires II (1980) adopts a slightly different approach. Its implemtnation is somewhat complex because of the number of variables.

The author is in the process of developing a total model which will be known as the P.M.S. Model. The latter has been outlined in at least three papers (MacLennan 1984, 1985a, 1985b). It is being developed as part of a research programme for the Australian Uniform Building Regulation Co-ordinating Council and also as part of MacLennan's doctoral programme at Portsmouth Polytechnic. The Model will also incorporate additional areas of research such as;

1.  Exit Affiliation   Sime (1984a, 1984b)
2.  Orientation and Wayfinding (Weisman Ozel, 1984)
3.  Aspects of Visual Access which could be adapted from the work of Archea (1984)

4. Implications of behavioural research that will permit confident generalisations to be made especially in terms of sequence Keating and Loftus (1984) and as described by Sime (1984b) re the contribution of his former colleague Breaux.
5. Importance and relevance of Perceived Time Needed and Perceived Time Available Sime (1984b).

The main problem of the behavioural component of the Model is one of sequence and the prediction of times. A possible answer will be presented in a later section of this paper.

Models Available

A number of egress models have already been developed viz;

1. Evacnet :            R.L. Francis of the University of Florida

2. Search and Rescue:  D.M. Alvard; National Bureau of Standards

3. B Fires II      :   F.I. Stahl formerly of the National Bureau of Standards

4. Descriptive Model:  E. Kendik
   for Movement

5. Effective Width
   Model          :    J.L. Pauls, National Research
                       Council of Canada.

The most powerful of these models in terms of movement is Evacnet. In the writer's opinion it only deals with part of the problem, seeing it does not really cater for human behaviour in terms of time. Human Behaviours such as perception of cues, investigation for the purpose of seeking information and general coping behaviour that may not involve actual movement towards exits or along egress routes are not really included. The time associated with these behaviours could be classified as start up time (ST). Once the decision has been made to evacuate an area then individuals or groups of individuals would move towards the exits. This time could be termed (MT). The latter could be defined as the time taken for occupants to reach an exit. There are however, still problems associated with;

(a) Orientation and wayfinding (Weisman/Ozel 1984)
(b) Exit affiliation (Movement towards the familiar)
    (Sime 1984a) and hence exit choice
(c) Perceived time needed and available that could affect both (ST)
    and (MT)

The time from initial perception to exit access could be summarised as being the clearance time for an area (CT). Evacnet could be modified to cater for this in certain occupancies such as assembly buildings. This is, however, an oversimplification of the problem. It does not allow for the cyclic relationship between behaviour and movement. Movement through exit systems has been well researched by Pauls, Fruin and others. The Search and Rescue Model has partly overcome the problem, but it is only really suitable at present for Board and Care Homes. It too has potential for other occupancy types, where there is an evacuation plan and team. B. Fires II is a simulation model but its main problem is complexity in terms of the number of variables. Kendik and Pauls' models only address movement through exits in the main. The writer would maintain that there is a need for a model that is based on a systems approach as developed in management science. The egress problem can be visualised as being a management exercise be it self management or group management. The impact of the environment at large together with the emergency environment must be catered for and identified.

582

Prediction of the Required Safe Time for Egress (RSET)

Available Time for Egress. Cooper (1983) postulated that the available
safe egress time (ASET) in enclosure fires is defined as the time between
fire detection and the onset of conditions which are hazardous to continued
human occupancy. The other side of the equation is RSET. Cooper (1983)
defines this as being the length of time, subsequent to alarm, which is
actually required for safe occupant egress from threatened spaces. This
definition is valid however in certain occupancy types such as assembly
building where the population density and the evacuation plan are such that
egress is almost entirely movement controlled eg. stadia. Sime (1984b) is
concerned with detection in terms of time. He discusses the problem of
perception in terms of the time that people think they have available PTA
and the time they think they need to evacuate the building safety (PTN).
These times can be related directly to cues especially in terms of whether
they are ambiguous or unambiguous. It is therefore imperative that RSET be
related to ASET in terms of cues. Cooper's model deals mainly with single
room or compartment fires.

There are other models that have been developed that go beyond ASET
such as FAST (Model for the Transport of Fire Smoke and Toxic Gases (Jones
1984). Jones maintains that it is necessary from a life safety and operational standpoint to be able to make accurate predictions of the spread of
fire, smoke and toxic gases. This will in turn provide valuable inputs for
an egress model in terms of time available in spatial terms. This can then
be related to the actual layout and/or configuration of various building
types together with the resultant egress behaviour. FAST also opens up
possibilities according to Jones for combatting the various problems that
may arise as well as taking preventive measures. Much of this can be
achieved through design intervention and "evacuation" planning. Other models
have been developed such the Multi Room Fire Spread Model by Tanaka (1983)
and a Two Room/Compartment Model by Zukowski and Kubota (1980). The Centre
for Fire Research is currently developing a "family" of models that will
eventually address most instances and provide the necessary input for the
various forms of egress/evacuation models in the form of constraints. The
simple concept of the designed safe egress criterion basically still stands
viz -

RSET ( $t_{hazard}$ - $t_{detection}$) < ASET

The problem is therefore one of confidently predicting RSET. The one
component of RSET namely human behaviour is the one that poses the problem
(Sime 1984b).

Generalisations about Human Behaviour that will permit Sequence and Time
Predictions. The main researchers into Human Behaviour have been Wood,
Canter, Bryan and Keating. The work of Wood (1972, 1980) and Bryan in
Project People and Project People II are perhaps the most widely read and
quoted. They provided the necessary start and in fact identified most of
the actions and/or responses. Sime, Canter and Breaux also made valuable
contributions (1980), and Sime demonstrated (1984b) the relevance of this
work in terms of RSET. Their "decomposition" diagrams (1980) provided a
valuable insight into the relationship between behaviours in emergencies.
Sime also raised problems associated with time and perception in terms
of egress.

Keating and Loftus (1984) have developed a new interviewing technique
known as the Behavioural Sequence Interview Technique. The main benefit
of this work is that it will provide real world answers and a data base
which will eventually be large enough to allow confident generalisations
about human reactions and behaviour during fires. When this technique is
coupled with the work carried out by Sime, Breaux and Canter (1980) and
Sime (1984a) together with work of Bryan and Wood then the behavioural

component will start to take shape. The writer intends to use this
technique in a modified form to design questionnaires for use in evacuation
drills that he will be carrying out as part of a research project (MacLennan
1984).

Movement. Movement of people along egress routes and through exits has
provided sufficient data for the prediction of the movement phase of an
egress model. Stahl and Archea 1977, Stahl, Crosson and Margulis (1982)
and Pauls (1984) have already appraised the literature. Kendik (1984) has
developed a model based on the work of Predtechenskii and Milinski (1978).
This "model" has vast potential for use within an integrated model. Kendik
will be also examining the Evacnet and Effective Width Models as well this
year, so that there could be further developments in this area. MacLennan
intends to incorporate the results of this research into his model.

Other perameters have to be catered for such as the problems of orientation and wayfinding, group behaviour perception of additional cues and
most important of all exit choice. The latter is a function of exit affiliation viz. movement towards the familiar (Sime 1984a). The study of
orientation and wayfinding (Weisman, Ozel 1984) is in its infancy, but
studies in non emergency conditions will (Passini) provide valuable information and data. Visual access is equally as important (Archea 1984) in terms
of what features of building a person can see from a certain station within
that building. When this is coupled with cognitive mapping and wayfinding
then the problem with the spatial form and distribution of exits in relation
to the normal circulation spaces could be more clearly stated.

Integrated Model for the Prediction of RSET. The author is currently
developing a model which will be known as the P.M.S. Model (Perception -
Movement Safety). Any model in this vein could be seen as being futile
where it failed to view egress/evacuation as a system. The model views the
occupants of a building as a form of organisation in event of an emergency.
Certain types of building occupancy such as shops are frequented by the
public who may not be familiar with the building. Occupants of office
buildings could quite well be familiar with the building. This poses
structural problems about the result "emergency organisation".

The environment at large is turbulent. "The environment" within a
building during an emergency such as fire is constantly changing so that it
too is turbulent. "Organisations" within the building can no longer be
isolated from each other and other external organisations which go to constitute the environmental world of interactions. The organisation must
therefore be studied as a system especially in terms of:

1. Its component parts (incl. interaction)
2. Reaction/Interaction with the Environment and the resultant change
3. Setting of Objectives
4. Input - Throughput - Output in terms of management effort to achieve
   objectives.

Organisations if they are to operate within a changing environment,
must interact positively with that environment and modify their modus
operandi as a result thereof. They must therefore be viewed as open
systems seeing they are also similar to living systems in that in order to
maintain themselves or "survive" (especially in emergencies) they must be
constantly exchanging material, energy and information with the environment
and modifying their goals (in part), behaviour, strategy, structure etc.
Goals are extremely important in terms of reaction to and with the environment. They must be compatible if the organisation is to survive and hence
the individuals within that organisation.

The occupants of buildings comprise individuals. The latter may exist in groups comprising organisations or as members of the public. There is no set ratio between the two other than a broad range that would relate to the occupancy type. Even the number of organisations occupying an office building could vary depending on the tenancy strategy of building owner. The latter may even occupy the whole building in the form of one organisation. Whatever the profile may be, this may be modified in event of an emergency.

In an emergency such as a fire the "environment" is immediately modified. The occupants of the building are therefore collectively "threatened". They receive certain cues, start to investigate them, seek information from others, carry out other coping actions, decide to evacuate, try to find their way to exits, encounter further cues, investigate further, assist others and the like. These behaviours may be structured depending on prior training and emergency management practices. The occupants in relating to one another and to the environment have common objectives;

1. To find out what is "going on"
2. To escape safely or be assured that they will be safe as result of following a set of instructions which they perceive to be correct.

The occupants are spatially related. They therefore comprise an organisation. The evacuation of or egress from a building involves its occupants either individually or in groups reacting to cues from an emergency generated environment. The main objective for each occupant or for each group and in fact for all groups is life safety. The achievement of such an objective is series of problem solving exercises. Problem solving is synonyuous with management practice. The writer is of the opinion that the systems approach can be used to accomplish this end i.e. organisations can be viewed as open systems.

The subsystems of an emergency organisation are seen as being;

1. Goals and Values: that of the "organisation" and its members related to emergencies which would be concerned with safety.

2. Structures: i.e. structuring of relationships, patterns of communication and information gathering plus the processing thereof, line of authority, level of responsibility and the like. Existing organisational hierarchies must therefore be considered.

3. Psychosocial: individual needs, motivation, behaviour, values, participation, perception of the environment, group dynamics and the like related to emergencies.

4. Technical: level of knowledge in the main. This also relates to safety education and prior evacuation training. Specialist advice may often therefore be required.

5. Managerial: need for management and coordination of the organisation to achieve objectives e.g. safety "Throughput" is therefore critical. See Figure 2a.

All of the above subsystems are part of the "whole" i.e. the organisation. Each subsystem is related to the other. Systems theory allows these relationships to be clearly identified and analysed. Emergency situations seldom occur and exist within the life of a building. The occupants of a building therefore have little opportunity of functioning as an emergency organisation in a condition of emergency. Training is therefore an essential part of the management of emergency organisations.

Another problem that exists with buildings is in the design thereof. Architects and their consultants may not be entirely aware of the makeup

of the "end users" of the building they are designing.  If they carry out
market research and determine their client's needs properly, then building
safety can be taken into account.  Building Owners and Designers often
consider that it is sufficient to merely comply with the provisions of the
various Codes.  This could hardly be seen as being "design intervention".
Each building is unique within itself together with its resultant impact
on the end user.

The results of research in the area of egress and fire provide valuable
input for any model concerned with life safety.  These highlight the need for
designers to cater for the needs of people.  Fire protection and detection
systems, smoke control systems, evacuation plans and procedures, location and
use of exits, internal subdivision need to be designed around the use of the
building and to satisfy the needs of the future occupants of the building.
The P.M.S. model is being developed to allow both the owner and the designer
to develop a full life safety strategy based on the analysis of a special
"organisation" functioning in an emergency environment.  It will also allow
for a similar type of analysis for existing buildings where life safety is
inadequate.

The PMS Model is shown in systems format in figure one.  It has been
based on a similar management model developed by Kast and Rosenzweig (1978).
The throughput has been analysed and summarised in a collapsed form in
figures 2a and 2b.  The operational flow is cyclic in nature as shown and
has been developed as a result of discussions held in Ottawa at the National
Research Council of Canada between Messrs. Pauls and MacLennan and Dr. Sime.
The components of the flow have been identified as Cues, Relatively Complex
Behaviour, Direct Movement to and through exit and Safety.  The diagram is
simplistic and is virtually self explanatory.  It will allow for the complete
integration of the results of research of egress and the fire scenario
within the various environmental constraints.  The results of research clearly
indicate the need for both management and responsible design intervention
strategies to cater for life safety.  There is a need for total involvement
of people.  If an organisation is seen as being dynamic then the open systems
approach can be used and the psychosocial subsystem catered for.  The Goals
of the emergency organisation can be aligned with those of its members via
an integrated Management approach such as "Management by Objectives".  The
"emergency organisation" would then be established and the individuals in
that organisation motivated to be concerned with safety via a safety committee.
Training via drills and education in the aspects of emergency scenarios such
as fire would then be invaluable.  Failure to adopt this approach could
result in confusion.  This would increase the overall time required for
safe egress to the point where it exceeded the time available.

Relatively complex behaviour as a component of the throughput includes
the detection of a cue, investigation and the seeking of further information, other forms of coping behaviour including fire fighting, the decision
to evacuate and assisting and warning others.  The Direct Movement to Exits
Component includes the movement aspect, orientation and wayfinding, exit
choice and affiliation, visual access, receipt of further cues which would
alter the flow as shown in figure 2a and eventually movement along egress
routes through exits to safety.  It also highlights the fact that evacuation
may not be the safest solution and that the objective of safety may be
accomplished by other means such as rescue.  The flow for any situation
could be set down in a predetermined safety or emergency plan.  The flow of
information within the organisation and hence through the building in spatial
terms is therefore vital.  The Model is therefore seen as being an integrated
approach for the achievement of the life safety objective and hence the prediction of RSET as applicable.

586

INPUT: FALLS INTO TWO CATEGORIES

1. FROM GENERAL ENVIRONMENT
 • KNOWLEDGE RE FIRE SCENARIO
 • PRIOR TRAINING
 • FIRE PROTECTION/DETECTION/SAFETY
 • OTHER

2. FROM EMERGENCY ENVIRONMENT
 REFERS TO ACTUAL INTERNAL ENVIRONMENT DURING
 THE EMERGENCY; CHANGES RAPIDLY ESPECIALLY IN
 EVENT OF FIRE; IT PROVIDES THE INITIAL CUES.

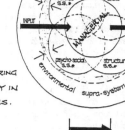

INPUT-OUTPUT →←
FLOW OF ENERGY,
INFORMATION AND
RESPONSE.-
EXTERNAL PRESSURES

# THE P.M.S. MODEL.
(PERCEPTION/MOVEMENT/SAFETY)

Figure 1

THROUGHPUT.

OUTPUT: ACHIEVEMENT OF SAFETY OBJECTIVE.
1. ESCAPE TO OPEN SPACE
2. ESCAPE TO AREA OF REFUGE
3. NO ACTION; EMERGENCY BROUGHT UNDER CONTROL
 OR RESCUE EFFECTED.

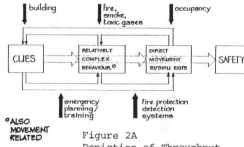

°ALSO MOVEMENT RELATED

Figure 2A
Depiction of Throughput
PMS Model

Figure 2b
Time Variability by Com-
ponent and Occupancy Type-
PMS Throughput Model.

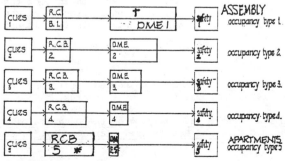

† DIRECT MOVEMENT TO/THRU EXITS
✻ RELATIVELY- COMPLEX BEHAVIOUR

587

The Estimation of Times for the RCB and DME Portions of the Throughput of the PMS Model. The current data available from Research into Human Behaviour cannot be used to make accurate generalised predictions. BSIT as developed by Loftus and Keating (1984) could overcome this problem when viewed in conjunction with work carried out by Canter Breaux and Sime (1980) (Sime 1984b). It identifies sequence hence permitting data derived from this technique to be used in making accurate generalised predictions re sequence. Activities could be "sequenced", but there will still be the problem with estimating times. The Australian Research study will utilise this approach in the design of evacuation drills. MacLennan is aware that these drills will not be under emergency conditions, but each building that is to be evacuated will be operated in "fire mode". Sime, Pauls and MacLennan are in the process of developing a questionnaire that will follow the concept of BSIT. The net result will be the development of data that can be analysed in association with other existing data and that can be used to estimate times for each generalised activity. These sequences will then be integrated to derive an overall time which will be matched against the appropriate fire growth or smoke production/filling/movement models.

This problem does not exist to the same extent with the DME portion of the throughput, as previously discussed. Effects of orientation and wayfinding, exit affiliation, visual access and the like will also be identified together with the effect of faulty design intervention strategies on the questionnaire. Direct observation techniques such as those used by Pauls (1980) will be used to reinforce and/or supplement the questionnaire.

The objectives of the throughput should be;

1.  The prediction of a total RSET.
2.  The comparison of RSET with "ASET" at strategic points (not necessarily Cooper's version).
3.  The evaluation of the degree of safety associated with each building analysed.
4.  The identification of areas requiring design intervention and adoption/modification of emergency management and training procedures.
5.  Adequacy of Building Codes re Life Safety.
6.  Framework for the improved design of egress systems and management.

CONCLUSION

Research will be constantly making progress with each year. Data will and is becoming more reliable especially in the area of human behaviour. Models will be able to utilise this data coupled with that of the movement phase of egress so that the prediction of a total egress time can be made. The P.M.S. Model will endeavour to achieve this end seeing it uses a systems approach.

REFERENCES

1.  Alvord, G.: Status Report of Escape and Rescue Model, NBS-GCR-83-432, Washington, June 1983, National Bureau of Standards.

2.  Bryan, J.L.: An Examination and Analysis of the Dynamics of Human Behaviour in the M.G.M. Grand Hotel Fire, Clark County, Nevada, November 21, 1980, National Fire Protection Association, Revised Edition, April, 1983.

3.  Bryan, J.L.: Smoke as a Determinant of Human Behaviour in Fire Situations (Project People), Center for Fire Research, National Bureau of Standards,

Program for Design Concepts, Grant No. 4 - 9027, 1977.

4. Bryan, J.L.: Implications for Codes and Behaviour Models from the Analysis of Behaviour Response Patterns in Fire Situations as selected from the Project People and Project People II Study Programs, NBS-GCR-83-425, Washington, National Bureau of Standards, March 1983.

5. Bryan, J.L.: The Determination of Behaviour Response Patterns in Fire Situations, Project People II, Final Report - Health Care, NBS-GCR-81-343, Washington, National Bureau of Standards, September 1981.

6. Canter, D., Breaux, J.J. and Sime J.D.: "Domestic Multiple Occupancy and Hospital Fires" in D. Canter (Ed.), Fires and Human Behaviour, Ch. 8, pp.117-136, Chichester New York, J. Wiley and Sons, 1980.

7. Collins, B.L. and Lerner, N.D.: An Evaluation of Exit Sign Visibility, Final Report, NBSIR 83-2675, Washington, National Bureau of Standards, April, 1983.

8. Cooper, L.Y.: "A Concept for Estimating Available Safe Egress Time in Fires" in Fire Safety Journal, Vol. 5, pp.135-144, 1983.

9. Fruin, J.J.: Pedestrian Planning and Design, Metropolitan Association of Urban Designers and Environmental Planners, New York, 1971, (currently under revision - 1985).

10. Herz, E., Edelman, P. and Bickman, L.: The Impact of Fire Emergency Training on Knowledge of Appropriate Behaviour in Fires, NBS-GCR-78-137, National Bureau of Standards, Washington, January 1979.

11. Hunt, J.W.: The Restless Organisation, John Wiley and Sons Ltd., Sydney, 1976.

12. Jones, W.J. and Quintiere, J.G.: "Prediction of Corridor Smoke Filling by Zone Models", in Combustion Science and Technology, 1983, Vol. 11, Gordon and Breach Science Publishers Inc., London, 1983.

13. Jones, W.J., A Model for the Transport of Fire, Smoke and Toxic Gases (FAST), NBSIR 84-2934, Gaithersburg, Maryland, National Bureau of Standards, September, 1984.

14. Kast, F.E. and Rosenzweig, J.E., Organisation and Management, Systems Approach, 3rd Edition, McGraw-Hill Kogakusha Ltd., Tokyo, 1978.

15. Keating, J.P. and Loftus, E.F., Post Fire Interviews: Development and Field Validation of the Behavioural Sequence Interview Technique, NBS-GCR-84-477, Gaithersburg, National Bureau of Standards, October 1984.

16. Kendik, E.: "Assessment of Escape Routes in Buildings - Discussion of a Design Method for Calculating Pedestrian Movement", Paper presented at a Fire Conference held at the Center for Fire Research, National Bureau of Standards, Gaithersburg, Maryland on the 18th October 1984, unpublished.

17. Levine, M.: "Cognitive Maps and You-are-here Maps", Article based on address presented to the A.P.A., Washington, D.C., 1982, State University of New York at New York.

18. MacLennan, H.A.: "Current Research Relating Time Required and Time Available for Egress", Paper presented at the Speakers' Session No. 4;

"Evacuation Time Criteria for Buildings - What do we know about the Time Needed and the Time Available?", Fall Meeting, National Fire Protection Association, San Diego, November, 1984.

19. MacLennan, H.A.: "The Problem with Estimating the Safe Time Required for Egress", Paper submitted and accepted for the ASHRAE National Meeting to be held in Honolulu, Hawaii, 23rd-27th June 1985. (1985a).

20. MacLennan, H.A.: Paper submitted and accepted for the International Conference on Building Use and Safety to be held in Los Angeles, 12th-14th March 1985, Building Use and Safety Institute and the National Institute of Building Sciences, Washington, D.C., (1985b).

21. MacLennan, H.A.: (On Management and Organisation Theory), unpublished Lecturer Notes, Building Degree Course, Faculty of Architecture and Building the New South Wales Institute of Technology, N.S.W., Australia, 1976-1978.

22. Pauls, J.L.: "The Movement of People In Buildings and Design Solutions for Means of Egress", in Fire Technology, Vol. 20, 1984, pp.27-47.

23. Pauls, J.L.: "Building Evacuation: Research Findings and Recommendations", in D. Canter (Ed.), Fires and Human Behaviour, Chichester/New York, Wiley, 1980, Ch. 4, pp.251-276.

24. Pauls J.L., Sime, J.D. and MacLennan, H.A.: Private Communication and Discussion, National Research Council of Canada, Ottawa, November, 1984.

25. Sime, J.D.: "Movement towards the familiar: Person and Place Affiliations in a Fire, Entrapment Setting", published in Duerk, D. and Campbell, D., (Eds.), The Challenge of Diversity: Proceedings of EDRA/15, 1984, Environmental Design and Research Organisation, Washington, D.C., pp.100-109,1984a.

26. Sime, J.D.: "Psychological and Social Factors Influencing and Evaluation", Paper presented at the Speakers' Session No. 4; "Evacuation Time Criteria for Buildings - What do we know about the time needed and the time available?", Fall Meeting, National Fire Protection Association, San Diego, November 1984b.

27. Stahl, F.I. and Archea, J.C.: An Assessment of the Technical Literature on Emergency Egress Behaviour during Fires, Calibration and Analysis, NBSIR 79-1713, National Bureau of Standards, Washington D.C., 1977.

28. Stahl, F.I., Crosson, J.J. and Marguus, S.T.: Time-based Capabilities of Occupants to Escape Fires in Public Buildings: A Review of Code Provisions and Technical Literature, NBSIR 82-2480, National Bureau of Standards, Washington D.C. 1982.

29. Stahl, F.I.: "BFIRES/VERSION 2: Documentation of Program Modifications", NBSIR 80-1982, 1980, Washington D.C., National Bureau of Standards.

30. Tanaka, T.: A Model of Multi Room Fire Spread, Published as a NBSIR, National Bureau of Standards, Washington, 1983.

- 31. Wiseman, G.D. and Ozel, F.: "Way Finding, Cognitive Mapping and Fire Safety: Some Directions for Research and Practice", paper submitted to the Environmental Design and Research Organisation, Annual Conference, San Luis Obispo, California, June-July 1984, University of Winsconsi, Milwaukee, 1984.

32. Wood, P.G.: "A Survey of Behaviour in Fires", in Canter D., (Ed.), Fires and Human Behaviour, John Wiley and Sons Ltd., Chichester/New York, 1980, pp.83-95.

33. Zukowski, E.E. and Kubota, T.: "Two Layer Modelling of Smoke Movement in Building Fires, Fire and Materials 4, 17, 1980.

# TRANSLATION OF RESEARCH INTO PRACTICE

Session Chair

**Prof. Robert W. Fitzgerald**
Worcester Polytechnic Institute
100 Institute Road
Worcester, Massachusetts 01609, USA

# Translation of Research into Practice

J. J. KEOUGH
Keough Consultants
5 Devon Street,
North Epping, NSW 2121, Australia

ABSTRACT

It is difficult to introduce new concepts to the building industry.
Innovation must be approved by:

                         the designer
                         the owner
                         the lending authority
                         the insuring body
                         the design code
                         the building regulations

The preferred approach to secure early acceptance of innovation is to
submit the concept to the professional committee responsible for the design
code.  Endorsement by that committee can be used to secure the several approvals
but incorporation in the appropriate design code can lead directly to incorporation in regulations and thus assure the other approvals.  Once the concept is
incorporated in a design code or the regulations it will almost certainly be
incorporated in the curriculum of educational establishments.

KEYWORDS

Innovation, lending authority, insuring body, design code, building regulations, high-speed sprinklers, steel, concrete, restraint, education, research.


INTRODUCTION

Research workers in all countries are often frustrated by what is seen as
an inordinately long time before the findings of successful research projects
are adopted in everyday practice.  It is probably true in most industries but
it is certainly true in the building industry where my experience lies.

Those who have given the matter only passing thought probably accept human
reluctance to accept change as the explanation.  Certainly the building industry
is not renowned for its rapid acceptance of innovation, no matter how well the
merit may have been demonstrated and proven.

Those who have given the matter more thought explain it as a failure in communication. The practitioner does not study scientific journals where research results are publicised and busy practitioners are not able to attend research seminars or to undertake post graduate extension courses.

It is my experience that whereas the above two explanations are partially correct, the real explanation lies elsewhere.

CONSTRAINTS WITHIN THE BUILDING INDUSTRY

No matter how eager a designer may be to employ a new technique he is obliged to work within certain constraints. The principal constraint is that he must inform the owner of any intention to be innovative and obtain the owner's consent. No matter how adventurous the owner may be, he in turn may have to obtain the consent of the lending authority through which he is funding the project and secure the approval of the insurance body which is or will be under-writing the project. After these approvals have been obtained local building regulations have to be complied with and these almost invariably require com-pliance with the relevant design codes. I can cite a classic example of the ease with which research can be translated into practice if some of the fore-going constraints can be eliminated.

In 1971 - 72 the Fire Research Station carried out full-scale tests on fire in high-racked storage at the research facility at Cardington. These experiments demonstrated that fire would spread vertically to involve material above a con-ventional sprinkler head mounted within the storage rack before that head would be brought into operation by heat from the fire. Further testing demonstrated that a linewire detector could be arranged in the rack to operate sprinkler heads electrically and selectively to confine the fire at its level of outbreak.[1] The Experimental Building Station in Australia is a national building research laboratory attached to the Department of Housing and Construction. This Depart-ment is responsible for all federal government building and is probably the largest design and construction authority in the southern hemisphere. The department has to secure its client department's approval before any innovation but it is not necessarily subject to the other constraints. In 1972 the depart-ment was designing a high-racked storage building for the Royal Australian Air Force. The strategic importance of the stores required a high level of security and fire safety. A small section of the proposed racking was erected at the EBS and by means of real fires the performance of in-rack sprinkler systems to the latest insurance codes and the new high-speed system were demonstrated. The superiority of the new system was established clearly. The new system, incorpo-rating a computer to monitor continuously the continuity of all electrical circuits and the status of all valves and the pressures in the hydraulic systems, was installed in the multi-million dollar store. Shortly afterward similar sys-tems were incorporated in three large stores for another Australian government instrumentality. After 10 years experience the operators of these four complexes have full confidence in the system but international insurance bodies have not given approval so no commercial enterprise requiring insurance cover has been able to adopt the system. Without support from the insurance industry it cannot be introduced into national codes as full consensus is normally required before existing codes are altered or new codes are adopted.

HOW RESEARCH CAN MEET THE CONSTRAINTS

The constraints upon the practitioner's introduction of change have been identified as -

1. Approval of the building owner
2. Approval of the lending authority
3. Approval of the insuring body
4. Acceptance within the appropriate design code
5. Incorporation within building regulations

1. Approval of the building owner

Few building owners welcome innovation for the sake of innovation and most adopt the attitude that they would prefer to follow up on someone else's success. The experiences of the De Haviland and Boeing companies with their Comet and 707 respectively are well known and understood. To secure the approval of the owner the practitioner has to produce a cost/benefit analysis showing a balance clearly in favour of the innovation. If the balance can be expressed in monetary terms it is more likely to succeed. Successful practitioners are experienced in such activities and normally do not require assistance from the research worker responsible for conceiving the innovation.

2. Approval of the lending authority

The lending authority is normally concerned about the retention of the monetary value of the asset, at least throughout the mortgage period. Accordingly, lending authorities have a conservative approach to innovation in any form. Here the practitioner may need to call upon the research worker for data accumulated during the development program to produce a forecast of the soundness and durability of the innovation to offset the lending authority's complete lack of past experience with the innovation. If the matter is highly technical the lending authority may refuse approval until the body responsible for the design code has examined and approved the innovation. That is, acceptance under constraint number two is frequently conditional upon compliance with constraint number four.

3. Approval of the insuring body

As insurance is based on past experience this industry's reaction to any proposed innovation is cautious in the extreme. Some of the larger industry groups such as the Fire Offices Committee and the Factory Mutual Corporation have their own research laboratories and thus have access to professional expertise in evaluating the merit of particular innovations. In many cases an innovator will have to accept increased insurance costs and in some cases coverage may only be granted if the innovative owner has sufficient other insurance business to offset the feared increase in risk.  In these latter cases compliance of the innovation with constraint number four makes the innovation more digestible but not necessarily more attractive or palatable.

4.  Acceptance within the appropriate design code

Although listed as number four, acceptance in a design code is probably the most important constraint to be met before an innovation will be widely accepted in practice.  No practitioner will risk his professional standing by employing an innovation that is forbidden by the design code but the more talented practitioner will be prepared to use an innovation which he approves personally but which is not specifically approved by the design code.  The less talented practitioner works within the provisions of the code as it embodies the procedures in which he was instructed in his undergraduate days.  Any subsequent modifications have been examined and approved by a committee comprising the brighter and more experienced of his fellow practitioners together with academics and other industry representatives.  Once an innovation has been incorporated in the design code it virtually ceases to be an innovation.  The average practitioner is not concerned with the history of its derivation and does not wish to study the matter in depth to develop a full understanding of its underlying theory.  He wishes to have the code nominate the technique and the bounds within which he may use the innovation in his practice.  Therein lies the biggest delay in the translation from research into practice.

Most research workers believe their task is completed once they have succeeded in having a paper describing a new research finding accepted for publication in a leading scientific journal.  In most cases this is all that their laboratory or institution requires of them.  Indeed it is on the basis of such publications that promotion and their future professional career is based.  If the research institute sees potential for exploitation by industry it may seek to establish a licensing arrangement to give a particular firm exclusive rights to develop the research finding into a marketable commodity.  Otherwise the finding may languish in the literature until an applied research worker at another institute or commercial research laboratory sees a potential for its application in particular problems.  The applied research worker explores the application both theoretically and experimentally to determine the factors that limit its safe application to particular problems.  Publication of this information eventually provides the basis for incorporation in the design code.  The contribution of the second worker is often more important than that of the primary worker and usually requires a much longer gestation period.

However, not all research follows the foregoing pattern.  I would like to summarize the history of a research undertaking that has evolved over some thirty years and in which many of us here at this First International Symposium on Fire Safety Science have played largely unco-ordinated roles of varying importance.

In the late 1950's several research workers became concerned about the significance of results that were emerging from tests on structural elements carried out at the Northbrook laboratories of Underwriters Laboratories Inc.  Studies to explain these results revealed a serious lack of information about the properties of structural steels at temperatures significantly above ambient.

Indeed in 1960 the definitive work on the relation between yield strength and temperature of structural steel was that of Lea[2] published in the early 1920's. To correct this a joint project was undertaken between the Buildings Research Establishment and the Fire Research Station in the United Kingdom[3] and a project was undertaken by the United States Steel Corporation in the U.S.A.[4] An EBS approach to the research laboratories of the steel industry in Australia at that time was rebuffed and suitable information on Australian steels did not become available until the 1970's[5].

The explanation for the test results from ULI was found to be the very high forces of restraint that could be generated during tests on specimens of steel and reinforced concrete held within the very strong specimen - containing frames that had been built for use at Northbrook when the facilities were transferred from Chicago. This highlighted the need for a better definition of the standard fire resistance test for use internationally and ISO/TC92 began work on what ultimately became ISO 834.[6] The Portland Cement Association developed a test facility at Skokie capable of controlling the magnitude and direction of forces of restraint at the boundaries of test specimens. The findings from this facility lead to ULI adopting the practice of publishing separate ratings for restrained and unrestrained structures.[7] Work at P.C.A. Laboratories[8] and at Universities in Sweden[9] and Germany[10] produced detailed information on material properties at high temperatures.

The Federation Internationale de la Precontrainte established a Commission on Fire Resistance which assembled the research information relevant to concrete structures and this panel of international experts drew up a set of guidelines for practising structural engineers designing concrete structures to possess nominated levels of fire resistance[11]. Later national committees produced guidance documents for the design of reinforced concrete structures in United Kingdom[12] and U.S.A.[13]

The foregoing is an example of the general recognition of an area of ignorance and the conduct of independent research programs to produce knowledge to overcome that ignorance. The position now achieved is that the professional associations have prepared the basis for national design codes for the design of fire resisting structures of steel and of concrete. In many countries the movement seems to have stalled at that point and difficulty is being encountered in coping with the next of the constraints.


5. Incorporation within building regulations

In Scandinavian countries there does not appear to be as much difficulty in securing legal acceptance of technical innovations as there is in other countries. Scandinavian countries have for some time accepted buildings that have been individually designed to match the severity of fire that the designing engineer has calculated to be probable for that building.

Most countries still adhere to the system of regulating fire safety in buildings by grouping buildings into classes according to usage and assuming that each class represents one level of potential fire severity. Buildings in each class are required to incorporate a common set of active and passive fire safety features that is varied according to the height and the floor area of the building. The writers of the building regulations have the option of incorporating detailed design requirements in the building regulation document or of specifying a level of performance and requiring compliance with the provisions of a separate nominated design code.

Fortunately in Australia the committee responsible for the drafting of the Australian Model Uniform Building Code has selected the second of these options. The AMUBC nominates the level of performance and requires compliance with the provisions of the appropriate national standard for design and installation produced by the Standards Association of Australia. The Association's standards are drafted by committees representing the appropriate technical and community interests and a representative of the building regulations committee serves on any standards committee responsible for drafting a standard that may be called up in the regulations. This assures appropriate liaison between the two groups. National standards produced by the SAA have no legal status unless they are specifically called up in a regulation.

The SAA Committees responsible for the steel structures code and for the concrete structures code have accepted that fire, like wind and earthquake, is a loading condition that the designer should consider in producing his design. Accordingly an appropriate section on design for fire will be incorporated in the 1985 editions of these standards and should thus become part of the AMUBC.

This system has committees of appropriate technical experts drafting technical standards and leaves the building regulations committee to concentrate on the moral, social and economic considerations that determine the desirable minimum quality of buildings. The system makes it possible for research findings to be considered by the relevant technical committee of SAA, for the national standard to be modified as appropriate and then become part of building regulations. The progress from SAA standard to building regulations is not automatic as the regulating committee has to approve the modified standard. Because of the system of representation on the standards committee a significant delay is unlikely provided that issue of the revised standard is harmonised with the review of regulations. One factor that has to be observed carefully is that standards must be expressed in mandatory rather than advisory terms if they are to be passed into law. While this may seem an editorial matter it does involve the technical committee in consideration of those design matters to be defined as obligatory and those to be expressed as desirable. The technical committee drafting the standard is best qualified to do this.

If the second option is followed the process becomes more involved. In February 1981 American Concrete Institute Committee 216 published a guide for determining the fire endurance of concrete elements.[13] The guide represents an excellent and detailed treatment of the subject but is not in a form suitable for adoption in building regulations. Had that ACI committee of experts co-opted representatives from building code authorities to prepare a companion document suitable for incorporation in building codes the matter would have been dealt with expeditiously and the U.S.A. would have had common rules for fire-resisting concrete structures. Instead the system requires each of the code authorities to form its own ad hoc committee to prepare its own rules. The duplication of the work undertaken by the Ad Hoc Committee on Calculated Fire Resistance of the Southern Building Code Congress[14] by each of the several other code authorities represents a major effort and very likely will have resulted in a lack of uniformity throughout the country.

INTRODUCING RESEARCH THROUGH EDUCATION

At post-graduate seminars, extension courses and conferences organised by fire protection associations, fire research workers are afforded the opportunity to outline progress with fire research to industry. Practitioners attending the seminars and courses are interested in learning techniques or other developments that can be applied in their practice. They are interested in learning the background and basis for changes that are being made in design codes or regulations but they are not interested greatly in research that has not yet reached that stage of advancement.

I have studied the professional backgrounds of attendants at regional and national conferences organised by the Australian Fire Protection Association. The majority of attendants are from the fire protection industry itself and the remainder are federal, state or local government officials or fire safety officers from large commercial or industrial undertakings. When questioned, the average architect or engineer will reply that he cannot spare the time to attend such functions but that he would like to be able to do so.

The Australian Model Uniform Building Code, wherever practical, calls up standards and codes of the SAA to express its requirements in the various building fields. Fire safety is a comparatively young science and because there is no SAA fire safety code the regulations devote a lot of attention to detailed fire safety provisions. Tertiary institutions such as the New South Wales Institute of Technology have introduced extension courses to acquaint building inspectors and practitioners with the intent of and background to these new regulations. Schools of Engineering and Architecture are including similar instruction in their undergraduate courses.

The overall picture shows clearly that the average practitioner and educationist becomes really interested when new concepts are introduced by design codes or by regulation. Until that stage is reached they are not interested.


CONCLUSION

Most industries, and certainly the building industry, are so structured that it is difficult for an individual designer to attempt to introduce a practice that is seen to be a major innovation. Consequently it is difficult for a major research advance to be introduced to practice by individual effort. The preferred approach is to submit the research findings to a national professional committee for appraisal. If the work can be incorporated in a design code it can then progress to incorporation in regulations. Once this is achieved it will be incorporated in the curriculum of educational institutions. Figure 1 is a schematice diagram of the procedure that must be followed with an individual attempt to be innovative and Figure 2 outlines the preferred approach through national design codes.

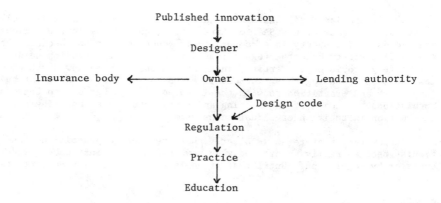

Figure 1    INDIVIDUAL EFFORT

---

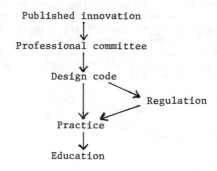

Figure 2    DESIGN CODE APPROACH

REFERENCES:

1. Nash, P., Bridge, N.W. and Young, R.A.: "The rapid extinctinction of fires in high-racked storages". Joint Fire Research Organisation Fire Research Note 944/1972.

2. Lea, F.C.: "The effect of temperature on some of the properties of metals". Proc.I.Mech.E., Vol.II, June 1922, pp 885-904.
   Lea, F.C.: "The effect of low and high temperatures on materials". Proc.I.Mech.E., Vol.II, Dec., 1923, pp 1053-1096.

3. Stevens, R.F.: "Discussion of 'Steel Reinforcement' by R.I. Lancaster in Structural Concrete, Vol. 3, No. 4, July/August 1966 pp 184-186.

4. "Steel for elevated temperature service", U.S. Steel Corporation, Pittsburgh, Pa.

5. Skinner, D.H.: "Determination of the high temperature properties of steel," BHP Tech. Bull. 16(2), Nov. 1972  pp 10-21.

6. Fire Resistance Tests - Elements of building construction, ISO 834 - 1975 (E), International Organisation for Standardisation, Switzerland.

7. Fire Resistance Index, Underwriters Laboratories, Inc., Northbrook, Jan. 1972.

8. Abrams, M.S.: "Behaviour of Inorganic Materials in Fire", Design of Buildings for Fire Safety, ASTM STP 685, E.E. Smith and T.Z. Harmathy, Eds, ASTM, 1979, pp 14-75.

9. Anderberg, V.: "Mechanical properties of reinforcing steel at elevated temperatures," in Swedish with English summary, Tekniska meddelande nr 36, Halmstad Jarnverk AB, Lund 1978.

10. Kordina, K. et al.: Sonderforschungsbereich 148, T.U. Braunschweig, 1977.

11. Guides to good practice: Recommendations for the design of reinforced and prestressed concrete structural members for Fire resistance. 1st ed. Wexford Springs 1975.

12. "Design and detailing of concrete structures for fire resistance," Interim guidance by a joint committee of The Institute of Structural Engineers and The Concrete Society, April 1978.

13. "Guide for determining the fire endurance of concrete elements", Report No. ACI 216R - 81, Concrete International Vol.3, No.2, February 1981, pp 13-47.

14. Gustaferro, A.H.,: Consulting Engineer - Private communication, September 1981.

# Translation of Research into Practice: Building Design

**MARGARET LAW**
Ove Arup Partnership, London

ABSTRACT

Progress towards 'fire engineering by design' will be slow unless
researchers can demonstrate to the public authorities and the engineering
community that the objectives of fire precautions can be achieved more
effectively than at present. Research can give methods of quantifying and
defining the requirements for public safety. It is then possible for the
building designer to adopt the fire protection measures appropriate to the use of
the building and its importance in terms of people and property. Practical
examples of applying fire engineering design methods are given.

INTRODUCTION

It is over ten years since I joined a large engineering practice, with the
broad brief to "provide fire engineering by design". The results of fire
research must be a major input, because most of our regulations and code of
practice cannot be applied directly to innovative buildings. Working with
creative architects and engineers to find a practical solution is a challenge.
The fire engineering advice must be soundly based and respect the needs and
aspirations of the designers and users of the building. Fortunately the fire
research is of a high standard, but the administrative procedures and
bureaucratic attitudes we must work with can be frustrating. It is in changing
these attitudes that I believe research should make a major impact. Except
perhaps in North America, the relatively small numbers of professional fire
engineers exert little influence at a national level; they are certainly not as
influential as the structural engineers, for example. In this paper I would like
to explore this theme but first, as the French say, a little history.

THE LAST FOUR DECADES

Forty years ago the UK government published a report on 'Fire Grading of
Buildings'(1). The objectives of fire precautions were stated to be 'to
safeguard life and property' and to this end fire precautions should aim 'to
reduce the number of outbreaks of fire' and, in relation to fire grading, should
'limit the development and spread of a fire in the event of an outbreak
and......provide for safe exit of occupants'. It was considered that a rational
and economic combination of passive and active defence measures was the ideal,
but was difficult to achieve in the absence of a coordinated body of fire
statistics.

Shortly after publication of the Report, the systematic collection of fire statistics was initiated.  In 1980 the UK Home Office published a consultative document on 'Future Fire Policy'(2) which summarised the results of a study started in 1976 to review and measure the total cost of fire to the community and submit proposals as to the range and level of measures which should desirably be taken to have the optimum effect in mitigating losses.  The fire statistics provided a major input.  Some brief extracts from the report are given here.

The current numbers of annual deaths and injuries in the UK are 900 and 6000 respectively and the casualty problem which exists 'derives essentially from a number of small fires, mostly in dwellings, affecting just one person, normally close to the source of ignition'.  In discussing Building Regulations and life safety the report suggests that 'because so high a proportion of fire casualties in dwellings results from personal incapacity or behavioural factors, there is little scope for additions to the Regulations which could further influence life-safety.  In the case of other occupied buildings, the number of casualties by type of building is too few to enable firm conclusions to be drawn'.  The report draws attention to the high fire protection costs in large buildings to which the public are admitted and suggests they are expensive compared with those for other areas of public safety.

This report confirms that the UK fire grading requirements of the Regulations have been broadly successful in relation to life safety and structural stability.  This is generally true for the other developed countries.  Whether the requirements are rational and economic is a topic which is not really addressed in the report although there are some interesting assessments of the economic benefits to the community of installing active fire protection measures in various sectors.


DEVELOPMENTS IN RESEARCH AND PRACTICE

Research and Regulations

With the background of buildings being relatively successful in surviving fires, research has been devoted more to the behaviour of the contents.  In relation to life safety, the areas of research interest have been domestic dwellings and institutions.  In relation to property safety, the major research interest has been in the contents of industrial and storage buildings.  However, in practical building design the major constraints and the increased expenditure on fire safety, needed to comply with regulations, are incurred in quite different areas.  For life safety, it is the buildings which contain large numbers of the public which are the major concern, while for property, the standards of protection, both active and passive, are being increased in buildings which are not particularly valuable, either in structure or in contents.

Why should this have happened?  One reason is that the building authorities perceive - quite rightly in my view - that the larger the number of people exposed to risk, which means the more serious the consequences of failure, the higher the standard of public safety should be.  However there is no agreed definition of this standard and no assessment of whether the measures required are effective and make the best use of resources.  The fire authorities are only just beginning to admit that such an assessment is desirable but they believe it is not possible to do one.  Research should show that it is possible and how it can be done.

The authorities also perceive that the public can become alarmed by a large property loss, even when there are no deaths or injuries. What is the critical size of loss and how does it vary with sector?

How much should public regulations mandate the safety of private property? Mandatory automatic sprinkler protection is increasing while at the same time the standards of structural fire protection, at least in the UK, are being raised.

The improvement in structural protection is being achieved not by changing the regulations, but by re-writing the rules for the design of elements to achieve a specified grade of fire resistance. Structural engineers protest that buildings designed to the old rules did not fall down in fires, so why is it necessary to increase the standard of protection? The grading should be reduced to maintain the status quo. The dilemma for the authorities is both political - if the required levels of grading are reduced it will appear that safety is being decreased - and technical - what reduction can be accepted and does it apply to all structural materials? If the beneficial effect of sprinklers can be recognised how is it quantified?

A further problem identified is the change in methods of structural design at normal temperatures: if there are smaller margins of safety in modern buildings, then past performance in fire cannot be relied on as a guide to their performance in the future. It is difficult to resolve these problems because although there have been many standard fire resistance tests carried out on single elements of construction, there has been much less study of the ways of assessing levels of safety or fire performance of building structures.

Research and Testing

Standard testing of materials and components is an important part of the procedure needed for compliance with the requirements of the authorities, but too many tests are administrative tools, not design tools. Tests become part of a regulation and are only incidentally fire safety measures. Any test method should at least define what is being measured and there should be an attempt to establish its relevance to practical conditions. The research people should be quite firm about this; if the authorities want a test they should say why it is wanted and what they hope to achieve. If they will not say this, then the research people should not be involved.

It is important that the authorities take into account the costs and benefits of the tests and regulations they impose. However, they should be clear about where their responsibilities end and the commercial market takes over. For example, our UK standard for fire resistance tests is currently undergoing extensive revision, in an attempt to obtain more uniform test results from the various commercial laboratories. The differences in results can be important in commercial terms to the manufacturers and the laboratories but if, as I suspect, they have little practical significance in relation to public safety then why involve the state-funded researchers and administrators? As far as I am aware, the authorities have not even asked the question - do these differences affect public safety?

FIRE ENGINEERING BY DESIGN - SOME PRACTICAL EXAMPLES

Covered Shopping Centres

The covered shopping development, where the old style open street is replaced by a covered pedestrian mall, is familiar around the world. When they began to appear in the UK, the authorities perceived that they would break the rules on compartment size and therefore automatic sprinklers were required in all the shops, as an alternative method of reducing the chance of a fire becoming large. The safety of people escaping was then considered and it was realised that smoke protection of the malls would be needed so that there would be time for the people to reach the open air outside the development. The amount of smoke generated by a fire during the escape phase would determine the amount of smoke venting needed, but this was difficult to quantify. Instead, it was observed from statistics of sprinkler operation that most fires in shops were controlled by very few sprinkler heads, and it was decided that the final fire size was not likely to exceed a 3m x 3m plan area. This size was therefore adopted for smoke design and assuming a heat output of about 0.5 MW/m$^2$, a total size of 5MW was formulated. It is now part of the UK mythology. The Fire Research Station produced design guidance for smoke venting the 9m$^2$, 5 MW fire, in a form which could be used by engineers; as a consequence, with new developments, the negotiations with the authorities are relatively straightforward.

As might be expected, the older shopping centres with open malls have become less popular and many are being refurbished and covered. With the temperate climate of the UK, all that may be required is a fairly basic shelter from the wind and the rain and some simple measures to alleviate the effects of solar gain.

The first such scheme we became involved in was Basildon Town Square. Here, shops border all four sides of a pedestrianized square which measures 200m long by 36m wide. A translucent fabric roof varying in height from 11.5-15m was proposed, to cover the whole Square. The enclosed space would not be conditioned, but would be naturally ventilated to be in balance with the exterior. With the Square being enclosed the authorities immediately asked for sprinklers to be installed in all the shops. They were not concerned about an increased risk of fire spread but about smoke being contained and hampering escape from the Square. Unfortunately the cost of the sprinkler systems and even more important, the costs of compensation for disruption of trading ruled out the sprinkler option.

The architects' view was that the enclosed space was so large that people could escape before smoke from a non-sprinklered fire became a problem. The national experts advised them however, that without sprinklers to control a fire it was not possible to establish the fire size as a basis for determining if the people would be safe. Not being convinced, the architects came to see me, and I suggested that we tried some scenarios, assuming a fire which was growing.

Most experimental studies of fire growth were of single artefacts burning in small rooms. These studies were the wrong scale for our problem. Fire statistics for actual fires in industrial buildings showed an exponential form of spread for fire-damaged area. There were no comparable data available for shops but we thought they were likely to be exponential too. Eventually, by using as a starting point the largest non-sprinklered shop permitted by regulations without smoke venting, we evolved a model which said that the fire area was initially 3m$^2$ and doubled in size every 4 minutes. (Recent analysis of statistics for shops indicates that 4 minutes is conservative).

The convective heat output per unit fire area was taken as 0.5 MW/m² as before and we assumed this all flowed into the Square. The entrainment into the plume and the subsequent smoke layering under the roof were calculated using the Fire Research Station methods, and we were able to estimate a time for the smoke to descend. This time had to be compared with the time taken for people to escape.

People needed time to move from a shop into the Square and then to walk across the Square to an exit. Some of the distances were quite large compared with those in our codes. Our escape codes are based on travel distances and exit widths which will give evacuation within a notional 2½ minutes. A direct distance on plan to an exit from a shop is limited to 30m, and the fire authority will therefore quote a walking speed of 12m/min. This speed actually relates to a crowd of people lining up in a corridor in order to move through an exit. In less crowded conditions people can move more quickly and can certainly cover a greater distance than 30m in 2½ minutes. We allowed for this in assessing the time needed to evacuate the people in the Square, and the people emerging from the surrounding shops. The fire authorities did not like this approach; they were, and are, very reluctant to accept increased travel distances, whatever the circumstances. In the end an extra exit was provided by sacrificing a shop unit. We did not think this extra exit was really needed but we were able to exploit it as a protected access point for the Fire Brigade.

Finally, we had to assess the effect of wind on the efficiency of the natural smoke venting. Mechanical extract was of course quite impractical because of the volumes involved. Wind tunnel tests were carried out to determine the location of the wind sensors and the sequence of operation of the vents.

Basildon Town Square was the first of many refurbishment projects we have worked on. Most of the shopping projects have been smaller than Basildon Town Square. Most of them are also without sprinklers, but for these we have not used the fire growth scenario. Instead, we have adopted an 18m² area 10MW fire for smoke venting design, on the assumption that a higher standard of smoke extract is needed where sprinklers are not installed. Statistics of fires in shops attended by fire brigades indicate that the chance that a fire in a sprinklered shop will exceed 9m² area is about the same as the chance that a fire in a non-sprinklered shop will exceed 18m². This comparison may not be directly relevant to the initial escape period of the fire but at least the risks appear comparable. On escape times we find that for a given population, in most circumstances the controlling factor is the width of the exits from the malls, rather than the distance travelled to reach the exits. We also think that automatic detection systems and a public address system are important safety features, if early evacuation is to be achieved. The smoke extract may be mechanical but more often natural venting is used. Wind tunnel tests may be needed if there are adjacent tall buildings.

We use what research data we can to solve our problems; clearly, a great deal of research information is available for direct use, if only it can be presented in the right way. What is difficult to present is a sense of proportion. It seems wasteful to apply the same standards to all covered shopping centres whatever their size. Some single storey centres are smaller in area than one floor of a department store. Others are multi-storey and hold many people. Hand in hand with our fire and escape scenarios there should be some more explicit relationship between the level of safety and the number of people at risk.

Transport Terminals

In the 19th century, the age of steam, large covered railway stations were built, with just enough vents in them to make them tolerable for the passengers. In the 20th century we build large air terminals, without the steam and without the vents.  These vast enclosures break the usual limits on compartment size in order to function efficiently.  However, they are specifically designed to move people and we ought to be able to exploit this aspect for escape purposes.  Our approach has been to categorise the public areas into two types of use.  These are first, the concourse and waiting areas, with very low fire loads, which are places of relative safety and second the shop and catering areas with significant fire loads, which must be protected selectively. This we can do with automatic sprinklers and powered smoke extract designed to prevent smoke entering the concourse.

For terminals, as for shopping centres, we think the code limits on travel distance to an exit are not realistic.  The assumption during normal working conditions in a terminal is that people move on average at about 1 m/s.  Why should it be slower in fire conditions?  We are obtaining measurements of the walking speed of people with baggage trolleys, as this is likely to be the critical value, and it does appear to be of the order of 1 m/s.

One terminal we are working on is a very large single-storey space, about 300m long by 180m wide by 12m high.  We would prefer people to escape directly to open air through exits on the perimeter, but this would mean their travelling longer distances than normal.  The authorities want people to go down into a tunnel in order to limit the distance to an exit.  We have therefore estimated the rate of smoke generation and movement in this large space and involved one of our environmental physicists to do the calculations using computer programs, in conjunction with the Fire Research Station at Borehamwood.  In large buildings such as this and in large atrium buildings a fire can be considered initially as a local injection of hot air into the building environment.  The effect of building physics becomes an important aspect of fire safety design.

Atriums

We have followed the fashion from the USA, and atriums are now very popular architectural features.  For fire safety reasons, our authorities have ensured that many atriums are rather sterile spaces, with not much more fire load than a few plants, and they are mainly confined to office buildings.  There is much confusion about the fire safety measures which should be adopted.  For acoustic reasons there is usually glazing between the atrium and the surrounding accommodation.  Drenchers may be required on the glazing as well as arbitrary smoke venting provisions.  Some people have adopted the mythical 5 MW fire as a basis for smoke design although there is no obvious reason why offices and covered shopping centres should be treated in the same way.  There is no explicit recognition that the rules should take into account the number of people at risk and the nature of the risk.  The rules that we may well be saddled with are likely to be dominated once again by the needs of the bureaucrats to pass or fail a building, and not by the need to design fire safety for buildings that people want.  We need the intellectual rigour of the researcher to question such rules and provide an alternative rational framework.

Structural Fire Protection

The use of the fire resistance test for the grading of structures is not
often questioned by engineers, although it is perceived as irrational.  They
would like an alternative 'compartment fire', approach to be available for those
occasions where it can be demonstrated that because of the nature and disposition
of the building contents the fire exposure is different from that normally
assumed.  The detailing of structural assemblies is perceived as an important
area which has not been studied sufficiently while the development and acceptance
of calculation methods for fire resistance proceeds rather too slowly.


WHAT THE DESIGNER NEEDS

For many simple, run-of-the-mill buildings, the architect and engineer will
prefer to have simple, if arbitrary, rules.  For more complex buildings, the
designers need to have flexibility if they are to provide a cost-effective
building which meets the needs of the client and the people who use the building.

There are many constraints on the design of a building; fire safety is one
important, but small aspect which cannot be allowed to dominate.  What is needed
is a choice of measures which in combination can achieve a required level of
safety.  Physical limitations may preclude the use of an extra escape route, or
an element of a given fire resistance rating.  What compensatory measures could
be acceptable?  We need a rational framework and agreed numbers.

I believe that we have enough information available to produce the rational
framework.  If we are successful then the practitioners and administrators will
start to ask the right questions - which will benefit us all.


REFERENCES

1.   Fire Grading of Buildings.  Part 1.  Post-war Building Studies No. 20.   The
     Ministry of Works, London, H M Stationery Office, 1946.

2.   Future Fire Policy.  A Consultative Document.  Home Office Scottish Home and
     Health Department, London, H M Stationery Office, 1980.

# Reliability Study on the Lawrence Livermore National Laboratory Water–Supply System

**HARRY K. HASEGAWA** and **HOWARD E. LAMBERT**
Hazards Control Department
Lawrence Livermore National Laboratory
P.O. Box 5505 (L-442)
Livermore, California 94550, USA

## ABSTRACT

We conducted a reliability analysis of the Mocho water–supply system for the Lawrence Livermore National Laboratory (LLNL) to determine if an adequate supply of water would be available in the event of a major fire. We used the digraph fault–tree approach for logic model generation. The initiating–enabling event interval reliability approach was used for a probabilistic evaluation of the Mocho system fault tree. In the event of a major fire, the Mocho system demand unavailability was calculated to be $3.6 \times 10^{-4}$. We identified 16 single–component failures that would cause failure of the control loop which monitors storage–tank level. These failures would go undetected by monitoring personnel at LLNL. Our recommended changes would provide a redundant measurement of the tank level, resulting in a decrease of the predicted system unavailability by about a factor of 50.

## INTRODUCTION

The Fire Science Group at Lawrence Livermore National Laboratory (LLNL) conducted extensive fault–tree analyses of both dry–pipe and wet–pipe sprinkler systems.[1] The fault trees for these sprinkler systems contained one common basic event: No matter how reliable sprinkler systems and fire fighters are, a fire cannot be extinguished without an adequate water supply. To complete the fault–tree analysis, we conducted a reliability study of the Mocho water supply system, LLNL's primary water supply. The Mocho water supply system contained many control loops that maintained the water level in the system storage tanks and also alerted LLNL personnel in the event of low water level in the tanks.

The digraph fault–tree methodology used in this study is particularly useful for fault analysis of control systems.[2] We constructed a fault tree with the Top Event of "Insufficient Supply of Water in Storage Tanks and No Detection of Same in Bldg. 511." The initiating–enabling event interval reliability approach is used to perform a probabilistic evaluation of the fault tree and to compute various systems reliability characteristics, such as the unavailability of water in the event of a major fire.[3]

## MOCHO WATER SUPPLY AND LLNL MONITORING SYSTEM

The main source of water to LLNL is the Hetch Hetchy Aqueduct, located 800 ft below ground at the Mocho pumping station, 8 mi south of LLNL. The water is first pumped

---

This work was performed under the auspices of the U.S. Department of Energy by Lawrence Livermore National Laboratory under contract No. W-7405-ENG-48.

Howard E. Lambert is an independent consultant, Oakland, California.

to the surface and into two standpipes. These standpipes have a capacity of 20,996 gallons each. The water flows by gravity from the standpipes to three main storage tanks located 1/2 mi south of LLNL on the hill above Sandia National Laboratories. The three storage tanks have a total capacity of 1,238,800 gallons and provide the head pressure necessary to supply LLNL. The tanks and standpipes are all at atmospheric pressure. The Central Control Room for the system is located on-site in Bldg. 511.

As an alternate or standby water supply, LLNL has water available from the Zone 7 water district. This water supply is used only during times when Hetch Hetchy water is unavailable due to tunnel maintenance, pump failure at the Mocho Pumping Station, or line failure between the Mocho standpipes and the storage tanks. The Zone 7 water supply must be activated manually on site.

Figure 1 is a simplified schematic of the LLNL water-control system. The system consists basically of two feedback subsystems: (1) the water-level control for the Mocho standpipes, and (2) the water-level control for the storage tanks. Any two of the three pumps at the Mocho pumping station control the water level in the standpipes. Water level in the storage tanks is controlled by opening a valve that causes the Mocho standpipes to drain; gravity feeds water as needed to LLNL from the storage tanks. Alarms, status indicators, and control signals are transmitted via frequency division multiplexed frequency-shift tone equipment. Selector switches, relays, water-level meters, and pilot lights display the data being transmitted and received on a control console in Bldg. 511. Manual commands from the control console can open or close valves at the water-storage tanks, and also start and stop pumps at the Mocho pumping station. The water level in the standpipes and storage tanks is continuously monitored in Bldg. 511. Any abnormal condition, such as a high or low water level in the storage tanks or in the standpipes, or any pump failure, initiates an audible and visible alarm at the control console.

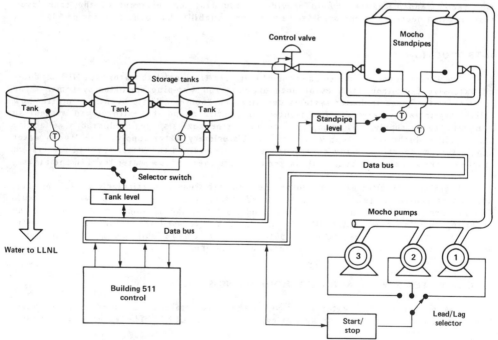

**Figure 1.** Simplified diagram of the Mocho water-supply system.

## ANALYSIS OF THE MOCHO SYSTEM

We first had to understand and model the components of the water–supply control system. Digraph fault–tree analysis uses steps common to traditional fault–tree analysis. In assessing the adequacy of the Mocho water–supply system in the event of a major fire, we found 16 single events that would cause failure of the control loop which monitors storage-tank level. These failures would go undetected in Bldg. 511 because the feedback loop performs both control and detection functions.

We will briefly describe the digraph fault–tree methodology used to arrive at our recommendations.

### Water Required to Fight a Major Fire

Based on firefighting experience in the chemical industry, LLNL's Fire Safety Division defined the amount of water necessary to extinguish a major fire on site: a continuous flow-rate of 3500 gallons per minute (gpm) for 4 hours, for a total of 840,000 gallons. In addition to the 1,283,800–gallon capacity of the three storage tanks, standpipes A and B contain 41,992 total gallons, and 180,180 gallons also sit in the line leading from the standpipes. We must also include the make–up capacity of the two Mocho pumps (1 lead and 1 lag) during this 4–hour period. The lead pump, No. 3, has a capacity of 1100 gpm; each of the lag pumps, No. 1 or 2, has a 500–gpm capacity (but only one pump can operate at a time). In 4 hours, the pumps have a total capacity of 1600 gpm, or 384,000 gallons. Therefore, the total capacity of the entire system plus make–up is 1,889,972 gallons. During 1981, the maximum daily water consumption was 950,000 gallons. Subtracting this amount from the Mocho system capacity leaves 939,972 gallons to fight a major fire, which exceeds the recommended 840,000 gallons. If the system is working, an adequate supply of water will be available for the postulated fire.

In the 15 years the Mocho system has been operating, however, the storage tanks have drained dry twice due to human error. Such errors can lead to an inadequate supply of water in the storage tanks, which no one in Bldg. 511 will detect. Since the total of the pump make–up plus the standpipes and the line leading from the standpipes is only 606,000 gallons, at least an additional 234,000 gallons must be in the storage tanks to meet the 840,000–gallon requirement. Therefore, if the level in the storage tanks drops below 234,000 gallons, there may not be enough water to extinguish the fire.

A normal low level in the tanks will generate an alarm in Bldg. 511, and accordingly LLNL personnel will take the appropriate measures, such as activating Zone 7 supplies.

### Understanding the Water–Supply System

By touring the entire water–supply system and interviewing LLNL personnel familiar with it, we identified the following information germane to the digraph fault–tree analysis.

The independent measurement of water level in the storage tanks was removed, leaving only one sensor for three tanks. Pump No. 3, an 1100–gpm pump, is the lead; a 500–gpm pump is the lag. In a No. 3 failure, either No. 1 or No. 2 becomes the lead pump. With No. 3 out of service, the water makeup to the standpipes takes longer but the system can be successfully operated in this mode. The main water valve will not open if either the stand-pipe is in a low–water–alarm condition or if the water tanks are in a high–water–alarm condition. A spurious signal for condition will cause the main water valve to close, and this results in the storage tanks standing at low level. In the event of a complete failure of the Mocho system, it takes approximately 15 min to cut in the Zone 7 water supply. The existing controls for the Mocho system include over 100 mechanical relays for logic and timing. Understanding the operational sequence of the control system was important to a detailed analysis of its reliability.

**Event sequence for storage–level control**   Water use by LLNL or neighboring Sandia

National Laboratories will cause the level in the storage tanks to drop. When the level in the storage tanks drops to approximately 12 ft–6 in., a water–pressure transducer will cause the automatic water valve to open, and water begins to flow from the Mocho standpipes. This water flows into the top of the No. 2 storage tank that is connected to No. 1 and No. 3 through service valves located at the bottom of the tanks. The automatic valve remains open until a water level of about 14 feet is reached in the storage tanks.

**Event sequence for standpipe–level control** As the automatic valve at the storage tanks open and water begins to flow, the level in the standpipes drops. When the level drops to 12 ft–6 in., the No. 3 pump starts pumping into the top of the No. 1 standpipe, connected by bottom piping to the No. 2 standpipe, from which water flows to the storage tanks. However, water flows through the automatic valve faster than No. 3 can pump, so the water level continues to drop. When the standpipe level reaches 8 ft–8 in., the lag pump automatically starts pumping. The combined output of both lead and lag pump is greater than the amount of water flow through the automatic valve, so the level will now rise again in the standpipe. When the level rises to 12 ft–10 in., the lag pump switches back off. When the automatic water valve at the storage tanks closes, the level in the standpipe rises until it reaches 13 ft–5 in., at which point the lead pump cuts off and everything comes to rest.

## Failure Modes and Effects Analysis

With input from LLNL personnel, we performed a detailed Failure Modes and Effects Analysis, FMEA, on the Mocho system. The results from this study provided much of the information to construct the digraph and the fault tree, and the probabilistic evaluation of the fault tree.

## SYSTEM DIGRAPH

A digraph is a multivalued logic model useful in constructing fault trees of control systems. The digraph consists of both nodes and edges (or arrows) connecting the nodes. A node represents a process variable and an edge represents the gain or the relationship between the nodes. A Top Event is defined as a deviation or disturbance and is the starting variable in the digraph. In this case, the Top Event is "Insufficient Supply of Water in Storage Tanks and No Detection of Same in Bldg. 511." The digraph is then constructed deductively, similar to constructing a fault tree. The limit of resolution in the digraph is equipment failure, human error, or environmental conditions.

The next step is to find the control loops in the digraph. A synthesis algorithm is devised to construct the fault tree from the digraph. Basically, the algorithm delineates how a control loop can cause or pass a disturbance, resulting in an occurrence of the Top Event.

The advantages of constructing a digraph are that the topology of the system variables is displayed and that the digraph resembles the system schematic; in addition, the digraph can consider multivalued logic and timing. By contrast, a fault tree bears no relationship to the schematic, and cannot effectively consider multivalued logic and timing.

### The Preliminary System Digraph

The preliminary system digraph (Fig. 2) shows the structure of the detailed digraph. The Top Event variable is "Flow rate to LLNL." The cycles in the digraph show the two basic feedback loops: (1) the storage–tank–level–control feedback loop, and (2) the standpipe–level–control feedback loop.

The sensed variable in both cases is static pressure and the manipulated variable is flow–rate through the main control valve (for the storage tanks) or through the Mocho pumps (for the standpipes). An arrow from one variable (the independent variable) to the other variable (the dependent variable) indicates that a change in the independent variable causes a change in the dependent variable.

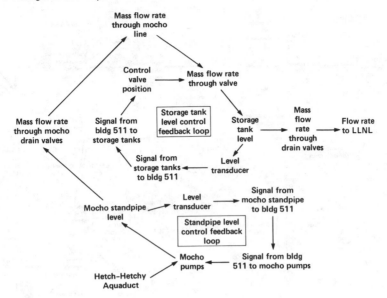

**Figure 2.** Preliminary system digraph.

## The Detailed Digraph

The detailed system digraph is segmented according to sites: the Mocho pumps, the Mocho standpipes, the storage tanks, and Bldg. 511. Each digraph was made using the control system schematic. Because of the magnitude of the detailed system digraphs, only the digraph of the Mocho standpipes is included as an example in Fig. 3.

Events which inactivate control loops as well as the information flow appear with the symbol "0:". In addition, system variables which deviate from their normal values appear in dashed circles. Inactivation events and system disturbances appear as basic events in the fault tree, which is discussed below.

## Fault–Tree Construction

Since the whole fault tree is 10 pages long, only the first page is displayed in Fig. 4. The causes are displayed for the Top Event:
   o  one or more storage–tank drain valves closed and no detection of low tank level;
   o  insufficient flow through control valve and no detection in Bldg 511.
Only once during the history of the system has a drain valve to a storage tank been closed and the selector switch not been changed to measure the water level in tank No. 1. This event caused the system to drain and, since a full tank was being monitored, simultaneously inactivated the control loop that operates the valve. Consequently, the remaining two tanks drained with no detection in Bldg. 511. The event — excessive system demand — has been included in the fault tree for completeness; however, as described earlier, this event has not occurred during the system life and will be excluded from further consideration.

It is important to note that if the storage–tank–level feedback loop is inactive or out of tolerance, then a Top Event will occur. The feedback loop is used simultaneously to control level in the storage tanks, and also to send a signal to Bldg. 511 in the event of a low level. If the control valve is open, and the loop fails out of tolerance, then the control

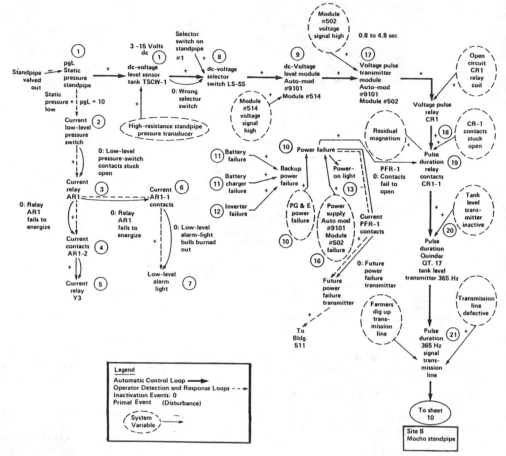

**Figure 3.** Detailed digraph of Mocho standpipes.

loop will command the valve to close for a longer period of time than desired, resulting in a low level in the storage tank. If the loop is inactive, the control valve will fail to open when it should, again resulting in a low level in the storage tanks.

A "close all valves" signal can result in the control valve being closed for too long. Two things can cause this: a spurious high storage–tank–level signal, or a low standpipe–level signal (which can be a spurious signal, or due to the feedback loop being inactive or out of tolerance, or due to failures that can cause an actual low standpipe level).

Another cause of a low tank level is an insufficient supply from the Mocho pumps. Note that we do not consider the case of drained Mocho standpipes. This is because a low standpipe level will cause the control valve to close, resulting in a low level in the storage tanks (which means that the fault–tree logic for draining the standpipes will generate non–minimal cut sets, since additional failures must occur to drain the standpipes). However, we do consider draining the pipe leading away from the standpipes, which can occur simply by closing the drain valve in standpipe No. 2. Since the standpipe level is constant in this case, the valve will continue to remain open until the pipe is drained. Note that throughout

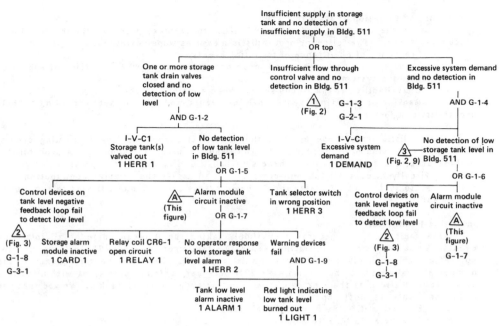

**Figure 4.** Fault tree of Mocho standpipes.

the fault tree, when a detection loop fails, an AND gate is generated. This is the result of the feed–forward operator described in Lapp and Powers.[2,3]

## QUALITATIVE FAULT–TREE EVALUATION

The fault tree contained 98 basic events, and an additional 640 minimal (min–cut) sets. Min–cut sets, also known as the system failure modes, are combinations of basic events which cause the Top Event to occur. The number of min–cut sets according to order is given below (order refers to the number of basic events in a min–cut set):

| ORDER | 1 | 2 | 3 | 4 | 5 | 6 | 7 |
|---|---|---|---|---|---|---|---|
| MIN–CUT SETS | 16 | 19 | 134 | 244 | 163 | 56 | 8 |

The 16 single–event min–cut sets are single failures of control devices on the storage-tank–level feedback loop as described earlier. Of the 98 total basic events, 72 basic events are initiating events, and 26 are enabling events. As described in Dunglinson et al., enabling events inactivate the system's mitigative or protective features but do not cause the Top Event to occur.[4] For example, an inactive alarm or burned–out light does not cause a low level but does fail the operator in the event of an alarm condition. For initiating events, we must compute the frequency of occurrence; for enabling events, we compute the demand unavailability when the initiating event occurs. Reliability data for the 98 basic events were obtained from actual system experience during the last 15 years.

For initiating events, we will assume a fault–duration time of 1 h. This includes the amount of time required for Bldg. 511 personnel to detect and diagnose the cause of failure and to take action after a failure of the Mocho system has occurred. For the Top Event to occur, we will assume no detection in Bldg. 511, although low water pressure will be detect-ed on site. The fault–duration time for enabling events, 0.13 years, corresponds to the average time a failure can exist with an inspection interval of 3 months.

617

## Probabilistic Evaluation of the Fault Tree

We used the computer code IMPORTANCE to evaluate probabilistically the Mocho System fault tree.[5] The following probabilistic measures were computed:

o    Frequency of occurrence of initiating events,

o    Frequency of occurrence and mean occurrence time when insufficient supply of water exists in the system,

o    Unavailability of the Mocho system when a major fire occurs, and

o    Ranking of initiating events, enabling events, and min–cut sets according to their probabilistic importance.

Probabilistic importance assesses the quantitative contribution of enabling events, initiating events, and min–cut sets to the Top Event occurrence frequency. A probabilistic ranking according to importance is necessary because it is virtually impossible for an analyst to visually inspect all the min–cut sets and to assess the relative contribution of a component to system failure (viz. the 98 basic events and 640 min–cut sets in the Mocho system fault tree).

## Initiating–Event Fault Tree

One fault tree was generated by simply taking the Boolean union of all initiating events. Since there were 72 initiating events, this fault tree generates 72 single–event min–cut sets. Table 1 displays the results of this fault tree. The Top–Event frequency — the number of challenges to the system — is 20.4 per year, which is consistent with historical data. Table 1 also lists the ranking of the initiating events through rank 9. We see that the following events are most important:

o    Electric utility (PG&E) power failure.

o    Oiler relay failure (to Mocho pumps).

o    Noise on transmission line.

We assume that the dominant failure cause of the "close all valves" transmitter is noise on the transmission line.

## RESULTS OF THE MOCHO SYSTEM FAULT–TREE ANALYSIS

Table 2 lists the system–reliability characteristics of the Mocho system fault tree. We predict that the Mocho system will have an inadequate supply of water to extinguish a major fire on the average of 3.1 times per year. It must be pointed out that this number corresponds only to a low level in the storage and not necessarily to a totally dry condition. Another measure of system adequacy is the demand unavailability of the system in the event of a major fire. The demand unavailability is calculated to be $3.6 \times 10^{-4}$. Comparing this number to historical data, we see that during the 15–year life of the system, it has been dry twice. If we assume that the system was unavailable for 4 hours each time, this results in a system unavailability of (8 hours x 1 year/8760 hours)(system lifetime/15 years) = $6.1 \times 10^{-5}$ year.

Since this number does not include other times in which the storage tanks might have been low instead of dry, we see that the calculated unavailability, $3.6 \times 10^{-4}$, agrees reasonably well with the historical data, $6.1 \times 10^{-5}$. Table 2 also ranks the initiating events through rank 4, which includes the basic events that are single–event min–cut sets. The one exception is PG&E failure, which requires failure of the backup–power supply system as well. These single events are failure of control devices on the tank–level feedback loop, which would go undetected in Bldg. 511.

Table 3 ranks the enabling events. We see that failure of components in the backup power supply system (the inverter, battery, and battery charger) are dominant enabling events. The next most important enabling event is failure of the operator in Bldg. 511 to respond to a low–level tank alarm. We assign a probability of 0.01 for this event, which is consistent with data given by Swain and Guttman.[6] Table 3 also ranks the most important min–cut sets. These min–cut sets include the important initiating and enabling events described above.

Table 1    Results from the initiating-event fault tree. The mission time is 15 yrs.

| Rank | Basic event description | Importance value | Failure rate per year | Mean fault duration (hr) |
|---|---|---|---|---|
| 1 | I-V-C1 PG&E power failure | 0.232 | 5.00 | 1.00 |
| 2 | I-P-A2 Oiler failure relay PCW #1 | 0.155 | 3.33 | 1.00 |
| 2 | I-P-A1 Oiler failure PCW #3 | 0.155 | 3.33 | 1.00 |
| 3 | I-V-D1 Close all valve transmitter failure on | 0.464 E-01 | 1.00 | 1.00 |
| 3 | I-V-D2 Noise on line to control valve transmitter | 0.464 E-01 | 1.00 | 1.00 |
| 3 | I-P-D1 Noise on line from standpipe to Bldg. 511 | 0.464 E-01 | 1.00 | 1.00 |
| 4 | I-P-A2 Control power contacts R3-1 transfer open | 0.218 E-01 | 0.47 | 1.00 |
| 4 | I-P-A1 Control power contacts R9-1 transfer open | 0.218 E-01 | 0.47 | 1.00 |
| 5 | I-V-C1 High resistance WLT #1 transducer | 0.186 E-01 | 0.40 | 1.00 |
| 5 | I-P-B1 High resistance standpipe pressure transducer | 0.186 E-01 | 0.40 | 1.00 |
| 6 | I-P-B1 Farmers dig up transmission line | 0.155 E-01 | 0.333 | 1.00 |
| 7 | I-P-B1 Power supply Auto-Mod #9109 failure off | 0.124 E-01 | 0.267 | 1.00 |
| 8 | I-V-C1 Tank level module voltage high | 0.617 E-02 | 0.133 | 1.00 |
| 8 | I-V-D2 Valve module contacts open | 0.617 E-02 | 0.133 | 1.00 |
| 8 | I-P-D1 Pump switch module voltage high | 0.617 E-02 | 0.133 | 1.00 |
| 8 | I-P-B1 Module #514 voltage signal high | 0.617 E-02 | 0.133 | 1.00 |
| 8 | I-P-B1 Module #502 voltage signal high | 0.617 E-02 | 0.133 | 1.00 |
| 8 | I-P-A1 PCW #3 motor fails to function | 0.617 E-02 | 0.133 | 1.00 |
| 8 | I-V-D2 TCW S-tank level rec card voltage high | 0.617 E-02 | 0.133 | 1.00 |
| 8 | I-V-D1 Standpipe level module failure (on) | 0.617 E-02 | 0.133 | 0.13 yr |
| 8 | I-V-D1 Tank level alarm module failure (on) | 0.617 E-02 | 0.133 | 0.13 yr |
| 9 | I-P-A2 Pressure switch out of tolerance | 0.311 E-02 | 0.670 E-01 | 0.13 yr |
| 9 | I-P-A1 Pressure switch out of tolerance | 0.311 E-02 | 0.670 E-01 | 0.13 yr |
| 9 | I-V-C2 2400-Hz receiver fails low | 0.311 E-02 | 0.670 E-01 | 1.00 |
| 9 | I-V-C2 Valve stem failure | 0.311 E-02 | 0.670 E-01 | 1.00 |
| 9 | I-V-C1 Residual magnetism relay CR1 | 0.311 E-02 | 0.670 E-01 | 1.00 |
| 9 | I-V-D2 Relay coil CR7 open circuit | 0.311 E-02 | 0.670 E-01 | 1.00 |
| 9 | I-V-C1 Open circuit relay CR1 | 0.311 E-02 | 0.670 E-01 | 1.00 |
| 9 | I-V-D1 CR5-3 contacts close | 0.311 E-02 | 0.670 E-01 | 1.00 |
| 9 | I-P-D1 TSCW level receiver voltage high | 0.311 E-02 | 0.670 E-01 | 1.00 |
| 9 | I-P-A1 Lead pump receiver failure off | 0.311 E-02 | 0.670 E-01 | 1.00 |
| 9 | I-V-C1 Level transmitter fails low | 0.311 E-02 | 0.670 E-01 | 1.00 |
| 9 | I-P-D1 CR3-2 contacts transfer open | 0.311 E-02 | 0.670 E-01 | 1.00 |
| 9 | I-P-D1 CR1A-1 contacts transfer open | 0.311 E-02 | 0.670 E-01 | 1.00 |
| 9 | I-V-C2 Agastat TDR-4 contacts transfer open | 0.311 E-02 | 0.670 E-01 | 1.00 |
| 9 | I-V-D2 Flow valve selector switch opens | 0.311 E-02 | 0.670 E-01 | 1.00 |
| 9 | I-P-A1 PSR6 contacts transfer open | 0.311 E-02 | 0.670 E-01 | 1.00 |
| 9 | I-V-D1 CR2A-1 contacts close | 0.311 E-02 | 0.670-01 | 1.00 |

Table 2.  Ranking of most important initiating events for the Top Event: "Insufficient level in storage tank and no detection."  Mean time to system failure = 2847.6 h (0.325 year).

| Rank | Basic Event Description | Importance Value | Failure Rate per Year |
|---|---|---|---|
| 1 | I-V-D2 Noise on line to control valve transmitter | 0.325 | 1.00 |
| 2 | I-V-C1 PG&E power failure | 0.163 | 5.00 |
| 3 | I-V-C1 High resistance to WLT #1 transducer | 0.130 | 0.400 |
| 4 | I-V-D2 Valve module contacts open | 0.0432 | 0.133 |
| 4 | I-V-C1 Tank-level module voltage high | 0.0432 | 0.133 |
| 4 | I-V-D2 TCW storage tank level rec card voltage high | 0.0432 | 0.133 |

**Table 3.** Ranking of most important enabling events for the Top Event.

| Rank | Basic Event Description | Importance Value | Failure Rate per Year | Mean Fault Duration (h) |
|------|-------------------------|------------------|-----------------------|-------------------------|
| 1 | Inverter failure, storage tank | 0.564 E-01 | 0.267 | 0.130 |
| 1 | Battery failure at storage tank | 0.564 E-01 | 0.267 | 0.130 |
| 1 | Battery charger failure, storage tank | 0.564 E-01 | 0.267 | 0.130 |
| 2 | No operator response to low tank level | 0.433 E-02 | 0.100 E-01* | -- |
| 3 | Relay coil CR6-1 open circuit | 0.377 E-02 | 0.670 E-01 | 0.130 |
| 3 | Storage alarm module inactive | 0.377 E-02 | 0.670 E-01 | 0.130 |
| 4 | No operator response to low standpipe level | 0.557 E-03 | 0.100 E-01* | -- |

*Represents a demand probability.

Based on our analysis, we recommend that each storage tank in the Mocho water-supply system has its own independent water-level sensor. If this is done, we estimate that the probability of system unavailability would decrease by a factor of 50. Restoring this independent measurement would result in no single-event minimal cut sets in which failures would go undetected.

In summary, we feel that the use of the digraph fault-tree methodology enabled us to perform a very detailed, complete, and accurate analysis of the water-supply system. The results seem reasonable because the failure-rate data was derived directly from the system's 15-year operating experience. This general methodology is ideal for the analysis of large and complex control systems.

## REFERENCES

1. H. K. Hasegawa, N. J. Alvares, A. E. Lipska, H. W. Ford, S. Priante, and D. G. Beason, "Fire Protection Research for Energy Technology Projects: FY 80 Year-End Report," Lawrence Livermore National Laboratory, Livermore, California, UCRL-53179 (1981).

2. S. A. Lapp and G. J. Powers, "Computer-Aided Synthesis of Fault-Trees," IEEE Trans. on Reliability, R-26 (April 1977, pp. 2-13.

3. S. A. Lapp and G. J. Powers, "The Synthesis of Fault Trees," in Nuclear Systems Reliability Engineering and Risk Assessment, J. B. Fussell and G. R. Burdick, eds., SIAM (1977).

4. C. Dunglinson and H. E. Lambert, "Interval Reliability for Initiating and Enabling Events," IEEE Transactions on Reliability, R-32, 2 (1983).

5. H. E. Lambert, "IMPORTANCE: The IMPORTANCE Computer Code," Lawrence Livermore National Laboratory, Livermore, California, UCRL-79529 (1977).

6. Swain and Guttman, "Handbook of Human Reliability Analysis with Emphasis on Nuclear Power Plant Applications," Nuclear Regulatory Commission, Washington, D.C., NUREG-1278 (1983).

# Decision Analysis for Risk Management of Firesafety Hazards

**FRANK NOONAN**
Worcester Polytechnic Institute
Worcester, Massachusetts 01609, USA

ABSTRACT

Applications of decision analysis in fire protection are reviewed for the purpose of trying to glean collectively from previous
studies a design framework for a general purpose decision support
system for risk management in fire protection.

I.    INTRODUCTION

This paper examines the use of decision analysis in the practice
of risk management for fire protection within business environments.
Risk management, for fire protection, concerns the management of a
firm's exposure to losses from fire where solution strategies cover
loss prevention, loss transfer and loss retention.  As a process
risk management has been depicted as involving five steps:

1.    identification of risks
2.    assessment/measurement of risks
3.    generation and evaluation of alternative
      solutions to risk
4.    application of the preferred solution
5.    monitoring performance of the solution

Here, only the second and third steps, risk assessment and solution
evaluation are of concern.

Decision analysis is a discipline for systematic evaluation of
alternative actions as a basis for choosing among them.  Decision
analysis models often include probabilities that quantify judgements
about uncertain future events as well as a utility function which
expresses the decision maker's attitudes, or the organization's
policies, as regards the assumption of risk (1).  The description
suggest a strong overlap between the decision analysis discipline
and steps 2 and 3 in the risk management process.  This paper reviews
the literature for applications of decision analysis in fire protection
with a view of trying to collectively glean from previous studies a
design framework for a general purpose decision support system (DSS)
for risk management in fire protection.

As a final introduction note, it is appropriate to delineate various categories of solution alternatives for fire protection and, also, the general categories of risk or loss that a business operation is exposed due to fire hazards. Loss prevention alternatives can include variations in building design, installing fire walls and automatic detection and suppression systems. Other solutions exist through contingency planning for reducing business interruption risk. There is also loss transfer (i.e. insurance) where a reasonable set of alternatives will examine various lower and upper limits on the amount of loss to be transferred. In measuring risk it is first necessary to estimate the certain or deterministic cost components associated with any solution strategy. These will include one time investment costs and costs which occur periodically over the planning horizon of interest. Second, it is necessary to have procedure for assessing the probabilistic components of loss. The latter, not only includes the direct losses associated with fire damage but, also, business interruption (BI) loss. BI loss can be further decomposed into insurable BI loss and non-insurable (but nonetheless real) market opportunity losses.

## II. LITERATURE REVIEW

In (2), Shpilberg and DeNeufville present an excellent decision analysis for evaluating investments into fire protection equipment and fire resistive construction vs. the procurement of insurance for both large and small airport facilities. Rather than simply evaluate alternatives in terms of expected loss, exponential utility theory is used and alternatives are compared on the basis of risk adjusted loss. A probability loss function is constructed for each alternative by using historical data on losses from airport fires. However, in using this procedure for risk assessment the authors recognize two significant limitations. First, the historical data base did not include business interruption losses, a fact which may explain the preference (within their study) for insurance over loss prevention investments. Second, they note that the lack of adequate loss statistics preclude examination of a larger, more reasonable, set of loss prevention alternatives.

There are two basic approaches to risk assessment; proability models can be based on subjective estimates or, as in the case described above, they can be estimated from an empirical data base. When the assessment process is subjective, uncertain events are systematically decomposed into conditional events such that expert judgement can be effectively used in assigning event probabilities (e.g. event or probability trees). In fire protection studies, the practice of risk assessment tends more often to be empirically based. Shpilberg and DeNeufville's study points out that this practice will restrict the number of risk reduction strategies (particularly innovative strategies) that can be evaluated due to the lack of historical statistics reflecting the causal impact of each solution strategy.

Cozzolino (3) has developed general formulations for risk
management in fire protection.  He considers loss prevention, loss
transfer and loss retention alternatives all in one integrated
framework.  Like (2), he applies exponential utility theory and
uses risk adjusted loss as the performance evaluation measure.
However, Cozzolino's work is general and does not really address
specific procedure for risk assessment.

Halpern (4) presents an application of decision analysis
for choosing between two ways of improving fire protection in a
community; he considers the installation of detection alarm systems
in dwellings vs. the addition of more fire companies in the
community.  While Halpern does not apply utility theory and only
uses expected loss to evaluate alternatives, his paper is of
interest for the procedures used in the assessment of risk.  Halpern
does use historical data for estimating the probability of loss
from a fire.  However, when statistics were not available to fully
reflect the causal impact of solution strategies, the author forms
a hybrid model by merging rational and subjective based models
with the empirical results on fire loss.

In (5) Fitzgerald describes the Engineering Method, a general
methodology for modeling smoke production and flame movement across
spaces and barriers.  The Engineering Method uses probability
trees to address site specific characteristics with respect to:

- fire growth hazard potential
- barriers
- detection
- automatic suppression
- manual suppression
- distribution of value

and generates a probability function on the percent of physical
loss.  Fitzgerald's work differs from the previous references in
that it describes a general purpose tool focused on the task of
risk assessment.  Its approach to risk assessment is also different
in that it systematically decomposes uncertain events into
conditional events which require subjective assessments from the
user.  In its current stage of development the Engineering Method
has a number of limitations.  It does not address business interruption loss and it doesn't consider insurance or contingency
planning strategies.  In its current form, the user interface
may involve too many inputs and, also, in its current development
there are no built in tools/resources to assist the user in
making subjective assessments (e.g. availability of empirical
statistics or Bayes Rule methodology).  However, the Engineering
Method does offer a general purpose tool and is certain to play
a role in designing any general DSS for risk management in fire
protection.

In (6) Helzer et al apply decision analysis to evaluate strategies for reducing upholstered furniture fire losses. Like the Engineering Method, the study utilizes probability trees and a subjective framework for risk assessment.

## III. DSS DESIGN GUIDELINES

In industry, decision analysis methodology, such as utility theory and subjective probability assessment, is not applied to risk management for fire protection. Risk assessment seems limited to what can be measured from historical results and the selection of a risk management strategy for a given facility is limited to the use of simple, one dimensional, decision rules. For example, "invest in loss prevention only if required by building code or insurance" or "establish contingency plans only if the maximum business interruption loss from a fire incident can exceed $X". While this type of decision making is attractive for its simplicity and ease of execution one has to question its overall cost-effectiveness and wonder whether the practice of risk management for fire protection is not indeed ready for utilizing a DSS which can be comprehensive, situation specific and, yet, reasonable in terms of implementation effort. Assuming this question warrants a positive response, a set of design guidelines for developing a DSS is listed below.

A.  Evaluation of Risk
    The evaluation module should formally recognize the existence of risk aversion by evaluating alternatives according to risk adjusted loss rather than expected loss; failure to incorporate risk aversion will lead to under-expenditure for loss prevention and loss transfer. As previously noted, (2) and (3) describe the use of exponential utility for computing risk adjusted loss; (7) provides an excellent review on how to assess a firm's utility function.

B.  Insurance Pricing Module
    An insurance module is necessary which defines insurance prices and risk reduction constraints, all as a function of any probability loss model on physical and business interruption loss.

C.  Risk Assessment
    If the DSS is to be both comprehensive and application specific, then the risk assessment process should fully utilize rational models and subjective probability assessment. Loss categories should include physical loss and both insurable and non-insurable business interruption loss. Significant correlation can exist between these categories and the assessment process must consider this coupling. The loss models must be generic with perhaps separate sub-modules for manufacturing, administrative and warehousing facilities.

When decision makers are required to assess values for specific
variables and conditional event probabilities, an expert
systems framework should exist where historical statistics
and Bayes Rule methodology is available to aid in making assessments. Expert system sub-modules should probably be available
on combustibility data and on cost and effectiveness data for
both manual and automatic, detection systems and suppression
systems. The human interface for assessing risk must be manage-
able. While subjective techniques are essential to a general
purpose DSS, the interface should not require hundreds of
inputs from the user. Sensitivity analysis can be employed
within the system design, in order to optimize the decomposition
of uncertain events and thereby keep assessment requirements
to a minimum.

## REFERENCES

(1) Brown, Rex V., Andres S. Kahr and Cameron Peterson.
Decision Analysis for the Manager. New York: Holt, Rinehart
and Winston, Inc., 1974.

(2) Shpilberg, David and Richard DeNeufville, "Use of Decision
Analysis for Optimizing Choice of Fire Protection and Insurance:
An Airport Study", Journal of Risk and Insurance, 1975,
pp. 133-149.

(3) Cozzolino, John M., "A Method for the Evaluation of Retained
Risk," Journal of Risk and Insurance, 1978, pp. 449-471.

(4) Halpern, Jonathan, "Fire Loss Reduction: Fire Detectors vs.
Fire Stations", Management Science, Vol. 25, No. 11, 1979.

(5) Fitzgerald, R.W., "Risk Analysis Using the Engineering Method
for Building Firesafety", Center for Firesafety Studies,
Worcester Polytechnic Institute, 1985.

(6) Helzer, S.G. et al, "Decision Analysis of Strategies for
Reducing Upholstered Furniture Fire Losses", NBS Technical
Note 1101, U.S. Department of Commerce, National Bureau of
Standards, 1979.

(7) Farquhar, P.H., "Utility Assessment Methods", Management
Science, Vol. 30, No. 11, 1984.

# Evaluation of the Risk Problem and the Selection of the Optimum Risk Management Solution

CRAIG VAN ANNE
The Hartford Steam Boiler Inspection and Insurance Company

## ABSTRACT

The determination of optimum fire protection engineering solutions at present is a predominantly subjective based process. An analytical technique is offered to evaluate risk and the manager's aversion toward risk to better employ the risk management options of loss transfer, loss absorption, and loss reduction.

An engineering method presently exists for numerically evaluating a relative level of risk in any building. The flame movement part of the method involves the determination of the probability of success in terminating a fire within each space of a building. The effectiveness of each barrier surrounding the space, also can be evaluated whether it be a floor, ceiling, wall, or an empty void.

An illustrative case study has been offered to demonstrate the incorporation of the engineering method into a decision analysis of possible alternatives based on the risk attitude of the decision maker. This case study illustrates how the three risk management options can be considered rationally and quantitatively.

## THE ENGINEERING METHOD AND RISK MANAGEMENT

There exists today a detailed engineering method (1) which, regardless of size or occupancy, can evaluate in a consistent manner any building. Through the application of this engineering method, a COPE (*) evaluation is performed which yields relative assessments of risk. Once the fire risk of a particular building is quantified, the appropriate parties can analyze possible risk management solutions. The evaluation of these components yields a probability value in the form of an L-curve. The L-curve is a graphical representaion of the cumulative probability (from 0.0 to 1.0) that a fire will be limited to the space being considered. Limited, in this sense, means that the fire will not propagate beyond the area which has been evaluated. Figure 1 shows typical L-curves.

This paper will concentrate its use of the engineering method on the Flame Movement Analysis. We wish to evaluate the likelihood that a fire will be limited to an area of a building. This likelihood is based on four engineering method parameters which evaluate COPE: an evaluation of the hazard present --the I curve; the automatic suppression system--the A-curve; manual fire fighting--the M-curve, and barrier effectiveness; either physical or spatial.

(*) COPE is acronym for the class of construction, occupancy, installed fire protection systems, and external exposures of a building.

An evaluation of a barrier is dependent on the specific type of fire that will attack the barrier, and of course barrier construction. A barrier can either be physical, or a space separation. A fire must penetrate the barrier to cause ignition in the next space.

RISK MANAGEMENT APPLICATIONS

By simply inserting a zero value for any L-curve component (I-, A-, M-,& B) into a network calculation, the effect that that component has in the L-curve can be negated. The ease of this "what if" situation capability is ideal for an effective comparison of various loss prevention alternatives and the resources required to implement them.

The extent of loss under different risk management situations is of primary importance if risk decisions are to be made. Loss estimates of the various risk management parameters can be easily assessed with the aid of the final product of the engineering method: the L-curve. This paper will address NLE and MFL risk management parameters using the L-curve..

NLE and MFL CURVES

The risk management parameters commonly used are Normal Loss expectancy (NLE) and Maximum Foreseeable Loss (MFL). Though these terms are used widely by those professionals dealing with risk management, definitions of each parameter are not widely consistent. Therefore, definitions of the risk management parameters as used in this paper will be given.

NLE Curve

The NLE is an evaluation of the normal loss to be expected when automatic suppression systems and public or private fire fighting assistance is present and fully functional. An engineering method generated NLE curve would then simply be the unaltered L-curve. In this situation the A-curve, M-curve, I-curve, and barrier performance could all have values greater than zero.

MFL Curve

The absolute worst risk management parameter is described by an MFL estimate. This estimate considers all automatic fire protection systems are impaired, and no manual fire fighting assistance is received. Any fire spread limitation offered by interior fire separation walls is completely ignored unless a MFL(*) fire wall is present. The engineering method MFL curve only considers the fire protection implications of loss due to the fire peril presented.

The engineering method generates an MFL-curve by assigning zero values to the A-curve, M-curve and barrier performance. Should an MFL wall be present, the values then assigned to Barrier Performance would be 1.0. Therfore, a network calculated MFL-curve can take only two shapes: that of the I-curve or that of the accumulated I-curve and barrier performance values. Figure 1 summarizes the risk management curves just discussed.

(*) MFL fire wall as defined by Loss Prevention Data Sheet 1-22, Factory Mutual Engineering Corporation.

BASIC CURVES

| SURVEY CURVE | A** | M | I | B | GENERAL L-CURVE SHAPE |
|---|---|---|---|---|---|
| NLE | >0 | >0 | >0 | >0 | 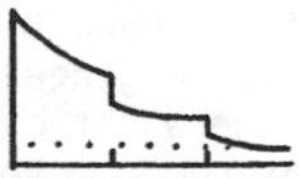 |
| MFL | 0 | 0 | >0 | 0 | 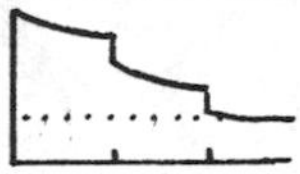 |
| MFL W/BARRIER (MFL WALL) | 0 | 0 | >0 | 1.0 | 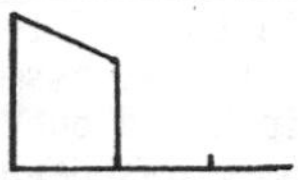 |

** BEYOND ROOM OF ORIGIN, AUTOMATIC SUPPRESSION ASSUMED
   INEFFECTIVE: A-CURVE = 0.0

Figure 1 NLE, MFL
RISK MANAGEMENT CURVES

DECISION ANALYSIS AND THE ATTRACTIVE ALTERNATIVE

Engineering method generated probabilities can be used together in a decision analysis model to determine what course of action to take for a simple "two room" building. In conjunction with this model, a major parameter in choosing an alternative is the decision maker's risk attitude.

The rooms in this example could be of any size that would meet the objectives of the analysis. Figure 2 is a simple plan of this example building. The objective of this illustration is to show the overall process of applying decision analysis using probabilities generated by the engineering method. Therefore, the details of the construction, occupancy and protection of this two room building are not imperative for illustrative purposes. The

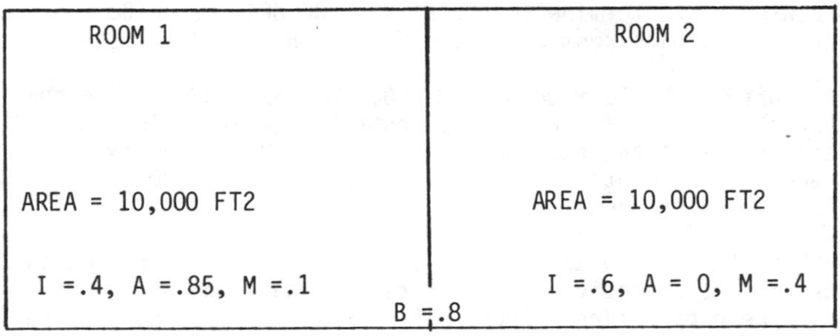

| ROOM 1 | ROOM 2 |
|---|---|
| AREA = 10,000 FT2 | AREA = 10,000 FT2 |
| I =.4, A =.85, M =.1 | I =.6, A = 0, M =.4 |

B =.8

Figure 2 Case Study Floor Plan

same analytic process used in this simplistic approach would be used in the more complicated risks found in actual practice. Room 2 of Figure 2 has been identified as the area of concern based on the highest loss potential in dollars of value. A fire is assumed to start in Room 1 and proceed from left to right. If the fire penetrates the barrier, it must propagate to Room 2. This is a simple condition for the engineering method. In more complicated situations, the engineering method's computer program (2) is capable of identifying the worst case scenario fire which would pose the greatest threat to any room or area of interest in the building.

## FIELD SURVEY

An evaluation of this building has provided "engineering method data" which yields the assessments of Table 1. This table represents the worst case scenario based on data supplied by the evaluating engineer. In this example, Room 1 has been identified as the ROOM OF ORIGIN because it has the lowest probability of a fire self-extinguishing: $P(I) = 0.4$. The computer program developed for the engineering method can easily determine the worst ROOM OF ORIGIN, or be instructed to accept a particular room of origin for a "what-if" analysis.

| ASSESSED SPACE | SURVEY ASSESSMENTS | | | | ASSESSED $ | % of BLDG. VALUE |
|---|---|---|---|---|---|---|
| | I | A | M | B | | |
| * ROOM 1 | .4 | .85 | .1 | .2 | 6.71M | 34.4% |
| ** ROOM 2 | .6 | .0 | .4 | .8 | 12.81M | 65.6% |
| | | | | TOTAL: | 19.52M | 100% |

*ROOM OF ORIGIN    **ROOM BEYOND

Table 1 SURVEY DATA

This table represents the worst case scenario based on data supplied by the evaluating engineer. In this example, Room 1 has been identified as the ROOM OF ORIGIN because it has the lowest probability of a fire self-extinguishing: $P(I) = 0.4$. The computer program developed for the engineering method can easily determine the worst ROOM OF ORIGIN, or be instructed to accept a particular room of origin for a "what-if" analysis.

Room 2 is a ROOM BEYOND THE ROOM OF ORIGIN (ROOM BEYOND) and is also an object of concern for this evaluation as it reprsents the greatest loss potential. Room 2 represents 66% of the exposed value of this building; anything from a warehousing area to a computer room. As above, a specific ROOM BEYOND can be assigned in the computer program.

The manner in which the survey assessments of Table 1 were determined is not important to the present objective. This assessment would have used a normal application of the engineering method. Therefore, in order to provide an illustration of the decision analysis process, the actual details of construction, occupancy, protection and exposure used in decomposed probability assessments (3) can be assumed to have been performed by the engineering analysis in order to provide an illustration of the decision analysis process.

The probability of limiting a fire at individual rooms is obtained from network diagrams of the engineering method. With this probability we can then determine the probability of limiting the fire to "Z" sq.ft. in space X, given the fire was <u>not</u> limited to "Y" sq. ft. in space X, where Z is greater than Y:

P(L1 @ Z / <u>L</u>1 @ Y)     Where <u>L</u> = not limited

For example,

$$P(L1 @ 3,000/NLE FIRE) = .919$$

$$P(L1 @ 6,000/\underline{L}1 @ 3000) = 8.1 \times 10\text{-}6 = 0$$

are obtained from an evaluation of Room 1 using the engineering method for the segmented floor areas. Likewise, this same evaluation would yield sector probabilites for Room 2. Sector areas are shown in Figure 3.

The seemingly unwarranted number of significant figures are included in Figure 3 to illustrate a conservative assumption. If a fire is limited to Room 1, the vast majority of that probability of being limited would fall within the 3000 sq. ft. floor area. Therefore, the succeeding sector probabilities would be much smaller; approaching zero at the overall room area.

| EVENT SECTOR | P(L) |
|---|---|
| P(L1 @ 3,000/NLE FIRE) | .9189919 |
| P(L1 @ 6,000/L1 @ 3,000) | $8.1 \times 10$ -6 |
| P(L1 @ / L1 @ 6,000) | 0.0 |
| P(<u>L</u>1 @ / <u>LI</u> @ 6,000) | .0810 |
| | SUM = 1.0 |

Figure 3 SECTOR PROBABILITIES

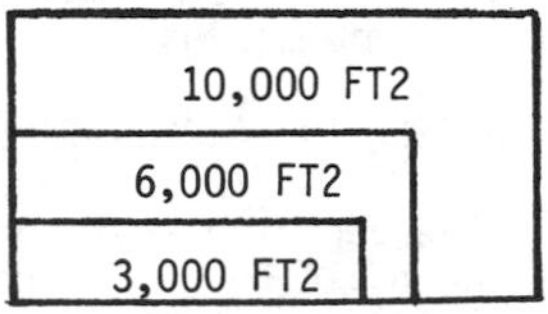

```
+---------------------------+
|                           |
|      10,000 FT2           |
|                   +-------+
|    6,000 FT2      |
|                   |
+-----------+-------+
| 3,000 FT2 |
+-----------+
```

SECTORS

DECISION MAKING AND THE ATTRACTIVE ALTERNATIVE

With the knowledge of the probabilities in Figure 3, a decision can now be made on which fire protection engineering alternative to implement. It can be shown that in every decision, there are three types of attitudes toward risk:

1) risk aversion
2) risk neutrality
3) risk seeking

When a decision involves uncertainty, a rational decision maker is not risk neutral. A decision maker in a business environment, again, if rational; is not risk seeking and therefore is risk averse. Depending upon corporate attitude toward risk, market conditions, and the size of the company, different degrees of risk aversion at different times will be evident. The risk attitude of the decision maker can be described by a risk constant, r, which enables a normalizing of all risk management alternatives.

Once all alternatives have been normalized, the most attractive alternative is identified based on maximizing gain or minimizing loss. The risk constant reflects the risk attitude of the decision maker. As the risk constant changes, the degree of the risk attitude exponentially changes. If r>0, the decision maker is risk averse; if r=0, risk neutral; and if r<0, risk seeking.

Weighing is done using the following exponential utility relationship to determine the certainty equivalent (CE) of an alternative:

(EQ. 1)          $CE = -1/r \ln[EV(\exp -rz)]$

where EV is the individual expected values of each alternative. Alternatives are normalized by determining a CE. Expected value (EV) is simply an outcome of all possible outcomes, losses or gains, weighed by their probabilities of occurrence. The risk attitude of the decision maker is not considered in the EV calculation. Therefore the CE of an alternative, which considers risk attitude, is desired.

Decision making based on expected value criteria shows the risk neutral attitude. But, risk management decisions must consider associated uncertainty. Given the risk constant of the decision maker, the basis on which a decision is made can now be rationally changed. Selection of the most attractive risk management alternative can now be made based on the incorporated risk aversion of the decision maker. The CE calculation incorporates the decision maker's attitude toward relative amounts of money, or in other words, risk. The new decision criteria is now that of maximizing gain/minimizing loss.

EVALUATING THE ALTERNATIVES

The decision tree for the STATUS QUO (field evaluation) situation for the building of Figure 2 consists of two possibilities. Given established burning (EB*) in the following example, either a NLE fire can occur or a MFL fire can occur. The decision tree reflecting value exposed in the STATUS QUO situation under a NLE fire is provided by Figure 4. The MFL analysis branching off this tree is shown in Figure 5. Note that this example only considers NLE and MFL fires.

The A-curve assessment of a MFL fire, P(MFL), is obtained from a decomposed probability technique by evaluating those events which can impair sprinkler system operation. In a similar manner, the complete L-curve is obtained, and the probability P(NLE) is simply

(EQ. 2)                $1 - P(MFL) = P(NLE)$

as these events are mutually exclusive. Therefore, the fire risk presented by the STATUS QUO situation is the accumulation of NLE and MFL fire risks. The respective values at the end of each branch are negative because they represent expenditures either as a fire loss or as the cost to carry out a fire protection engineering recommendation, as the case may be.

(*) EB is the size fire which threatens the building; usually taken as 250mm. EB can change depending on the occupancy and construction of the building being addressed.

Resulting from the original engineering survey which generated Figures 4
and 5, three engineering solutions have been offered to reduce the loss
potential of this risk.  Each alternative can be implemented either alone
or together, at some respective cost.  How does the fire risk manager
evaluate the cost effectiveness of each alternative and thus quantify the
resulting reduction in risk?

A cost effectiveness analysis for an alternative and the STATUS QOU
situation would be done exactly the same.  The post-survey engineering
analysis of the property results in engineering recommendations. These must
be evaluated to determine which offers the most increased protection,
decreased insurance costs, favorable affect on risk acceptability, etc; for
the cost of implementing the recommendation.  These recomendations result in
the following modified input data for the engineering method:

ALTERNATIVE 1

ROOM 1:   I = 0.6

The fuel loading (stored combustibles) of Room 1 is either eliminated
or rearranged such that flame spread is hampered or eliminated.  COST TO
IMPLEMENT:  $ 0

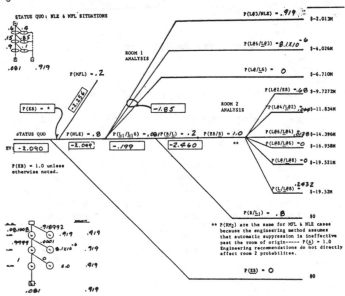

Figure 4 NLE Analysis For STATUS QUO

ALTERNATIVE 2

ROOM 1:   I = 0.6 & M = 0.2

In addition to the above, improvement in detection is made by the installation
of smoke and flame detectors.  COST TO IMPLEMENT: $50,000.

633

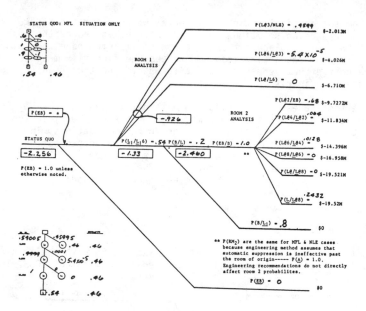

Figure 5 MFL Analysis For STATUS QUO

## ALTERNATIVE 3

P(MFL) = 0.05; ROOM 1: I = 0.6. M = 0.2

Sprinkler valve maintenance program is implemented and complete electronic valve monitoring is acheived by installation of a new system.  COST TO IMPLEMENT: $25,000.

These modifications result in a change in network probability calculations of the engineering method.  The resulting new probabilites are inserted into decision tress of the respective alternatives.  The trees are rolled back to obtain an expected valve at the front of each tree for respective MNL & MML cases of the above three engineering alternatives (4).  Note, a modified NLE is a MNL; a modified MFL, is a MML.

An expected value ranking of these three engineering options and the STATUS QUO is:

|   | EV RANKING | |
|---|---|---|
| 1) | ALTERNATIVE 1: | $-2.0616M |
| 2) | ALTERNATIVE 3: | $-2.0639M |
| 3) | STATUA QUO   : | $-2.0904M |
| 4) | ALTERNATIVE 2: | $-2.1046M |

Table 6 EXPECTED VALUE RANKING

If the decision maker is neutral towards risk, r=0, the most attractive option of the four alternatives is Alternative 1 because it shows the least loss.  However, as previously discussed, if uncertainty is invovled a rational decision maker is not risk neutral. The EV ranking of Table 6 does not consider the aversion to the uncertainty associated with loss potentials.

Risk aversion of the decision maker (corporation/risk manager) must now be included in the evaluation of the four alternatives. Given the range of loss potentials for the building of Figure 2 are,

$-19.52M: total amount subject to loss

and

$-50,000: greatest expenditure of any one alternative,

two different risk constants, r, are determined:

$$r(1) = 1.25 \times 10^{-7}$$
$$r(2) = 1.00 \times 10^{-5}$$

Substituting the EV values for the four alternatives into equation 1, based on respective risk constants, the four risk management options under consideration now fall into new rankings of attractiveness:

r = 1.25 x 10-7: LESS CONSERVATIVE          r = 1.00 x 10-5: MORE CONSERVATIVE

1) ALTERNATIVE 1: CE = $-2.0617M          1) ALTERNATIVE 3: CE = $-2.069M
2) ALTERNATIVE 3: CE = $-2.0697M          2) ALTERNATIVE 1: CE = $-2.075M
3) STATUS QUO    : CE = $-2.0908M          3) ALTERNATIVE 2: CE = $-2.116M
4) ALTERNATIVE 2: CE = $-2.1057M          4) STATUS QUO    : CE = $-2.136M

Additional r values could yield yet different rankings of these alternatives. If the risk constant accurately describes the risk attitude of the decision maker, then the best engineering alternative is identified.

SUMMARY

An engineering method exists today for numerically evaluating a relative level of risk in any building. The flame movement part of the method involves the determination of the probability of success in terminating a fire within each space of a building. The effectiveness of each barrier surrounding the space, whether it be a floor, ceiling, wall, or an empty void; is also evaluated. A computer model exists which will describe any or all possible fire propagation paths from any specified room of origin and quantify the threat to any space along the path. The model then will calculate coordinates of the L-curve in time and space for each fire propagation path.

Once the probabilistic description of how the building in the example would react in a fire was determined, a decision analysis of possible fire protection alternatives was performed. This decision analysis evaluated the effectiveness of each fire protection alternative, and its cost to implement, against the status quo by incorporating the decision maker's (corporation/individual) willingness to accept risk associated with minimum and maximum loss potentials.

The case study has been offered to demonstrate the incorporation of the engineering method into a decision analysis of possible alternatives based on the risk management attitude of the decision maker. This case study offers how the three risk management options of loss transfer, loss absorption or loss reduction can be considered rationally and analytically.

REFERENCES

(1) Fitzgerald, Robert. Risk Analysis Using the Engineering Method for Building Analysis. Worcester Polytechnic Institute, 1985.

(2) A computer program currently being developed by Professor Robert Fitzgerald, Worcester Polytechnic Institute; Worcester, Massachusetts.

(3) & (4) Van Anne, Craig. Managing Risk Through Decision Analysis to Evaluate Firesafety Hazards and Extent of Probable Loss by Use of an Engineering Method. Worcester Polytechnic Institute, 1985.

# Simple and Not So Simple Models for Compartment Fires

**M. R. CURTAT and X. E. BODART**
Groupe Recherche Service Feu
Centre Scientifique et Technique du Bâtiment
84, Av. Jean Jaurès, Champs s/Marne 77420, France

## Abstract

Simple predictive models can be very useful for engineering applications or education in fire safety. Two aspects of compartment fires are addressed here : the growing fire (risk of flashover) and the fully developped fire (fire resistance problems). For each aspect, two predictive methods are presented : a simple model needing small calculation equipment and a computer code based on a zone model. Some results are given as examples showing the possibilities and limits of these methods.

Keywords : Compartment fire - Fire modelling - Fire safety engineering.

## INTRODUCTION

Different types of models have been written during the last decade, that can be used as prediction tools for compartment fires: stochastic models, deterministic field or zone models or correlation formulas from empirical or computer results. Field models, based on local equations, lead to codes demanding large computer capabilities with the theoretical advantage of a more rigorous physical description. The concept of zone model for compartment fire has been used to predict the evolution of several variables in the fire growth process. The computer codes based on zone models need smaller memory size and CPU time than the codes based on field models, with the theoretical drawback of more approximate physical description. Correlation formulas using dimensionnal analysis and/or simple equations can give approximate predictions with little computation work. Stochastic models are of a different type but could be used with deterministic models (if not too big) in coupled deterministic and stochastic modelling (10).

For short term applications in Fire Safety, a compromise has to be made between required precision, computation capabilities and cost, and size of a global model and code for Fire Safety evaluation, in which compartment fire is an element among others. The continuous progress in computer performances and the desired evolution of an integration of the different elements of Fire Safety are in favor of an increasing use of tools such as the simple - or not very complex - models presented in this paper.

combustible linings or walls (8). The simple version of ISBA uses a numerical
approach giving a very quick integration of the set of differential equations
corresponding to the equations of conservation of mass and energy in upper and
lower gaseous zones and the discretized expressions of the equation of heat diffusion (4). The main variables that are calculated in function of time are: rate
of heat release, temperatures of upper and lower zones, height of the thermal
interface, concentration of oxygen and unburnt fuel in upper layer. The pyrolysis rate evolution has to be given as input data. The typical CPU time is about
1/10 of the duration of the fire on a VAX-VMS 11/780 system.

To illustrate the predictive capabilities of ISBA, a comparison of TISBA
(Computed temperature of upper layer) and TEXP (averaged measured temperature of
upper layer) is shown in Figure 2 for three tests. The references of the tests
are given in Table 1. A very good agreement is observed for the sofa (●) and the
wood crib (□) fires. The agreement is poorer for the foam slab fire (Δ) : the
computation was made with a rough (linearized) description of the equivalent
radius of the pyrolyzing surface for this simple version of ISBA though a more
complex description is necessary (5).

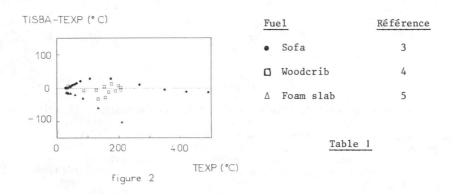

| Fuel | | Référence |
|------|------|-----------|
| ● | Sofa | 3 |
| □ | Woodcrib | 4 |
| Δ | Foam slab | 5 |

Table 1

figure 2

In order to present a certain number of results from (a), prediction (TCOR) were
gathered, versus TEMPER several, TEMPER being TEXP tests values (from ref.(2),
(3), (4), (6), (7)) or TISBA predictions from some examined fire situations (see
Figure 3 giving two evolutions of $\dot{m}_p$ and Tables 2 and 3 showing the input data).
We chose the interval 400-600 °C as a definition for a critical temperature range
(9) and modified the results from (a) by introducing a factor 0.8. The results
are given in Figure 4: the standard deviation in about 100 °C in the critical
interval, if we except the highest TCOR values corresponding to expanded polystyrene walls. Deriving general conclusions would of course need a much broader
set of experimental results and a deeper analysis. Nevertheless the present work
gives a new light on the possibilities of the two predictive approaches retained,
on giving very quick but approximate results, the other based on more physical
basis and needing a numerical solution.

FIRE GROWTH IN A ROOM

The prediction of an average upper gas temperature – among several possible variables describing the fire – is not sufficient for a precise description of the growth of a fire but can give precious information about the risk of igni- tion of items exposed to fire and then the risk of flashover. A model, as simple as possible, for predicting the evolution of this temperature could be used by any person who needs approximate answers one cannot get from conventional tests results. An analytical expression of the evolution of the upper gas temperature during the growth period of a fire was derived by Mc CAFFREY (1) and QUINTIERE (2). The variables $X_1$ and $X_2$ used in this correlation were obtained by writing the conservation of energy equation under a simple way where, in particular, the heat exchange between gas in upper layer and walls (including ceiling and floor) is linearized in :

$$h_{cond} \cdot (T_u - T_0)$$

where $h_{cond}$ is a conductance term :

$$(K\rho c/t)^{1/2} \text{ for } t < t_p$$

$$K/e \text{ for } t > t_p$$

$$\text{with } t_p = (\rho c/K) \cdot (\rho/2)^2$$

An algebra was defined in ref.(2) to take into account the case of different thermal properties of the walls of a given room.

$$\text{With } X_1 = \dot{Q}/(g^{1/2} \cdot C_p \cdot \rho_0 \cdot T_0 \cdot A_{op} \cdot H_{op}^{1/2})$$

$$\text{and } X_2 = h_{cond} \cdot A(g^{1/2} \cdot C_p \cdot \rho_0 \cdot A_{op} \cdot H_{op}^{1/2})$$

TEXP-TO (°C)

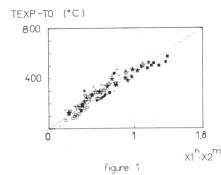

figure 1

$$(TEXP - TO)/TO = 1.6 (X_1)^m (X_2)^n \tag{a}$$

$$n = 2/3, \quad m = -1/3$$

The results of a comparison between TEXP – TO and the predictions from (a) are shown in Figure 1 for 8 data sets (2).

The Fire Research Group of CSTB has been developping since 1979 zone models and computer codes of compartment fires. A recent version of ISBA code (Incendie Simulé dans un BAtiment) was simplified in order to reduce CPU time: the main modifications consisted of a simpler description of the pyrolysis rate and a new writing of the differential equations. The basic physical contents of the model are almost similar to what can be found in the HARVARD model 1978 (5). This ISBA version can only address fires caused by burning items located in the lower part of the room. An other version, more complex, can consider the contribution of

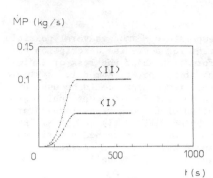

MP (kg/s)

0.15

0.1

⟨II⟩

⟨I⟩

0        500      1000

t (s)

figure 3

$H_c = 15.10^6 \ Jkg^{-1}$

Room size :

(s)    4 m x 3 m x 2.5 m

(lr)   8 m x 6 m x 2.5 m

Opening :

(d)  : 0.8 m x 2 m

(dd) : 1.6 m x 2 m

(w)  : 1 m x 1 m, at 1 m from floor.

| WALL MATERIAL | $m_p$ | ROOM | OPENING |
|---|---|---|---|
| C | I | S | D |
| C | II | S | D |
| AC | I | S | D |
| AC | II | S | D |
| EPS | I | S | D |
| EPS | II | S | D |
| WO | II | S | D |
| WO | I | S | D |
| AC | I | S | W |
| AC | I | S | DD |
| AC | I | LR | D |
| C | I | S | W |
| C | I | S | DD |
| C | I | LR | D |

Wall material: concrete (c), aerated concrete (ac), expanded polystyrene (eps), wood (wo)

TABLE 2

| MATERIAL | e | k | $\rho$ | C | $\varepsilon$ |
|---|---|---|---|---|---|
| Concrete | 0.1 | 1.6 | 2400 | 750 | 1 |
| Aerated concrete | 0.1 | 0.26 | 500 | 960 | 1 |
| Wood | 0.1 | 0.12 | 540 | 2500 | 1 |
| Expanded polystyrene | 0.1 | 0.034 | 20 | 1500 | 1 |

Thermal propertie of walls (S.I. units)

TABLE 3

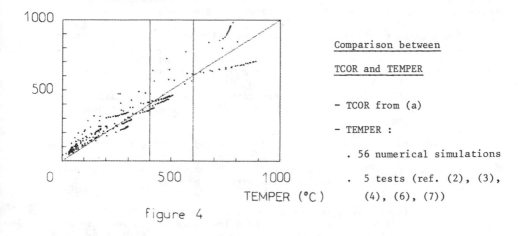

TCOR (°C)

Comparison between

TCOR and TEMPER

- TCOR from (a)

- TEMPER :

. 56 numerical simulations

. 5 tests (ref. (2), (3),

(4), (6), (7))

figure 4

## POST - FLASHOVER FIRE

During the last twenty years, several numerical approaches were developped for the prediction of post-flashover fire temperatures. KAWAGOE and SEKINE (11) presented a graphical technique for obtaining approximate solutions in 1963. More recently, computer was used to solve the coupled set of differential (and/ or algebraic) equations derived from heat and mass balances in the gaseous internal zone (assumed to be well-stirred), and the equation of heat diffusion through the walls - References (12) and (13) are examples of more recent works.

Babrauskas proposed in 1981 (14) an approximate calculation method for determining post-flashover fire temperatures in a compartment containing a single ventilation opening. A theoretical model (15) and a computer code (13) were assumed appropriate and adequate : from computed results a closed-form approximation was derived by curve-fitting. The choice was made to write the average gas temperature as :

$$T_g = T_0 + (T^* - T_0) \cdot \theta_1 \cdot \theta_2 \cdot \theta_3 \ldots \tag{b}$$

where $T^* = 1725 \, °C$ is the value of an "adiabatic" temperature giving the best agreement in the range of interest and where the $\theta$ factors represent the influence of burning rate stoichiometry, thermal losses (steady or transient) at the walls, opening height effect, and combustion efficiency (see (14) for expression of $\theta$ factors). The agreement between the results from this closed-form expression and the computer code calculation results was found to be within +-3 %. An approximate fire temperature can be calculated very quickly through expression (b) on a small microcomputer and form an useful input for the prediction of the thermostructural response of fire-exposed elements.

The evolution of the rate of fuel release has of course to be given in this calculation method (this is a limitation for all the present methods), as the description of the room, wall properties. Then the input data needed are approximately the same as for more complex computerized methods.

We developped recently at CSTB a theoretical model and a computer code of post-flashover fire based on the same basic assumptions as in (13). The main features of CSTB model are described at references (16 and 17). The typical CPU time on a VAX-VMS 11/780 is about 10 sec. for a simulation of a 30 min fire.

An example of prediction of gas temperature is shown at Figure 5. The experimental values are obtained by averaging 20 thermocouples outputs. The test is n° 8 in Ref. (18).

The closed-form expression of Ref. (14), called here CF, and the CSTB model Z1 are then two possible tools for an engineering approach of some aspects of Fire Safety. As the CPU times are short for both computer codes, they could be introduced in a coupled deterministic-stochastic approach.

Some comparisons were made between the predictions from CF and Z1. The fire situations are described in Table 4. Two mass loss rate evolutions were examined (Figure 6). The results on gas temperature are given on Figure 7 under the form TCF versus TZ1. A very good agreement is observed for concrete and expanded polystyrene walls. The agreement is worse for aerated concrete and wooden walls. The mean value of TCF-TZ1 is −5 °C and the standard deviation is about 150 °C. A complete analysis of the observed differences between these two calculated results would necessitate a critical examination of the predictive capability of expression (b) as a deeper evaluation of Z1 model, on the basis of a comparison with a larger number of experimental results.

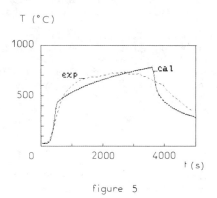

figure 5

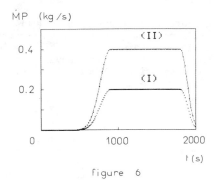

figure 6

642

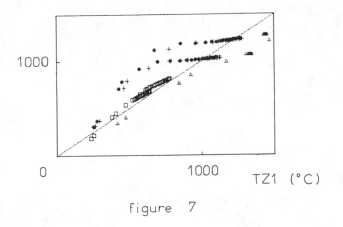

TCF (°C)

1000

0          1000     TZ1 (°C)

figure  7

COMPARISON

Between TCF and TZ1 :

- mass loss rate
  is given at figure 6

- Room :
  4.5 m x 4.5 m x 2.5 m

- Opening :
  2.2 m x 0.95 m

- Thermal properties of
  walls(see table  3)

TABLE 4

CONCLUSIONS and REMARKS

. Fire models for short term engineering applications should be simple to use
and  demanding  short computer times.

. Four fire models of this type are presented here aiming  at the prediction of
temperature evolution in a compartment fire.

. The models should not consume computer time for the calculation of variables
that are not necessary to answer a given question. In the design of such models
one has then to choose which variables are to be calculated (Wall temperatures
for Fire Resistance, Critical delay for flashover prediction, egress time for
people  evacuation,...

. The physical description of fire phenomena is more or less approximate. To
represent the influence of both the unknown data and the poorly described
phenomena, stochastic approaches are useful in association with deterministic
models.

. The capabilities of fire models for Fire Engineering applications should be
evaluated through the quality of the predictions from general Fire Safety
models in which they would be included.

NOMENCLATURE

| | |
|---|---|
| A | surface area of enclosure |
| $A_{op}$ | area of opening |
| C | specific heat of solid |
| $C_p$ | specific heat at constant pressure |

| | |
|---|---|
| e | thickness (walls) |
| $\varepsilon$ | surface emissivity |
| g | gravitationnal acceleration |
| $H_c$ | heat of combustion of fuel |
| $H_{op}$ | vertical opening dimension |
| k | thermal conductivity |
| $\dot{m}_p$ | mass rate of fuel supply |
| $\dot{Q}$ | rate of heat release |
| $\rho$ | density of gas |
| $\rho_0$ | density of ambient air |
| $T_0$ | initial or ambient air temperature |
| $T_u$ | temperature of upper gas layer |
| $T_g$ | post-flashover gas temperature |
| $\Delta T$ | temperature rise of hot gas |
| t | time |
| $t_p$ | thermal penetration time |

REFERENCES

1. B.H. McCAFFREY, J.G. QUINTIERE and M.F. HARKLEROAD, "Estimating Room Temperatures and the Likelihood of Flashover using Fire Test Data Correlations", Fire Technol., 17, 2, May 1981.

2. J.G. QUINTIERE, "A Simple Correlation for Predicting Temperature in a Room Fire", Nat. Bur. Stand., NBSIR 83-2712.

3. J. BLOQWVIST and B. ANDERSSON, "Modelling of Furniture Experiments with Zone Models", Res. Report from Division of Building Fire Safety and Technology - LUND Institue of Technology, presented at the 16th meeting of CIB Commission W14, Borehamwood, UK, May 14-18 1984.

4. M. CURTAT and X. BODART, "Point sur le modèle CSTB de développement du feu dans une pièce unique", rapport CSTB Octobre 1983.

5. H.W. EMMONS, H.E. MITLER and L.N. TREFHETEN, "Computer Fire Code III", Harvard University, Division of Applied Sciences, Home Fire Project Technical Report no 25, Jan. 1978.

6. P.A. CROCE (ed.), "A Study of Room Fire Development : the 2nd Fire Test July 1973", Factory Mutual Research, FMRC Sec. no 21011.4, June 1975.

7. R.L. ALPERT et al., "Influence of Enclosure on Fire Growth, Vol.1, Test 4 : Open Door and Window", Factory Mutual Research, FMRC Job I.D. no 01052.BU-4, July 1977.

8. M. CURTAT, "Modélisation du Feu à l'Intérieur d'Un Local : Participation du Combustible en Paroi Verticale", Rapport CSTB Déc 1984, Marché numéro 82.21.286 du Ministère de l'Intérieur et de la Décentralisation, Direction de la Sécurité Civile.

9. P.H. THOMAS, "Testing Products and Materials for their Contribution to Flashover in Rooms", Fire and Materials, vol. 5 no 3, Sept. 1981.

10. R.B. WILLIAMSON, "Coupling Deterministic and Stochastic Modeling to Unwanted Fires", Fire Safety Journal, 3 (1980-81), 243-259.

11. K. KAWAGOE and T. SEKINE, "Estimation of Fire Temperature-Time Curve in Rooms", BRI Occasionnal Report no 11, Building Research Institute, Tokyo, 1963.

12. S.E. MAGNUSSON and THELE ANDERSSON, "Temperature–Time Curve of Complete Precess of Fire Development", LUND Institute of Technology, SWEDEN, Bull. no 16, 1970.

13. V. BABRAUSKAS, "COMPF2, A Program for Calculating Post–Flashover Fire Temperatures", Nat. Bur. Stand., Tech. Note 991, 1979 UCB–FRG 716–16, Nov. 1976.

14. V. BABRAUSKAS, "A Closed Form Approximation for Post–Flashover Compartment Fire Températures", Fire Safety Journal, no 4, 1981.

15. V. BABRAUSKAS and R.B. WILLIAMSON, "Post–Flashover Compartment Fires : Basis of a Theoretical Model", Fire Mater., 2, p. 39, 1978.

16. P. FROMY, Thesis report under preparation, CSTB, 1985.

17. M. CURTAT and P. FROMY, "Modèle Z1 de feu développé dans une pièce", Rapport CSTB, Mai 1985.

18. J.C. TOURRETTE, "Compte-Rendu sur la deuxième et la troisième série d'essais CSTB", Convention no 77.61.154 avec le Ministère de l'Urbanisme et du Logement, Rapport CSTB, 1980.

# Assessment of Extent and Degree of Thermal Damage to Polymeric Materials in the Three Mile Island Unit 2 Reactor Building

**N. J. ALVARES**
Lawrence Livermore National Laboratory, USA

ABSTRACT

This paper describes assumptions and procedures used to perform thermal damage analysis caused by post loss-of-coolant-accident (LOCA) hydrogen deflagration at Three Mile Island Unit 2 Reactor. Examination of available photographic evidence yields data on the extent and range of thermal and burn damage. Thermal damage to susceptible material in accessible regions of the reactor building was distributed in non-uniform patterns. No clear explanation for non-uniformity was found in examined evidence, e.g., burned materials were adjacent to materials that appear similar but were not burned. Because these items were in proximity to vertical openings that extend the height of the reactor building, we assume the unburned materials preferentially absorbed water vapor during periods of high, local steam concentration. A control pendant from the polar crane located in the top of the reactor building sustained asymmetric burn damage of decreasing degree from top to bottom. Evidence suggests the polar-crane pendant side that experienced heaviest damage was exposed to intense radiant energy from a transient fire plume in the reactor containment volume. Simple hydrogen-fire-exposure tests and heat transfer calculations appoximate the degree of damage found on inspected materials from the containment building and support for an estimated 8% pre-fire hydrogen.

INTRODUCTION AND BACKGROUND

About 10 hours after the 28 March 1979 loss-of-coolant accident began at the Three Mile Island Unit 2 Reactor Building, a hydrogen deflagration of undetermined extent occurred inside the reactor building. Hydrogen was generated as a result of reaction between zirconium nuclear fuel rod cladding and steam produced as the reactor core was uncovered. Ignition of the hydrogen-and-air mixture release after the breach of the reactor-coolant drain tank (RCDT) rupture disk resulted in nominal thermal and overpressure damage to susceptible materials in all accessible regions of TMI-2. Initiation of burn and subsequent termination of induced fires are indicated by data from a variety of pressure and temperature sensors located throughout the containment

This work was performed under the auspices of the U. S. Department of Energy by Lawrence Livermore National Laboratory under contract No. W-7405-ENG-48 and sections of this paper were originally published in GEND-INF-023, Vol. VI, U.S. Nuclear Regulatory Commission, Washington, D.C. (1983) under DOE Contract No. DE-AC07-76ID01570.

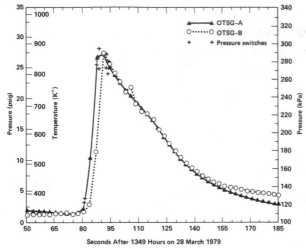

Fig. 1. Pressures recorded during the burn from OTSG (once-through steam generator) pressure transmitters and pressure switch actuation times. Corresponding average temperature via procedure described in Ref. 1 added to psig scale.

volume. Activation of the building spray system is defined by inflection and increase in the negative slope of interior-pressure-reduction curves (Fig. 1).[1]

The hydrogen-in-air concentration [$H_2$] was estimated to be approximately 6 to 8%. At this concentration range, propagation of flame is possible upward and horizontally in quiescent conditions, but not downward; however, turbulent conditions, established circulation patterns, and the ambient absolute humidity of the mixture can perturb propagation patterns in ways that are only qualitatively understood.[2,3] Assuming uniform mixing of 8% hydrogen-in-air concentration and induction of adequate turbulence in internal circulation flows, flame speeds to 5 m/sec (16 ft/sec) are possible -- even in the presence of saturated steam environments.[4]

A cross section of the reactor building (Fig. 2) and plan view of the main (347-ft) operations level (Fig. 3) show the regions of thermal and burn damage. Given that few operational ignition sources were available in the reactor building above the 305-ft level, the time delay to achieve peak overpressure is consistent with an ignition location in the basement. The potential electrical shorting of electrical control systems caused by basement water spillage and the frequency of steam release from the reactor coolant drain-tank pressure-release system supports this assumption.

Thermal damage to fine fuels* indicates general exposure of all susceptible interior surfaces to fire with the exception of random materials including fabric ties of unknown composition, 2 x 4 framing lumber on both the 305-ft and 347-ft levels, and various polymeric materials. These unburned items are evident in photographic and video surveys, and were visually reconfirmed by various entry participants. This pattern is reported in several preliminary reports.[5,6] Possible mechanisms to prevent thermal

---

*Fine fuel is defined as a flammable material with high surface-to-volume ratio.

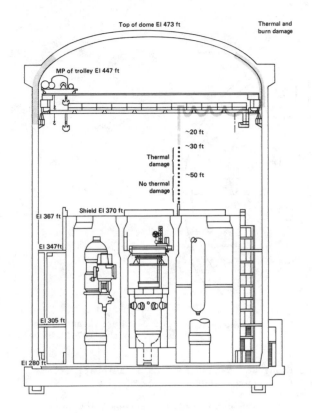

Fig. 2. Cross section of the TMI-2 reactor containment building.

damage to these items include:

1. Preferential absorption of water from saturated atmosphere, requiring greater thermal exposure to produce thermal damage.
2. Direct exposure to high-concentration steam and water vapor, requiring greater thermal exposure to produce thermal damage.
3. Shielding from thermal radiation by position or geometric obscuration.
4. Shielding from the expanding flame front or convectively driven hot gases by physical obstruction.

Although photographic surveys of internal reactor building vistas, ensembles, items, and surfaces were abundant (approx 600 photos from 29 entries), clarity of burn detail in most photographs was not adequate for diagnostic purposes. However, the extent of thermal damage was defined (Figs. 2 and 3) as regions where thermally degraded materials were located, photographed, and, in some cases, extracted from the reactor building for further examination.

Ignition of a uniformly distributed near-lower-limit mixture of hydrogen in air, spreading from basement ignition sources to the top of the reactor building dome by turbulent propagation modes, occurred in the time period

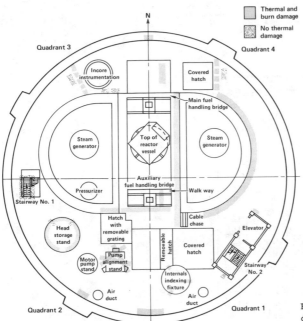

Fig. 3. Thermal and burn
damage on the 347-ft level.

indicated in Fig. 1.  The flame front would have been at an adiabatic flame
temperature of about 700°C to 800°C (approx 1000°K).

Exact paths of flame propagation are undefined.  Because of low hydrogen
concentration, preferential flame spread was upward in quiescent atmosphere;
however, air motion produced by reactor building coolers, steam/hydrogen
release from the rupture disk line of the RCDT and natural convection processes
caused turbulent flow conditions which greatly modify flame spread rates.  The
source of major hydrogen release was located near the west open stairway on the
undersurface of the 305-ft level plane.  Henri and Postma[1] conclude that the
primary entry path of the hydrogen-and-steam mixture to the total reactor
building above the basement (282 ft) level was through this stairwell.  How
these gases from the rupture disk line interacted with total ventilation patterns is not known.  This may be a moot point since, by the time ignition
occurred, hydrogen in the reactor building was undoubtedly uniformly mixed.

Identification of a specific ignition source is not possible from available documentation; however, two potential basement source-types are considered.
(1) Several circuit boxes, instrument racks, meters, and control components
were at various locations around D-rings and containment walls at undefined
(as built) heights above the basement floor.  Thus, failure of circuit
components may have been caused by immersion in water.  (2) Plant operators
who control core and reactor building conditions may have produced ignition
arcs from control components perturbed by thermal or mechanical effects of
reactor excursion.[7]  The inner perimeter of the reactor building basement
had no obstructions to block or blind flow of gases outside of the D-ring.
Approximately 10% of the cooled gases from the cooling system plenum
(25,000 ft$^3$/min) was distributed to the basement (outside of the D-ring)

through committed ducting. The only exit paths for these gases were the 4-in. seismic gaps (a space that physically separates each floor level from the reactor building) and the open stairwell that extended from the basement space to the 347-level without barrier. If ignition occurred at sources away from the open stairwell, the preferred flame propagation would be upward through the seismic gap, and above the 305-ft level, through the grating in the 347-ft level floor. Horizontal spread would occur, but at a slower rate, even during turbulent propagation conditions. Ample evidence exists on the 347-ft level to confirm flame propagation through the seismic gap regions and the floor grating.

At the peak pressure rise of about 28 psig during the hydrogen burn, the adiabatic temperature rise during combustion of 6 to 8% hydrogen-in-air mixture is about 1000°K. At this temperature, calculated exposure radiative and convective flux ($\dot{q}_t$) from an optically thick combustion plume is

$$2.2 \text{ W/cm}^2 < \dot{q}_t < 4.5 \text{ W/cm}^2 . \tag{a}$$

For calculational purpose assume emittance of 0.5, then $\dot{q}_r = \varepsilon \sigma T^4 = 2.8 \text{ W/cm}^2$. This range is approximate because we assume values for combustion plume emittance ($\varepsilon$) at the limits of the range $0.2 < \varepsilon < 0.8$. It is quite possible that $\varepsilon$ could be larger for optically thick hydrogen combustion plumes.[8] Heat transfer coefficient for minimum and maximum convective heat transfer is based on gas velocity ($u_g$) at the limits of the range:

$$3 \text{ m/sec} < u_g < 12 \text{ m/sec} . \tag{b}$$

At $u_g = 12$ m/sec and $L = 1$ m, $\bar{h} = 1.2 \times 10^3 \text{ K(Pr)}^{1/3} \text{ Re}^{1/2} = 2.33 \times 10^{-3} \text{ W/cm}^2\text{K}°$, $\dot{q}_c = \bar{h}A (T_s - T_\infty) = 1.6 \text{ W/cm}^2$. Total heat transfer to surface: $\dot{q}_t = \dot{q}_r + \dot{q}_c = 4.5 \text{ W/cm}^2$.

## EXAMINATION OF TMI MATERIALS

To estimate the intensity of thermal exposure to damaged materials and to analyze thermal damage patterns, it is necessary to examine their condition and to determine their composition. Photographic evidence is inadequate for such appraisal. We examined materials removed from the reactor building, and recommended removal of additional materials for analysis. We examined the following materials:*

| Level 305 | Level 347 | Polar Crane |
|---|---|---|
| 1. Polypropylene bucket | 1. Plywood board | 1. Fire extinguisher |
| | 2. Wood from tool box | 2. Hypalon polar-crane pendant jacket control box |
| | 3. Two radiation signs, probably polyethylene | |
| | 4. Hemp and polypropylene rope | |
| | 5. Catalog remains | |
| | 6. Telephone and associated wire | |

---

*Available July 1983.

(a) Bell telephone at TMI

(b) Charred manual on electrical box

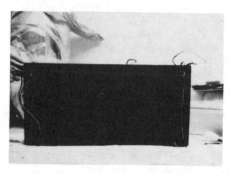

(c) Plywood panel (back)

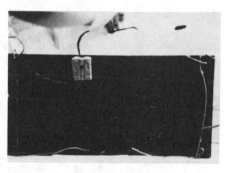

(d) Plywood panel (front)

Fig. 4. Hydrogen-burned in-containment materials extracted from TMI-2.

These materials retain residual radioactive contamination. Consequently, all examinations were performed under radiologically-safe conditions. Chemical or physical analytical procedures could only be done with contaminated or easily decontaminated instruments. We were unable to locate expendable diagnostic equipment; therefore, our examination of extracted materials was limited to detailed photography and macroscopic observations.

Figure 4 shows photographs of plywood on the reactor building south wall and remains of an instruction or maintenance manual located on the reactor building north wall, both ignited by fire propagation through the seismic gap and/or radiant exposure from combustion gases in reactor building free volume. In Fig. 4(a) note the wires along the wall also exhibit burn trauma. Figures 4(c) and 4(d) show the front and the rear surface of the plywood panel after it was extracted from the south wall of the reactor building, over the seismic gap. Both sides are charred, as are edges and holes through which wire ties penetrate. Surface char condition indicates the panel ignited to flaming combustion for a short period before self-extinguishing or being quenched by the reactor spray system. Regardless of the ignition source location, it is apparent that a hydrogen-and-air flame front traversed most of the reactor building volume above (and probably below) the 305-ft level. Duration of this

propagation was about 12 sec. Slow temperature decay before operation of the
building spray system ensured thermal exposure to combustible or thermally
sensitive surfaces was sufficient to produce thermal damage and/or ignition of
these materials, particularly in regions where volume of the combustion plume
was optically thick.

THERMAL MEASUREMENTS ON EXEMPLAR MATERIALS

To augment this analysis, we located exemplar materials generally similar
to those removed from the reactor building. Response properties of the
exemplar materials were measured in a thermal gravimetric analyzer (TGA) to
ascertain the temperature range of thermal degradation and weight-loss rates.
Figure 5 shows TGA patterns for ABS (acrylonitrile butadiene styrene), a
standard material from the National Bureau of Standards (NBS) used as a
control for smoke tests. ABS is similar to telephone body material.

Thermograms are obtained by isothermally heating milligram-sized samples
of materials, supported on a micro balance, at a constant temperature rate.
Weight loss with temperature indicates thermal degradation mode and mechanism.
The temperature range of maximum weight loss indicates critical conditions for
producing potentially ignitable pyrolyzates. Figure 5 shows that NBS-ABS
flammable pyrolyzates are produced in the temperature range of 370° to 500°C,
leaving about 20% inert materials as residue. These pyrolyzates are flammable
which, with an external ignition source, will ignite within this range.

The temperature corresponding to the median of weight loss during the
first major weight-loss experience in any polymer can be used to estimate the
condition where the rate of thermal destruction is maximum, as in the case of
pyrolyzate production. Thus, we can use this temperature to define the time
when subject materials are most susceptible to ignition.

Using standard solutions for transient heat conduction in semi-infinite
solids with constant thermal properties, it is possible to calculate the time
at which a material's surface will attain a specific temperature upon exposure
to constant thermal flux levels. Adjustments should be made to account for
re-radiation heat losses from exposure surfaces and latent heat processes
required to produce pyrolyzates from polymers. With specific surface temperature, exposure heat flux, and defined thermal constants, the time required to
reach this temperature is determined by solution of the differential equation
for transport heat flow in a semi-infinite solid:

$$t = \left(\frac{\pi T_s}{2\dot{q}_t}\right)^2 k\rho c_p \ , \tag{1}$$

where

$\dot{q}_t$ = total thermal exposure,

$T_s$ = surface temperature,

$k\rho c_p$ = material thermal constants.

Polymeric materials present in most items in common use have thickness of
order of 0.2 cm. At this thickness Biot number is less than 1. In distributed
systems, thermal penetration time for such materials is $\Delta x^2/\alpha$ and for
properties of PVC or PMMA, this time is approx 30 sec.
Times calculated using this equation should be short relative to those

653

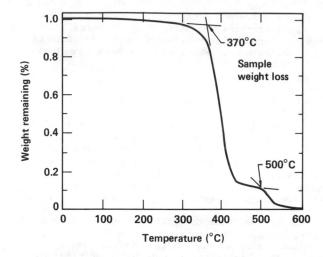

Fig. 5. Thermogram of NBS-ABS. In air, 20 C per min heating rate.

for real materials which experience both thermal and mass convection heat losses. To account for these losses, we adjust $\dot{q}_t$ by subtracting from it the surface radiation energy at the specified critical surface temperature and the mass convection losses (the product of surface mass loss and latent heat of pyrolysis). The resultant effective energy exposure rate $q_e$ replaces $q_t$ in Eq. (1), giving a longer time to attain the critical temperature level. Values for time obtained by using both $\dot{q}_t$ and $\dot{q}_e$ in Eq. (1) bound the time range between exposure of an inert solid and a solid experiencing both re-radiation and latent heat losses. Critical temperature for the three materials is estimated to be 600°K, and thermal exposure energy is the high value calculated from both convective and radiative exposure during combustion of 8% hydrogen in air ($\dot{q}_t$ = 4.5 W/cm$^2$).

These materials and times to critical weight-loss are

| Material | $t_e(\dot{q}_t)$ | $t_c(\dot{q}_e)$ |
|---|---|---|
| Pine wood | 5.3 sec | 9.4 sec |
| PVC | 32.0 sec | 54.7 sec |
| Acrylic | 40.0 sec | 68.0 sec. |

Times to attain critical temperature conditions in these materials are the same order of duration as those recorded during the hydrogen burn in free volumes of the reactor building. Thus, all susceptible materials exposed to this energy should (and did) experience thermal degradation and/or flaming ignition.

HYDROGEN-FIRE-EXPOSURE TESTS

Thermal constants of most polymeric materials are defined only for virgin compounds. It is virtually impossible to calculate thermal response properties of commercially available polymers because additives, retardants, and fillers modify fundamental properties; however, simple hydrogen-fire-exposure tests may indicate accident exposure conditions. To assess this possibility, we conducted selected exposure tests on our exemplar materials using a Meeker burner

adjusted to a fully pre-mixed burning mode.[8] Size and thickness of these
samples were as we found them. The tests were crude and no attempts were made
to conform to condition other than to confine time of exposure to the range
indicated by TMI-2 pressure measurements. Flow was adjusted to produce a
measured flame temperature of 833°K (note: during measurement, the 20-mil
thermocouple was incandescent, so measured temperature was substantially lower
than actual flame temperature). A simple-copper-slug calorimeter measurement
of total thermal flux indicated an exposure flux of 6 W/cm². This level of
flame temperature and thermal flux was within reasonable limits of projected
TMI-2 accident measurements and estimated reactor exposure conditions. Thus,
resulting data trends should be similar to thermal response variations of
materials that suffered hydrogen-flame exposure in the TMI-2 reactor building.

Similarity of thermal damage sustained by materials from the reactor
building and those used in the small-scale test were encouraging. Both duration and intensity of test thermal exposure is in the range of estimated thermal fluxes extant during the reactor building burn. Note that these are very
simplistic tests. No attempt was made to refine temperature or thermal energy
measurement. We have no illusion as to the distribution of convective or
radiative contribution from the test burner; however, the results give data
trends which are intuitively acceptable.


CONCLUSIONS

On the basis of

1.  Photographic and video surveys of the TMI-2 reactor building interior,
2.  Visual and photographic analysis of materials extracted from the reactor
    building,
3.  Macro- and micro-experiments with materials of composition generically
    similar to that of extracted TMI samples, and
4.  Calculations using proposed physical conditons and assumed material
    properties,

the following conclusions are posed:

1.  Hydrogen concentration in the reactor building prior to burn is
    confirmed to be about 8%, as calculated by analyzers of TMI-2
    pressure and temperature records.
2.  No defined path for hydrogen propagation has been established.
3.  The most probable ignition site for the hydrogen burn is in the
    basement volume outside of the D-ring: radial location is not defined.
4.  Thermal degradation of most susceptible materials on all levels is
    consistent with direct flame contact from hydrogen fire.
5.  The directional character of damage to lower pendant length suggests
    potential geometric limitation of the hydrogen-fire plume.
6.  The total burn pattern of the plywood tack board for the south-wall
    telephone on the 347-ft level indicates flame propagation through the
    seismic gap.
7.  Lack of thermal degradation of random, thermally susceptible materials
    may result from preferential moisture absorption. Because of the
    random nature of this evidence, it is not likely that undamaged
    materials resulted from selective shadowing.
8.  Burn patterns in the reactor building indicate that the dome region
    above the 406-ft level was uniformly exposed to direct hydrogen flame;
    the region between the 406-ft level and the top of the D-ring was partially
    exposed to hydrogen flame (most likely in the south and east quadrants);

and, the damage on the 305-ft level was geometrically similar to that above the 347-ft level but less severe.

REFERENCES

1.  J. O. Henri and A. K. Postma, Analysis of the Three Mile Island (TMI-2) Hydrogen Burn, Rockwell International, Rockwell Hanford Operations, Energy Systems Group, Richland, WA, RHO-RE-SA-8, 1982.

2.  M. Hertzberg, Flammability Limits and Pressure Development in $H_2$-Air Mixtures, Pittsburgh Research Center, Pittsburgh, PA, PRC Report No. 4305, 1981.

3.  Flame and Detonation Initiation and Propagation in Various Hydrogen-Air Mixtures, With and Without Water Spray, Rockwell International, Atomics International Division, Energy Systems Group, Canoga Park, CA, Al-73-29.

4.  W. E. Lowry, B. R. Bowman, and B. W. Davis, Final Results of the Hydrogen Igniter Experimental Program, Lawrence Livermore National Laboratory, Livermore, CA, UCRL-53036; U. S. Nuclear Regulatory Commission, NUREG/CR-2486.

5.  G. R. Eidem and J. T. Horan, Color Photographs of the Three Mile Island Unit 2 Reactor Containment Building: Vol. 1--Entries 1, 2, 4, 6, U. S. Nuclear Regulator Commission, Washington, DC, GEND 006, 1981.

6.  N. J. Alvares, D. G. Beason, and G. R. Eidem, Investigation of Hydrogen Burn Damage in the Three Mile Island Unit 2 Reactor Building, U. S. Nuclear Regulatory Commission, Washington, DC, GEND-INF-023 Vol. I, 1982.

7.  M. Hertzberg, A. L. Johnson, J. M. Kuchta, and A. L. Furno, "The Spectral Radiance Growth, Flame Temperatures, and Flammability Behavior of Large-Scale, Spherical, Combustion Waves," Proceedings of the Sixteenth Symposium (International) on Combustion, The Combustion Institute, Pittsburgh, PA, 1976.

8.  B. Lewis and G. von Elbe, Combustion, Flames and Explosions of Gases, 2nd ed., 490, Academic Press, New York, 1961.

# Exponential Model of Fire Growth

**G. RAMACHANDRAN**
Fire Research Station
Borehamwood, Herts, WD6 2BL, United Kingdom

ABSTRACT

This paper discusses a statistical model for estimating the rate of fire growth and doubling time from data on actual fires attended by fire brigades. The model is based on the assumption that the area damaged in a fire has an exponential relationship with duration of burning. Results of application are presented for a few groups of industrial buildings, three major areas of fire origin and two materials ignited first.

Keywords: Exponential Model, Fire, Growth, Actual Fires, Statistical Analysis.

INTRODUCTION

Rate of fire growth plays a central role in planning the evacuation of a building in the event of a fire and determining effective methods of detection and suppression. According to scientific theories supported by experimental evidence, the heat output of a fire increases as an exponential function of time. This implies that the area damaged by direct burning has an exponential relationship with duration of burning. Assuming this relationship, it is shown in this paper how the rate of fire growth can be estimated by a logarithmic regression analysis of data on actual fires attended by fire brigades. Applying this statistical method fire growth rates and doubling times in terms of area destroyed were estimated for eight groups of industrial buildings, three areas of fire origin and two (early and later) periods of fire development.

For four groups of industrial buildings, growth rates for the early period were also estimated for two materials ignited first. Using the doubling time in terms of area destroyed, doubling times in terms of volume destroyed were calculated under three assumptions regarding the relationship between horizontal and vertical rates of fire spread. The expected value and confidence limits of these parameters would provide a framework for comparing growth rates based on actual fires with rates determined from experimental fires.

THE MODEL

According to a simple deterministic model[1]

$$q = q_o \exp{(at)} \tag{1}$$

where

q  = heat output of fire at time t
$q_o$ = heat output of fire at time zero
a  = growth factor

This model has been postulated for fire growth in a building with radiation as the dominant mode of heat transfer. Inside an enclosure there will be re-radiation from building walls, smoke as well as from flame under the ceiling. The exponential model in equation (1) is supported by experiments on spread of fire in a building Labes[2], for example, has discussed this model for home fires in terms of volume involved in fire.

It is, therefore, reasonable to assume that

$$A(T) = A(o) \exp (\beta T) \tag{2}$$

where

$A(T)$ = floor area damaged in T minutes since ignition
$A(o)$ = floor area initially ignited
  $\beta$  = fire growth parameter

Equation (2) is for flame damaged area only and does not include smoke and water damage.

The duration of burning T, can be divided into the following five periods:

$T_1$ - ignition to detection or discovery of fire
$T_2$ - detection to calling fire brigade
$T_3$ - call to arrival of brigade at the scene of fire
    (attendance time)
$T_4$ - arrival to the time when fire is brought under control by brigade
    (control time)
$T_5$ - control to extinction

$T_4$ denotes the duration up to the time when the fire has been effectively controlled or surrounded by the brigade and a message is sent to the fire station to stop the despatch of further reinforcements. The growth of fire is practically negligible during $T_5$ and hence only the periods $T_1$ to $T_4$ need to be considered in a statistical investigation. It must be noted that A(T) is not the area damaged or burning during the $T^{th}$ minute; it is the area damaged in T minutes.

A fire detected soon after ignition will be in its early stage of growth when fire fighting by brigade commences and hence can be controlled quickly. A reduced detection time($T_1$) would therefore shorten the control time ($T_4$) as well, thus reducing the fire duration T and the damage to an appreciable extent. An automatic detection system is designed to achieve this saving. In addition the call time ($T_2$) can be reduced by linking the detector directly to the local fire brigade. The average attendance time ($T_3$) of a fire brigade can be reduced by establishing more fire stations or relocating some of the existing stations. The control time ($T_4$) can be reduced by adopting an efficient fire fighting strategy. Other time periods can be introduced into the model in equation (2) by considering the time for the commencement of first-aid fire fighting methods such as extinguishers, the time of operation of sprinklers and the time when the structural fire protection (walls, floors and ceilings) will be expected to play its part. Also, as noted by Labes[2], the basic time-growth curves may be affected by events such as roof venting, the collapse of a floor or ceiling or any other change that alters the compartmentation. These events can occur strictly as a result of fire development.

The five time periods, $T_1$ to $T_5$ mentioned above are consistent with the information given in the reports (FDR1) on fires furnished by the fire brigades in the United Kingdom. In this report the fire brigade is required to estimate for each fire the detection time according to the following classification.

(i)   discovered at ignition
(ii)  discovered under 5 minutes after ignition
(iii) discovered between 5 and 30 minutes after ignition
(iv)  discovered more than 30 minutes after ignition.

In the pilot investigation[3] on the economic value of automatic fire detectors classes (ii) to (iv) were considered as a single group. In a subsequent detailed investigation[4] average values of 2, 17 and 45 minutes were used for the three classes (ii), (iii) and (iv) respectively. In this later study, the expanded model

$$A(T) = A(o) \exp \, \Sigma_i \beta_i T_i \tag{3}$$

was employed to allow for different growth rates during different periods; $\beta_i$ is the rate of fire growth during the $i^{th}$ period of duration $T_i$ and $T = \Sigma_i T_i^1$.

For statistical estimation equation (2) may be turned into the simple regression model.

$$\text{Log } A(T) = \text{Log } A(o) + \beta T \tag{4}$$

and equation (3) into the multiple regression model

$$\text{Log } A(T) = \text{Log } A(o) + \Sigma_i \beta_i T_i \tag{5}$$

The $i^{th}$ period may be expected to make a contribution of $\beta_i T_i$ towards Log $A(T)$.

The three periods $T_1$, $T_2$ and $T_3$ before the commencement of fire fighting by the brigade are conceptually independent but are all correlated with the control time $T_4$. These interactions with $T_4$ and the ranges of values for the four periods should be taken into consideration in a detailed investigation. The additive model in equation (5) should be regarded as a simple approximation.

The rate at which fire grows can be expected to vary depending on the part of the building where fire starts. Three major areas of fire origin can be identified for an industrial building - production area, storage area and 'other' areas. The rate would also depend on the presence or absence of first-aid fire fighting before the arrival of the fire brigade. All these factors were taken into consideration in the detailed study[4] (mentioned earlier). Since this study was on the economic value of automatic fire detectors, total damage including smoke and water damage was used for the variable $A(T)$.

APPLICATION

Using data furnished by the fire brigades in the United Kingdom an investigation of growth rates of fires in buildings without sprinklers was carried out by Fire Research Station in collaboration with Swedish Fire Protection Association. The presence or absence of first-aid fire fighting was not included as a factor affecting rate of fire growth. The growth parameter was obtained for area damaged by direct burning as mentioned in equation (2). The analysis was based on data on fires which occurred during 1979 and 1980. The number of fires (Sample sizes) ranged from 78 for the storage area of clothing, footwear, leather and fur to 1103 for production area of metal manufacture.

659

The results are given in Table 1 for eight groups of industrial building where $\beta_A$ and $\beta_B$ are the growth rates for the following two periods

$t_A$ = time of ignition to the time of fire brigade arrival at the scene of fire $(T_1 + T_2 + T_3)$.

$t_B$ = arrival time to the time when the fire is brought under control by the brigade $(T_4)$

The model employed in this analysis may be written as

$$A(T) = A(o) \exp \left[ \beta_A t_A + \beta_B t_B \right] \qquad (6)$$

where $A(o)$ is the area initially ignited. The growth parameters $\beta_A$ and $\beta_B$ were highly significant (1 per cent level) or significant (5 per cent level) in almost all cases.

Conceptually, $A(T) = 0$ at time $T = 0$. However, the best engineering application might be made by using the values of parameters giving the best fit for equation (6) rather than values of a modified curve passing through the origin. For any industrial group and area of fire origin the $A(o)$ given in Table 1 is an average value based on all fires in the sample analysed. For a particular building and area in this group, if considered necessary, a distribution of values for $A(o)$ can be obtained by carrying out a fire load survey. The average value or any other parameter of this distribution can be used for 'design' purposes.

The growth parameters given in Table 1 are also expected or average values based on samples of fires analysed. It is difficult to identify the characteristics of the 'average building' to which these results would apply. Information required for this purpose is not available from the fire brigade reports. However, using deterministic models, the growth rates in Table 1 can be modified to take into account room geometry, fuel loading, arrangement of objects, ventilation and other such factors affecting the rates.

DOUBLING TIME

'Doubling time' is the parameter generally used for characterising rate of fire growth. This is the time for fire to double in size and is a constant for the exponential model of fire growth. For example, if it takes 5 minutes for the area damaged to increase from 20 $m^2$ to 40 $m^2$ it will also take 5 minutes for the damage to increase from 40 $m^2$ to 80 $m^2$ amd 80 $m^2$ to 160 $m^2$ and so on. For the model in equation (2), the doubling time in terms of total floor area damaged is given by

$$d = (1/\beta) \log_e 2 \qquad (7)$$

Using equation (7) the doubling times $d_A$ and $d_B$ corresponding to the growth rates $\beta_A$ and $\beta_B$ are given in Table 2.

According to figures in Table 2, $d_A$ varied from 14.44 minutes to 40.77 minutes for the early period and from 9.24 minutes to 53.32 minutes for the later period.

Equation (7) is applicable in terms of horizontal area destroyed. If it is assumed that fire spreads uniformly in all directions (horizontal and vertical) the doubling time in terms of volume destroyed would be approximately

$$d_v = (\tfrac{2}{3}) d \qquad (8)$$

TABLE 1   Fire Growth Parameters

| Industry | Production A(0) (Sq.mtrs) | $\beta_A$ | $\beta_B$ | Storage A(0) (Sq.mtrs) | $\beta_A$ | $\beta_B$ | Other A(0) (Sq.mtrs) | $\beta_A$ | $\beta_B$ |
|---|---|---|---|---|---|---|---|---|---|
| Food, drink, tobacco | 0.504 | 0.020 | 0.013 | 0.694 | 0.017 | 0.049 | 0.327 | 0.042 | 0.026 |
| Chemicals and allied | 0.225 | 0.038 | 0.033 | 0.628 | 0.048 | 0.035 | 0.218 | 0.027 | 0.044 |
| Metal manufacture | 0.341 | 0.033 | 0.026 | 1.160 | 0.017 | 0.045 | 0.425 | 0.032 | 0.041 |
| Mechanical, instrument and electrical engineering | 0.248 | 0.038 | 0.038 | 0.619 | 0.018 | 0.072 | 0.225 | 0.042 | 0.045 |
| Textiles | 0.304 | 0.047 | 0.029 | 1.793 | 0.037 | 0.037 | 0.215 | 0.032 | 0.053 |
| Clothing, footwear, leather and fur | 0.723 | 0.038 | 0.064 | 1.346 | 0.025 | 0.039 | 0.315 | 0.028 | 0.075 |
| Timber, furniture, etc. | 0.485 | 0.046 | 0.046 | 0.949 | 0.037 | 0.052 | 0.566 | 0.030 | 0.037 |
| Paper, printing and publishing | 0.213 | 0.044 | 0.052 | 0.985 | 0.027 | 0.044 | 0.235 | 0.023 | 0.060 |

TABLE 2   Doubling Time (mts)

| Industry | Production $t_A$ | Production $t_B$ | Storage $t_A$ | Storage $t_B$ | Other $t_A$ | Other $t_B$ |
|---|---|---|---|---|---|---|
| Food, drink, tobacco | 34.66 | 53.32 | 40.77 | 14.15 | 16.50 | 26.66 |
| Chemicals and allied | 18.24 | 21.00 | 14.44 | 19.80 | 25.67 | 15.75 |
| Metal Manufacture | 21.00 | 26.66 | 40.77 | 15.40 | 21.66 | 16.91 |
| Mechanical, instrument and electrical engineering | 18.24 | 18.24 | 38.50 | 9.63 | 16.50 | 15.40 |
| Textiles | 14.75 | 23.90 | 18.73 | 18.73 | 21.66 | 13.08 |
| Clothing, footwear, leather and fur | 18.24 | 10.83 | 27.73 | 17.77 | 24.76 | 9.24 |
| Timber, furniture, etc. | 15.07 | 15.07 | 18.73 | 13.33 | 23.10 | 18.73 |
| Paper, printing and publishing | 15.75 | 13.33 | 25.67 | 15.75 | 30.14 | 11.55 |

The doubling time in terms of volume destroyed would be

$$d^1_v = (\tfrac{1}{2})\ d \tag{9}$$

if the vertical rate of spread is twice the horizontal rate and

$$d^{11}_v = (2/5)\ d \tag{10}$$

if the vertical rate is three times the horizontal rate.

For some data quoted by Thomas[5] the doubling time, apparently in terms of volume destroyed, ranged from 1.4 minutes to 13.9 minutes. Rasbash[16] has also quoted a doubling time of 10 minutes for fires in rooms with traditional furniture. At the Factory Mutual Laboratories[7] the growth rates of a series of rather spreading fires were measured by continuous weighing of combustible materials. Data from these tests when forced to fit the exponential growth law indicated doubling times ranging from 21 seconds to 4 minutes.

CROSS VALIDATION

The growth rates derived from statistics of real fires (Table 1) attended by fire brigades involve a number of materials and structural elements of buildings. But the growth rates estimated from experimental fires pertain to individual materials and experimental conditions. There is a need to combine the experimental growth rates of different objects and estimate a composite growth rate for a room or a building. Such an exercise will provide a mechanism for cross validating results of experiments and statistical analysis of real fires. A cross-validation, however, is possible only in probabilistic terms. A real fire involving some material may not continue to burn or spread to another material; this depends on the arrangement of materials in the room, ventilation and environmental conditions.

A cross validation of results for the early stage of fire growth can be attempted by comparing experimental and statistical growth rates for different materials or objects. The difference between the expected (average) values of growth rates for the two cases should be tested statistically for its significance. An alternative test would be to judge whether the experimental growth rate falls within the confidence limits of the statistical growth rate. In Table 3, the expected values of statistical growth rates based on fire area together with their confidence limits are given for each area of fire origin and for two materials for which large numbers of observations were available. The probability of exceeding the upper limit or falling short of the lower limit is 0.025. In three cases only one material with sufficient data was available. The doubling times were calculated according to equations (7) to (10). The figures in Table 3 relate to the early period ($t_A$) and buildings without sprinklers. Fires during the years 1978 to 1980 provided the necessary data. This pilot exercise for four groups of industrial buildings can be extended to other materials and groups of industrial and commercial buildings.

In problems related to fire safety the expected value of the growth rate may not be the appropriate parameter. There is a fifty per cent chance that the growth rate in a real fire will be greater than the expected value. There is only a 2.5 per cent chance that the rate in a fire would exceed the upper confidence limit. For determining fire protection requirements for a building one could choose a growth rate corresponding to an acceptable level of risk (probability).

TABLE 3   Material ignited first – fire growth rate and doubling time

| Industry/Area | Material | Parameter | β | d (mts) | $d_v$ (mts) | $d_v^1$ (mts) | $d_v^{11}$ (mts) |
|---|---|---|---|---|---|---|---|
| **Timber, furniture,** | | | | | | | |
| Production | Dust powder, flour | (a) | 0.052 | 13.33 | 8.89 | 6.67 | 5.33 |
| | | (b) | 0.069 | 10.05 | 6.70 | 5.02 | 4.02 |
| | | (c) | 0.035 | 19.80 | 13.20 | 9.90 | 7.92 |
| | Raw materials | (a) | 0.038 | 18.24 | 12.16 | 9.12 | 7.30 |
| | | (b) | 0.069 | 10.05 | 6.70 | 5.02 | 4.02 |
| | | (c) | 0.006 | 115.53 | 77.02 | 57.76 | 46.21 |
| Storage | Dust powder, flour | (a) | 0.017 | 40.77 | 27.18 | 20.39 | 16.31 |
| | | (b) | 0.057 | 12.16 | 8.11 | 6.08 | 4.86 |
| | | (c) | – | – | – | – | – |
| | Raw materials | (a) | 0.150 | 4.62 | 3.08 | 2.31 | 1.85 |
| | | (b) | 0.246 | 2.82 | 1.88 | 1.41 | 1.13 |
| | | (c) | 0.054 | 12.84 | 8.56 | 6.42 | 5.13 |
| Other | Dust powder, flour | (a) | 0.037 | 18.73 | 12.49 | 9.37 | 7.49 |
| | | (b) | 0.064 | 10.83 | 7.22 | 5.42 | 4.33 |
| | | (c) | 0.010 | 69.32 | 46.21 | 34.66 | 27.73 |
| **Textiles** | | | | | | | |
| Production | Textiles | (a) | 0.045 | 15.40 | 10.27 | 7.70 | 6.16 |
| | | (b) | 0.068 | 10.19 | 6.80 | 5.10 | 4.08 |
| | | (c) | 0.022 | 31.51 | 21.01 | 15.75 | 12.60 |
| | Raw materials | (a) | 0.095 | 7.30 | 4.86 | 3.65 | 2.92 |
| | | (b) | 0.183 | 3.79 | 2.53 | 1.89 | 1.52 |
| | | (c) | 0.007 | 99.02 | 66.01 | 49.51 | 39.61 |
| Storage | Packaging | (a) | 0.010 | 69.32 | 46.21 | 34.66 | 27.73 |
| | | (b) | 0.105 | 6.60 | 4.40 | 3.30 | 2.64 |
| | | (c) | – | – | – | – | – |
| Other | Textiles | (a) | 0.007 | 99.02 | 66.01 | 49.51 | 39.61 |
| | | (b) | 0.064 | 10.83 | 7.22 | 5.42 | 4.33 |
| | | (c) | – | – | – | – | – |
| | Electrical | (a) | 0.094 | 7.37 | 4.92 | 3.69 | 2.95 |
| | | (b) | 0.145 | 4.78 | 3.19 | 2.39 | 1.91 |
| | | (c) | 0.043 | 16.12 | 10.75 | 8.06 | 6.45 |

(a) Expected value;   (b) Upper confidence limit;   (c) Lower confidence limit;   – negative value (inadmissible)

TABLE 3   Material ignited first – fire growth rate and doubling time   (cont'd)

| Industry/Area | Material | Parameter | β | d (mts) | $d_v$ (mts) | $d_v^1$ (mts) | $d_v^{11}$ (mts) |
|---|---|---|---|---|---|---|---|
| **Chemical**<br>Production | Raw materials | (a) | 0.052 | 13.33 | 8.89 | 6.67 | 5.33 |
| | | (b) | 0.079 | 8.77 | 5.85 | 4.39 | 3.51 |
| | | (c) | 0.025 | 27.73 | 18.48 | 13.86 | 11.09 |
| | Electrical | (a) | 0.071 | 9.76 | 6.51 | 4.88 | 3.91 |
| | | (b) | 0.111 | 6.25 | 4.16 | 3.12 | 2.50 |
| | | (c) | 0.031 | 22.36 | 14.91 | 11.18 | 8.94 |
| Storage | Packaging | (a) | 0.019 | 36.48 | 24.32 | 18.24 | 14.59 |
| | | (b) | 0.050 | 13.86 | 9.24 | 6.93 | 5.55 |
| | | (c) | – | – | – | – | – |
| | Raw materials | (a) | 0.031 | 22.36 | 14.91 | 11.18 | 8.94 |
| | | (b) | 0.073 | 9.50 | 6.33 | 4.75 | 3.80 |
| | | (c) | – | – | – | – | – |
| Other | Electrical | (a) | 0.023 | 30.14 | 20.09 | 15.07 | 12.06 |
| | | (b) | 0.060 | 11.55 | 7.70 | 5.78 | 4.62 |
| | | (c) | – | – | – | – | – |
| | Raw materials | (a) | 0.039 | 17.77 | 11.85 | 8.89 | 7.11 |
| | | (b) | 0.074 | 9.37 | 6.24 | 4.68 | 3.75 |
| | | (c) | 0.004 | 173.29 | 115.53 | 86.64 | 69.32 |
| **Paper**<br>Production | Dust powder, flour | (a) | 0.076 | 9.12 | 6.08 | 4.56 | 3.65 |
| | | (b) | 0.152 | 4.56 | 3.04 | 2.28 | 1.82 |
| | | (c) | – | – | – | – | – |
| | Raw materials | (a) | 0.062 | 11.18 | 7.45 | 5.59 | 4.47 |
| | | (b) | 0.121 | 5.73 | 3.82 | 2.86 | 2.29 |
| | | (c) | 0.003 | 231.05 | 154.03 | 115.53 | 92.42 |
| Storage | Raw materials | (a) | 0.066 | 10.50 | 7.00 | 5.25 | 4.20 |
| | | (b) | 0.174 | 3.98 | 2.66 | 1.99 | 1.59 |
| | | (c) | – | – | – | – | – |
| Other | Electrical | (a) | 0.041 | 16.91 | 11.27 | 8.45 | 6.76 |
| | | (b) | 0.080 | 8.66 | 5.78 | 4.33 | 3.47 |
| | | (c) | 0.002 | 346.57 | 231.05 | 173.29 | 138.63 |
| | Dust powder, flour | (a) | 0.032 | 21.66 | 14.44 | 10.83 | 8.66 |
| | | (b) | 0.110 | 6.30 | 4.20 | 3.15 | 2.52 |
| | | (c) | – | – | – | – | – |

ACKNOWLEDGEMENTS

The author wishes to acknowledge the assistance of Mr S E Chandler of Fire Research Station in carrying out the statistical analysis.

This study was carried out in collaboration with Mr S Bengtson of the Swedish Fire Protection Association and was partly financed by the Swedish Fire Research Board. It formed part of the work of the Fire Research Station, Building Research Establishment, Department of the Environment. It is contributed by permission of the Director, BRE.

REFERENCES

1. Thomas, P H. Fires in model rooms : CIB Research Programmes. Current Paper CP32/74. February 1974. Building Research Establishment, Fire Research Station, Borehamwood, Hertfordshire, England.

2. Labes, W G. The Ellis Parkway and Gary dwelling burns. Fire Technology, Vol 2 (4), 1966, 287-297.

3. Ramachandran, G. Economic value of automatic fire detectors. Information Paper IP27/80. November 1980. Building Research Establishment, Fire Research Station, Borehamwood, Hertfordshire, England.

4. Ramachandran, G and Chandler, S E. The economic value of fire detectors. Fire Surveyor, 13 (2), April 1984, 8-14.

5. Thomas, P H. Fire modelling and fire behaviour in rooms. Eighteenth Symposium (International) on combustion. The Combustion Institute, 1981, 503-518.

6. Rasbash, D J. The time factor in fire safety. Short Course on Appraisal and Measurement of Fire Safety. University of Edinburgh, October 1978.

7. Friedman, R. Quantification of threat from a rapidly growing fire in terms of relative material properties. Fire and Materials, Vol 2, No 1, 1978, 27-33.

# The Use of a Zone Model in Fire Engineering Application

**STAFFAN BENGTSON**
The Swedish Fire Protection Association, Sweden

**BENGT HÄGGLUND**
National Defence Research Institute, Sweden

SUMMARY

In this paper is shown the possibility to modernize fire protection engineering in a way that it will be on the same level as other parts of the
design of a building. A fire simulation model will be presented shortly.
The model predicts the time for smoke-logging, flash-over and detection
events for a given geometry of the building. The input parameter is the fire
growth curve. In a special section it is shown how the fire growth can be
predicted. Sometimes the fire area is known, but more often statistical
information has to be used. Finally some engineering applications are
indicated, where the consequences of a fire in a bus garage and a textile
factory are discussed.

## 1. INTRODUCTION

During the last few decades extensive efforts have been made to create
analytical models by which it is possible to judge the safety level objec-
tively. There are three main reasons for this interest:

a)  As the costs of fire safety measures are increasing the cost
    effectiveness is of great interest.

b)  Architects and consulting engineers have found that rigid praxis,
    inflexible building codes and insurance recommendations very often
    are a hindrance to an effective design of a building.

c)  By quantitative analyzis you wish to obtain information on the effec-
    tiveness of new types of measure systems without waiting for expe-
    rience from real fires.

In spite of the fact that there now exist analytical models to describe a
fire and the threat from it, few attempts have been made to use them as
an engineering tool. In the following will be demonstrated how it is
possible with the knowledge of today to predict with good accuracy the
threat from fire and the effect of various fire safety measures. Similar
analyzis have been made by Bengtson, Hägglund and Ramachandran (1-5).

## 2. THE SIMULATION MODEL

The threat from fire is calculated by a zone model which is described in
(6). Its physical outlay is described in Figure 1. The enclosure is divided into two homogeneous layers, a layer next to the ceiling which contains the hot combustion products and one next to the floor which con-

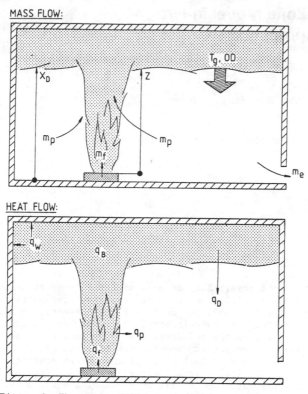

Figure 1. The smoke filling simulation model

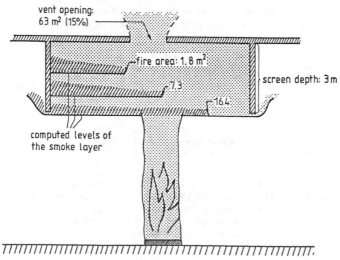

Figure 2. Computed levels of the smoke layer for kerosene fires of area 1.8, 7.3 and 16.4 m².

tains fresh air. The interface moves down at a velocity determined by the rate of air entrained into the fire plume ($m_p$) and by the rate of mass flow through leaks close to the floor ($m_e$). By means of a mass balance for the lower cool region it is possible to calculate this drop velocity.

The burning item releases heat at a rate $q_f$. The heat is lost by radiation and convection to the enclosing structures of the upper gas layer ($q_w$). The upper layer also loses heat by radiation to the lower cool region of the enclosure ($q_D$). Other loss terms are the radiant heat from the fire plume ($q_p$) and the heat to rise the temperature of the upper layer ($q_B$). By means of a heat balance for the upper gas layer, including the fire plume, the temperature of the descending smoke layer is computed.

In addition to the closed-single-space model, a model which accounts for roof-venting is inserted into the model. For given values of the vent area and the depth of screens, extending downwards from the ceiling, the model predicts the depth and temperature of the smoke layer.

## COMPARISON WITH EXPERIMENTAL FIRES

In the following is shown the agreement between calculated results and the measured data from two experimental series with roof-venting.

In the first series, experimental fires were conducted in a central test cell of an aircraft hangar (7). The floor area of the test cell was 20x21 meters and the height approximately 15 meters. Trays of aviation kerosene were burnt with the areas of 1.8, 7.3 and 16,4 $m^2$. The duration of the steady-state burning was 3-4 minutes and the linear burning rate was approximately 4 mm/min. The effective discharge area of the vents was 1.125 %, 4.5 % or 9 % of the floor area. The depth of the screens was 0, 1.5 or 3 meters.

Figures 2 and 3 display the results of the model in comparison with the measured data. In Figure 2 the computed levels of the smoke layer are displayed for the roof vent of the largest area (63 $m^2$) and the screen depth of 3 meter. For the biggest fire (16.4 $m^2$) the computed layer thickness was a few cm below the screens. The result agrees well with what is stated from the experimental observations: "In the experiments involving the greatest outlet venting and the 3-m draught curtains very little spillover of smoke into the roof beyond the test cell occurred". Figure 3 shows the average values of the measured temperature rise of the hot gases below the roof in the test cell as a function of fire size and the distance below the roof – and for all combinations of the test variables in the experimental series. The computed values – based on an upper homogeneous and isothermal zone – are in good agreement with the measured data. The distribution of the filled points in the figure refers to different test conditions.

In the second series, experimental fires were conducted in an industrial building (8). The floor area of the building was 1000 $m^2$ and the height 9.5 meter. Various numbers of wood cribs were burnt and various areas of fire ventilation were available. In Figure 4 the computed height of the smoke layer from one experiment is compared with the one observed. The agreement is fairly good.

## 3. THE DESIGN FIRE

The design fire can either be described by the horizontal fire area or the heat output as a function of time. The scenario can be chosen in at least

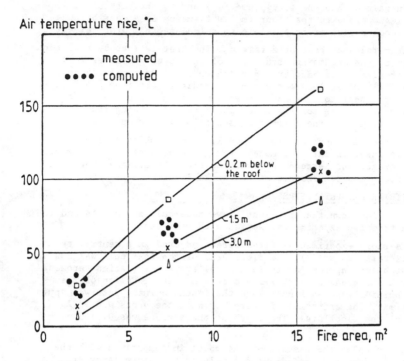

Figure 3. Measured and computed values of the temperatures near the ceiling for kerosene fires of area 1.8, 7.3 and 16.4 m².

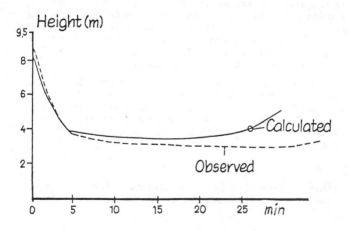

Figure 4. Measured and computed values of the smoke layer for a wood crib fire in an industrial building

four different ways:

a)  Known fire area

b)  Experiences from fire tests

c)  Experiences from real fires using sprinkler statistics

d)  Experiences from real fires using statistical information
    on fire area

## 3.1 Known fire area

Very seldom the fire area is exactly known. An exception may be, however,
if the expected fire will be in an oil tray or a similar object, or in a
container for refuse. Experiences from fires with only one burning item
can also be used, which may be the case in a bus garage, for example, where
you know depending on how the fire is expected to start and on the material
inside the bus - when the bus windows are breaking and the whole bus is burning. You also have a good knowledge of how long it will take before the
next bus catches fire.

## 3.2 Experiences from fire tests

Hägglund has used information from tests at the Factory Mutual laboratories
to predict the fire growth as a function of time (2). The fire growth was
assumed to be an exponential function of time obtained in a series of full-
scale tests (9). Doubling times between 1 and 5 minutes were reported which
can be representative of stored goods under similar conditions like for instance in the tests. Bengtson and Hägglund have in (3) used data on burning
furniture items to predict the times for smoke-filling and temperatures in
single enclosures like patient rooms, assembly halls and atrium buildings.

## 3.3 Experiences from real fires using sprinkler statistics

This method has been used by Bengtson in (1) and in the preliminary work
for the Swedish Fire Ventilation Design Guide (10). The fire area in these
works has been related to the hazard groups as defined in the sprinkler
regulation. This has been done by using information from fires where sprinklers have operated. In (10) the fire area at operation time was assumed to
be the same as the sprinkler area in the hazard group. It is evident, however, that it is a mistake to give the same value of the fire area as the
sprinkler area at operation time for sprinklers. Bengtson therefore made
some modifications of the fire growth rates as presented in (1). As there
were some rough assumptions which had to be made a cooperation between the
Swedish Fire Protection Association and the Fire Research Station was initiated to develop better fire growth models. This work is presented below.

## 3.4 Experiences from real fires using statistical information on fire area

In United Kingdom information is collected from almost any fire to which
the fire brigade is called. The information can be analyzed in different
ways. The area damaged after ignition is reported, for example, and also
the time period between an estimated fire-start and the control of the
fire. The number of jets is also reported. All information is categorized
depending on the activity in the building. For the various groups the information can be related to where the fire started: production areas,
storage areas and "other" areas. The statistical information, which is
gathered and administered by the Fire Research Station, has been used for
various applications (4-5).

# 4. ENGINEERING APPLICATIONS

Computer-based analyzis by the model in the solution of two fire protection engineering problems is demonstrated below. The consequences of a fire in a bus garage and in a textile factory are discussed.

## 4.1 Bus fire in a garage

In the garages for the transportation companies there are many buses parked during the night, which means that a fire can produce extensive communication problems for the passengers the day after a fire. The question has therefore been raised how to protect the buses against fire. For this engineering work the simulation model presented in this paper has been used.

In Figure 5 the calculated smoke-filling times are presented for a garage with floor area of 4000 m$^2$ and of a height of 5 meters. Also the temperatures of the smoke layer are displayed in the figure. The calculations are based on two heat output scenarios, 20 and 40 MW. 20 MW corresponds to one bus burning and 40 MW to two buses burning. The time given in the figure is the time after flash-over of the bus. Flash-over of the bus is assumed to occur 10 minutes after the fire start. At that stage the bus windows collapse and smoke and hot gases flow out into the garage. That means that the heat output is zero until 10 minutes after the fire starts and then grows to 20 MW. The fire area is constant until the next bus, which is parked close to the first one, is ignited, which will take only 2-3 minutes.

As can be noticed from the figure the garage - with one bus burning - can be smoke-filled down to the level of 3 meter about 15 minutes after the fire starts in the bus. The temperatur of the smoke layer at that time is about 150 $^\circ$C.

The consequences of the fire.

If sprinkler is installed, the first bus will be totally damaged, the neighbouring buses probably only damaged by heat and perhaps 10-20 buses only damaged by smoke.

If there is no sprinkler system installed or if it is out of work the fire brigade must check the fire. To give them any chance at all a fire detection system must be installed.

If the attendance time is 10 minutes this means that the fire brigade can start fighting the fire about 21 minutes after the fire start. Flash-over in the garage has not yet occurred but the smoke is down to 1.5-2 m from the floor and at least 2 buses are burning. This means that the fire is extremely difficult to fight, especially as there are only 0,5 meters between the buses, which makes it impossible to send in any fire men. A total damage is a probable result. If a fire ventilation system correctly designed, is provided together with a fire detection system fire fighting may be possible and the result will be that only 2 buses are damaged.

## 4.2 Fire in a textile factory

Data on fire areas from textile factories were used as input in the fire simulation model. As the statistics give you the area damaged after the fire is extinguished, but the equation for fire growth describes the fire size at a specific time, you probably get some exaggeration of the threat. It is not certain that the whole area is burning at the same time. In future works this problem must be analyzed. The rate of fire growth presented in (4) was used, defined by eq 1

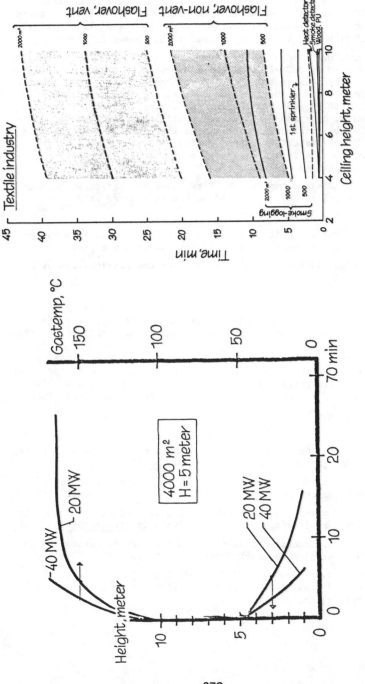

Figure 6. Flashover time, smoke filling and detection events in a textile industry

Figure 5. Computed values of the smoke layer and temperature for a bus fire in a garage

673

$$A = 4.7 \cdot e^{0.063 \cdot t} \qquad (1)$$

where A denotes the area damaged in t minutes. The calculated result is presented in Figure 6 for single enclosures of various sizes and ventilation conditions. The floor area ranges from 500 to 2000 $m^2$ and the ceiling height from 4 to 10 meters. Two ventilation conditions are considered, NON-VENT and VENT. Under the ventilation condition NON-VENT the enclosure is assumed to be closed except for openings close to the floor. Under the ventilation condition VENT vents are assumed to be opened in the roof.

The critical events when smoke-logging and flash-over occur in the single enlosures are denoted by SMOK-LOGGING and FLASH-OVER in the figure. Flash-over is assumed to occur when the computed gas temperature for the entire volume reaches 300 $^\circ$C as an average. The detection events are denoted by SPRINKLER, HEAT DETECTOR and SMOKE DETECTOR - representing the time when the first sprinkler, heat and smoke detector will operate. The sprinkler-head is assumed to be located.2.5 meters from the plume axis of the fire and the heat and smoke detector 4 and 7 meters, respectively, from the axis. The criteria of the events are based on works by Alpert (11) and by Heskestad and Delichatsios (12).

The number of hose reel-jets reported to Fire Research Station which must be used to fight the fire may be related to the fire area, eq 2

$$\text{nojet} = 4 \cdot \ln A_F + 2.71 \qquad (2)$$

When giving this correlation it is expected that 5 hose-reel-jets correspond to 1 main-jet. The equation is valid for production areas.

Comments on the computed results will be given at the presentation of this paper.

## 5. FUTURE WORK

In the future the theories presented in this paper have to be verified in various ways. The rates of fire growth for various industrial groups should be determined to be used as input in analytical models. The rate of fire growth at the beginning of the fire should be studied in more detail as it has a great influence especially on the detection times for heat and smoke detectors. The computed results and predictions should be compared with real fires as regards, for example, detection events and observed times to flash-over.

## 6. REFERENCES

1.  Bengtson, S., The effect of different protection measures with regard to fire damage and personal safety. FoU-Brand (1978):1

2.  Hägglund, B., Hazardous conditions in single enclosures subjected to fire, a parameter study. Stockholm: FOA C 20524-D6. Dec. 1983

3.  Hägglund, B & Bengtson, S., A smoke filling simulation model and its engineering applications. To be published in Fire Technology,1985

4.  Ramachandran, G., Economic value of automatic fire detectors. Bore-hamwood: Building Research Establishment. BRE IP 27/80. 1980

5.  Ramachandran, G., Exponential model of fire growth. The First International Symposium on Fire Safety Science. Gaithersburg, Md, USA, October 1985

6.  Hägglund, B., Simulating the smoke filling in single enclosures.
    Stockholm: FOA C 20513-D6. 1983

7.  Keough, J J, Venting fires through roofs, (Experimental fires in
    an aircraft hangar). Chatswood, Australia: Commonwealth Ex-
    perimental Building Station. Report 344. 1972

8.  Kruppa, J & Lamboley, G, Contribution á l'etude des incendies dans
    les batiments de grand volume realises en construction metalli-
    que. Centre Technique Industriel de la Construction Metallique.
    September 1983.

9.  Friedman, R., Quantification of threat from a rapidly growing fire
    in terms of relátive material properties. Fire and Materials
    2(1978):7

10. Brandventilation för industri- och lagerbyggnader. Sthlm: SBF.
    SBFs rekommendationer 5:3 1982

11. Alpert, R L, Calculation of response time of ceiling-mounted fire
    detectors. Fire technology 8(1972):3, pp 181 ff.

12. Heskestad, G & Delichatsios, M A, Environments of fire detectors
    phase I: Effect of fire size, ceiling height and materials,
    Vol. 2. Analyzis. Wash., D.C.: Natl Bur of Stand. (Factory
    Mutual Research Corp). NBS-GCR-77-95. 1977

# DETECTION

Session Chair

**Prof. Takashi Handa**
Center for Fire Science and Technology
Science University of Tokyo
Kagurazaka, Shinjuku-ku
Tokyo 162, Japan

# Overview on Fire Detection in Japan

**AKIO WATANABE and HIROAKI SASAKI**
Fire Research Institute, Fire Defence Agency
3-14-1, Nakahara, Mitaka, Tokyo 181, Japan

**JUZO UNOKI**
Nohmi Bosai Kogyo Co., Ltd.
4-7-3, Kudan-Minami, Chiyoda-ku, Tokyo 102, Japan

ABSTRACT

Some of unexpected fires may self-extinguish due to insufficient heat re-lease or air supply and some may be discovered and controlled by occupants. Accordingly an automatic fire detection system plays an important role in cases where effective human detection measures couldn't be expected. In Japan, auto-matic fire detection systems have been obligatorily installed in buildings for specific uses having a floor space above a certain value as prescribed by the Fire Service Law. Automatic fire detection systems seem to be considerably effective but there are many cases where the bell of the control unit does not operate because the custodians of the buildings keep the bell switched off in order to avoid false alarms. The probabilities of giving false alarm per year per detector are 6.5% in smoke detectors and 0.8% in heat detectors. There seem to be two current trends in the solution of false alarm problems. One is "right detector in the right place" and the other is to increase the informa-tion content on which discrimination between genuine fires and fire simulating phenomena can be made. Several methods have already been put into practice to greatly reduce the probability of false alarm occurrence to 1%/year·detector, and much more sophisticated fire detection systems will be adopted in near future.

INTRODUCTION

Fire detection technique has made great progress for the past 25 years, and the use of automatic fire detection systems is felt to be closer to accept-ance by the general public. In Japan, the annual output of fire detectors has increased from 80,000 in 1954 to 5 million in 1984, and they have been obliga-torily installed in buildings for specific uses having a floor space above a certain value as prescribed by the Fire Service Law.

At present, the reliability of detectors can be easily confirmed by vari-ous testing methods and life prediction methods. However, to pursue more ef-fective fire detection methods, the following should be considered;

a)  from ignition to extinction, there are many fire-developing stages and corresponding systems, each of which consists of many sub-systems and components and achieves the mission under different conditions, and,

b)  even if the reliability of a hardware system is high, the system ef-fectiveness will be considerably affected by the human attitude to-wards the system and fire.

The present paper describes the role of human and hardware systems in fire detection and touches on the improvement of the system effectiveness.

ROLE OF HUMAN AND HARDWARE SYSTEMS IN FIRE DETECTION

Most of the unexpected fires are caused by human activities at their living place. So, there may be frequently somebody near the fire origin. If he is the person involved directly in the ignition, it's natural that he may well discover the fire first. FIGURE 1 (a) illustrates the first persons who detected the fires inside building between 1979 and 1983 in Tokyo. The figure also gives a transition probability of fire spread from "small fire" to

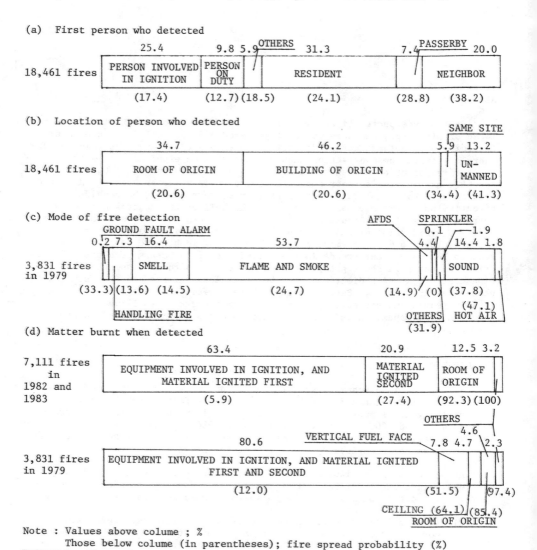

Note : Values above colume ; %
    Those below colume (in parentheses); fire spread probability (%)
FIGURE 1.  Human detectors

"partially-destroyed fire" as one of the indexes showing the effect of detection methods. In this paper, "small fire" means that the burnt floor space is below 3.3 sq. m and the ratio of fire loss to the assessed value of the building is below 10%, while "partially-destroyed fire" means that the loss is over that of small fire.

Other questions would be where, how and when the person discovers the fire. The answers are shown in FIGURE 1 (b) - (d). From these figures, it may be concluded;

a) both fires during people's absence and the fires discovered by a guard are increased by the recent reduction of the persons on night duty in Japan,

b) lesser loss does not always link with earlier detection, and seems to depend on locating the fire spot and human activities after detection, and,

c) the largest fire spread probability is associated with detection by sound and hot air, and a relatively smaller one can be expected with automatic detection; it may be noted that detection by smell leads to slight damage.

After unexpected ignition, some of fires may self-extinguish due to insufficient heat release or air supply, and some may be discovered by the occupant and extinguished with his hands and feet before an automatic fire detection system operates. Statistical analysis has been done for the fires in buildings in which an automatic detection system should be installed by the law, as shown in FIGURE 2. Since fires are often controlled by occupants before the detection system operates as in FIGURE 2, the system plays an important role in cases where effective human detection measures couldn't be expected.

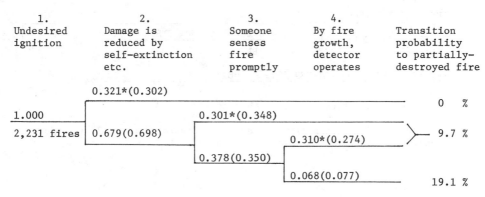

| 1.<br>Undesired<br>ignition | 2.<br>Damage is<br>reduced by<br>self-extinction<br>etc. | 3.<br>Someone<br>senses<br>fire<br>promptly | 4.<br>By fire<br>growth,<br>detector<br>operates | Transition<br>probability<br>to partially-<br>destroyed fire |
|---|---|---|---|---|

0.321*(0.302) ———————————————— 0 %

1.000

2,231 fires | 0.679(0.698)

0.301*(0.348)

0.310*(0.274) ⟩— 9.7 %

0.378(0.350)

0.068(0.077) ———— 19.1 %

Note : * ; probability of success
       Number in parentheses ; obtained by the last investigation

FIGURE 2. Probabilities of fire detection

EFFECTIVENESS OF AUTOMATIC FIRE DETECTION SYSTEM

FIGURE 3 shows the result of the above-mentioned statistical analysis. It is seen from this figure that the fire spread probability is a little over 7% for the buildings fully protected by the automatic fire detection system and 24% for the buildings not yet equipped, that is, when the system is installed, the fire spread probability would be reduced by a factor of 3.4 as against that of non-protected buildings.

The Tokyo Fire Department have surveyed 2006 fires which occurred during 6 years from 1976 to 1981 in buildings where automatic fire detection systems are installed. According to this report, 726 fires of the surveyed fires were detected first by the automatic fire detection systems and 797 fires were alarmed by these systems although they were detected first by human. Therefore, 1523 fires, that is, 76% of total fires were successfully detected by automatic fire detection systems. In the remainder, 309 fires, 15.4% of total fires occurred where no fire detector was installed, and 174 fires, 8.7% of total fires could not be detected by the automatic fire detection systems. 76% of these 174 unsuccessful cases happened because the bell of the control unit did not operate due to fault of management (56% were caused by stopping the bell and 20% by cutting off the power supply). This fact indicates that false alarms were caused so frequently that the custodian of the building kept the bell switched off in order to avoid false alarms.

As to reliability of equipments, Japan Fire Equipment Inspection Institute made investigation on the degree of changes in function of 670 thermal and smoke detectors in operation under severe environmental condition for more than five years. According to this result a large number of detectors were found to show considerable changes in their function particularly among those installed in spa if classified by districts of their installation, and those in hotels and factories if classified by occupancies of buildings.

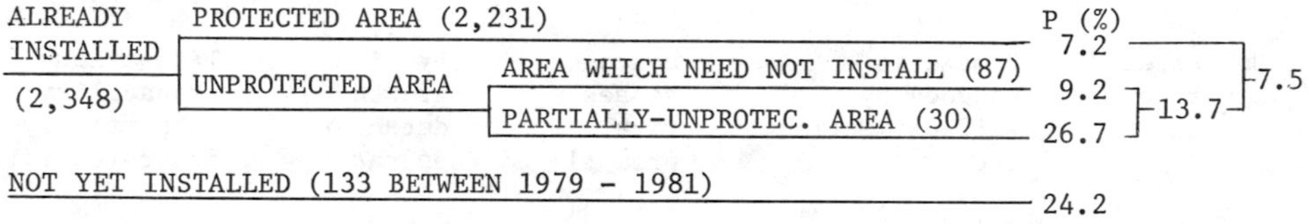

```
                                                                        P (%)
ALREADY      PROTECTED AREA (2,231)                                      7.2
INSTALLED    ┌──────────────────────────────────────────────────────┐        ┐7.5
─────────    │               AREA WHICH NEED NOT INSTALL (87)         │  9.2   │
(2,348)      │UNPROTECTED AREA├──────────────────────────────────────┤        ├13.7
             │               PARTIALLY-UNPROTEC. AREA (30)            │  26.7 ─┘

NOT YET INSTALLED (133 BETWEEN 1979 - 1981)                              24.2
```

Note : Number in parentheses ; the number of fires

FIGURE 3. Transition probability to partially-destroyed fire; P, in buildings where automatic fire detection system should be installed

ACTUAL CIRCUMSTANCES OF FALSE ALARM

At the second half of 1960's, the ratios of false alarms to genuine fire alarms were 3.9 : 1 in Japan and 11.1 : 1 in England. The ratio in England was 23.8 : 1 in 1971 according to the British Fire Protection Systems Association. The reasons for world-wide increasing tendency of the ratio may be frequent use of smoke detectors for fire-safety, and the adoption of powerful air conditioning units and airtight building structures.

In grasping the causes of false alarm, we meet with the following difficulties.

a) it's difficult to decide on the detector giving false alarm, as plural detectors with different principles are connected in one circuit,

b) it's very rare to give false alarm at a specified place, and false alarm is frequently caused by a very quick phenomenon, and,

c) the users are apt to complain of the operation by heat or smoke produced by everyday life, as false alarm, especially in case of the obligatory installation.

The recent statistical analysis on false alarm were reported (a) on 2,787 false alarms of smoke detectors by the Association of Fire Alarm of Japan and the Tokyo Fire Alarm Maintenance Corporative Association in 1977, (b) on 553 false alarms by the Fire Equipment Safety Center of Japan, and on 7,469 false alarms in 1,500 buildings by the Tokyo Fire Department in 1982. Although some of the results were already presented in the English language, the main findings are summarized as follows;

a) the number of false alarms, classified by month, shows the highest value between July and September.

b) the number, classified by hour, shows the highest value at 7 to 9 o'clock in the morning, and then human activities seem to affect the false alarm.

c) though half to 1/3 of false alarms could't be traced to the causes, the causes mentioned were strong wind, maintenance, high humidity, artificial smoke etc. in a report, while they were cooking, smoking, exhause gases etc. in a recent report, putting the causes in order of the frequency.

d) the probabilities of giving false alarm per year per detector are summarized in TABLE 1.

The above findings suggest that,

a) most of false alarms from smoke detectors are caused by normal combustion and heating in daily life such as smoking and cooking. False alarm of fixed temperature type detectors can be explained as they are installed under severe conditions,

b) the probabilities in the place of living and kitchen were maximum 5.6%/year·detector and 41.6%/year·detector, respectively,

c) the delay type smoke detectors could be expected to give less false alarms.

TABLE 1.  Probability (%), of giving alarm per year·detector

| Type / Alarm given by | Probability of giving alarm per year detector (%) | | | | |
| | Smoke | | Heat | | |
| | Ioni-zation | Photo-electric | Fixed temp. | Rate-of-rise | |
| | | | | Spot | Line |
|---|---|---|---|---|---|
| Genuine fire | 0.08 (0) | 0.12 (0) | 0 | 0.06 | 0 |
| Combustion or heating in daily life | 4.36 (0.47) | 0.87 (0) | 2.46 | 0.03 | 0 |
| Environmental factors having no connection with comb. or heating | 0.63 (0.23) | 0.56 (0.27) | 0 | 0.06 | 0 |
| Device/system defect or bad maintenance | 0.04 (0.12) | 0.19 (0) | 0.30 | 0.06 | 0 |
| Unknown cause | 3.11 (0.70) | 1.68 (0) | 1.49 | 0.67 | 0 |
| TOTAL EXCEPT THOSE BY GENUINE FIRE | 8.14 (1.52) | 3.30 (0.27) | 4.25 | 0.82 | 0 |

No. of detectors; 12,980 (Smoke ones 65.4%)

No. of false alarms; 553 (during a year)

| TOKYO FIRE DEPT's FALSE ALARM DATA | 6.5 | 0.8 |
|---|---|---|

No. of detectors; 262,152 (smoke ones 37.5%)

No. of false alarms; 7.469 (during a year)

## MUCH BETTER SYSTEM

As mentioned above, there are a wide range of possible causes of false alarm. Therefore, the problem of the false alarm may be difficult to solve by a single method. There seem to be two current trends in its solutions.

One is "right detector in the right place" by using interchangeable mounting-base. This is quite a realistic solution. The other is to increase the information content as to complicated fire situations and occupants' activities. Most of the present detectors are of one-parameter and one-threshold value. They give alarm when a fire characteristics parameter exceeds the predetermined threshold value. Indeed, the fire is accompanied with heat, smoke, radiation, gas etc., but fire is not a unique source for the sensor output. Several misleading phenomena could be the source giving alarm, too. In this way it is possible for detector to give false alarm, and it is necessary to increase the information content on which discrimination between genuine fires and fire simulating phenomena is made.

Up to now, from the viewpoint of the maintenance of detectors and the signal processing, there have been proposed several methods as follows.

### Adoption of multiple confirmation or integral circuit

This system checks whether the output of sensor is temporary or continuous. If the former is the case, the alarm is not given. For example, it is generally said that the use of time delay type detector is expected to decrease the false alarms by 45%.

### Detector with individual address

To identify the detector giving alarm, the addition of individual detector address has been proposed in recent years. The outputs of devices with ability of transmitting such an address in 1984 were about 2,000 detectors and 17,000 relaying devices having encoder. The most sophisticated system is able to remotely check the detector sensitivity; replacing the defective detectors resulted in probability of the false alarm occurrence of 0.15%/year.detector.

### Sensitivity-floating detection systems

The sensitivity level is automatically set corresponding to ambient conditions such as the start of strong heating and air conditioning system.

### Line type or areal type detection systems

This example is the smoke detection method measuring light-extinction over long paths. Its independence of localized heat or smoke would decrease the false alarm.

### Multi-stage detector

This can give several stages of alarms corresponding to the progress of fire.

Combined or compensation type detector

This detector is exemplified by the composite of 2 kinds of sensors so that the threshold level of smoke detector become more sensitive with temperature rise.

Application of pattern recognition method

The processes of this method are as follows.

1. Storage of data on fire characteristic values in fires and normal conditions.
2. Extraction of fire features, i.e. grasp of time variation and/or spatial distribution of fire characteristics as patterns.
3. Comparison between patterns of fire characteristic values and of measured quantity.
4. Discrimination between genuine fires and deceptive phenomena.

Thus, the system can avoid giving the false alarm. But the fire pattern is dynamic in actual fires, and so real time recognition is desirable. In this point this method is different from the ordinary pattern recognition, which is static. Citing the present examples, the pattern recognition type fire detection systems have been developed by the Japan National Railway since the most of train fires occur in specific places such as overheat of the devices under the floor in Japan. The other relates to the visual organ of intelligent fire fighting robots.

CONCLUSION

Some of above-mentioned methods have already been put into practice to greatly reduce the probability of false alarm occurrence to 1%/year.detector. Even if the probability of individual detector is low, when more than one thousand detectors are installed in one building, the owner may suffer from false alarm. Also, if all automatic fire detection systems in a municipality are directly connected to a fire department, the department is driven by necessity for the solution of false alarm. In these two cases, the accumulated number of false alarms in the system cannot be generously tolerable despite of the low probability.

In these years we are experiencing drastic reduction in cost of semiconductor components as well as great progress of information analysis technology, which will make it possible to adopt much more sophisticated fire detection systems as solution for the false alarm problem (cf. REFERENCE 11 ∿ 17).

Both pertinent human actions after detection of fire and the maintenance of the system are indispensable to effective application of the system. Therefore, the important future problem is how the system assists human beings. In other words, the goal of the system must be foolproof or fail-safe. Remembering the fire detection system is of us, by us and for us; we must get along with this honest device for our safety life.

ACKNOWLEDGEMENTS

The authors are grateful to the Tokyo Fire Department and the Fire Equipment Safety Center of Japan for their admissions of the use of the data.

REFERENCES

1. Tokyo Fire Department : <u>Actual Condition of Fires</u>, 1980ed., 1981ed., 1982ed., 1983ed. and 1984ed. (in Japanese).

2. Watanabe, A. : <u>Effectiveness of Active Fire Protection Systems</u>, CIB Symposium on Systems Approach to Fire Safety in Buildings (TSUKUBA), Aug. 1979.

3. Katsuno, J. : "Maintenance and Management of Fire Protection and Equipments" in <u>Actual State of the Effectiveness of Fire Protection Facilities and Equipments</u>, Journal of Architecture and Building Science, Vol.98, No.1210, 1983, Architectual Institute of Japan. (in Japanese).

4. Ohkubo, I. : <u>Outline of Follow-up Study on Detection Performance Changes with the Passage of Time in Fire Alarm Systems</u>, Journal of Japanese Association of Fire Science and Engineering, Vol.32, NO.5, 1982. (in Japanese).

5. Gupta, Y. P. and Hart, S. J. : <u>Some Aspects of Capability and Reliability of Automatic Fire Detection System</u>; Second National Reliability Conference, March 1979.

6. Fry, J. F. and Eveleigh, C. : "False Alarm" in <u>Automatic Fire Detection</u>, FPA Journal, Oct. 1970.

7. Gomizawa, M. : <u>False Alarm and Failed Alarm of Fire Detection Systems</u>, Committee on Fire Alarm, Illuminating Engng. Inst. of Japan, Report No. HO-412 and 413, 1968. (in Japanese).

8. Nagata, K. : <u>Actual Circumstances of False Alarm of Smoke of Detectors</u>, Committee on Fire Alarm, Report No. 77-9, 1977. (in Japanese).

9. <u>Interim Report for the Fiscal Year 1980</u>, Research Committee on Fire Protection System, Fire Protec. Equipment Safety Center of Japan, June, 1981. (in Japanese).

10. Tokyo Fire Department : <u>Investigation Findings on False Alarm of Automatic Fire Detection System</u>, Tokyo Fire Dept. Oct. 1982. (in Japanese).

11. Takahashi, N. : <u>Selection between Fire Alarms and Non-genuine Alarms of Ionization Smoke Sensors by μP</u>, Proceeding of Annual Meeting of Japanese Association of Fire Science and Engineering, May 1983. (in Japanese).

12. Takahashi, N. et al. : <u>Filtering of Ionization Smoke Sensor Output and Detection of Smoldering and Burning Fire</u>, ibid., May 1984. (in Japanese).

13. Takahashi, N. et al. : <u>Algorithm drawing Qualitative Information from Quantitative Output of Sensor and Result of Simulation</u>, ibid., May 1985. (in Japanese).

14. Ono, T. et al. : <u>Incipient Fire Pattern based on Sensor Response</u>, ibid., May 1983. (in Japanese)

15. Ono, T. et al. : <u>Algorithm on Judgement of Incipient Fire by Sensor Response</u>, ibid., May 1984. (in Japanese).

16. Ono, T. et al. : <u>Basic Research on giving Intelligence to Fire Alarm System -Frequency Response of System and Compensation for Room Figure</u>, ibid., May

1985. (in Japanese).

17. Takeuchi, K. and Ishii, H. : <u>Intelligent Systems : Present Design, What they are and What they might achieve</u>, Seminar paper reprints of IFSEC'85, London, April 1985.

# Attenuation of Smoke Detector Alarm Signals in Residential Buildings

**R. E. HALLIWELL and M. A. SULTAN**
Division of Building Research
National Research Council of Canada
Ottawa, Canada K1A 0R6

ABSTRACT

The propagation of sound from smoke detector alarms through residential buildings has been investigated with respect to the effect of furnishings, type of heating system, and number of closed doors. A simple model representing the propagation path as a series of linked rooms, each modifying the sound level, is described and compared with measured attenuations in 11 houses. The model can be used to determine the optimum number and location of smoke detector alarms.

INTRODUCTION

It is estimated[1] that 40 to 50% of the people killed in fires each year could be saved if adequate early-warning fire detection devices were installed. A study by Jones[2] of multiple death fires in the U.S. indicates that 81.4% of fires occur between 8:00 p.m. and 8:00 a.m., with the largest number (40.5%) between midnight and 4:00 a.m.

Smoke detectors are generally considered to be more effective than people in detecting fire aerosols, and because they can be used to monitor unfrequented areas they are effective early-warning devices for fire. It is important to remember that in many cases the sound of the smoke detector is the only means of alerting a sleeping person to the existence of fire. Detectors can save lives only if people hear them.

The fire detector has two main components: a combustion aerosol detector that determines the existence of fire, and an alarm device to alert the occupants. A breakdown in either component will prevent a fire warning.

In the last two decades aerosol detection has received world-wide attention and is now reasonably well understood, but the problem of producing a sufficiently loud alarm signal has received scant consideration. Defining the alarm level distribution throughout a residential home is important, however, if the self-contained detector alarm is to be an effective warning device.

The sound power output of a smoke detector must be such that after attenuation as it propagates through a building the sound level is still sufficiently high to waken a sleeping occupant. The Underwriters Laboratories standard UL217[3] requires that such a device must provide an A-weighted sound pressure level of at least 85 dB at 3.05 m when mounted near to two reflecting surfaces. Nober et al.[4] concluded that an alarm level of 55 dBA at the ear position in a quiet background is adequate to waken a college-age person with normal hearing, but that 70 dB is required if a window air conditioner is operating in the bedroom. In another study, Kahn[5] suggested that 70 dB is the minimum level required at the ear in a quiet background to waken college-age persons with normal hearing. Clearly, the minimum level to waken people is not well defined and requires further study. Although it is beyond the scope of the present investigation, the question requires attention if the results of this study are to be used effectively.

Once the alarm sound level has been established, there is still the question of where to locate an alarm so as to provide maximum benefit. The answer to this question requires a model that can be used to calculate the attenuation of the alarm signal as it propagates through a building. The model would permit one to determine the optimum location for an alarm to achieve the required signal level at any location in the building. It is the purpose of the present study to develop such a model.

To assess the attenuation of the alarm signal from smoke detectors in residential buildings it was necessary to make measurements in a number of buildings and from the data to develop a general model to be applied for any residential building. Eleven buildings were studied, constituting a reasonable cross-section of the common types of dwelling: bungalows, split-level, and two-story houses. Included in the study were both furnished and unfurnished homes.

MEASUREMENT PROCEDURE

Measurements were made using a smoke detector (modified to operate continuously) as a source of alarm signal. It was mounted on a stand 2.1 m in height so as to simulate a ceiling-mounted detector and placed in a number of locations in each dwelling: in the basement near the furnace room, in the main hallway near the kitchen, and in the hallway near the bedrooms. From each source location the attenuation of noise was measured to every other room. This was done first with all doors in the propagation path open, then with them closed successively until all doors in the path were closed.

To determine the attenuation along each path, the sound level was measured simultaneously near the source and in the receiving room. The source microphone was in a fixed position 1 m from the smoke detector, while the receiving room microphone was moved about the room to provide an average sound level for the room. A two-channel FFT analyser collected data from the two microphones simultaneously. Sixty-four spectra were averaged and the resultant spectra for each microphone were stored for subsequent analysis. A calibration signal was recorded on each microphone at the beginning and end of each measurement period.

As acoustical data are usually provided in third-octave bands, the narrow-band spectra provided by the FFT analyser were converted to third-octave spectra by summing the energy within the standard third-octave bands. This was done by assuming a realistic filter shape and corrected using the calibration signal to obtain the absolute sound levels in third-octave bands. The attenuation was then calculated as the difference between source and receiver levels for each third-octave.

Sound power measurements were made for two smoke detectors of each of seven models in accordance with ANSI S1.31-1980[6] in the reverberation chamber at the Division of Building Research, National Research Council of Canada.

## DISCUSSION OF RESULTS

The reduction in sound level that is provided by walls, doors, etc., within a building and the sound absorption provided by furnishings increase with increasing frequency. To be most effective it would thus be reasonable for a smoke detector to have most of its acoustical output at low frequencies, say below 500 Hz. The human ear, on the other hand, is most acute in the 2000- to 5000-Hz range. It is also more economical to produce an inexpensive alarm operating in a higher frequency range, but this means that such alarms must operate at a higher sound power if they are to be adequate as warning devices.

Sound power measurements on a number of smoke detectors are listed in Table 1, which shows that most smoke detectors only provide noise output in a few bands, the two dominant ones being the 3150- and 4000-Hz bands. For the purpose of this study only the 3150-Hz band has been used to develop a propagation model. The higher frequency band, which was not present for all smoke detectors, will tend to be attenuated more and thus will be less useful in alerting occupants. Where there is energy in lower frequency bands, the model will predict too little attenuation and thus provide an extra margin of safety.

Two different models were considered for predicting the attenuation of noise from the alarms. The first was based on a model proposed by Berry.[7] Its most attractive feature is its simplicity, the basic attenuation being assumed to be a function of the straight-line horizontal distance between source and the mid-point of the receiving room, without regard for changes in elevation. Added to this basic attenuation are three corrections, one for the number of floor changes in elevation, another for each closed door along the propagation path, and a third for each open doorway along the propagation path.

TABLE 1.  Maximum sound power output of smoke detectors

| Detector[1] | Duty Cycle[2] ($\tau$) | 1/3 Octave Frequency Band, Hz | | | | | | | | | | |
|---|---|---|---|---|---|---|---|---|---|---|---|---|
| | | 500 | 630 | 800 | 1000 | 1250 | 1600 | 2000 | 2500 | 3150 | 4000 | 5000 |
| A1 | 0.203 | 38 | 39 | 39 | 39 | 63 | 57 | 73 | 96 | 84 | 63 | 50 |
| A2 | 0.230 | 37 | 38 | 38 | 38 | 44 | 56 | 70 | 98 | 92 | 67 | 56 |
| B1 | 0.877 | 82 | 82 | 60 | 71 | 74 | 81 | 79 | 95 | 95 | 95 | 88 |
| B2 | 0.870 | 79 | 81 | 66 | 72 | 76 | 81 | 77 | 93 | 94 | 96 | 92 |
| C1 | 0.986 | 44 | 44 | 44 | 45 | 45 | 50 | 61 | 79 | 102 | 90 | 69 |
| C2 | 0.989 | 44 | 44 | 44 | 45 | 45 | 50 | 62 | 79 | 102 | 91 | 70 |
| D1 | 0.844 | 46 | 46 | 46 | 46 | 47 | 52 | 63 | 80 | 103 | 93 | 71 |
| D2 | 0.845 | 44 | 44 | 44 | 45 | 45 | 50 | 62 | 80 | 102 | 88 | 68 |
| E1 | 1.0 | 84 | 70 | 69 | 85 | 76 | 92 | 88 | 96 | 92 | 91 | 80 |
| E2 | 1.0 | 76 | 83 | 63 | 69 | 80 | 87 | 85 | 97 | 100 | 91 | 89 |
| F1 | 1.0 | 61 | 60 | 72 | 70 | 70 | 74 | 86 | 75 | 83 | 90 | 82 |
| F2 | 1.0 | 58 | 61 | 69 | 70 | 72 | 77 | 90 | 81 | 82 | 89 | 82 |
| G1 | 0.643 | 37 | 37 | 37 | 38 | 39 | 50 | 63 | 88 | 95 | 69 | 55 |
| G2 | 0.667 | 38 | 38 | 38 | 38 | 39 | 48 | 61 | 84 | 95 | 71 | 56 |

[1]Detectors with the same letter designation are identical models.
[2]The duty cycle is the fraction of time during which the alarm is operating.
10 $\log(1/\tau)$ was added to the measured mean sound power level to give the maximum sound power level.

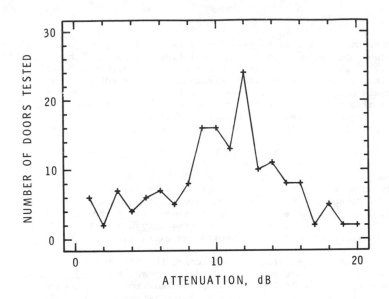

Figure 1. Histogram of sound attenuation due to closure of single door in propagation path.


Figure 1 shows a histogram of the increase in attenuation provided by closing a single door in a propagation path. The wide range in attenuation is a result of the wide variation in fit among doors, from doors with large gaps beneath to those carefully weather-stripped. The mean value is 10 dB, and this was used in the model as the correction for closed doors. At first glance this result appears to be in conflict with NFPA[8] work, which suggests 15 dB, and with that of Bradley and Wheeler[9] and of Nober,[4] who found 16.4 dB and 15 dB, respectively. This is, however, the change in attenuation with a door closure rather than an overall transmission loss of the wall-door system.

The best fit to the data was found with the following corrections: 10 dB for each floor between the source and receiver, 3 dB for each open doorway along the path, and 10 dB more for each closed door along the path. Figure 2 shows the measured attenuation minus the three corrections plotted as a function of distance from the source for the best fit values for these corrections. The slope of the least squares fit line drawn through the points gives the dependence on distance for this model. There is a very large range in the scatter, making this a less than satisfactory model for sound attenuation. Some of the scatter is a result of the range in measured attenuation of doors, but it is insufficient to explain all of the scatter. The major weakness of the model is its inability to make allowances for differing sound absorption in different rooms.

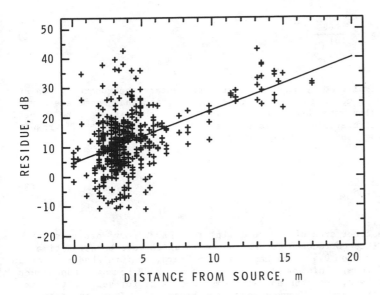

Figure 2.  Measured attenuation minus corrections for open doors, closed doors, and number of floors between alarm and receiver.

The second model considered takes a slightly different approach.  In it, the propagation path is viewed as a series of linked rooms, each of which modifies the sound level.  The path to be used is the most direct path as would be traversed by a person walking from the source to the receiver.  Each space enclosed by walls or partitions, including hallways, is counted as a room provided that the opening leading from the previous room is a doorway or equivalent.  For the purpose of this model, it is assumed that little, if any, sound is transmitted through the partitions or floors.  From reverberation room theory, the sound level in a room due to transmission of sound through an opening or partition into the room is given by[10]

$$L_R = L_s - R + 10 \log \left[ \frac{S \, T_{60} \, C \, (1.086)}{60 \, V_R} \right] \tag{1}$$

where R = transmission loss of partition,
  $L_S$ = sound pressure level in source room,
  $L_R$ = sound pressure level in receiving room,
  S = area of partition ($m^2$),
  $T_{60}$ = reverberation time,
  C = speed of sound (m/s),
  $V_R$ = volume of receiving room ($m^3$).

At room temperature, this reduces to

$$L_R = L_s - R + 10 \log \left[ \frac{S \, T_{60}}{0.161 \, V_R} \right] \qquad (2)$$

It may be further simplified by assuming that sound enters the room only via an open doorway of area 2 m$^2$ with zero transmission loss, and that rooms are always 2.4 m high. A normally furnished room of average size, that is one with carpet and furniture, has a reverberation time of about 0.4 s at 3150 Hz.* The result is that the receiving room level is given by

$$L_R = L_s - \left[ 10 \log \left( \frac{area}{2.08} \right) + corr \right] \qquad (3)$$

where 'area' is the floor area of the receiving room and 'corr' provides a means of adjusting this correction for instances in which reverberation time differs substantially from 0.4 s, as happens in a "hard," unfurnished room or in an extremely "soft" room. This would have a value of −2 dB for hard rooms such as bathrooms or kitchens, zero for normal rooms, and +2 dB for very soft rooms such as a bedroom with carpet, heavy drapes, and bedspread. Thus, the term

$$10 \log \left( \frac{area}{2.08} \right) + corr \qquad (4)$$

may be viewed as a correction to the sound level due to absorption within the room or, alternatively, as the room attenuation. Attenuation due to absorption can thus be calculated for each room in the house, independent of where source and receiver are located. The overall attenuation of the detector alarm is thus the sum of the attenuations for all rooms in the propagation path plus 10 dB for each closed door.

The derivation of Eq. (1) is based on the assumption that there is a diffuse sound field in both source and receiving rooms, a condition very unlikely to occur in a residential building. Similarly, the assumption of zero transmission loss through the open doorway is an over-simplification because it ignores any edge or interference effects of the doorway. A comparison of the sound attenuation predicted by this model with the measured attenuation indicates that an additional 5 dB attenuation needs to be added for each room in the propagation path. This may be viewed as the transmission loss associated with an open doorway. Note that the addition of transmission loss for the open doorway to the 10 dB insertion loss of the closed door gives a total attenuation of 15 dB, which compares favourably with values quoted by other workers.

It is well established from field studies of transmission loss of walls and floors that heating ducts can provide a flanking path that will short-circuit a partition and result in lower noise reductions than would otherwise be obtained. This was borne out in the present study; it was found that buildings that do not

*J.S. Bradley. Unpublished data.

have forced-air heating provide an additional 6 dB attenuation for each room in the propagation path.

These corrections can all be summarized in the following expression:

$$\text{Atten} = [\sum_{r=1}^{n} \{10 \log (\frac{\text{area}_r}{2.08}) + 5 + \text{corr}_r + K\}] + 10 \text{ (door)} \tag{5}$$

where area = floor area of room 'r' ($m^2$),
      door = number of closed doors in path,
      corr = -2 for hard rooms (kitchen, bath),
            = 0 for normal rooms,
            = 2 for soft rooms (rugs, draperies),
        K = 0 for forced air heating,
           = 6 for electric or hot water heat,
        n = number of rooms in path from smoke detector to point of interest, not counting room containing smoke detector.

Figure 3 shows the attenuation calculated using this model plotted against measured attenuations for all source-receiver configurations in the 11 houses studied. The solid line is the least-squares fit to the data, with a standard deviation of 7.5 dB. Although some of this scatter will be due to variation in the attenuation for closed doors (shown in Fig. 1), much of it appears to be associated with measurement of the source room sound level. The source room sound pressure level was measured at a single position rather than with a moving microphone, as was done in the receiving room. The measured sound level will be

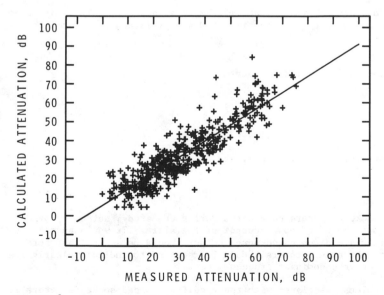

Figure 3. Comparison of calculated and measured attenuations.

more representative of the near field of the alarm, as modified by adjacent reflecting surfaces, rather than the mean sound level in the room.

Obviously there are many other transmission paths in real buildings which could be included in a more detailed calculation. The inclusion of such extra details would require extremely complicated calculations and is unlikely to provide a better fit to the measured data than the semi-empirical method described above.

The attenuation calculated by Eq. (5), when subtracted from the initial sound level provided by the alarm signal, gives the alarm signal level at the point of interest. The initial sound level provided by the alarm signal can be obtained in one of two ways. The most direct is to measure the mean sound level in the room containing the smoke detector alarm. This is not always practical, however, especially if one is trying to ascertain the best room in which to locate the alarm. Thus, the second method is to calculate the initial sound level from the sound power output of the alarm,[6] using the expression

$$L_s = P - 10 \log \left[ \left( \frac{V_s}{T_{60}} \right) \left( 1 + \frac{S_s \lambda}{8 V_s} \right) \right] + 14 \tag{6}$$

where  $L_s$  = mean sound pressure level in source room,
 $\quad\quad P$  = sound power output of alarm (dB),
 $\quad\quad V_s$  = volume of source room ($m^3$),
 $\quad\quad T_{60}$  = reverberation time,
 $\quad\quad S_s$  = surface area of source room ($m^2$),
 $\quad\quad \lambda$  = wave length of sound (m).

For an alarm operating primarily at 3150 Hz in an approximately square room with 2.4-m ceiling and a reverberation time of 0.4 s, this can be reduced to

$$L_s = P + 14 - 10 \log \left[ 6.06A + .3\sqrt{A} \right] - corr \tag{7}$$

where A = area of room,
 $\quad$ corr = -2 for hard rooms (kitchen, bath),
 $\quad\quad\quad$ = 0 for normal rooms,
 $\quad\quad\quad$ = 2 for soft rooms (rugs, drapes, etc.).

RECOMMENDATIONS

Of the seven smoke detectors considered during the study, not one provided any information about the sound power output of the alarm. It would seem to be desirable that either a minimum sound power output should be established for smoke detectors or that manufacturers should be required to indicate clearly the sound power output and the dominant frequency band.

One of the problems associated with these self-contained smoke detectors is that the optimum location for early detection of fires is not the same as the optimum location to ensure that the alarm is heard. This problem can be

overcome in several ways, one being to make the alarms louder. It may increase
the cost of the unit, and may also risk hearing damage for people close to the
alarm when it is activated. Another is to have a number of smoke detectors
interconnected so that detection of a fire by any one would trigger all the
alarms. A third method would be to separate the detector and the alarm so that
each could be placed in its optimum location. It is recommended that the last
two options be incorporated in building regulations and practices for fire
safety design.

CONCLUSION

A simple expression has been developed to calculate the attenuation of the
alarm signal from a smoke detector as it propagates through a residential
building, with the path viewed as a series of connected rooms. Attenuation
depends on floor area and type of furnishings in each room. Corrections are
applied if the house does not have forced air heating or if a number of doors
are closed. The expression can be used to determine the optimum location for
alarms. As the best location for an alarm is not necessarily the best location
for a smoke detector, it is recommended that interconnected multiple
detector/alarm systems be used or that detector and alarm be separated.

REFERENCES

1.  Bright, R.G.: "Recent Advances in Residential Smoke Detectors," _Fire
    Journal_, 68:6, 69-77, 1974.

2.  Jones, J.C.: "1982 Multiple-Death Fires in the United States," _Fire
    Journal_, 77:4, 10-25, 1983.

3.  Underwriters Laboratories Inc. Standard for Safety UL217: "Single and
    Multiple Station Smoke Detectors."

4.  Nober, E.H., Peirce, H., Well, A., and Johnson, C.C.: "Waking
    Effectiveness of Household Smoke and Fire Detection Devices,"
    NBS-GCR-80-284, Washington, DC, 20234, 1980.

5.  Kahn, M.J.: "Detection Times to Fire-Related Stimuli by Sleeping
    Subjects." NBS-GCR-83-435, Washington, DC, 20234, 1983.

6.  ANSI S1.31-1980.: "American National Standard Precision Method for the
    Determination of Sound Power Levels of Broad-Band Noise Sources in
    Reverberation Rooms." American Institute of Physics, New York.

7.  Berry, C.H.: "Will Your Smoke Detector Wake You?" _Fire Journal_, 72:4,
    105-108, 1978.

8.  American National Fire Codes, National Fire Protection Association,
    NFPA 74, Batterymarch Park, Quincy, MA, 02269, 1974.

9.  Bradley, H.L. and Wheeler, W.P.: "The Analysis of the Audible Signal From
    Single-Station Heat and Smoke Detectors." Unpublished paper for ENFP 416,
    Fire Protection Engineering Department, University of Maryland, College
    Park, MD, 1977.

10. ASTM E90-81: "Standard Method for Laboratory Measurement of Airborne Sound
    Transmission Loss of Building Partitions."

# Detection of Smoldering Fire in Electrical Equipment with High Internal Air Flow

**HIROBUMI HOTTA and SATORU HORIUCHI**
Nohmi Bosai Kogyo Co., Ltd.
4-7-3, Kudan-Minami, Ciyoda-ku, Tokyo 102, Japan

ABSTRACT

A series of fire tests of smoke detectors was conducted with a view to seeking a method for early detection of electrical component failures, e.g. smoldering of printed circuit boards and cables, burst of capacitor, in highly integrated (using LSI's) electrical equipment as represented by electronic computers. As a result, it has been found that, in equipment where there is high air flow for cooling, the ionization detector shows better response compared with the photoelectric detector if the air velocity exceeds about 1m/s. It has been a general practice to avoid installation of the ionization detector in a place where it is subjected to high air flow because the output level changes due to air flow characteristics. However, a small ionization detector which is intended for use in equipment such as computer and which is stable against air flow has been developed. This paper describes this newly developed detector as well.

INTRODUCTION

As proved by introduction of computers with capacities of 64KB, 256KB and 1MB-RAM on the market, development of computers is being directed toward higher integration using VLSI's and speeding up. As a result of this, there is a greater danger that a large amount of heat is generated in limited spaces, thus it has become one of the important questions to adopt adequate measures (cooling) against heat as well as fire retardation of electronic components. In general, electronic components are provided with radiative fins for air cooling to maintain the equipment below a predetermined temperature. In some types of equipment, air velocity exceeds 10m/s.

However, there are such cases where printed circuit boards and cables burn due to defective components, overload etc. Although some equipment is provided with heat detectors for detection of such fires, it is difficult to detect them at the smoldering stage. A possible alternative is the smoke detector. As smoke detectors, the ionization detector and light scattering detector are known. Characteristics of those detectors are described in detail in papers[1,2] by T.G.K. Lee, R.W. Bukowski, and G.W. Mulholland. It is generally known from their characteristics that the former is highly sensitive to flaming fire like open wood fire, and the latter is highly sensitive to smoldering fire of PVC insulated cables and beds. However, according to the above paper[1], there is a difference in the smoke particle size distribution between the cases where a punk is burning in still air and in moving air as shown in FIGURE 1, and there-

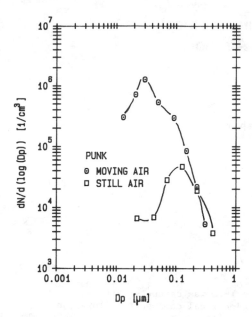

FIGURE 1  Effect of air movement (2m/s) on particle size[1]

fore the light scattering detector may not always have a good response to smoldering fires.  Besides, punk sticks continue self-combustion and especially tend to keep glowing if air is supplied.  Since electronic components such as printed circuit boards and cables, discussed in this paper are flame retardant and hardly burn unless electric power is continuously supplied at the initial stage of combustion, their burning characteristics are different from that of the punk stick.  Therefore, this paper describes experiments which were carried out with a view to studying combustion of electrical components especially printed circuit boards, cables, and capacitors in equipment with internal flow of cooling air, and the sensor to detect fires from these components.

In this paper, "detector" and "sensor" are distinguished in that a "detector" sends out discrete ON-OFF output and a "sensor" sends out analog output.

TEST METHOD

The apparatus used for the tests was a smoke tunnel with an open path of 7.0m long to obtain conditions close to those of electrical equipment with air cooling fans.  FIGURE 2 shows the tunnel structure.  Air velocity in the tunnel was controlled by using two fans: one for freely changing the air velocity at the fire point, and the other for setting the air velocity at the measuring point constant regardless of the air velocity at the fire point.  A constant air velocity was set within a range 0.5 to 8m/s by means of voltage regulators according to the indication given by hot wire anemometers provided at each measuring point.  The reasons why air velocity at the sensors' location was kept constant were to reduce the dilution effect by air flow and to maintain the normal level and response of the sensors constant.

For heating up the electrical components to smolder, a 3KW hot plate (300mm in diameter, 200V AC) was used. Temperature control was made by voltage regulator with reference to the reading of type K thermocouple (elemental wire AWG #30). However, combustion of capacitors was initiated by applying a voltage exceeding the breakdown voltage of each capacitor. Smoke densities were measured by means of analog output type ionization sensor (type FDS 221) and a light scattering sensor (type FDK 224) manufactured by Nohmi Bosai Kogyo Co., Ltd. At the same time, the smoke densities were also measured by two kinds of smoke densitometers having a light path length of 28cm. One of them is the same type as that used for the approval testing of smoke detectors in Japan and has a sensitivity range within the visible region (typical wavelength of 570nm) and almost the same characteristics as the densitometer used for the test prescribed in UL-268[3]. The other is sensitive in the near-infrared region (typical wavelength 950nm) conforming to the standard prescribed in EN54 Part 7[4]. The following are three types of electrical components subjected to the tests and their combustion conditions.

(i)  Printed Circuit Board
     20 x 20 x 1.6 mm
     Fiber-reinforced epoxide

Combustion in the range
300 to 500°C on the hot plate.
Typical temperature 450°C.

(ii) Cable
     PVC insulated flat cable
     used for computer wiring

Combustion in the range 300
to 450°C on the hot plate.
Typical temperature 350°C.

(iii) Aluminum electrolytic
      capacitor
      Withstand voltage 16V,
      Capacity 2200 μF

Breakdown at an applied
voltage of 16V AC or 35V DC.

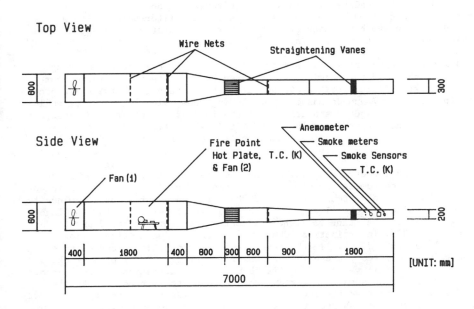

FIGURE 2   Smoke test tunnel

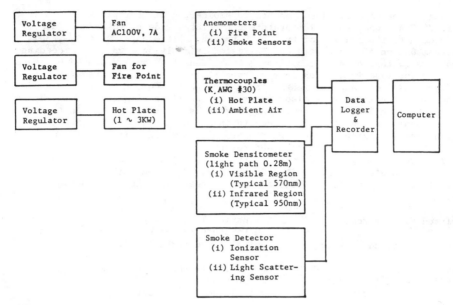

FIGURE 3   Block diagram

SENSOR CALIBRATION

The ionization sensor and light scattering sensor were calibrated in accordance with Japanese technical standards[6] for fire alarm systems.  The test
chamber used for this calibration is a smoke circulating box.  Smoke for the
calibration was generated by smoldering of a sheet of filter paper which was
sandwiched between two hot plates of 75mm diameter and heated up to 400°C.  The
circulating air velocity was 0.2m/s.

The ionization sensor is calibrated with a parallel electrode measuring
ionization chamber and the density is indicated by the rate of ionization current change ($x=(Io-I)/Io$).  FIGURE 4 indicates the correlation between MIC prescribed in EN54 Part 7 and Japanese MIC including the scale of Y value as well
as x Value[5,7].  The standard sensitivity of the ionization detector in Japan is
$x = 0.24$ and the corresponding value in EN-MIC is about 1.2 in Y value.  The
value obtained with the ionization sensor in the test is expressed in Y value,
which is obtained by calibrating the sensor output with this calibration value.

The smoke densitometer for calibrating the light scattering sensor is an
extinction type densitometer sensitive in the visible region.  It uses an incandescent lamp with a color temperature of 2800K for light source and a luminosity
-corrected selenium photo-cell with a typical wavelength of 570nm for the light
receiving element.  The standard light scattering detector used in Japan has a
sensitivity of 10%/m in extinction rate.  The corresponding m value defined in
EN 54 Part 7 is 0.46 (dB/m).  In the test results for the light scattering sensor, the value indicated represents the sensor output for filter paper smoke and
converted into m (dB/m) value.  Here, it should be noted that the selenium photo
-cell differs in light sensitive region from the densitometer used in EN standards and that the light scattering sensor used has a light emitting cycle of 0.2
second which is 20 times shorter than that for ordinary detectors.

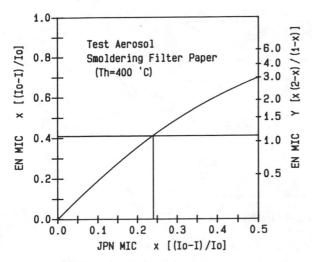

FIGURE 4   Correlation between EN-MIC and Japanese-MIC

["MIC": Measuring Ionization Chamber]


TEST RESULTS AND CONSIDERATION

     FIGURE 5 shows typical results obtained with the ionization and light scat-
tering sensors in smoldering fire of a printed circuit board.  The conditions
under which these data were taken are as follows.

° Amount of combustion:   Glass-fiber reinforced epoxide printed
                          circuit boards
                          3 sheets (initial weight 3.6g)

° Hot plate temperature:  450°C

° Duration of combustion: 3 minutes

° Air velocity at         1m/s (constant)
  location of sensor:

° Air velocity at fire    0.5 to 8m/s
  point:

     The result shows that ionization sensor output increases and light scat-
tering sensor output decreases with increasing air velocity at the fire point.
This result also indicates a decrease of smoke particle size and increase of
particle concentration (z).  To prove this, the extinction type smoke densities
... m values (dB/m) both in the visible region and near-infrared region are compared in FIGURE 6.  The ratio of m values obtained in 0.8 for an air velocity at
the fire point of 0.5m/s, but falls to 0.35 for an air velocity of 4.0m/s.
FIGURE 7 shows the comparison of response of the ionization sensor with that of
the light scattering sensor, and includes results obtained by changing quantities of the samples in such a way that they show weight loss of 0.2g, 0.4g, and
0.6g during three minutes' combustion under five different air velocities, from
0.5m/s to 8m/s at the fire point.  The plotted data show the maximum values of

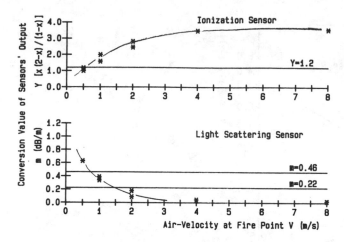

FIGURE 5  Correlation between air velocity at fire poing and output of smoke sensor

Burning material: printed circuit board

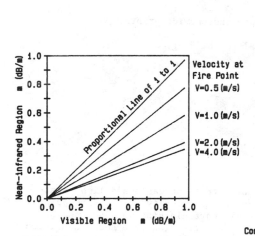

FIGURE 6  Correlation between smoke densities in visible region and near infrared region with air velocity at fire point

Burning material: printed circuit board

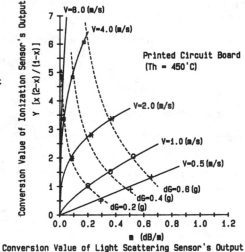

FIGURE 7  Comparison of response of the ionization sensor with that of the light scattering sensor

the sensor outputs in each test. The broken lines show the curves at the reduction of the corresponding weight and solid lines are isolines of air velocity.
For a velocity of 0.5m/s, the relationship between m and Y values is approximately linear but as the velocity increases, the solid lines become more apart
from the straight line forming curves. This suggests coagulation in a shorter
time or increase of particle size following increase of the smoke generation
rate.

FIGURE 8a shows the rate of weight loss versus air flow during 3 minutes
required for operation of the ionization detector with threshold level Y=1.2 and
the light scattering detector with threshold level m=0.22 and m=0.46. While the
weight loss increased with increase of air velocity at fire point, in the case
of operation of the light scattering detector, there was decrease in the weight
loss in the case of the ionzation detector. This indicates that the ionization
detector has higher sensitivity. Although the air velocity at the sensors' location was kept constant, it practically increased with increase of air velocity
at the fire point. Therefore, dilution by air flow must be taken into consideration. If the particle diameter is constant, the weight loss at the time of operation of detector is the function of air flow rate. FIGURE 8b shows the rate
of weight loss with consideration of dilution. It shows that the sensitivity of
the ionization detector is almost constant, but that of the light scattering
detector is extremely decreased. It is varied by the setting level of the
threshold of the detector, but the turnning piont is 0.5 to 1.0 m/s.

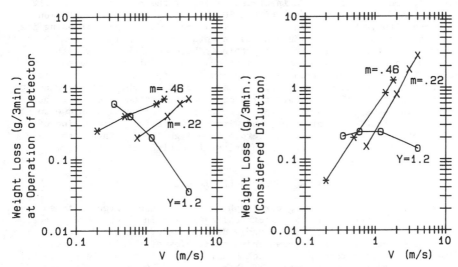

FIGURES 8a(left), 8b(right)  Weight loss required for operation of the
smoke detector

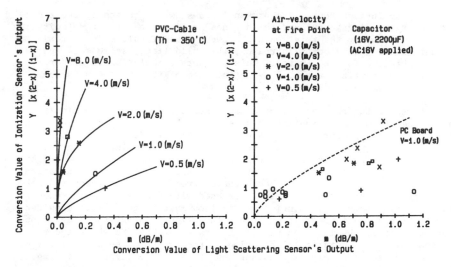

FIGURES 9(left), 10(right)  Comparison of response of the ionization
sensor with that of the light scattering sensor

FIGURE 9 shows data for smoldering of a PVC insulated cable at 350°C, indicating a trend similar to the example for the printed circuit board.  This indicates that when commercially available ionization detectors and light scattering
detectors are compared, the former show better response to smoldering fire in
the printed circuit board and PVC insulated cable, if the air velocity at the
location of the fire exceeds 1m/s.  FIGURE 10 shows data obtained from combustion of a capacitor, which was caused by application of a voltage exceeding its
breakdown voltage in air flow.  In these data we see no significant trend by air
velocities at the fire point.  Possibly, this is because smoke grows within the
capacitor, and almost stabilized smoke is discharged out of the capacitor.
Therefore, it shows the same trend as smoldering fire in the printed circuit
board at low air velocities, and it can be said that the light scattering sensor
shows better response.  However, attention must be paid to the fact thet the
light scattering sensor emits light in pulses to prevent misoperation due to
background noise from external light, and with consideration of reduction of
power consumption, LED life, etc.  Therefore, the light scattering sensor may
fail to detect transient smoke from a capacitor in a high air velocity.


DEVELOPMENT OF IONIZATION SENSOR HOUSED IN ELECTRICAL EQUIPMENT

As described in the preceding paragraph, the ionization detector is suitable for early detection of fire in electrical equipment in which there is a flow
of cooling air.  An ionization sensor newly developed for use within such equipment is shown below.  FIGURE 11 shows an external view of this compact sensor
with dimensions of less than 40mm on each side.  FIGURE 12 shows the horizontal
and vertical air flow characteristics of this sensor, in which the maximum and
minimum values are given to show output fluctuation ranges in deviation from the
air velocity of 0m/s when angles of the horizontal and vertical air flow are
varied.  Threshold levels in the figure are standard values, the outputs at each
air velocity of 10m/s being +20% or less of these values.  This sensor has less
possibility of false operation of delayed detection which is caused by influence
of air flow and which is inherent in the conventional ionization detector.  The

706

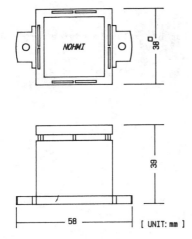

FIGURE 11   Ionization sensor
(FDS765C)

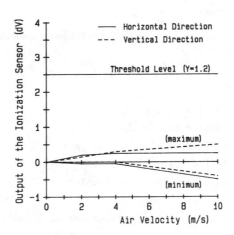

FIGURE 12   Air flow characteristics of
the ionization sensor

response time to smoke in air flow exceeding 1m/s is within 15 seconds.  This
analog ionization sensor, connected to a control unit, is capable of monitoring
smoke down to low output level.  It has also been possible to provide the sensor
with a trouble detecting function, multistage alarm function, and remote operation test function.  Therefore, this sensor is suitable for early detection of
equipment fire.


CONCLUSION

Generally, the ordinary smoke detectors are arranged in such a way that
each of them supervises a large area.  Therefore, even in the case of a flaming
fire to which the ionization detector shows good response, it takes a certain
time for smoke to reach the detector if the fire occurs remote from the detector,
and thus detection may be delayed due to smoke coagulation.

However, even in the case of smoldering fire which is not favorable to the
ionization detector, it has proved to show good response if smoldering fire occurs within the equipment where there is an air flow.  It has also been found
that the ionization detector shows better response than the light scattering detector to smoldering fire in printed circuit boards and cables, if air velocity
exceeds about 1m/s.  There is an exceptional case in which the light scattering
detector shows better resopnse to smoldering fire in the capacitor, but might
fail to give an alarm depending on the light emitting pulse cycle if there is
high air flow.  Therefore, the ionization detector is preferable for early detection of fire in equipment with a cooling fan.  It is expected that the small
ionization sensor reported in this paper will demonstrate its ability to detect
fire within electrical equipment.

ACKNOWLEDGEMENT

The author wishes to thank Professor T. Handa for his valuable advice in preparing this paper.

NOMENCLATURE

| | | |
|---|---|---|
| dG | weight loss | [g] |
| I | ionization current | [pA] |
| Io | initial ionization current | [pA] |
| m | optical density | [dB/m] |
| Th | surface temperature of hotplate | [°C] |
| V | air velocity | [m/s] |
| x | rate of ionization current change | [(Io-I)/Io] |
| Y | smoke concentration | [x(2-x)/(1-x)] |

REFERENCE

1.  Lee, T.G.K. and Mulholland, G.W.: "Physical Properties of Smokes Pertinent to Smoke Detector Technology," NBS, NBSIR 77-1312, 1977.

2.  Bukowski, R.W. and Mulholland, G.W.: "Smoke Detector Design and Smoke Properties," NBS, TN973, 1978.

3.  "Smoke Detectors for Fire Protective Signaling Systems," Standard for Safety, UL268-1981, Underwriters Laboratories Inc., 1983.

4.  "Components of Automatic Fire Detection Systems, Part 7, Specification for Point-type Smoke Detectors using Scattered Light, Transmitted Light or Ionization," EN54-Part 7.

5.  "Components of Automatic Fire Detection Systems, Part 9, Methods of Test of Sensitivity to Fire," EN54-Part 9.

6.  "Ministerial Ordinance for Technical Standards for Fire Alarm System (Japan)," Minitry of Home Affairs, 1977.

7.  Scheidweiler A.: "The Ionization Chamber as Smoke Dependent Resistance," Fire Technology, 2, 12, 1976.

8.  Takemoto A. and Watanabe A.: "The Particle Size Distribution Products and Thier Effects on the Response of Smoke Detectors," Report of Fire Research Institute of Japan, 38, 1974.

9.  Watanabe A. and Takemoto A.: "Response Characteristics of Smoke Detectors," Bulletin of the Fire Prevention Society of Japan, 2, 21, 1972.

10. Helsper C., Fissan H.J., Muggli J. and Scheidweiler A.: "Particle Number Distributions of Aerosols from Test Fires," J. Aerosol Sci., 11, 1980, (UK).

# Global Soot Growth Model

**G. W. MULHOLLAND**
Center for Fire Research
National Bureau of Standards
Gaithersburg, Maryland 20899, USA

ABSTRACT

Analytical results for soot concentration, average particle size, and $\sigma$ of the size distribution are obtained for a free radical soot growth model which includes a constant nucleation source, growth, and coagulation. Results are obtained with and without coagulation included and for a size independent growth rate as well as growth rate proportional to the surface area. Neither this model nor a nucleation pulse model is able to account for all the results on soot formation for shock tube and pyrolysis experiments.

INTRODUCTION

Soot produced in fires is a major concern. The emission of radiation from soot plays a dominant role in fire spread. The reduced visibility caused by soot is a significant impediment to persons escaping from fires. The deposition of soot in the respiratory tract is a potential health hazard especially in the case of repeated exposures by firemen. On the other hand, a low concentration of soot is adequate to activate a smoke detector alarm and thus provide a warning of a fire.

While there has been a large amount of experimental work concerning soot formation under a wide variety of conditions, as evidenced by some 388 references in a recent review article by Haynes and Wagner [1], there has been relatively little theoretical effort to integrate the various processes leading to soot formation and growth into a model capable of predicting soot concentration and particle size. Jensen [2] has developed such a model for predicting soot generation from methane fuel in an exhaust jet. The approach taken here is similar in spirit to Jensen's method though with less structure in the individual processes. The advantage of our model is that it can be solved analytically, and this allows a more general understanding of the effect of the various growth processes on the final particle size.

The model includes the formation of soot nuclei, surface growth, and coagulation. The nucleation process is modeled as a thermal pyrolysis leading to a free radical species. The growth stage is first treated in a manner analogous to chain polymerization. Growth proportional to the surface area is also considered. For the combined growth model, the soot concentration, the average particle size, and the width of the size distributions are determined.

The model assumes a homogeneous isothermal system and for that reason the most direct validation is with shock tube experiments and pyrolysis experiments rather than with flame data. The model results are compared with the available data.

## FREE RADICAL SOOT GROWTH MODEL

This model is very similar to what is called polymerization without termination in the polymer literature. The first step involves the thermal pyrolysis of the fuel to yield free radical species.

$$F \xrightarrow{k_1} R_1 + R_a$$

where $R_1$ indicates the incipient soot nuclei and $R_a$ indicates a gaseous radical such as H atoms. The rate constant $k_1$ is taken to have an activation energy, $E_a$. Of course, the chemistry leading to the soot precursor in an actual system is not a simple elementary reaction so this is a global expression of the kinetics.

The free radical $R_1$ initiates the chain polymerization.

$$R_1 + F \xrightarrow{k_2} R_2$$

$$R_j + F \xrightarrow{k_2} R_{j+1}$$

The symbols, $F$, $R_1$, $R_j$, etc. represent species in the equations above; we use the same symbols to represent concentration in the kinetic equations developed below. The rate constant $k_2$ is taken to be a constant independent of particle size. It is assumed that the fuel adds directly to the radical species. The polymer species $R_j$ are considered to be soot particles. We denote by N the total number concentration of soot.

$$N = \sum_{j=1}^{\infty} R_j \tag{1}$$

The rate of change of N depends only on the pyrolysis reaction.

$$dN/dt = k_1 F \tag{2}$$

The rate of change of the fuel concentration is obtained by adding up all the loss terms,

$$dF/dt = -k_1 F - k_2 FN \tag{3}$$

In the small time limit, we replace F with $F_o$ in eqs. (2) and (3) and obtain

$$N = k_1 F_o t \tag{4}$$

$$F_o - F = k_1 F_o t + 1/2\, k_1 k_2 \left(F_o t\right)^2 \tag{5}$$

The fuel depletion, $F_o - F$, is also proportional to the soot volume concentration for this model, because there are only two types of species--fuel and soot. Equations (2) and (3) have also been solved exactly and the results plotted in figure 1 for the following conditions: initial fuel concentration of $2.5 \times 10^{17}$ cm$^{-3}$, $k_2 = 10^{-11}$ cm$^3$/s, and $k_1 = 10^{-3}$ s$^{-1}$. The linear portion of the curves correspond to eqs. (4) and (5). These expressions remain valid until the fuel becomes depleted by 10-20%. The time dependence of the average volume can be determined from eqs. (4) and (5).

$$\bar{v} = v_o \left(F_o - F\right)/N = v_o + 1/2\, v_o k_2 F_o t \tag{6}$$

where $v_o$ is the volume of a fuel molecule.

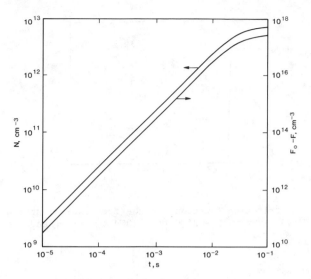

FIGURE 1. Number concentration, $N$, and fuel depletion, $F_0 - F$, are plotted versus time for the free radical growth model.

For this model, the size distribution function, $R_j$, can also be obtained. The rate equations for the species $R_j$ are given by

$$dR_1/dZ = k_1/k_2 - R_1$$

$$dR_j/dZ = R_{j-1} - R_j$$

$$(7)$$

where $dZ = k_2 F dt$. Solving this set of linear equations sequentially, we obtain

$$R_j = \frac{k_1}{k_2}\left[1 - \sum_{i=0}^{j-1} \frac{Z^i}{i!} e^{-Z}\right] \qquad (8)$$

As an example, the size distribution function is plotted in figure 2 for the same set of conditions as described above. The value of $j$ at which $R_j$ drops by a factor of two, denoted by $j_1$, is given by $j_1 \approx Z$. The value of $R_j$ changes rapidly about the value $j_1$ changing from $0.99\ k_1/k_2$ to $0.01\ k_1/k_2$ as $j$ increases from $j_1 - 2\sqrt{j_1}$ to $j_1 + 2\sqrt{j_1}$.

The average particle volume, $\bar{v}$, and $\sigma$, defined by $\left(\overline{v^2} - \bar{v}^2\right)^{1/2}$ for this size distribution are found to be,

$$\bar{v} = v_1/2 \qquad (9)$$

$$\sigma = \bar{v}/\sqrt{3} \qquad (10)$$

where $v_1$ corresponds to a particle with $j_1$ monomer units. The abrupt change in $R_j$ at $j = j_1$ is approximate as a step change in deriving eqs. (9) and (10). So, as is apparent from figure 2, the size distribution is broad with the width proportional to the mean volume.

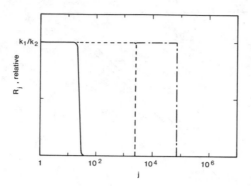

FIGURE 2. Size distribution function for the free radical growth model for
$Z = 25$ $(t = 10^{-5}$ s) ————, $Z = 2500$ $(t = 10^{-3}$ s) -----, and $Z = 70,000$
$(t = 0.1$ s) ——— - ———.


GROWTH PLUS COAGULATION

At the very high particle concentration typical of a flame environment, particles
will collide and stick together in the coagulation process. The rate of decrease
of the total number concentration, $N$, is proportional to the square of the number
concentration with coagulation coefficient designated by $\Gamma$.

$$dN/dt\big|_{coag} = - \Gamma N^2$$

This represents a first order treatment of coagulation; a more detailed treatment
would include the dependence of the coagulation coefficient on particle size such
as the analysis by Dobbins and Mulholland [4]. Including coagulation, eq. (2)
becomes

$$dN/dt = k_1 F - \Gamma N^2 \tag{11}$$

In the following analysis, we assume that the fuel is only slightly depleted so
that $F$ may be replaced by $F_0$ in eq. (11). Integrating eq. (11), we obtain

$$N = N_{ss} \tanh (t/\tau), \tag{12}$$

where 
$$N_{ss} = \left(k_1 F_0 / \Gamma\right)^{1/2} \tag{13}$$

$$\tau = \left(k_1 F_0 \Gamma\right)^{-1/2} \tag{14}$$

Substituting from eq. (12) into eq. (3) with $F$ replaced with $F_0$ on the right hand
side and integrating, we obtain

$$F_0 - F = k_1 F_0 t + \frac{k_2 F_0}{\Gamma} \log [\cosh(t/\tau)] \tag{15}$$

The number concentration, $N$, and fuel depletion, $F_0 - F$, for small time have the
same form as the case of no coagulation. For long time, the number concentration
approaches a steady state as the rate of formation is balanced by the loss rate
from coagulation. It is convenient to use the following approximate form for the
time dependence of $N$:

$$N = N_{ss} \, t/\tau \qquad t < \tau$$
$$N = N_{ss} \qquad\qquad t > \tau \tag{16}$$

Using this form for N in eq. (3) leads to the following expression for the fuel depletion:

$$F_o - F = k_1 F_o t + 1/2 \, k_1 k_2 F_o^2 t^2 \qquad t < \tau$$
$$F_o - F = k_1 F_o t + k_2 N_{ss} F_o (t - \tau/2) \qquad t > \tau \tag{17}$$

As seen in figure 3, the approximate expressions are valid in the limits of short and long time and, at worst, are about 15% greater than the correct value.

The rate equation for the species $R_j$ are given by

$$dR_1/dt = k_1 F_o - k_2 R_1 - 2 \, \Gamma \, R_1 N$$
$$dR_j/dt = k_2 F_o R_{j-1} - k_2 F_o R_j + \Gamma \sum_{i+k=j} R_i R_k - 2 \, \Gamma \, R_j N \tag{18}$$

The kinetic equations for the first three moments of the distribution function, N, V, and $V_2$, are readily derived from eq. (18).

$$dN/dt = \Sigma R_j = k_1 F_o - \Gamma \, N^2$$
$$dV/dt = v_o \Sigma_j R_j = k_1 F_o v_o + k_2 F_o N v_o \tag{19}$$

$$dV_2/dt = v_o^2 \, \Sigma j^2 R_j = k_1 F_o v_o^2 + 2 \, \Gamma \, V^2 + 2 \, k_2 F_o v_o V$$

The volume concentration V is simply $v_o(F_o - F)$. In integrating eq. (19) to obtain $V_2$, we approximate N with eq. (16) and V using the second term in eq. (17), which dominates the first for most cases of interest. For a similar reason, the first term in eq. (19) is dropped. The intensive quantities $\bar{v}$ and $\sigma$, derived from N, V, and $V_2$, are given by

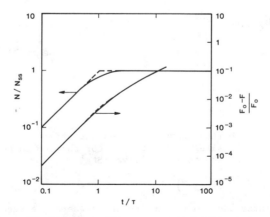

FIGURE 3. Reduced number concentration and reduced fuel depletion are plotted versus time for the free radical growth model with coagulation included. The dashed curves correspond to the approximations given by eqs. (16) and (17).

713

$$\bar{v} = \frac{v_o k_2 F_o \tau}{2} \ (t/\tau) \qquad\qquad t < \tau$$

$$\bar{v} = v_o k_2 F_o \tau \ (t/\tau - 1/2) \qquad\qquad t > \tau \qquad\qquad (20)$$

$$\sigma = \frac{(v_o k_2 F_o \tau)}{2 \sqrt{3}} \ (t/\tau) = \bar{v}/\sqrt{3} \qquad\qquad t < \tau$$

$$\sigma = (v_o k_2 F_o \tau) \ \sqrt{2/3} \ (t/\tau)^{3/2} = \sqrt{2/3} \ (t/\tau)^{1/2} \ \bar{v} \qquad t > \tau \qquad (21)$$

The presence of coagulation causes the ratio $\sigma/\bar{v}$ to grow with time. Without coagulation, the ratio $\sigma/\bar{v}$ is constant. It can be seen from the study of Dobbins and Mulholland [4] that, with a constant nucleation source and no growth except by coagulation, one obtains $\sigma/\bar{v} \sim t^{1/2}$ for large time, just as in eq. (21) above. Boisdron and Brock [5] have analyzed eq. (18) for the initial value problem where, at time t equal zero, there are only species $R_1$ plus the growth species. The nucleation term, $k_1 F_o$, is not present in their analysis. Under this condition, the number concentration decreases with respect to time rather than increasing to a steady state value and $\sigma$ becomes equal to the average particle volume in the long time limit. We shall consider these results in relation to experimental results below.

SURFACE GROWTH

In the free radical growth model, it is assumed that there is only a single active site on each growing soot particle. There is experimental evidence for premixed flames [6] that the growth rate is proportional to the surface area. To consider this possibility, we take the growth rate for $R_j$ to be proportional to $j^{2/3}$. The model equations then become

$$dR_j/dt = k_2 \ (j-1)^{2/3} \ R_{j-1} - k_2 \ (j)^{2/3} \ R_j \qquad\qquad j > 1$$

$$dR_1/dt = - k_1 F_o - k_2 R_1 \qquad\qquad (22)$$

The rate of change of the fuel concentration in the limit of small fuel depletion is given by

$$dF/dt = - k_1 F_o - k_2 F_o \ \left[\sum_j R_j j^{2/3}\right] \qquad\qquad (23)$$

The presence of the fractional exponent complicates the problem. We approximate the sum in eq. (23) as follows:

$$\sum_j R_j \ (j^{2/3}) = N \left(\frac{F_o - F}{N}\right)^{2/3} \qquad\qquad (24)$$

The change in the growth rate expression does not affect the expression for dN/dt; eq. (2) is still valid. Substituting for the sum in eq. (23) and then taking the ratio of eq. (23) to eq. (2), we obtain

$$dF/dN = - 1 - k_2/k_1 \ N^{1/3} \ (F_o - F)^{2/3} \qquad\qquad (25)$$

We are primarily interested in the growth stage where the second term on the right hand side is large compared to the first. So dropping the first term and integrating, we find

$$F_o - F = \left(1/4 \; \frac{k_2}{k_1}\right)^3 N^4,$$ (26)

or

$$F_o - F = (1/4)^3 \left(k_2 F_o\right)^3 k_1 F_o t^4$$ (27)

The average volume is given by

$$\bar{v} = v_o \; (1/4)^3 \left(k_2 F_o t\right)^3.$$ (28)

We see that the particle growth is much more rapid than in the case of the free radical growth.

The steady state size distribution function obtained from eq. (22) is given by

$$R_j = k_1/k_2 \; F_o j^{-2/3}$$ (29)

In the case of free radical growth treated in Section 2, the steady state concentration is independent of size. Using eq. (28) and assuming an upper size cutoff at j=n, we obtain the following expressions for $\bar{v}$ and $\sigma$:

$$\bar{v} = v_o/4 \; n$$ (30)

$$\sigma = 12/\sqrt{112} \; \bar{v} = 1.13 \; \bar{v}$$ (31)

As in the free radical growth, $\sigma$ is proportional to $\bar{v}$ though, in this case, the coefficient is about twice as large as for free radical growth.

The effect of coagulation is considered by approximating the sum in eq. (23) using eq. (24) and then substituting the expression given by eq. (16) for N. Integrating, we obtain

$$
\begin{aligned}
F_o - F &= (1/4)^3 \left(k_2 F_o \tau\right)^3 N_{ss} \; (t/\tau)^4 & t < \tau \\
&= (1/3)^3 \left(k_2 F_o \tau\right)^3 N_{ss} \; (t/\tau - 1/4)^3 & t > \tau
\end{aligned}
$$ (32)

The corresponding expressions for $\bar{v}$ are given by

$$
\begin{aligned}
\bar{v} &= v_o \; (1/4)^3 \left(k_2 F_o \tau\right)^3 (t/\tau)^3 & t < \tau \\
&= v_o \; (1/3)^3 \left(k_2 F \tau\right)^3 (t/\tau - 1/4)^3 & t > \tau
\end{aligned}
$$ (33)

The fuel depletion is less rapid with coagulation included as a result of the smaller number of growth sites, but the average particle size is about two times larger. We are not able to solve for the $\sigma$ of the size distribution with the coagulation included.

COMPARISON WITH EXPERIMENTS

Flames are not homogeneous with respect to concentration and temperature; in addition, there are both growth and burnout processes taking place. The most direct experimental tests of the models are with shock tube and high temperature pyrolysis experiments.

In a shock tube study by Frenklach et al. [7], soot formation was studied as a function of fuel concentration, temperature, and time. The results for acetylene for T < 2000 k can be expressed in the form

$$F_o - F \sim e^{-E_a/RT} F_o^a t^b. \tag{34}$$

From the data in series D for times of 0.5, 1.0, 1.5, and 2.0 ms, we obtain a value of about 2.7 for b. From data at carbon atom concentrations of about $2.0 \times 10^{17}$ cm$^{-3}$ and $5.0 \times 10^{17}$ cm$^{-3}$, we obtain a value of 1.5 for a with a large uncertainty, $\pm 0.5$, because of the limited range of data. Prado and Lahaye [8] measured the dependence of the soot volume fraction on the concentration of benzene in nitrogen at a temperature of 1383 k for a 0.5 second residence time. Measurements were made at eight concentrations over the range 0.25 to 10% with a value of 1.5 for a. The free radical growth model without coagulation is the best compromise for the exponents a and b with values of 2.0 for both versus the experimental values of 1.5 and 2.7. The dependence of $F_o - F$ on $F_o$ and t for each model is given in Table 1.

TABLE 1. Summary of Model Results[a]

| | F.R.G. | F.R.G. + Coag. | S.G. | S.G. + Coag. |
|---|---|---|---|---|
| Constant Nucleation Source[b] | | | | |
| N | $k_1 F_o t$ | $(k_1 F_o)^{1/2}$ | $k_1 F_o t$ | $(k_1 F_o)^{1/2}$ |
| $F_o - F$ | $k_1 F_o^2 t^2$ | $k_1^{1/2} F_o^{3/2}(t-\tau/2)$ | $k_1 F_o^4 t^4$ | $k_1^{1/2} F_o^{7/2}(t-\tau/4)^3$ |
| $\bar{v}$ | $F_o t$ | $F_o(t-\tau/2)$ | $F_o^3 t^3$ | $F_o^3(t-\tau/4)^3$ |
| $\bar{v}/\sigma$ | $\sqrt{3}$ | $\sqrt{3/2}\,(\tau/t)^{1/2}$ | $1/1.13$ | -- |
| Nucleation Pulse | | | | |
| N | $N_o$ | $t^{-1}$ | $N_o$ | $t^{-1}$ |
| $F_o - F$ | $F_o t$ | $F_o \log t$ | $F_o^{3/2} t^3$ | $F_o^{3/2} t^2$ |
| $\bar{v}$ | $F_o t$ | $F_o t \log t$ | $F_o^{3/2} t^3$ | $F_o^{3/2} t^3$ |
| $\bar{v}/\sigma$ | $\sqrt{\bar{v}}$ | 1 | -- | -- |

[a]Numerical factors are not included for N and $F_o - F$, and the temperature independent constant $k_2$ is not included.

[b]Abbreviations: F.R.G. - free radical growth; S.G. - surface growth; Coag. - coagulation.

The pyrolysis experiments of Prado and Lahaye indicate that the number concentration is independent of residence time. This is consistent with the steady state result obtained in the model as a result of a balance between nucleation and coagulation. Prado and Lahaye interpret their results differently by stating that the constancy of the number concentration is a result of both nucleation and coagulation being completed by 0.1 seconds when their first sample is taken. For the conditions of pyrolysis experiment with a temperature of about 1400 K and a particle size of 50 nm, we calculate a free molecular coagulation coefficient [9] of $3.4 \times 10^{-9}$ $cm^3/s$. For an initial number concentration of $10^{10}$ $cm^{-3}$, the concentration would decrease by a factor of 70 over the two second residence time in the pyrolysis tube as a result of coagulation.

Prado and Lahaye found that while the average particle diameter changed by a factor of five as the fuel concentration was increased 40 fold, the ratio $\bar{D}/\sigma_D$ had a constant value of about 5.5. To compare with our results in Table 1, we compute $\bar{v}/\sigma_v$ based on the assumption that the size distribution is Gaussian with respect to particle diameter. We obtain

$$\frac{\bar{v}}{\sigma_v} = \frac{k \, (1 + 3/k^2)}{(9 + 36 \, k^{-2} + 15 \, k^{-4})^{1/2}} \tag{35}$$

where $k = \bar{D}/\sigma_D$. For a value of 5.5 for $k$, we obtain $\bar{D}/\sigma_v$ equal 1.89. Again, the free radical growth model without coagulation gives the best agreement with this experimental result.

The growth species in the models presented is assumed to be the fuel molecule. If, instead, the growth species were acetylene, then for fuels other than acetylene, the pyrolysis reactions leading to the growth species must be included in the model.

The activation energy, $E_a$, determined by Frenklach et al. [7] for acetylene was about 126 kJ/mol. This is to be compared with a value of about 500 kJ/mol for the free radical growth model. Including coagulation reduces the model prediction by a factor of two, but, as pointed out above, the other aspects of the model agree better with the free radical growth without coagulation.

DISCUSSION

The free radical growth model contains a chemical nucleation with an Arrhenius type kinetics. For constant temperature and small fuel depletion, the nucleation rate is constant. This differs from the short nucleation pulse predicted by homogeneous nucleation theory [10]; in this case, the nucleation is turned off by a slight decrease in the concentration of the condensing vapor. As a limiting case, the nucleation is assumed to be completed before the surface growth and coagulation begin. In Table 1 we have included the predictions for such a case which corresponds to a fixed initial number concentration. A similar model was described by Boisdron and Brock [5]. The best agreement between the model and the shock tube and pyrolysis data for soot concentration is for the surface growth case. This is in contrast to the constant nulceation source where the best agreement is for free radical growth (rate constant $k_2$ independent of particle size). The narrower size distribution predicted for the nucleation pulse case appears to agree better with the available data [8], though the relatively constant number concentration observed by Prado and Lahaye [8] agrees better with the prediction of the constant nucleation source with coagulation included. At the high concentration typically observed, coagulation is surely important unless there is a very small sticking coefficient for colliding particles.

717

So it is seen that no one of these models is consistent with all the experimental observations. Major improvements in our understanding of soot growth will require innovative experimental approaches coupled with theoretical analyses. There is a need for data on the time dependence of the nucleation process. The nucleation rate and duration probably have the largest effect on the amount of soot produced and its size distribution. In order to better understand the growth processes, information is needed in regard to the sticking probability for gas-particle and particle-particle collisions as a function of particle size. The viscosity of the growing particles is required to model the transition from coalescence to agglomeration.

REFERENCES

[1]   Haynes, B.S. and Wagner, H. Gg., Soot formation, Prog. Energy Combust. Sci, 7, 229 (1981).

[2]   Jensen, D.E., Prediction of soot formation rates: a new approach, Proc. R. Soc. Lond. A. 338, 375 (1974).

[3]   Allen, P.E.M. and Patrick, C.R., Kinetics and Mechanisms of Polymerization Reactions, 162, Wiley, NY (1974).

[4]   Dobbins, R.A. and Mulholland, G.W., Interpretation of optical measurements of flame generated particles, Combustion Science and Technology, 40, 175 (1984).

[5]   Boisdron, Y. and Brock, J.R., Particle growth processes and initial particle size distributions, Assessment of Airborne Particles (Mercer, T.T., Morrow, P.E. and Stöber, W.) 129, Thomas, Springfield, IL (1972).

[6]   Harris, S.J. and Weiner, A.M., Surface growth of soot particles in premixed ethylene/air flames, Combustion Science and Technology 31, 155 (1983).

[7]   Frenklach, M., Taki, S., Durgaprasad, M.B. and Matula, R.A., Soot formation in shock-tube pyrolysis of acetylene, allene, and 1,3-butadiene, Combust. Flame 54, 81 (1983).

[8]   Prado, G. and Lahaye, J., Physical aspects of nucleation and growth of soot particles, Particulate Carbon Formation During Combustion (edited Siegla, D.C. and Smith, G.W.) 143, Plenum Press, NY (1981).

[9]   Friedlander, S.K., Smoke, Dust and Haze, 179, Wiley, NY (1977).

[10]  Frenkel, J., Kinetic Theory of Liquids, Dover, NY (1955).

# Numerical Simulations of Smoke Movement and Coagulation

YUKIO YAMAUCHI
R & D Center, Hochiki Corp.
246 Tsuruma, Machida-shi, Tokyo 194, Japan

ABSTRACT

For the purpose of understanding the smoke detector's response to enclosure fires, a computational model for predicting the evolution of the local concentration and size distribution of smoke aerosol is presented. The model utilizes the characteristic of smoke aerosol that the large size part of the size distribution is not significantly affected by coagulation and that its size distribution takes a virtually time-independent reduced form. By numerically solving the conservation equations of total particle volume and total particle number, the size distribution of the smoke aerosol is determined by the reduced expression of the size distribution where the total particle volume and the total particle number are the parameters. The results of sample calculations and relevant experiments showed a reasonable agreement.

Key words: Aerosols; enclosure fires; detector sensitivity; fire detectors; modeling; particle size; room fires; simulation; smoke detectors.

## INTRODUCTION

Smoke detectors are an effective means for early detection of fires in enclosures. Light scattering type smoke detectors and ionization type smoke detectors are mostly in use. However, the response of these detectors is strongly dependent on the particle size of the smoke aerosol and its concentration [1,2,3]. Since the particle size distribution and the concentration of smoke aerosol vary during the evolution of the fire, fire tests have been the only tools for evaluating the detector response [4].

The variation in the size distribution of smoke aerosol is explained as a result of the phenomenon of particle coagulation due to the Brownian collisions [5]. In order to predict the detector response to a given fire in an enclosure it is necessary to evaluate the effect of coagulation as well as the large scale smoke movement.

The purpose of this report is to present a computational procedure for predicting the evolution of the local concentration and size distribution of smoke aerosol in an enclosure containing a fire source. The procedure utilizes the findings of Mulholland, et al. [6] that the large size part of the size distribution of smoke aerosol is not greatly affected by coagulation and that its size distribution takes a virtually time-independent reduced form:

$$\zeta = 0.1 \, (\eta + 0.1)^{-2}, \qquad\qquad\qquad (1)$$

where $\zeta$ and $\eta$ are the reduced number distribution and the reduced particle volume respectively. The corresponding unreduced size distribution is

$$n(v,t) = 0.1V[v + 0.1V/N(t)]^{-2}. \tag{2}$$

This suggests that if the total particle volume V and the total particle number N are given the number distribution n as a function of particle volume v is determined by Eq. (2).

In this paper a computational model for predicting the total particle volume V and the total particle number N locally and thus predicting the local size distribution and concentration of smoke aerosol is described. In order to validate the theoretical model the predictions were compared with some experimental data.

THEORETICAL MODEL

Smoke Movement

Smoke aerosol generated at the heat/smoke source is transported by the velocity field of buoyant convection. Thermal molecular diffusion also plays a small role. Regarding the smoke concentration (mass, volume or number) as a passively-transported scalar quantity in the field, its time averaged conservation equation for the constant density flows takes the form:

$$\frac{\delta C}{\delta t} + \frac{\delta C u_j}{\delta x_j} = \frac{\delta}{\delta x_j}\left(D^*\frac{\delta C}{\delta x_j}\right) + S, \tag{3}$$

where C is the smoke concentration: $u_j$ (i,j = 1,2,3) is the component velocity in the direction $x_j$; $D^*$ is the effective diffusion coefficient; S is the source/sink term. The effective diffusion coefficient $D^*$ is the combined laminar and turbulent diffusions defined by

$$D^* = D + K \tag{4}$$

where D and K respectively are the laminar and turbulent diffusion coefficient.

Assuming that the wall adsorption and settling of aersol are insignificant, the source/sink term in the conservation equation of total particle volume can be neglected except at the point of heat/smoke source:

$$\frac{\delta V}{\delta t} + \frac{\delta V u_j}{\delta x_j} = \frac{\delta}{\delta x_j}\left(D^*\frac{\delta V}{\delta x_j}\right) \tag{5}$$

where $V = V(x_i,t)$ is the total particle volume of smoke aerosol within a unit volume at the point $x_i$ at at the time t.

Smoke Coagulation

Smoke particles moving in the enclosure collide and adhere with each other as a result of the Brownian motion. Through this process the average particle size increases while the total number of particles decreases. The coagulation frequency is closely related to the temperature and the size of the aerosol particles and its concentration.

The basic equation for the variation of the number of aerosol particles versus time due to thermal coagulation takes the form:

$$\left[\frac{\delta\,n\,(x_i\,,v\,,t)}{\delta\,t}\right]_{int} = \int_0^v \Gamma\,(v-v'\,,v')\;n(x_i\,,v-v'\,,t)\;n(x_i\,,v'\,,t)\;dv$$

$$- 2\;n(x_i\,,v\,,t)\int_0^\infty \Gamma\,(v\,,v')\;n(x_i\,,v'\,,t)\;dv' \qquad (8)$$

where $n(x_i\,,v\,,t)dv$ means the number of aerosol particles at the point $x_i$ at the time t, whose volume lies between the values of v and v + dv; $\Gamma\;(v\,,v')$ is the coagulation coefficient between aerosol particles of volumes of v and v'. Subscript 'int' denotes that only the internal process within the idealized spatial volume element is concerned here.

Eq. (6) is so complicated that an analytical solution is very difficult to find. Trying to get numerical solution of the equation is impractical because of the wide range of particle size (roughly 0.01 to 10 $\mu$m in diameter) concerned.

Assuming that the coagulation coefficient ($\Gamma$) is not affected by the flow fields and assuming $\Gamma$ is independent of particle size, then we find by integrating over particle volume:

$$\left[\frac{\delta\,N}{\delta\,t}\right]_{int} = -\,\Gamma_0\;N^2 \qquad (7)$$

where $N = N(x_i\,,t)$ is the total number of particles within a unit volume at the point $x_i$ at the time t. As the size distribution can be estimated using Eq. (2), this assumption is quite useful.

Building this relation into the source/sink term of Eq. (3) we get a conservation equation for the total particle number:

$$\frac{\delta\,N}{\delta\,t} + \frac{\delta\,Nu_j}{\delta\,x_j} = \frac{\delta}{\delta\,x_j}\left(D^*\frac{\delta\,N}{\delta\,x_j}\right) - \Gamma_0 N^2. \qquad (8)$$

Given the velocity field as a function of time Eqs. (5) and (8) can be solved numerically by giving the generation term as part of the boundary conditions.

Natural Convection

Our next problem is the prediction of flow fields in the enclosure. For our sample calculations, which were two-dimensional, the following set of time-averaged equations for constant density flows were used. The equation of vorticity transport was derived from the Navier-Stokes' type equation of motion by eliminating the pressure term. In representing the influence of buoyancy, the Boussinesq approximation is incorporated. The buoyancy term is included in the equation of vorticity transport as $g\beta\,\delta\,\theta\,/\,\delta\,x$.

Equation of energy

$$\frac{\delta\,\theta}{\delta\,t} + u_x\frac{\delta\,\theta}{\delta\,x} + u_y\frac{\delta\,\theta}{\delta\,y} = \alpha^*\left(\frac{\delta^2\theta}{\delta\,x^2} + \frac{\delta^2\theta}{\delta\,y^2}\right) \qquad (9)$$

Equation of stream function

$$u_x = \frac{\delta \psi}{\delta y},$$

$$u_y = -\frac{\delta \psi}{\delta x},$$

(10)

Equation of scalar vorticity

$$\omega = \frac{\delta u_y}{\delta x} - \frac{\delta u_x}{\delta y} = -\left(\frac{\delta^2 \psi}{\delta x^2} + \frac{\delta^2 \psi}{\delta y^2}\right),$$

(11)

Equation of vorticity transport

$$\frac{\delta \omega}{\delta y} + u_x \frac{\delta \omega}{\delta x} + u_y \frac{\delta \omega}{\delta y} = \nu^*\left(\frac{\delta^2 \omega}{\delta x^2} + \frac{\delta^2 \omega}{\delta y^2}\right) + g\beta \frac{\delta \theta}{\delta x}.$$

(12)

Here $\theta$ is the temperature; $\psi$ is the stream function; $\omega$ is the scalar vorticity; $\alpha^*$ is the effective heat diffusivity; $\nu^*$ is the effective kinematic viscosity; g is the gravitational acceleration; $\beta$ is the volumetric expansion coefficient.

## COMPUTATIONAL AND EXPERIMENTAL PROCEDURE

In order to validate the theoretical model, we experimented using a rectangular enclosure. Numerical simulations were applied to the two-dimensional enclosure having the same dimensions. Both results were compared.

### Smoke Source Evaluation

Cotton wicks were used for both the heat and smoke source. This material is used by Underwriters' Laboratories, Inc. (UL) for testing smoke detectors in Standard No. 268. Prior to the enclosure experiments and calculations, the size distribution and the generation rates of smoke aerosol were evaluated using a smoke box of the size 1.6 x 0.8 x 0.5 m.

Three cotton wicks were set glowing then held inside the smoke box for 60 seconds. After allowing the smoke to become well mixed (1 min) by a fan positioned inside the box, the size distribution was measured as a function of time using an optical particle counter (Particle Measuring Systems, Inc. model LAS-X). In order to keep the concentration of aerosol within the measuring range, a 100 to 1 or 200 to 1 diluter was used in the sampling line of the optical particle counter. It took about 40 seconds for measuring the size range of 0.09 - 3.0 $\mu$m. Light extinction by the smoke aerosol inside the box was monitored using an extinction meter equivalent to UL 268 specifications.

The measured size distribution of cotton wick smoke is plotted in Fig. 1, where the readings of the optical particle counter is converted into values of size disribution, dn/dv, by assuming the particle shape as spherical. The observed light extinction coefficient for this size distribution was about 2.5 %/m. The effect of coagulation was insignificant in this level of aerosol concentration.

The measured particle size distribution showed a good agreement with the estimated size distribution by Eq. (2). Reversely by assuming that the size distribution follows Eq. (2) even outside the measured range, the generation rates of smoke aerosol in total number N and total volume V were calculated. The calculated aersol generation rates of a single glowing cotton wick are

722

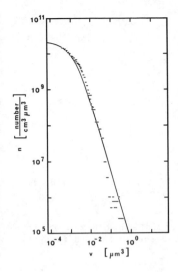

FIGURE 1. The measured size distribution of smoke aerosol from glowing cotton wicks. The dashes (-) show the data points. The solid line shows the assumed size distribution used for the calculations.

$dN/dt = 1.2 \times 10^{11}$ particles/sec and $dV/dt = 1.9 \times 10^{9} \mu m^{3}/$ sec. These values give the average particle volume of $\bar{v} = 1.6 \times 10^{-2} \mu m^{3}$.

Experimental Setup

The size of the enclosure is 1.8 m W. x 1.6 m H. x 0.9 m D. with a hung wall of 0.4 m down at the open end (Fig. 2). The interior surface of the enclosure was made of polystyrene foam thermal insulator. 20 glowing cotton wicks were positioned in line embedded at the center of the floor. The scalar velocity was measured at three different places in order to check the flow reproducibility. A thermal anemometer (Nihon Kagaku Kogyo Co. Ltd. model 6161) was used for measuring the velocity. The same optical particle counter and diluter described above were used for measuring the aerosol size distribution.

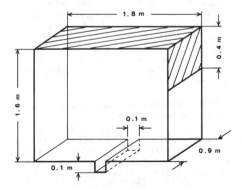

FIGURE 2. The experimental enclosure.

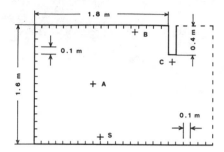

FIGURE 3.   The computational domain: S shows the heat/smoke source. A, B and C show the points where comparison with the experiments was made.


     The flow state in the enclosure became steady in about 3 min after beginning the experiment. The data taken at 4 min (240 sec) were used for the comparison.


Numerical Calculations

     Fig. 3 shows the computational domain of the simulation. For computational purposes the domain was extended to the free boundary region outside the doorway. The letter S shows the point of heat/smoke source. A, B and C show the points where comparison with the experiments was made.

     Computational scheme. The finite difference method was used for the numerical calculations. In deriving the finite difference equations the two-point backward implicit scheme was employed for all the time derivatives. For the spatial derivatives, the up-wind difference and central difference schemes were taken for the convective and non-convective terms respectively.

     Boundary and initial conditions. The two types of boundary, solid or free, have to be considered. On the solid boundaries the non-slip condition was employed on the velocity components, thus the stream function on the interior walls was set to be constant. For the temperature equation, adiabatic side walls and adiabatic ceiling were used. The temperature of the floor was kept constant (ambient) except at the nearest point to the heat/smoke source where adiabatic conditions were taken. For the smoke concentration equations, adiabatic conditions were used for all the interior walls. The vorticity on the solid boundary was calculated by assuming the mirror points in the stream function. On the free boundaries, the derivatives of stream function and vorticity normal to the free surfaces were equated to zero. The temperature and the smoke concentration conditions were separated by the flow directions. At points of in-flow, ambient temperature and zero concentration were specified, at points of out-flow the derivatives of each variable normal to the surface were equated to zero.

     As the intial conditions, the fluid in the enclosure was set motionless, uniform ambient temperature (293 K) and zero smoke concentration.

     Computational details. A 17 x 26 grid system was used for the calculations. The computational time step was set as 0.1 sec. The physical constants used were as follows: the effective heat diffusivity $\alpha^* = 6.0 \times 10^{-3}$ m$^2$/sec, the effective kinematic viscosity $\nu^* = 6.0 \times 10^{-3}$ m$^2$/sec, the effective diffusion coefficient $D^* = 6.0 \times 10^{-3}$ m$^2$/sec, the coagulation coefficient $\Gamma_0 = 4.0 \times 10^{-10}$ cm$^3$/sec, the volumetric expansion coefficient $\beta = 1/273$ deg$^{-1}$, and the gravitational acceleration g = 9.8 m/sec$^2$.

$\alpha^*$, $\nu^*$ and $D^*$ were set to be  contant and have the  same value of 400 $\nu$. For the value of $\Gamma_0$, the coagulation coefficient,  the suggested value for the punk smoke by Lee and Mulholland [5] was used.

Constant heat and smoke generation were assumed and simulated by injecting additional quantities at each time step into the idealized mesh cell at the heat/smoke source. The experimentally determined values described in the previous section were used for the smoke generation. The temperature rise rate of 10 deg/sec was assumed from observations and used for the heat generation.

It took about 30 min computational run time for the calculation up t = 240 sec using IBM 370 computer system.

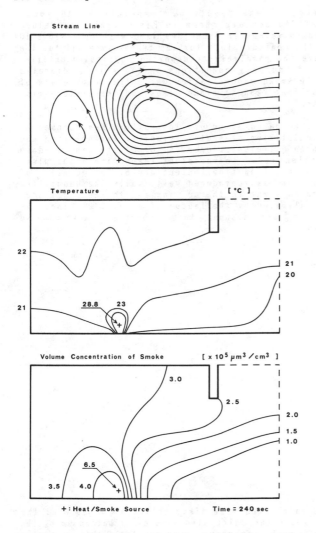

Stream Line

Temperature                    [ °C ]

22

21

28.8    23

21

Volume Concentration of Smoke        [ x 10$^5$ μm$^3$/cm$^3$ ]

3.0

2.5

2.0

1.5

1.0

6.5

3.5    4.0

+ : Heat/Smoke Source          Time = 240 sec

FIGURE 4.   Contours of calculated stream function, temperature and smoke concentration.

TABLE 1.  The observed and calculated quantities of scalar velocity.

|                                  | Point A | Point B | Point C |
|----------------------------------|---------|---------|---------|
| Observed  velocity ( cm / sec )  | 8.0     | 4.0     | 6.0     |
| Calculated  velocity ( cm / sec )| 7.2     | 1.8     | 3.6     |

RESULTS AND DISCUSSIONS

Contours of calculated stream function, temperature and smoke concentration at the time of 240 sec are shown in Fig. 4. All the values represent the spatially averaged values in each idealized mesh cell (10 x 10 cm). The difference between the contour plots for the temperature and for the smoke concentration - both are the passively-transported physical quantities - came from the difference of boundary conditions on the floor. Adiabatic conditions were taken for the smoke concentration while the temperature of the floor was fixed to be ambient (293 K) since almost no temperature rise was observed in the experiments.

The reproducibility of the calculation of flow pattern was checked by comparing with the observed quantities of scalar velocity. The observed and calculated quantities at three different places (A, B, C) are tabulated in Tab. 1. These points were selected as to represent respectively the plume flow, the ceiling flow and the out-flow from the enclosure. The quantity of the velocity in the plume region was reproduced well, while the quantities of the velocity at the points adjacent to the ceiling and to the under edge of the hung wall were under-estimated by the factor of about 1/2. Since the effective kinematic viscosity $\nu^*$ was assumed to be constant, more precise agreement would not be expected.

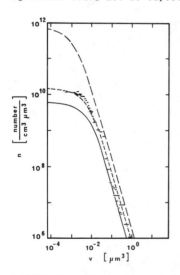

FIGURE 5.  The measured and calculated size distribution at point C at the time 240 sec. The solid line shows the calculated size distribution using the experimental data. The short-dashed line show the calculated result using the modified generation rates and coagulation coefficient. For comparison the size distribution at the heat/smoke source is plotted by long dashes.

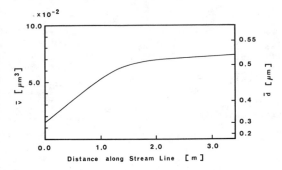

FIGURE 6. The variation of the calculated average particle volume v̄ versus travel distance along the stream line from the heat/smoke source.

Fig. 5 shows the measured and calculated size distribution of smoke aerosol at point C at the time 240 sec. Although the calculated size distribution using the experimentally determined aerosol generation rates gave a good reproduction in the profile of the size distribution, the volume concentration was 2.7 times under-estimated compared with the experimental data. Calculations with modified generation rates and modified coagulation coefficients were made and one of the results is also plotted in Fig. 5. The data set used in this case were $dV/dt = 3.24 \times 10^{11}$ particles/sec, $dN/dt = 5.13 \times 10^{9} \mu m^{3}/sec$ and $\Gamma_0 = 1.5 \times 10^{-10} cm^{3}/sec$.

Fig. 6 shows the variation of the calculated average particle volume v̄ versus travel distance along the stream line from the heat/smoke source. This result suggests that the coagulation of aerosol occurs mostly in the plume region and becomes insignificant in the ceiling flow.

The reason why we had to adjust the values of the smoke generation rates and the coagulation coefficient have to be discussed here. Not only these parameters but also the diffusion coefficient ($D^*$) have the effect of varying the coagulation frequency. Smaller $D^*$ increases the smoke concentration in the plume region and in effect increases the average particle volume (v̄) at the ceiling. Thus only the realistic values of the smoke generation rates, the coagulation coefficient and the diffusion coefficient should give a good reproduction of the phenomena. Since the exact values for these parameters were unknown, the values of the smoke generation rates and the coagulation coefficient were adjusted. As far as the detector response is concerned, this adjustment would not reduce the utility of this model. Baum et al. reported a somewhat similar study of simulating the fire induced flow and smoke coagulation [7]. In their model the smoke movement was simulated by the particle tracking method which corresponds to D* equal zero in Eqs. (5) and (8). We conducted calculations with D* = 0, which gave larger average particle volume at the ceiling.

It has become clear that a deterministic approach to understand the detector's response to enclosed fire is possible. This will allow us to utilize the computational model as a development tool in the software for signal processing of 'intelligent' fire detection systems such as described in [8,9].

ACKNOWLEDGEMENTS

The author is grateful to Professor Takashi Handa of Science University of Tokyo for continuous advice and encouragement. The author also wish to thank Dr. Masahiro Morita of Science University of Tokyo for his invaluable advice on the computational schemes.

## NOMENCLATURE

| | |
|---|---|
| C , | smoke concentration; |
| D , | diffusion coefficient (laminar); |
| $D^*$ , | effective diffusion coefficient; |
| g , | gravitational acceleration; |
| K , | turbulent diffusion coefficient; |
| N , | total particle number; |
| n , | particle number distribution; |
| S , | source or sink for the smoke concentration; |
| t , | time; |
| $u_i$ , | velocity component in $x_i$ axis (i = 1,2,3); |
| V , | total particle volume; |
| v , | particle volume; |
| $\bar{v}$ | average particle volume; |
| $x_i$ , | cartesian coordinates (i = 1,2,3); |
| x , y , | horizontal and vertical coordinates; |

Greek letters

| | |
|---|---|
| $\alpha^*$ , | effective heat diffusivity; |
| $\beta$ , | volumetric expansion coefficient; |
| $\Gamma$ , | coagulation coefficient; |
| $\zeta$ , | reduced number distribution; |
| $\eta$ , | reduced particle volume; |
| $\theta$ , | absolute temperature; |
| $\nu^*$ , | effective kinematic viscosity; |
| $\psi$ , | stream function; |
| $\omega$ , | scalar vorticity. |

## REFERENCES

1. Welker, R. W. and Wagner, J. P.: "Particle Size and Mass Distributions of Selected Smokes. Effect on Ionization Detector Response," J. fire & Flamability, 8 : 26 - 37, 1977.

2. Bukowski, R. W. and Mulholland, G. W.: "Smoke Detector Design and Smoke Properties," Nat. Bur. Stand. (U.S.), Tech. Note 973, 1978.

3. Mulholland, G. W. and Liu, B. Y. H.: "Response of Smoke Detectors to Monodisperse Aerosols," J. Research Nat. Bur. Stand., 85: 3, 223 - 238, 1980.

4. Bukowski, R. W. and Bright, R. G.: "Results of Full-Scale Fire Tests with Photoelectric Smoke Detectors," Nat. Bur. Stand. (U.S.), NBSIR 75-700, 1975.

5. Lee. T. G. K., and Mulholland. G. W.: "Physical Properties of Smokes Pertinent to Smoke Detector Technology," Nat. Bur. Stand. (U.S.), NBSIR 77-1312, 1977.

6. Mulholland, G. W., Lee, T. G. and Baum, H. R.: "The Coagulation of Aerosols with Broad Initial Size Distributions," J. Collid and Interface Science, 62: 3, 406 - 420, 1977.

7. Baum H. R., Rehm R. G. and Mulholland G. W.: "Computation of Fire Induced Flow and Smoke Coagulation," 19th Symposium (International) on Combustion/The Combustion Inst., 921 - 931, 1982.

8. Scheidweiler, A.: "The Distribution of Intelligence in Future Fire Detection Systems, " Fire Safety J., 6 : 3, 209 - 214, 1983.

9. Tomkewitsch, R.: "Fire Detector Systems with 'Distributed Intelligence'- The Pulse Polling System," Fire Safety J., 6 : 3, 225 - 232, 1983.

# Dynamic Performance of Pneumatic Tube Type Heat Sensitive Fire Detectors

HEINZ O. LUCK and NORBERT DEFFTE
Duisburg University
Electrical and Electronic Engineering
Dept. of Communication Technology
Bismarckstr. 81; D-4100 Duisburg, FRG

## ABSTRACT

Pneumatic tube type heat sensitive fire detectors are used in several different applications, e.g. the protection of road tunnels. Some quantitative
knowledge in the dynamic performance of these systems is essential for the
prediction of response times under different operating conditions or for the
calculation of a suitable response threshold setting.

The paper discusses an approximative method to calculate the dynamic performance that takes into account the inherent nonlinearities of some system
elements. The method is based on some simplifying assumptions and uses models
for all system elements. The simplification results in an approximative overall model for the pneumatic system which can be described by a fairly simple
nonlinear differential equation. This equation can be solved by using a suitable Runge-Kutta method and yields the pressure difference $\Delta p$ at the pneumatic
switch as a function of time. It is shown furthermore that this solution
cannot be considered as valid in general but can be used as a calculation tool
for response times and response behaviour of the system in the range of
normally used response pressure thresholds with a suitable accuracy. So the
method is simple enough and usable for practical purposes.

Keywords: Pneumatic tube type fire detectors, heat sensitive fire detectors,
pneumatic models, nonlinear systems.

## 1. INTRODUCTION

Pneumatic tube type heat sensitive fire detectors are frequently applied for
road tunnel protection. Several other applications are known /1/. The practical advantages of this type of fire detecting system are a robust installation
that is insensitive against electromagnetic interference and a fairly good
fire detection capability.

Several attempts have been made to calculate the dynamic performance of the
underlying pneumatic system by using approximative models that apply methods
known from linear electrical network analysis /1, 2/. Although the response
time figures calculated from these models seem to be fairly close to those
measured in some experiments the models themselves suffer from the fact that
all the elements involved have to be modelled as linear elements. This is
obviously not consistent with physical reality.

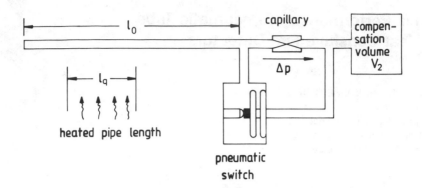

Fig. 1   Schematic sketch of a tube type fire detection system
         (SECURITON AG, Zollikofen, Switzerland)


The  following article presents a method to calculate the dynamic  performance
that  is approximative as well but takes nonlinearities into account where  it
seems to be necessary. The method has been developed using the characteristics
and  the  essential parameters of the TRANSAFE fire detection system for  road
tunnels manufactured by the Securiton AG (Zollikofen, Switzerland).

## 2. BASIC MODELS FOR SYSTEM ELEMENTS

Fig.  1  shows a schematic sketch of the configuration used in the above  mentioned tube type fire detection system. If a part of the pipe is heated a time
varying pressure difference $\Delta p$ will occur that operates a pneumatic switch.

The  basic  model that has been used in this study to  calculate  the  dynamic
performance is based on the following conditions:

(1.) The  pressures  and the fluid velocities within the pneumatic  systems  -
     except in the heated part of the sensor pipe - are small enough to assume
     that the specific air density $\rho$ be a constant.
(2.) The  flow resistances within the system - except that associated with the
     resistive capillary - are negligibly small.
(3.) The rate of change of all pneumatic values in the system is small  enough
     to  assume  that  dispersion within the system  can  be  neglected.  This
     assumption  results in a model that considers a "system with concentrated
     elements"  and there is no need to go back to the basic equations of  gas
     dynamics (e.g. /3/ p. 182 or /4/).

Under the above conditions pressure variations in pneumatic subsystems can  be
described  by  associated  variations  in the mass flow to or out of  the  sub-
systems. Furthermore the model system itself can be described by using the so-
called Kirchhoff-laws known from electrical network theory.  The following  is
based on the equation for ideal gases

$$p \ V = \frac{R}{m} \ M \ T = \beta \ M \ T \tag{2.1}$$

where p is the pressure and V an airfilled volume which e.g.  is a part of the
pneumatic sensor pipe. M is the total mass, T the absolute temperature and the

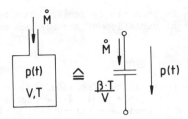

Fig. 2  Pneumatic capacity

constant $\beta$ incorporates the gas-constant R and the molecular weight of the gas (air in this particular case). Introducing the specific air density $\rho = \frac{M_O}{V_O}$ at a known initial temperature $T_O$ in the system, $\beta$ can easily be calculated as follows using (2.1):

$$p_O = \beta \frac{M_O}{V_O} T_O = \beta \rho T_O \Rightarrow \beta = \frac{p_O}{\rho T_O}$$

In the following the system elements are discussed using the basic assumptions.

Airfilled Volumes at Constant Temperatures and with Time Varying Mass Flows

This model describes the constant temperature part of the sensor pipe of the length $l_O - l_q$ and the compensation vessel volume $V_2$ (see Fig. 1). Introducing the mass flow

$$\overset{\circ}{M} = \frac{dM}{dt} \quad \text{we get}$$

$$\overset{\circ}{p} V = \beta T \overset{\circ}{M} \quad \text{or} \quad \overset{\circ}{p} = \frac{\beta T}{V} \overset{\circ}{M} \tag{2.2}$$

where $\frac{\beta T}{V}$ is a constant. (2.2) can be interpreted as a description of a pneumatic capacity, i.e. an element for potential energy storage, as shown in Fig. 2. From Fig. 1 it can easily be seen that the whole not heated part of the sensor pipe can be concentrated to one pneumatic capacity. Different elements of this kind can easily be connected in series as shown in Fig's. 3 and 4.

The Resistive Capillary

The capillary shown in Fig. 1 works as a flow resistive element. If friction within the flow is neglected and the flow is a laminar one the Bernoulli-law leads to

$$\Delta p = \text{const.} \ \overset{\circ}{M}_2 |\overset{\circ}{M}_2|$$

to describe the resistive behaviour of this element. Experimental results show on the other hand that the capillary flow is not laminar in the device under consideration and the best description for $\Delta p(M_2)$ is /5/

$$\Delta p = a\overset{\circ}{M}_2 + b\overset{\circ}{M}_2^c \tag{2.3}$$

where  a,  $b_o$  and c are constants that can be derived from these  experimental
results   $p(M_2)$  by using a slight variation of the minimum mean  square   error
method /5/.

## Model for the Heated Part of Sensor Pipe

The most difficult part of the whole system for modelling is the heated length
of  the  sensor pipe (see Fig.  1).  This part is the source for  all  dynamic
changes in the system. In order to produce a fairly well solvable differential
equation  (see eq.  3.7) the calculation model is based on the idea to replace
the heated air volume by a volume of the same size $V_q$ at constant  temperature
$T_o$ and an additional time varying air flow $\overset{o}{M}_q(t)$ as indicated in Fig. 3.  The
left hand part of Fig. 3 a) and b) models the source while the right hand part
(volume V) is a general pneumatic load that represents the remaining pneumatic
system.  This procedure follows the idea of equivalent-source-technique  which
is very common in electrical network theory.  It is not quite correct to model
the  remaining  pneumatic system simply by a volume V because V  represents  a
purely "capacitive" load  while the system contains a resistive  element  as
well. Numerical  calculations  show  that this error has an  acceptable  small
influence on the final result at least for the system under consideration.

Assume that the temperature in the heated volume is rising according to

$$T(t) = T_o \, f(t) \quad \text{and} \quad f(t) = \begin{cases} 1 & \text{for } t < 0 \\ f(t) & \text{for } t \geq 0 \end{cases} \tag{2.4}$$

where $f(t)$ is some suitable function of time.

The system in Fig. 3 a) is obviously described by

$$P_q = p \qquad\qquad \Rightarrow \overset{o}{P}_q = \overset{o}{p} \tag{2.5}$$

$$M_q + M = \text{const.} \quad \Rightarrow -\overset{o}{M}_q = \overset{o}{M} \tag{2.6}$$

$$P_q V_q = \beta \, T \, M_q = \beta \int \frac{\partial}{\partial t} (T \, M_q) \, dt \tag{2.7}$$

$$p \, V = \beta \, T_o \, M = \beta \, T_o \int \frac{\partial}{\partial t} M \, dt \tag{2.8}$$

Rewriting (2.7) and (2.8) and using (2.4) we get

$$\overset{o}{P}_q = \frac{\beta \, T_o}{V_q} (\overset{o}{f} \, M_q + f \, \overset{o}{M}_q) \quad \text{and} \tag{2.9}$$

$$\overset{o}{p} = \frac{\beta \, T_o}{V} \overset{o}{M}. \tag{2.10}$$

Using (2.5) and (2.6) the result after some minor calculations is

$$\overset{o}{M}(t) = M_{qo} \left(1 + \frac{V}{V_q}\right) \frac{\frac{V}{V_q} \overset{o}{f}(t)}{\left(1 + \frac{V}{V_q} f(t)\right)^2} \tag{2.11}$$

where

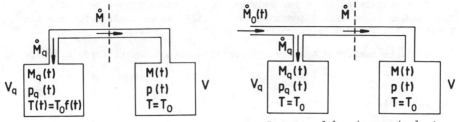

**a.)** Physical source model           **b.)** Source model using equivalent source technique

Fig. 3 Approximative models of pneumatic source

$$M_{qo} = M_q(t)\big|_{t=0} = \rho\, V_q \tag{2.12}$$

In order to be equivalent the model according to Fig. 3 b) has to deliver the same flow $\overset{\circ}{M}$ to the load V.

From Fig. 3 b) we get

$$p_q = p \qquad\qquad \Rightarrow \overset{\circ}{p}_q = \overset{\circ}{p} \tag{2.13}$$

$$\overset{\circ}{M}_o = \overset{\circ}{M}_q + \overset{\circ}{M} \tag{2.14}$$

$$p_q V_q = \beta\, T_o\, M_q \qquad \Rightarrow \overset{\circ}{p}_q = \frac{\beta\, T_o}{V_q} \overset{\circ}{M}_q \tag{2.15}$$

$$p\, V = \beta\, T_o\, M \qquad \Rightarrow \overset{\circ}{p} = \frac{\beta\, T_o}{V} \overset{\circ}{M} \tag{2.16}$$

Following the same line as above the introduced mass flow $\overset{\circ}{M}_o$ can be written as

$$\overset{\circ}{M}_o = \frac{V_q}{V} \overset{\circ}{M} + \overset{\circ}{M} = \frac{V + V_q}{V} \overset{\circ}{M}$$

or using the result from (2.11)

$$\overset{\circ}{M}_o(t) = \rho\, \frac{(V + V_q)^2}{V_q}\; \frac{\overset{\circ}{f}(t)}{\left(1 + \frac{V}{V_q} f(t)\right)^2} \tag{2.17}$$

Having obtained this result it is necessary to consider the following items:

-   The introduced source mass flow $\overset{\circ}{M}_o$ depends on the load V. In other words it models the system including this particular capacitive load.

-   Thus this way to proceed seems to be somewhat artificial and yields only approximative results if one predetermined source is applied to several pneumatic networks different from that used to determine it. On the other hand it can be easily shown that the resulting differential equation (3.7) becomes unsuitably sophisticated for practical use if the system in Fig. 4

would be calculated using the original source model according to Fig. 3a).
The results in section 3 indicate that (2.17) can be applied for practical
purposes to obtain a fairly suitable accuracy.

It should be mentioned that (2.17) contains both pneumatic open circuit

$$V_o = 0 \implies \overset{o}{M}_o(t) = \rho \, V_q \, \overset{o}{f}(t)$$

and pneumatic short circuit

$$V \to \infty \implies \overset{o}{M}_o(t) = \rho \, V_q \, \frac{\overset{o}{f}(t)}{f^2(t)}$$

as well. So it covers the whole possible load range.

## 3. A CALCULATION MODEL FOR TUBE TYPE FIRE DETECTION SYSTEMS

Using the element models discussed in section 2 the tube type fire detection
system according to Fig. 1 can be represented by an approximative model as
shown in Fig. 4. $V_q$ represents the heated part of the sensor pipe, $V_2$ is the
compensation volume, $V_q + V_1 = V_o$ stands for the sensor pipe volume and
$V_q + V_1 + V_2 = V_{ges}$ represents the whole volume of the pneumatic system.

The following equations can easily be derived from Fig. 4:

$$P_q = P_1 \qquad\qquad \implies \overset{o}{P}_q = \overset{o}{P}_1 \tag{3.1}$$

$$P_1 = \Delta p + P_2 \qquad \implies \overset{o}{P}_1 = \Delta\overset{o}{p} + \overset{o}{P}_2 \tag{3.2}$$

$$\overset{o}{M}_o = \overset{o}{M}_q + \overset{o}{M} \tag{3.3}$$

$$\overset{o}{M} = \overset{o}{M}_1 + \overset{o}{M}_2 \tag{3.4}$$

The expression (2.3) in sec. 2 can be written in a more general form

$$\Delta p = F(\overset{o}{M}_2) \tag{2.3}$$

and yields

$$\Delta\overset{o}{p} = \frac{d}{dt} F(\overset{o}{M}_2) = \frac{\partial F}{\partial M_2} \overset{oo}{M}_2 \tag{3.5}$$

and using (2.2) we have

$$\overset{o}{P}_q = \frac{\beta T_O}{V_q} \overset{o}{M}_q; \qquad \overset{o}{P}_1 = \frac{\beta T_O}{V_1} \overset{o}{M}_1; \qquad \overset{o}{P}_2 = \frac{\beta T_O}{V_2} \overset{o}{M}_2 \tag{3.6}$$

Combining (3.1) through (3.6) we arrive, after some simple calculations, at a
differential equation for $M_2$:

$$\frac{\partial F}{\partial M_2} \overset{oo}{M}_2 + \frac{\beta T_o V_{ges}}{V_o V_2} \overset{o}{M}_2 = \rho \, \beta \, T_o \, \frac{V_q}{V_o} \left(\frac{V}{V_q} + 1\right)^2 \frac{\overset{o}{f}(t)}{\left(1 + \frac{V}{V_q} f(t)\right)^2} \tag{3.7}$$

with the initial condition $\overset{o}{M}_2(0) = 0$.

734

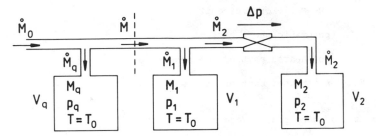

Fig. 4   Approximative model of a tube type fire detection system
(SECURITON AG, Zollikofen, Switzerland)

After solving this for $\overset{\circ}{M}_2$ (2.3) can be used to calculate $\Delta p$ as a function of time. In general (3.7) cannot be solved analytically and some numerical methods have to be applied. There are only very special parameter sets that result in an easy analytical solution, e.g. F being a quadratic function of $M_2$, f a linear function of t and $V \to 0$ (see /5/).

So a computer program based on a suitable Runge-Kutta method has been developed and carefully tested that is based on

$$\Delta p = F(\overset{\circ}{M}_2) = a\overset{\circ}{M}_2 + b\overset{\circ}{M}_2^c \tag{2.3}$$

with a wide range for the parameters a, b and c and on four different functions f(t) for the temperature rise in the heated part of the sensor vessel.

$$f_1(t) = 1 + \alpha_1 t \qquad \text{a linear temperature rise} \tag{3.8}$$

$$f_2(t) = e^{\alpha_4 t} \qquad \text{an exponential temperature rise} \tag{3.9}$$

$$f_3(t) = f_{max} - (f_{max}-1)\, e^{-\alpha_3 t} \tag{3.10}$$

an exponential temperature rise with a
predetermined limit $f_{max}$

$$f_4(t) = \frac{f_{max}\, e^{\alpha_4 t}}{(f_{max}-1) + e^{\alpha_4 t}} \tag{3.11}$$

a mixture of (3.9) and (3.10) that is likely to
model practical cases quite well.

The only parameter in (3.7) which is not directly derived from the detection system under consideration is the volume V. V had been used to derive the model for the driving source and models the not heated part of the pneumatic system (see Fig. 3). Therefore it is a free parameter which may have an unsuitable influence on the result. In order to check this influence, some calculations have been made using the actual parameters of the above mentioned system (TRANSAFE, Securiton AG, Zollikofen, Switzerland) with four different choices for the volume V:

V = 0                    pneumatic open circuit

$V \rightarrow \infty$                    pneumatic short circuit

$V = V_1$                    not heated part of the sensor vessel

$V = V_1 + V_2$          not heated part of the whole system
                          volume

and a linear temperature rising according to (3.8).

The result is shown in Fig. 5. Very similar results can be obtained with other types of temperature rises. They indicate considerable differences which disqualify the outlined calculation method for general application. On the other hand it has to be taken into account that the alarm pressure threshold $\Delta p_a$ for the pneumatic switch normally is in the range of 150 Pa to 400 Pa. In this range no significant differences are produced in the whole range possible for V by the calculation method discussed above. This is shown in Fig. 6 in some more detail. It justifies the conclusion that the above mentioned method may be applied successfully to calculate the response time of tube type heat sensitive fire detection systems in practice. In addition it can very well be applied to calculate the minimum rate of rise of temperature in the heated part of the sensor that is necessary to trigger an alarm signal at all if the response pressure threshold is fixed. Or it can be used to calculate the minimum length of the sensor vessel to be heated that is necessary to produce an alarm. Or it can be used to calculate the temperature variations within the protected premises which are tolerated by the system without a false alarm.

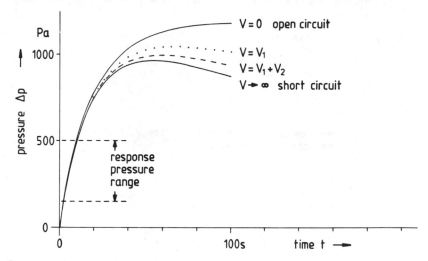

Fig. 5   Pressure $\Delta p$ at the pneumatic switch as a function of time

## 4. SUMMARY

An approximative method has been presented for the calculation of the dynamic performance of tube type heat sensitive fire detection systems. It is based on some basic assumptions which simplifiy the calculation procedure to make it

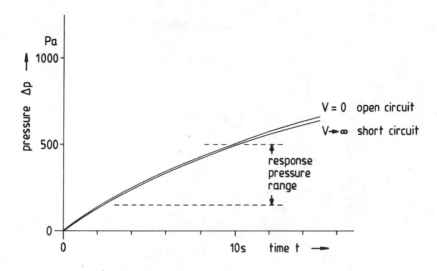

Fig. 6   Pressure Δp at the pneumatic switch

applicable in practical cases. On the other hand it is limited to a time range of approximately 50 s and a response pressure threshold up to 400 Pa for the pneumatic switch. Within this range the method can be used to calculate e.g. response times or suitable response pressure thresholds under different system parameters.

## REFERENCES

/1/   S. Nakauchi and Y. Tsutsui, "A Study on the Pneumatic Tube Type Fire Detector", Report of Fire Research Institute of Japan, Vol. 7 No. 1 - 2, Dec. 1956

/2/   D. Iseli, TRANSAFE, Brandmeldesystem für Straßentunnels, Proceedings of AUBE '82, 8. International Conference on Problems in Automatic Fire Detection, Duisburg University, Oct. 1982

/3/   G. Joos, Lehrbuch der theoretischen Physik, Akademische Verlagsges., Frankfurt/M. 1959 (10. Auflage)

/4/   R. Haug, Pneumatik, Teubner Verlag, Stuttgart 1980

/5/   N. Deffte, Berechnung des dynamischen Verhaltens eines nichtlinearen Netzwerks als Analogon für pneumatisch arbeitende Brandmeldesysteme, Diploma thesis, Duisburg University, Electrical and Electronic Dept. 1984

# Installation and Reliability of a Free Smoke Detector

STAFFAN HYGGE
The National Swedish Institute for Building Research

ABSTRACT

The present study reports 832 telephone interviews during the spring of 1984 with persons from two insurance companies in the same geographical area. In the fall of 1982 one of the insurance companies started to distribute a free smoke detector to all of its customers. The other insurance company served as a control and comparison group for studying the effects of a free smoke detector on installation and reliability.

The population in each insurance company was stratifed into two insurance policy groups, and crossed with three age groups. Random samples of approx equal sizes were drawn. Around 94 percent of the persons in the samples were actually reached, and the interview about installation and reliability ended with a request to press the test button of the smoke detector.

The net effect of sending out a free smoke detector is more pronounced with household policy holders, although home policy holders more often have a smoke detector installed.

There is no general tendency that people who have bought a smoke detector take better care of it, maintain it better, install it in the right place more often, check it more often, or make it operate properly more often than those who have got a free smoke detector. However, for old people the free smoke detector has been out of working order more often than for young people.

Any such effect of a free smoke detector is not more pronounced with those who have a household insurance policy than with those who have a home insurance policy. Nor is there any such effect of age group.

It should be noted that around 16 percent of the smoke detectors that are reported by their owners to be in working condition, do not respond to the test-button being pressed.

## INTRODUCTION

In the fall of 1982, a local insurance company in Sweden (Länsförsäkringar, Gästrikland, LFG) began distribution of a free smoke detector to all its customers with a policy including fire liability. Around 21.000 policy holders with a household (rental units or condominiums) or home (privately owned single-family homes) insurance policy received their detectors by mail before Christmas 1982, and some hundreds more after that time.

Very little is known about the installation and reliability of smoke detectors in general in Sweden. SMI [1] reported from a telephone survey in February 1984, that the average proportion of detectors had tripled (from 16 to 45 percent) from 1981 to 1984, and that 88 percent of those who had a detector were sure that it was in working order. The same study also reported the proportion of smoke detectors in apartments or small homes and across a few age strata.

Two mail surveys, which among other things addressed the installation and reliability of the free smoke detectors, have been carried out among the LFG policy holders (Blom et al, [2]; Eriksson et al, [3]). In the first of these two studies it was reported that on the average 79 percent had installed the free smoke detector, and that the proportion varied with age and kind of insurance policy. The second study reported that 86 percent of the free smoke detectors were said to work.

None of the above three studies have deliberately been designed to test hypotheses about installation and reliability of smoke detectors in various layers of the population. None of the two surveys on the free LFG smoke detectors made any attempt to evaluate whether there was any difference in these respects between free smoke detectors and smoke detectors that people have paid for. None of the three studies made any attempt to find a more valid measure of whether the detectors were in working order, than merely a report from the person interviewed.

The figures for the U.S. are higher than in Sweden for smoke detector installation. Gancarski and Timoney [4] reported that by 1982 two out of three U.S. homes had detector coverage. Their paper references three survey reports [5, 6, 7] and four field research projects [8, 9, 10, 11, 12] adressed to the effectiveness of home smoke detectors. It is interesting to note that one of surveys [7] gathered data by means of actual home visits rather than telephone interviews. It is also noteworthy that in [7] 30 percent reported never having tested their detectors, and in [12] only 38 percent of the detectors had been tested during the last month.

The study presented here tested a number of explicit hypotheses related to the fact the the LFG smoke detectors were free, and what effect this had on where the detector was put up, how it was maintained and checked. Of particular interest was to find out whether there was any variation in reliability and maintenance due to type of policy and age of policy holder. Comparisons were therefore made with another insurance company (Folksam, FO) in the same geographical area.

Another objective of the study was to find out whether or not the smoke detectors actually responded to the test button. This was a strong reason for employing telephone interviews.

The more general hypotheses can be phrased in the following way:

I. Those who have received a free smoke detector do not take as good care of it, do not maintain it, do not install it in the right place, and do not make it operate properly to the same extent as those who bought it themselves.

Ia. This effect of a free smoke detector is more pronounced with those who have a household insurance policy than with those who have a home insurance policy.

Ib.  This effect of a free smoke detector is more pronounced with older people than with young and middle aged people.

With regard to the installation of free detectors the following hypotheses were formulated:

II.  Those who have been sent a free smoke detector more often have an installed smoke detector than those who have bought one themselves.

IIa.  This effect of a free smoke detector is less pronounced with those who have a household insurance policy than with those who have a home insurance policy.

IIb.  This effect of a free smoke detector is less pronounced with older people than with young and middle aged people.

Note that Ia, Ib, IIa, and IIb are formulated as the interactions of the effects of a free smokedetector with insurance policy and with age group. That is, the effects of policy form and age groups in the company that did not give out a free smoke detector will be compared to the effects observed within the company who did.

## METHOD

A total of twelve random samples of approx 75 persons each were taken from the two insurance companies.  Stratifying variables were age (61+, 42-60, -41) and type of insurance policy (household, home).  In March 1984 all of the persons in the sample were mailed a letter, saying they would be contacted over the phone for an interview.  Nothing was however mentioned about smoke detectors being the subject of the interview.

From March through August 1984 842 telephone interviews were made, reaching 93.8 percent of the original net sample of 898 persons (see Table 1). Persons with non listed phone numbers were contacted by mail up to three times and asked to call back.

All of the interviewed were asked about whether they had or had had a smoke or fire detector installed, when they installed it, whether it was operative or not, in what room and where in the room it was put up, whether it had been in order all the time, how often and in what way they tested the smoke detector, whether it had given an alarm signal at any other time than during testing and the cause for that.  At the end of the interview all those who said they had a smoke detector installed were asked to press the test button so the interviewer could hear the signal over the phone.  Those who tried but did not succeed the first time were asked to try again and to keep the button pressed for 20 sec.  Arguments for refusing were recorded, and later scored into "good" and "bad" arguments.  Only arguments stating lack of will, but not lack of capability were scored as "bad".

## RESULTS

In order to test the above hypotheses a saturated logit anlysis (SPSS-X) was made, followed by successive eliminations of non-significant terms, employing the likelihood ratio chi square statistic for testing the remaining model against observed frequencies.  (See [13,14] for more information about the statistical methods)

TABLE 1. The sample divided by insurance policy and age groups

|  | Household | | Home | |
|---|---|---|---|---|
|  | FO | LFG | FO | LFG |
| **61+** | | | | |
| Net sample | 70 | 76 | 76 | 72 |
| Not reached | 3 | 4 | 2 | 4 |
| Interviewed | 67 | 72 | 74 | 68 |
| **42-60** | | | | |
|  | 73 | 76 | 74 | 73 |
|  | 9 | 13 | 3 | 0 |
|  | 64 | 63 | 71 | 73 |
| **-41** | | | | |
|  | 81 | 75 | 75 | 77 |
|  | 5 | 7 | 3 | 3 |
|  | 76 | 68 | 72 | 74 |

**Percent Smoke Detectors Installed Across Years**

Figure 1 shows the cumulative percent of smoke detectors across years. There is no difference between the two insurance companies within the two policy forms before the fall of 1982 when LFG started to distribute the free detector.

Also, there is virtually no difference between age groups in adoption of smoke detectors.

The proportions of policy holders in the various sub-strata who report they had a smoke detector installed at the time of the interview are shown in Figure 2. For FO the average is 32 percent and for LFG 71 percent, a highly significant difference.

The significant effects of company (C) (z= -11.01) and the interaction C x policy (P) (z= -2.05), means that the effect of a free detector is different for the two policy forms. As can be seen in Figure 2 (row 3, col 1), there is a greater difference between the companies for the household insurance, which is contrarary to hypothesis IIa.

742

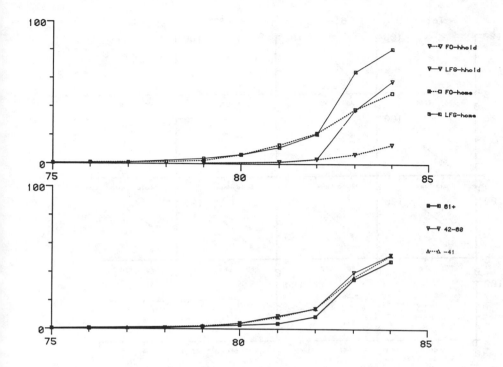

FIGURE 1. Cumulative percentage smoke detectors by year

The interaction C x age group (A) (z= -2.50, 2.11) indicates different age effects for the two insurance companies. In Figure 2 it can be seen that the difference is largest for the oldest group, which is contrary to hypothesis IIb. The significant interaction P x A ( z= .86, 2.41) means that for the youngest group, there is the largest difference between policy forms.

Thus, the following conclusions can be drawn with regard to the hypotheses:

II. Yes, sending out a free smoke detector has increased the number of installed smoke detectors.

IIa. No, the effect of a free smoke detector is not less pronounced with those who have a household insurance policy than with those who have a home insurance policy.

IIb. No, the effect of a free smoke detector is not less pronounced with older people than with young and middle-aged people.

## Operating All the Time

The percent of respondents with a smoke detector who said "yes" to the question whether their smoke detector had been operating all the time, are shown in Figure 3.

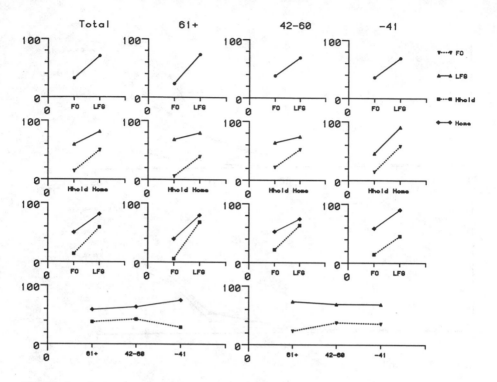

FIGURE 2.   Percentage installed smoke detectors

Hypothesis I is supported, since there is a significant average difference
between companies (z= 7.37).

The expectation inherent in hypothesis Ia, that a difference between the
companies should be more prounounced for household policy holders than for
home policy holders, is not borne out, since there is no significant
interaction C x P.

Concerning hypothesis Ib about the effect of age on the difference between the
companies, there is an interaction C x A (z= 12.97, -).  This means that the
smoke detector has more often been in working order for FO than for LFG for
older and middle ges people, while the reverse is true for young people.
Hypopthesis Ib therefore receives partial support.

## Placement, Checks and False Alarms

In what room or where in the room the smoke detector is put up, whether the
detector is checked at least once a month, and the number of false alarms did
not significantly differ between groups in the way hypotheses I - Ib suggest.

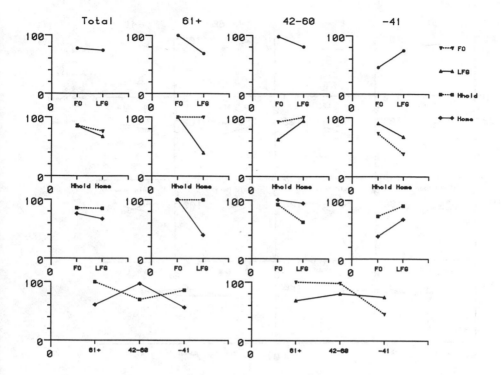

FIGURE 3.  The smoke detector has been in working order all the time

## Is the Smoke Detector in Operation?; Verbal Response

On the average 91 percent of the installed smoke detectors are verbally
reported to be in working order.  This figure does not significantly differ
between subgroups in the ways stated by the hypotheses.

## Is the Smoke Detector in Operation?; Button Press

Of those who reported that they had an operating smoke detector, and who
agreed to press the test button (approx 17 percent did not press the test
button for good or bad reasons), 84 percent of the smoke detectors responded.
Thus, as shown if Figure 4, on the average 16 percent
of the smoke detectors which are reported to be in working order did not react
to pressing the test button.  (If it is assumed that all of those who refused
to press the button for bad reasons, have an installed but non-responding
detector, this figure is doubled.)

In terms of hypotheses I - Ib there is no indication that the differences
shown in Figure 4 reach significance.

Thus, hypotheses I - Ib are not supported.

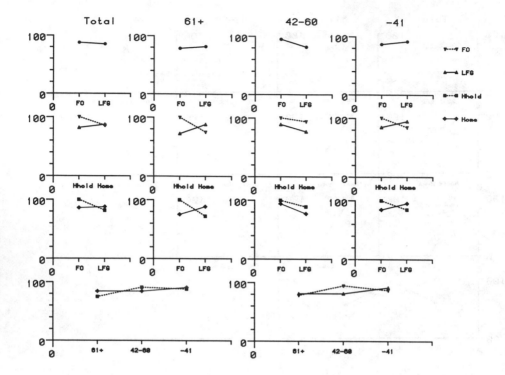

FIGURE 4. Smoke detector's response to pressing the test button

## Refusals To Press the Test Button

The percentages who refused to press the test button are shown in Figure 5. There is no indication of an average difference between companies. and hypothesis I is not supported.

Also, there is no support for hypothesis Ia and Ib, since there are no indications of an interation C x P or C x A.

There is an effect of age (z= -2.77, -2.17) and of policy form (z= -3.06).

## DISCUSSION AND CONCLUSIONS

Household policy holder do install their free smoke detector to a larger extent than do home policy holders. However, this may be a ceiling effect attributable to the fact that there was a higher pecent smoke detectors in homes than in households before the free smoke detector was given out.

Older people do not differ from other age groups in how often they install a free smoke detector.

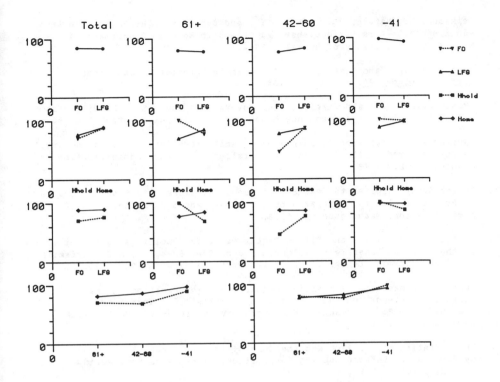

FIGURE 5.  Pressed vs refused to press test button

There is no general tendency that people who have bought a smoke detector take
better care of it, maintain it better, install it in the right place more
often, check it more often, or make it operate properly moref often than those
who have got a free smoke detector.

Nor is any such effect of a free smoke detector generally more pronounced with
those who have a household insurance policy than with those who have a home
insurance policy, or for any age-group.

It should be noted that approx 16 percent of the smoke detectors that are
reported by their owners to be in working condition, do not respond to
pressing the test-button.

REFERENCES

1.  Skandinaviska Marknadsinstitutet, SMI.  "Telefonintervjuer - brandkunskap
    hos allmänheten," (Telephone Interviews - The Public's Knowledge About
    Fire).  Skandinaviska Marknadsinstitutet, Stockholm, 1984.

2.  Blom, R., Hallberg, K., Johansson, L., and Maikulangara, V.  "Ingen rök
    utan eld - en undersökning av brandvarnare i Gästrikland," (No Smoke
    Without a Fire - A Study of Smoke Detectors in Gästrikland).  Department
    of Statistics, University of Uppsala, 1983.

3.  Eriksson, M., Natvig, H., Jansson, T., and Berge, K. "Brandsläckare - en vän i nöden," (Fire Extinguisher - a Friend in Need). Department of Statistics, University of Uppsala, 1984.

4.  Gancarski J.L. and Timoney, T. "Home Smoke Detector Effectiveness," Fire Technology, 20: 4, 57-62, 1984

5.  "Residential Smoke and Fire Detector Coverage in the U.S.: Findings from a 1982 Study," Federal Emergency Managament Agency, Washington, DC, 1982.

6.  "Survey and Analysis of Occupant-Installable Smoke Detectors: A Summary," Report prepared by the Aerospace Corporation, U.S. Fire Administration, Washington, DC, 1978.

7.  "Pilot Study Desgined to Test Effectiveness of Smoke Detection Devices in Private Dwellings," prepared by L.N. Moyer and S.E. Miller, U.S. Fire Administration, Washington DC, 1978.

8.  "Detector Sensitivity and Siting Requirements for Dwellings (a report of the "Indiana Dunes Tests"),." prepared by R.W. Bukowski, W.J. Christian and T.E. Waterman, Center for Fire Research, Washington, DC, 1975.

9.  "Detector Sensitivity and Siting Requirements for Dwellings: Phase 2 (part 2 of a report on the "Indiana Dunes Tests"),." prepared by S.W. Harpe, T.E. Waterman and W.J. Christie, Center for Fire Research, Washington, DC, 1976.

10. "Fire Statistics for 1980 and the Role of Smoe Detectors," prepared by J.H. Darge, Ontario Housing Corporation, Maintenance Engineering Branch, Toronto, Ont., Can., 1981.

11. Moore, D.A. "Remote Detection and Alarm for Residences: The Woodlands System," Fire Journal, 74, 1, 1980.

12. "An Evaluation of Residential Smoke Detectors under Actual Field Conditions: Final Report," prepared by the International Association of Fire Chiefs Foundation, U.S. Fire Administration, Washington, DC, 1982.

13. SPSS-X. User's Guide. New York: McGraw Hill 1983.

14. Haberman, S.J. Analysis of Qualitative Data. Volume 1. New York: Academic Press 1978.

# Correlation Filters for Automatic Fire Detection Systems

**HEINZ O. LUCK**
Duisburg University
Electrical and Electronic Engineering
Dept. of Communications
Bismarckstr. 81, D-4100 Duisburg, FRG

## ABSTRACT

The technical development in automatic fire detection is mainly governed by the rapidly increasing development of electronic components including microprocessors that opens a variety of new signal processing tools. Most applications used in automatic fire detection systems are aimed at the improvement for system handling in the alarm or in the fault signal situation or at the improvement of automatic system monitoring. Only rather few attempts have been made to use modern electronic tools including software solutions to improve the detection capability. This article deals with a method to develop signal detection algorithms which can be used in automatic fire detection systems and which, in addition, can very well be realized as software controlled electronic circuits, e.g. as software program operated microcomputers or as special VLSI circuits. The method is based on a simple and fairly general model for the physically measurable signal in the vicinity of a developing fire and is therefore not restricted to a special kind of fire detector.

Keywords:
Automatic fire detection, detection capability, fire detection algorithm, correlation filters.

## 1. INTRODUCTION

Any extinguishing measure in a developing fire situation requires a preceeding detection procedure. From this point of view fire detection is a most important part in the chain of fire prevention measures and far too little attention has been payed to the associated techniques in the past compared to other important fields. The situation can be very roughly characterized as follows:
- Only rather few characteristic fire parameters physically measurable in the vicinity of a developing fire are used for detection purposes. These parameters are mainly temperature rise, smoke concentration and flame radiation intensity in the IR- and UV-wavelength range /1/.
- As far as the detection procedure itself is concerned rather poor signal processing methods have been applied in the past. The alarm signal mainly is created within the detector head by a simple threshold comparison. Signal and data transmission require no sophisticated techniques. Only few detectors are grouped together to protect small areas in the building so that an alarm from a detector group instantaneously indicates the location of the developing fire in the building in a very simple manner.
- The important electrical connection lines within fire detection systems e.g. between detectors and the control and indicating equipment or to

remote control stations (fire-brigade) are monitored automatically and fault signals are indicated differently from the alarm signals on the indicating panel.

In some special cases a more sophisticated signal and data processing procedure was required and therefore it has been introduced. If the false alarm problem became too incomfortable some coincidence techniques between two different detector groups or a suitable time delay for the alarm signal generation with the check whether or not the alarm criterion holds within the delay time were introduced. Furthermore it was necessary to provide flame detectors with more intelligent signal processing tools in order to bring them into operation at all.

This short review indicates that the classical fire detection technique does not require very challenging signal detection tools. The advantages associated to this classical detection technique are:
- Its performance is easy to understand.
- It is inexpensive (e.g. low developing costs are involved).
- It results in robust systems and installations.
- The reliability is quite satisfactory and the lifetime for components and installations can fairly well be estimated.
- The efficiency has obviously been satisfactory too (the main figure is in the range of one false alarm per year and 100 detector heads).

But there are some severe shortcomings which are the main reason for a rather bad image associated with this technique.
- The associated efficiency cannot be increased any more. For large installations with several hundred detector heads or large areas with large numbers of installations that have to be supervised by one fire brigade the small figure of "one false alarm per year and 100 detectors" is no longer satisfactory. In addition it is a matter of fact that the absolute number of false alarms considerably exceeds the number of real fire alarms /2/.
- Automatic monitoring and control for correct performance is mainly restricted to the power supply components and some electrical signal lines.
- The actual state of the important system components is not automatically monitored. Partial control of the system is mainly provided by some maintenance procedures which are performed in some regular time intervals.

The application of microprocessor techniques introduces the possibility to realize important system properties by software solutions in a very flexible manner. Microprocessors and software solutions in the development of fire detection systems are aimed at the following goals:
(a.) Improvement and facilitation for system handling in the case of alarm or in fault signal situations, where some measures have to be taken.
(b.) Improvement of automatic system monitoring which may result in a higher reliability level with the consequence that the maintenance requirements be reduced.
(c.) Improvement of the detection capability.

In all the three areas considerable progress is possible if a suitable use is made of new signal and data processing facilities. Many attempts have been made already concerning the first two points /5, 6/. But rather few work concerning the improvement of detectivity is known to the author so far /3, 4/. If the term detection capability is defined as the capability of a fire detection system to safely distinguish between the situations "Fire" and "Not Fire" it is obviously the essential parameter in this context at all. Therefore this article deals with item (c.) in the following.

## 2. THE BACKGROUND FOR SIGNAL DETECTION FILTERS IN AUTOMATIC FIRE DETECTION SYSTEMS

Fig. 1 shows the basic procedure in an automatic fire detection installation. The following paper only deals with the signal processing part $m(t) \rightarrow y(k)$ and

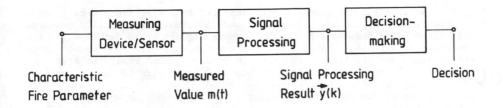

| Measuring Device/Sensor | Signal Processing | Decision-making |
|---|---|---|

Characteristic      Measured      Signal Processing      Decision
Fire Parameter      Value m(t)      Result $\vec{y}(k)$

Fig. 1  Signal Processing in Autmatic Fire Detectors

does not consider the type of sensor which may be used.
The most simple mathematical model for m(t) is given by

$$m(t) = \begin{cases} s_j(t) + n(t) & \text{if a fire signal is present} \\ n(t) & \text{in normal conditions} \end{cases} \qquad (2.1)$$

$\{s_j(t)\}$ is the set of different signal components in m(t) which may be produced by different fire conditions. n(t) represents the measured value without any fire influence.

In automatic fire detection systems an extremely high efficiency and reliability is required.  Compared with a "normal" signal detection situation the  following aspects are of mayor importance in this context:

(1.) The  signal detection has to be as fast as possible because the  associated costfunction depends very much on the detection time /7/.

(2.) The  false  alarm  rate must be extremely low  because  any  countermeasure initiated in an alarm situation may cause unexceptable costs.

(3.) The  signals  $\{s_j(t)\}$  to be detected are not known in  all  their  details because they are determined by a very large number of different parameters.

(4.) There  is only few information available about the noise n(t) that  has  to be taken into account.

In  this  situation the classical detection filter theory based on  the  matched filter  concept is no longer applicable.  Some other techniques,  e.g.  nonpara- metric signal detection, has to be used. In the following the methods to be used are discussed more systematically.

- All signal components are considered as signal vectors with a finite number n of components:

$$m(t) \rightarrow \vec{m} = \begin{bmatrix} m_1 \\ \cdot \\ m_i \\ \cdot \\ m_n \end{bmatrix} \text{ and } s(t) \rightarrow \vec{s}_j = \begin{bmatrix} s_1 \\ \cdot \\ s_{ji} \\ \cdot \\ s_n \end{bmatrix} \text{ and } n(t) \rightarrow \vec{n} = \begin{bmatrix} n_1 \\ \cdot \\ n_i \\ \cdot \\ n_n \end{bmatrix}$$

$$\vec{m} = \begin{cases} \vec{s}_j + \vec{n} & \text{if a fire is present} \\ \vec{n} & \text{under normal conditions} \end{cases} \qquad (2.2)$$

The  limited number of signal vector components can be derived by a  suitable sampling  procedure  if  an observation interval of limited  duration  $T_o$  is considered.

It  is not known in advance which of the possible signals  $s_j(t)$  will  occur in  a  real fire case.  Therefore it is necessary to match the detection  algo- rithm  to  those  properties of the signals  $s_j(t)$,  that are  common  to  all elements of the set of possible fire signals $\{s_j(t)\}$.  So the component  $\vec{s}_j$  is considered  as an element  $\vec{s}_j$ of a set of signals $\{\vec{s}_j\}$ where all the  elements in  the set have some common properties which are significant for this set of signals. This may be shown by a simple example.

751

Fig. 2 shows some typical signals $s_j(t)$ measured in a fire experiment where the noise $n(t)$ is suppressed. There are two different smoke density measurements taken simultaneously and both serve as signals to be detected in the same dangerous situation. They are different but they have some common properties which are significant for the situation. It can easily be seen that in all possible observation periods of length $T_o$ both signals have a positive or "increasing" trend.
This property can be indicated by using a ranking technique where each measured signal vector

$$\vec{s}_j = \begin{bmatrix} s_{j1} \\ \cdot \\ s_{ji} \\ \cdot \\ s_{jn} \end{bmatrix} \Rightarrow \vec{r}_{sj} = \begin{bmatrix} r_{sj1} \\ \cdot \\ r_{sji} \\ \cdot \\ r_{sjn} \end{bmatrix} \tag{2.3}$$

is associated with a rank vector $\vec{r}_{sj}$.
The components $r_{sji}$ are the ranks of the associated signal components $s_{ji}$ within the signal vector $\vec{s}_j$ according to their magnitude. Using this notation every type of trend within the signal vector easily can be indicated. In particular all monotonously increasing signals have the same rank vector

$$\vec{r}_{nat} = \begin{bmatrix} 1 \\ \cdot \\ i \\ \cdot \\ n \end{bmatrix} \tag{2.4}$$

A positive trend can now be defined by having the rank vector $\vec{r}_{sj}$ in some neighbourhood of $\vec{r}_{nat}$. In other words: In this example the set of signals $\{s_j\}$ to be detected can be <u>defined</u> by the significant and common signal property of a positive trend in the observation period $T_o$.
As a general result <u>we are looking for detection filters for the detection of predetermined signal properties</u> which are associated with a set of otherwise different signals.

## 3. GENERALIZED MATCHED FILTER CONCEPT /8/

In order to get a mathematical (or calculable) formulation it is necessary to have suitable definitions for
- the general term "signal properties" and
- a quality measure associated with a criterion of optimization.
Definitions:
For convenience the subscript j in $\{s_j\}$ will be dropped so that $\vec{s}_j = \vec{s}$.
(1.) The modified sign function:

$$u(\xi) = \begin{cases} +1 \text{ for } \xi > 0 \\ 0 \text{ for } \xi \leq 0 \end{cases}$$

(2.) The term "signal property" is defined by any transformation of $\vec{m}$ or $\vec{s}$ resp. of the form:

$$\vec{f}_i(\vec{m}) = \begin{bmatrix} f_{i1}(\vec{m}) \\ \cdot \\ f_{ij}(\vec{m}) \\ \cdot \\ f_{in}(\vec{m}) \end{bmatrix} \quad \text{or} \quad \vec{g}_i(\vec{s}) = \begin{bmatrix} g_{i1}(\vec{s}) \\ \cdot \\ g_{ij}(\vec{s}) \\ \cdot \\ g_{in}(\vec{s}) \end{bmatrix}$$

where $\vec{f}_i$ and $\vec{g}_i$ in general are vector functions of $\vec{m}$ or $\vec{s}$.
We briefly discuss an example, which afterwards will be used:

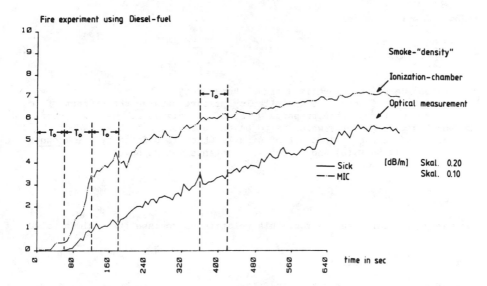

Fig. 2  Typical Fire Signals

Example:

The "signal property" is the order of a single signal sample $s_i$ within the signal vector $\vec{s}$ according to the signal sample magnitudes and can be indicated by

$$\vec{g}_i(\vec{s}) = \begin{bmatrix} u(s_1 - s_i) \\ . \\ u(s_j - s_i) \\ . \\ u(s_n - s_i) \end{bmatrix}$$

Other "signal properties" do not require a notation of such complexity. In these cases the vector function $\vec{f}_i$ or $\vec{g}_i$ resp. are reduced to simple scalar values. The conclusion which can be drawn from this kind of definition is that every transformation or function of $\vec{s}$ that can be written in the form of a vector $\vec{f}_i(\vec{s})$ can serve as mathematical model for a signal property. A convenient "quality measure" is a generalized correlation coefficient introduced by Kendall /9/ which can be written in the following form ($f^T$ indicates the transposed matrix):

$$C = \frac{\sum\limits_{i=1}^{n} \vec{f}_i^T(\vec{m}) \ \vec{g}_i(\vec{s})}{\sqrt{\sum\limits_{i=1}^{n} \vec{f}_i^T(\vec{m}) \ \vec{f}_i(\vec{m})} \ \sqrt{\sum\limits_{i=1}^{n} \vec{g}_i^T(\vec{s}) \ \vec{g}_i(\vec{s})}} \qquad (3.1)$$

Using the Cauchy-Schwartz inequality we have

$$\sum_{i=1}^{n} \vec{f}_i^T(\vec{m}) \ \vec{g}_i(\vec{s}) \leq \sqrt{\sum_{i=1}^{n} \vec{f}_i^T \ \vec{f}_i} \ \sqrt{\sum_{i=1}^{n} \vec{g}_i^T \ \vec{g}_i} = \sqrt{E_m} \ \sqrt{E_s}$$

and $\quad C \leqslant 1$. The terms

$$E_m = \sum_{i=1}^{n} \vec{f}_i^{T}(\vec{m})\ \vec{f}_i(\vec{m}) \quad \text{and} \quad E_s = \sum_{i=1}^{n} \vec{g}_i^{T}(\vec{s})\ \vec{g}_i(\vec{s}) \tag{3.2}$$

may be considered as generalized signal "energies".
The coefficient $C$ serves as a similarity measure between the property $\vec{f}_i(\vec{m})$ of the filter input $\vec{m}$ and the property $\vec{g}_i(\vec{s})$ of the signal $\vec{s}$. We now regard $\vec{g}_i(\vec{s})$ as that signal property which is to be detected or, in other words, $\{\vec{s}\}^i$ is regarded as a predetermined set of signals which is defined by the known signal property $\vec{g}_i(\vec{s})$. So $\vec{g}_i(\vec{s})$ and $E_s$ become constants and we have:

$$\frac{1}{\sqrt{E_m}} \sum_{i=1}^{n} \vec{f}_i^{T}(\vec{m})\ \vec{g}_i(\vec{s}) \leqslant \sqrt{E_s}$$

Obviously the left hand part of this relation approaches its maximum value $\sqrt{E_s}$ if

$$\vec{f}_i(\vec{m}) = \vec{g}_i(\vec{s}) \quad \text{for all } i$$

Following this line the detection filter can be based on Kendall's correlation coefficient choosing $\vec{f}_i = \vec{g}_i$

$$C_T = \frac{1}{\sqrt{E_m}} \sum_{i=1}^{n} \vec{g}_i^{T}(\vec{m})\ \vec{g}_i(\vec{s}) \leqslant \sqrt{E_s} \tag{3.3}$$

where $C_T$ serves as a quality measure and the above relation is the associated criterion of optimization, i.e. $C_T$ has to be made as large as possible at most equal to $\sqrt{E_s}$.
In the absence of noise, i.e. $\vec{m} = \vec{s}$ the filter output produces its maximum value if the filter input $\vec{m}$ belongs to the set $\{\vec{s}\}$ of signals defined by $\vec{g}_i(\vec{s})$. From this point of view we have a detection filter "matched" to all signals $\vec{s}$ belonging to $\{\vec{s}\}$ or, in other word, "matched" to all signals $\vec{s}$ with the signal property $\vec{g}_i(\vec{s})$. Such a detector may be called a generalized matched filter. Necessarily the special form of $\vec{g}_i(...)$ has to be known.
The filter behaviour in a noisy background cannot be treated in such a general way. Furthermore it is beyond the scope of this article. In the following some examples shall be discussed.

Example 1:
   If the signal property to be detected is the signal trend it can be based on
   a suitable indication of the individual ranks for all signal components
   according to their numerical values.

$$\vec{g}_i(\vec{s}) = \begin{bmatrix} u(s_1 - s_i) \\ \cdot \\ u(s_j - s_i) \\ \cdot \\ u(s_n - s_i) \end{bmatrix} \quad \text{and} \quad \vec{g}_i(\vec{m}) = \begin{bmatrix} u(m_1 - m_i) \\ \cdot \\ u(m_j - m_i) \\ \cdot \\ u(m_n - m_i) \end{bmatrix}$$

and Kendall's generalized correlation coefficient (3.3) yields

$$y(T_o) = C_T = \frac{1}{\sqrt{E_m}} \sum_{i=1}^{n} \sum_{i=1}^{n} u(m_j - m_i)\ u(s_j - s_i) \leqslant \sqrt{E_s}$$

with

$$E_m = \sum_{i=1}^{n} \sum_{j=1}^{n} u(m_j - m_i), \quad E_s = \sum_{i=1}^{n} \sum_{j=1}^{n} u(s_j - s_i)$$

Consider now the property of a monotonously increasing signal $\vec{s}$ which has a strictly increasing trend and

$$u(s_j - s_i) = u(j-i) = \begin{cases} 1 & \text{for } j > i \\ 0 & \text{for } j \leq i \end{cases}$$

so that

$$E_s = \sum_{i=1}^{n} \sum_{j=1}^{n} u(j-i) = \frac{n}{2}(n-1); \quad E_m \leq \frac{n}{2}(n-1)$$

$$y(T_o) = \sum_{i=1}^{n} \sum_{j=1}^{n} u(m_j - m_i) \leq \frac{n}{2}(n-1)$$

This is the well known (nonlinear) nonparametric Kendall-$\tau$-detector for trend. It has a considerable advantage because it can be rewritten in a recursive form

$$y(k) = y(k-1) - \sum_{i=1}^{n} u[m(k-(n-i)) - m(k-n)] + \sum_{i=1}^{n} u[m(k\ ) - m(k-(n-i))]$$

An associated filter structure can very easily be derived as is shown in Fig. 3.

Example 2:
Many applications in fire detection yield the following signal property: For all signals $s(t)$ it is possible to define a generalized (time dependent) threshold function $m_o(t)$ within the observation period $T_o$ with

$$s(t) > m_o(t) \quad \text{for all} \quad t\varepsilon[0,T_o]$$

as can be seen from Fig. 2.
If we introduce a suitable threshold $m_o(t)$ as a standard signal

$$m_o(t) \rightarrow \vec{m}_o = \begin{bmatrix} m_{o1} \\ \cdot \\ m_{oi} \\ \cdot \\ m_{on} \end{bmatrix}$$

and consider the difference $\vec{s} - \vec{m}_o > 0$ i.e. $s_i > m_{oi}$ for all i, as that signal property to be detected, we get

$$\left. \begin{aligned} \vec{g}_i(\vec{s}) &= g_i(\vec{s}) = g_i(s_i) = u(s_i - m_{oi}) = 1 \\ \vec{g}_i(\vec{m}) &= g_i(\vec{m}) = g_i(m_i) = u(m_i - m_{oi}) \end{aligned} \right\} \quad \text{for all } i$$

and

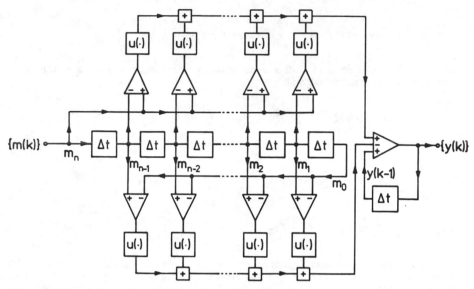

Fig. 3  Trend Detection Filter (Kendall-τ Type)

$$E_s = \sum_{i=1}^{n} u(s_i - m_{oi}) = n; \quad E_m = \sum_{i=1}^{n} u(m_i - m_{oi}) \lessgtr n$$

$$y(T_o) = C_T = \sum_{i=1}^{n} u(m_i - m_{oi}) \lessgtr n \quad \text{or written as a recursive algorithm}$$

$$y(k) = \sum_{i=1}^{n} u\left[ m(k-(n-i)) - m_{oi} \right] \lessgtr n$$

Example 3:

In some important fire detection cases the signals to be detected very rapidly rise to a rather high signal energy.  A simple signal energy detector can be derived by using the method indicated above. This results in

$$y(T_o) = C_T = \sum_{i=1}^{n} m_i^2 \lessgtr n$$

or written in a recursive form

$$y(k) = \sum_{i=1}^{n} m^2(k-(n-i))$$

which is a normalized signal energy detector.

## 4.  APPLICATIONS

In  practical  applications  mostly a combination of  several  different  signal properties  define  a set of signals $\{ \vec{s} \}$ which are to be detected,  or in  other

756

words, the detection filter has to be matched to a set of signal properties. Following the theory outlined above a fire detector is a combination of several single detection filters each of them matched to a special property. A logic connection has then to be made according to the detection situation which is to be solved. Fig. 4 shows an automatic fire detection system based on the signal properties outlined above.
A prefiltering shown in Fig. 4 is to reduce the influence of highly correlated noise and has not been discussed in this presentation /10/.

## 5. FIRE DETECTION CAPABILITY

Up to this point the interference by the noise component has been neglected. It has not been incorporated in the method of the detection filter design. On the other hand it has a considerable influence upon the efficiency of the fire detector performance.
The filter behaviour in a noisy background cannot be treated in such a general way mainly because no general information about the noise is available. Nevertheless there is some evidence from field experiments for the assumption that the noise n(t) consists of at least two different components

$$n(t) = n_o(t) + n_t(t)$$

where $n_o(t)$ is a low pass approximative stationary signal so that the associated sample-vector $\vec{n}_o$ contains statistically dependent components. $n_t(t)$ is a rather rare and transient signal which is very hard to measure in the field.
Under this condition it is hardly possible to calculate the detector performance efficiency in terms of false alarm or detection probability or any other statistical measure.
The only way to overcome this problem is to use some simulation technique.
In order to study the behaviour and the efficiency of detector types discussed in this article a special simulation and measuring device has been used /4/. It is computer-controlled and generates the input data with suitable accuracy and simulates the detector algorithm. In addition, the important and interesting statistical parameters of both input data as well as detector output can be measured and plotted in a suitable form. The basic configuration of the device is discussed elsewhere /6/. In this context the following items are of major importance.
(1)  It is important to use a simulation method which generates reliable results much faster than they can be drawn from measurements in the field. This is mainly because false alarms are very rare events in practical cases if one single detector device or even a group of several detectors is considered. On the other hand, it is obviously impossible to draw any reliable conclusion from a statistical measurement without having observed at least $10^3$ false alarms.
(2)  The main difficulty in this context is the fact that a signal generator is needed to produce a variety of simulated input signals for the detector device, which allows to determine both the distribution function or probability density $p_m(m)$ of the generated signal {m} and its autocorrelation function $R_{mm}(k)$ independently from each other. $R_{mm}(k)$ is a measure for the two-dimensional dependency of the signal samples m(k) and is therefore closely linked to the two-dimensional probability density function. So a practical solution cannot comply with the above requirement in principle. On the other hand, there is - within some mathematical limits - an approximate solution which cannot be outlined in this article. Some details are given in /4/.
Results from several simulation experiments using the method indicated above show that fire detection systems based on detection filters shown in Fig. 4 very well may be able to improve the fire detection capability, i.e. reliabi-

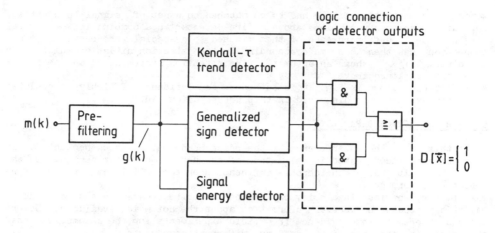

Fig. 4   Fire Detector Matched to a Combination of Signal and Noise Properties

lity of detection as well as false alarm rate, to a considerable amount. Algorithms like those mentioned above can easily be implemented by using modern microcomputer means as software programs or they can be introduced as specific VLSI-circuits both without an unsuitable increase in system prices. Future development will, probably, increasingly be directed towards an improvement of fire detection capability using modern electronic means including software solutions. This, on the other hand, is necessarily going to change the testing methods and associated international standards.

## REFERENCES

/1/ H. Luck, "Neuere Entwicklungen und Entwicklungstrends in der Technik der automatischen Brandentdeckung", Tagungsbericht Fachtagung des VBSF und des BVD, Nov. 1983, Bern, Schweiz

/2/ J.F. Fry, "The Problem of False Alarms from Fire Detection Systems", Proceedings of 6th International Seminar on 'Probleme der automatischen Brandentdeckung', Institut für Elektrische Nachrichtentechnik, RWTH Aachen, Oct. 1971, pp. 37-53

/3/ R. v. Tomkewitsch, "Fire Detector Systems with 'Distributed Intelligence' - The Pulse Polling System", Fire Safety Journal 6 No. 3 (1983)

/4/ H. Luck and K.-R. Hase, "Signal Detection Aspects in Automatic Fire Detection", Fire Safety Journal 6 No. 3 (1983)

/5/ Fire Safety Journal 6 No. 3 (1983), (Special Issue on Automatic Fire Detection)

/6/ H. Luck (ed.), "Proceedings of AUBE '82 - 8. Internationales Vortragsseminar über Probleme der automatischen Brandentdeckung", Universität Duisburg, Oktober 1982

/7/ H. Luck, "Wirksamkeitsmaße für spezielle Signaldetektoren einfacher Struktur", AEÜ 26 (1972) pp. 267-275.

/8/ H. Luck, "Signal Detection Using a General Matched Filter Concept", AEÜ 36 (1982) pp. 217-222

/9/ M. Kendall, "Rank Correlation Methods", Griffin, 4th edn. London 1975

/10/ H. Luck, "Signal Detection in Automatic Security Systems", Proceedings of EUSIPCO-83 (2nd European Conference on Signal Processing), H.W. Schüssler (ed.), Elsevier Science Publishers B.V. (North-Holland)(1983).

# SPECIALIZED FIRE PROBLEMS

Session Chair

**Prof. Yoichi Uehara**
Yokohama National University
156 Tokiwadai, Hodogaya-ku
Yokohama 240, Japan

# Historical Aspects of Fires, after Impact, in Vehicles of War

W. JOHNSON
School of Industrial Engineering
Purdue University
West Lafayette, Indiana 47907, USA

Key words:   Fire    Vehicles   Historical
             War     Impact     Damage

## ABSTRACT

Bellcose, destructive engagements between land, sea and air vehicles throughout history are reviewed, where fire has been the primary destructive agent or has been a powerful secondary consequence of the mechanical impact of projectiles.

## INTRODUCTION

The subject of fire in vehicles of war was strikingly brought to the attention of the people of the U.K., (and indeed elsewhere), by the Falklands War of 1982. News-reel coverage of incidents in which ships were hit and set on fire by Exocet missiles was extensive and the public in Britian engaged in much discussion about the reasons for the vulnerability of their present day naval craft to fire, especially after missile attack.

Studying the effects of fire in, or caused by, vehicles of war, brings about a realisation that fire has long been a deliberately offensive weapon and sometimes an unforeseen horrendous secondary consequence of the use of missiles.

This short paper endeavours to draw together some of the more general but conspicuous features of fire as encountered historically in vehicles of war. There are, necessarily, lacunae in this treatment of the subject because of want of information about weapons' effects and because the topic has not, apparently, been greatly researched in extenso.

The review begins with brief historical facts and references to pre-Christian and Roman naval operations in which fire was employed and these are followed by notes, mostly about Greek fire and its use predominantly by the Byzantines, with some interspersed remarks drawn from Needham's writings on firearms and gunpowder in ancient China. Little has been found about fire in vehicles of war at the time of the Middle Ages and indeed up to the relatively modern post-1600 A.D. era. From simple reflection it is easily concluded that until the 20th. century, it is naval vehicles that must be considered the principal topic of interest, since the two large categories, automotive vehicles and aircraft, previously had no existence, whilst railways or locomotives have been operating for but a century and a half. For a survey of the subjects of fires after impact in non-warlike vehicles, see ref. (1).

## FIRE IN WAR VEHICLES

### I. Pre-Christian Naval Engagements Using Fire

The maximum speed of pre-Christian warships has been estimated at between 6 and 13 knots,[2] it tending to increase in proportion to the number of super-imposed banks of oars. Phoenician biremes, and Greek triremes and penteconters are known of from several centuries B.C.[3], and many illustrations of them appear on ancient vases.

For penetrating the hulls of the ships of their adversaries beneath the water, solidly pointed underwater rams[3] were employed. Besides these it was common to throw fire and shoot flammable arrows and the like from towers specially built on warships.

Chapter 2 in refs. (4), notes several conspicuous pre-Christian warlike encounters in which the deliberate promotion of on-board fire was attempted, thus,

1. In the battle of Samalis, 480 B.C., the Greeks placed red-hot coals in kettles on the end of long spars and tipped them into enemy (Persian) ships.
2. In the battle of Actium in 31 B.C., heavy weights and balls of fire were thrown on to Roman ships from wooden towers built on to the vessels in Mark Anthony's fleet.
3. It is recorded that Archelaus in a war (of 87 B.C.) against the Romans, washed a wooden tower with a solution of 'alum' (sic) to render it fire resistant. Apparently however, this alum seems not to have been used to protect ships' timbers against fire and in any case was not present-day alum (see the discussion in ref. (5)). However, Partington[9] does attribute the use of real alum to Greeks for fire-proofing timber and to Romans for fire-proofing seige engines in 296 A.D.
4. Recall that Archimedes was supposed to have set fire to Roman ships using metal mirrors.

## II. Principally on Greek Fire

To convey a sense of proportion about the true importance of fire in the whole of early warfare, it is well to recall an excerpt from Oman's[6] 'The Art of War in the Middle Ages, A.D. 378-1515.'

".... Greek fire .... though its importance in poliorcetics (siege warfare) and naval fighting was considerable .... was a 'minor engine of war' and not comparable as a cause of Byzantine success to their excellent strategical and tactical system ...."

However, the impression which Greek Fire could have on troops is well instanced by the following extract from a Russian chronicle,[7]

"The Greeks have something which is like lightning from heaven, and, discharging it, they set us on fire; that is why we did not defeat them."

The above quotation refers to a naval attack on the Russians by the Byzantine navy in 941, when a Russian armada was virtually annihilated by the use of Greek fire. Of the year 971 A.D., there is a report of Byzantine fire-shooting ships "capable of turning the very stones to cinders". Greek fire was a notorious 'secret weapon' for which nations from Eastern Europe sought 'samples' of this 'priceless commodity' from the Byzantine Empire. One early 13th. century manuscript[7] shows an artist's idea of such an encounter. Early occasions as in the 7th. century when "fire ships" were fitted with siphons are noted in ref. (9).

Scoffern in 1858, gave an early and relatively lengthy discussion of the origins of Greek fire, whilst a shorter different survey on Inflammables and Explosives, is to be found in the book published in 1906 by Cowper.[9] The most significant work of scholarship on this subject is undoubtedly that from Partington[9] in 1960, entitled "A History of Greek Fire and Gunpowder."

The ingredients of Greek fire, variously stated to be mixtures of any or all of sulphur, resin, quick-lime, pitch and turpentine, or naptha,[†] Needham categorically states to have been used from the classical period of Greece up to the 14th. century A.D., by which time gun-powder had been developed or invented. This fire was projected by bellows or pumps (from ships or even land vehicles) on to enemy vessels which took fire and caused black smoke which could seldom be suppressed. The projected incendiary material, say the naptha, was likely to have been thickened by resinous materials and sulphur since the jet would otherwise have been dissipated too quickly by the air.

---

[†] From "liquid petroleum from oil wells in Iraq", p. 28, ref. (9).

The common idea that the Chinese used gun-powder only for fireworks, Needham holds, is quite false.[11] That it was developed solely by artisans he maintains is also untenable; he claims it was the outcome of 'systematic' explorations by Taoist alchemists.

The name 'wet fire' or 'wild fire' has been used to describe the 'fire' ejected from siphons; and yet another variant seems to have been 'sea fire' which probably included salt-petre;[10] the latter went out of use in 1200 A.D.

Needham,[11] has described a 'fire lance' of the 10th. century which was used to destroy rigging and woodwork; when first used it was held manually by fire-weapon soldiers. Later it was made of bamboo, (he claims it to be the archetypal gun barrel) and was called an eruptor; it is indicated to have been long and nearly one foot in diameter, being mounted on a framework of legs and even wheels for mobility. This device is supposed to have shot flaming projectiles. Edward Gibbon's Chap. LII of his Decline and Fall of the Roman Empire,[8] contains an interesting reference to the use of copper tubes in the bows of ships, for projecting Greek fire in assaults on Constantinople in the late 8th. century.

Table 1 from ref. (1) seems to show that knowledge of gun-powders, allegedly first used as a 'match' for flame-throwers in 919 A.D., was brought to Europe from Chinese lands in the second half of the 13th. century by Moslem and Christian merchants and adventurers and by proselytizing Nestorian priests. The later forms of gun-powder were brisant with the nitrate proportion in compositions increased to give destructive explosions.

Greek fire was employed in Western Europe in the Middle Ages for burning towns, the very common roofs of straw or shingle being highly vulnerable to fire. In 1379 at Oudenarde, inhabitants covered their houses with earth to resist fire hurled into the basse-cour.

Generally, ancient fire extinguishers were water, sand, dry or moist earth and manure or urine (which contain phosphates) and probably salty vinegar.

The same applied in respect of bridges and shipping. A Brabanter wooden bridge over the Meuse was attacked by the Guerlois with 'engins feu' in 1388 and caused to collapse into the river. At Breteuil in 1356, those besieged were provided with 'canons jetant feu...' to deter attackers on advancing towers.[12]

If we include Middle Ages seige engines as early vehicles in our review then notice may be taken of a full description of the use of the Sow or Cat (the ancient Vinea) in the seige of Jerusalem in 1099 which was given by William of Malmesbury. The Sow is essentially a low, mobile covered timber construction protecting advancing engineers engaged in mining the foundations of walls. Mines prepared in about 1200 A.D., were caverns in which pillars of wood, supporting incumbent walls, were smeared with pitch, surrounded with combustibles, and then fired to allow collapse. Also, Berefreids or moderate-sized towers were employed as a stage for attacking soldiers contesting defenders on ramparts. The defenders it is reported, 'trusted their whole security by pouring down boiling grease and oil upon the Tower'. This was replied to by the Franks by throwing 'faggots flaming with oil on to an adjoining wall tower'. Wet animal hides were frequently drapped over wooden walls to protect structures.

Cowper[8] notes the wide and early use of inflammable propellants for projecting rockets and fire pots from tubes in 'early India' by the Siamese, at Delhi by Mahoud V against Timur Beg and their throwing by catapult in 1290 by Ala-ed-din. He quotes too, examples of projected ignited moss on to burning villages by South American Indians. Cowper's summary that "inflammables for war derive from Assyria which bequeathed it as Median fire to Byzantium - the Arabs stole (sic!) it from the Greeks - and the Europeans learned (sic!) it in the Crusades", may well be as true as Needham's line of suggestion in imputing to it a unique origin in China. Partington,[9] again, devotes whole Chapters to this subject. It is not all improbable that many pyro-jetting practices were developed in several societies independently of one another, much as is sometimes the case with scientific discoveries.

Partington devotes a chapter to incendiaries of war and especially mentions several Muslim fire books which show among other things recipe proportions for gunpowder and the use of naptha.

## III. Fire Ships and Hot Shot

Wooden naval ships have always feared incendiary attack - an increasing threat onward from the time of the introduction of Greek fire. Cowper says that Procopius in his history of the wars of the Goths of Africa describes the use of Median fire and Genseric is said to have used fire ships against the Greeks in the 5th. century A.D.

Fireships were used, successfully, against Drake at Cadiz in 1490 by the Spaniards, and the Dutch filled ships with explosives to cause explosions in the Duke of Parma's fleet which was beseiging Antwerp in 1585. In battles against the British in the latter half of the 17th. century, the Dutch always had on hand a number of specially combustible craft. Fire was used to great advantage in a Dutch raid (on English ships) up the river Medway in 1667.*

Fire ships consisting of fishing smacks of faggots and pitch were apparently to be used by the English to attack the Spanish armada in 1492 in the Calais Roads but as they could not be assembled rapidly enough, eight small ships of the fleet, probably victualers, were prepared. A Spanish screen of protective patrol boats mostly fled as tide and wind bore the fired ships into the assembled armada. These fire ships had 'shotted' guns and their explosions added to the panic induced. The starting of ship fires in the armada apparently failed completely but they were successful in as far as they induced the 120 Spanish ships to slip or cut their cables to escape in terror.

It has often been observed that, certainly in the late 18th. century, it was rare for ships-of-the-line to be sunk by gun fire: burning or capture† was the rule. In a Dutch-English naval battle of 1794, the Dutch had at least seven and the British two ships burnt or blown-up. In the battle of Trafalgar no ship was sunk on either side as a direct consequence of enemy action.[14] Though fire ships were included in most large fleets there is little evidence that they directly caused fire disasters. But fires in wooden ships were a terrible threat and their menace generally led to the British Navy taking thorough precautions against them.

Fire ships were used as recently as the first quarter of the 19th. century, for example in 1809 by an English fleet under Lord Cochrane against French ships at anchor in the Basque Roads, and in the War of Greek Independence the Greek fireships used successfully against the Ottoman navy led to their command of the sea.

Notwithstanding the panic which tended to be induced by fire ships hampering the movements and dispersal of a fleet, they could be easily sunk by enemy fire or towed away by enemy boats. Also, experience showed that premature explosion by those putting fire ships into position could often be fatal. Typically, fire ships of later times consisted of building 'a fire chamber' between decks from the forecastle to a bulkhead constructed abaft of the mainmast. This was fulfilled with resin, pitch, tallow and tar and with gunpowder in iron vessels. The gunpowder and the combustibles were connected by powder trains and by bundles of brushwood ('bavins'). Service with these fire ships was so dangerous that the reward of £100, or in lieu of it a gold chain, was given by the British Navy!

It was long the practice when wooden naval ships were in vogue to exchange hot shot or cannon balls in the hope that fires would be started or ammunition caused to explode on the enemy's ships. It was infrequent for shot to penetrate the sides of ships

---

* Interestingly, it is recorded that Newton heard the noise of this raid from as far away as Cambridge and was able to pinpoint the time of day and deduce that serious losses had been incurred.

† Firing a ricochet to demast ships and thus leave them to be boarded later or at the mercy of the weather was a serious 18th./19th. century tactic, of the British Navy. See W. Johnson and S.R. Reid, 'Ricochet of Spheres Off Water', J. Mech. Eng. Sci., 17, 71-81, 1975.

of a fleet which were often three feet thick in hard woods such as oak, teak and mahogany. In ref. (13) it is implied that it took a close range 32 lb. shot from a 10 ft. long cannon to do so. In a long seige of Gibraltar by wooden French ships in the 1780's, red hot shot was apparently fired from the land fortress.[14] Many reports exist of ships exploding after gunpowder carried in magazines had been heated by unintentional ship fires or penetration by missile fragments.

The invention of red hot shot, in 1579, is attributed by Brigadier O.F.G. Hogg in his foreward (page v) to the book, Great Art of Artillery[†] (Artis Magnoe Artillerioe) by Polish General C. Simienowicz, to the King of Poland. The book itself is a fascinating text book for would-be pyrobolists rather than artillerists; it was originally published in Latin in 1650, then translated into French and from French into English in 1729. The variety of fire projectiles treated by him exceeds at least twenty. Simienowicz says in his Chap. XVII that red hot shot was "far from being modern" and refers to Diodorus Siculus testifying that the Tyrians projected it into the "works of Alexander the Great". Also an occasion is mentioned when in a war of 1598 a red hot ball penetrated a tower, fell into a barrel of gun powder and destroyed the entire structure. Gibbon (see above) in his Chap. LII refers to the launching of red hot balls of stone and iron in the 8th. century as well as to inflammable oil deposited in fire ships.

Under this heading brief mention needs to be made of incendiary shells first introduced in 1460 and variously developed since. In the main these were spherical carcasses or shells containing ignitable compositions which were vented radially to allow the egress of flame. Martin's shell (c. 1860) was filled with molten iron and used against shipping by the British whilst another design due to Hodgkinson (c. 1914) was used against Zeppelins (See O.F.G. Hogg, Artillery, Archon Books, 1970, p. 171).

Firing fleets at anchor as described above merely underlines the perennial risk from fire in any gathering together of vehicles, e.g. of buses and coaches, see ref. (23). We may mention that the biggest naval holocaust of the 19th. century occurred in the U.S.A. at the Norfolk Naval Base when Southern sympathizers torched warships on the occasion of Virginia's secession from the Union in 1861.

The effects on wooden ships of a lightning strike, namely 'rigging set on fire, masts split and severed to pieces...' was reported in writing in 1775 by Bathae.[3] The reason for the latter was probably due in part to the anisotropic character or the grain of the wood; its moisture when instantaneously vapourised 'explodes' the wood whilst axiality in the flow of electricity leads to severing or 'chivering'. Early references to this (end of the 17th. century), and closely related work on fulgerite[*] formation is given in ref. (24).

Lightning conductors had been introduced and were nearly universal in British naval vessels by the beginning of the 19th. century.

## IV. Armoured Fighting Vehicles and Fire in the 20th. Century

In all military fighting vehicles fire poses a huge and constant threat. Not only does it destroy expensive structures and equipment but the personnel, aside from purely humane considerations, are frequently expensively equipped and costly to train. The flame-thrower has already been mentioned in an historical context[‡] as an ancient

---

[†] Republished by S.R. Publishers Ltd., 1971, Wakefield, England and Scholar Press Ltd., Menston, Yorkshire, U.K.

[*] Fulgerites: tubes of vitrified sand found in the earth after a lightning strike, so-called after the Roman god of lightning, Fulger.

[‡] Compare Gibbon (Chap. LIII) who refers to the use in (probably) the 8th. century of a "mischievous engine that discharged a torrent of the Greek fire, the feu Gregois ..." In a footnote he says that Joinville described something as "like a winged long-tailed dragon...with the report of thunder and the velocity of lightning..." "It was shot with a javelin from an engine that acted like a sling", seemingly a mangonel, but more likely a balista.

naptha projector which was certainly in use according to Needham[†] in 919 A.D., when gun-powder of a kind was first identified as being a 'match' for naptha. A British origin and development for napalm has been claimed in ref. (15). Flame Fougasse as a one-shot weapon was apparently developed in about 1940 to project onto advancing tanks about one ton of burning petroleum.[†] The mobile Churchill flame-throwing tank was later developed to assist in clearing operations and a slightly fictionalised account of this Crocodile as used operationally in World War II is given in 'Flame Thrower' by A. Wilson.[16] Some details as outlined by him follow.

The flame-projector was mounted in the front of a Churchill Mark VII tank with the flame-gunner alongside the driver. Four hundred gallons of heavy viscous flame fuel, (of petrol, naptha and rubber - a crude form of napalm) was carried in a seven-ton trailer connected to the tank by an armoured link containing the fuel pipe. The fuel was projected from the nozzle gun by means of 350 p.s.i. of nitrogen, - actually reduced from 5000 p.s.i., as stored in five steel 'bottles' carried in the trailer. The 'rod' of fuel was ignited by a jet of burning petrol that passed between two electrodes. The gun had a range of 90 yards and could flame continuously for two minutes.

In 'The Secret War' by Gerald Pawle,[17] a 'Cockatrice' is described in chapter 4 as designed mainly by the Lagonda Car Company. It fired diesel oil and used 8 gallons of fuel per second. Later models fired 200 yards; these were developed by the Army and the Anglo-Iranian Oil Company. The Cockatrice consisted of a $2^{1}/_{2}$ ton lorry, "invulnerable" to fire, possessing a tank holding 2 tons of fuel with the flame thrower itself mounted in a turret behind the cab and operated by a gunner in the turret. The all-up weight was 12 tons.

It was tried at sea after installing it on a trawler, by firing it vertically against low-flying aircraft: the pillar of flame was over 400 ft. high, its height being enhanced by the heat of the fuel. However, in tests with even up to one half an aircraft wing being driven through the flame, no serious damage to the plane ensued. A few naval flame-throwers were produced and installed on sea-coasters but their maintenance was difficult; very high pressures had to be maintained and unless well handled, ships and their crews were liable to be smothered in tar and oil.

It seems that German experience after testing a somewhat similar device which projected lighted fuel from a pipe and nozzle which ran up a ship's mast, showed it to be a failure.

Tests of the ability of assault landing-craft to enter a harbour with the infantry passing through the fire of any flame-thrower mounted on a breakwater, showed they needed no canopy. To fire a flame along a straight trajectory when slightly depressed was said to be very difficult. Tests showed that dummies, paper and mice in the bottom of craft, emerged without scorching or poisoning.

Impact leading to fire and/or explosion is promoted when some modern tank armours are penetrated by copper-lined shaped-charges; the temperature of local target material is raised but particularly it has been noted that hot spall particles are ejected through tank compartments. If they are small (mainly aluminium), they may burn so rapidly that effectively an explosion occurs and they may have a lethal effect on a tank crew. Suppression is provided by plastic liners.

A shaped-charge passing through a partially filled container of gasoline, or volatile hydraulic fluid, renders it rapidly combustible when heated and atomised. Diesel fuel thus tends to be used and its tank container located outside the crew compartment armour. Shaped-charge jets may also penetrate ammunition, but steps are now taken to quench explosion propagated through propellant charges by surrounding them with an extinguishing agent.

---

[†] Hogg, op. cit., p. 174, refers to an engine tested for projecting liquid fire at Woolwich, near London, in 1709.

## V. Airships

Fire sources in airships and catastrophic experiences with them are outlined in refs. 18 and 19; also many references are quoted therein which detail well known disasters, for example Deighton and Schwartzmans' 'Airshipwreck'.[20] Because the risk of fire aboard airships has been so heavily reduced, mainly due to ceasing to use hydrogen as the lifting gas - and for several other reasons[21] - many governments have become interested in developing them for off-shore patrolling work such as protecting oil installations and fisheries, etc., and as having anti-submarine capability. They have been thought of as capable of carrying air-to-air and other missiles and torpedoes, all both cheaper and faster (at up to about 100 knots) than conventional surface ships, able to have light armour and being easily maintained with a small crew. It has been stated that such airships have a lower radar profile and little infra red and noise emission, which makes them less attractive targets for homing missiles.

Stepping back into history, the Britannica Encyclopaedia (1947) gives a detailed historical account of air balloon development in which instances of disastrous fires occur, many similar in result to contemporary disasters typified by a 1983 report in the Sunday Times of a 140,000 ft$^3$ hot air balloon being projected 200 ft in the air after a propane cylinder had exploded.

By contrast, 'My Airships' is an interesting but little known book,[24] by A. Santos-Dumont (1873-1937) about his airship development and flying activities between 1900 and 1905. Among many of his observations, this pioneer thought that the first practical use of airships would be found in war, (p. 80). He recorded that he had no fear of fire (p. 45), and that he could see but one dangerous possibility of it (p. 48), namely, "that of the petroleum reservoir taking fire by a retour de flamme from the motor" - (a sucking back of flame). This innovator, contrary to the subsequent experience of others, appears to have suffered no fire disasters despite his large number of journeys.

## VI. 20th. Century Naval Ship Fire Safety [25]

"Naval ships are often floating ammunition farms and fuel dumps housing multi-million dollar (equipments)... and densely packed with personnel..."
"The most common feature of modern naval warfare is internal fire caused by shells bursting among combustibles and explosives."

This topic is reputed to have received much attention from about the time of the Russo-Japanese war (1904-1905) when fire, smoke and burning paint on the Russian ships, together with acrid gas from Japanese gun powder ('shimosa'), had devastating effects.

In about 1880 the coal bunkers of battleships were deliberately placed between the outer skin of a vessel and its vital organs. It was noted from tests in 1878 that 2 ft of coal had the resisting power of 1 inch of iron (ref. 26, p. 412) and was not set on fire by shell (i.e. gun fire) explosion. However, tests in 1904 using torpedoes showed that similar protection was not available against them.

From World War II to the mid-1970's, major fire disasters aboard naval ships and aircraft carriers especially, hardly diminished.

H.M.S. Hood, a 41,000 ton British capital ship, (having a deck only 3 inches thick and turret roofs only 5 inches), apparently never modified for the containment of powder flash from burning powder magazines after experience in World War I, suffered a direct hit from The Bismark which caused a great fire that quickly led to the explosion of the magazines. The German ship suffered nearly 400 direct shell hits and 6 torpedoes and only sank after being scuttled. Oscar Parkes[26] in his huge and authoritative volume on British Battleships, 1860-1950, notes that propellant cordite which exploded when exposed to flash in the magazines, explains the loss of battle cruisers at The Battle of Jutland. A different treatment of cordite (separation into two sections) by the Germans did not give rise to explosion despite being fired. Parker notes that among the German battle cruisers associated with nine penetrations there were eight fires but no explosions (p. 641 and p. 679).

World War II made clear the vulnerability to conflagration from oil, aviation gasoline and explosives of many Japanese and American aircraft carriers which had only lightly armoured decks; many ships were lost due to these, and after attack by bomb and Kamikaze attacks. This reflects the fact that a new, essential and far reaching lesson for sea power (after the battle for Midway) was that centuries of dependance on big ships carrying big guns was ended and their place was to be taken by aircraft carriers. Many battery explosions and electrical fires in U.S. submarines occurred between 1949-1955, whilst in the late 1960's, three carriers were the subjects of conflagrations due to 'burning pyrotechnics' and deck fires fed by aircraft ordnance and fuel ignitions.

The 1970's saw in fires on the U.S.S. Forrestal and U.S.S. Saratoga, impassable passage-ways affording fire spread and fire propagation along wire and cable-ways. Most traumatic was the fire resulting from the collision of cruiser U.S.S. Belknap and carrier U.S.S. J.F. Kennedy in 1975. The mast and aluminium superstructure of the former were sheared off by the overhang of the Kennedy's angle deck sponson; most damage resulted from the melting of aluminium structure due to fire,[†] fed early on by

JP-5 fuel pouring from the sheared flight-deck fuel risers on the U.S.S. Kennedy.

Hydro-carbon fires are combated best by excluding air and reducing the vapourisation rate of applied liquid Aqueous Film-Forming Foam (AFFF) concentrate (mixed with sea water to give 6% solution). Halon gases in machinery spaces extinguish fires by interfering with the combustion chain reaction.

Fire protection design against present-day fire threats to ships are outlined in ref. 4.

Oil fire extinguishment it is held, may be possible by using pressurised containers of water and very fine mist nozzles; the water spray droplets become steam which absorbs thermal energy and cools the fire.

A Flight-Deck Washdown System was introduced after the unfortunate experiences with the U.S.S. Forrestal (1967) and U.S.S. Enterprise (1969). The system can deluge all or part of a flight deck with AFFF through spray nozzles on a flight deck or along a deck edge.

Items for upgraded fire protection on naval ships include, among other things, luminescent markers to make clear escape routes to weather decks, refractory felt for the protection of aluminium superstructure, improved magazine sprinkler systems and fire spots for cable ways.

For detailed papers about missile magazine protection, see refs. (29) and (30); Halon Expansion foam fire protection systems for ships, are discussed in the Naval Engineers Journal of recent years.

## VII. Mainly About the Use of Aluminium in Naval Vessels [31,32]

Aluminium has wide use in modern naval craft, especially in the hull and super-structure in order to reduce weight, improve stability and to help conserve fuel. Unfortunately, the relatively low temperatures vis-a-vis steel at which aluminium loses its strength, renders fire threat of enhanced importance. Thus, alongside the use of aluminium it has become necessary to develop light weight thermal mitigation systems. This leads to trade-offs as between fire resistance (or time to structural failure under fire conditions) and weight, in the design of high performance craft.

---

[†] The ability of metals when sliding, cutting or dragged under pressure against another surface to reach a high flash temperature and thus to be an energy source sufficiently high to ignite available vapourised fuel is empirically addressed in ref. (27). Ref. (28) seems to be the most recent scientific paper in which junction growth and heat due to the work of plastic deformation, are incorporated into the heat-transfer equations of the Jaegar-Archard theory and provide results which compare well with those of experiment for estimating flash temperatures.

Aluminium is about one third the density of steel but has about the same ultimate strength, which impressively contrasts the strength of these two materials over a large range of temperatures. Whilst the various aluminium alloys do perform differently, they do not do so significantly over the temperature range of interest; certainly at 700°F all aluminium strength has effectively disappeared. When active systems are inadequate, malfunction or by human error are not released, then passive systems are relied on. Particularly this means minimising serious fire threat where aluminium is concerned; time is purchased prior to active fire protection systems being operated, by adding thermal insulation. Fire prevention, early fire detection and rapid fire extinguishment are essential active features, especially on aluminium craft. Examples exist of aluminium bulk heads, aluminium hulls and indeed whole ships being melted down; in contrast these items in steel do not.

Fire insulation materials currently used are mineral wood, refractory felt or blanket polyisocynurate foams, polyimide foam and fibre glass, all with densities in the range 2.5 to 7 lb/ft$^3$. With a temperature of 450°F as a failure level, one hour of protection is given by about $1^3/_4$ inches of 4 lb/ft$^3$ refractory felt. Note however that heated and ignited isocyanurate foams give off large amounts of HCN gas.

## VIII. Naval Vessels in the Falklands' War[*] of 1982 [21,19]

Fire and its consequences in fighting ships in the Falklands' War highlighted the subject's enormous importance and underlined the cost of taking, or neglecting to take, its likelihood seriously enough. Especially, a lesson of the Falklands' War of 1982 was the need to reduce smoke in battle damaged warships; this was made clear when the destroyer, HMS Sheffield was hit at 8-15 ft. above the water-line, and set on fire by a Super-Etendard 600 mph Exocet missile of 658 kg weight, which carried a 164 kg warhead. It is reported that within minutes of it striking the ship, smoke poured from deck openings at 100 ft and more from the point of impact. The ship was set on fire and burning solid propellant was scattered, as well as the missile's own fuel; these in part created the massive smoke pall. Doubtlessly the ship's ventilating systems contributed to the rapid spread of fire as well as the ship's own materials and the availability of cable ways.[#] Paint on the ship's side came off around the zone near to the missile and close to the region of penetration the hull glowed red. Though fought for several hours it was eventually abandoned out of fear that the magazine might explode.

H.M.S. Coventry was hit by several bombs which caused flooding beyond the design limits of the ship. Smoke rendered the operations room unusable, as much as anything because of disorientation effects.

Of torpedoes which helped sink the Argentine troop carrier Belgrano, one was reported by the Captain to have created 'a fireball', (compare and see the references in (34)), and another 'a cloud of dirty smoke'. A huge hole was created in her bottom, amidships, and her bow broke off.

H.M.S. Antelope suffered from an unexploded bomb that detonated when defusing was in progress. The ship burned for many hours apparently with the aluminium superstructure remaining intact and not being responsible for her loss.

The Altantic Conveyor received one or two Exocet missiles and burned for two to three days, but the superstructure on this occasion distorted without collapsing.

---

[*] Evidence and sources for many remarks in this section will be found intersperced among two references stated.

[#] Remarks concerning pre-routing ventilation systems to permit smoke removal and for improving ventilation fans for de-smoking have been made. Existing fans apparently cannot remove hot gases since the fan motors cannot withstand the temperatures. Eductors from exhaust systems have been recommended for investigation.

The New Scientist catalogued the following explosion-fire 'errors' as contributing to the H.M.S. Sheffield 'fire trap', though the £85 million ship (commissioned in 1980) had the usual sprinkler systems, fireproof doors and hatches, etc.

1. Generators providing power to fire-fighting pumps were out of action: the missile knocked-out the aft main generator and the forward one failed minutes later.
2. Fire-fighting pumps did not work because vital parts were missing: the ship was deficient in small fire extinguishers since only a small number was available and not rechargeable.
3. In some cases, breathing equipment containing compressed air was almost empty.
4. Foam mattresses burnt easily, giving off clouds of toxic smoke. It is reported that in the latest U.K. battleships, these have been dispensed with in favour of horse hair, which only smoulders and gives off smoke less rapidly than synthetic materials.*
5. Hydraulic fluid sprayed uncontrollably from burst pipes feeding the fires.
6. Sailors were wearing polyester[+] uniforms that melted on to their skins. Anti-flash personnel protective equipment, e.g. hoods and gloves, was a major factor in preventing and limiting injuries from flash and blast.[#] Multiple layers of any clothing served better than one layer; polyester or other synthetic fabrics were counter-productive.

Reports described missile blasts as going "upwards and outwards" - upwards as frequently happens in buildings where ceilings are usually the weakest of a room's containing surfaces - rupturing H.M.S. Sheffield's deck and almost one third of the ship from galley to damage control headquarters, causing it to burst into flames with thick toxic smoke; "...the gaping hole in the ship's side and engines fed the flames, which rapidly turned into an inferno".

Early allegations that in the damaged H.M.Ss Invincible, Sheffield and Ardent, 'aluminium burnt' and contributed to loss have been refuted.* In a recent leaflet, the (British) Aluminium Federation felt it necessary to state that aluminium does not burn under natural conditions, but that it does melt and vent a fire. However, that there was much flammable material aboard ships at the Falklands (and as unnecessary furnishings) was conceded and that it would burn when some critical temperature was reached was unavoidable; but most of it was "fuel and ammunition followed by electrical cabling, deck and bulkhead linings and furnishings". Apparently, 175 miles of cable were then to be found in a British destroyer. With H.M.S. Sheffield it is acknowledged that the smoke and fire were chiefly the products of unspent rocket fuel from the attacking missile motor, and oil from the ship's fuel tanks. Noteworthy, is the little known fact that H.M.S. Glamorgan suffered a hit from a land-based Exocet but that the major fire was extinguished by a large hole in the deck being plated-over and other major equipment repairs being performed.

A possible useful conclusion, valuable for future ship design that has been noted is that control functions (which includes damage control) might be prudently separated from those of internal command. Also that the location of generators and similarly

---

* Plans to do away with oil for potato chip making are afoot too!

[+] Suits now provided are flame resistant. Partington[9] (p. 207) mentions fire-suits for men and horses, of felt and compositions of talc and brick dust etc., being used in about the 14th. century by the Arabs!

[#] Interestingly, many World War II bombers had extensive interior hydraulic lines which blew up or burned rapidly when damanged. In prisoner-of-war camps it was observed that the type of aircraft that had been flown by certain in-coming prisoners, (B-24 crews), was known by the kind of bandages for burns which they wore.

* An instance in which metal probably did burn occurred in the Le Mans (French) Grand Prix of 1955, when a Mercedes car containing an engine with a high fraction of magnesium, swerved at 180 mph during the race and ran up an earth safety bank, somersaulted, bounced and "exploded into white-hot component parts", amid spectators, killing 82. ['Great Disasters', Treasure Press, (1976), p. 61.]

'soft' vital auxiliaries might be more thoughtfully distributed around ships in view of likely hits from smart munitions: and that loss of air conditioning, low pressure air and centrally-supplied services can disable weapons and command positions.

In one discussion on the latter topic, it was pointed out that modern engine rooms are in a kind of deep pit with very hot machinery at the bottom and with volatile and combustible fuel oil about. Steam drench was thought preferable for putting out machinery space fires but as noted above, "the ultimate is large quantities of salt water with AFFF foam or some equivalent". Inert gases are generally used in weapon electronics compartments in gas turbine modules and in the engine rooms of gas turbine ships. Nitrogen is used in submarines though halogens are favoured; lower over-pressure is required for fire extinguishers and personnel survive it better. Fire has to be contained for inert gases to be effective and large missile holes in ships preclude this.

The U.S. Navy aircraft carrier U.S.S. Forrestal (1966), already referred to above, was almost totally destroyed by fire after a missile on a plane exploded on the flight deck and many died of carbon monoxide poisoning. A consequence was that a Survival Support Device (S.S.D.) was developed, but regretably too few of these were available to British Naval personnel in the Falklands' war. (Apparently, emergency life-support apparatus holding air for 8 minutes has since been ordered). About 120,000 such units were with the U.S. fleet in 1978 and now an Emergency Escape Breathing Device (E.E.B.D.), has been developed to provide a 15 minute air supply for the U.S. fleet.

## CONCLUSION

It is hoped that some general appreciation of the significance of fire associated with fighting vehicles in a primary or secondary role over the centuries has been conveyed and that in particular some of their vulnerable features have been usefully noted. This significance will be seen to have grown with time and vessel system sophistication.

## ACKNOWLEDGMENTS

The author would like to thank the Leverhulme Trust which supported his diversions into this field.

He also wishes to thank his wife, Mrs. S. Purlan, and Ms. Denise Evans for typing the original script and subsequent drafts of this paper.

## REFERENCES

1. Johnson, W.: Vehicle Impact and Fire Hazards, Structural Impact and Crashworthiness Conf., Vol. I Keynote Lectures, Elsevier Appl. Sci. Pubs., 75-113, London, 1984.

2. Landels, J.G.: Engineering in the Ancient World, University of California Press, 1978.

3. a) Morrison, J.S. and Williams, R.T.: Greek Oared Battleships, Cambridge University Press, 1968.

   b) Torr, C.: Ancient Ships, Argonaut Inc. Pubs., Chicago, MCMLXIV (1964).

4. Rushton, F.: Fire Aboard, 2nd Edition, Brown, Sons and Ferguson, Ltd., pp. 638, Glasgow, 1973.

5. Beckmann, J.: A History of Inventions, R. and J.E. Taylor, London, pp. 184, 1846.

6. Oman, C.W.C.: The Art of War in the Middle Ages, A.D. 378-1515, Cornell University Press, 1953.

7. Obolensky, D.: The Byzantine Commonwealth, Sphere Books, pp. 246, 1974.

8.  a) Scoffern, J.: <u>Projectile Weapons of War,</u> Richmond Publishing Co., 50-63, 1858.

b) Cowper, H.S.: <u>The Art of Attack,</u> W. Holmes Ltd., Ulverston, 1906, and reprinted by E.P. Publishing Ltd., 280-295, 1977.

c) Gibbon, E.: <u>Decline and Fall of the Roman Empire,</u> Vol. V, 1802.

9.  Partington, J.R.: <u>A History of Greek Fire and Gunpowder,</u> W. Heffer and Sons Ltd., Cambridge, 1960.

10. _____: <u>Brittanica Encyclopaedia:</u> Greek Fire, 16th Edn., Vol. 10, 1946.

11. Needham, J.: <u>Science in Traditional China,</u> Harvard University Press, Cambridge, Mass., Ch. 2, 1981.

12. Hewitt, J.: <u>Ancient Armour and Weapons in Europe,</u> Vol. I, J.H. and J. Parker, Oxford, 1860.

13. Harbon, J.D.: <u>The Spanish Ship of the Line,</u> Scientific American, December 1984.

14. Browne, D.G.: <u>The Floating Bulwark:</u> The Story of the Fighting Ship: 1514-1945, St. Martin's Press, N.Y., pp. 134, 1963.

15. Lloyd, G.: Letter to <u>The Times,</u> 18 June, 1982.

16. Wilson, A.: <u>Flame Thrower,</u> Bantam Books, 1984.

17. Pawle, G.: <u>The Secret War,</u> Transworld Pubns., 1959.

18. Johnson, W. and Walton, A.C.: Airshipwreck: Analysis of Safety in Rigid Airships in terms of Impact Mechanics and Fire, <u>5th. Sym. on Engineering Applications of Mechanics,</u> Ottawa, Canada, 237-241, 1980.

19. Johnson, W.: <u>Structural Crashworthiness,</u> Eds. N. Jones and T. Wierzbicki, Analysis and Structural Damage in Airship and Rolling Stock Collisions, Butterworths, London, 417-439, 1983.

20. Deighton, L. and Schwartzman, A.: <u>Airshipwreck,</u> Jonathan Cape, London, 74, 1978.

21. Johnson, W.: <u>Crashworthiness, Vol. I,</u> Ed. G.A.O. Davies, (Conf. London, 1984), Chapter 3, Vehicle Impact and fire Hazards, Elsevier, 74-114, 1984.

22. Santos-Dumont, S.: <u>My Airships,</u> Dover Pubns., N.Y. 1973, (originally printed in 1904).

23. Johnson, W. and Walton, A.C.: Fires in Public Service Vehicles, (Buses and coaches), in the U.K., <u>Int. Veh. Des., 2</u> (3), 322-334, 1981.

24. Williams, D.J. and Johnson, W.: A Note on the Formation of Fulgerites, <u>Geological Magazine, 3,</u> 293-260, 1981.

25. a) Pohler, C.H., McVoy, J.L., Carhart, H.W.,Leonard, J.T. and Pride, J.S.: Fire Safety of Naval Ships: an Open Challenge, <u>Naval Engineers' Jnl.,</u> 21-30, April, 1978.

b) Gustafson, R.E. and Schab, H.W.: General Concepts for a Combined Halon-High Expansion Foam Fire Protection System for Shipboard Application, Ibid. 24-32, February 1973.

26. Parkes, O.: British Battleships, 1860-1950, Archon Books, 1972, pp. 701.

27. Aircraft Crash Survival Design Guide Vol. III - Aircraft Structural Crashworthiness. D.H. Laananen, et.al., U.S. Department of Commerce, ADA 08914, 1980.

28. Lingard, S.: Estimation of Flash Temperatures in Dry Sliding, Proc. Inst. Mech. Engrs., Vol. 198c, No. 8, 91-97, 1984.

29. Seeger, B.F. and Lapp, R.H.: Water Injection: A New Protective System for Magazines, Naval Engineers' Jnl., 719-722, November, 1959.

30. Marcus, S.W.: Protection of Deep Stowage Missile Magazines, Naval Engineers' Jnl., 55-259, February, 1972.

31. Ventriglio, D.R.: Fire Safe Materials for Navy Ships, Naval Engineers' Jnl., 65-74, October, 1982.

32. Winter, A. and Butler, F.: Passive Fire Protection for Aluminium Structures, Naval Engineers' Jnl. 59-66, December, 1975.

33. Walker, J.W., Capt.: Lessons of the Falklands, Naval Engineers' Jnl., 129-132, July, 1983.

34. Williamson, B.R. and Mann, L.R.B.: Thermal Hazards from Propane (L.P.G.) Fireballs, Comb. Sci. and Tech., 25. 141-145, 1981.

# The Thermal Response of Aircraft Cabin Ceiling Materials during a Post-Crash, External Fuel-Spill, Fire Scenario

LEONARD Y. COOPER
Center for Fire Research
National Bureau of Standards
Gaithersburg, Maryland 20899, USA

ABSTRACT

An algorithm is developed to predict the thermal response of aircraft ceiling materials during a post-crash fire scenario. The scenario involves an aircraft's emergency exit doorway which opens onto the flames of a fuel-spill fire which engulfs the fuselage. Data of near-ceiling temperatures acquired during full-scale, post-crash test simulations provide indirect validation of the algorithm. The post-crash time-to-ceiling-ignition is proposed as a measure of cabin fire safety. This measure would be used as a surrogate for the post-crash time available for passengers to safely evacuate the cabin. In this sense, the algorithm is exercised in an example evaluation of the fire safety of a candidate honeycomb ceiling material used together in cabin systems involving polyurethane cushion seating.

INTRODUCTION

The purpose of this investigation is to analyze aircraft cabin ceiling surface temperature data recently acquired during full-scale test simulations of post-crash fires. The analysis is carried out with a view toward the development of a procedure for estimating the temperature histories of overhead aircraft cabin materials subsequent to the ignition of exterior, fuel-spill fires. With such a capability it would be possible to estimate the time for such materials to reach ignition temperatures. This would result in a rational means of ranking the fire safety of candidate overhead aircraft cabin materials.

All tests described here were carried out by the U.S. Department of Transportation Federal Aviation Administration (FAA), Atlantic City, New Jersey.

DESCRIPTION OF THE TESTS

The experiments simulated a wide-body aircraft cabin post-crash fire, similar to those reported previously[1]. The scenario involved a fuselage with two open doorways where one of these is engulfed by an external fuel spill fire. The fire is simulated by a burning 2.44 m x 3.05 m pan of jet fuel (JP-4). The threat to the cabin by this test fire has been shown[1] to be representative of the threat by real, external fuel-spill fires. No-wind conditions were simulated. The test article was a surplus U.S. Air Force C133A cargo aircraft.

The ceiling of the test cabin was made up of 0.0127 m thick rigid Kaowool® ceramic fiber board, where $k = 0.045$ W/mK; $\alpha = 2.67$ $(10^{-7})$ $m^2/s$. A mockup seat made of cushions on a steel frame was placed in the cabin in front of the open doorway exposed to the fire. The study involved eight tests. The only parameter which varied from test to test was the seat cushion construction. Test 111 is designated as the background test since it involved the seat frame with no cushioning. Data from Test 111 were available for 240 s after ignition. Data from all other tests were only available for 120 s. A schematic of the test setup is presented in Fig. 1.

During the tests, the radiant heat flux near the doorway, and 0.30 m and 0.91 m above the floor, was measured with fluxmeters facing outward toward the fire. Throughout each test, and from one test to another these two fluxes were substantially similar. It will be assumed that this flux, $\dot{q}''_{rad-door}$, is uniform and isotropic across the entire doorway, and that it can be approximated by the lower flux measured in Test 111 (see Fig. 2).

This study considers near-ceiling temperatures measured by three thermocouples placed in the line traversing the width of the cabin, and directly above the center of the doorway as shown in Fig. 1. The thermocouples were of 24 gage (0.000584 m diameter) chromel/alumel wire. The wire was supported several centimeters from its bead, and there was an attempt to position the bead close to

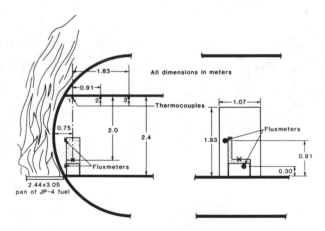

FIGURE 1. A schematic of the test setup.

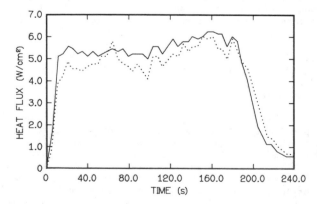

FIGURE 2. Measured doorway heat flux vs time (lower fluxmeter ....; upper fluxmeter ——); test 111 (background).

776

the ceiling surface so that the bead temperatures would be substantially similar to the nearby, ceiling temperatures. The bead-to-ceiling distances were probably of the order of 0.001 m.

Up to 120 s after ignition, the measured temperatures at each of the three positions and for all eight tests were substantially similar[2]. It is, therefore, reasonable to assume that, for the threat scenario being simulated and up to the 120 s, fire development in a single, mockup seat would not add significantly to the ceiling surface fire threat. Thus, it is assumed to be adequate to study the thermal response of the ceiling only during Test 111. Plots of the measured near-ceiling thermocouple temperatures during this test are presented in Fig. 3.

AN ANALYSIS OF THE THERMAL RESPONSE OF THE CABIN CEILING MATERIAL

Two major phenomena can lead to relatively prompt lower surface heating of the cabin ceiling. The first involves the thick flames and copious products of combustion which engulf the exterior of the fuselage near the exposed, open, doorway. These lead to radiative and convective heat flux to the cabin ceiling.

The convection is from the hot, buoyant gases of the fire which are captured by the open doorway. Upon entering the cabin, these gases are driven upward toward the ceiling, forming an outward (i.e., away from the doorway and toward the cabin interior) moving ceiling jet. After spreading radially from the doorway, this ceiling jet is redirected away from the general location of the doorway and toward the front and rear of the cabin. Eventually the hot, captured, products of combustion start to fill the cabin. They then participate in venting from the second open doorway and in complicated entrainment processes which develop at the fire-exposed, open doorway itself. An analysis of the external fire and the captured flow under rather general wind conditions has been presented previously[3].

The second phenomenon leading to ceiling heating involves the fire which spreads in the seating. Here, the single-seat scenario of the present tests results in only marginally important levels of ceiling heat flux. Yet, fire spread in a fully outfitted cabin could lead to a significant additional threat to the cabin ceiling. The seating fire leads to both radiative and convective heating of the ceiling. The radiation would be primarily from the fire's combustion zone, and the convection from the fire's plume-driven ceiling jet. This ceiling jet would augment the previously mentioned, captured-gas-driven ceiling jet.

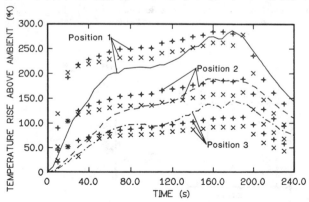

FIGURE 3. Computed Test 111 ceiling temperature (+:β=0.; x:β=3.0 m²), and corresponding measured near-ceiling temperatures at positions 1, 2, and 3.

Other components of heat flux to/from both the upper and lower ceiling surfaces are radiation from relatively cool, far-field surfaces and reradiation from the ceiling surfaces themselves. In an analysis of the ceiling heating it is reasonable to account for natural convection cooling of the ceiling's upper surface, and to adopt the relatively simple geometry of Fig. 4. Estimates for the components of ceiling heat transfer are developed below. Using these, the problem for the thermal response of the ceiling is then formulated and solved.

Radiation from Doorway to Ceiling

The radiant flux through the door, and to the ceiling is taken to be

$$\dot{q}''_{door-ceiling} = \dot{q}''_{rad-door} F_{A-dA} \tag{1}$$

where $F_{A-dA}$ is the viewfactor[4] given in Fig. 5.

Captured External Fire Product Gases — An Equivalent Buoyant Source

The "captured gas" doorway plume is modeled by a nonradiating, equivalent, point source of buoyancy located at the center of the horizontal surface of the mockup seat (see Fig. 4). The strength of the equivalent source, $\dot{Q}_{equiv}$, is assumed to be directly proportional to $\dot{q}''_{rad-door}$. Thus

$$\dot{Q}_{equiv} = \beta \dot{q}''_{rad-door} \quad (\beta \text{ in } m^2) \tag{2}$$

Radiation and Convection for the Seating Fire

During the first 120 s of the fire, ceiling heat transfer from the burning single mockup cabin seat was not significant. However, in fully outfitted cabins, it is anticipated that this situation would be changed, especially after the first minute or two subsequent to ignition. By these times, fires in multiple-seat configurations have been observed to grow and spread beyond single seat involvement. Since the present analysis will be extended to fully outfitted cabin scenarios, ceiling heat transfer contributions from the seating fire will be included at the outset.

The seating fire is simulated by a time-dependent point source of energy release rate, $\dot{Q}_{seat}$, assumed to be located with the nonradiating source, $\dot{Q}_{equiv}$, at the center of the horizontal surface of the outer, exposed, doorway seat. A fraction, $\lambda_{r,seat}$, of $\dot{Q}_{seat}$ is assumed to be radiated uniformly over a sphere to the far field. The remaining energy release rate, $(1-\lambda_{r,seat})\dot{Q}_{seat}$, drives the buoyant fire plume upward. Thus, the radiation from the seating fire to the ceiling is assumed to be

$$\dot{q}''_{rad-seat} = \lambda_{r,seat} \dot{Q}_{seat} / [4\pi H^2 (1+r^2/H^2)^{3/2}] \tag{3}$$

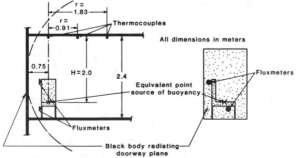

FIGURE 4.   A simplified version of the post-crash fire scenario.

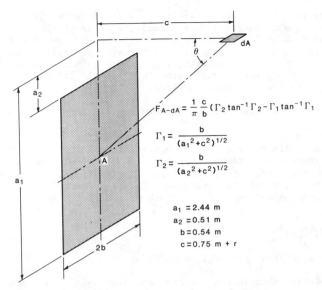

$$F_{A-dA} = \frac{1}{\pi}\frac{c}{b}\left(\Gamma_2 \tan^{-1}\Gamma_2 - \Gamma_1 \tan^{-1}\Gamma_1\right)$$

$$\Gamma_1 = \frac{b}{(a_1^2 + c^2)^{1/2}}$$

$$\Gamma_2 = \frac{b}{(a_2^2 + c^2)^{1/2}}$$

$a_1 = 2.44$ m
$a_2 = 0.51$ m
$b = 0.54$ m
$c = 0.75$ m $+ r$

FIGURE 5.   The viewfactor between the doorway and a ceiling element.

$\dot{Q}_{seat}$ would vary from one seat cushion construction to another. $\dot{Q}_{seat}$ would typically have to be estimated from test data, and then specified in the present analysis. $\lambda_{r,seat}$ would also vary somewhat from one construction to another, although it is reasonable to choose the value 0.35, a value which characterizes the radiation from flaming combustion zones of many practical fuel assemblies[5]. This value is adopted here.

Convective Heat Transfer from a Combined, Equivalent Source of Buoyancy

$\dot{Q}$, is the combined enthalpy flux of the upward moving combustion gases.   Thus

$$\dot{Q} = \dot{Q}_{equiv} + (1 - \lambda_{r,seat})\dot{Q}_{seat} \tag{4}$$

All convective heat transfer to the cabin ceiling is from the $\dot{Q}$-generated, plume-driven, ceiling jet, and is estimated by[6,7]

$$\dot{q}''_{conv,L} = h_L\left(T_{ad} - T_{s,L}\right) \tag{5}$$

$$\frac{\left(T_{ad} - T_{amb}\right)}{T_{amb}\dot{Q}*^{2/3}} = \begin{cases} 10.22\exp(-1.77 r/H), & 0 \leq r/H \leq 0.75 \\ 2.10(r/H)^{-0.88}, & 0.75 \leq r/H \end{cases} \tag{6}$$

$$h_L/\tilde{h} = \begin{cases} 7.75\mathrm{Re}^{-0.5}\left[1 - (5.0 - 0.390\mathrm{Re}^{0.2})(r/H)\right], & 0 \leq r/H \leq 0.2 \\ 0.213\mathrm{Re}^{-0.3}(r/H)^{-0.65}, & 0.2 \leq r/H \leq 1.03 \\ 0.217\mathrm{Re}^{-0.3}(r/H)^{-1.2}, & 1.03 \leq r/H \end{cases} \tag{7}$$

$$\dot{Q}* = \dot{Q}/\left(\rho_{amb}C_p T_{amb}g^{1/2}H^{3/2}\right); \quad \tilde{h} = \rho_{amb}C_p g^{1/2}H^{1/2}\dot{Q}*^{1/3}; \quad \mathrm{Re} = g^{1/2}H^{3/2}\dot{Q}*^{1/3}/\nu \tag{8}$$

The above algorithm is for heat transfer to unconfined ceilings. In using it here, two major assumptions are made; namely, effects of the upper smoke layer are relatively weak during the early times of interest, and the interactions of the ceiling jet and lateral cabin wall surfaces, especially surfaces immediate to the doorway side of the plume-ceiling impingement point, will not lead to total heat transfer flux amplitudes which are significantly larger than peak values that will be estimated with their neglect.

## Radiation Between the Lower Ceiling Surface and the Far-Field Cabin Surfaces

The lower ceiling surface is assumed to radiate diffusely to the illuminated surfaces of the cabin and its furnishings. Responding to this, the temperatures of those surfaces also increase with time. However, for times of interest here, it is assumed that these latter temperature increases are always relatively small compared to the characteristic increases of $T_{s,L}$. Accordingly, the net radiation exchange between the ceiling and the nonburning surfaces below can be approximated by

$$\dot{q}''_{rerad,L} = \varepsilon_L \sigma (T^4_{s,L} - T^4_{amb}) \tag{9}$$

## Heat Transfer from the Upper Ceiling Surface

Heat is transferred through the ceiling, and eventually the temperature of its upper surface, which is also assumed to be exposed to a constant $T_{amb}$ environment, begins to rise. Heat transfer from this surface has convective and radiative components. These are estimated by

$$\dot{q}''_{conv,U} = h_U (T_{s,U} - T_{amb}); \quad \dot{q}''_{rerad,U} = \varepsilon_U \sigma (T^4_{s,U} - T^4_{amb}) \tag{10}$$

$$\text{where}[8] \quad h_U = 1.675 |T_{s,U} - T_{amb}|^{1/3} \ W/m^2 \ (T \ in \ K) \tag{11}$$

## The Boundary Value Problem for the Ceiling, and the Method of Its Solution

The temperature field of the ceiling is assumed to be governed by the Fourier heat conduction equation. Initially, the ceiling is at temperature, $T_{amb}$. The rates of heat transfer to the lower and upper surfaces, are

$$\dot{q}''_L = \dot{q}''_{door-ceiling} + \dot{q}''_{rad-seat} + \dot{q}''_{conv,L} - \dot{q}''_{rerad,L}; \quad \dot{q}''_U = -\dot{q}''_{conv,U} - \dot{q}''_{rerad,U} \tag{12}$$

Radial gradients of variables of the problem are assumed to be small enough so that conduction in the ceiling is quasi-one dimensional in space. An illustration of the idealized, fire scenario is presented in Fig. 6.

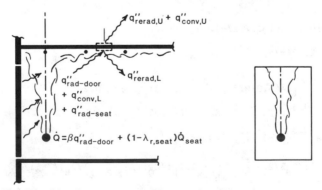

FIGURE 6. The idealized post-crash fire scenario.

A computer program for solving the above problem was developed. The solution to the heat conduction equation for the ceiling at every radial position of interest is by finite differences[10],[11]. For a given calculation, $N \leq 20$ equally spaced points are positioned at the surfaces and through the thickness of the ceiling. The spacing of these, $\delta Z$, is selected to be large enough to insure stability of the calculation. The change in time for all time steps is made small enough so that, at a given lower surface node, the temperature increases from time step to time step never exceed one percent of the current value of T.

CALCULATION OF THE RESPONSE OF THE CEILING IN THE POST-CRASH TEST SIMULATION

The algorithm was used to predict the response of the Kaowool® ceiling during the first 240 s of Test 111. Here and in the next section all surfaces are assumed to radiate and absorb as black bodies. $\dot{Q}_{seat}$ was taken to zero, and $\dot{q}''_{rad-door}$ as identical to the Test 111, underseat flux measurement. Ceiling temperatures at positions 1, 2 and 3 were computed for different $\beta$'s in the range $0 \leq \beta \leq 6.0m^2$. (This range of $\beta$ leads to the approximate $\dot{Q}_{equiv}$ range $0 \leq \dot{Q}_{equiv} \leq 300$ kW.) The computed lower ceiling histories for $\beta = 0.$ and $3.0$ $m^2$ are plotted in Fig. 3.

The Importance of $\dot{Q}_{equiv}$

If convective ceiling heating from doorway-captured products of combustion is equivalent to that from a seat fire of the order of a few hundred kW, then the calculated results plotted in Fig. 3 indicate that such heating is not significant compared to doorway radiation. (Except for the very earliest few seconds, convection from the relatively weak source associated with $\beta=3.0m^2$ is seen to lead to net cooling of the strongly irradiated ceiling surface.) This result is consistent with earlier observations where variations in single seat cushion construction (peak energy release rates likely never exceeding the few hundred kW level) did not lead to significant differences in near-ceiling temperatures.

Comparisons Between Computed and Measured Temperatures

Per Fig. 3 the peak computed values of ceiling temperature compare favorably with the corresponding peak temperatures measured by the near-ceiling thermocouples. However, the basic qualitative characteristics of the computed and measured transient thermal responses are significanlty different. Namely, the measured temperatures do not have the same type of rapid response which the solution properly predicts for the ceiling surface temperatures. Also, the close tracking of the position 2 and 3 thermocouples at early times does not compare favorably with a like tracking of the computed temperatures.

Two conclusions result from these observations: the thermocouples are not at the temperature of the ceiling surface, and, therefore, data to validate the analysis are not evident. As a result of these conclusions, an analysis of the response of the thermocouples was carried out in order to explain the measured thermocouple responses, and with the hope of obtaining a measure of experimental validation, albeit indirect, for the predicted ceiling response.

AN ANALYSIS OF THE THERMAL RESPONSE OF THE NEAR-CEILING THERMOCOUPLES

The objective of the present analysis is to predict the thermal response of the thermocouples when placed near, but not touching the ceiling. The procedure for positioning these devices prior to testing was such that the thermocouple wires were essentially parallel to the lower ceiling surface and at a distance, d, of the order of 0.001 m. The actual orientation of the wire relative to the doorway plane is unknown. As depicted in Fig. 7, the analysis will consider two extreme configurations for the wire, viz., normal and parallel to the doorway.

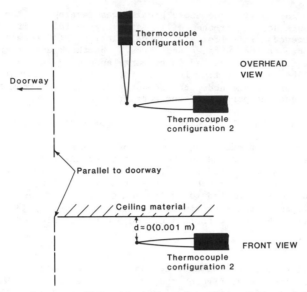

FIGURE 7. Two extreme configurations for placement of the near-ceiling thermocouples.

The characteristic time for conductive heat transfer through the wire thickness is of the order of tenths of a second. It will therefore be assumed that the wire is spatially uniform in temperature. Properties of the chromel/alumel wire will be taken as those of Nickel, viz., $\rho = 8800 kg/m^3$, $C_p = 460 Ws/(kgK)$.

From the literature[5,11,12] it is estimated that the thickness of the ceiling jet within which the thermocouples are submerged are of the order of several centimeters. With a characteristic d, of the order of 0.001 m, it is therefore reasonable to assume that gas velocities local to the thermocouple wire are so small that forced convection vs radiative heating of the wire is negligible. Also, the characteristic Grashof numbers would be relatively small, and any natural convection would be reduced to a conduction limit. This would be dependent on the unknown distance d.

At early times radiation from the doorway drives the temperature increase of the thermocouple. Also, a steady-state analysis which balances doorway heating and radiation exchanges between thermocouple, ceiling and ambient (i.e., which ignores conduction) leads to a result which is consistent with late-time, Fig. 3, measured and computed temperatures of thermocouple and ceiling, respectively.

The thermal analysis which emerges from the above discussion leads to the following equation for the temperature, $T_w$, of the thermocouple wire

$$\frac{\pi}{4}\rho C_p D^2 \frac{dT_w}{dt} = \dot{Q}'_{door-wire} + \dot{Q}'_{ceiling-wire} + \dot{Q}'_{amb-wire} - \dot{Q}'_{wire} \tag{13}$$

$$\dot{Q}'_{wire} = \pi D \sigma T_w^4; \quad \dot{Q}'_{amb-wire} = \frac{\pi D}{2}\sigma T_{amb}^4$$

$$\dot{Q}'_{ceiling-wire} = \frac{\pi D}{2}\sigma T_{s,L}^4; \quad \dot{Q}'_{door-wire} = \alpha D \dot{q}''_{door-ceiling} \tag{14}$$

$$\alpha = \begin{cases} 1 \text{ for configuration 2 of Fig. 7} \\ 1/\sin\theta \text{ (see Fig. 5) for configuration 1 of Fig. 7} \end{cases} \tag{19}$$

To obtain $T_w$ one would specify $\alpha$ and $T_{s,L}$, use the measured values of $\dot{q}''_{\text{rad-door}}$ to obtain $\dot{q}''_{\text{door-ceiling}}$, and solve Eq. (13) subject to $T_w(t=0)=T_{amb}$.

Solutions for $T_W$ in the Test 111 Scenario

The above procedure was applied to the Test 111 scenario. The analysis was carried out numerically for a thermocouple in position 1, 2 or 3 and in configuration 1 or 2. In each case, $T_{s,L}$ was taken from the ceiling temperature calculations described earlier.

$T_w$ calculations were carried out for $\beta$ values of 2.0 $m^2$, 3.0 $m^2$, and 4.0 $m^2$. $\beta=3.0$ $m^2$ results are presented in Fig. 8, which includes the measured $T_w$ of Fig. 3.

Comparison Between Computed and Measured Temperatures - A Choice for $\beta$

Perhaps of greatest significance in Fig. 8 is the early-time thermocouple temperature predictions, which were of particular concern in the ceiling vs thermocouple temperature comparisons of Fig. 3. Here, the simulations of the early, near-linear responses of the thermocouples are noteworthy.

Of further significance is the fact that the calculations reveal a possible explanation for the close tracking of the response of the thermocouples at positions 2 and 3. Namely, such behavior is predicted if the thermocouple wire at position 2 was normal to the door plane (configuration 2), and the thermocouple wire at position 3 was parallel to the door plane (configuration 1).

Fig. 8-type plots provide a basis for selecting the "best" value for $\beta$. The $\beta$ predicting a ceiling response which, in turn, yields the most favorable comparisons between calculated and measured values of $T_w$ would be the obvious choice. Calculations reveal that the $T_w$ predictions are not very sensitive to $\beta$ variations in the appropriate range 2.0-4.0 $m^2$. Furthermore, of the values $\beta=2.0$ $m^2$, 3.0 $m^2$, and 4.0 $m^2$, all yielded reasonable $T_w$ predictions, and no one of these values clearly yields more favorable $T_w$ predictions than the others. $\beta=3.0$ $m^2$ will be chosen as the "best" value.

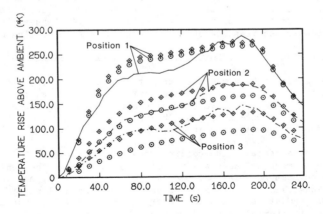

FIGURE 8. Predicted and measured test 111 thermocouple temperatures ($\beta=3.0$ $m^2$). <•>: Predicted $T_w$, configuration 1; O: Predicted $T_w$, configuration 2; ____, __ __ __, __ • __: Measured $T_w$

PREDICTING THE POST-CRASH TIME-TO-IGNITION OF CEILING CONSTRUCTIONS IN A FULLY
SEATED CABIN

The above results provide some confidence in the ceiling thermal response
algorithm. To use it to simulate the post-crash fire exposure in a fully-seated
cabin, effects of fire spread in an array of seating must be included. This
would be done by inputing appropriate, nonzero, $\lambda_{r,seat}$ and $\dot{Q}_{seat}$ terms in Eqns.
(3) and (4). Then, using the k and $\alpha$ of a candidate ceiling material, the algorithm would calculate the ceiling's time-dependent, post-crash, thermal response.

In the most likely case of a combustible ceiling material, one could, for example,
predict the time for the lower surface to reach a characteristic ignition temperature. Results of a previous FAA program indicate that away from the combustion
zone tenable conditions are maintained throughout the cabin prior to ceiling
ignition. The time-to-ceiling ignition would therefore provide a reasonable
measure of post-crash cabin fire safety, viz., the minimum time available for
passengers to evacuate the cabin or the Available Safe Egress Time (ASET)[13].
Hopefully, evaluations of practical cabin ceiling material candidates would lead
to associated ignition times, or minimum ASET's, which exceed the time required
for cabin evacuation. In any event, the greater the time-to-ignition of a
material the better.

In the case of a noncombustible ceiling, time-to-ignition in the above discussion
would be replaced by time to reach some agreed upon ceiling temperature, e.g.,
600°C, which is often associated with cabin flashover.

Estimates of Post-Crash Fire Growth in Arrays of Cabin Seats - An Example

Estimates of the energy release rate of post-crash fires spreading through arrays
of seats were obtained previously[14]. Based on FAA, full-scale, 21 seat tests which
were similar to Tests 104-111, estimates of fire growth in two types of seat construction were obtained. The first type of seats, designated as "regular" seats,
were made of fire retarded polyurethane foam covered with wool-nylon fabric. The
second seat construction was similar to the first, except that it included a blocking layer constructed of a 0.0048 m thick sheet of neoprene with a polyester scrim.

The estimates of $\dot{Q}_{seat}$ for the two types of seats are plotted in Fig. 9. The
plots terminate at 140 s and 185 s, at which times video-tape recordings of the
tests indicated the initiation of either flashover (140 s) or of rapid development of total obscuration (185 s). These estimates will be used below to
evaluate the post-crash response of a specific, honeycomb ceiling material.

POST-CRASH RESPONSE OF A HONEYCOMB CEILING MATERIAL - ESTIMATES OF TIME-TO-IGNITION

The algorithm developed here was used to estimate the post-crash thermal response
of a 0.0254 m thick, honeycomb composite, aircraft lining material with an epoxy
fiberite covering. The effective thermal properties of the composite were
measured, and found to be[15] $k=5.9(10^{-5})kW/(mK)$; $\rho=110.kg/m^3$; $\alpha=4.8(10^{-7})m^2/s$;
$C_p=1.11kJ/(kgK)$. $\dot{Q}_{seat}$ was simulated by the plots of Fig. 9.

The predicted temperature of the ceiling above the doorway seat is plotted in
Fig. 10 for both "regular" seating and "blocked" seating. The ignition temperature of the honeycomb material had been measured previously, and was found to
be[16] 536°C. Thus, results of Fig. 10 predict onset of ceiling ignition at 148
and 204 s for "regular" and "blocked" seating, respectively. For cabin ceilings
of this honeycomb material, blocked rather than unblocked seating would lead to a
56 s advantage in ASET.

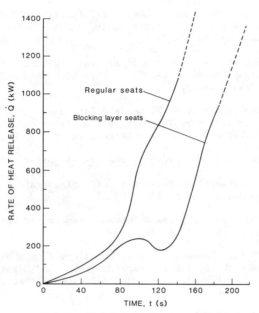

FIGURE 9. Estimate for $\dot{Q}_{seat}$ for arrays of polyurethane seats with and without blocking layers [13] (_ _ _ _ extrapolated from curves of [13]).

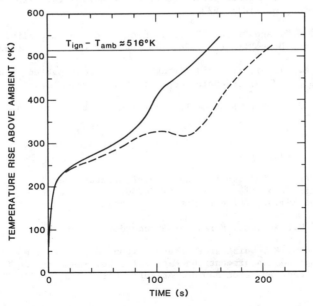

FIGURE 10. Predicated temperature of the honeycomb ceiling material in a cabin with polyurethane seats: with (_ _ _ _) and without (____) blocking layers.

REFERENCES

1.  Hill, R.G., Johnson, G.R. and Sarkos, C.P., Postcrash Fuel Fire Hazard Measurements in a Wide-Body Aircraft Cabin, FAA-NA-79-42, Fed. Aviation Admin., Atlantic City, NJ, 1979.

2.  Cooper, L.Y., The Thermal Response of Aircraft Cabin Ceiling Materials During a Post-Crash External Fuel-Spill Fire Scenario, NBSIR 84-2912, Nat. Bur. of Stand. Gaithersburg, MD, 1984.

3.  Emmons, H.W., The Ingestion of Flames and Fire Gases Into a Hole in an Aircraft Cabin for Arbitrary Tilt Angles and Wind Speed, Home Fire Proj. Rpt. 52, Harvard Univ. Div. Appl. Sciences, Cambridge, MA, 1982.

4.  Eckert, E.R.G. and Drake, R.M., Heat and Mass Transfer, McGraw-Hill, 1959.

5.  Cooper, L.Y., A Mathematical Model for Estimating Available Safe Egress Time in Fires, Fire and Mat., 6, p. 135, 1982.

6.  Cooper, L.Y., Heat Transfer from a Buoyant Plume to an Unconfined Ceiling, J. Heat Trans., 104, p. 446, 1982.

7.  Cooper, L.Y., Thermal Response of Unconfined Ceilings Above Growing Fires and the Importance of Convective Heat Transfer, 22nd Nat'l. Heat Transfer Conf., ASME Paper 84-HT-105, 1984 and NBSIR 84-2856, Nat. Bur. Stand., Gaithersburg, MD, 1984.

8.  Yousef, W.W., Tarasuk, J.D. and McKeen, W.J., Free Convection Heat Transfer from Upward-Facing, Isothermal, Horizontal Surfaces, J. Heat Trans., 104, p. 493, 1982.

9.  Emmons, H.W., The Prediction of Fires in Buildings, 17th Symp. (Inter.) on Combustion, p. 1101, 1979.

10. Mitler, H.E. and Emmons, H.W., Documentation for the Fifth Harvard Computer Fire Code, Home Fire Proj. Rpt. 45, Harvard Univ., Cambridge, MA, 1981.

11. Poreh, M., Tsuei, Y.G. and Cermak, J.E., Investigation of a Turbulent Radial Wall Jet, ASME J. of Appl. Mech., p. 457, 1967.

12. Alpert, R.L., Turbulent Ceiling-Jet Induced by Large-Scale Fires, Comb. Sci. and Tech., Vol. 11, p. 197, 1975.

13. Cooper, L.Y., A Concept of Estimating Safe Available Egress Time, Fire Safety Journal, Vol. 5, p. 135, 1983.

14. Steckler, K., Chapter 1: The Role of Aircraft Panel Materials in Cabin Fire and Their Properties, DOT-FAA CT 84/30, Nat. Bur. Stand. rpt. to Fed. Aviation Admin., Atlantic City, NJ, 1985.

15. Parker, W., National Bureau of Standards, private communication.

16. Harkleroad, M., Quintiere, J. and Walton, W., Radiative Ignition and Opposed Flame Spread Measurements on Materials, DOT/FAA/CT-83/28 (Nat. Bur. Stand. rpt. to Fed. Aviation Admin., Atlantic City, NJ, 1983.

NOMENCLATURE

| | |
|---|---|
| $a_1$, $a_2$, b, c | dimensions, Fig. 5 |
| $C_p$ | specific heat |
| D | wire diameter |
| d | thermocouple-to-ceiling separation distance |
| $F_{A-dA}$ | view factor, Eq. (1), Fig. 5 |
| g | acceleration of gravity |
| H | seat fire-to-ceiling distance |
| $h_L$, $h_U$ | lower/upper surface heat transfer coefficient |
| $\tilde{h}$ | characteristic heat transfer coefficient, Eq. (8) |
| k | thermal conductivity |
| N | number of grid points in ceiling analysis |
| $\dot{Q}$, $\dot{Q}*$ | enthalpy flux in plume, Eq. (4), dimensionless $\dot{Q}$, Eq. (8) |
| $\dot{Q}_{equiv}$, $\dot{Q}_{seat}$ | equivalent fire strength, strength of seat fire |
| $\dot{Q}'_{amb-wire}$, $\dot{Q}'_{ceiling-wire}$ | radiation: ambient to wire, ceiling to wire per unit length |
| $\dot{Q}'_{door-wire}$, $\dot{Q}'_{wire}$ | radiation: doorway to wire, from wire per unit length |
| $\dot{q}''_{conv,U}$, $q''_{conv,L}$ | convection to upper/lower ceiling |
| $\dot{q}''_{door-ceiling}$ | radiation from doorway to ceiling |
| $\dot{q}''_{rad-door}$, $\dot{q}''_{rad-seat}$ | radiation from doorway, from seat fire to ceiling |
| $\dot{q}''_{rerad,U}$, $\dot{q}''_{rerad,L}$ | radiation from upper/lower ceiling |
| $\dot{q}''_U$, $\dot{q}''_L$ | net heat transfer to upper/lower ceiling |
| Re | Reynold's number, Eq. (8) |
| r | distance from plume impingement point |
| $T_{ad}$; $T_{amb}$ | adiabatic ceiling temperature, Eq. (6); ambient temperature |
| $T_{s,U}$, $T_{s,L}$ | upper/lower surface ceiling temperature |
| $T_w$ | thermocouple wire temperature |
| t | time |
| Z | indepth ceiling coordinate |
| $\alpha$ | thermal diffusivity/wire configuration constant, Eq. (15) |
| $\beta$ | a constant |
| $\Gamma_1$, $\Gamma_2$ | constants, Fig. 5 |
| $\delta Z$ | indepth spacing of ceiling grid points |
| $\varepsilon_U$, $\varepsilon_L$ | lower/upper ceiling emissivity |
| $\theta$ | configuration angle, Fig. 5 |
| $\lambda_{r,seat}$ | fraction of $\dot{Q}_{seat}$ radiated |
| $\nu$ | kinematic viscosity of ambient air |
| $\rho$, $\rho_{amb}$ | density, density of ambient |
| $\sigma$ | Stefan-Boltzmann constant |

# Evaluation of Aircraft Interior Panels under Full-scale Cabin Fire Test Conditions

CONSTANTINE P. SARKOS and RICHARD G. HILL
Federal Aviation Administration
FAA Technical Center
Atlantic City Airport, New Jersey 08405, USA

ABSTRACT

Realistic full-scale fire tests demonstrated the potential safety benefits of advanced interior panels in transport aircraft, and displayed the characteristics of cabin fire hazards. The tests were conducted in a C-133 airplane, modified to resemble a wide-body interior, under postcrash and in-flight fire scenarios. The safety benefit of the advanced panel ranged from a 2-minute delay in the onset of flashover when the cabin fire was initiated by a fuel fire adjacent to a fuselage rupture, to the elimination of flashover when the fuel fire was adjacent to a door opening or when an in-flight fire was started from a seat drenched in gasoline. Analysis of the cabin hazards measured during postcrash fire tests indicated that the greatest threat to passenger survival was cabin flashover, and that toxic gases did not reach hazardous levels unless flashover occurred.

## INTRODUCTION

### Objective

The primary objective of this paper is to describe the safety benefits of advanced interior panels under realistic full-scale aircraft cabin fire test conditions. A secondary objective is to characterize and analyze the hazards affecting occupant survivability in cabin fires.

### Background

Although the accident record of the airline industry is excellent, on rare occasions accidents do occur with grave consequences. For the United States (U.S.) airline industry, an average of 32 fatalities per year are attributable to fire. [1] All of these fatalities have occurred in crash accidents which are usually accompanied by the spillage and ignition of jet fuel. In spite of the intensity and apparent dominance of a jet fuel fire, under certain accident conditions, the survivability of cabin occupants will be established by the hazards of burnin interior materials. [2] The Federal Aviation Administration (FAA) is supporting and conducting research, testing and development to minimize the hazards of burning interior materials in the postcrash fire environment. [3] Also, the in-flight fire problem is now receiving more

attention because of this type of accident experience with foreign carriers;
e.g., Air Canada DC9 accident in Cincinnati. [4]

Improvements for two important types of cabin interior materials have been
investigated — seat cushions and panels. Because of the flammable nature
of urethane foam cushions, a fire blocking layer concept was developed that
provides significant safety benefits for both postcrash and in-flight cabin
fires. [5] The FAA has proposed more stringent flammability regulations for
seat cushions. [6] The current emphasis by FAA is to develop improved test
requirements and materials for interior panels, which constitute the sidewalls, ceiling, stowage bins, and partitions of a contemporary transport cabin
interior. The importance of panels during a cabin fire stems from their large
surface area and location in the upper cabin (ceiling, stowage bins) where
fire temperatures are highest.

Full-scale fire tests are necessary to determine the potential benefits of
safety improvements in real fires and to corroborate the trends indicated by
small-scale test results. During full-scale tests, important real-world
conditions such as fire source, geometry, and scale are reasonably simulated.
Another important application of full-scale fire tests, is for the analysis of
the hazards affecting survivability during a cabin fire. These hazards are
grouped into three categories: heat, smoke (visibility), and toxic gases.
What is the relative importance of each of these hazards? What are the
effects of different types of fire scenarios on the significance of each
hazard category? Realistic full-scale tests can provide information which, at
the very least, give insight for answering these complex and far-reaching
questions.


DISCUSSION

Interior Panel Materials

Figure 1 describes the advanced and inservice panels evaluated in this paper.

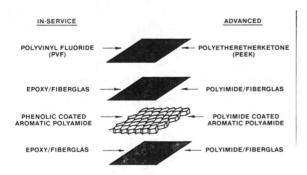

Figure 1.  Composition of Composite Panels


The in-service panel contained epoxy/fiberglass facings and represented the
type of panel design employed in the earliest wide-body jet interiors.
Polyimide was selected in the advanced panel design for resin and core coating
because of its high degradation temperature. Polyetheretherketone (PEEK) was

the decorative film in the advanced panel design, primarily to eliminate the production of hydrogen fluoride during thermal decomposition of the polyvinyl fluoride film commonly used in contemporary panels. The superior thermal stability of the advanced panel was evidenced alone by its cure temperature; viz., 500° F for 16 hours vs. 350° F for approximately 2 hours for the in-service panel.

Test Article

The full-scale test article was a C-133 aircraft, modified to resemble a wide-body cabin interior, as shown in figure 2 and reference 2. It was utilized to compare the performance of the advanced and in-service panels installed in a representative cabin interior layout as sidewalls, stowage bins, ceiling and partitions, under simulated postcrash and inflight fire conditions. Under the postcrash scenarios, the interior was subjected to an external fuel fire adjacent to an opening (door or fuselage rupture) in the forward part of the fuselage (figure 2). For the in-flight fire scenario, the fuselage openings were covered and a perforated ducting system simulated the ceiling discharge of air into the cabin as occurs with the cabin environmental control system (ECS).

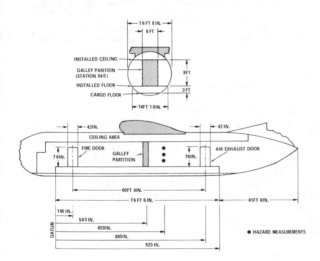

Figure 2. Schematic of C-133 Wide-Body Cabin Fire Test Article

Fire Scenarios

Table 1 outlines the fire scenarios utilized in this study. In the postcrash scenario, previous work had demonstrated that the size of the C-133 external fuel fire produced 80 percent of the radiant heat flux into the interior expected from an infinite fire. [7] Thus, the experimental fuel fire gave a reasonable simulation of a large pool of burning fuel. With an unfurnished C-133 interior and zero wind, there is virtually no accumulation of fuel fire hazards (temperature, smoke, and gases) inside the test article. [7] For this reason, the cabin hazards measured with interior materials installed and a zero wind fuel fire are attributed to burning materials, although fuel fire flames are drawn into the interior as the materials begin to ignite and burn.[8] The main role of the fuel fire is to subject the interior materials to intense radiant heat.

Table 1.  Fire Scenarios

| NO. | DESIGNATION | TYPE | IGNITION SOURCE | FUSELAGE CONFIGURATION | VENTILATION |
|---|---|---|---|---|---|
| 1 | FUEL FIRE/ RUPTURE | POSTCRASH | FUEL FIRE | INTACT, TWO OPENINGS: RUPTURE (FIRE) DOOR (AFT) | NATURAL ZERO WIND |
| 2 | FUEL FIRE/ OPEN DOOR | POSTCRASH | FUEL FIRE | INTACT, TWO OPENINGS: DOOR (FIRE) DOOR (AFT) | NATURAL ZERO WIND |
| 3 | GASOLINE/ SEAT | IN-FLIGHT | SPILLED GASOLINE ON SEAT | CLOSED | CONTROLLED |

The in-flight fire scenario consisted of the ignition of a passenger seat doused with one quart of gasoline.  It probably represented the most intense in-flight fire that is likely to occur out in the open (in contrast to a fire in a concealed area).  The use of forced ventilation in a closed fuselage for the in-flight scenario was expected to affect the fire characteristics, compared to the postcrash case with fuselage openings and natural ventilation.

TEST RESULTS AND ANALYSIS

Postcrash Fuel Fire and Fuselage Rupture Scenario

The arrangement of materials with the postcrash fire scenario with a fuselage rupture adjacent to the fuel fire is shown in figure 3.  Basically, a small area of the interior in the vicinity of the fuselage rupture was lined with the panels being examined and furnished with seats and carpet.  The seats were surplus aircraft passenger seats protected with cushion fire blocking layers and the carpet was new, aircraft grade wool/nylon carpet.  The quantity of materials employed was more than adequate to produce non-survivable conditions in the event of ignition and adequate fire growth.  By using seats and carpet in addition to the panels being evaluated, the effect of panel flammability on the ignition and burning of other cabin materials used in large quantities, and vice-versa, was taken into consideration.

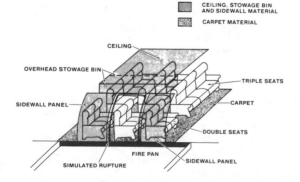

Figure 3.  Postcrash Fuel Fire/Rupture Scenario

The postcrash fuel fire scenario with a fuselage rupture was the most severe fire condition used, primarily because a seat was centered in the rupture and exposed to high levels of radiant heat. A flashover — defined in this paper as the sudden and rapid uncontrolled growth of the fire from an area in the immediate vicinity of the fuel fire to the remaining materials — occurred with both types of panels. However, the time to flashover was much earlier in the test with in-service panels than in the test with advanced panels. As shown in figure 4, the difference in flashover times was approximately 140 seconds. Since the occurrence of flashover is the event in a postcrash cabin fire that creates nonsurvivable conditions, as discussed later in this paper and in an earlier study (reference 2), the advanced panels also resulted in 140 seconds of additional time available for evacuation. This difference in available evacuation time was clearly a significant benefit to be gained from the advanced panels.

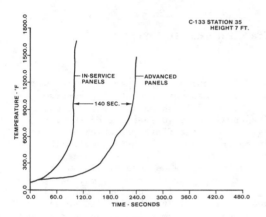

Figure 4. Benefit of Advanced Composite Panels-External Fuel/Fuselage Rupture Scenario

Postcrash Fuel Fire and Open Door Scenario

The arrangement of materials with the postcrash fire scenario with an opened door adjacent to the fuel fire was similar to the fuselage rupture scenario except that the center row of seats was eliminated and a box-like structure representing a galley was installed. The resultant fire condition was less severe than with the fuselage rupture scenario because of the removal of the passenger seat next to the opening.

The superior fire performance of the advanced panels was even more evident with the fuel fire/open door scenario. Under this scenario, the usage of advanced panels eliminated flashover. This result is demonstrated in figure 5, which compares the temperature history inside the test article for both types of panels. With in-service panels, flashover occurred in aproximately 2 1/2 minutes; however, with advanced panels, flashover did not occur over the 7-minute test duration. A comparison of the results with both types of postcrash scenarios (see figures 4 and 5) demonstrates the consistency of the data and illustrates that the rate of development of a cabin fire is largely dependent on fire scenario.

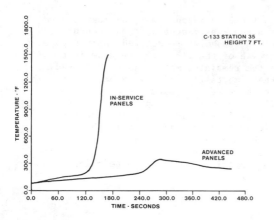

Figure 5.  Benefit of Advanced Composite - External Fuel Fire/Open
Door Scenario


An analysis of the cabin hazards measured in the fuel fire/open door test with
in-service panels revealed the importance of flashover in dictating survivability during a postcrash cabin fire.  This data is shown in figure 6, which
contains the hazard histories measured approximately 40 feet aft of the fire
door at an elevation of 5 feet 6 inches.  Before the flashover which occurred
at approximately 150 seconds, the cabin environment was clearly survivable;
after flashover, the conditions very suddenly deteriorated to such a degree
that survival would have been highly unlikely.  The suddeness of flashover,
and perhaps the fact that it occurs without any apparent warning, may make
passengers unaware of the imminent dangers that they face during a cabin fire.
For example, within 30 seconds, as shown in figure 6, visibility decreased
from about 30 feet to 3 feet, temperature measured from slightly above ambient
to over 400° F, CO increased from zero to over 2500 ppm, and oxygen decreased
from ambient to 16 percent.  Therefore, it was concluded that improvements in
postcrash cabin fire safety, when burning interior materials are the dominant
factor, can be best attained by delaying the onset of flashover.  If material
selection is on the basis of state-of-the-art small-scale fire tests, then
the use of an appropriate flammability test would seem to be far more beneficial than the use of either smoke or toxicity tests.

Why were the hazards measured 40 feet aft of the fire door at an elevation of
5 feet 6 inches virtually zero for over 2 minutes in the fuel fire/open door
test with in-service panels?  There are two likely reasons for this result.

First, the small mass burning rate before flashover and the large cabin volume
(13,200 cubic feet) made dilution and wall loss effects (heat transfer,
adsorption) dominant.  Secondly, the hazards that are produced before flashover are largely contained in the hot "smoke layer" which clings to the
ceiling, above the measurement location and probably above the head of most
passengers.  Previous C-133 tests [2], and the photographic/ video coverage
from the tests described in this paper, document the significant stratification during a postcrash cabin fire with natural ventilation; i.e., with no
forced ventilation.

The in-flight fire scenario was the least severe of the three scenarios
studied. Figure 8 compares the temperature history near the fire source for
the in-service and advanced panels. As in the fuel fire/open door test,
flashover did not occur with the advanced panels. The fire resistance of the
more flammable in-service panels was also sufficient to delay the onset of
flashover until 8 minutes. From a practical viewpoint, an in-flight fire of
this kind with inservice panels would, under most circumstances, have been
extinguished by crewmembers utilizing hand-held extinguishers before the fire
became out ot control.

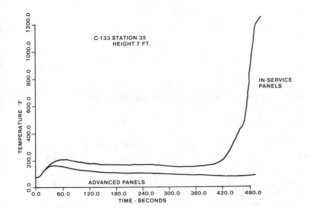

Figure 8.  Benefit of Advanced Composite Panels - In-Flight Fire Scenario

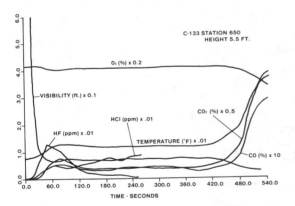

Figure 9.  Hazard Time Profiles With In-Service Composite Panels - Flight
Fire Scenario

The controlled ventilation in the in-flight scenario tended to distribute the
seat fire hazards throughout the airplane. Figure 9 presents the measured
hazard histories for the in-service panel test. Each of the measured hazards
was detected before the onset of flashover, apparently because of the mixing
action associated with the controlled ventilation. In contrast, for the
postcrash tests where the cabin was ventilated naturally through fuselage
openings, the hazards were primarily contained in the ceiling smoke layer, and
remained virtually undetected at the 5-foot 6-inch sampling height until the

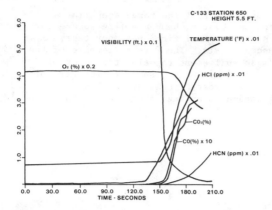

Figure 6. Hazard Time Profiles With In-Service Composite Panel - External
Fuel Fire/Open Door Scenario

Figure 7 also demonstrates that the hazards over this 7-minute test were
clearly survivable. At 7 minutes, the temperature had only increased by 20° F
over ambient, the concentration of $CO_2$ was 2000 ppm, the concentration of
$O_2$ remained at ambient, and visibility had decreased to 50 feet. The toxic
gases CO, HCl, HCN, and HF were not detected. This data also supports the
conclusion that in a postcrash cabin fire, the hazards effecting survival are
created by a flashover. Also, smoke and toxic gas hazards affecting survivability did not materialize as a consequence of flashover being prevented.

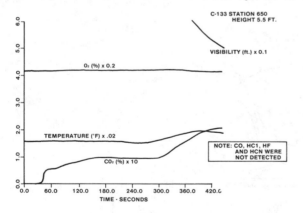

Figure 7. Hazard Time Profiles With Advanced Composite Panels External
Fuel Fire/Open Door Scenario

In-Flight Fire Scenario

For the in-flight scenario, the fuselage openings were covered and a perforated duct simulated air discharge from the cabin ECS. The seat next to the
covered door, doused with one quart of gasoline, served as the fire source.
This type of seat fire will burn for 2 mnutes, with a peak burning rate at
40 seconds before self-extinguishing because of the fire blocking layer. [9]

cabin flashover (e.g., see figure 6). For the in-flight test, however, each measured hazard before flashover was well below its estimated incapacitation level. For example, at 8 minutes the calculated dose of CO was approximately 4000 ppm-minutes, which is significantly below the estimated human escape impairment dose of 30,000-40,000 ppm-minutes. [10] Also, the measured concentration of HCl, which was less than 100 ppm, would have been easily tolerated by passengers, based on recent primate studies. [10] The main peril before flashover was the dramatic loss in visibility due to smoke (calculated visibility was less than 10 feet at 30 seconds). It is interesting to note that significant smoke obscuration can occur without hazardous levels of toxic gases or elevated temperatures.

## SUMMARY OF SIGNIFICANT FINDINGS

Based on the realistic, full-scale cabin fire tests and analysis described in this paper, and on the composite panel materials evaluated and the types of fire scenarios employed, the following are the significant findings:

1. Advanced interior panels can provide a significant safety improvement during postcrash and in-flight cabin fires.

2. The greatest threat to passenger survival during postcrash cabin fires dominated by burning interior materials, is cabin flashover.

3. Toxic gases produced during postcrash cabin fires consisting of a fuel fire adjacent to a fuselage opening or in-flight fires initiated by a gasoline-drenched seat fire do not reach hazardous levels unless flashover occurs.

4. During an in-flight fire, the cabin environmental control system has a major effect on the distribution and dissipation of hazards.

## REFERENCES

1. Final Report of the Special Aviation Fire and Explosion Reduction (SAFER) Advisory Committee, Federal Aviation Administration, Volume I, Report FAA-ASF-80-4, June 26, 1980.

2. Sarkos, C. P., Hill, R. G., and Howell, W. D., The Development and Application of a Full-Scale Wide-Body Test Article to Study the Behavior of Interior Materials During a Postcrash Fuel Fire, AGARD Lecture Series No. 123 on Aircraft Fire Safety, AGARD-LS-123, June 1982.

3. Engineering and Development Program Plan Aircraft Cabin Fire Safety, Federal Aviation Administration, Report FAA-ED-18-7, revised February 1983.

4. Aircraft Accident Report: Air Canada Flight 797, McDonell-Douglas DC-9-32, C-FTLU, Greater Cincinnati International Airport, Covington, KY, June 2, 1983, National Transportation Safety Board, Report NTSB/AAR-84/09, August 8, 1984

5. Sarkos, C. P., and Hill, R. G., Effectiveness of Seat Cushion Blocking Layer Materials Against Cabin Fires, SAE Technical Paper No. 821484, presentd at Aerospace Congress and Exposition, October 25-28, 1982.

6. Flammability Requirements for Aircraft Seat Cushions; Notice of Proposed Rulemaking, DOT/FAA, Federal Register, Volume 48, No. 197, p. 46251, October 11, 1983.

7.  Hill, R. G., Johnson, G. R., and Sarkos, C. P., Postcrash Fuel Fire Hazard Measurements in a Wide-Body Aircraft Cabin, Federal Aviation Administration, Report FAA-NA-79-42, December 1979.

8.  Quintiere, J. G., and Tanaka, T., An Assessment of Correlations Between Laboratory and Full-Scale Experiments for the FAA Aircraft Fire Safety Program, Part 5: Some Analyses of the Postcrash Fire Scenario, National Bureau of Standards, Report NBSIR 82-2537, July 1982.

9.  Hill, R. G., Brown, L. J., Speitel, L., Johnson, G. R., and Sarkos, C. P., Aircraft Seat Fire Blocking Layers: Effectiveness and Benefits Under Various Scenarios, Federal Aviation Administration, Report DOT/FAA/CT-83/43, February 1984.

10. Kaplan, H. L., Grand, A. F., Rogers, W. R., Switzer, W. G., and Hartzell, G. E., A Research Study of the Assessment of Escape Impairment by Irritant Combustion Gases in Postcrash Aircraft Fires, Federal Aviation Administration, Report DOT/FAA/ CT-84/16, September 1984.

# Preliminary Test for Full Scale Compartment Fire Test (Lubricant Oil Fire Test: Part 1)

**T. TANAKA**
Toshiba Corporation
8, Shinsugita, Isogo-ku, Yokohama, Japan

**Y. KABASAWA**
Chubu Electric Power Company, Inc.
1, Toushincho, Higashi-ku, Nagoya, Japan

**Y. SOUTOME**
Hitachi Works of Hitachi, Ltd.
3-1-1, Saiwaicho, Hitachi, Ibaraki-Pref., Japan

**M. FUJIZUKA**
Mitsubishi Heavy Industries, Ltd.
(Mitsubishi Atomic Power Industries, Inc.)
2-4-1, Shibakouen, Minato-ku, Tokyo, Japan

ABSTRACT

This study reports on two kinds of fire tests: one kind intended for the purpose of understanding the burning characteristics of turbine oil in a free space and a second kind intended for elucidating the fire characteristics in a compartment. The latter were carried out using methanol and investigated how the ventilation pattern and the position of the fire source affect fire behavior.

In the former, five different sized oil pans (0.1 - 4.0m$^2$) were utilized and the burning rate, radiation heat flux, etc. were measured. Test results showed no difference in characteristics for the two turbine oils VG32 and VG56. Empirical equations for the burning rate and radiant emittance of turbine oil were obtained from the data.

In the compartment tests, the burning rate, room temperature and radiation heat flux were measured in a 6x7x5m compartment constructed of fire resistant board. They were determined for a total of nine different combinations of three ventilation patterns and three positions of the fire source, using methanol in an 0.5m$^2$ oil pan. The test results showed that the room temperature in a fire is related to the ventilation pattern caused by the presence of openings such as doors, while the burning rate and radiation heat flux are not significantly related to the ventilation pattern or the location of the fire source.

INTRODUCTION

Although turbine oil is used as lubricant in many industrial plants, its burning characteristics, especially in a confined space or a compartment, are not well-known. To clarify these burning characteristics, tests were conducted in oil pans of five different sizes which were filled with several liters of turbine oil (VG32 and/or VG56) as a fire source to measure the burning rate, flame height, radiation heat flux, etc. Compartment fire tests were also carried out to investigate how ventilation patterns (i.e. locations of air intake and exit) and position of the fire source affect fire behavior. Nine cases were tested in a fire resistant board compartment using methanol in a burning oil pan as the fire source.

BURNING CHARACTERISTICS OF TURBINE OIL IN A FREE SPACE

Test Conditions

To obtain the burning characteristics of the turbine oil, fire sources were made using several liters of turbine oil in steel pans of five different sizes (0.1, 0.5, 1.0, 2.0 and 4.0m$^2$). The tests were carried out in a laboratory to eliminate wind effects. Light oil and methanol were also used as fuel sources.

Test Results and Evaluation

Table 1 compares the individual burning characteristics. Since the characteristics of turbine oil VG32 in the 0.5m$^2$ pan were the same as that of turbine oil VG56, further tests were carried out with VG32. All the measured values increase with increases in the oil pan size.

Estimation of turbine oil VG32 burning rates. The burning rate of combustible liquid varies with the oil pan diameter. With larger diameters, the burning rate is controlled by the radiation heat flux and approximately given by the following equations.[1]

$$V = V_\infty (1 - e^{kD}) \tag{1}$$

where  V  : Burning rate of "D" meter oil pan (mm/min)
       $V_\infty$ : Burning rate with no effect from oil pan size. (Burning rate of a sufficiently large oil pan) (mm/min)
       D  : Oil pan diameter (m)
       k  : Constant (m$^{-1}$)

and

$$V_\infty = \frac{0.076 \ Hc}{Hv + \int_{t}^{t_B} Cp \cdot dt} = \frac{0.076 \ Hc}{Hv + \overline{Cp} \cdot \Delta t} \tag{2}$$

Table 1    Summary of Results

| Fuel Source | Oil Pan size (m$^2$) | Burning Rate kg/m$^2$·min (mm/min) | Radiation Heat Flux (kw/m$^2$) at 3m | at 5m | Flame Height (m) | Flame Width (m) | Calorific Heat (kw/m$^2$) |
|---|---|---|---|---|---|---|---|
| Turbine Oil VG32 | 0.1 (φ0.36m) | 0.5 (0.6) | 0.18 | 0.058 | 0.8 | 0.3 | 370 |
| | 0.5 (φ0.80m) | 1.2 (1.4) | 1.75 | 0.76 | 2.0 | 0.8 | 870 |
| | 1.0 (φ1.13m) | 1.7 (1.9) | 4.07 | 1.98 | 2.5 | 1.0 | 1160 |
| | 2.0 (φ1.60m) | 2.0 (2.3) | 10.1 | 4.07 | 3.8 | 1.3 | 1400 |
| | 4.0 (φ2.26m) | 2.2 (2.5) | 17.2 | 6.86 | 4.3 | 2.1 | 1500 |
| Turbine Oil VG56 | 0.5 (φ0.80m) | 1.2 (1.4) | 1.75 | 0.76 | 2.0 | 0.8 | 870 |
| Light Oil | 0.5 (φ0.80m) | 1.5 (1.8) | 1.98 | 0.81 | 1.8 | 0.8 | 1070 |
| | 2.0 (φ1.60m) | 2.2 (2.6) | 10.1 | 4.07 | 3.8 | 1.7 | 1500 |
| Methanol | 0.5 (φ0.80m) | 1.4 (1.6) | 0.41 | 0.18 | 1.3 | 0.6 | 440 |

where  $V_\infty$ : Burning rate of a sufficiently large oil pan (mm/min)
       $H_c$ : Calorific Power (kJ/kg) = 43,000 (kJ/kg)
       $H_v$ : Latent heat (kJ/kg) = 128 (kJ/kg)
       $C_p$ : Specific heat (kJ/kg·K)
       $t_B$ : Boiling point (K) = 645(K)
       $t$  : Initial oil temperature (K) = 293(K)
       $\overline{C_p}$ : Average specific heat in the temperature range between the initial
            oil temperature and the boiling point. (kJ/kg·K) = 2.52 (KJ/kg·K)
       $\Delta t$ : Temperature difference between oil temperature and boiling
            point. (deg) = 352 (deg).

As the burning rate of turbine oil VG32 of the sufficiently large pan ($V_\infty$) is
estimated to be 3.22 mm/min from the above equations, the burning rate of
turbine oil can be expressed as Equation (3).

$$V = 3.22 \left(1 - e^{-0.714D}\right) \tag{3}$$

A comparison of the burning rate between Equation (3) and observed data is made
in Fig. 1.

   Estimation of the radiant emittance.  If the flame is considered as a
gray body, the radiation heat flux at any point is expressed by the equation
given below.

$$E = \phi \cdot \varepsilon \cdot \sigma \cdot T^4 \tag{4}$$

where  $\phi$ :  Shape factor of flame
       $\varepsilon$ :  Emissivity of flame
       $\sigma$ :  Stefan-Boltzmann Constant (5.77 x $10^{-11}$ kW/m$^2$·K$^4$)
       $T$ :  Flame temperature (K)

Assuming the shape of the flame is cylinder, the shape factor $\phi$ is expressed by
the following equation.[2]

$$\phi = \frac{1}{\pi Y}\tan^{-1}\cdot\left(\frac{X}{\sqrt{Y^2-1}}\right) +$$

$$\frac{X}{\pi}\cdot\left[\frac{(A-2Y)}{Y\cdot\sqrt{AB}}\cdot\tan^{-1}\left(\sqrt{\frac{A}{B}\cdot\frac{(Y-1)}{(Y+1)}}\right) - \right.$$

$$\left.\frac{1}{Y}\tan^{-1}\left(\sqrt{\frac{(Y-1)}{(Y+1)}}\right)\right] \tag{5}$$

where  A = $(1+y)^2+X^2$
       B = $(1-y)^2+X^2$
       X = 2H/D
       Y = 2L/D
       H = Flame height (m)
       D = Flame Diameter (m)
       L = Any point from the center of
           the flame (m)

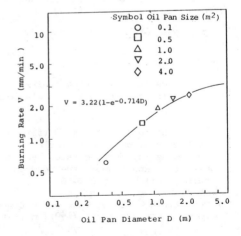

Fig. 1

Comparison of Burning Rate between
Observed Data and Equation (3)

The calculated results of the shape factor obtained from Equation (5) are shown in Fig. 2.

The radiant emittance Rf can be calculated by using the radiation heat flux and the shape factor as follows.

$$Rf = \frac{E}{\phi} \qquad (6)$$

The radiant emittance is estimated by the following equation.[3]

$$Rf = K \cdot V \cdot Hc$$

Where K : Constant
      V : Burning rate (kg/m$^2$·h)
      Hc : Calorific Power (kWh/kg)

The constant K is 0.02 according to Reference 3).

Then a comparison of emittance calculated from the experimental radiation heat flux data with that calculated from the experimental burning rate is made as shown in Fig. 3. It is clear that the emittance calculated from the burning rate (experimental data and calculated results) is smaller than that calculated from the experimental radiation heat flux data. As the constant K is determined to be 0.0384, the equation for the radiant emittance is written as follows.

$$Rf = 0.0384 \ V \cdot Hc \qquad (7)$$

In Summary

As a result of the nine fuel source burning tests, the following conclusions were reached.

(1) There was no significant difference in the burning characteristics including the burning rate, radiation heat flux and flame shape between turbine oil VG32 and VG56.

(2) The burning rate of VG32 was estimated to be $V = 3.22 \ (1-e^{-0.714D})$ (mm/min).

(3) There was no significant difference between the turbine oil (VG32 and VG56) and light oil in terms of the flame height.

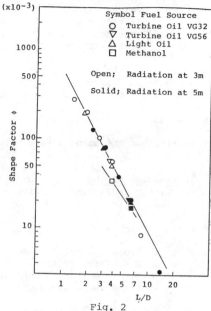

Fig. 2
Shape Factor $\phi$ VS. L/D

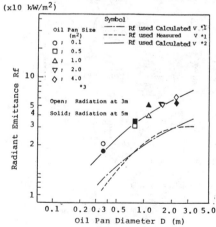

*1   Rf = 0.02 V · Hc
*2   Rf = 0.0384 V · Hc
*3   Calculated Rf by Radiation Heat Flux

Fig. 3
Radiant Emittance Rf VS. Oil Pan Diameter

(4) The turbine oil VG32 radiant emittance empirical equation is:

$$Rf = 0.0384 \ V \cdot Hc \ (kW/m^2)$$

ON COMBUSTION WHEN CHANGING VENTILATION PATTERN AND LOCATION OF FIRE SOURCE IN A COMPARTMENT

Test Conditions

   Test conditions, that is, the shape of the fire source, the kind of fuel, the combinations of ventilation patterns and the locations of fire source were selected as described below.

   Shape of fire source. The round oil pan shown below was used as fire source in the test.

D  =  0.798 m
H  =  0.2 m
t  =  3.2 mm

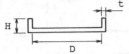

   Kind of fuel. Methanol was used as the fuel because it produces little smoke, which helps in determination of the burning rate and radiation heat flux and therefore, identification of the direction of air movements induced by fire.

   Test compartment. The outside dimensions of the test compartment were 6m wide, 7m long and 5m high. It was built by installing an angle steel frame on a concrete floor and attaching calcium silicate boards (t = 25 mm) to it.

   Ventilation patterns. The ventilation patterns were assumed to influence the fire behavior. While paying particular attention to the location of air intake, the following three cases were examined.

   Pattern I; Air is drawn naturally through the door opening (2m$^2$, 1m wide x 2m high) and exhausted forcibly in the vicinity of the ceiling on a diagonal line.

   Pattern II; Without any opening in the walls, air is forcibly supplied through a duct at an elevation of 3m and forcibly exhausted in the vicinity of the ceiling on a diagonal line.

   Pattern III; Air is drawn naturally through the door opening and supplied forcibly in the vicinity of the ceiling, and exhausted in the vicinity of the ceiling on a diagonal line.

   Locations of the fire source. Three locations were selected for the tests.

   Center; at the center of the test compartment

   Near wall (1); on the longitudinal center line, 1.5m from the wall on the intake side.

   Near wall (2); at a corner, 1.5m from the wall on the intake side and 1.5m from its right wall.

Rough sketches of the ventilation patterns and locations of the fire source are shown in Fig. 4.

| Ventilation Pattern / Factors (mm) | I | II | III |
|---|---|---|---|
| 1. Duct size | 350x350 | | |
| 2. Exhaust size | 300x200x3 | | |
| 3. Intake size | — | 300x300x3 | |
| 4. Doorway size | 1,000x2,000 | — | 1,000x2,000 |

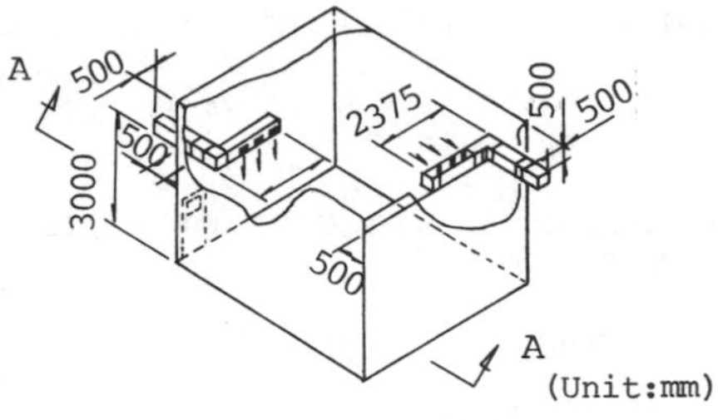

(Unit:mm)

| Fire source Location(m) | Center | Near Wall (1) | Near Wall (2) |
|---|---|---|---|
| L | 3.5 | 1.5 | |
| W | 3 | | 1.5 |

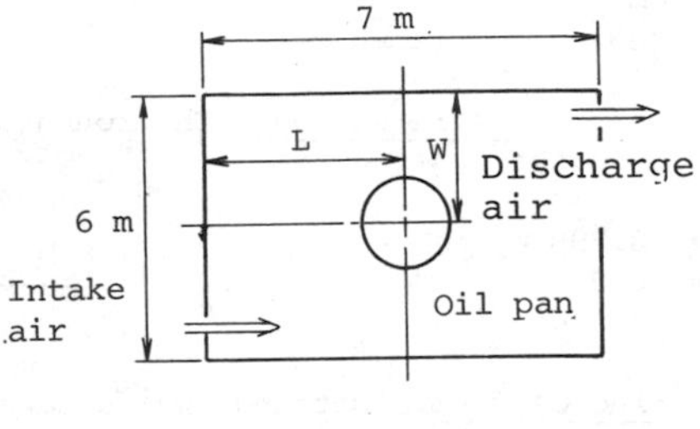

A-A section

Fig. 4  Ventilation Patterns and Locations of Fire Source

Instrument layout.  For instrument layout, the test room is longitudinally divided into six sections to identify spatial distributions of each measurement site in the test compartment.  Sensors are arranged on a mesh in each section as shown in Fig. 5.

Data processing.  For speedy processing, the data taken at a total of 130 measuring points are centrally routed to a sensor terminal for processing by a computing on data logger and personal computer.  Graphs were automatically drafted by a X-Y plotter.  Tests were monitored on a TV screen and burning conditions were recorded by a VTR.

Test Results

Burning rate.  The burning rate was determined from weight loss of fuel using loadmeters.  The results for each case are shown in Table 2.

Radiation heat flux.  Figs. 6 through 8 show radiation heat flux at each position for various ratios of L/D, the linear distance between the oil pan and radiometer L to the diameter of oil pan D.  The results for each case are summarized in Table 2.

Temperature and air flow velocity distributions and air flow characteristics.  Fig. 9 shows the temperature and air flow distributions on the central longitudinal section for each ventilation pattern.  Comparisons of temperature right above on flame and the mean temperature within the section for each case are also shown in Table 2.

804

(Unit : mm)

(a) Instrumentation

| Symbol | Instrument | Total |
|--------|------------|-------|
| • | Thermocouple | 104 sets |
| △ | Radiometer | 9 sets |
| ○ | Anemometer | 12 sets |
| ⊢ | Smoke density meter | 2 sets |
| ■ | Loadmeter | 3 sets on 50 kg range |

(b) Items and purposes of measurements

| Measuring Item | Purpose | Instrument |
|----------------|---------|------------|
| Temperature | To study temperature distribution in room | CA (chromelalumel) thermocouples |
| Air flow Velocity | To study the flow in room induced by burning | Pito-tube anemometer |
| Radiation Heat Flux | To understand the effect of radiation heat flux from burning flame on the room boundary | Radiometer |
| Burning rate | To investigate the difference in burning rate of oil (methanol) due to oil pan location and ventilation pattern | Loadmeter |
| Air flow direction | To study air flow direction caused by fire | Pyrotechnics smoke, using VTR, camera and visual observation |

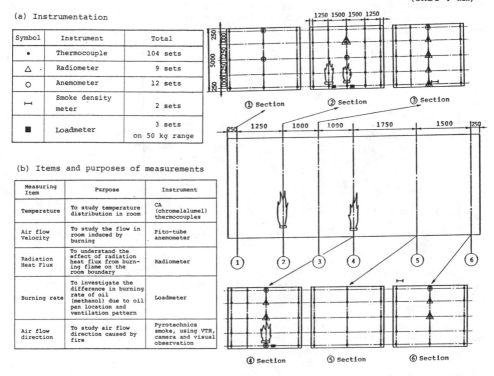

(c) Compartment longitudinal sections

Fig. 5  Instrument Layout

Table 2  Test Results

| | Ventilation pattern / Fire source location | I | II | III |
|--|--|--|--|--|
| Burning rate (kg/m²) | Center | 1.38 | 1.30 | 1.37 |
| | Near Wall (1) | 1.42 | 1.32 | 1.45 |
| | Near Wall (2) | 1.36 | 1.24 | 1.39 |
| Radiation heat flux (Kw/m²) | Center | 0.98 | 1.19 | 1.20 |
| | Near Wall (1) | 1.22 | 1.37 | 1.52 |
| | Near Wall (2) | 1.23 | 1.01 | 1.42 |
| Temperature at a typical point* (K) | Center | 432 | 449 | 436 |
| | Near Wall (1) | 430 | 466 | 440 |
| | Near Wall (2) | 423 | 448 | 432 |
| Mean Temperature** (K) | Center | 417 | 429 | 414 |
| | Near Wall (1) | 416 | 441 | 422 |
| | Near Wall (2) | 407 | 427 | 412 |
| Height of neutral zone from floor (m) | Center | 1.3 | 1.5 | 1.65 |
| | Near Wall (1) | 0.5 | 0.5 - 0.6 | - |
| | Near Wall (2) | 1.5 - 1.6 | 1.5 - 1.6 | 1.5 - 1.6 |

\* Temperature at a point 250 mm under the ceiling directly above the flame, taken 15 minutes after ignition.

\*\* Mean temperatures of the upper 18 thermocouples (2.5m above floor) which lie on the central longitudinal section, taken 15 minutes after ignition.

805

Oil pan - 0.5m$^2$ (D = 0.798m)

5 times/hour of ventilation

Building height - 5m

L: Linear distance between radiometer and fire source (m)

D: Diameter of oil pan (m)

Mean value from 14 min. 32 sec. to 15 min. 20 sec. after ignition

| Symbol | Location |
|---|---|
| —·—O—·— | Center |
| ——△—— | Near Wall (1) |
| —·—□—·— | Near Wall (2) |

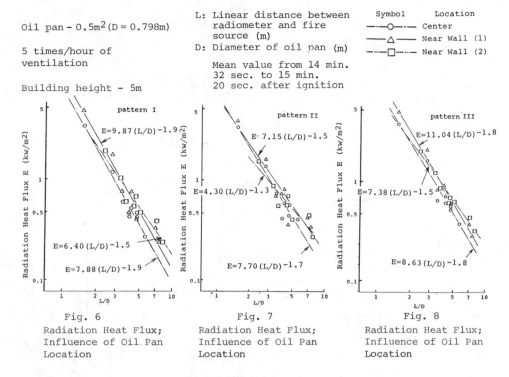

Fig. 6
Radiation Heat Flux; Influence of Oil Pan Location

Fig. 7
Radiation Heat Flux; Influence of Oil Pan Location

Fig. 8
Radiation Heat Flux; Influence of Oil Pan Location

Height of neutral zone. All ventilation patterns become steady several minutes after ignition. The height of the neutral zone for each case is shown in Table 2.

In Summary

Based on the experimental data, the following conclusions can be derived on the effect of ventilation patterns and locations of fire source on firing status.

(1) The burning rate, radiation heat flux and temperature right above the oil pan were found to have no significant correlation with ventilation patterns and locations of fire source.

(2) However, it might well be said that the burning rate and radiation heat flux were the severest for ventilation pattern III and location of fire source near wall (1). Room air temperature was the highest in the case of ventilation pattern II and location of the fire source near wall (1). In ventilation pattern III, the blow-off of burnt hot air and the flow-in of ambient cool air through the opening appear to help reduce the room air temperature, in spite of the high burning rate.

806

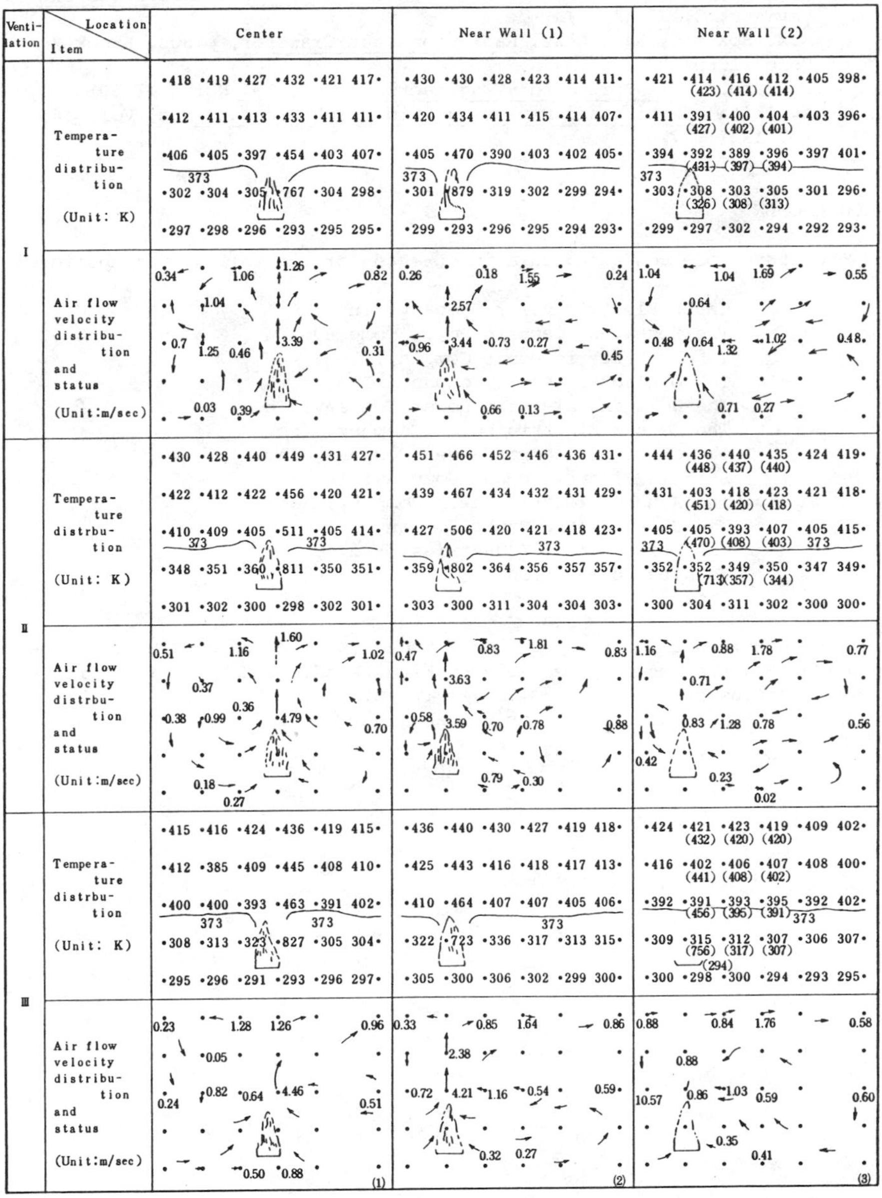

| Venti-lation | Location Item | Center | Near Wall (1) | Near Wall (2) |
|---|---|---|---|---|
| I | Tempera-ture distribu-tion (Unit: K) | ·418 ·419 ·427 ·432 ·421 417·<br>·412 ·411 ·413 ·433 ·411 411·<br>·406 ·405 ·397 ·454 ·403 407·<br>373<br>·302 ·304 ·305 ·767 ·304 298·<br>·297 ·298 ·296 ·293 ·295 295· | ·430 ·430 ·428 ·423 ·414 411·<br>·420 ·434 ·411 ·415 ·414 407·<br>·405 ·470 ·390 ·403 ·402 405·<br>373<br>·301 ·879 ·319 ·302 ·299 294·<br>·299 ·293 ·296 ·295 ·294 293· | ·421 ·414 ·416 ·412 ·405 398·<br>(423) (414) (414)<br>·411 ·391 ·400 ·404 ·403 396·<br>(427) (402) (401)<br>·394 ·392 ·389 ·396 ·397 401·<br>(431) (397) (394)<br>373<br>·303 ·308 ·303 ·305 ·301 296·<br>(326) (308) (313)<br>·299 ·297 ·302 ·294 ·292 293· |
| | Air flow velocity distribu-tion and status (Unit:m/sec) | 0.34 1.06 1.26 0.82<br>1.04<br>0.7 1.25 0.46 3.39 0.31<br>0.03 0.39 | 0.26 0.18 1.55 0.24<br>·2.57<br>·0.96 ·3.44 ·0.73 ·0.27 0.45<br>0.66 0.13 | 1.04 1.04 1.69 0.55<br>·0.64<br>·0.48 ·0.64 ·1.02 0.48·<br>1.32<br>0.71 0.27 |
| II | Tempera-ture distrbu-tion (Unit: K) | ·430 ·428 ·440 ·449 ·431 427·<br>·422 ·412 ·422 ·456 ·420 421·<br>·410 ·409 ·405 ·511 ·405 414·<br>373 373<br>·348 ·351 ·360 ·811 ·350 351·<br>·301 ·302 ·300 ·298 ·302 301· | ·451 ·466 ·452 ·446 ·436 431·<br>·439 ·467 ·434 ·432 ·431 429·<br>·427 ·506 ·420 ·421 ·418 423·<br>373<br>·359 ·802 ·364 ·356 ·357 357·<br>·303 ·300 ·311 ·304 ·304 303· | ·444 ·436 ·440 ·435 ·424 419·<br>(448) (437) (440)<br>·431 ·403 ·418 ·423 ·421 418·<br>(451) (420) (418)<br>·405 ·405 ·393 ·407 ·405 415·<br>373 (470) (408) (403) 373<br>·352 ·352 ·349 ·350 ·347 349·<br>(713)(357) (344)<br>·300 ·304 ·311 ·302 ·300 300· |
| | Air flow velocity distrbu-tion and status (Unit:m/sec) | 0.51 1.16 1.60 1.02<br>0.37<br>0.38 0.99 0.36 4.79 0.70<br>0.18 0.27 | 0.47 0.83 1.81 0.83<br>·3.63<br>·0.58 ·3.59 0.70 0.78 0.88<br>0.79 0.30 | 1.16 0.88 1.78 0.77<br>·0.71<br>0.83 1.28 0.78 0.56<br>0.42 0.23 0.02 |
| III | Tempera-ture distrbu-tion (Unit: K) | ·415 ·416 ·424 ·436 ·419 415·<br>·412 ·385 ·409 ·445 ·408 410·<br>·400 ·400 ·393 ·463 ·391 402·<br>373 373<br>·308 ·313 ·323 ·827 ·305 304·<br>·295 ·296 ·291 ·293 ·296 297· | ·436 ·440 ·430 ·427 ·419 418·<br>·425 ·443 ·416 ·418 ·417 413·<br>·410 ·464 ·407 ·407 ·405 406·<br>373<br>·322 ·723 ·336 ·317 ·313 315·<br>·305 ·300 ·306 ·302 ·299 300· | ·424 ·421 ·423 ·419 ·409 402·<br>(432) (420) (420)<br>·416 ·402 ·406 ·407 ·408 400·<br>(441) (408) (402)<br>·392 ·391 ·393 ·395 ·392 402·<br>(456) (395) (391) 373<br>·309 ·315 ·312 ·307 ·306 307·<br>(756) (317) (307)<br>(294)<br>·300 ·298 ·300 ·294 ·293 295· |
| | Air flow velocity distribu-tion and status (Unit:m/sec) | 0.23 1.28 1.26 0.96<br>·0.05<br>0.24 0.82 0.64 4.46 0.51<br>0.50 0.88<br>(1) | 0.33 0.85 1.64 0.86<br>2.38<br>·0.72 4.21 1.16 0.54 0.59<br>0.32 0.27<br>(2) | 0.88 0.84 1.76 0.58<br>0.88<br>10.57 0.86 1.03 0.59 0.60<br>0.35 0.41<br>(3) |

Temperature and flow velocity measured
15 minutes after ignition

Values in parentheses are those
taken on the section passing
through the fire source

Fig. 9   Temperature Distribution, Air Flow Velocity
Distribution, and Air Flow Status

REFERENCES

1) D.S. Burgess, A. Strasser, J. Grumer, <u>Fire Research Abstracts and Reviews</u>, 3,177 (1961)
2) E.M. Sparrow, R.D. Cess, <u>Radiation Heat Transfer</u>, p.302, Brooks/Cole, 1966
3) T. Yumoto, <u>Oil Fire Safety Engineering</u>, Vol. 19, No. 6 (1980)
4) K. Akita, <u>Burning and Explosion Guide High Pressure Gas</u>, Vol. 14, No. 12 (1977)

ACKNOWLEDGMENT

The tests were conducted in a joint study of the following companies.

Chubu Electric Power Company, Inc.
The Hokkaido Electric Power Company, Inc.
Tohoku Electric Power Company, Inc.
Tokyo Electric Power Company, Inc.
The Hokuriku Electric Power Company, Inc.
The Kansai Electric Power Company, Inc.
The Chugoku Electric Power Company, Inc.
Shikoku Electric Power Company, Inc.
The Kyushu Electric Power Company, Inc.
The Japan Atomic Power Company
Mitsubishi Heavy Industries, Ltd.
Toshiba Corporation
Hitachi, Ltd.

The authors wish to thank the many technical and engineering staff members for conducting the tests, especially Emeritus Prof. S. Hoshino (University of Tokyo), Emeritus Prof. H. Saito (University of Tokyo) and Prof. Y. Uehara (Yokohama National University) for valuable advice.

# Full Scale Compartment Fire Test with Lubricant Oil (Lubricant Oil Fire Test: Part 2)

**M. FUJIZUKA**
Mitsubishi Heavy Industries, Ltd.
(Mitsubishi Atomic Power Industries, Inc.)
4-1, Shibakouen 2-chome, Minato-ku, Tokyo, Japan

**Y. KABASAWA**
Chubu Electric Power Company, Inc.
1, Toushincho, Higashi-ku, Nagoya, Japan

**Y. SOUTOME**
Hitachi Works of Hitachi, Ltd.
3-1-1, Saiwaicho, Hitachi, Ibaraki-Pref., Japan

**J. MORITA**
Toshiba Corporation
8, Shinsugita, Isogo-ku, Yokohama, Japan

ABSTRACT

The objective of the tests was to provide the data for use in evaluating the environmental effects of lubricant oil fires in the compartment. Fifty-one tests were conducted to research the environmental effects (atmospheric temperature, radiation heat flux and so forth) on the surroundings with the various size fuel pans and compartments and the various air exchange rates. The concentration of the smoke and CO were also measured in the tests. Test results showed that the radiation heat flux from the high temperature gas and soots was greater than that from the flame because the later was interrupted by the smoke. The gradients of atmospheric temperature were hardly observed in the horizontal temperature distribution but were observed in the vertical temperature distribution in the compartment. It appeared that the concentrations of smoke and CO were able to be estimated from the ratio of burning rate and air supply rate. From the test results, we have developed a conventional and conservative method to evaluate the environmental effects of lubricant oil fire in the compartment.

INTRODUCTION

It seems that the lubricant oil fires may at times occur in an industrial plant facilities. Therefore, fire protection on lubricant oil fire shall be taken into consideration in planning the fire protection program of the industrial plant. In this consideration, we conducted the full scale compartment fire tests with turbine oil. The objective of the tests was to research the thermal influences (ex; radiation heat flux, atmospheric temperature and so forth) to the surroundings and the atmospheric condition (ex; smoke and CO concentration) in the compartment. Preliminary fire tests preceding this test were conducted to get the fundamental data on lubricant oil fire and to obtain necessary data to decide the test condition. (See Reference 1.)

TEST METHOD

Test Condition

Fuel Source    Fuel sources was lubricant oil (Turbine oil VG32) contained in circular pans. The size of the fuel pans were 0.1m$^2$, 0.3m$^2$, 0.5m$^2$, 1.0m$^2$ and 2.0m$^2$.

Test Models    Rectangular parallel-piped compartment were used in these tests. The internal dimensions of the compartment were 6.0m width, 5.0m height. The length was variable to 3.5m, 7.0m, 10.5m and 14.0m. For the additional case, the compartment with the internal dimensions 20m length, 8.0m width and 5.0m height was also used. The walls of the compartment were constructed with fire resistant insulation board (calcium silicate plate) of 25mm thickness. Test models are shown in Table-1.

Table-1   List of the test models                    Total:  51 models

| Pattern No. | Compartment size & Fire location | Opening | Air exchange rate ($h^{-1}$) | Fuel pan size 0.1m² | 0.3m² | 0.5m² | 1.0m² | 2.0m² | Remarks |
|---|---|---|---|---|---|---|---|---|---|
| 1 | 20m × 8m (o) | None | 0 | - | - | - | - | ○ | Duration of the instrumentation |
| | | | 2 | - | - | △ | ○ | ○ | |
| | | | 10 | - | - | △ | ○ | ○ | ○:20 min. |
| | | 2m² | 0 | - | - | - | - | ○ | △:30 min. |
| | | | 2 | - | - | - | - | ○ | □:40 min. |
| | | | 10 | - | - | - | - | ○ | |
| 2 | 14m × 6m (x, o) | None | 0 | □ | △ | ○ | - | - | Fire location |
| | | | 2 | - | - | x:○ o:△ | ○ | - | |
| | | | 5 | □ | △ | △ | ○ | - | o |
| | | | 10 | - | - | △ | ○ | - | x |
| 3 | 10.5m × 6m (x, o) | None | 0 | □ | △ | ○ | - | - | |
| | | | 2 | - | - | x:○ o:△ | - | - | |
| | | | 5 | □ | △ | △ | ○ | - | |
| | | | 10 | - | - | △ | ○ | - | |
| 4 | 7m × 6m (x, o) | None | 0 | ○ | △ | ○ | - | - | |
| | | | 2 | □ | - | x:○ o:△ | - | - | |
| | | | 5 | □ | □ | △ | - | - | |
| | | | 10 | - | - | △ | ○ | - | |
| 5 | 3.5m × 6m (o) | None | 0 | □ | △ | - | - | - | |
| | | | 2 | □ | - | - | - | - | |
| | | | 5 | □ | △ | ○ | - | - | |
| | | | 10 | □ | - | - | - | - | |

Ventilation    Forced ventilation system (push-pull type) was provided to the compartment as shown in Fig. -1.

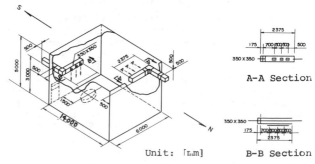

Unit: [mm]

A-A Section

B-B Section

Fig. -1   Configuration of the Compartment model

Fire location   The center of the oil pan was located on the north-south center-line of the room and 1.5m away from the south wall.

Location of the instrumentation   The location of the instrumentation are shown in Fig. -2.

| ○ T.C 11 | ○ T.C 25 | ○ T.C 19 | ○ T.C 19 | ○ T.C 15 | ○ T.C 15 |
|---|---|---|---|---|---|
| ○ V 2 | ○ V 2 | ○ V 3 | ○ V 3 | ○ V 0 | ○ V 2 |
| | △ Rb 1 | △ Ra 0 | △ Ra 3 | △ Ra 3 | △ Ra 3 |
| | | | | □ CO 2 | ⊢ Cs 1 |

Exhaust Dust

| T.C | 2 |
|---|---|
| V | 1 |
| Cs | 1 |

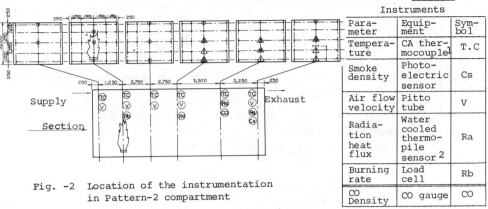

Fig. -2   Location of the instrumentation in Pattern-2 compartment

Instruments

| Para-meter | Equip-ment | Sym-bol |
|---|---|---|
| Tempera-ture | CA ther-mocouple[1] | T.C |
| Smoke density | Photo-electric sensor | Cs |
| Air flow velocity | Pitto tube | V |
| Radia-tion heat flux | Water cooled thermo-pile sensor[2] | Ra |
| Burning rate | Load cell | Rb |
| CO Density | CO gauge | CO |

## Test Procedure

Pan was filled with turbine oil to the adequate depth according to the duration of an experiment and the gasoline was floated on the surface.  It was set on fire by a little gun powder.  After ignited, atmospheric temperature, smoke density, air flow velocity, radiant heat flux, CO density and so forth were measured.

## TEST RESULTS AND CONSIDERATIONS

### Burning Rate

As shown in Fig. -3, lubrication oil burning rate in the compartment fire was nearly constant not relating to the values of parameter "β"[3] derived from the compartment volume and air exchange rates.  Fig. -4 is the comparison of the burning rates of closed space fires with that of free burning fires (open space fires).  (See Reference 1.)  It appears that the variability of the burning rates of the compartment fire is similar to that of the open space fires and the burning rates is correlative to the diameter of fuel pan.  Considering the above, we reached to the conclusion that we could estimate the burning rates of the compartment fire based on those of open space fire.

---

[1]Fine thermocouples were used to restrict the ratio of the radiation heat flux in the measurement to a low degree
[2]Sensors with water jacket were used to restrict the ratio of the convective heat flux in the measurement to a low degree.
[3]$\beta = V(n+1)$   β:  Air supply rate (m³/h)
   V:  Compartment volume (m³)
   n:  Air exchange rate (h⁻¹)

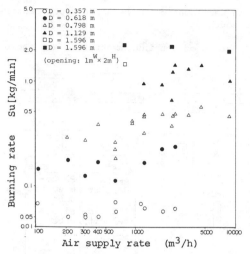

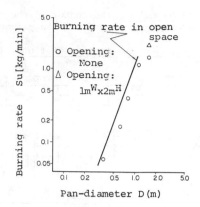

Fig. -4  Dependence of burning rate
on pan-diameter

Fig. -3  Dependence of burning rate on air volume

## Radiation Heat Flux

As shown in Fig. -5, received radiation heat flux in the compartment fire
was greater than that in the open space fire represented by theory locus curves
and the decrease of the received radiation heat flux on the distance was not
observed except in the proximity of the fuel pan.  This might be because the
radiation heat flux from the flame was interrupted by the smoke (gas & soots)
and that from the gas was rather greater than that from the flame.

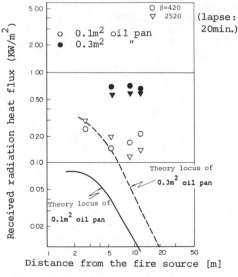

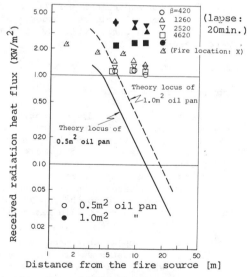

Fig. -5-1  Experimental and
predicted received radiation
heat flux (Compartment pattern-2
$0.1m^2$ & $0.3m^2$ oil pan)

Fig. -5-2  Experimental and
predicted received radiation
heat flux (Compartment pattern-2,
$0.5m^2$ & $1.0m^2$ oil pan)

Fig. -6 shows that the radiation heat flux increases with the increase of gas layer thickness (distance from south wall to receiver) and gas temperature which rises with the lapse of time. In case of small fuel pan, however, radiation heat flux from the flame was greater than the others. This might be because the smoke concentration and gas temperature was low. Considering the above, we studied the radiation heat flux as a function of the gas temperature. The correlation of radiation heat flux and gas temperature is shown in Fig. -7 and -8.

Experimental estimation formulas were developed from the test data as shown in Table-2. In this table, constant "A" means the product of gas emissivity and Stefan-Boltzmann Constant as follows:

$$A = \sigma(1-e^{-K \cdot L})$$

R: Coefficient of correlation

σ: Stefan-Boltzmann Constant
K: Gas absorption factor
L: Gas layer thickness

In spite of the formulas being developed, we reached to the conclusion that the radiation heat flux from gas might be estimated from the following formula because of three reasons that follows:

1. The difference of the radiation heat flux due to the distance was small.

2. The measured values were the limited one because radiation heat flux sensor could not catch the heat flux beyond the angle of elevation (=2/3 radian).

3. We estimated gas emissivity to the maximum (=1.0). For, we could not estimate the actual gas emissivity because we had not analyzed gas composition in tests.

|  | Distance from south wall (m) | $Ax(10^{-11})$ | Remarks |
|---|---|---|---|
| 1 | 2.50 | 3.670 | R= 0.116 |
| 2 | 3.375 | 4.310 | R= -0.756 |
| 3 | 3.50 | 3.310 | R= 0.988 |
| 4 | 4.25 | 4.017 | R= -0.655 |
| 5 | 5.25 | 3.313 | R= 0.976 |
| 6 | 5.75 | 3.083 | R= 0.991 |
| 7 | 6.75 | 2.894 | R= 0.950 |
| 8 | 7.00 | 3.509 | R= 0.981 |
| 9 | 7.875 | 3.710 | R= 0.987 |
| 10 | 10.0 | 2.807 | R= 0.973 |
| 11 | 10.25 | 3.395 | R= 0.983 |
| 12 | 10.4 | 3.455 | R= 0.984 |
| 13 | 13.75 | 3.514 | R= 0.990 |
| 14 | 15.0 | 2.663 | R= 0.967 |
| 15 | 19.75 | 2.858 | R= 0.980 |

$$Qr=A((T)^4 - (T_o)^4)$$

Table-2 Experimental estimation formulas on received radiation heat flux

$$Qr = \sigma(T^4 - T_0^4) \ (KW/m^2)$$

$$\sigma = 5.67 \times 10^{-11} \ (KW/°K^4 \cdot m^2) \ \text{(Stefan-Boltzmann Constant)}$$

Qr: Received gas radiation heat flux ($KW/m^2$)
T: Atmospheric temperature (°K)
$T_0$: Receiver temperature (°K)

In the end, we reached to the conclusion that the radiation heat flux in the compartment fire could be estimated conservatively by summing the heat flux values of gas radiation and flame radiation.

Atmospheric Temperature

As shown in Fig. -9, the gradients of atmospheric temperature were observed in the vertical temperature distribution and were on the increase depending on the increase of fuel pan diameter or the decrease of the air exchange rate.

813

Oil pan size
○ 0.1 m²
● 0.3 m²
△ 0.5 m²
▲ 1.0 m²

Air exchange rate
○ n=0
◔ n=2
◑ n=5
◕ n=10

Oil pan size
○ 0.1 m²
● 0.3 m²
△ 0.5 m²
▲ 1.0 m²

Air exchange rate
○ n=0
◔ n=2
◑ n=5
◕ n=10

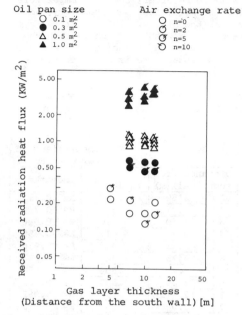

Gas layer thickness
(Distance from the south wall) [m]

Fig. -6-1 Dependence of received radiation heat flux on gas layer thickness (lapse: 10min.)

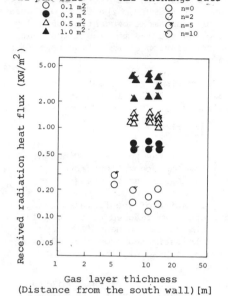

Gas layer thichness
(Distance from the south wall) [m]

Fig. -6-2 Dependence of received radiation heat flux on gas layer thickness (lapse: 20min.)

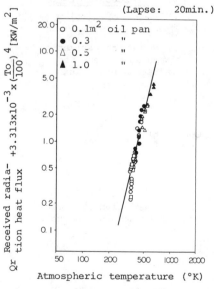

(Lapse: 20min.)

○ 0.1m² oil pan
● 0.3       "
△ 0.5       "
▲ 1.0       "

Atmospheric temperature (°K)

Fig. -7 Radiation heat flux vs. gas temperature
(at 6.75m against south wall)

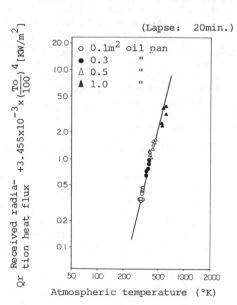

(Lapse: 20min.)

○ 0.1m² oil pan
● 0.3       "
△ 0.5       "
▲ 1.0       "

Atmospheric temperature (°K)

Fig. -8 Radiation heat flux vs. gas temperature
(at 13.75m against south wall)

814

And the difference due to the fire location was not observed. The gradients of atmospheric temperature were also not observed in the horizontal distribution except just above the fuel pan. That is to say, atmospheric temperature was constant everywhere at the same elevation in the compartment. The correlation of the maximum atmospheric temperature at the same elevation and the parameter "α"[1] is shown in Fig. -10.

$$^1\alpha = \frac{V(1+nt)}{S}$$

α: Radio of burning fuel weight to total air supply
V: compartment volume ($m^3$)
n: Air exchange rate ($h^{-1}$)
t: lapse (h)
S: burning fuel weight (kg)

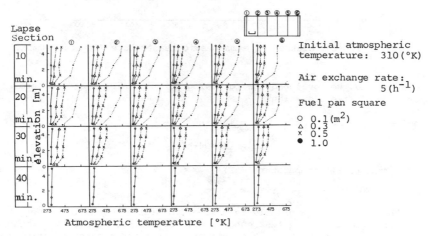

Fig. -9-1   Atmospheric temperature distribution in pattern-2 compartment (Effects of fuel pan size)

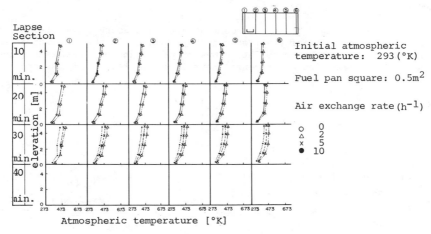

Fig. -9-2   Atmospheric temperature distribution in pattern-2 compartment (Effects of air exchange rate)

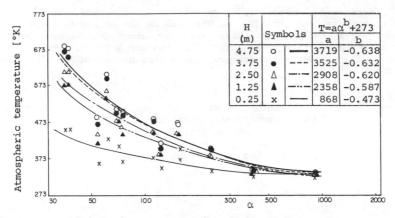

Fig. -10   Correlation of atmospheric temperature (Lapse:   20min.)


In the end, we developed the following formula to estimate the atmospheric temperature of free elevation.

$$T = a\alpha^b + 273 \qquad a = 1344 \cdot h^{0.354}$$

$$b = -0.494 \cdot h^{0.0745}$$

T:   Atmospheric temperature (°K)

h:   Elevation (m)

NOTE:   This formula is available in case of initial atmospheric temperature being about 293°K (=test condition).  Difference in initial atmospheric temperature should be taken into consideration in case of estimating the atmospheric temperature in other initial condition.

Smoke

As shown in Fig. -11, the concentration of the smoke became beyond the measuring limit (Cs=5) of the instrument by about 5 minutes.  It was observed that the speed of smoke accumulation was very high in the compartment fire.  We developed the following experimental formula to estimate the smoke concentration.

$$Cs = 40.57 \cdot \alpha^{-0.485}$$

Extinct coefficient:  Cs (1/m)

$$I = I_0 \cdot e^{-Cs \cdot \ell}$$

$I_0$:   Source illuminance (Lux)

I:   Receiver illuminance (Lux)

$\ell$:   Distance from the source to receiver (m)

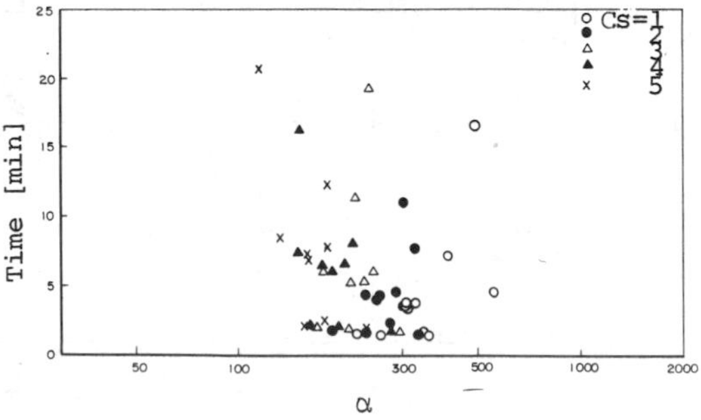

Fig. -11   Dependence of smoke concentration on α

816

CO Concentration

As shown in Fig. -12, the values of CO concentration were on the decrease with the increase of parameter "α". It became to the dangerous concentration for long stay (approximately, 1000 - 3000ppm) in case of parameter "α" being under 100.

We developed the following experimental formula to estimate the CO concentration.

EL: 1.25M        $CO = 3.33 \times 10^5 \cdot \alpha^{-1.282}$

EL: 2.50M        $CO = 7.16 \times 10^5 \cdot \alpha^{-1.482}$

    CO:  CO concentration (ppm)

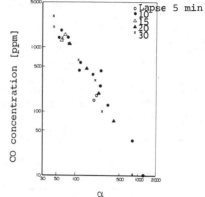

Fig. -12  Dependence of CO concentration on α (H=1.25m)

Air Flow Velocity

As shown in Fig. -13, the values of air flow velocity were on the decrease with the increase of convection distance.

We developed the following experimental formula to estimate the air flow velocity.

$V = a \cdot x^{-b}$

$a = 6.73 \cdot Su^{0.164}$

$b = 0.880 \cdot Su^{-0.0607}$

    V:  Air flow velocity (m/S)

    Su: Burning rate (kg/min.)

    x:  Convection distance (m)

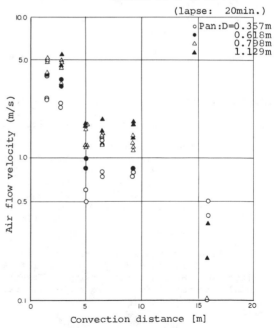

Fig. -13  Dependence of air flow velocity on convection distance

817

CONCLUSION

We developed the method to estimate the compartment fire behavior conservatively as follows.

(1) The burning rate in the compartment fire can be estimated at that in the open space fire.

(2) The radiation heat flux can be estimated at the sum of heat flux values of gas radiation and flame radiation.

(3) Atmospheric temperature can be estimated depending on the following formula.

$$T = a \cdot \alpha^b + 273 \ (°K) \qquad a = 1344 \cdot h^{0.354}$$
$$b = -0.494 \cdot h^{0.0745}$$

(4) Concentration of the smoke can be estimated depending on the following formula

$$Cs = 40.47 \cdot \alpha^{-0.485} \ (1/m)$$

(5) Concentration of the CO can be estimated depending on the following formula.

$$CO = 3.33 \times 10^5 \cdot \alpha^{-1.282} \quad (ppm) \quad (EL: \ 1.25M)$$
$$CO = 7.16 \times 10^5 \cdot \alpha^{1.482} \quad (ppm) \quad (EL: \ 2.50M)$$

(6) Air velocity of convection flow can be estimated depending on the following formula.

$$V = a \cdot X^{-b} \ (m/s) \qquad a = 6.73 \cdot Su^{0.164}$$
$$b = 0.88 \cdot Su^{-0.0607}$$

ACKNOWLEDGEMENT

The tests were conducted by the joint study of the following companies.

Chubu Electric Power Company, Inc.; The Hokkaido Electric Power Company, Inc.; Tohoku Electric Power Company, Inc.; Tokyo Electric Power Company, Inc.; The Hokuriku Electric Power Company, Inc.; The Kansai Electric Power Company, Inc.; The Chugoku Electric Power Company, Inc.; Shikoku Electric Power Company, Inc.; The Kyushu Electric Power Company, Inc.; The Japan Atomic Power Company; Mitsubishi Heavy Industries, Ltd.; Toshiba Corporation; Hitachi, Ltd.

The authors wish to thank many technical and engineering staff in conducting the tests, and especially emeritus Prof. S. Hoshino (University of Tokyo), emeritus Prof. H. Saito (University of Tokyo) and Prof. Y. Uehara (Yokohama National University) for the valuable advices.

REFERENCE

1. T. Tanaka, Y. Kobasawa, Y. Soutome and M. Fujizuka: "Preliminary Test for Full Scale Compartment Fire Test (Lubricant Oil Fire Test; Part-1)," First International Symposium on Fire Safety Science, 1985

# Fire Safety Research and Measures in Schools in Belgium

**A. F. VAN BOGAERT**
State School Building Fund
Ministry of Education
B-1040 Brussels, Belgium

ABSTRACT

   Starting from an analysis of the fire safety concept, this paper describes the scope, philosophy, methods and results of a radical fire safety research done by the Belgian School Building Fund in the larger framework of an overall school building research to meet the new requirements resulting from important educational evolutions.

   While covering three main groups of items : fire preventive, fire restrictive and fire protective measures, it stresses the need
- to take the human behaviour as a starting-point,
- to pre-calculate the evacuation times from the design stage onwards,
- to care for handicapped pupils in regular schools,
- of particular safety concern in special schools and homes for disabled children,
- to take safety arrangements for impaired visitors to the school as a community centre during offhours.

   The paper points to some near future developments in school life risks and responding measures.

   It is noteworthy that in 1982 the results of this research were translated into a national Belgian norm (NBN) : S 21-204, that was given force of law by Royal Act.

   The conclusion says that we build schools in a much safer way than we live in them, owing to shortcomings in education itself, and thus ironically wonders why we should build schools.

                         *  *  *

Key words : Belgian norm; boarding schools; building design, disabled pupils; evacuation; fire prevention; fire protection, fire restriction; fire safety; safety concept; schools; special schools.

INTRODUCTION

   Up to 1974, the ministry of Public Works was responsible for school building and maintenance.  Owing to a shift in departmental tasks, this charge was transferred to the School Building Fund (SBF) of the Education department.  Facing the burden of future new qualifications and responsibilities, the SBF felt the need to rethink the entire school building problem, going down to, and then starting from, the very essence of functions, needs and conditions in the educational process.

In this research work, fire safety took an important sector closely linked with building layout, with equipment policy and with daily educational activities. The down to the roots research produced an

# 1. ANALYSIS OF THE FIRE SAFETY CONCEPT

Fire safety in a school is a complex notion because it depends on some constant and many alterable factors :
- the design and construction of the building (constant factor)
- the contents of the building (variable factor)
- the occupants' behaviour (highly variable factor)

1.1. The users' daily avocations breed a multicellular mosaic of hazards which are constantly challenging safety in the building. Prospect and caution, prevention habits, self discipline, drilled preparedness and trained co-operation are the positive poles of sound safety behaviour. Age, mental and physical condition, education, assimilated information and applied experience together draw the daily safety diagram of a community.

1.2. The contents in their turn relate to the occupants' activities and to their mental and physical abilities. These activities often require apparatus, machines, equipment, furniture and stocks that may hide unsuspected fire and associated risks such as smoke and toxic gas generation. Thorough maintenance and regular checking are preventive measures against these dangers.

1.3. Finally the building, as a solid environment, holds and protects its contents and the human activities. In its design and construction it should closely match all the dimensions of these functions, including those which aim at fire safety. Research of these functions should start with measuring hazards originated by the daily activities in the premises and with the prospect of human abilities and failures on the threshold of a disaster. According to these risks, fire preventive, fire restrictive and fire protective settings should be included in the building and be considered from the very start of design operations.

1.4. Diagram 1 shows an analysis of the three main components into their respective factors, their interactions, and their final impact on the safety level of a school. It clearly indicates that all fire safety research and development work should start from the human factor, because the users' abilities and shortcomings must be matched by the physical setting (Ref. 1). This conclusion became the basic philosophy for the research work on firesafe school building.

# 2. SCOPE OF THE RESEARCH

As said before, this research was part of an overall study in view of an important shift in departmental qualifications. At the same time it was connected with the renovation trends in educational objectives and structures and in didactic means and methods. General new characteristics gradually appeared in school life : increased movement of pupils due to frequent alterations in group size and composition; intensified use, both by teachers and pupils, of electric and electronic apparatus; longer students' self governed occupancies;

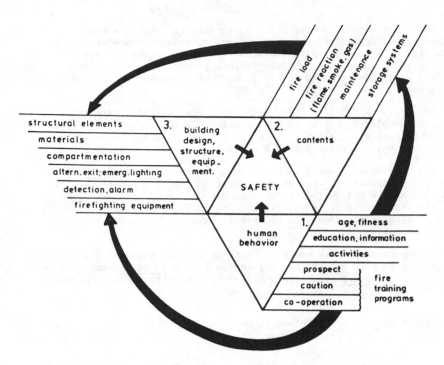

Diagram 1 : Components of the SAFETY notion; logical growth
     philosophy of fire prevention norms.


use of screens for isolating team work; introduction of lighter furniture
allowing immediate re-arrangements for varying activities.

     This evolution required new adequate approaches to school building.  These
were : more diversity in the capacity of rooms, varying from 6 to more than
100 pupils; larger teaching and smaller circulation areas, more compact layout,
requiring mechanical ventilation in the core; folding partitions allowing flexibility in space use, increased storage, multiplication of rooms needing gas
and electric connexions etc.  (Ref. 1 and 2).  It was clear that all these features would entail
     higher fire-load,
     increased energy-risks,
     intenser potentiality of smoke generation and
     more complicated evacuation.


3. RESEARCH AND RESULTS

     The research covered three main groups of items : fire preventive, fire
restrictive and fire protective measures.


     3.1. PREVENTION mainly focussed on the general layout. In fact, prevention
became an integrated part of the design work avoiding to locate risky spaces
in the core of the building, but pushing towards the skin of the building, or
even outside in separate premises such spaces as the boiler room, the school

kitchen, important storage rooms, tension transformer cabins, workshops. As far as the operational context allows, gas and power equipped rooms are grouped, thus reducing the spread of special equipment risks. If this cluster is conceived in vertical sense, the laboratories are put on the highest floor, in order to limit expansion to other floors in case of explosion or of fire followed by explosions.

Another prevention matter lies in the choice of building materials. Not only their fire related properties are important, but also the way they are applied. For, properly speaking, there are rather few dangerous building materials, but many materials are used or applied in hazardous manners.

A third prevention measure aims at limiting risks during offhours : all workshops and laboratories have, next to their entrance, a general gas stop cock and an electric key switch, both commanding all conducts inside. Parallel to this, the teacher's desk in laboratories is equipped with devices enabling him, at the slightest incident, to cut off all gas and electricity conducts used by the students.

The same prevention idea prevails in rooms without constant human presence and holding inflammable goods. They are controlled by smoke detectors (in most cases) and heat detectors (in boiler houses). In the latter case they command automatic extinguishing devices.

3.2. The FIRE RESTRICTIVE research mainly concentrated on three items : fire resistance of structural elements, fire reaction of materials, compartmentation.

Fire resistance : In case of fire, the structural elements should maintain their loadbearing or space separating function long enough to allow complete evacuation and adequate firefighting without a threat of crash. The fire resistance requirements are determined by the nature, function and position of the structural elements. Thus the longest fire resistance is required for three types of elements,
- the 1st being the skeleton and the floors, because of their impact on the stability of the building;
- the 2nd being the compartmentation walls, the stairs and their encasing, on behalf of their life protective importance;
- the 3rd being the enclosures of risky places like boiler rooms and kitchens.

A second factor influencing the fire resistance requirements is the height of the building, and a third factor is related to the building's nature : on behalf of their night occupancy, boarding schools and students' homes require a higher fire resistance degree from their structural elements than equally-sized schools.

Thus, fundamental fire resistance values of structural elements in school buildings vary from 1/2 h to 2 h.

A second fire restrictive measure concerned materials for floor, ceiling and wall finishes, particularly in relation to their flame spread rate and their potential gas and smoke generation. This was probably the most daunting part of our research work : on one hand we wanted to get rid of the harsh, severe, impersonal interiors of our own schooldays. We were eager to introduce gayer colours, cosier floors, more homelike finishings. On the other hand at that moment the tapestry, carpeting and wall paint markets were flooded with new materials, preponderantly, the fire reactions of which were largely divergent, unsufficiently known and over-advertised in optimistic terms. As we didnot want to buy heavens of nocious gases in pretty pink boxes, a long term contract was made with a specialized fire control organism charged to collect laboratory-tested information on all present and future floor, wall and ceiling coverings on the Belgian market, and to conclude, for each material, whether it was fit or not to be applied in schools and students' homes. Later on, the study was extended to textile, with a stress on synthetics as used in bedroom equipment.

It was an expensive operation, but the price was well worth the purchase. Not only it provided us with reliable information, but the manufacturers have become conscious that safety research and application should be indispensable and integrated parts of their production.

As to compartmentation, the main fire restrictive measure, a conflict came up with the need of a wide diversity in room spaces and a flexible use of the building. It was felt that permanent evolution in the nature of education would probably call for later serious interior remodelling works. So, restricted compartments would go against adaptability of the building, leaving it physically sound, but functionally obsolete after a couple of decades (Ref. 1 & 2). Extensive compartments might increase risks but were indispensable to ensure future adaptability to the ever changing educational needs. Once again, moveable education, requiring a kinetic architecture, conflicted with fire restrictive options (as do open doors in smokefree evacuation routes). A compromise was adopted : the provision of generous exitways from each compartment was to compensate the extension of its size. Backed by the remedy of a short evacuation time, the maximum compartment surface was raised to 3500 m2, which seems to be a record in European school building.

As boarding-schools and students' homes are not subject to ulterior major alteration in space distribution, the utmost size of their compartments was limited to 2000 m2.


3.3. FIRE PROTECTION is above all concerned with the users' security.
In this sector three major problems were dealt with : warning and alarm systems, the time needed for evacuation, the escape routes.

Obviously warning and alarm techniques vary according to the size of the school, from simple means to extensive systems, from the school bell, rang in a special way, or a portable siren, to the detector operated system showing the fire location on a synoptic panel, and in some cases combined with an automatic warning transmission to the fire station. In boarding schools each tutor(being in charge of 21 pupils) and the administrator can warn the fire station at once via a phone in their respective bedrooms. Fire alarm boxes are placed in the corridors, near each of the tutors' rooms and at an utmost distance of 60 m from each other; it was found practical to locate the fire-extinguishers next to these alarm-boxes.

Evacuation times : The central idea of safety concern in school building being : "Save the people and then, if possible, the building" prompted to sharp concentration on the evacuation problem.

What we wanted was a reliable computation method for future, for planned buildings, - properly speaking, a method capable to precast the evacuation time from the first design stage onwards, when it is still possible to remedy short-comings by increasing the exiting capacity.

The major factor taken into account in our computations was the evacuation flow rate in the different escape route segments, itself being composed of several subfactors, such as the density of the evacuation stream, the width of the different escape route segments and the normal values of horizontal velocity, vertical downward velocity and vertical upward velocity. Thus we found

three flow ratings per passage unit of 60 cm, per second :
horizontal rate :                1,5   )
vertical downward rate :        1,1   )  persons/passage unit/second (Ref.
vertical upward rate :          0,9   )                              1 and 2)
These values, combined with the relevant interior environment factors, made it
possible to compose the computing method needed to predict the evacuation time
from the building design stage onwards.  Moreover it includes prediction of the
origin, location and duration of traffic congestions which inevitably occur at
places where wider segments flow into narrower ones or at moments when the
downstream evacuees are still occupying staircase segments when upstream people
arrive behind them.

    A description of this method would go beyond the scope of this lecture, but
it should be mentioned that later on, it has been enlarged to all sorts of
buildings, meeting the total agreement of the Belgian Firefighting Techniques
Institute (ref. 4).

    Another noteable feature was the research after what time evacuation beco-
mes unsafe or impossible.  The analysis of numerous tests and real fires seemed
to show that there is no general answer.  Yet we found a series of 5 co-effi-
cients (total surface area, number of floors, degree of space partitioning,den-
sity of occupation, equipment or production risks), the combined impact of
which brought us to recommend evacuation of a school building in less than 5,
and a boarding school or studenthome in less than 7 minutes.  Up to now, in fire
drills, evacuation times have not exceeded 3 1/2 and 5 1/2 minutes respectively.

    Evidently, the third fire protection concern, escape routes, was closely
linked to, and based on the evacuation time conclusions.  The basic rule to
ensure at least one alternative wayout in all circumstances requires skilful
choice and combination of length, width, number and location of the exit ways.
Thus the overall school building layout is co-governed by some dominating evacuation principles :
- The every-day traffic routes are also the escape routes.
- Avoid the need to build emergency stairs.
- Divide the building, and even each compartment, into evacuation sectors, i.e.
  the areas to be discharged by each staircase and each outdoor exit.
  Throughout the building, these sectors should be balanced in importance
  according to the prognosticated number of their users.
- Avoid mixing stairway exits with those that discharge the groundfloor.  This
  means that wherever possible, the staircase-bottom should have a doorway
  direct to open air.
- All stairways are completely encaged and closed by walls and doors of 1 to 2
  h fire resistance, according to the height.
- The wider you plan the stairs, the fewer you have to provide, the more you
  further cumbersome concentrations of evacuees, the fewer are the chances to
  spread the alternative exit ways.
- So, whenever possible, use a stairwidth of 2 passage units with railings on
  both sides.  Stairs of 3 passage units are dangerous, the central user having
  no rail protection.  In case of 4 passage units, the stairs have two side and
  one central railings.

    3.4. All essential conclusions for renewed school conception, inclusive
safety, were ready in 1970, but two more years were needed for testing, verify-
ing and refining several newly made opinions, methods and measures. The complete
results were issued in 1972 (Ref. 1) and 74 (Ref. 2) as a self-service guideline
for the School Building Fund.

    But meanwhile the scope of the task had widened, calling for additional
research on a new type of school that had sprung up in the 60-ies for special

education of handicapped children.

4. DISABLED PUPILS

Let us clearly make the difference between physically handicapped children or students who attend a regular school, and communities of disabled children suffering from various handicaps and educated in special schools.

4.1. To meet the needs of the former, who are extremely rare (+ 0,3 %), primary schools do not cause major difficulties, since these are generally single-storey-buildings. If they are multi-storey, all common and special rooms are always on ground floor level, as well as a number of home rooms. This space distribution always allows to keep groups that contain handicapped pupils on the evacuation level.

As such combinations are impossible in the multistorey buildings of second-ary schools, these have been or are being equipped with lifts that can hold wheelchairs. But as lifts are not to be used in case of fire, the periodical evacuation drills include precise behaviour lines and exercises for the respon-sible teacher and some appointed volunteers to carry the handicapped student into safety.

The permanent character of daily school life organization warrants that these measures will do. But the growing community use of schools has imported a new problem, not so widely known. It is caused by the possible presence of handicapped off-hours adolescents and adults. Our lifts can bring them to any floor. But who gets them down and out in case of fire? Who, moreover, knows about their presence and where they are?

These are our provisional measures : All handicapped are welcome, whether alone or in company, on the ground floor. But movement or sensorial impaired visitors, wanting to use parts of the premises above the evacuation level, should be accompanied by at least one person, not impaired, and should report their presence on entering and withdraw it on leaving. (Ref. 3).

4.2. In specific schools and boarding schools for handicapped children,the situation called for an analysis of the pupils' shortcomings as to perception, response and mobility, the negative results of which should be met by special arrangements in the layout, construction and equipment of the building. We soon discovered that it was dangerously erroneous to base on the school structure qualifications of mental or physical deficiencies, because there are many bodi-ly handicapped among the mentally or emotionally disturbed. What really matters is to know how many pupils in each impairment category are able to escape, without help, from a building under fire.

A close investigation in 51 special schools and boarding schools showed alarming percentages of non-ambulatory pupils among both groups of mentally and physically handicapped. This statement became the keynote of our conclusions (Ref. 4) :
a. Design, construction and equipment of buildings for handicapped children and adolescents must be matched to the presence of non-ambulatory impaired among both mentally or emotionally and bodily disturbed pupils.
b. Compartments in special schools should be smaller (max. 1000 m2) and more numerous than in normal schools. They should be still smaller (max.500 m2) and more numerous in special boarding schools, so as to limit smoke and fire spread and to procure sheltering spaces that offer quick but provisional sa-fety by way of horizontal egress. This evacuation method corresponds with the minimal disruption system recommended for hospitals (Ref. 6).
c. As the evacuation process takes more time, and as the stay in a shelter

compartment may be prolonged, the fire resistance of structural and
compartmentation elements should attain 1 or 2 h, according to the number of storeys.
d. Limitations of risks call for arrangements in time and space, whereby handicapped should be located on, or not far from, the evacuation level (E)
which is generally the ground floor (O), as marked in the tables below.

These tables clearly reflect the need to limit the height of special
schools and boarding schools to two floors above the ground floor or evacuation
level. A strong preference goes towards a single-storey design or to buildings,
all levels of which can be evacuated horizontally to the outside.
e. In these institutes, fire drills should more than anything else aim at preparing staff members to coordinate their exiting assistance and to manage
the egress procedure.
f. Finally, basing on several nightly evacuation drills, it was proposed that
the boarding personnel to pupils ratio should be raised to 1-8, whereby at
least one staff member should not be groupbound.

5. A NATIONAL NORM ON FIRE SAFETY IN BUILDINGS FOR SCHOOLS, BOARDING SCHOOLS
AND STUDENTS' HOMES.

Wishing to officialize the fire safety measures as found and applied
by the SBF in regular and special educational institutions, the Belgian Institute of Normalization took them as a firm base and framework for a national
norm, issued in 1982 as NBN S 21-204 (ref. 7). It was followed by a Royal Act
making its application compulsory for all new school buildings, by whomever
they be erected : the State, regional or local authorities and private organizations. The rather exceptional legal statute of this norm stresses the
recognition of its importance.

Yet, like all norms, the S 21-204 deals with forthcoming, to be built
situations. About existing buildings it says that their safety conditions

| | Disability \ Level | - 1 | 0 or E | + 1 | + 2 |
|---|---|---|---|---|---|
| In day-time | Non-ambulatory | – | ✳ | ✳ | – |
| | Ambulatory | ✳ | ✳ | ✳ | ✳ |
| | Visual deficiency | – | ✳ | – | – |
| | Auditive deficiency | ✳ | ✳ | ✳ | ✳ |

| | Disability \ Level | - 1 | 0 or E | + 1 | + 2 |
|---|---|---|---|---|---|
| At night | Non-ambulatory | – | ✳ | – | – |
| | Ambulatory | – | ✳ | ✳ | ✳ |
| | Visual deficiency | – | ✳ | – | – |
| | Auditive deficiency | – | ✳ | ✳ | ✳ |

should approach the norm's prescription in the best possible way and degree.
Among the older constructions, those that were not originally built for school
use prove to be the most unsafe. As it is almost impossible to improve their
fire resistance and reaction, the systematic approach to raising their safety
level consists of a threefold measure :
- ensuring a rapid evacuation by multiplying the exit ways,
- accelerating the alarm by means of a general detection system,
- delaying fire spread by means of sprinklers in all unoccupied spaces.

## 6. NEAR FUTURE

6.1. Multi-purpose rooms, skill and handicraft training opportunities,
space for multi-tuitonal activities that partake a considerable amount of floor
area, direct interaction between practice rooms, intenser mobility within learning departments are all in growing demand. As a combined effect of these new
characteristics, considerable segments of the horizontal escape ways will soon
no more be enclosed. This calls for the forthcoming need at least to clearly
mark off those specific floor areas which should be rigidly kept clear from
hindrances in order to ensure their exiting function.

6.2. In a number of specialized higher technical and artistic institutes,
computer rooms, broadcasting and T.V. studios are in rapid expansion. These
environments with highly concentrated and expensive electronic equipment require
dry extinguishing methods, using halon gas, on account of its characteristics
(non-destructive; penetrating into inaccessible spaces; almost harmless to
people).

6.3. We are not satisfied at all with the general lack of safety education
in our schools. Clear and complete instructions, dating from 1975 (Ref. 8),may
have brought some insight in prevention and protection, but apart from periodical fire-drills they have scarcely altered the caution pattern in daily school
life. Yet, you need not be an educator to know that it is perfectly possible
to introduce items of safety motivated education into the curriculum of physical
training, chemistry, physics, mathematics, morals.

6.4. In relation to this shortcoming, the SBF would also like to be followed in an earlier suggestion, to create series of slides or short sequence films
staged in actual schools and showing safe and unsafe behaviour and consequences.
The viewing should be followed by discussions.

6.5. There is a more promising prospect of regular relations between
schools and fire brigades, - a movement that has already started and that consists of pupils' visits to the fire-station, inclusive their interviewing the
fire-men; or fire officers being invited to schools, to talk on their job,
evolving into a discussion on safety behaviour in the very premises.

6.6. In the technical sector, forthcoming strivings go to the extension of
the direct fire brigade warning systems commanded by detection. Priority will
be given, in this order, to special boarding schools and special schools, to
regular boarding schools in old premises, to technical schools with chemical or
nuclear sciences curriculum.

## 7. CONCLUSION

Although the aims described in this paper are common knowledge, some particular methods and results probably deserved to be stressed: the human behaviour as a starting-point; the pre-calculation of evacuation times; the utmost
care given to escape routes, inclusive the complete enclosure of all staircases;
the particular concern for disabled children; the edition of a national specific
norm, with legal status, on firesafe school building. Summarizing, the whole

bulk might be reduced to three questions :

Do we build our schools safely? Undoubtedly yes, and most probably they rank among the safest on the European continent, for a range of details and refinements going beyond norms and standards ensure a better control in emergency cases. Still, constant alertness to educational evolutions is indispensable.

But - do we safely live in our school buildings? A tiptoe answer : the hardware of our safety system is of real good quality; the software is psychologically underdeveloped. For laziness to break with wrong habits,daily inadvertance,unbelief in fire, are tremendous shortcomings in education itself.

Hence this third, somewhat vindictive, question : Why the deuce do we build schools?

REFERENCES

1. Van Bogaert, A.F.: Logica en Actie in de Scholenbouw,Ed. Stevin,Brussel, 1972.

2. Van Bogaert, A.F.: Prospective dans la construction scolaire, Ed. Vander, Brussel and Cesson (France), 1974.

3. SBF Directorate : IDG 80.040 : Naschools gebruik van normale-school-gebouwen door gehandicapten, Brussel, 1980.

4. School Building Fund : Fire prevention in Schools and Boarding Schools for Handicapped, Brussels, 1978 and NBSIR, Washington DC, 1980, n° 2070.

5. Van Bogaert, A.F.: Fire and Evacuation Times, Evacuatietijden bij Brand, Ed. Story-Scientia, Gent, 1978.

6. Marchant, E.W. : Escape Route Design, Edinburgh, 1975.

7. B.I.N. (Belgian Institute of Normalization) : NBN S 21-204 : Fire Safety in Schools, Boarding Schools and Students' Homes, Brussel, 1982.

8. Departmental Regulations : I.M. 74.120 : Brandveiligheid in Internaten, Brussel, 1975.
I.M. 75.040 : Brandveiligheid in Scholen, Brussel, 1975.

# Fire Spread along Roofs—
# Some Experimental Studies

**KAI ÖDEEN**
Swedish Fire Protection Association

ABSTRACT

In the paper is described a number of test series where the influence of the
wind velocity on the fire spread along the external surface of a roof has
been studied. Different types of roof coverings and insulating materials
within a wide range of combustibility properties have been studied. The
results show that the spread of flame is fairly independent of these properties if the covering material is glued to the roof over its entire surface.
For point-wise (mechanically) fixed roof covering materials, however, the
fire spread is rapid and the damages extensive. The reason seems to be that
the fire can take place and spread also in the space under the covering
material.

Key words: fire spread, roofing materials, heat insulating materials, wind
effects.

INTRODUCTION

During the last years the Swedish Fire Protection Association has been
carrying out investigations with the aim to demonstrate the fire technical
properties of light weight roof structures in different respects and to give
the base for testing and judging these components from the points of view of
the fire insurers. As a result of these studies a proposal for a fire
testing and classification procedure has been presented by the author /Ödeen,
1985/. This procedure is at present implemented  in the routine activities
of the Swedish fire insurance companies.

Concerning the total fire technical behaviour of a roof structure the spread
of a fire along the external surface creates an important phenomenon which has
turned out to be subject to very few studies. In this paper some Swedish experimental tests are presented and the results are briefly discussed.

HISTORY

The fire technical properties of the light weight roof structures in general
have been subject to discussions and investigations since they were introduced
in the fifties. Due to the famous Livonis fire Factory Mutual in the USA an
extensive research work  was  started   resulting in a half-scale test method
"Construction Materials Calorimeter" /Factory Mutual Research Corporation,
1980/. This method is at present used for insurance classification purposes
in the USA (in combination with the "tunnel test"). The method was essentially
new      and reflected when it was introduced in many respect a new way of
thinking in fire testing. Especially its ability to measure the fuel contribution of the test specimen is of great value. The method was based on comparisons with about 50 "full scale tests" in a test building with the horizontal measures 100 x 20 ft.

As this type of roof structures were introduced in Scandinavia the fire authorities as well as the fire insurance companies were faced to a number of
intricate questions concerning the fire technical behaviour of the roofs. To
find some answers the Norwegian and Swedish national testing institutes performed some test series in the early sixties. The aim of these tests was to
give the base for regulations and guidelines in the national building codes.
A comprehensive description of these tests is given in a report from the
Swedish National Testing Institute /Statens provningsanstalt, 1966/.

In practice the problem of external spread of flame along roofs with combustible thermal insulation or combustible surface material is tackled in different
ways. In e.g. Norway and the FRG a roof insulation of this type is divided
into fields of a certain maximum area by strips of non-combustible material,
in most cases mineral wool. Some preliminary tests in the sixties indicated
this to be a useful method to prevent the ignition of larger roof areas. However, it must be emphasized that these tests were performed without taking
into account the effect of a wind along the roof surface.

830

Experimental investigations of the mechanism at external fire spread along
the roof surface in combination with wind have previously been carried out in
Sweden. Of special interest are some studies in 1966 by the National Swedish
Testing Institute in cooperation with some industries where the influence on
the fire spread of the wind velocity was studied. This test series was later
followed up by similar tests in 1973 /Bengtson 1973/.

The 1966- tests were carried out in a specially designed test rig with test
specimens 70 x 500 cm, slightly inclined and with a fan arrangement which
could be controlled up to a wind velocity along the surface of maximum 5
meters per sec. In total 8 tests were performed. 7 test specimens were built
up of steel tin profiles and heat insulation of polystyrene-foam and cork and
one as a conventional wood panel roof. In all cases the roofs were covered
with normal bitumenous roofing felt.

Even if these tests must be regarded as preliminary some general conclusions
can be drawn from the observations. One is that the speed of the fire spread
along the roof seems to increase with increasing wind velocity, however, only
up to a maximum value. At further increase of the wind velocity a tendency to
decrease of the fire spread can be observed. The results are not very clear
but in any case it can be stated that the speed of the fire spread does not
increase over a certain wind velocity (cf figure 1).

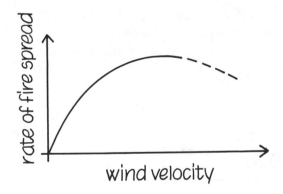

FIGURE 1. Rate of fire spread vs wind velocity.

The 1973-tests were performed with larger test specimens but following the
same concept. The test specimens had the width 100 cm and the length 500 cm
(or in some cases 1000 cm). The fan arrangement could give a wind velocity
of up to 10 meters per sec.

At the tests was clearly demonstrated that the method of separating a combustible roof insulation with non-combustible mineral wool strips gave an almost neglectable effect as soon as the influence of wind was taken into account. A separation of this type did only give considerable effect if it was combined with other arrangements e.g. a non-combustible screen on the roof and the tests gave some indications to how this should be designed. The very small effect of the traditional separation metod turned out to depend on the fact that the fire spread took place essentially in the roofing felt and its adhesive and that the combustible insulation material (polystyrene foam) contributed only to a minor degree to the fire development.

Later on complementary tests have been performed with entirely non-combustible heat insulation as well as with light-weight concrete and all the results indicate that the spread of flame is almost unaffected by the combustibility and thermal inertia of the underlaying material. However, this preliminary statement has certain important exceptions which are further commented later on /Ahlen - Ödeen, 1982/.

In the following sections are described some tests series with modern industrial roof structures with different types of insulating and covering material.

TEST ARRANGEMENT

The tests were performed in a surplus industrial building mainly built of light weight and normal concrete. The test arrangement was similar to that of the previous tests and is shown in figure 2. The test specimens were 2 x 8 meters. In total 6 tests were carried out (A - F).

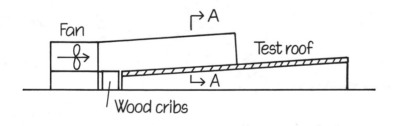

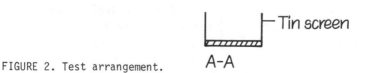

FIGURE 2. Test arrangement.  A-A

The test roofs were of a conventional steel-tin type (tests A - E) or light-weight concrete (test F). The details are shown in figure 3.

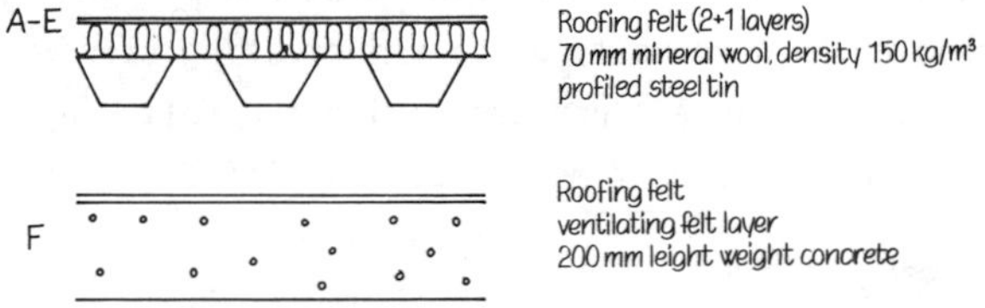

A-E    Roofing felt (2+1 layers)
       70 mm mineral wool, density 150 kg/m³
       profiled steel tin

F      Roofing felt
       ventilating felt layer
       200 mm leight weight concrete

FIGURE 3. Test specimens

The roofing felt was glued to the underlaying material with a bitumenous glue. The amounts of glue were determined to 4.5 - 6.1 kg per sq meter (test A - E) and 2.6 kg per sq meter (test F). These figures do not include the bitumen in the roofing felt itself. The source of ignition was in the two first tests (A and B) two wooden cribs with a total weight of 150 kg. Observations at these tests raised doubts whether this gave a realistic picture of a fire attack. Therefore three cribs with a total weight of 225 kg were used as ignition source in the rest of the test series. The net calorimetric heat value can be estimated to appr. 2500 MJ for two cribs and 3700 MJ for three cribs. The cribs were built up of sawn pieces 50 x 50 x 500 mm. On a first layer of 4 pieces another 12 layers with 6 pieces each were laid cross-wise. The very upper layer with 4 pieces was kept at the same level as the nearest short edge of the test specimen (cf figure 1). The cribs were ignited simultaneously with 2 liters of alcohol in small tin containers under each crib.

At two  preliminary tests (A and B) the wind velocity was chosen  2 and 5 meters per sec. At these tests no significant fire spread took place and the damages were restricted to the area immediately adjacent to the ignition source. Therefore these low wind velocities were judged to be too small and of no relevance for the rest of the test series.

After extensive discussions test C was performed with widely varying wind in order to give guidance as far as possible for chosing a proper profile for a test procedure. Observations at this test indicated that maximum impact and fire spread could be achieved at a velocity profile increasing step-wise from 0 during the first 3 minutes to 5 meters per sec during the following 5 minutes and then up to a final value of 7 meters per sec (cf figure 4).

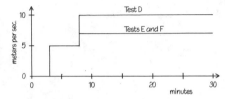

FIGURE 4. Wind velocity profiles.

The tests started with the ignition of the cribs. After about 3 minutes they were entirely ignited and the flames were 1.5 to 2 meters high. At this time the fan was started and the fire in cribs and roof was allowed to develop freely. The tests were terminated after about 30 minutes when the cribs had almost burnt down.

The fire courses were basically the same in all tests with only minor damages to the test roofs. It should be observed that these two types of roofs at present are classified in the best class according to Swedish insurance rules.

## INFLUENCE OF INSULATION MATERIAL AND OF METHOD FOR FIXING THE COVERING MATERIAL

The testing method described above has been applied to a number of roof struc-tures manufactured by Swedish manufacturers. The scope of this test series was primarily to show the influence of the thermal and combustibility properties of the insulating material and furthermore to demonstrate the difference - if any - of different methods of fixing the covering material (roofing felt or similar) to the roof. Tests were performed with the following insulating materials.

834

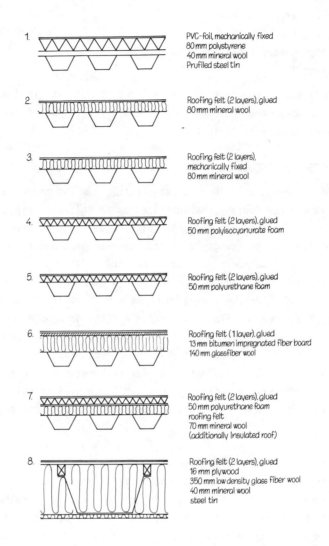

1. PVC-foil, mechanically fixed
   80 mm polystyrene
   40 mm mineral wool
   Prufiled steel tin

2. Roofing felt (2 layers), glued
   80 mm mineral wool

3. Roofing felt (2 layers),
   mechanically fixed
   80 mm mineral wool

4. Roofing felt (2 layers), glued
   50 mm polyisocyanurate foam

5. Roofing felt (2 layers), glued
   50 mm polyurethane foam

6. Roofing felt (1 layer), glued
   13 mm bitumen impregnated fiber board
   140 mm glassfiber wool

7. Roofing felt (2 layers), glued
   50 mm polyurethane foam
   roofing felt
   70 mm mineral wool
   (additionally insulated roof)

8. Roofing felt (2 layers), glued
   16 mm plywood
   350 mm low density glass fiber wool
   40 mm mineral wool
   steel tin

Figure 5. Compositions of the test specimens.

835

- Polystyrene foam
- Mineral wool
- Polyisocyanurate foam (PIR)
- Polyurethane foam (PUR)
- Bitumen impregnated fiber board
- Wood (plywood)

Covering materials glued over the entire area as well as fixed point-wise with mechanical devices were included in the test series. A brief description of the test specimens is given in figure 5.

TESTS RESULTS AND DISCUSSION

The fire developments were in most cases fairly similar and were - for the roofing felt covered roofs - characterized by an initially rapid spread of flames along the upper surface of the felt. However, this spread decreased after some time and the damages were limited even in those cases were the insulating material was more or less combustible. It is important to state that this did only hold for the surface layers glued to the insulation over its entire surface. For the layers which were mechanically, point-wise attached to the roof the spread of flames was rapid and the damages extensive. The reason is obviously that in these cases the combustion takes place not only along the upper surface of the layer but also in the space between the surface layer and the insulating material. This type of fire behaviour is essentially the same as has previously been clearly observed for e.g. internal wall coverings tested in the American Tunnel test, the Scandinavian Box test or in full scale room tests.

A conclusion of all test results seems to be that the fire spread in a combustible roofing layer is fairly independent of the combustibility properties of the under-laying material. However, this conclusion must be restricted to surface layers which are glued to the roof over its entire area whereas roofing systems with mechanically fixed surface layers increase the risk of rapid spread of fire over the roof surface. One consequence of this conclusion - drawn by Swedish insurance companies - is that the latter type of roof cannot be approved in the highest (best) class without further testing proving acceptable fire behaviour of the roof.

Finally it can be noted that there has been developed a small scale test based on the test results described above. This test is a modification of the test method for external roof surfaces referred to in the Scandinavian building codes (NORD-TEST, Test Method NT Fire oo6).

REFERENCES

Ahlen B - Ödeen K.: Utvändig brandspridning längs tak - Kompletterande studium av vindhastighetens inverkan (External spread of flame along roofs - Supplementary studies of the effect of wind velocity), Swedish Fire Protection Association, Stockholm 1982.

Bengtson, S.: Flamspridning längs utvändigt isolerade plåttak, rapport 1, 2 och 3 (Spread of flame along externally insulated tin roofs), The National Swedish Testing Institute, Stockholm 1973.

Factory Mutual Research Corporation: Construction Materials Calorimeter. January 1970.

Statens provningsanstalt: Rapport angående brandförsök med takkonstruktioner av stålplåt med värmeisolering (Report from fire tests with roof structures of steel tin with heat insulation), Stockholm 1966.

Ödeen, K.: Brandtekniska egenskaper hos industritak (Fire Technical Properties of Industrial Roofs), Swedish Fire Protection Association, Stockholm 1985.

# Evaluation of Garment Flammability using Thermal Mannequins

Y. UEHARA and M. UMEZAWA
Department of Safety Engineering
Division of Materials Science and Chemical Engineering
Faculty of Engineering
Yokohama National University
156 Tokiwadai, Hodogaya-ku, Yokohama 240, Japan

ABSTRACT

The flammability of garments was studied using thermal mannequins with sensors which were dressed in 73 sets of clothes commonly available in Japan. The results showed that all of them, except for 18 sets, were highly flammable and indicated a possibility of serious fire injuries within a short time. The maximum temperature and the maximum heat flux obtained on the body surface were 437°C and 335 kJ/m2 s, respectively. To evaluate the hazards, method of evaluation based on the burning rate, size and degree of fire injury plus a combination of these factors were proposed to be useful.

INTRODUCTION

Clothing, together with food and housing, is the most basic factor in our daily life and indispensable in maintaining protection of the body against cold weather and outer hazards. However, some garments are so flammable that a large number of casualties are the result of garment fires every year. Being aware of the hazards, several countries in the world have already established standards of flame proof garment which are centered on children's nightclothes, or restrictions have been placed on garments.

This paper intends to explain garment fires in Japan and to report on the classification of garment flammability based on tests using thermal mannequins dressed in everyday clothes.

GARMENT FIRES IN JAPAN

Of the 1,332 total deaths, except for suicidal arson, 130(9.8%) were caused by garment fires in 1978, according to the "Process to Death" in the White Paper on Fire Service (1). The death toll by garment fires accounts for 180(13.8%) of 1,301 deaths in 1979 and 141(11.4%) of 1,238 in 1980 (143(13.0%) of 1,096 in 1982 and 153(13.3%) of 1,152 in 1983). Of the 491 total deaths, except for suicidal arson, 50 are considered to have been caused by garment fires within the jurisdiction of the Tokyo Fire Department Agency during the five years from 1975 through 1979. This rate is almost equal to that in the White Paper on Fire Service (2). The rates consist of only the cases where clothes caught fire directly from a flame. If those cases of fires transferred by intermediaries are included, such as where a cigarette smoked in bed ignited bedding which then transferred to night clothes, or where leaked oil caught fire and spread to garments when an oil stove was accidentally overturned, the death toll by garment fires amounts to 128 out of 491 deaths.

Of those total deaths, people over 61 and children under 10 account for 50.6% and 12.6%, respectively. Girls under 5 account for a remarkably high percentage among children.

EXPERIMENTS

Experimental Apparatus (Thermal Mannequins)

The mannequins employed in this study were those on the market and manufactured by FRP. Their surface was covered with an asbestos and cement mixture with a thickness of 5 to 10 mm to provide them with heat resistance. Heat conductivity of asbestos-cement is 1.67kJ/m h deg at $90^{\circ}C$, and is close to 1.80 to 2.72kJ/m h deg, which is the heat conductivity of human skin obtained at 23 to $25^{\circ}C$. Three types of thermal mannequins were designed: one with a height of 175cm for a male, one with a height of 166cm for a female and another with a height of 113cm for a child. To measure heat flux, a 0.3 mm chromel-alumel thermocouple was welded onto the center of 15 mm diameter and 1 mm thick copper-plates. The surface of these sensors was coated with black heat resistant paint to increase heat absorption. Eight such thermoplates were installed on the mannequin and their positions are shown in Fig.1. These thermoplates were calibrated using a black body radiation surface. Conducting wires of the thermoplates were taken out from the wrists and heels of the mannequins.

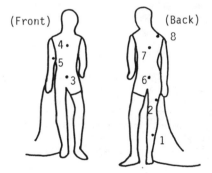

(Front)          (Back)

Fig. 1   A Thermal Mannequin (Male) and Position of Sensors

Samples

Generally, in burning tests the results are not always consistent even though the same samples are employed. Thus, tests are usually performed several times under the same conditions. However, since there are many types of garments, it is not possible to test all of them. This explain why relatively simple styled garments, such as nightclothes and A-line, one piece dresses, had mostly been used in past thermal mannequin tests (3-6,15). In the present tests, samples which cover as many types of garments as possible, ranging from nightclothes to suits including underwear, were selected and examined under the same conditions used every day. Detailed descriptions of the samples are omitted here due to the limitation of papers. However, care was taken to ensure that main fabrics, such as cotton, polyester, nylon, silk, wool and acrylic fiber, were covered.    The samples were brand-new and no special pretreatment was carried out.

Test Method and Items

The tests were carried out in a laboratory room with a floor space of 55m$^2$ and 7.85m high.    Air flow was generated by ventilating fans at a velocity of 0.2 to 0.3m/s at a position 1m above the floor from the left front to the right of a mannequin during the tests.    Ignition was made on contact with a diffused flame for 10 seconds using a Bunsen burner charged town gas (46,000kJ/m3) at a velocity of 325ml/min.    Ignition place were in principle: in trousers, the front bottom of the leg on which a sensor was installed; in skirts and kimonos, the front hem.    The burnings were recorded by VTR and photographed. Outputs of the thermoplates were also continuously recorded during the tests. The following were obtained from the temperature curve based on the output of the thermoplates:(1) the time required from contact with a flame to the beginning of combustion;(2) the time required from the beginning of combustion to the peak;(3) the peak temperature;(4) the maximum heat flux;(5) the time required from contact with a flame to causing a second degree burn.

Here, a second degree burn means a burn which injured depth is 100 micrometers or which has blisters and some broken skin. Henrique, Stoll, Greene, Chienta and others have the relation between a second degree burn and heat flux (7-11). The time required for a second degree burn was determined by measuring heat flux at various times.    The maximum heat flux was obtained from the output of the thermoplates that provided the sharpest slope during the time from the beginning of combustion to its peak.

RESULTS

Comparison of Flammability

Tests were conducted on 73 sets of general everyday garments, of which men's, ladies', boys' and girls' were numbered 18, 27, 14 and 14, respectively.    Of all the samples, 18 sets of garments were not ignited by a 10-second contact with a flame of a bunsen burner or the fire self-extinguished after a small part was burnt.    The remaining 55 sets were readily ignited and blaze up.

Flammability of Various Garments

   Yukata dresses (cotton kimonos for summer wear).    Nightclothes generally showed high flammability and yukata dresses in particular were highly combustible.    In a test using a female mannequin dressed in a pure cotton yukata and nylon underwear, flames reached near the face in 8 seconds after ignition, then flared up to 50cm over the head.    The yukata itself was almost totally consumed in a minute and 10 seconds.    At the back waist, the maximum temperature and the maximum heat flux were 300°C and 138kJ/m2 s, respectively; it took 21 seconds to produce a second degree burn.    At the jaw, the maximum temperature was 108°C and the maximum heat flux 25.1kJ/m2 s.    The time required for a second degree burn was 43 seconds.

   Pajamas.    A test of male mannequin wearing a cotton, short-sleeved pajama jacket and trousers, a cotton short-sleeved undershirt with a U shaped neck, and cotton briefs with a 15% polyester blend, demonstrated that flames extended to the face 20 seconds after ignition on contact with a gas flame. The garment almost burnt up in one minute and 40 seconds. Figure 2 shows the relationship between the time and the temperature and heat flux at measuring points. The maximum temperature of $149\,^{\circ}$C and the maximum heat flux of 110kJ/m2 s were obtained at the waist and hips, respectively. In the latter, it took 18 seconds to produce a second degree burn.

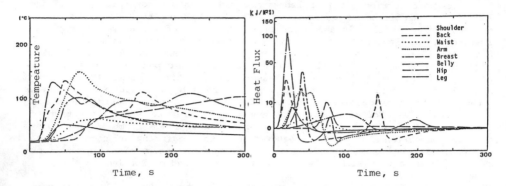

Fig. 2  Time Dependence of Temperature and Heat Flux (M1)

Negligees.  A test was conducted using a female mannequin with a cotton short-sleeved negligee, a nylon of a polyurethane mix and cotton panties. Flames reached the face 12 seconds after contact with a gas flame and the garment was consumed within one minute and 30 seconds. The maximum heat flux at the waist and that at the left arm were 139.7kJ/m2 s and 58.6kJ/m2 s, respectively.

Suits.  In a test of woolen suits, a small flame was raised by a 10-second contact with a gas flame on the bottom of the right leg. However, it extinguished immediately when the gas flame was removed. Then a 25-second contact with a flame was made on the back bottom of the jacket. The lininig (55% cuprammonium rayon and 45% nylon) caught fire and spread to the shirt and the combustion reached its peak one minute and 20 seconds later. Eight minutes were required until the suits were completely consumed. A similar situation occurred with polyester suits with 45% wool. They were ignited when touched with a flame on the front bottom of the jacket for 15 seconds. A fire extended to the face in one minute and 20 seconds afterward, and the combustion reached its peak 5 minutes and 30 seconds later. It took 16 minutes until most of the suites were consumed.

These two types of suits were relatively flame proof. They began burning only when the lining was ignited. However, cotton corduroy suits exhibited a rapid combustion owing to the fluffy surface. Flames reached the face in one minute and 30 seconds and it took 2 minutes until the burning reached its peak. The maximum temperature and the maximum heat flux were 208 C and 46.0kJ/m2 s, respectively, at the foot. It took 27 seconds to produce a second degree burn.

Jeans.  Cotton jeans were ignited after being touched with a flame for 10 seconds. Since they fit the body tightly, their burning rate is low; it took 5 minutes before the flame reached the femur. In two types of jeans, the fire self-extinguished at the femur. Although modes of combustion were similar, big flames occurred in the other type and a fire transferred to a polo shirt in one minute and 30 seconds,or an acrylic sweater in 2 minutes.

Combination of sweaters and skirts or trousers(including pantaloons).
It is difficult to generalize the flammability of these combinations because
their modes of combustion are greatly affected by combination.  For example,
the bottom portion may be dressed in flammable skirt or trousers, and if the
upper portion may be dressed in a sweater with high wool content it will not
catch fire easily.  On the contrary, the more acrylic fiber contained in the
garment, the more flammable it becomes.  An example is given by a female
mannequin wearing an acrylic sweater and an acrylic skirt with 25% polyester
and 5% wool.  In this test, flames reached the jaw 30 seconds after ignition,
where the maximum temperature of $344^{\circ}C$ and the maximum heat flux of 334.7kJ/m2
s were obtained.  It is considered that these high values were obtained because flames attacked the sensor directly.

Combination of shirts (blouses) and skirts or trousers.  In a test of a
female mannequin with a T-shirt and a skirt, the latter exhibited fierce
combustion.  The face was covered in flames 30 seconds later and the skirt was
burnt out in one minute and 30 seconds to 3 minutes.  On the contrary, the T-
shirt was not severely damaged.  In a combination of a blouse and a polyester
skirt, the latter was not ignited but melted in drops.  However, fierce combustion was observed in the case where underwear caught fire which spread to the
blouse.

Tests of boys' T-shirts and short pants presented low burning rates with small
damage to the T-shirts, regardless of the pants being cotton or polyester.A
comparison between a combination of a T-shirt with a skirt and that of a T-
shirt with short pants proved that the former was greatly hazardous.

One-Piece dresses.  Most of the one-piece dresses were readily ignited.
Flames covered the face 20 to 30 seconds after a gas flame touched the dress
and the garment burnt up about 2 minutes and 30 seconds later.  An example is
given by a test of a sleeveless, polyester, one-piece dress with a 35% cotton
mix.  The maximum temperature and the maximum heat flux at the abdomen were
$172^{\circ}C$ and 68.5kJ/m2 s, respectively.

Kimono dresses.  A woolen men's kimono ensemble (a kimono dress and a
short coat) with 15% silk and 5% nylon blended, was ignited after a 10-seconds
contact with a gas flame.  However, the fire extinguished 40 seconds later.
Contact with a gas flame was again made on the bottom of the right sleeve for
20 seconds.  The polyester lininig caught fire and flames leapt up to the face
one minute and 30 seconds later.  Two minutes and 30 seconds were required
until the combustion reached its peak and 9 minutes were required until most
of the clothes burnt out.  A maximum temperature of $212^{\circ}C$ and a maximum heat
flux of 20.1kJ/m2 s were obtained at the waist and the back, respectively.

A silk ladies' kimono was not ignited, although a flame was in contact for 10
seconds.  The fire extinguished burning a small part of the lining only.
Then, contact with a flame was made on the bottom of the sleeve for 20 seconds; however, the kimono did not ignite and only the ignition place was
carbonized.  Lastly, an obi (a sash belt, the outer side of which is polyester
and the lininig rayon) was ignited after a 10-second contact with a flame and
exhibited an extremely dull combustion.  A maximum temperature of $437^{\circ}C$ was
measured at the arm approximately 5 minutes 50 seconds later.  The maximum
heat flux of 25.1kJ/m2 s was observed at the abdomen.

Working clothes.  A polyester working garment with a 20% cotton blend was
ignited after being touched with a flame for 10 seconds.  Although the combus-

843

tion was slow, flames spread over the clothes. The maximum temperature was obtained at the back 7 minutes later and the maximum heat flux indicated 3.3kJ/m2 s. Another one, made of aromatic polyamide, did not exhibit a flame at all and only the part which was directly touched with a gas flame was carbonized although it was exposed to a gas flame for a long period.

Sports wear. Polyester sports wear was tested and when ignited, they melted fiercely, falling down in drops. The part which were burnt peeled off. Therefore, they exhibited a low apparent flame propagating rate and the temperature of the body did not increase. Therefore, good results were obtained from the tests. However, in one of them, a sudden rise in temperature was observed at the abdomen 3 minutes later and it exceeded 250$^{\circ}$C 4 minutes and 30 seconds later.

Coats. Three types of coats, a polyester one blended with 40% wool, a polyester one with 35% cotton and a 100% cotton one, were tested. In all tests, flames spread to the face one minute to one minute and 20 seconds after coming into contact with a gas flame. The combustion reached its peak one minute and 30 second later. However, in the case of coats, clothes put on under a coat, such as suits, greatly affected the combustion; the heat flux on the body surface became small because of the suits. Thus, a higher safety mark was given to the coats than one would assume from the appearance of combustion.

Comparison with past tests

In the present tests, the heat flux of 4 to 84kJ/m2 s is frequently obtained. A heat flux of 335kJ/m2 s, the maximum of all measured, is obtained once. On the other hand, the maximum temperature generally does not exceed 300 C, although the temperature of 437$^{\circ}$C is observed once.

Finley and others (3-5) recorded the maximum heat flux of 11kJ/m2 s and the maximum temperature of 204$^{\circ}$C in a test of a cotton A-shaped one-piece dress, in which the time taken to reach the peak of combustion was 37 seconds.

In the tests of flammability of yukata dresses designed by the Ministry of International Trade and Industry of Japan, the maximum heat flux was 4.2 to 13.4kJ/m2 s and it took about 35 to 55 seconds to obtain the highest value(12).

The temperature and time given by these two tests nearly coincide with those of the present tests. However, the heat flux seems to be a little lower in their tests.

On the contrary, in Ohya's experiments (13), the heat flux of 250kJ/m2 s was frequently observed, and that of 355kJ/m2 s was also reported. The time to the peak of combustion is 30 to 50 seconds which is a little longer than the value measured in the present tests. Hence, it can be said that the values shown in the present tests were proved to be appropriate.

Summary of Garment Flammability Tests

Flammable garments. The results of fire tests are summarized as follows:
(1) Clothes made of flammable materials are hazardous enough to make flames reach the face in 10 to 90 seconds. Heat flux was generally 4 to 84 kJ/m2 s but 335kJ/m2 s was observed as the maximum flux once.
(2) It is difficult to make a conclusion owing to various factors; however,

cotton, acrylic clothes, those of a mixture of cotton and polyester and those of a mixture of cotton and acryl are relatively flammable. Silk, woolen, nylon and aromatic polyamide clothes are relatively flame proof.
(3) In cotton clothes, thick textures, such as jeans, exhibit a very low burning rate. On the contrary, a mixture of cotton and polyester and that of cotton and acryl are relatively flammable regardless of their thickness.
(4) In clothes which are made of cotton and polyester and fit the body relatively well, the fibers contract and the clothes shrink to the surface of a mannequin when heated. A fire self-extinguished because of the shortage of air in most cases. It is difficult to classify this type of case because it appears to be flame proof. On the other hand, the heat contraction of fibers is small in clothes of a mixture of cotton and acryl, therefore, fire spreads.
(5) Generally, girls' wear is flammable in both materials and forms. More attention should be paid to this.

Nonflammable garments. As was mentioned above, 18 sets of all the garments were not ignited by the first contact with a gas flame for 10 seconds. The reasons of self-extinguishment of almost the garments are due to melt, drip or shrink of tex-tiles, or the fitness of clothes. After all, only the garments made of wool, silk, mixture of wool and nylon and aromatic polyamide were found to be substantially flame proof.

PROPOSAL OF A NEW TEST METHOD FOR THE CLASSIFICATION OF GARMENT FLAMMABILITY USING THERMAL MANNEQUINS

Classification Method

To classify garment flammability using thermal mannequins, three measurements, such as burning rate, size of degree of fire injuries were introduced.

Burning rate. After fire tests were performed, the mean times required to cause a second degree burn at various parts were calculated. These values were classified into four classifications as shown in the following. The percentages are also listed for each group for the 73 sets of garments.

| Classification | Type | Rate (%) |
|---|---|---|
| A | 180 seconds or more | 34 |
| B | from 120 to 179 seconds | 25 |
| C | from 60 to 119 seconds | 25 |
| D | 59 seconds or less | 16 |

Size of fire injuries. Classification was made by the total number of second degree burns as follows:

| Classification | Type | Rate (%) |
|---|---|---|
| A | 1 - 2 places | 12 |
| B | 3 - 4 places | 18 |
| C | 5 - 6 places | 44 |
| D | 7 - 8 places | 26 |

Degree of fire injuries. The mean maximum heat flux was calculated for portions where the maximum heat flux was larger than 2.9kJ/m2 s, designed as the degree of fire injury and classified as follows:

| Classification | Type | Rate(%) |
|---|---|---|
| A | 10kJ/m2 s or less | 12 |
| B | 10 - 20 kJ/m2 s | 44 |
| C | 21 - 30 kJ/m2 s | 22 |
| D | 30kJ/m2 s or more | 22 |

General classification. Using the above three evaluation standards, measurements, A, B, C and D were given 4, 3, 2 and 1 point, respectively, and the general classification of garment flammability was evaluated by counting the points. The results are follows:

| Classification | Type | Rate(%) |
|---|---|---|
| A | 11 - 12 points | 11 |
| B | 8 - 10 points | 34 |
| C | 5 - 7 points | 41 |
| D | 3 - 4 points | 14 |

Discussion on the General Classification

Classification results are shown in Table 1. Judging from the general classification, working cloth made of fire resistant fiber provided good results without doubt. Classified into class A are a combination of a woolen sweater and trousers, that of a woolen jumper and trousers with a nylon blend, polyester sports wear, woolen suits, suits of wool and nylon, a combination of a cotton T-shirt and trousers blended with polyester. These clothes are flame proof in material and fit the body relatively well. Good results were obtained in polyester sports wear because they melted fiercely and fell in drops when ignited. However, they should be re-examined.

Clothes which come under class B in the general classification are a combination of a T-shirt and trousers, that of a T-shirt and jeans, kimono dresses and a combination of a shirt and trousers, arranged according to superiority in other classifications. They are relatively difficult to burn except for kimono dresses. Therefore, good results were yielded although many of them were cotton. Kimono dresses made of wool or silk were difficult to burn. Listed in a lower rank of class B are two-piece dresses, a combination of a polo shirt and pantaloons, that of a sweater and a skirt and cotton suits. Their materials are flammable and many of them are ladies' wear.

Coupled within a high rank of class C are a combination of a shirt and trousers, pajamas and one-piece dresses. Those listed in a lower rank are a combination of a T-shirt and a skirt, one-piece dresses, negligees and so on. Most of them are ladies' wear and some girls' wear is included.

Those classified into class D, the lowest group, are all kinds of nightclothes such as yukata dresses and pajamas, sports wear and a combination of a T-shirt and skirt. This classification evidently shows how flammable nightclothes are.

The results of the general classification obtained after those three evaluation standards are completely identical with experimental results. Besides, points are given according to the level of safety, and it is clear that the proposed method allows a useful and quantitative evaluations of the safety of garments from fire.

Table 1    Classification of Garments According to the Proposed Method

| No *1 | Garment | Composition *2 | Weight (g) | Burning rate | Size of burn | Degree of burn | General *3 | No. of contact with a flame |
|---|---|---|---|---|---|---|---|---|
| B 1 | T-shirt & trousers | CT70,PE30/CT100 | 148 | A | B 4 | A 6 | A | |
| B 2 | T-shirt & trousers | CT100/CT95,PU5 | 173 | B | A 2 | A 3 | A | |
| B 3 | T-shirt & trousers | PE65,CT35/PE50CT50 | 220 | B | C 1 | B 7 | B | |
| B 4 | T-shirt & trousers | CT100/PE80,CT20 | 216 | A | C 5 | B 5 | B | |
| B 5 | T-shirt & trousers | PE65,CT35/PE100 | 154 | A | B 4 | B 4 | B | |
| B 6 | Yukata dress | CT100 | 186 | C | D 8 | B 8 | C | |
| B 7 | Yukata dress | CT100 | 298 | D | D 7 | D 7 | D | |
| B 8 | Pajama(half sleeve) | CT100 | 143 | B | D 7 | B 8 | C | |
| B 9 | Pajama(half sleeve) | CT100 | 143 | B | D 8 | B 8 | C | |
| B 10 | Yukata dress | CT100 | 204 | D | D 7 | D 8 | D | |
| B 11 | Pajama(half sleeve) | CT100 | 160 | B | D 8 | B 8 | C | |
| B 12 | Training wear | PE100 | 461 | B | A 1 | B 1 | B | 2 |
| B 13 | Sweater & jeans | AN80,WL20/CT80PE15 | 235 | A | C 6 | B 7 | B | |
| B 14 | Sweater & trousers | WL100/CT100 | 162 | A | A 1 | A 2 | A | |
| G 1 | T-shirt & skirt | CT100/PE65,CT35 | 138 | B | B 4 | B 7 | C | |
| G 2 | T-shirt & skirt | CT100/PE65,CT35 | 224 | C | D 8 | C 8 | C | |
| G 3 | T-shirt & skirt | CT50,PE50/CT100 | 135 | D | C 6 | C 6 | C | |
| G 4 | T-shirt & skirt | CT50,PE50/CT100 | 146 | C | D 7 | D 7 | D | |
| G 5 | Yukata dress | CT100 | 286 | D | C 6 | D 7 | D | |
| G 6 | Pajama(half sleeve) | CT100 | 133 | D | D 8 | D 8 | D | |
| G 7 | Pajama(half sleeve) | CT75,PE25 | 179 | C | D 8 | C 8 | C | |
| G 8 | One-piece dress | CT80,PE20 | 130 | C | C 5 | B 6 | C | |
| G 9 | One-piece dress | CT100 | 148 | C | C 6 | B 6 | C | |
| G 10 | One-piece dress | CT50,PE50 | 70 | C | A 1 | D 1 | C | |
| G 11 | One-piece dress | PE100 | 150 | C | C 5 | B 5 | C | Drip |
| G 12 | Sweater & skirt | AN70,WL30/PE65RY35 | 161 | C | B 3 | C 2 | C | |
| G 13 | Sweater & skirt | AN80,PE20/CT100 | 175 | B | C 6 | B 7 | B | |
| G 14 | Sweater & skirt | AN100/AN75PE20WL5 | 100 | C | C 6 | D 6 | C | 2 |
| L 1 | One-piece dress | PE100 | 185 | A | C 5 | B 6 | B | Drip |
| L 2 | Negligee | CT100 | 140 | D | C 5 | D 7 | D | |
| L 3 | Pajama | CT100 | 170 | C | D 7 | C 8 | C | |
| L 4 | Negligee,half sleeve | CT80,PE20 | 171 | C | C 6 | C 7 | C | |
| L 5 | Blouse & skirt | PE65,CT35/PE100 | 216 | A | C 6 | C 6 | B | 2 Drip |
| L 6 | One-piece dress | CT100 | 253 | D | C 5 | B 5 | C | |
| L 7 | One piece dress | PE65,CT35 | 274 | C | C 6 | C 6 | C | |
| L 8 | Blouse & skirt | PE100/PE100 | 377 | B | B 3 | D 3 | B | 3 Drip |
| L 9 | Polo & Pantaloon | CT50,PE50/CT50PE50 | 492 | A | C 5 | B 6 | B | |
| L 10 | Blouse & skirt | CT80,RY20/PE100 | 419 | A | C 6 | B 6 | B | |
| L 11 | Pajama(half sleeve) | CT70,PE30 | 217 | B | C 6 | C 6 | C | 2 |
| L 12 | Cardigan & pantaloon | WL100/CT50,PE50 | 256 | B | B 4 | C 5 | B | |
| L 13 | Polo & pantalon | PE52,CT48/CT50PE50 | 529 | B | C 6 | B 7 | B | |
| L 14 | T-shirt & jeans | CT50,PE50/CT100 | 539 | A | A 1 | D 1 | B | 2 |
| L 15 | Yukata dress | CT100 | 325 | C | C 5 | B 7 | C | |
| L 16 | T-shirts & jeans | CT100/CT100 | 577 | A | B 3 | B 4 | B | |
| L 17 | Yukata dress | PE65,CT35 | 495 | C | D 7 | C 7 | C | |
| L 18 | Polo & jeans | PE65,CT35/CT100 | 600 | A | C 5 | C 6 | B | 2 |
| L 19 | Negligee,half sleeve | WL100 | 175 | D | A 1 | D 2 | C | 2 |
| L 20 | Training wear | PE100 | 681 | A | A 2 | B 5 | A | 2 |
| L 21 | Pajama(half sleeve) | CT70,PE30 | 217 | B | C 5 | C 6 | C | |
| L 22 | One-piece dress | AN94,WL6 | 712 | B | D 7 | B 7 | C | |
| L 23 | Two-piece dress | WL90,NL10 | 838 | C | B 3 | B 4 | B | 4 |
| L 24 | Sweater & skirt | AN90,WL10/WL70AN20 | 446 | B | C 6 | C 6 | C | |
| L 25 | Coat | PE65,CT35 | 340 | A | C 5 | B 5 | B | |
| L 26 | Coat | CT100 | 745 | A | C 5 | B 5 | B | 3 |
| L 27 | Kimono dress | SILK100 | 727 | A | B 4 | B 5 | B | |
| M 1 | Pajama(half sleeve) | CT100 | 316 | D | C 6 | C 8 | D | |
| M 2 | Pajama | CT70,PE30 | 475 | B | D 7 | B 7 | C | |
| M 3 | Pajama | CT70,AN30 | 453 | C | D 7 | A 8 | C | |
| M 4 | Training wear | CT100/PE65,CT35 | 785 | B | D 8 | D 8 | D | |
| M 5 | T-shirt & trousers | PE65,CT35/CT100 | 700 | A | C 5 | D 6 | C | |
| M 6 | Bath robe | CT100 | 330 | D | D 7 | C 7 | D | |
| M 7 | Yukata dress | CT100 | 390 | D | C 6 | D 6 | D | |
| M 8 | Yukata dress | PE65,CT35 | 480 | D | D 7 | D 7 | D | |
| M 9 | Jumper & trousers | NL100/PE95,CT5 | 600 | A | D 7 | B 7 | B | 3 |
| M 10 | Suits | WL100 | 1200 | A | B 3 | A 7 | A | 3 |
| M 11 | Jumper & trousers | AN90,PE10/PE65RY35 | 993 | A | B 4 | D 4 | B | 2 |
| M 12 | Suits | PE55,WL45 | 737 | A | B 4 | A 4 | A | 3 |
| M 13 | Working cloth | PE65,CT35 | 826 | A | C 6 | A 7 | B | |
| M 14 | Kimono dress | WL100 | 1400 | A | D 7 | B 7 | B | 2 Drip |
| M 15 | Suits | CT100 | 1400 | B | C 6 | B 7 | B | |
| M 16 | Jumper & trousers | NL100/WL100 | 924 | A | A 2 | B 2 | A | 3 |
| M 17 | Coat | PE60,WL40 | 740 | A | C 5 | A 4 | B | |
| M 18 | Working cloth | AA80,NL20 | 900 | A | A 0 | A 0 | A | |

Note:
*1  B:Boy, G:Girl, L:Lady and M:Man
*2  CT:Cotton, WL:Wool, PE:Polyester, AN:Polyacrylonitrile, NL:Nylon
    PU:Polyurethane, RY:Rayon, AA:Aromatic polyamide
*3  Numerical value indicates the number of thermoplates counted.

CONCLUSION

The fire tests using thermal mannequins dressed in 73 sets of everyday garments showed that all of them, except for 18 sets, exhibited fierce combustion for 10-second contact with a gas flame and also indicated a possibility of causing serious fire injuries to the body within a short time. On the other hand, some clothes were proved to be fire proofed and these were classified into three groups: one group is that the material itself is fire resistant, such as aromatic polyamide, wool and silk; another is that the fiber melts and falls in drops, such as polyester and nylon; and the third group is that the material is thick or fits the body like jeans, though the material itself is flammable, such as cotton. Standards of classification, such as burning rate, size and degree of fire injuries and a combination of these factors were proposed to be viable.

ACKNOWLEDGEMENT

We are deeply grateful to Mr. Hideo Terasaki, Chief of the Research Division of the Japan Fire Retardant Association for a great amount of information for this study. We also extend our thanks to Mr. Akira Kubota, a student of this department, for his assistance in our tests, Messrs. Kazuhiko Tamura and Hiroshi Morijiri, research students, Mr. Gouichiro Kuwaki and Miss Kaoru Shimogawara, students of this department, for helping with data processing.

REFERENCES

1. Fire Defence Agency of Japan, Annual White Paper on Fire Service
2. Data from the Tokyo Fire Department Agency (1980)
3. Finley, E.L.: "Ignition System used as a Testing Technique for Studying Garment Flammability," J. Fire and Flammability, 1: 166-174 (1970)
4. Finley, E. L. and Carter, W. H.: "Temperature and Heat-Flux Measurements on Life-Size Garments ignited by Flame Contact," ibid., 2: 298-320 (1971)
5. Finley, E. L. and Butts, C. T.: "Garment Conformation on a Mannequin changes Flammable Characteristics," ibid., 4: 145-155 (1973)
6. Chouinard, M. P., Knodel, D. C. and Arnold, H. W.: "Heat Transfer from Flammable Fabrics," Textile Research Journal, 431: March 1973
7. Seaman, R. E.: Bull. N.Y. Acad. Med., 43: 8, 649 (1967)
8. Henriques, F. C., Jr.: Archives of Pathology, 43: 489-502 (1947)
9. Stoll, A. M. and Greene, L. C.: J. Applied Physiology, 14: 373-382 (1959)
10. Stoll, A. M. and Chianta, M. A.: "Burn Production and Prevention in Convective and Radiant Heat Transfer," Aerospace Med., 39: 1097-1100 (1968)
11. Moritz, A. R. and Henriques, F. C.,Jr.: American Journal of Pathology, 23: 695 (1947)
12. Ministry of International Trade and Industry: "On the Results of Tests of Flammability of Children's Yukata Dresses" (1980)
13. Ohya, H.: Garment Flammability and Heat Injury, 60, Maruzen, Tokyo (1983)
14. Braun, E., Cobble, V. B., Helzer, S., Krasny, J. F., Peacock, R. D. and Stratton, A. K.: "Back-Up Report for the Proposed Standard for the Flammability of General Wearing Apparel," NBSIR 76-1072, National Bureau of Standards, Washington, D. C. (1976)
15. Cooperative Industry Program on General Apparel Flammability (1977)

848

# Burning Rate of Upholstered Chairs in the Center, alongside a Wall and in the Corner of a Compartment

T. MIZUNO and K. KAWAGOE
Center for Fire Science and Technology
Science University of Tokyo
2641, Yamasaki, Noda-shi
Chiba-ken, 278, Japan

ABSTRACT

This paper describes three series of fire tests of an upholstered chair for the investigation of their burning behavior and for the comparison of results between the three series, which were (1) C-series in which the test specimen was located at the center of a room, (2) W-series in which it was located alongside a wall, and (3) K-series in which it was located in the corner of walls. Measurements were made of the following items: mass burning rate, radiative heat transfer to the surroundings, heat transfer to the wall, optical smoke density, etc. In comparing the three series, no obvious difference between mass burning rates could be seen. This tendency were verified by the burning of wood cribs and urethane foams. In this paper only the comparison of mass burning rate is described.

keywords: burning rate, upholstered chair, full-scale burning test, burning position

## 1. INTRODUCTION

A study of the burning behavior of furniture in pre-flashover period was presented by Mizuno and Kawagoe in Ref.[1], in which the test item was located at the center of a test room. An engineering model of burning rate for upholstered chairs was proposed, too. Usually, furniture is placed not only in the center of a room but also alongside a wall or in the corner of the room. Saito[2] pointed out from the result of kerosene pool fire that the wall had the effects on the burning item within the distance of the wall two times burning source diameter. Walls and corners may have little effect unless the flames touch to the walls. In Saito's result, when kerosene pan was placed close to the walls a flame touched to the walls, and a mass burning rate was larger than free burning. In the case of upholstered chairs, the chair can be roughly divided two parts of the seat cushion and the seat back. The seat cushion would not be placed close to a wall but the seat back can. Therefore, the presence of the wall would influence the burning rate, to say the least of it, of the seat back. However, it is not clear what differences of burning behavior, specially, burning rate, would be found due to the position of furniture in a room, because there have been a little studies to investigate the burning behavior of furniture located alongside a wall or in the corner.

In our earlier study[1], fire tests of wood cribs and chairs were conducted at the center of a room and the burning rate, smoke generation and radiant heat flux from flame were reported. Also it was pointed out that the mass burning rate is one of the key factors which describe the burning behavior. Then an engineering model for the mass burning rate was proposed for furniture on the basis of the results of wood crib tests.

Fig.1 shows the mass burning rate curve of a wood crib in Ref.[1]. It reveals that a peak value appears at a few minutes after ignition, and it gradually decreases after the peak. It can be assumed that the burning rate after the peak is proportional to the residual weight. Rockett, also, points out in Ref.[3] that the burning rate of urethane foam after the peak is proportional to the residual weight using the test results by Babrauskas[4]. The assumption is expressed by the following equation;

$$\dot{m}=-dm/dt=Am \qquad (1)$$

where $\dot{m}$ is the mass burning rate,
   $m$ is the residual weight, and
   $A$ is the burning coefficient.

Then, integration of equation (1) gives

$$m=C\cdot\exp(-At) \qquad (2)$$

where $C$ is the constant of integration. Let Tp denote the time to reach the peak mass burning rate and r denote the rate of residual weight to the initial weight, mo, at the peak, then equation (2) becomes

$$m=r\cdot mo\cdot\exp(-A(t-Tp)) \qquad (t>Tp) \qquad (3)$$
$$RW=m/mo=r\cdot\exp(-A(t-Tp)) \qquad (4)$$

Therefore equation (1) becomes

$$\dot{m}=A\cdot r\cdot mo\cdot\exp(-A(t-Tp)) \qquad (t>Tp) \qquad (5)$$

Equations (3) and (5) include four characteristic burning parameters of r, A, Tp and mo. Equation (4) can be rewritten to;

$$\ln(RW)=-At+\ln(r)+A\cdot Tp$$

By plotting a graph of RW vs. time, t, on a semi-log graph, the line of RW becomes a straight line and its slope is equivalent to

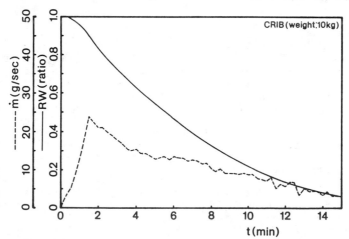

Fig.1 Burning Rate and Residual Weight Ratio of Wood Crib

the value of A which indicates the intensity to burn.
    In the case of chairs, by defining the modified initial
weight, $mo^*$, as the difference between the initial and final
chair weight, and the modified residual weight, $m^*$, as the
difference between the residual weight and the final chair
weight, and adopting the modified residual weight ratio, $RW^*$,
instead of RW, it may be possible to apply the model described
before for upholstered chairs.

## 2. EXPERIMENT AND RESULTS

    Three series of burning tests of upholstered chairs, wood
cribs and urethane foams were conducted in a test room which has
a volume of about 175 $m^3$, floor dimensions of approximately 6m by
6m, a height of 4.85m, and one doorway opening of 0.9m width by
2.0m height. The three series were (1) C-series: the test
specimen was located at the center of a room, (2) W-series:
alongside a wall, and (3) K-series: in the corner of walls. The
wall or corner erected in the test room was made of ALC (
Autoclaved Lightweight Concrete ) boards of 120 mm thickness. The
ceiling height of the test room was higher than a typical
residential room, but this would not give any effect on the
burning rate[6][7].
    The ignition sources for the chairs and the wood cribs
consisted of the two chips of insulating fiber-board (2x20x1 cm)
soaked with about 100 cc of methanol which were placed on the
seat of the chairs or under the crib, and for the urethane foams
the ignition source was a Methenamine pill placed on the surface
center of the foam. The test item was placed on a load-platform
and their mass loss was measured with a set of load-cells at 15
second intervals. In the case of upholstered chairs, since foams
and fabric generally burn first and the wood frame very slow,
these tests were stopped when the remaining foam and fabric
became very small.

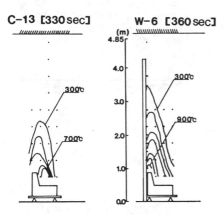

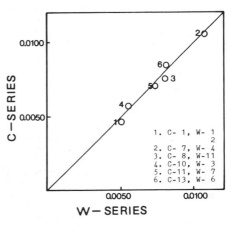

Fig.2 Typical Temperature Contour
Map of Burning Chairs ( 100°C of
interval of lines )

Fig.3 Comparison of Parameter A
Between C and W series

## 2.1 Fire Tests of Chairs: Part 1.  C-series and W-series

Upholstered chairs of 46 in total [5] were burned at two positions in the room, and the following items were measured; flame height, mass burning rate, air temperature over the chair, room temperature, smoke density, etc.  The most of chairs tested were used items including many kinds and shapes.  In this study picking out the results of six sets of the same chairs on materials and chair shape, the comparison of mass burning rate between C-series and W-series is made.

Fig.2 shows the typical temperature contour map of the same type of chairs, C-13 and W-6, at the time of the peak of burning. The mass burning rates of these chairs at the peak were the same value of about 25 g/sec.  In the case of wall side test, W-6, it reveals high temperature near the wall surface.  Comparing 300°C lines shows that the flame in the case of wall side test leans against to the wall.  However, Fig.3 shows that there is no obvious difference in the parameter A in equation (3) and (5). It was found that, when the upholstered chairs burn individually in the pre-flashover period, the burning rate may be assumed to be the same value whether the position of burning item is in the center of a room or alongside a wall.  But, based on these results alone, it was difficult to affirm that.

## 2.2 Fire Tests of Chairs: Part 2. C, W and K-series

To explore further the former experimental results, two sets of 3 upholstered chairs of the same materials and the same shape were purchased and burned each in the center of the test room, along the side of a wall erected in the test room, and in a corner erected in the test room.  The test specimens were commercial goods and very cheap ones, so they were consisted simply of only urethane foam and a cover fabric supported by a wood frame.  The chairs of type A, CC-1, WC-1 and KC-7, were about 8.9kg in weight.  The chairs of type B, CC-3, WC-2 and KC-9, were about 7.7kg.  Materials and shapes of type A and type B chairs were almost the same kind of ones.  CC-1 and CC-3 were tested in the center of the room, WC-1 and WC-2 along side the wall, and KC-7 and KC-9 in the corner.

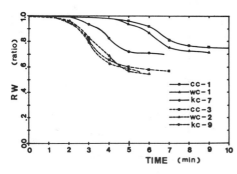

Fig.4 Residual Weight Ratios of Upholstered Chairs

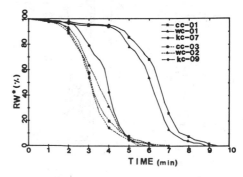

Fig.5 Modified Residual Weight Ratios of Upholstered Chairs

Fig.4 shows the residual weight ratios of the burning chairs. These upholstered chairs burned very fast, much as the urethane foams because they were simple ones. The upholstered chairs described in section 2.1 burned more slowly than in this section because they consisted not only of a filling foam and a fabric cover but also an interlining, a suspension and/or a cotton batting, which slowed the burning rate of the chair.

The residual weight ratios of the chairs, when the tests were stopped, were almost the same value for each type. In the case of type A the ratios were about 70 % and in the case of type B they were about 60%. The residual weights were nearly equal to the weights of frames; the upholstered chairs leave substantial residual mass due to unburnt parts such as the wood frame after the active burning period. Fig.5 shows the modified residual weight ratios of the burning chairs. In the case of Fig.4, the times when the curves begin to drop are somewhat scattered, but the slopes of downward curves are roughly the same. Fig.5 clearly shows this point.

The burning rate ratios of the chairs are shown in Fig.6. The plotting curves of the burning rate of chairs which burn suddenly, as shown in Fig.6, indicate an evident peak. Comparing Fig.5 and Fig.6 shows that residual weight ratios at the peak are about 40 %. This tends to be a smaller value than found in section 2.1. It was observed that the flame of the seat back touched to the walls in the case of the corner tests, and in the case of type A chairs the peak value of the burning rate of KC-7 is larger than in the other chairs. So, in the corner test there might be a bit effect of the wall. However, the overall burning rate ratios within the three burning locations show no obvious differences. Thus, when the upholstered chairs burn individually

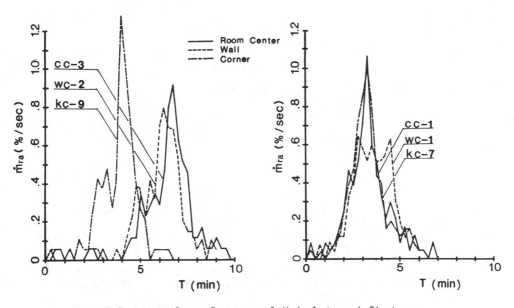

Fig.6 Burning Rate Ratios of Upholstered Chairs

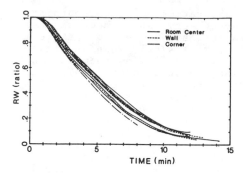

Fig.7 Residual Weight Ratios
of Wood Cribs

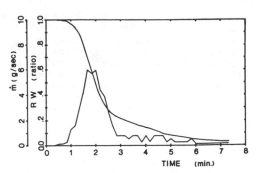

Fig.8 Typical Residual Weight
Ratio and Burning Rate of
Urethane Foam

the burning rate may be assumed, from an engineering point of
view, to be the same value whether the position of burning item
is at the center of a room, along the side of a wall, or in the
corner of a room.


## 2.3 Verification Tests by Wood Cribs and Urethane Foams

In the second step, the burning tests using wood cribs and
urethane foams were carried out to verify the results obtained
from the burning of upholstered chairs.
The test setup for the wood cribs was the same as for the
upholstered chairs. The wood cribs consisted of sticks of
Japanese cedar, 4.5 cm by 3.5 cm with a length of 45 cm. The
number of sticks per layer was 4 and the number of layers 10;
total weight of a wood crib was about 9 kg.
Wood cribs burn very smoothly. Fig.1 reveals a peak value at
a few minutes after ignition, followed by a gradually decreases
after reaching the peak. The time to reach the peak mass burning
rate was about 120 seconds after ignition. In the corner tests
it was observed that a flame touched to walls, and in the wall
side tests it was not obvious. Thus, there might be a little
effect of walls on the burning rate in the corner test.
Fig.7 shows the residual weight ratios of the wood cribs.
The curves in Fig.7 are closely coincident, and they indicate no
obvious differences between the three burning positions. It could
be concluded that walls have little effect on the burning rate.
Burning tests of urethane foams were, also, carried out with
the same setup as for the upholstered chair tests. The urethane
foam slabs had dimensions of 52 cm by 53 cm, thickness of 6 cm,
and weight of approximately 250 g. The density of the urethane
foam was about 0.015 $g/cm^3$. Two types of tests, single slab tests
and double slab tests, were conducted. In the double slab tests,
slabs were piled up, so the burning area was the same within the
two types of tests.
Fig.8 shows the typical residual weight ratio and mass
burning rate of the urethane foam in the corner of the test room.
The curves of the burning rate as shown in Fig.8 indicate a
shape like a mountain, similar to that of the upholstered chairs.

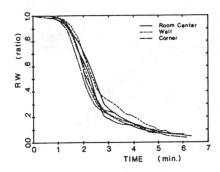

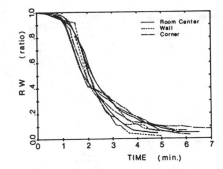

Fig.9 Residual Weight Ratio of
Urethane Foams (one slab)

Fig.10 Residual Weight Ratio of
Urethane Foams (two slab)

In the corner tests it was observed, as same as in wood cribs
tests, that a flame touched to walls, and in the wall side tests
it was not obvious. Fig.9 and Fig.10 show the residual weight
ratios of urethane foam. The curves of the residual weight
ratio are closely coincident but in the case of double slab the
slope of the curves of the corner tests may be a few greater than
the others. However, in the roughly speaking, the curves of the
residual weight ratio in Fig.9 and Fig.10 indicate no obvious
difference between the three burning positions. Therefore, it
could be also concluded that walls have little effect on the
burning rate.

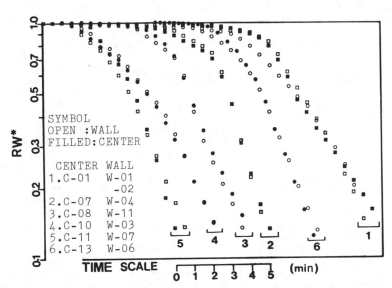

Fig.11 Comparison of Modified Residual Weight Ratio

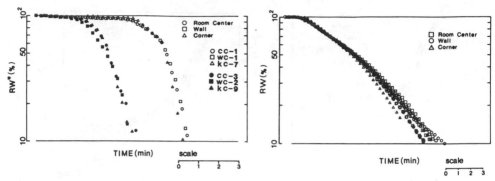

Fig.12 Modified Residual Weight
Ratios of Chairs in Semi-log

Fig.13 Residual Weight Ratios
of Wood Cribs in Semi-log

## 3. DISCUSSION

The burning rate of a fire source in a room is one of the
key factors which describe the burning behavior. The radiant
heat flux, the flame height, the smoke generation, etc. can be
considered as a function of the mass burning rate. The
engineering model of the mass burning rate for upholstered chairs
proposed in Ref.[1] and Ref.[5] is given by equation (5). Then
the residual weight ratio, RW, is expressed by the following
equation:

$RW=m/mo=r \cdot \exp(-A(t-Tp))$

Therefore, by plotting RW and t on a semi-log graph, the line of
RW becomes straight line.

Fig.11 shows the modified residual weight ratios of the
burning chairs of section 2.1 in the semi-log graph. The plots of
six sets of chairs are coincident and give essentially straight
lines after the peak. Fig.12 shows the modified residual weight
ratios of the burning chairs of section 2.2 in the semi-log
graph. Also, the plots of both type of chairs for the these
plotting are coincident and, after the peak, give essentially
straight lines.

Fig.13 shows the residual weight ratios of the wood cribs in
the semi-log graph, and, the plots of the residual weight ratios
of wood cribs are similar for the three burning locations. And
after the peak they may coincide in almost straight lines.
Fig.14 and Fig.15 show the residual weight ratio of urethane foam
in the semi-log graph. There is some scatter in the latter
period, but it is possible to neglect this; the plots may give
essentially straight lines. There is no obvious difference in
the burning rate whether the burning location of a burning item
in a room is the room center or not.

Walls and corners may have little effect unless the flames
touch to the walls. In the case of chairs, it was observed that
the flame of the seat back touched to the walls in the wall side
and the corner tests. And in the corner test of wood cribs and
urethane foams the flame touched to the walls.

When burning items are located closely along the side of a
wall or in the corner of a room their burning rate must be
affected somewhat by the adjacent wall. For example, the area

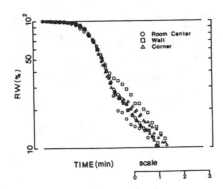

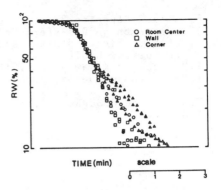

Fig.14 Residual Weight Ratios
of Urethane Foams (one slab)

Fig.15 Residual Weight Ratios
of Urethane Foams (two slab)

for air supply to the burning region decreases, and the diffusion
of the heat is disturbed by a wall. The decomposition of charred
materials may be influenced mainly by the radiation inside
materials. Any way, as the burning rate ought to depend on many
factors, it is not obvious how the burning rate depends on any
factors.

Strictly speaking, the results of the upholstered chairs,
wood cribs and urethane foams show, that the burning rate ratio
in the case of the corner test may be a bit larger than in the
case of the room center test, and that in the case of wall side
test, it is not clear whether the burning rate ratio is larger or
not. Thus, walls may have a bit effect in the case of the corner.
But, from an engineering point of view, one could assume that the
burning rate values of upholstered chairs are the same value
whether the position of the burning item is at the center of a
room, along the side of a wall, or in the corner of walls.

## 4. CONCLUSION

It was verified that the burning rate of upholstered chairs,
when they burn individually in the preflashover period, is
assumed to be the same value whether the position of burning item
is in the center of a room or not. On the basis of the present
and former[1] studies, equation (5), $\dot{m}=A \cdot r \cdot mo \cdot \exp(-A(t-Tp))$,
could estimate the burning rate of upholstered chairs of which
burning position is in the center of a room or not.

## 5. ACKNOWLEDGEMENT

The study was basically supported by Science and Technology
Agency of Japan. The author wishes to thank undergraduate
students of the Science University of Tokyo for their assistance
in the experiments.

## REFERENCES

1. T.Mizuno and K.Kawagoe, Fire Science and Technology, Vol.4,
   No.1, 1984.

2.  F.Saito, "Burning Behavior of wooden materials ( in Japanese )", Research on Disaster, Vol.1, 1980
3.  J.Rockett, Fire and Materials, Vol.6, No.2, 1982
4.  V.Babrauskas, NBSIR 77-1290, 1977
5.  T.Mizuno and K.Kawagoe, INTERFLAM'85., Guildford,1985
6.  T.Handa and O.Sugawa, J. of Fire and Flammability, Vol.12, 1981
7.  T.Handa and O.Sugawa, J. of Fire and Flammability, Vol.13, 1982

# Cable-Fire Tests under the Raised Floor of Data-Processing Installations

**KLAUS MARTIN**
Allianz Versicherungs-AG
Munich, West Germany

ABSTRACT

Fire tests were performed for various cable types and layouts under practical conditions. Surveys were carried out in a number of data-processing centers to determine a cable layout that meets the requirements of data-processing installations. The following tests were carried out: Flaming of cables outside the raised floor, flaming of cables under the raised floor and flaming of a cable support with cables under the raised floor. One litre of n-hexane, which was electrically ignited, served as ignition source. Thermocouples were used to measure the temperature course at many measuring points and the course of the fire was documented photographically. When the cables were flamed outside the space under the raised floor, the fire did not spread. When the cables were flamed under the raised floor, fire damage to the cables were more extensive, the cables spread the fire from the space under the raised floor into the test room. When only coaxial cables with improved fire-resistant characteristics were used, the extent of the damage was restricted to the area of the ignition vessel. The conclusions are briefly discussed with respect to fire protection measures.

INTRODUCTION

Automatic data-processing is more and more becoming part and parcel of the every-day routine in a great many areas of our economy. Company data is entered into electronic data-processing systems and controlled by these systems; virtually all departments take advantage of data-processing. Consequently, the question pertaining to the prevention of damage to data-processing systems, as well as that with regard to system availability have become more important than ever.

The most significant hazards that threaten data-processing installations are unusual environmental hazards resulting from water, heat, smoke and corrosive gases that may occur as a result of accidents, fires or other potential sources of damage. Protecting DP installations from damage caused by water is primarily a matter of taking the necessary precautions during the building phase of the premises. Such precautions, as well as technical protective devices help to reduce or prevent damage in case of fire. A great many DP installations are nowadays protected by burglar alarm systems and highly sophisticated access-control systems prevent unauthorized persons from entering certain areas.

Data-processing installations with their costly equipment represent an extraordinary risk both to the company that owns them and to the insurer. Even if the damage to equipment is only slight, corrosive fumes may result in extensive damage when they lead to interruption of work.


OBJECTIVES

In general, the behaviour of cables in raised floors of data-processing installations is not quite clear when it comes to a fire; consequently, the same is true with regard to fire-protection measures in respect of such floors. Despite the fact that this matter has been discussed for many years, it has been impossible to find satisfactory answers to the various questions, mainly due to the lack of fire tests. Furthermore, two conflicting opinions prevail:

-   The raised floor must be protected by a smoke detector and equipped with automatic fire extinguishers as the damage under the raised floor would otherwise be considerable of a fire were to break out.
-   If special fire-resistant cables are used, neither a smoke detector, nor automatic fire extinguishers are required since the fire cannot spread via the cables.

Fire tests carried out under realistic conditions in the fire-test room of Allianz insurance company served the purpose of clarifying the existing uncertainties. Results from such tests may help to come to a better assessment of the risk that is run by departments whose contribution to the overall success of a company is vital.


CABLE CONFIGURATION AND LAYOUT

In order to determine a cable layout that meets the requirements of data-processing installations, surveys were carried out in a number of data-processing centers. They were selected from various areas, such as business services, administration, research, production control, etc. Apart from the configuration, the nature of the cables, their cross sections and layout under the raised floor was also of importance since this can have a decisive effect on their behavior in case of fire. A further distinction was made between system-internal cables, having improved fire-resistant characteristics, and conventional commercially available cables. Altogether some 20 surveys of existing DP installations were made. The surveys revealed that cables under raised floors are not normally grouped in bundles in accordance with their cross sections. Invariably combinations of different types of cable and different cross sections were found, with heavy concentrations of cables in certain areas and virtually no cables in others.

The surveys revealed the following average configuration:

75 % system-internal cables
25 % commercially available cables

The outer diameters of the system-internal cables ranged from 10 mm to 40 mm and those of the commercially-available cables from 0.5 mm to 20 mm.


SETUP OF THE TEST ENVIRONMENT

Raised floor

A raised floor measuring 4.2 m x 4.2 m (size of the floor panels 60 cm x 60 cm) was installed in the test room in such a way that it was enclosed at two sides by two of the walls of the room. The two open sides were closed off by means of plasterboard. A plexiglass window in the plasterboard permitted observation of the fire under the raised floor. The clearance under the raised floor was 500 mm.

The floor panels rested on steel uprights with pressurecast aluminum supporting platforms. The floor panels themselves were made of highly-compressed chipboard with a zinc-plated steel lining glued to the bottom and carpet at the top. Two of the floor panels showed a cutout measuring 300 x 150 mm through which the cables were run. 30 % of the floor panels were ventilation panels (figure 1).

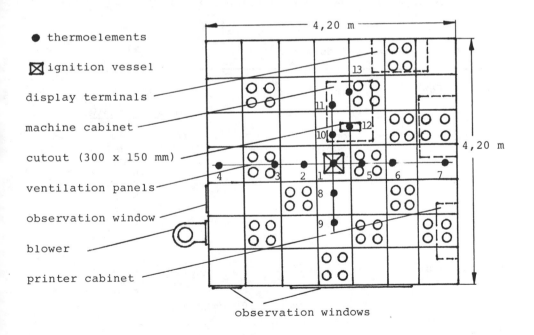

figure 1: Setup of the test environment

Two display terminals and the cabinet of a printer were placed on the raised floor in order to be able to determine the impact of chloride contained in the fumes. A cabinet placed over one of the cutouts served to simulate actual positioning of equipment and cables in the second test.

## Cables

In accordance with our findings in the field, an average cable configuration was used under the raised floor. There was an unsystematic arrangement of the cables above the ignition vessel in the first and second test. In the third test, single coaxial cables were laid on a metal cable support (22 x 6 x 300 cm). The cable support was parallel with the observation window, and its distance to the lower floor was 220 mm. Cable configuration for all tests see table 1.

In tests 1 and 2, ten system-internal cables and one commercially-available power cable emerged from the cutout in the floor; in test 3 they were replaced with 75 coaxial cables.

## Ventilation

In the majority of all the DP installations that were surveyed, ventilation occurs via the raised floor. The air is blown into the room through openings in the floor, and extracted via the ceiling. For this reason, the tests were carried out only in ventilated raised floors.

TABLE 1    Cable configuration

---

TEST 1 AND 2

| System-internal cables: | outer diameter |
|---|---|
| 13 standard interface cables | 23 / 28 mm |
| 3 standard interface cables, shielded | 23 mm |
| 2 modem attachment cables, shielded | 10 mm |
| 2 control cables, shielded | 10 mm |
| 2 power-supply cables | 10 mm |
| 1 main control cable | 15 mm |
| 1 power cable | 40 mm |
| 1 signal cable, shielded | 20 mm |

| External cables: | |
|---|---|
| 4 power-supply cables, shielded | 20 mm |
| 4 power-supply cables | 13 mm |
| 3 PTT cables | 5 mm |
| 1 plastic cable support (20 x 6 x 300 cm) | |

TEST 3

| 228 coaxial cables | 6 mm |
|---|---|

---

The air that was blown into the raised floor by a radial blower corresponded with an air exchange of 25 times.

The fumes were permanently extracted from the fire-test room and replaced with fresh air. The air was let out of the raised floor through 14 ventilation panels. Each of these ventilation panels had 4 openings with a diameter of 120 mm. Plastic dust traps in the form of tapered baskets were placed in each of the openings. The openings themselves were covered with air-outlet covers (of the whirl type), made of cast aluminum.

Ignition source

In test 1 the ignition source was placed on the raised floor under the cables emerging from the cutout. In tests 2 and 3 it was placed on the lower floor under the stacked cables and under the cable support, respectively. The ignition vessel was a porcelain dish, 35 cm in diameter, and 9 cm high. Prior to each test 1 liter of n-hexane ($C_6H_{14}$) was poured into the dish. The hexane was ignited electrically.

Measuring the temperature

The temperature was measured by means of Ni-CrNi jacketed thermo-elements with a diameter of 3 mm. These were installed under the raised floor at a distance of 400 mm from the lower floor. The distance between the thermo-elements and the ignition vessel was 50 cm. Further thermo-elements were placed at the point where the cables emerged from the raised floor and at the end of the emerging cables. The temperature was continuously recorded during the tests.

·TEST PROCEDURE

Ignitial tests

The time during which 1 liter of n-hexane burned in the ignition vessel that was used in the tests was determined at 6 minutes.

Laboratory tests revealed the behavior of the cables in a fire.

Main tests

The following tests were carried out:

1. Flaming of the cables outside the raised floor.
2. Flaming of the cables under the raised floor.
3. Flaming of a cable support with cables under the raised floor.

A chronological summary of the events that occurred during the above-mentioned tests was made.

863

TEST RESULTS

Flaming of the cables outside the raised floor (test 1)

After the hexane had been ignited, it burned with a flame approximately 50 cm tall. After one minute, dark smoke developed. Smoke development increased further and within five minutes the entire fire-test room was filled with smoke.

When the hexane was burnt, the cable insulation continued to burn. The flames spread beyond the ignition vessel. In this test, however, there was hardly any evidence of burnt cables beyond the range of action of the flames in the ignition vessel. There was no dripping of flames from the cable insulation. The fire did not spread into the space under the raised floor, not did the temperature under the raised floor increase. The temperature of the copper conductors at the end of the cables increased to approximately 144 °C as a result of thermal conduction.
27 minutes after the ignition the last burning cable extinguished.


Flaming of the cables under the raised floor (test 2)

After the hexane had been ignited, the flames reached the underside of the floor panels. After 30 seconds black smoke emerged from the raised floor. The intensity of the fire was increased as a result of the fact that the flames bounced off the underside of the raised floor. After one minute, the first few flames emerged from the ventilation opening; the plastic dust traps had melted. The smoke continued to develop rapidly and thickly. After some three minutes, pitch-black smoke had filled the fire-test room. After that the smoke continued to increase. After six minutes the light of the halogen photo lamp was no longer visible. This thick smoke lasted until the 10th minute.

After some 20 minutes there was a real fire under the raised floor of the fire-test room, although the raised floor was still an uninterrupted surface. After the 23rd minute, the temperature over the ignition source, which had first fallen, began to rise again (figure 2: thermoelement measuring points 1 - 8, figure 3: thermoelement measuring points 9 - 13).

It is likely that at that time the floor boards began to burn. It appeared later that the metal sheeting on the underside had become detached. In the 30th minute some of the floor panels collapsed. One supporting platform of the steel uprights had melted completely, two others had disintegrated partly. In an area of 1.62 $m^2$ the cables were burned completely. In the direction of the air stream the damage was greater. The flames reached the cabinet over the cutout; it burned out completely and beyond repair. The cables emerging from the raised floor burned completely (figure 4).

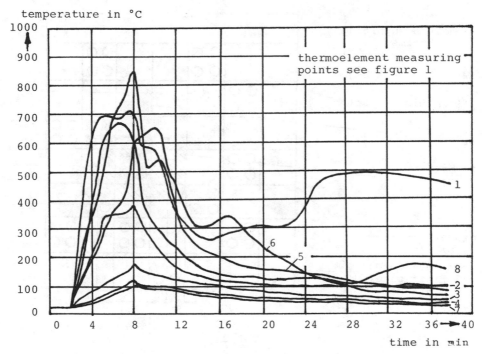

figure 2: Temperature-time-recordings (test 2, measuring points 1-8)

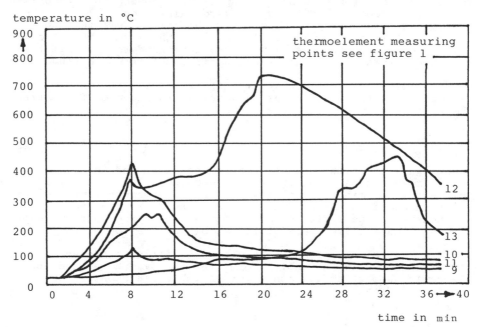

figure 3: Temperature-time-recordings (test 2, measuring points 9-13)

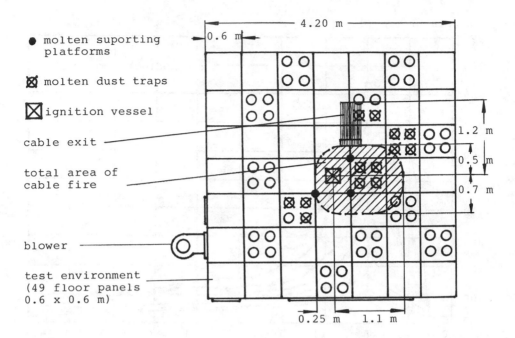

figure 4: Extent of damage in test 2

The evaluation of the recorded temperatures showed that temperatures ranging from 650 °C to 800 °C were maintained until all the hexane had burned away; subsequently the temperature fell up to the 15th minute to approximately 300 °C. From the moment when the floor panels began to burn, the temperature increased to approximately 500 °C and remained at that level until the fire was extinguished.

After 35 minutes the first attempt to extinguish the fire with manually operated $CO_2$ fire extinguishers was unsuccessful. After the 40th minute, the fire was extinguished with water.

Flaming of an occupied cable support under the raised floor (test 3)

After ignition of the hexane, the flames enveloped the cable support. Smoke begins to emerge from the raised floor after 1.15 minutes. After two minutes the first few flames emerged from the ventilation openings. After four minutes, the smoke had become thicker. The hexane in the ignition vessel burned for approximately seven minutes. Small flames were still visible at the top of the cables.

Cindered cables were found only where the flames had been able to get at them: under the cable support at the oval holes and above the cable support at the top layer of cables. 82 coaxial cables remained undamaged.

CONCLUSIONS

The following conclusions may be drawn from the initial tests:

- All raised floors in DP installations that were surveyed contained a mixture of system-internal cables and external (commercially available) cables.
- Invariably, cables of different outer diameters were found.
- Each time when an installation is extended or replaced, new cables are supplied with the systems. Because only in very few cases the old cables are removed, system extension generally means more damage in case of fire.
- The cable layout under the raised floor was generally unsystematic and in layers.
- All raised floors that were inspected were ventilated.
- The design and the materials of the raised floor used in the fire test conformed with standard practice.
- Laboratory tests showed that the system-internal cables with improved fire-resistant characteristics have these characteristics only with regard to the outer sheating. The sheated internal wires showed a normal behavior in the fires.
- The smoke produced by system-internal cables with improved fire-resistant characteristic varies, and can be extraordinarily intensive.

The following conclusions may be drawn from the main tests:

- When the cables were flamed outside the space under the raised floor, the fire did not spread. After the hexane was burned up, the cables continue to burn for another 20 minutes or so until they extinguished. The extent of the damage was restricted to the local area, and the cables did not spread the fire into the space under the raised floor.

- When the cables were flamed under the raised floor, fire damage to the cables was more extensive in the direction of the air flow (maximum damage at a distance of 1.1 m from the ignition vessel!). Outside this area, the cables under the raised floor were undamaged. The test was stopped after 40 minutes and the fire was extinguished with water.

- The cables spread the fire from the space under the raised floor into the test room; the cables burned out completely, and the cabinet that served to simulate a piece of DP equipment was roasted. We must therefore assume that - in the event of a fire - the equipment in the vicinity of the seat of the fire is bound to be damaged by the fire.

- When only coaxial cables with improved fire-resistant characteristics were used, the extent of the damage was restricted to the area of the ignition vessel. The flames extinguished of their own accord after approximately 10 minutes. The fire did not spread into the room.

- If a fire breaks out under a raised floor, an extremely fast development of thick smoke must be expected (no visibility after 6 minutes).

- An examination of the equipment used showed that the impact of chloride ions as a result of the fumes amounted to 10 µg Cl$^-$ per cm$^2$; this exceeds the limit beyond which repair is impossible.

- Premature detachment of the metal lining glued to the underside of the floor panels resulted in spreading of the fire. The smoldering floor panels were extinguished with water; the use of $CO_2$ was unsuccessful.

- The floor over the ignition vessel collapsed prematurely because the supporting aluminum platforms for the floor panels melted away.

- Burning drops of cable insulation were not observed.

SUMMARY

The tests showed that the fire protection of raised floors in data-processing installations requires at least the use of smoke detectors. Their number and position should be governed by the influence of the ventilation, and the display system must permit fast location of the source of the smoke. Damage resulting from corrosive fumes and intensive smoke must be expected after a relatively short time, and spreading of the fire to the equipment via the ventilation openings can be prevented only by an automatic fire extinguishing system. The tests have also shown that mechanical fastening of the metal lining on the underside of the floor panels will prevent the lining from becoming detached prematurely and the fire from spreading. The supporting platforms for the raised floor should be made of steel, because melting of the aluminum platforms causes the floor to collapse, which in turn can cause the fire to spread and constitutes a hazard to the firemen.

# Spray Fire Tests with Hydraulic Fluids

GÖRAN HOLMSTEDT
Department of Physics, Lund Institute of Technology
P.O. Box 118, S-221 00 Lund, Sweden

HENRY PERSSON
Division of Fire Technology, National Testing Institute
P.O. Box 857, S-501 15 Borås, Sweden

ABSTRACT

Two test series simulating the hazards associated with the accidental release of hydraulic fluid near to a source of ignition have
been carried out with six hydraulic fluids; mineral oil, organic
ester, phosphate ester, water in oil emulsion and two polyglycols
in water solution.
In one of the test series the fluids were sprayed (1-4 kg min$^{-1}$)
through different nozzles at various hydraulic pressures into a
diffusion flame under a semi-open hood which collected all the
combustion gases; thus the rates of generation of smoke and gases
($O_2$, $CO$, $CO_2$) could readily be measured.
In the other test series the fluids were sprayed (7-30 kg min$^{-1}$)
through various nozzles at various hydraulic pressures into a
diffusion flame or against a hot metal plate in a large fire hall.
The flame length, temperature and radiation and the auto-ignition
temperature were measured.
The correlation between the two test series regarding rates of heat
release between 1 and 20 MW was very good. As a result a test
method is proposed. In this test method the flammability hazard of
hydraulic fluid spray fires is measured in terms of their combustion efficiency, net heat of combustion, radiant fraction and
smoke and toxic rate of production.

Keywords: Hydraulic fluids, Spray combustion, Combustion effi-
          ciency, Smoke, Toxic gases, Auto-ignition temperature,
          RHR-measurements.

INTRODUCTION

Hydraulic fluids have been the subject of concern due to the fire
hazards associated with them  especially in industries where "hot
processes" are used, such as  steel-making plants, steel-rolling
mills, forge workshops etc. In Sweden, for example, a number of
fires have been caused by leaking hydraulic fluid which has led to
very expensive damage [1].
The most common source of leakage in hydraulic systems is from
fittings, valves, steel reinforced rubber hoses and steel and
copper pipes. The high pressure in the hydraulic system leads to
leakage in the form of very fine sprays. Hydraulic fluids which,

when held in bulk are not particularly flamable, can therefore through leakage, in the form of fine sprays, atmoize and apon ignition lead to very large flames.

The risk of hydraulic fluid fires can be minimized in many ways; engineering, through the introduction of warning- and fluid-stop systems, hose break valves, double pipes, sprinklers etc., and chemically, by changing the composition nature of the hydraulic fluid. A fluid spray's tendency to combust is influenced by many factors such as droplet size distribution, the fluid's flash-point, auto-ignition temperature, heat of combustion etc. The droplet size can increase, for example, through the addition of certain polymers, which have high molecular weights. These polymers, which reduce the tendency to formation of mists, are on the other hand very sensitive to shear forces and degrade quickly in the hydraulic system due to the wear caused by the moving parts of pumps, filters and valves [2].

A number of hydraulic fluids, which have a higher flash-point and auto-ignition temperature and a lower heat of combustion than mineral oil, have been developed and are normally called "fire-resistant hydraulic fluids". These include oil in water emulsions, water in oil emulsions, polyglycols in water solutions, phosphate esters, halogenated hydrocarbons and organic esters. It is then necessary to introduce modifications into the hydraulic system for many of these fluids, in order to make seals and metals compatible with the fluid. The problem of toxic substances can also arise with the fluids and their combustion products; this is especially the case with fluids containing phosphates and halogenated hydro-carbons.

The concept "fire-resistant hydraulic fluids" is a diffuse concept which relates to manufacturers' or institutions' classification rules based on small-scale tests [2-12]. The tests, which are normally carried out with a limited quantity of fluid (0.05-0.5 kg min$^{-1}$), very often give no quantitative information about any combustion property. Instead the fluid is "passed" or "failed". Fluids which, according to such tests have been classified as fire-resistent, however, have caused severe damage in industrial environments and have under other tests [11, 13] given results which indicate that the difference between their combustion properties and the combustion properties of mineral oil is not so large.

Two test series have been carried out with six hydraulic fluids. In these tests the combustion efficiency, radiant fraction, water content, auto-ignition temperature and rate of production of smoke and toxic gases were determined.


EXPERIMENTAL METHOD

Test equipment

Two sets of test equipment have been used. In one of the test series the hydraulic fluid was sprayed (1-4 kg min$^{-1}$) through different nozzles at various hydraulic pressures into a diffusion flame under a semi-open hood which collected all the combustion gases, Fig. 1. The hood was connected to an evacuation system through an exhaust duct and equipped with shields to collect the combustion gases. The openings on both sides of the spray were in total 10 m$^2$ and the entrance opening of the hood 9 m$^2$. During the

870

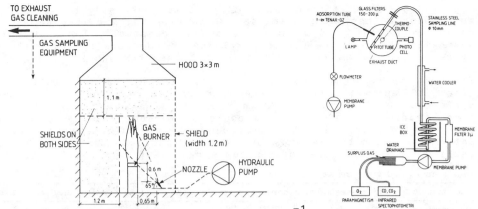

Fig.1 Experimental equipment (1-4 kg min$^{-1}$)
    a) semi-open hood                    b) analytical equipment

test air was evacuated at a rate of approximately 4 Nm$^3$s$^{-1}$, giving
an average speed of the cold gas so low (<0.5 ms$^{-1}$) that it did not
interfere with the spray flame. In the exhaust duct, where the
gases were well mixed, mass flow (pilot tube, thermocouple, velocity profile), optical smoke density and gas composition ($O_2$, CO,
$CO_2$) were measured as shown in Fig. 1. The test equipment is
presented in more detail in ref. 14.
In the other test series the fluids were sprayed (7-30 kg min$^{-1}$)
through different nozzles at various hydraulic pressures through a
diffusion flame or against a hot metal plate in a large fire hall
(18 x 22 x 20 m), Fig. 2. The flame length, temperature and radiation were measured.


Fluid spray

The fluid flow to the spray nozzle was provided by a hydraulic
pump. The pressure in the system could be adjusted between 0 and 30
MPa using a pressure limiting valve, and the temperature between 10
and 90 °C using a thermostatically controlled electrical heater
mounted on a 160 l hydraulic reservoir. The hydraulic pressure in
the nozzle was measured using a pressure transducer of strain gauge

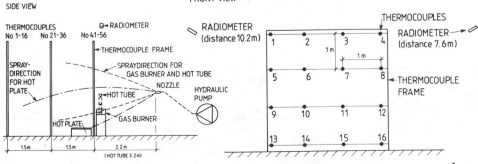

Fig.2 Experimental equipment in a large fire hall(7-30 kg min$^{-1}$).

type (34.5 MPa, 0.25% accuracy, 0.1 ms time constant), the temperature using a thermocouple and the fluid flow using a turbine wheel gauge (30 l min$^{-1}$, 1% accuracy, 1 s time constant). The flowmeter was calibrated for the different fluids at the working temperature by measuring the amount of fluid which flowed through it in a given time.

To simulate real leakages which can give rise to many different types of sprays, four different hydraulic pressures and four different nozzles were used. In this way the number density and size distribution of the drops could be varied in order to avoid that the results being be too dependent on a single spray form. Information about the nozzles used,( manufactured by Spraying Systems Co.) which give homogeneous sprays is given in Table 1.

Table 1. Data regarding the nozzles

| test equipment | nozzle | coneangle,deg. | hydraulic pressure,MPa | fluid flow,l min$^{-1}$ | med.vol.diam.,$\mu$m($H_2O$,21$^{\circ}$C) |
|---|---|---|---|---|---|
| semi-open hood | Tg 0.4 | 60 | 5, 10, 15, 25 | 1 – 2.5 | 70 – 50 |
| semi-open hood | Tg 0.7 | 60 | 5, 10, 15, 25 | 2 – 4 | 130 – 90 |
| large fire hall | D3/31 | 65 | 15, 25 | 7 – 9 | 110 – 90 |
| large fire hall | D5/56 | 30 | 15, 25 | 23 – 27 | 200 – 150 |

## Ignition sources

The sprays were ignited by a diffusion flame from a porous propane burner (0.3 x 0.3 m) with a power of 200 kW. In addition a hot steel plate was used as the ignition source in some tests in the large fire-hall. The plate (0.65 x 0.75 x 0.02 m) was heated from below with a thermostatically controlled electrical heater. Due to radiation losses the steel plate was insulated during the heating period. Just before the test the insulation was withdrawn whereupon the surface temperature slowly started to decrease. During a test the surface temperature, as indicated by three thermocouples fell by less than 25 $^{\circ}$C.

## Hydraulic fluids

The results of the tests were used to compare the combustion properties of the different hydraulic fluids with the combustion properties of mineral oil. To complement the manufacturers' data, the water content and the heat of combustion were determined and are given together with other values in Table 2.

Table 2. Tested hydraulic fluids

| | min.oil | org.ester | phosph.ester | water in oil em. | pol.glyc.I in w.s. | pol.glyc.II in w.s. |
|---|---|---|---|---|---|---|
| density,kgm$^{-3}$ + | 892 | 920 | 1125 | 964 | 1055 | 1064 |
| viscosity,cSt (40$^{\circ}$C) + | 66.8 | 40.0 | 43.0 | 110 (38$^{\circ}$C) | 43 (38$^{\circ}$C) | 45 |
| flash-point,$^{\circ}$C + | 212 | 280 | 245 | data missing | – | – |
| auto-ign.temp.,$^{\circ}$C + | 350 | 460 | 545 | data missing | – | – |
| water cont.,% of weight | 0 | 0 | 0 | 38 | 35 | 35.2 |
| heat of comb.,kJg$^{-1}$ ($H_2O$ liq.) | 45.0 | 39.8 | 32.4 | 29.2 | 16.9 | 14.6 |
| heat of comb.,kJg$^{-1}$ ($H_2O$ gas ) | 42.3 | 37.7 | 30.3 | 26.6 | 14.7 | 12.5 |

+ manufacturers´data

## Test procedure

The temperature of the hydraulic fluid was kept constant at 37 ± 2 °C. The pressure was adjusted to the desired value and the fluid was allowed to circulate back to the reservoir. For the experiments under the semi-open hood the ignition source was allowed to burn for 30 s before the spray was turned on by shifting the fluid flow from recirculating to the spray nozzle with an electrically controlled three-way valve. The spray was allowed to burn for another 30 s with the ignition source on. For the experiments in the large fire-hall the corresponding times were 5 and 20 s. The experiments were repeated using different pressures, nozzles, hydraulic fluids and ignition sources. In all about 150 tests were carried out.


## CALCULATIONS

### Rate of heat release and combustion efficiency

In the experiments under the semi-open hood the rate of oxygen consumption was measured. This is a way of estimating the rate of heat release as the heat of combustion per unit of oxygen consumed is approximately the same for most fuels [15]. The rate of heat release, q, is given by [16]:

$$q = 17.2(X^O_{02} - X^S_{02})\dot{V}_s/\alpha \qquad \text{where}$$

$q$ = rate of heat release, MW
$X^O_{02}$ = volume % of oxygen in the incoming air
$X^S_{02}$ = volume % of oxygen in the exhaust gases
$\dot{V}_s$ = volume flow of the exhaust gases, $m^3s^{-1}$ (25 °C, 0.1 MPa).
$\alpha$ = expansion factor for the fraction of air that is depleted of its oxygen ($\approx 1.1$)

The total inaccuracy, including the factors which have been neglected in the formula above (CO and $H_2O$ content) is estimated to be 25 kW or ± 10% of the calculated value. Starting from the calculated rate of heat release the combustion efficiency can be derived from:

$$\emptyset = (q-q_{ign})/(\dot{m} \cdot Q) \qquad \text{where}$$

$\emptyset$ = combustion efficiency
$q_{ign}$ = power of the ignition source = 0.2 MW
$\dot{m}$ = mass flow of the hydraulic fluid, kg s$^{-1}$
$Q$ = heat of combustion of the hydraulic fluid, MJ kg$^{-1}$ ($H_2O$ gas)


### Production of smoke and toxic gases

In the experiments under the semi-open hood the volume flow, light obstruction, and the CO- and $CO_2$ concentrations in the exhaust gases were measured. The smoke potential is given by [17]:

$$D_O = 10 \log (I_O/I) \cdot \dot{V}_t/(\dot{m} \cdot L) \qquad \text{where}$$

$D_O$ = smoke potential, ob $m^3g^{-1}$
$\dot{V}_t$ = volume flow exhaust gases, $m^3s^{-1}$ (at exh. gastemp. and press).
$\dot{m}$ = mass flow of the hydraulic fluid, gs$^{-1}$
$L$ = diameter of the exhaust duct, m

In a similar manner the CO- and $CO_2$ potential, l g$^{-1}$, can be calculated. The total inaccuracy is estimated to be ±10% of the calculated value.

Radiant fraction

The radiation was measured using two Medtherm radiometers (10 kWm$^{-2}$, 3% accuracy) placed at different distances on either side of the flame axis. It is difficult to calculate the total radiated power from only two point measurements. The flames from nozzles Tg 0.4, Tg 0.7 and D3/31, however, were approximately spherical and small compared with the distance, r, to the radiometers, so that the power could be estimated by the measured radiation multiplied by $4\pi r^2$. The flames from nozzle D5/56 were comparable in length to the distance to the radiometers. In this case the total radiation was calculated by regarding the flame as a cylinder which radiated with uniform intensity and with a length equal to the flame length. The difference in calculated power from the two radiometers was less than 10%.


RESULTS

Rate of heat release and combustion efficiency

Tests were carried out under the semi-open hood with six hydraulic fluids at four pressures using two different nozzles. The rate of heat release and the generated heat per gram sprayed organic ester as a function of time are shown in Figs. 3a and b, respectively. When the ignition source is on the rate of heat release stabilizes at 0.2 MW. Thirty seconds later when the spray is applied the rate of heat release increases and stabilizes at a new level. The generated heat per gram fluid, and consequently the combustion

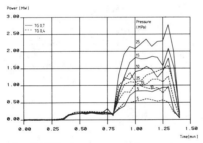 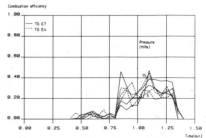

Fig.3 a) RHR for organic ester as a function of time.   b) Generated heat per gram sprayed org.ester as a function of time.

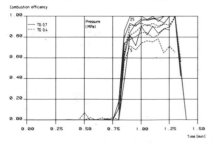

Fig.4 a) Combustion eff. for min.oil as a function of time.   b) Combustion eff. for polyglycols I in water solution as a function of time.

874

efficiency, is, as shown in Figs. 3b, 4a and 4b, relatively independent of the hydraulic pressure and the nozzle used. The results are given in more detail in ref. 18. In Table 3 the combustion efficiencies are given for the hydraulic fluids tested.

## Production of smoke and toxic gases

The smoke-, CO- and $CO_2$ potential were also found to be relatively independent of the hydraulic pressure and nozzle used. In Figs. 5a and b the CO- and $CO_2$ potential are shown as a function of time and in Table 3 the potentials for the tested hydraulic fluids are given.

Table 3. Test results

|  | min.oil | org.ester | phosph.ester | water in oil em. | pol.glyc.I in w.s. | pol.glyc.II in w.s. |
|---|---|---|---|---|---|---|
| combust.efficiency,% | 90 | 80 | 70 | 60 | 25 | 30 |
| radiant fraction,% | 32 | 20 | 32 | 20 | 10 | 10 |
| smoke potential,$cbm\,g^{-1}$ | 0.2 | 0.2 | 2-6 | 0.4 | 0.1 | 0.1 |
| CO pot.,$l\,g^{-1}$(25°C,0.1 MPa) | 0.03 | 0.035 | 0.1 | 0.035 | 0.02 | 0.02 |
| $CO_2$pot.,$l\,g^{-1}$(25°C,0.1 MPa) | 1.7 | 1.3 | 0.8 | 0.6 | 0.2 | 0.2 |

## Radiant fraction

It was also found that the radiant fraction was relatively independent of the hydraulic pressure and the nozzle used. In Figs. 6a and b the measured radiation level and the radiation energy per gram sprayed fluid are shown. In Table 3 the radiant fraction is given for the fluids tested.

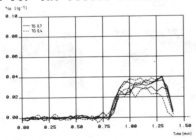
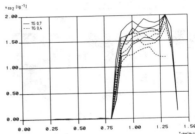

Fig.5 a) CO-potential for w.i.oil emulsion as a function of time.  b) $CO_2$-potential for mineral oil as a function of time.

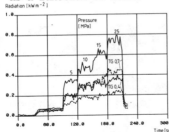
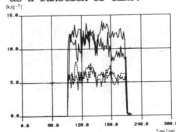

Fig.6 a) Radiation level from flames of min.oil(——) and org.ester (- -)as a function of time.  b) Radiation per gram fluid as a function of time(Fig 6a divided by the mass flow).

Flame length

The length of the flames in the tests carried out in the large fire-hall was estimated from pictures and video recordings. The geometry of the flames were somewhat different depending on whether the fluid was sprayed freely in the air, which was the case with a diffusion flame as the ignition source, or if the fluid was sprayed close to the floor which was the case with the hot plate. In Figs. 7a and b some typical pictures are shown from comparable tests with the diffusion flame. In Fig. 7c comparable tests are shown with the hot plate. In the tests using the hot plate, the mineral oil and organic ester started a heavy pool fire in the fluid which hit the floor. The pool fire stabilized the flames which increased in intensity and gave rise to a strong turbulence rumble and a heavy smoke generation. Under the same test conditions phosphate ester and water-in-oil emulsion only gave rise to a small pool fire on the floor and the two polyglycols in water solution produced only a small local fire at the ignition source.

Mineral oil 2 m    Organic ester 2 m    Water in oil emulsion 1.5 m    Polyglycols in water solution 0.75 m

Fig.7 a)Flame lengths from free sprays with nozzle D3/31 ( 7 kg min⁻¹ ) at 15 MPa:s pressure.

Mineral oil 5.5 m    Organic ester 5 m    Water in oil emulsion 3.5 m    Polyglycols in water solution 1.25 m

   b)Flame lengths from free sprays with nozzle D5/56 ( 23 kg min⁻¹ ) at 15 MPa:s pressure.

Mineral oil 7 m    Organic ester 5 m    Water in oil emulsion 3 m    Polyglycols in water solution 0 m

   c)Flame lengths from sprays close to the floor with nozzle D5/56 ( 23 kg min⁻¹ ) at 15 MPa:s pressure.

## Auto-ignition temperature

The auto-ignition temperature of the fluids was determined by testing the fluid sprays against a hot surface at various temperatures. In Table 4 the highest surface temperature at which ignition did not occurred and the lowest temperature at which ignition did occur are shown together with the auto-ignition temperature given by the manufacturers. The results are in good agreement since a difference of 50 to 100 $^{\circ}$C between different test methods is not unusual [19].

Table 4. Auto-ignition temperature

| | mineral oil no ign. / ign. $^{\circ}$C | | organic ester no ign. / ign. $^{\circ}$C | | phosphate ester no ign. / ign. $^{\circ}$C | | water in oil em. no ign. / ign. $^{\circ}$C | | pol.glyc.I in w.s. no ign. / ign. $^{\circ}$C | | pol.glyc.II in w.s. no ign. / ign. $^{\circ}$C | |
|---|---|---|---|---|---|---|---|---|---|---|---|---|
| hot plate | 300 | 350 | 400 | 450 | 500 | 550 | 350 | 400 | >900 | | >900 | |
| manuf.data | | 350 | | 468 | | 545 | missing | | - | | - | |

## DISCUSSION

The risks of injury on personnel and damage to property which arise from hydraulic fluid fires are caused by spray flames, secondary fires and the presence of smoke and toxic gases. Flame length, rate of heat release, radiation and the generation of smoke and toxic gases are the most important factors which it is necessary to determine to be able to assess the hazards.

The flame length is dependent on the extent of the leakage as well as on the combustion properties of the fluid. In the aerosol the droplets are not uniformly distributed with respect to size and number. The smaller droplets are easily retarded and cause the spray to mix with the surrounding air [20]. If the droplets combustion time is short the flame length depends on the quantity of air which must be entrained to give complete combustion. A long flame is therefore correlated with a large combustion air requirement which in turn is correlated with a large rate of heat release. There are difficulties involved in measuring the flame length because of the turbulent nature of the flame and possible obstruction by smoke. An alternative to measuring the flame length is therefore to measure the rate of heat release [11]. This apparent correlation between the flame length and the rate of heat release is shown for comparable tests in Table 3 and Figure 7.

The results from the two test series show that sprays from hydraulic fluids can burn when they hit an ignition source, in spite of this they are sometimes classified as fire-resistant in small-scale tests. The combustion efficiency is influenced by a number of properties such as the size- and number distribution of the droplets, and the type of hydraulic fluid etc. Comparable tests in the two test series have shown that the combustion efficiency for hydraulic fluid spray fires differs with the hydraulic fluid used but is relatively independent of hydraulic fluid pressure, nozzle size and spray angle. The same conditions seem to be valid for the radiant fraction and smoke- and toxic gas potential. In Figure 8 the total and radiant heat output per gram sprayed fluid are shown as a function of the fluids net heat of combustion. The net heat of combustion of a hydraulic fluid seems to be the most important parameter which controls the combustion efficiency in spite of differences in the auto-ignition temperature of a few hundred degrees C.

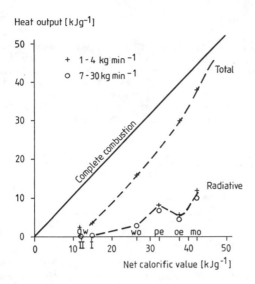

Fig.8 Total and radiation output from burning hydraulic fluid as a function of the fluids net heat of combustion

The tests showed that a spray fire close to the floor often led to a pool fire and that thin thermocouple wires through the spray led to small fires which stabilized the spray fire. These test series have therefore been performed with the ignition source on during the whole test. Tests which assess the ability of free fluid sprays to stabilize themselves when the ignition source is withdrawn are often dependent on the choice of spray nozzle. The size of the ignition source does not seem to have a pronounced influence on the combustion efficiency of the fluids. The combustion efficiency was hardly affected when fluid flow rates varying from 1 to 30 kg min$^{-1}$ were sprayed against a 200 kW propane burner as well as when a constant flow of 2 kg min$^{-1}$ mineral oil was sprayed against four different diffusion flames varying from a cigarett lighter to 200 kW [21].

CONCLUSION

The results from the two test series show that sprays of hydraulic fluids can burn when they hit an ignition source in spite of the fact that they have been characterized as fire-resistant on the basis of small scale tests. To assess the fire hazards associated with hydraulic fluids, the testing method should determine the combustion efficiency, net heat of combustion, radiant fraction and smoke- and toxic gas potential of the fluid.

ACKNOWLEDGMENT

This work was supported by the Swedish Fire Research Board and National Testing Institute, Sweden.

# REFERENCES

1. Bäckström, H., Skandia (PM 811216).
2. Romans, J.B. and Little, R.C., Fire Safety Journal 5 (1983), p 115.
3. Test Procedure: Less Hazardous Hydraulic Fluid, Factory Mutual Research, June, 1975.
4. High Temperature - High Spray Ignition, Federal Test method Standard 791, Method 6052T.
5. Low Pressure Spray Ignition, AMS - 3150 C.
6. Loftus, J.J., et al, Flammability Measurements on fourteen different hydraulic fluids using a temperature pressure spray ignition test, NBSIR 81 - 2247, U.S. Department of commerce.
7. Loftus, J.J., MSHA wick test for hydraulic fluids: A preliminary evaluation, NBSIR 81 - 2312, U.S. Department of commerce.
8. Loftus, J.J., An evaluation of the MSHA temperature-pressure spray ignition test of hydraulic fluids, NBSIR 81 - 2373, U.S. Department of commerce.
9. Loftus, J.J., An assessment of three different fire resistance tests for hydraulic fluids, NBSIR 81 - 2395, U.S. Department of commerce.
10. Flammability spray test for hydraulic fluids, Draft for Development 61, British Standard Institution, London (1979).
11. Roberts, A.F. and Brookes, F.R., Fire and Materials 5 (1981), p 87.
12. CETOP RP 77 H 760301, (Comité Européen des Transmissions Gléo Hydrauliques et Pneumatiques).
13. Råback, E., Prov för att iaktta några olika brinnande vätskors spridningsförmåga, Fagersta AB (1979).
14. Wickström, U., Sundström, B. and Holmstedt, G.S., Fire Safety Journal 5 (1983), p 191.
15. Huggett, C., Fire and Materials 4 (1980), p. 61.
16. Parker, W.J., Calculation of the Heat Release Rate for Oxygen Consumption for Various Applications, NBSIR 81 - 2427 - 1, U.S. Department of commerce.
17. Rasbash, D.J. and Pratt, B.T., Fire Safety Journal, 2 (1979/80), 23-37.
18. Holmstedt, G.S., Persson, H. and Rydermann, A., Teknisk Rapport SP-RAPP 1983:29 ISSN 0280-2503.
19. Hilado, C.J. and Clark, S.W., Fire Technology 8 (1972), p 218.
20. Faeth, G.M., Prog. Energy Combust. Sci. 8 (1977), p 191.
21. Persson, H., Teknisk Rapport SP-RAPP 1983:50 ISSN 0280-2503.

# A Study on the Fire Spread Model of Wooden Buildings in Japan

**YOSHIRO NAMBA** and **KENJIRO YASUNO**
Japanese Association of Fire Science and Engineering
Department of Architecture, Faculty of Engineering, Kinki University
1000 Hiromachi, Kure-city, Hiroshima, Japan

ABSTRACT

This paper deals with the fire spread within and between wooden buildings in Japan. The relation of the burnt area and the time draws approximately an S-curve. As former theories fail to estimate the real fire spread condition correctly, we have examined a polynomial curve and a logistic one as the most appropriate substitutes. According to the method of MAICE proposed by Dr. Akaike (1973), the optimal degree of the polynomial is shown to be three. But the polynomial does not fit the real fire spread phenomenon because it does not always follow a monotone function. Thus we have suggested that the logistic curve is good enough to explain the fire spread phenomenon. But our model exhibits a symmetry at the point of inflection. As the fire spread phenomenon is not guaranteed to have a symmetry, we have examined a more general type of logistic curve which discribes the fire spread.

In case of determining the parameters of the logistic curve, we should strictly use the nonlinear least squares method as we have already reported in our former studies. By analyzing the data of real fire experiment by this method, we have obtained a very significant result, and also wish to refer to the method of improving the determination method of the parameters of the fire spread model.

## 1. INTRODUCTION

In most Japanese cities and towns, the percentage of wooden buildings and the building-to-land ratio are relatively high. Therefore once a fire breaks out, there is a danger of being burnt down together with neibouring area. Many conflagrations have thus happened in Japan till about 1960's. Consequently, many fire spread formulas have been developed as mentioned in the next section. They were useful for planning measures of the fire prevention, the fire fighting and so forth. However, they were based on the data of conflagrations till about 1945 and the fire spread velocity by their formulas seems to have become out-dated, for present conditions of cities, buildings and fire fighting measures have profoundly changed since 1945. Today in Japan, small fires often break out and they catch several buildings at most. However, this facts does not mean that there is no more danger of confraglations. For example, it is said that there can still be in Japan a danger of conflagrations when the fire fighting ability is lost in such as a case of great earthquakes.

Recently, the authors have developed a model of fire spread by
analyzing the fire data of real experiment in 1980 by Fire Research
Institute, Ministry of Home affaires, Government of Japan(1982,1983).
These data show that the relationship between the burnt area and the
time discribes approximately an S-curve. This means that former formulas
fail to estimate the state of the real fire spread correctly. As a result,
we have examined the polynomial and logistic curves in our studies,
by assuming them better substitutes. According to the method of MAICE
proposed by Dr. Akaike (1973), the optimal degree of polynomial is
shown to be three. But the polynomial does not fit the real fire spread
phenomenon because it is not always a monotonial. On the other hand,
it appears that a logistic curve function is good enough to explain
the fire spread phenomenon.

But our model exhibits a symmetry about the point of inflection.
As it is difficult to assume that a fire spread phenomenon would be
symmetrical, we have examined mathematically a more general type of
logistic curve of the fire spread model.

This paper discusses the best type of function to discribe the growth
of fire in wooden buildings and the best way to determine the required
constants using the data from a typical fire. This is a mathematical
model because most of the real fires show in Japan a similar tendency to
the data.

## 2. REVIEWS OF FIRE SPREAD FORMULAS AND MODELS OF WOODEN BUILDINGS IN JAPAN

### 2.1 A Brief Review of Literature

Empirical formulas of fire spreading speed have been proposed
by Suzuki and Kinbara (1940), the Study team of Fire Prevention of
Tokyo (1942), Tosabayashi (1947), Hamada (1951), Hishida (1954), and
Horiuchi(1961). Further, Fujita (1975) and Sakai (1983) developed
simulation models of the fire-spreading based on Hamada's formula.
These formulas and simulation models have been developed on the basis
of fire data before 1945. Therefore, it is doubtful to apply those
formulas to modern urban area where the structure of houses and buildings
have greatly changed since 1945. In fact, it is pointed out by Sasaki
(1976) and Yamashita (1977) that Hamada's and Horiuchi's formulas tend
to overestimate the fire-spreading speed. Moreover these formulas do not
follow an S-curve.

### 2.2 Progress of authors' studies

We have developed recently a model of fire spread by analyzing
the fire data of real experiments by Fire Research Institute, Government
of Japan. It is obvious at a glance of the data that the relationship
between the burnt area and the time constitutes approximately an S-curve
as shown in Fig. 1. The dots in Fig.1 are the observed values. The
smaller curves are indicated the state of fire spread for each building.
The large curve is the state of fire spread of four buildings as a whole.
Each building is similar to each other, and the conditions of experiment
are in brief as follows:

(1) Structure : Wooden and one-story
(2) The average of building area : 132 $m^2$
(3) The average distance between buildings : 4.5 m
(4) The average of wind velocity : 1.8m/sec

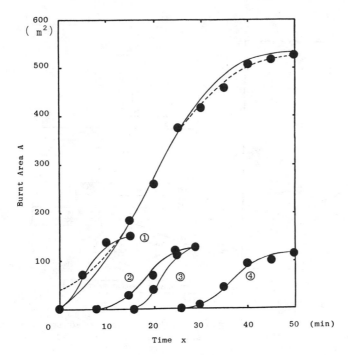

Fig.1 The relation of the Burnt Area A and the Time x : ( After the
fire experiment data at Saganoseki by Fire Research Institute,
Government of Japan ; The dots are observed values.; ① - ④
are the building numbers.)


(5) The method of ignition : Dividing the house into two equal parts,
the fire was started at about the center of the two parts each.


　　Former theories as reviewed in foregoing section fail to explaine
the real fire spread condition correctly.  As the most appropriate
expression, we have examined a polynominal curve and a logistic one.
By the method of MAICE (Minimum Akaike Information Criterion Estimation)
proposed by Dr. Akaike,  the optimal degree of polynomial can be
determined.  Fig.2 shows the relationship between the value of AIC (
Akaike Information Criterion) and degree of polynomial.
　　The value of AIC is given by

$$AIC = n \cdot \log(2\pi) + n \cdot \log(S) + n + 2 \cdot (m+2) \tag{1}$$

where n is the number of data and m is the degree of polynomial.
S is sum of squares residual and log is natural logarithm.
　　As the optimal degree is the case when the value of AIC is minimum,
the optimal degree is shown to be three from Fig.2.  But the polynomial
does not fit the real fire spread phenomenon because it is not always
a monotonial.  On the other hand,  the function of logistic curve is
good enough to explain the fire spread phenomenon.
　　The function of logistic curve given by

(×10²)

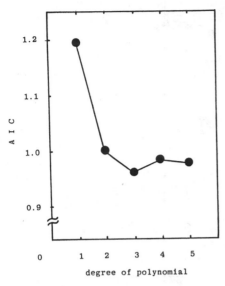

Fig.2 The relation of AIC and
and degree of polynomial
as a fire spread model

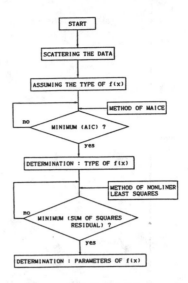

Fig.3 Schematic diagram of determining the parameters of
f(x) : In this case, f(x)
is considerd polynomial

$$A = \frac{G}{1 + \exp(-a(x-c))} \qquad (2)$$

where A ($m^2$) is burnt area at the time x (min) . G ($m^2$) is the maximum
area of wooden buildings that is vulnerable to fire spread. The values
of a and c are parameters to be determined statistically by the data.

By using above data, parameters of a and c is estimated to be
0.130 and 19.3 respectively by the nonlinear least squares method. This
curve is drawn with dotted line in Fig.1 . It is not convenient that
A does not equal zero if x=0 and also A does not equal G if $x=x_i$. $x_i$
is the burning-out time of building area G ($m^2$). Thus we modified the
equation (2) as follows:

$$A = \frac{G + d_1}{1 + \exp(-a(x-c))} - \frac{G + d_1}{1 + \exp(ac)} \qquad (3)$$

where $d_1$ is a modified coefficient given by

$$d_1 = \frac{1 + 2\cdot\exp(-ax_i + ac)+\exp(-ax_i+2\cdot ac)}{\exp(ac) - \exp(-ax_i + ac)} \qquad (4)$$

In this case, $d_1$ is about 46.2 and this curve is drawn with the solid
lines in Fig.1.

## 3. ANALYSIS OF THE NEW MODEL : THEORETICAL REMARKS

Our fire spread model formulated by equation (2) has a symmetry about the point of inflection. As the fire spread phenomenon does not always exhibit a symmetry, we have investigated a more generalized model of fire spread.

Equation (2) has a property of symmetry diagrammatically because a part of exponent is a linear function of x . If we choose another function, described f(x) as arbitrary function of x, the generalized equation is given by

$$A = \frac{G}{1 + \exp(f(x))} \qquad (5)$$

or

$$\log ( G/A - 1 ) = f(x) \qquad (6)$$

What kind of function is suitable to f(x), depends upon the data. If y = log( G/A -1 ), then this is no more than a problem of curve fitting to be determined statistically by the data set of x and y. In the case of equation (2), f(x) is considered a monomial. It is possible to consider that f(x) is a polynomial. If f(x) is asummed as polynomial of m degree, then

$$f(x) = b_0 + b_1 \cdot x + b_2 \cdot x^2 \cdots\cdots + b_m \cdot x^m \qquad (7)$$

The optimal degree of polynomial can be estimated by the method of MAICE which is already mentioned in the previous section.

There is no difficulty in being regarded the error of A as normal distribution statistically because A is the value of observation. But error of y=log(G/A-1) is not considered normal distribution because the error of A is considered of normal distribution. It is impertinent to determine the parameters of polynomial by the method of least squares to equation (6), because the basic assumption of the method of least squares is that the variables of data are considered normal distribution. Equation (6) is used for the purpose of determining the type of function f(x) only. In order to estimate the parameters of f(x) strictly, we should apply to the method of nonlinear least squares for equation (5). This methodology is indicated by Fig.3.

Here we would like to submit our opinion about the conditions of model formulation, summarized as follows:

(1) The property of the curve must be constituted with observed data. For example, it is necessary to satisfy with initial and boundary conditions as strictly as possible.
(2) The tendency of the curve is not contradictory to the phenomenon. For example, if the phenomenon has a property of monotone increment, the curve should not vibrate.
(3) The difference between the observed values and predicated ones must be as small as possible.

In order to make the equation (6) satisfy the above condition (1), we modified the equation (6) as follows:

$$A = \frac{G + d_2}{1 + \exp(f(x))} - \frac{G + d_2}{1 + \exp(f(0))} \qquad (8)$$

where

$$d_2 = \frac{1 + 2 \cdot \exp(f(x_i)) + \exp(f(x_i) + f(0))}{\exp(f(0)) - \exp(f(x_i))} \cdot G \qquad (9)$$

## 4. RESULTS AND DISCUSSION

In Fig.4, the relationship between $y=(G/A-1)$ and x in Eq.(6) is plotted. The points are totalled nine that are obtained from the whole fire of four buildings. If the function of this relationship is asummed polynomial, the optimal degree can be determined by the method of MAICE as described in the previous section. The optimal case is defined that the value of AIC given by Eq.(1) is minimum. The value of AIC were calculated for several cases and then the relationship between AIC and the degree of polynomial is shown in Fig.5. From this, the optimal function is known as trinomial. This is written by

$$f(x) = b_0 + b_1 \cdot x + b_2 \cdot x^2 + b_3 \cdot x^3 \qquad (10)$$

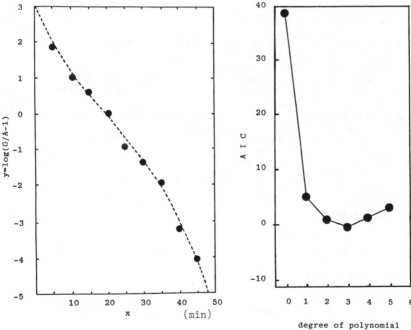

Fig.4 The relation of $y=\log(G/A-1)$ and x in equation (6)

Fig.5 The relation of AIC and degree of polynomial as $f(x)$ in equation (6)

where the unit of time for x is hour in view of circumstancies of the
calculation. In order to estimate the parameters of f(x) strictly,
we would submit that the method of nonlinear least squares should be
applied for Eq.(5). But conventional methods such as the log-transformed
type of Eq.(6) applying the method of ordinary least squares, though
simple and convenient, is impertinent. In recent years, the computer
and its argorithms have made rapid progress and the computer programs
of the method of nonlinear least squares have been developed such as
NOLLS1 by Mr. Tanabe (1981). The difference of the estimate parameters
between both methods are indicated by Table 1. From this, the ratios
of each parameter range from 0.96 to 1.87 and the nonlinear method of
sum of squares residual is smaller than the log-transformed one.
This means that the nonlinear method conforms to the data better.

Table 2 showns the sum of squares residual of the symmetric models
( Eq.(2) and Eq(3) ) and the asymmetrical models ( Eq.(5) and Eq.(8) ).
Eq.(2) and Eq.(5) are the prototype of these models. Eq.(3) and Eq.(8)
are the modified equations in order to satisfy the condition (1) of model
formulation. The sum of squares residual of each fire and whole one are
indicated in table 2. From this, it is known that the sum of squares
residual of the symmetric model is larger than the one of asymmetric
model. This means that the symmetric model dose not conform to the
data in comparison with the asymmetrical one.

Comparing the prototype and the modified one, neither of the two
is better at all studies. But modified one is superior to the prototype
in view of satisfication of the condition (1) of model formulation.
In Fig.6, the full line corresponds to the model of Eq.(8) and the
dotted line corresponds to the model of Eq.(5) about the whole fire
of four buildings .

Synthesizing the results of each paragraph of this paper, it
becomes clear that Eq.(8) is a more general model for the fire spread
of wooden buildings in Japan.

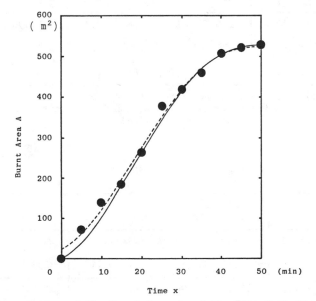

Fig.6 The relation of the Burnt Area A and the Time x as asymmetric
model of Eq.(5) and Eq.(8) about the whole fire

Table 1　Estimated parameters and sum of squares residual of trinomial

| | | $b_0$ | $b_1$ | $b_2$ | $b_3$ | S |
|---|---|---|---|---|---|---|
| Building NO.1 | ⓐ log-trns. | 2.714 | −39.57 | 121.8 | −291.7 | 446.477 |
| | ⓑ nonlinear | 4.525 | −69.75 | 227.2 | −280.2 | 13.217 |
| | ⓑ/ⓐ | 1.67 | 1.76 | 1.87 | 0.96 | 0.0296 |
| Building NO.2 | ⓐ log-trns. | 4.696 | −28.34 | 59.33 | −103.6 | 64.763 |
| | ⓑ nonlinear | 4.646 | −29.96 | 79.42 | −148.8 | 38.296 |
| | ⓑ/ⓐ | 0.99 | 1.06 | 1.34 | 1.44 | 0.591 |
| Building NO.3 | ⓐ log-trns. | 3.034 | −35.26 | 106.5 | −340.6 | 324.554 |
| | ⓑ nonlinear | 4.685 | −59.69 | 187.9 | −364.3 | 174.483 |
| | ⓑ/ⓐ | 1.54 | 1.69 | 1.76 | 1.07 | 0.538 |
| Building NO.4 | ⓐ log-trns. | 5.040 | −31.87 | 28.95 | −11.83 | 44.142 |
| | ⓑ nonlinear | 5.995 | −42.02 | 51.91 | −16.31 | 17.453 |
| | ⓑ/ⓐ | 1.19 | 1.32 | 1.79 | 1.38 | 0.395 |
| The whole fire | ⓐ log-trns. | 2.636 | −10.55 | 11.38 | −12.43 | 2487.96 |
| | ⓑ nonlinear | 3.070 | −13.89 | 19.11 | −17.97 | 2067.22 |
| | ⓑ/ⓐ | 1.16 | 1.32 | 1.68 | 1.45 | 0.831 |

(Abbreviations)　log-trns.　: the method of log-transformed type
nonlinear　: the method of nonlinear least squares
S　:　sum of squares residual

Table 2　Sum of squares residual of fire spread models

| | | Fire for each building | | | | The whole fire |
|---|---|---|---|---|---|---|
| | | No.1 | No.2 | No.3 | No.4 | |
| symmetric model | Eq.(2) | 65.118 | 147.405 | 241.421 | 112.429 | 2942.35 |
| | Eq.(3) | 61.731 | 130.428 | 230.522 | 103.122 | 4913.71 |
| asymmetric model | Eq.(5) | 13.217 | 38.296 | 174.483 | 17.453 | 2067.22 |
| | Eq.(8) | 39.597 | 31.241 | 221.433 | 50.445 | 3358.76 |

# 5. CONCLUDING REMARKS

The authors have reviewed former mathematical and statistical
fire spread formulas and models of wooden buildings in Japan. As a
result, it is found out that former theories fail to explain real fire
spread condition in present urban areas. The authors have therefore
deveroped a new fire spread model. This paper consolidates the reports
hitherto published by authors (1982,1983) as well as newly obtained
findings. A summary of our conclusion about the results is as follows:

(1) A polynomial model for the fire spread phenomenon would be best
    explained in a trinomial form. But this does not fit the real fire
    spread because it does not always constitute a monotone function.
(2) The logistic curve of Eq.(2) is good enough to explain the fire
    spread phenomenon consistently. As the modified model of Eq.(3)
    satisfies the condition (1) of model formulation, this is better
    than the prototype of Eq.(2) .
(3) But the above model exhibits a symmetry about the point of
    inflection. As it is difficult to assume that the fire spread
    phenomenon is symmetrical, we have examined a more general model.
    Synthesizing the results of each section of this paper, it becomes
    clear that Eq.(8) is a more general model for the fire spread of
    wooden buildings in Japan .
(4) Concerning the determination of parameters of nonlinear function,
    we should apply the method of nonlinear least squares and should
    not apply such method like the log-transformed method. In this
    study, the ratios of the difference of parameters between the
    nonlinear method and the log-transformed one range from 0.96 to 1.87.

Because this study is the basic one on the fire spread model,
we should make these fruitful conclusions apply the practical problems.
For example, Dr. Kuroda and authers have recently applied these findings
for the planning of aseismic fire cisterns in Japan (1985). Needless
to say, the present study is as yet a very fundamental one, so that
a number of questions remain to be further investigated for improving
fire prevention planning.

# 6. AKNOWLEGEMENTS

The authors express their gratitude to the Fire Research Institute,
Government of Japan, for providing useful data and are greatly indebted
to Dr. Tone for his helpful advice and to Mr. Tanabe for providing the
computer program NOLLS1 of nonlinear least squares method. The authors
also wish to thank Kinki University for the partial financial support
for the study.

# NOMENCLATURE

$A$ : burnt area of floor($m^2$)
AIC : Akaike Information Criterion given by Eq.(1)
$a$ : coefficient of Eq.(2)
$b_1, b_2 \cdots b_m$: coefficient of polynomial
$c$ : coefficient of Eq.(2)
$d_1$ : modified coefficient of Eq.(4)
$d_2$ : modified coefficient of Eq.(8)
$G$ : building area ($m^2$) of floor that a fire is vulnerable to spread

m : degree of polynomial
n : number of data
S : Sum of squares residual
x : elapsed time from the fire-break-out time (min or h)
$x_i$ : burning-out time (min or h) of building area G


REFERENCES

1. Akaike, H. : Information Theory and An Extention of The Maximum
   Likelihood Principle, 2nd Inter. Symp. on Information Theory
   ( Petrov, B. N. and Csaki, F. eds. ), Akademiai Kiado, Budapest,
   1973 (in English)
2. Yasuno, K. et al. : A basic study on fire spread of buildings,
   Bull. of JAFSE, Vol.31, No.1, 1982.a (in Japanese)
3. Yasuno , K. et al. : A basic study on fire spread formula of houses
   used by logistic curve, Trans. of AIJ, NO.311, 1982.b (in Japanese)
4. Yasuno, K. et al. : A basic study on effectiveness of suppressing
   building fire by water application, Bull. of JAFSE, Vol.31, No.2,
   1982.c (in Japanese)
5. Yasuno, K. et al. : A basic study of fire spread formula (logistic
   curve) of wooden buildings by finite difference graphical method ,
   Trans. of AIJ, NO.311, 1983 (in Japanese)
6. Fire Research Institute, Ministry of Home Affaires, Government of
   Japan : Experimental Report on the Coutermeasures againt Large-Scale
   Fire by Using the Actual Buildings at Saganoseki, Fire Research
   Institute, 1980 (in Japanese)
7. Suzuki, S. and Kinbara, T. : On Fire spreading of a conflagration,
   Applied Physics , vol.9 ,No.10 ,1940 (in Japanese)
8. Study Team of Fire Prevention of Tokyo : Fire spread speed, Marin
   and Fire Insurance Association of Japan, 1942 (in Japanese)
9. Tosabayashi, T. : On Conflagration of Muramatsu, Marine and Fire
   Insurance Association of Japan ,1947 (in Japanese)
10. Hishida, K. : Arithmetic of fire risk level, Marine and Fire
    Insurance Association of Japan ,1947 (in Japanese)
11. Horiuchi, S. : Study on fire prevention facilities in urban area ,
    a Dissertation submitted to Kyoto University, 1961 (in Japanese)
12. Fujita, T. : Model for fire spread and simulation, Research on
    Disaster, Vol.8, 1975 (in Japanese)
13. Sakai, K. : Study on Hamada's formula in view of burnt area and
    movement of fire spreading, City Planning Review Vol.18, 1975
    (in Japanese)
14. Sasaki, H. : Road width and fire spread, Trans. of Scientific
    Lecture meeting of AIJ, 1983 (in Japanese)
15. Yamashita, K. et al. : Report of fire spreading conditions of
    Sakata conflagration (I) - (IV), Fire, Vol.72, No.2-6, 1977
    (in Japanese)
16. Tanabe, K. et al. : NOLLS1, A FORTRAN subroutine for nonlinear
    least squares by quasi-Newton method, Computer science monographs,
    The Institute of Statistical Mathematics, Tokyo, 1981 (in English)
17. Kuroda, K. et al. : Decision theoretic approch to location planning
    of aseismic fire cisterns, Proceedings of JSCE, No.353, 1985
    (in English)

Abbreviations
  AIJ   : Architecutural Institute of Japan
  JAFSE : Japanese Association of Fire Science and Engineering
  JSCE  : Japan Society of Civil Engineers

# Full Scale Test of Smoke Leakage from Doors of a Highrise Apartment

**OSAMI SUGAWA and IICHI OGAHARA**
Center for Fire Science and Technology
Science University of Tokyo
2641 Yamasaki Noda-shi, 278 Chiba, Japan

**KAZUO OZAKI**
Mitsui Fudousan (Mitsui Real Estate) Co., Ltd.
2-1-1 Nihonbashi Muromachi Chuoku 103 Tokyo, Japan

**HIROOMI SATO**
Building Engineering Department, Kajima Corporation
2-19-1 Tobitakyu, Chofu-shi 182 Tokyo, Japan

**ISAO HASEGAWA**
Research Division of Mitsui Construction Co., Ltd.
518-1 Komagi Nagareyama, 270-01 Chiba, Japan

ABSTRACT

Check of smoke leakage of an entrance door,class A fire door, for highrise apartment was carried out in a full scale model using a model fire source which was designed to smolder 1 hour and then to flame. The door openable inward and outward with and without air tight material were used. A total 13 types of experimental conditions was carried out with major variables of door situation (open or close) and of pressure difference between fire room and corridor. Concentrations of smoke, gas, and smoke particles, pressure, temperatures,and weight of fire source were measured. No difference in smoke leakage performance between doors openable inward and outward was obtained. Smoke and combustion gas in corridor were hardly observed when the entrance door was closed, therefore it suggests clearly that the middle corridor is safe enough as an evacuation route when the door was closed.

KEY WORDS; FULL SCALE TEST, SMOKE LEAKAGE FROM DOOR, CLASS A FIRE TIGHT DOOR, SMOKE PARTICLES, FIRE DETECTOR, SMOLDERING

1. Introduction and Objective of this Work

The highrise apartment for this study has a star shape arrangement with middle corridor and veranda as two evacuation routes. There is danger of hot smoky gas filling in the middle corridor as if combustion gas comes out through/around the entrance door. It is very plausible that the residents would take the middle corridor as an evacuation route because they know it well. Therefore, it is necessary and important to keep the middle corridor free from fire products.

The objectives of this experimental study is to check the doors performance against smoke leakage under the conditions of pressure difference given between the fire room and corridor with using an early stage of a fire source which grows from smoldering to flaming. Fire detectors performance is also examined.

## 2. Fire scenario and Model fire source

The fire statistics shows that about 26% of fires is caused by a cigarette and about 29% of first item of fire development is occupied by a bedding and a sofa. The death rate in midnight is higher than the other period for resident fire. Concern with these situations, the fire scenario to follow is set. A lighted cigarette fell on a sofa and makes smoldering combustion first, then flaming which develops to involve the wallpaper and finally to flashover. The first two stages of the fire, about 1 hour smoldering and 10 min flaming, are adopted for this study as the model of an early development of fire.

### (a) mockup cushion

Babrauskas(ref.1) reported that the optical smoke concentration (Cs) of 1/m was generated from a standard chair in 28.3 $m^3$ volume of fire room. Kawagoe and Mizuno(ref.2) also reported that a cushion gave 1-1.2/m of Cs. A sofa and two loungers, are like a bedding, were adopted as the first item of a fire. The smoke concentration from them is about 4-5 times greater than it from a chair. The volume of fire room is about 42.5 $m^3$ then the smoke load is expected to be about 200 $m^{-1} \cdot m^3$. Several kinds of mockup cushions fire had been tried in preliminary tests, and its smoke evolution repeatability was especially tested. The fire source adopted finally was consists of 60 cm X 60 cm X 6 cm(T) of polyurethane foam of about 240 g, cotton batting of 160 g, and thick cotton cloth of 270 g. It gave the smoldering duration of about 60 min and Cs V of about 70 $m^{-1} \cdot m^3$. Thus, three cushions were used simultaneously to give the designed smoke load. The average mass loss rate of a chair is about 6-8 g/sec for its first 10 min (ref.1,2) flaming. Therefore, the heat release rate of about 100 Kw for 10 min was planned to give by the alcohol burning.

### (b) Full size fire source

A sofa of about 22.7 kg, two loungers of 9.3 and 8.8 kg, a wooden side table of about 8 kg, a wooden magazine rack of 0.5 kg, a 14' TV set with wood frame, carpet of 4 m X 4.5 m, wall- and ceiling paper, and a book stand of 90 cm width and 120 cm height with about 150 kg of books were adopted.

## 3. Experimental

Monthly average wind velocities in Tokyo area are about 2.9, 3.3, 3.0, and 2,8 m/sec for Jan., May, Aug., and Oct.. Thus 3 m/sec was adopted as an average wind velocity for estimation of the pressure difference between fire room and middle corridor. The corridor is semi-enclosed, therefore, coefficients of upwind and downwind pressures are reduced to half of them, compared with those of ordinary case. About 0.5 mmAq of pressure difference was given between fire room and middle corridor. The plan and vertical views of the facility is shown in Fig. 1. The total ventilating openings was controlled to adjust to the respective value of the real one by sealing off with tapes and putty.

Door 1 was a wooden partition door. Door 2 (inward openable) and door 3 (outward openable) are class A fire door. Leakage characteristics of the door was tested preliminary, and gave the performance of $Q = 0.22 \cdot P^{0.76}$ (0.5 $\leq$ P< 20), where Q is ventilation air volume ($m^3$/h) and P is positive pressure difference ($kg/m^2$) between inside and outside.

In order to measure the smoke movement and concentration, extinction beams were set in as partly illustrated in Fig. 2. Temperature were measure by K-type thermocouples. The sample gas was introduced about 10 cm below the ceiling and was sent into the analysers. CO, CO2 and O2 gas concentrations were measured. The sample gas was then returned back to the respective section. Two kinds of smoke detectors, ionization (I.S.D.) and scattering type (L.S.D.) which were modified to give analog outputs, and also two kinds of heat detectors, rate of rise heat and fixed temperature type were employed. Simple weighing system was used which consists of load cells with a water jacket and of a bed or a platform. Outputs from many sensors were recorded every 2 min. The smoke filling and movements were also observed and recorded by camera and two sets of video system.

Experimental conditions of door openings, fire source, and pressure difference are shown in Table 1.

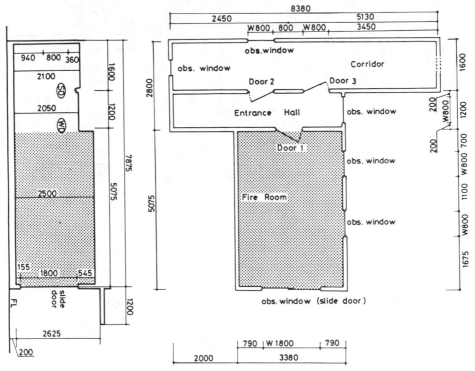

Figure 1    The plan view and vertical view of the facility for this experimental study. The mm unit is adopted for scale.

4. Results and Discussion
(a) Mockup cushions Fire
    A pill of METHENAMINE with an electric heater which activated for 10-15 sec was used for ignition method.
    The smoldering area on the mockup cushion increased circular-

893

ly at the rate of about 1 cm/min in radius, and about 30 min after, burning zone developed into the foam. The typical results for weight loss versus time are shown in Fig.3. Each plot shows a remarkable tendency to cluster the curve. This indicates that every fire source gave almost same amounts of smoke, gas, and heat at the almost same release rates. Almost steady state of smoldering combustion at the rate of 15.5 g/min of three cushions was observed for 30-65 min. After 60 min, the alcohol was ignited and gave the heat release rate of about 107 kW for 10 min.

Fire Room
The average smoke filling rate was about 0.3 m/min. After 30 min, the smoke concentration increased rapidly based on the maximum weight loss rate and smoke concentration (Cs) of 4-5/m was observed for 50-60 min.

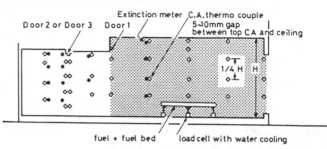

Figure 2    Some locations of measurements system for temperature and smoke.

CO and $CO_2$ gas concentration gradually increased to about 0.2% and 1% at around the end of smoldering duration, and $O_2$ decreased to 20.3%. In this period the ratio of $CO/CO_2$ were almost constant of 0.2. At around the end of smoldering duration, average temperature rise in fire room was about 8 K with very flat temperature distribution along the vertical direction. This suggests that the fire detection applying temperature rise, both fixed type and rate of rise heat type, may fail in detection even in fire room for a smoldering fire. After 60 min, hot layer of 1 m thick under the ceiling indicated 70-80 K temperature rise above initial temperature and which drove the smoke and gas away to the entrance hall and gave the decreases in concentrations of smoke and gas. Little

Table 1    EXPERIMENTAL CONDITION

| Exp. No. | Press diff. | Fire source | DOOR1 (1) | DOOR2 (2) | DOOR3 (3) | Slide door |
|---|---|---|---|---|---|---|
| PRE-1 | 0.5mmAq | REAL | CLOSE | OPEN | — | CLOSE |
| A-2 | 0.5mmAq | CUSHION | CLOSE | OPEN | — | CLOSE |
| A-3 | NO | CUSHION | CLOSE | OPEN | — | CLOSE |
| B-1 | 0.5mmAq | REAL | OPEN | CLOSE | — | CLOSE |
| B-2 | 0.5mmAq | CUSHION | CLOSE | CLOSE | — | CLOSE |
| B-3 | NO | CUSHION | CLOSE | CLOSE | — | CLOSE |
| B-4 | NO | CUSHION | OPEN | CLOSE | — | CLOSE |
| B-5 | 0.5mmAq | CUSHION | OPEN | CLOSE | — | CLOSE |
| C-1 | 0.5mmAq | CUSHION | CLOSE | — | CLOSE | CLOSE |
| C-2 | NO | CUSHION | CLOSE | — | CLOSE | CLOSE |
| D-1(4) | 0.5mmAq | CUSHION | CLOSE | — | CLOSE | CLOSE |
| E-1 | 0.5mmAq | CUSHION | OPEN | CLOSE | — | OPEN |
| ADD-1 | 0.5mmAq | CUSHION | CLOSE | | CLOSE | CLOSE |
| ADD-2 | 0.5mmAq | CUSHION | CLOSE | CLOSE | — | CLOSE |

wooden door(1)    inward openable(2)    outward openable(3)
without air tight material (4)

difference in smoke and gas concentrations was found independent of door 1 opened or closed. Table 2 and 3 show the concentrations of smoke and CO, CO2 and O2 gas in fire room, entrance hall, and middle corridor.

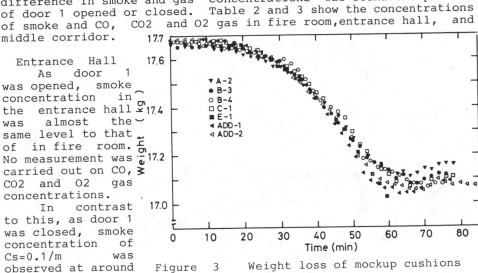

Figure 3 Weight loss of mockup cushions versus time.

Entrance Hall
As door 1 was opened, smoke concentration in the entrance hall was almost the same level to that of in fire room. No measurement was carried out on CO, CO2 and O2 gas concentrations.
In contrast to this, as door 1 was closed, smoke concentration of Cs=0.1/m was observed at around 120 cm which almost corresponds to the height of eyes and nose and maximum temperature rise of 3-5 K was obtained at the end of flaming duration. However, thicker smoke of about 1/m was obtained under the ceiling and above the floor. CO and CO2 gas concentration were almost 1/5 to the ones in fire room in smoldering duration. After 60 min, an increase of smoke concentration of 0.7-1.2/m was observed, which was almost 1/3 - 1/2 of the concentration in fire room. Concentrations of CO and CO2 increased to 0.063-0.064% and to 0.7-0.9%, respectively. The apparent dilution ratio between fire room and entrance hall depend on gas concentration change was about 1/3 which is almost equal ratio to the one estimated by Cs change. 3-5 K of temperature rise was observed at the end of smoldering duration, and in flaming duration about 40 K was obtained under the ceiling indicating the triangle shape distribution with about 0.2 K/cm along the vertical direction. If there were no apparent natural convection and flow in an entrance hall, it is almost certain that

**Table 2  SMOKE CONCENTRATION (Cs; m⁻¹)**

| FIRE ROOM | DOOR 1 | ENTRANCE HALL | DOOR 2 or 3 | CORRIDOR |
|---|---|---|---|---|
| Cs = 4~5 ↓ 60 min after ↓ Cs = 2~2.4 | O P E N | Cs = 4~5 ↓ 60 min after ↓ Cs = 2~2.4 | C L O S E | Cs = 0 |
| Cs = 4~5 ↓ 60 min after ↓ Cs = 2~2.4 | C L O S E | Cs = ~0.1 (max. 0.7~1.2) ↓ 60 min after ↓ Cs = 0.7~1.2 | C L O S E | Cs = 0 |
| Cs = 4~5 ↓ 60 min after ↓ Cs = 2~2.4 | C L O S E | Cs = ~0.1 (max. 0.7~1.2) ↓ 60 min after ↓ Cs = 0.7~1.2 | O P E N | Cs = 0.1~0.2 ↓ 60 min after ↓ Cs = 0.7~0.8 |

895

the hall which is partitioned even by a simple wooden door is safe enough to use as an evacuation route.

Middle Corridor

The smoke leakage from the entrance door was hardly measurable even in the flaming duration which gave about 5 mmAq pressure difference between fire room and corridor.

In contrast to this, when entrance door was opened, smoke concentration at the front of entrance door was about 0.1-0.2/m and peak concentrations of CO gas was 0.011-0.013% and of CO2 was 0.053-0.06% which were about 1/3 to the ones in entrance hall, and the ratio of CO/CO2 of 0.2 was conserved in smoldering duration. After 60 min, smoke concentration increased to 0.7-0.8/m with time delay of about 2 min. Little dilution on smoke concentration was observed with it parts when it was driven to corridor from entrance hall. Smoke moved so slowly with drifting and conserved its distribution like a smoke-cloud. This kind of smoke movement was also reported as the movement in a long corridor (ref.3). Table 2 shows the almost the same smoke concentrations were observed in entrance hall and in corridor after flaming, and these are almost 1/3-1/2 to the one in fire room. The upper limit of smoke concentration which begins to give serious emotional fluctuation to residents is about Cs of 0.1-0.15/m (ref. 4), and Cs of 0.7-0.8/m gives walking speed of 0.3-0.7 m/sec (ref. 5). This tells that there is a strong fear of smoke blocking against the evacuation in middle corridor when door was opened. Closing of door 2, independent of door 1 opened or closed, produced about 1/30 CO gas concentration relative to one in fire room. As can be seen from Table 3, this concentration level may give an enough time to evacuation before toxic gases rise up to serious levels. As the closed entrance door without air tight material gave the smoke concentration of about 0.3-0.5/m about 10 min before the flaming period. The smoke contaminated region was found at least 4 m for both side from the entrance door. Little temperature rise was measured in corridor even in the flaming duration. Therefore, these phenomena strongly suggested that it is necessary to set the auto door closer and air tight material to the entrance door.

**Table 3    GAS CONCENTRATIONS (%)**

| FIRE ROOM | DOOR 1 | ENTRANCE HALL | DOOR 2 or 3 | CORRIDOR |
|---|---|---|---|---|
| CO = 0.2<br>CO2 = 1.0<br>O2 = 20.3<br>60 min after ↓<br>CO = 0.2<br>CO2 = 2.0<br>O2 = 18.5 | CLOSE | CO = 0.037 ~ 0.047<br>CO2 = 0.20<br>O2 = 20.7<br>60 min after ↓ ~ 20.8<br>CO = 0.063 ~ 0.064<br>CO2 = 0.7~0.9<br>O2 = 20.03 | CLOSE | CO = 0.0048<br>CO2 = 0.048<br>O2 = 20.81<br>60 min after ↓<br>CO = 0.0056<br>CO2 = 0.069<br>O2 = 20.78 |
| CO = 0.2<br>CO2 = 1.0<br>O2 = 20.3<br>60 min after ↓<br>CO = 0.2<br>CO2 = 2.0<br>O2 = 18.5 | CLOSE | CO = 0.037 ~ 0.047<br>CO2 = 0.20<br>O2 = 20.7<br>60 min after ↓<br>CO = 0.063 ~ 0.064<br>CO2 = 0.7~0.9<br>O2 = 20.03 | OPEN | CO = 0.011 ~ 0.013<br>CO2 = 0.053 ~ 0.060<br>O2 = 20.9<br>60 min after ↓<br>CO = 0.023<br>CO2 = 0.11~0.12<br>O2 = 20.83 |
| CO = 0.2<br>CO2 = 1.0<br>O2 = 20.3<br>60 min after ↓<br>CO = 0.2<br>CO2 = 2.0<br>O2 = 18.5 | OPEN | NOT MEASURED | CLOSE | CO = 0.0048<br>CO2 = 0.048<br>O2 = 20.81<br>60 min after ↓<br>CO = 0.0056<br>CO2 = 0.069<br>O2 = 20.78 |

Pressure change and air ventilation rate
        In the smoldering duration,  very little pressure difference
was  obtained,   but  pressure jumped up to about 5 mmAq  based  on
alcohol  burning.   This   high  pressure  difference   resulted   in   a

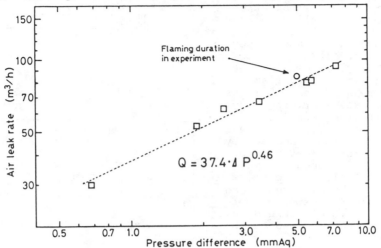

Figure   4      Logarithmic   plots   of   air   leak   rate   against
              pressure difference between fire room and corridor.

greater ventilation rate through   many   small  openings, then negative   pressure difference of about -10 mmAq was given to the   fire
room depend on O2 consumption. Air exchange rate during smoldering
and  flaming  was monitored using a tracer gas.  The air  exchange
rate  estimated by the tracer gas concentration change  was  1.1/h
for  smoldering  and  1.8/h for flaming for  this   facility.    The
relationship between pressure difference and air volume of leakage
was   tested preliminarily with using a fan with controller   and   a
Venturi tube. This gave the relationship of Q = 37.4 P  ,as shown
in  Fig.4.   Air exchange rate in flaming duration is shown with a
circle in Fig. 4.
        No  difference  was found in the leakage  behavior  of  smoke
through/around  the   entrance  door with or without  the  pressure
difference which was given to estimate the natural wind effect.

   Working time of fire detectors
        The working times of smoke fire detectors estimated based  on
their  outputs and are shown in  Fig.  5.   These figures tell the
apparent tendency that L.S.D.  worked a little earlier than I.S.D.
did in fire room,  and roughly the opposite tendency was  obtained
in  entrance hall.  And in corridor,  the I.S.D.  worked in a  few
cases  of entrance door opened,  and of the door without air tight
material. These working behaviors  may come from the difference in
size  distribution  of smoke particles which was  induced  by  the
difference  of diffusive characteristics depend on particle  size.
Number  concentration and mobility of larger size smoke  particles
is small relative to smaller ones (ref.  6).  And for smaller size
particles,  I.S.D.  can  work earlier relative to L.S.D.(ref.  7).
It  is plausible that these situation gave the  earlier  detecting

time to I.S.D. relative to L.S.D. In this series of experiments, observation windows which connects to outside were closed except E-1. An attention should be called on that the ambient air in fire room, in entrance hall, and in corridor was very still, no apparent air flow was given except the flows induced by flame or by pressure difference. These caused no excess driving force when particles gets into the labyrinth of L.S.D. In real case, air flow induced by natural and/or forced convection give the adequate movement to particles. Therefore, no significant difference on working times is expected between both types of smoke detector when they mounted to the ceiling of entrance hall which connects to fire room.

No heat detector worked in smoldering duration even in fire room. It was apparent that smoke detectors are advantageous to heat detector on a smoldering fire.

(b) Real Furniture Fire

In smoldering duration, the changes of temperature, smoke concentration, gas concentration depend on growth of fire were almost as same as ones which were observed for mockup cushions fire. At a-

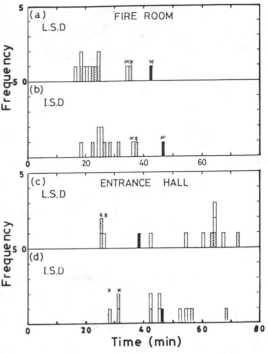

Figure 5 Frequency of smoke detector of light scattering type (L.S.D) and ionization type (I.S.D) in fire room and in entrance hall. Histogram with * is door 2 open, with mesh screen is pressure difference given, and with rigid one is real furniture fire source.

round 50 min, smoldering zone developed into the foam of back part of the sofa, then after past 60 min, it reached to a small amount of alcohol which was set preliminarily to get a flaming combustion. Flaming combustion grew rapidly inside the back space of the sofa and at about 65 min developed to wallpaper and to other combustible materials. Rapid temperature rise was observed and which gave pressure difference of over 5 mmAq, which was beyond the range of our system. Hot dense smoky gas exhausted giving hiss through the gaps of aluminum sash of slide door. Combustible hot gas which drifted under ceiling burned toward down drawing a layer of 30 - 50 cm thick. This burning phenomena were observed at least twice in about 1 min before break down of slide door. Temperature, smoke concentration, and number concentration of particles covering 0.2-2.0 μ by 5 steps of channel isolation were monitored in corridor. Increase of temperature and smoke concentration which were measured by employed system were hardly observed both in flaming duration and after flashover. Fig. 6 shows the number concentration of particles versus time. When entrance door was open, it is clear of smoke coming into corridor as shown in Fig. 6-(a). In contrast to this, as the entrance door closed, the number concentration of particles increased little till about 30 min, and slightly increased for 30-40 min. And at around 50 min and after, apparent increasing were obtained in each channels as shown in Fig.6-(b),(c). The measurable upper limit of the number concentration for covering range of 0.2-0.5 μ, was about 2 x $10^{11}$/$m^3$ depend on coincidence loss (ref. 8). Therefore, it is better to use the number concentration of channel 4 and channel 5 to compare the leakage performance. For example, the case of entrance door open, number concentration of 1.0-2.0 u particles was 10 times greater than one in case of entrance door closed.

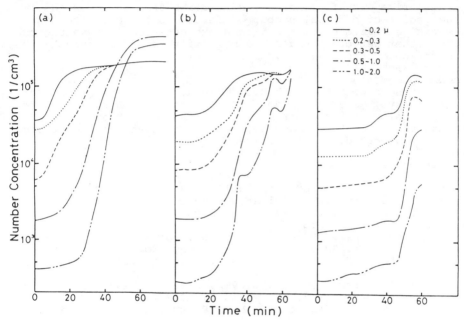

Figure 6    Time histories of number concentration of smoke particles, (a) Exp. A-2, (b) Exp. B-2 for mockup cushions and (c) Exp. B-1 of real furniture fire.


5. Conclusion
     In the middle corridor, smoke leakage from the entrance door(class A fire door with air tight material) was hardly observed even the fire room was pressurized as high as 5 mmAq or more by fire. However, the door without air tight material permitted the smoke leakage of over 0.1/m of which concentration must give the disorientation to residents. Therefore, the key point of successful evacuation using a middle corridor depends on closing of the entrance door and installation of an auto door close and an air tight material.

No difference was obtained with respect to the smoke leakage performance between inward and outward openable door with the pressure difference given between fire room and corridor. If the entrance door satisfy the smoke leakage performance as mentioned above, a criterion of door selection whether it opens inward or outward primarily depends on security performance and on the matter of convenience for usual use. However, it is preferable to adopt a outward panic door which installed in a middle corridor to confirm the compartmentation and evacuation.

It was hard to expect the early detecting of smoldering fire by heat detector. In the fire room, light scattering smoke detector worked earlier relative to ionization smoke detector , however the difference in the working time between them were not serious. Considering a development of fire growth from smoldering to flaming and a fire with flaming at its starting and also considering on the reduction of total number of detectors to be mounted to a residential compartment, ionization smoke detector is advantageous to the other detector as a residential fire detector except the consideration on the frequency of false alarm.

6. Acknowledgments

The authors wish to thank Mr. Takeda, chief official of Sci. Univ. of Tokyo, and Mr. Endo, a member of Nohmi Bosai Ltd., for their kind help for the experiment. We thank Mr. Nishimoto, Mr. Yamanaka, and students of Kawagoe, Handa, and Ogahara laboratories for fully support of this experiment.

7. References

1) Babrauskas, V., NBS TECHNICAL NOTE 1103 Aug. (1979)
2) Kawagoe, k., Mizuno, T., Private Communication
3) Handa, T., Hamada ,T., Fukaya, H., Sugawa, O., Kaneko, K.,
   Furukawa, Y ., and Endo, K.,
   Bull. Jpn. Asso. Fire Sci. and Eng. vol.28, No.2 (1978)
4) Jin, T., J. Jpn. Asso. Fire Sci. and Eng. No.97 p44 (1975)
5) Jin, T., Bull. Jpn. Asso. Fire Sci. and Eng. vol.30,
   No.1 p1 (1980)
6) Handa, T., and Sugawa, O., CIB Symp. III-5, TSUKUBA Aug.(1977)
7) Lee, T., and Mulholland, G., NBSIR 77-1312 (1977)
8) Handa, T., and Nagashima, T., Fire Flamm. vol.1 p265 (1977)

# External Radiation at a Full Scale Fire Experiment

**ISAO TSUKAGOSHI and EIICHI ITOIGAWA**
Building Research Institute
Oho, Tsukuba, Ibaraki 305 Japan

ABSTRACT

For the vast wooden house built-up areas in Japanese cities, Firebreak Belts (F.B.B.) are thought practical and effective against a post-earthquake urban fire. When a wooden house area is guarded by a vertical type of F.B.B. such as elevated traffic ways, a row of fireproof buildings, etc., this area might be exposed to the radiation mainly from the upper part of flame. The existing calculation method which assumes the homogeneous distribution within an imaginary plane of heat source should be reconsidered and the difference between upper and lower parts of flame in radiation intensity must be discussed theoretically and empirically. In this paper[1], after the descriptions on the background and purpose of study, the outline of a full scale fire experiment are explained and the data of its external radiation are analyzed. Finally, a calculation model of the radiation from a big urban fire is proposed for the evaluation of effect of F.B.B..

INTRODUCTION

Background of Study

Major Japanese cities such as Tokyo and Osaka are quite vulnerable to urban fires because of the existence of a great number of combustible buildings. Specially in case of a great earthquake, this condition of urban area might cause the wide spread of multiple and simultaneous fires, which will grow up into a great urban fire as sizable as a whole built up area.

In order to cope with this problem, a system of Urban Fire Prevention Units is proposed. (Outline of the system is introduced in Ref. 3) The basic concept of this system is the compartmentalization of a vast combustible area, being divided into a number of small combustible zones by means of a effective allocation of linear non-combustible zones called Firebreak Belt (F.B.B.). The realization of this system would require a specific urban planning and some administrative institutions and, at the same time, it is quite necessary to establish a practical design method for F.B.B. including a quantitative evaluation method for effect of F.B.B..

---

[1]This work is a revision of Ref. 1 and Ref. 2.

 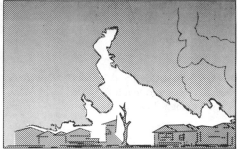

Figure 1.   Shizuoka big fire
(reproduced by the author
from the ilustration in Ref. 5.)

Figure 2.   Ohdate big fire
(reproduced by the auther
from the photo in Ref. 6.)

Shape of Urban Fire Flame

To evaluate practically the effect of F.B.B. against an urban fire, it is
necessary to estimate the radiation intensity from a big merged flame by means
of a simplified model of complicated shapes of flames.   In the Ref. 4, Dr.
Hamada has proposed a model of radiation source as a rectangular plane inclined
downwind and regarded the radiation distribution within this plane as
homogeneous.

There are two types of F.B.B.; one composed of flat elements like surface
traffic ways, flat open spaces, etc., and the other the combination of vertical
elements such as elevated traffic ways, densely planted trees and a row of fire
proof buildings.

The radiation received behind the flat type of F.B.B. can be calculated
almost correctly even through the model of homogeneous heat source since a
heat receiving point is facing against the whole plane of heat source.   In case
of vertical type, however, this method is not necessarily reasonable because the
vertical walls interrupt the radiation from the lower part of flame and only the
upper part of flame above the walls is important. Therefore, we must discuss
the distribution of radiation to the vertical direction along the surface of flame.

An urban scale fire is different from ordinary fire mainly in its size of
flame.   The height of flame for a two storied wooden house is approximately 15
meters in the maximum, but in case of an urban scale big fire, it is believed to
become much higher with the effect of merge of flames.

Figure 1. and 2. show the shapes of flames at the Shizuoka big fire in
1940 and at the Ohodate big fire in 1955 respectively.   They indicate that the
flames of urban scale fire are shaped like a series of pyramids.   For this kind
of shape, the heat source plane can be assumed as multiple triangles, but this
assumption leads to a meaningless complication at the calculation of the view
factor. (See Figure 3.)   So that, we have assumed, like Hamada's method, a
rectangular plane where the radiation intensity depends upon the vertical
coordinate of heat source.   The flame shape illustrated in Figure 3. makes us
anticipate a linear reduction of heat flux according to the height of heat
source.   To verify this relation quantitatively, a full scale fire experiment was
executed.

902

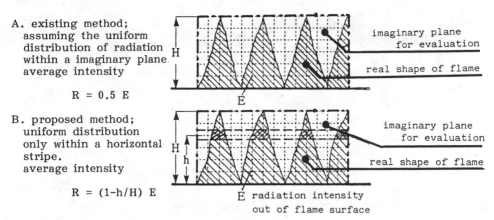

A. existing method;
   assuming the uniform
   distribution of radiation
   within a imaginary plane
   average intensity

   R = 0.5 E

B. proposed method;
   uniform distribution
   only within a horizontal
   stripe.
   average intensity

   R = (1-h/H) E

imaginary plane
   for evaluation

real shape of flame

imaginary plane
   for evaluation

real shape of flame

E radiation intensity
out of flame surface

Figure 3.   Real shape of flame and imaginary plane of heat source

EXPERIMENT

Outline of Experiment

   For the purpose of this study, it was quite desirable to realize the
simultaneous burning of a group of houses but in view of the safety condition
of experiment it was difficult.   And among the aborted buildings in the
Tachikawa base of U.S.army, a two storied wooden house for four families were
selected as the object of experiment.   The outline of this building is shown in
Table 1. and in Figure 4. and other conditions of the experiment are indicated
in Table 2..

   The experiment was performed on October 1st in 1980.   In order to get a
maximum size of flame, the burning were initiated at the four points indicated
in Figure 5. using four wooden cribs, under the condition of all the windows

Table 1.   Outline of test building

| | |
|---|---|
| place | natl. park of Syouwa-Kinen Kohen, Tachikawa, Tokyo; |
| use | residential |
| floor area | ground fl.   334.48 $m^2$<br>total fl.   428.01 $m^2$ |
| structure | wooden; partly 2 storied |
| roof | asbestos cement roof tile |
| ext.wall | mortal and paint |
| ceiling | plaster bd. and paint |
| partition | plaster bd. and paint |
| floor | p.v.c. tile |

Table 2.   Experiment condition

| | |
|---|---|
| date | Oct. 1, 1980,   11:00 |
| climate | fair, 23°c, 42%,<br>av. wind 2.7 m/s N-NE |
| loaded combust. | timber of destructed house 57kg/$m^2$ uniformly set on the floor,   (water 23.3%) (fixed load   143 kg/fl.$m^2$) |
| total fuel | 200 kg/$m^2$ per total fl.<br>256 kg/$m^2$ per ground fl. |
| ignit. | 4 points same time at the north entrances, using wood crib |
| windows & doors | all opened to the full |

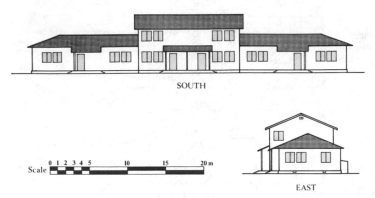

Figure 4. South and east side views of test building

and doors remained open. Immediately after ignition, the fires were fully developed all over the initial rooms. About three minutes after, they reached to the all the rooms on the ground floor and 10 or 11 minutes later through the staircases to the second floor. After the roof of the second floor was broken (about 20 minutes after ignition), the heights of flames became higher as is shown in Figure 6.

Observation Method

Following equipments were used for the observation in regard to the radiation;

1. Thermoviewer (Infrared thermometer)   x 1,
2. Video camera                          x 2,
3. Radiation meter                       x 4.

Measuring conditions of the equipments are shown in Table 3. and the allocation of them in Figure 5.

Table 3. Measuring condition

| equipment & type | record interval | recording method |
|---|---|---|
| thermoviewer (Nihon Densi co.ltd.) | 1 sec | on 1/2" video tape by Sony SLO-330. |
| video camera (Sony DXC-2000) | 1/30 sec | NTSC composite signal on 3/4" tape by Sony VO-2950 |
| radiation-meter | 30 sec (R1,R4) 120 sec (R2,R3) | YODAC thermal recorder (Yokokawa Denki co,ltd) |

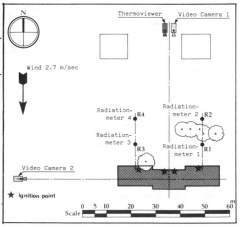

Figure 5. Location of equipments

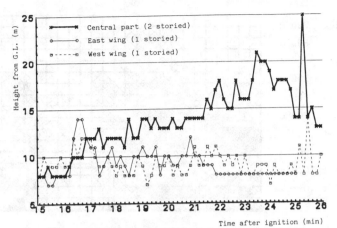

Figure 6. Time history of flame height

With a scanning system, the thermoviewer has a function to output the temperature distribution of the object whose emissivity can be considered the same as the setting value on the equipment. In this case, the emissivity was set up to 1.0. Therefore, the indicated value for the object of low emissivity doesn't mean the exact value of temperature but it corresponds to the intensity of radiation flux including the effect of emissivity.

Time History of Flame Height and Radiation

The time histories of the maximum flame heights (above the ground level) for the one and two story parts were obtained with two video cameras as are shown in Figure 6.. From these data, it can be known that the most active burning period for the two story part would be from 23 to 25 minutes after ignition. Figure 7. is the time histories of radiation intensities observed at four radiation meters and shows the most active burning period from 21 to 23

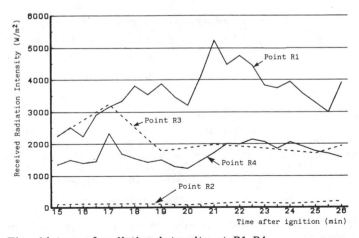

Figure 7. Time history of radiation intensity at R1-R4

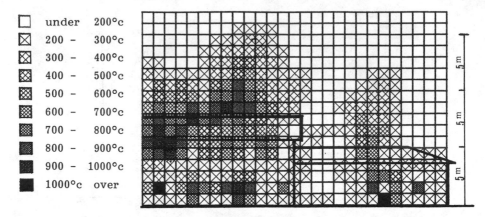

| | |
|---|---|
| ☐ | under 200°c |
| ⊠ | 200 - 300°c |
| ⊠ | 300 - 400°c |
| ⊠ | 400 - 500°c |
| ⊠ | 500 - 600°c |
| ▨ | 600 - 700°c |
| ▨ | 700 - 800°c |
| ▨ | 800 - 900°c |
| ▨ | 900 - 1000°c |
| ■ | 1000°c over |

Figure 8. Distribution of average temperature from 21 to 25 min. after ignition

minutes. After all, we have judged that the most active and stable burning period of this experiment was from 21 to 25 minutes after ignition.

ANALYSES

Distribution of Radiation Intensity

The thermoviewer analogue data recorded on a magnetic tape were processed to digital data in accordance with the divided meshes (one meter square for each) on the vertical plane assumed at the north side of the building. Processing time interval was every 30 minutes.

With these data, an average temperature for each mesh can be calculated according to the following formula;

$$T_{ij} = \sqrt[4]{T_{ijt}^{4}/p} \tag{1}$$

where

$T_{ij}$ : average temperature for mesh (i,j)  (K)

$T_{ijt}$ : observed temperature for mesh (i,j) at time t  (K)

p : number of data for a mesh (t=1 to p)

i : horizontal position of mesh (i=1 to m)

j ; vertical position of mesh (j=1 to n).

Figure 8. is the illustration of average temperatures ($T_{ij}$) in the period from 21 to 25 minutes. Within the plane of heat source, the difference of temperature might be caused by the fluctuations of the flame shape above the roof line and by the allocation of openings on solid mortal walls below the roof line. However, we can recognize a total tendency that the average value of temperature in the upper part is less than in the lower part.

Comparison between Observed and Estimated Radiation

For the validation of the function of thermoviewer, estimated radiation values corresponding to the thermoviewer temperature were compared with the

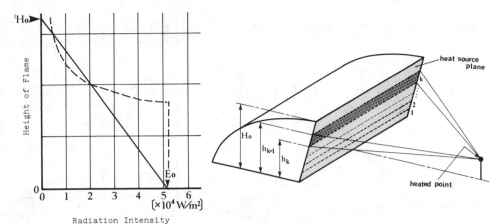

Figure 11.  Simplified profile
of radiation on a heat source
plane

Figure 12.  Heat source for calculation of
radiation

$$R_{ij} = f_{ij} \cdot s \cdot T_{ij}^{4} \tag{4}$$

where $T_{ot}$ in the formula (3) is relatively small and has been neglected.


Vertical Profile of Radiation

An average value of different radiation intensities for the same vertical
position of a heat source mesh was calculated by the following formulae;

$$E_{j} = ( \sum_{i} \sum_{t} E_{ijt} ) / (p \cdot m) \tag{5}$$

$$E_{ijt} = s \cdot (T_{ijt})^{4} \tag{6}$$

where    $E_{j}$    : average value of different radiation intensities for the same
                vertical position      $(W/m^{2})$

         $E_{ijt}$   : radiation intensity for mesh (i,j) at time t      $(W/m^{2})$

With the observed data for the two story part, the vertical profile of
radiation was calculated and is indicated in Figure 10.   It is clear that above
the roof line, the average radiation decreases according to the increase in
height of a heat source.    Below the roof line, however, the large scatter in
radiation flux has been caused by the influence of the openings located on the
solid mortal wall.


CONCLUSION

If the results of this experiment can be generalized, the dotted line in
Figure 11. should be proposed as a model of vertical profile in radiation flux.
However, to adopt this model for an actual urban area, this must be well
confirmed through other validation studies; for different constructions of
building and for different fire conditions.

907

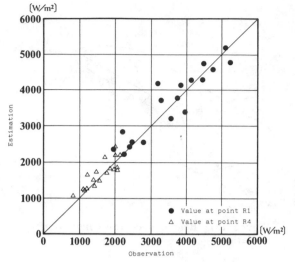

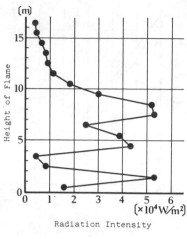

Figure 9. Comparison of observed & estimated radiations

Figure 10. Vertical profile of radiation on the flame

actual radiation values measured by the radiation meters of R1 and R4. The method of estimation is as follows;

$$R_{ot} = \sum_i \sum_j R_{ijt} \qquad (2)$$

$$R_{ijt} = f_{ij} \cdot e_{ij} \cdot s \cdot (T_{ijt}^{4} - T_{ot}^{4}) \qquad (3)$$

where
$R_{ot}$ : total radiation at receiving point at time t  (W/m$^2$)

$R_{ijt}$ : radiation flux from mesh (i,j) at time t  (W/m$^2$)

$f_{ij}$ : view factor between mesh (i,j) and a receiving point  (-)

$e_{ij}$ : (emissivity for mesh (i,j) )x(absorptivity for a receiving point)  (-)

$s$ : constant of Stefan Boltzmann  $5.67 \times 10^{-8}$ (W/m$^2$/K$^4$)

$T_{ot}$ : temperature at a receiving point at time t  (K)

The observed and estimated values of received radiation at the positions of radiation meters R1 and R4 are plotted on Figure 9, where $f_{ij}$ can be calculated by the geometric relation between heat source meshes and a heat receiving point, and $e_{ij}$ can be assumed as 1.0 because the emissivity of a flame is included in the observed value of temperature as the thermoviewer's function which has been explained in the previous section. Under these conditions, a good correlation between two values has been obtained as is shown in Figure 9.

Therefore, the thermoviewer is allowed to work as a kind of scanning radiation meter, instead of a thermometer and for this purpose, only the conversion of data by the next formula is required;

908

So that, in this stage of study, we can only propose, as the simplest model which will present the overall tendency, a linear profile like the solid line in Figure 11.

This linear model can be described as follows (see Figure 12.);

$$R_k \doteqdot f_k \cdot E_o \cdot (1-(h_k+h_{k+1})/2H_o) \tag{7}$$

where
    $R_k$ : radiation from the k-th horizontal stripe   $(W/m^2)$

    $f_k$ : view factor for the k-th horizontal stripe   (-)

    $E_o$ : maximum radiation intensity  $= 5.12 \times 10^4$  $(W/m^2)$

    $h_k$ : height of lower bound of the k-th stripe

    $h_{k+1}$ : height of upper bound of the k-th stripe

    $H_o$ : height of flame  (L)

    k   ; vertical position of horizontal stripe on heat source plane.

Predicted value for the radiation below the roof line might be different to some extent from the actual value but, when the model is applied to the large flame in an urban fire, the difference would become relatively small. For a design of flat type of firebreak, only the total amount of radiation flux is important and for a vertical type of firebreak, the radiation out of the heat source below the roof line could be concealed.

With these reasons, we have determined to adopt the simple linear model for a design of firebreak but, for a more strict argument, additional experiments and further studies are necessary.


REFERENCE

1.  Itoigawa, E., Tsukagoshi, I. : "Distribution of Radiation Intensity along the Vertical Direction of Flame (Estimation of Radiation Intensity in case of Seismic City Fire, part 1)"; Summaries of Technical Papers on Annual Meetings of Architectural Institute of Japan, Sept.,1983.

2.  Tsukagoshi, I., Itoigawa,E. : "Evaluation of Fire Blocking Belt in the Planning of Urban Fire Prevention Unit, part 1 Estimation of fire load to fire blocking belt" : Transaction of Architectural Institute of Japan, No. 340 June 1984.

3.  Tsukagoshi, I., Tanahashi, I., Itoigawa, E., Iwakawa, N., Kumagai, Y., Murosaki, M, Ogawa, Y., Sato, T : "A STudy on Urban Fire Prevention in case of Big Earthquake" : 8th World Conference on Earthquake Engineering - San Francisco, July 1984.

4.  Hamada, M., Suzuki,Y. : " Taishin Kasai-ji no Shoushitu han-i no Suikei" (Prediction for burning area in case of post-earthquake urban fire) : Summaries of Technical Papers on Annual Meetings of Architectural Institute of Japan, 1633-1636, Sept. 1974.

5.  Fujita, K.: "Fire spread caused by radiation and its prevention" in Kasai no Kenkyuu, Sagami-Shoboh, Tokyo, 1951.

6.  Kamei, K. : "A report on Ohdate big fire" in Journal of Japanese Association of Fire Science and Engineering, vol.3. 127, 1955.

# Oil Pool Fire Experiment

**TAKAAKI YAMAGUCHI and KENJI WAKASA**
Petroleum Association of Japan
1-9-4, Ohtemachi, Chiyoda-ku
Tokyo, 100 Japan

ABSTRACT

It is important to know the behavior of a large-scale oil pool fire, so a large-scale oil pool fire experiment was carried on in JAPAN on May 30, 1981. It was carried out by filling excavated storage tanks measuring 30 m, 50 m and 80 m in diameter each with water of about 200 mm in depth, floating kerosene to a depth of about 20 mm above the water and burning the kerosene.

Measurements were made on the flame temperature, radiant heat, partial radiant heat, burning rate of a hydrocarbon pool fire, composition of gaseous combustion products, naturally-induced airflow velocity and flame shape, seven items in all.

The experiment results clearly indicate that the results of a small-scale experiment on a storage tank whose diameter is less than 10 m cannot be extrapolated to a large-scale oil pool fire.

INTRODUCTION

The oil pool fire experiment conducted this time was mainly aimed at clarifying the behavior of a large-scale oil pool fire with emphasis on the analogy due to the sizes of storage tanks, and at examining the validity of the conventional danger evaluation based on the extrapolation of the results of a small-scale experiment.

The experiment was conducted at the Higashi-Fuji Exercise Field of the Ground Self-Defense Force in Gotenba City, Shizuoka Prefecture on May 30, 1981 with the participation of the Petroleum Association of Japan and Petrochemical Industry Association and the cooperation of Shizuoka Prefecture.

Formulation of the plan, guidance to execution and analysis of results of the experiment were commissioned to the Safety Engineering Association. The Association established an Oil Pool Fire Experiment Committee consisting of 13 experts to accomplish the task.

Preparation

The experiment was carried out using excavated storage tanks of 30 m, 50 m and 80 m in diameter and by burning kerosene. The specifications and arrangement of the fuel tanks are shown in Fig. 1. Since the fuel was floated on the water in the tank, the bottom of each fuel tank was covered with polyethylene sheeting to prevent water leakage.

Kerosene was used as fuel for the following reasons: If crude oil with a wide boiling-point range had been used, it would have caused changes in fluid composition due to evaporation during combustion, and no regular combustion would be obtainable, resulting in difficulty of comparison of combustion characteristics. It was desirable to have a thicker layer of fuel in order to prolong the steady state of combustion, but it was determined to use a fuel layer of about 20 mm due to various restrictions, and this layer of fuel was floated on the surface of water of about 200 mm in depth.

1. Fuel tank specifications

| Tank description | Average diameter | Tank height | Free board |
|------------------|------------------|-------------|------------|
| 30 m tank | 30.04 m | 0.45 m | 0.25 m |
| 50 m tank | 50.02 m | 0.45 m | 0.25 m |
| 80 m tank | 80.07 m | 0.45 m | 0.25 m |

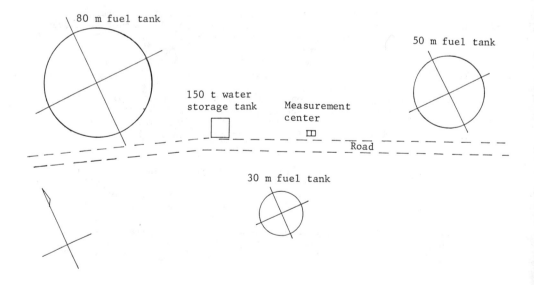

Sectional view of fuel tank

2. Fuel tank layout

80 m fuel tank

50 m fuel tank

150 t water
storage tank

Measurement
center

Road

30 m fuel tank

FIGURE 1.  Specifications and layout of fuel tanks

Ignition

Due to the reasons that kerosene with a higher ignition point was used as fuel and it was necessary to cause the ignited fire to grow into a complete conflagration within as short a time as possible, small amounts of naphtha were made to flow out of several points on the oil surface and ignited electrically using ignition balls.

1.  Ignition position

Since the thickness of kerosene was 20 mm and thin, it was necessary to wholly burn kerosene for a short time after ignition so that the whole burning time would be made long.

After pre-test results, it was found that liquid surface falling speed was about 2.1 mm/min and fire propagation speed was 3 to 4 m/min.

The fire became a conflagration in 3 minutes after ignition.

Kerosene consumption until then was thought to be about 2 mm on the average for all tanks. Kerosene of 18 mm or above was consumed when a conflagration occurs.

Since fire propagation speed was 3 to 4 m/m, it was assumed that fire would expand to 9 to 12 m in 3 minutes.

The position and number of ignition points were determined, so that any parts of the tank can be within about 10 m from the ignition position. As a result, it was determined that 50 m tank should ignite at 4 points and 80 m tank would ignite at 11 points. It was determined that the 30 m tank should ignite at central 1 point in order to measure the fire propagation speed.

2.  Ignition facilities

Since it was difficult to ignite directly to kerosene, it was determined that ignition should be made by a small quantity of naphtha and ignition balls to the extent of which influence would not given to the burning properties of kerosene.

The head tank of ignition naphtha was installed at the place 5 to 10 m away from the edge of each tank. The pre-test was made so that the naphtha quantity, which flowed out of respective outlets, would be about 500 ml/min.

The height of the head tank was set so that its bottom would be 1 m higher from the outlet. Outlets, to which naphtha would flow from the head tank, were installed to the ignition equipment. They were connected with the PVC tube. Outlets were set facing upward direction at about 5 cm above the liquid.

The circumference of outlets was covered with cotton cloth hung from above. When the stop valve opened, which had been installed to the head tank containing naphtha, naphtha flowed out of outlets. Part of naphtha was impregnated into cotton cloth and remaining naphtha was mixed with kerosene in the tank. Ignition balls were set into cotton cloth so that naphtha would not be directly placed. The switch of ignition balls was installed at D/4 outside the tank and ignition balls of respective tanks were designed to be operated by setting one switch. Naphtha of 2500 ml per outlet was placed into the head tank so that naphtha of 500 ml for each minute per outlet would flow for 5 minutes.

The number of head tanks was 30. One head tank each was installed for 50 m tank and 2 head tanks were installed for 80 m tank.

3. Ignition procedures

PVC tubes were made full of naphtha beforehand so that naphtha would flow out from outlets at the same time when the valve was opened. The valve of the head tank was opened 1 minute before ignition. One minute after naphtha was discharged, the switch of ignition balls was turned ON and ignition balls were caused to generate.

According to this operation, ignition balls would be ignited, naphtha, which was impregnated into cotton cloth, would burn, kerosene mixed with naphtha would be lit and, further, fire would propagate to kerosene of the whole tank.


Measurement

Measurements were made on many items including the flame temperature, radiant heat, partial radiant heat, burning rate of the hydrocarbon pool fire, composition of gaseous combustion products, naturally-induced airflow velocity and flame shape.

1. Flame temperature measurement

Temperatures in the liquid and on and above the liquid surface during combustion were measured by installing sheath-type thermocouples at six locations in the center of each tank and four locations at positions of D/8, D/4, and D/2 (D: tank diameter) from the center, 18 locations in all.

2. Radiant heat measurement

Radiant heat from the flames was measured outside the fuel tank and on the liquid surface in the fuel tank. Measurements were made by installing radiant heat meters at positions of 1D, 2D, 3D, and 4D from the center in each tank for measuring outside the fuel tank, and by installing liquid-level radiant heat meters at three points in the 80 m tank and at one point each in 50 m and 30 m tanks for measuring on the liquid surface.

3. Partial radiant heat measurement

To find out the magnitudes of radiant heat from various portions of flames, partial radiant heat was measured by installing radiant heat meters, whose heat receiving surface was partially shielded, at a position of 2D from the center of each tank.

4. Measurement of burning rate of hydrocarbon pool fire

The burning rate of fuel was measured by the liquid level descending rate. The average descending rate of the entire liquid level was measured at one point in each tank, while regional liquid level descending rates were measured at four points in the 80 m tank, two points in the 50 m tank and one point in the 30 m tank.

Further, a video camera was installed at the center of the 50 m tank to monitor the liquid level condition and liquid level descending rate during combustion.

5. Measurement of composition of gaseous combustion products

To find out the detailed burning condition during an oil conflagration, gases on the liquid surface and in the flames were sucked in and collected from seven points in each tank, and concentrations of carbon monoxide, carbon dioxide, hydrocarbon and oxygen contained therein were analyzed by gas chromatograph. In addition, oxygen concentration was continuously measured using an oxygen concentration meter separately from the gas chromatograph.

6. Measurement of naturally-induced airflow velocity

The velocity of the naturally induced airflow generated by the oil fire was monitored by propeller and bilam type anemometers installed at eight points each in each tank to measure the wind direction and velocity near the edge of each tank.

7. Flame shape measurement

To clarify the behavior of an oil conflagration including the height and shape of flames, the burning state in each tank was recorded by a 16 mm movie camera and a video camera.

Results

Experiments with the 30 m tank (Experiment No. 1) and 50 m tank (Experiment No. 2) were carried out as planned under conditions of near calm, but the experiment with the 80 m tank (Experiment No. 3) was conducted under windy conditions, and some points which were slightly different from the original plan such as deviation of the oil layer on the water surface occurred.

The present combustion experiments used unprecedently large-scale storage tanks, and measurements also employed new techniques such as observation of liquid level variation by a submerged video camera, and measurements of the regional liquid level descending rate and partial radiant heat of flames, thereby obtaining many valuable experiment results. Major findings from the experiments are enumerated below.

1. In the 1st experiment, naphtha burned up at the same time when ignition balls switch was turned ON, and just after that, the liquid surface began burning. In the 2nd experiment, cotton naphtha at all 4 points burned up and the liquid surface began burning simultaneous when ignition balls switch was turned ON. In the 3rd experiment, cotton cloth naphtha was ignited at only 6 points due to strong wind.

Three points out of them were set on the water surface where there was no oil, because the oil surface inclined to part of the burning tank due to the wind; hense fire propagation did not occur to the liquid surface and the fire was put out immediately after the flow of naphtha. Where the points were set on the liquid surface, the liquid surface was ignited immediately. In the 1st and 2nd experiments, a conflagration was made in about 3 minutes as scheduled. In the 3rd experiment, propagation down the wind side was in 2 to 3 minutes. The whole surface on the wind side where there was kerosene burned in about 5 minutes after ignition.

2. As analysis results of burning gas constituents, only 1 time of the result concerning the 80 m tank was shown. This was due to the fact that since 1) kerosene in the 80 m tank was blown away towards the wind side by the wind, 2) about 1/2 of the kerosene surface only burned, 3) the kerosene surface severely inclines, 4) the sample introduction section completely was exposed in the air, measurement became meaningless; hence this part was omitted.

In the continuously recorded results of oxygen concentration in respective tanks, it was observed that turbulence was caused by cock operation for sampling. From the positional relationship between the fire condition at nearly the time when sampling was made and the sample introduction section, it can be thought that in the 30 and 50 m tanks, gas constituents in the flames can be sampled, but in the 80 m tank, the sample introduction section is exposed very much in the air. When we consider the positional relation between the condition of the fire nearly at the time, when sampling is made, and the sample introduction section, we find that we have probably been able to sample gas constituents in the fire for at 30-m and 50-m tanks, but the sample introduction section is quite extensively exposed to air at the 80 m tank.

3. Light emission of the flames was not dependent upon the sizes of storage tanks, but became the maximum at a height 0.8 to 1.0 times the tank diameter length. Such flames developed a pattern of respiration with a period of several seconds, and the period became longer as the storage tank size increased.

4. The maximum temperature of the flames was obtained near the center of the tank and reached 1,400 to 1,800°C. This value did not vary with the sizes of tanks, but was slightly higher than that of a crude oil fire.

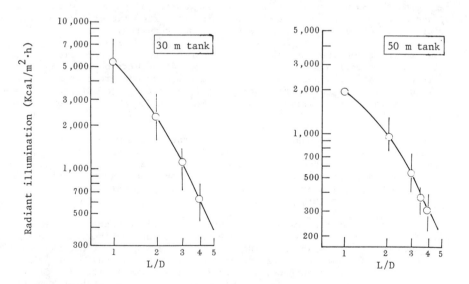

FIGURE 2. Relation between radiant heat and dimensionless distance under calm conditions (L is the distance from tank center to radiation meter)

5.  The receiving quantity of radiant heat along the ground surface--when com-
    pared with dimensionless distance L/D which was obtained by dividing the
    distance from the tank center L by storage tank diameter D--suddenly
    dropped as L/D increased, and the heat receiving quantity, the dimensionless
    distance being equal, did not collate with the value obtained by conven-
    tional predictive calculation, and sharply decreased as the size of the tank
    increases.

6.  Oil level descending rate due to combustion was about 4.7 mm/min on the
    average of the entire oil surface, and did not appear to be dependent on
    tank sizes.  On the other hand, the regional liquid level descents varied
    with locations and time within the range of 2 to 7 mm/min, and were related
    to the movements of flames near the oil surface.

7.  Naturally-induced airflow velocity necessary for burning was greater at the
    position higher than the ground surface and indicated the value of several
    m/s.  This rate dropped as the tank size increased and showed cyclic varia-
    tion as the respiration of the flames.  Further, the oil surface developed
    ripples even during calm owing to this naturally-induced airflow by combus-
    tion.

8.  Near the liquid level at the tank center during a conflagration, oxygen
    supply was minimal, and oxygen concentration became higher at a position
    high above the liquid level.  As a result, oxygen deficiency occurred at
    the center portion, and steam which had evaporated from the liquid surface
    developed pyrolysis and generated low-molecular-weight combustible gases
    such as methane.

9.  During combustion, heat transfer from flames to the liquid surface was
    mainly performed by radiant heat transfer, and convective heat transfer was
    only about 10% of the radiant heat transfer.

10. Slanting of flames due to winds was at about 30° from horizontal at the
    lower part of the flames in Experiment No. 3 in which wind velocity varied
    between 5 to 10 m/s.  This slanting angle increased to about 70° at the
    upper part of the flames.

11. The wind did not greatly affect the maximum temperature of flames.  Radi-
    ant heat from flames when it was windy showed a great difference depending
    upon the wind receiving direction.  It became the maximum in the right-
    angle direction with the wind direction and the minimum on the windward
    side.

General Conclusions

     These results will be analyzed in detail in the future.  Some items of the
experiment results are important, because they suggest that the results of the
small-scale experiment with a tank having a diameter of less than 10 m, as they
are, cannot be extrapolated to an oil conflagration.  In particular, the finding concerning radiant heat in Item 3. indicates that since the heat receiving
quantity decreases as the storage tank size increases, the dimensionless distance being equal, it is not appropriate to compare radiation heat danger uniformly by the dimensionless distance and that the calculation of radiant heat
by the conventionally used danger prediction calculation method requires re-
assessment, because such calculation gives over-estimation of several times as
much as the measured value, thereby suggesting that the said finding has an
important meaning in practice.

The measured values of radiant heat obtained by the present experiment are compared with calculated values obtained by the conventional calculation method as follows:

Measured and calculated values of radiant heat intensity

| Burnt tank diameter | Distance from burnt tank center | Radiant heat intensity | | Ratio Em/Ec % |
|---|---|---|---|---|
| | | Measured value Em* Kcal/m²h | Calculated value Ec Kcal/m²h | |
| 30 m | 1D | 5,400 | 10,540 | 51 |
| | 2D | 2,300 | 4,300 | 53 |
| | 3D | 1,100 | 2,180 | 50 |
| | 4D | 620 | 1,290 | 48 |
| 50 m | 1D | 1,900 | 10,540 | 19 |
| | 2D | 960 | 4,300 | 22 |
| | 3D | 540 | 2,180 | 25 |
| | 4D | 310 | 1,290 | 24 |

\* Average during calm

The present experiment is expected to increase in value as research progress in this field, and actual application of the experiment results will play important roles in the practical and research aspects in the future.

REFERENCES

1. Japan Society for Safety Engineering: The Report of the Oil Pool Fire Experiment, J.S.S.E., Yokohama, Japan, 1981.

2. Yumoto, T.: Journal of Japan Society for Safety Engineering, J.S.S.E., Yokohama, Japan, 10: 3, 143, 1971.

3. Yumoto, T.: Journal of Japan Society for Safety Engineering, J.S.S.E., Yokohama, Japan, 16: 1, 58, 1977.

4. Yumoto, T., Nakagawa, N., and Sato, K.: Journal of Japan Society for Safety Engineering, J.S.S.E., Yokohama, Japan, 21: 1, 30, 1982.

5. Yamaguchi, T., Konishi, S., and Yamamoto, Y.: Journal of Japan Society for Safety Engineering, J.S.S.E., Yokohama, Japan, 21: 4, 215, 1982.

6. Kashio, T., and Akita, K.: Proceedings of the 13th Symposium of Safety Engineering, Japan Society for Safety Engineering, Japan, 81, 1983.

# The Behaviour of Heavy Gas and Particulate Clouds

R. J. BETTIS, G. M. MAKHVILADZE, and P. F. NOLAN
Department of Chemical Engineering
Polytechnic of the South Bank
London SE1 0AA, United Kingdom

ABSTRACT

Models have been applied to stages of two phase releases following the failure of pressurised vessels. The initial expansion stage has been modelled using an experimental apparatus involving the measurement of pressure histories and of Freon-11 aerosol droplet sizes and velocities. The experimental data combined with a thermodynamic analysis has allowed estimations of the time dependent processes. The analytical description of the critical pressure decrease on the opening of the vessel is presented.

The later evolution stage has been mathematically modelled by assuming that the two phases have separate velocities. The cloud motion characteristics have been shown to depend upon the degree of hydrodynamic interactions between the particles via the gaseous phase. For small particles' fractional volumes, this interaction is small and each particle behaves as a single particle corresponding to a "filtration" regime. However, for large concentrations the air between the particles becomes entrained due to the particle motion and the cloud velocity exceeds that of the single particle. In this "entrainment" regime large scale vortex motion occurs.

## INTRODUCTION

Despite continuous efforts (1-9), problems still remain with regard to the prevention of explosions following the failure of pressurised storage vessels Substances, such as liquid petroleum gas (e.g. propane) require the maintenance of an increased pressure in the vessel simply because their boiling points at normal pressure are lower than ambient temperatures (e.g. boiling point of propane is - 45°C (228K) at 1 bar). In order to store propane as a liquid at, say 20°C (293K) a pressure of 10 bar must be applied. Such storage under pressure increases the risk of explosion due to vessel failure (1,2,5,7).

The failure of a vessel, is accompanied by an abrupt decrease in pressure, the liquid becomes superheated, rapidly vaporises, and a mixture of vapour and liquid is released from the vessel. The cloud formed as this release mixes with the surrounding air is highly dangerous, both from the potential environmental pollution effects and also from the possibility of a flammable mixture being ignited. Both gas dynamics and hydrodynamics are involved in a liquid petroleum gas (LPG) release.

The present address for G. M. Makhviladze is Institute for Problems in Mechanics, USSR Academy of Sciences, prosp. Vernadskogo, 101, Moscow, 117256.

EXPERIMENTAL WORK - The Initial Expansion

The experimental work modelled the first phase of a pressurised liquid release, during which the pressure falls to that of the surroundings. The model material was Freon-11, which has a normal atmospheric boiling point of C. 23°C (296K).

The release was produced by "failing" a fully instrumented vessel consisting of two hemispheres held together pneumatically and containing the model material maintained at measured internal pressure, temperature and fill-level. The assembled vessel is 114mm in diameter, and has a maximum fill of 1.1 kg. of Freon-11. A heater in the vessel allows energy to be put into the closed system, increasing the temperature and pressure of the Freon-11. When the appropriate release conditions had been established, the pressure on the pneumatic system holding the vessel together was reversed, and the two halves pulled apart.

The resulting release consisted of three parts:

(i) Vapour; produced during the rapid boiling as the pressure was reduced.

(ii) Aerosol; the liquid broken up during boiling and entrained into the expanding vapour, which carried it away from the release point. The size of the droplets which form the release was measured using a laser diffraction system, and flash photography. The profile of the spray, and its overall velocity was examined using video and high speed cine photography.

(iii) Bulk liquid; any liquid which remained in the immediate vicinity of the release point. This was collected on an instrumented tray below the vessel.

Preliminary results in Table 1 represent average values from several experiments.

The droplet sizes (the mean sizes of the best log.- normal distribution) were measured at a point level with the centre-line of the vessel, at a distance of 0.5m from the centre. The velocity estimates were average values over the first two metres of travel away from the vessel. The time given as release time was the time for equalisation of vessel internal pressure with the atmosphere. For simplicity, in calculations of the later stage of evolution the mean droplet size was taken as 100 $\mu$m and the characteristic velocity of the spray as 10 m s$^{-1}$.

TABLE 1. Results from experimental work

| Pressure kPa | Fill-level kg | Release time te $10^{-3}$ s | Droplets | | Residual liquid kg |
|---|---|---|---|---|---|
| | | | Size $\mu$m | Velocity ms | |
| 310.3 | 1.00 | 255 | 80 | | 0.289 |
| 310.3 | 0.75 | 222 | 82 | | 0.246 |
| 310.3 | 0.50 | | 85 | 6.7 | |
| 310.3 | 0.30 | 97 | 90 | | 0.178 |
| 413.7 | 0.50 | | | | 0.070 |
| 413.7 | 0.30 | | | | 0.045 |
| 482.7 | 0.50 | | | 12.8 | |
| 517.2 | 1.00 | | | | 0.103 |
| 206.9 | 0.50 | | | 4.5 | |

THERMODYNAMIC ANALYSIS - The Initial Expansion

Following the example set by Hardee and Lee (1), the initial expansion can be considered isentropic

thus $S_t = X_t S_{Vt} + (1 - X_t) S_{Lt} =$ a constant $\qquad$ (1)

hence $S_o = X_o S_{Vo} + (1 - X_o) S_{Lo} = X_t S_{Vt} + (1 - X_t) S_{Lt}$ $\qquad$ (2)

i.e. $X_t = [X_0(S_{Vo}-S_{Lo}) + (S_{Lo}-S_{Lt})]/(S_{Vt}-S_{Lt})$ $\qquad$ (3)

On the assumptions that the vapour behaves as an ideal gas

$$\rho_{vt} = P_t \, \varphi \, /(R \, T_{Lt})$$ $\qquad$ (4)

and that the vapour temperature is a constant and equal to the room temperature, the vapour volume is given by

$$V_{vt} = x_{vt} \, W_v \, / \, \rho_{vt}$$ $\qquad$ (5)

Thus, if the relationships between entropy and pressure are known, then the pressure dependencies of quality, vapour density and vapour volume can be calculated. Pressure-entropy relationships were taken from curves fitted to data from standard tables (10).

The calculated data for quality is presented in Figure 1. This figure shows that between 10 and 45% of Freon-11 is vaporised during the expansion. This agrees with the results from the experimental work and that of others (1,3). However, it is the time dependence of these parameters, not pressure, which is important. Hardee and Lee (1) produced a theoretical time dependence for the pressure.

A typical experimental pressure history trace is given in Figure 2 (P = 310.3 kPa, W = 1 kg). Point (a) corresponds to the beginning of vessel failure. This is followed by a rapid fall in pressure, taking 15 to 20 x $10^{-3}$s. It is thought that this is due to the exit of vapour from the vessel, and occurs before the flash boiling begins. Occasionally there is then a small rise in apparent pressure lasting for a similar time. This pressure is probably due to the impact of liquid on the pressure sensor resulting from the onset of flash boiling. After this "momentum peak" has subsided the pressure falls smoothly to atmospheric, although at a much slower rate than in the initial decay. It is during this period that the two phase release occurs, with the expanding vapour entraining the liquid and carrying it away from the vessel.

It is possible to model the changes, which occur in the pressure history prior to the onset of flash boiling at Point (b).

Integrating the continuity equation, over the volume $V_V$, corresponding to the volume occupied by gas in the vessel, leads to

$$\frac{\partial}{\partial t} \int_{V_v} \rho_v \, dV + \int_{V_v} \text{div} \, \rho_v \, \vec{u}_v \, dV = 0$$

$\qquad$ (6)

921

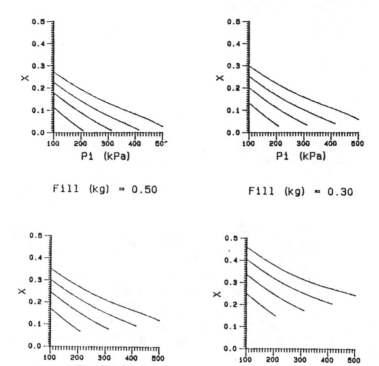

Fill (kg) = 1.00

Fill (kg) = 0.75

Fill (kg) = 0.50

Fill (kg) = 0.30

Figure 1.    Theoretical quality curves for 2,3,4 and 5 atm
             initial pressure.

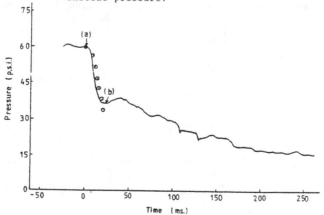

Figure 2.    Pressure versus time during release.

By applying Gaussian theory to the second term, and assuming that the gas velocity at the vessel walls is zero, and that the annular velocity, $u$, is constant, then

$$\int_{V_v} \text{div } \rho_v \vec{u}_v \, dV = \rho_v u'_v A \tag{7}$$

and

$$\rho_v = Z P_v \quad \text{where} \quad Z = W_v^*/(R T_v) = \text{constant} \tag{8}$$

The pressure of the gas in volume $V_v$ can be considered constant but will change rapidly on the failure of the vessel

$$V_v \, dP_v/dt = - P_v u'_v A \tag{9}$$

By taking a hydraulic approximation to the momentum balance, it is possible to introduce a quasi-stationary relationship for the out-flow of the jet

$$u'_v = [(P_v - P_A)/\rho_v]^{1/2} \tag{10}$$

The annular area, $A$, changes during the process from zero to the value $A_e$ at the time of pressure equalisation. For simplicity, this will be described by a power law

$$A = A_e (t/t_e)^n \tag{11}$$

By substitution of equations (10) and (11) into (9)

$$\partial p'/\partial t' = [p'(p' - 1)]^{1/2} (t')^n \tag{12}$$

where $p' = P_v/P_A \quad t' = t/t*$

and $t* = [Z^{1/2} v_v/(A_e t_e)]^{1/(n+1)} t_e \tag{13}$

Equation (12) describes the rate of the pressure drop. Integration of equation (12) with the initial condition of

$$p' = p'_o \quad \text{where} \quad p'_o = P_o/P_A$$

leads to

$$\frac{(t')^{n+1}}{n + 1} = \ln\left( \frac{2 p' - 1 + [(2 p' - 1)^2 - 1]^{1/2}}{2 p'_o - 1 + [(2 p'_o - 1)^2 - 1]^{1/2}} \right) \tag{14}$$

This equation is only valid for the initial period and cannot be applied for the whole transient process. The value of t* in equation (13) determines the characteristic time of the pressure drop.

Figure 2 also illustrates the correlation between this theoretical approach and the experimental pressure drop for the following conditions

$V_V = 0.97 \times 10^{-3}$ m$^3$ $W_V = 137.37 \times 10^{-3}$ $T_V = 300K$ $t_e = 0.229$ s.

It was assumed that

$n = 1$ $A_e = 0.59 \times 10^{-3}$ (corresponding to a final opening of 1.5mm)

Between points (a) and (b) in Figure 2 good agreement was found but, as expected, deviations occurred beyond this region.

MATHEMATICAL MODEL - The Later Evolution

The final stage of cloud evolution is assumed to start when the particles have reached their maximum height. Here they can be described as a cloud of stationary, mono-sized, spherical particles suspended above a horizontal surface. The particles then start to move downwards under gravity, giving rise to gas movement.

The main assumption of multi-phase mechanics is that the diameter of the particles and the mean distance between the particles are much smaller than the distance over which there is a significant change in the macroscopic parameters (11). The system can be considered as two intermixed, interacting media, the gas and the particles. The diameter of the particles is much larger than the molecular scales.

Brownian motion, evaporation and break-up of the particles are neglected, although it is recognised that these may be important, particularly evaporation.

Assuming $\Theta_L \ll \Theta_v$ , $\rho_v / \rho_L \ll 1$ and constant dynamic viscosity.

The evolution of the cloud can be considered to be an isothermal process, since the heating of the system due to the viscous energy dissipation is small and both gas and air temperature are held at the initial temperature $T_o$.

Both planar (see reference (12)) and axi-symmetrical cases can be considered. For the planar case, it is assumed that the cloud size is greater in one horizontal direction than the other. The force for interaction at the interface is defined by

$$\vec{F} = \frac{3}{4} \frac{\epsilon}{d} C_d \ Re_p \frac{\rho_v \rho_L}{1 - \Theta_{Lo}} \ | \vec{u}_v - \vec{u}_L | \ ( \vec{u}_v - \vec{u}_L )$$  (15)

where $C_d = (1 + 0.158 \ Re_p^{0.5}) \ 24/Re_p$

$Re_p = Re_{po} \ | \vec{u}_v - \vec{u}_L | \ \rho_v /\Theta_v$

The origin for the co-ordinates is situated under the cloud mass centre, the x-coordinate being along the surface and the y-coordinate along the plane of symmetry for the planar case and along the axis of symmetry for the axi-symmetrical case.

924

Initially, the stationary gas is in a static equilibrium with the cloud of stationary particles and a Gaussian distribution of particle concentration about the centre is applicable. The boundary conditions take into account the symmetry about the plane x = 0; and the equilibrium state at infinity.

Particles which reach the surface play no further part in the process.

In the above example, the following values for the parameters are based on data from the initial expansion stage

$$k = 10^3 \text{ cm}^{-3} \qquad\qquad \Theta_{Lo} = 10^{-3} \qquad\qquad M^2 = 0.72 \times 10^{-3}$$
$$\gamma = 1.4 \qquad\qquad\qquad \epsilon = 10^{-3}$$

and the terms which depend on particle size have ranges

$$d = 3.3 \times 10^{-6} \rightarrow 2 \times 10^{-4} \qquad Re = 7 \rightarrow 60 \quad \Theta_{Lo} \simeq 10^{-3} \rightarrow 10^{-2} \qquad\qquad h = 1.5 \rightarrow 14$$
$$Re_{po} = 6500000 \ d = 21.4 \rightarrow 1300$$

The external Reynolds' numbers are large and indicate turbulent motion and are estimated using an effective turbulent viscosity. The particles are very small in comparison with the scale of turbulence and the Reynolds' number for the particles uses a molecular viscosity. Numerical methods have been developed (13) and two regimes of cloud evolution can be considered.

The interaction with the horizontal surface begins when the height of the cloud centre is equal to or approaching twice the radius of the cloud.

The stationary particles start to move downwards under gravity. The entrainment of the surrounding gas due to friction is described by an empirical law (equation (15)). The effect of entrainment increases with decreasing distance between the particles. In this case hydrodynamic interaction occurs between the particles throughout the cloud as a result of large scale motion of the air. The velocity of the cloud increases continuously and exceeds that of a single particle. This is called the "entrainment" regime.

If the distance between the particles is sufficient, each particle behaves as a single free-falling sphere because it is unaffected by gas motion caused by the presence of others. This is called the "filtration" regime; the gas filters through the particles.

The boundary between the two regimes in the planar case is described by

$$\ln \Theta_{Lo} = (0.825 \times 10^4) \ d - 4.56 \qquad\qquad\qquad\qquad (16)$$

with 7% accuracy for the ranges

$$3.3 \ 10^{-5} \le d \le 2 \ 10^{-4} \quad \text{and} \quad 10^{-5} \le \Theta_{Lo} \le 10^{-2}$$

At a fixed value of d, the filtration regime occurs if the parameter $\Theta_{Lo}$ is small. In the filtration regime, as every particle moves individually, the initial cloud shape does not change. In the entrainment regime the cloud shape changes continuously. At first, the lines of equal particle concentration form concentric circles. The cloud is similar to a "drop" moving through a liquid i.e. the particles move downwards in the centre of the cloud and move upwards near the edge. With time and an increasing cloud velocity, the "drop" is

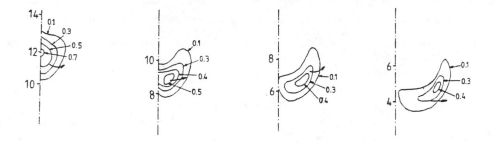

Figure 3.   Cloud evolution with high particle concentration.

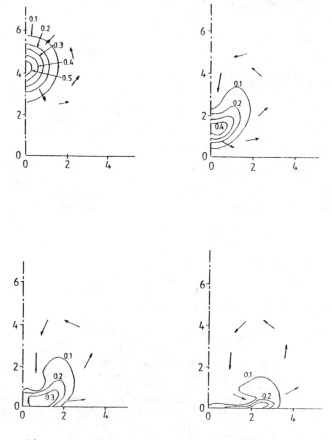

Figure 4.   Characteristics of cloud sedimentation.

deformed. A "dimple" forms in the bottom of the "drop". Gradually the drop transforms into the "cap" described previously (14, 15, 16). The gas zone is increasing continuously. Large scale vortex motion, causing lateral displacement of the particle, is formed. The zones with maximum particle concentration are removed from the plane x = 0 (or at axis x = 0, in the axisymmetrical case). As a result, the lines of equal concentration breakoff at a definite time.

Two symmetrical cylindrical vortices with the particles in their core are formed in the planar case (or toroidal vortex in the axi-symmetrical case). The evolution of the cloud is shown in Figure 3. The flow field and lines of equal concentration are presented for $\Theta_{Lo}$ = $10^{-3}$ and d = 6.67 x $10^{-5}$ at various times i.e. (a) 4.16 (b) 6.94 (c) 9.71 (d) 12.5.

The process of evolution of the cloud can be followed experimentally by noting the changes in shape.

## CHARACTERISTICS OF SEDIMENTATION

In the filtration regime all the particles are descending vertically on direct line trajectories. The interaction obviously does not depend on the initial cloud height.

The distribution of the particle concentration falling on the surface can be determined analytically by means of integration of the initial particle distribution on coordinate y.

In the entrainment regime the large scale vortex motion leads to lateral displacement of the particles above the surface. Therefore, the particles are scattered across the surface. The scattering increases with increasing initial cloud height.

Cloud sedimentation, using $\Theta_{Lo}$ = $10^{-3}$, d = 6.67 x $10^{-5}$ and h = 4.8 at various times are shown in Figure 4. It is possible to identify the formation of

(a) the vortex flow structure
(b) the "cap"
(c) the deformation of cloud shape during sedimentation.
(d) gas propagation across the surface and the scattering of particles.

Initially the particles land near the line of symmetry. Then the fall-out takes place at a certain distance. The final distribution gives a maximum located at a certain distance from the central line and a considerable number of particles can fall on the surface outside the initial projection of the cloud.

This effect can be estimated by introducing a coefficient for cloud scattering onto the surface. This coefficint is equal to the fraction of the particles which land outside the initial geomtrical shadow of the cloud on the surface.

$$e = 1 - N_s/N \tag{17}$$

where    N = total number of particles
         $N_s$ = number of particles landing outside the initial cloud projection.

In the "filtration" regime, the coefficient is near to zero.   In the "entrainment" regime, the coefficient depends on several parameters e.g. h, d and $\Theta_{Lo}$.   With increasing initial cloud height the coefficient increases, since the vortex motion takes place in a larger gas volume, and the particles move considerable distances from the line, x = 0.   For constant h and $\Theta_{Lo}$, the scattering coefficient diminishes with increasing particle size.   If $\Theta_{Lo}$ is increased, keeping the other parameters constant, the coefficient "e" increases to mark the transformation from the "filtration" regime to the "entrainment" regime.   It then declines because of the large vertical component of the particle velocity.   The particles fall too rapidly to move from the line of symmetry.

An isothermal mixture has been assumed, since the boiling point of Freon-11 at normal pressure is not significantly different to ambient temperature.   However, for other hydrocarbons the temperature difference in liquid and gaseous phases may be significant.

CONCLUSIONS

Physical and mathematical modelling has been used to investigate the separate stages of the release of a two-phase mixture from a pressurised vessel.

In order to study the first stages of such processes, experiments have been conducted on the "failure" of a vessel filled with liquid Freon-11.   Data was obtained of the size and velocity of particles, pressure equalisation time and the time dependence of the pressure in the vessel.   An analysis of the experimental data combined with thermodynamic data has allowed a definition of the time dependence of the process.

The assumption that the process is isentropic, despite only approximating to the actual nature of the process, provides satisfactory results.   The results obtained can be used to estimate the parameters of the cloud formed during the failure of a pressurised vessel.

Further work will involve the investigation of

(a) flow patterns and temperature distributions set up during the release, for both the particles and the vapour.

(b) the cloud formation with simultaneous combustion inside the vessel.

The mathematical formulation of the cloud evolution has been based on the two phases having separate velocities.

The character of the cloud motion has been shown to depend upon the extent of the hydrodynamic interaction which occurs between the particles via the gaseous phase.

For a low concentration of particles this interaction is small, and each particle behaves as a single particle (the "filtration" regime).   If the concentration of the particles is high enough the air between the particles is entrained by the particle motion; as a result the cloud velocity exceeds that of a single particle.   In this "entrainment" regime large scale vortex motion arises: the cloud is transformed into two cylindrical vortices (in the case of planar symmetry).   The regime of cloud motion defines the features of the sedimentation of the particles on the horizontal surface.   In the "entrainment" regime, the particles entrained into the vortex motion move laterally, therefore they sediment over some distance.

928

Further developments will consider

(a) The evaporation of the particles, and the inclusion of the temperature difference between the gas and the particles.

(b) The combustion of the cloud above the vessel. Methods exist to enable non-isothermal conditions to be considered (17, 18).

## NOMENCLATURE

| | | |
|---|---|---|
| A | = | exit area |
| d | = | relative diameter of particles |
| e | = | scattering coefficient |
| h | = | height above surface |
| k | = | particle concentration |
| n | = | power term for vessel failure |
| P | = | pressure |
| R | = | gas constant |
| s | = | entropy |
| T | = | temperature |
| t | = | time |
| u,u' | = | velocity |
| V | = | volume |
| W | = | total mass |
| X | = | quality (vapour mass fraction) |
| x,y | = | geometrical coordinates |
| Z | = | release characteristic |
| $\varepsilon$ | = | density ratio |
| $\varphi$ | = | molecular weight |
| $\rho$ | = | density |
| $\gamma$ | = | specific heat ratio |
| $\Theta$ | = | volume fraction of particles |

Groups

| | | | | | |
|---|---|---|---|---|---|
| $C_d$ | = | drag coefficient |
| F | = | drag force |
| M | = | Mach number |
| p' | = | non-dimensional pressure |
| Re | = | Reynolds number |
| t' | = | non-dimensional time |
| t* | = | characteristic equalisation time |

Subscripts

| | | |
|---|---|---|
| A | = | of the atmosphere |
| e | = | at pressure equalisation |
| o | = | at the initial state |
| s | = | on the surface |
| t | = | at time, t |
| V | = | vapour phase |
| L | = | liquid phase |

A vector quantity is indicated with the symbol "$\rightarrow$" above it e.g. $\vec{u}$ and div. is a vector operation.

## REFERENCES

1. H.C. Hardee, D.O. Lee; "Expansion of Clouds from Pressurised Liquids" Accident analysis and Prevention, 7, 91-102, 1975

2. J.A. Barton, P.F. Nolan; "Runaway Reactions in Batch Reactors" I.Chem.E. Symp. Series No. 85, The Protection of Exothermic Reactors and Pressurised Storage Vessels 13-22, Chester, 1984

3.  H. Giesbrecht, K. Hess, et al; "Explosion Hazard Analysis of Inflammable Gas Released Spontaneously into the Atmosphere", Chemie Ingenieur Technik 52, 2, 114-122, Feb. 1980

4.  K. Sato, K. Hasegawa; "Study on the Fireball following Steam Explosion of of n-Pentane". 2nd Int. Symp. on Loss Prevention and Safety Promotion in the Process Industries, 1977.

5.  R.E. Britter, R.F. Griffiths; "The Role of Dense Gases in the Assessment of Industrial Hazards", Journal of Hazardous Materials 6, 3-12, 1982

6.  P. Field; "Dust Explosion Protection", Journal of Hazardous Materials 8, 223-238, 1984

7.  F.P. Lees; "Loss Prevention in the Process Industries", Butterworths, 1980

8.  D.R. Blackmore, M.N. Herman, J.L. Woodward; "Heavy Gas Dispersion Models", Journal of Hazardous Materials, 6, 107-128, 1982

9.  J.L Woodward, J.A. Havens, W.C. McBridge, J.R. Taft; "A Comparison with Experimental Data of Several Models for Dispersion of Heavy Vapour Clouds", Journal of Hazardous Materials, 6, 161-180, 1982

10. Kuzman Rasnjevic; "A Handbook of Thermodynamic Tables and Charts", 1980

11. R.I. Nigmatulin; "Osnowy Mechaviki Getterogennych Sred. M. Nauka", 1978

12. G.M. Makhviladze, O.I. Melichov; "On the Cloud Motion and Sedimentation under Gravity above a Flat Horizontal Surface" Izvestia of USSR Academy of Sciences, Liquid and Gas Mechanics, 6, 64-73, 1982

13. G.M. Makhviladze, S.B. Tscherbak; "Numerical Method for the Investigation of Non-stationary Spatial Motions of Pressurised Gas " Inginere Physical Journal, 38, 3, 528-537, 1980

14. G.W. Slack; "Sedimentation of Compact Clusters of Uniform Spheres", Nature, 200, 4905, 466-467, 1963

15. G.W. Slack; "Sedimentation of a large Number of Particles as a Cluster in Air", Nature 200, 4913, 1306, 1963

16. K. Adachi, S. Kiriyama, N. Yoshioka; "The Behaviour of a Swarm of Particles Moving in a Viscous Fluid", Chemical Engineering Science, 33, 1, 115-121, 1978

17. G.M. Makhviladze, O.I. Melichov; "On the Motion and Evolution of Cloud with Initially Hot Particles", Doklady USSR Academy of Sciences, 267, 4, 844-847, 1982

18. G.M. Makhviladze, O.I. Melichov; "Combustion of Aerosol Cloud above a Flat Horizontal Surface", Chemical Physics, 7, 991-998, 1983

# An Event Tree Model for Estimation of Fire Outbreak Risks in Case of Large-Scale Earthquake

**HIDEKI KAJI**
University of Tsukuba, Japan

**TETSUYA KOMURA**
Nippon Univac Kaisha, Ltd., Japan

ABSTRACT

This paper aims to develop a new method for estimating fire outbreak risks in the case of a large-scale earthquake using computer simulation technique, in which causal relationships concerning fire outbreak are modeled as an event tree structure. As a case study, kitchens for commercial use were taken, and about 5,000 kitchens of different types of business were surveyed particularly in terms of the fire appliances used therein. An event tree structure was designed on the basis of practical causal relations of 55 fires which broke out in past earthquakes. Monte Carlo simulation technique was applied for estimating fire outbreak risks by business type. Although many questions remain unanswered, the usefullness and high applicability of this method are clearly shown.

INTRODUCTION

In Japan a number of earthquakes have taught us that damage caused by seismic city fire is more severe than that caused by the actual collapse of houses and other urban facilities when a large-scale earthquake hits the Tokyo Metropolitan Area. Prevention of seismic city fire is therefore considered the most important mitigation policy in Tokyo.

According to the Tokyo Disaster Prevention Council, it is estimated that fires would break out simultaneously at about 300 places: 208 fires could be extinguished by Metropolitan Fire Services, but 92 fires would spread uncontrollably over the city.

Obviously, this kind of estimation is essential for formulating an appropriate earthquake prevention plan. Thus, many efforts to estimate the number of seismic fires have been made. However, a truly reliable method has not yet been developed. The most popular and well-known method is the Kawazumi's formula, on which the above mentioned estimate is based.

However, Kawazumi's formula is criticized because it depends too much on data which is basically drawned from the Kanto earthquake of 1923. Fire appliances and life styles have changed radically since then, therefore Kawazumi's statistical model can not be directly applied today.

This paper aims to develop a new method for estimating fire outbreak risks in case of a large-scale earthquake, by computer simulation technique in which causal relationships concerning fire outbreak are modeled as an event tree structure.

ASSUMPTIONS AND DATA COLLECTION

In this study, it is assumed that the magnitude of the earthquake is 7.9, which is the same as that of the Kanto earthquake. This means that the acceleration at the ground level is about 250 to 400 gal, and the response acceleration of a building is about 980 gal.

The basic structure of the model is designed by formulating the interactive behavior between some fire appliances and its surrounding combustible materials when the buildings are quaked with such acceleration.

In general, fire appliances vary in kind, e.g., for cooking, for heating, for manufacturing, and they are used in different places and situations. It would be too much to develop a general model that covers each different case.

In this paper, a commercial kitchen and the fire appliances used therein are taken as a case study. This is the most dangerous case from the viewpoint of fire outbreak risks and from casualty resulting from the quake, since the commercial district is where many non-resident customers gather during the day time.

About 5,000 kitchens from different types of business were surveyed, particularly in terms of the fire appliances used there. The business were classified into eight types, they are as follows:

1. Nightclub and bar
2. Japanese restaurant
3. General restaurant
4. Department store
5. Retail store
6. Hotel
7. Hospital and clinic
8. Kindergarten

With regard to the fire appliances, it was observed that more than 20 different kinds of fire appliances were used in the kitchens. In this study, however, three major fire appliances normally used in every kitchen are counted as a fire source. They are; gas heater, gas table and gas range. The average number of fire appliances used in a kitchen was about 1.5 units.

Based on this obervation, simulation cases by business type, which will be discussed later, were set up. This paper focuses mainly on how to design and operate the event tree model (Figure 1).

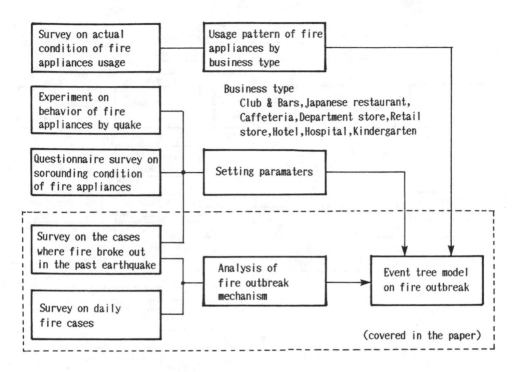

FIGURE 1. Diagram of the study


DESIGN OF EVENT TREE STRUCTURE OF FIRE OUTBREAK PROCESS

As the first step for designing the event tree structure of
fire outbreak process, three major earthquakes that occurred during
the past 20 years were carefully examined.  They are:  Niigata
earthquake (1964), Tokachioki earthquake (1966), and Miyagiken-oki
earthquake (1978).  According to the reports, 55 fires broke out
in these three earthquakes.  Figure 2 and 3 show main causal
relationship of events which led to the fire.  Based on this
analysis, referring causal factor of the daily fire case, a general
cause and effect structure of fire outbreak was designed (Figure 4).
As an exogenous condition for applying this model, typical usage
patterns of fire appliances by business type were also set up.

Causal relationships as seen in figure 4 can normally be
described as a stochastic process using probability of how frequently
each event happens.  This causal relation model, however, consists
of multiple feedback loops which make it difficult to describe by
simple stochastic equations of the conditional probability.  Thus,
the Monte Carlo Simulation Method that random numbers are generated
according to the occurrence ratio exogenously assigned is applied.

The structure of the model is basically divided into three parts.
They are; description of human behavior, dynamic movement of fire

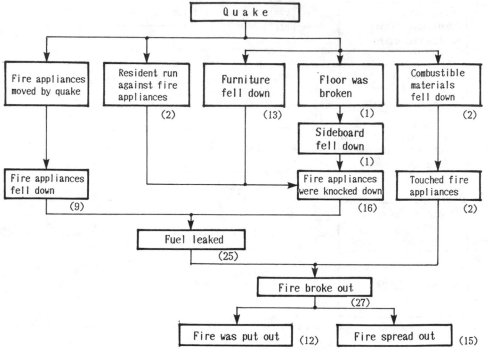

FIGURE 2. Fire outbreak process by liquid fuel of oil stove and oil heater

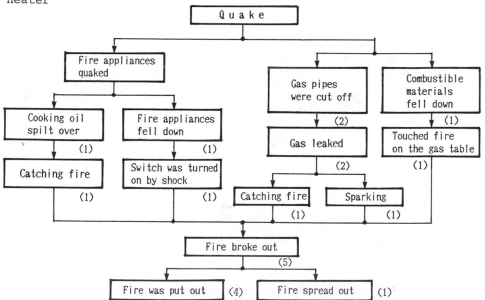

FIGURE 3. Fire outbreak process by gas fuel of cooking gas table

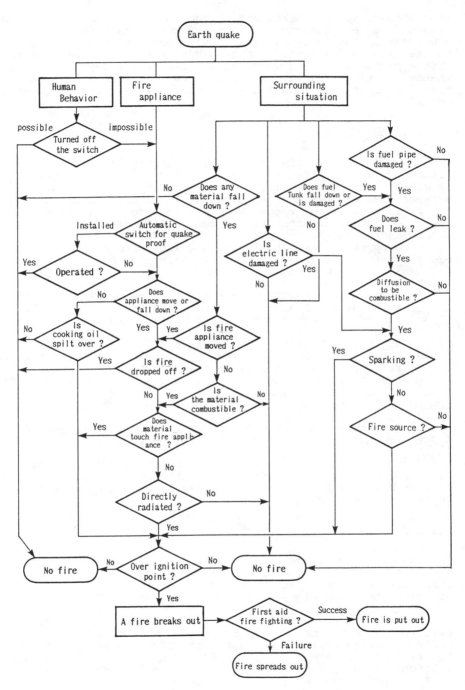

FIGURE 4. Event tree structure

appliance caused by quake, and change of circumstance. As seen in figure 4, an event happening in one part causes another event in the other part. The model describes this interaction between three parts on the time series basis. It should also be noted that the chance that a particular event would occur or not, may depend totally on whether suitable conditions for the event is formed at that particular time. A bit of time lag plays an important role for an event to occur.

SPECIFICATION OF PARAMETER

In order to operate this computer simulation model, several parameters which control the occurrence time and frequency have to be established exogenously. Not so many parameters, however, can be specified with a reasonable amount of accuracy based on the experimental results or past earthquakes, because the Kanto earthquake is too old to be of accurate reference and recent earthquakes are not as big as the one being assumed in the study. Thus, some parameters which we specified cannot always be justified practically. The basic strategy for parameter specification in this study is as follows:

a. To utilize the knowledge from experiments, if any. (Such as the behavior of appliances by quake, leak of fuel or spilling over of the cooking oil by quake, and the flashing point of oil by overheating.)
b. To apply the theoretical model in the engineering field. (Such as the computation of the response acceleration of a building.)
c. To refer the experience of the past earthquake. (Such as the breakdown ratio of the gas piping and the possibility of fire fighting at an early stage.)
d. To utilize the experience of daily fire cases. (Such as time until ignition or flash point.)
e. To keep the balance between parameters through intuitive insight based on inspection by experts on fire-proofing.

Some parameters which are difficult to specify are set tentatively first and modified later through sensitivity analysis, so that outcomes are reasonably stable in a common sense for experts in fire fighting (Figure 5).

ESTIMATION OF FIRE OUTBREAK RISKS

In order to estimate fire outbreak risks by applying the model described earlier, a pattern of fire appliance usage which is characterized by type and number of fire appliances, number of operators and operation place, has to be specified.

These characteristics vary with the type of business mentioned earlier and therefore, estimation of fire outbreak risks by a usage pattern of fire appliances can be directly translated into that by the type of business. In figure 6, as an example, the results from a kitchen in a typical restaurant with the following usage pattern of fire appliances is shown (Table 1).

936

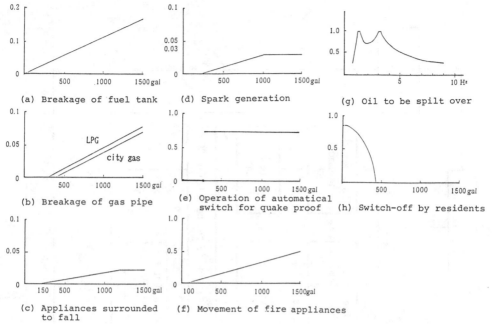

(a) Breakage of fuel tank

(b) Breakage of gas pipe

(c) Appliances surrounded to fall

(d) Spark generation

(e) Operation of automatical switch for quake proof

(f) Movement of fire appliances

(g) Oil to be spilt over

(h) Switch-off by residents

FIGURE 5. Occurrence probability of each event

TABLE 1.   Usage Pattern of Fire Appliances

| | | |
|---|---|---|
| Response acceleration of the building | | 482 gal |
| Fire appliances | gas ring | 1 piece |
| | gas range | 1 piece |
| | frying machine | 1 piece |
| Operators (cooks) | | 2 persons |

The digits seen in figure 6 show the average frequency of occurrence of each event obtained through ten times of computation, each of which generates random number 5,000 times.

First it was judged whether the switch of the fire appliances could be turned off by the cooks working in the kitchen.  In this result, the switches were turned off only 709 times out of 15,000 attempts on three appliances.  This means that on average only one fire appliance out of three can be switched off by people.  The next step is to compute how the appliances on fire would react.  Some materials which were kept nearby fell down 214 times out of 5,000 times and they touched and moved fire appliances 48 times.  As a total, fire appliances moved 528 times.

Fire of the fire appliances was put out naturally 1,337 times by movement, materials falling down and other reasons.  When materials kept nearby fell down, they could touch fire directly, and fire would break out if these were combustible.  This event occurred 66 times in this computation.  Frying machine was kept on fire 3,789 times and as a result, oil contained was overheated, and 36 fires broke out by this process.  In some cases oil was spilt over by

937

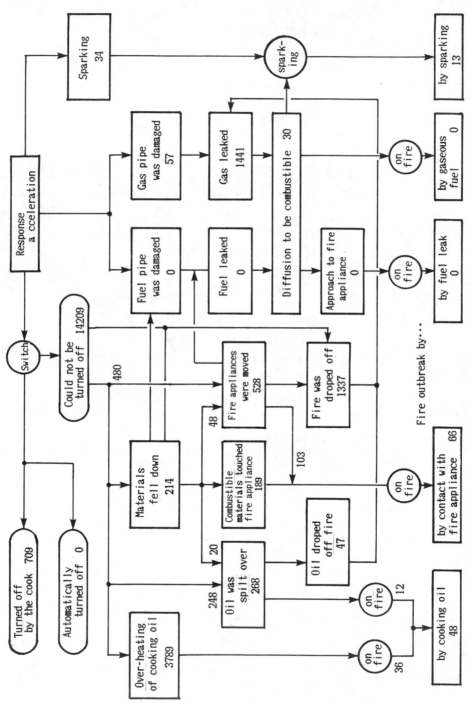

FIGURE 6. An example of outcome of fire outbreak estimation

quake, 12 fires broke out through burning oil.  As a result,a total of 48 fires were caused by oil.  The total number of fire outbreak was 127 in this case.  Therefore, probability of fire outbreak can be computed as 0.0254.

In order to compute the average probability of fire outbreak by business type based on this simulation result, the following formula was applied.

$$P = (P_o * U_o + P_m) * U_f * F_e * C_m$$

Where     $P$: Probability of fire outbreak by business type
            $P_o$: Simulation result of fire outbreak by oil
            $U_o$: Usage ratio of frying pan
            $P_m$: Simulation result of fire outbreak by other fire source
            $U_f$: Fire appliance usage ratio
            $F_e$: Failure of fire extinguishing operation multiplier
            $C_m$: Combustible materials multiplier

Table 2 shows the average probability of fire outbreak by business type computed by the above formula.  The results are different on the ground condition.  In case of the soft ground (clay), the probability is about 50 per cent over that of the hard ground (loamy soil).  The kitchen of a restaurant is the most dangerous. This means that the fire outbreak risks is high in the commercial district where these facilities are concentrated.

TABLE 2.  Fire Outbreak Risks by Business Type

| Type of Business | Fire Outbreak Risks (%) | |
| --- | --- | --- |
| | Hard ground (loamy soil) | Soft ground (clay) |
| Nightclub and bar | 0.29 | 0.41 |
| Japanese restaurant | 0.30 | 0.40 |
| General restaurant | 0.32 | 0.46 |
| Department store | 0.17 | 0.20 |
| Retail store | 0.12 | 0.17 |
| Hotel | 0.19 | 0.25 |
| Hostel and clinic | 0.06 | 0.08 |
| Kindergarten | 0.10 | 0.13 |

Major causal routes to fire outbreak cases are as follows:
a.  Fire appliances on fire $\longrightarrow$ Overheating of cooking oil $\longrightarrow$ Fire outbreak
b.  Kitchen appliances moved or fell down $\longrightarrow$ Contacted with combustible materials $\longrightarrow$ Fire outbreak
c.  Cooking oil spilt $\longrightarrow$ Contacted with fire source $\longrightarrow$ Fire outbreak
d.  Combustible appliances moved $\longrightarrow$ Damaged gas pipe $\longrightarrow$ Gas leaked $\longrightarrow$ Sparking $\longrightarrow$ Fire outbreak
e.  Boiling soup spilt over $\longrightarrow$ Flame of gas stove put out $\longrightarrow$ Gas leaked $\longrightarrow$ Sparking $\longrightarrow$ Fire outbreak

Although cases of d and e have not occurred yet in the past earthquakes, possibility of occurrence in the future is very high in areas where liquid propane gas is used.

VALIDITY OF THE MODEL

Validity of the model should be tested by reference or comparison with practical data in a past earthquake or experimental results.  In this model, however, there are neither large earthquakes, except the Kanto earthquake, nor any real scale experiment to which the model could be referred.  Therefore, instead of applying the normal validity test, intuitive insight and the expertise of skilled inspectors of fire-proofing were used to check the results of the estimation from the following viewpoints:

1.  Whether they felt that the relative occurrence ratio and order of each phenomena were reasonable.
2.  Whether they felt that the estimate of fire outbreak risks matched their intuition.
3.  Whether they felt that the occurrence ratio of fire by business type was reasonable.

The results of the estimation was thought to be acceptable by the inspectors.  The estimated sum of the number of fire outbreaks in the Tokyo Metropolitan area, obtained by muliplying the fire occurrence ratio by the total number of business, comes to 2,000.  This figure appears to be too large for the number of fires occurring in the kitchens of the special types of business.  Apart from these, other sources such as housing, chemical factories etc. must be taken into consideration.  In which case the number of fires in Tokyo would double.  This seemed to be unimaginable.

Thus it is suggested that this model be examined further from the following viewpoints.

1.  The season or time when the earthquake would take place should be considered and introduced to the model so that the usage ratio of fire appliances can be reduced.
2.  Fire source would not be left so long because people may put out the fire after the first strong quake finishes.  In which case the fire outbreak caused by overheating of cooking oil or other similar sources may be avoided.
3.  Parameters should be examined more  carefully.

REFERENCES

1.  Tokyo Fire Department, The estimation of fire outbreak risks in commercial buildings and emergency preparedness for large scale earthquake, 1983
2.  Bureau of Fire Defense, Research report on fire caused by Niigata earthquake, 1965
3.  Tokyo Fire Department, Report on Miyagiken-oki earthquake of 1978, 1978
4.  Bureau of Science and Technology, Human behavior depending on intensity of quake, 1981

# STATISTICS, RISK, AND SYSTEM ANALYSIS

Session Chair

**Dr. Marita Kersken-Bradley**
Vilshofenerstrasse 6
8000 München 80, Federal Republic of Germany

# Towards a Systemic Approach to Fire Safety

**ALAN N. BEARD**
Unit of Fire Safety Engineering, University of Edinburgh
King's Buildings
Edinburgh, EH9 3JL, United Kingdom

ABSTRACT

In order to understand fully the nature of fire safety, which encompasses social values and engineering hardware, it is necessary to consider it as a part of a 'dynamic whole'. Fire may be regarded as a failure of a system. A methodology which may be of help in attempting to approach fire safety from this point of view is suggested.

INTRODUCTION

Fire safety is often considered in a fragmentary way. That is to say, the elements which combine to produce fire and possible loss of life or injury are often effectively regarded as independent of each other. Such a disjointed approach must inevitably lead to a superficial appreciation of the problems. In order to gain a deep and comprehensive awareness of the nature of the risk in a particular situation it is necessary to attempt to consider all aspects of the problem in a coherent way. As a part of this an elucidation of the factors involved and the pertinent relationships is vital. The fire risk in a given situation is a result of the interaction of a number of 'parts'. That is, fire safety is a characteristic of an entire system and in order to understand fire safety it is necessary to understand the system.

SYSTEMIC APPROACH

The word 'system' has been used in many different ways and it is suitable here to adopt the broad definition that a system is any entity, conceptual or physical, which consists of inter-dependent 'parts'. The word 'parts' has been put in inverted commas as there is discussion as to just what a 'part' is. (See, for example, ref. 1). However, such considerations will not be pursued further here and it will be assumed that a 'part' may be fairly easily understood. A closed system is such that no interaction takes place with elements outside the system. Ultimately there is only one closed system i.e. the Universe. Smaller systems will be open to a greater or lesser extent.

A system may be considered as 'failed' if there are aspects of the system which are regarded as undesirable by one or more people involved. Whether or not something represents a 'failure' depends upon one's point of view and position within the entire system. With this in mind it is possible to think of fire as a failure of a system.

In order to gain a full understanding of the fire situation it is necessary to consider the systems involved and to look beyond the immediate horizon. There is a need for a 'systemic approach' to the fire problem and a systemic approach is not the same as a systematic approach. The word 'systematic' may be thought of as implying 'methodical' or 'tidy', but 'systemic' implies something else. A systemic approach is to see the 'dynamic wholeness' in a situation. It is a way of looking at things which should help one to see pattern and inter-relationship within a complex whole. A mode of thinking may be systematic and yet not be systemic. The significance of the concept of the 'whole' has been graphically and simply illustrated by M'Pherson (2). He takes the example of a swarm of gnats as a 'whole' and points to "the fact that each and every gnat turns back towards the centre of the swarm whenever it finds itself at the edge, which is a behavioural property that cannot be understood by only counting the gnats and tracking their motion".

The realization that it is necessary to look upon things as a dynamic whole is not new; it goes back to at least 500 B.C. At about that time Heraclitus put forward his idea that everything is in a state of perpetual change. The 'essence of Being is becoming' and all is part of the 'Universal flux':

> "One cannot step into the same river twice,
> nor touch substance twice in the same state ....
> Into the same rivers we step and we do not step."

It is interesting that Heraclitus regarded all bodies as transformations of just one element - fire. If anything, it is more important to have this active, all embracing, view of things today than it was 2,500 years ago. A problem needs to be seen in its context and not in isolation.

In order to carry out a 'systemic study' it is necessary to have an idea of the objectives of the study and an appropriate methodology. Which methodology is suitable depends upon the study objectives and the nature of the systems involved. It is useful to consider systems as either 'hard' or 'soft' (3). A 'hard' system is one in which the parts and relationships are well defined and quantified, such as an engineering system. A 'soft' system is one in which not all the parts and relationships are easily defined and quantified. All systems in which human beings play a large part are essentially 'soft'. Also, it may not be possible to give exact expression to the objectives of a study of a soft system, at least at the beginning.


METHODOLOGY FOR FIRE SAFETY

A suggested methodology for fire safety, taking into account its systemic setting, is shown in figure 1.

Although this methodology is written as a series of steps it should not be regarded as a 'sausage machine' which, when completed, produces a 'correct answer'. The need will probably arise to return to earlier steps and cycle round. It will almost certainly be necessary to return to the stage 'Formulation of the problem' more than once. The methodology should be regarded as dynamic and not static.

The 'problem' may become very complex and usually there cannot be a single simple 'solution'.

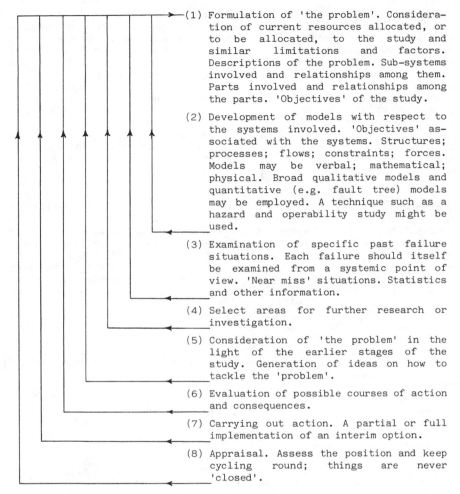

(1) Formulation of 'the problem'. Consideration of current resources allocated, or
to be allocated, to the study and
similar limitations and factors.
Descriptions of the problem. Sub-systems
involved and relationships among them.
Parts involved and relationships among
the parts. 'Objectives' of the study.

(2) Development of models with respect to
the systems involved. 'Objectives' associated with the systems. Structures;
processes; flows; constraints; forces.
Models may be verbal; mathematical;
physical. Broad qualitative models and
quantitative (e.g. fault tree) models
may be employed. A technique such as a
hazard and operability study might be
used.

(3) Examination of specific past failure
situations. Each failure should itself
be examined from a systemic point of
view. 'Near miss' situations. Statistics
and other information.

(4) Select areas for further research or
investigation.

(5) Consideration of 'the problem' in the
light of the earlier stages of the
study. Generation of ideas on how to
tackle the 'problem'.

(6) Evaluation of possible courses of action
and consequences.

(7) Carrying out action. A partial or full
implementation of an interim option.

(8) Appraisal. Assess the position and keep
cycling round; things are never
'closed'.

FIGURE 1: A suggested methodology for a systemic study in relation to fire
risk.

    A problem will, in general, require cycling through the stages of the
methodology indefinitely. That is, in principle, a never-ending series of
iterations may be required. The statement of 'the problem', and understanding
of it, will change through time. Intermediate working solutions may be
formulated at different points in this process. Although the difficulties are
great it is suggested that the methdology does provide a guiding structure
for tackling a situation. Within the structure techniques may be used which
may be 'sausage machines' to a greater or lesser degree.

    Both theoretical and experimental models might be used. For example, a
theoretical model might be constructed which aimed at providing the
probability of fatalities given ignition. Another model might be of a purely
verbal kind relating, for example, bed-sores in a hospital ward to the type
of mattress or the staffing level. Such knowledge might be put forward by

nursing staff. Different forms of model should be considered in an attempt to throw light on different facets of the overall system. It must be stressed that a model does not constitute a systemic approach itself. However, models may be used as part of a systemic approach; a model would attempt to answer a relatively limited question within a broader consideration. Within its structure a technique may bring in many different factors. However, a distinction must be drawn between a model which brings in many different factors as part of an attempt to answer a particular question and a systemic approach itself. (More is said about models later in this paper). In short, a systemic approach should not be dogmatic and prescriptive but open and flexible. The tentative methodology outlined here may be thought of as an initial attempt to provide a green field within which people may fruitfully graze rather than an algorithm for the forced feeding of a goose. Some comments on the stages are in order:

(1) Formulation of 'The Problem'. Why is the study being carried out? What has prompted it? What are the initial objectives of the study? In the case of hospitals the 'problem' might be rather tentatively stated as 'How to increase fire safety in hospitals'. However, the exact statement of the 'problem' might alter as the study progressed. For example, it might improve fire safety to reduce the number of electrical appliances in a hospital, but then the absence of a particular electrical instrument might have a very damaging effect with respect to some other aspect of the system. There will be conflicting objectives. People with different points of view need to be brought in. Also it will usually be the case that limitations of current resources allocated, or to be allocated, to the study (together with similar limitations and factors, e.g. temporal) constitute part of the problem itself.

Considering the hospital fire safety problem further, some of the relevant sub-systems might be:

1. Patients.
2. Nursing and medical staff.
3. General public.
4. National Health Service.
5. Department of Health and Social Security.
6. Home Office
7. 'Design system' for construction of hospitals.
8. 'Fire safety design system' for construction of hospitals.
9. Hospital 'fire safety system'. (After construction.)
10. Fire brigade.
11. Ambulance service.
12. Local authority.
13. Fire research system.
14. Technical systems.
15. Systems directly associated with the chemistry and physics of fire processes.
16. Manufacturers of fire safety and other equipment used in hospitals.
17. British Government.
18. British socio-economic system.
19. International socio-economic system.

Most of these sub-systems overlap. The order given above is not meant to imply any kind of 'order of importance'; it is simply a list of some of the systems which might be pertinent to a study. Other systems may also be relevant.

(2) Development of Models with Respect to the Systems Involved. A model is a representation of an aspect of 'reality' and expresses, amongst other things, the point of view of the person constructing it. The construction of models will already have started in stage 1. There are many different types of model, ranging from broad verbal statements to deterministic mathematical models. One may also consider physical models. In addition to models which are intended to represent a situation in a general manner, there are also 'simulation' models which are based on the generation of specific cases. Mathematical probabilistic models of this type might use, for example, the Monte Carlo technique. An example of a physical simulation model would be the representation of the flow of a river and its tributaries by flow of electrical current through wires.

Most of the systems we encounter have 'objectives' associated with them. That is, there may be aims and expectations which people have with respect to a system. A person or a group of people may want one or more things from a system. However, the objectives of a person or group may conflict. Also, that which a person or group wants from a system may conflict with that which another person or group wants from the system. For example, the objectives of a manufacturer of motor cars will be different from the objectives of a buyer of a motor car. In general, for any complex system, there will almost certainly be conflicts of objectives. Clarification and understanding of objectives and how they arise is necessary.

The structures, processes and forces existing within systems and crossing system boundaries need to be investigated. Flows of information and material both within systems and between systems need to be understood. There may also be constraints for a given system. For example, the amount of money received by a local authority from government may be fixed.

The models should help us to understand better the relationships within the system as a whole.

However, a word of warning is necessary regarding the use of models. All models have limitations and it is vital to be aware of what those limitations are. The assumptions, both explicit and implicit, in each model must be clearly realized. It is as important to know what a model cannot say as to know what it can say. This is true for both quantitative and qualitative models. In particular, for quantitative models, it is important not to attach an unjustified significance to numbers which result. At an obvious level there are uncertainties which will be associated with models. For numerically quantitative models these may be expressed as, for example, 'errors', 'confidence limits' etc. Numerical results should be seen in their context. According to the mathematician Gauss a lack of appreciation of the value of mathematics is "nowhere revealed so clearly as by meaningless precision in numerical studies".

Another issue raised in the application of specific methods is that models, and techniques in general, may on occasion be applied to situations for which they are not appropriate. It has been said that this is the most common form of mis-use (3).

In addition to these 'overt' points there may also be mis-construction at a deeper level. A simplistic approach can sometimes lead one to effectively ascribe to models powers which they do not possess. (One might almost call this a 'fetishism' which may be associated with models.) For example, if one wishes to compare risk situations then a relevant quantity to take into account would be the probability of fatality associated with each risk. It would, however, be foolish to pretend that a comparison of the probabilities

947

of fatality would represent a comparison of the situations which may give rise to fatality. Such over-simpliicty is sometimes an implied assumption if not an explicit one. There are many different dimensions involved in each risk and not all of these dimensions may be quantifiable; mathematical models cannot provide a 'correct answer'. A comparison of the risk associated with cardio-vascular disease and that associated with road accidents is not straight forward. Looking at the probabilities of fatality is to consider only one aspect of these risk situations. In general, qualitatively different facets should not be collapsed onto a one-dimensional scale.

Further, it may well not be possible in practice to associate numbers with many of the characteristics involved even if it may be possible to do so in principle. If we effectively insist, say, that a hazard only really exists if plausible numbers can be associated with it we shall be limiting our conceptualization dramatically. Also, considering things in purely quantitative terms may lead to a false sense of security because if numbers cannot be attached to a characteristic then there may be a tendency to ignore it; conceptual features may be lost. More generally, we must realize that all explanation is interpretive and context-dependent and, as Thomas has pointed out (4), how we measure fire safety is itself a value-judgement.

Having sounded these cautionary notes it must be said that the appropriate use of models may be of great help in enabling us to comprehend things. Both deterministic and probabilistic models may aid us and one probabilistic technique is seen in the stochastic model which has been developed by the author. Many of the assumptions contained in that model are not firmly based and it needs to be improved. Clarification of the assumptions and limitations is contained in the references. The purpose of that technique is to afford an idea of the likely number of fatalities given that a fire starts in a hospital ward. An estimate may thereby be found for the changes in the likely number of fatalities which would be expected to result from changes in the ward sub-system. It is described further in the Appendix.

(3) <u>Examination of Specific Past Failure Situations</u>. Studies of past fires are obviously of crucial importance. Each past fire should itself be examined from an overall systemic point of view. 'Near miss' occasions should be studied i.e. occasions which very nearly could have produced a failure but in fact did not. ('Failure' might be taken to mean, for example, 'injury or death due to fire'.)

Information, both statistical and non-statistical, should be considered. In the case of hospitals information afforded by hospital staff could prove to be very useful.

(4) <u>Select Areas for Further Research or Investigation</u>. Both experimental and/or non-experimental work might be pursued. For example, it might be decided that it would be useful to carry out some specific tests. Different kinds of survey could be considered. In the hospital case it might be decided to ask staff to answer survey questions.

(5) <u>Consideration of 'The Problem' in the Light of the Earlier Stages. Generation of Ideas on how to Tackle 'The Problem'</u>. Radical changes in the systems may be necessary. Given the understanding of the situation reached so far, what possible changes seem suitable as a means of eliminating or reducing the 'problem'? Alternative ideas may be tentatively suggested.

(6) <u>Evaluation of Possible Courses of Action and Consequences</u>. Each of the alternatives to emerge in the previous stage should be considered in detail.

An attempt must be made to elucidate the full implications of each alternative.

(7) Carrying Out Action

(8) Appraisal. Whatever actions are carried out the study is never 'finished'. It will always be necessary to re-think old ideas.

Before leaving this discussion of the stages of the methodology it is important to make the general point that the investigators themselves are also part of the system under study and this must be realized (5).

CONCLUSION

Fire safety needs to be examined from a systemic point of view and not in a disjointed or narrow fashion. Fire is a product of a system and it is necessary to understand the 'dynamic wholeness' of things in order to understand the failure.

To carry out a study from a systemic point of view it is necessary to have a guiding structure (or methodology) which is dynamic and not static. Within a methodology specific techniques may be employed which have specific tasks. It must be emphasised, however, that a particular technique would have a relatively limited remit and would only consitute a part of a systemic approach. A distinction must be drawn between a theoretical model which attempts to acount for different factors in trying to answer a fairly narrow question and a systemic consideration of the problem as a whole. A possible outline for such a methodology has been suggested and it is hoped that such a framework would help in the task of exploring as many aspects of the problem as possible.

Essentially, the central theme of this paper has been the need to consider things in their entirety. Concern about fire safety has been the 'starting point' for this consideration. However, as has been mentioned, in any complex situation involving human beings there will be conflicting objectives and in any action it is necessary to attempt to deal with all of these objectives. We need to try to develop a system within which conflicts and contradictions do not arise, as far as we possibly can.

Fire safety has been the starting point in this paper but if we view things as an entirety then in principle all aspects of a system should enter into consideration no matter what is the starting point.

In short, one can rarely, if ever, make a straight forward 'fire safety decision'. Generally, one can only make a 'decision' which has 'fire safety consequences', amongst others. Usually a decision will have numerous ramifications, implications and facets. Many people, bringing in different points of view, need to be involved in the process.

To close, a line from the New England poet Emily Dickinson puts things rather well:

"In broken mathematics we estimate our prize,
Vast in its fading ratio to our penurious eyes."

We need to be able to develop eyes which are no longer penurious but are capable of seeing the nature, complexity and inter-relationship of the world we live within. In such a context mathematical models may have a truly valuable contribution to make.

APPENDIX : MODELLING THE NUMBER OF FATALITIES IN A FIRE

This Appendix concentrates on a brief description of a type of mathematical model which may be used as a part of a guiding methodology. It serves
as an example of a general class of techniques but it is not meant to be
implied that this kind of technique is necessarily 'better' than other
models. As stressed in the main text, many different kinds of model should be
considered and all techniques should be employed with caution and circumspec-
tion.

This Appendix does not describe an application of the methodology
suggested. It attempts to answer a specific question and considers different
causal factors as parts of the model. Such a technique would form only a
small part of a systemic approach.

Before describing the stochastic model reference may be made to three
logic-trees which have been constructed for three past fire disasters. Each
tree was used to assess the reduction in the likelihood of the occurrence of
the disaster if things had been different. The three fires considered were
the Coldharbour Hospital fire (6), the Fairfield Home fire (7) and the St.
Crispin Hospital fire (8).

In addition to these specific logic-trees a stochastic model has been
developed which is intended to give an idea of the likely number of deaths
resulting from a fire in a given space. The development of the fire in terms
of heat, smoke and toxic gases is considered as well as the response of those
present and evacuation characteristics. The model has been constructed for
hospital wards and has been applied to the ignition of a bed having a
polyurethane mattress. Two types of ward have been looked at, the first being
of 'Nightingale' design containing thirty, bed-bound, non-ambulant patients
and the second being a ward consisting of six-bedded bays containing similar
patients (9,10). Further considerations, including sensitivity studies, have
also been carried out (11).

A stochastic framework has been devised by considering a number of
'critical events' which the fire may pass through and the times between
critical events. Particular rates of heat release are associated with the
critical events and these are given in table 1. The probability of a fire
going through a critical event given that it has gone through an earlier one
is represented by a 'transition probability', $P_{ij}$. Probability density
functions have been assumed for the times between critical events and fires
have been simulated using the Monte Carlo method. Each fire passes through
one of six chains with number 1 corresponding to the most trivial fire and
number 6 the most serious. Table 2 shows the relative frequency of each chain
and the numbers of fires corresponding to given numbers of fatalities for
each chain. This is for the Nightingale ward calculation.

For the bay calculation it was found that the likelihood of multiple
fatalities resulting is higher than for the Nightingale arrangement but that
the mean number of fatalities for 500 simulations is approximately half that
for the Nightingale ward.

Sensitivity studies have been conducted for the Nightingale calculation
by looking at the effects of changes in the assumptions on the mean number of
deaths. The most striking change is seen upon alteration of each of the first
two transition probabilities, all else remaining the same, (Fig. 2). It has
also been found that the calculated mean number of fatalities is approximately halved if the smoke and gases are assumed to stratify rather than fill the
ward uniformly.

TABLE 1 – Critical Heat Event (CHE) Definitions.
         (CE1 refers to the first critical event).

| | |
|---|---|
| CE1 | Ignition |
| CHE2U | Fire passes through 2 kw on way up |
| CHE3U | Fire passes through 50 kw on way up |
| CHE4U | Fire passes through 1000 kw on way up |
| CHE5U | Fire passes through 10,000 kw on way up |
| CHE6U | Fire passes through a level $\dot{q}_m$ (upwards) |

TABLE 2 – The relative frequency of each chain and the numbers of fires corresponding to given numbers of fatalities for each chain.

| Chain | Relative Frequencies | Number of fires with X fatalities | | | |
|---|---|---|---|---|---|
| | | X=0 | X=28 | X=29 | X=30 |
| 1 | 0.718 | 359 | 0 | 0 | 0 |
| 2 | 0.188 | 94 | 0 | 0 | 0 |
| 3 | 0.058 | 8 | 7 | 13 | 1 |
| 4 | 0.016 | 0 | 0 | 8 | 0 |
| 5 | 0.006 | 0 | 0 | 3 | 0 |
| 6 | 0.014 | 0 | 0 | 7 | 0 |

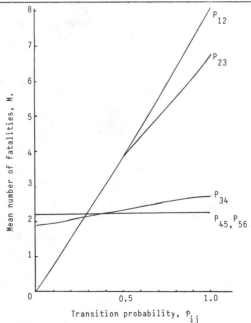

FIGURE 2 – Variation in mean number of fatalities, M, with each transition probability, $P_{ij}$.

(Unless otherwise: $P_{12} = P_{23} = 0.3$; $P_{34} = 0.4$; $P_{45} = 0.5$; $P_{56} = 0.6$)

REFERENCES

1.  Beer, S., 'Below the Twilight Arch-A Mythology of Systems'; In Proceed-
    ings of the First Systems Symposium, Case Institute of Technology, 1961.

2.  M'Pherson, P.K., 'A Perspective on Systems Science & Systems Philosophy'
    Futures, Vol. 6, no. 3, pp 219-239, June 1974.

3.  'Systems Performance-Human Factors and System Failures', The Open
    University Press.

4.  Thomas, P.H., 'Fire Safety: Some General Aspects of Research, Regulation
    and Design'; In conference entitled "Systems Approach to Fire Safety in
    Buildings", Tsukuba, Japan, 29-30 August, 1979.

5.  Boulding, K.E., 'General Systems as a Point of View'; In proceedings of
    the Second Systems Symposium, Case Institute of Technology, 1964.

6.  Beard, A.N., 'Applying Fault-Tree Analysis to the Coldharbour Hospital
    Fire', "Fire", vol. 71, pp 517-519, 1979.

7.  Beard, A.N., 'A Logic-Tree Approach to the Fairfield Home Fire'; "Fire
    Technology", vol. 17, no. 1, pp 25-38, February 1981.

8.  Beard, A.N., 'A Logic-Tree Approach to the St. Crispin Hospital Fire',
    "Fire Technology", vol. 19, no. 2, pp 90-102, May 1983.

9.  Beard, A.N., 'A Stochastic Model for the Number of Deaths in a Fire';
    "Fire Safety Journal", vol. 4, pp 169-184, 1981/82.

10. Beard, A.N., 'A Stochastic Model for the Number of Deaths Resulting from
    a Fire in a Bay in a Hospital Ward'; "Fire Safety Journal", vol. 6, pp
    121-128, 1983.

11. Beard, A.N., 'A Stochastic Model for the Number of Deaths in a Fire:
    Further Considerations', "Fire Safety Journal", vol. 8, pp 201-226,
    1985.

ACKNOWLEDGEMENTS

    I would like to thank Professor David Rasbash for his comments on this
work. Also, I am grateful to the Fire Protection Association, London, for
permission to use material from the article "Towards a Rational Approach to
Fire Safety" which was published in 'Fire Prevention Science and Technology',
No. 22, pp 16-20, 1979. This work was supported financially by the Science &
Engineering Research Council and the Department of Health & Social Security.

# The Use of Probabilistic Networks for Analysis of Smoke Spread and the Egress of People in Buildings

**WAI-CHING TERESA LING**
Department of Mathematics and Science
City Polytechnic of Hong Kong

**ROBERT BRADY WILLIAMSON**
Department of Civil Engineering, University of California
Berkeley, California 94720, USA

## ABSTRACT

This paper focuses on the use of a network analysis approach to solve the fire safety problem associated with the spread of smoke and the probability of escape by the occupants before the fire and/or smoke blocks their path. Many random factors affecting smoke production and spread can be accounted for by coupling each smoke spread network to a given fire spread network. Smoke spread is examined for different fire scenarios, and the occupants' egress problem is treated as a dynamic network flow problem for a given fire scenario. The time to detection and to untenable conditions, as calculated from the smoke spread network, determine the time period of the dynamic network under consideration.

## INTRODUCTION

Determining the spread of fire and smoke between rooms in buildings and other structures is one of the unsolved problems of fire protection engineering. Extensive research has been carried out in the area of fire growth and smoke production within the "room-of-origin", which has contributed to a better understanding of one aspect of the fire and smoke spread problem between rooms. The other aspect of this problem is strongly probabilistic, and although deterministic modeling can be extended from the room-of-origin models, the solution to this problem will depend on a quantitative treatment of the stochastic aspects of the process.

The application of networks to solve different fire protection problems has been suggested by several writers. Elms, Buchanan and Dusing [1] represented possible paths by defining the rooms as nodes and the links between these nodes as possible paths for room to room fire spread. Berlin, Dutt and Gupta [2] also depicted a building as a network; however, their nodes represented the location of the occupants, and their links the possible exit paths for these occupants in the case of a fire. Ling and Williamson [3,4] used Elms et al.'s approach for representing the fire spread, and they associated an element of time as well as a probability with each link. This paper focuses on the use of a network analysis approach to solve the fire safety problem associated with the spread of smoke and the probability of escape by the occupants before the fire and/or smoke blocks their path.

In recent years, most research efforts dealing with smoke problems have been concentrating on the measurement of smoke, the reduction of smoke potential of materials, and active smoke control within buildings. High rise smoke simulation experiments have been conducted to quantitatively measure the smoke movement under controlled and uncontrolled conditions [5,6,7]. It should be noted that all these experiments were conducted for vertical smoke movements.

In their analysis of fire spread networks, Ling and Williamson [3,4] calculated the probability of occurrence for each fire scenario by means of an equivalent fire spread network using Mirchandani's algorithm [8]. The same approach is used in this paper for representing both the spread of smoke and the egress of occupants in buildings. Many random factors affecting smoke production and spread can be accounted for by coupling each smoke spread network to a given fire spread network. Smoke spread is thus examined for different fire scenarios, and in a similar fashion, the occupants' egress problem is treated as a dynamic network flow problem for a given fire scenario. The time to detection and to untenable conditions, as calculated from the smoke spread network, determine the time period of the dynamic network under consideration. With additional data input into the network regarding the number of occupants in each room and the degree of mobility of the occupants, the question "are all the occupants able to escape without injury in the time allowed" can be answered.

## CONSTRUCTION OF A PROBABILISTIC HORIZONTAL SMOKE SPREAD NETWORK

The framework for the construction of a smoke spread network will be laid out in this section. Nodes will represent space as well as the smoke conditions of that space. Links will represent possible movement of smoke from space to space, or the transition of different smoke conditions within that space. Smoke conditions of special interest with respect to detection and exiting of occupants will be discussed below.

If there are operative smoke detectors, they sound the alarm when smoke reaches a threshold quantity in the vicinity of the detector. In the probabilistic smoke spread network, this possible detection state will be depicted by a star-shaped node ✶ .

Cooper [9] has suggested that a hazardous condition for safe egress exists when either the smoke/toxic layer has dropped down to .91 m (3 ft) above the floor, or the average temperature of the upper layer has reached $183^{\circ}$C (corresponding to .25 W/cm$^2$). An untenable condition in the corridor is defined here as the time when the smoke layer drops to .91 m above the floor of such a corridor. This condition is depicted by ■. Thus ■$_C$ represents an untenable condition in the corridor.

The method of constructing the fire spread network can also be applied to the construction of the smoke spread network. Associated with each link in the smoke spread network is a pair of numbers $(p_r, t_r)$ where $p_r$ represents the probability that the rate of smoke discharge in the room of origin is r, r being a material property which can be given in terms of mass of smoke particles divided by the mass of material burned per unit time, and $t_r$ represents the time it takes for the smoke to travel the link, given that the rate of discharge is r. If r is multiplied by the combustion rate of the material in the room of origin, it becomes a "source term" in terms of total mass of smoke particles generated per unit time. Certain smoke detectors are activated when the air passing through them has a density of smoke particles (products of combustion) exceeding a specific limit. The time $t_r$ is obtained by dividing the total number of particles needed for such a volume by the smoke particle generating rate.

Once the probability distribution for $t_r$ associated with each link is known, Mirchandani's algorithm can be applied to construct an equivalent network for smoke spread. The reliability and the expected values can then be computed. Since an untenable condition in the corridor signals the end of safe exiting for the occupants remaining on such a floor, the horizontal smoke spread network presented in this paper considers the time it takes for smoke/toxic gases to spread from the room of fire origin to cause an untenable condition in the corridor.

The construction of a horizontal smoke spread network is presented by way of examples. The numerical evaluation of such a network is contingent upon the availability of data which, at this time, is still mostly lacking. However, the methodology is introduced in this paper to illustrate its usefulness and to suggest a direction for future data collection.

Assumptions:

To facilitate the construction of the smoke network the following general smoke behavior is assumed:
1.  Movement – During the preflashover state of a fire in the room of origin, smoke will start filling up the space from the ceiling of the room down to the soffit above the door header. If the door is open, smoke will then spread to the adjacent space assumed to be the corridor. If the door is closed, smoke will first fill the whole room of origin. When a sufficient built-up of pressure is reached in this room, the smoke will begin to escape into the corridor through the door assemblies or air duct.
2.  Materials – It is assumed here that the materials in the room of origin are either cellulosic or plastic. For cellulosic materials it is assumed that throughout the preflashover period the smoke/toxic gases are only a threat to the occupants in the room of origin. After flashover, particularly during the ventilation control period, it is assumed that there is a major threat to the occupants beyond the room of origin with respect to both toxic effects and visibility. In the case of plastic materials, it is assumed that they generate large quantities of thick, black smoke which, if the door is open, may cause low visibility conditions outside the room of origin even before flashover occurs. It should be noted that many plastic materials are capable of creating very toxic gases after the fire flashes over, and they continue to pose a threat to life even after the fire has gone into the fuel control period.
3.  Detection – It is assumed that detection of the cellulosic smoke by means of an automatic detector (POC) or by human olfactory senses may be possible outside the room of origin at $\alpha\, t_f$, $0 < \alpha < 1$, where $t_f$ = time to flashover, if the door is open. If the door is closed, detection in this area is possible only after flashover. In the case of plastic materials, even if the door to the corridor in the room of origin is closed, it is assumed that there is sufficient leakage before flashover, making detection in the corridor possible at $1/2\, t_f$. And if the door to the corridor in the room of origin is open, it is assumed that detection will be accomplished at $0.1\, t_f$.
4.  Untenable Condition – It is assumed that if the door to the corridor in the room of origin is open when flasheover occurs, the corridor will become untenable instantaneously for both plastic and cellulosic materials. If the corridor door with a fire endurance of $t_b$ minutes is closed in the room of origin, it is assumed that in the case of cellulosic fires, the corridor will become untenable when the fire breaches the door. For plastic fires under similar conditions, it assumed that the corridor will become untenable in $1/2\, t_b$ minutes after flash-over.
5.  Smoke Contribution – It is assumed that there is no smoke contribution from the materials in the corridor.

Smoke Spread Network

   The floor plan shown in Fig. 1 and the basic module shown in Fig 2. will be used to demonstrate the construction of a smoke spread network. Since it is assumed that there is no header separating the ceiling of the whole corridor $C_1$ through $C_7$, the corridor is represented by one node C.

   In this sample illustration of the construction of a smoke spread network, the following fire scenario will be used. The fire starts in Room 1, flashes over, spreads into the corridor segment $C_1$ via an open door and then spreads into the corridor segment $C_2$. The path of the chosen scenario is shown in Fig. 3(a) by means of the fire spread network. It should be noted that this represents only one of many possible fire spread paths, as described in the previous papers by Ling and Williamson [3,4]. When applying this methodology, one would first analyze the potential fire scenarios and their probability using Ling and Williamson's method [3,4] and then construct the smoke spread networks for each viable scenario. Using the notations from the fire spread network, and the new smoke symbols $\star_C$ and $\blacksquare_C$, a probabilistic network of smoke spread from Rm 1 to the untenable condition in the corridor is shown in Fig. 3(b).

   Note that the uncertainty of fire spread was taken care of by the choice of a specific fire scenario. In other words, smoke spread is considered with the assumption that the chosen fire scenario will occur. Thus every one of the probabilities of spread time associated with the links taken directly from the

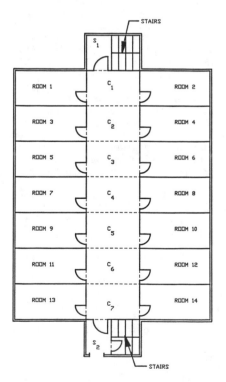

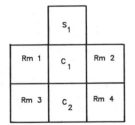

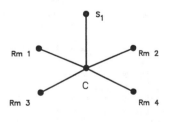

FIGURE 1. Floor plan of a typical office building, apartment or hotel.

FIGURE 2. Diagram showing the transformation of a floor plan of a module (top) into a smoke spread graph (bottom).

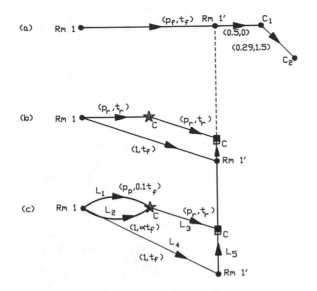

FIGURE 3. Diagram of the construction of a smoke spread network of a fire starting in Room 1, flashing over, spreading into the corridor Segment $C_1$ via an open door and then spreading into the corridor segment $C_2$. (a) Fire spread network (b) Smoke spread probabilistic network (c) Smoke spread equivalent network.

fire spread network is put to unity. The equivalent network for the chosen fire scenario is constructed using the above assumptions, as shown in Fig. 3(c). The probability $p_p$ in $L_1$ represents the probability that some of the burning contents or lining materials in the room of origin are plastic. If the burning materials are not plastic, then by our assumption, they must be cellulosic causing the smoke detector to be activated at $\alpha t_f$ where $0 < \alpha < 1$. Link $L_3$ in Fig. 3(c) represents the situation in which a very slow growing but smoky fire with plastic materials generates enough smoke to make the corridor untenable before the room of origin flashes over. Link $L_4$ is a part of the fire scenario and thus the probability is 1, and the $t_f$ is dictated by the fire scenario. Link $L_5$ corresponds to the assumptions that the corridor door in the room of origin is open and that the corridor will become untenable instantaneously when flashover occurs for either plastic or cellulosic fires.

Figure 4 depicts the situation in which a fire flashes over in the room of origin, breaches a closed door with fire endurance $t_b$ and then spreads to the corridor. With these stated assumptions, the only remaining uncertainty in the smoke spread network is the probability of the room contents and linings to be of plastic materials. This scenario is thus represented in Fig. 4 where the smoke spread network is shown below the fire spread network links.

The smoke spread for the scenario in which a fire flashes over in the room of origin whose door is closed, breaches the wall and spreads to the next room, Rm 3, is shown in Fig. 5. The fire then flashes over in Rm 3 and spreads into the corridor via an open door. In this fire scenario, even though the fire does not breach the fire door in Rm 1, the room of origin, smoke penetrates through the fire door once Rm 1 flashes over. Thus Fig. 5 shows the links $(p_p, 0.5t_d)$ and $(1, t_d)$ between the detection node and the untenable node. Note that $t_d$ is the fire endurance of the corridor door in Rm 1; it is larger than the endurance of the wall between Rm 1 and Rm 3.

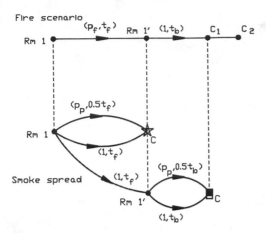

Fire scenario

FIGURE 4. Illustration of the scenario in which the fire flashes over in the room of origin, breaches a closed door with a fire endurance $t_b$ and then spreads into the corridor. Here the smoke spread network is shown below the fire spread network links.

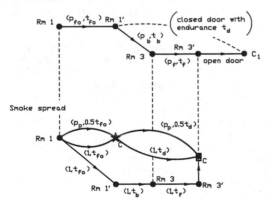

Figure 5. Diagram of the smoke spread equivalent network of the scenario in which a fire flashes over in the room of origin whose door is closed, breaches the wall and spreads into the next room, Rm 3. The fire then flashes over in Rm 3 and spreads into the corridor via an open door. This Fig. shows the links $(p_p, 0.5t_d)$ and $(1, t_d)$ between the detection node and the untenable node.

After the formation of the smoke spread equivalent network, the recursive formulae shown by Ling and Williamson [4] can be applied to calculate the expected shortest spread time from ignition to the possible detection state . Similarly, the expected shortest spread time from ignition to the untenable node $\blacksquare_C$ can be calculated. The difference between these two values represents the time available for safe egress under a given fire scenario. It is interesting to note that although fire spreads more quickly with an open door, because of possible early detection made via such an open door, the available egress time is actually longer with an open door than with a closed, unrated door.

## OCCUPANT EGRESS

Different methods for calculating the time needed for egress in case of fire are mentioned in the literature. The classic work in the field was published by Predtechenskii and Milinskii, and is now available in an English translation [10]. Their methods were used by Kendik [11] to calculate the time required to evacuate three high rise buildings. She then compared the predictions with actual evacuation tests, and estimated the "projected area factor" to be between 0.12 and 0.15 $m^2$. Berlin, Dutt, and Gupta used a simulation method to evaluate the egress network for estimating the time needed for the elderly to be evacuated from rest homes. Pauls [12] reported the results of evacuation drills in Ottawa high rise office buildings. The results are given for the total time required for evacuating the whole building and the descending speed in terms of number of people per 22" width stairwells. Chalmet, Francis and Saunders [13] used the procedure of Ford and Fulkerson [14] to transform a static transshipment problem into a dynamic network flow model representing the evacuation of a building as it evolves over time, where time is represented discretely by consecutive time periods. The dynamic model, while directly minimizing the average amount of time each occupant takes to exit the building, simultaneously maximizes the total number of people to be evacuated from the building during periods 1 through p for all values of p, and minimizes the time taken by the last evacuee to exit the building. This triple optimization result is mathematically proven by Jarvis and Ratliff [15].

The model for the occupants' egress is based on the framework that had been devised for the fire and smoke spread. For a given fire scenario, there is no uncertainty regarding the fire growth pattern, the detection time, or the time allowed for safe egress. Other uncertainties, such as the number and the degree of mobility of the occupants in the room at the time the fire breaks out, still remain. Once the type of occupancy for a particular building is identified, the building code sets guidelines on the maximum number of people that can be expected in each room. Although the actual number of occupants at the time of a fire is highly stochastic, the maximum number allowed by the building code can be considered as a reasonable estimate. Thus the capacity for occupancy set by the building code will be used as the number of people who need to be evacuated in a given fire situation. Given the type of occupancy, the travel time required by the occupants for exiting can now be determined from empirical data, such as those collected by Pauls [12]. In the event of mixed occupancy, the longest needed travel time should prevail. This is similar to the PERT type network in which the longest path determines the shortest project completion time. Based on the above discussed ground rules, the method for applying the dynamic network flow to the evacuation of people egress is presented in the following section.

## Dynamic Network Flow Model

The first step in the construction of a dynamic network flow model is to build a static model with a floor plan in which the rooms, corridors and stairs are represented by nodes. The dynamic model is then constructed to represent the movement of occupants over time. The total time period to be considered extends from the start of the fire in the room of origin until the onset of untenable conditions blocking the last exit path. For the individual room, this time period is from the time the fire is detected until the time the path leading to safety from this room is blocked by smoke. The duration of this time period is determined from the calculation of the smoke spread network for a particular fire scenario. If the total time is divided into K equal time periods, then there will be (K+1) copies of the static nodes in the dynamic model.

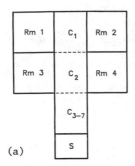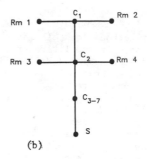

Figure 6. Illustration of the occupants' egress model demonstrating use of the dynamic flow method. (a) Floor plan. (b) Graph representation


The floor plan shown in Fig. 6(a) with a graph representation shown in Fig. 6(b) will be used to demonstrate this dynamic flow method. This static model has four nodes for Rm 1 through Rm 4; three nodes $C_1$, $C_2$, and $C_{3-7}$ for the corridor segments; one node S for the stairs; and four nodes Rm 1 through Rm 4* representing the state when the occupants in the room have detected the fire. Thus there is a total of twelve nodes. Note that once Rm i* node has been reached, the situation is irreversible and there is no need to go back to the corresponding Rm i node. Thus after detection of the fire, there are only eight static nodes to work with. For demonstration purposes, the following fire scenario was chosen. A cellulosic fire started in Rm 1 and spreads to the corridor via an open door. Under this fire scenario and using the assumptions described earlier, smoke will cause the whole corridor ($C_1$, $C_2$ and $C_{3-7}$) to become untenable as Rm 1 flashes over. This marks the end of safe egress for all the occcupants still remaining in the rooms. It is assumed that the length of time is divided into ten equal periods and the detection time for all the rooms, except the room of origin, is one time period. There are eleven copies of the static nodes. Assume the number of occupants in Rm i to be $a_i$, for i=1, 2, 3, and 4. They are source inputs into the dynamic network flow. For those nodes that do not have any occupants at the start of the fire, such as the corridor nodes, assign a capacity of zero. Construct a link between every copy of a node to the adjacent copy of the same node without any restrictions on the capacity for these holdover links.

The next step is to consider the links which indicate the possible movements of occupants from one area of the floor to an adjacent area. The capacity associated with each of these movement links is the maximum flow rate between nodes in terms of number of people per time period. It is assumed that the link capacity for the link leaving Rm i* is represented by $f_i$ for i=1, 2, 3, 4. Those between $C_1$ and $C_2$, $C_2$ and $C_{3-7}$ are represented by $f_5$ and $f_6$, respectively. Finally, the capacities for the links that go from $C_{3-7}$ into the stair node S are represented by $f_7$.

For the complete dynamic network, the holdover links and the movement links are to be combined. To simplify the network, all nonessential nodes and links are eliminated; i.e. the links and nodes not part of any path going from source nodes to exit nodes. In the sample network here, the source nodes are nodes Rm 1 through Rm 4 in time period zero, and the exit nodes are S nodes in all the time periods. The complete dynamic network is shown in Fig. 7. The dynamic network depicting the fire scenario in which a fire starts in a vacant Rm 1 with a closed five-minute door is shown in Fig. 8. The difference between an early and a tardily detected fire can be viewed by comparing Figs. 7 and 8.


960

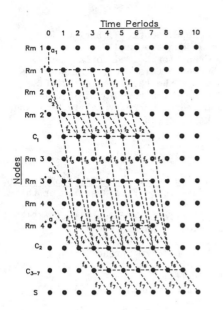

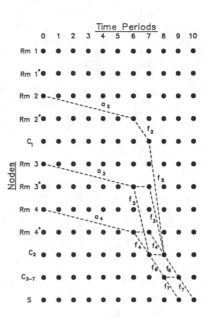

Figure 7. Diagram of a complete, dynamic network of a fire in a room
whose door is open. To simplify the
network, all nenessential nodes and
links are eliminated.

Figure 8. Complete dynamic network of a
fire starting in the vacant Rm 1 which
has a closed five-minute rated door. To
simplify the network, all nonessential
nodes and lings are eliminated.

In these examples, each link capacity is kept constant over all time periods. If data become available reflecting conditions of egress becoming more
difficicult with time as the generation of smoke creates a visibility problem,
even before the onset of untenable conditions, then the link capacity may be
decreased for later time periods. This represents just one of the modifications that can easily be incorporated into the model whenever new relevant
data become available.

Once this dynamic network is constructed, it is solved as a maximal flow
problem with the flow rate limit specified by the link capacities. By the
Maximum Flow-Minimal Cut Theorem, the value of the maximal flow will be at
most the sum of the $a_i$'s which is the total number of occupants on the floor.
If the maximal flow equals the sum of the $a_i$'s, the floor may be considered
fire safe and all the occupants can be evacuated within the time allowed.

If the results of the dynamic flow network indicate that the egress of the
occupants is not assured, a further look can be taken at the probability of
occurrence for such a happening. Then the decision must be taken as to whether
the risk is deemed acceptable or whether an improvement in the fire protection,
such as replacing the five minute unrated corridor doors by twenty minute fire
rated doors, should be undertaken. In this way, the model facilitates an evaluation and comparison of design changes. Though simplistic, the numerical
examples on the four room module illustrated how the method works, and gave a
good indication of the effectiveness of different fire protection measures.

ACKNOWLEDGEMENT
The authors are greatly indebted to Professors E. Scott, R. Barlow, R. Sawyer and J. Marsden. We are also grateful to J. Hagstrom and Prof. Mirchandani for their suggestions. We thank C. Grant for her editorial assistance, and all the other colleagues who contributed to this work. The work was presented in partial fulfillment of requirements for the Ph.D. degree at UCB, and it was supported by the Nuclear Regulatory Commission through the Sandia National Laboratory in Albuquerque, NM, at the Lawrence Berkeley Laboratory, and the National Bureau of Standards, Center for Fire Research.

REFERENCES
1. Elms, D. G., Buchanan, A. H. and Dusing, J. W.: Fire Technology, 20, 1, 11-19, 1984.
2. Berlin, G. N., Dutt, A. and Gupta, S. M.: Modeling Emergency Evacuation from Group Homes, a paper presented at the Annual Conference on Fire Research at the National Bureau of Standards, Oct. 1980.
3. Ling, W. C. T. and Williamson, R. B.: "Using Fire Tests for Quantitative Risk Analysis", in ASTM Special Publication STP 762 ed.G.T. Castino and T.Z. Harmathy, American Society for Testing and Materials, Philadelphia, 1982.
4. Ling, W. C. T., and Williamson, R. B.: "Modeling of Fire Spread Through Probabilistic Networks". To be published in Fire Safety Journal, 9, #3.
5. Fung, F. C. W.: "Evaluation of a Pressurized Stairwell Smoke Control System for a 12 Story Apartment Building," National Bureau of Standards, NBSIR 73-277, Washington, June 1973.
6. Fung, F. C. W., and Zile, R. H.: "Evaluation of Smoke Proof Stair Towers and Smoke Detector Performance," National Bureau of Standards, NBSIR 75-701, Washington, September 1975.
7. Tamura, G. T. and McGuire, J. H.: "The Pressurized Building Method of Controlling Smoke in High Rise Buildings," National Research Council of Canada, NRCC 13365, Ottawa, September 1973.
8. Mirchandani, P. B.: Computers and Operations Research, 3, 3447-355, 1976.
9. Cooper, L: "Estimating Safe Available Egress Time from Fires," National Bureau of Standards, NBSIR 80-2172, Washington, February 1981.
10.Predtechenskii, V. M., and Milinskii,A. I.: Planning for Foot Traffic Flow in Buildings, Tr. from the Russian, Pub. for NBS and NSF by Amerind Publishing Co., New Delhi, 1978.
11.Kendik, E.:Fire Safety Journal, 5, 3&4, 223-232, 1983.
12.Pauls, J. L.: "Movement of People in Building Evacuations," in Human Response to Tall Buildings, 34, Dowden, Hutchinson & Ross, Community Development Series, Stransburg, PA, 1977.
13.Chalmet, L. G., Francis, R. L., and Saunders, P. B.:"Network Models for Building Evacuation," Management Science, 28, 1, 1982.
14.Ford, L. R. Jr., and Fulkerson, D. R.: Flows in Networks, Princeton University Press, Princeton, 1962.
15.Jarvis, J. J., and Ratliff, H. D.:"Some Equivalent Objectives for Dynamic Network Flow Problems, Management Science, 28, 1, 1982.

NOMENCLATURE

| | |
|---|---|
| $\bigstar_C$ | detection state in the corridor |
| $\blacksquare_C$ | untenable state in the corridor |
| $(Pr, tr)$ | smoke travels the link in tr minutes with probability Pr |
| $t_f$ | time to flashover in a compartment |
| $t_b$ | fire endurance of door or wall |
| $P_p$ | probability that some burning contents/lining materials are plastic |
| $t_d$ | fire endurance of door |
| $t_{fo}$ | time to flashover in the room of origin |
| $Rm_i$ | the state when the occupants in Room i have detected the fire |
| $a_i$ | number of occupants in Room i at the start of the fire |
| $f_i$ | link capacity associated with movement links |

# Reliability and Maintainability for Fire Protection Systems

HOWARD D. BOYD
Fire Protection, USA

CHARLES A. LOCURTO
Safety, R & M, and Quality Engineering, USA

ABSTRACT

Reliability and Maintainability (R&M) issues are discussed which pertain to Fire Protection (FP) systems. Although R&M technology has been developed primarily for military and electronic systems, the philosophies and methodologies are applicable to the field of Fire Protection. To illustrate this, the reliability of a representative system for a high rise building is analyzed. Chance of failure and MTBF (Mean Time Between Failure) indices are calculated for a sub-set of equipment which delivers electrical power to a pump automatically. The system power depends upon redundant Diesel Generators (DG's) which back-up the electric Utility Line. Actual failure data is derived for each equipment. The results raise concern over the appreciable chance of failure of such systems. The case where only one back-up DG is used (typically) represents a condition which is worse. Several Maintainability concepts are presented. These include specification design requirements, fault detection, isolation, built-in test, automatic test equipment, and schemes utilizing Computer Management Systems for periodic exercise and monitoring of FP equipment. Overviews of several adaptable contemporary R&M programs are provided. These include the treatment of purely mechanical equipment. The authors highly recommend incorporation of modern R&M technology into the Science of Fire Safety. This should reduce the present lag in R&M applications for FP systems.

INTRODUCTION

There is a significant lag in the application of R&M technology to the field of Fire Protection compared to fields such as military weapon systems. The original impetus for development of R&M programs was spurred by Congress in the 1940's to improve the faltering availability of American weapon systems during, and after World War II. R&M programs developed in the last 40 years deal primarily with military systems and their associated electrical and electronic equipment. However, the philosophies and methodologies developed in these programs have been applied to sundry other fields of endeavor. Amongst these, to name but a few, are the nuclear power industry, commercial aircraft, automobiles, and household equipment. We take the view, as others have done, that these R&M methodologies are also applicable to FP systems. There is a paucity of data on FP peculiar hardware compared to highly developed failure rate data and specifications for electronic components. Nevertheless, in our judgment there is sufficient data and information to support the full application of modern R&M methodologies to the field of FP and its associated systems.

To illustrate this, we have chosen to discuss R&M issues which pertain to the FP system of a high rise building. It is beyond the scope of this paper to present a full discussion of the many R&M methods and procedures which can apply. There is not sufficient room allocated here to analyze an existing high rise building. However, our main purpose is to discuss enough R&M issues to enhance the use of this technology in FP systems. Accordingly we have limited the scope of this paper to a discussion of some of the cogent R&M issues. For ease of presentation we have employed a simplified FP system which is representative and which does not necessarily include all the equipment hardware found in a specific, existing high rise building.

The discussion is divided into three parts. First the Reliability issues are presented; second the Maintainability issues, and the third part presents an overview of R&M program elements which can be adapted to FP systems.

RELIABILITY

Figure 1 relates to a representative or notional FP system for a high rise building. It includes some of the more important equipment existing in real cases. The basic features include a single electric water pump backed up by redundant Diesel Generators, an automatic controller with manual override, transfer switching, redundant standpipes, and other features typically found in a simple system.

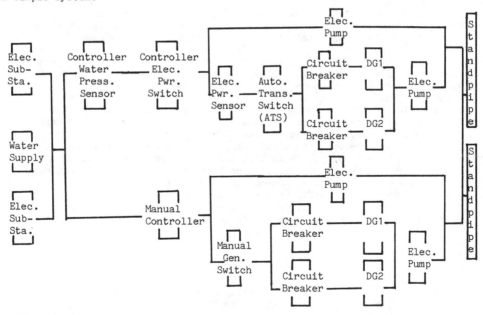

FIGURE 1. National Fire Protection System for a High Rise Building.

It should be noted that Figure 1 is a reliability block diagram. Accordingly, it shows which equipment <u>must</u> operate (i.e., in series), and those equipments which may work as either one or the other, to perform the required function (i.e., in parallel). This approach helps to visualize the concepts of single failure points, and the idea of redundancy. A single failure point exists in any mode of operation wherein a failure of a specific unit of hardware

964

will completely fail the total protection system. The following summaries present several reliability issues pertaining to the system represented in Figure 1.

- In the fully automatic mode of operation, before there is time for the arrival of the Fire Department with auxiliary water (pumped in through siamese or "Y" connections) there are four single failure points. One is the water supply, and others are the controller water pressure sensor, the controller electrical power switch and the electric pump. These must operate for the system to be successful.

- There are three single failure points in the manual override mode of operation. These are the water supply, the manual controller switch and the pump.

- It is obvious that single failure points are the weak links. One has to assess the risk of allowing them to dominate the reliability of the system. The recommended practice is to design them out of the system by using redundant, or stand-by techniques.

- This system includes a redundant standpipe architecture. It is desirable because if one fails the other may work. However, reliance on redundancy must always be wary of equipment susceptability to failure modes whose physics of failure are such that the redundancy does not really exist. In other words, both units may fail because of equal susceptability to aging, cracks (particularly in cast iron), freeze-ups and rust. Experience tells us, for example, that one of the overall weakest reliability characteristics of the system in Figure 1 could be the pipe water distribution system, in spite of redundancy.

- One of the difficulties associated with analyses of Fire Protection Systems is the lack of historical failure data on hardware equipments which comprise the system. However, some progress can be achieved. To illustrate this, let us consider a limited portion of the hardware elements in Figure 1. We limit our investigation to a portion which supplies electric power to the pump automatically. Also to make it easier we will modify the system arrangement and use an elementary fault tree type of illustration of the logic involved. Accordingly this subsystem fails to supply electrical power, as shown in Figure 2.

- The sources of the failure data for each equipment in Figure 2 are as follows:

  DIESEL GENERATOR: Fails while running. The failure rate, $\lambda = 6 \times 10^{-4}$/HR. (Reference 1, P. 1211). The $\lambda$ is for emergency standby DG's. This data was gathered by IEEE team visits to nine operating nuclear power plants. The failure rate data are also substantiated by the report Diesel Generator Reliability at Nuclear Power Plants, Data and Preliminary Analysis. Electrical Power Research Institute, EPRI NP-2433, Interim Report, June 1982.

  CIRCUIT BREAKER: Fails open. $\lambda = 0.14 \times 10^{-6}$/HR. (MTBF = 815 years) (Reference 2). The failure data is obtained from the MIL handbook equipment failure summaries for typical electrical circuit breakers.

  AUTOMATIC TRANSFER SWITCH (ATS): Fails to transfer to Utility line. $\lambda = 0.14 \times 10^{-6}$/HR. (Reference 1, p. 120). The data are for indoor ac circuit breakers for all modes of failure. This is a conservative estimate for ATS switch failure (the data is not readily available).

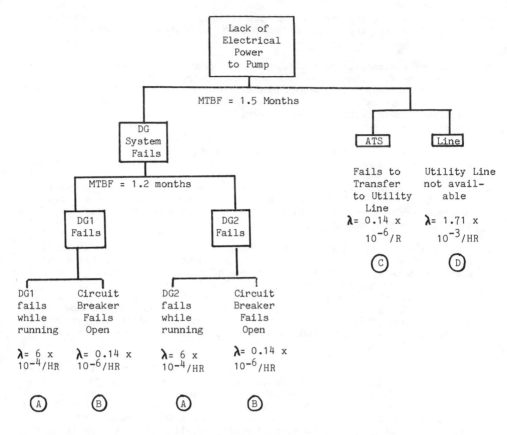

FIGURE 2. Elementary Fault Tree for Power Interruption to Pump.

UTILITY LINE NOT AVAILABLE: Failure is either unacceptable interruption or lack of power when needed. (From reference 3 the MRBF = 200 HRS. Utility interruption = 43.8/year. From reference 4 Utility power interruption = 0.6/month. MRBF = 1200 HRS. Utility interruption = 7.3/year.) After reviewing these data with persons who have some experience with power interruptions we settled on an estimate of 15 Utility interruptions per year, or $\lambda$ = 1.71 x $10^{-3}$/HR.

- From Probability Theory[1] the failure rate of the redundant generator subsystem is:

  $\lambda$DG1 + $\lambda$DG2 = 2 (6 x $10^{-4}$) = 12 x $10^{-4}$/HR. The MRBF = 1.2 months.

- The reliability ($R_s$) and chance of failure of the system in Figure 2, in a one-month period, are determined as follows:
  From Probability Theory:
  $$R_s = R_A^2 R_B^2 + R_C R_D - R_A^2 R_B^2 R_C R_D \qquad (1)$$

---

[1]See also MIL-STD-756B Reliability Modeling and Prediction.

by substitution:

$$R_S = (.64920)^2(.99993)^2 + (.9993)(.29194) - (.42146)(.99979)^3(.29194)$$

$$R_S = 0.59032 = 59\%$$

The chance of failure in one month = 41%

- The Mean Time Between Failure for the above is obtained by integration of the time (t) dependent equation 1, determined as follows:

$$MTBF = \int_0^\infty Rs(t)\,dt \tag{2}$$

By integration:

$$MTBF = \frac{1}{2\lambda_A + 2\lambda_B} + \frac{1}{\lambda_C + \lambda_D} - \frac{1}{2\lambda_A + 2\lambda_B + \lambda_C + \lambda_D} \tag{3}$$

By substitution:

$$MTBF = \frac{1}{.0012} + \frac{1}{.00171} + \frac{1}{.00291}$$

MTBF:   1074.5 HRS.

MTBF:   1074.5/720 = 1.5 months

- The analysis outlined above is not intended to be exhaustive. It is intended to illustrate the value of utilizing this type of methodology. Caution is advised when applying these limited results because a number of enhancing and detracting factors should also be considered. Some of these are as follows:

  - only random, independent failures are considered.
  - failure rates are assumed to be constant; i.e. no aging of equipment was considered.
  - uncertainty of derived failure rates exists.
  - major components have not been analyzed in detail.
  - probable faults in the electrical substations should also be considered.
  - Many Fire Codes in the United States require emergency power for Life Safety-related subsystems such as smoke control and sprinkler systems. Unfortunately in many existing buildings these systems are rife with single failure points resulting from using the same wiring throughout the system. The analysis should be extended to include Life Safety subsystems.

- The foregoing numerical results are not fully complete because of the assumptions and limitations outlined above. Nonetheless, the results of this simple analysis cannot be dismissed from the truth. Therefore, the managers of real fire protection systems which depend on electrical back-up generators, especially those cases which depend on a single back-up generator, would be well advised to review the reliability of their FP system. The chance of failure is appreciable, with a low MTBF.

- There are many other reliability analytical and management procedures which can and should be applied in a typical FP system for a high rise building. However, the foregoing may be enough to illustrate the plausibility of such analyses, and the ability to develop rational issues in spite of the lack of sophisticated data.

## MAINTAINABILITY

Maintainability plays an important role in the long term average failure rate of any real Fire Protection System. Experience has shown that with thorough viligance for maintenance, the long term reliability and the availability of the system is improved. The following presents issues which arise on this subject:

- It has been shown mathematically (reference 5), that the force of mortality (system failure characteristic) is diminished by the practice of ideal maintenance philosophies. This means that a FP system which is poorly maintained and exercised will have a higher long term average failure rate than has the <u>same</u> system when it is maintained with a dedicated program, and exercised periodically.

- Exercises such as those where the Fire Department comes to measure pump output pressure periodically are fundamental to good maintainability.

- Maintainability parameters can be used in design. The probability of detection of a failure, the probability of isolation (failures in one unit of a redundant system, for example, are difficult to isolate) and the probability of restoration (involves difficulty and time required to restore a downed system to full operational status) are each predictable, specifiable, and measurable hardware design parameters.

- Detections of failure are enhanced not only by periodic visual inspections, but also by the efficacy of built-in-test (BIT) and automatic test equipment (ATE). Important detection parameters to be monitored in high rise building applications are water column pressure, electrical power in, pump starting characteristics, and the maintenance of critical system water pressure. Some advanced cases exist which utilize computer systems with programs so arranged that they can automatically exercise every generator and also measure the protection system parameters. Automatic exercise periods are typically one-half hour each week. (Refer, for example, to the Honeywell Computer Management System, Delta Program 1000).

- As always, nature is a jealous mistress. The return on investment in sophisticated BIT and ATE schemes is limited by the effectiveness of the BIT equipment itself. The BIT equipment will require outstanding reliability characteristics which are at least an order of magnitude better than the FP system hardware which it monitors.

- Again, much progress has been made in maintainability design for military weapon system development. This technology can be directly tailored for use in FP fault detection and isolation design. We have covered only a smattering of the available concepts here.

## R & M PROGRAM ELEMENTS

The following presents an overview of several R&M programs and practices which exist in other fields, and which should be adapted to FP systems.

- The program elements included in such highly developed procedures as Relia-bility Program for Systems and Equipment Development and Production MIL-STD-785B, and Maintainability Program Requirements MIL-STD-470 are directly applicable.

- A military publication called: Application of Reliability-Centered Mainten-ance to Naval Aircraft, Weapon Systems and Support Equipment MIL-HDBK-266 (AS) deals with an advanced concept in maintenance. Its detailed procedures assist in determination of the significant items to be maintained, the method of partitioning the system to a workable level, the method for evaluation of failure consequences and methods for how to schedule maintenance tasks. Al-though this is slanted towards Naval equipment the genre of hardware is much the same as it is for commercial Fire Safety systems.

- Typical R&M program plans include techniques which would be useful in re-search, development, and test and evaluation phases of FP system design. Two special techniques will be mentioned here:

  One is a Reliability Allocation, Analysis and Assessment (RAAA) study. This consists of an initial and periodic analytical assessment which com-pares reliability estimates with measurement of critical components as part of the technical evaluation phase of a given program. Included are numerical apportionment of requirements in hardware procurement specifica-tions, assessment of designs by stress analysis, and measurement of the impact of systematic improvements of failure and suspected failure mech-anisms through specific corrective actions. A proposed military standard entitled Procedures for Performing a Reliability Stress Analysis of Mech-anical Equipment is close to publication. Preliminary copies are avail-able.

  Another analytical technique is known as a Maintainability Allocation, Analysis and Assessment (MAAA) study. This consists of an initial and periodic assessment which compares maintainability estimates with measure-ment and achievement, as a part of the program technical evaluation of progress. Included are numerical apportionments of maintainability re-quirements, such as allowable detection and isolation characteristics, assessment of built-in-test effectiveness by progressive measurements, optimization of inspection periods, evaluation of ease of maintenance, and progressive analysis of design parameters.

- Some additional techniques which are available as part of a well-determined R&M Program are listed, briefly, below. The scope of this paper does not allow a discussion of each at this time:

  - Failure Modes and Effects Criticality Analysis
  - Electrical and Mechanical Reliability Stress Analysis
  - Quality Assurance (much work is being done on assurance of quality and reliability. The technical literature is available). This should be particularly important for inclusion in specifications and contract requirements for the acquisition and procurement of critical FP hardware.

- Design Guidelines for Prevention and Control of Avionic Corrosion NAVMAT P 4855-2, June 1983. Department of the Navy. Although this guide addresses the general subject of avionic equipment, it also presents many highly de-veloped concepts which are directly applicable to mechanical as well as electrical equipment. This guide is intended to direct the designer's attention to the use of techniques and methods which will eliminate, prevent and reduce corrosion. It is not the last work on corrosion prevention, but rather a document which can direct the attention of a FP system designer to a complex problem.

● There are many other valuable R&M program elements and military standards and handbooks, which could be adapted to the design and evaluation of FP systems. We could only touch upon a few at this time.


CONCLUSION

The discussion presented in this paper is intended to engender interest in the efficacy of applying existing R&M technology to the design and evaluation of Fire Protection systems. Initial development of military R&M methodologies has become the cornerstone for design of tailored R&M programs for many commercial and military scientific applications. Accordingly, we highly recommend the foregoing concepts, and others available in the R&M technical literature, for full incorporation into the Science of Fire Safety. This will help to reduce the lag in application of this technology to the design and development of Fire Protection systems such as those for high rise buildings.


REFERENCES

1.  IEEE Guide to the Collection and Presentation of Electrical, Electronic, and Sensing Component Reliability Data for Nuclear-Power Generating Stations. IEEE Std. 500-1984.

2.  Reliability Prediction of Electronic Equipment, MIL-HDBK-271D, 15 January 1982, Rome Air Development Center, Griffiss Air Force Base, New York.

3.  Locurto, C. A. and several application engineers (represents an extreme case opinion of Utility Line failure rate). Unpublished paper.

4.  Kesterson, A. and Maker, P. Computer Power-Problems and Solutions. EC&M, December 1982, p. 67.

5.  Pieruschka, E. Principles of Reliability. Prentice-Hall, Inc., 1963.

# Fire Following Earthquake

**CHARLES SCAWTHORN**
Dames & Moore
500 Sansome Street
San Francisco, California 94111, USA

ABSTRACT

Fires following earthquakes have caused the largest single losses due to earthquakes in the United States and Japan. The problem is very seriously regarded in Japan, but not very seriously considered in the US or other earthquake-prone countries. Yet, the potential for future conflagrations following earthquakes is substantial. Earthquakes in the US in 1983 and 1984 have recently highlighted this problem. The scenario for post-earthquake fire must consider structural and non-structural damage, initial and spreading fires, wind, building density, water supply functionality and emergency response. Each of these factors is reviewed in this paper, and an analytical model and preliminary results for San Francisco, California are presented.

INTRODUCTION

Although many aspects of fire and earthquakes have been investigated in recent years, one aspect that has seen very little treatment has been the subject of fire spread in an urban region following an earthquake, herein termed post-earthquake fire. This problem is important in cities in Japan, the US and other countries which have a large building stock composed primarily of wood. This is true of all Japanese cities, and most cities in the US, especially the seismically active West (e.g., Los Angeles, San Francisco). This hazard also exists with regard to industrial installations, such as oil refineries, large factories, chemical plants and installations dealing with hazardous materials. The problem is an especially complex and challenging one, since an earthquake has the potential for initiating a chain of events involving damage to structures and to lifelines such as water supply, gas, electricity, transportation and communications systems, that can turn a moderately damaging seismic event into a conflagration of disastrous proportions.

In Japan, the problem is very seriously regarded, which is appropriate given the holocaust that occurred in Tokyo on Sept. 1, 1923, as well as Japan's history of conflagrations in general. Indeed, in Japan earthquakes are feared equally for post-earthquake fire, on the one hand, and for shaking, tsunami, liquefaction and landslide damage on the other.
However, in the US and most other earthquake-prone countries, the problem is largely ignored, which is strange since in the US as well as in

Japan the single most damaging earthquake of the twentieth century has actually been a post-earthquake fire. This refers, of course, to San Francisco 1906 and Tokyo 1923. In 1906, 80% of the damage in San Francisco was due to fire, amounting to a burnt area of 12.2 sq. km., 28,000 buildings in all. At today's prices, this would be about $3 billion. Little appreciated today is that fires occurred throughout the heavily shaken portions of California, and that the central business district of the city of Santa Rosa was also destroyed by fire in 1906. Tokyo's conflagration in 1923 was much worse, burning 38.3 sq. km. and reaching firestorm proportions, with a tragic death toll of about 140,000.

POST-EARTHQUAKE FIRE CONSIDERATIONS

The post-earthquake fire problem is complex and involves many diverse elements (Figure 1). It begins with the occurrence of the earthquake, which causes structural and non-structural damage to buildings, lifelines, fire stations, communications networks, etc. Structural damage results in the loss of integrity of many of the fire safeguards we rely upon, such as firewalls/stops/doors, etc., fireproof wall and roof coverings (i.e., loss of stucco, brick, etc., wall coverings), sprinkler systems, fire alarms, etc. Additionally, damage occurs to urban lifelines such as water supply, transportation, communications, etc.

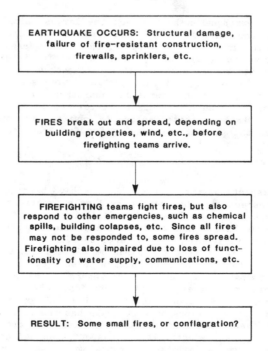

Figure 1. Simplified Scenario for the Post-Earthquake Fire Problem

Beyond physical damage, the shaking causes fires to break out, due to a variety of causes including overturning of open flames (candles, fireplaces, water heaters), electrical short circuits, arcing of power lines, hot equipment coming into contact with debris, friction or sparking of articles during the shaking itself, etc. Many of these fires can quickly and easily be put out by citizens if the citizens are uninjured, aware of the fire, can get to it and have the means with which to fight it. Experience shows that citizens are capable of doing this in a good percentage of the initial fires (Ref. 1). However, because of people not being aware of some of these fires in the initial stage, or not being able to put a fire out, some fraction of initial fires spread to the stage where they require well-equipped firefighting teams. This fire spread is initially within the compartment, then the building. Fire spreading to neighboring buildings may be gradual or rapid, depending on the spacing and exposure between buildings, the materials of the buildings and their contents, the damage to these materials, the wind and the emergency response. Wind is an especially critical factor since inter-building fire spreading velocity increases exponentially with wind speed.

Historically, this experience has been borne out by several earthquakes:

| Earthquake and year | | Number of Initial Fire Outbreaks |
|---|---|---|
| San Francisco | 1906 | 50 |
| Tokyo (Japan) | 1923 | 129 |
| Fukui (Japan) | 1948 | 24 |
| San Fernando | 1971 | 109 |
| Coalinga | 1983 | 19* |

*15 grass fires; 4 buildings

Most recently, the 1983 Coalinga and 1984 Morgan Hill, California earthquakes have pointed out the potential for major post-earthquake fires (Refs. 10, 11). These two events highlighted several lessons, including:

-   Fire departments functioned well but were inundated with numerous demands involving not only fire but structural damage, search and rescue, hazardous material incidents and medical aid.

-   Communications were seen to be extremely vital but highly vulnerable, especially with regard to reporting initial fires to the fire departments.

-   Fire departmental response can be retarded, due to the following factors:

    1.  delays in reporting the fires, due to telephone overload
    2.  delays in proceeding to the fire, due to rubble-blocked roads, landslides, downed wires, traffic jams, etc.
    3.  problems at the scene, including downed wires, collapsed buildings, and especially, lack of water due to damaged mains or, less frequently, insufficient pressure.

Restoration of utilities (gas and electricity) can result in delayed fires, at the time of the service restoration. This can be hours to days after the initial disaster. Restoration needs careful thought as to how and when to reconnect an area. Consideration might be given to not restoring service before individuals are present in every structure (with public officials authorized to enter those structures whose owners are not available in in order to check for fire or gas leaks, and for standby fire units to be in place in the area at the time of utility restoration.

The diversity, significance and complexities of fires following earthquakes are evident. This complexity requires simulation modeling, which has rarely been applied to this problem (Refs. 2-4).

ANALYTICAL MODELING

This section reviews some previous analytical modeling of the author, performed in Japan, and presents some preliminary results of ongoing work.

Scawthorn et al. (Refs. 4,9) presented an analytical methodology for the probabilistic estimation of losses due to fire following earthquake in an urban setting, Figure 2. That work, which was conducted in Japan and utilized the city of Osaka as a case study area, showed that fire following earthquake was related to patterns of overall seismic damage and, under certain circumstances, could be more significant than damage due to general shaking (as indeed it had proven several times historically). Table 1 for example indicates total expected fire and shaking losses for the city of Osaka, Japan under two hypothetical but realistic earthquake scenarios. These losses are for direct structural material losses only, not including any human casualty equivalents or indirect social losses.

TABLE 1:  Comparison of Post-earthquake Fire Spreading and Other Losses for Osaka, Japan* (after Ref. 4)

| Damage Agent (low-rise bldgs. only) | Losses (US $ millions) | | |
|---|---|---|---|
| | Annual | M 6.5 @ 40 km. | M 8.0 @ 160 km. |
| Fire | 45 | 188 | 13 |
| Shaking | 80 | 158 | 17 |
| Liquefaction | 20 | 3.4 | 0 |

*(Osaka has a population of 2.8 million people, and an area of 208 sq. km.)

More recently, the author has extended this methodology and is presently employing it to examine the effects of a major earthquake on the city of San Francisco (population 700,000, area 125 sq. km.). Figure 3 shows San Francisco and the distribution of fire stations (symbols) and engine companies (numbers), while Figure 4 shows seismic intensity in the Modified Mercalli Intensity scale (MMI) for San Francisco, given a magnitude 8.3 earthquake on the San Andreas fault similar to the 1906 event. Seismic intensity is determined according to the interaction of intensity of ground shaking (decreasing with distance from an earthquake fault) and local site geologic effects, which may increase or decrease seismic intensity. Note that the city generally sustains about MMI VII+ (significant structural damage) although MMI IX (heavy damage, some collapse) is exceeded, especially in the eastern portion of the city where old marshy areas will experience ground failure.

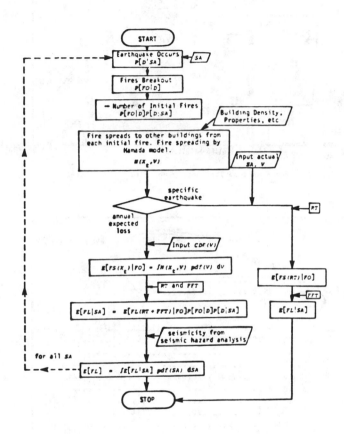

Figure 2   Schematic Diagram of Analytical Methodology (Reference 9)

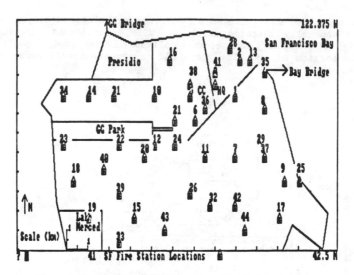

Figure 3   San Francisco Fire Department Fire Station Locations

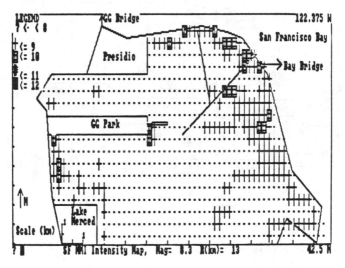

Figure 4   MMI Intensities for San Francisco for a M8.3 on the San Anreas Fault

Initial fire outbreaks are next determined, as a function of seismic intensity, building structural characteristics and occupancy. Initial fire outbreaks are due to a variety of sources, including gas pipe breakage (e.g., overturning water heaters), open flames (e.g., candles), electrical

976

malfunctions (e.g., damaged wires, malfunctioning appliances), reactions of spilled chemicals, etc. We consider fire outbreaks which are not extinguished by local citizens. That is, we only consider fires which grow to considerable size, and require trained firefighting personnel and equipment. While deterministic analyses for initial fire outbreak rate are conceptually possible, present data are insufficient. Instead, based on past US and Japanese earthquake experience, regressions have been performed to determine initial fire outbreak as a function of seismic intensity and occupancy. Employing these correlations together with building and seismic intensity distributions, we see in Figure 5 that about 27 fires will occur. This initiation of fire occurrence is based on a random Poisson process, and is not unique as to specific fire location or total number. Numerous trials have shown that the total number of fires is usually in the range of 25-35, with similar patterns of occurrence.

Each of these fires requires response by at least one of San Francisco's 41 fire engines. There are insufficient engines to respond to each fire under standard procedures and, if we assume that at least one engine responds to each fire, we can estimate engine arrival time at each fire under post-earthquake conditions, Table 2. We see that several of the fires can be suppressed but that many of the fires are multi-alarm in nature, requiring several or more engines and other resources. Alternatively, if normal multi-engine response procedures are employed, then some fires cannot be responded to. Note that ordinary mutual-aid assumptions will be inappropriate in this situation, since neighboring jurisdictions will have their own problems. This example is for a 10 mph wind from the west. Under such a scenario we can see that one or more large spreading fires are likely. These results are preliminary, and studies are being performed to simulate subsequent fire department response, fire development, and water supply damage.

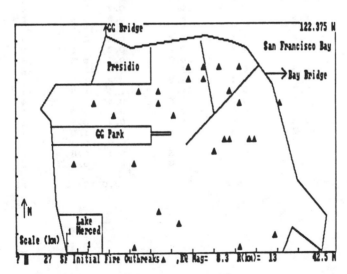

Figure 5   Initial Outbreak of Fires for MMI Distribution in Figure 4

977

Table 2  Magnitude 8.3 San Andreas Event Scenario,
San Francisco:  Initial Fire Department Response and Demand

Wind velocity  10  mph from azimuth 270     INITIAL RESPONSE

| Fire | T(arrival) (mins) | Fire Vel. (ft/min) | A(Burnt) (acres) | No. Burnt Bldgs | Loss ($mil) | FFlow (Mgpm) | No.Engs. Reqd. | No.Pers. Reqd. |
|------|------|------|------|------|------|------|------|------|
| 1  | 10.5 | 9.2  | 0.2 | 2.2  | 0.2 | 1.3  | 2  | 11  |
| 2  | 6.2  | 7.6  | 0.1 | 1.6  | 0.1 | 1.0  | 1  | 8   |
| 3  | 8.1  | 8.4  | 0.1 | 2.8  | 0.2 | 1.7  | 2  | 14  |
| 4  | 7.4  | 8.1  | 0.1 | 1.1  | 0.1 | 0.6  | 1  | 5   |
| 5  | 8.5  | 8.5  | 0.1 | 1.3  | 0.1 | 0.8  | 1  | 7   |
| 6  | 19.9 | 11.4 | 0.8 | 18.7 | 1.4 | 11.1 | 15 | 93  |
| 7  | 21.8 | 11.7 | 1.0 | 22.9 | 1.7 | 13.6 | 18 | 113 |
| 8  | 6.2  | 7.6  | 0.1 | 1.6  | 0.1 | 1.0  | 1  | 8   |
| 9  | 15.2 | 10.6 | 0.4 | 3.8  | 0.3 | 2.3  | 3  | 19  |
| 10 | 8.1  | 8.4  | 0.1 | 2.8  | 0.2 | 1.7  | 2  | 14  |
| 11 | 11.6 | 9.6  | 0.3 | 2.4  | 0.2 | 1.4  | 2  | 12  |
| 12 | 9.7  | 9.0  | 0.2 | 10.0 | 0.7 | 5.9  | 8  | 49  |
| 13 | 8.1  | 8.5  | 0.1 | 6.4  | 0.5 | 3.8  | 5  | 32  |
| 14 | 15.6 | 10.4 | 0.6 | 26.0 | 2.0 | 15.5 | 21 | 129 |
| 15 | 7.4  | 8.2  | 0.1 | 5.0  | 0.4 | 2.9  | 4  | 25  |
| 16 | 10.5 | 9.2  | 0.2 | 2.2  | 0.2 | 1.3  | 2  | 11  |
| 17 | 10.5 | 9.2  | 0.2 | 10.0 | 0.7 | 5.9  | 8  | 49  |
| 18 | 15.2 | 10.3 | 0.5 | 22.8 | 1.7 | 13.5 | 18 | 113 |
| 19 | 8.5  | 4.9  | 0.1 | 2.5  | 0.2 | 1.5  | 2  | 12  |
| 20 | 12.4 | 9.7  | 0.3 | 14.4 | 1.1 | 8.6  | 11 | 71  |
| 21 | 5.0  | 6.2  | 0.1 | 5.6  | 0.4 | 3.3  | 4  | 28  |
| 22 | 6.2  | 4.7  | 0.0 | 1.5  | 0.1 | 0.9  | 1  | 8   |
| 23 | 9.3  | 8.8  | 0.1 | 1.3  | 0.1 | 0.8  | 1  | 7   |
| 24 | 14.7 | 10.7 | 0.3 | 1.6  | 0.1 | 1.0  | 1  | 8   |
| 25 | 9.3  | 8.7  | 0.1 | 0.6  | 0.0 | 0.3  | 0  | 3   |
| 26 | 9.3  | 8.8  | 0.1 | 1.3  | 0.1 | 0.8  | 1  | 7   |
| 27 | 11.6 | 9.7  | 0.2 | 1.0  | 0.1 | 0.6  | 1  | 5   |
|    |      |      | 6.6 | 173.5 | 13.0 | 103.04 | 137 | 859 |

## CONCLUDING REMARKS

Large post-earthquake fires are a low probability, high consequence event. The largest fires in the US and Japan in this century have been due to post-earthquake fire. Adequate preparation and response for fire following earthquake is extremely difficult but is aided by analytical modeling as presented herein. This modeling shows that, on average , post earthquake fire spreading is less of a problem than direct structural damage due to shaking, but that there exist situations where this may be reversed.

## ACKNOWLEDGEMENTS

This work was initiated in Japan under a grant from the Japanese Ministry of Education, and is continuing under a grant from the US National Science Foundation, program managers Drs. G. Albright, F. Krimgold and W. Anderson. Numerous individuals and institutions have aided the author, including Drs. M. Kobayashi, I. Oppenheim, P.H. Thomas, Chief E. Condon (SFFD), R. Reitherman, the Danish Fire Prevention Committee and the San Francisco, Oakland, San Jose, Morgan Hill and Coalinga Fire Departments. Many others have been omitted due to lack of space.

## REFERENCES

1.  Kobayashi, M., Urban Post-Earthquake Fires in Japan, 8th World Conference on Earthquake Engineering, San Francisco, 1984.
2.  Mizuno, H., On Outbreak of Fires in Earthquakes, Dissertation, Kyoto Univ., Kyoto, Japan, 1978 (in Japanese).
3.  Kobayashi, M., A Systems Approach to Urban Disaster Planning, Dissertation, Kyoto Univ., Kyoto, Japan, 1979 (in Japanese).
4.  Scawthorn, C., Lifeline Effects on Post-Earthquake Fire Risk, Proc., Lifeline Earthquake Engineering, Am. Soc. Civil Engineers, Oakland, 1981.
5.  Namba, S., Fire Estimation, Urban Disasters and Land Use Planning, Dissertation, Kyoto Univ., Kyoto, Japan, 1983 (in Japanese).
6.  California Div. Mines and Geology, Earthquake Planning Scenario for a Magnitude 8.3 Earthquake on the San Andreas Fault in the San Francisco Bay Area, Special Publication 61, Sacramento, 1982.
7.  Tokyo Metropolitan Government, A Report on Estimated Earthquake Damage in Tokyo's Wards, Tokyo Disaster Prevention Committee, Tokyo, 1978 (in Japanese).
8.  Oppenheim, I.J., Modeling Earthquake-Induced Fire Loss, 8th World Conference on Earthquake Engineering, San Francisco, 1984.
9.  Scawthorn, C., Yamada, Y. and Iemura, H., A Model for Urban Post-Earthquake Fire Hazard, DISASTERS, The International Journal of Disaster Studies and Practice, v. 5, no. 2, pp. 125-132, London, 1981.
10. Scawthorn, C. and Donelan, J., Fire-related Aspects of the Coalinga Earthquake, in Reconnaissance Report, Coalinga Earthquake of May 2, 1983, Earthquake Engineering Research Institute, Berkeley, 1984.
11. Scawthorn, C., Bureau, G., Jessup, C. and Delgado, B., Fire-related Aspects of 14 April 1984 Morgan Hill Earthquake, in The 1984 Morgan Hill, California Earthquake, Special Publication of California Department of Conservation, Division of Mines and Geology, 1984.

# CIB-Concept for Probability Based Structural Fire Safety Design

**L. TWILT and A. VROUWENVELDER**
Institute TNO for Building Materials and Building Structures
Rijswijk, The Netherlands

## 1. INTRODUCTION

In most countries, the verification of the fire protection requirements
gets much more attention than the requirements themselves. It is felt that
this renders to a heavily unbalanced fire safety design. In view of this
discrepancy, it has been decided within the Fire Commission of the Conseil
International du Bâtiment (CIB-W14) to prepare a design concept which
covers both afore mentioned aspects in an integrated way and which may
serve as a framework for national design guides in this field.

Such an improved design concept should be based on clearly specified fire
safety objectives and should recognize the contribution of structural
design provisions to these objectives. Therefore structural requirements
should be functional – i.e. refer to the expected performance of the
structure and its members in an actual fire – so that verification of
compliance can be done by an engineering design, comparable to the design
for non-accidental situations. This calls for the use of analytical models
for verification, as an alternative to experimental verification.
Moreover, structural requirements should allow for a certain equivalency
of different design solutions comprising structural (passive) and operational (active) fire protection measures. This involves an assessment of
the inherent degree of reliability, calling for a probabilistic approach.

Since, in 1980, this work was started by a CIB-W14 Workshop, based on
earlier work, e.g. [1], [2], the following documents have been prepared:

- A Conceptional Approach Towards a Probability Based Design Guide on
  Structural Fire Safety [3]
- Design Guide on Structural Fire Safety [4].

In [3] the framework for a probabilistic design method on structural fire
safety is outlined and a state of the art review of calculation models for
verification is given. On basis of this report, in [4] operational rules
for the assessment of buildings with respect to structural fire safety are
presented. An important feature of this design procedure is, that not only
one but a variety of assessment methods, with different levels of refinement is offered, allowing for an optimal balance between the verification
method on the one hand and the accuracy of available input data and the
relevance of structural performance on the other hand.

This paper decribes some of the principles of the CIB-design concept as well as its main limitations. A comparison will be made with a more traditional design concept for structural fire safety. To illustrate the benefit of the CIB design concept, a practical situation will be evaluated.

2. PRINCIPLES AND LIMITATIONS OF THE CIB DESIGN CONCEPT

Objective of fire protection is to limit:

- individual life risk and societal risk
- neighbouring property risk
- directly exposed property risk

to a level which is acceptable by society. Evaluation of directly exposed property risk is based on economic considerations only and should thus be the client's decision. More specificly, the aim of the design procedure is to confine fully developed fires within a compartment and to prevent local failure, leading to failure of the whole structural system (progressive collapse).

In order to achieve these aims, the design concept provides functional requirements for an adequate load bearing capacity of the structure and an adequate separating function of the structural components in case of a fire, severe enough to cause structural damage. The related limit states are:

(1) load bearing capacity
(2) thermal insulation
(3) integrity.

For verification, heat exposure models (H) as well as structural response models (S) are necessary. With respect to their level of refinement, different models of each type are presented:

Heat Exposure Models

(H1) a rise of temperature versus time according to ISO 834, the duration of which is equal to the "required time of fire duration" expressed in building regulations and codes for the particular use of the building or fire compartment;

(H2) a rise of temperature versus time according to ISO 834, the duration of which is approximated on the basis of the combustion and thermal conditions expected to prevail in the particular fire compartment;

(H3) a rise of temperature versus time expected in a compartment fire, directly related to the combustion and thermal conditions expected to prevail in the particular fire compartment.

Structural Response Models

(S1) the structure is considered as a number of individual structural members with simplified support and restraint conditions - the model can either be experimental or analytical;

(S2) the structure is considered as a number of sub-assemblies - analytical models prevail;

(S3) the structure is analyzed as a whole, assuming fire exposure through-out the structure or only within an individual compartment.

Each combination of heat exposure model and structural response model, as an element of the matrix in Fig. 1, represents a particular design proce-dure. It is evident that not all models can be used in all possible combi-nations. The rule should be to provide a sensible relation in the levels of advancement of both models. In the text in Fig. 1, reference is made to this aspect [5].

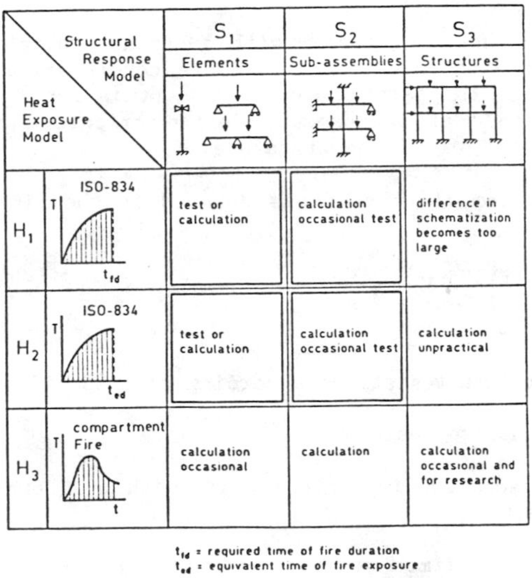

| Structural Response Model — Heat Exposure Model | $S_1$ Elements | $S_2$ Sub-assemblies | $S_3$ Structures |
|---|---|---|---|
| $H_1$ ISO-834 | test or calculation | calculation occasional test | difference in schematization becomes too large |
| $H_2$ ISO-834 | test or calculation | calculation occasional test | calculation unpractical |
| $H_3$ compartment Fire | calculation occasional | calculation | calculation occasional and for research |

$t_{fd}$ = required time of fire duration
$t_{ed}$ = equivalent time of fire exposure

FIGURE 1: Matrix of heat exposure and structural response models in sequence of improved schematization.

Taking the three above mentioned heat exposure models as a starting point, the CIB design concept distinguishes between the following assessment methods:

- Assessment Method 1: Method on the basis of ISO standard fire exposure. The design criterion is that the fire resistance, determined either by experiments or analytically, is equal to or exceeds the time of fire duration required by building regulations or codes. Reference for application: Model combinations H1-S1 or H1-S2.
- Assessment Method 2: Method on the basis of a standard fire exposure. The design criterion is that the fire resistance, determined either by experiments or analytically, is equal to or exceeds the equivalent time of fire exposure, a quantity which relates compartment (non-standard) fire exposure to the ISO standard fire. Reference for application: Model combination H2-S1 or H2-S2.
- Assessment Method 3: Method characterized by a direct analytical design on the basis of compartment (non-standard) fire exposure. Reference for application: Model combination H3-S1, H3-S2 or H3-S3.

In addition to the heat exposure models and the structural response models, also reliability models are to be defined, comprising - in prin-ciple - aspects such as:

983

- Intrinsic randomness of design parameters and properties.
- Model uncertainties of the models for heat exposure and structural response.
- Assessment of frequency, such as the probability of occurrence of a large fire, the effect of fire brigade actions, the reliability of sprinklers.
- Safety considerations from both human and economic point of view such as height, volume and occupancy of the building, availability of escape routes and rescue facilities as well as consequences of violating a limit state.

Verification requires the proof that - with a certain reliability and for a certain application - no relevant limit state conditions are failed. As a general design format the partial safety concept is used. For the three above introduced methods of assessment, design criteria are formulated dependent on their level of refinement/nature.

Assessment Method 1: The limit state is defined in the time domain. The design criterion reads:

$$[\text{req. } t_f] - [\text{eval. } t_f] \leqslant 0 \tag{1}$$

wherein:

$[\text{req. } t_f]$ = required fire resistance according to a specified fire safety class;
$[\text{eval. } t_f]$ = evaluated fire resistance.

The probabilistic aspects are implicitly dealt with by a proper choice of the fire safety class.

Assessment Method 2: The limit state is defined in the time domain. The design criterion reads:

$$t_f/\gamma_f - \gamma_e \, t_e \, \gamma_n \leqslant 0 \tag{2$^a$}$$

wherein:

$t_f$ = evaluated (characteristic) value of the fire resistance;
$t_e$ = (characteristic) value of the equivalent time of fire exposure calculated on basis of the fire load density (= $q_f$) the ventilation conditions (= w) and the thermal properties of the surrounding thermal properties (= c); $t_e = c \cdot w \cdot q_f$;
$\gamma_f$ = partial safety factor related to the fire resistance and for average reliability requirements;
$\gamma_e$ = partial safety factor related to the equivalent time of fire exposure and for average reliability requirements;
$\gamma_n$ = differentiation factor accounting for safety differentiation ($\gamma_{n1}$; e.g. height of building, number of persons involved) and frequency differentiation ($\gamma_{n2}$; e.g. envisaged alarm and sprinkler systems, force of fire brigade); $\gamma_n = \gamma_{n1} \cdot \gamma_{n2}$.

with $\bar{\gamma} = \gamma_e \cdot \gamma_f$ equ. (2$^a$) can be rewritten as:

$$t_f > \bar{\gamma} \cdot t_e \cdot \gamma_n \tag{2$^b$}$$

wherein:

984

$\bar{\gamma}$ = global safety factor for average reliability requirements.

It is seen that in Assessment Method 2, the probabilistic aspects are accounted for explicitly.

Assessment method 3: The limit state is defined in either the mechanical strength domain (load bearing structures) or the temperature domain (separating structures). The design criteria are formulated in a way, analogue to the one specified for Assessment Method 2, on the understanding that $t_f/\gamma_f$ is replaced by the minimum design value of the ultimate load bearing capacity (resp.: maximum temperature at the unexposed side of the structure, acceptable with respect to insulation) and $t_e \cdot \gamma_e$ is replaced by the design load (resp.: the highest design temperature at the unexposed side of the structure). Design values, again, may be calculated using the partial safety factor concept.

A most interesting feature of the CIB-approach is, that the partial safety factors and the differentiation factors, introduced in Assessment Method 2 and 3 are derived from target reliabilities and/or occurrence rates of fires over a probabilistic analysis, thus allowing for a differentiated and functional design. Therefore, in this paper, Assessment Methods 2 and 3 will be denoted as "advanced".

With regard to practical application, the CIB-concept for structural fire safety design is subject to some limitations:

- All heat exposure models refer to a situation of post-flashover fires and a more or less uniform temperature distribution within the fire compartment. For very large fire compartments this assumption may be questionable.

- The verification of the limit state of integrity requires an experimental analysis. Especially in the case of Assessment Method 3, which is, due to its very nature focussed on analytical models, this may lead to some inconsistency.

- Regarding the probabilistic models, there is a lack of reliable data. This especially holds for the relations which describe the occurrence rates of fires. Also the target failure probabilities are only determined in a global way, pending the availability of more extensive data and analysis of risk perception.

Despite these reservations, it is felt that the design guide should be used for practical applications, preferably together with more conventional design concepts. Thus the benefits and possibly the shortcomings of the new design concept will become more explicit.

3. COMPARISON WITH A TRADITIONAL DESIGN CONCEPT

As an example, the directives for structural fire safety design which are currently used in The Netherlands [6] will be reviewed and be compared with the CIB-concept.

In The Netherlands the fire safety requirements are primarily directed towards the limitation of individual life risk, societal risk and neighbouring property risk. To limit directly exposed property risk is principally outside the scope of the directives; consequently no explicit guidance in this respect is provided. The CIB-concept does give such guidance, by differentiating with respect to target reliabilities.

Effectively, the Dutch concept for structural fire safety is to be considered as a rating system for building components. There are, practically spoken, four classes, defined by a minimum required fire resistance according to ISO-834: 30,60, 90 and 120 minutes. Classification is as follows: To start with, the effective time of fire exposure according to the ISO-standard fire, $t_{eff}$, is determined by:

$$t_{eff} = q_t \qquad\qquad [min] \qquad\qquad\qquad (3)$$

wherein:

$q_t$ = the total fire load density due to building components and contents in the fire compartment [kg wood/$m^2$ floor area].

The fire load density is determined as a representative value for the building type and occupancy under consideration. An individual assessment of the fire load is - as opposed to the suggestion in the CIB-concept - normally not accepted. The effective fire duration (= $t_{eff}$) is transformed to the "required fire safety class" according to:

$0 < t_{eff} \leqslant 30$ : min. fire resistance : 30 min.
$30 < t_{eff} \leqslant 60$ :   "     "      "    : 60 min.

etc.

The CIB-approach is more flexible since - for Assessment Method 2 and 3 - in principle continuous limit state functions are specified.

In the Dutch design concept as well as in the CIB approach, safety and frequency differentitation is dealt with separately.
As far as frequency differentiation is concerned, the Dutch method only gives explicit guidance for the risk reducing effect of the public fire brigade. It is assumed that the public fire brigade all over the country meets the same standard and that fires will be under control within 60 minutes after flash-over. This means that the effective fire duration according to equ. (3) will be limited to 60 minutes, irrespective of the actual fire load. The CIB-method differentiates with respect to public and residential fire brigade and/or sprinkler installation, on basis of the anticipated reduction of fire occurrence.
Regarding safety differentiation, the building height and the function of the structural components under consideration are main parameters in the Dutch design method. Specific rules are only given for a limited number of building categories, such as appartment buildings and hotels. In these cases, the required fire resistance following from the relevant value of the effective fire duration is increased by 30 minutes (= one safety class); if the height of the upper floor is more than a certain value (= 13 m) above street level, the increase amounts 60 minutes (= two safety classes). The additional fire resistance is only required for the limit state of load bearing capacity and for main structural elements, i.e. those elements the failure of which may lead to progressive collapse. These aspects are also recognized in the CIB-concept, but are here - more consistently - expressed in terms of target failure probabilities.

Verification in the Dutch directives is on the level of Assessment Method 1 according to the CIB design concept (cf. equ. 1).

It follows from the above comparison that the traditional (Dutch) design concept has some important features in common with (advanced) CIB-Assessment Method 2, e.g. limit states, idea of equivalent time of fire duration, idea of safety and frequency differentiation.

However, in the CIB-concept these aspects are evaluated in a consequent and deducible manner, whereas the traditional method gives global rules, mainly based on practical experience. Moreover, in the traditional method, a proper probabilistic basis is missing, reason why this method is essentially to be identified as a CIB-Assessment Method 1 approach.

In the next chapter some practical implementations of the use of either the traditional (Dutch) or the advanced CIB-methods will be reviewed.

4. IMPLEMENTATIONS FOR PRACTICAL DESIGN - A CASE STUDY

In Rotterdam, The Netherlands, a series of 4 similar police offices is under design. One of the options is a cube shape concept with dimensions of $12 \times 12 \times 12$ m$^3$. In this "cube", three floor levels are planned. Since there are no compartmentation walls inside the building, each floor is to be considered as one fire compartment with a floor area of $12 \times 12 = 144$ m$^2$. Per storey, in each of the four outer walls, three windows with dimensions $1.8 \times 1.8$ m$^2$ are situated as is shown in Fig. 2.

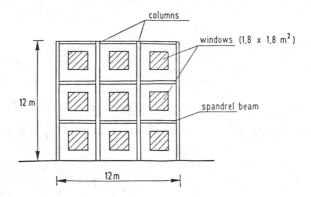

FIG. 2: Front view of the facade

For the main load bearing structure a three dimensional steel frame is planned, consisting of columns, spandrel beams and floor beams. All columns are outside the facade and between the windows. The same holds for the spandrel beams. Thus, the facade acts as a heat shield to protect the external steel work from a fire inside the building. The floors consist of reinforced concrete slabs, cased in situ and carried by the floor and spandrel beams. Apart from the window openings, the facades are of masonry.

A key element in the architectural concept is the application of bare external structural steel. Protection of the internal steelwork - if necessary - is acceptable. For this reason and because the concrete floors and masonry facades are not critical, the following discussion will be focussed on the fire safety of the steel structure.

In the traditional (Dutch) design concept, the actual fire conditions are ignored, and requirements – if any – are expressed solely in terms of fire resistance according to ISO 834 (standard fire conditions). The level of required fire resistance depends with equ. (3) on the total fire load density $q_t$. According to the Dutch regulations, the representative fire load density for offices amounts 60 kg/m$^2$ . This value is based on Dutch statistics and corresponds roughly to the 80% fractile. Taking into account the additional 30 minutes for safety differentiation, it follows for the required fire resistance of the steel frame: 60 + 30 = 90 minutes. The fire resistance of unprotected steel elements under common design loads is in the order of magnitude of 10 to 30 minutes. So, without detailed analysis it is clear that with bare (external) steel, the specified requirement cannot be met. As a direct consequence the whole design concept fails.

In view of this extreme important consequence, it is at hand to re-evaluate the design on basis of one of the advanced and thus more differentiated CIB-methods.

Assessment Method 2 seems to be an appropriate method for evaluating the requirement for the internal steelwork, since the related concept of equivalent fire duration is derived for ventilation controlled compartment fires in not too large compartments.
An evaluation on basis of Assessment Method 2, taking into account the same fire load statistics as in the afore mentioned analysis, renders a required fire resistance of 90 minutes under the following (main) conditions:

- tolerable annual failure probability ($p_f$) = 2 x $10^{-6}$;
- annual probability of initial fires (p): 0.5 . $10^{-6}$ . A, where A is the floor area of the fire compartment;
- reduction of the probability of a severe fire due to standard public fire brigade only ($p_1$): $10^{-1}$;
- safety differentiation factor to allow for the relative importance of the steel frame ($\gamma_{nl}$) : 2.1;
- reduction of the variable part of the fire load (m): 0.4, to allow for the effect of partial protection.

It is noted that the adopted rate of fire occurrence, p, as well as the reduction effect of the fire brigade action, $p_1$, are in line with European data. The applied reduction factor m on the variable fire load is obtained for conditions, which are representative for offices [7]. Calculation of the partial safety factors is based on 80% (20%) fractiles for $t_f$ and $t_e$ while variation coefficients as suggested in [4] are taken into account. The resulting average reliability level may, in view of the choice of $p_f$, be associated with a situation in which structural collapse involves low personal risk and medium economic loss [8]. By introducing the safety differentiation factor $\gamma_{nl}$ = 2.1, the average reliability is increased by roughly one to two orders of magnitude.
Thus – a posteriori – a functional motivation is provided for the 90 minutes requirement, which follows from the traditional (Dutch) design method. It will be clear, that the above results do not proof whether the 90 minutes requirement is "right" or "wrong". They give however a basis for rational discussion. Other options can easily be evaluated. As far as the internal steelwork in the considered design is concerned, such a discussion is not of much help: Also a decrease of the required fire resistance to, say, 60 minutes, would mean that the steel is to be protected.

For the external steelwork, however, the 90 minutes requirement should not be analysed on basis of Assessment Method 2 due to the specific type of fire exposure, which is not accounted for in the concept of equivalent fire duration. In this case an Assessment Method 3 analysis is necessary. Such an analysis is carried out, using the calculation model for fire exposed bare external steelwork, described in [9]. In this model, rules are given to calculate the heat transfer to the steel coming from both the internal fire, radiating through the windows and the emerging flames. As a result, steady state steel temperatures are determined, which may be considered as upper bound values of the steel temperatures which will be attained during a real fire. Geometry of the fire compartment, wind conditions and fire load density are main parameters.

For the limit state expressed in the temperature domain, design verification is for instance accomplished by ensuring that:

$$T_c/\gamma_c - T_s \ (\gamma_s \ \gamma_n \ q_f) \geqslant 0 \tag{4}$$

wherein:

$T_c$ = representative value for the critical steel temperature
$T_s$ = calculated steel temperature, given the fire load and ventilation
$\gamma_c, \gamma_s$ = appropriate partial safety factors
$\gamma_n$ = differentation factor as defined for equ. 2
$q_f$ = representative value for the fire load.

The CIB design guide does not provide fully operational guidance for the evaluation of equ. (4). For practical application the following reasoning will, however, suffice: For a certain situation defined by building geometry, type of wind conditions (viz. through draught or no through draught) etc., the fire load density $q_f$ is the main variable for $T_s$. The relation between $q_f$ and $T_s$ is - for the considered building design - presented in Fig. 3 for both the columns and the spandrel beams.

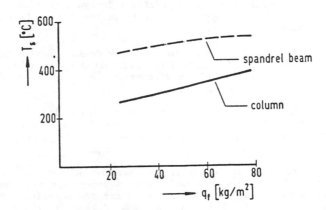

FIG 3: Relation between the fire load density ($q_f$) and the calculated steel temperature ($T_s$) of the columns and the spandrel beams.

989

Note that the influence of $q_f$ on the maximum attained steel temperature is rather insignificant. When $T_s$ is determined over a certain design range of $q_f$, the choice of this design value (= $q_{fd}$) will, therefore, not be very critical. It is suggested to choose, as a rough approximation:
$q_{fd} = q_f \, \gamma_n \, \gamma_s$ with $\gamma_s$, $q_f$, $\gamma_n$ as in the Assessment Method 2 evaluation.

This renders $q_{fd} \sim 80$ kg/m$^2$ (= 1520 MJ/m$^2$) . The corresponding values for the steel temperature can be read from Fig. 3 and amount 400°C and 530°C for the columns and the spandrel beams respectively.
The (characteristic) values of the critical steel temperature $T_c$ can be calculated on basis of e.g. [10]. Due attention must be paid to a proper load combination for accidental design, including partial safety factors and combination coefficients. In the CIB design guide, this is left to the national authorities.Therefore, in the present analysis, current (conservative) Dutch rules are followed. It can be shown then that design values for the critical temperature $T_{cd} = T_c/\gamma_c$ of 500 and 850°C hold for the columns and spandrel beams respectively. Thus the design criterion according to equ. (4) is met. This means that – with appropriate reliability – the steel frame may be assumed to withstand the anticipated fire, so that bare external steelwork is acceptable.


4.  CONCLUSIONS

Structural fire safety requirements may render important consequences, not only in terms of building costs but, occasionally, also for the whole building concept. In traditional methods for structural fire safety design the requirements are conventional and based on global experience, which may lead to unjustified and/or uneconomic decisions.
The presentation of functional requirements which are consistent with the level of the applied verification models, is considered as the main advantage of the advanced CIB-methods for structural fire safety design, viz. CIB-Assessment Methods 2 and 3.

This paper explains some of the principles of these methods and exemplifies their application for a practical design situation, with reference to a more traditional approach. More in particular it follows from the case study that:

-   the advanced CIB-design concepts provide the possibility for a functional analysis of given requirements; already in their present form they may be a useful tool in discussions with building officials on fire safety requirements;

-   verification on basis of Assessment Method 3 is – e.g. regarding the probabilistic aspects – not fully operational; in situations which are more complicated than the reviewed one, this may render practical problems.

More in general, it is emphasized that the performance of systematic case studies of the kind reviewed in this paper, is very useful because it constitutes an important means to explore the benefits and possible shortcomings of the advanced design methods. Others are therefore invited to do similar exercises and to exchange the results.

ACKNOWLEDGEMENT

The CIB-design concept, reviewed in this paper has been developed within a workshop of the Fire Commission of the Conseil International du Bâtiment (CIB W14).

REFERENCES

[1]   Magnusson, S.E. and Pettersson, O.: "Rational design methodology for fire exposed load bearing structures", Fire Safety Journal, 3 (1981)

[2]   DIN 18230: "Baulicher Brandschutz im Industriebau", Vornorm, Berlin, 1982.

[3]   CIB/W14: "A Conceptional Approach Towards a Probability Based Design Guide on Structural Fire Safety", Fire Safety Journal, Vol. 6, no. 1, 1983.

[4]   CIB/W14: "Design Guide for Structural Safety", to be published in Fire Safety Journal, 2nd half 1985.

[5]   Witteveen, J.: "A Systematic Approach Towards Improved Methods of Structural Fire Engineering Design", Proceedings 6th International Fire Protection Seminar organized by VFDB, Karlsruhe, 1982.

[6]   Twilt, L.: "New Approach to Fire Protection in The Netherlands", Fire Safety in Buildings with Steel, 1st International Symposium of the ECCS, The Hague, 1974.

[7]   Bryl, S.: "Fire Load in Office Buildings - Fire Safety in Constructional Steelwork", Part III, ECCS-III-74-2-D.

[8]   CEB/FIP: "Model Code for Concrete Structures", Vol. 1, Joint Committee on Structural Safety, Comité Euro-international du Béton, Bureau de Paris.

[9]   Law, M. and O'Brien, T.: "Fire Safety of Bare External Steel", Constrado, May 1981.

[10]  ECCS-Technical Committee 3: "Fire Safety of Steel Structures", European Convention for Constructional Steelwork, Elsevier Scientific Publishing Company, Amsterdam, 1983.

# Risk Analysis Using the Engineering Method for Building Firesafety

ROBERT W. FITZGERALD
Worcester Polytechnic Institute
Worcester, Massachusetts 01609, USA

ABSTRACT

The analytical techniques incorporated in the engineering method for building firesafety make it possible to evaluate a comparative level of firesafety risk today. At the same time, the structure of the method serves as a bridge between research and practice to facilitate technology translation and communication between those two groups. This paper describes briefly the parts of the engineering method, the quantifications and development activities, and some applications that have been addressed.

## I. INTRODUCTION

In the early 1970's, Harold E. Nelson developed a method for evaluating relative levels of building fire risk. This method was described at that time as a "Goal Oriented Systems Approach to Building Firesafety" (1). The descriptor of risk was a cumulative curve of the probability of success in limiting fire involvement to selected floor areas within a building. The "L-Curve", as this cumulative success curve was called, was obtained by calculation using fault tree techniques.

Since that time, considerable work has been done at Worcester Poly-technic Institute toward extending and adapting Nelson's original concepts into a rigorous, disciplined, and practical method by which consistent results can be achieved in building evaluations. This work has focussed on two major areas. The first is the theoretical foundation. Much effort has been devoted toward identifying the complex interactions of fire and buildings. Major attention (2) has been given to the logic and structure of these interactions by utilizing classical mathematical and systems procedures that have proven to be effective in other engineering fields. The second area of focus has been directed toward practical application of the method. The evolution of the theoretical treatment has always been accompanied by testing it in field applications.

The goal of this work is to develop a method of analysis and design for building firesafety for use by professional engineering practitioners. That procedure, called the "engineering method" in this paper, is envisioned to function in a manner analogous to structural and mechanical engineering methods. Although the building firesafety method has reached a level of maturity where it can be used for engineering applications today, it has evolved only part of the way toward the complete engineering procedure. Four stages are envisioned to complete the method.

The first, and possibly the most difficult stage, is the identification of the major components of the system and the basis for their evaluation. This was done by Nelson (1) in 1972. The second stage is the identification of the anatomy of a system that links the complex interactions in a manner that can be utilized in general engineering practice. Temporary quantification techniques and calculation procedures have been developed to determine measures of performance.

The third stage is the quantification of the components and their linkages by the development of relatively simple equations that enable the behavior or performance of components to be predicted within normal engineering accuracy. The final stage is the establishment of a rational, mathematically based design method suitable for routine engineering practice. Incorporation of recognized factors of safety is an integral part of this work.

At the present time some components are in Stage 3, most are at Stage 2, and a few are still at Stage 1. Nevertheless, the method can be used for a variety of practical applications at its present state of development. More importantly, the tasks needed to reach the goal are recognizable.

## II. PARTS OF THE ENGINEERING METHOD

The complete engineering method involves eight major parts that can be grouped into three categories (3):

A. Performance Identification and Needs
   1. Establish Performance Criteria
B. Building Analyses
   2. Prevent Established Burning Analysis
   3. Flame Movement Analysis
   4. Smoke Movement Analysis
   5. Structural Frame Analysis
C. Engineering Design
   6. People Protection
   7. Property Protection
   8. Continuity of Operations Protection

The first category addresses the single aspect of identifying needs and establishing performance criteria for a building. These firesafety criteria describe the following:

   a. Life safety. Identify the acceptable time interval between when the occupant is alerted to a fire and the time when the critical rooms or segments of a building are expected to be untenable with regard to products of combustion. Anticipated occupant needs, as well as the level of tenability and its measurement scale, are incorporated within this requirement.

   b. Property. Identify an acceptable value or level for property damage in a building fire. Property may include the contents or the building structure itself. Often specific items or rooms are identified as requiring special attention.

c. Continuity of operations. Identify an acceptable period of operational downtime that an owner or an occupant may tolerate after a building fire. Again certain spaces normally are more sensitive to operational needs and require special attention.

Building analyses described as parts 2 - 5 above involve engineering procedures to predict the performance of an existing or a proposed building and its firesafety system. Established Burning is the demarkation between prevention activities and the building's firesafety performance. Established Burning (EB) is the size of fire the engineer identifies as the start of the threat of the building. For most buildings a flame 250 mm high is a practical and logically convenient flame size to define as EB.

Parts 6 - 8 involve engineering solutions for the firesafety problem areas. Here, the engineer compares the results of the technical analyses to the performance criteria and to any special needs the owner, an occupant, an insurance carrier, a lender, or an authority having jurisdiction. When the building does not meet a performance criteria, the engineer identifies alternative solutions, their costs, and their effectiveness. Solutions are tailored to meet specific needs, and may involve one or a combination of operational, economic, or technical features.

## III.    FLAME MOVEMENT ANALYSIS

Flame movement analysis is the central focus of the engineering method because the building response to the products of combustion is time and fire size dependent. From a flame movement perspective, every building can be described as a grouping of spaces and barriers. The probability of success in limiting a fire is evaluated within each space and at each barrier. Large, open buildings, such as warehouses and industrial buildings, can be subdivided with "zero strength barriers." This technique enables large areas to be evaluated in a manner consistent with compartmented buildings.

At the present time, the quantifiable descriptor of building firesafety performance is the L-Curve. The L-Curve describes the cumulative probability of success in limiting flame movement within sequential rooms along a specific path of possible fire propagation. Consequently, the L-Curve can be used to identify a numerical level of risk for all parts of a building, such as within a specified area, in a room, at a barrier, or within n connecting rooms, as well as for the entire building or group of buildings.

Figure 1 shows the format of the L-Curve. Note that the origin is located at the top of the graph. The curved lines describe the probability of limiting the flames to the space. The vertical lines describe the probability that a barrier will prevent extension of flames into the next space. A short vertical line indicates a weak barrier, while a long line describes a strong barrier.

Because each building is unique, the L-Curve can be used to compare different buildings, as shown in Figure 2(a). In addition, different design alternatives for the same building can be compared, as shown in Figure 2(b).

995

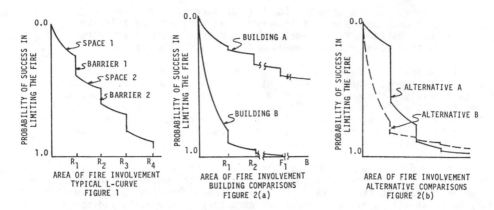

AREA OF FIRE INVOLVEMENT
TYPICAL L-CURVE
FIGURE 1

AREA OF FIRE INVOLVEMENT
BUILDING COMPARISONS
FIGURE 2(a)

AREA OF FIRE INVOLVEMENT
ALTERNATIVE COMPARISONS
FIGURE 2(b)

Building analysis and design for fire is an extremely complex process. Not only are there many human and phenomenological components that influence events and their outcomes, but also their interactions are often conditional. In order to describe the logic of this complex system, an organized structure of networks has been developed to identify the interrelationships. The networks serve a dual role. First, they allow a probabilistic firesafety analysis today for any building regardless of size, construction, use, or occupancy. Second, they serve to guide the future development of a deterministic procedure to analyze the performance of the components.

The flame movement analysis first evaluates a room as a room of origin. Then, it evaluates the performance of the barriers surrounding that room and the succeeding spaces and barriers along any path of fire propagation. Within any space, a fire can be limited by self-termination of the fire itself, by automatic suppression from a sprinkler system, or by manual extinguishment by the fire department. These events are given the symbols I, A, and M respectively. Figure 3 shows the network that is used to evaluate the limit for any area within a room of origin. The evaluation of I, A, and M are described by cumulative probability curves that show the liklihood of success in terminating a fire within a room. For convenience of notation, the conditional terms $A/I$ and $M/I$ are simply called the A-Curve and the M-Curve. The conditionality is understood to exist.

After the limit (L) for the room of origin is determined, the expected barrier performance is assessed. Given full room involvement, $(\bar{L})$, the barriers are evaluated both for the occurrence of small, localized penetration failures $(\bar{T})$ and for the existence of large, massive failures, $(\bar{D})$. The analysis evaluates field performance of the barriers. Consequently, construction features, loading conditions, barrier openings, and automatic barrier protection are incorporated into the evaluation.

When these components are evaluated probabilistically, the evaluation can consider first the probability that full room involvement occurs, represented by $P(\bar{L})$, and then the conditional probability of failure, given full room involvement. This allows the isolation of the components as,

$$P(\bar{D}) = P(\bar{L})P(\bar{D}/\bar{L}) \qquad (1)$$
$$P(\bar{T}) + P(\bar{L})P(\bar{T}/\bar{L}) \qquad (2)$$

The values of $P(\overline{D}/\overline{L})$ and $P(\overline{T}/\overline{L})$ are time dependent. Their values will change for each increment of time beyond full room involvement. Therefore, as the fire continues to burn without extinguishment, the type of barrier deterioration becomes an integral part of the analysis of the next room.

The type of barrier failure influences the success in terminating a fire in the adjacent space. Therefore, the evaluation of the I-, A-, and M- values for the adjacent space is conditional on the type of barrier failure. Figure 4 shows the network for the combined barrier-space module for any segment beyond the room or origin.

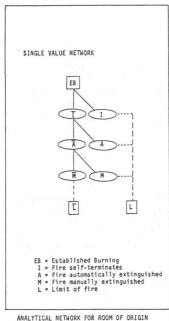

SINGLE VALUE NETWORK

EB = Established Burning
I = Fire self-terminates
A = Fire automatically extinguished
M = Fire manually extinguished
L = Limit of fire

ANALYTICAL NETWORK FOR ROOM OF ORIGIN
FIGURE 3

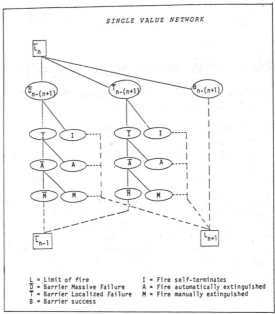

SINGLE VALUE NETWORK

L = Limit of fire
$\overline{D}$ = Barrier Massive Failure
$\overline{T}$ = Barrier Localized Failure
B = Barrier success

I = Fire self-terminates
A = Fire automatically extinguished
M = Fire manually extinguished

ANALYTICAL NETWORK MODULE FOR ROOMS BEYOND THE ROOM OF ORIGIN
FIGURE 4

The L-Curve analysis evaluates any or all possible space-barrier paths of propagation by combining the networks of Figures 3 and 4 for the sequence of rooms emanating from any room of origin. For any given time increment, hand calculations could be used to calculate values of L and $\overline{L}$. However, this becomes tedious for a continuum since the values of L and $\overline{L}$ vary with time due to the barrier deterioration as the fire continues, and because of the introduction of intervention methods, such as automatic sprinkler effectiveness and fire department suppression, which change as time progresses. Consequently, a computer model has been constructed that calculates the L-Curve for any or all paths of fire propagation in time and space. With this computer analysis it is possible to construct an envelope of L-Curves, such as that shown in Figure 5, for any room of origin or for an entire building.

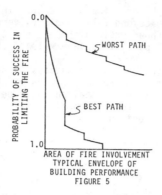

PROBABILITY OF SUCCESS IN LIMITING THE FIRE

0.0

WORST PATH

BEST PATH

1.0

AREA OF FIRE INVOLVEMENT
TYPICAL ENVELOPE OF
BUILDING PERFORMANCE
FIGURE 5

## IV. QUANTIFICATION OF FLAME MOVEMENT ANALYSIS COMPONENTS

The quantification of the engineering method today utilizes a knowledge base including such factors as

a.  Physical and chemical phenomena
b.  Fire test results
c.  Code and standards experience
d.  Building analyses
e.  Computer models

Engineering judgment provides the link between the available scientific and experiential knowledge and the expected fire performance of the unique building being evaluated. The expected performance of the components is expressed in terms of their probability of success. Other than in the expression of results, the present application of the engineering method is similar in approach to those used in traditional structural and mechanical design methods. The anatomy of the method keeps the complex interactions of the fire and the building in order so that the components may be integrated into the complete system.

The fundamental organization of the hierarchial networks and framework has been structured in a manner that allows new knowledge to be incorporated without disruption or change to the basic analytical framework. This allows each component to be evaluated in a manner analogous to a free body analysis of structural mechanics. That is, each component can be evaluated in isolation and then returned to the framework with confidence that its interactions will be addressed. Consequently, all new findings can serve to increase the confidence in the results. The basic analytical structure need not change. This feature will allow a transition from the probabilistic judgmental engineering evaluations of today to deterministic equations as the evolution continues.

The quantification "tomorrow" will involve the use of relatively simple deterministic equations to evaluate the major components. When the deterministic equations are developed, the probabilistic descriptors used today will become obsolete. The probability assessments then assume an entirely different role. They change character and become reliability evaluations of the equations, their interactions, and the ancillary design requirements that are the parts of the engineering method.

The design method must incorporate a factor of safety or a safety index. This may be achieved by adjusting the deterministic equations and design requirements by a safety index that incorporates an acceptable probability of failure. The acceptable probability of failure will reflect the level of risk and the related cost of protection that society is willing to accept. An assessment of existing code complying buildings will provide a useful base to assist in identifying the safety level. Techniques for doing this are a regular part of contemporary engineering literature. Magnusson (4) described procedures for structural fire resistances in 1974.

To illustrate this technique, consider the load, S, and the resistance, R, functions shown in Figure 6(a). In flame movement analysis, the load, S, would be the design fire. The resistance, R, could be the automatic sprinkler capability; or the fire department capability; or the barrier effectiveness. The probability of failure can be described as

$$P_f = P \ (R < S) \tag{3}$$

(where $P_f$ is the probability of failure)

When we combine the distributions of R and S, as shown in Figure 6(b), the probability of failure becomes,

$$P_f = P(R-S) < 0.0 \tag{4}$$

The safety index, $\beta$, times the mean, $\mu$, will define the level of safety achieved by the design equations and procedures.

Obviously, each component has many variables and conditions that will influence its behavior. However, since we are interested only in the probability of (R-S) < 0, many of the variables which are important influences for the fire or the suppression behavior may become insignificant, depending on the form of the design equation. This allows a reliability analysis to be conducted by load and resistence evaluations similar to those described by Galambos (5).

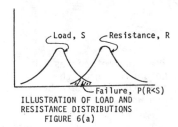

ILLUSTRATION OF LOAD AND
RESISTANCE DISTRIBUTIONS
FIGURE 6(a)

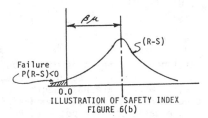

ILLUSTRATION OF SAFETY INDEX
FIGURE 6(b)

## V. SMOKE, STRUCTURAL, AND ESTABLISHED BURNING ANALYSES

The smoke movement analysis currently evaluates the time and probability that a selected space will remain tenable. At the present time, tenability evaluations are described by obscuration levels.

The evaluation is done in two parts. The first evaluates for each increment of time the probability that the space will remain tenable given that no suppression occurs, P (tenable/$\bar{L}$). The second part incorporates the probability that the fire is not limited, P($\bar{L}$), as evaluated by the flame movement analysis. The S-Curve is then obtained from the conditional evaluations of

$$P \text{ (tenable)} = P \text{ (tenable/}\bar{L}) \; P(\bar{L}) \qquad (5)$$

Structural fire protection has received considerable attention over the years. The European fire community in particular has assumed such a leadership role that quantification and design procedures are well advanced.

The evaluation of the probability of failure using safety index procedures becomes feasible because one can separate the structural situation after full room involvement, P(structural failure/$\bar{L}$), from the probability of full room involvement, P($\bar{L}$). In this way, the conditionality allows the two parts to be evaluated as separate entities and then combined as

$$P(\text{structural failure}) = P(\text{structural failure/}\bar{L}) \; P(\bar{L}) \qquad (6)$$

Established Burning (EB) is an important concept in building analysis. As described in the "Anatomy of Building Firesafety" (2) the likelihood of ignition, fire growth, occupant extinguishment, and special hazards automatic suppression is incorporated into this analysis conveniently. This demarkation of EB becomes quite useful with regard to consistency of engineering analysis and design. It also allows the free body concept, described earlier, to be applied more readily to individual buildings.

Established Burning is composed of two parts. One is the probability of ignition, P(IG). The second is the probability of reaching the fire size defined as Established Burning, given ignition. Equation (7) shows these parts. As noted earlier, the free body concept allows these components to be evaluated separately and then combined later into a complete risk analysis of the entire building, if it is of value.

$$P(EB) = P(IG) \; P(EB/IG) \qquad (7)$$


## VI. APPLICATIONS

The development of the present status of the engineering method has involved the integration of several parts. One is the unfolding of the method itself. Great care has been taken to ensure that the structure has a sound theoretical and logical foundation. While much work remains to be completed, the basis still appears to be correct.

A second part has involved techniques of quantification. Temporary quantification measures have utilized research reports, basic theory, experiences, and engineering judgment. Reports from the fire research community have been integrated into the quantification wherever possible. More importantly, the general form and type of questions and results needed to translate research results into general use are more clearly understood.

The third major part of the process has included practical applications. The actual use of the method to solve real firesafety problems has been an essential ingredient.

The simultaneous attention to all three parts has been essential to the present state of the evolution of a disciplined engineering analysis and design process for building firesafety. The bridge between research and practice has been the engineering method. The method has provided both a framework to facilitate technology translation from research into practical tools, and also a conduit to enhance communication between the two paths. The range of application activities has been broad. Up to this point all activities have been unsponsored. Consequently, practice has been limited to specific problem areas. Frequently, when feasibility has been proved or the practicality has been demonstrated, the project is halted.

The following areas illustrate the type of applications that have been undertaken in conjunction with the engineering method. Listing does not imply completed techniques. Listing means that the application has been studied to a depth whereby one could recognize that further research on the topic could result in the refinement of procedures that would have a high probability of a project success.

1. Building firesafety analyses
   a) Risk assessments
   b) Cost-benefit analyses
2. Management science decision analyses
   a) Insurance underwriting
   b) Corporate fire risk management
3. Building code anlayses
4. Selected fire standards analyses
5. Building regulatory studies
6. Fire department management
   a) Indexing of buildings
   b) Prefire planning and training
7. Indexing of ships


VII. CONCLUSION

It is possible to evaluate the comparative level of risk today by the analytical techniques described in the engineering method for building firesafety. The numerical values are, of course, comparative and not absolute. Nevertheless, the engineering method does permit a relative consistency in building evaluations because of the structure of the analysis.

The method can incorporate new research findings without disruption of the basic framework and structure. If guidelines were prepared for numerical evaluation of the components, collection of statistical data on an ongoing basis, incorporated into a Bayesian updating would eventually produce an absolute level of risk.

The current approach is not so much to develop an absolute risk analysis for a building, even though it is possible to do so, but rather to integrate building analysis experiences with research results to develop deterministic equations that eventually can be used for office calculation in engineering practice. These deterministic equations will be subjected to realiability analyses to gain an insight into the important variables and incorporated into the equations to form a basis for a fire design code that is analogous to a structural design code.

# REFERENCES

(1) Nelson, Harold E., "Building Firesafety Criteria, Appendix D" General Services Administration, 1972.

(2) Fitzgerald, Robert W., "The Anatomy of Building Firesafety", Center for Safety Studies, Worcester Polytechnic Institure, 1982

(3) Fitzgerald, Robert W., "An Engineering Method for Building Firesafety Analysis", SFPE Symposium for Computer Application in Fire Protection, 1984.

(4) Magnusson, Sven Erik, "Probabilistic Analysis of Fire Exposed Steel Structures", Lund Institute of Technology, Lund, Sweden, 1974.

(5) Ravindra, M. K. & Galambos, T.V. "Load and Resistance Factor Design for Steel", Journal of the Structural Division, ASCE (1978)

# The Development and Use of the United Kingdom Home Office Fire Cover Model

GEOFFREY H. DESSENT and JOHN A. HARWOOD
Scientific Research and Development Branch, Home Office
United Kingdom

ABSTRACT

The paper describes the further development of the United Kingdom Home Office Fire Cover Model and its use in providing advice to the Joint Committee on the Standards of Fire Cover. The Committee was tasked with reviewing the arrangements made by the Local Authority Fire (County) Brigades for providing fire appliance attendance to fires and other emergency incidents. The Home Office wished to consider the effect on resources required if various possible changes were made in the current arrangements. The model was used to assess the effects in a sample of four brigades.

## INTRODUCTION

There are three main functions of the fire service in the United Kingdom (1):

   i) the extinction of fires and protection of property and life in case of fire

   ii) special services

   iii) fire prevention

The arrangements made by fire brigades to provide the rapid response necessary in i) and ii) are known as "fire cover". The planning of fire cover is based on categorising areas by fire risk and recommending a specified minimum level of brigade attendance for each risk area. There are four categories of risk: A,B,C,D; corresponding approximately to commercial and industrial city complexes, centres of large towns, built-up areas of towns and rural areas. Especially high risks and remote rural areas are treated as special cases. For each risk area there is a recommended first attendance. For A risk, it is two pumps (1st attendance fire appliance) within a maximum period of five minutes, and one further pump within eight (for B risk, 1 in 5 plus a second in 8; for C, 1 in 8 to 10; and for D, 1 in 20). For special services, such as road traffic accidents and for known small fires such as those on waste ground or in derelict buildings, brigades use their discretion.

The fire cover arrangements described above followed recommendations made in the report by the Joint Committee on Standards of Fire Cover (JCSFC) in 1958 (2). The Committee was reconvened in 1980 to consider whether there was a need to modify the arrangements recommended by their predecessors. As part of their information gathering, they requested the Operational Research Division of the Scientific Research and Development Branch (SRDB) of the Home Office to advise on the likely outcome of various policy options. SRDB had a long tradition of working in the Fire Cover area (formerly as the Scientific Advisory Branch, SAB) and a number of models had been developed so the task was undertaken with some confidence.

One type of model calculated fire brigade attendance times from pump dispositions and fire occurrance. A second type calculated fire losses from attendance times (loss/attendance). By combining the two models a relationship between fire cover strategy and fire losses was obtained. This was then minimised to obtain the most cost/effective fire brigade arrangements. So far only the first type of model has achieved full acceptability.

In an early model, described by Hogg in 1973 (3), fire was modelled as existing in states (confined to object, confined to room, beyond room and beyond building) with time independent probabilities of transition determined from the data collected. It took account of the first pump at each station and the first at each fire and provided output in terms of arrival times of the first pump.

A further model looked into the relationship between the amount of fire spread after the arrival of the first pump and the arrival time of the second and subsequent pumps (4). As the work progressed the difficulties of formulating realistic loss/attendance relationships became more evident, and the emphasis of the work turned to improving these. However the inherent problems and the difficulty of obtaining sufficient and reliable data left doubts in the minds of researchers and potential users. In 1980 reseach responsibilities in the Home Office were reorganised and SRDB took over responsibility for this area of work. Although it was at that time apparant that loss/attendance models had limited credibility, this aspect was not necessary to adequately cope with the committee's request.


DEVELOPMENT OF THE MODEL

In the SAB model queuing theory was used to model the process of a sequence of calls (the customers) being attended by the nearest available pumps (multi-servers). This process is complex since the initial attendance to calls depends on the nature of the call and risk category (as outlined above).

The model took into consideration the variable call rate throughout the day by splitting the day into six four-hour periods (during which a constant call rate was assumed). A road network of the principle roads used by pumps responding to incidents was set up for brigade areas by defining "nodes" at major road junctions and fire stations. From the lattice road system, travel times were assessed and fed into the model. By calculating the likelihood of the nearest appliance being available it was possible to compute attendance times for a year's calls. The final part of the model attempted to convert the fire brigade attendance times into direct financial loss - the loss/attendance relationship.

In view of the objective of assessing alternative fire cover standards considerable development was considered necessary. In particular, the loss-attendance, as mentioned, was not essential, and was removed. The resultant response time model had the advantage that the results could be easily validated.

The original SAB model used population statistics to estimate the likely number of fires. Although some statistical evidence was put forward to suggest that population was a good predictor for fire incidence, it was far from conclusive. A Statistical survey carried out of 14 U.K. fire brigades for the JCSFC for 6 months provided details of the location of all incidents, which were used to feed the SRDB model.

The SAB model "ran" on a computer in "batch" mode and produced a large amount of hard copy output. When the model had been successfully transferred to SRDB's VAX 11/780, the program was made fully interactive, allowing the user to make changes at the keyboard to the number of fire appliances and type of manning at a station and various other parameters. The output was reduced dramatically and now simply provides a recapitulation of the input data and the attendance times by risk category as follows:

Output from the SRDB Fire Cover Model
Percentage of Occasions when standards broken

| Risk Category | First Pump | Second Pump | Third Pump |
|---|---|---|---|
| A | 6% | 12% | 2% |
| B | 0% | 5% | - |
| C | 1% | - | - |
| D | 0% | - | - |

This shows typical results from the model. In this theoretical example, 6% of all A Risk first appliances failed to arrive within the standard of 5 minutes. Further, in this example all D Risk attendances and the first pump in B Risk areas attended incidents within the standard.

In addition, the model produces a further table which has the same format as the above, but shows the "average time by which the standard is broken" (for those occasions on which the standard is broken).

MODEL USAGE

Since it was not practicable to study all brigades (there are 54 brigades in England and Wales), it was agreed that a small sample of "typical" brigades should be examined with the model. In order to make use of the detailed information provided by the Statistical Exercise on incident location, the brigades studied were chosen from among the fourteen in that survey. Further, the original SAB model was tested out by consultants (from the Local Government Operational Research Unit) in 1976 on 11 fire brigrades. Therefore LGORU (who were contracted to carry out the data collection) were able to provide travel time matrices (which needed a little modification) for these 11 brigades.

The final choice of four brigades was simply the overlap in brigades studied previously by LGORU and those which took part in the statisical survey. The four brigades studied were :-

(1) Cleveland

(2) Greater London

(3) Greater Manchester

(4) Hertfordshire

These brigades were very different in many respects and had quite different risk maps. Cleveland is unusual (5) as it contains much C and D risk along with the largest area of Special Risk in the UK made up for the most part of 22 Major Chemical Plants. Greater London is mostly A and B risk. Greater Manchester exhibited a very high incident rate, has a good deal of C risk and some B risk but much less A risk than London. Hertfordshire contains no A risk, very litle B, some C but is predominently D risk.

The model was validated against the "actual" by means of the "percentage of occasions the standards were broken". In all cases the figures were within 5% of those actually recorded. Initial validation of Cleveland, the first study, showed that the nodal density should be linked to risk category, which proved to be vital in final validation.

The model was used to assess the resource implications of alternative standards of fire cover. For example, one question was "Would sending a second appliance to arrive within 10 minutes to C risk areas require further appliances (and men)?".

The model showed that it would but that it was virtually impossible to achieve this aim in some areas without a considerable station building programme. Another question was "Would a reduction from 3 to 2 appliances in A risk areas mean that current resources would result in overprovision?". The model showed that this was not so, as in general the third appliance had a first attendance role in an adjacent station's area. These and other questions were examined by the model providing a useful insight into the question "what would happen if?".

CONCLUSIONS

Use of the Fire Cover Computer Model has demonstrated that the resource implications of different Fire Cover Standards vary from brigade to brigade, depending on two major factors:-

    i.   the "busyness factor"

    ii.  the "geographical factor"

The former simply reflects the rate of incidents, whilst the latter pertains to road configurations, risk map definition, and station and pump location. It seems likely that brigades which reflect comparable "busyness" and "geographical make-up" will exhibit similar resource implication characteristics. However, many brigades may be considered unique for the following reasons:-

    i.   Brigade Area has costal border.

    ii.  Many Special Risks in brigade area.

    iii. Brigade contains many natural barriers (eg. rivers).

It has also demonstrated the usefulness of an operational research approach and a computer model in providing information to a policy committee. The success of the work has lead to a request being to us to investigate the potential for developing a loss/attendance model. This is currently in progress.

Further, the computer model is well suited to tackle fire cover problems at a more local level, enabling fire brigades to " get the best from the resources available ". It has been suggested that not only could fire brigades consider the allocation of major pumping appliances to stations but could also examine the disposition of special appliances, such as turntable ladders, emergency tenders and the like.

REFERENCES

1.  Review of Fire Policy, Home Office, 1980

2.  Report of the Joint Committee on Standards of Fire Cover, HMSO, 1958.

3.  Losses in Relation to the Fire Brigades Attendance Times, J. Hogg SAB Report 5/73.

4.  A Model of the Growth of Fought Fires, R.Rutstein and R.J. Barnes SAB Report 1/75

5.  Cleveland County Research and Intelligence Report: Fire Brigade Statistics- Initial results from a six months study of incidents, D. Peace

# National Fire Costs—A Wasteful Past but a Better Future

**TOM WILMOT**
World Fire Statistics Centre
Geneva, Switzerland

ABSTRACT

Fires are costing many countries around 1% of Gross Domestic Product, but compared with other aspects of national waste such as Road Safety, Crime Prevention and Industrial Safety, Fire Prevention ranks low in political priorities.

Countries have a responsibility to develop a fire strategy aimed at reducing fire costs. For measuring the success of the strategy, national fire cost statistics are needed for comparison with other countries. These statistics need to cover both fire losses and fire protection costs.

In 1983 the World Fire Statistics Centre was formed under the auspices of the Geneva Association and with headquarters in Geneva. Its object is to encourage better world fire statistics and to encourage politicians to rank fire protection higher in the list of political priorities.

Major Centre activities have included cooperation with the United Nations in a fire statistical scheme under which 14 countries have submitted annual fire cost statistics and with the Commission of the European Communities (EEC) in the International Fire Symposium held in Luxembourg in 1984. Details of national fire cost statistics are included in this paper, together with some suggestions for future progress, including a plea for a World Fire Research Council.

Address all correspondence to 12 Kylestrome House, Cundy Street, London SW1W 9JT.

NATIONAL FIRE COSTS – A WASTEFUL PAST BUT A BETTER FUTURE

Political interest in fire prevention

Although fires are costing many countries around 1% of Gross Domestic Product, fire prevention generally ranks low in national political priorities. This fact was dramatically illustrated in a study published by the Geneva Association in 1979 – European Fire Costs – The Wasteful statistical gap. The study contained tentative estimates for 6 key items of fire costs for 12 European countries covering the years 1970-1975. If all 12 countries had produced annual statistics for each of the 6 items, a regular European collection of 72 valuable fire statistics would have been available for practical fire use. In fact only around 30 of these 72 statistics appeared to be published. Not surprisingly the study went on to comment "This suggests that little political will exists in Europe to reduce fire waste".

Why is this lack of statistics so wasteful? The simple answer is that in order to monitor the success of national fire strategy, Governments need to survey the annual trends in their own national fire costs and then to measure them against trends in countries abroad – without statistics, measurements are impossible and decision-making remains speculative.

Yet other types of national waste are tackled impressively, both at national and international level. Road Safety, Industrial Safety and Crime Prevention all receive top-level political attention. Where political will exists to reduce waste, spectacular results can be achieved as any survey of the achievements of the World Health Organisation dramatically illustrates.

Progress on fire statistics

There are encouraging signs of increased political interest in fire safety, both at international and at national level. In 1981 the United Nations started a fire statistical scheme and in 1984 the Commission of the European Communities (EEC) held a most successful International Fire Symposium. With the prime object of encouraging better world fire statistics the World Fire Statistics Centre was formed in 1983 under the auspices of the Geneva Association and with headquarters in Geneva.

United Nations fire statistical scheme

In 1981, The United Nations Economic and Social Council, through its Working Party on Building, agreed to support a proposal for better fire statistics. The proposal came from the meeting held by the International Council for Building Research Studies and Documentation (CIB) in Athens in 1980 at which a spirited discussion had taken place on the need for better fire statistics. As a result of the CIB initiative, the United Nations agreed that the fire statistical problem needed tackling at international level and commissioned a pilot study under which 4 countries provided data for the years 1978/9 covering 6 items of fire costs – direct fire losses, Indirect fire losses, Human losses, plus the costs of fire brigades, of fire insurance and of protecting buildings against fire.

The study involved stimulating challenges to ensure that the statistics should be as accurate as possible and that the data should be produced on a uniform basis. As there is no uniform European system for collecting fire statistics, a method had to be evolved to bring together a widely varying set of national data. The method adopted was to draft a questionnaire asking countries to submit published data for each of the 6 items of fire costs. A list of possible

adjustments for each item was then included in the questionnaire and fire
experts in each country were asked to estimate additions or deductions to the
published figures, in order to bring the final figures  onto a standardised
European basis.  In some cases these adjustments were far-reaching, for example
the direct loss calculation allowed for 8 possible additions and for 3 possible
deductions.

The following year, the UN studied the report on the pilot study and felt that
this method of standardising national data was practical.  13 countries then
agreed to join an extended scheme under which they would provide data for the
years 1979/80.  A year later, all 13 countries agreed to update their figures
and the Canadian Fire Commissioner added Canada to the scheme.  The data is analysed annually by the World Fire Statistics Centre in a report for presentation
to the United Nations.  In this report, which covers around 12 pages, estimates
are given of the national fire costs covering the 6 items, which are then compared with the Gross Domestic Product (GDP) or the population of the country
concerned.  Some indication of these estimates is given below.

1.  Direct fire losses

Direct fire losses are, of course, a major item of national fire costs and frequently cover around 30% of total fire costs.  The figures for the countries
submitting figures are:-

TABLE 1.  -  COST OF DIRECT FIRE LOSSES

| Country | 1979 | 1980 | Percentage of Gross Domestic Product |
|---|---|---|---|
| Hungary | Ft 600 m | FT 750 m | 0.10 |
| Japan | Yen 240,000 m | Yen 276,000 m | 0.12 |
| Spain | Ptas 21,900 m | Ptas 26,700 m | 0.18 |
| Austria | AS 2,550 m | AS 1,500 m | 0.21 |
| Netherlands | Dfl. 650 m | Dfl. 725 m | 0.21 |
| United Kingdom | £375 m | £500 m | 0.21 |
| Finland | Fmk 350 m | Fmk 525 m | 0.25 |
| United States | US $6,300 m | US $6,600 m | 0.26 |
| Sweden | SKr 1,300 m | SKr 1,325 m | 0.30 |
| France | F7,750 m | F8,750 m | 0.32 |
| Denmark | DKr 1,100 m | DKr 1,500 m | 0.36 |
| Norway | NKr 925 m | NKr 1,025 m | 0.37 |

The volume of direct losses can, of course, vary considerably from year to year,
so too much cannot be read into two years results.  But it is interesting to see
that both the Eastern European countries (Hungary and Austria) enjoyed good
results.  Japan has long been respected for its low fire losses, helped by their
scientific approach to fire problems.  At the lower end, the Scandanavian
countries clearly suffer from climatic conditions.

2.  Indirect fire losses

Indirect fire losses were defined as "the losses falling on the national economy
as the result of consequential fire losses".  A few years ago, it was frequently
suggested that these indirect losses were as high, or higher than direct losses,
but it is now becoming clear that indirect losses are less than 10% of total
fire costs.

Precise calculations are very difficult to make and it might be misleading to
publish a Table of the tentative estimates.  Where a country enjoys a thriving

business interruption insurance market, the insured losses provide a valuable guide to the total national indirect losses. In other cases, the industrial fire losses often provide the best indication of the total indirect losses. In Holland the fire brigades produce statistics. For all countries, it was assumed that only 50% of the actual indirect losses affect the national economy (the balance being taken up by gains to trade competitors etc).

## 3. Human losses

Fire deaths and fire injuries are to most people the most tragic aspects of the fire problem. So it is not surprising that figures for fire deaths are more reliable than most fire statistics. Not only do fire brigades keep records in most countries for fire deaths to which they respond, but the World Health Organisation publish Government figures for deaths from fires and flames. The following Table is based on the higher of these 2 sets of figures and frequently includes an addition for unreported fire deaths.

Table 2. - POPULATION COMPARISONS FOR FIRE DEATHS (1979/1980)

| Country | Deaths per 100,000 persons |
| --- | --- |
| Switzerland | 0.64 |
| Netherlands | 0.64 |
| Austria | 1.00 |
| Spain | 1.25 |
| Denmark | 1.33 |
| Sweden | 1.80 |
| Japan | 1.80 |
| France | 1.90 |
| Norway | 2.27 |
| United Kingdom | 2.39 |
| Finland | 2.56 |
| Hungary | 2.64 |
| United States | 3.63 |

It is interesting to compare this Table with Table 1 (Direct losses) and to notice that Netherlands and Austria continue to show up well. Spain too shows low losses, perhaps helped by the warm climate. The high death toll in the USA has of course, been recognised for many years and energetic steps are being taken to reduce the numbers.

Many countries also publish figures of fire injuries, but as the definition of "fire injury" varies so widely, I have not included any statistics. It may, however, be a reasonable supposition to expect the national pattern of fire injuries to be similar to that for fire deaths.

## 4. Fire-fighting organisations

So far, the Tables have related to fire losses. The remaining Tables deal with the cost of protection. Until recently, international Fire Tables tended to concentrate on losses rather than protection, but figures for both classes are needed to assess national fire problems.

It is rather surprising that although the cost of fire fighting organisations is running at the rate of around 15% of total national fire costs, many countries publish no annual figures for fire brigade costs. The major part of the cost is, naturally, related to the public fire brigades, but additions need to be made for the cost of such organisations as industrial and volunteer fire brigades.

1012

TABLE 3.    COSTS OF FIRE-FIGHTING ORGANISATIONS

Average percentage of Gross Domestic Produce (1979/1980)

Country

| | |
|---|---|
| Denmark | 0.09 |
| Netherlands | 0.16 |
| Finland | 0.20 |
| Norway | 0.21 |
| United Kingdom | 0.24 |
| United States | 0.27 |
| Japan | 0.31 |
| Sweden | 0.33 |

Possibly a factor in the low cost of Danish fire brigades is the strength of the Falck privately owned system.  The high Japanese figure may come as a surprise to some fire experts – it may well be part of the price that Japan is paying in order to achieve its low fire losses.

Sadly no figures were forthcoming for the cost of the Austrian fire brigades since their situation is a most interesting one.  With a population of 7½ million, there are around 2000 firemen in the public brigades and a staggering 250,000 in the volunteer brigades.  The strength of the volunteer brigades may be the key to Austria's fine fire record, since not only are the fire brigade costs low, but the expert training of the volunteer firemen and their families must contribute towards the cause of the low Austrian fire losses.

5.  Fire insurance

The cost of administering fire insurance is somewhat similar to that of running the fire brigades and is running at the rate of around 15% of total national fire costs.  Surprisingly, it is rare for countries to publish the annual cost of administering fire insurance, since the fire figures tend to be merged with other classes of insurance business.  However, with the help of insurance experts, most countries were able to answer this part of the questionnaire.

TABLE 4.  -  COSTS OF FIRE INSURANCE ADMINISTRATION

Average percentage of Gross Domestic Product (1979/1980)

Country

| | |
|---|---|
| Hungary | 0.01 |
| Finland | 0.07 |
| Spain | 0.08 |
| Sweden | 0.08 |
| Denmark | 0.13 |
| Austria | 0.14 |
| Norway | 0.14 |
| United Kingdom | 0.15 |
| Japan | 0.15 |
| France | 0.16 |
| United States | 0.22 |

The low Hungarian figure is due to the State monopoly.  A remarkably low expense ratio accounts for the low Finnish figure.

If any further study into fire insurance were contemplated, then it would best be carried out by some such body as the Fire Group of the Comite European des Assurances who have carried out valuable studies into fire insurance problems.

## 6. Building protection

The cost of fire protection to buildings ranks with direct losses as the most costly section of national fire costs - around 30%.

No annual statistics for this item are published and an accurate calculation is not easy. An important pioneering work in this area was published in 1967 by Alan Silcock of the UK Fire Research Station - Protecting buildings against fire - (Architects Journal Information Library 13.12.1967). In this paper, Silcock estimated that the average fire protection cost as a percentage of total building costs varied from 1% (Housing) to 7% (Industry). By applying these percentage costs to the government figures for new construction, he arrived at a figure of around 2½% of the total building costs in 1965. Today, the more stringent nature of UK Building Regulations means that some figures are probably slightly higher, although annual patterns depend on the relationship between domestic and industrial volumes of building.

Several countries put forward estimates for the cost of fire protection to buildings:-

TABLE 5. - FIRE PROTECTION TO BUILDINGS

Average percentage of Gross Domestic Products (1979/1980)

Country

| | | |
|---|---|---|
| France | 0.16 | (1978/1979) |
| Denmark | 0.18 | (1978/1979) |
| United Kingdom | 0.18 | |
| Netherlands | 0.19 | |
| Japan | 0.32 | |
| United States | 0.32 | |
| Sweden | 0.38 | |
| Switzerland | 0.44 | |
| Hungary | 0.45 | |
| Norway | 0.65 | |

Too much should not be read into these figures since a great deal of further research is needed before any reliable international Table can be produced. This expensive aspect of national fire costs is perhaps the one with most potential for rewarding research and it is encouraging to know that an increasing amount of interest is being shown in it - one recent example being the Japanese study by Mr H Nokamura Analysis of the Fire Protection Index Appendix to minutes of World Fire Statistics Centre Seminar March 23/4 1983. This study involved computerised studies of the fire protection costs of 1300 buildings.

## Value of United Nations statistics

The United Nations Committee are under no illusions about the difficulties in evaluating and comparing fire statistical data with other countries. Such international comparisons are fraught with danger, even in such well developed areas as company financial statistics. Governments have been passing laws on company accounting for over a century, but even today investment analysts find extreme difficulties in making international comparisons of multi-national companies using their annual published accounts. But nevertheless, the United

Nations statistics provide the best indications available of the relative success that each country enjoys in its fire strategy.

There are encouraging signs for the future with new countries considering joining the scheme including W Germany and New Zealand. Another indication of increased Government interest in fire costs was the major initiative of the Commission of the European Communities (EEC) in holding an International Fire Symposium "to consider the measures that can be taken to cut fire costs and to reduce fire risks". The Symposium ended with a spirited discussion on possible EEC future initiatives including a suggestion that the EEC goal should be to bring about a substantial reduction in the cost of fire by the end of the century. In listing some suggestions for fruitful international action, I have drawn heavily on this EEC discussion.

Suggestions for a better fire future

Earlier I have mentioned that the object of the World Fire Statistics Centre is to encourage the production of better world fire statistics and to encourage politicians to rank fire protection higher in the list of political priorities. But these two aims are merely steps towards the ultimate goal which is "To reduce international fire costs". If these costs are to be reduced, then there is scope for action in both the statistical and political field. Given steady progress in these two areas, specific areas for cost-reduction can be singled out - Arson and the high cost of the fire protection parts of Building Regulations being two possible targets.

Taking these problems of statistics, political involvement and targets for savings in turn, I will attempt to suggest a few possible ways of moving forward. In doing so, I should like to emphasise that these proposals are not put forward as a model method for cutting fire costs; many participants in the Symposium are infinitely more knowledgeable on fire matters than I am and better qualified to suggest the ideal way to move forward. But if this paper succeeds in provoking discussion on how better to cut international fire costs, then its purpose will have been achieved.

1. Scope for better fire statistics

I hope that I have convinced readers of this paper, that without adequate fire statistics, monitoring of national fire strategy is impossible and therefore decision-making becomes unduly speculative.

The obvious two areas for better statistics are:- 1. The national fire statistics for the 6 key-items of national fire costs, and 2. National statistics covering causes and locations of fire. The top priority for the national fire cost statistics is, I think, to improve their quality rather than their quantity. A few months spent on an international study of the costs of fire protection to buildings could be particularly rewarding. So far as statistics for causes and locations of fires are concerned, the scope for improvement is immense. Many important countries produce no annual statistics at all for this crucial problem and of those countries which do publish figures, frequently the value of the figures is drastically reduced by the lack of figures for monetary losses in the Tables. One important exception to this situation is the Annual Report of the Canadian Fire Commissioner - Fire Losses in Canada which packs a mass of essential information into less than 50 pages and is well worthy of study by any country planning to produce their own national figures. An unfortunate constraint on any international comparison of national statistics for "causes and locations" is the widely varying sets of classifications used in different countries. However, there are in Europe, a limited number of statistics produced on a uniform basis - the analysis of large fires carried out by the Comité

Europeén des Assurances – and closer cooperation between fire experts and the CEA could produce worthwhile results.

## 2. Scope for greater political involvement in fire cost-cutting

As I have mentioned earlier, the progress recently in this field has been most encouraging, so some say that it is greedy to ask for more. Nevertheless, a little more involvement would be invaluable and could yield savings infinitely greater than costs. The obvious logical step forward is the formation of a United Nations Fire Research Council which would coordinate national efforts and take international initiatives. The UN is holding this year its 7th International Congress on Crime Prevention – a conference which takes place at five-yearly intervals. The organisation of a similar Congress on Fire Prevention (with the main aim of cutting fire costs) could be an early priority for the new Fire Council. Another valuable role could be the provision of small sums of money to bodies such as the Conference of Fire Protection Associations for their international work. Such organisations lack funds to employ staff able to spare much time on international affairs, which can only be given at the expense of handling domestic national day-to-day problems. So a small grant could enable one or two people to be employed with specific international responsibilities, who could help in such activities as the organisation of the proposed international congress, or international studies into current fire problems.

## 3. Scope for identifying specific areas for cutting fire costs

Individual lists of the most fruitful area for international cooperation will, clearly, vary greatly – the following suggestions are not intended to comprise a definitive list – rather are they put forward in order to indicate the scope for worthwhile international cooperation, which I believe exists.

## a) Direct and Indirect fire losses

Many fire experts regard arson as the major area warranting urgent action. The announcement of an international will to reduce arson losses, coupled with the provision of a small budget to help the Conference of Fire Protection Association in their international anti-arson work would be a practical step forward – formation of an international anti-arson Task Force could speed up progress. It goes without saying that better statistics covering arson would be needed to monitor progress.

Another early priority might be the growing problem of warehouse fires. An attempt to encourage safer building and furnishing material could be cost-effective. A study could review recent international improvements, review the cost-benefit equation and put forward proposals for better pooling of international knowledge in this rapidly-developing field. Another possible study might cover the scope for better fire detection, including the progress being made on greater reliability, the scope for cutting costs and an attempt to discover new ideas for anti-arson detectors.

## b) Fire Brigade costs

Considering that this item represents around 15% of total national fire costs, surprisingly little international cooperation exists. In fact several leading countries show so little interest in the problem that they do not even produce national figures.

The scope for international study, therefore, appears to be almost unlimited. But such study cannot be carried out in isolation from a study of total national fire costs. Some countries might well be spending too little (not too much) on their fire brigades – the example of Japan shows how an expensive, but high-quality, fire brigade network can contribute towards a low national fire loss.

## c) Cost of fire protection to buildings

This item ranks with Direct fire losses as the largest item of national fire costs (around 30%). There is a feeling in several countries that Building Regulations now involve too high implementation costs. An international Government-inspired study might produce startling ideas for savings.

Other possible cost-effective areas for international cooperation include acceleration of harmonisation progress and the introduction of National Certification systems such as the recently formed British Approval of Fire Equipments (BAFE).

## d) Publicity

One area where greater expenditure might well be justified is the field of publicity for national fire prevention efforts. Success and failure in this area differs widely from country to country. An international review identifying outstanding successful publicity campaigns could make very stimulating reading for fire experts disheartened by the paucity of press and television interest in their own country. A particularly interesting international study could be conducted into the best method of bringing up children to become powerful fire prevention allies. A sideline of such a study might include the effect of volunteer fire brigades on youth training – one cannot help suspecting that the low Austrian fire losses are influenced by the astonishing success of the volunteer fire brigades, who number 1 out of every 20 Austrians amongst their membership of around 250,000. Another Central European country with a large volunteer fire brigade movement is Switzerland with over 200,000 volunteer firemen.

## Conclusion

If I have been guilty of pessimism in sketching a wasteful past history of international efforts to cut fire costs, I hope that I have restored the balance by proper recognition of current progress and by an optimistic view of the future. If such optimism is to be justified, it seems to me that a powerful international Fire Research Council needs to be formed receiving strong support from several leading national Governments. The Council's Governing body of fire experts could then identify priorities, allot budgets, monitor progress using statistics and finally cut fire costs.

# Investment Model of Fire Protection Equipment for Office Building

**HIROYUKI NAKAMURA**
Institute of Technology, Shimizu Construction Co., Ltd.
No 4-17 Etchujima 3-chome, Koto-ku
Tokyo 135, Japan

## ABSTRACT

This report presents (1) a canonical correlation analysis that describes the relations between characteristics of a building and cost indices of fire protection equipment, and (2) how to apply the analysis for a standard and a trade-off for investment model of fire protection equipment when designers would decide to invest in fire protection equipment.

The results of the analysis show that the variate pairs between characteristics of a building and cost indices of fire protection equipment are highly related, and also are available for a standard and a trade-off for investment model of fire protection equipment. Based on the canonical loadings, designers evaluate the installation of sprinkler systems against widths within the buildings, and decide to emphasize the installation of dry risers or sprinkler systems by comparison between the width and service cost index of the buildings.

## 1. INTRODUCTION

There has been a misunderstanding that costs relating to fire protection are usually unduly high in comparison with the entire construction cost of a building. On the other hand, detailed research into the exact cost of fire protection does not seem to have been conducted. There are difficulties involved in conducting such research; estimated data usually are not recorded systematically. Moreover, many fire protection items are not merely installed for the single purpose of fire fighting, but are for multifunctional use. For example, fire walls, staircases, safety zones, and so on.

In my prior reports[1,2,3], the "fire protection costs" included in past estimates from the records of estimation and discussions with a view to determining the facts on their investment are described. Furthermore, taking into account the results of such analyses, the office buildings were chosen for further detailed research.

This report presents a canonical correlation analysis which describes the relations between a set of predictor variables (characteristics of a building) and a set of criterion variables (cost indices of fire protection equipment). And by using the results, also discussed are a standard and a trade-off for investment model of fire protection equipment when designers would decide to invest in fire protection equipment.

## 2. DATA BASE AND ITEMS TO BE ANALYZED

### 2.1 Data Base

Shimizu Construction Co Ltd has kept systematic records of estimates on every building construction job it has undertaken. For each building the estimators recorded their estimation results on coding sheets according to an entry manual. A set of coding sheets contains about 2,000 different items, for which 12,350 bytes are computerized as a random-access file. This data base system was, however, suspended for the purpose of reviewing its recording procedures and items. The period of time such estimates cover ranges from March 1970 through December 1983.

### 2.2 Limitations and Conditions of Analyses

In planning fire safety, there are many fire protection items to be considered. In the data base, most of their quantities are counted in terms of number or area. It is not always possible to pick up the exact cost of each and every item because of certain limitations[1,2,3] in the data base system.

On the other hand, however, it is relatively easy to determine costs relating to fire protection where "electrical and mechanical" works are recorded as separate items under the limitations. In the analysis that follows, the ratio of the costs, and also building construction job, to estimated entire construction cost will be called cost index. The records of 359 office building, with 27 characteristics and 12 cost indices of fire protection equipment of a building, are chosen for the analysis from the system in which are entered 1,592 statistics on buildings.

## 3. CANONICAL CORRELATION ANALYSIS

The study of relations between a set of predictor variables and a set of criterion variables is know as Canonical Correlation Analysis (CCA). CCA is the most general of the multivariate techniques. In fact, the other procedures - multiple regression, discriminant function analysis, and MANOVA - are all special cases of it. CCA should be used in simultaneously analyzing several predictor variables and criterion variables. It is particularly appropriate when the criterion variables are themselves correlated. The Canonical Correlation Model (CCM), formulated from the CCA, is employed for two reasons: (1) to find a linear combination of the original predictor variables that best explain variation in the criterion variables and (2) to investigate relations between the two sets of variables by duly considering the canonical weights or canonical loadings. The canonical weights express the importance of a variable from one set with regard to the other set in obtaining a maximum correlation between sets. Thus, they are comparable with multiple regression weights.[4] The canonical loading gives the ordinary product-moment correlation of the original variable and its respective canonical variate. Thus, it reflects the degree to which a variable is represented by a canonical variate.[4]

Let "m" be the number of predictors and "p" be the number of criterion variables. The CCM is described as a set of predictor variables-$X_m^*$ (characteristics of buildings) and a set of criterion variables-$Y_p^*$ (fire protection equipment) as shown in Table 1. This can be expressed as --

$$Y_p^* = b_1 X_m^* + b_0 \tag{1}$$

where $b_0$ and $b_1$ are defined as regression coefficients.

$X_m^*$ and $Y_p^*$ of CCA can be defined as the following two-linear combinations of the "m" predictors and the "p" criterion variables:

$$X_m^* = c_1x_1 + c_2x_2 + \ldots + c_mX_m$$

$$Y_p^* = d_1y_1 + d_2y_2 + \ldots + d_pY_p$$

(2)

and

$$d_1y_1 + d_2y_2 + \ldots + d_pY_p = c_1x_1 + c_2x_2 + \ldots + c_mX_m$$

(3)

where $d_i$ (i = 1 ... 27) and $c_j$ (j = 1 ... 12) are defined as canonical weights.

The size or scale of canonical weights is influenced according to those of predictor and criterion variables because the variables under study possess different and arbitrary units and scales. To avoid such an influence, an arbitrary normalization of $x_{zi}$ (i = 1 ... 27) and $y_{zj}$ (j = 1 ... 12) are calculated according to the following:

$$x_{zi} = (x_i - \bar{x}_i) / S_{xi} \text{ (i = 1 ... 27)}$$

$$y_{zj} = (y_j - \bar{y}_j) / S_{yj} \text{ (j = 1 ... 12)}$$

(4)

where $x_{zi}$ and $y_{zj}$ are defined as normalized score (Z-score), $\bar{x}_i$ and $\bar{y}_j$ are mean, and $S_{xi}$ and $S_{yj}$ are standard deviation for the predictor and criterion variables. Therefore, $X_m^*$ and $Y_p^*$ have unit variance, that is mean $(X_m^*) = 0$ and similarly mean $(Y_p^*) = 0$; and $b_1$ in equation (1) is defined as a correlation coefficient between $X_m^*$ and $Y_p^*$; then $b_0 = 0$.

Following the previous data transformation, the canonical weights do not depend on the original scale of measurement, and are expressed in standardized form, given by --

$$d_1y_1 + d_2y_2 + \ldots + d_{12}y_{12} = c_1x_1 + c_2x_2 + \ldots + c_{27}x_{27}$$

(5)

## 4. RESULTS OF CANONICAL CORRELATION ANALYSIS

Table 1 shows the means, standard deviations and Box-and-Whisker plots of 27 characteristic variables, predictor variables, and 12 items of fire protection equipment, criterion variables, of a building. The Box-and-Whisker[5] plot is designed by J W Tukey to summarize the location of the bulk of the data with a box that covers the central 50% of the points, extending from the first to the third quartile. In addition, a "*" in the box points the median, and "whiskers" extend to the extreme points. The box shows the "body" of the data, and the whiskers portray the "tails" with suitably less visual impact.[6]

Table 2 shows that the canonical correlations are large (0.834 and 0.717), which implies that the canonical variate pairs are highly related. Figure 1 might show spuriously high canonical correlation because of the dot "α". Then additional analysis was conducted which eliminated a dot "α" and samples of which buildings were not required to install even one item of fire protection equipment listed in Table 1 according to relevant laws and regulations. The canonical correlations in the above-mentioned analysis are 0.814 and 0.638. Therefore, the canonical weights shown in Table 2 will be employed as a

Table 1. Means, Standard deviations and Box-and-Whisker plots of 39 variables of a building.

| variables | | means | standard deviations | Box-and-Whisker plots* |
|---|---|---|---|---|
| $x_1$ number of stories/basement | | 0.20 | 0.476 | |
| $x_2$ number of stories | | 3.98 | 2.378 | |
| $x_3$ building area | $(m^2)$ | 467.22 | 478.674 | |
| $x_4$ architectural area | $(m^2)$ | 1853.24 | 3033.737 | |
| $x_5$ cost index/building work | (%) | 26.73 | 4.860 | |
| $x_6$ /exterior finishing work | (%) | 14.55 | 6.962 | |
| $x_7$ /interior finishing work | (%) | 15.31 | 6.472 | |
| $x_8$ /other work | (%) | 1.59 | 3.486 | |
| $x_9$ /services | (%) | 25.30 | 7.864 | |
| $x_{10}$ estimation date ** | | -4.18 | 2.106 | |
| $x_{11}$ unit cost | ( 1000Yen/$m^2$) | 120.10 | 35.710 | |
| $x_{12}$ area index/fire escapes | (%) | 0.92 | 2.155 | |
| $x_{13}$ /balcony | (%) | 0.63 | 1.920 | |
| $x_{14}$ /public | (%) | 17.60 | 8.276 | |
| $x_{15}$ /control | (%) | 1.28 | 4.815 | |
| $x_{16}$ /services | (%) | 4.24 | 4.315 | |
| $x_{17}$ number of spans/ridge direction | | 4.00 | 2.106 | |
| $x_{18}$ /bay direction | | 2.53 | 1.372 | |
| $x_{19}$ typical span length/ridge direction(m) | | 6.56 | 1.954 | |
| $x_{20}$ /bay direction | | 8.34 | 2.693 | |
| $x_{21}$ number of structural bays | | 9.08 | 8.681 | |

1022

Table 1. (Continued)

| variables | | means | standard deviations | Box-and-Whisker plots* |
|---|---|---|---|---|
| $x_{22}$typical structural bay spacing | ($m^2$) | 51.46 | 18.147 | |
| $x_{23}$number of columns/external | | 53.42 | 49.427 | |
| $x_{24}$ /internal | | 14.04 | 26.308 | |
| $x_{25}$total length/external wall | (m) | 292.35 | 198.908 | |
| $x_{26}$ /girders | (m) | 809.12 | 946.430 | |
| $x_{27}$ /beam | (m) | 457.52 | 764.771 | |
| $y_1$ dry riser | | 0.17 | 0.344 | |
| $y_2$ sprinkler system | | 0.02 | 0.109 | |
| $y_3$ foam extinguishing system | | 0.06 | 0.365 | |
| $y_4$ Halon system | | 0.04 | 0.256 | |
| $y_5$ carbon dioxide extinguishing system | | 0.00 | 0.039 | |
| $y_6$ smoke ventilation system | | 0.07 | 0.251 | |
| $y_7$ independent electric power source | | 0.16 | 0.574 | |
| $y_8$ emergency power source | | 0.06 | 0.200 | |
| $y_9$ fire alarm indicating equipment | | 0.08 | 0.139 | |
| $y_{10}$automatic fire alarm system | | 0.29 | 0.406 | |
| $y_{11}$interlocking device by smoke detector | | 0.13 | 0.205 | |
| $y_{12}$fire detection and alarm warning | | 0.07 | 0.149 | |

** $x_{10}$=(Year-1977)+(Month-1)/12

* Box-and-Whisker plots

"H-spread"=difference between values of hinges.
"step"=1.5 times H-spread.
"inner fences" are 1 step outside hinges.
"outer fences" are 2 step outside hinges.(and thus 1 step outside of inner fences)
values beyond outer fences are "far out".

Figure 1. Scatter plot between the first canonical score X* (characteristics of a building) and Y* (cost indices of fire protection equipment).

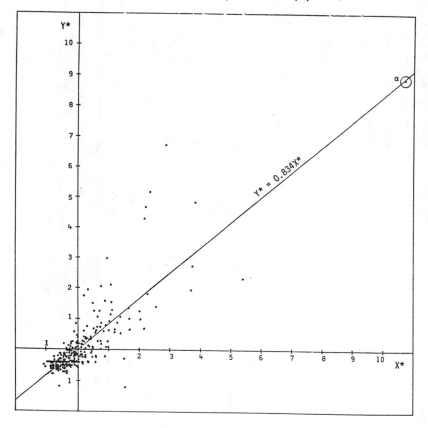

Table 2.  Canonical correlation analysis of relations between characteristics of a building and fire protection equipment.

| | | Variate 1 | | Variate 2 | |
|---|---|---|---|---|---|
| Variables | | Canonical loadings | Canonical weights | Canonical loadings | Canonical weights |
| **Predictor set     X*** | | | | | |
| $x_1$ number of stories/basement | | 0.71 | 0.24 | -0.20 | -0.37 |
| $x_2$ number of stories | | 0.61 | 0.07 | -0.26 | -0.38 |
| $x_3$ building area | (m²) | 0.68 | -0.03 | -0.27 | -0.71 |
| $x_4$ architectural area | (m²) | 0.92 | 0.02 | 0.03 | 0.24 |
| $x_5$ cost index/building work | (%) | 0.01 | 0.01 | 0.14 | -0.08 |
| $x_6$ /exterior finishing work | (%) | -0.19 | 0.03 | 0.06 | -0.06 |
| $x_7$ /interior finishing work | (%) | -0.22 | -0.04 | 0.31 | -0.00 |
| $x_8$ /other work | (%) | -0.02 | 0.04 | -0.10 | -0.10 |
| $x_9$ /services | (%) | 0.43 | 0.11 | -0.44 | -0.24 |
| $x_{10}$ estimation date | | -0.08 | -0.03 | 0.10 | 0.09 |
| $x_{11}$ unit cost | (×1000Yen/m²) | 0.19 | -0.09 | -0.14 | -0.08 |
| $x_{12}$ area index/fire escapes | (%) | 0.01 | -0.02 | -0.14 | 0.03 |
| $x_{13}$ /balcony | (%) | -0.03 | -0.01 | 0.04 | 0.07 |
| $x_{14}$ /public | (%) | -0.07 | -0.02 | -0.03 | -0.04 |
| $x_{15}$ /control | (%) | -0.06 | -0.01 | 0.06 | -0.01 |
| $x_{16}$ /services | (%) | 0.48 | 0.08 | -0.23 | 0.08 |
| $x_{17}$ number of spans/ridge direction | | 0.54 | 0.13 | -0.16 | 0.07 |
| $x_{18}$ /bay direction | | 0.46 | 0.06 | 0.07 | 0.04 |
| $x_{19}$ typical span length/ridge direction(m) | | 0.13 | 0.04 | -0.07 | 0.05 |
| $x_{20}$ /bay direction | | 0.20 | 0.01 | -0.38 | -0.12 |
| $x_{21}$ number of structural bays | | 0.70 | -0.03 | 0.11 | -0.09 |
| $x_{22}$ typical structural bay spacing | (m²) | 0.30 | 0.04 | -0.40 | -0.17 |
| $x_{23}$ number of columns/external | | 0.91 | 0.17 | -0.06 | 0.08 |
| $x_{24}$ /internal | | 0.87 | 0.06 | 0.03 | -0.08 |
| $x_{25}$ total length/external wall | (m) | 0.40 | -0.04 | -0.57 | -0.34 |
| $x_{26}$ /girders | (m) | 0.70 | -0.19 | -0.32 | -0.16 |
| $x_{27}$ /beam | (m) | 0.95 | 0.59 | 0.17 | 1.21 |
| Explained variance | | 25.9% | | 6.2% | |
| **Criterion set (cost index(%)) Y*** | | | | | |
| $y_1$ dry riser | | 0.39 | 0.18 | -0.49 | -0.34 |
| $y_2$ sprinkler system | | 0.84 | 0.66 | 0.44 | 0.86 |
| $y_3$ foam extinguishing system | | 0.30 | 0.19 | -0.10 | -0.13 |
| $y_4$ Halon system | | 0.29 | 0.02 | -0.40 | -0.44 |
| $y_5$ carbon dioxide extinguishing system | | 0.28 | -0.08 | -0.06 | -0.30 |
| $y_6$ smoke ventilation system | | 0.64 | 0.25 | -0.18 | -0.22 |
| $y_7$ independent electric power source | | 0.34 | 0.07 | -0.25 | -0.24 |
| $y_8$ emergency power source | | 0.50 | 0.17 | -0.32 | -0.07 |
| $y_9$ fire alarm indicating equipment | | 0.15 | 0.03 | -0.33 | -0.20 |
| $y_{10}$ automatic fire alarm system | | 0.27 | 0.09 | -0.29 | -0.06 |
| $y_{11}$ interlocking device by smoke detector | | 0.21 | 0.02 | -0.17 | -0.02 |
| $y_{12}$ fire detection and alarm warning | | -0.28 | -0.08 | 0.32 | 0.12 |
| Explained variance | | 18.0% | | 3.4% | |
| Canonical correlation | | 0.83 | | 0.72 | |
| Redundancy | | 0.18 | | 0.03 | |

criterion  when  designers  determine  which  fire  protection  equipment  will  be installed and how many items should be assigned to the building that they design.

Table 2 however, shows that the percentage of explained variance -- 25.9 and 6.2 percent for the criterion variables, and 18.1 and 3.4 percent for the predictor variables - are relatively small.    Moreover, only about 21 percent (18.0+3.2) of the variation in the $Y_{12}^*$-set is accounted for by the $X_{27}^*$-set variate.    Further, there are a number of algebraic sign reversals, and the rank ordering of variables varies substantially depending on whether the canonical weights or loadings are used.

Though canonical correlations suggest strong relations, and the structure of canonical loadings in these results demonstrates some similarity with those of the canonical weights, there are important differences due to multicollinearity.    Thus, it is difficult for the canonical weights and loadings to be employed to determine the structure or relation between the 27 characteristics of a building and the 12 cost indices of fire protection equipment.

Based on these canonical loadings, the following results might be offered: designers who evaluate the installation of sprinkler systems $(y_2)$ against the width within the buildings, total length of beams $(x_{27})$, architectural area $(x_4)$ and number of external columns $(x_{23})$ [variate 1] , and who decide to emphasise the installation of dry riser $(y_1)$ or a sprinkler $(y_2)$ by comparison between the width within the building, total length of external wall $(x_{25})$, and services cost index of the building $(x_9)$ [variate 2] .

## 5.  DISCUSSIONS

### 5.1 A Standard for Fire Protection Investment

In this section is interpreted application of the CCM to the design standard of how to apply the cost of fire protection equipment.

Table 3 shows characteristics of building "A" for the applied calculation and the result by utilizing the CCM.    Therefore, a standard for fire protection equipment is given by --

$$Y^*_{A12} = d_1 y_{z1} + d_2 y_{z2} + \ldots + d_{12} y_{z12} = 0.834 \times (-0.041) \tag{6}$$

The stability of the CCM, however, is questionable because of the extrapolation: estimation date $(x_{10})$ is the variable for price fluctuations.    As previously mentioned, the period of time that this estimation system covers ranges from March 1970 through December 1983, and calculation of a standard is extrapolated from 1984 retrogressively by the CCM.

Table 3.    Characteristics of buildings for applied calculation and the result by utilizing the CCM.

| Variables | | Buildings "A" | Buildings "B" | Z-score ·x Canonical weight "A" | Z-score ·x Canonical weight "B" |
|---|---|---|---|---|---|
| $x_1$ number of stories/basement | | 0.00 | 0.00 | -0.101 | -0.101 |
| $x_2$ number of stories | | 7.00 | 8.00 | 0.089 | 0.118 |
| $x_3$ building area | (m²) | 642.00 | 588.00 | -0.012 | -0.007 |
| $x_4$ architectural area | (m²) | 4704.00 | 4704.00 | 0.019 | 0.019 |
| $x_5$ cost index/building work | (%) | 26.73 | 26.73 | 0 | 0 |
| $x_6$ /exterior finishing work | (%) | 14.55 | 14.55 | 0 | 0 |
| $x_7$ /interior finishing work | (%) | 16.31 | 16.31 | 0 | 0 |
| $x_8$ /other work | (%) | 1.59 | 1.59 | 0 | 0 |
| $x_9$ /services | (%) | 25.30 | 25.30 | 0 | 0 |
| $x_{10}$ estimation date | | 5.00 | 5.00 | -0.131 | -0.131 |
| $x_{11}$ unit cost | ( 1000Yen/m²) | 120.10 | 120.10 | 0 | 0 |
| $x_{12}$ area index/fire escapes | (%) | 0.92 | 0.92 | 0 | 0 |
| $x_{13}$ /balcony | (%) | 0.63 | 0.63 | 0 | 0 |
| $x_{14}$ /public | (%) | 17.60 | 17.60 | 0 | 0 |
| $x_{15}$ /control | (%) | 1.28 | 1.28 | 0 | 0 |
| $x_{16}$ /services | (%) | 4.24 | 4.24 | 0 | 0 |
| $x_{17}$ number of spans/ridge direction | | 4.00 | 4.00 | 0 | 0 |
| $x_{18}$ /bay direction | | 3.00 | 3.00 | 0.021 | 0.021 |
| $x_{19}$ typical span length/ridge direction(m) | | 7.00 | 7.00 | 0.009 | 0.009 |
| $x_{20}$ /bay direction | | 8.00 | 7.00 | -0.001 | -0.005 |
| $x_{21}$ number of structural bays | | 12.00 | 12.00 | -0.010 | -0.010 |
| $x_{22}$ typical structural bay spacing | (m²) | 56.00 | 49.00 | 0.010 | -0.005 |
| $x_{23}$ number of columns/external | | 98.00 | 112.00 | 0.153 | 0.201 |
| $x_{24}$ /internal | | 42.00 | 48.00 | 0.064 | 0.077 |
| $x_{25}$ total length/external wall | (m) | 728.00 | 784.00 | -0.088 | -0.099 |
| $x_{26}$ /girders | (m) | 1624.00 | 1736.00 | -0.164 | -0.186 |
| $x_{27}$ /beam | (m) | 588.00 | 588.00 | 0.101 | 0.101 |
| 0.834*X* | | | | -0.034 | 0.002 |

## 5.2 A Trade-off for Fire Protection Investment

In this section is explained how application of the CCM to the trade-off between characteristics of the building and the fire protection equipment should be installed.

Table 3 also shows the characteristics of another building "B" for trade-off as compared with those of building "A". In this case, a standard for fire protection equipment is given by --

$$Y_{B27}^* = d1y_{z1} + d2y_{z2} + \dots + d12y_{z12} = 0.834 \times (0.002) \tag{7}$$

Let $y_{z10}$ (automatic fire alarm system) and $y_{z12}$ (fire detection and alarm warning) be the fire protection equipment for the trade-off, and assume that both items of fire protection equipment have similar effectiveness against fire protection, while other equipment is installed at the mean values. The following trade-off relations between building "A" and "B" can be defined --

$$0.091y_{z10} - 0.084y_{z12} = -0.034 \tag{8}$$

$$0.091y_{z10} - 0.084y_{z12} = 0.002 \tag{9}$$

where equation (8) is for building "A" and equation (9) is for building "B" in Table 3.

Now, to simplify this problem, either equipment of $y_{z10}$ or $y_{z12}$ should be selected. Under the above assumption, Table 5 shows the calculated cost indices of each item of fire protection equipment in case of one of above-mentioned two items is selected. The combination of building "A" and $y_{z12}$ (fire detection and alarm warning) would be a better solution in this case.

Table 5. Calculated cost indices for fire protection equipment.

| Buildings | Fire protection equipment | |
|---|---|---|
| | $y_{10}$ | $y_{12}$ |
| "A" | -0.036 | 0.014 |
| "B" | 0.124 | -0.050 |

## 6. CONCLUSIONS

The results of CCA are as follows:

1) The variate pairs between characteristics of a building and cost indices of fire protection equipment are highly related, and also are available for a standard and a trade-off for investment model of fire protection equipment.

2) Based on the canonical loadings, designers evaluate the installation of sprinkler systems against the width within the buildings, and decide to emphasize the installation of dry risers or sprinkler systems by comparison between the width and service cost index of the buildings.

REFERENCES

1.  Nakamura, H: "Analysis of the Fire Protection Cost Index", Geneva Association Seminar, 23/24, March 1983.

2.  Nakamura, H and Yashiro, Y: "Analyses of the Fire Protection Cost Index", 7th UJNR, Oct 1983.

3.  Nakamura, H: "Analyses of the Costs of Fire Protection in the Electrical and Mechanical Works, and Occupancy Clustering", Fire Sci and Tech vol 4, No 1 1984.

4.  Dillon, W and Goldsmith, M: Multivariate Analysis, John Wiley and Sons.

5.  Tukey, J W: Exploratory Data Analysis, Addison-Wesley Pub Co.

6.  "Statistical Data Analysis", Proc of Symposia in Applied Mathematics, American Mathematical Society.

7.  Langdon-Thomas, G L: Fire Safety in Building, Adam & Charles Black.

8.  Egan, M D: Concepts in Building Firesafety, John Wiley & Sons.

9.  Joedicke, J: Office Buildings, Crosby Lockwood & Sons Ltd.

10. Glossary of Terms Associated with Fire, BS4422 Part 4  1975, British Standards Institution.

# Risk Management Application of Fire Risk Analysis

**MARDYROS KAZARIANS and NATHAN SIU**
Pickard, Lowe and Garrick, Inc.
2260 University Avenue
Newport Beach, California, USA

**GEORGE APOSTOLAKIS**
School of Engineering and Applied Science
University of California
Los Angeles, California 90024, USA

ABSTRACT

Probabilistic risk assessment (PRA) has shown that the contribution of fires to the frequency of core damage and radionuclide release in some nuclear power plants can be significant. This article discusses the use of PRA results in fire risk management. The decomposition of these results leads to the identification of the most important contributors to the risk and, thus, allows for the identification of potential modifications that can have the greatest impact on risk. This paper discusses the process of generating these options and offers several insights that have been gained from an actual study.

Keywords: risk management, fire risk, nuclear power plants.

## 1. INTRODUCTION

Probabilistic risk assessment is a systematic approach to the quantification of the risk from complex industrial facilities and the identification of the major accident scenarios. PRA provides the means by which this risk, which may be economic as well as health-related, may be reduced. The decision to reduce this risk, and the process employed to achieve the desired reduction, fall within the realm of risk management. PRA, therefore, is a valuable source of information in terms of identifying and quantifying the impacts of various alternatives for the risk management process.

Recently, a number of PRAs have been performed for a variety of nuclear plants (Reference 1). In these PRAs, the risk is typically quantified in terms of the frequency of severe core damage and the frequencies of several public health effects; e.g., latent cancers. The conduct of a PRA requires the construction of a plant model; i.e., a logical representation of the plant that, using fault tree and event tree methods quantifies the response of the plant to a large number of disturbances (initiating events), including earthquakes, hardware failures, operator actions, and fires.

As a result of these studies, it has been shown, albeit with large uncertainties, that fires in a number of nuclear power plants can contribute significantly to those plants' total risk (e.g., References 2 and 3). More importantly, from a risk management standpoint, these studies indicate which fire scenarios are important and, therefore, what measures can be taken to decrease their importance in a verifiable manner. The ability to verify the impact of a candidate risk-reducing measure is an important risk management consideration, and will be further discussed in Section 2.

In the case of interest, one such plant-specific risk study showed that the mean frequency of fire-initiated accidents leading to core damage was about 20% of the total core damage accident frequency. Similarly, the same study showed that fires in that plant contribute roughly 50% to the total frequency of accidents leading to large-scale releases of radionuclides from the plant. The methodology employed to estimate the fire risk involves the evaluation of the frequency of each specific fire scenario in terms of the fire location, initial fire severity, fuel bed characteristics, suppression characteristics, and availability of additional plant systems that can mitigate the effect of the scenario on the plant's functioning (References 4 through 6).

From the standpoint of facility management, the results of PRAs are of interest when applied to the following questions:

- What possible options are there for reducing risk?
- How effective (and believable) are these options in reducing the risk?
- How desirable are these options?

This paper addresses these questions. It does not, however, address the selection of an optimal alternative; this decision is the province of management, which must weigh the benefits and disadvantages of each alternative with respect to economic, regulatory, operational, and risk considerations.

2. METHODOLOGY

2.1 Major Contributors to Risk

The different layers of a nuclear plant PRA's results are shown in Figure 1 (adapted from Reference 7). The decomposition process starts from the top (level 1), the final result, or risk curve level. Level 2 reveals the important release type (which characterizes how the radionuclides are released over time) where importance is measured in terms of degree of contribution to the risk curves. Level 2 also indicates the important degraded plant conditions (i.e., the plant damage states that lead to these release types), and the initiating events (e.g., a fire in a cable spreading room) that lead to these damage states. Level 3 identifies the important sequences of events (the accident scenarios) that lead to the various plant damage states. In levels 4 through 6, the important system and component failures, the causes of these failures, and the data supporting the quantified analysis of the frequencies of these causes and failures are respectively identified.

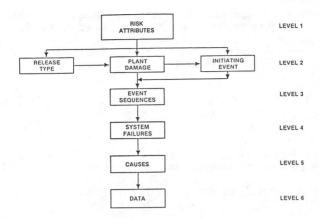

FIGURE 1. A graphical representation of the risk decomposition process (adapted from Reference 7)

In the following section, the fire risk analysis methodology employed in the base case PRA is described, giving attention to those factors (and groupings of factors) that may affect our choice of risk management options.

2.2 Fire Risk Analysis Methodology

The methodology for analyzing fire risk in nuclear power plants has been developed specifically to handle the fire-related characteristics of those plants. These plants typically consist of several large concrete buildings that are subdivided into rooms (fire zones) with relatively thick concrete walls and floors. Almost all fire zones contain numerous electrical cables in trays and conduits. In addition to cables, a zone may contain pumps, electrical switching gear, control panels, batteries, and/or piping and valves. The combustible loading of a typical zone is relatively low when compared to that of other commercial structures, such as office buildings.

For protection against the occurrence and consequences of a fire, nuclear plants must be designed and operated according to several strict regulations. The most important fire-related regulation is the Appendix R to Code Federal Regulation Title 10, Part 50 (Reference 8). As a result of these safety considerations and the low combustible loadings in a plant, fires have the potential to initiate serious accident sequences only in a small number of plant locations.

As described in References 4 through 6, the fire risk analysis methodology for these plants proceeds as follows. First, the potentially important fire scenarios are identified by establishing the exact locations of the components whose simultaneous failure may have a severe impact on the plant. For each such location, different fire scenarios (involving fires of varying severities at or near the location) are postulated and their frequencies of occurrence are quantified using both statistical data and judgment. The fraction of fires that damage the components is established by modeling the physical effects of fires. This involves identifying fire propagation patterns, modeling the components' thermal responses, and computing the likelihood of fire suppression before the important components are damaged. Finally, to assess the impact of the fire on the plant, the likelihood of the failure of components outside of the fire zone due to other causes, such as maintenance, mechanical failure, and human actions, is assessed. Thus, the frequency of plant damage of type X from a fire scenario can be written as

$$\lambda_X = \sum_j \lambda_j Q_{d|j} Q_{X|d,j} \tag{1}$$

where

$\lambda_j$      = annual frequency of fires of class j, where the classes are determined both by the location and initial severity of the fire.

$Q_{d|j}$    = fraction of class j fires which lead to damage to a specified set of components.

$Q_{X|d,j}$ = fraction of class j fires causing damage to the specified set of components that lead to plant damage type X.

The fraction of fires that lead to damage, $Q_{d|j}$, is the fraction of fire scenarios where component damage from fire growth occurs prior to fire suppression. This can be written as (Reference 6)

$$Q_{d|j} = Fr\{T_G < T_H\} \tag{2}$$

where Fr{A} denotes the frequency of occurrences of event A, $T_G$ is the time it takes for the fire to grow and damage the important components, and $T_H$ is the total time required to detect and suppress the fire.

The third parameter of Equation (1), $Q_{X|d,j}$, is generally a function of numerous system and component unavailabilities, as well as operator action frequencies; indeed, the major portion of a nuclear plant PRA is dedicated to establishing these relationships. A very simple representation of these complex equations is

$$Q_{X|d,j} = Q_{CD|d,j} \; Q_{X|CD,d,j} \tag{3}$$

where

$Q_{CD|d,j}$    = fractions of class j fires that include additional component failures that lead to core damage.

$Q_{X|CD,d,j}$ = fraction of those fires that include additional failures that lead to damage state X.

An example for an event characterized by $Q_{X|CD,d,j}$ is the failure of containment cooling-related equipment. This failure does not influence the likelihood of core damage, but has a profound impact on the severity and the type of radionuclide release.

2.3 Development of Fire Risk Management Options

From the preceding discussion, it can be seen that the fire risk analysis methodology employed allows a decomposition of the risk down to the bottom level of Figure 1. Equation (1) represents a summation of risk contributions of different fire-initiated accident sequences (level 3); each product in Equation 1 contains terms that represent the level 4 system and component failures due to fire and to other causes, (level 5), and the analysis procedure used to quantify each term identifies the level 6 contribution to risk. Thus, the primary contributors to risk at each level can be identified, and the impacts of alternatives intended to reduce the risk at each level can be evaluated.

To reduce the risk from a fire of specified initial severity and location, options can be chosen to reduce any one or more of the factors shown in Equation (1). For example, the likelihood of fire occurrence may be decreased by increasing administrative controls over the movement of combustibles and the performance of maintenance-related activities in the location of interest. The likelihood of component damage can be addressed by slowing down the fire growth rate, say by reducing combustible loadings or by installing fire barriers, or by speeding up the rate of detection and suppression, e.g., by installing automatic sprinklers. The likelihood of further component failures, given a certain number of fire-induced failures, can be decreased by increasing the redundancy of important equipment in areas independent of the location of interest.

A somewhat different risk management option is suggested by the fact that the exact value of each term in Equation (1) is not known with certainty. If the uncertainties in any term of a dominant product in Equation (1) are very large, a potential risk-reducing measure may be a more detailed analysis of those terms characterized by large uncertainties. Thus, if the risk from a scenario involving an uncertain fire damage threshold for electrical cables is large, it may be more efficient to test the cables under appropriate conditions than to actually make changes in the plant. Of course, it is not certain that a reduction in risk will result from further analysis; thus, this measure may be less desirable than others that guarantee some degree of reduction.

It is important to realize that the uncertainties in the quantification of the effectiveness of each risk management option must be included as an integral part of the analysis. If the expected reduction in risk from a particular alternative is small with respect to the uncertainty bands about the original value, or the new value, there clearly will be doubts as to the actual effectiveness of the alternative.

3. CASE STUDY

3.1 Base Case Decomposition

Only three potentially significant fire scenarios were identified and explicitly analyzed in the base case. Because of this, the event sequence structure of Equation (1), the decomposition of the base case results from level 1 (where fires, as a group of events, are recognized to contribute significantly to the total risk) to level 4 (the "system failure" level) is straightforward.

In the base case study, two of the three fire scenarios analyzed contributed significantly to the total plant risk. We refer to the two fire zones housing these contributing scenarios as zones 1 and 2; the total contributions from these two zones as well as the breakdowns of these contributions, are given in Table 1. It can be seen that, in both cases, the state-of-knowledge uncertainties in the component frequencies can be very large.

1033

TABLE 1. Summary of the case study decomposition results

| Zone Designator/ Scenario | Percentile | Frequency, Events Per Year | | $\lambda_j$ | $Q_{d\mid j}$ | $Q_{CD\mid d,j}$ | $Q_{R\mid CD,d,j}$ |
|---|---|---|---|---|---|---|---|
| | | $\lambda_{CD}$* | $\lambda_R$** | | | | |
| 1. Fire Under Cables Damaging Switch-Gears and Power Cables to Component Cooling and Safety Injection Pumps | 5th 50th 95th Mean | 4.6-8 7.9-6 4.2-4 7.1-5 | 4.6-8 7.9-6 4.2-4 7.1-5 | 1.1-7 1.3-5 3.7-4 1.2-4 | 0.32 0.62 0.90 0.62 | 1.0 | 1.0 |
| 2. Fire in the Aisle Damaging Power Cables to Component Cooling and Safety Injection Pumps | 5th 50th 95th Mean | 5.5-8 4.7-6 1.0-4 2.4-5 | 5.5-8 4.7-6 1.0-4 2.4-5 | 1.2-7 8.4-6 1.6-4 4.2-5 | 0.20 0.55 0.87 0.57 | 1.0 | 1.0 |
| 3. Fire on the Floor Damaging Control Cables 10 Feet Above the Floor and Failing All Control and Instrumentation Capability | 5th 50th 95th Mean | 3.0-10 7.3-8 3.3-6 1.9-6 | <1.0-10 7.3-9 5.9-7 3.0-7 | 5.3-7 3.3-5 5.0-4 1.5-4 | 0.12 0.45 0.80 0.48 | 2.5-4 5.0-3 1.0-1 2.6-2 | 0.02 0.1 0.5 0.16 |

*Core damage frequency; $\lambda_{CD} = \lambda_j Q_{d\mid j} Q_{CD\mid d,j}$.

**Radionuclude release frequency; $\lambda_R = \lambda_j Q_{d\mid j} Q_{CD\mid d,j} Q_{R\mid CD,d,j}$.

NOTE: Exponential notation is indicated in abbreviated form; i.e., 4.6-8 = $4.6 \times 10^{-8}$.

The contributions to risk from fires in other zones (other than the three that were analyzed) were judged to be much smaller, due to a variety of reasons (e.g., independent critical equipment lie outside of the zone of interest). Although the contributions from these latter fire scenarios are not important when quantifying the base case risk, they do become visible when the risk from the dominant scenarios is reduced. In other words, as the magnitude of one particular problem is lowered, other, formerly less important problems, become more noticeable. This interesting observation indicates that PRAs cannot be completely bottom-line oriented if they are to be used in risk management.

A resulting task from this observation, therefore, is the quantification of the risk contributors from the "next level" of fire scenarios. The total mean frequency of core damage ($\lambda_{CD}$) for the next 17 fire zones is estimated to be $6.0 \times 10^{-6}$ per reactor year; the corresponding release frequency ($\lambda_R$) is $8.9 \times 10^{-7}$ per reactor year.

3.2 Identification of Potential Options

Four general categories of the fire risk management options can be identified, based on the discussion of Section 2.3. They are options to:

1. Improve the models employed in the original analysis.

2. Reduce the frequency of occurrence and the potential severity of fires in the critical location of interest.

3. Reduce the likelihood of important component damage, given a fire.

4. Reduce the likelihood of subsequent plant system failures, given the loss of important components.

The first option stems from the recognition that the original PRA was performed under specific time, budget, and state-of-knowledge constraints. It is conceivable that further investigation of the critical fire scenarios will result in the identification and elimination of conservatisms in the original analysis. On the other hand, it is also conceivable that further study will corroborate the earlier results or may even identify some optimistic assumptions. However, because of the uncertainty in the outcome of this option, and because the then current state of knowledge in fire risk modeling did not allow any simple improvements in the analysis (in other words, a major research project would have to have been undertaken), this option was not pursued.

The frequency of fire in a particular zone ($\lambda_j$) may be reduced by changing the design of equipment in the zone or by improving administrative procedures. For example, zone 1 of our case study contains oil-lubricated compressor sets. If the system is changed such that oil need not be brought into the room, the likelihood of a severe fire would be decreased. Another option would be to introduce a permanent fire watch into the zone. None of these options was pursued in this study, primarily because significant decreases in the fire frequency could not be verified. We note that current administrative procedures for fire prevention in nuclear power plants are already quite strict, and that the long-term effectiveness of using a fire watch to prevent the occurrence of very rare fires (see Table 1) is dubious.

From Equation (2), it can be seen that the likelihood of component damage can be reduced by increasing the growth time, $T_G$, or by reducing the hazard time (the sum of the detection and suppression times), $T_H$. The installation of protective barriers serves the former purpose and is relatively cheap. This option will be further discussed in Section 3.3.

Improvements to the existing automatic detection and suppression system would reduce $T_H$, but are fairly expensive. The possibility of spurious suppression system actuation also reduces the desirability of this option from an operations and maintenance point of view. Furthermore, detailed detection and suppression time models needed to formally analyze the improvement in $T_H$ (Reference 5) were not available at the time the analysis was performed. This option, therefore, was not investigated further.

The last set of potential modifications concentrates on improving the plant system response to the loss of critical components due to fire; i.e., on reducing $Q_{x|d,j}$. Since the essential characteristic of the dominant fire scenarios is that a single fire damages a large number of important components and thereby renders unavailable many safety systems, a natural solution is to ensure that one or more critical safety systems are entirely independent of the dominant fire zone. In this case study, zones 1 and 2 contain the power cables for the "component cooling" and the "safety injection" pumps. If these cables are damaged by a fire, these pumps would lose all power, the "charging pumps," which are cooled by the component cooling system, would be lost, and severe core damage would eventually result. Two relatively efficient plant modifications to mitigate the effects of a severe fire in either zone are (1) the installation of an independently cooled and powered charging pump that does not depend on the component cooling system and (2) the provision of an alternate electrical power source for the component cooling system.

## 3.3 Fire Barriers

The installation of fire barriers in zones 1 and 2 is intended to perform essentially the same function as the plant system modifications discussed in Sections 3.4 and 3.5; the barriers are to render the selected power cables for the component cooling and safety injection pumps independent (or nearly so) of the remainder of the two zones. However, because the barriers are supposed to prevent, rather than mitigate, component damage, their effectiveness is modeled in the analysis of $Q_{d|j}$ instead of $Q_{X|d,j}$.

The fire barriers considered for this option are thermal insulating boards composed of noncombustible material, and are about 1.3 cm thick. These barriers would enclose the cable tray holding the power cables to one of the three pumps of both the component cooling and the safety injection systems. The same type of barriers would also separate redundant switchgear cabinets. In zone 3, the barriers are to extend the length of the room and enclose two sets of three cable trays. This would protect the power cables for two of three component cooling and safety injection pumps.

To evaluate the impact of these barriers, the same procedure as used for the original study is employed. The thermal calculations underlying the analysis of $Q_{d|j}$ indicate that not only is the time to damage longer with the installation of the barriers, but the initial severity of the initiating fire must be greater as well. Thus, this modification also leads to a reduction in $\lambda_j$. The results of this analysis are given in Table 2.

TABLE 2. Summary of the case study final results

| Option | Description | Percentile | $\lambda_{CD}$ Events Per Year | Reduction Factor | $\lambda_R$ Events Per Year | Reduction Factor |
|--------|-------------|------------|-------------------------------|------------------|-----------------------------|------------------|
| 0 | Base Case | 5th | $2.2 \times 10^{-6}$ | | $1.5 \times 10^{-6}$ | |
| | | 50th | $3.0 \times 10^{-5}$ | | $2.6 \times 10^{-5}$ | |
| | | 95th | $1.1 \times 10^{-3}$ | | $9.7 \times 10^{-4}$ | |
| | | Mean | $1.0 \times 10^{-4}$ | 1.0 | $9.6 \times 10^{-5}$ | 1.0 |
| 1 | Fire Barriers | 5th | $5.9 \times 10^{-7}$ | | $2.1 \times 10^{-7}$ | |
| | | 50th | $9.1 \times 10^{-6}$ | | $7.4 \times 10^{-6}$ | |
| | | 95th | $2.3 \times 10^{-4}$ | | $2.1 \times 10^{-4}$ | |
| | | Mean | $3.9 \times 10^{-5}$ | 2.6 | $3.3 \times 10^{-5}$ | 2.9 |
| 2 | Self-Contained Charging Pump | 5th | $1.6 \times 10^{-6}$ | | $3.6 \times 10^{-7}$ | |
| | | 50th | $8.8 \times 10^{-6}$ | | $3.4 \times 10^{-6}$ | |
| | | 95th | $9.9 \times 10^{-5}$ | | $9.2 \times 10^{-5}$ | |
| | | Mean | $1.9 \times 10^{-5}$ | 5.3 | $1.2 \times 10^{-5}$ | 8.0 |
| 3 | Alternate Power Source | 5th | $1.7 \times 10^{-6}$ | | $5.7 \times 10^{-7}$ | |
| | | 50th | $7.1 \times 10^{-6}$ | | $3.0 \times 10^{-6}$ | |
| | | 95th | $4.8 \times 10^{-5}$ | | $2.6 \times 10^{-5}$ | |
| | | Mean | $1.4 \times 10^{-5}$ | 7.1 | $6.9 \times 10^{-6}$ | 14.0 |

## 3.4 Self-Contained Charging Pump

A diesel engine-driven charging pump that does not require any external plant systems for motive power or cooling is proposed as an addition to the original charging system. The pump must be located in an area that does not contain any portions of the component cooling system. The new pump would provide cooling to crucial plant components in the case of fire damage to the component cooling system.

To analyze the risk reduction of this modification, the unavailability of this pump is calculated from data for existing diesel engine-driven pumps. This unavailability is then multiplied with the original value of $Q_{X|d,j}$ for those zones where total loss of component cooling due to a fire is possible. The final results are summarized in Table 2.

## 3.5 Alternate Power Source

The main purpose of this modification is to provide an alternate source of electrical power to some of the important pumps that can be affected by fires in zones 1 and 2. The modification can be implemented using an existing switchgear in an independent zone as the source of power; new power cables are to be routed from this switchgear, through areas outside of zones 1 and 2, to one component cooling pump and one safety injection pump. The unavailability of this new power source is computed by taking into consideration possible equipment failures and the potential errors that the personnel could make during the hookup. Also accounted for is the time window available to them to correct their errors. Personnel errors are the main contributors to the unavailability of the alternative power source. The risk impact of this modification is shown in Table 2.

## 3.6 Discussion of Options

As can be seen from Table 2, all three options provide a measurable reduction in risk. It can also be seen that the alternate power source option is the most effective of the three. The fire barrier option is less effective, primarily because of the analysis uncertainties in quantifying the frequency of very severe fires. The diesel-driven charging pump option is more effective than the fire barrier option, but it has less impact than the alternate power source option because the unavailability of the diesl engine is somewhat greater than the operator error rate in attaching the alternate feed cables. Coincidentally, the alternate power source option also was the most desirable alternative for both the plant operations and licensing personnel. This option was eventually chosen for implementation by the plant management.

## 4. CONCLUDING REMARKS

In this paper, some of the desired features of a PRA that enhance its use in a fire risk management study have been addressed. These features include the ease and extent to which the results can be disassembled to determine the principal contributors to risk, the inclusion of sufficient detail to allow analysis of the effectiveness of various risk-reducing options, and the complete treatment of uncertainties to express our confidence in the results. These uncertainties can significantly affect the rankings of a number of alternatives. If the uncertainties in the risk-reducing benefits of a particular modification (e.g., administrative improvements in the control of combustibles) are sufficiently large, the decision makers may opt for a somewhat more expensive change (e.g., hardware additions), whose benefit is more clear cut.

The case study identification and analysis of fire risk management options from the results of a nuclear power plant PRA serves to underline these arguments. For example, the necessity to quantify risk sequences of secondary importance indicates that purely bottom-line oriented studies may require additional work before they may be used to evaluate the effectiveness of the various options.

5. REFERENCES

1. Garrick, B. J., "Recent Case Studies and Advancements in Probabilistic Risk Assessment," Risk Analysis, Vol. 4, pp. 267-279, March 1984.

2. Pickard, Lowe and Garrick, Inc., Westinghouse Electric Corporation, and Fauske & Associates, Inc., "Indian Point Probabilistic Safety Study," prepared for the Power Authority of the State of New York and Consolidated Edison Company of New York, Inc., March 1982.

3. Wood-Leaver and Associates, Inc., "Probabilistic Risk Assessment, Big Rock Point Plant," prepared for Consumers Power Company, Jackson, Michigan, March 1981.

4. Kazarians, M., N. Siu, and G. Apostolakis, "Fire Risk Analysis for Nuclear Power Plants: Methodological Developments and Applications," Risk Analysis, Vol. 5, No. 1, pp. 33-51, 1985.

5. Siu, N., and G. Apostolakis, "Models for the Detection and Suppression of Fires in Nuclear Power Plants and Some Related Statistical Problems," Draft Report UCLA-ENG-8440, December 1984.

6. Apostolakis, G., "Some Probabilistic Aspects of Fire Risk Analysis for Nuclear Power Plants," presented at the First International Symposium on Fire Safety Science National Bureau of Standards, Gaithersburg, Maryland, October 9-11, 1985.

7. Garrick, B. J., "Examining the Realities of Risk Management," Society for Risk Analysis, 1984 Annual Meeting, Knoxville, Tennessee, September 30-October 3, 1984.

8. U.S. Nuclear Regulatory Commission, "Title 10, Part 50, Code of Federal Regulations," Appendix R, February 17, 1981.

# Some Probabilistic Aspects of Fire Risk Analysis for Nuclear Power Plants

GEORGE APOSTOLAKIS
School of Engineering and Applied Science
University of California
Los Angeles, California 90024, USA

ABSTRACT

The development of a methodology for the quantification of the risks associated with fires in nuclear power plants requires the use of judgment, deterministic models and statistical evidence. Several issues that arise are discussed including: 1. The impact of design changes, plant-to-plant variation and other qualitative factors on the assessment of the frequency of fires in specific locations; 2. The importance of distinguishing among statistical, parameter and modeling uncertainties, when fire propagation is studied; 3. How parameter and modeling uncertainties can be assessed; and 4. How the probability distribution of detection and suppression times can be determined.

INTRODUCTION

The evaluation of the risk associated with fires in nuclear power plants has attracted considerable attention in the last few years. While it was believed earlier, on the basis of crude estimates, that this risk was very small and, therefore, dominated by other risks, more detailed analysis showed that the risks from fires could be significant. For example, fires constitute the second (after seismic) most important contribution to core melt frequency at Indian Point 2 (1). These fires (occurring at specific locations) could damage vital power cables, thus causing a small loss of coolant accident through the failure of the reactor coolant pump seals. At the same time, several other power cables could fail incapacitating vital safety components (safety injection pumps, containment spray pumps and fan coolers). The mean frequency of core melt due to fires was found to be $2.0 \times 10^{-4}$ per reactor year, while the 5th and 95th percentiles were $6.0 \times 10^{-6}$ and $7.6 \times 10^{-4}$ per reactor year.

The preceding example is fairly typical of the findings of a fire risk analysis for nuclear power plants and illustrates several of its salient features. The methodology is focused on the effects of fires on control and power cables and traces the impact of their failure on the plant. Furthermore, the scenarios that are identified have very low frequencies and special care must be taken to display the uncertainties. We have found that the Bayesian interpretation of probability (2-3) provides a convenient framework for the explicit and rigorous treatment of the uncertainties and the subjective judgments that are necessary.

The purpose of this paper is to discuss several of the statistical issues that arise when a fire risk analysis is carried out. Since the basic methodology has been presented elsewhere (4-5), its details will not be presented here. While this methodology has been developed for nuclear power plants, the issues that we discuss are relevant to any effort to quantify the risks associated with fires in any large and complex technological system. This belief is based on the realization that these issues are the consequences of two features that all such systems share: 1. Fires must occur at the "right" location to seriously threaten the safety of the system, and 2. The events of interest are rare, therefore, conventional statistical evidence is very weak and the analysts must devise methods that extract the maximum possible amount of information from this evidence.

OVERVIEW OF THE METHODOLOGY

The methodology is centered around the following three tasks:

1.   The identification of the "critical" locations and the assessment of the frequency of fires.

2.   The estimation of fire growth times and the competing detection and suppression times.

3.   The response of the plant.

A location is declared as "critical", when the occurrence of a fire there has the potential of creating an abnormal condition in the plant and damaging the engineered safety functions (ESF). Thus, tasks 1 and 3 are related. As stated earlier, only fires that affect control and power cables are considered. Figure 1 shows a situation, where the cable trays are elevated. The plant analysis has already determined that trays A, B and C carry cables, which, if failed, would lead to undesirable consequences. The occurrence of a large fire on the floor, as shown, could fail these cables, therefore, this location merits a detailed analysis, and is "critical". In other situations the cable trays are closer to the floor and one worries about a fire starting on one tray and propagating to the others.

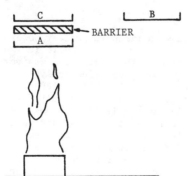

Figure 1.   Sample Room Configuration.

In the second step the time $T_G$ that it takes for the fire to propagate and damage the cables of the trays is estimated using the computer code COMPBRN (6-7). This code uses empirical correlations for such quantities as the fire's mass burning rate and flame height to characterize the fire as a heat source. It also employs heat and mass balances to determine the extent and temperature of the layer of hot gases accumulating near the room ceiling. The heat transfered to objects within the room and the room boundaries is then calculated and the thermal response of these receivers determined. Damage is assumed to occur when the surface temperature of an object exceeds a predefined limit.

The detection and suppression times are representd by the "hazard" time $T_H$, i.e, the fire actually burns for a time $T_H(8)$. The potential occurrence of fire damage to components is modeled as the outcome of a competition between fire growth and fire suppression processess over time. The conditional frequencey of damage, given that the fire occurs, is

$$Q = \mathrm{Fr}\{T_G < T_H \mid \mathrm{fire}\} \tag{1}$$

where $\mathrm{Fr}\{A/C\}$ denotes the frequency of occurrence of event A conditioned on the occurrence of event C. Eq. (1) simply says that the frequency of damage, given a fire, equals the frequency by which the growth time is smaller than a hazard time, i.e, the time it takes to detect and suppress the fire.

The last step of the analysis (plant response) is highly plant specific and requires intimate knowledge of the plant. It essentially entails a detailed investigation (using, usually, fault and event trees) of whether the damaged cables can prevent the successful termination of the neutron chain reaction and whether the heat that is produced in the core after the shutdown (the "decay" heat) is successfully removed, thus avoiding a melt down. We note that the plant response is affected not only by the components damaged by the fire, but also by the components which are unavailable due to other causes (e.g., maintenance).

The analysis described above is carried out for all the critical locations. The total frequency of core melt due to fires is, then, the sum of the frequency contribution from each location, i.e.,

$$\lambda_{CM\text{-}Fire} = \sum_j \lambda_j Q_{d \mid j} Q_{CM \mid d,j} \tag{2}$$

where

$\lambda_j$ = frequency of fires of class j, where the classes are determined both by the location and initial severity of the fire (step 1)

$Q_{d \mid j}$ = fraction of class j fires which lead to damage to a specified set of components (step 2, see Eq. (1)), and

$Q_{CM \mid d,j}$ = fraction of class j fires causing damage to the specified set of components which lead to core melt (step 3).

THE FREQUENCY OF FIRES

Modeling Needs

The first factor in eq. (2) to be estimated is the frequency of fires. The difficulties of doing so become evident, when we go back to Fig. 1 and realize the $\lambda_j$ represents fires that occur in the "right" location, i.e., beneath the cable trays or thereabout, and are of sufficient magnitude to cause damage to the cables. Each such situation is unique and it is impossible to find statistical evidence which is directly applicable. However, evidence regarding "similar" rooms or equipment in other plants is available and must be utilized. Of course, what are similar rooms can not be defined precisely. The model which has been used considers $\lambda_j$ as the product of several factors, as follows:

$$\lambda_j = \lambda_{C_j} \; f_{C_j,L} \; f_{C_j,L,S} \tag{3}$$

where

$\lambda_{C_j}$      = annual frequency of fires in class $C_j$ of rooms or equipment

$f_{C_j,L}$      = fraction of these fires that occur in the appropriate location.

$f_{C_j,L,S}$    = fraction of these fires that have the appropriate severity.

Each of these quantities is discussed below.

## The Frequencies of Fires in Classes of Rooms or Equipment

The classes $C_j$ are obtained by dividing the plant into generic buildings, rooms, or equipment. Reference 9 presents one such division in which the frequencies of fires in auxiliary buildings, turbine buildings, control rooms, and cable spreading rooms are evaluated. It also presents the frequency of fires in reactor coolant pumps ( in a pressurized water reactor) per year of operation and the unavailability of a diesel generator unit due to fires. The general approach is to develop these distributions from Bayes' theorem (3), in which the prior distribution is noninformative and the likelihood function is determined by the statistical evidence. The latter creates problems of both interpretation and use.

Design and procedure changes may create questions about the applicability of some of the operating experience. For example, one plant's cable spreading rooms have experienced three relay fires within 5 months in the motor control center (MCC) cabinets that are installed there (9). These fires were due to a generic defect in certain relays. Corrective actions have since been taken in those units and in other plants that had the same kind of relays. Thus, we suspect that those fires were unique to that station at that time and their likelihood of recurrence has been substantially reduced. The question is, then, how many fires we should include in our data when we estimate the frequency of fires in these rooms. In this case the three incidents were counted as one, i.e., only the cause of these fires was counted. Of course, the design changes have, presumably, eliminated this cause, however, one can look at this situation from the point of view that there is a class of failure causes that may occur and elimination of one specific cause does not eliminate the class as a whole.

In addition to the questions of interpretation of the evidence, another issue is whether all rooms or equipment in a class should be treated as being identical, when, in fact, there may be significant plant-to-plant variation. The plant-to-plant variability of the occurrence rates can be very important, especially when the event of interest has occurred in some plants but not in the majority of plants. Formal methods for developing generic distributions which explicitly model this variability have been developed (10). An example that shows how significantly different the results may be deals with the occurrence of fires in control rooms (9). The statistical evidence is of the form $\{k_i, T_i\}$, where $k_i$ is the number of fires over $T_i$ control room years at plant i.

$T_i$ is in the range of 1 to 20 years, while $k_i$ is zero for all but one $i$. Using formal methods we can derive a generic distribution for the frequency of fires which reflects the plant-to-plant variability and has characteristic values:

5th percentile:  $1.3 \times 10^{-7}$ per control room year

50th percentile:  $3.2 \times 10^{-4}$ per control room year

95th percentile:  $1.5 \times 10^{-2}$ per control room year

mean:  $4.9 \times 10^{-3}$ per control room year

If, on the other hand we ignore the variablity of the rate of fires over plants and we say that the evidence is $k = \sum_i k_i = 1$ fire over $T = \sum_i T_i = 453$ control room years, then the resulting characteristic values are:

5th percentile:  $1.2 \times 10^{-4}$ per control room year

50th percentile:  $1.5 \times 10^{-3}$ per control room year

95th percentile:  $6.6 \times 10^{-3}$ per control room year

mean:  $2.2 \times 10^{-3}$ per control room year

This distribution is much narrower than the first one. The plant-to-plant curve, being broader, has a higher 95th percentile ($1.5 \times 10^{-2}$) and a mean value which is more than twice that of the second distribution. While the first model is more realistic (in the sense that it allows for differences among plants), it is the second model that is used by most analysts, mainly because of lack of appreciation of the difference that the plant-to-plant variability can make.

Specialization Factors

The factors $f_{C_j,L}$ and $f_{C_j,L,S}$ of eq. (3) have to be assessed using judgment, since they are highly situation specific and statistical evidence does not exist. The ability of analysts to quantify their judgment in terms of probabilities has been studied extensively (for some observations pertinent to risk analysis see (3)). While it has been found that people, in general, are not very good at it, it has also been determined that the use of formal methods and training can improve this ability to a significant degree. Essentially, the assessors must try to develop numerical estimates, which fit "coherently" (11) in their body of knowledge and beliefs. Several considerations which generally help in the present context are:

1.  The number of rooms that the buildings of class $C_j$ have. While all rooms are not necessarily equally likely to have the fire occur there, this number is a useful input.

2.  The contents of the rooms. For example, rooms containing oil or machinery with moving parts are more likely to have fires.

3.  Even within a room a fire must occur in specific locations to cause damage, e.g., in Fig. 1, the fire must occur beneath the cable trays. This argument is, of course, intimately related to the types and quantities of fuel that could be reasonably assumed to be introduced into the room. This is a particularly thorny issue, since the amounts of oil, for instance, that have to be assumed present are not found, under normal conditions, in these rooms. On the other hand, there have been instances in which several gallons of oil have been left inadvertently in controlled areas.

4. The frequency of visits by plant personnel.

Even though considerations of this kind do help the analysts, the fact remains that this is a "soft" and controversial part of the analysis.

FIRE GROWTH

As stated earlier, the computer code COMPBRN is used to calculate the time to damage a specified component, $T_G$. This time we call the deterministic reference model (DRM) estimate. The question, now, is what kinds of uncertainties are associated with this number.

The uncertainties associated with modeling of physical processes can be thought of as being of two kinds: the statistical uncertainty and the state-of-knowledge uncertainty. The statistical uncertainties stem from the random nature of fire; if we perform a particular experiment a large number of times under 'identical' conditions and measure $T_G$, we will obtain a frequency distribution. Our physical model cannot predict the environmental fluctuations which lead to this distribution. These uncertainties are inherent in the nature of fire. Even if our knowledge of fire were to increase dramatically, we would not be able to reduce these uncertainties significantly. Such an increase in knowledge, however, would markedly reduce our state-of-knowledge uncertainties.

There are two kinds of state-of-knowledge uncertainty: the parameter uncertainty and the modeling uncertainty. The parameter uncertainty is due to insufficient knowledge about what the input to the code should be. The modeling uncertainty is due to simplifying assumptions and the fact that the models used may not accurately model the true physical process.

The DRM is a collection of approximate models which are valid under certain conditions. The modeling uncertainty is due not only to our uncertainty in the accuracy of each model's predictions under the conditions they were developed for, but also to our uncertainty in the synthesis of these independent models. One of our primary concerns is whether or not the synthesis contains enough component models (i.e., if all important phenomena have been modeled).

The three types of uncertainty that we have discussed, i.e., statistical, parameter and modeling uncertainty, must be expressed in terms of probability distributions, which will be used in the risk assessment. The development of these distributions requires, again, substantial judgment, since conventional statistical evidence is lacking. It is judged that the variation of the estimated propagation time is dominated by state-of-knowledge uncertainties. This means that the uncertainties in the models and the parameter values are considered to overwhelm the statistical uncertainty, an assumption which is reasonable in the light of the current state of the art.

Most of the parameters that must be used as input to COMPBRN are either fixed by the problem, e.g., its geometry, or are known to a satisfactory degree. On the other hand, significant uncertainties (at least to the extent that they must be represented by probability distributions) exist regarding the thermal and combustion properties of cable insulation and jacket material. The principal sources of uncertainty are the lack of information regarding the composition of these materials and of experimental evidence.

To propagate the parameter uncertainties, either Monte Carlo methods or response surface methods can be used. The purpose of the response surface is to replace the expensive computer code with a much simpler analytical expression, that is, to represent the output of a computer code by a simple function of its inputs. The form of the specific response surface depends on the specific application. The coefficients in the response surface are estimated by using the data from a few runs of the computer code.

To treat the modeling uncertainties, Ref. 6 proposes to use an "error" or "uncertainty" factor $E_\tau$, which is multiplied with the code prediction to yield the actual growth time. Because the exact value of $E_\tau$ is unknown for an arbitrary scenario, we work with its probability distribution. In Reference 6, $E_\tau$ is assumed to be lognormally distributed with a 5th percentile of 0.8 and a 95th percentile of 4.0, thus recognizing that COMPBRN yields generally conservative results. These percentiles are loosely based upon comparison of early simulations of cable tray fires with experimental values and, mainly, expert opinion.

The modeling uncertainty must, of course, be revised as the code is updated or as more evidence becomes available. For example, a recent modification of COMPBRN consists of the inclusion of heat losses due to radiation and convection, at a component's surface (these losses were neglected in the first version of the code), and the relaxation of the assumption that components are damaged only when they are ignited (12). Consequently, a new distribution for the uncertainty factor must be used in connection with this new version of the code.

The presence of both parameter and modeling uncertainties has an additional consequence. When experimental results become available, it is not clear how the code can be used to simulate them. Ideally, of course, we would like to know all the exact values of the parameters of the experiments, in which case the modeling approximations of the code can be tested. In practice, however, such is not the case and how exactly one should proceed is not well understood.

DETECTION AND SUPPRESSION

The issues that arise in developing the distribution of the hazard time $T_H$ of eq. (1) are, to a large extent, similar to those that we have discussed. This, of course, is not surprising, since the root causes are the same, i.e., the data are sparse and vague. Often, however, we can infer qualitative levels of a few characteristics. For example, we may be able to roughly judge the size of a fire (at detection) based upon what type of fuel (which determines the intensity) and what type of component (which determines the burning area) was involved. Similarly, the growth rate of the fire can be assessed by determining the fuel involved and the ignition source.

CONCLUDING REMARKS

We have discussed several of the issues that arise when the risk associated with fires in nuclear power plants is quantified. These issues occur naturally, when one attempts to develop a methodology, which combines deterministic models, statistical evidence, and judgment. The methodology is relatively new and has not had the benefit of extensive applications and peer reviews. As such, it is continually evolving; nevertheless, its basic elements appear to be sound and its use provides useful insights, which can be used in decision making concerning plant modifications (13).

ACKNOWLEDGMENT

This work has been sponsored by the Division of Risk Analysis of the U.S. Nuclear Regulatory Commission. I thank B. Buchbinder for his continuing support. Two of my former graduate students, M. Kazarians and N. Siu, who participated in the early stages of the project and made substantial contributions, deserve special mention.

REFERENCES

1.  Indian Point Probabilistic Safety Study, Consolidated Edison Company of New York, Inc. and Power Authority of the State of New York, May 1982.

2.  Kaplan, S., and Garrick, B.J., "On the Quantitative Definition of Risk," Risk Analysis, 1:11-37, 1981.

3.  Apostolakis, G., "Data Analysis in Risk Assessment", Nuclear Engineering and Design, 71: 375-381, 1982.

4.  Apostolakis, G., Kazarians, M., and Bley, D.C., "Methodology for Assessing the Risk from Cable Fires," Nuclear Safety, 23: 391-407, 1982

5.  Kazarians, M., Siu, N.O., and Apostolakis, G., "Fire Risk Analysis for Nuclear Power Plants: Methodological Developments and Applications," Risk Analysis, 5: 33-51, 1985

6.  Siu, N.O., and Apostolakis, G., "Probabilistic Models for Cable Tray Fires," Reliability Engineering, 3: 213-227, 1982.

7.  Siu, N.O., "Physical Models for Compartment Fires," Reliability Engineering, 3: 229-252, 1982.

8.  Siu, N. and Apostolakis, G., "Modeling the Detection and Suppression of Fires in Nuclear Power Plants," Presented at the American Nuclear Society/European Nuclear Society Topical Meeting on Probabilistic Safety Methods and Applications, San Francisco, California, February 24 - March 1, 1985.

9.  Kazarians, M. and Apostolakis, G., "Modeling Rare Events: The Frequencies of Fires in Nuclear Power Plants," in: Low-Probability/High-Consequence Risk Analysis, Waller, R.A., and V.T. Covello, Editors, Plenum Press, New York, 1984.

10.  Kaplan, S., "On a 'Two-Stage' Bayesian Procedure for Determining Failure Rates from Experimental Data", IEEE Transactions on Power Apparatus and Systems, PAS-102: 195-202, 1983.

11.  de Finetti, B., Theory of Probability, Vols. 1 and 2, John Wiley & Sons, New York, 1974.

12.  Chung, G., Siu, N., and Apostolakis, G., "Improvements in Compartment Fire Modeling and Simulation of Experiments," Nuclear Technology, 69: 14-26, 1985

13.  Kazarians, M., Siu, N.O., and Apostolakis, G., "Application of Nuclear Power Plant Fire Risk Analysis to Risk Management," Presented at the First International Symposium on Fire Safety Science National Bureau of Standards, Gaithersburg, Maryland, October 9-11, 1985.

# A Probabilistic Method for Optimization of Fire Safety in Nuclear Power Plants

DIETMAR HOSSER and WOLFGANG SPREY
König und Heunisch, Consulting Engineers
Oskar-Sommer-Strasse 15-17, D-6000 Frankfurt/Main 70, FRG

ABSTRACT

As part of a comprehensive fire safety study for German Nuclear Power Plants*) a probabilistic method for the analysis and optimization of fire safety has been developed. It follows the general line of the American fire hazard analysis, with more or less important modifications in detail. At first, fire event trees in selected critical plant areas are established taking into account active and passive fire protection measures and safety systems endangered by the fire. Failure models for fire protection measures and safety systems are formulated depending on common parameters like time after ignition and fire effects. These dependences are properly taken into account in the analysis of the fire event trees with the help of first-order system reliability theory. In addition to frequencies of fire-induced safety system failures relative weights of event paths, fire protection measures within these paths and parameters of the failure models are calculated as functions of time. Based on these information optimization of fire safety is achieved by modifying primarily event paths, fire protection measures and parameters with the greatest relative weights. This procedure is illustrated using as an example a German 1300 MW PWR reference plant. It is shown that the recommended modifications also reduce the risk to plant personnel and fire damage.

INTRODUCTION

From 1982 to 1984 a comprehensive theoretical and experimental study on fire safety in nuclear power plants /1/ was conducted by several German research institutes. The work was sponsored by the Federal Minister of the Interior (BMI) and was coordinated by the Gesellschaft für Reaktorsicherheit (GRS).

One of the main aims of the study was the development of a method for analysing quantitatively fire hazards in critical plant areas in order to

- compare the fire risk with the risk due to other internal or external events
- detect weak points in fire safety concepts
- reduce fire risk by more efficient combinations of fire safety measures
- make fire safety measures more efficient by influencing the most important parameters.

*) Optimization of fire safety measures and quality control in nuclear power plants. Study SR 144/1, sponsored by the German Federal Minister of the Interior, 1982 - 1984

At the beginning, American methods for fire hazard analysis /2/ and fire risk analysis (e. g. /3, 4 /) were studied. These methods seemed to be less appropriate for German nuclear power plants because

- the German fire safety concept is mainly based on physical separation of systems and less on fire suppression measures
- the fire effects on fire protection measures and safety systems are not explicitly taken into account
- the dependences between single failures due to the time-dependent fire effects are not clearly treated in the event tree analyses.

Therefore, a somewhat modified methodology based on first-order reliability theory was developed consisting of:

- the assessment of time-dependent fire event trees
- the definition of simplified failure models for fire protection measures and safety systems to be used in reliability analyses
- the analysis of the fire event trees with the help of first-order system reliability methods
- the optimization of fire protection measures based on the results of the event tree analyses.

The latter two steps will be illustrated using as an example a German 1300 MW PWR reference plant.

TIME-DEPENDENT FIRE EVENT TREES

The risk-orientated investigations in /1/ started with the selection of areas in a typical German PWR plant, in which potential fire hazards could endanger safety systems or plant personnel. For these areas event sequences induced by the occurrence of an initial fire were established. Similar to /2/ different protective measures are provided to detect and suppress a fire or to limit the effects of a fire on safety sytems and personnel to the compartment affected (Fig. 1). From experience the most probable times of actuation (after ignition) with lower and upper bounds can be estimated for all active fire protection measures.

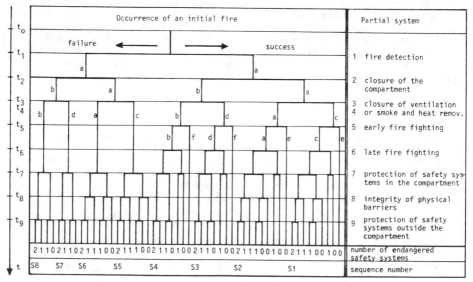

Fig. 1   Time-dependent fire event tree

Depending on success or failure of the active fire protection measures different time-histories of fire effects are expected. In Fig. 2 a set of temperature-time-histories is shown with the following boundary conditions:

curve a - normal conditions, compartment closed, fixed forced ventilation rate, no fire suppression
curve b - at least one door open (higher ventilation rate), no fire suppression
curve c - like a, but ventilation stopped at time $t_3$
curve d - like b, but ventilation stopped at time $t_3$
curve e - like a, but fire suppression started at time $t_5$
curve f - like b, but fire suppression started at time $t_5$.

If fire suppression measures are properly designed and actuated in due time the temperature decrease is so fast that curves e and f can be neglected in the analysis of consequences.

One of the above mentioned temperature-time-histories is assigned to each branch of the event tree in Fig. 1. Depending on the respective temperature at the time of demand failures of fire protection measures or safety systems due to fire can occur. Therefore, the consequences of a fire in a plant area depend on time, too. In the analysis of the event tree the frequencies of critical consequences, e. g. failure of one redundancy of safety systems in the fire compartment and failure of a physical barrier between two compartments and failure of a second redundancy in the adjacent compartment, are checked at varying time steps t*.

FAILURE MODELS FOR FIRE PROTECTION MEASURES AND SAFETY SYSTEMS

In order to account for dependences due to the time-dependent fire effects the single failures are described with the help of simplified mechanical models. The models are constructed as follows:

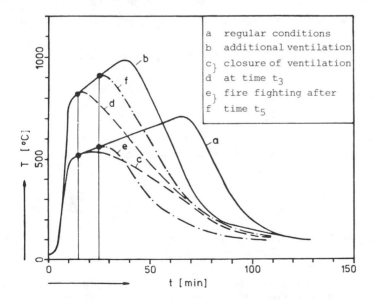

Fig. 2  Typical temperature-time-histories in NPP compartments

The active fire protection measures are divided in 6 "partial systems" as indicated in Fig. 1:

- fire detection and alarm
- closure of openings in the compartment boundary
- closure or shut down of compartment ventilation
- removal of smoke and heat
- early fire fighting inside the compartment
- late fire fighting from outside the compartment.

These partial systems are composed of "components" which act in parallel or series arrangements. The failure frequencies of the partial systems can be derived from failure rates of the components using fault tree models (e.g. Fig. 3).

Failure rates of the "components" are only partly known from statistical data. Especially, the portion of failures due to fire or late actuation is not sufficiently well covered by data. Therefore, simplified limit-state models are formulated and treated with the help of first-order reliability theory. For the partial system "early fire fighting" shown in Fig. 3, the limit-state functions are as follows:

$$P_{51} = P \{Z_{51} \leq - \beta_{51}\}$$

$\beta_{51}$ = standardized Gaussian variable calibrated with statistical data for $P_{51}$

$$P_{52} = P \{Z_{52} \leq - \beta_{52}\}$$

$\beta_{52}$ = analogical to $\beta_{51}$

$$P_{53} = P \{Z_{53} \leq T_{RM} - T (t_5)\}$$

$T_{RM}$ = ultimate temperature (°C) for manual fire suppression

$T(t_5)$ = gas temperature (°C) at time $t_5$

$t_5$ = $t_1 + \Delta t_5$
= time of fire detection + delay from detection to arrival of fire emissary

$$P_{54} = P \{Z_{54} \leq - \beta_{54}\}$$
$\beta_{54}$ = analogical to $\beta_{51}$

$$P_{55} = P \{Z_{55} \leq T_{RL} - T (t_5)\}$$
$T_{RL}$ = ultimate temperature (°C) for fire suppression system

$$P_{56} = P \{Z_{56} \leq - \beta_{56}\}$$
$\beta_{56}$ = analogical to $\beta_{51}$

$$P_{57} = P \{Z_{57} \leq t^* - t_5 - \Delta t_5^*\}$$
$t^*$ = varying time for checking the consequences

$\Delta t_5^*$ = duration of fire fighting until success.

The failure frequencies of passive fire protection measures (physical barriers) and safety systems depend strongly on fire effects, especially on gas temperature, which are functions of the time after occurrence of the initial

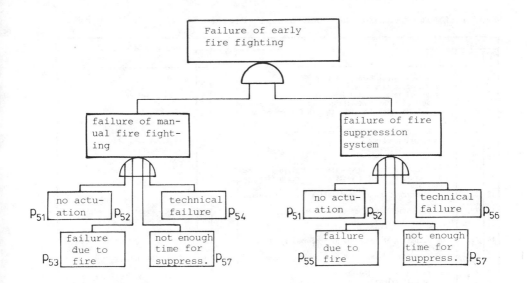

Fig. 3  Fault tree for failure of early fire fighting within the compartment

fire. For safety systems (mechanical and electrical components) ultimate gas temperatures have to be specified during design, based on fire test or experience. Passive fire protection measures are usually tested in standard fire tests, e. g. in Germany according to DIN 4102 /6/. The fire resistance of these measures does not only depend on the gas temperature but also on the time of action of the temperature; therefore, the time-integral of the standard fire curve up to the fire resistance time is taken as ultimate limit of fire resistance. In /1/ it was shown that this ultimate limit is valid not only for standard fires but also for natural fires in the compartments under consideration (cf. /7/).

All limit-state definitions used for the "components" of active fire protection measures as well as for passive fire protection measures and safety systems are summarized in Tab. 1. The parameters influencing the limit-states are random variables which are described by distribution parameters (cf. Tab. 2). The limit-states are dependent, due to common parameters.

SYSTEM RELIABILITY ANALYSIS

The fire event tree in Fig. 1 can be treated like a technical system consisting of the different event sequences with the same consequence in series arrangement. Within each event sequence the partial systems according to the preceding section are arranged in parallel. Finally, the partial systems act as parallel or series system with several "components". The state of the overall system can be formulated with the help of Boolean algebra. Alternatively, the system state can be directly related to the states of the individual "components" or it can be determined indirectly using intermediate systems (i. e. event sequences) or partial systems as a kind of macro-components.

In the following, mainly the second "subsystem method" is used because of its advantages with respect to calculation effort and interpretation of the results. Because of the above mentioned dependences between some of the single components the classical methods for fault tree and event tree analysis are not

Tab. 1   Limit-states of the event tree in Fig. 1

| no. | name | failure for $Z_i \leq 0$ | fire curve | no. | function |
|---|---|---|---|---|---|
| 1 | $Z_{11}$ | no manual direct fire detection | | | |
| 2 | $Z_{12}$ | no automatic fire detection | | | |
| 3 | $Z_{13}$ | no manual indirect fire detection | | 1 | fire detection |
| 4 | $Z_{14}$ | no automatic indirect fire detection | | | |
| 5 | $Z_{15}$ | no indirect fire detection through failure of components | | | |
| 6 | $Z_{21}$ | Opening in physical separation not closed | | | |
| 7 | $Z_{22}$ | no automatic closure | | 2 | physical separation of |
| 8 | $Z_{23}$ | technical failure | | | the compartment |
| 9 | $Z_{24}$ | no manual closure | | | |
| 10 | $Z_{31}$ | no closure of the air ventilation through personnel | | | |
| 11 | $Z_{32}$ | no closure of the air ventilation through fire emissary | | | |
| 12 | $Z_{33}$ | no closure of the air ventilation from the control room | | | |
| 13 | $Z_{34a}$ | no automatic closure of the air ventilation | a | 3 | closure of the air ventilation |
| 14 | $Z_{34b}$ | | b | | |
| 15 | $Z_{35}$ | technical failure | | | |
| 16 | $Z_{41}$ | no actuation of smoke and heat removal system by personnel | | | |
| 17 | $Z_{42}$ | no actuation of smoke and heat removal system by fire emissary | | 4 | smoke and heat removal |
| 18 | $Z_{43}$ | no switch over of the air ventilation | | | |
| 19 | $Z_{44}$ | technical failure | | | |
| 20 | $Z_{51}$ | no early fire fighting through personnel | | | |
| 21 | $Z_{52}$ | no early fire fighting through fire emissary | | | |
| 22 | $Z_{53a}$ | failure of manual fire fighting due to fire | a | | |
| 23 | $Z_{53b}$ | | b | | |
| 24 | $Z_{53c}$ | | c | | |
| 25 | $Z_{53d}$ | | d | 5 | early fire fighting in the compartment |
| 26 | $Z_{54}$ | technical failure of the manual fire fighting | | | |
| 27 | $Z_{55a}$ | failure of the fire suppression system due to fire | a | | |
| 28 | $Z_{55b}$ | | b | | |
| 29 | $Z_{55c}$ | | c | | |
| 30 | $Z_{55d}$ | | d | | |
| 31 | $Z_{56}$ | technical failure of the fire suppression system | | | |
| 32 | $Z_{57}$ | not enough time for fire suppression | | | |
| 33 | $Z_{61}$ | not enough time for late fire fighting | | 6 | late fire fighting from |
| 34 | $Z_{64}$ | technical failure | | | outside the compartment |
| 35 | $Z_{71a}$ | failure of safety systems in the compartment | a | | |
| 36 | $Z_{71b}$ | due to fire effects | b | 7 | protection of safety systems |
| 37 | $Z_{71c}$ | | c | | in the compartment |
| 38 | $Z_{71d}$ | | d | | |
| 39 | $Z_{81a}$ | failure of physical barriers due to fire effects | a | | |
| 40 | $Z_{81b}$ | | b | 8 | integrity of physical |
| 41 | $Z_{81c}$ | | c | | barriers |
| 42 | $Z_{81d}$ | | d | | |
| 43 | $Z_{91a}$ | failure of safety systems in an adjacent compartment due to fire effects | a | | |
| 44 | $Z_{91b}$ | | b | 9 | protection of safety systems |
| 45 | $Z_{91c}$ | | c | | in an adjacent compartment |
| 46 | $Z_{91d}$ | | d | | |
| 47 | $Z_{92}$ | late fire spread to an adjacent compartment or redundancy | | | |

applicable; i. e. the frequencies of the overall system states cannot be calculated by multiplying (for intersections) or summing up (for unions) the component or macro-component state frequencies. Therefore, first-order system reliability methods are used which are based on proposals in /7-10/. Only very few aspects of these methods can be discussed here.

As shown before, the states of all single components are described by state functions $Z_i$ (cf. Tab. 1) where

$Z_i \leq 0$: failure of the component

$Z_i > 0$: success of the component.

Tab. 2 Random basic variables for the limit-states of Tab. 1

| no. | name | Random variables significations |
|-----|------|------------------------------|
| 1 | $p_{11}$ | failure probability of personnel in the compartment |
| 2 | $p_{12}$ | failure probability of automatic alarm in the compartment |
| 3 | $p_{13}$ | failure probability of personnel in adjacent compartments |
| 4 | $p_{14}$ | failure probability of automatic alarm in adjacent compartments |
| 5 | $p_{15}$ | probability of not recognizing component failures |
| 6 | $p_{21}$ | probability of physical separations being not closed |
| 7 | $p_{22}$ | failure probability of automatic closure |
| 8 | $p_{23}$ | probability of a technical failure |
| 9 | $p_{32}$ | probability of air ventilation being not closed by the fire emissary |
| 10 | $p_{33}$ | failure probability of actuation from the control room |
| 11 | $p_{35}$ | probability of smoke and heat removal system not being actuated by personnel |
| 12 | $p_{42}$ | probability of smoke and heat removal system not being actuated by f. emissary |
| 13 | $p_{43}$ | probability of air ventilation not being switched over |
| 14 | $p_{44}$ | probability of technical failure |
| 15 | $p_{54}$ | failure probability of manual fire fighting equipment |
| 16 | $p_{56}$ | failure probability of fire suppression system |
| 17 | $p_{62}$ | failure probability of equipment for indirect fire suppression |
| 18 | $T_S$ | actuation temperature of solder |
| 19 | $\Delta T_A$ | temperature difference between compartment and exhaust air duct |
| 20 | $\Delta T_Z$ | temperature difference between compartment and supply air duct |
| 21 | $\Delta T_{KN}$ | temperature difference between compartment and safety system in adjacent area |
| 22 | $T_{RM}$ | ultimate temperature of the manual fire fighting |
| 23 | $T_{RL}$ | ultimate temperature of the fire suppression system |
| 24 | $T_{RK}$ | ultimate temperature of safety systems in the compartment |
| 25 | $T_{RKN}$ | ultimate temperature of safety systems in the adjacent compartment |
| 26 | $\int T\,dt$ | temperature capacity of the physical barriers |
| 27 | $T_{10}$ | |
| 28 | $a_1$ | |
| 29 | $a_{A1}$ | parameters to describe the temperature-time-history a |
| 30 | $\alpha_{A1}$ | |
| 31 | $t_{A1}$ | |
| 32 | $T_{20}$ | |
| 33 | $a_2$ | |
| 34 | $a_{A2}$ | parameters to describe the temperature-time-history b |
| 35 | $\alpha_{A2}$ | |
| 36 | $t_{A2}$ | |
| 37 | $a_{A3}$ | - " - c |
| 38 | $\alpha_{A3}$ | |
| 39 | $a_{A4}$ | - " - d |
| 40 | $\alpha_{A4}$ | |
| 41 | $t_1$ | time of fire alarm |
| 42 | $\Delta t_3$ | delay from alarm to closure of air ventilation (arrival of fire emissary) |
| 43 | $\Delta t_4$ | delay from alarm to actuation of the smoke and heat removal system |
| 44 | $\Delta t_5$ | delay from alarm to actuation of fire fighting |
| 45 | $\Delta t_5$ | duration of fire fighting until successful suppression |
| 46 | $\Delta t_6$ | delay from direct to indirect fire fighting |
| 47 | $\Delta t_6^*$ | duration of indirect fire fighting |
| 48 | $t_W$ | delay from initial fire to fire spread into an adjacent area |
| 49 | $t^*$ | varying time for checking the consequences of the fire |

If $Z_i$ is a function of a parameter vector $\underline{X}$ according to Fig. 4 and each parameter is known with its probability distribution, then e. g. the probability of component failure

$$p_{fi} = P\,(Z_i\,(\underline{X}) \leq 0) \tag{1}$$

can be calculated by a first-order reliability method. The basic principle of the applied method is to transform the limit-state $Z_i$ into a linear function of uncorrelated standardized Gaussian variables. Then the probability distribution $\Phi_{Zi}$ is standardized Gaussian, too and can easily be determined; the probability of failure is $\Phi_{Zi}\,(Z_i = -\beta_i)$ where $\beta_i$ is the so-called safety index. The contributions of the random variations of the parameters $\underline{X}_i$ to the safety index $\beta_i$ are given by so-called weighting factors $\alpha_{Xi}$ which are calculated during linearization of the limit-state following an idea in /8/.

The weighting factors $\alpha_{\underline{X}_i}$ are an appropriate means for identifying the relative importance of the parameters $\underline{X}_i$ for a limit-state under consideration. They help also to evaluate the degree of correlation between two limit-states $Z_i$ and $Z_j$ with commom parameters $\underline{X}$ because the correlation coefficient $\rho_{ij}$ is simply

$$\rho_{ij} = \sum_{k=1}^{n} \alpha_{ik} \circ \alpha_{jk} \qquad (2)$$

Now, the conditions for system analysis are as follows:

- All components of the system are described by state functions $Z_i$.
- The safety indices $\beta_i$ and the weighting factors $\alpha_{\underline{X}_i}$ have been calculated separately for each limit-state
- The correlation coefficients $\rho_{ij}$ for each couple of two limit-states $Z_i$ and $Z_j$ are determined with Eq. (3)
- The states of the partial systems with components in parallel and series arrangement have to be analyzed as intersections and unions of correlated component states, e. g. for failure $F_5$ of partial system no. 5 "early fire fighting" according to Fig. 3:

$$F_5 = \{(Z_{51} \leq 0) \cup (Z_{52} \leq 0) \cup (Z_{53} \leq 0) \cup (Z_{54} \leq 0) \cup (Z_{57} \leq 0)\}$$

$$\cap \{(Z_{51} \leq 0) \cup (Z_{52} \leq 0) \cup (Z_{55} \leq 0) \cup (Z_{56} \leq 0) \cup (Z_{57} \leq 0)\}$$

- The state of the overall system has to be evaluated as intersection of the states of different event sequences defined as unions of the states of the correlated partial systems, e. g. for the consequence "loss of two redundancies of safety systems" according to Fig. 1:

$$F = \{S1 \cup S2 \cup S3 \cup \ldots \cup S8\}$$

with

$$S1 = \{\bar{F}_1 \cap \bar{F}_2 \cap \bar{F}_3 \cap F_5 \cap F_6 \cap F_7 \cap F_8 \cap F_9\}$$

To analyze the state probabilities approximate solutions of the multinormal probability integral on the basis of /9/ for intersections and /10/ for unions are applied. By using equivalent linearizations according to /7/ for partial systems, intermediate systems and the overall system , equivalent safety indices and equivalent weighting factors can be evaluated for all these systems. These values are very helpful for interpreting the results of such complex system analyses.

## OPTIMIZATION OF FIRE PROTECTION MEASURES

The optimization of fire protection measures and quality controls in nuclear power plants can have different aims, e. g.:

- minimization of the total of construction cost, control and maintenance cost and damage cost for a given fire safety level
- minimization of the frequency of fire-induced consequences for given total cost
- reduction of the frequency of fire-induced consequences with the help of more effective fire protection measures.

Since the information on the different cost contributions was very poor the more pragmatic third aim was chosen for the optimization in /1/. A good basis for the assessment of fire protection measures are the results of the system reliability analyses. They show clearly

- which plant area is critical with respect to the consequences of a fire for reactor safety, plant personnel or plant operation

- which protective measures really reduce the frequency of the consequences or limit the damage cost
- which parameter has the greatest influence on the efficiency of the most important protective measures.

The fire safety level in uncritical plant areas should be chosen according to conventional requirements. In critical areas a higher fire safety level seems to be reasonable in order to minimize the consequences of a fire. Fire protection measures which are expensive but unreliable should be avoided. Also protective measures without any influence on frequency or extend of consequences are unreasonable. The best way to increase the efficiency of fire protection measures is by variation of parameters with the greatest relative weight.

APPLICATION TO A REFERENCE PLANT

The methods described in the preceding sections were applied in /1/ to a German 1300 MW PWR reference plant in order i) to demonstrate the efficiency of the methodology, ii) to check the completeness of the available input data and to study the influence of uncertain data, iii) to assess the fire safety concept and identify relative weak points and iv) to derive recommendations for the optimization of fire protection measures and related quality controls.

For all selected plant areas the frequencies $p_f$ of critical fire-induced consequences were calculated as functions of time after occurrence of the initial fire. In most cases exists a maximum of $p_f$ indicating the most critical situation during a fire. The decrease of $p_f$ after the maximum results either from the cooling phase of the fire or from the effect of fire suppression; the closure of the air ventilation has no influence because of the unreliable actuation. Beside the frequency $p_f$ also the time-dependent squared weighting factors $\alpha_i^2$ (equivalent values related to the overall system) were determined. For illustration, the frequency $p_f$ and weighting factors $\alpha_i^2$ from the analysis of the area of the main cooling pumps in the reactor building containment are depicted in Fig. 4. The main impact of the fire on safety systems comes from fire-induced failures of electrical equipment. Only the "regular" temperature-time-history a is of interest. The most critical situation is reached in an early stage of the fire when fire fighting by the fire suppression system is not yet manually actuated.

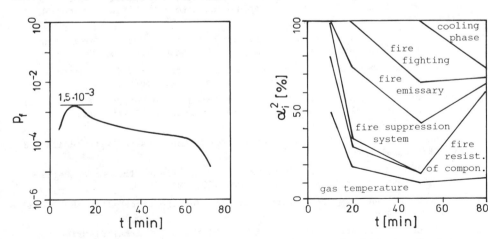

Fig. 4  Time-dependent frequency $p_f$ and squared weighting factors $\alpha_i^2$ of parameters for the event "fire-induced loss of the main cooling pumps"

Important parameters at this stage are the gas temperature and the ultimate temperature of the electrical equipment in this area. About 10 min. after the occurrence of the initial fire the early fire fighting by the suppression system becomes effective and $p_f$ is reduced. Further reductions of $p_f$ come from the effect of the late fire fighting and the beginning of the cooling phase of the fire. As the failure of the electrical equipment of all main cooling pumps is to be expected at an early time, the only way to reduce the failure frequency $p_f$ is to actuate immediately the fire suppression system, either manually from the control room after checking the situation by TV monitors or automatically by fire detectors. Such a modification would also reduce the risk to plant personnel and limit the damage due to spreading corrosive smoke.

CONCLUSIONS

A probabilistic method for the quantitative evaluation of fire hazards in nuclear power installations has been developed. It is based on fire event sequences which depend on success or failure of different active or passive fire protection measures. Single failures of these measures and of safety systems endangered by a fire are dependent events due to the common influence of the fire effects. With the help of a first-order reliability method the dependences can be modelled and properly taken into account in the event tree analysis. Beside frequencies of undesired consequences of event sequences, relative weights of event sequences, fire protection measures and parameters influencing the measures are determined. Based on such information weak points in fire safety concepts can easily be identified and optimal combinations of fire protection measures for a required fire safety level can be recommended.

REFERENCES

/1/ ABK/GRS: Optimization of fire safety measures and quality control in nuclear power plants. Final report of the research project SR 144/1 sponsored by the Federal Minister of the Interior. December 1984 (in German).

/2/ Berry, D. L. and E. E. Minor: Nuclear Power Plant Fire Protection - Fire Hazard Analysis (Subsystems Study Task 4). NUREG/CR-0654, September 1979.

/3/ Fleming, K. N., W. J. Houghton, F. P. Scarletta: A Methodology for Risk Assessment of Major Fires and its Application to a HTGR Plant. General Atomic Company, July 1979.

/4/ Gallucci, R. and R. Hockenbury: Fire-induced Loss of Nuclear Power Plant Safety Functions. Nuclear Engineering and Design 64 (1981), 135 - 147.

/5/ DIN 4102: Brandverhalten von Baustoffen und Bauteilen; Teil 2: Bauteile, Begriffe, Anforderungen und Prüfungen. Ausgabe September 1977.

/6/ Schneider, U. und D. Hosser: Reliability based Design of Structural Members. First International Symposium on Fire Safety Science. Gaithersburg, MD (USA), October 9 - 11, 1985.

/7/ Hohenbichler, M. and R. Rackwitz: First-Order Concepts in System Reliability. Structural Safety 1 (1983).

/8/ Hasofer, A. M. and N. C. Lind: An Exact and Invariant First-Order Reliability Format. Journal Eng. Mech., ASCE, 100, EM1, 1974.

/9/ Breitung, K.: An Asymptotic Formula for the Failure Probability. EUROMECH 155, Kopenhagen, 15 - 17 Juni 1982.

/10/ Hohenbichler, M.: Approximate Evaluation of the Multinormal Distribution Function. Berichte zur Zuverlässigkeitstheorie der Bauwerke, Technische Universität München, Heft 58, 1981.

# SMOKE TOXICITY AND TOXIC HAZARD

Session Chair

**Dr. Jack E. Snell**
Center for Fire Research
National Bureau of Standards
Gaithersburg, Maryland 20899, USA

# Mathematical Modeling of Toxicological Effects of Fire Gases

**G. E. HARTZELL, D. N. PRIEST, and W. G. SWITZER**
Department of Fire Technology
Southwest Research Institute
P.O. Drawer 28510
San Antonio, Texas 78284, USA

ABSTRACT

Research in combustion toxicology over the past few years has led to a reasonable understanding and even quantification of some of the effects of fire effluent toxicants and, with the availability of a modest amount of both non-human primate and human exposure data, the combustion toxicologist is gaining increasing capability to assess and predict the toxicological effects of smoke inhalation.

This paper presents a mathematical approach, based on experimental data for CO, HCN and HCl, for the prediction of both incapacitating and lethal effects on rats exposed to these toxicants. Elementary examples are given for computer simulation of the development of toxic hazards in fires and comparisons are made with actual experimental results. These comparisons show that computer-predicted times to toxicological effects lie within the standard deviation of experimental mean values.

INTRODUCTION

From the wide variety of fire gases that may be generated, the toxicant gases may be separated into three basic classes: the asphyxiants or narcosis-producing toxicants; the irritants, which may be sensory or pulmonary; and those toxicants exhibiting other and unusual specific toxicities.

Although many asphyxiants may be produced by the combustion of materials, only carbon monoxide and hydrogen cyanide have been measured in fire effluents in sufficient concentrations to cause significant acute toxic effects. The toxicity of carbon monoxide is primarily due to its affinity for the hemoglobin in blood. Even partial conversion of hemoglobin to carboxyhemoglobin (COHb) reduces the oxygen-transport capability of the blood (anemic hypoxia), thereby resulting in a decreased supply of oxygen to body tissues. Hydrogen cyanide, a very rapidly acting toxicant, does not combine appreciably with hemoglobin, but does bind with the trivalent ion of cytochrome oxidase in cellular mitochondria. The result is inhibition of the utilization of oxygen by cells (histotoxic hypoxia). Effects of the asphyxiant or narcosis-producing toxicants are dependent on the accumulated dose, i.e., both concentration and time of exposure, and increase in severity with increasing doses.

This work was supported under U.S. National Bureau of Standards Grant No. NB83NADA4015 and under Southwest Research Institute Internal Research Project No. 01-9316.

From an extensive review of methodologies for assessment of the incapacitating effects of the narcotic fire gases, both with rats and with non-human primates, it has been concluded that rats appear to be sensitive to approximately the same range of accumulated doses as may be deemed potentially hazardous to human subjects [3]. Thus, for both carbon monoxide and hydrogen cyanide, the rat is expected to be a reasonably appropriate model for the development of methodology for estimating toxicological effects on humans.

Irritant effects, produced in essentially all fire gas atmospheres, are normally considered by combustion toxicologists as being of two types. These are sensory irritation, including irritation both of the eyes and of the upper respiratory tract, and pulmonary irritation. Most irritants produce signs and symptoms characteristic of both sensory and pulmonary irritation, however.

Airborne irritants enter the upper respiratory tract, where nerve receptors are stimulated causing characteristic physiological responses, including burning sensations in the nose, mouth and throat, along with secretion of mucus. Senosry effects are primarily related to the concentration of the irritant and do not normally increase in severity as the exposure time is increased. There is no evidence that sensory irritation, per se, is physically incapacitating, either to rodents or to primates. On the contrary, recent studies involving exposure of baboons to hydrogen chloride have shown that even massive concentrations (up to 17,000 ppm for 5 minutes) are not physically incapacitating and do not impair escape [4].

Of potentially greater importance with both rodents and primates is that, following signs of initial sensory irritance, significant amounts of inhaled irritants are quickly taken into the lungs with the symptoms of pulmonary or lung irritation being exhibited. Tissue inflammation and damage, pulmonary edema and subsequent death often follow exposure to high concentrations, usually after 6 to 48 hours. Unlike sensory irritation, the effects of pulmonary irritation are related both to the concentration of the irritant and to the duration of the exposure.

In view of the importance of the common asphyxiants and also of pulmonary irritation, this study focused on the development of models for predicting the effects of CO, HCN and HCl on rats. Only the post-exposure lethality of rats was felt significant for the HCl part of this study; however, both incapacitation (leg-flexion shock-avoidance) and within-exposure lethality were employed for CO and HCN.

MATHEMATICAL MODELING

The basic concepts for toxicological modeling developed here involve the use of concentration-time relationships as a quantification of the "exposure dose" required to produce a given effect. From fundamental principles of toxicology, it has been demonstrated that concentration, mean exposure time to effect a given response, and percent of subjects responding are interrelated mathematically [6]. For example, a plot of mean exposure time required to effect a response as a function of concentration (Ct plot) also represents the $EC_{50}$ (or $LC_{50}$) as a function of exposure time, or conversely, the $ET_{50}$ (or $LT_{50}$) as a function of concentration.

Typical concentration-time relationships are illustrated in Figure 1 for incapacitation and lethality of rats from exposure to carbon monoxide. From this figure, the "exposure dose" required to cause incapacitation can, for example, be obtained by multiplying the values of any pair of coordinates; e.g., 2000 ppm x 20 min = 40,000 ppm-min "exposure dose." The "exposure dose" required to cause an effect for a given toxicant is relatively constant (Haber's Rule) over the curved portion of the plot. For many purposes, the relative constancy of the Ct product can be used as a first approximation for estimation of the time-to-effect for a desired exposure concentration. However, for purposes of developing mathematical models having predictive value, further refinements must be made to take into account the fact that the Ct product "exposure dose" decreases with increasing concentration and may actually cover a two- to three-fold range.

**CARBON MONOXIDE INTOXICATION OF RATS**

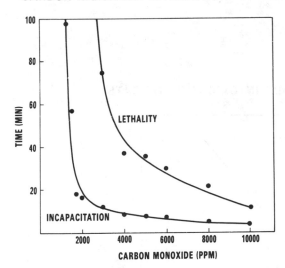

FIGURE 1.

Specific Ct "exposure doses" required to produce a given toxicological effect for a particular concentration may be determined as follows:

From the expression for Haber's Rule,

$$Ct = K, \tag{1}$$

simple rearrangement of terms produces

$$C = K \left(\frac{1}{t}\right). \tag{2}$$

This is a linear equation, indicating that a plot of concentration against the reciprocal of time should be a straight line with the slope being K. Figures 2 and 3, using mean times-to-effect for exposure of rats to carbon monoxide, show that the relationship is, indeed, linear. Similar linear relationships between concentration and the reciprocal of time-to-effect are shown in Figures 4 and 5 for incapacitation and death of rats due to exposure to hydrogen cyanide.

In the case of lethality of rats from hydrogen chloride, time of exposure required to effect 50 percent post-exposure lethality, rather than time to death, must be used. Therefore, $LC_{50}$ values, determined for a series of different exposure times, are plotted as a function of the reciprocal of exposure time as shown in Figure 6. In contrast to Figures 2 through 5, which show mean times-to-effect for CO and HCN exposures, Figure 6 reflects the time of exposure required to cause 50-percent post-exposure lethality for any concentration of HCl.

**CARBON MONOXIDE INTOXICATION OF RATS**

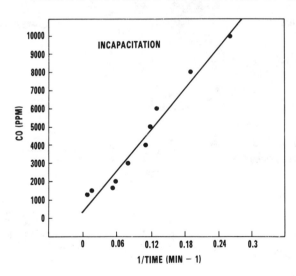

FIGURE 2.

## CARBON MONOXIDE INTOXICATION OF RATS

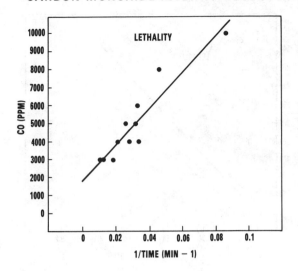

FIGURE 3.

## HYDROGEN CYANIDE INTOXICATION OF RATS

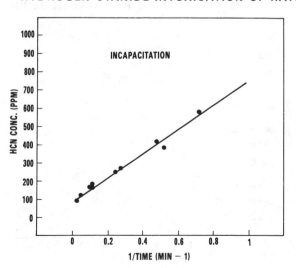

FIGURE 4.

## HYDROGEN CYANIDE INTOXICATION OF RATS

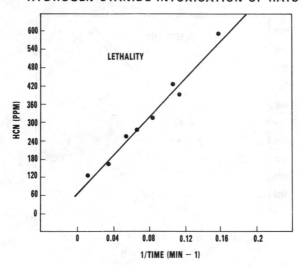

FIGURE 5.

## POST-EXPOSURE LETHALITY OF RATS
## FROM EXPOSURE TO HYDROGEN CHLORIDE

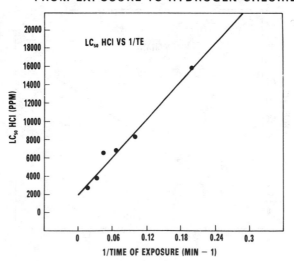

FIGURE 6.

Since the plots of 1/t against concentration do not pass through the origin, equation (2) must be modified to include b, the y-axis intercept.

$$C = K \left(\frac{1}{t}\right) + b \tag{3}$$

Equation (3) can be used directly for calculation of the $EC_{50}$ or $LC_{50}$ for any desired time of exposure once the constants have been determined. From equation (3) can also be derived an expression for the Ct product "exposure dose" as a function of the concentration,

$$Ct = K \left(\frac{C}{C-b}\right), \tag{4}$$

where K and b are determined from the linear plot of experimental data. Equation (4) enables calculation of the Ct product "exposure dose" of the toxicant required to produce a given effect when present at any concentration. The slope K of the plot of concentration vs. 1/t represents the minimum value of the "exposure dose" required to effect a 50-percent response at sufficiently high concentrations such that the term (C/C-b) approaches unity. Under these conditions, Haber's Rule (i.e., Ct = K) would be valid. The y-axis intercept, b, would thus appear to be a measure of deviation from Haber's Rule at relatively low concentrations of a toxicant.

In Table I are presented the values of K and b for both incapacitation and lethality for CO and HCN, and post-exposure lethality for HCl. Examples of calculated $EC_{50}$'s and $E(Ct)_{50}$'s for 15- and 30-minute exposures of rats are also shown. Data in Table I were derived from populations of 138 and 78 rats for incapacitation and lethality due to CO, populations of 70 and 46 rats for incapacitation and lethality due to HCN and a population of 180 rats for post-exposure lethality due to HCl exposures.

TABLE I.  Tabulation and use of modeling constants

|  | K | b | $EC_{50}$ 15-min (ppm) | $EC_{50}$ 30-min (ppm) | $E(Ct)_{50}$ 15-min (ppm-min) | $E(Ct)_{50}$ 30-min (ppm-min) |
|---|---|---|---|---|---|---|
| | | | INCAPACITATION | | | |
| CO | 36,509 | 233 | 2,667 | 1,450 | 40,004 | 43,499 |
| HCN | 698 | 92 | 139 | 115 | 2,078 | 3,458 |
| | | | LETHALITY | | | |
| CO | 102,874 | 1,778 | 8,636 | 5,207 | 129,544 | 156,214 |
| HCN | 3,130 | 66 | 275 | 170 | 4,120 | 5,110 |
| HCl (post-exposure) | 69,299 | 1,935 | 6,555 | 4,245 | 98,324 | 127,349 |

A restriction on the direct application of simple concentration-time relationships is that they are appropriate only for modeling constant concentration or "square wave" toxicant exposures such as those from which they were obtained experimentally. Fires do not produce constant toxicant concentrations, but rather, produce changing (usually increasing) concentrations. Mathematical models to predict effects from fire-generated toxicants must be able to accommodate varying concentrations which, in turn, involve varying Ct "exposure doses" required to effect incapacitation and/or death.

In Figure 7 is illustrated conceptually how this may be accomplished. From a plot of toxicant concentration as a function of time, incremental "doses" (C x Δt) are calculated and related to the specific Ct "exposure dose" required to produce the given toxicological effect at the particular incremental concentration. Thus, a "fractional effective dose" (FED) is calculated for each small time interval. Continuous summation of these "fractional effective doses" is carried out and the time at which this sum becomes unity represents the time of greatest probability of occurrence of the toxicological effect in 50 percent of the exposed subjects. (As with all statistical data, there is a probability distribution for the predicted time-to-effect.)

# MODELING OF TOXICOLOGICAL EFFECTS OF FIRE GASES

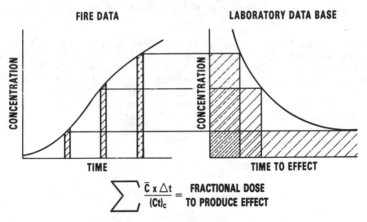

$$\sum_i \frac{\bar{C} \times \Delta t}{(Ct)_c} = \text{FRACTIONAL DOSE TO PRODUCE EFFECT}$$

EFFECT OCCURS AT TIME t WHEN Σ FRACTIONAL DOSES = 1

FIGURE 7.

The laboratory data Ct curve illustrated in Figure 7 is presented for conceptual purposes only. In practice, the specific Ct "exposure dose" corresponding to each incremental concentration is calculated from Equation (4).

The mathematical modeling concept illustrated in Figure 7 was tested for both incapacitation and lethality of rats exposed to increasing concentrations of carbon monoxide as might occur in a fire. The carbon monoxide concentration was ramped from 0 to about 9500 ppm over a period of about 10 minutes, followed by maintaining the concentration at the maximum level. By continuous summation of fractional effective doses for each 15 seconds of exposure, computer projections estimated incapacitation of exposed rats should occur at 10.0 minutes, with death expected at 20.8 minutes. Actual exposure of six rats to the ramped conditions yielded mean times of 10.2 ± 1.9 and 22.8 ± 3.5 minutes for incapacitation and death, respectively. The predicted times to both incapacitation and death were well within the standard deviations for the experimentally determined values.

Another test of the fractional effective dose model for carbon monoxide was conducted which employed a slower ramping profile for generation of the toxicant. The maximum CO concentration was attained at about 30 minutes. The computer-predicted time to incapacitation of 16.5 minutes was in excellent agreement with the experimental time of 17.2 ± 1.5 minutes determined using five rats. Death of the rats was predicted at 32.5 minutes. The experimentally determined mean time-to-death was 43.9 ± 13.9 minutes, with two of the five rats dead by 33.0 minutes.

The model was also tested for hydrogen cyanide, with exposure of six rats to HCN ramped from 0 to 195 ppm over about 12 minutes, followed by maintaining that concentration throughout the remainder of the exposure. Incapacitation and death were predicted by computer for 13.8 minutes and 32.2 minutes, respectively. Experimentally, mean times-to-incapacitation and death were 13.8 ± 2.3 minutes and 36.8 ± 11.9 minutes, respectively. Three of the six rats had died by 32 minutes.

A second test of the FED model for hydrogen cyanide was conducted with the concentration of HCN ramped from 0 to 270 ppm over 30 minutes. The model predicted incapacitation and death for 22.8 minutes and 34.0 minutes, respectively. Mean times-to-incapacitation and death of the six rats exposed in the ramped test were 20.8 ± 2.8 minutes and 34.2 ± 2.1 minutes, respectively.

A further refinement of the FED model enabled estimation of the error expected in the projection of times to incapacitation and death for CO and HCN. This involved determination of the K and b modeling constants for concentration vs. $1/t$ plots representing standard deviations of the original time-to-effect data. Analogous to the summation of fractional effective doses for the projection of mean times-to-effect, summations were also carried out corresponding to Ct products representing standard deviations of the original laboratory data. A comparison of computer-projected and experimental data, including predicted errors and/or standard deviations is presented in Tables II and III. It can be seen that predicted and experimental ranges for times to effect are sufficiently comparable as to suggest that model predictions are as accurate as experimental determinations.

TABLE II. Comparison of FED model with experimental data for carbon monoxide

|  | Incapacitation (Minutes) | | Lethality (Minutes) | |
|---|---|---|---|---|
|  | Mean | Range (S.D.) | Mean | Range (S.D.) |
| FED Model | 10.0 | 7.5 − 11.0 | 20.8 | 17.0 − 24.0 |
| Experimental | 10.2 | 8.3 − 12.1 | 22.8 | 19.3 − 26.3 |
| FED Model | 16.5 | 12.2 − 18.2 | 32.5 | 26.8 − 37.2 |
| Experimental | 17.2 | 15.7 − 18.7 | 43.9 | 30.0 − 57.8 |

Table III. Comparison of FED model with experimental data for hydrogen cyanide

|  | Incapacitation (Minutes) | | Lethality (Minutes) | |
| --- | --- | --- | --- | --- |
|  | Mean | Range (S.D.) | Mean | Range (S.D.) |
| FED Model | 13.8 | 12.0 - 15.8 | 32.2 | 25.2 - 35.2 |
| Experimental | 13.8 | 11.5 - 16.1 | 36.8 | 24.9 - 48.7 |
| FED Model | 22.8 | 21.0 - 24.2 | 34.0 | 30.5 - 37.0 |
| Experimental | 20.8 | 18.0 - 23.6 | 34.2 | 32.1 - 36.3 |

Validation of the fractional effective dose model for exposure of rats to HCl would be expected to be somewhat more difficult, since lethality occurs up to 14 days post-exposure and is due to indirect causes. Hydrogen chloride was ramped stepwise to a maximum concentration of about 6400 ppm. The FED summation reached a value of 0.87 at 28.5 minutes. The six exposed rats were then withdrawn and observed for up to 14 days, with the expectation that half should survive. Although all the subjects died within the 14-day period, this result is not unusual for a single experiment. The confidence limits for experimental post-exposure $LC_{50}$'s of HCl are quite broad, due, in part, to wide animal response variability with post-exposure effects.

The validity of the FED model has thus far been tested only for exposure of rats to single toxicants. However, with input data from appropriate laboratory experiments, the model would be expected to be quite appropriate for combinations of toxicants. The initial success of this methodology is anticipated to open the way to the combustion toxicity testing of materials without the necessity of using laboratory animals on a routine basis. Furthermore, the estimation of toxicological effects of smoke inhalation by humans exposed in a fire would also appear to be feasible.

REFERENCES

1. Kaplan, H. L., Grand, A. F. and Hartzell, G. E., Combustion Toxicology: Principles and Test Methods, Technomic Publishing Company, Incorporated, Lancaster, Pennsylvania, 1-20, 1983.

2. Hartzell, G. E., Packham, S. C. and Switzer, W. G. "Toxic Products from Fires," Am. Ind. Hyg. Assoc. J., 44 (4), 248-255, 1983.

3. Kaplan, H. L. and Hartzell, G. E. "Modeling of Toxicological Effects of Fire Gases: I. Incapacitating Effects of Narcotic Fire Gases," J. Fire Sciences, Vol. 2, No. 4, 286-305, 1984.

4. Kaplan, H. L. et al., "A Research Study of the Assessment Escape Impairment by Irritant Combustion Gases in Postcrash Aircraft Fires," Southwest Research Institute, Final Report, DOT/FAA/CT-84/16, U. S. Department of Transportation, Federal Aviation Administration, Atlantic City, New Jersey, 1984.

5. Kaplan, H. L., Southwest Research Institute, Unpublished Work.

6. Packham, S. C. and Hartzell, G. E., "Fundamentals of Combustion Toxicology in Fire Hazard Assessment," Journal of Testing and Evaluation, Volume 9, No. 6, 341-347, November 1982.

# Toxicity Testing of Fire Effluents in Japan: State of the Art Review

**FUMIHARU SAITO**
Testing Laboratory of Center for Better Living
2 Tatehara, Oho-machi, Tsukuba-gun, Ibaraki-ken, Japan

**SHYUITSU YUSA**
Building Research Institute, Ministry of Construction
1 Tatehara, Oho-machi, Tsukuba-gun, Ibaraki-ken, Japan

ABSTRACT

The present state of research on toxicities of fire effluents in Japan is described. This review centers on the status of a joint study project which has been under way in recent years, participated in by research institutions in the country. The research setup in the country, the method of applying conditions of actual fires to testing apparatus of laboratory scale, the role of $CO_2$ mixed with pure gases, and the results of square wave exposure of fire effluents in a newly developed testing apparatus are described.

INTRODUCTION

Research in Japan on toxicities of fire effluents has been mainly carried out at universities and by national research institutes and testing facilities of various government ministries and agencies. Studies to be carried out at the research institutions of the ministries and agencies are based on the administrative needs of the individual government organs. Therefore, the concepts of toxicities of fire effluents of these research organizations differ in degree and are being studied by means of different techniques. The Science and Technology Agency of the Japanese government has set up a system for allocating funds, taking up cases of joint studies by a plural number of research institutions as special research projects. A 5-years joint research program concerning toxicities of fire effluents was started 3 years ago with the cooperation of the United States and Canada (hereafter called "Project Research") under this budgeting system. In Japan, five national research organs, two universities, and one public testing organ are cooperating and sharing in the research work. This research is being done by the three working groups on establishment of burning conditions, development of testing apparatus, and development of an evaluation method, with three or four institutions participating in each working group. This is the first time that such a research organization has been set up in the field of fire safety science. So far, satisfactory results have been obtained. Here, the studies in Japan in recent years of toxicities of fire effluents will be described including the abovementioned results.

BURNING CONDITIONS WHEN EVALUATING COMBUSTION TOXICITY

The composition of gas emitted during combustion of a material is affected by burning conditions such as heating temperature, partial pressure of oxygen, and configuration of the combustion chamber, and by the size and orientation of the sample. Consequently, it is clear that the potential for toxicity of effluents

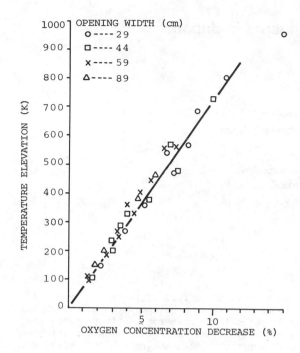

FIGURE 1. Relatiohship between the temperature rise and the oxygen concentration in the burn room (1).

in an actual fire varies depending on conditions and factors such as the space in the building, configurations and sizes of openings, and the quantities and shapes of combustibles. It is required that for toxicity testing of fire effluents these conditions and factors are reproducible on a laboratory scale. However, since it is difficult from the standpoint of manpower and cost to carry out these studies on a routine basis, the individual researchers have up to now arbitrarily set up only one or at most a few conditions of fire in testing. And it had not been made very clear whether these conditions simulated actual fires. In Project Research, with regard to heating conditions to which materials may be subjected during actual fires, a series of full scale steady state fire experiments were conducted with a propane burner as the fire source in a burn room (3.45 x 3.55 x 2.17 meters) and a strong correlation has been recognized between temperature and oxygen concentration of the atmosphere as shown in Fig. 1 (1). As a result, it is beginning to be shown that this correlation is useful as an aid in removing unreasonable test conditions in case of applying combinations of the two to toxicity tests.

EVALUATION

Test Animals

During a fire, a reduction in $O_2$ concentration occurs along with generation of $CO_2$ and CO; possibly along with gases such as HCN, HCl, and acrolein, depending on the chemical composition of the material burning. The mechanism and tolerance when any of these gases acts on the human body are specific to each individual gas, but during a fire the condition is that a plural number of these component gases are

normally mixed together, and it will be necessary to evaluate the toxicity of fire
effluents as the comprehensive effect of these mixed gases. In this case, the factors to cause impediments to oxygen supply capacities of bronchi, lungs, and blood
of the human body differ depending on the component gas, and it is highly questionable whether the toxicity of the mixed gas can be considered to be the sum of the
toxicities of the individual component gases. In order to clarify this point it
will be necessary to use test animals, but this will again be accompanied by problems such as the correlation between test animals and human beings, and errors in
evaluations using animals. To obtain correlation with human beings it is necessary
to use primates as the test animals, but in Japan at present, mice are generally
used from the fact that many mice can be tested at once and from the standpoints of
cost, and ease of handling. The rotary cage method using mice has been adopted for
Project Research, also.

Studies Using Pure Gases

When attempting toxicity evaluations based on chemical analysis values of fire
effluents, it is first necessary to ascertain their actions exposing animals to
pure gases. In the past, studies in Japan had been based on mortality of mice (2),
but in Project Research, mice in rotary cages are exposed to pure gases of CO, $CO_2$,
$NH_3$, $NO_2$, HCl, and HCN individually and in mixtures, and toxicities of various mixtures of the gases are evaluated based on incapaciation of the mice(3). Fig. 2
shows the results in separate exposures to the individual component gases. Here,
the relationship between gas concentration at which mice are incapacited in 5 to
30 minutes and time to incapaciation is expressed by the following equation:

$$D = \frac{a}{T} + b \qquad\qquad (1)$$

where, D: gas concentration (% or ppm)
       T: time to incapacitation of mice

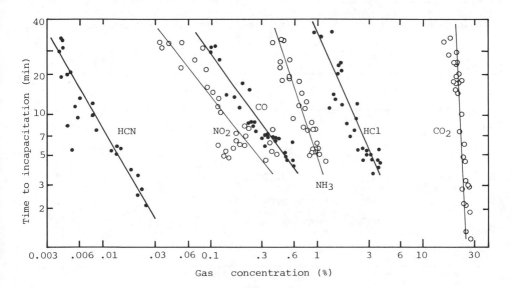

FIGURE 2. Incapacitation data for single gas.

a, b: constant specific to the gas

The constants 'a' and 'b' in equation (1) were determined using the method of least squares. The values of constants in the study are shown in Table 1. The regression lines for the relationships between the incapacitation times and the gas concentration are shown in Fig. 2.

In the discussion of mixed gases, a non-dimensional gas concentration index (Ig), which is defined by the following equation, was used.

$$Ig = \frac{Dm - b}{D_5 - b} \tag{2}$$

where, Dm: measured gas concentration
$\quad\quad$ $D_5$: gas concentration corresponding to five minutes of incapacitation of mice in equation (1)
$\quad\quad$ b : constant in equation (1)

Fig. 3 shows the results of mixed gases using the gas concentration indexes. The curves in Fig. 3 are ideal ones (Ig = 5/T) which correspond an additive effect between the individual gas. With regard to mixed gases, antagonism were seen in gases of CO mixed with $NH_3$, and HCl while with a mixed gas of CO and HCN, it was confirmed that there is an additive action. However, there was antagonism when CO and HCN of low concentration were mixed. As for when $CO_2$ was further added to a number of these mixed gases, complex actions were seen to occur. In effect, the three kinds of mixed gases containing NH3, HCl and HCN were recognized to have their toxicities increased when $CO_2$ was added compared with mixed only with CO. However, it may need further investigation for the influence of the mixture of individual gases below the asymptotic concentration b.

Heat and moisture are generated simultaneously with combustion gas during a fire, and therefore, an environment in which these factors are also added is produced in a fire. It may expected that the increase in temperature causes the volume of respiration to increase, and it is thought that the physiological action of toxic gases is accelerated by an increase in the volume respiration. The results

TABLE 1.  Calculated values of constants 'a' and 'b'.

| Gas | a | b | $\sigma^1$ | Unit of D |
|---|---|---|---|---|
| CO | 2.34 | 0.016 | 0.037 | % |
| $CO_2$ | 9.97 | 20 | 1.12 | % |
| $NH_3$ | 3.15 | 0.378 | 0.064 | % |
| $NO_2$ | 1.14 | 0.029 | 0.053 | % |
| HCl | 10.34 | 0.904 | 0.156 | % |
| HCN | 491 | 25.3 | 13.8 | ppm |

$^1\sigma = \sqrt{\sum_{}^{N} (Dm - D)^2 / N}$

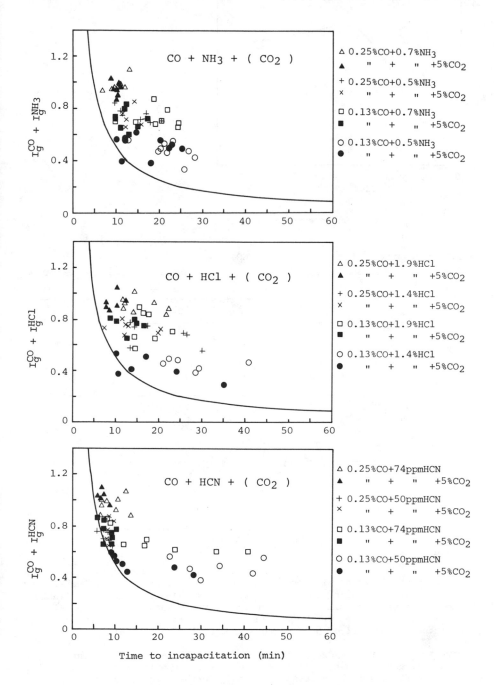

FIGURE 3. Relationship between the total gas concentration index and time to incapacitation of mice.

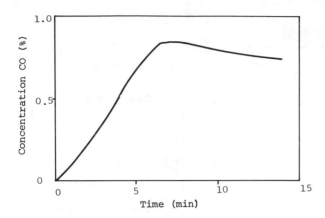

FIGURE 4. CO concentration in exposure chamber.

TABLE 2. Concentration time products for CO.

| Temperature (°C) | | humidity (%) | | |
|---|---|---|---|---|
| | | 70 | 80 | 90 |
| 30 | $Ct^1$ | 3.45 | 2.82 | 2.67 |
| | $s^2$ | 1.03 | 0.91 | 0.78 |
| | $n^3$ | 22 | 27 | 40 |
| 40 | Ct | 3.08 | 2.83 | 2.72 |
| | S | 0.86 | 0.64 | 0.59 |
| | n | 60 | 43 | 39 |
| 50 | Ct | 2.13 | 0.77 | 0.39 |
| | S | 0.58 | 0.35 | 0.15 |
| | n | 30 | 38 | 30 |

[1] Average concentration time products for CO
[2] Standard deviation
[3] Number of mice

of the experiments in which mice are exposed to some combinations of temperature
and humidity under increasing CO concentration (Fig. 4) are indicated in Table 2
which shows Concentration time products (Ct) of CO for incapacitation of mice. Although it is not clear whether the influence of humidity is significant or not, it
seems that the temperature of 50°C affects greatly to incapacitation of mice.

There are cases when hydrophiric gases such as $NH_3$ and HCl are contained in
fire effluents, and although it is known that there are antagonisms between these
gases in dry air, it will be necessary to study further regarding their biological
effects when they exist in environments of high humidity.

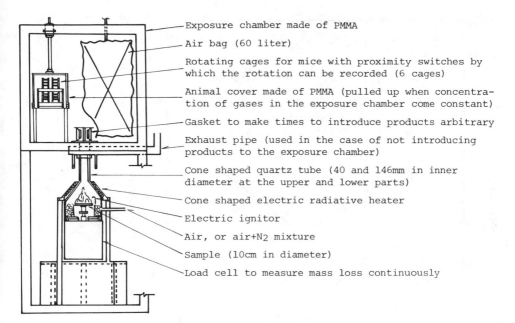

- Exposure chamber made of PMMA
- Air bag (60 liter)
- Rotating cages for mice with proximity switches by which the rotation can be recorded (6 cages)
- Animal cover made of PMMA (pulled up when concentration of gases in the exposure chamber come constant)
- Gasket to make times to introduce products arbitrary
- Exhaust pipe (used in the case of not introducing products to the exposure chamber)
- Cone shaped quartz tube (40 and 146mm in inner diameter at the upper and lower parts)
- Cone shaped electric radiative heater
- Electric ignitor
- Air, or air+N$_2$ mixture
- Sample (10cm in diameter)
- Load cell to measure mass loss continuously

FIGURE 5. Toxicity test apparatus developed at Building Research Institute.

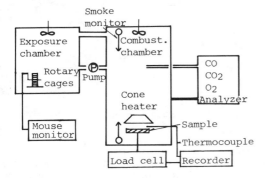

FIGURE 6. Toxicity test apparatus at Research Institute for Polymers and Textiles.

## Toxicity Evaluation of Fire Effluents

The research in Japan on toxicities of fire effluents from materials, broadly deviced, consists of that based on the surface area of the material heated and that based on weight. In case of the former, research based on Ministry of Construction Proclamation 1231, the test method prescribed for control of fire-retardant materials used in high-rise buildings in Japan, is main. In this method, the material is burned in a furnace which is a modified version of the fire propagation test apparatus in Part 6 of BS 476, and the time of incapaciation of a mouse is compared with that using wood (red lauan) which is the reference materials. In the latter, it is aimed to discern the toxicity par unit weight of material burned. There are versions such as a method with the volume of the test animal exposure chamber as

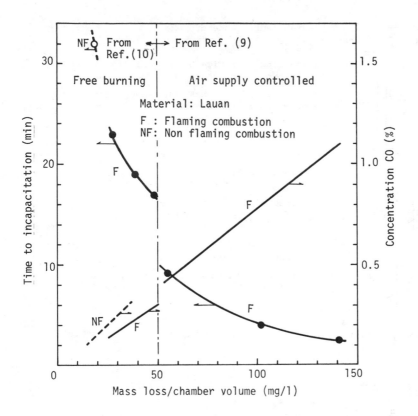

FIGURE 7. Animal test results for lauan at Building Research Institute and Research Institute for Polymers and Textiles.

the variable(5), a method with quantity of material varied(6), and a method with exposure time constant(7).

   Two new type of test apparatus are presently under development in Project Research. One is an apparatus with which partial pressure of oxygen and air supply volume are variable(8)(9)(Fig. 5), while the other is a closed type with $O_2$ concentration in the combustion chamber variable(10)(Fig. 6). Both use cone-type furnaces and continuous mass loss measurements by load cell and the animals can be subjected to square wave exposure to combustion products in both methods.

   With regard to the animal exposure experiments, the results obtained with the two test apparatus so far on the same material (lauan) are shown in Fig. 7. The experiments of the former method (right part of the Fig. 7) were carried out under the conditions of 2.5 W/cm² of heat flux, 4.6 1/sec/cm² of supplying air flow rate in flaming mode, and with the latter method (left part of the figure), the conditions were 2.0 W/cm² for smoldering, and 4.2 W/cm² for flaming. As CO evolution depends on burning condition (in this case controlled air supply rate or free burning), times to incapacitation of mice are different for the two method. However, it needs further study for lower range of the mass loading less than 50 mg/l in the former method. It is planned to accumulate more detailed data hereafter varying the different parameters.

CONCLUSION

As mentioned at the beginning of the paper, studies on toxicities of fire effluents in Japan have in the past been carried out on relatively individual bases; however, with the initiation of Project Research, cooperated in by the United States and Canada, studies recently are in a more systematic manner. The number of researchers to cope with the problems that must be solved is still extremely small. Delay in resolving the problem of fire effluent toxicity means that many more human lives will be lost, with lack of more paper control of materials. It will be necessary for a systematically stronger research organization, including a setup for international research, to be established to achieve further advances in this field of study.

REFERENCES

1. Tanaka, T., Nakaya, I., Yoshida, M.: "Full scale experiments for determing the burning condition to be applied to toxicity test," in the Paper to be present at the first international symposium on fire safety science, Nat. Bur. Stand. Gaithersburg, MD, 1985.

2. Kishitani, K., Nakamura, K.: "Research on evaluation of toxicities of combustion gases generated during fires," in Proceedings of 3rd Joint Conference of UJNR Fire Research and Safety, NBS SP 540, 485-519, 1978.

3. Sakurai, T.: "Toxic gas test by the several pure and mixture gases," 3rd conference of Canada-Japan-USA Trilateral Study Group on Fire Gas Toxicity, NRCC, Ottawa, Canada, Paper 20, 1984.

4. Saito, F.: Unpublished data arranged from "Factors accelerating the gas toxicity in a fire," in Proceeding of 3rd Joint Conference of UJNR Fire Research and Safety, NBS SP 540, 528-548, 1978.

5. Saito, F.: "Evaluation of toxicity of combustion products," in J. Comb. Tox. Vol. 4, p. 32, 1977.

6. Kishitani, K., Yusa, S.: "Study on evaluation of relative toxicities of combustion products of various materials," in J. Fac. Univ. of Tokyo (B), Vol. 35, No. 1, 1979.

7. Report on the Fire Prevention Systems for Rolling Stocks, Japan Association of Rolling Stock Industries; March 1978 (in Japanese).

8. Yusa, S.: "Development of laboratory test apparatus for evaluation of toxicity of combustion products of materials in fire." in Proceedings of 7th Joint Panel Meeting of the UJNR Panel on Fire Research and Safety, NBSIR 85-3118, 472-487, 1985.

9. Yusa, S.: "Animal exposure test by new apparatus," 3rd Conference of Canada-Japan-USA Trilateral Cooperative Study on Fire Gas Toxicity, NRCC, Ottawa, Canada, Paper 13, 1984.

10. Furuya, M.: "Combustion condition of plastic materials and evaluation of toxicity of combustion products," 3rd Conference of Canada-Japan-USA Trilateral Cooperative Study on Fire Gas Toxicity, NRCC, Ottawa, Canada, Paper 14, 1984.

# Thermal Decomposition of Poly(vinyl chloride): Kinetics of Generation and Decay of Hydrogen Chloride in Large and Small Systems and the Effect of Humidity

C. A. BERTELO, W. F. CARROL, JR., M. M. HIRSCHLER,
and G. F. SMITH
The Vinyl Institute
355 Lexington Avenue
New York, New York 10017, USA

ABSTRACT

Experiments in which PVC wire insulation was decomposed by an electrical overload in a plenum or in a 200 $\ell$ PMMA box showed that the highest concentration of HCl in the atmosphere was always less than 40% of the theoretical amount of chlorine in the original sample. Furthermore, this concentration quickly decays to a level that is dependent on humidity, but never more than 4% of the theoretical maximum. A mathematical model was used to determine a number of parameters that describe the system. The same model applies to both large and small scale tests. This treatment showed that less than 48% of the chlorine in the wire reached the atmosphere. It also showed that the rate of decay was primarily dependent on the rate of transport in the system, producing a half life of 5-6 min without external agitation. Furthermore, the final HCl concentration was found to be dependent on relative humidity, and on surface composition (painted gypsum or PMMA).

INTRODUCTION

Large scale experiments have recently been conducted[1] in order to confirm earlier observations from small scale work[2-7] that hydrogen chloride (HCl) is not indefinitely stable in the atmosphere, but rather that it "decays." With the objective in mind of developing a mathematical model of the airborne HCl concentrations observed, which can be used for a variety of scenarios, smaller scale experiments were designed to facilitate quantification of the effects of relative humidity (RH) and air recirculation on the transport and decay of HCl.

EXPERIMENTAL

The large scale experiments consisted of the decomposition of 30 ft of commercially available, PVC-insulated wire typical of that used to carry electric power in a home or office[1]. The decomposition was performed in an 8 ft x 12 ft x 12 ft enclosure divided into an 8 ft high room and a 4 ft plenum. Decomposition was afforded by current overload of 180 or 250 A.

In addition, the apparatus included a rudimentary air recirculation system which withdrew air from the plenum in the center of the 4 ft x 8 ft wall, filtered it and returned it through a duct which extended from the center of the 4 ft x 12 ft plenum wall to the center of the ceiling and down into the room. This was run at eight room-air-changes per hour in Experiment 3 and not used in the others.  In some of the experiments (2 and 3) an 8 in by 24 in grating was placed in the ceiling centered along and within a foot of the 12 ft wall.  This grating became a part of the air recirculation system in Experiment 3, serving as the return air vent.

The apparatus was instrumented with continuous analyzers for CO, $CO_2$ and hydrocarbons.  HCl was analyzed by drawing the atmosphere for a prescribed period of time through 3 mm tubes packed with 20-30 mesh soda lime.  Hereafter, these are called soda lime tubes or SLTs.  Additional detail on the apparatus and methods of analysis is documented elsewhere.[1]

Four variations were conducted in the large scale apparatus.  The first three tests were designed to investigate flow conditions, with similar sample configuration and decomposition, at 180 A.  For the first test the recirculation system was off and the ceiling was solid (no grate).  For the second experiment, the grate was in place.  For the third experiment, both the recirculation system and the grate were utilized.  The fourth experiment used a large decomposition current, 250 A, but was configured as in the first.  Heating ceased in this experiment after 5 minutes when the conductor fused open.

For mathematical treatment of the large scale experiments, plenum HCl values were averaged, and are presented in Table I.  For Experiments 1, 2 and 4, this is an average of three sites.  For Experiment 3 (where the recirculation system is utilized) the SLT site over the grate was not included as the final HCl concentrations recorded at this site were significantly different from the others.  They are noted separately in Table I as is the room site with the

TABLE I.  HCl Concentrations for Large Tests (ppm in Atmosphere).

| Time, min | Test 1 Plnm Avgs | Room Ste 7 | Test 2 Plnm Avgs | Room Ste 7 | Test 3 Plenum Ste 1&2 | Ste 3 | Room Ste 7 | Time, min | Test 4 Plnm Avgs | Room Ste 7 |
|---|---|---|---|---|---|---|---|---|---|---|
| 3-7 | 326 | 10 | 370 | <10 | 280 | 220 | <10 | 0-2 | 97 | <10 |
| 7-11 | 2553 | <10 | 2593 | <10 | 1635 | 1680 | 100 | 2-4 | 2103 | <10 |
| 11-15 | 2260 | <10 | 2163 | <10 | 935 | 1310 | 200 | 4-7 | 2897 | 10 |
| 15-20 | 1463 | <10 | 1410 | <10 | 370 | 440 | 150 | 7-11 | 1767 | 10 |
| 20-25 | 950 | <10 | 877 | 10 | 150 | 210 | 60 | 11-15 | 1110 | 10 |
| 25-30 | 603 | 10 | 640 | <10 | 65 | 210 | 40 | 15-20 | 663 | 10 |
| 30-37 | 436 | <10 | 430 | <10 | 30 | 410 | 10 | 20-25 | 400 | 10 |
| 37-45 | 250 | <10 | 277 | 10 | 20 | 290 | 10 | 25-30 | 243 | 10 |

highest readings.  Note that forced recirculation (Test 3) causes a significant amount of HCl to be transferred to the room.  Also, in Test 2 there is a large difference in HCl concentration between the sites directly above and below the grate which are separated by only 2 ft.

After the first large scale test, a sample of ceiling tile was analyzed for chlorine.  Extrapolation of excess chlorine recovered from this sample to the entire ceiling surface area accounted for 17 per cent of the chlorine originally present in the insulation.  A crude estimate of total chlorine deposition can be obtained by further extrapolation to the entire plenum wall and ceiling surface.  This would account for 62 per cent of the chlorine originally present.

The small scale experiments were conducted in a 47.25 in x 16.5 in x 14.125 in poly(methyl methacrylate) (PMMA) box, similar to the chamber used in the NBS toxicity protocol.[8] This is a direct scale down of the larger tests based on mass PVC per plenum unit volume. Mixing was accomplished by a small fan fitted at one end of the box.

The wire samples consisted of 6 in of PVC insulation left in the center of a 2 ft section of conductor, laid on a small Marinite® board, elevated two inches by placement on an inverted Pyrex dish. Decomposition was accomplished by 180 A current overload. Instrumentation for sampling the same gases as in the large scale experiments was incorporated into the box.

Three HCl sampling devices were situated at different heights to test vertical stratification. In addition to three thermocouples situated vertically through the box at the wire sample end, one thermocouple was attached directly to the wire with a standard potting ceramic. Wall deposition of HCl was measured in these tests by analysis of evenly spaced 3 in x 3 in PMMA coupons placed on the floor of the box. Also, after each experiment the atmosphere was pumped through a large soda lime tube. This was combined with a similar "filter" on the continuous sampler throughout the experiment and analyzed in the same manner as the other SLTs.

Ten small scale tests (numbered 11 through 20) were conducted, all at 180 A. The recirculation fan was on for 6 of the these 10 tests although the recirculation rate could not be set totally reliably. The relative humidity (RH) levels used were 80, 50 and 5%. The order of experimentation was obviously influenced somewhat by the weather.

Data for the small scale tests are presented in Tables II and III. HCl concentrations represent averages of all the sampling sites. For some of the Fan On experiments, however, only the top two sets of SLTs were analyzed (Tests 11, 14, 17 and 18), when it was determined that the fan truly produced a homogeneous atmosphere. Final HCl concentrations are very low--<300 ppm in any event--representing less than 4% of the chlorine in the original sample. Chlorine analyses for the char, filters and coupons (assuming the coupons represent a uniform coating on the interior surfaces of the test chamber) are also included.

SLTs provide HCl values averaged over the time the sample was taken; however, for the mathematical treatment, distinct time/concentration pairs are necessary. To accomplish this the time intervals presented in Tables I and II were represented by the median value. The shape of these curves, shown in Figure I, suggest exponential phenomena for both production and decay of the HCl concentrations. In addition, a term must be included to account for the final, non-zero, HCl concentrations. While this seems at first to be a complex problem, equations of this form have been used to explain the kinetics of various chemical systems. A general set of equations that describes a group of species, the concentrations of which are interrelated by first-order phenomena, has been described by Moore and Pearson.[9]

Use of these mathematical techniques is by no means limited to chemistry. A scheme that represents the physical system in a manner consistent with this approach is shown in Figure II, along with equations adapted to the model. In addition to the three constants, $k_{12}$, $k_{23}$ and $k_{32}$, an induction period, $t_0$, and the total amount of HCl released, $A_0$, must also be estimated.

TABLE II.  HCl Concentrations for Small Scale Tests.[1]

| Time, min | Test 11 | Test 12 | Test 13 | Test 14 | Test 15 | Test 16 | Test 17 | Test 18 | Test 19 | Test 20 |
|---|---|---|---|---|---|---|---|---|---|---|
| 0-2 | 0 | 53 | 87 | 30 | 60 | 0 | 40 | 0 | 0 | 0 |
| 2-4 | 0 | 83 | 170 | 40 | 123 | 87 | 65 | 70 | 120 | 113 |
| 4-6 | 130 | 490 | 413 | 310 | 495 | 1270 | 545 | 440 | 1407 | 1147 |
| 6-8 | 255 | 547 | 503 | 1115[2] | 867 | 1743 | 1060 | 920 | 2283 | 1850 |
| 8-10 | 265 | 447 | 330 | 430[3] | 545 | 1513 | 730 | 805 | 2020 | 1500 |
| 10-14 | 185 | 273 | 213 | 245 | 260 | 1000 | 335 | 500 | 1340 | 1017 |
| 14-19 | 155 | 203 | 137 | 170 | 117 | 617 | 125 | 255 | 733 | 503 |
| 19-25 | 115 | 177 | 137 | 115 | 73 | 430 | 95 | 150 | 400 | 257 |
| 25-30 | 95 | 133 | 110 | 100 | 47 | 207 | 55 | 70 | 207 | 143 |
| 30-35 | 80 | 127 | 83 | 100 | 47 | 143 | 50 | 40 | 117 | 90 |
| 35-40 | 60 | 110 | 80 | 95 | 37 | 107 | 40 | 40 | 83 | 57 |
| 40-45 | 50 | 97 | 80 | 60 | 37 | 80 | 35 | 30 | 57 | 47 |

[1] Parts per million in atmosphere
[2] Time for this sample is 6-7 min
[3] Time for this sample is 8.33-10 min

FIGURE I.  Full Scale Test 1, Average Plenum Concentrations.

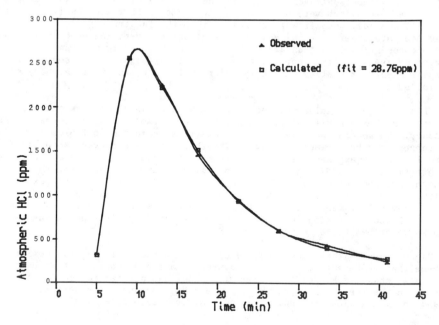

TABLE III.  Wet Analyses for Chlorine.[1]

| | Test 11 | Test 12 | Test 13 | Test 14 | Test 15 | Test 16 | Test 17 | Test 18 | Test 19 | Test 20 |
|---|---|---|---|---|---|---|---|---|---|---|
| Char | 31.5 | 40.7 | 30.3 | 42.1 | 55.6 | 39.1 | 46.4 | 32.7 | 21.5 | 45.5 |
| Surface[2] | 4.2 | 6.5 | 4.1 | 4.2 | 4.0 | 8.9 | 3.1 | 2.6 | 5.8 | 4.6 |
| Filters | 1.4 | 5.9 | 1.5 | 1.4 | 1.0 | 8.6 | 2.7 | 2.7 | 8.9 | 9.0 |

[1]Per cent of total theoretical (2.08 g Cl/sample)
[2]Extrapolated from coupons

The simplex method[10] was chosen as the mathematical optimization protocol.
This system investigates a local area of the goodness-of-fit response surface and
uses the slope of this portion to estimate a direction and magnitude for the
adjustment of the parameters.  Graphically, this results in a dimension for each
of the parameters to be optimized plus one for the goodness-of-fit, thus
producing a six-dimensional problem.  Two computer programs were independently
developed for the IBM Personal Computer to incorporate the methods outlined by
Deming and Morgan and each gave essentially the same results.  Goodness-of-fit
was found using a simple standard deviation with n-1 weighting.

FIGURE II.  Proposed Deposition Process and Mathematical Treatment.

$$\text{Potential HCl} \xrightarrow{k_{12}} \text{Atmospheric HCl} \underset{k_{32}}{\overset{k_{23}}{\rightleftharpoons}} \text{Deposited HCl}$$

$$[A] \qquad\qquad [B] \qquad\qquad [C]$$

$$[B] = A_0 \left[ \frac{k_{12}k_{32}}{l_2 l_3} + \frac{k_{12}(k_{32}-l_2)}{l_3(l_2-l_3)} e^{-l_2 t} + \frac{k_{12}(l_3-k_{32})}{l_3(l_2-l_3)} e^{-l_3 t} \right]$$

where

$$l_2 = \frac{(p+q)}{2} \qquad l_3 = \frac{(p+q)}{2}$$

and

$$p = (k_{12} + k_{23} + k_{32}) \qquad q = [p^2 - 4(k_{12}k_{23} + k_{12}k_{32})]^{1/2}$$

A three-dimensional presentation of the goodness-of-fit response surface is presented in Figure III. For the three dimensions not shown, the induction period $t_0$ was fixed at 4.8 min, $k_{32}$ was fixed at 0.0087 $min^{-1}$, and $k_{23}$ was described empirically as a function of $k_{12}$ and $A_0$ by the equations:

$$k_{23} = k_{12} - 0.5231 + 0.7319x - 1.332x^2 + 0.6696x^3$$

with $x = (A_0-4000)/7000$.

The simplex program was modified so as to hold one of the parameters constant while optimizing the others and, after all the points of the simplex produced similar goodness-of-fit (in this case +/- 0.01), to step the first parameter to a new point in order to investigate better the area at the bottom of the response surface "valley". This approach details the bottom of the valley, and the result is presented in Figure IV, together with the production and decay rate-constants corresponding to these points. This graph clearly shows the same general form noted for the valley in Figure III and further reveals a distinct minimum at low values of $A_0$. The exact minimum was found using a fully optimized simplex routine and the optimum values for the parameters were confirmed by reproduction from a different starting point.


RESULTS AND DISCUSSION

The results of the modeling calculations for the large scale tests are presented in Table IV. All tests produced similar values for the total amount of HCl released, $A_0$, representing 53% of the theoretical maximum, based on the chlorine in the PVC (the somewhat smaller value for Test 3 could be due to less

TABLE IV.  Summarized Large Scale Test Results - Fully Optimized Constants

| | Starting Points ($A_0$) ppm HCl | $A_0$, ppm HCl | $t_0$, min | $k_{12}$; $min^{-1}$ | $k_{23}$; $min^{-1}$ | $k_{32}$; $min^{-1}$ | Goodness of Fit, ppm HCl |
|---|---|---|---|---|---|---|---|
| Test 1 | 4000 | 4551.8 | 4.78 | 0.338 | 0.120 | 0.0059 | 28.76 |
|        | 5000 | 4569.3 | 4.78 | 0.336 | 0.121 | 0.0059 | 28.79 |
| Test 2 | 3500 | 4294.0 | 4.77 | 0.401 | 0.116 | 0.0070 | 23.42 |
|        | 5000 | 4258.5 | 4.77 | 0.406 | 0.114 | 0.0067 | 23.32 |
| Test 3 | 2500 | 3687.7 | 4.80 | 0.402 | 0.244 | 0.0017 | 7.51 |
| Test 4 | 3500 | 4002.7 | 2.61 | 2.010 | 0.138 | 0.0053 | 43.68 |
|        | 4500 | 3996.8 | 2.67 | 2.350 | 0.137 | 0.0052 | 43.30 |
| Average | | 4194.4[1] | 4.78[2] | 0.377[2] | 0.124[3] | 0.0060[3] | |
| Std Dev | | 296.4 | 0.01 | 0.032 | 0.008 | 0.0006 | |
| Percent Variation | | 7.1 | 0.2 | 8.6 | 6.5 | 10.7 | |

1)  All tests
2)  Tests 1, 2 & 3 only
3)  Tests 1, 2 & 4 only

FIGURE III. Response Surface Analysis, Large Scale Test 1.

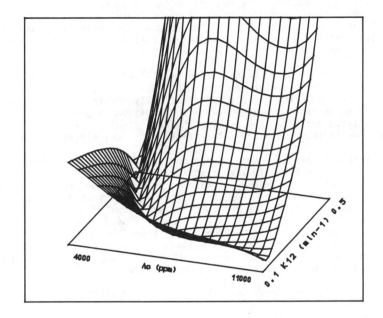

FIGURE IV. Response Surface Cross-Section, Large Scale Test 1.

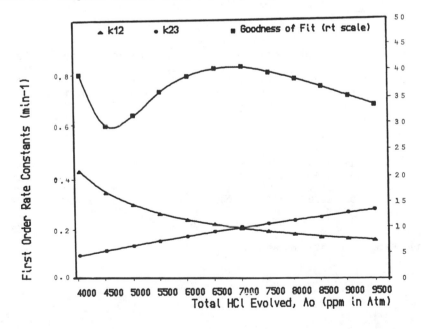

complete decomposition). In Tests 1, 2 and 3, the values of the $k_{12}$ and $t_0$ are the same while in Tests 1, 2 and 4 the values of $k_{23}$ are the same. This shows that $k_{12}$ and $t_0$ are related to production of HCl as caused by the use of current and are not affected by the recirculation fan. The fan, however, affects the rate of transport of HCl to the sampling site, shown by the higher value of $k_{23}$ in Test 3. Two further points of interest are that the presence of the grate does not affect rates (compare Tests 1 and 2) and that $k_{23}$ in Test 3 is virtually identical to the recirculation rate (0.26 $min^{-1}$), confirming that $k_{23}$ closely tracks the rate of transport. In all static tests, $k_{23}$ assumes similar values but is much smaller in the dynamic test.

No relationship was found in the small scale tests between chlorine in the char and RH. Recirculation decreased the HCl found on the coupons and on the filters. Although this result seems initially surprising, it is conceivable that the assumption of homogeneous HCl deposition is poor, and that greater decomposition occurs on the ceiling than the floor in "Fan Off" experiments. HCl concentration on filters and surfaces increased slightly with RH.

Since the total recovery of HCl from char, filters and extrapolated surfaces was significantly less than 100%, experiments were performed to determine other possible deposition sites. These experiments were carried out in a similar PMMA box and they showed that no less than 48% of the theoretical chlorine ultimately becomes attached to the Maranite® on the floor of the box. Significantly more may reside in the Maranite® used to support the sample. HCl also condenses on metallic surfaces. Additionally, sorptive surfaces other than Maranite exhibited similar behavior when tested (e.g. painted drywall, ceiling tile and cement block). The rate of HCl decay and maximum HCl concentration were lower in the presence of these sorptive surfaces.[11]

TABLE V.  Summarized Small Scale Tests – Fully Optimized Constants

| Expt | RH, % | Wt Lost, gm | $A_0$, ppm | $t_0$, min | $k_{12}$, $min^{-1}$ | $k_{23}$, $min^{-1}$ | $k_{32}$, $min^{-1}$ | Std Dev, ppm | Cl, Moles[1] | Pct Theo |
|------|------|------|------|------|------|------|------|------|------|------|
| | | | | | FAN ON | | | | | |
| 13 | 80 | 3.67 | 1250 | 2.49 | 0.315 | 0.293 | 0.0217 | 40.79 | 0.0087 | 14.8% |
| 14 | 80 | 3.30 | 1882 | 4.89 | 1.620 | 0.425 | 0.0254 | 30.16 | 0.0131 | 22.3% |
| 15 | 50 | 2.51 | 1363 | 4.75 | 1.800 | 0.257 | 0.0086 | 41.79 | 0.0095 | 16.2% |
| 17 | 50 | 2.64 | 1930 | 4.57 | 0.838 | 0.303 | 0.0082 | 25.61 | 0.0134 | 22.8% |
| 18 | 5 | 3.76 | 1457 | 4.48 | 0.729 | 0.178 | 0.0048 | 25.32 | 0.0101 | 17.2% |
| | | | | | FAN OFF | | | | | |
| 12 | 80 | 2.17 | 1158 | 2.85 | 0.645 | 0.182 | 0.0232 | 27.85 | 0.0080 | 13.7% |
| 16 | 50 | 3.50 | 2321 | 4.20 | 1.077 | 0.118 | 0.0036 | 40.13 | 0.0161 | 27.4% |
| 19 | 5 | 4.04 | 3337 | 4.25 | 0.786 | 0.141 | 0.0022 | 40.06 | 0.0232 | 39.6% |
| 20 | 5 | 3.25 | 2550 | 4.40 | 1.067 | 0.143 | 0.0020 | 35.14 | 0.0177 | 30.2% |
| | | | Fan Off Avg. | | 0.146 | | | | Avg. | 22.7% |

(1)  Based on 6.94E-6 moles/ppm

The results of applying the mathematical model to the small scale tests (except Test 11) are presented in Table V; they show graphical patterns similar to those of Figure III. The total HCl released ($A_0$) and the production rate constant ($k_{12}$) do not seem to be affected by RH. The induction periods for all experiments were similar except 12 and 13. No other data (wire temperature, hydrocarbon production, etc.) suggest a different effective starting time for these tests; however, the SLT sample times were recorded on a clock separate from that used for the other data. The decay rate constant, $k_{23}$, shows a slight positive relationship with humidity for the Fan On experiments but little or no effect without forced circulation. Comparing $k_{23}$ for the small-scale and large-scale Fan Off tests reveals that within the accuracy of these calculations all values are the same. This again supports the hypothesis that $k_{23}$ is related to the rate of transport.

The final parameter, $k_{32}$, increases strongly with RH. The values vary over a full order of magnitude (0.02 to 0.002), are very constant from test to test and are not dependent on circulation. Curiously, they also vary directly with residual airborne HCl, suggesting that $k_{32}$ may be a measure of "non-decayable" HCl, perhaps present as an aerosol. Therefore, if in two identical experiments the chamber volume were to be changed, the values of $k_{32}$ would change in the same way. In fact, in the large scale tests, $k_{32}$ decreased by a factor of three when the recirculation system was on. This factor corresponds to the ratio of the volume of the entire apparatus to that of the plenum. Assuming the room is not involved in the experiments without recirculation, while the entire volume of the apparatus is involved with recirculation, the change in $k_{32}$ could be viewed simply as a dilution factor, and as another indication that the constant in some way tracks residual HCl.


CONCLUSIONS

Clearly this work confirms earlier observations that HCl, generated during the thermal decomposition of PVC, decays. The peak concentrations of HCl were less than 45 of the maximum theoretical concentration and quickly decreased from this level. Decay continued until after about thirty minutes the concentration of HCl reached 300 ppm for the large scale and 30-100 ppm for the small scale tests.

A mathematical model facilitated calculation of a number of parameters that describe the atmospheric HCl concentration observed and lead to an understanding of the system. Decay rates found suggest half-lives for HCl of 5-6 min without external agitation. This rate is primarily dependent upon the rate of transport in the system, but also varies with the type of surface.

There are no indications that the concentration of HCl rises above 50% of the theoretical concentration based upon total release of chlorine. Even though decomposition approaches 100%, about 25% of the chlorine remains bound in the char as inorganic salts. The observation of low $A_0$ values may indicate that 20 to 30% of the HCl is deposited very quickly and never gets further than a few inches from the decomposition site. This may not, however, be valid for every possible experimental design. An additional parameter, used to account for the final HCl concentration, suggests that these values are humidity dependent; however, a complete understanding of the significance of the parameters awaits further study.

REFERENCES

1.  J. J. Beitel, C. A. Bertelo, W. F. Carroll, Jr., R. O. Gardner, A. F. Grand, M. M. Hirschler and G. F. Smith, "Hydrogen Chloride Transport and Decay in Large Apparatus I. Decomposition of Poly(Vinyl Chloride) Wire Insulation in a Plenum by Current Overload," J. Fire Sci., in press.

2.  M. M. O'Mara, Pure & Appl. Chem., 49, 649-660 (1977).

3.  Y. Tsuchiya and K. Sumi, J. Appl. Chem., 17, 364-366 (1967).

4.  C. Boudene, J. M. Jouany, R. Truhaut, J. Macromol. Sci., Chem., A11, 1529-1545 (1977).

5.  P. G. Edgerley, K. Pettett, Plastics and Rubber, Proc. and Applic., 1, 133-137 (1981).

6.  W. D. Woolley, J. Macromol. Sci., Chem., A11, 1509-1517 (1977).

7.  J. C. Spurgeon, "A Preliminary Comparison of Laboratory Methods for Assigning a Relative Toxicity Ranking to Aircraft Interior Materials," FAA Report No. FAA-RD-73-37 (1975).

8.  B. C. Levin, A. J. Fowell, M. M. Birky, M. Paabo, A. Stutte, D. Malek; NBSIR 82-2532, National Bureau of Standards, 1982.

9.  J. W. Moore and R. G. Pearson, "Kinetics and Mechanism, Third Edition," Wiley-Interscience, New York, 1981, p. 299.

10. S. N. Deming and S. L. Morgan, Anal. Chem., 45, 278A-283A (1973).

11. J. J. Beitel, C. A. Bertelo, W. F. Carroll, Jr., R. O. Gardner, A. F. Grand, M. M. Hirschler and G. F. Smith, J. Fire Sci., manuscript in preparation.

# Quantitative Determination of Smoke Toxicity Hazard—A Practical Approach for Current Use

RICHARD W. BUKOWSKI
Hazard Analysis
Center for Fire Research
National Bureau of Standards
Gaithersburg, Maryland 20899, USA

ABSTRACT

The concepts of fire hazard assessment are discussed. The development of these concepts into the framework for a hazard assessment model is described. This model, which is actually a group of interacting models, is presented in terms of the component functions and the interactions necessary to accomplish a hazard analysis. The most critical research issues which must be resolved in order to use this hazard analysis model for practical problems are identified. Preliminary results of experiments to assess the predictive accuracy of the multi-compartment transport model used within the hazard model are presented. A simple, engineering approach to toxicity evaluation included in the current model is also discussed.

INTRODUCTION

Over the past decade, the field of fire modeling has progressed to the point that quantitative predictions of fire in buildings can be made to an accuracy which is useful for engineering purposes. Over the past two years, the Center for Fire Research (CFR), National Bureau of Standards (NBS), has been working on the application of fire modeling techniques to the prediction of the hazard to occupants from building fires. Hazard assessment is a logical extension of fire modeling. Its development has been driven primarily by the need to evaluate the role of combustion product toxicity in relation to other hazards associated with fire.

This paper presents a framework for using fire models for hazard assessment and focuses on the progress made with two critical components: the assessment of the transport model's (FAST) predictive accuracy and the prediction of occupant response to toxic combustion products.

HAZARD MODELING

Fire models consist of sets of equations which describe the physical processes associated with fires in buildings. They describe the evolution and distribution over time, of energy and mass released by a burning material. Thus, with the proper model or combination of models, the environment in each compartment in the building is described in terms of the temperature, smoke density and gas concentrations. These time-varying conditions represent the exposure of the building occupants as a function of their location during the fire. Hazard analysis requires that this predicted exposure be evaluated in terms of the expected response of the occupants to it. Thus, exposure-response is translated into the consequences of the exposure in terms of incapacitation, injury, or death. The ability to assess the exposure-response represents a critical step in moving from fire models to hazard models.

But there are many scenario dependent factors which influence the exposure and the likelihood of an injury or fatality. In the simplest terms, these factors all relate to one common denominator - time. As so aptly put by Cooper [1], we need to know whether the time available for occupant escape is greater than the time needed. The former encompasses all the environmental conditions (e.g., temperature, radiant flux, smoke obscuration) which may delay or prevent successful escape and the latter includes the occupants' awareness and physical ability to reach a refuge or exit the building.

FRAMEWORK

Figure 1 presents a block diagram of the major components of a fire hazard analysis method. Each block represents a model or calculation method which describes a general process, or a data input which is specific to the scenario being evaluated. Within this modular framework the most appropriate model or data is used for each element and improved techniques can be easily substituted as they develop. The calculation begins with a specified fire or combustion model. The specified fire is described in the form of total energy and mass release rates deduced from data from small or large scale calorimeter measurements or experiments. The combustion model (a limited version is contained in the Harvard Computer Fire Code [2]) predicts the total energy and mass release rates taking into account the thermochemical properties of the burning material and the thermophysical properties of the enclosure. If a suppression system is present, its effect on the total release rates of energy and mass is taken into account. A transport model, such as the computer code FAST [3], then takes these total release rates and predicts the distribution of temperature and combustion products throughout the structure, accounting for the influences of building geometry, construction materials, HVAC (including any smoke control system), stack effect and wind. This results in a description of the time varying environment within each compartment to which any occupants therein would be exposed. This information is also used to predict detection or suppression system actuation.

The response of each occupant to this exposure based on a set of prescribed tenability limits is then evaluated (see example later) as a function of location and time of exposure. Location is provided by an evacuation model which begins with a specified, initial occupant distribution and predicts their movement, accounting for notification delays and their capabilities, either inherent or as they might change as a result of exposure to fire products. The eventual result is a prediction of the expected consequences of the fire scenario in terms of injuries or fatalities, and the time, location, and factor(s) related to each.

CALCULATING SMOKE TOXICITY HAZARD

Determining the Exposure-Dose

The procedure for estimating smoke toxicity hazard involves four steps: (1) determining fuel mass loss rate, (2) calculating mass concentration, (3) calculating exposure-dose, and (4) fixing the time at which the exposure begins. Consider an arbitrary compartment containing a growing fire and an occupant exposed to the combustion products. The measured heat release rate of an item of upholstered furniture can be approximated by a triangle as shown in

INTERRELATIONSHIPS OF MAJOR COMPONENTS OF A FIRE HAZARD MODEL

Figure 1 - Hazard Model Framework

Figure 2a using a method described by Babrauskas [4]. By assuming that the effective heat of combustion is a constant, the mass loss rate of the fuel (burning item) can be calculated.

By also assuming that all of the fuel mass lost goes to "combustion products", and that all of these combustion products are contained within the occupied volume, then the increase over time of the mass concentration of combustion products in the space and the resulting exposure to the occupants can be calculated as shown below. Note that both the concentrations and the resultant exposures (Figures 2b and c) vary, depending on whether it is assumed that the combustion products entering the occupied volume are fully mixed or stratified.

For the fully mixed case, the concentration and exposure as a function of time are simply given by equations (1) and (2) below.

$$C(t) = \frac{1}{V} \int_{o}^{t} \dot{m} dt \tag{1}$$

$$E(t) = \int_{o}^{t} C(t) dt \tag{2}$$

where:  C(t) is the combustion products mass concentration
        V is the volume of the space
        $\dot{m}$ is the fuel mass loss rate
        E(t) is the exposure dose [5]

For the stratified case, the volume into which the combustion products are distributed (the upper layer) changes with time, and the exposure only begins when the occupant is immersed in the upper layer. Thus, for this case, equations (3) and (4) would apply.

$$C(t) = \frac{\int_{o}^{t} \dot{m} dt}{AD(t)} \tag{3}$$

$$E(t) = \int_{t'}^{t} C(t) dt \tag{4}$$

where:  D(t) is the upper layer depth at a time t
        A is the compartment floor area
        t' is the time that the layer interface reaches the level of the occupant

For equation (4), the quantity t', i.e., the time at which the interface reaches the level of the occupant must be estimated. Simple filling models such as ASET [6] can be used to determine a value for t'. Otherwise, when the difference in height between the assumed occupant position and the level (base) of the fire is less than about 10% of the compartment ceiling height, a crude approximation for t' can be made from a family of curves (Fig. 2(d)) based on

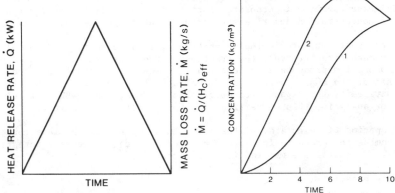

(a) Triangular approximation for upholstered furniture heat release rate

(b) Concentration of "combustion products" vs. time, assuming (1) fully mixed and (2) stratified (linear filling rate)

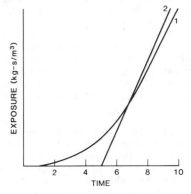

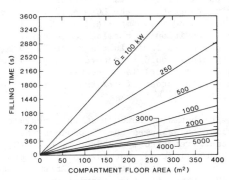

(c) Exposure of occupant, assuming (1) fully mixed and (2) stratified (linear filling rate)

(d) Time to fill to the level of the fire

Figure 2 – Procedure For Hazard Estimation

the work of Cooper [7]. These curves give estimated time to fill an enclosure to the level of the fire as a function of compartment floor area for various levels of heat release rate. This time is approximately t'.

The above described procedure for determining exposure-dose for the stratified case has been incorporated into FAST version 17 [3] by treating the product of concentration and time as a transportable species called CT.

It represents an interim implementation of a smoke toxicity dose calculation. For each fire interval, the fraction of fuel mass which is converted to "toxic" combustion products is calculated. In the NBS protocol [8], the lethal concentration, the $LC_{50}$, is defined as the total fuel mass loaded into the furnace divided by the exposure chamber volume. Thus, where NBS protocol data is used for analysis, this conversion fraction is defined as unity.

The species CT calculated by the model then represents the mass concentration of combustion products in the upper layer of each compartment integrated over time. The units are mg-min/liter = gram-min/m$^3$.

Quantifying Hazard

This method enables easy correction of the predicted value of CT to allow for the actual time to start of exposure (t'). If, for example, it is assumed that the exposure begins when the interface reaches 5 feet (1.5 m) from the floor (nose level of a standing person), it is only necessary to determine the value of CT at this time, and subtract this value from all subsequent values of CT to provide the corrected results. To determine a critical value for CT (called CT*), the $LC_{50}$ for the fuel material is multiplied by the exposure time over which the $LC_{50}$ was determined. For example, if the fuel is PVC undergoing flaming combustion, the $LC_{50}$ = 17 mg/ℓ for a 30 min. exposure. Thus, CT* = 17 x 30 = 510 mg-min/ℓ = 510 g-min/m$^3$. When CT = CT* for the fuel, a lethal condition is considered to exist. Note that, since the 30 min. $LC_{50}$ for most common fuels is in the range 20-40 mg/ℓ, a CT* value of approximately 900 mg-min/ℓ could be generally applied for estimating purposes where a specific value for the fuel is unknown. Likewise, since CT* values for incapacitation are often of the order of 1/2 the value for lethality [9], a value of 450 mg-min/ℓ might be used.

It should be noted that this evaluation procedure (but not the model calculation) assumes the CT product which causes a biological effect is a constant (referred to as Haber's Law [10]). Recent data indicate that this is not generally true, but is the best approximation which can currently be made with available toxicity data. If $LC_{50}$ data are available for different exposure times for the material in question, the Fractional Effective Dose (FED) procedure described by Hartzell, et al. [11], can be used to correct the CT* estimate.

Where the fuel (burning item) consists of a mixture of materials for which the $LC_{50}$ data are available for each, an effective $\overline{LC}_{50}$ (and thus an effective CT*) can be estimated by the following equation [12].

$$\frac{1}{\overline{LC}_{50}} = \sum_i \frac{f_i}{LC_{50_i}}$$

where $f_i$ is the fraction of total fuel mass represented by material i and, $LC_{50_i}$ is the $LC_{50}$ (generally for a 30 min. exposure) of material i

Then $\overline{CT}^* = \overline{LC}_{50} \times 30$ min

If $IC_{50}$ (concentration necessary to incapacitate), or $EC_{50}$ (concentration necessary to produce any specified effect) data are available, they would be used in exactly the same way to produce a CT* and predict time to incapacitation or other effect.

Progress on Assessing the Accuracy of FAST

Initial work on the development of techniques for assessing the accuracy of fire models has used the transport model FAST an an example; but the techniques are intended to be generally applicable to any model. This work has shown the value of coordinated experiments both to establish the statistical accuracy of predicted quantities and to identify the sensitivity of results to model assumptions.

The criterion used for comparison of model predictions and experiments was that, for a given parameter, the model prediction must lie within the normal variation for a set of replicate experiments. Thus, a set of experiments (generally 5) was conducted and the derived parameters were statistically analyzed to produce an envelope of ± one standard deviation about the mean value for the set. This requires that the prediction be within a range covered by 68% of the experimental values. The preliminary results of this exercise with FAST version 17 [3] are presented in the next section.

Experiments

The facility constructed for the study consisted of a 2.4 m (8 ft) cubic burn room at one end of a 10 m (30 ft) corridor. For some experiments, a target room of the same size as the burn room was included near the far end of the corridor. The fire source was a natural gas burner where about 15% acetylene was mixed with the natural gas to produce visible smoke. The door at the far end of the corridor from the burn room was below a collector hood equipped for oxygen consumption measurements as a check on heat release rate in the effluent flow. In some experiments, this door was closed except for a 20 mm undercut. Complete details on the facility and experimental results will be published in a separate report by Peacock, et al. [13].

Some of the results for two sets of experiments (five open door and five closed door) with a 100 kW fire strength are presented in Figures 3 and 4. In each case, the experimental results are presented as two curves, representing plus and minus one standard deviation for a series of five experiments. The predictions by the model treating the facility as a two room arrangement are shown. On some plots, the effect of including the short corridor-like space between the burn room and corridor as a third compartment in the model is also shown. The level of agreement varies from excellent to fair, although the reasons for disagreement are not always indicative of problems with the model. This is demonstrated by the comparisons with corridor interface height.

The interface position is derived from experimental data in two ways: a temperature method similar to one proposed by Cooper, et al. [14] and a new smoke method. The temperature method estimates the temperature at the interface at 10% to 20% of the temperature rise on the top thermocouple in a tree (15% was used for the data presented). The vertical location of this tempera-

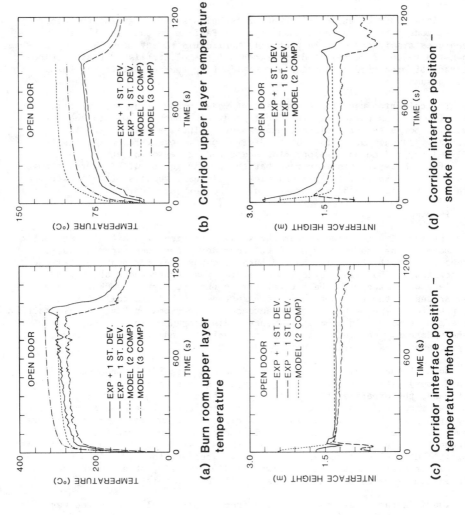

(a) Burn room upper layer
    temperature

(b) Corridor upper layer temperature

(c) Corridor interface position –
    temperature method

(d) Corridor interface position –
    smoke method

Figure 3 – Comparison Of Experiments To Model – Open Door

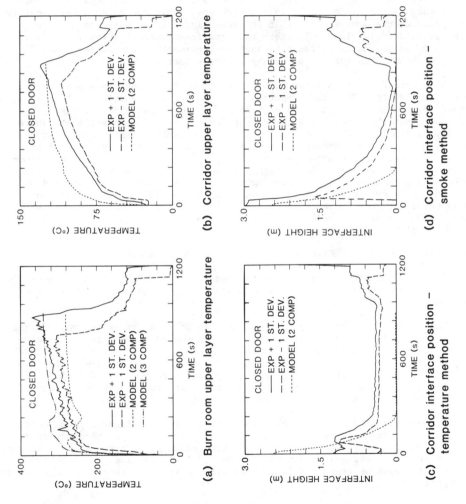

(a) Burn room upper layer temperature

(b) Corridor upper layer temperature

(c) Corridor interface position – temperature method

(d) Corridor interface position – smoke method

Figure 4 – Comparison of Experiments To Model – Closed Door

ture is then calculated by linear interpolation between the thermocouples in the tree. This method is thus bounded by the physical location of the top and bottom thermocouples - in this case 15 cm (6 in) below the ceiling and above the floor.

The smoke method employs a horizontal smoke meter located near the ceiling and a vertical smoke meter from the floor to ceiling (actually built into the floor and ceiling so that the path length is full room height). From the horizontal meter, the optical density (OD) per unit path length in the upper layer is obtained. This value is then set equal to the OD per unit path for the vertical meter, and the "effective path length" which produces this result is obtained. This is the upper layer thickness. This technique makes the same assumption as the zone model - that the upper layer is uniform and the lower layer is relatively clean. If there is significant contamination in the lower layer, it can be accounted for with another horizontal meter in the lower layer; although this was not necessary in these tests.

Both methods were employed to measure the corridor interface position in these experiments. Only the temperature method was used within the burn room. Comparison of the two methods to each other and to visual observations indicates that the results are comparable for the open door experiments where the layer stabilized at about mid level. Both showed the same major features including the rise in interface position after the burner was turned off at 900 s. The temperature method shows the layer beginning below the top thermocouple at zero time due to a small thermal layer from the burner pilot. The smoke method shows the layer beginning at the ceiling since the pilot did not have the acetylene feed. For the closed door tests, however, the smoke method shows the layer goes to the floor which agrees with visual observations, while the temperature method shows it stopping at the bottom thermocouple for the reasons explained previously.

IMPLICATIONS OF EXPERIMENTAL RESULTS FOR ASSUMPTIONS IN FAST

The data from the experiments were used to identify potential improvements in FAST. These improvements relate to conduction through layered walls, the door jet, and leakage.

The experimental facility surfaces are kaowool over brick in the burn room and calcium silicate board over gypsum board in the corridor. In the model an attempt was made to treat this multilayered wall as a single layer of material with composite properties. Poorer agreement between the model prediction and experiments and the difficulty of estimating composite properties indicate this approximation should not be used. Thus a conduction routine which accounts for up to three layers in a wall has been developed.

As most models do, FAST assumes the door jet is a horizontal plume, with a circular cross section; when, in fact, the cross section is rectangular due to flow through the door opening. This different geometry effects the entrainment. The effect of a rectangular geometry in the model was examined and some improvement in agreement was seen. This feature will be included in the next generation transport model.

The effect of leakage at the wall/ceiling junction was also investigated. In the experiments, visual observations identified that leaks occurred, and the mass flows at the corridor door showed more flow in than out. If this difference in corridor door mass flow rates were included in the model as a leak from the upper layer, the agreement with experimental conditions should improve substantially. Thus, care must be taken to either eliminate leaks in the test, or to include them in the model.

SUMMARY

A framework for fire hazard assessment which describes the major components and their interactions was presented. Specific models or techniques which might be used within this framework may vary depending on the level of accuracy required by the application.

An engineering approach to smoke toxicity evaluation was then presented within the context of this framework. The easily used approach utilizes data produced by currently available toxicity test methods.

Finally, an example of the use of validation experiments to provide an understanding of the predictive accuracy of the transport model FAST was described. The importance of some of the assumptions used in FAST was indicated by the experiments. Considerably more work has been completed than is reported here, and a complete report is in preparation [13].

REFERENCES

[1]  Cooper, L.Y. and Nelson, H.E., Life Safety Through Designed Safe Egress – A Framework for Research, Development and Implementation, Proceedings of the 6th UJNR, Tokyo, Japan, 1982.

[2]  Mitler, H.E. and Emmons, H.W., Documentation for CFC V, The Fifth Harvard Computer Fire Code, National Bureau of Standards, NBS-GCR-81-344, Oct. 1981.

[3]  Jones, W.W., A Multicompartment Model for the Spread of Fire, Smoke, and Toxic Gases, Fire Safety Journal, Vol. 19, Nos. 1 and 2, May/July 1985.

[4]  Babrauskas, V. and Krasny, J.F., Fire Behavior of Upholstered Furniture, NBS Monograph MN-173, in press.

[5]  Huggett, C., Combustion Conditions and Exposure Conditions for Combustion Product Toxicity Testing, J. of Fire Sciences, 2, 328-347, Sept/Oct 1984.

[6]  Cooper, L.Y., A Mathematical Model for Estimating Available Safe Egress Time in Fires, Fire and Materials, Vol. 16, Nos. 3 and 4, 135-144, 1983.

[7]  Cooper, L.Y., The Development of Hazardous Conditions in Enclosures with Growing Fires, NBSIR 82-2622, NBS, Gaithersburg, MD 20899, December 1982.

[8]  Levin, B.C., Fowell, A.J., Birky, M.M., Paabo, J., Stolte, A., and Malek, D., Further Development of a Test Method for the Assessment of the Acute Inhalation Toxicity of Combustion Products, NBSIR 82-2532, NBS, Gaithersburg, MD 20899, June 1982.

[9]  Kaplan, H.L. and Hartzell, G.E., Modeling of Toxicological Effects of Fire Gases: I. Incapacitating Effects of Narcotic Fire Gases, J. of Fire Sciences, Vol. 2, No. 4, 286-305 (1984).

[10] Kaplan, H.L., Grand, A.F., and Hartzell, G.E., Combustion Toxicology Principles and Methods, Technomic Publishing Co., Inc., Lancaster, Pa., p. 25.

[11] Hartzell, G.E., Priest, D.N., and Switzer, W.G., Mathematical Modeling of Toxicological Effects of Fire Gases, Proceedings of the 8th UJNR, Tokyo, Japan, in press.

[12] Bukowski, R.W., Evaluation of Furniture Fire Hazard Using a Hazard Assessment Computer Model, Fire and Materials, in press.

[13] Peacock, R.D., Davis, S., and Lee, B.T., Experimental Data for Model Validation, NBSIR, in press.

[14] Cooper, L.Y., Harkleroad, M., Quintiere, J., and Rinkinen, W., An Experimental Study of Upper Hot Layer Stratification in Full-Scale Multiroom Fire Scenarios, J. Heat Transfer, 104, 741-749.

# The Effects of Fire Products on Escape Capability in Primates and Human Fire Victims

DAVID A. PURSER
Huntingdon Research Centre Ltd.
Huntingdon, Cambridgeshire, England

ABSTRACT

Animal studies of incapacitation by thermal decomposition products indicate that the common asphyxiant gases, CO, HCN, low $O_2$ and $CO_2$ are almost certainly responsible for the severe narcosis and death of fire victims overcome by smoke. Eye and upper respiratory tract irritation also probably impair escape capability but to an unknown degree. Time to narcotic incapacitation in man should be predictable if the fire profile in terms of the above gases is known. Incapacitation of victims of smouldering and post-flashover fires can be explained in these terms, but victims should be able to escape from early flaming fires. It is suggested that the high incidence of victims in the room of fire origin may be partly due to sleeping victims being intoxicated by CO during the smouldering phase.

INTRODUCTION

Statistical surveys of fire casualties in the UK during the mid 1970's revealed that not only were a large proportion of fatal and non-fatal fire casualties being reported in the category "overcome by smoke and toxic gases" rather than by heat and burns, but that there was a four-fold increase in this category between 1955 and 1971[1]. This increasing trend has continued over the last decade so that now approximately half of all fatal casualties and a third of all non-fatal casualties of fires in dwellings, the majority caused by fires in furniture and bedding, are reported as being overcome by smoke and toxic gases[2]. This has occurred despite the fact that the total annual numbers of fires have remained approximately constant over this period of time.

There is therefore good evidence that smoke and toxic gases are a considerable problem in fires and in general two rather different approaches have been used to evaluate the toxicity associated with burning materials. Once consists basically of a 'black box' approach, whereby materials are decomposed in a standard small scale apparatus and materials are ranked on the basis of rodent $LC_{50}s$[3]. The problems with this approach are that little attempt is usually made to identify the decomposition products responsible for the toxicity and that laboratory scale furnaces cannot recreate the complex temperature/time/product profiles occurring in large scale fires. The other approach is to measure the conditions in large scale fire tests in terms of temperature, smoke density and the profiles of the major toxic products with time (usually without animal exposures) and then to attempt to predict the likely effects on human fire victims from a knowledge of the toxicity of specific fire gases. This approach suffers from an inadequate knowledge of the important toxic fire products and exactly how they cause incapacitation and death individually and in combination[4].

However, if it could be demonstrated that toxic effects in fires are caused by a small number of identifiable products, then by studying the effects of these gases individually and in combination it should be possible, by measurements of the concentration/time profiles of the common fire gases, to predict how a person confronted with a particular situation would be incapacitated and how their ability to escape would be affected. The main function of small scale toxicity tests would then be to confirm that the toxicity associated with particular burning materials was due to the common toxic fire products, by means of chemical atmosphere analysis and animal exposures, and to identify those cases where unusual toxic effects occurred(5).

At Huntingdon Research Centre we have been studying the effects of combustion atmospheres on animals, mainly primates, at sublethal levels to examine the mechanisms whereby people become incapacitated in fires(6).

INCAPACITATION BY COMBUSTION PRODUCTS

As a result of the animal exposures to combustion product atmospheres from a wide range of materials it was found that despite the great complexity in chemical composition of the products the basic toxic effects on the animals were relatively simple, and for each individual smoke atmosphere the toxicity was always dominated by a narcotic gas or by irritants(6). Interactions between individual narcotic gases, or between narcotics and irritants, were found to be minor, so that a reasonably good predictive model of incapacitation could be developed by considering each of a small number of individual toxic gases as acting separately. From the results of these animal studies and available human data a model has been developed for predicting the incapacitating effects of fire products on victims and the time during exposure when they should occur, to determine likely effects on escape capability. This is achieved by consideration of 2 sources of information:

1.  An extrapolation from animal data and such human data as is available to determine probable time to and nature of incapacitation for common fire products;

2.  Consideration of the concentration/time profiles of the fire products in large scale experimental fires of different types.

These predictions have then been compared with data on real fire victims, derived from the Strathclyde fire victim pathology studies(7). Home Office Statistics(2) and a small number of case reports, in an attempt to test the predictability of the model and to determine if this approach helps to explain how people are affected physiologically by fire and how some mitigation of toxic hazard in fires might be achieved.

NARCOSIS BY FIRE GASES

Narcotic gases cause incapacitation mainly by effects upon the central nervous system, and to some extent the cardiovascular system(8). In general time to incapacitation and its severity are predictable in that there is usually a relatively sharp cut off between a near normal state and one of severe incapacitation(6,9). Most narcotic fire gases produce their effects by causing brain tissue hypoxia(8,10), and since the body posesses powerful adaptive mechanisms designed to maximize oxygen delivery to the brain it is usually possible to maintain normal body function up to a certain concentration or dose of narcotic. However once the point is reached where normal function can no longer be maintained deterioration is rapid and severe, beginning with signs similar to the effects of alcohol, consisting of lethargy or euphoria with poor physical co-ordination followed rapidly by unconsciousness and death if exposure continues(9,10). The major narcotic gases found in fires are CO, HCN, low $O_2$ and high $CO_2$(6,11).

CO

The most important narcotic fire gas is CO which presents a serious hazard at concentrations of approximately 100 ppm and above. In the Strathclyde pathology study(7) lethal levels (>50% carboxyhaemoglobin) were found in 54% of all fatalities, while some 69% of victims had carboxyheamoglobin levels capable of causing incapacitation (>30% carboxyhaemoglobin). Incapacitating levels of carboxyhaemoglobin were also common in victims surviving the immediate fire, so that CO is evidently important as a cause of both incapacitation and death. CO uptake and intoxication are extremely insidious. During the early stages, as the carboxyhaemoglobin levels build up gradually in the blood, the effects are minimal, and in low level exposures in man(12) the first symptoms were of a headache at 15-20% carboxyhaemoglobin, and objective tests at these levels show only minor performance deficits. When significant effects do occur their onset is sudden and rapid(9), and the degree of incapacitation is severe, so that by the time a victim is aware that he is affected he is probably unable to take effective action. These findings may explain why deaths from CO derived from defective heating appliances are so common. Survivors of such situations often report that they, or other victims that died, experienced headaches or nausea, but had no idea of the cause, so they did not attempt to leave the area until overcome by fumes[13].

During the early stages of incapacitation the main effects appear to be on motivation and psychomotor ability, with a tendency for the victim to sleep if left undisturbed[9]. Under these conditions one might expect a subject, if alerted by a sudden noise such as of breaking glass (often reported by fire survivors) to 'sober up' and rouse himself sufficiently to make an escape attempt. However such a victim is likely to fail for 3 reasons:

Firstly because this stage is rapidly followed by unconsciousness and coma;

Secondly because active subjects are seriously affected at carboxyhaemoglobin concentrations which have only minor effects on sedentary subjects. Thus whereas sedentary animals were often unaffected at carboxyhaemoglobin levels of up to 40%, those engaged in light activity were seriously affected at carboxyhaemoglobin levels in the 25-35% range[9]. Similarly, in one human study, although a sedentary subject could perform tasks such as writing, even at the exceptionally high level of 55% carboxyhaemoglobin, the subject collapsed and became unconscious immediately when he attempted to get up and walk[14]. Thus if a victim in a bed or chair did attempt to escape not only would he be in danger of a rapid collapse due to continued CO uptake, but even if no further uptake occurred, the ability to perform even light work or exercise would be severely impaired, and even the simple act of rising from a horizontal to an upright position could precipate loss of consciousness;

The third important feature is that the rate of uptake of CO depends on the respiration (respiratory minute volume - RMV) and hence the activity of the subject.

A MODEL FOR THE PREDICTION OF TIME TO INCAPACITATION BY CO IN FIRES

FIGURE 1

TIME TO INCAPACITATION BY CARBON MONOXIDE FOR A 70 kg MAN
AT DIFFERENT LEVELS OF ACTIVITY

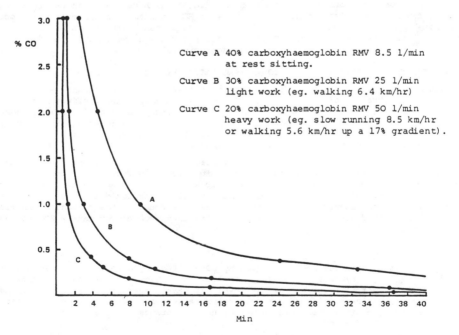

Curve A 40% carboxyhaemoglobin RMV 8.5 l/min at rest sitting.

Curve B 30% carboxyhaemoglobin RMV 25 l/min light work (eg. walking 6.4 km/hr)

Curve C 20% carboxyhaemoglobin RMV 50 l/min heavy work (eg. slow running 8.5 km/hr or walking 5.6 km/hr up a 17% gradient).

Incapacitation by CO depends upon a dose accumulated over a period of time until a carboxyheamoglobin concentration is reached where compensatory mechanisms fail and collapse occurs. In order to predict time to incapacitation of fire victims due to CO it is necessary to know the carboxyhaemoglobin concentrations at which incapacitation is likely to occur and the rate of uptake of CO so that the time to achieve this concentration can be calculated. The carboxyhaemoglobin concentrations likely to cause incapacitation depend upon the activity of the victim and should be similar to those measured in the primate experiments[9]. The rate of CO uptake in man can be calculated using the Coburn Foster Kane equation[15]. From these data I have constructed Figure 1, which shows the probable time to incapacitation (loss of consciousness) for a 70 kg man exposed to different CO concentrations at 3 levels of activity. The figure shows that the degree of activity can have a major effect on time to incapacitation.

HCN

Cyanide has been measured in the blood of both fatal[16] and non-fatal[17] fire victims. However in the Strathclyde fire fatality study high concentrations of cyanide in the blood of victims were usually associated with lethal levels of carboxyhaemoglobin, so that the role of cyanide as a cause of incapacitation was difficult to determine[16]. It is also difficult to relate blood cyanide levels from samples collected after a fire to likely HCN exposure, since the dynamics of HCN uptake and removal from the blood are poorly understood[18].

The pattern of incapacitation for HCN is somewhat different from that produced by CO in that the effects occur more rapidly, as unlike CO, HCN is not held almost exclusively in the blood, but carried rapidly to the brain[19]. With HCN although the accumulation of a dose is one factor, the most important determinant of incapacitation appears to be the rate of uptake of HCN, which in turn depends upon the HCN concentration in the smoke and the subjects' respiration. Thus in the animal experiments[8,18], it was found that at HCN concentrations below approximately 80 ppm the effects were minor over periods of up to 1 hour, with a mild background hyperventilation. At concentrations above 80 ppm up to approximately 180 ppm an episode of hyperventilation with subsequent unconsciousness occurred at some time during a 30 minute period and there was a loose linear relationship between HCN concentration and time to incapacitation while above 180 ppm the hyperventilatory episode began immediately with unconsciousness occurring within a few minutes. Data on human exposures to HCN is limited but Kimmerle[20] does quote some approximate data showing a similar effect in man, with incapacitation occurring after 20-30 minutes at 100 ppm HCN and after 2 minutes at 200 ppm, death occurring rapidly at concentrations of above approximately 300 ppm.

A MODEL FOR THE PREDICTION OF TIME TO INCAPACITATION BY HCN IN FIRES

From these results it is possible to predict that HCN concentrations below a threshold concentration of approximately 80 ppm will have only minor effect over periods of up to 1 hour. From 80-180 ppm the time to incapacitation (unconsciousness) will be between 2 and 30 minutes according approximately to the relationship: Time to incapacitation = (ppm HCN - 185)/-4.4. For concentrations above approximately 180 ppm incapacitation will occur very rapidly (o-2 minutes).

HCN could be particularly dangerous in fires due to its rapid knock down effect, and low HCN levels in the 100-200 ppm range could cause fire victims to loose consciousness rapidly and remain in the fire to die later as a result of accumulation of CO or some other factor. Also a small change in HCN concentration could cause a large decrease in time to incapacitation so that for example doubling the concentration from 100 to 200 ppm could bring the incapacitation time down from approximately 20 minutes to approximately 2 minutes[18,20].

HYPOXIA

Apart from the tissue hypoxia caused by CO and HCN, hypoxia in fires can also be caused by low oxygen exposure. Due to physiological compensatory mechanisms hypoxia has little effect down to 15% $O_2$, but as the level decreases towards 10% $O_2$ these compensatory mechanisms begin to fail and narcotic intoxication with lethargy, and sometimes euphoria rapidly occurs. This is followed by unconsciousness, and death at concentrations below 7% $O_2$[8,10,20].

CO$_2$

Carbon dioxide, like carbon monoxide, is universally present in fires. It is not toxic at concentrations up to 5%. However at 5% breathing is strongly stimulated (by a factor of 3) [8], and this hyperventilation, apart from being stressful, can increase the rate at which other toxic fire products (such as CO) are taken up. However at concentrations above 5% CO$_2$ is itself narcotic, at 7-10% unconsciousness unconsciousness occurs in man after a few minutes [21,22,23].

## PREDICTION OF TIME TO INCAPACITATION BY LOW O$_2$ or CO$_2$

For these two gases the degree of incapacitation does not appear to increase significantly with time once equilibrium has been established, over periods of up to 30 minutes [8,10], but narcosis is likely to occur over periods of upto 5 minutes below a threshold concentration of approximately 12% O$_2$ or above approximately 10% CO$_2$. Also at either of these concentrations it is likely that exercise tolerance will be severely reduced [10].

## INTERACTIONS BETWEEN TOXIC FIRE GASES

The effects of interactions between combinations of these gases on time to incapacitation in fires is an area that requires further investigation as little information is available. The most important effect is that hyperventilation due to CO$_2$ exposure is likely to increase the rate of uptake of other toxic gases and thus decrease time to incapacitation in proportion to the increase in breathing. For other combinations CO/low O$_2$/HCN, available data suggest that only minor interactive effects are likely [24], so that if allowance is made for hyperventilation due to CO$_2$, errors in estimation of time to incapacitation will probably not be great (< 20%) if narcotic gases are assumed to act independently at concentrations existing in practical fire situations.

## IRRITANT FIRE PRODUCTS

Unlike the incapacitating effects of narcotics, which are clear cut and well understood, the incapacitative effects of irritants are much more difficult to determine. Irritant fire products produce incapacitation by their painful effects upon the eyes and upper respiratory tract, and to some extent also the lungs. They can also be dangerous to victims surviving the immediate fire exposure, producing a pulmonary irritant response consisting of oedema and inflammation which can cause respiratory difficulties and even death 6-24 hours after exposure [25,26]. The effects do not show the sharp cut off of narcosis, but lie on a continuum from mild eye irritation to severe pain, depending upon the concentration of the irritant and its potency [27,28,29]. The effects do not depend upon an accumulated dose but occur immediately upon exposure, and usually lessen somewhat if exposure continues [6,29].

The effects of low concentrations of irritants can best be considered as adding to the obscurational effects of smoke by producing mild eye and upper respiratory tract irritation. In this situation irritants may have some effect by impairing the speed of movement through a building, as would visual obscuration. The limitation of escape capability may not be simply limited to direct physiological effects, but also to physchological and behavioural effects such as the willingness of an individual to enter a smoke filled corridor [30].

At the other end of the scale when irritants are present at high concentrations there is some disagreement about the likely degree of incapacitation. Some investigators believe that the painful effects on the eyes and upper respiratory tract would be severely incapacitating, so that for example escape from a building would be rendered extremely difficult [31]. Others believe that the effects peak out at moderate concentrations, and that although the effects may be very unpleasant they would not significantly impair the ability to escape from a building and that they would provide a strong stimulus to escape that might almost be beneficial [32].

The most extensive studies of the effects of severe irritancy in man have been performed on volunteers exposed to riot control agents such as CS (o-chlorobenzylidine malonitrole) or CN (a-chloroacetophenone). Even these studies do not really show how the ability to escape from a building might be effected but they do to some extent convey the severity of the effects. The effects of CS, which are probably similar to those of any severe sensory irritant, have been described by Beswick et al(29). They consist of an almost instantaneous severe inflammation of the eyes accompanied by pain, excessive lacrimation and blepharospasm (spasm of the eyelids). There is irritation and running of the nose with a burning sensation in the nose, mouth and throat and a feeling of intense discomfort during which these subjects cough, often violently. If the exposure continues, the discomfort spreads to the chest and there is difficulty in breathing, and many subjects describe a tightness of the chest or pain as the worst symptom. At this stage most individuals are acutely apprehensive and highly motivated to escape from the smoke. However if exposure continues there is some remission of signs and symptoms.

Among fire victims reports are conflicting. Some people say they went through dense smoke without experiencing any great discomfort, while others say that respiratory difficulties prevented them from entering smoke filled areas(33). Anyone who has had bonfire smoke in their eyes will know how painful the experience can be. However, the effects can be mitigated by blinking or shutting the eyes, and the effects on the nose can be mitigated by mouth breathing and breath holding. Also, it is known that people are often unaware of painful stimuli when in emergency situations(34). It is therefore likely that irritant smoke products do have some effects on the escape capability of fire victims, but it is not possible at present to determine the degree of incapacitation.

FIRE SCENARIOS AND VICTIM INCAPACITATION

From the point of view both of product composition and toxic hazard, it is possible to distinguish three basic types of fire situations:

1. Smouldering fires where the victim may be in the room of origin of the fire or a remote location.

2. Early flaming fires where the victim is in the room of origin.

3. Fully developed or post-flashover fires where the victim is remote from the fire.

In the UK 80% of fire deaths and injuries occur in domestic dwellings, and in most cases the casualties occur in the compartment of origin of the fire. This class of fire is responsible for the highest incidence of deaths (60%) and a high incidence of injuries (39%), and these fires occur mostly in living rooms or bedrooms, and in upholstery or bedding(2). In these cases fire is often confined to the material first ignited. The toxic hazard in such fires depends upon whether there is a long period of smouldering, or whether there is a rapidly growing flaming fire.

With smouldering fires the decomposition temperatures are relatively low (~400°C) and materials are decomposed into a mixture of pyrolysis and oxidation fragments containing mixtures of narcotic and irritant gases and particulates. Under these conditions the highest yields of a great variety of products are formed, many of which are irritant. Also incomplete oxidation is favoured and $CO_2/CO$ ratios approach unity, so that CO is likely to be an important toxic factor. The formation of high yields of HCN is however not normally favoured(11). Although toxic products are formed under these conditions the rate of evolution is slow, smoke is seldom dense and room temperatures are relatively low. A potential victim therefore has ample time to escape if alerted sufficiently early, but may be overcome by fumes after a long period of time if unaware of the danger, particularly if asleep. Here the main danger is almost certainly narcosis by CO, with possibly a small contribution from low $O_2$ if the victim is in a room with a poor air supply(6,35,36). It is not possible from fire statistics to determine how common this type of fire is, since in many cases smouldering fires become flaming fires before they are detected. However, it is likely that fires which are estimated to have burned for 30 minutes or more before discovery have involved long term smouldering, and it may be relevant that deaths are twenty times more likely in this situation, than for fires discovered within 5 minutes of ignition, which are often rapidly growing flaming fires(2).

For flaming fires where the victim is in the room of origin the hazard relates to the early stages of fire growth. Such fires often grow quickly, but even the most rapidly growing flaming fires take approximately 3 minutes to reach levels of heat and gases hazardous to life(37), which should allow ample time to escape from a room, and of course most people do escape. As Figure 2 shows the hazard in this situation relates to a number of factors all of which may reach life threatening levels simultaneously as the fire reaches the rapid phase of exponential growth. In the high temperature, well

FIGURE 2

## SMOKE, HEAT AND GASES DURING SINGLE CHAIR ROOM BURN

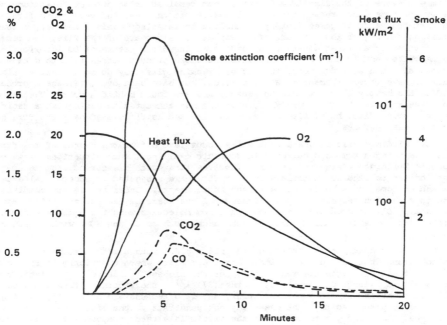

Armchair - Polystyrene with polyurethane cushions and covers.
Room      - 29 m$^3$, open doorway. Gases measured in doorway at 2.1 m height.
FROM BABRUSKAS 1979

oxygenated flames of early flaming fires most thermal decomposition products are consumed to form simple comparatively innocuous products such as $CO_2$ and water, the $CO_2/CO$ ratios being in the 200-1000 range initially.  Since CO is approximately 10-50 times as toxic as $CO_2$ it is thus concievable that in this type of fire $CO_2$ could present more of a narcotic hazard than CO.  However as the $CO_2$ concentration in the fire compartment approaches 5% and the $O_2$ concentration decreases towards 15% the combustion becomes less efficient and the $CO_2/CO$ ratios decrease to the region of 50-100 and CO tends to become a more important factor.  Nevertheless as the armchair burn in Figure 2 shows the atmospheres obtained in a rapidly growing fire can contain narcotic concentrations of $CO_2$ (>5%), CO (>1000 ppm) and low oxygen (<15% $O_2$).  In addition some of the pyrolysis products escape the flame zone giving rise to potentially irritant smoke.  A victim in this situation is therefore likely to be confronted simultaneously by high temperatures and heat radiation, smoke and high concentrations of CO and $CO_2$ accompanied by low $O_2$, any one of which could incapacitate a victim and prevent escape.

The inability of victims to escape from such fires seems to depend upon a number of factors.  Casualties include a higher proportion of young children and old people than does the general population (for over 65s fatalities in bedding fires 7 times that expected from population distribution - 1978) [38], and people who are incapacitated by a previous period of smouldering or by some other infirmity (such as a physical disability, alcohol or drug intoxication) are obviously more at risk [7].  However there seem to be two other factors of importance, the behaviour of the victim and the exponential rate of fire development.

In many cases the victim has a short period in which to carry out the correct actions enabling escape, after which he is rapidly trapped. Some victims may be asleep during this critical escape 'window', but there are also reports of situations where the victim was aware of the fire from ignition, but remained in an attempt to extinguish the fire or for some other reason failed to leave before the phase of very rapid fire growth when heat and narcotic gases rapidly reach life threatening levels. Another, perhaps surprising finding is that victims often appear to be unaware of the fire, and remain to be discovered in a burned out chair or bed. The insidious nature of CO intoxication has been described and it also seems that irritant smoke products often fail to wake sleeping victims, although a sudden noise such as of breaking glass may do so. Other victims appear to have roused themselves at some stage, but have been overcome, again probably by CO or HCN before they are able to escape, and are found behind a door. There are also cases reported by survivors where a victim has attempted to extinguish a rapidly growing flaming fire, but failed to leave in time and is discovered near the fire having been overcome by fumes[35,36].

The third scenario is where casualties occur remote from the source of the fire. Apart from being a common occurrence in domestic dwellings, such situations often occur in public buildings where the situation involves fire which has spread from the material first ignited to others. Materials in such fires are subjected to substantial external flux and in some cases to oxygen deficient environments. Under the severe conditions found in such high temperature post-flashover fires where oxygen concentrations are low, the basic pyrolysis products break down into low molecular weight fragments and can contain high concentrations of narctoic substances such as CO and HCN, with $CO_2/CO$ ratios of <10[11].

Under such conditions a building can fill rapidly with a lethal smoke capable of causing incapacitation and death within minutes. Fires where the victim is remote from the compartment of origin are responsible for the highest incidence of non-fatal casualties (48%) and a large proportion of deaths (37%)[2]. The victim is five times more likely to be killed by smoke than by burns, and is often unaware of the fire during the crucial early phase, so that the gases may not penetrate to the victim until the fire has reached its rapid growth phase and the victim is already trapped. The major causes of incapacitation and death in this type of fire are almost certainly narcotic gases, particularly CO, which can build up rapidly to high concentrations, although the role of irritants in causing incapacitation and impeding escape attempts may be crucial.

COMMENT

The severe narcotic incapacitation and subsequent deaths of many fire victims is almost certainly due to the common narcotic gases.

However, the importance of irritants in impeding escape is unknown, and from narcotic gas profiles it is not obvious why so many fatalities occur in the room of fire origin. Useful information may be obtainable from survivors who have experienced exposure to dense, irritant smokes and from case studies of 'room of origin' fires.

POSSIBLE ROUTES TO MITIGATION OF TOXIC HAZARD

For smouldering fires it would be an advantage if materials were designed to self extinguish, and if the formation of products other than CO during decomposition could be encouraged (such as oxidised hydrocarbon fragments or $CO_2$). Early audible warning by smoke alarms may be particularly advantageous as sound often appears to alert victims where the presence of irritant smoke or heat fails.

For early flaming fires where the victim is in the room or origin, any measure which limits the rate of growth once ignition has occurred will give a victim more time to extinguish a small fire or escape from a growing one.

For fully developed fires where the victim is remote from the point of origin the most important mitigating factors are probably early warning and containment of the fire and gases within the original fire compartment.

REFERENCES

1.  Bowes, P.C., "Smoke and Toxicity Hazards of Plastics in Fires", Annals of Occupational Hygiene, 17 : 143-157, 1974.

2.  United Kingdom Fire Statistics 1980. London, Home Office, Published annually.

3.  Anderson, R.A., Croce, P.A., Feeley, F.G. and Sakura, J.D., "Study to Assess the Feasibility of Incorporating Combustion Toxicity Requirements into Building Material and Furnishing Codes of New York State", A.D. Little, Reference 88712, 1983.

4.  Sarkos, C.P., Hill, R.G. and Howell, W.D., "The Development and Application of a Full-scale Wide Body Test Article to Study the Behaviour of Interior Materials During a Postcrash Fuel Fire", AGARD-LS-123, 601-621, 1982.

5.  Purser, D.A., "Combustion Toxicology Research and Animal Models", Society of the Plastics Industry meeting. Hilton Head S.C. U.S.A., August 9th, 1983.

6.  Purser, D.A. and Woolley, W.D., "Biological Studies of Combustion Atmospheres", Journal of Fire Sciences, 1: 118-144, 1983.

7.  Anderson, R.A., Watson, A.A. and Harland, W.A., "Fire Deaths in the Glasgow Area: 1. General Conclusions and Pathology", Medicine, Science and Law, 21: 175-183, 1981.

8.  Purser, D.A., "A Bioassay Model for Testing the Incapacitating Effects of Exposure to Combustion Product Atmospheres using Cynomolgus Monkeys", Journal of fire Sciences, 2 : 20-36, 1984.

9.  Purser, D.A. and Berrill, K.R., "Effects of Carbon Monoxide on Behaviour in Monkeys in Relation to Human Fire Hazard", Archives of Environmental Health 38 : 308-315, 1983.

10. Luft, U.C., "Aviation Physiology - The Effects of Altitude", in "Handbook of Physiology. Section 3 : Respiration", eds. W.O. Fenn and H. Rahn, 1110, American Physiological Society, Washington D.C., 1965.

11. Woolley, W.D. and Fardell, P.J., "Basic Aspects of Combustion Toxicology", Fire Safety Journal, 5 : 29-48, 1982.

12. Stewart, R.D., Peterson, J.E., Baretta, E.D., Dodd, H.C. and Herrmann, A.A., "Experimental Human Exposure to Carbon Monoxide", Archives of Environmental Health 21 : 154-164, 1970.

13. "Carbon Monoxide Poisoning Due to Faulty Gas Water Heaters in British Holidaymakers on the Algarve". "Sunday Times", 1983.

14. Von Leggenhager, K., "New Data on the Mechanisms of Carbon Monoxide Poisoning", Acta Medica Scandanavica, 196/suppl: 563, 1-47, 1974.

15. Peterson, J.E. and Stewart, R.D., "Predicting the Carboxyhaemoglobin Levels Resulting From Carbon Monoxide Exposure", Journal of Applied Physiology, 39 : 633-638, 1975.

16. Anderson, R.A., Thompson, I. and Harland, W.A., "The importance of Cyanide and Organic Nitriles in Fire Fatalities", Fire and Materials, 3 : 91-99, 1979.

17. Clark, C.J., Campbell, D. and Reid, W.H., "Blood Carboxyhaemoglobin and Cyanide Levels in Fire Survivors", Lancet, June 20th : 1332-1335, 1981.

18. Purser, D.A., Grimshaw, P. and Berrill, K.R., "The Role of Hydrogen Cyanide in the Acute Toxicity of the Pyrolysis Products of Polycarylonitrile", Archives of Environmental Health, 39 : 394-400, 1984.

19. Ballantyne, B., "Artifacts in the Definition of Toxicity by Cyanides and Cyanogens", Fundamental and Applied Toxicology, 3 : 400-408, 1983.

20. Kimmerle, G., "Aspects and Methodology for the Evaluation of Toxicological Parameters During Fire Exposure", Journal of Combustion Toxicology, 1 : 4-50, 1974.

21. Haldane, J.S., "Respiration", New Haven : Yale University Press, 1922.

22. Schulte, J.E., "Sealed Environments in Health and Disease", Archives of Environmental Health, 8 : 427-452, 1964.

23. Documentation of the Threshold Limit Values for Substances in Workroom Air. American Council of Governmental Industrial Hygienists, 1980.

24. Purser, D.A., "The Role of Narcosis by Common Fire Toxicants", in Toxic Hazards From Fire, ed. GE Hartzell, Technomic, Lancaster P.A., 1984.

25. Purser, D.A. and Buckley, P., "Lung Irritance and Inflammation During and After Exposure to Thermal Decomposition Products From Polymeric Materials", Medicine, Science and Law, 23 : 142-150, 1983.

26. Campbell, D., "Respiratory Tract Trauma in Burned Patients", Research Colloquim "People and Fire", Fire Research Station, Borehamwood, 1985.

27. Kane, L., Barrow, C.S. and Alarie, Y., "A Short-term Test to Predict Acceptable Levels of Exposure to Airborne Sensory Irritants", American Industrial Hygiene Association Journal 40 : 207-229, 1979.

28. Punte, C.L., Owens, E.J. and Gutentag, P.J., "Exposure to Ortho-chlorobenzylidene Malonitrile", Archives of Environmental Health, 6 : 366-374, 1963.

29. Beswick, F.W., Holland, P. and Kemp, K.H., "Acute Effects of Exposure to Orthochlorbenzylidene malonitrile (CS) and the development of tolerence", British Journal of Industrial Medicine, 29 : 298-306, 1972.

30. Jin, T., "Studies of Emotional Instability in Smoke from Fires", Journal of Fire and Flammability, 12 : 130-142, 1981.

31. Alarie, Y., "Sensory Irritation by Airborne Chemicals", CRC Critical Reviews of Toxicology, 2 : 299-363, 1973.

32. Kaplan, H.L., Grand, A.F., Rogers, W.R., Switzer, W.G. and Hartzell, G.C., "A Research Study of the Assessment of Escape Impairment by Irritant Combustion Gases in Postcrash Aircraft Fires", Report No. DOT/FAA/CT-84/16, 1984.

33. Canter, D., "Studies of Human Behaviour in Fire : Empirical Results and their Implications for Education and Design", University of Surrey, June, 1983.

34. Melzack, R. and Wall, P.D., "Pain Mechanisms : A New Theory", Science, 150 : 971-979, 1965.

35, Silcock, A., Robinson, D. and Savage, N.P., "Fires in Dwellings - An investigation of Actual Fires. Part II; Hazards from Ground-floor Fires, Part III; Physiological effects of fire", Building Research Establishment Current Paper CP80/78, 1978.

36. Purser, D.A., Unpublished data obtained from case reports, fire investigations and personal communications involving the Fire Research Station, Home Office and UK Fire Brigades, 1982-1985.

37. Babrauskas, V., "Full-scale Burning Behaviour of Upholstered Chairs", National Bureau of Standards Technical Note. 1103, 1979.

38. "A Statistical Review of Fire Starting in Textiles or Funiture in Dwellings in the United Kingdom", Home Office S3 Division, October, 1980.

# Toxicity of the Combustion Products from a Flexible Polyurethane Foam and a Polyester Fabric Evaluated Separately and Together by the NBS Toxicity Test Method

BARBARA C. LEVIN, MAYA PAABO, CHERYL S. BAILEY,
and STEVEN E. HARRIS
Center for Fire Research
National Bureau of Standards
Gaithersburg, Maryland 20899, USA

ABSTRACT

Representative specimens of two materials, a flexible polyurethane foam and a
polyester, were thermally decomposed separately and together in order to compare
the toxicological effects of the combustion products from the combined materials
with those from the single homogeneous materials. Gas concentrations (CO, $CO_2$,
$O_2$ and HCN), blood carboxyhemoglobin, and $LC_{50}$ values [the concentration of
material necessary to kill 50% of the test animals (Fischer 344 male rats) during
a 30 minute exposure and a 14 day post-exposure observation period] were determined for the separate and combined materials under both flaming and non-flaming
conditions. The results of the combined experiments indicated that under non-
flaming conditions, both materials contributed in an additive manner to the
concentration of the combustion products. However, under flaming conditions,
the generation of HCN and CO is greater than that predicted from the addition of
the maximum amounts produced by the materials separately.

INTRODUCTION

In the industrialized world, the United States is second only to Canada in
the number of fire deaths per capita [1]. The fire scenario which produces
the most fire deaths in the U.S. begins with an inadvertently dropped cigarette
in an upholstered piece of furniture. Since the majority of commercially
available upholstered furniture today contains some formulation of flexible
polyurethane foam as a filling material and a covering fabric which is either a
cellulosic or a thermoplastic such as polyester, these two materials were chosen
for this study. Many small-scale laboratory studies have examined the toxicity
of the combustion products from flexible polyurethane foams [2] or polyesters
[3]. There have also been numerous large-scale room burns of chairs, multiple
materials, or composite materials which included these materials. Alarie et al.
compared the toxicity of individual materials (determined in small-scale tests)
with the toxicity of multiple combined materials (determined in large-scale chair
burns) [4]. However, the objective of their study was to compare the toxicity of
the major components of the chairs (flexible polyurethane foam, polyester, and

---

cotton fiber) with their individual toxicity in small-scale tests. They did not study the toxicity of the combined components in the small-scale tests.

This study was designed to examine and to compare the toxicological effects from the combustion products of a flexible polyurethane foam and a polyester fabric in order to determine the contribution of the combustion products from each material to the overall toxicity of the mixture. Two separate aspects of this problem were considered: (1) Would the toxicity be affected merely because the increased mass of the combined materials increase the concentrations of the pyrolysis or combustion products or does some unexpected toxicological interaction occur? (2) Would the types or yields of toxicants be affected?

MATERIALS AND METHODS

The materials studied, polyester fabric and a flexible polyurethane foam, were generically classified, i.e., the specific chemical formulations were unknown. Both the polyester upholstery fabric (100% polyester, scoured and dyed dark blue) and the flexible polyurethane foam were obtained from the Consumer Product Safety Commission, Washington, DC 20207. The results of a previous toxicological study on this polyurethane foam, designated CPSC #13, have been published [5].

The acute inhalation toxicity of the combustion products from these materials was evaluated according to the NBS Toxicity Test Method [6]. Each material was examined at 25°C above and below its autoignition temperature (Tables 1 & 2). In addition, polyester was tested at the non-flaming temperature of the polyurethane foam (375°C), and the flexible polyurethane foam was examined at the flaming temperature of the polyester (525°C). Combinations of the two materials were thermally decomposed in the cup furnace at a non-flaming temperature of 375°C (which was the highest possible non-flaming temperature, since the polyurethane foam would flame at higher temperatures) and a flaming temperature of 525°C (the temperature at which both the polyurethane and polyester would undergo flaming combustion, if tested separately). In all cases, the amount of material consumed was determined by weighing the residue.

Carbon monoxide (CO) and carbon dioxide ($CO_2$) were measured continuously by non-dispersive infrared spectroscopy. Oxygen concentrations were measured continuously by a galvanic cell or a paramagnetic analyzer. The HCN generated from the polyurethane foam was sampled with a gas-tight syringe approximately every three minutes and analyzed with a gas chromatograph equipped with a thermionic detector [7].

Fischer 344 male rats, weighing 200-300 grams, were obtained from the Harlan Sprague-Dawley Company (Walkersville, Maryland) or Taconic Farms (Germantown, New York) and were allowed to acclimate to our laboratory conditions for 10 days prior to experimentation. Animal care and maintenance were performed in accordance with the procedures outlined in the National Institutes of Health's "Guide for the Care and Use of Laboratory Animals" [8].

Six animals were exposed in the head-only mode in each experiment. Exposures were for 30 minutes, during which blood for carboxyhemoglobin (COHb) analysis was taken at 0 time, approximately 15 minutes and just before the end of the experiment from cannulated animals (one or two animals per exposure were surgically prepared with a femoral arterial cannulae 24 hours before experiments [9]). The number of animals that died at each mass loading of material was plotted to produce a concentration-response curve from which an $LC_{50}$ value was calculated [10]. The $LC_{50}$, in this case, is defined as the mass loading of material per

Table 1

| Material | AIT (°C) | Mode | Initial Temp. of Expt. (°C) | Type of Expt. | Mass Chamber Loaded (mg/ℓ) | Mass Vol. Consumed (mg/ℓ) | Average Gas Concentration[1] CO (ppm) | $CO_2$ (ppm) | $O_2$ (%) | HCN (ppm) | Max. CO (ppm) | Highest % COHb (30 min) | No. died / No. tested Within Exp. | Within & Post | Latest Day of Death | $LC_{50}$ 30 min. + 14 days (mg/ℓ) |
|---|---|---|---|---|---|---|---|---|---|---|---|---|---|---|---|---|
| Poly-urethane #13 | 400 | NF | 374 | A | 10.0 | 9.3 | 420 | 470 | 20.8 | 8 | 540 | NA | NA | NA | NA | 37.0 (29.8-46.0)* |
| | | | 375 | R | 19.8 | 17.9 | 620 | 2370 | 20.4 | ND | 800 | ND | 0/6 | 1/6 | 14 | |
| | | | 375 | R | 30.0 | 26.6 | 700 | 2690 | 20.4 | 4 | 1000 | ND | 0/6 | 1/6 | 11 | |
| | | | 375 | R | 32.0 | 27.4 | 600 | 2800 | 20.3 | 5 | 1100 | 26.3 | 0/6 | 1/5 | 7 | |
| | | | 377 | R | 35.0 | 31.1 | ND | ND | ND | 9 | ND | 47.0 | 0/6 | 4/4 | 11 | |
| | | | 375 | R | 40.0 | 34.3 | 740 | 2500 | ND | 1 | 1320 | ND | 0/6 | 3/5 | 12 | |
| Poly-urethane #13 | 400 | F | 425 | A | 10.0 | 9.7 | 170 | 8700 | 19.9 | 19 | 210 | NA | NA | NA | NA | >40 |
| | | | 425 | R | 20.0 | 19.9 | 320 | 21400 | 18.2 | 17 | 370 | ND | 0/6 | 0/6 | NA | |
| | | | 425 | R | 30.0 | 29.8 | 520 | 28400 | 17.4 | 24 | 590 | ND | 0/6 | 0/6 | NA | |
| | | | 425 | R | 40.0 | 39.1 | 840 | 33500 | 16.7 | 27 | 940 | 46.5 | 0/6 | 0/4 | NA | |
| | | | 524 | A | 20.0 | 19.9 | 630 | 10700 | 19.5 | 51 | 720 | NA | NA | NA | NA | |
| | | | 528 | A | 20.0 | 19.9 | 390 | 19100 | 18.7 | 37 | 430 | NA | NA | NA | NA | |

Legend:

1. Average gas concentration = $\dfrac{\text{integrated area under instrument response curve for 30 minutes}}{30 \text{ minutes}}$ = $\dfrac{\text{ppm-min}}{30 \text{ min}}$

A. Analytical experiment
R. Rat experiment
ND. Not determined
NA. Not applicable
F. Flaming
NF. Non-flaming
AIT. Auto-ignition temperature
*. 95% confidence limits
FPU 13. Flexible polyurethane #13
PE. Polyester
Comb. Combined weight of FPU 13 + PE

Table 2

| Material | AIT (°C) | Mode | Initial Temp. of Expt. (°C) | Type of Expt. | Mass Chamber Loaded (mg/ℓ) | Mass Vol. Consumed (mg/ℓ) | Average Gas Concentration[1] CO (ppm) | $CO_2$ (ppm) | $O_2$ (%) | HCN (ppm) | Max. CO (ppm) | Highest % COHb (30 min) | No. died / No. tested Within Exp. | Within & Post | Latest Day of Death | $LC_{50}$ 30 min. + 14 days (mg/ℓ) |
|---|---|---|---|---|---|---|---|---|---|---|---|---|---|---|---|---|
| Polyester | 500 | NF | 474 | A | 20.0 | 16.4 | 970 | 1700 | 20.7 | NA | 1270 | NA | NA | NA | NA |  |
|  |  |  | 474 | A | 25.0 | 21.3 | 1670 | 2550 | 20.6 | NA | 2330 | NA | NA | NA | NA |  |
|  |  |  | 473 | R | 35.0 | 30.6 | 2440 | 4770 | 20.5 | NA | 3460 | 81.6 | 0/6 | 0/4 | NA |  |
|  |  |  | 473 | R | 36.5 | 30.9 | 2330 | 4910 | 20.4 | NA | 3140 | 67.8 | 0/6 | 0/4 | NA | 39.0 (38.4-39.5)* |
|  |  |  | 476 | R | 37.1 | 32.5 | 2810 | 5590 | 20.3 | NA | 3900 | 83.7 | 0/6 | 0/4 | NA |  |
|  |  |  | 475 | R | 37.5 | 32.4 | 2570 | 5350 | 20.4 | NA | 3590 | 79.3 | 1/6 | 5/5 | 2 |  |
|  |  |  | 475 | R | 38.5 | 34.3 | 2910 | 5920 | 20.3 | NA | 4090 | 82.7 | 2/6 | 3/6 | 12 |  |
|  |  |  | 473 | R | 39.2 | 33.2 | 2650 | 5300 | 20.4 | NA | 3730 | 75.0 | 2/6 | 3/5 | 2 |  |
|  |  |  | 475 | R | 40.0 | 34.9 | 2660 | 4990 | 20.4 | NA | 3830 | 81.0 | 4/6 | 5/6 | 1 |  |
|  |  |  | 377 | A | 40.0 | 4.3 | 50 | 600 | 20.9 | NA | 160 | NA | NA | NA | NA |  |
|  |  |  | 376 | R | 40.0 | 21.9 | 370 | 2980 | 20.5 | NA | 770 | 13.5 | 0/6 | 0/4 | NA | >50 |
|  |  |  | 375 | R | 50.0 | 20.4 | 410 | 2710 | 20.6 | NA | 840 | 15.2 | 0/6 | 0/5 | NA |  |
| Polyester | 500 | F | 524 | R | 30.0 | 28.7 | 2220 | 25200 | 18.4 | NA | 2990 | 82.0 | 0/6 | 0/4 | NA |  |
|  |  |  | 525 | R | 35.0 | 33.7 | 2290 | 27300 | 18.2 | NA | 3400 | 83.0 | 1/6 | 2/5 | 0 | 37.5 (35.3-39.8)* |
|  |  |  | 523 | R | 37.5 | 36.1 | 2640 | 28500 | 18.0 | NA | 3860 | 83.2 | 3/6 | 3/6 | NA |  |
|  |  |  | 524 | R | 40.0 | 38.8 | 2990 | 30500 | 17.8 | NA | 4310 | 84.6 | 4/6 | 4/5 | NA |  |

For Legend, see Table 1

1114

unit chamber volume (mg/ℓ) which caused 50% of the animals to die during the 30 minute exposure plus the 14 day post-exposure observation period. (Animals that were still losing weight on day 14 were kept until they died or recovered as indicated by three days of successive weight gain. All deaths were included in the $LC_{50}$ calculation. Surviving cannulated animals were sacrificed following the test and only counted in the determination of the $LC_{50}$ if they died during the exposure.) If no deaths occurred at the highest concentration tested, the $LC_{50}$ is listed as greater than that concentration.

RESULTS AND DISCUSSION

Flexible Polyurethane Foam

The chemical and toxicological data obtained from the flexible polyurethane foam thermally decomposed under non-flaming (375°C) and flaming (425°C and 525°C) conditions are presented in table 1. Similar to other non-fire retarded flexible polyurethane foams tested in this laboratory, no animal deaths occurred during the 30 minute exposures to concentrations up to 40 mg/ℓ regardless of the mode of decomposition [5,6]. Post-exposure deaths only occurred following the non-flaming experiments. The $LC_{50}$ value for the non-flaming mode was 37.0 mg/ℓ with 95% confidence limits of 29.8-46.0 mg/ℓ, whereas, the $LC_{50}$ value for the flaming mode was greater than 40 mg/ℓ, i.e., no animal deaths were noted from any of the concentrations tested up to 40 mg/ℓ.

Recent results [11] from this laboratory on the toxicity of CO, $CO_2$ and HCN alone and in various combinations have shown that the 30 minute $LC_{50}$ for CO in air was 4600 ppm. No animals died below 4100 ppm or post-exposure. The 30 minute $LC_{50}$ for $CO_2$ in air was greater than 18% (1% = 10,000 ppm). However, when CO and $CO_2$ were combined, the presence of 5% $CO_2$ increased the toxicity of CO such that animals died from 30 minute exposures to 2500 ppm. Some of these deaths were within 24 hours. The combination of CO and HCN (30 minute HCN $LC_{50}$ = 160 ppm) showed the following additive effect:

$$\text{If } \frac{[CO]}{LC_{50} \text{ CO}} + \frac{[HCN]}{LC_{50} \text{ HCN}} \geq 1, \text{ the animals died.}$$

When this formula equalled less than 1, the animals lived. Again deaths were observed up to 24 hours post-exposure.

Comparison of the gas concentrations generated from the polyurethane experiments to the pure gas experiments quoted above showed that lethal amounts were not produced in any of the tests (Table 1). Therefore, the deaths, which occurred as late as 14 days in the non-flaming mode, were due to other toxic combustion products or undetermined factor(s).

Polyester

All the chemical and toxicological data collected from the thermal degradation of polyester are shown in table 2. In the non-flaming mode at 475°C, the $LC_{50}$ value of the polyester was 39.0 mg/ℓ with 95% confidence limits of 38.4 - 39.5 mg/ℓ. Animal deaths were noted both during and following the 30 minute exposures. At 375°C, however, no animal deaths were observed up to concentrations of 50 mg/ℓ. However, it is important to note that at the lower temperature (375°C), only 22-55% of the original sample was consumed; whereas, at 475°C, approximately 85% of the sample was consumed. Based upon a comparison of the toxicological effects at the actual masses consumed at 475°C, deaths would not be expected at the masses consumed at the lower temperature.

Examination and comparison of the gas concentrations that were generated during these non-flaming experiments with our pure and combined gas toxicity experiments discussed above indicate that the average CO levels are 37-80% lower than that necessary to cause death by CO alone. In many experiments, the CO levels did not plateau but continued to rise throughout the exposures reaching a maximum at 30 minutes. This maximum value, however, was still lower than the 30 minute $LC_{50}$ value for CO in air (4600 ppm) (Table 2). The average $CO_2$ present is about 10% of that necessary to increase the susceptibility of the rats to lower levels of CO. However, the maximum COHb levels at the end of these 30 minute lethal exposures are relatively high, 75-83%. These results indicate that CO, although low, is contributing to the within-exposure deaths, but other toxic or irritant gas(es) are also acting in conjunction with or to potentiate the effects of the CO. The cause of the late post-exposure deaths are unexplained.

In the flaming mode, the $LC_{50}$ value for the polyester was 37.5 mg/ℓ with 95% confidence levels of 35.3 - 39.8 mg/ℓ (Table 2). In the lethal experiments, the COHb levels ranged from 83-85% and the rats died within exposure or shortly thereafter. These factors would implicate CO as the main toxicant. However, the average CO is approximately 50-65% of the lethal concentration determined for CO alone. Even considering the effect of $CO_2$ on CO, the average values of CO and $CO_2$ from flaming polyester are still too low to account for the deaths that occurred during these 30 minute exposures. Only if one considers the maximum CO levels along with the $CO_2$ concentrations would the deaths be predictable.

Combined Flexible Polyurethane Foam and Polyester

Non-flaming experiments. The thermal decomposition of both flexible polyurethane foam and polyester in the non-flaming mode was studied at 375°C which was 25°C below the autoignition temperatures of the polyurethane. In these experiments, the polyester fabric was folded and dropped into the cup furnace immediately preceding the polyurethane foam. Upon heating, the samples collapsed in less than one minute and formed a black ball in approximately two minutes.

Since the polyester, by itself, was not toxic at 375°C[2] even at the highest loading tested (50 mg/ℓ), a sublethal amount of polyester (20 mg/ℓ) was chosen to test whether this addition would increase the toxicity (lethality) of the polyurethane foam in the combination experiments. If the polyester component has no effect at this temperatures, then the addition of 20 mg/ℓ of polyester to the $LC_{50}$ value of the polyurethane would increase the $LC_{50}$ value of the mixture by 20 mg/ℓ; that is, the $LC_{50}$ value of the polyurethane, 37 mg/ℓ, would increase to approximately 57 mg/ℓ. The results, however, showed that the $LC_{50}$ value of the combined materials only increased to 47.5 mg/ℓ, an indication that the polyester was not inert but contributed to the toxicity by about 10 mg/ℓ (Table 3). The total amount of polyurethane in the combined $LC_{50}$ is only 27.5 mg/ℓ, which is outside the 95% confidence limits of the $LC_{50}$ for polyurethane alone.

Since a significant proportion of the polyester is not decomposed at 375°C, these data were also analyzed on the basis of mass consumed/chamber volume. The experiments on the polyester alone at 375°C showed that when 3.88 grams (20 mg/ℓ) were loaded into the cup furnace, 78% remained as residue and only 4.3 mg/ℓ

---

[2]This lower toxicity is probably due to the large fraction (more than 45% of the initial mass loading) of the polyester which is not consumed at the lower temperature.

Table 3

| Material | AIT (°C) | Mode | Type of Expt. | Mass Chamber Loaded (mg/ℓ) | | | Volume Consumed (mg/ℓ) | Average Gas Concentration[1] | | | | Max. CO (ppm) | Highest % COHb (30 min) | No. died / No. tested Within Exp. | Within & Post | Latest Day of Death | LC$_{50}$ 30 min. + 14 days (mg/ℓ) |
|---|---|---|---|---|---|---|---|---|---|---|---|---|---|---|---|---|---|
| | | | | FPU 13 | PE | Comb. | | CO (ppm) | CO$_2$ (ppm) | O$_2$ (%) | HCN (ppm) | | | | | | |
| Polyester plus poly- urethane #13 | 500 | NF | R | 20.0 | 20.0 | 40.0 | 21.0 | 690 | 2600 | 20.5 | ND | 890 | 31.9 | 0/6 | 0/4 | NA | |
| | 400 | | R | 27.5 | 20.0 | 47.5 | 26.2 | 670 | 3120 | 20.5 | ND | 960 | 26.4 | 0/6 | 2/4 | 8 | 47.5 (43.0-52.5)* |
| | | | R | 30.0 | 20.0 | 50.0 | 31.2 | 850 | 3400 | 20.5 | 5 | 1270 | 35.2 | 0/6 | 3/4 | 7 | |
| | | | R | 32.5 | 20.0 | 52.5 | 33.4 | 1130 | 3920 | 20.3 | ND | 1420 | 43.9 | 0/6 | 3/4 | 2 | |
| | | | R | 35.0 | 20.0 | 55.0 | 35.4 | 1300 | 3810 | 20.4 | ND | 1710 | 35.4 | 0/6 | 4/4 | 16 | |
| | | | R | 37.5 | 20.0 | 57.5 | 37.3 | 1390 | 3850 | 20.4 | ND | 1860 | 54.9 | 0/6 | 4/4 | 2 | |
| | | | R | 40.0 | 20.0 | 60.0 | 43.1 | 1160 | 4090 | 20.4 | ND | 1550 | 40.9 | 0/6 | 4/4 | 2 | |
| Polyester plus poly- urethane #13 | 500 | F | R | 20.0 | 15.0 | 35.0 | 34.0 | 1870 | 30200 | 17.5 | 62 | 2370 | 76.2 | 1/6 | 1/5 | NA | |
| | 400 | | R | 20.0 | 20.0 | 40.0 | 38.9 | 2270 | 33600 | 17.1 | 63 | 3120 | 80.2 | 3/6 | 3/5 | NA | 39.0 (36.0-42.2)* |
| | | | R | 20.0 | 22.5 | 42.5 | 41.8 | 2410 | 31100 | 17.4 | 50 | 3420 | 78.7 | 5/6 | 5/6 | NA | |
| | | | R | 20.0 | 25.0 | 45.0 | 44.0 | 2780 | 34400 | 16.8 | 48 | 3930 | ND | 6/6 | 6/6 | NA | |
| | | | R | 20.0 | 30.0 | 50.0 | 49.0 | 3070 | 34700 | 17.1 | 59 | 4500 | ND | 6/6 | 6/6 | NA | |
| | 425 | | R | 20.0 | 30.0 | 50.0 | 43.7 | 1750 | 31500 | 17.4 | 13 | 1950 | 69.1 | 1/6 | 1/5 | NA | |

For legend, see Table 1

1117

were actually consumed. Using the same null hypothesis as before, that is, the polyester at this temperature has no effect on the combined toxicity, then one would expect the $LC_{50}$ of the polyurethane (31.9 mg/ℓ, consumed weight) should increase by 4.3 mg/ℓ producing a combined $LC_{50}$ of 36.2 mg/ℓ. However, the $LC_{50}$ of the combination is only 26.2 mg/ℓ, consumed weight, indicating that the polyester increases the toxicity by about 10 mg/ℓ; this is the same value calculated when the mass loaded, rather than mass consumed, was considered.

In these non-flaming experiments in which 20 mg/ℓ of polyester were added to different loadings of flexible polyurethane foam, all deaths occurred during the post-exposure period. These results are more characteristic of the polyurethane experiments, decomposed by itself, and different from those seen with the polyester alone. The concentrations of measured gases (CO, $CO_2$, HCN) were not responsible for the post-exposure deaths that occurred.

The average concentration of the primary gases (CO, $CO_2$, HCN) generated from the thermal decomposition of the mixture of the materials appear to be approximately equal to the sum of the average concentrations generated from the individual materials under non-flaming conditions (Table 4 and Figure 1). Therefore, if the concentrations of the primary gases from the thermal decomposition of the individual components are known, then a reasonable prediction of the gas concentrations from the mixture decomposed under the same conditions can be made.

Flaming experiments. The experiments in which the flexible polyurethane foam and polyester were combined and tested in the flaming mode were conducted at 525°C (25°C above the autoignition temperature of the polyester) to ensure that both materials would flame. In these experiments, the mass concentration of polyurethane was kept constant at 20 mg/ℓ and only that of the polyester was varied (Table 3). The reason for this approach was to see if a non-lethal amount of the less toxic material (in this case, the polyurethane foam) would increase the toxicity of the polyester whose $LC_{50}$ could be measured. The polyurethane foam when tested by itself in the flaming mode at 425°C had produced no deaths either during or post-exposure at concentrations up to 40 mg/ℓ, whereas, the polyester fabric when decomposed by itself in the flaming mode at 525°C had produced both within and post-exposure deaths. The $LC_{50}$ value for the flaming polyester fabric by itself was 37.5 mg/ℓ. Therefore, if the polyurethane was toxicologically inert, the addition of 20 mg/ℓ of polyurethane should have raised the $LC_{50}$ value to 57.5 mg/ℓ. In actuality, the 30 minute and 14 day $LC_{50}$ value calculated for the combined exposures was 39.0 mg/ℓ with 95% confidence limits of 36.0 - 42.2 mg/ℓ. These results, showing that the $LC_{50}$ value for the combined materials was lower than expected by almost the exact amount of polyurethane added to the system, are an indication that the polyurethane and the polyester are both contributing in an additive manner to the toxicity. In other words, the combination of 19 mg/ℓ of polyester and 20 mg/ℓ of polyurethane produced the $LC_{50}$; whereas, 20 mg/ℓ of the polyurethane foam decomposed by itself in the flaming mode produced no deaths (Table 1) and the polyester decomposed by itself did not produce any deaths below a concentration of 35 mg/ℓ (Table 2). Thus individual sublethal concentrations of this polyurethane foam and polyester fabric are adding up to a concentration which is lethal.

Examination of the average gas concentrations of CO, $CO_2$, and HCN which were generated during these exposures and comparison of these gas values with our pure gas toxicological studies shows that the concentrations of these gases were sufficient to account for the deaths that occurred. Table 3 also shows HCN levels higher than those seen in the flaming exposures of polyurethane alone at 525°C which, in turn, were greater than at 425°C (Table 1). Figure 2 shows the generation of HCN from 20 mg/ℓ of the flexible polyurethane when decomposed alone or combined with the polyester under various flaming conditions.  Polyurethane decomposed alone at 425°C produced an average HCN concentration of 19 ppm; whereas, at 525°C, it produced an average of 37 ppm in one experiment and 51 ppm in another.  The flaming decomposition of various amounts of polyester with 20 mg/ℓ of polyurethane at 525°C produced greater concentrations of HCN than in any of the experiments on the polyurethane alone (Tables 1,3 and Fig. 2).  This result was unexpected since polyester contains no nitrogen and should not contribute to the HCN generation.  Figure 2 also shows that the HCN generation over time from 20 mg/ℓ of flaming polyurethane foam alone tends to plateau during the 30 minute test, whereas, in the combination studies of this polyurethane (20 mg/ℓ) and polyester, the HCN continues to increase throughout the experiments.  The reason for this increased level of HCN is unexplained at this time.

Table 4

Gas Concentrations from the Thermal decomposition of
Polyurethane Foam and Polyester Alone and in Combination

| Mode | Temp. (°C) | Material | Mass Loaded Chamber Volume (mg/ℓ) | Average Gas Concentration[1] | | |
|------|------|----------|------|------|------|------|
| | | | | CO (ppm) | $CO_2$ (ppm) | HCN (ppm) |
| Flaming | 525 | Polyurethane | 20[2] | 510 ($\pm$ 120) | 14900 ($\pm$ 4200) | 44 ($\pm$ 7) |
| | | Polyester | 30 | 2220 | 25200 | -- |
| | | Total | | 2730 | 40100 | 44 |
| | | Polyurethane + Polyester | 20 + 30 | 3070 | 34700 | 59 |
| Non-Flaming | 375 | Polyurethane | 30 | 700 | 2690 | 4 |
| | | Polyester | 20[3] | 50 | 600 | -- |
| | | Total | | 750 | 3290 | 4 |
| | | Polyurethane + Polyester | 20 + 30 | 850 | 3400 | 5 |

1  Average gas concentration:

$$\frac{\text{integrated area under instrument response curve for 30 minutes}}{30 \text{ minutes}} = \frac{\text{ppm-min}}{30 \text{ min}}$$

2  Results are average $\pm$ range of two analytical experiments
3  Analytical experiment (no animals)

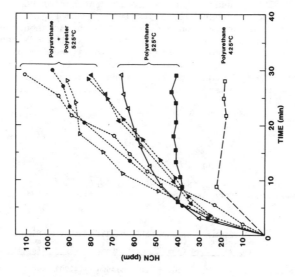

Figure 2. Generation of HCN from 20 mg/ℓ of Polyurethane Decomposed Alone and with Different Amounts of Polyester Under Flaming Conditions.

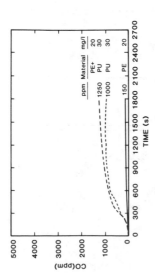

Figure 1. CO Generation from Flexible Polyurethane Foam and Polyester Decomposed Alone and Together. Non-Flaming 375°C.

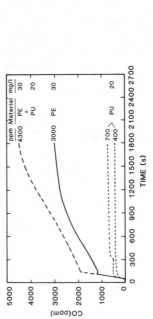

Figure 3. CO Generation from Flexible Polyurethane Foam and Polyester Decomposed Alone and Together. Flaming 525°C.

In the flaming mode, the concentrations of CO from combined materials was also greater than the sum of the CO concentrations from the individual materials (Figure 3). This was more apparent from the graphic representation of the actual generation of CO than from the tabular depiction of the average concentrations (Table 4).

CONCLUSIONS

Individual Materials

- The decomposition products of the flexible polyurethane foam produced no deaths during exposure and only caused post-exposure deaths in the non-flaming mode.

- The polyester when decomposed 25°C above or below its autoignition temperature caused deaths both during and following exposures.

- Comparison of the CO, $CO_2$, and HCN concentrations generated from the individual materials with pure gas toxicity experiments (performed with single and multiple gases) indicated:

  1. the deaths from flexible polyurethane could not be explained by the concentrations of these gases,

  2. non-flaming polyester produced relatively high COHb (75-83%) levels, but lower than lethal average or maximum CO concentrations. Even when CO was considered with $CO_2$ (which potentiates the toxicity of CO), the combination was not sufficient to account for the deaths, and

  3. the deaths from exposures to flaming polyester products were probably due to CO since COHb values were 83-85%. In this case, the maximum (not the average) concentrations of CO plus $CO_2$ were sufficient to predict the deaths.

Combined Materials

- Depending on the amount thermally decomposed, both materials contributed to the combined toxicity. In the flaming mode, the contribution was additive.

- Similar to the polyurethane results, the non-flaming combined experiments only produced post-exposure deaths which were not attributable to the generated CO, $CO_2$, and HCN concentrations.

- The deaths observed from the flaming combined experiments were explainable based on the concentrations of CO, $CO_2$, and HCN.

- Comparison of the gas concentrations from the combined materials to those from the individual materials indicated:

  1. The non-flaming generation of CO, $CO_2$, and HCN appear to be approximately equal to the sum of the concentrations from the single materials.

1121

2. The flaming generations of CO and HCN were greater than the sum of those from the single materials.

ACKNOWLEDGEMENTS

This work was supported in part by the Consumer Product Safety Commission, Washington, DC, Dr. Rita Orzel, Project Officer. The conclusions are those of the authors and not the Consumer Product Safety Commission.

REFERENCES

1. Fire in the United States, Second Edition, Federal Emergency Management Agency, FEMA-22, July, 1982.

2. "Literature Review of the Combustion Toxicity of Flexible Polyurethane Foam," Status Report on Fire Combustion Toxicity, TAB-B. Consumer Product Safety Commission, Washington, DC 20207, May 31, 1984, 103 p.

3. Braun, E. and Levin, B.C.: "Polyesters: A Review of the Literature on Products of Combustion and Toxicity," Nat. Bur. Stand. (U.S.) Gaithersburg, MD, NBSIR 85-3139, June 1985.

4. Alarie, Y., Stock, M.F., Matijak-Schaper, M., and Birky, M.M.: "Toxicity of Smoke During Chair Smoldering Tests and Small Scale Tests Using the Same Materials," Fund. & Appl. Tox. 3:619-626, 1983.

5. Levin, B.C., Paabo, M., Fultz, M.L., Bailey, C., Yin, W., and Harris, S.E.: "An Acute Inhalation Toxicological Evaluation of Combustion Products from Fire Retarded and Non-Fire Retarded Flexible Polyurethane Foam and Polyester," Nat. Bur. Stand. (U.S.) Gaithersburg, MD, NBSIR 83-2791, November, 1983.

6. Levin, B.C., Fowell, A.J., Birky, M.M., Paabo, M., Stolte, A., and Malek, D.: "Further Development of a Test Method for the Assessment of the Acute Inhalation Toxicity of Combustion Products," Nat. Bur. Stand. (U.S.) Gaithersburg, MD, NBSIR 82-2532, June 1982.

7. Paabo, M., Birky, M.M., and Womble, S.E.: "Analysis of Hydrogen Cyanide in Fire Environments," J. Comb. Tox. 6:99-108, 1979.

8. Committee on Care and Use of Laboratory Animals: "Guide for the Care and Use of Laboratory Animals," DHEW publication No. (NIH) 78-23. U.S. Dept. of HEW, Public Health Service, National Institutes of Health, 1978.

9. Packham, S.C., Frens, D.B., McCandless, J.B., Patajan, J.H., and Birky, M.M.: "A Chronic Intra-arterial Cannula and Rapid Technique for Carboxyhemoglobin Determination," J. Comb. Tox. 3:471-478, 1976.

10. Litchfield, J.T., Jr., and Wilcoxon, F.: "A Simplified Method of Evaluating Dose-effect Experiments," J. Pharmacol. and Exp. Therapeut. 96:99-113, 1949.

11. Levin, B.C., Paabo, M., Gurman, J.L., Harris, S.E., and Bailey, C.S.: "Toxicological Effects of the Interactions of Fire Gases and Their Use in a Hazard Assessment Computer Model," The Toxicologist, 5:127, 1985.

# Calculation of Smoke Movement in Building in Case of Fire

**TAKAYUKI MATSUSHITA, HIROSHI FUKAI, and TOSHIO TERAI**
Department of Architecture
Kyoto University, Japan

Calculation methods of smoke movement by using graph theory are presented. If both the routes of smoke movement and of evacuation are to be represented by the same method, the analysis of interaction is very simplified.

The main features of this programs are :

1. Flow circuit is expressed by incidence matrix, or loop matrix. If the data such as the incidence matrix, the geometry of branch and the initial conditions are given, no other modification is necessary to the program.

2. In order to facilitate analysis of the interaction between evacuation and smoke flow, methods are proposed to select the tree which embeds the evacuation route into part of the smoke flow tree, and to number the nodes and branches to simplify the incidence matrix.

3. The flow rate assuming method has been shown to be more efficient than the pressure assuming method .

KEYWORDS : graph theory, incidence matrix, loop matrix, smoke movement, evacuation, pressure assuming method, flow rate assuming method, Newton's method

## 1. INTRODUCTION

The safety of evacuation is usually estimated without taking into account smoke movement. When doors of staircases are opened during evacuation, it is possible that smoke flow through them may trap those evacuating. These effects have not been considered.

In large buildings, the longer the evacuation time, the higher is the chance for smoke to flow through evacuation routes. To analyze interaction between smoke flow and evacuation, it is convenient to formulate both routes by the same method using graph theory[1] . Mathematically there are many possibility to fix the graph, but when taking into account the safety, tree of the graph should be coincident with the route of the evacuation, therefore the tree is fixed.

In addition to the evacuation routes which are the main route causing the smoke movement, other main smoke flow routes, e.g. elevator shafts and ducts, have to be selected as a part of the tree in the incidence matrix. After routes of evacuation and smoke movement are represented by identical mathematical expression, the analysis of the system is very simplified. The incidence matrix is easily transformed to the loop matrix, both matrices could be used to calculate smoke flow. The one uses pressure as independent variables and uses incidence matrix, and the other loop flow rate and loop matrix. In this paper, two methods of formulation and the advantages of latter formulation are explained.

In order to compute smoke movement, there are some published programs based on the same method as in ventilation which assumes uniform mixing in nodes. Wakamatsu [2] has used regula-falsi method from node to node relaxation. This program has to be changed with network. Klote and Fothergill [3] improved the input of network data, but they also used regula-falsi method. Yoshida et al.[4] solved simultaneous linear equations by using one dimentional Newton's method from equation to equation. The method constructing equations is not flexible to other networks, as Wakamatsu's. In this report graph theoretical formulation is used. To solve the system iteratively, all nodes are relaxed simultaneously by using multi-dimensional modified Newton's method.[5,6] The regula-falsi method sometimes shows poor convergence for the building consisted of small and large openings, however modified Newton's method showed faster and better convergence.

## 2. SYMBOLS

[ ] : matrix
{ } : column vector

Upper case letters represent variables at nodes.

$A$ : floor area
$C_p$ : specific heat
$H$ : height to ceiling
$P$ : room pressure at floor level
$Q$ : strength of heat source in node
$T$ : node temperature
$W$ : strength of mass flow source in node
$Y$ : smoke boundary height from floor
$\Delta P$ : error of node pressure
$\Delta W$ : error of mass flux (hypothetical source in node)
$\Delta V$ : error of node volume
$\Gamma$ : density in node

Lower case letters represent variables at branches.

$p$ : branch pressure difference
$p_b$ : pressure source in branch
$w^+,(w^-)$ : flow rate to specified (opposite) directions
$w$ : net mass flow rate, $w = w^+ - w^-$
$\bar{w}$ : mass flow rate of loop (mass flow rate of the co-tree is the same as that of the loop, while mass flow rate of the branch of the tree is the sum of that of the loops relating to this branch.)

$\Delta p$ : error of branch pressure (hypothetical source in branch)
$\Delta \bar{p}$ : error of pressure around loop (hypothetical pressure source in branch of co-tree)
$\Delta w$ : error of mass flow rate (hypothetical source in branch)
$\Delta \bar{w}$ : error of mass flow rate around loop (hypothetical source in branch of co-tree)
$\gamma$ : branch density difference

Suffix

Lower case roman letters are used to specify the node, while lower case greek letters to branch. When the flow direction specified to branch $\lambda$ is from $i$ to $j$ , then
$p_\lambda = P_i - P_j$, $w_\lambda = w_\lambda^+ - w_\lambda^-$ and $\gamma_\lambda = \Gamma_i - \Gamma_j$.

$s$ : smoke
$a$ : air
$t$ : tree
$l$ : loop (co-tree)

$[I] = [I_t, I_l]$ : reduced incidence matrix, order $(n-1) \times \beta$
where $n$ is total number of node and $\beta$ is that of branch.
$[L] = [L_t, E]$ : loop(tie-set) matrix, order $(\beta - n + 1) \times \beta$
From $[I][L']=[0]$, we get $[L_t] = -[I_l'] [I_t']^{-1}$,
where prime shows transpose of matrix.
$[E]$ : unit matrix

## 3. FORMULATION OF GRAPH OF SMOKE MOVEMENT AND EVACUATION ROUTES

Usually it is convenient to select open air (final egress place) as the reference node (root).

It is better to number the nodes and branches so as to simplify the incidence matrix. This can be done if
(1) the number of nodes should increase from the root to the end of branches,
(2) the branches should have the flow direction from nodes with greater number to smaller number, that is, the same direction to escape,
(3) the number of a branch of the tree should be equal to the upstream node number,
but the numbering of branches on the corresponding co-tree is arbitrary.

It is easy to check the incidence matrix, because for each column the sum of elements must be zero. $[I_t]$ thus obtained is upper triangular matrix.

The difference between the method in calculating the ventilation and the smoke flow is that the former supposes temperatures in all nodes as uniform and steady, and the latter as different and unsteady. When the temperature in node $i$ is different from node $j$ , a pair of fluxes in opposite direction, $w_\lambda^+$ and $w_\lambda^-$,

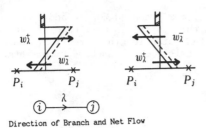

Direction of Branch and Net Flow

Fig.3.1 Relation between direction of branch and net flow rate.

would exist between these nodes as shown in Fig.3.1. Because it is sufficient to consider the net flow rate to evaluate conservation of flow in nodes, only one specified direction between nodes would be sufficient. When the direction of branch is from $i$ to $j$ , $p_\lambda = P_i - P_j$ ,$w_\lambda = w_\lambda^+ - w_\lambda^-$. $p_\lambda$ and $w_\lambda$ could be positive or negative, but derivative $\partial w_\lambda / \partial p_\lambda$ is always positive.

## 4. COMPUTATIONAL METHOD

### 4.1 Steady Flow (perfect mixing flow)

1) Pressure Assuming Method (PAM)
   a) mass conservation in node
$$[I]\,\{w\} \;-\; \{W\} \;=\; \{0\} \tag{4.1}$$
   b) relation between branch pressure and node pressure
$$\{p+p_o\} = [I']\,\{P\} \tag{4.2}$$
   c) relation between mass flow rate and pressure in branch
$$w = f(p, \Gamma, opening\ geometries) \tag{4.3}$$
      There are many formulas to eq.(4.3) according to their precision. The most simple and common formulation is shown in Appendix A .
   d) relation between errors of mass conservation and of pressure in node

$$( [I]\, [\tfrac{\partial f}{\partial p}]\, [I']\, )\, \{\varDelta P\} = \{\varDelta W\} \ . \tag{4.4}$$

2) Flow rate Assuming Method (FAM)
   a) tree mass flow rate, from eq.(4.1) by using the relation $[I][L']=[0]$
$$\{w_t\} = [L_t']\,\{\tilde{w}\} \;+\; [I_t]^{-1}\,\{W\} \tag{4.1'}$$
   b) relation to zero loop pressure, from eq.(4.2)
$$[L]\,\{p+p_o\} = \{0\} \tag{4.2'}$$
   c) inverse relation to eq.(4.3)
$$p = g(w, \Gamma, opening\ geometries) \tag{4.3'}$$
      This is usually not expressed explicity. Instead, eq.(4.3) should be solved iteratively or be substituted by approximate relations.
   d) relation between errors of loop pressure and of loop mass flow rate

$$( [L]\, [\tfrac{\partial g}{\partial w}]\, [L']\, )\, \{\varDelta \tilde{w}\} = \{\varDelta \tilde{p}\} \tag{4.5}$$

     where $\partial g / \partial w = 1/(\partial f / \partial p)$. $\hfill (4.6)$
      The derivative eq.(4.6) can be calculated as the reciprocal of the derivative of eq.(4.3), therefore it is not necessary to define eq.(4.3') explicitly.

      PAM and FAM are both iterative methods. The PAM starts from initial guess of node pressures and iterates by solving eq.(4.4) until eqs.(4.1),(4.2) and (4.3) are satisfied within prescribed error, and the FAM takes the same process with loop flow rates and eqs. (4.1'), (4.2'), (4.3') and (4.5). In addition to PAM, FAM is tested by using Newton's Method in this paper. In most cases iteration by using modified Newton's method with constant step length correction, could give a stable result, even though the circuit consists of branches with large and small flow coefficients.

## 4.2 Unsteady Flow

1) Common relations to PAM and FAM
  A) in case of perfect mixing flow
    a) change of temperature

$$\{AHΓ\frac{dT}{dt}\} = - [I] \{\overline{wT}\} + \{\frac{Q}{C_p}\} \tag{4.7}$$

$$\text{where} \quad \overline{wT} = \begin{cases} w^-(T_i-T_j) & for \ (i) \ node \\ w^+(T_i-T_j) & for \ (j) \ node. \end{cases} \tag{4.8}$$

  B) in case of two layers flow
    a) change of temperature in smoke zone

$$\{A(H-Y)Γ_s\frac{dT_s}{dt}\} = - [I] \{\overline{u_s}\overline{T_s}\} + \{\frac{Q_s}{C_p}\} - \{W_{as}(T_s-T_a)\} \tag{4.7'}$$

where $W_{as}$ is the flow rate from air zone to smoke zone .
    b) change of smoke boundary height

$$\{AΓ_a\frac{dY}{dt}\} = - [I] \{w_a\} - \{W_{as}\} . \tag{4.9}$$

Considering the meaning of the net flow in branch, it is not difficult to extend from perfect mixing to two layers.

2) Pressure Assuming Method (PAM)
  a) relation of volume flow in node (By taking into account the volume change in node, conservation of volume flow rate is relevant.)

$$[I] \{(\frac{w}{Γ})\} - \{\frac{Q}{C_pΓT}\} = \{0\} \tag{4.10}$$

where $(w/Γ)$ is the net volume flow rate in branch, $(w/Γ) = w^+/Γ_i - w^-/Γ_j$.
  For two layers flow, $w^+/Γ_i = w_s^+/Γ_{si}+w_a^+/Γ_{ai}$ (positive total volume flow of smoke and air) and $w^-/Γ_j = w_s^-/Γ_{sj}+w_a^-/Γ_{aj}$ (negative total volume flow of smoke and air).
  b) relation between branch pressure and node pressure
    $\{p+p_o\} = [I'] \{P\}$ $\qquad\qquad\qquad\qquad\qquad$ (4.2)
  c) relation between volume flow rate and pressure in branch

$$(\frac{w}{Γ}) = \overline{f}(p,Γ,Y,opening \ geometries) \tag{4.11}$$

  d) relation between errors of volume conservation and of pressure in node

$$([I] [\frac{\partial \overline{f}}{\partial p}] [I'] ) \{\varDelta P\} = \{\varDelta V\} . \tag{4.12}$$

3) Flow rate Assuming Method (FAM)
  a) from eq. (4.10)

$$\{(\frac{w}{Γ})_t\} = [L_t'] \{(\frac{\tilde{w}}{Γ})\} + [I_t]^{-1} \{\frac{Q}{C_pΓT}\} \tag{4.10'}$$

  b) relation to zero loop pressure
    $[L] \{p+p_o\} = \{0\}$ $\qquad\qquad\qquad\qquad\qquad\qquad$ (4.2')

c) inverse relation to eq.(4.11)

$$p = g((\frac{w}{\Gamma}),\Gamma,Y,opening\ geometries).$$ (4.11')

d) relation between errors of loop pressure and of loop volume flow
   rate

$$( [L]\ [\frac{\partial \bar{g}}{\partial\ (w/\Gamma)}]\ [L']\ )\ \{ \varDelta\ (\frac{\tilde{w}}{\Gamma})\} = \{ \varDelta\tilde{p}\}$$ (4.13)

   where in the same as eq.(4.6), elements of diagonal matrix in eq.(4.13),
$\partial\bar{g}/\partial\ (w/\Gamma)$  , are the reciprocal of $\partial\bar{f}/\partial p$ .

   In unsteady case, at each time step eq.(4.10) should  be  used  instead  of
eq.(4.1)  for  PAM,  or  eq.(4.10')  instead  of eq.(4.1') for FAM. Procedure of
iteration is the same as that of the steady case. This implies the  conservation
of  volume  flow  should  be  satisfied at each time step even in the calculation of
unsteady case. After the norm of errors becomes  smaller  than  the  convergence
limit  at  each time step, temperatures in nodes for next time step are obtained
by solving eq.(4.7) for perfect mixing in node. In  case  of  two  layers  flow,
temteratures  in  smoke  zone and smoke boundary heights are calculated by using
eq.(4.7') and  (4.9).


## 5.  COMPARISON BETWEEN LOOP FLOW RATE ASSUMING METHOD AND
##     NODE PRESSURE ASSUMING METHOD

   The computing times of flow rate assuming  method  (FAM)  and  of  pressure
assuming method (PAM) are compared. The main computing time is used to solve the
systems of equations repeatedly. Therefore it depends mainly on  the  dimensions
of  the  variables.  Dimensions  of  eq.(4.1)  or  (4.10) for PAM is $(n-1)$ and of
eq.(4.1') or (4.10') for FAM is $(\beta-n+1)$. Generally in tall building,  there  are
many cases that $(n-1)>(\beta-n+1)$.
   The  conservation  of  flow in node is satisfied always for FAM, but has an
error for PAM. This means that FAM is more favorable in stability than PAM  when
the temperature or concentration in nodes are calculated, because they are based
on the conservation of flow in nodes. In case of PAM,  pressures  are  given  by
eq.(4.2)  and  the  flow  rates of branches are calculated by eq.(4.3) or (4.11)
explicitly, while in FAM, the pressure of branch is expressed by eq.(4.3') or
(4.11'),  which  is  the  inverse relation to eq.(4.3) or (4.11), and can not be
calculated explicitly. But this difference of the computing time between PAM and
FAM is not significant compared to time to solve the systems of equations.
   In  order  to combine the evacuation to smoke flow, the change of effective
area $\alpha A$ of door in each time  should  be  taken  into  account.  This  could  be
implemented  to  PAM as well as FAM by assigning appropriate small $\alpha A$ for closed
door, and need not change incidence matrix and loop matrix.
   Programs by PAM and FAM in steady case are shown in Appendix B.
   Example of calculation in steady flow is shown in Fig.5.1  and  Table  5.1.
This  building  has  $n=100$  nodes and $\beta=153$ branches. Therefore numbers of loops
$\beta-n+1=54$ for FAM and numbers of reduced nodes $n-1=99$ for PAM. Computing time  is
0.24  sec  for  FAM  and  1.19  sec  for PAM by FACOM-M382. The calculations are
carried at Kyoto Univ. Data Processing Center.


## 6.  DISCUSSION

   In order to consider interaction between smoke movement and evacuation, the
method of calculation is proposed. When graph theory is used, the formulation is
simple and no modification of program is necessary.

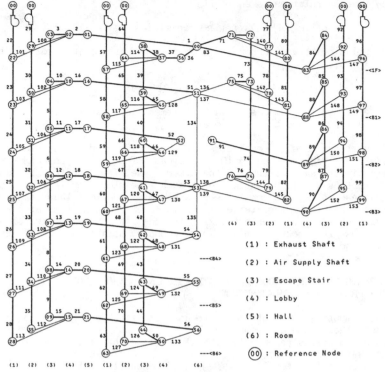

Fig.5.1 Graph of building. This is underground part
of a building (B6-6F). Thick lines show the TREE.

Table 5.1 Data and results of example calculation.

| number of nodes | $n$ | 100 |
|---|---|---|
| number of branches | $\beta$ | 153 |
| number of loops | $\beta - n + 1$ | 54 |
| dimension of | FAM | 54 |
| variables | PAM | 99 |
| computational | FAM | 0.24 |
| time (sec) | PAM | 1.19 |

FAM : flow rate assuming method
PAM : pressure assuming method

(1) : Exhaust Shaft

(2) : Air Supply Shaft

(3) : Escape Stair

(4) : Lobby

(5) : Hall

(6) : Room

(00) : Reference Node

This program is written taking into account the sparse properties of $[I]$ and $[L]$. This is the more efficient than the programs proposed in references[2,3,4].

The advantage of flow rate assuming method over the pressure assuming method is mainly a reduced computing times for solving of system of equations .

Net flow rates have some errors in PAM, while no errors in FAM. Therefore FAM is advantageous than PAM to calculate unsteady heat and concentration in its stability .

## Reference

1. Busacker,R.G. and Saaty,T.L., Finite Graphs and Networks, McGraw-Hill, 1965
2. Klote,J.H. and Fothergill,J.W.Jr., Design of Smoke Control System for Buildings, ASHRAE and NBS, Sep. 1983
3. Wakamatsu,T., Calculation of Smoke Movement in Buildings, BRI Research Paper No.34,1968
4. Yoshida,H.,Shaw,C.Y. and Tamura,G.T.,A FORTRAN IV Program to calculate Smoke Concentrations in a Multi-Story Buildings, NRCC, DBR Computer Program No.45, June 1979
5. Terai,T., Calculation Method of Ventilation to apply Smoke Exhaust, Report of Sub-Committee of Architectural Institute of Japan, Kinki Branch, May 1971
6. Terai,T.,Matsushita,T. and Fukai,H., Effect of Pressure Change with Time on Smoke Movement and Difference between Pressure Assuming and Flow Assuming Methods, Trans. of Architectural Institute of Japan, Kinki Branch, Vol.24, June 1984

# Appendix A

When the direction of branch is from $i$ to $j$, the relations corresponding to eq.(4.3) are

$$w_\lambda = \text{sgn.}(p_\lambda)\,\alpha b(h_2-h_1)\sqrt{2g\Gamma\,|\,p_\lambda\,|}$$

for isothermal case $(\Gamma = \Gamma_i = \Gamma_j)$

$$\bar{w} = |\,|\,1-\bar{h}\,|^{3/2} - |\,\bar{h}\,|^{3/2}\,| \qquad \text{if } \gamma_\lambda < 0, \bar{h} < 0 \text{ or } \gamma_\lambda > 0, \bar{h} > 1$$

$$\bar{w} = -\alpha\,|\,|\,1-\bar{h}\,|^{3/2} - |\,\bar{h}\,|^{3/2}\,| \qquad \text{if } \gamma_\lambda < 0, \bar{h} > 1 \text{ or } \gamma_\lambda > 0, \bar{h} < 0$$

$$\bar{w} = (1-\bar{h})^{3/2} - \alpha(\bar{h})^{3/2} \qquad \text{if } \gamma_\lambda < 0, 0 \le \bar{h} \le 1$$

$$\bar{w} = (\bar{h})^{3/2} - \alpha(1-\bar{h})^{3/2} \qquad \text{if } \gamma_\lambda > 0, 0 \le \bar{h} \le 1$$

for non-isothermal case
where

$$\alpha = \sqrt{\Gamma_j/\Gamma_i} \;,\; \bar{h} = \frac{h_n - h_1}{h_2 - h_1} \;,\; \bar{w} = \frac{w_\lambda}{w_0^\dagger} \;,$$

$$w_0^\dagger = \frac{2}{3}\alpha b\sqrt{2g\Gamma\,|\,\gamma_\lambda\,|}\,(h_2-h_1)^{3/2} \;,\; p_\lambda = h_n\gamma_\lambda$$

and $h_1, h_2, h_n$ are defined as showing in Fig.A.1, $\alpha$ is flow coefficient and $b$ is width of opening.

# Appendix B

**===== LIST OF VARIABLES =====**

(1) PRESSURE ASSUMING METHOD
```
  IP(J)   : POSITIVE NODE OF BRANCH J
  IM(J)   : NEGATIVE NODE OF BRANCH J
            INCIDENCE MATRIX (IP(J),J) = 1.0
            INCIDENCE MATRIX (IM(J),J) =-1.0
  IB(J)   : IDENTIFICATION OF BRANCH
            ABS(IB(J))= 0  HORIZONTAL OPENING BRANCH
            ABS(IB(J))= 1  VERTICAL OPENING BRANCH
            ABS(IB(J))= 2  FAN BRANCH
            ABS(IB(J))= 3  FAN BRANCH
            ABS(IB(J))= 4  WIND-PRESSURED BRANCH
            ABS(IB(J))= 5  WIND-PRESSURED BRANCH
  PP(I)   : NODE PRESSURE ( mmAq )
  GG(I)   : NODE AIR SPECIFIC GRAVITY ( AIR DENSITY ) ( Kg/m**3 )
  TT(I)   : NODE ABSOLUTE AIR TEMPERATURE ( K )
  TH(I)   : NODE AIR TEMPERATURE ( C )
            TH(I)=TT(I)-273.16
  P(J)    : BRANCH PRESSURE DIFFERENCE ( mmAq )
  PS(J)   : BRANCH PRESSURE SOURCE ( mmAq )
  G(J)    : BRANCH AIR SPECIFIC GRAVITY DIFFERENCE ( Kg/m**3 )
  GBP(J)  : BRANCH AIR FLOW SPECIFIC GRAVITY ( POSI-DIRECT. )
            ( Kg/m**3 )
  GBM(J)  : BRANCH AIR FLOW SPECIFIC GRAVITY ( NEGA-DIRECT. )
            ( Kg/m**3 )
  WP(J)   : BRANCH AIR MASS FLOW RATE ( POSI-DIRECT. ) ( Kg/sec )
  WM(J)   : BRANCH AIR MASS FLOW RATE ( NEGA-DIRECT. ) ( Kg/sec )
  VP(J)   : BRANCH AIR VOLUME FLOW RATE ( POSI-DIRECT. ) ( CMH )
  VM(J)   : BRANCH AIR VOLUME FLOW RATE ( NEGA-DIRECT. ) ( CMH )
  WPP(J)  : CHANGE OF BRANCH AIR FLOW RATE WITH PRESSURE DIFFERENCE
            ( POSI-DIRECT. )
  WMP(J)  : CHANGE OF BRANCH AIR FLOW RATE WITH PRESSURE DIFFERENCE
            ( NEGA-DIRECT. )
  DWDP(J) : CHANGE OF TOTAL BRANCH AIR FLOW RATE WITH PRESSURE
            DIFFERENCE
            DWDP(J)=WPP(J)+WMP(J)
  ON(J)   : NUMBER OF OPENINGS
  ALF(J)  : FLOW COEFFICIENT
  B(J)    : BREADTH OF OPENING ( m )
  HH(J)   : TOP HEIGHT OF OPENING FROM FLOOR LEVEL ( m )
  HL(J)   : BOTTOM HEIGHT OF OPENING FROM FLOOR LEVEL ( m )
  HS(J)   : FLOOR LEVEL DIFFERENCE ( m )
  DWW(I)  : NODE FLOW BALLANCE ERROR ( Kg/sec )
  DWWMAX  : MAXIMUM OF DWW(I) ( Kg/sec )
  ERR     : MAXIMUM OF RELATIVE ERRORS
  DPP(I)  : NODE PRESSURE CORRECTING VALUE ( mmAq )
  AJ(I,J) : COEFFICIENT MATRIX
  PFAC    : CORRECTIVE FACTOR
  VO      : OUTSIDE WIND VELOCITY ( m/sec )
  IMAX    : TOTAL NUMBER OF NODES
  IMAXS   : NUMBER OF NODES EXCEPT BASE NODE
            IMAXS=IMAX-1
  JMAX    : TOTAL NUMBER OF BRANCHES
  IFIRE   : NODE NO. OF FIRE ROOM
  ISTEP   : ITERATION TIMES
  ERRLIM  : CRITICAL VALUE OF RELATIVE ERROR
  ISTLIM  : ALLOWABLE ITERATION TIMES
  VW(I)   : WORK SPACE
  IPP(I)  : WORK SPACE
  SUM(I)  : WORK SPACE
(2) FLOW RATE ASSUMING METHOD
  AL(L)   : VALUE OF Lth ELEMENT OF LOOP MATRIX
  LL(L)   : NO. OF LOOP OF Lth ELEMENT
  LB(J+1)-LB(J) : NUMBER OF LOOP RELATED TO Jth BRANCH
  W(J)    : BRANCH NET AIR MASS FLOW RATE ( Kg/sec )
  DPDW(J) : CHANGE OF BRANCH NET PRESSURE DIFFERENCE WITH NET AIR
            FLOW RATE
  DPL(L)  : LOOP PRESSURE BALLANCE ERROR ( mmAq )
  DPLMAX  : MAXIMUM OF DPL(L) ( mmAq )
  ERR     : MAXIMUM OF RELATIVE ERRORS
  DWL(L)  : LOOP FLOW RATE CORRECTING VALUE ( Kg/sec )
  AJ(L,L) : COEFFICIENT MATRIX
  WFAC    : CORRECTIVE FACTOR
  LMAXS   : NUMBER OF INDEPENDENT LOOP
            LMAXS=JMAX-IMAX+1 (=JMAX-IMAXS)
  VW(L)   : WORK SPACE
```

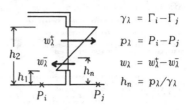

$$\gamma_\lambda = \Gamma_i - \Gamma_j$$
$$p_\lambda = P_i - P_j$$
$$w_\lambda = w_\lambda^\dagger - w_\lambda^-$$
$$h_n = p_\lambda/\gamma_\lambda$$

**Fig.A.1 Definitions of $h_1$, $h_2$ and $h_n$.**

```
  IPP(L)  : WORK SPACE
  ABSP(J) : WORK SPACE
```

## Main Routine of Pressure Assuming Method

```
*****************************************************************
*                                                               *
*   CALCULATION OF SMOKE MOVEMENT IN BUILDING FIRE              *
*                                                               *
*   UNIFORMLY MIXED STEADY STATE , PRESSURE ASSUMING METHOD     *
*   USING REDUCED INCIDENCE MATRIX & NEWTON'S METHOD            *
*                                                               *
*                                                               *
*****************************************************************
      CHARACTER MON*10,COM(4)*72
      DATA NO,VER,MON,ID,IY,COM/1,4.1,' JANUARY,',13,1985
     * ,'UNIFORMLY MIXED STEADY STATE , PRESSURE ASSUMING METHOD'
     * ,'USING REDUCED INCIDENCE MATRIX & NEWTON''S METHOD'
     * ,' ',' ',' '/
* * * * * * * * * * * * * * * * * * * * * * * * * * * * * * * * *
      PARAMETER (NN=210,NB=380)
      DIMENSION IP(NB),IM(NB),IB(NB)
     * ,PP(NN),GG(NN),TT(NN),TH(NN)
     * ,P(NB),PS(NB),G(NB),GBP(NB),GBM(NB)
     * ,WP(NB),WM(NB),WPP(NB),WMP(NB),DWDP(NB),VP(NB),VM(NB)
     * ,ON(NB),ALF(NB),B(NB),HH(NB),HL(NB),HS(NB)
     * ,DWW(NN),DPP(NN),AJ(NN,NN),VW(NN),IPP(NN),SUM(NN)
      DATA EPS,PFAC,ISTEP,IX/1.E-12,0.6,2*0/

*....... READ CONDITIONS & DATA FOR CALCULATION ..................
      CALL RCOND1(NW,IPRD,ERRLIM,ISTLIM)
      CALL RDATA1(COM(4),TT,TH,IP,IM,IB,ON,ALF,B,HH,HL,HS
     *          ,IMAX,IMAXS,JMAX,IFIRE,VO,NN,NB)

*....... PREPARATION FOR CALCULATION ..............................
      CALL MGSET (1,6)
      CALL TITLE (NO,VER,MON,ID,IY,COM,4,NW)

*....... INITIALIZATION OF INDOOR PRESSURES ......................
      CALL RANU2 (IX,PP,IMAXS,ICON)
      IF(ICON.EQ.30000) STOP 'CONDITION ERROR OCCURS IN RANU2'
      PP(IMAX)=0.0

*....... CALCULATION OF SPECIFIC GRAVITY & PRESSURE SOURCE ........
      CALL CALGPS(TT,GG,GBP,GBM,G,HS,PS,IP,IM,IB,VO,IMAX,JMAX)

*::::::: PRINTOUT OF THE DATA ::::::::::::::::::::::::::::::::::::::
      CALL WDATA1(TH,GG,IP,IM,IB,ON,ALF,B,HH,HL,G,PS,IMAXS,IMAX,JMAX
     *          ,IFIRE,VO,IPRD,NW)
*:::::::::::::::::::::::::::::::::::::::::::::::::::::::::::::::::::::

*========================================= ITERATION LOOP ======
1000 CONTINUE
      ISTEP=ISTEP+1

*....... CALCULATION OF DWW .......................................
      CALL CALFLO(PP,P,PS,WP,WM,WPP,WMP,VP,VM,GBP,GBM,G,ON,ALF,B
     *           ,HH,HL,IP,IM,IB,IMAX,JMAX)
      CALL CALWP (WPP,WMP,DWDP,JMAX)
      CALL CALDWZ(WP,WM,DWW,DWWMAX,ERR,IP,IM,IMAX,JMAX,SUM)

*---------------------------------------- JUDGEMENT OF DWW ------
      IF(ERR.LT.ERRLIM) GO TO 1100
*-----------------------------------------------------------------

*....... CORRECTION OF INDOOR PRESSURES ..........................
      CALL SETX (DWW,DPP,IMAXS)
      CALL JAC2 (DWDP,AJ,IP,IM,NN,IMAX,JMAX)
      CALL LAX  (AJ,NN,IMAXS,DPP,EPS,1,IS,VW,IPP,ICON)
      IF(ICON.GE.20000) STOP 'CONDITION ERROR OCCURS IN LAX'
      CALL CORP (PP,DPP,PFAC,IMAXS)

*---------------------------------------- JUDGEMENT OF ISTEP -----
      IF(ISTEP.GE.ISTLIM) GO TO 1110
*-----------------------------------------------------------------

      GO TO 1000
*========================================= END OF ITERATION LOOP ======
*....... PRINTOUT OF THE RESULT ..................................
1100 CALL WRELT1(PP,WP,WM,VP,VM,IMAXS,JMAX,ISTEP,ISTLIM,ERR,ERRLIM
     *          ,DWWMAX,IPRD,NW)
      STOP
1110 CALL WMSG (NW,IPRD,1)
      CALL WRELT1(PP,WP,WM,VP,VM,IMAXS,JMAX,ISTEP,ISTLIM,ERR,ERRLIM
     *          ,DWWMAX,0,NW)
      STOP 'TIMES OF STEP OVER THE MAXIMUM COUNT'
      END
```

## Main Routine of Flow Rate Assuming Method

```
******************************************************************
*                                                                *
*   CALCULATION OF SMOKE MOVEMENT IN BUILDING FIRE               *
*                                                                *
*   UNIFORMLY MIXED STEADY STATE , FLOW ASSUMING METHOD          *
*   USING REDUCED LOOP MATRIX & NEWTON'S METHOD                  *
*                                                                *
*                                                                *
******************************************************************
      CHARACTER MON*10,COM(4)*72
      DATA NO,VER,MON.ID.IY.COM/2,4.1,' JANUARY,',13,1985
     *  ,'UNIFORMLY MIXED STEADY STATE , FLOW ASSUMING METHOD'
     *  ,'USING REDUCED LOOP MATRIX & NEWTON''S METHOD'
     *  ,' ',' ',' '/
* * * * * * * * * * * * * * * * * * * * * * * * * * * * * * * *
      PARAMETER (NN=210,NB=380,NL=182)
      PARAMETER (NNL=NN*NL)
      DIMENSION IP(NB),IM(NB),IB(NB),AL(NNL),LL(NNL),LB(NB)
     *  ,GG(NN),TT(NN),TH(NN)
     *  ,P(NB),PS(NB),G(NB),GBP(NB),GBM(NB)
     *  ,W(NB),WP(NB),WM(NB),DPDW(NB),VP(NB),VM(NB)
     *  ,ON(NB),ALF(NB),B(NB),HH(NB),HL(NB),HS(NB)
     *  ,DPL(NL),DWL(NL),AJ(NL,NL),VW(NL),ABSP(NL)
     *  ,AW(NN,NN),IPP(NN)
      DATA EPS,WFAC,ISTEP,IX/1.E-12,1.0,2*0/

*...... READ CONDITIONS & DATA FOR CALCULATION ..........
      CALL RCOND1(NW,IPRD,ERRLIM,ISTLIM)
      CALL RDATA1(COM(4),TT,TH,IP,IM,IB,ON,ALF,B,HH,HL,HS
     *  ,IMAX,IMAXS,JMAX,IFIRE,VO,NN,NB)
      LMAXS=JMAX-IMAXS

*...... PREPARATION FOR CALCULATION ..................
      CALL MGSET (1,6)
      CALL LOOP21(IP,IM,AL,LL,LB,NNL,NN,LMAXS,IMAXS,JMAX,AW)
      CALL TITLE (NO,VER,MON,ID,IY,COM,4,NW)

*...... INITIALIZATION OF LOOP FLOW RATE ................
      CALL RANU2 (IX,W(IMAX),LMAXS,ICON)
      IF(ICON.EQ.30000) STOP 'CONDITION ERROR OCCURS IN RANU2'

*...... CALCULATION OF SPECIFIC GRAVITY & PRESSURE SOURCE ..........
      CALL CALGPS(TT,GG,GBP,GBM,G,HS,PS,IP,IM,IB,VO,IMAX,JMAX)

*::::::: PRINTOUT OF THE INPUT DATA ::::::::::::::::::::::::::::::
      CALL WDATA1(TH,GG,IP,IM,IB,ON,ALF,B,HH,HL,G,PS,IMAXS,IMAX,JMAX
     *  ,IFIRE,VO,IPRD,NW)
*::::::::::::::::::::::::::::::::::::::::::::::::::::::::::::::::::

================================================== ITERATION LOOP ======
1000 CONTINUE
      ISTEP=ISTEP+1

*...... CALCULATION OF PRESSURE BALLANCE ERROR ................
      CALL CALW21(W,AL,LL,LB,NNL,IMAXS,JMAX)
      CALL CALPRE(W,WP,WM,VP,VM,P,DPDW,GBP,GBM,G,ON,ALF,B,HH,HL
     *  ,IB,JMAX)
      CALL CDPL21(P,PS,DPL,DPLMAX,ERR,AL,LL,LB,NNL,LMAXS,IMAXS,JMAX
     *  ,ABSP)

*================================================= JUDGEMENT OF ERR ******
      IF(ERR.LT.ERRLIM) GO TO 1100
*

*...... CORRECTION OF LOOP FLOW RATE ................
      CALL SETX  (DPL,DWL,LMAXS)
      CALL JACW21(DPDW,AJ,AL,LL,LB,NNL,NL,LMAXS,JMAX)
      CALL LAX   (AJ,NL,LMAXS,DWL,EPS,1,IS,VW,IPP,ICON)
      IF(ICON.GE.20000) STOP 'CONDITION ERROR OCCURS IN LAX'
      CALL CORW  (W(IMAX),DWL,WFAC,LMAXS)

*================================================= JUDGEMENT OF ISTEP ******
      IF(ISTEP.GE.ISTLIM) GO TO 1110
*

      GO TO 1000
================================================= END OF ITERATION LOOP ======
*...... PRINTOUT OF THE RESULT ..................
1100 CALL WRELT2(W,WP,WM,VP,VM,JMAX,ISTEP,ISTLIM,ERR,ERRLIM
     *  ,DPLMAX,IPRD,NW)
      STOP
1110 CALL WMSG  (NW,IPRD,1)
      CALL WRELT2(W,WP,WM,VP,VM,JMAX,ISTEP,ISTLIM,ERR,ERRLIM
     *  ,DPLMAX,0,NW)
      STOP 'TIMES OF STEP OVER THE MAXIMUM COUNT'
      END

      SUBROUTINE RCOND1(NW,IPRD,ERRLIM,ISTLIM)

      WRITE (6,900)
      READ (5, * ) IANS
      WRITE (6,600)
      IF(IANS.EQ.1) WRITE (6,910)
      READ (5, * ) NW,IPRD,ERRLIM,ISTLIM
      RETURN

600 FORMAT(' ',T5,'INPUT (NW),(IPRD),(ERRLIM),(ISTLIM)')
900 FORMAT(' ',T5,'DO YOU NEED EXPLANATION OF FOLLOWING INPUT ? '
     *  ,'(1)=YES OR (0)=NO')
910 FORMAT(' ',T8,'(NW)     : OUTPUT UNIT IDENTIFIER'
     *  /' ',T8,'(IPRD)   : PRINTOUT INPUT DATA (1)=YES OR (0)=NO'
     *  /' ',T8,'(ERRLIM) : LIMIT OF RELATIVE ERROR'
     *  /' ',T8,'(ISTLIM) : LIMIT OF ITERATIVE TIMES')
      END
```

```
      SUBROUTINE RDATA1(COM,TT,TH,IP,IM,IB,ON,ALF,B,HH,HL,HS,M,MS,N,IF
     *  ,VO,NN,NB)
      CHARACTER*(*) COM
      DIMENSION    TT(NN),TH(NN)
     *  ,IP(NB),IM(NB),IB(NB),ON(NB),ALF(NB),B(NB)
     *  ,HH(NB),HL(NB),HS(NB)

      READ(50,500) M,IF,VO
      READ(50,510) (TH(I),I=1,M)

      READ(51,520) COM
      READ(51,530) N
      READ(51,540) (IP(J),IM(J),IB(J),ON(J),ALF(J),B(J),HH(J),HL(J)
     *  ,HS(J),J=1,N)

      DO 100 I=1,M
      TT(I)=TH(I)+273.16
100 CONTINUE

      MS=M-1
      RETURN

500 FORMAT(2I4,F4.0)
510 FORMAT(F8.1)
520 FORMAT(A)
530 FORMAT(I4)
540 FORMAT(3I4,F4.0,4F8.4,F8.3)

      END


      SUBROUTINE LOOP21(IP,IM,AL,LL,LB,KL,KM,LS,MS,N,A)
      DIMENSION IP(N),IM(N),AL(KL),LL(KL),LB(N+1),A(KM,MS)

      DO 100 I=1,MS
      DO 101 J=1,MS
101 A(I,J)=0.
      A(I,I)=1.
100 CONTINUE
      DO 110 I=MS,2,-1
      DO 110 J=1,MS
      A(IM(I),J)=A(IM(I),J)+A(I,J)
110 CONTINUE

      LB(1)=1
      LT=0
      DO 120 J=1,MS
      DO 121 L=1,LS
      B=0.
      I=MS+L
      IF(IP(I).LE.MS) B=B-A(J,IP(I))
      IF(IM(I).LE.MS) B=B+A(J,IM(I))
      IF(B.EQ.0.) GO TO 121
      LT=LT+1
      LL(LT)=L
      AL(LT)=B
121 CONTINUE
      LB(J+1)=LT+1
120 CONTINUE

      DO 130 J=MS+1,N
      L=J-MS
      LT=LT+1
      LL(LT)=L
      AL(LT)=1.
      LB(J+1)=LT+1
130 CONTINUE

      RETURN
      END

      SUBROUTINE CALW21(W,AL,LL,LB,KL,MS,N)
      DIMENSION W(N),AL(KL),LL(KL),LB(N+1)

      DO 100 J=1,MS
      W(J)=0.
      DO 100 L=LB(J),LB(J+1)-1
      W(J)=W(J)+AL(L)*W(MS+LL(L))
100 CONTINUE

      RETURN
      END


      SUBROUTINE CALGPS(TT,GG,GBP,GBM,G,HS,PS,IP,IM,IB,VO,M,N)
      DIMENSION TT(M),GG(M),GBP(N),GBM(N),G(N),HS(N),PS(N)
     *  ,IP(N),IM(N),IB(N)

*...... CALCULATION OF NODE SPECIFIC GRAVITY ..................
      DO 100 I=1,M
      GG(I)=353.25/TT(I)
100 CONTINUE

      DO 110 J=1,N

*...... CALCULATION OF BRANCH SPECIFIC GRAVITY ................
      GBP(J)=GG(IP(J))
      GBM(J)=GG(IM(J))
      G(J)=GBP(J)-GBM(J)

*...... CALCULATION OF PRESSURE SOURCE OF BUOYANCY ........
      IF(HS(J)) 1,2,3
1     PS(J)=(GBM(J)-GG(M))*HS(J)
      GO TO 10
2     PS(J)=0.0
      GO TO 10
3     PS(J)=(GBP(J)-GG(M))*HS(J)

*...... CALCULATION OF PRESSURE SOURCE OF WIND ..................
10    IF(IB(J).EQ.-4) PS(J)=PS(J)-0.7*GG(M)*VO*VO/19.61
      IF(IB(J).EQ. 4) PS(J)=PS(J)+0.7*GG(M)*VO*VO/19.61
```

```
      IF(IB(J).EQ.-5) PS(J)=PS(J)+0.4*GG(M)*V0*V0/19.61
      IF(IB(J).EQ. 5) PS(J)=PS(J)-0.4*GG(M)*V0*V0/19.61
  110 CONTINUE
      RETURN
      END

      SUBROUTINE CDPL21(P,PS,DPL,DPLMAX,ERR,AL,LL,LB,KL,LS,MS,N,ABSP)
      DIMENSION P(N),PS(N),DPL(LS),AL(KL),LL(KL),LB(N+1),ABSP(LS)
      DPLMAX=0.
      ERR   =0.

      DO 100 L=1,LS
      DPL(L)=0.
      ABSP(L)=ABS(P(MS+L))
  100 CONTINUE
      DO 110 J=1,N
      DO 110 L=LB(J),LB(J+1)-1
      DPL(LL(L))=DPL(LL(L))+AL(L)*(P(J)+PS(J))
  110 CONTINUE
      DO 120 L=1,LS
      IF(ABSP(L).LE.1.) ABSP(L)=1.
      DPLMAX=AMAX1(DPLMAX,ABS(DPL(L)))
      ERR=AMAX1(ERR,ABS(DPL(L))/ABSP(L))
  120 CONTINUE

      RETURN
      END

      SUBROUTINE JACW21(DX,AJ,AL,LL,LB,KL,KN,LS,N)
      DIMENSION DX(N),AJ(KN,LS),AL(KL),LL(KL),LB(N+1)

      DO 100 L1=1,LS
      DO 100 L2=1,LS
      AJ(L1,L2)=0.
  100 CONTINUE
      DO 110 J=1,N
      I1=LB(J)
      I2=LB(J+1)-1
      DO 110 L1=I1,I2
      DO 110 L2=I1,I2
      AJ(LL(L1),LL(L2))=AJ(LL(L1),LL(L2))+AL(L1)*DX(J)*AL(L2)
  110 CONTINUE

      RETURN
      END

      SUBROUTINE JAC2(DX,AJ,IP,IM,K,M,N)
      DIMENSION DX(N),AJ(K,M),IP(N),IM(N)

      DO 100 I2=1,M
      DO 100 I1=1,M
      AJ(I1,I2)=0.
  100 CONTINUE
      DO 110 J=1,N
      AJ(IP(J),IP(J))=AJ(IP(J),IP(J))+DX(J)
      AJ(IP(J),IM(J))=AJ(IP(J),IM(J))-DX(J)
      AJ(IM(J),IP(J))=AJ(IM(J),IP(J))-DX(J)
      AJ(IM(J),IM(J))=AJ(IM(J),IM(J))+DX(J)
  110 CONTINUE

      RETURN
      END

      SUBROUTINE CALWP(WPP,WMP,DWDP,N)
      DIMENSION WPP(N),WMP(N),DWDP(N)

      DO 100 J=1,N
      DWDP(J)=WPP(J)+WMP(J)
  100 CONTINUE

      RETURN
      END

      SUBROUTINE CALPRE(W,WP,WM,VP,VM,P,DPDW,GBP,GBM,G,ON,ALF,B,HH,HL
     *                 ,IB,N)
      DIMENSION W(N),WP(N),WM(N),VP(N),VM(N),P(N),DPDW(N),GBP(N),GBM(N)
     *         ,G(N),ON(N),ALF(N),B(N),HH(N),HL(N),IB(N)

      DO 100 J=1,N
      WP(J)=0.0
      WM(J)=0.0
      VP(J)=0.0
      VM(J)=0.0
      IABSIB=IABS(IB(J))

*...... CALCULATION OF FAN PRESSURE ....................
      IF(IABSIB.EQ.2) THEN
      C1=1./3.
      C2=1./6.
      C3=-1.5
      P0=ON(J)
      Q0=ALF(J)

      IF(IB(J).LT.0) THEN
      WM(J)=-W(J)
      VM(J)=WM(J)*3600./GBM(J)
      X=VM(J)/Q0
      Y=(C1*X+C2)*X+C3
      DYDX=2.*C1*X+C2
      P(J)=-P0*Y
      DPDW(J)=P0*DYDX*3600./(Q0*GBM(J))
      ELSE
      WP(J)=W(J)
      VP(J)=WP(J)*3600./GBP(J)
      X=VP(J)/Q0
      Y=(C1*X+C2)*X+C3
      DYDX=2.*C1*X+C2
      P(J)=P0*Y
```

```
      DPDW(J)=P0*DYDX*3600./(Q0*GBP(J))
      END IF

      ELSE IF(IABSIB.EQ.3) THEN
      P0=ON(J)
      Q0=ALF(J)

      IF(IB(J).LT.0) THEN
      WM(J)=-W(J)
      VM(J)=WM(J)*3600./GBM(J)
      X=VM(J)/Q0
      IF(X.GE.0.6) THEN
      C1= 25./18.
      C2=-C1
      C3=-1.
      ELSE
      C1=-25./18.
      C2= 35./18.
      C3=-2.
      END IF
      Y=(C1*X+C2)*X+C3
      DYDX=2.*C1*X+C2
      P(J)=-P0*Y

      DPDW(J)=P0*DYDX*3600./(Q0*GBM(J))
      ELSE
      WP(J)=W(J)
      VP(J)=WP(J)*3600./GBP(J)
      X=VP(J)/Q0
      IF(X.GE.0.6) THEN
      C1= 25./18.
      C2=-C1
      C3=-1.
      ELSE
      C1=-25./18.
      C2= 35./18.
      C3=-2.
      END IF
      Y=(C1*X+C2)*X+C3
      DYDX=2.*C1*X+C2
      P(J)=P0*Y
      DPDW(J)=P0*DYDX*3600./(Q0*GBP(J))
      END IF

*....... ISOTHERMAL CASE ..........................  .....
      ELSE IF(IABSIB.EQ.1.OR.G(J).EQ.0.) THEN
      CC=19.61*(ON(J)*ALF(J)*B(J)*(HH(J)-HL(J)))**2

      IF(W(J).LT.0.0) THEN
      WM(J)=-W(J)
      VM(J)=WM(J)*3600./GBM(J)
      P(J)=-WM(J)*WM(J)/(CC*GBM(J))
      DPDW(J)=2.*WM(J)/(CC*GBM(J))
      ELSE
      WP(J)=W(J)
      VP(J)=WP(J)*3600./GBP(J)
      P(J)=WP(J)*WP(J)/(CC*GBP(J))
      DPDW(J)=2.*WP(J)/(CC*GBP(J))
      END IF

*....... NON-ISOTHERMAL CASE ...................:.....  .....
      ELSE
      ABSG=ABS(G(J))
      HHL=HH(J)-HL(J)
      CC=ON(J)*ALF(J)*B(J)*SQRT(19.61*ABSG)
      C=2./3.*CC*HHL**1.5
      WBP=C*SQRT(GBP(J))
      WBM=C*SQRT(GBM(J))
      GMP=SQRT(GBM(J)/GBP(J))
      GPM=1./GMP

      IF(W(J).GT.WBP) THEN
      WP(J)=W(J)
      VP(J)=WP(J)*3600./GBP(J)
      CC=9.*ABSG*WP(J)/(CC*HHL)**2/GBP(J)
      P(J)=CC/8.*WP(J)+G(J)*(HH(J)+HL(J))/2.
      DPDW(J)=CC/4.

      ELSE IF(W(J).LT.-WBM) THEN
      WM(J)=-W(J)
      VM(J)=WM(J)*3600./GBM(J)
      CC=9.*ABSG*WM(J)/(CC*HHL)**2/GBM(J)
      P(J)=-CC/8.*WM(J)+G(J)*(HH(J)+HL(J))/2.
      DPDW(J)=CC/4.

      ELSE IF(G(J).GT.0.) THEN
      CP=CC*SQRT(GBP(J))
      CM=CC*SQRT(GBM(J))

      CC=3.*ABSG/(CC*SQRT(GBP(J)*HHL))
      AA=2.*(1.+GMP)
      P(J)=(CC*W(J)+G(J)*(GMP*HH(J)+HL(J)))/2./AA
      DPDW(J)=CC/AA
      WP(J)=2./3.*CP*ABS(HL(J)-P(J)/G(J))**1.5
      WM(J)=2./3.*CM*ABS(HH(J)-P(J)/G(J))**1.5
      VP(J)=WP(J)*3600./GBP(J)
      VM(J)=WM(J)*3600./GBM(J)

      ELSE
      CP=CC*SQRT(GBP(J))
      CM=CC*SQRT(GBM(J))
      CC=3.*ABSG/(CC*SQRT(GBM(J)*HHL))
      AA=2.*(1.+GPM)
      P(J)=(CC*W(J)+G(J)*(GPM*HH(J)+HL(J)))/2./AA
      DPDW(J)=CC/AA
      WP(J)=2./3.*CP*ABS(HH(J)-P(J)/G(J))**1.5
      WM(J)=2./3.*CM*ABS(HL(J)-P(J)/G(J))**1.5
      VP(J)=WP(J)*3600./GBP(J)
      VM(J)=WM(J)*3600./GBM(J)
      END IF

      END IF
  100 CONTINUE
```

```
                RETURN                                              100 CONTINUE
                END                                                     RETURN
                                                                        END
                SUBROUTINE CALFLO(PP,P,PS,WP,WM,WPP,WMP,VP,VM,GBP,GBM,G,ON,ALF
              *               ,B,HH,HL,IP,IM,IB,M,N)                 *....... ISOTHERMAL CASE ............................. .........
                DIMENSION PP(M),P(N),PS(N),WP(N),WM(N),WPP(N),VP(N),VM(N)    ELSE IF(IABSIB.EQ.1.OR.G(J).EQ.0) THEN
              *         ,GBP(N),GBM(N),G(N),ON(N),ALF(N),BCN(N),HH(N),HL(N)    IF(P(J).LT.0.) THEN
              *         ,IP(N),IM(N),IB(N)                                     C=ON(J)*ALF(J)*B(J)*(HH(J)-HL(J))*SQRT(19.61*GBM(J))
                DO 100 J=1,N                                                   SQP=SQRT(-P(J))
                                                                              WM(J)=C*SQP
*....... CALCULATION OF BRANCH PRESSURE ....................                  VM(J)=WM(J)/GBM(J)*3600.
                P(J)=PP(IP(J))-PP(IM(J))-PS(J)                                 WMP(J)=0.5*C/SQP
                                                                              ELSE IF(P(J).GT.0.) THEN
*....... CALCULATION OF FLOW RATE ..........................                  C=ON(J)*ALF(J)*B(J)*(HH(J)-HL(J))*SQRT(19.61*GBP(J))
                WP(J)=0.0                                                     SQP=SQRT(P(J))
                WM(J)=0.0                                                     WP(J)=C*SQP
                VP(J)=0.0                                                     VP(J)=WP(J)/GBP(J)*3600.
                VM(J)=0.0                                                     WPP(J)=0.5*C/SQP
                WPP(J)=0.0                                                    ELSE
                WMP(J)=0.0                                                    WPP(J)=1.E5
                IABSIB=IABS(IB(J))                                            WMP(J)=1.E5
                                                                              END IF
*....... CALCULATION OF FAN FLOW RATE ......................
                IF(IABSIB.EQ.2) THEN                                *....... NON-ISOTHERMAL CASE ......................... .........
                C1=1./3.                                                    ELSE
                C2=1./6.                                                    CP=ON(J)*ALF(J)*B(J)*SQRT(19.61*GBP(J)*ABS(G(J)))
                C3=-1.5                                                     CM=ON(J)*ALF(J)*B(J)*SQRT(19.61*GBM(J)*ABS(G(J)))
                P0=ON(J)                                                    HN=P(J)/G(J)
                Q0=ALF(J)                                                   HHN=ABS(HH(J)-HN)
                                                                            HLN=ABS(HL(J)-HN)
                IF(IB(J).LT.0) THEN                                         SQHHN=SQRT(HHN)
                D=AMAX1(0.0,C2*C2-4.*C1*(C3+P(J)/P0))                       SQHLN=SQRT(HLN)
                SQD=SQRT(D)
                VM(J)=Q0*(SQD-C2)/(2.*C1)                                   IF(HN.LT.HL(J)) THEN
                WM(J)=VM(J)*GBM(J)/3600.                                    IF(G(J).LT.0.) THEN
                IF(D.EQ.0.) WMP(J)=2.E5                                     WP(J)=2./3.*CP*ABS(HHN*SQHHN-HLN*SQHLN)
                IF(D.NE.0.) WMP(J)=GBM(J)*Q0/(3600.*P0*SQD)                 VP(J)=WP(J)/GBP(J)*3600.
                ELSE                                                        WPP(J)=CP/ABS(G(J))*ABS(SQHHN-SQHLN)
                D=AMAX1(0.0,C2*C2-4.*C1*(C3-P(J)/P0))
                SQD=SQRT(D)                                                 SUBROUTINE CALDW2(WP,WM,DWW,DWWMAX,ERR,IP,IM,M,N,SUM)
                VP(J)=Q0*(SQD-C2)/(2.*C1)                                   DIMENSION WP(N),WM(N),DWW(N),IP(N),IM(N),SUM(M)
                WP(J)=VP(J)*GBP(J)/3600.                                    DWWMAX=0.
                IF(D.EQ.0.) WPP(J)=2.E5                                     ERR   =0.
                IF(D.NE.0.) WPP(J)=GBP(J)*Q0/(3600.*P0*SQD)
                END IF                                                      DO 100 I=1,M
                                                                            DWW(I)=0.
                ELSE IF(IABSIB.EQ.3) THEN                                   SUM(I)=0.
                P0=ON(J)                                             100 CONTINUE
                Q0=ALF(J)                                                   DO 110 J=1,N
                X=P(J)/P0                                                   DWW(IP(J))=DWW(IP(J))+WP(J)-WM(J)
                IF(IB(J).LT.0) X=-X                                         DWW(IM(J))=DWW(IM(J))-WP(J)+WM(J)
                                                                            SUM(IP(J))=SUM(IP(J))+WM(J)
                IF(X.GE.-4./3.) THEN                                        SUM(IM(J))=SUM(IM(J))+WP(J)
                C1=25./18.                                           110 CONTINUE
                C2=-C1                                                      DO 120 I=1,M-1
                C3=-1.                                                      IF(SUM(I).LE.1.) SUM(I)=1.
                D=AMAX1(0.0,C2*C2-4.*C1*(C3-X))                             DWWMAX=AMAX1(DWWMAX,ABS(DWW(I)))
                SQD=SQRT(D)                                                 ERR=AMAX1(ERR,ABS(DWW(I)/SUM(I)))
                ELSE                                                 120 CONTINUE
                C1=-25./18.
                C2=35./18.                                                  RETURN
                C3=-2.                                                      END
                D=AMAX1(0.0,C2*C2-4.*C1*(C3-X))
                SQD=-SQRT(D)                                                SUBROUTINE SETX(Y,X,M)
                END IF                                                      DIMENSION Y(M),X(M)

                IF(IB(J).LT.0) THEN                                         DO 100 I=1,M
                VM(J)=Q0*(SQD-C2)/(2.*C1)                                   X(I)=Y(I)
                WM(J)=VM(J)*GBM(J)/3600.                              100 CONTINUE
                IF(D.EQ.0.) WMP(J)=2.E5
                IF(D.NE.0.) WMP(J)=GBM(J)*Q0/(3600.*P0*SQD)                 RETURN
                ELSE                                                        END
                VP(J)=Q0*(SQD-C2)/(2.*C1)
                WP(J)=VP(J)*GBP(J)/3600.                                    SUBROUTINE CORW(WL,DWL,WFAC,LS)
                IF(D.EQ.0.) WPP(J)=2.E5                                     DIMENSION WL(LS),DWL(LS)
                IF(D.NE.0.) WPP(J)=GBP(J)*Q0/(3600.*P0*SQD)
                END IF                                                      DO 100 L=1,LS
                                                                            WL(L)=WL(L)-DWL(L)*WFAC
                ELSE                                                 100 CONTINUE
                WM(J)=2./3.*CM*ABS(HHN*SQHHN-HLN*SQHLN)
                VM(J)=WM(J)/GBM(J)*3600.                                    RETURN
                WMP(J)=CM/ABS(G(J))*ABS(SQHHN-SQHLN)                        END
                END IF
                                                                            SUBROUTINE CORP(PP,DPP,PFAC,M)
                ELSE IF(HN.GT.HH(J)) THEN                                   DIMENSION PP(M),DPP(M)
                IF(G(J).GT.0.) THEN                                         DO 100 I=1,M
                WP(J)=2./3.*CP*ABS(HHN*SQHHN-HLN*SQHLN)                     PP(I)=PP(I)-DPP(I)*PFAC
                VP(J)=WP(J)/GBP(J)*3600.                              100 CONTINUE
                WPP(J)=CP/ABS(G(J))*ABS(SQHHN-SQHLN)                        RETURN
                ELSE                                                        END
                WM(J)=2./3.*CM*ABS(HHN*SQHHN-HLN*SQHLN)
                VM(J)=WM(J)/GBM(J)*3600.
                WMP(J)=CM/ABS(G(J))*ABS(SQHHN-SQHLN)                *****  Omitted Subroutine *****
                END IF
                                                                            SUBROUTINE ; TITLE
                ELSE IF(G(J).LT.0.) THEN                                    --- Subroutine to print title
                WP(J)=2./3.*CP*HHN*SQHHN                                    SUBROUTINE ; WDATA1
                WM(J)=2./3.*CM*HLN*SQHLN                                    --- Subroutine to print input data
                VP(J)=WP(J)/GBP(J)*3600.                                    SUBROUTINE ; WRELT1, WRELT2
                VM(J)=WM(J)/GBM(J)*3600.                                    --- Subroutine to print result
                WPP(J)=-CP/G(J)*SQHHN                                       SUBROUTINE ; WMSG
                WMP(J)=-CM/G(J)*SQHLN                                       --- Subroutine to print error message
                ELSE                                                        SUBROUTINE ; RANU2
                WP(J)=2./3.*CP*HLN*SQHLN                                    --- Subroutine to create random number
                WM(J)=2./3.*CM*HHN*SQHHN                                    SUBROUTINE ; LAX
                VP(J)=WP(J)/GBP(J)*3600.                                    --- Subroutine to solve system of equations by
                VM(J)=WM(J)/GBM(J)*3600.                                        LU decomposition algorithm
                WPP(J)=CP/G(J)*SQHLN
                WMP(J)=CM/G(J)*SQHHN
                END IF
                END IF
```

# Effects of Combustion Gases on Escape Performance of the Baboon and the Rat

HAROLD L. KAPLAN
Department of Fire Technology
Southwest Research Institute
P.O. Drawer 28510
San Antonio, Texas 78284, USA

ABSTRACT

In postcrash aircraft fires, only a few minutes are often available for egress. To assess the potential of combustion gases to impair human escape, a signalled avoidance task was developed for use with the juvenile baboon. After a 5-minute exposure, the animal was required to select and depress the correct lever to open an escape door and exit into the adjacent compartment of a shuttlebox. With CO, the $EC_{50}$ for escape failure was 6850 ppm. Acrolein (12 to 2780 ppm) neither prevented escape nor affected escape times, despite irritant effects at all concentrations. Similar results were obtained with HCl (190 to 17,200 ppm) in that, all animals successfully performed the escape task, even at concentrations that produced severe post-exposure effects and lethality. With a comparable shuttlebox and escape paradigm for rats, the $EC_{50}$ of CO was 6780 ppm. Five-minute exposures to HCl (11,800 to 76,730 ppm) did not prevent escape but severe post-exposure respiratory effects and lethality occurred at 15,000 ppm and higher. In both species, HCl did not affect escape time but the number of intertrial responses was significantly related to concentration. The results indicate that the rat and the baboon have a comparable tolerance to CO and irritant gases and that laboratory test methods of incapacitation of rodents may be useful in evaluating the potential of combustion gas atmospheres containing CO and irritant gases to prevent human escape.

## INTRODUCTION

All commercial aircraft contain a wide variety of interior polymeric materials which, when combusted, evolve toxic decomposition products. The most prevalent product is carbon monoxide (CO), but other toxicants may be formed, depending on the chemical structure of the material and the combustion conditions. The fire gases are often classified into two major classes, the asphyxiants or hypoxia-producing toxicants and the irritants.

For passengers to survive a postcrash aircraft fire, their escape capability must not be severely impaired by toxic combustion gases during the few minutes available for egress. Our knowledge of the potential for these gases to impair human escape performance is very limited. In studies of the hypoxia-producing toxicants (CO and hydrogen cyanide [HCN]), laboratory test methods generally have utilized loss of gross locomotor function or of shock-avoidance response to measure the incapacitating effects of these gases (1). Although the mechanisms of action of both CO and HCN appear comparable in the rodent and man, the correlation between incapacitation of rodents and loss of escape capability in humans has not been established. As for the irritant combustion gases, studies with

This research study was supported under Federal Aviation Administration Contract No. DTFA03-81-00065.

rodents have shown that sensory irritants cause a reflex inhibition of respiratory rate whereas a temporary increase in respiratory rate occurs upon inhalation of pulmonary irritants (2). The relevance of these respiratory effects in the rodent to human escape capability also has not been established.

These studies were conducted in order to assess the potential of asphyxiant and irritant combustion gases to impair human escape performance, using a nonhuman primate model and an operant escape task. A secondary objective was to evaluate the usefulness of presently used laboratory methods with rodents to predict the potential of these gases to prevent human escape.

METHODS

Effects of Gases on Escape Performance of Nonhuman Primates

Animal subjects. Male juvenile (ages 2 to 3 years) baboons (Superspecies Papio cynocephalus) were obtained from a breeding colony at the Southwest Foundation for Biomedical Research, San Antonio, Texas. All animals were healthy and free of respiratory problems at the beginning of the study.

Exposure and escape performance test system. The system for exposure of animals and measurement of escape performance (Figure 1) basically consisted of three elements: (1) a gas mixing chamber and associated ducts and valves; (2) a gas bypass loop; and (3) a primate escape performance test apparatus. Exposure atmospheres were premixed in the gas mixing chamber and recirculated through the gas bypass loop until the desired concentrations were obtained. For exposure of an animal, activation of appropriate valves removed the gas bypass loop from the system and the atmosphere was recirculated between the gas mixing chamber and one chamber of the primate escape performance test apparatus.

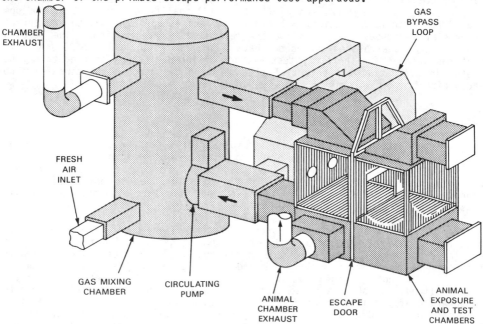

FIGURE 1. Gas mixing/exposure system and escape performance test apparatus for primate tests

The primate escape performance test apparatus was a shuttlebox consisting of two identical chambers separated by a vertically sliding escape door. Within each chamber was a cage constructed of aluminum bars to which constant current shockers could provide 4 to 8 milliamps of current. Two response levers were mounted on the wall opposite the escape door in each chamber and, above each lever, were two cue lights, one white and one red. During a trial, the white light was activated over one lever and the red light was activated over the other lever, with the sequence randomized by a computer control system. Depression of the lever over which the white light was "on" caused the escape door to open; depression of the lever when its red light was "on" had no effect.

Escape performance paradigm. The behavioral paradigm for measurement of escape performance was a signalled avoidance task. An audio cue (Sonalert buzzer) signalled the start of a trial and, simultaneously, the white light over one lever and the red light over the other lever were turned on to indicate to the subject which lever was the correct response, i.e., would open the escape door. Ten seconds later, an electric shock was applied to the bars of the cage and maintained for 20 seconds. If the subject pressed the correct lever and moved through the door opening into the adjacent chamber within 10 seconds, the response was designated an "avoidance." If the animal pressed the correct lever and exited after 10 seconds, but within 30 seconds, the response was termed an "escape." The response was a "failure" if the subject did not exit the exposure chamber or exited after 30 seconds. Both avoidance and escape responses were considered escapes or successful performance in the treatment of the data.

A behavioral control system, consisting of a Data General Nova 3 minicomputer equipped with a BRS/LVE Corporation INTERACT System, was used to program and control the escape performance equipment and paradigm and to record performance data. Performance data included: (1) time to first lever press; (2) time to first correct lever press; (3) time to chamber exit; (4) number of correct and incorrect lever presses; (5) number of intertrial lever presses (ITIs), i.e., number of presses made prior to initiation of a trial; (6) cumulative number of avoidances and escapes; and (7) number of shock pulses delivered.

Exposures. The effects on escape performance of a five-minute exposure to each of three combustion gases, CO, acrolein and hydrogen chloride (HCl), were investigated. In each experiment, the exposure of a subject was initiated after the desired concentration of the gas was equilibrated in the gas mixing chamber, thereby enabling the atmosphere to rapidly reach a stable concentration in the exposure chamber. With CO, each of six subjects was exposed to each of four concentrations in order to evaluate concentration-response relationships for the escape performance measures and to derive an $EC_{50}$ value for escape failure. With acrolein and HCl, usually concentrations were increased in successive experiments until post-exposure lethalities occurred, in an effort to determine a threshold concentration for escape failure. Some animals exposed to lower concentrations were exposed to these gases a second time, with weeks to months intervening and provided the subjects were asymptomatic.

Generation and analysis of exposure atmospheres. Compressed gas cylinders of pure CO were used to generate CO atmospheres, by metering the gas through a calibrated flowmeter into the gas recirculation system. Continuous analysis of these atmospheres was accomplished by means of a Beckman 865 non-dispersive infrared analyzer. For the experiments with acrolein, predetermined quantities of liquid acrolein were injected directly into the recirculation system to obtain the desired concentrations of exposure atmospheres. Continuous analysis of these atmospheres was accomplished by means of a hydrocarbon analyzer calibrated by gas chromatographic analyses of frequent syringe samples withdrawn from sampling ports in the exposure chamber. Hydrogen chloride atmospheres were generated by the same procedure as for CO, using a compressed gas cylinder of pure

HCl. During each exposure, five one-minute samples of the atmosphere were obtained using soda-lime aborption tubes; these samples were desorbed with water and analyzed by titration with mercuric nitrate (3).

## EFFECTS OF GASES ON ESCAPE PERFORMANCE OF THE RAT

Animal subjects. Male Sprague-Dawley rats (Timco Breeding Laboratories, Houston, Texas), weighing between 340 and 500 grams, were used in these studies.

Escape performance test apparatus and paradigm. A modified commercially available shuttlebox (Lafayette Instrument Company, Model No. 85103), consisting of two identical chambers separated by a vertically sliding partition driven by an electric motor, was used to measure escape performance. Two levers were mounted on each side of the partition in each chamber and a white cue light was mounted above each lever. The floor and top of each chamber consisted of a grid of electrically isolated bars for the administration of an electric shock to the subjects. The escape performance paradigm and behavioral control system were the same as for the primates except that the rat was not required to discriminate between red and white cue lights because of the lengthy training time required. Each trial was initiated with a tone (Sonalert buzzer) and the lighting of the white light over each of the two levers. Pressing of either lever by the rat opened the partition and allowed the animal to escape into the adjacent chamber. Responses were designated "avoidance," "escape" or "failure" according to the same criteria as in the primate studies.

Exposures. The effects on escape performance of five-minute exposures to CO and HCl were investigated in the rat. With CO, each of four or six animals was exposed to each of four concentrations in order to evaluate concentration-response relationships for the escape performance measures and to derive an $EC_{50}$ value for escape failure. For these exposures, the rodent shuttlebox was placed inside the exposure chamber of the primate test system and connected to the computer control system. For the HCl exposures, the rodent shuttlebox was interfaced with a 300-liter acrylic exposure chamber in a manner that allowed the exposure cage to be located within the acrylic chamber and the escape cage outside of this chamber. Each animal was exposed only once to HCl. The procedures for the generation and analyses of CO and HCl exposure atmospheres were basically the same as for the primate studies.

## RESULTS

Effects of carbon monoxide on escape performance of the baboon and the rat. The results of the experiments with CO in Table 1 (baboons) and in Table 2 (rats) show a decreasing percentage of escaping animals, both baboons and rats, with increasing average concentrations of CO. Linear regression analysis of performance data for the baboon did not show a statistically significant relationship between CO concentration and escape time or other performance parameters of those subjects that successfully performed the escape task. In rats, however, escape time increased as CO concentration increased ($p<0.05$). None of the other performance parameters of rats was affected by exposure to CO. The concentration-response curves for escape performance of the two species (Figure 2) were obtained by plotting percentage of failures against the logarithm of the mean CO concentration and deriving the best fitting line by the probit method of Finney (4). From these curves, $EC_{50}$ values for escape failure were determined to be 6850 ppm (95-percent confidence limits: 6043-7773) for the juvenile baboon and 6780 ppm (95 percent confidence limits: 6367-7271) for the rat.

TABLE 1. Effects of Carbon Monoxide on Escape Performance of the Baboon

| AVERAGE CO CONCENTRATION (PPM) | 6120 ± 47 | 6840 ± 30 | 7010 ± 114 | 7520 ± 86 |
|---|---|---|---|---|
| NUMBER OF AVOID/ESCAPE/FAIL RESPONSES | 4 AVOID 2 FAIL | 2 AVOID 1 ESCAPE 3 FAIL | 2 AVOID 1 ESCAPE 3 FAIL | 2 AVOID 4 FAIL |
| AVOID/ESCAPE TIME (SEC)[a] | | | | |
| PRE-EXPOSURE | 7.4 ± 1.6 | 6.2 ± 0.5 | 6.6 ± 0.8 | 7.9 ± 1.8 |
| TEST | 4.2 ± 1.5 | 8.5 ± 3.6 | 9.4 ± 3.1 | 5.0 ± 1.2 |
| TIME TO FIRST LEVER PRESS (SEC)[a] | | | | |
| PRE-EXPOSURE | 3.0 ± 0.7 | 2.4 ± 0.4 | 2.2 ± 1.0 | 2.9 ± 0.9 |
| TEST | 1.9 ± 1.3 | 2.6 ± 1.8 | 2.4 ± 1.1 | 1.7 ± 0.4 |
| TIME TO FIRST CORRECT LEVER PRESS (SEC)[a] | | | | |
| PRE-EXPOSURE | 4.6 ± 1.3 | 3.1 ± 0.3 | 3.5 ± 1.0 | 3.6 ± 0.2 |
| TEST | 1.9 ± 1.3 | 2.6 ± 1.8 | 5.4 ± 4.6 | 1.7 ± 0.4 |
| NUMBER OF INCORRECT LEVER PRESSES[a] | | | | |
| PRE-EXPOSURE | 0.5 ± 0.3 | 0.4 ± 0.3 | 0.9 ± 0.3 | 0.6 ± 0.6 |
| TEST | 0.0 ± 0.0 | 0.3 ± 0.6 | 2.0 ± 1.0 | 0.0 ± 0.0 |
| NUMBER OF LEVER PRESSES/MIN DURING ITI[a,b] | | | | |
| PRE-EXPOSURE | 1.0 ± 0.6 | 1.0 ± 0.8 | 1.7 ± 1.2 | 1.5 ± 1.6 |
| TEST | 2.5 ± 1.7 | 1.8 ± 1.9 | 2.4 ± 1.7 | 2.1 ± 1.7 |

[a]VALUES REPRESENT MEAN ± S.D. (N = NO. OF AVOID/ESCAPE ANIMALS)
[b]ITI = INTERTRIAL INTERVAL.


TABLE 2. Effects of Carbon Monoxide on Escape Performance of the Rat

| AVERAGE CO CONCENTRATION (PPM) | 6150 ± 48 | 6730 ± 46 | 6990 ± 47 | 7220 ± 195 |
|---|---|---|---|---|
| NUMBER OF AVOID/ESCAPE/FAIL RESPONSES | 3 ESCAPE 1 FAIL | 4 ESCAPE 2 FAIL | 1 AVOID 1 ESCAPE 4 FAIL | 1 ESCAPE 3 FAIL |
| AVOID/ESCAPE TIME (SEC)[a] | | | | |
| PRE-EXPOSURE | 6.4 ± 4.3 | 6.3 ± 1.8 | 3.3 ± 1.6 | 11.3 ± 5.4 |
| TEST[b] | 12.2 ± 1.2 | 18.6 ± 5.5 | 8.7 ± 1.6 | 18.9 |
| TIME TO FIRST LEVER PRESS (SEC)[a] | | | | |
| PRE-EXPOSURE | c | 4.8 ± 1.9 | 2.2 ± 1.5 | c |
| TEST | c | 14.4 ± 3.5 | 6.3 ± 5.4 | c |
| NUMBER OF LEVER PRESSES[a] | | | | |
| PRE-EXPOSURE | 1.2 ± 0.1 | 1.0 ± 0.1 | 1.0 ± 0.0 | 1.2 ± 0.4 |
| TEST | 2.0 ± 1.1 | 1.0 ± 0.0 | 1.5 ± 0.7 | 1.0 |
| NUMBER OF LEVER PRESSES/MIN DURING ITI[a,d] | | | | |
| PRE-EXPOSURE | 0.2 ± 0.3 | 0.1 ± 0.1 | 0.3 ± 0.6 | 0.1 ± 0.1 |
| TEST | 0.6 ± 0.6 | 1.2 ± 1.1 | 0.9 ± 0.6 | 1.6 ± 1.0 |

[a]VALUES EXPRESSED AS MEAN ± S.D. (N = NO. OF AVOID/ESCAPE ANIMALS).
[b]TEST VALUES SIGNIFICANTLY ($p < 0.05$) RELATED TO CO CONCENTRATION.
[c]DATA NOT AVAILABLE DUE TO EQUIPMENT MALFUNCTION.
[d]ITI = INTERTRIAL INTERVAL.

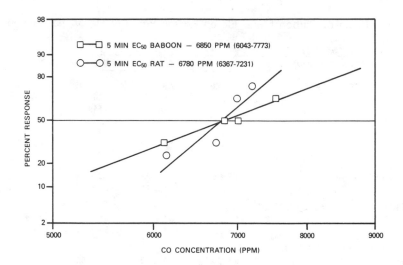

FIGURE 2. Concentration-response curves and $EC_{50}$ values for escape failure of the baboon and the rat by carbon monoxide

Effects of acrolein on escape performance of the baboon. In nine experiments in which baboons were exposed for five minutes to acrolein at average concentrations of 12 to 2780 ppm (Table 3), all subjects except one made avoidance responses. In the one exception (95 ppm), the subject was mobile and did not exhibit signs of incapacitation but did not perform the escape task. The result is not considered meaningful, however, because the shock stimulus did not operate in this experiment and, in addition, animals exposed to much higher concentrations were able to perform the task. In most experiments, test escape times were less than pre-exposure mean escape times, suggesting that the animals were attempting to escape from the irritant environment, but the difference was not statistically significant. Nor did statistical analyses of the data indicate any effect of acrolein on other performance parameters. Although escape performance was not impaired, irritant effects were evident at all concentrations of acrolein, increasing in severity from blinking and closure of the eyes and rubbing of the eyes/nose at lower concentrations to salivation, nasal discharge, violent shaking of the head and nausea at higher concentrations. The animals exposed to the two highest concentrations (1025 and 2780 ppm) expired at 24 and 1.5 hours, respectively, following exposure, with severe pulmonary edema and hemorrhage as the most significant histopathologic findings.

Effects of hydrogen chloride on escape performance of the baboon and the rat. In eight experiments in which baboons were exposed for five minutes to average HCl concentrations of 190 to 17,290 ppm (Table 4), all animals performed the escape task, with avoidance responses by six subjects and escape responses by two subjects. Statistical analyses of the data did not indicate a significant relationship between acrolein concentration and escape time or other performance parameters except for intertrial responses, which showed a significant increase with increasing concentrations ($p < 0.05$). Irritant effects were observed at all concentrations except the lowest, increasing in severity from coughing and frothing at the mouth at lower concentrations to profuse saliva-

TABLE 3. Effects of acrolein on escape performance of the baboon

| Average Acrolein Conc. (ppm) | Test Response | Avoid/Escape Time (sec)[a] Pre-Exposure/Test | Time to 1st Lever Press (sec)[a] Pre-Exposure/Test | Time to 1st Correct Lever Press (sec)[a] Pre-Exposure/Test | No. of Incorrect Lever Presses[a] Pre-Exposure/Test | No. Lever Presses/ Min During ITI[a,b] Pre-Exposure/Test |
|---|---|---|---|---|---|---|
| 12 | Avoid | 6.3 ± 2.4 / 4.9 | 1.7 ± 0.9 / 1.1 | 3.0 ± 2.4 / 1.5 | 1.3 ± 1.7 / 1 | 1.4 ± 1.4 / 2.6 |
| 25 | Avoid | 5.0 ± 0.4 / 6.5 | 1.8 ± 0.4 / 3.2 | 1.8 ± 0.4 / 3.1 | 0.0 ± 0.0 / 0 | 0.0 ± 0.0 / 0 |
| 95[d] | Fail | c / c | c / c | c / c | c / c | c / c |
| 100 | Avoid | 5.9 ± 1.2 / 5.8 | 2.7 ± 1.3 / 3.1 | 2.7 ± 1.3 / 3.1 | 0.0 ± 0.0 / 0 | 0.8 ± 1.1 / 0.6 |
| 250 | Avoid | 5.8 ± 1.4 / 5.2 | 2.3 ± 1.4 / 2.1 | 2.3 ± 1.4 / 2.1 | 0.0 ± 0.0 / 0 | 0.0 ± 0.0 / 0 |
| 505 | Avoid | 6.8 ± 1.8 / 5.1 | 2.6 ± 1.5 / 1.7 | 3.7 ± 1.9 / 1.7 | 1.3 ± 1.7 / 0 | 0.0 ± 0.2 / 0.4 |
| 505[d] | Avoid | 6.6 ± 1.7 / 4.7 | 1.2 ± 0.6 / 0.9 | 1.7 ± 1.4 / 0.9 | 1.0 ± 0.9 / 0 | 1.9 ± 1.7 / 3.6 |
| 1025[d] | Avoid | 5.9 ± 1.7 / 4.0 | 1.7 ± 0.9 / 1.1 | 2.6 ± 1.3 / 1.1 | 0.9 ± 0.9 / 0 | 2.8 ± 0.8 / 4 |
| 2780[e] | Avoid | 6.6 ± 1.7 / 4.3 | 2.4 ± 1.8 / 1.8 | 2.4 ± 1.8 / 1.8 | 0.0 ± 0.0 / 1 | 1.5 ± 2.3 / 1.8 |
| Mean ± S.D. | | 6.1 ± 0.6/5.1 ± 0.8 | 2.0 ± 0.5/1.9 ± 0.9 | 2.5 ± 0.6/1.9 ± 0.9 | 0.6 ± 0.6/0.3 ± 0.5 | 1.0 ± 1.0/1.4 ± 1.6 |

[a] Pre-exposure value for each concentration expressed as mean ± S.D. (N = 10-12); test value (N = 1).

[b] ITI = Intertrial Interval.

[c] Equipment malfunction, data not included.

[d] Subjects not previously exposed to acrolein.

[e] Animal exposed to 100 ppm acrolein 5 weeks earlier.


TABLE 4. Effects of hydrogen chloride on escape performance of the baboon

| Average HCl Conc. (ppm) | Test Response | Avoid/Escape Time (sec)[a] Pre-Exposure/Test | Time to 1st Lever Press (sec)[a] Pre-Exposure/Test | Time to 1st Correct Lever Press (sec)[a] Pre-Exposure/Test | No. of Incorrect Lever Presses[a] Pre-Exposure/Test | No. Lever Presses/ Min During ITI[a,b] Pre-Exposure/Test[c] |
|---|---|---|---|---|---|---|
| 190 | Avoid | 5.4 ± 1.4 / 4.5 | 1.5 ± 0.9 / 0.9 | 1.8 ± 1.2 / 0.9 | 0.8 ± 0.7 / 1 | 3.0 ± 3.0 / 2.8 |
| 810[d] | Avoid | 5.1 ± 1.0 / 4.4 | 1.5 ± 0.2 / 1.5 | 1.5 ± 0.2 / 1.5 | 0.0 ± 0.0 / 0 | 0.0 ± 0.0 / 1.8 |
| 890 | Avoid | 7.7 ± 1.7 / 5.2 | 2.6 ± 1.1 / 1.8 | 3.5 ± 1.4 / 1.8 | 0.9 ± 1.2 / 2 | 1.0 ± 2.1 / 2.6 |
| 940[d] | Avoid | 5.0 ± 2.4 / 4.2 | 1.4 ± 1.4 / 1.4 | 2.2 ± 1.4 / 1.4 | 0.9 ± 0.9 / 1 | 2.6 ± 2.2 / 1.2 |
| 2,780[d] | Avoid | 6.0 ± 1.3 / 7.8 | 2.0 ± 1.3 / 3.5 | 2.8 ± 1.2 / 3.7 | 1.4 ± 1.0 / 1 | 2.8 ± 1.3 / 9.2 |
| 11,400[d] | Escape | 6.7 ± 2.1 / 16.3 | 2.1 ± 1.0 / 10.0 | 3.9 ± 2.1 / 13.4 | 1.3 ± 1.3 / 6 | 8.2 ± 5.4 / 11 |
| 16,570[e] | Avoid | 8.3 ± 3.6 / 5.9 | 1.4 ± 0.4 / 2.4 | 3.9 ± 3.8 / 2.4 | 2.4 ± 2.4 / 0 | 0.0 ± 0.1 / 1.4 |
| 17,290[d] | Escape | 8.2 ± 3.5 / 10.9 | 4.2 ± 2.0 / 1.9 | 5.4 ± 3.4 / 7.1 | 2.2 ± 2.6 / 1 | 0.3 ± 0.7 / 9.6 |
| Mean ± S.D. | | 6.5 ± 1.4/7.4 ± 4.3 | 2.1 ± 1.0/2.9 ± 3.0 | 3.1 ± 1.3/4.0 ± 4.3 | 1.3 ± 0.8/1.5 ± 1.9 | 2.2 ± 2.7/5.0 ± 4.2 |

[a] Pre-exposure value for each concentration expressed as mean ± S.D. (N = 10-12); test value (N=1).

[b] ITI = Intertrial Interval.

[c] Test values significantly (p<0.05) increased with increasing concentrations.

[d] Subjects exposed to HCl one time.

[e] Subject not previously exposed to HCl but exposed to 95 ppm acrolein several months earlier.

tion, blinking/rubbing of eyes, and shaking of head at higher concentrations. In animals exposed to the two highest concentrations (16,570 and 17,290 ppm), severe dyspnea persisted after exposure, with death at 18 and 76 days, respectively, post exposure. Histopathologic examination revealed pneumonia, pulmonary edema and tracheitis with epithelial erosion.

In twelve experiments in which rats were exposed for 5 minutes to HCl at concentrations of 11,800 to 87,660 ppm (Table 5), all animals performed the escape task, except at the highest concentration which resulted in death during exposure. Statistical analyses of the data did not indicate any effect of HCl on escape time or other performance parameters, except for the number of intertrial responses which significantly (p<0.05) increased with increasing concentration. All concentrations produced signs of severe irritation (respiratory tract and/or eyes), with persistent respiratory effects and post-exposure lethality produced by exposure to concentrations of 15,250 ppm and greater. The times at which post-exposure deaths occurred ranged from 3 minutes to 13 days following exposure, and were correlated inversely with HCl concentration.

TABLE 5. Effects of hydrogen chloride on escape performance of the rat

| Average HCl Conc. (ppm) | Test Response | Escape Time (sec)[a] Pre-Exposure/Test | Time to First Lever Press (sec)[a] Pre-Exposure/Test | Number of Lever Presses[a] Pre-Exposure/Test | No. Lever Presses/ Min During ITI[a,b] Pre-Exposure/Test[c] |
|---|---|---|---|---|---|
| 11,800 | Escape | 15.5 ± 5.1 / 25.7 | 11.6 ± 1.3 / 14.4 | 1.1 ± 0.4 / 1 | 0.0 ± 0.0 / 0.6 |
| 14,410 | Escape | 13.3 ± 1.2 / 13.9 | 11.3 ± 1.6 / 13.1 | 1.2 ± 0.4 / 2 | 0.8 ± 1.0 / 2.8 |
| 15,250 | Escape | 13.5 ± 1.7 / 15.3 | 10.8 ± 1.2 / 10.0 | 1.4 ± 0.5 / 6 | 0.0 ± 0.0 / 5.2 |
| 18,430 | Avoid | 10.0 ± 3.7 / 7.0 | 9.5 ± 3.8 / 6.3 | 1.0 ± 0.0 / 1 | 0.5 ± 1.0 / 4.6 |
| 22,260 | Escape | 13.5 ± 5.3 / 21.5 | 13.7 ± 5.5 / 13.0 | 1.1 ± 0.3 / 2 | 0.0 ± 0.0 / 2.4 |
| 25,300 | Escape | 13.6 ± 4.5 / 16.1 | 13.1 ± 4.5 / 7.9 | 1.0 ± 0.0 / 1 | 0.2 ± 0.6 / 3.2 |
| 25,850 | Escape | 15.1 ± 5.0 / 14.0 | 4.5 ± 5.2 / 13.4 | 1.5 ± 0.9 / 1 | 0.1 ± 0.3 / 0.2 |
| 27,690 | Escape | 16.8 ± 4.6 / 16.6 | 12.5 ± 3.8 / 10.2 | 1.5 ± 0.8 / 13 | 0.3 ± 0.8 / 3.4 |
| 50,910 | Escape | 12.2 ± 2.5 / 14.7 | 10.9 ± 1.1 / 14.0 | 1.5 ± 0.7 / 1 | 0.1 ± 0.3 / 7.0 |
| 53,900 | Avoid | 13.3 ± 2.3 / 2.9 | 12.7 ± 2.2 / 2.4 | 1.1 ± 0.3 / 1 | 0.0 ± 0.0 / 2.8 |
| 76,730 | Escape | 12.1 ± 3.9 / 21.4 | 10.6 ± 2.9 / 10.2 | 1.0 ± 0.0 / 11 | 0.2 ± 0.4 / 5.8 |
| 87,660 | Died | -- | -- | -- | -- |
| Mean ± S.D. | | 13.5 ± 1.8/15.4 ± 6.4 | 11.0 ± 2.5/10.5 ± 3.7 | 1.2 ± 0.2/3.6 ± 4.4 | 0.2 ± 0.2/4.2 ± 3.3 |

[a] Pre-exposure value expressed as mean ± S.D. (N = 10-12); test value (N = 1).

[b] ITI = Intertrial Interval

[c] Test values significantly (p<0.05) increased with increasing concentration.

DISCUSSION

The similarity of 5-minute $EC_{50}$ values for CO-induced escape failure in the baboon and rat suggests that the rat may be a useful model for evaluating the potential of CO atmospheres to cause loss of escape capability in nonhuman primates and, possibly, in man. Additional insight into the utility of the rat as a model for man may be obtained from Kimmerle's (5) human toxicity data which indicate that symptoms produced in some individuals at between 30 and 40 percent carboxyhemoglobin (COHb) may be sufficiently severe as to prevent escape. Using the Stewart-Peterson equation (6) with a respiratory minute volume (RMV) of 20 liters (light activity), this range of COHb saturation may be anticipated in man after a 5-minute exposure to from 7,000 to 9,000 ppm CO. These exposures equate to a concentration-time product (Ct), i.e., accumulated dose, of 35,000 to 45,000 ppm-minutes. The 5-minute $EC_{50}$ Ct values obtained for the baboon (34,250 ppm-min) and the rat (33,900 ppm-min) are near this range, providing support for the usefulness of the the the rat and baboon (with the shuttlebox) for estimating the potential of CO atmospheres to prevent human escape. Comparable Ct values reported for incapacitation of rats by CO using other methods, such as the rota-rod (7), the motor-driven exercise wheel (8) and leg-flexion shock avoidance (9), indicate that these methods, also, may have utility in estimating the effects of CO-containing atmospheres on human escape capability.

In the experiments with acrolein, a five-minute exposure did not prevent the baboon from performing the escape task, even at concentrations (1025 and 2780 ppm) that cause post-exposure lethality. These results do not appear to be consistent with claims that the $RD_{50}$ concentration of an irritant gas will incapacitate man within a few minutes (10, 11), in view of the reported $RD_{50}$ concentration of only 1.7 ppm of acrolein in mice (12). The effects of acrolein on escape performance of the rat were not investigated in these studies nor are experimental data for acrolein using other methods for measuring incapacitation of rodents available in the literature. Unpublished data provided by the FAA/CAMI laboratory, however, indicate that 5-minute exposures to approximately 5000 to 10,000 ppm of acrolein incapacitate the rat in the motor-driven exercise wheel and cause post-exposure lethality (13). It is possible that these concentrations would produce sufficient pulmonary damage in the baboon within 5 minutes to prevent performance of the escape task. Thus, both the rat and the baboon appear to be highly tolerant of the incapacitating effects of acrolein.

The results obtained with HCl in baboons were similar to those with acrolein in that 5-minute exposures to HCl did not prevent the baboon from performing the escape task, even at concentrations (approximately 17,000 ppm) that caused severe post-exposure respiratory effects and lethality. These results also do not appear to be consistent with the prediction that man will be incapacitated within a few minutes by the $RD_{50}$ concentration (309 ppm in mice [12]) of HCl. Rats, also, tolerated high concentrations of HCl (11,800 to 76,730 ppm) without loss of performance of the escape task, although post-exposure lethalities occurred at concentrations above approximately 15,000 ppm. These results are in accord with those of experiments conducted by the FAA/CAMI laboratory (13) in which approximately 5 to 7 minutes of exposure to 65,000 to 100,000 ppm incapacitated rats in the motor-driven exercise wheel, with subsequent lethality.

From the results of these studies and those of Crane (13), the rat and baboon appear to be capable of enduring high concentrations of acrolein or HCl without incapacitation or loss of escape capability. These results are not consistent with the conclusions of Henderson and Haggard (14) that a concentration of 24 ppm of acrolein is unbearable and of 10 ppm and above is lethal to man in a short time and that inhalation of 1000 ppm of HCl is dangerous. It is possible that this inconsistency is due to species differences in sensitivity to irritant gases. However, it is also possible that man can tolerate higher concentrations of irritant gases than anticipated, when unavoidable, without complete loss of escape capability.


ACKNOWLEDGEMENTS

The author wishes to acknowledge the significant contributions to this study by the co-investigators of the Department of Fire Technology and the Department of Bioengineering of the Southwest Research Institute, the invaluable data and technical recommendations provided by Dr. Charles R. Crane of the FAA/CAMI Laboratory and the guidance and advice given by Mr. Constantine P. Sarkos of the Federal Aviation Administration.

REFERENCES

1. Kaplan, H. L., Grand, A. F. and Hartzell, G. E., Combustion Toxicology: Principles and Test Methods, Technomic Publishing Company, Incorporated, Lancaster, Pennsylvania (1983).

2. Alarie, Y., "Sensory Irritation by Airborne Chemicals," CRC Critical Reviews in Toxicology, Volume 2, pp. 299-362 (1972).

3. Grand, A. F., Kaplan, H. L., Beitel, J. J., III, Switzer, W. G., and Hartzell, G. E., "An Evaluation of Toxic Hazards from Full-Scale Furnished Room Fire Studies," Fire Safety: Science and Engineering, ASTM STP 882, T. Z. Harmathy, Ed., American Society for Testing and Materials, Philadelphia, in press (1985).

4. Finney, D. J., "Probit Analysis" (3rd ed.), Cambridge University Press (1971).

5. Kimmerle, G., "Aspects and Methodology for the Evaluation of Toxicological Parameters During Fire Exposure," The Journal of Fire and Flammability Combustion Toxicology Supplement, Vol. 1 (February 1974).

6. Stewart, R. D., Peterson, J. E., Fisher, T. N., Hosko, M. J., Baretta, E. D., Dodd, H. C., and Herrmann, A. A., "Experimental Human Exposure to High Concentrations of Carbon Monoxide," Archives of Environmental Health, Volume 26, pp. 1-7 (January 1973).

7. Southwest Research Institute, Unpublished Work.

8. Crane, C. R., et al., "Inhalation Toxicology: I. Design of a Small-Animal Test System; II. Determination of the Relative Toxic Hazards of 75 Aircraft Cabin Materials," Report No. FAA-AM-77-9 Prepared for Office of Aviation Medicine, Federal Aviation Administration (March 1977).

9. Packham, S. C., and Hartzell, G. E., "Fundamentals of Combustion Toxicology in Fire Hazard Assessment," J. of Testing and Evaluation, Volume 9, No. 6, p. 341 (1981).

10. Barrow, C. S., Lucia, H., Stock, M. F. and Alarie, Y., "Development of Methodologies to Assess the Relative Hazards from Thermal Decomposition Products of Polymeric Materials," Amer. Ind. Hyg. Assoc. J., Vol. 40, pp. 408-423 (May 1979).

11. Kane, L. E., Barrow, C. S. and Alarie, Y., "A Short-Term Test to Predict Acceptable Levels of Exposure to Airborne Sensory Irritants," Amer. Ind. Hyg. Assoc. J., Vol. 40, pp. 207-229 (1979).

12. Alarie, Y., "Toxicological Evaluation of Airborne Chemical Irritants and Allergens Using Respiratory Reflex Reactions," Proceedings of the Inhalation Toxicology and Technology Symposium, sponsored by The Upjohn Company, Kalamazoo, Michigan (October 23-24, 1980).

13. Crane, C. R., Unpublished Data.

14. Henderson, Y. and Haggard, H. W. Noxious Gases, 2nd Edition, New York, Reinhold (1943).

# SUPPRESSION

Session Chair

**Prof. David J. Rasbash**
Unit of Fire Safety Engineering
University of Edinburgh
Edinburgh EH9 3JL, United Kingdom

# The Extinction of Fire with Plain Water: A Review

**D. J. RASBASH**
University of Edinburgh, Unit of Fire Safety Engineering
Edinburgh, EH9 3JL, United Kingdom

ABSTRACT

Research into the extinction of fire of solid and liquid fuels by plain water has been reviewed. Properties of water sprays and of the fires themselves that influence the performance of sprays as extinguishing agents are outlined. The results of basic investigations which have yielded data which are of assistance to quantifying extinction processes and of empirical investigations designed to provide information directly applicable to practical problems have been briefly summarised. Simple quantitative approaches to two forms of extinction for which basic data exists i.e. extinction of the flames by spray and cooling the fuel to the firepoint are outlined. The approach to extinction by cooling also points to an approach for deciding under which conditions stable flaming may be formed on the whole of a fuel surface in the absence of spray. Areas for future research are indicated.

INTRODUCTION

Fire and the way it is affected by water in the form of rain is a force that has helped fashion the very nature of our living environment. It is probably safe to say that since mankind first made use of fire, they made use of water to control it. Apart from rhetorical quotations, very little has come down to us from these aeons of time, on just how much water is needed to control fires of different kinds. It has undoubtedly long been found useful to have a few buckets of water to hand since they are very good at extinguishing fires before they become really dangerous. Apart from this there has been the tendency, for some fire authorities to base demands for waters supplies on instinct, with an eye perhaps more on what can be made available rather than what might be really needed.

Water although nominally quite cheap in most places, becomes quite expensive if it has to be handled and stored in large quantities for emergency purposes. The important practical problem also exists of making the best out of limited supplies of water that may be available in the early stages of fires of all kinds. These demands call for a fund of structured information on the subject. A significant amount of research has been carried out which has gone some way to filling this gap. It is the aim of this paper to summarise this information, to point to where coherent patterns lie within it, and to indicate where scope lies for further work.

# HEAT ABSORPTION PROPERTIES OF WATER

The major useful property of water as an extinguishing agent is its capacity to cool burning fuels to a temperature below which they cease to burn. In general this capacity substantially exceeds that of other extinguishing agents, including carbon dioxide and nitrogen, as indicated in Table 1. The exception is the inability of water to cool fuels that can burn near or below normal ambient temperatures, particularly low flash point liquids like gasoline.

TABLE 1 - Maximum cooling capacity of water and other agents for extinction processes, expressed as J/g agent.

| Agent | Agent temperature | Enthalpy change relevant to cooling solid and liquid fuels to: | | | Enthalpy change relevant to cooling reaction zone in flames to 1300°C | |
|---|---|---|---|---|---|---|
| | °C | 0°C | 100°C | 250°C | In gas phase and condensed phase | In gas phase alone |
| Water | 15 | 0 | 2612 | 2900 | 5317 | 2704 |
| Carbon dioxide (solid) | -78 | 637 | 734 | 872 | 2158 | 1585 |
| Nitrogen (liquid) | -196 | 405 | 509 | 652 | 1885 | 1685 |

# RELEVANT PROPERTIES OF JETS, SPRAYS AND FIRES

Water is normally applied to a fire in the form of a jet or a spray. A jet allows the water to reach the fire area more easily, but has a very limited capacity for rapid removal of heat whereas a spray can remove heat rapidly. However when a jet reaches a solid surface it can break up into a spray or form a film of water over the surface which improves its capacity to remove heat. It can also be moved manually to similar effect.

The reach of a jet depends on the pressure and the flow rate at the nozzle and its angle of elevation. It also depends on the nozzle design. Recent improvements in nozzle design have led to a significant increase in the throw of jets (1). The reach of sprays depends also on the cone angle of the spray; for a given pressure and flow rate, its reach is substantially less than a jet. The following formula for the throw Z(m) of sprays was arrived at by regression analysis (2).

$$Z = 1.1F^{0.36}P^{0.28}/(\tan \theta/4)^{0.57}$$
F = flow rate (0.7-33 l/s)
P = nozzle pressure (3-11 bar)
$\theta$ = cone angle (30°-90°)

The following properties of water sprays have been found relevant to their action on fires.

1. Mean flow rate per unit area in the region of the fire.
2. Distribution of the flow rate at and about the fire area.
3. Direction of application towards the fire area, particularly whether the spray is being applied downwards against the flow or with a substantial sideways component.
4. Drop size of the spray and drop size distribution.
5. The velocity of the current of air entrained by the spray.
6. The velocity of the drops related to (a) the entrained air current, (b) the flame the drops move through, and (c) the fuel which is impacted.

The size and velocity of the spray drops mainly affects the heat transfer to flame and fuel but they also affect the tendency of drops to enhance the burning rates of liquids by causing splashing and sputtering at the fuel surface. Also together with the entrained air current they control the ability of the spray drops to penetrate an upward moving flame. All practical sprays have a wide distribution of drop size and care has to be exercised in deciding on the relevant representative drop size to use since drop size enters into the physical laws that influence the effects of sprays on fires in widely different ways. In this paper the mass or volume median drop size will be used as a representative drop size. In general, it lies between the "volume mean" drop size ($\Sigma n d^4 / \Sigma n d^3$) and the "area mean" or Sauter Mean drop size ($\Sigma n d^3 / \Sigma n d^2$). However the "mean volume" drop size $\left( \sqrt[3]{\dfrac{\Sigma n d^3}{\Sigma n}} \right)$ which is considerably smaller, has also been used by some workers.

The nature of a fire also has an influence on the ease with which water can extinguish it. The temperature at which the fuel may and does burn, the geometrical distribution of the fuel surfaces and the thermal properties of the fuel and any char that forms are, of course, major influences. The size and upward velocity of the flames as well as their temperature and intensity of combustion influence the ease with which the flame may be extinguished and the ability of the spray to reach the fuel.

There follows a summary of research investigations of the extinction of fire with plain water. These will be divided into two areas; basic research when a quantitative approach to interpreting the results is aimed at, and empirical full scale tests.

BASIC RESEARCH ON EXTINCTION OF FIRE

Liquid Fires. Work by the author in the 1950's in which the majority of the above factors were monitored revealed the complexity of the interaction between sprays and liquid fires. The work was carried out with two forms of apparatus:

(a) Sprays of a range of drop size (0.2-2.0mm) produced from a battery of hypodermic needles, were allowed to fall at different rates onto a kerosine fire 11cm diameter, situated 120cm below the battery. The object of the hypodermic needles was to obtain sprays of uniform drop size (3). This was reasonably successful for drop sizes greater than 0.6mm.

(b) Sprays of a range of drop size and flow rate produced at different pressures from batteries of pairs of impinging jets projected downwards on fires burning in a vessel 30cm diameter 175cm below the jets. Most tests were carried out on a 6cm deep layer of burning kerosine or gasoline but other liquids were used as well (4,5).

The main observations in the tests were:

1. If the kerosine fire was not extinguished, it stabilised in a few minutes after application of the spray with a temperature near the surface which was usually lower, and a flame size and mean rate of burning which was usually less than the values obtained in tests without spraying. The temperature distribution within the fuel layer was also made more uniform. These results are exemplified in figures 1a, b, c and d.

2. There was a most effective drop size for the kerosine fire for reducing burning rate and cooling the fuel, which in series (a) was 0.4mm and in series (b) decreased from 0.7mm to 0.35mm as the pressure at the spray battery increased from 0.35 bar to 2.1 bar. At the most effective drop size 40-50% of the spray penetrated to below the burning fuel. Minimum stable burning rates occurred at this drop size, which were about 2-3 $g/m^2 s$, compared with 10 $g/m^2 s$ and 14 $g/m^2 s$ for the (a) and (b) fires respectively. At coarser drop sizes (more than 1.7 times most effective drop size) a considerable amount of splashing of the fuel into the flames took place, at finer drop sizes (less than about 0.7 of most effective drop size) only some 10-20% of water penetrated the fuel layer, and fuel was sputtered into the flame by the fine drops evaporating as they hit the hot kerosine. Both these mechanisms increased the burning rate. For series (b) it was possible to correlate the downward thrust of the entrained air stream, the drop size and the fraction of spray that penetrated the flame (6).

3. The kerosine fire was extinguished in two distinct ways:-
   (a) by cooling the liquid temperature to the neighbourhood of the fire point as measured by the Cleveland open cup (58-68°C). In this case the flames reduced to a small size before extinction and reignition did not take place immediately on the application of an ignition source following extinction. The flow rates of water required to bring about extinction increased from 100 $g/m^2 s$ to 300 $g/m^2 s$ as the drop size increased from 0.4mm to 1.2mm. The closed cup flash points of the unburned kerosines used were in the range 40-45°C.
   (b) the flame was itself extinguished when of substantial size. The fuel surface temperature at extinction was well in excess of the fire point. Usually substantial momentary partial clearance of flame took place prior to extinction. After extinction application of an ignition source brought about immediate reignition. Extinction times were usually much shorter than when the liquid was cooled to the fire point.

4. The gasoline fire was only extinguished by the second of the above mechanisms. At a spray pressure of 5.9 bar, extinction did not take place at a drop size greater than 0.6mm nor a flow rate less than 130 $g/m^2 s$. As long as the spray could penetrate the flames in sufficient quantity the extinction time was proportional to the fourth power of the drop size and inversely to the second power of the flow rate (4).

5. Certain sprays which could extinguish the gasoline fires readily (2-12 seconds extinction time) when the preburn time (i.e. time between ignition and spray application) was 15 seconds or more, could not extinguish the fire when the preburn time was less than 10 seconds. The spray caused the flames to be pushed into a horizontal shape burning in the entrained air of the spray. A similar phenomenon occurred with benzole fire. With the kerosine fire this phenomenon would give extinction by cooling to the fire point but in a much longer time. As a result for a given spray acting on a kerosine fire a bimodal distriubtion of extinction times as a function of preburn times occurred (4).

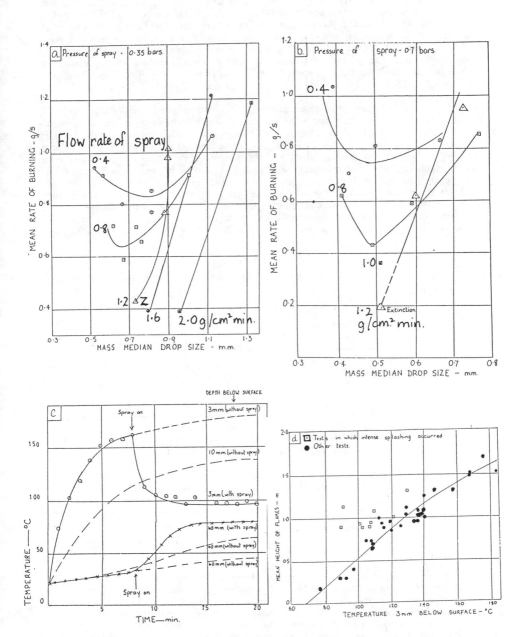

Fig. 1 - <u>Effect of water sprays on a kerosine fire.</u> Properties of fire: 300mm diameter, 20mm ullage, burning rate 1.0 g/s. (a) and (b) Effect of spray flow rate and dropsize on burning rate, (c) Effect of a spray (Z in (a)) on temperature below kerosine surface, (d) Relation between flame height and temperature 3mm below surface.

6.  For sprays of flow rate 267g/m² s and drop size 0.28mm, 0.38mm, 0.49mm and
    fires in diesel oil (fire point 104-115°C) and transformer oil (fire point
    175-180°C) there was a large upsurge of flame in the first 1-2 seconds.
    They nevertheless were usually extinguished a few seconds later and more
    quickly than a kerosine fire. The coarser spray was rather more effective
    than the finer spray. The two coarser sprays extinguished an alcohol fire
    by diluting the surface layers but the finest spray extinguished the
    flames of this fire in a few seconds.

Solid Fires.  As long as the water flow can reach the surfaces of the burning
fuel then flow rates of the order of only a few g/m² s have been found capable
of extinguishing many solid fires. Moreover the flow rate needed is not
particularly sensitive to the drop size of the spray. Thus Bryan (7) used a
moving jet of water to extinguish a rotating crib fire. The minimum flow rate
for extinction in his experiments was about 1.6 g/m² s. Magee and Reitz (8)
applied water spray to vertical and horizontal burning plastic materials,
exposed to different levels of radiation up to 11Kw/m² . Table 2 gives critical
water flow rates for extinction when there was no external radiation. For
delrin and PMMA the heat of vaporisation of the extra water flow needed when
the sample was exposed to external radiation balanced the external radiation.

TABLE 2 - Critical water flow rates for extinction with water spray. After Magee and Reitz. (No external radiation)

| Polymer | Surface vertical g/m² s | Surface Horizontal g/m² s |
|---|---|---|
| PMMA | < 2.7(~ 1.7) | < 1.7 (~ 1.3) |
| DELRIN | 1.9 | < 2.7 (~ 2.0) |
| PS | | < 4.1 (~ 3.0) |
| PE | | < 6.0 (~ 4.3) |

TABLE 3 - Critical fuel flow rates at extinction with water spray; after Magee and Reitz. Radiation 0 - 0.4 cal./cm² s. Various water flow rates.

| | Surface Vertical g/m² s | Surface Horizontal g/m² s |
|---|---|---|
| PMMA | 3.4-4.0 | 6.0 |
| DELRIN | 6 | 4 - 5.5 |
| PS | | 9 |
| PE | | 11 |

This suggests that water is capable of being more or less completely vaporised
at a solid burning surface. However, for polystyrene and polyethylene rather
higher flow rates were required in the absence of external radiation. Moreover
the extra flow needed for the irradiated sample was more than would balance
the extra radiation. When water spray was applied to these two polymers fuel
was sputtered into the flame. This is indicated in Table 3, which gives the
critical flow rates of fuel at extinction with water spray, indicating
substantially higher flow rates for these than for delrin or for PMMA. With
the former materials a coarse spray was found to be more effective than a

finer one. The measured fire point of polystyrene is 350°C and polyethylene 330°C (9), whereas the melting points are 240°C and 138°C (10). Thus a horizontal surface of these materials would be covered by a significant layer of very hot liquid. A vertical surface would burn as a running liquid fire.

It is characteristic of solid fires that burning surfaces may be obscured to a varying extent from the direct action of the spray i.e. there is obstruction. A review carried out by Heskestad (11), indicated a range of critical flow rates per unit area for wood in various geometrical forms from slabs to cribs varying from 1.3 to 3.0$g/m^2$s; the higher rates seemed to apply to the more densely packed cribs. Of course the total exposed fuel surface area on which burning takes place in a crib may substantially exceed the plan area of the fuel. Experiments were carried out by O'Dogherty et al (12), in which the spray was applied downwards to fire in cribs of the order of 0.9m wide and 1m deep, made up of 2.5cm sticks and with a total exposed stick area of about 48$m^2/m^2$ of top surface. These indicated that as long as the flames were beaten down to allow water spray to reach the burning fuel, then a flow rate of 187$g/m^2$s reaching the top surface of the crib, could extinguish the fire in about 10 minutes. This corresponds to about 4$g/m^2$ of fuel surface area per second. O'Dogherty also developed a standard test for extinguishers using wood cribs of varying lengths (13). His results indicate that an extinguisher delivering about 6$g/m^2$s could extinguish a fairly open crib fire in about 1 minute; comparing this with information in ref.7, suggests a critical rate of about 2.5$g/m^2$s.

Bhagat (14) found however that to extinguish glowing charcoal with spray of drop size 60-120$\mu$ carried in an air stream at 10.4m/s a flow rate of about 10 $g/m^2$s was needed. Lower flow rates tended to increase the burning rate of the charcoal above what they were without spray by removing ash from the surface.

QUANTITATIVE APPROACH TO EXTINCTION

The experiments on liquid fires show that extinction can occur either by extinction of flame or cooling the fuel. Evidence in the experiments on fire in solids, particularly the balancing of extra radiation with extra water flow needed, indicates also that cooling the fuel is the major aspect of extinction of these fires. Actually it will be seen in the discussion below that extinction by cooling the fuel is a special case of extinction of the flame.

Extinction of the Flames. It has been found that the amount of diluents such as nitrogen, carbon dioxide or water vapour needed to extinguish premixed flames is sufficient to reduce the adiabatic flame temperature to about that of the lower flammability limit. Indeed this limit may be regarded to arise from diluting the flame with air (oxygen and nitrogen) above that which is needed for combustion. A lower limit adiabatic flame temperature of about 1580K or 1300°C emerges for most hydrocarbons and other organic vapours, from these considerations. This may be regarded as a minimum temperature where a sufficient rate of reaction can be sustained in the limit flame relevant to the particular aerodynamic conditions under which the flame is propagating. The ratio of lower limit concentration to stoichiometric concentration for the vapours concerned is between 50-60%; the extra diluent is therefore responsible for absorbing about 45% of the heat of combustion at the limit temperature. It may be argued therefore that any heat transfer mechanism and particularly heat transfer between spray drops and flames which may be responsible for absorbing this fraction of the heat of combustion from the reaction zone should prevent the flame propagating. Table 1 also indicates that if all the heat transfer results in steam, the heat absorption needed to bring about extinction could be halved.

Experiments with diffusion flames have indicated that these flames are often extinguished with a lesser requirement of diluent than premixed flames (15). Thus 30 and 60cm diameter diesel oil fires were found to be extinguished when burning in a diluent gas based mainly on water vapour and nitrogen when the oxygen concentration was 14.5-15%, whereas the composition of the gas was such as to indicate that a reduction to 12.2% would be needed to prevent flame propagation in a premixed gas. Flame on a wick (dimension about 5mm) was extinguished at a concentration of 16.7% $O_2$. A flame about 3cm wide and 10cm long travelling vertically up a group of thin wooden sticks was extinguished with 13.2% of oxygen in nitrogen, whereas flame propagation through premixed vapours was estimated as requiring a reduction to 11.3%. The implication is that for diffusion flames of dimension 5-60cm, removal of about 30-35% of the heat of combustion from within the reaction zone would cause the flame to be extinguished. The probable reason is that the diffusion flames near extinction are losing substantially more heat by radiation from the reaction zone than premixed flame near the limit and also combustion is less complete. The latter would certainly be implied by the existence of luminous, thus sooty flames prior to extinction. A very small flame required a heat removal factor of only about 0.2 for extinction and there may be a partial blow out effect associated with this small value.

It was found that for the 30cm diameter liquid fuel fires tested with sprays, the intensity of combustion in the flames without spray was about 2W/cm³ and independent of the fuel. The mean temperatures across the flames were measured as 921°C (benzole to 1218°C (alcohol). While these are less than the 1300°C limit temperature mentioned above, it must be remembered that there was a substantial unmixedness factor for the turbulent flames concerned, these being non-uniform mixtures of unburned fuel vapours, unburned air, reaction zone and combustion products. Proceeding on the assumption that the capability of the spray to cool any of these items will contribute to the cooling of the reaction zone, indicates that there may be a correlation between the capability of the sprays to extract heat from the flames and the extinction of the flame. This was examined (16) and found to be the case. For extinction of the flame to take place it was found that the spray when entering the flame needed to have an ability to extract 0.4 of the heat produced in unit volume of flame. The downward force of the spray also needed to be greater than about 0.7 of the upward force of the flames, presumably to allow water spray to enter the lower parts of the spray, but one could argue that the downward entrained air current was forcing combustion products back into the flame. The difficulty in extinguishing flames when the preburn time was less than 10 seconds was probably caused by flames being made to burn stably but more intensely in the air current of the spray, as a horizontal sheet above the fuel surface.

Cooling the Fuel. The concept of the fire point is that it is the minimum fuel temperature at which a stable flame may be formed across the whole surface of the fuel. The observations of flame size as fire point conditions are approached (Fig. 1d) suggest that, apart from intense splash fires, heat transfer is probably dominated by convection from the flame which will approach a maximum at the fire point, rather than by radiation. It is possible for fuel surfaces to maintain continuous flashing flames at temperatures below the fire point if vapours evolved can produce concentrations within the limits and at thicknesses greater than about half the relevant quenching distance. However, heat transfer to the fuel will be significantly less than at the fire point, since at any instant only part of the surface is covered by flame. Moreover such flames are very vulnerable to wind and the entrained current of sprays. They are unlikely to be of importance during extinction with water sprays.

The author has approached extinction, by cooling the fuel on the basis of a heat balance at the fuel surface with separation of convective and radiant heat transfer (6,17,18,19). The spray will cool the fuel if it can cause removal of sensible heat from the surface at a rate per unit area S given by:

$$S = (H_f - \lambda_f)\dot{m}'' + R_a - R_s \qquad \ldots\ldots(1)$$

$\dot{m}''$ = rate of burning per unit area

$H_f$ = convective heat transfer from flames associated with unit mass of fuel entering the flame

$\lambda_f$ = heat imparted to the body of the fuel which is necessarily associated with the production of unit mass of fuel vapour; under steady state conditions this may be equated to the heat required to produce the volatiles,

$R_a$ = other forms of heat transfer to unit area of fuel surface including radiation from the flames, from the outside environment, and by conduction from hot metal etc.

$R_s$ = heat loss from surface not included in $\lambda_f$ especially heat loss by radiation.

To cool the fuel to its fire point, the spray has to be capable of removing heat at the relevant rate S throughout the whole cooling process from application of spray to the fire point. The capacity of water spray to remove heat from the fuel may remain relatively constant for a solid fire but could for a kerosine fire be lowered significantly as the temperature of the fuel is reduced. As the fuel is cooled the contribution of radiation from the flames to $R_a$ will be reduced since a reduction of $\dot{m}''$ will cause a reduction in flame size (Fig. 1d). However the convective heat transfer $H_f\dot{m}''$ will increase somewhat because of the reduction of the blocking effect caused by reduced fuel vapour flow. Extinction by cooling will be expected to occur if the cooling rate is greater than S throughout the cooling process and at the special condition at the fire point as well as covered by equation 2.

$$0 > (H_{fc} - \lambda_f)\dot{m}''_c + R_{ac} - R_{sc} - S_c \qquad \ldots\ldots(2)$$

The subscript c in equation 2 implies conditions at the fire point. $H_{fc}$ is a maximum value of $H_f$ and may be expressed as a fraction $\emptyset$ of the heat of combustion H. Also $H_f$ and $\dot{m}''$ may be related, and as the fire point condition approaches, $H_{fc}$ and $\dot{m}''_c$ may perhaps more tentatively be related, through Spalding's B number (20) as follows:-

$$\dot{m}''_c = f(\emptyset, H, m_{og}, T_g, T_s, h, c \text{ etc.}) \approx \frac{h}{c} \ln\left(1 + \frac{A}{\emptyset H}\right) \qquad \ldots\ldots(3)$$

$A = m_{og} H/r + C(T_g - T_s)$; $A/\emptyset H$ = B number at the fire point

Equation 3 may also be taken as a condition for establishing a stable flame that covers the whole surface area of the fuel.

As indicated above small flames were noted as the fire point condition was approached. Moreover there is a lack of sensitivity to fire size of critical flow rates for extinction by cooling of liquid fires for pools of diameter 0.11 to 2.43m (6). These observations provide some justification for ignoring radiation from the flame to the surface, as a factor influencing critical flow rates needed for extinction by cooling. Experiments by Spalding (20) on cooling kerosine fires on a small sphere with excess fuel and other experiments indicate that $\emptyset$ is about 0.2-0.4 for normal organic fuels of composition ranging between hydrocarbons - cellulosics, when burning in air at ambient temperatures, a mean value of 0.3 for $\emptyset$ has been suggested (17). This may be compared with a heat removal factor of about 0.33 for large radiant diffusion flames and 0.45 for premixed propagation flame. Estimates that then may be made using equation 2 of rate of burning under fire point conditions, and water flow rate requirements. are reasonably in agreement with rates mentioned above for kerosine, wood and PMMA. The equation however cannot cover conditions of intense sputtering or splashing.

Equations (2) and (3) do not explicitly take account of the influence that the flow of water vapour ($\dot{m}''_{wv}$) produced at the fuel surface may be having on cooling the flame. This is unlikely to have a major effect if this cooling, as estimated from the last column of Table 1, is small compared with the convective heat the flame is capable of losing without extinction ($\emptyset H \dot{m}''_c$). However this effect of steam can still be included in Eq. 2 and 3 if $\emptyset$ and H are both reduced to take account of the presence of the steam and $\dot{m}''_c$ is regarded as including $\dot{m}''_{wv}$.

A very similar set of concepts including the use of equation 3 has been used by Tamanini to account for the flow rate of water needed to extinguish wooden slabs burning on both sides (21). Tamanini focussed on the water requirement needed to reduce the adiabatic temperature of the limit flame to 1300°C. This implies the use of a value of $\emptyset$ of about 0.45 prior to its modification by steam that accompanies the fuel vapour.

By associating the $\emptyset$ factor of convective cooling of the flame at the fire point to the cooling of the reaction zone associated with flammable limits (18), it is possible to extend the use of equation 2 to a wider range of ambient temperatures, fuel chemistry and oxygen concentrations. However certain assumptions inherent in equation 3 will cease to operate for high values of $\emptyset$ which may be associated with high oxygen concentration, particularly when it is to be expected that there will be significant amounts of unburned oxygen at the surface when the fire point condition is approached.

The heat removal factor by extinguishant S is but one of a number of factors in equations 1 and 2. The fire will be expected to go out on its own if the heat balance is such as to cause the fuel to be cooled to below the fire point. Although a continuous flashing fire regime in which flames cover only part of the surface at any given moment is more likely to become apparent in the absence of a spray, a heat balance which causes a fire to cool through the fire point is unlikely to allow it to survive long in the flashing regime. It should also be mentioned in passing that because of the way the buoyant rising column of flame and hot gases dominate the aerodynamic conditions near the fuel surface, the value of the heat transfer coefficient h that needs to be used in applying equations 2 and 3 to upward and downward facing surfaces is the reverse of values appropriate to normal heat transfer calculations.

The above treatment separates processes of heat transfer in the flame leading to flame extinction, and in the fuel leading to extinction by cooling to the fire point. This of course is major simplification. A general approach to extinction would require the combination of both aspects. This is particularly important if the intention is to predict extinction time rather than critical flow rates. An attempt to do this for open forest fires has been made by Corlett and Williams (22), although they were concerned rather with the breakdown of flame coherence of large areas of fire. This they related to the combined effect of heat loss in the flames and the reduced fuel input into the flames caused by application of extinguishant.


FULL SCALE FIRE TESTS

While over the ages tests may well have been carried out to indicate how much water may be needed for fire extinction, no systematic published information on experimental work has been available until comparatively recently. However the practice has certainly now developed of carrying out large scale tests to give realistic answers to practical problems. Attention, for the most part, has focussed on flow rate and quantities of water needed for specific types of fire, with specific types of nozzle. In general the practical difficulties in full scale work has precluded obtaining the extensive information on the properties of jets, sprays and fires listed earlier, necessary for fundamental appraisal of the effects of water on the fire.

Protective Installations - Solid Fires. Sprinkler systems were introduced more than a century ago. The water flow rates needed for these systems were developed by trial and error in practice, and remained fairly constant at 0.1 gallon/ft$^2$min. or 5mm/minute or 82g/m$^2$s flow over all parts of the plan area, until the late fifties. However it then became clear that for many practical situations, substantially larger flow rates were required. This was due in the main to the introduction of new materials, particularly foam plastics, and high storage. These higher flow rates have found expression in FOC and NFPA rules (23,24). The advent particularly of automated high rack storage and the necessity for intermediate rack sprinklers to which these give rise, has prompted extensive research investigations which are continuing at the present time. It is not possible here to summarise this work adequately; much information is available from organisations such as the Fire Research Station in the U.K. and the Factory Mutual Corporation in the U.S.

It is appropriate to comment on one point. Traditionally the general design of a sprinkler was for it to throw about 50-60% of its water upwards to hit the ceiling, as a ring of water about 1.5m diameter. This water then fell to the ground as large drops. The rest of the water from the sprinkler was sprayed outwards. During the 1950's a different design of sprinkler was introduced which became the standard sprinkler in the U.S., which threw all its water outwards and downwards. Tests which for the most part appear to have been carried out with fires situated between the sprinklers showed the new type as significantly superior to the old type (25). However the design never found similar favour in the U.K., the "old type" sprinkler remained the standard while the new type was recognised as the "spray sprinkler". The reason for the old type sprinkler remaining in favour was that the large drops from the ceiling was thought to be able to penetrate flames moving upwards in the region of the sprinkler. Experience on the effectiveness of the sprinklers in practice has not shown either of the types to be markedly superior to the other. In recent years workers at the Factory Mutual Laboratory have designed coarse drop size sprinklers for specific use in situations where there may be a strong up-draft (26).

Protective Installations - Liquid Fires. Water spray systems were introduced for use against high flash point liquid fires in the U.K. in the thirties, particularly for risks like power stations. Flow rates of about 0.4 gal/ft$^2$min (330g/m$^2$s) of directional spray produced at pressures exceeding 2.7 bar were regarded as necessary for the risks concerned. The question arose in the fifties as to whether a fine spray which was distributed around the risk and which could be pulled in by the upward moving flames, could be regarded as a viable alternative. The requirement of the system was that it should bring about certain extinction of fire that might be expected in a power station within one minute of operation. Experiments (27,28) were carried out under semi-open conditions, on pan fires up to 1.2m diameter and with a rig of 50mm diameter empty tubes down which burning oil (mostly transformer oil) flowed at a rate up to 0.76 l/s. Extinction of the pan oil fires was found to be by cooling to the fire point and was found to be related to the flow rate and drop size. However with directional sprays at a pressure of 1.7 bar, it was found that stable splash fires were established in spite of the fact that the liquid fuel was cooled to well below the fire point. These fires became extinguished very quickly or reduced to small flames burning at the edge of the metal vessel used when the spray was turned off.

The tube rig experiment was very demanding with regard to water requirements because of the high temperatures which could be reached by the tubes in a very short time (550°C in 2 minutes and 700–900°C in 4 minutes) and because of the masking of flames behind the tubes and supporting struts. As the temperature of the tubes increased from 100–500°C, a marked increase was needed in the flow rate to the envelope area of the tube rig from 0.3 to 1 imp. gal./ft$^2$ min (250–850 g/m$^2$ s) to bring about certain extinction in 45 seconds. However a further increase of tube temperature from 500–900°C only required a modest increase from 850 to 980g/m$^2$ s. This suggests that a major difficulty encountered by the water spray was its inability because of film boiling to evaporate efficiently on a hot metal surface, under conditions where oil with a much higher boiling point (about 350°C) could evaporate and produce fuel vapour.

Again the effectiveness was regulated by the amount of water spray which could reach the burning oil on the tubes. The tendency of flames to draw in fine spray was generally frustrated by the comparatively low flow rates needed actually to reduce flame size, which in its turn would reduce the inward pull of air. An experiment (25) with gasoline fires sprayed upward on metal surfaces showed that fine spray introduced laterally was substantially more effective than sprinklers spraying downwards in reducing temperatures reached by metal in the fire. As at present, extinction of fires of low flash point liquids by plain water is regarded as unreliable and is in any case frowned upon because the production of heavy flammable vapours remain unsupressed. Water sprays are used predominantly for cooling neighbouring risks with this type of fire (29).

Fire in Compartments. A number of workers have carried out investigations on the extinction of fires in compartments by hand held water sprays and jets (30-33) and a theoretical model has been put forward to cover this type of problem (34). In the fifties an extensive series of tests on post flashover fires in a room 18m$^2$ area were carried out at the U.K. Fire Research Station (30), to explore possible benefits that the use of high pressure sprays may bring to the extinction of fires of this kind. They found that with water sprays at a range of rates of flow of 23-114 litres/min. produced at pressures of 5.6-35 bar, there was little significant difference in the amount of water needed to control or extinguish the fire. They also found there was no significant difference when jets were used rather than sprays.

Table 4 gives a summary of conditions and results of tests on compartment fires carried out by different workers. All flow rates used in all tests were capable of extinguishing the fire. In spite of a wide difference in ventilation and consequent rate of burning between the tests, the amount of water used to control the fire and the amount of water evaporated was

TABLE 4 (part 1) - Extinction tests on room fires. Properties of fires.

| Test origins and references | Borehamwood tests (mock furniture) Ref. 31 | University of Karlsruhe tests (wood cribs) Ref. 32 | University of Karlsruhe tests (real furniture) Ref. 32 | Salzberg et al (real furniture) Ref. 33 | Fire Technology Laboratory Finland Ref.34 |
|---|---|---|---|---|---|
| Room Area m$^2$ | 18 | 12.8 | 12.8 | 13.4 | 8.6 |
| Dimen- Ht. m | 2.8 | 2.8 | 2.8 | 2.4 | 2.4 |
| sions Vol. m$^3$ | 50 | 36 | 36 | 32 | 21 |
| Ventilation A$\sqrt{H}$ m$^{5/2}$ | 7.13 | 1.0 | 1.0 | 4.0 + door | 2.3 |
| Mass of fuel kg. | 360 | 380 | 380 | 300 | 20 + walls & ceiling |
| Surface area of fuel m$^2$ | 67.3 | 112 | 40 | ? | 43 |
| Rate of burning 6.5A $\sqrt{H}$ kg/min | 46.4 | 6.5 | 6.5 | 26 | 15 |
| Actual weight loss rate after flashover kg/min | 72(approx) | 10.4 | 11.3 | ? | 4.5 MW 15.9 kg/min. |
| Rate/unit area of fuel g/m$^2$ s | 17.8 | 1.55 | 4.7 | ? | 6.2 |
| Time to flash-over min. | 5-10 | 28-35 | 5-25 | ? | 4-5 |

much the same. However extra water was required in the Karlsruhe tests which were carried out under conditions of low ventilation and high preburn time, than in the other series which had high ventilation and low preburn time. This extra water was, in the main, needed to proceed from control to complete extinction and a substantial part of it was not evaporated.

A question concerning the extinction of post-flashover fires in compartments is whether they are extinguished by steam produced by heat transfer between hot gases or hot surfaces and water drops or whether they are extinguished by cooling the fuel. Certainly large quantities of steam are seen to be produced when water is used against a post-flashover fire. However one would expect that if the steam were the mechanism, then extinction would be made more difficult by increasing the vent size and if cooling, by increasing the preburn time. Thomas (35) obtained a significant increase of both critical rate of water application and total water needed for a very small fire (0.13m$^3$) as ventilation increased. However this was for a condition in which the spray used was fixed in position so that it could not impinge on all fuel surfaces and in any case for flow rates of water (approx. 7g/m$^2$ s), substantially in excess of critical flow rates for the fuel used. Thomas however also estimated that as the scale increased, cooling the fuel became more important than the production of steam, mainly because of an increase in volume to surface ratio. This tends to be bourne out by the information in Table 4. In particular the difference between the Borehamwood and Karlsruhe tests suggests that cooling the fuel is the mechanism of extinction. It will be noted in table 4 that the least rate of water used to bring about extinction, not necessarily a critical rate, was about 6g/m$^2$ of fuel area per second, which may be compared with a critical

TABLE 4 (part 2) - Extinction tests on room fires. Water application used and results. J = Jet; S = Spray

| Test origins and references | Borehamwood tests (mock furniture) Ref. 31 | University of Karlsruhe tests (wood cribs) Ref. 32 | University of Karlsruhe tests (real furniture) Ref. 32 | | Salzberg et al (real furniture) Ref. 33 | Fire Technology Laboratory Finland Ref. 34 | |
|---|---|---|---|---|---|---|---|
| Flow rate used l/min | 22.7-113.7 | 50-100 | 100 | 15-25 | 25-112 | 46(J) 18(S) | 47(S) |
| No. of tests | 40S 10J | 3(s) | 2(s) | 1(s) | 17(s) | 1 | 2 |
| Pressure bars | 5.6-35 | 5 and 35 | 5 | 5 | 17 | 2 | |
| Time after flashover before application of agent min. | 2 | 12 | 13 | 11 | 0.5-2 | 1-2 | |
| Amount burnt prior to application % | 43 | 43 | 42 | | ? | ? | |
| Water used to control l. | 32(mean) | (25-50?) | (50?) | 12 | 23-58 | 7-40 | |
| Water used to extinguish completely l. | 76 (mean) | 162 | 225 | 152 | 50-80 | 25-43 | |
| Amount collected l. | Nominally 0 | 64 | 126 | 50 | ? | ? | |
| Amount evaporated l. | Nominally 76 | 98 | 99 | 102 | ? | ? | |
| Min. flow rate/ unit fuel area g/m² s | 5.6 | 7.4 | 6.25 | | ? | 6.9 | |

rate for large wood crib fires mentioned earlier, of about $4g/m^2 s$, and a general flow rate of $1.3-3g/m^2 s$, summarised by Heskestad. It is difficult to be convinced that if a spray is aimed at fuel consisting of furniture any special effect of steam due to enclosure of the fire by the room is present unless flow rates less than about $6g/m^2$ of fuel area per second are used and the fire is controlled in well under one minute.

Experiments with an inert gas generator (36) indicated that to extinguish the flames of a post flashover fire in a room with a $1.3m^2$ opening, a flow rate of about $1.6m^3/s$ of partly inerted gas containing 10% oxygen was required. If this were pure steam this flow rate could be equivalent to 1000g/s or 60l/minute of water input. However one could argue that a third of this flow rate would be needed if the inerting capability of the steam, rather than reversal of air ingress at the window, controlled the flame extinction process. Flow rates of 20-60l /minute are in line with the range of flow rates used in the tests described in Table 4. One cannot therefore rule out flame extinction by steam formation, since steam can be formed in impingement of spray on any hot surface or in the flames themselves. Indeed in the tests carried out by Salzberg the water was stated as having been applied indirectly.

WATER USED IN ACTUAL FIRE FIGHTING

Thomas in the U.K. and Labes in the U.S. have carried out statistical analyses on the amount of water used in fire fighting for a range of fires of plan area 5-60000$m^2$. Both investigations indicate that the flow rate of water increases in proportion to the perimeter of the fire with approximately one jet in use for every 10 metres. The total amount of water used is proportional to the 1-1.2 power of the fire area. The actual relationships found are as follows:

Thomas(37)  $J = 0.33\sqrt{a}$,  $t = 3.3 \sqrt{a}$  (for fires mostly with a>200 $m^2$)

Labes (analysed by Baldwin) (38)  $W = 1.24 a^{0.664}$, $t = 1.66a^{0.554}$

J = number of jets used
t = time in minutes
W = flow rate of water used (l/s)
a = area of fire ($m^2$)

The two relationships are reasonably coincident if it is assumed that the flow rate of water/jet is about 500-600 l/minute.

For an area of 15$m^2$ the Labes relationships imply a flow rate of 450 l/minute and a quantity of water used of about 3,500 litres. These are about 10-20 times greater than flow rates and quantities used for rooms of similar area indicated in Table 4. This may be because at a post flashover fire, a fire brigade would tend to use a branch with a flow of 500 l/minute or more, rather than a hose reel jet of order 50 l/minute. Moreover the learning factor is less in a repeated series of controlled experiments on fires of one type, carried out in a short period, than it would be in a variety of different fires stretched over a long period. Thus it was found that trained firemen reduced the water required to extinguish a post flashover room fire by a factor of 2-3, once they learned, after a few tests carried out in 3 days, to vary their techniques so as to stop spraying as soon as large quantities of steam were seen to emerge from windows (39). Fry (40) showed that in 90% of fires in dwellings less than 450 litres of water was used which is much nearer the quantities shown in Table 4. However only about 10% of domestic fires are expected to be post flashover fires.

CONCLUSIONS

There is an increasing amount of information becoming available on the amounts of water that are needed to extinguish fires of different kinds. There is also the beginnings of a structure becoming apparent on the way these requirements may be estimated from a fundamental basis. As in many other areas of fire research this embryonic structure is capable of massive improvement. However it is difficult to see how this can be done without much more experimental work aimed at improving our understanding of the phenomena involved, and checking on the applicability of the structure outlined rather than providing empirical ad-hoc information for presssing practical problems. Of course we need the latter as well but perhaps we can assume that funds for this type of investigation will always be available if an urgent need arises.

Our knowledge of the capability of water sprays to extinguish flames can be improved by measuring the effects of sprays on different flames from both gases and volatile liquids burning at a range of intensities in different turbulent regimes. Investigations should at least cover turbulence that might be produced by wind and projected sprays themselves both with and without obstacles being present. Quantification of extinction by cooling could be helped by obtaining more information on the component items in equation 2. One item here still in need of major improvement is the heat of gasification of woods of all kinds with different moisture content. It is desirable also to explore to what extent experiments on pilot ignition may be invoked to give information on both $\emptyset$ and $\dot{m}''_c$ and how they are related (19). Williams has stressed the importance of the Damkohler number in the extinction process (41). This number which is essentially the ratio of the time that is needed for the flame reaction to the time that is available, is embedded in the various heat loss parameters for flame extinction, including $\emptyset$, that have been used here. The time available for reaction depends on aerodynamic conditions and indeed $\emptyset$ must be reduced almost to zero in a situation where a volatile liquid fire is blown out by air without extinguishing agent added. However blow out velocities are quite large for fires of practical size and they increase with fire dimension. This aspect of the effect of wind is therefore unlikely to be of major practical importance except perhaps in the extinction of solid fuel sticks in a high wind. Nevertheless it is very desirable to integrate this aspect of extinction with those dealt in this paper, since blowout experiments provide a neat method of defining the kinetics of the combustion chemistry. This can be exploited to obtain estimates of $\emptyset$, and to explore how $\emptyset$ might vary with such variations of aerodynamic conditions that occur in practice, particularly fuel dimensions and air velocity.

Sibulkin (42) has used finite chemistry based on William's experiment to estimate fuel flow rates at extinction of a vertical slab of PMMA burning under laminar conditions at different oxygen concentrations and external radiation. His results are similar to those that may be predicted using equations 2 and 3. Other calculations by Sibulkin (43) also indicated that reaction kinetics over the relevant range that leads to extinction did not influence the burning rate greatly nor, for a mass fraction of oxygen of 0.18, did it indicate the presence of oxygen at or near the fuel surface. This gives support to the use of equation 3 to cover conditions at the fire point, albeit at oxygen concentration of 0.236.

There are also certain other aspects of quantification of the extinction process which have hardly been touched upon. Estimation of extinction time is important, particularly for extinction by fuel cooling. Extinction of glowing may be more difficult than the suppression of flaming of solids and needs to be integrated into a quantitative approach. The conditions which can give rise to enhancement of fire by splashing or sputtering and the way this may seriously affect extinction by cooling the fuel has not yet been addressed. This is likely to be a highly complicated question to sort out quantitatively.

Finally wherein lies the road for improvement of fire fighting practice with water? There appears to be some gap - a factor of about 10-20 between what is needed and what is actually used and the reason for this would repay study. There is a degree of incompatibility between the necessity of delivering as high a fraction as possible of the water that is projected from a nozzle into a fire space, yet once there for the water to be dispersed so that it can reach all surfaces of the burning fuel. The task of harmonizing these two requirements could also repay research. There could be circumstances which could be pinpointed where the production of steam other than by cooling the burning fuel could be beneficial, for example, in extinguishing a fire in the upper part of a high room which has become masked by smoke and where steam formed would remain in place or when the burning fuel is particularly inacessible. However it would be essential to follow up by cooling the fuel to avoid the risk of flashbacks and even explosions once the steam formed has been diluted sufficiently with air. For volatile liquid fires in the open this risk is still present and although it is probably practicable to devise sprays which could extinguish flames of moderate size, their use would be limited. Extinction of large gas jet flames with water spray is less likely to suffer from this disadvantage because after extinction the jet tends to dilute itself in the atmosphere to below the limit by turbulent mixing. It is interesting to note there has been encouraging activity in this area (44).

REFERENCES

1.   Theobald, C.R., "The Design of a General Purpose Fire-Fighting Jet and Spray Branch, "Fire Safety Journal, 7: 2, 177-190, 1984.
2.   Thomas, P.H. and Smart, P.M.T., "The Throw of Water Sprays", Fire Research Note 168, 1955, Fire Research Station, Borehamwood, Herts., U.K.
3.   Rasbash, D.J., "The Production of Water Spray of Uniform Drop Size by a Battery of Hypodermic Needles", Journal of Scientific Instruments 30, 189-192, 1953.
4.   Rasbash, D.J., Rogowski, Z.W. and Stark, G.W.V., "Mechanisms of Extinction of Liquid Fires with Water Sprays", Combustion and Flame 4, 223-334, 1960.
5.   Rasbash, D.J., Rogowski, Z.W. and Stark, G.W.V., "Properties of Fires of Liquids", Fuel 35, 94-106, 1956.
6.   Rasbash, D.J., "The Extinction of Fires by Water Sprays", Fire Research Abstracts and Reviews 4: 1 and 2, 28-52, 1962.
7.   Bryan, J., "The Effect of Chemicals in Water Solution on Fire Extinction", Engineering 159, 457, 1945.
8.   Magee, R.S. and Reitz, R.D., "Extinguishment of Radiation Augmented Plastic Fires by Water Sprays", Factory Mutual Research Technical Report FMRC No. 22357-1, 1974.
9.   Drysdale, D.D. and Thomson, H.E., University of Edinburgh, Private Communication.
10.  Bradrup, J. and Mammergat, Polymer Handbook, 2nd ed., John Wiley 1975.
11.  Heskestad, G., "The Role of Water in Suppression of Fire", Fire and Flammability 11, 254-259, 1980.
12.  O'Dogherty, M.J., Nash, P., and Young, R.A., "A Study of the Performance of Automatic Sprinkler Systems", Fire Research Technical Paper No. 17, HMSO, 1967.
13.  O'Dogherty, M.J., Young, R.A. and Lange, A., "The Performance of Water-Type Extinguishers on Experimental Class A Fires", Fire Research Note No. 731, 1968, See Ref. 2.
14.  Bhagat, B.H., "The Extinguishment of Burning Wood Charcoal Surfaces", Fire Safety Journal 3, 47-53, 1980/81.
15.  Rasbash, D.J. and Langford, B., "Burning of Wood in Atmospheres of Reduced Oxygen Concentration", Combustion and Flame 12, 1, 33-40, 1968.
16.  Rasbash, D.J., "Heat Transfer Between Water Sprays and Flames of Freely Burning Fires", Proc. Symp. on the Interaction Between Fluids and Particles, I.Chem.E., 217, 1962.
17.  Rasbash, D.J., "Relevance of Fire Point Theory to the Assessment of Fire Behaviour of Combustible Materials", University of Edinburgh, 169, 1975.
18.  Rasbash, D.J., "Theory in the Evaluation of Fire Properties of Combustible Materials", VFDB, 5th International Fire Protection Seminar, Karlsruhe, 113, 1976.

19. Rasbash, D.J., Drysdale, D.D. and Deepak, D., "Critical Heat and Mass Transfer at Pilot Ignition and Extinction of a Material", The American Soc. of Mech. Eng. Symposium, Boston, 1983.
20. Spalding, D.B., Some Fundamentals of Combustion, Gas Turbine series, ed. J. Hodge, pp 62,63,126, Butterworth Scientific Publications, 1955, London.
21. Tamanini, F., "A Study of the Extinguishment of Vertical Wood Slabs in Self-Sustained Burning by Water Spray Application", Combustion Science and Technology 14, 1-15, 1975.
22. Corlett, R.C. and Williams, F.A., "Modelling Direct Suppression of Open Fires", Fire Research 1, 6, 323-337, 1979.
23. Rules for Automatic Sprinkler Installations, 29th ed., F.O.C. London, 1968.
24. NFPA: National Fire Codes, No. 13, 1973.
25. Thompson, N.J., Fire Behaviour and Sprinklers, NFPA, 1964.
26. Dundas, P.H., "Cooling and Penetration Study", FMRC Serial No. 18792. May 1974.
27. Rasbash, D.J. and Rogowski, Z.W., "Extinction of Fires in Liquids by Cooling with Water Sprays", Combustion and Flame 1, 4, 453-466, 1957.
28. Rasbash, D.J. and Stark, G.W.V., "Extinction of Running Oil Fires", The Engineer, 862, 1959.
29. NFPA, National Fire Codes, No. 15, 1982.
30. Hird, D., Pickard, R.W., Fittes, D.W. and Nash, P., "The Use of High and Low Pressure Water Sprays against Fully Developed Room Fires", Fire Research Note 388, 1959, See Ref. 2.
31. Fuchs, P., "Arbeitsgemeinschaft Feuerschutz", (AGF), Berichte 29, University of Karlsruhe, 1975.
32. Salzberg, F. Vodvarka, F.J. and Maatman, G.L., "Minimum Water Requirements for Suppression of Room Fires", Fire Technology 6, 22, 1970.
33. Kokkala, M., "Extinguishment of Compartment Fires using Portable Chemical Extinguishers and Water", Fire Technology Laboratory, Finland, private communication 1985.
34. Ball, J.A. and Pietrzak, L.M., Fire Research 1, 291, 1979.
35. Thomas, P.H. and Smart, P.M.T., "The Extinction of Fires in Enclosed Spaces", Fire Research Note 86, 1954, see Ref. 2.
36. Stark, G.W.V. and Card, J.F., "Control of Fires in Large Spaces with Inert Gas and Foam Produced by a Turbo-Jet Engine", Fire Research Note no. 550, 1964, see Ref. 2.
37. Thomas, P.H., "Use of Water in the Extinction of Large Fires", Quart. Instn. Fire Engineers 19, 35, 130-132, 1959.
38. Baldwin, R., "Use of Water in the Extinction of Fires by Brigades", Instn. Fire Engineers Q. 31, 82, 163-168, 1971.
39. Thomas, P.H. and Smart, P.M.T., "Fire Extinction Tests in Rooms", Fire Research Note 121, 1954, see Ref. 2.
40. Fry, J. and Lustig, "Water Used in Firefighting", Fire Research Note 492, 1963, see Ref. 2.
41. Williams, F.A., "A Review of Flame Extinction", Fire Safety Journal 3, 2-4, 163-175, 1981.
42. Sibulkin, M. and Gale, T., "The Effects of External Radiation on Solid fuel Diffusion Flames", J. Fire Science 2, 70, 1984.
43. Sibulkin, M., Kulkarni, A.K. and Annamalai, K., "Burning on a Vertical Fuel Surface with Finite Chemical Reaction Rate", Combustion and Flame 44, 187-199, 1982.
44. McCaffrey, "Jet Diffusion Flame Suppression using Water Sprays - an Interim Report", Combustion Science and Technology 40, 107-136, 1984.

LIST OF SYMBOLS

A — numerator of the B number = $m_{og}H/r + c(T_g - T_s)$

B — Spalding's number = $A/H_f$

F — total flow rate from nozzle

J — number of jets used for large fires

H — heat of combustion of fuel

$H_f$ — convective heat transfer from flames to surface associated with unit mass of fuel entering the flame

P — nozzle pressure

$R_a$ — heat transfer to unit area of fuel surface other than by convection from flame

$R_s$ — heat loss from surface including radiation per unit area

S — sensible heat removed by water spray per unit area fuel surface

T — temperature

W — water flow rate used for large fires

Z — throw of spray

a — area of fire

c — specific heat of gas

d — mean diameter of spray drops within a small part $\Delta d$ of the total drop size range

h — convective heat transfer coefficient in the absence of mass transfer;

$\dot{m}''$ — mass flow rate of fuel vapour per unit area fuel surface

$m''_{wv}$ — mass flow rate of water vapour from fuel surface to flame

$m_{og}$ — mass fraction of oxygen in the ambiennt atmosphere

n — number of particles within a small part $\Delta d$ of total drop size range

r — stoichiometric ratio

t — time to extinguish fire

$\lambda f$ — the heat that needs to be imparted to the body of the fuel to produce unit mass of fuel vapour

$\Theta$ — cone angle of spray

$\emptyset$ — fraction of heat of combustion of fuel removed by convective transfer to surface under fire point conditions


SUBSCRIPTS

c — at the fire point condition

g — in the ambient atmosphere

s — at the fuel surface.

# Investigation of Spray Patterns of Selected Sprinklers with the FMRC Drop Size Measuring System

HONG–ZENG YOU
Factory Mutual Research Corporation
1151 Boston-Providence Turnpike
Norwood, Massachusetts 02062, USA

ABSTRACT

Characteristics of single sprinkler sprays for three selected upright sprinklers were investigated with the FMRC drop size measuring system. The sprays of sprinklers with nominal orifice sizes of 16.3, 13.5, and 12.7 mm were mapped out at water pressures of 206 and 393 kPa and at 3.05 and 6.10 m below the sprinklers in a measuring sector whose bisector was perpendicular to the sprinkler pipe.

Correlations of gross drop size distributions were derived based on the measurements. The gross drop size distributions at the 3.05 m level could be reproduced with the data measured at the 6.10 m level. The gross drop size distributions of the tested sprinklers can be represented by a composite of a log–normal distribution and a Rosin-Rammler distribution. An economical test protocol was also developed to quantify the spray characteristics of commercial sprinklers.

## I. INTRODUCTION

Until recently, investigation of spray characteristics remained hindered by the lack of an efficient technique for counting and sizing drops within sprinkler sprays. Two methods have been employed in the past decade to measure drop sizes of sprinkler sprays: 1) freezing the falling liquid drops with liquid nitrogen; and 2) drop analysis by still–photography. Both freezing and photographic methods are time consuming in counting and sizing drops. Furthermore, the photographic method may bias the results due to incorrect focusing, while the freezing method is likely to produce erroneous results caused by breaking large particles and coalescing small ones.

In 1979 FMRC acquired a computerized high–speed drop sizing and counting system. This instrument is a laser–illuminated optical array imaging device originally designed to be carried by aircraft to measure the particle size spectra of clouds and precipitation[1]. Successive modification of the probe system from its original design, both in hardware and software, was performed to accommodate measurements under sprinkler spray[2]. The maximum effective sampling area of the probe is 61 mm x 6.4 mm, and the probe was designed to accommodate drop sizes ranging from 100 μm to 6400 μm. The probe system was calibrated with monodisperse and polydisperse samples of solid spheres[2,3]. The calibration tests with monodisperse samples demonstrated that the probe system sized the particles in the range 100–6000 μm within 30 μm and measured their velocities within 10% of the actual velocities. The calibration tests with polydisperse samples showed that the system could satisfactorily report the particle size distribution of a sample[3]. The detailed description of the current FMRC probe system is given in References 2, 3, and 4.

The probe system was not calibrated with water drops. However, the probe system software screened out those drops whose images were skewed from the circular image beyond a predetermined tolerance[2]. Therefore, the adverse effect of water drop deformation on the accuracy of the measurements reported in this study was minimized.

Detailed spray mapping is generally needed to adequately quantify a spray, because of the drastic variation in distributions within the sprinkler spray. It was found that, for tested commercial sprinklers[2], a 5-min period was required to achieve good measurements for certain locations within the spray. Thus, detailed spray mapping was not economical even with the high speed drop sizing and counting system. Therefore, before conducting a large number of tests to quantify spray characteristics of commercial sprinklers, an economical test protocol capable of generating reliable information should be established.

The objectives of the present investigation were: 1) Detail mapping of the variation of spray characteristics for selected upright sprinklers at selected heights and water pressures; 2) Correlation of the spray characteristics for the selected sprinkler sprays; 3) Use of the gathered data to develop a test protocol to quantify sprinkler sprays economically and accurately.

## II. EXPERIMENTS

In the present study, seven single sprinkler sprays were mapped out at elevations of 3.05 or 6.10 m below the sprinkler. The tested sprinklers were limited to three selected upright sprinklers illustrated photographically in Figure 1. A summary of the test variables is presented in Table I.

TABLE I.  Test Variables

| Case | Water Pressure (kPa) | Water Discharge Rate (liters/min) | Nozzle Dia. (mm) | Vertical Distance Between Probe and Sprinkler (m) |
|------|------|------|------|------|
| 1 | 206 | 227 | 16.3 | 3.05 |
| 2 | 393 | 314 | 16.3 | 3.05 |
| 3 | 206 | 227 | 16.3 | 6.10 |
| 4 | 393 | 314 | 16.3 | 6.10 |
| 5 | 206 | 163 | 13.5 | 3.05 |
| 6 | 206 | 163 | 13.5 | 6.10 |
| 7 | 206 | 117 | 12.7 | 3.05 |

The experiments were conducted in the FMRC fire test building at West Glocester, Rhode Island. Figure 2 schematically depicts the floor plan of the test setup. The test setup was located at about the floor center within a test site of 26.60 m x 26.60 m x 18.28 m high. The ceiling was 3.66 m x 3.66 m square and fixed at 7.62 m above the floor. The tested upright sprinkler was installed on a pipe of 25.4 mm pipe size and centered under the ceiling; the pipe was about 0.15 m below the ceiling. The sprinkler was so oriented that its two supporting arms were in alignment with the pipe.

A large-scale, remote-controlled, probe-traversing apparatus was constructed to traverse the probe within the circular sector shown in Figure 2. The radius of the sector was 6.1 m long and the angle encompassed by the arc was 100 degrees. The traversing apparatus rested on the floor and the probe elevation could be adjusted from 1 m to 6.1 m above the floor. Different vertical distances between the probe and ceiling were achieved by adjusting the probe elevation.

The water flux was measured separately by automatic water collectors[6] at the same locations relative to the sprinkler as with the drop size measurements. Each water collector covered a floor area of 500 mm x 500 mm. The design and operation of the collector is given in Reference 6.

III. DATA ANALYSIS AND RESULTS

In each series of the measurements, the probe was traversed azimuthally from 5 degrees to 95 degrees in increments of 10 degrees. Along each radius, measurements were conducted at six locations about equally spaced. The outermost location in each radius was determined visually to coincide with the edge of the spray.

FIGURE 1.  Tested upright sprinklers.

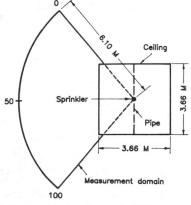

FIGURE 2.  Floor plan of test setup.

## 3.1 Water Flux Distribution

Detailed data listings of all tested cases are presented in Reference 5. Figure 3 shows the water flux distribution in the measuring sector for Case 5, to illustrate the drastic variation of water flux distributions within the sprays of real-world sprinklers.

We may divide the sector into a number of zones such that each zone contains a measuring location. If we assume that the average water flux in each zone can be represented by the measurement at the respective measuring location, the overall water delivery rate in the sector can be calculated by

$$\dot{V} = \sum_{k=1}^{K} \sum_{k=1}^{P} A_p \dot{v}''_{kp} \tag{1}$$

where P is the number of measurements along a radius; K is the number of azimuthal measurements; $A_p$ is the area of the p-th zone in a radial direction; and $\dot{v}''_{kp}$ is the flux at the location of the p-th radial zone and the k-th azimuthal angle. The delimitation of the divided zones along a radius was set at the center between the adjacent measuring positions. The same rule is employed to determine the delimiation along an arc. The edge of the outermost zone for each radius was delimited by the outer edge of the water collector.

Table II presents a comparison of the measured overall water delivery rates in the sector and the water delivery rates in the same area with axisymmetrical sprinkler sprays. In spite of the seemingly drastic variation of water flux from the bisector to the two sides of the sector, the water delivery rate based on axisymmetry is close to (better than 84%) the rate obtained by Eq (1).

TABLE II. Overall Water Delivery Rates, Median Drop Sizes, and C Constants for Eq. (4) in the 100-Degree Sector

| Case | Actual Rates in the Sector (liters/min) | Rates in a 100° Sector for Axisymmetrical Sprays (liters/min) | $d_m$ (mm) | C |
|------|------|------|------|------|
| 1 | 66.4 | 63.1 | 1.66 | 4.30 |
| 2 | 90.8 | 87.3 | 1.37 | 4.41 |
| 3 | 68.7 | 63.1 | 1.61 | 4.17 |
| 4 | 97.5 | 87.3 | 1.39 | 4.48 |
| 5 | 53.6 | 45.2 | 0.96 | 2.86 |
| 6 | 51.1 | 45.2 | 1.00 | 2.97 |
| 7 | 38.2 | 32.6 | 0.86 | 2.33 |

## 3.2 Local and Gross Drop Size Distribution

Figures 4 and 5 illustrate the typical radial and azimuthal variations of drop size distribution measured in the present study, respectively, using

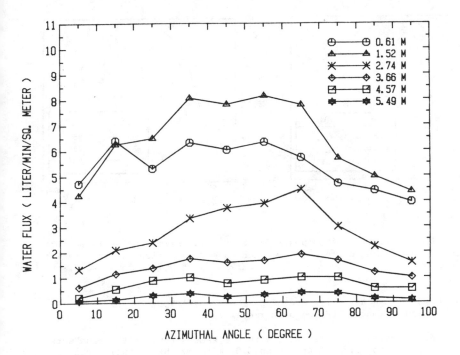

FIGURE 3. Spatial water flux distribution for Case 5.

Case 5 for examples. The measurements of drop size distribution of all seven tested cases are documented in Reference 5. The ordinate of Figures 4 and 5 represents the accumulative percents by volume (APV) below a drop size. Figure 4 shows that increasingly larger drops occur in the distribution with greater radial distance from the sprinkler. Figure 5 shows that small drop-lets tend to concentrate in the middle of the sector; correspondingly, increasingly larger drops exist toward the two sides of the sector. The actual factors causing drop size and water flux to systematically vary from the central region to the two sides of the sector are not fully understood. However, it is suspected that the effect of the pipe and the sprinkler's two arms on the initial spray formation might be responsible for producing the spray pattern found in this study.

Gross drop size distribution within the 100-degree measuring sector can be derived from the local water flux and drop size distributions measured in this area. The APV below a specified drop size can be written as:

$$APV_{\ell} = \frac{\sum\limits_{k=1}^{K} \sum\limits_{p=1}^{P} A_p \, \dot{v}''_{kp} \, APV_{kp\ell}}{\dot{V}} \qquad (2)$$

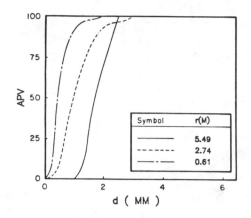

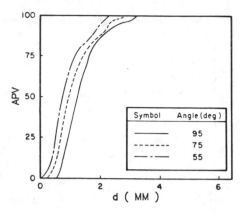

FIGURE 4. Radial variation of drop size distribution along the 45-degree radius for Case 5.

FIGURE 5. Azimuthal variation of drop size distribution at a radial distance of 2.74 m for Case 5.

where $APV_\ell$ is the gross APV below a drop size $d_\ell$; and $APV_{kp\ell}$ is the local APV at the k-th azimuthal angle and the p-th radial zone.

The present results indicate that, with the same sprinkler type and water pressure, the gross drop size distribution 3.05 m below the sprinkler can be reasonably reproduced at a vertical distance of 6.10 m below the sprinkler. Figure 6 demonstrates this finding by the gross drop size distributions of Cases 2 and 4, i.e., with a nozzle size of 16.3 mm and water pressure of 393 kPa. The gross drop size distributions of all seven tested cases are tabulated in Reference 5. The good agreement of gross drop size distributions obtained at the above two elevations implies that under the present test conditions, the process of drop formation is completed at 3.05 m below the sprinklers, and the net effects of drop breakup and coalescence, if any, is negligible below 3.05 m.

For geometrically similar nozzles, Heskestad[7] found that at room temperature the characteristic drop size can be related to the orifice size D and water pressure $\Delta P$ by the following expression:

$$d \sim \Delta P^{-1/3} D^{2/3} . \tag{3}$$

Equation (3) can be written in the following nondimensional expression by introducing the Weber number, $We = \rho_w U^2 D/\sigma_w$:

$$\frac{d_m}{D} = C/We^{1/3} \tag{4}$$

where $d_m$ is the median drop size, U is the water discharge velocity, and $\rho_w$ and $\sigma_w$ are water density and surface tension, respectively.

Table II tabulates the volumetric median drop sizes and associated C constants for all tested cases. The table indicates that, for a specific sprinkler, i.e., Cases 1, 2, 3, and 4, the C values are about the same for different water pressures. Based on the present limited data, the pressure effect on drop size indicated in Eq (3) appears to be valid for sprays of commercial sprinklers.

3.3 Correlations of Gross Drop Size Distribution

It was found that the gross drop size distribution obtained from the present study could not be fitted entirely by either log-normal or Rosin-Rammler distribution[8] as postulated in many previous applications of sprinkler sprays. To substantiate this finding, it was decided to fit the present data with both log-normal and Rosin-Rammler distributions in their applicable size ranges.

The equation of log-normal distribution is

$$ y = \frac{1}{d\sigma \, (2\pi)^{1/2}} \, \exp \frac{-[\ell n \, (d/d_m)]^2}{2 \, \sigma^2} \tag{5} $$

where y is the occurring probability at drop size d, and $\sigma$ is the variance of the distribution. The units of d, $d_m$, and $\sigma$ are in millimeters. The APV below a drop size could not be obtained analytically from Eq (5) but by numerical integration. The accumulative percents of a log-normal distribution should be in a straight line if plotted on log-probability graph paper.

The Rosin-Rammler equation is

$$ y = \gamma k d^{\gamma-1} \, Exp \, [-kd^{\gamma}] \tag{6} $$

where k is a constant, and $\gamma$ is the constant which depends on the characteristics of a spray nozzle.

Integrating Eq (5), we obtain:

$$ APV_\ell = 1 - Exp \, [-\phi \, (d_\ell/d_m)^{\gamma}] \tag{7} $$

where $\phi = k d_m^{\gamma}$.

The data correlations for Case 2 are shown in Figure 6. The data in the linear portion were fitted with Eq (5), while those in the curved position were correlated with Eq (7). Both the fitted straight line and curve are extended to the whole range of the measured drop sizes, to substantiate the deviations of the above two distribution patterns from the portions of data where they cannot be applied.

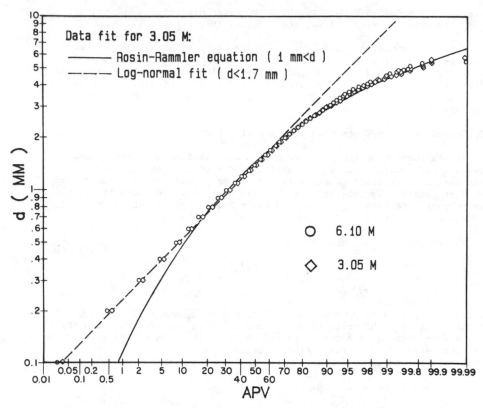

FIGURE 6.  Gross drop size distributions for Cases 2 and 4 and data fit for
Case 2.


Table III presents the parameters of σ, φ, and γ of Eqs (5) and (7), and
their respective applicable drop size ranges.


TABLE III.  The Values of σ, φ, γ

| Case | σ (mm) | Range (mm) | φ | γ | Range (mm) |
|------|--------|------------|------|------|------------|
| 1 | 0.76 | 0.1 - 2.0 | 0.66 | 1.78 | 1.0 - 5.8 |
| 2 | 0.78 | 0.1 - 1.7 | 0.67 | 1.67 | 1.0 - 5.7 |
| 3 | 0.73 | 0.1 - 2.0 | 0.61 | 1.69 | 1.0 - 5.9 |
| 4 | 0.75 | 0.1 - 1.7 | 0.64 | 1.68 | 1.0 - 5.4 |
| 5 | 0.58 | 0.1 - 1.7 | 0.70 | 1.69 | 1.0 - 3.9 |
| 6 | 0.56 | 0.1 - 1.4 | 0.66 | 1.73 | 1.0 - 4.1 |
| 7 | 0.62 | 0.1 - 1.5 | 0.68 | 1.54 | 1.0 - 3.9 |

## 3.4  A Proposed Tentative Test Procedure to Estimate Gross Drop Size Distribution with Few Measurements

From the economic standpoint, it is desirable that the gross drop size distribution in an area of interest be closely estimated by conducting only a few measurements at strategic locations instead of mapping out the entire region.

On our present domain of measurements, the drop size was found to increase progressively with radial distance; drops in the bisector region were generally smaller than those at the two sides of the measuring sector.  If we employ only the data along the bisector radius, the resulting distribution will be biased toward the smaller size regime.  On the other hand, if we employ measurements along either one of the two sides of the sector, the distribution will then be weighted toward the larger size regime.  However, a method may be developed by properly weighting the extreme spray properties at selected locations (i.e., along the bisector radius and either side of the present measuring sector) with their respective water flux.  The water flux and local drop size distribution were generally not exactly symmetrical with respect to the bisector[6]; therefore, the present proposed scheme is to weight the spray properties of the two sides with their water fluxes at corresponding radial locations; i.e.:

$$\dot{v}''_{e,p} = \dot{v}''_{5,p} \frac{\dot{v}''_{5,p}}{\dot{v}''_{5,p} + \dot{v}''_{95,p}} + \dot{v}''_{95,p} \frac{\dot{v}''_{95,p}}{\dot{v}''_{5,p} + \dot{v}''_{95,p}} \qquad (8)$$

$$APV_{e,p\ell} = (APV_{5,p\ell}\, \dot{v}''_{5,p} + APV_{95,p\ell}\, \dot{v}''_{95,p})/(\dot{v}''_{5,p} + \dot{v}''_{95,p}) \qquad (9)$$

where 5 and 95 represent the 5-degree and 95-degree radii respectively.

The gross drop size distribution can now be calculated with the data from the 50-degree radius (the bisector) and the combined properties of the two sides by:

$$APV_\ell = \sum_{p=1}^{P} A_p (APV_{50,p\ell}\, \dot{v}''_{50,p} + APV_{e,p\ell}\, \dot{v}''_{e,p})/\sum_{p=1}^{p} A_p (\dot{v}''_{50,p} + \dot{v}''_{e,p}) \qquad (10)$$

Figure 7 illustrates the gross drop size distributions for Case 7 derived from detailed measurements within the sector, as well as the distribution from the 50-degree radius only, and the combined result from the 5-, 50-, and 95-degree radii.  Comparisons for the other tested cases are shown in Reference 5.  In general, the combined results closely represent those obtained from detailed measurements and apparently are better than the estimation using data only from the 50-degree radius (the bisector).

## IV.  CONCLUSIONS

Characteristics of single sprinkler sprays for three selected upright sprinklers were investigated with FMRC drop size measuring system.  Sprinklers with nominal orifice sizes of 16.3, 13.5, and 12.7 mm were tested at water pressures of 206 and 393 kPa.  Distributions of water flux and drop size distribution were mapped out in sectors at vertical distances of 3.05 and

6.10 m below the sprinkler. In the present study, the bisector of the sector was perpendicular to the imaginary plane containing the two sprinkler arms which were in alignment with the supporting pipe. A tentative test protocol was developed to obtain gross drop size distribution without mapping out the entire spray region. Several observations and findings are drawn from the present investigation:

1) Under the measuring conditions of this study, the water flux is roughly symmetrical to the bisector of the measurement domain. The water flux tends to decrease from the bisector to the two sides of the sector.

2) Large drops gradually increase their proportional contribution to the water flux as the radial distance is farther from the sprinkler. Azimuthally, the same trend is observed moving from the bisector toward the two sides of the sector.

3) The gross drop size distribution at 3.05 m below the tested sprinklers can be reproduced by measurements at the 6.10 m level.

4) Based on present results, the gross drop size distributions of the tested upright sprinklers correspond to a composite of two different distributions: the log-normal distribution for small drops and the Rosin-Rammler distribution for large drops (above the median size).

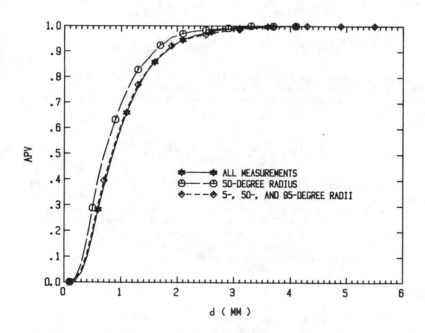

FIGURE 7. Comparison of three derived gross drop size distributions for Case 7 calculated from three different combinations of data.

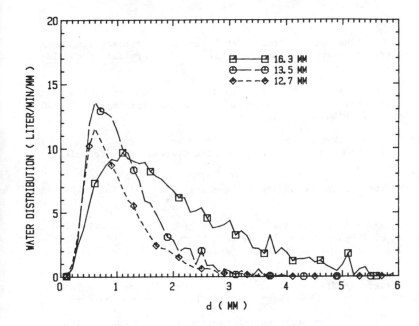

FIGURE 8. Water distribution as a function of drop size at 3.05 m below the sprinklers, with water pressure of 206 kPa.

5)     Under the same test conditions, sprinklers with larger orifice produce larger drops than sprinklers with smaller orifice, regardless of difference in sprinkler design. Therefore, if greater penetrability of drops into the fire plume is desired, sprinklers with large orifice should be considered. On the other hand, if cooling ability is emphasized (small drops), sprinklers with small orifice should be used. Figure 8 demonstrates the trend of drop size distribution for the present three tested sprinklers at a water pressure of 206 kPa and at 3.05 m below the sprinklers.

6)     Based on the present limited data, the median drop size of a sprinkler appears to vary inversely proportional to the one-third power of water pressure.

ACKNOWLEDGMENTS

The author gratefully acknowledges the support and encouragement of Mr. C. Yao and Dr. H. C. Kung during the course of this work. Special thanks are extended to: Mr. A. P. Symonds for providing the maintenance service for the FMRC Drop Size measuring system; Mr. W. R. Brown for designing and fabricating the probe traversing apparatus; Mr. E. E. Hill for assistance in constructing the probe traversing apparatus; and Mr. D. E. Charlebois and the Factory Mutual Test Center staff for assembling the test apparatus and for assistance in conducting the tests.

REFERENCES

1. Knollenberg, R. G.: "The Optical Array: An Alternative to Scattering on Extinction for Airborne Particle Size Determination," J. Applied Meteorology, 9, 86-103, 1970.

2. Croce, P.A., You, H.Z., Khan, M.M., and Symonds, A.P.: "Calibration of, and Preliminary Measurements with, FMRC's PMS Drop-Size Measuring System," FMRC Technical Report, J.I. OF0E2.RA, 1981.

3. You, H.Z. and Symonds, A.P.: "Sprinkler Drop-Size Measurements, Part I: An Investigation of the FMRC PMS Drop-Size Measuring System," FMRC Technical Report, J.I. OG1E7.RA, 1982.

4. Khan, M.M.: "Particle Measuring Systems: Operation Manual," FMRC Internal Memo, 1981.

5. You, H.Z.: "Sprinkler Drop-Size Measurements, Part II: An Investigation of the Spray Patterns of Selected Commercial Sprinklers with the FMRC PMS Droplet Measuring System," FMRC Technical Report, J.I. OG1E7.RA 1983.

6. Goodfellow, D.: "Apparatus and Procedure for Measuring the Actual Delivered Density of Sprinklers in Simulated Rack Storage Fires, Volume I - Apparatus and Procedure Development," FMRC Technical Report, J.I. OK0J4.RR, 1984.

7. Heskestad, G.: "Proposal for Studying Interaction of Water Sprays with Plume in Sprinkler Optimization Program," FMRC Interoffice Correspondence, 1972.

8. Rosin, P. and Rammler, E.: "The Laws Governing the Fineness of Powdered Coal," The Institute of Fuel, 29-36, October, 1933.

# Extinguishment of Rack Storage Fires of Corrugated Cartons Using Water

JAMES L. LEE
Factory Mutual Research Corporation
1151 Boston-Providence Turnpike
Norwood, Massachusetts 02062, USA

ABSTRACT

A series of large-scale fire tests were conducted on the extinguishment of corrugated cartons stored on metal racks. Three different storage heights (3.0 m, 4.5 m, and 6.0 m) were investigated. A specially designed water applicator, supported at a close distance from the top of the test array, was used to deliver a uniform water application density directly onto the array. The applicator could be actuated at any stage of the fire development process to simulate sprinkler response under different fire scenarios. Products of combustion from the fire were collected by a large-capacity calorimeter for the determination of heat release rate for the entire test duration. The effects of water application rate, fire size at the time of water application, and storage height were examined. A single empirical correlation of the extinguishment data was established between the fuel consumed during extinguishment normalized by the fire consumable mass left at the time of water application, $M_{ext}/M_{o,w}$, and the water application rate normalized by the mass burning rate at water application, $\dot{M}_w/\dot{M}_{b,w}$. A power law relationship exists between $M_{ext}/M_{o,w}$ and $\dot{M}_w/\dot{M}_{b,w}$ with the power being $-1.55$. The same kind of correlation was also obtained in a laboratory-scale extinguishment study on wood cribs and wood pallets of different heights. Based on the correlation, a critical water application rate (per unit exposed surface area) for rack storage array of corrugated cartons was determined to be $3.0$ $g/m^2s$ which is very close to the values reported for wood arranged in other geometries such as crib, slab, and pallet.

## I    INTRODUCTION

In the industrial fire loss history for the past decade, warehouse fires have been reported to be among the most costly. These fires present a severe challenge to existing sprinkler protection systems because of the configuration of the storage systems in the warehouses as well as the fire hazard potential of the new generation of storage materials. A popular approach in warehouse storage today is to store palletized goods on metal racks; this arrangement presents an ideal setup for accelerated flame spread among the storage goods upon ignition. Traditionally, the protection needs for this type of storage system were determined through full-scale fire tests; however, such tests are usually very costly and the data collected are primarily applicable to the particular test conditions chosen. In order to reduce the cost of fire testing and have data useful for a wider range of test conditions, a basic understanding of the suppressive action of water on these rack storage fires is essential.

In the past 20 years, a number of studies were conducted on the extinguishment of three-dimensional fuel structures using water alone[1-6]. The fuel examined was wood and the structures used were in the form of a crib or stack of pallets. Some interesting results were obtained in terms of a critical water application rate and the power law relationship between the fuel consumption during extinguishment and the water application rate. This study was intended to compare the results of this work with those from the laboratory-scale studies, particularly those by Kung and Hill[3], to advance our understanding of the extinguishment behavior of large-scale rack storage fires of corrugated cartons such as encountered in actual warehouses.

## II  EXPERIMENT

A series of 32 rack storage fire extinguishment tests were conducted at the Factory Mutual Test Center in West Glocester, Rhode Island. The fuel selected for the study was double triwall corrugated cartons with metal liners inside. Each carton measured 107 cm x 107 cm x 105 cm externally and weighed 38 kg (excluding the weight of the metal liner). A picture of the carton is shown in Figure 1. Each carton was placed on a wood pallet and stored on a double-row metal rack which allowed a two-pallet-load wide by two-pallet-load deep storage. Two-, three-, and four-tier high arrays were tested which had overall storage heights of 3.0 m, 4.5 m, and 6.0 m, respectively. Within each test array, flue spaces were maintained of 0.15 m between the vertical surfaces of the pallet loads, and of 0.33 m between the bottom of the pallets and the top of the commodity. A schematic showing the 4.5-m high array is presented in Figure 2.

FIGURE 1.  Metal-lined double triwall carton

The entire test array was erected on a 3,600-kg load platform which continuously monitored the weight loss of the fuel during the test. A specially designed water applicator (see Figure 3), consisting of eight

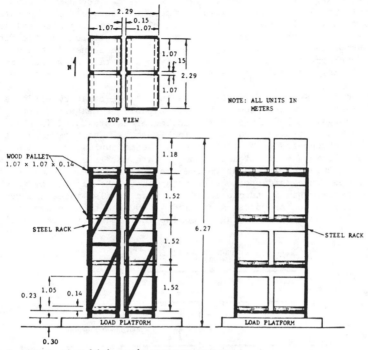

FIGURE 2. Four-tier high rack storage array.

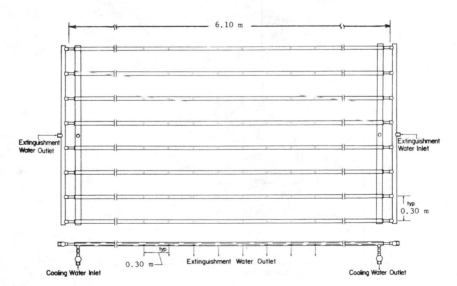

FIGURE 3. Water applicator.

parallel steel pipes fitted with eight spray nozzles (solid cone type) along each pipe, was supported at 0.3 m above the top of the test array. The nozzles were spaced 0.3 m apart to provide a uniform coverage over the top surface of the test array which measured 2.44 m x 2.44 m. Because of the proximity of the applicator to the fuel surface, the sprays from the nozzles were assumed to have 100% penetration and the effects of drop size and spray momentum under this situation were considered to be negligible on the fire extinguishment process. The water applicator could be actuated at any point of the fire development stage to deliver a given flow rate of water directly onto the test array for extinguishment.

The entire test setup was located beneath a large-capacity calorimeter called the Fire Products Collector, which is capable of measuring fires into the megawatts range. A schematic of the collector is presented in Figure 4 and a photograph of a typical test setup in Figure 5. The Fire Products Collector was instrumented with pitot probes, thermocouples, and a gas sampling probe at the instrumentation station (located 8.7 m or 5.7 duct diameters from the entrance orifice above the cone) for the measurement of the total flow, temperature, and specie concentration of the gas stream in the duct. These measurements in turn allowed the heat release rates and generation rates of combustion products, such as $CO_2$, CO, and total hydrocarbons, to be calculated. Oxygen depletion rate was also included in the calculations.

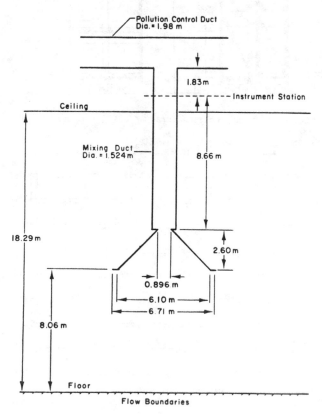

FIGURE 4.  The Fire Products Collector.

FIGURE 5.  Setup for rack storage fire extinguishment test.

More details on the Fire Products Collector and the algorithms for calculating
heat release rates and combustion products generation rates can be found in
reference 7.
    The ignition of the test array was by means of four cellucotton rolls,
7.6-cm dia x 7.6-cm long, each soaked with 120 ml of gasoline.  Each roll was
placed within a plastic bag and taped to a wooden stick.  The ignitors were
placed in the center flue space of the array next to the four inner corners of
the lowest pallet loads.  This arrangement is illustrated in Figure 6.

    At the beginning of each test, before the commodities were loaded onto
the rack, the water applicator was checked out and calibrated with a given
water flow rate to ensure free passage through the nozzles.  The commodities
were then loaded and ignitors put in place.  When ignition commenced, the
combustion products from the fire were collected by the Fire Products Collector which continuously monitored the fire behavior of the test array.  At a
predetermined level of heat release rate, the water applicator was actuated,
simulating the response of sprinklers in a given fire scenario, to deliver a
given water application rate onto the top surface of the test array.  The heat
release rate from the test fire during water application was recorded and
analyzed for the characterization of extinguishment behavior of the rack
storage arrays.


III  TEST RESULTS AND DISCUSSION

    The Fire Products Collector provided measurements for the calculation of
convective heat release rate and total heat release rate (based on the oxygen
depletion method) from the fire both before and during water application.  A
graph showing the convective heat release rate history measured for a given
test configuration under three different water application rates is presented
in Figure 7.

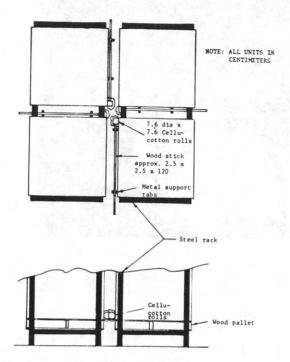

FIGURE 6. Ignition scheme.

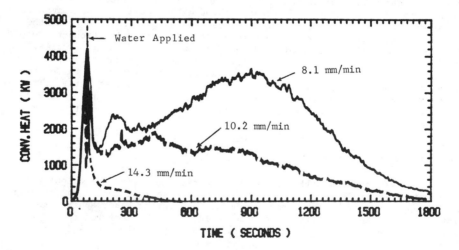

FIGURE 7. Convective heat release rate histories as a function of water application rate.

The variables of interest in this study were the storage height of the array, the heat release rate or burning rate at water application, and the water application rate.

The parameter chosen to represent the extinguishment behavior of the test array was $E_w$, the total energy released from the fire during water application. $E_w$ is simply the integrated value of the total heat release rate of the fire from the time of water application to the end of the extinguishment process. This parameter can be converted to a more practical unit, $M_{ext}$, the mass of fuel consumed, by dividing the total energy release by the actual heat of combustion of corrugated paper. A correlation between the extinguishment results and the test variables was obtained by using a nondimensional fuel consumption parameter and a nondimensional water application rate parameter. The first parameter consisted of the ratio of mass consumed during water application to fire-consumable mass left at the moment of water application, or $M_{ext}/M_{o,w}$. The second parameter consisted of the ratio of water application rate to mass burning rate at the moment of water application, or $\dot{M}_w/\dot{M}_{b,w}$.

Figure 8 is a plot of the nondimensional fuel consumption parameter versus the nondimensional water application rate parameter for all the extinguishment test results on rack storage fires. This correlation can be expressed as:

$$M_{ext}/M_{o,w} = 0.350 \, (\dot{M}_w/\dot{M}_{b,w})^{-1.55} \, . \tag{1}$$

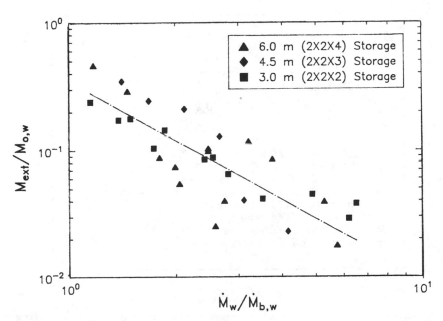

FIGURE 8. Correlation between fuel consumed during extinguishment and water application rate for rack storage array of cartons.

In their extinction studies of wood cribs and wood pallets[3], Kung and Hill also used a water application scheme similar to that used in this study to deliver different flow rates of water onto wood cribs (in three different crib heights) and wood pallets (in 1.22-m and 2.44-m high stacks) after different preburn periods (5-20% of the fuel's initial weight). They correlated the extinguishment results using the same nondimensional fuel consumption parameter and the nondimensional water application rate parameter given in eq (1). However, they actually measured the mass of fuel consumed after each test instead of converting it from an energy release term. In their study, they reported a correlation of:

$$M_{ext}/M_{o,w} = 0.312 \ (\dot{M}_w/\dot{M}_{b,w})^{-1.55} \tag{2}$$

for the wood crib tests and

$$M_{ext}/M_{o,w} = 0.150 \ (\dot{M}_w/\dot{M}_{b,w})^{-1.55} \tag{3}$$

for the wood pallet tests. Figure 9 is a plot of the extinguishment data of the three different fuel structures with their correlations. It is interesting that, for all three fuel structures, 1) a power law relationship holds between the fuel consumption and the water application rate, and 2) a common power of −1.55 was obtained from their correlations. The primary difference is in the proportionality constants which are 0.150, 0.312 and 0.350 for wood pallets, wood cribs, and rack storage of cartons, respectively. The −1.55 power appears to be characteristic of the extinguishment process of loosely packed assemblies as suggested by Kung and Hill[3]. The variation of the proportionality constant among the three correlations implies that, for the

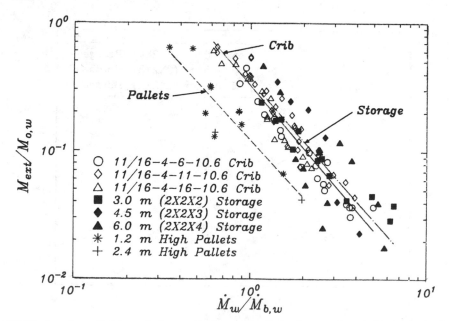

FIGURE 9. Correlations between fuel consumed during extinguishment and water application rate for wood cribs, wood pallet stacks, and rack storage array of cartons.

1184

same percentage of mass consumed, $M_{ext}/M_{o,w}$, the water application rate needed to suppress the rack storage fires of cartons is slightly higher than that for wood crib and even higher than that for wood pallets. This conclusion is not unreasonable in that it was much easier for the flame to penetrate through corrugated cardboard than through solid wood sticks and planks, resulting in a burning surface which is more difficult to extinguish.

Another point of interest is the critical water application rates calculated for these different fuel geometries based on their correlations. The critical water application rate is defined as the water application rate below which all burnable mass is consumed. Assuming the burning rate at water application is half the maximum burning rate of the given array, the critical water application rates per unit total exposed surface area calculated were 2.1, 1.9, and 3.0 $g/m^2s$ for wood cribs, wood pallets, and corrugated cartons, respectively. These values are very close to those reported by other researchers on wood cribs[8] arranged in different configurations using different modes of water application as well as for wood slab[8]. This result seems to imply that the critical water application rate is not dependent on the scale and geometry of the fuel array or the mode of water application. However, it was found to be slightly dependent on the heat release rate or burning rate at the time of water application.

IV  CONCLUSIONS

In conclusion, this work shows a single correlation of the extinction data on different storage heights of rack storage arrays of corrugated cartons. A power law behavior is exhibited similar to that obtained for wood cribs and wood pallets in laboratory-scale extinction studies. Moreover, the critical water application rate obtained for cartons stored on racks is similar to those reported for wood arranged in other geometries such as crib, slab, and pallet. This finding suggests that the critical water application rate of a given material is independent of the size and geometry with which the material is arranged.

REFERENCES

1. D.J. Rasbash, "The Extinction of Fires by Water Sprays," Fire Research Abstracts and Reviews, 4, p. 28-53 (1962).

2. H. Kida, "Extinction of Fire of Small Wood Crib with Sprays of Water and Some Solutions," Report of Fire Research Institute of Japan No. 36, p. 1 (1973).

3. H.C. Kung and J.P. Hill, "Extinction of Wood Crib and Pallet Fires," Combustion and Flame, 24, p. 305, (1975).

4. F. Tamanini, "The Application of Water Sprays to the Extinguishment of Crib Fires," Combustion Science and Technology, 14, p. 17 (1976).

5. M.J. O'Dogherty and R.A. Young, "The Performance of Automatic Sprinkler Systems: Part III - The Effect of Water Application Rate and Fire Size on the Extinction of Wooden Crib Fires," Joint Fire Research Organization, Fire Research Note No. 603 (1965).

6.  M. Stolp, "The Extinction of Small Wood Crib Fires by Water," 5th International Fire Protection Seminar, Karlsrule, p. 127 (September 1976).

7.  G. Heskestad, "A Fire Products Collector for Calorimetry into the MW Range," FMRC J.I. 0C2E1.RA, Factory Mutual Research Corporation, Norwood, MA (June 1981).

8.  G. Heskestad, "The Role of Water in Suppression of Fire:  A Review," Journal of Fire and Flammability, 11, p. 254 (1980).

NOMENCLATURE

$E_w$      total energy released from the fire during water application, kJ

$M_{ext}$   total mass of fuel consumed during extinguishment period, kg

$M_{o,w}$   burnable mass available at the time of water application, kg

$\dot{M}_w$      water application rate, g/s

$\dot{M}_{b,w}$   mass burning rate of the fuel at the time of water application, g/s

# Fire Extinguishing Time by Sprinkler

JUZO UNOKI
Nohmi Bosai Kogyo Co., Ltd.
4-7-3, Kudan-Minami, Chiyoda-ku, Tokyo 102, Japan

ABSTRACT

Sprinkler actuation time has been calculated for various cases, but there
has been little study on extinguishing time after sprinkler actuation.  Therefore, we have tried to draw an equation for the prediction of extinguishing
time in the compartment fires where the ceiling is not high, and burning rate
and water discharge rate are comparatively low.  It is considered that the fire
extinguishing performance depends on the interaction of sprinkler sprays with
buoyant plumes, and cooling effect on a fire source by water discharge.  The
ratio of the updraft gas velocity in the plume to the velocity of water drops
is considered as the former factor, and the ratio of the burning rate to the
water discharge rate as the latter.  We have obtained a generalized equation
for the extinguishing time after sprinkler actuation by dimensional analysis,
using these factors.  We also attempted numerical calculation of extinguishing
time for an example by this equation.  This is the case of a crib fire with a
constant burning rate.  We consider that the extinguishing time for sprinklers
will become one of the important factors on extinguishing performance of sprinklers.

## INTRODUCTION

Calculations of the time it takes for sprinklers to operate after their
detection of fires occurring in compartments have been made for various cases
[1], [2].  However, there have been published very few generalized equations
for the prediction of time from actuation of sprinklers to extinguishment of
the fires.  In this paper we assumed fires in compartments of residential
buildings and obtained an equation for the prediction of extinguishing time by
dimensional analysis and determined constants by experiment for the case that
the ceiling is not high, burning rate and water discharge rate are comparatively low.  We also attempted numerical calculations of extinguishing time for a
certain case, using the equations.

## EQUATIONS FOR THE PREDICTION OF EXTINGUISHING TIME

It is considered that the fire extinguishing effectiveness of sprinklers
depends on the following conditions.

1. To what extent spray water drops can penetrate through the updraft of the
   fire source.

2. At what discharge rate water must be delivered to the fire source in order to suppress heat generation from the fire source and to stop continued combustion.

As a dimensionless factor which governs the condition 1., the ratio of the updraft gas velocity in the plume, Up, to the velocity of water drops, Ut, is considered where the interaction between the fire plume and spray water drops is mainly governed by gravity because of comparatively low water discharge pressure [3]. As a dimensionless factor governing the condition 2., the ratio of the mass burning rate at the start of water discharge, Q, to the water discharge rate onto the upper surface of the fuel (in the case of no updraft of the gas), Qw, is considered [4]. On the other hand, as a dimensionless factor indicating the fire extinguishing effectiveness, the ratio of the time from the sprinkler actuation to the fire extinguishment, T, to the time from the outbreak of fire to the sprinkler actuation, To, is considered. Hence, the following relation is established among these dimensionless factors.

$$T/To = f(Up/Ut, \ Q/Qw) \qquad (1)$$

The maximum gas velocity of the updraft in the plume, Upm, is expressed by the following equation [2].

$$\frac{Upm}{\sqrt{gH}} = 3.16 \ \frac{\dot{Q}^{1/3}}{H^{5/6}} \ [Cp \ \gamma o \ to \ g^{1/2}]^{-1/3} \qquad (2)$$

where
H : distance from upper surface of crib to ceiling
$\dot{Q}$ : heat release rate from fuel
Cp: isopiestic specific heat of air
$\gamma$o: density of air under standard ambient condition
      (20°C, 1 atm.)
to: temperature of air under standard ambient condition
      (20°C, 1 atm.)
g : gravitational acceleration

Since there would not be substantial change in diameter and velocity of the plume extending from the upper surface of the fuel to the ceiling, whole Up may be represented by Upm. The heat release rate from the fuel is calculated from $\dot{Q} = Qh$, where h is the heat release rate per unit weight of the fuel and Q is the mass burning rate of the fuel. Consequently

$$Up \propto ( \ \frac{Q}{H} \cdot \frac{h \ g}{Cp \ \gamma o \ to} \ )^{1/3} \qquad (3)$$

Water drops having radius of approximately $2 \times 10^{-2}$ cm or larger and falling in the air are subjected to resistance as defined by Newton's law, and their velocity, Ut, is expressed by the following equation [5].

$$Ut = 2.11 \sqrt{\frac{g(\gamma w - \gamma)}{\gamma}} \ \sqrt{\frac{d}{2}} \qquad (4)$$

where
$\gamma$w: density of water
$\gamma$ : density of gas in plume
d : diameter of spray drop

Since $\gamma$ is very small compared to $\gamma w$, $\gamma w/\gamma$ may be used in lieu of $(\gamma w - \gamma)/\gamma$. If sprinkler models are specified, distribution of spray drop diameters is characterized for each sprinkler model, and therefore the mean diameter of the drops, dM, may be used in lieu of d. Accordingly

$$Ut \propto (\frac{g \gamma w\, dM}{\gamma})^{1/2} \qquad\qquad (5)$$

From the equations (3) and (5)

$$\frac{Up}{Ut} \propto (\frac{h}{Cp\, \gamma o\, to\, \gamma w^{3/2}\, g^{1/2}})^{1/3} (\frac{Q}{H})^{1/3} (\frac{\gamma}{dM})^{1/2} \qquad\qquad (6)$$

From the equations (1) and (6)

$$\frac{T}{To} = k(\frac{Q}{Qw})^m [A \cdot \frac{Q}{H} \cdot (\frac{\gamma}{dM})^{3/2}]^{\ell/3}$$

Where

$$A = \frac{h}{Cp\, \gamma o\, to\, \gamma w^{3/2}\, g^{1/2}}$$

With $\ell/3 = n$, the above equation becomes:

$$\frac{T}{To} = k(\frac{Q}{Qw})^m [A \cdot \frac{Q}{H} \cdot (\frac{\gamma}{dM})^{3/2}]^n \qquad\qquad (7)$$

Then, values for k, m and n are obtained by experiment.


EXPERIMENTS AND RESULTS

As shown in FIGURE 1 three sprinkler heads were installed on the ceiling with a height of 3.5m. They were arranged with the distance from the fire source, L, varied, so that water drops may fall upon the fire source at different delivery rates. For the fire source, cribs of cedar wood sticks, each sized

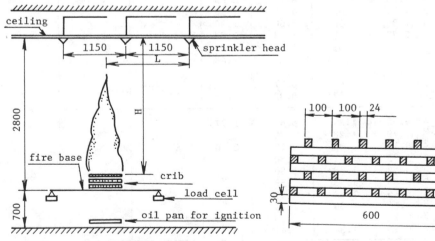

FIGURE 1  Arrangement of fire extinguishing experiment

FIGURE 2  Crib (example in the case of 8 layers)

24mm x 30mm x 600mm with moisture content of 9 ∿ 12% and stacked in six, eight and ten layers as shown in FIGURE 2 were used. To ignite them an oil pan (60cm x 60cm) containing 600cc of gasoline was placed under the crib. It was removed from the position in one minute after ignition. Water discharge was started when the wood cribs showed a weight loss by 30%, but the burning rates from the ignition to water discharge (60, 90, 130 kcal/s) were considered to be almost constant because of ignition by an oil pan (heat release rate:190 kcal/s). Disappearance of flame on the crib was regarded as conclusion of extinguishment. TABLE 1 shows measured values in the experiments.

TABLE 1.  Result of extinguishing test with sprinklers

| Test No. | | Weight of crib | Time from ignition to water discharge | Time from water discharge to extinguishment | Water discharge rate on crib | Mass burning rate just before water discharge | Distance from upper surface of crib to ceiling | Density of air in plume |
|---|---|---|---|---|---|---|---|---|
| Group | No. | W (kg) | To (min.) | T (min.) | QW (kg/min. 0.36m$^2$) | Q=0.3W/To (kg/min.) | H (m) | (kg/m$^3$) |
| A | 1 | 5.55 | 1.76 | 8.50 | 0.48 | | | |
|   | 2 | 5.95 | 1.88 | 7.66 | " | | | |
|   | 3 | 5.57 | 1.76 | 2.12 | 0.85 | | | |
|   | 4 | 5.70 | 1.80 | 1.75 | " | 0.95 | 2.62 | 0.82 |
|   | 5 | 5.90 | 1.87 | 1.10 | 1.35 | | | |
|   | 6 | 5.55 | 1.76 | 1.03 | " | | | |
|   | 7 | 5.45 | 1.72 | 0.55 | 1.77 | | | |
|   | 8 | 5.55 | 1.76 | 0.50 | " | | | |
| B | 1 | 7.70 | 1.65 | 7.08 | 0.76 | | | |
|   | 2 | 7.20 | 1.55 | 6.76 | 0.85 | | | |
|   | 3 | 7.75 | 1.66 | 4.50 | 1.19 | | | |
|   | 4 | 7.38 | 1.58 | 3.60 | 1.30 | | | |
|   | 5 | 7.27 | 1.56 | 1.53 | 1.77 | 1.40 | 2.56 | 0.74 |
|   | 6 | 7.14 | 1.53 | 1.42 | " | | | |
|   | 7 | 7.88 | 1.69 | 1.12 | 2.48 | | | |
|   | 8 | 7.95 | 1.71 | 1.02 | 2.97 | | | |
|   | 9 | 7.53 | 1.61 | 0.87 | 3.17 | | | |
|   | 10 | 7.30 | 1.57 | 0.83 | 3.29 | | | |
| C | 1 | 9.55 | 1.39 | 6.63 | 1.77 | | | |
|   | 2 | 8.79 | 1.28 | 5.75 | " | | | |
|   | 3 | 9.14 | 1.33 | 4.72 | 2.40 | | | |
|   | 4 | 9.75 | 1.42 | 3.80 | " | | | |
|   | 5 | 9.52 | 1.38 | 1.80 | 3.29 | 2.06 | 2.50 | 0.65 |
|   | 6 | 9.85 | 1.43 | 1.58 | " | | | |
|   | 7 | 9.15 | 1.33 | 1.35 | 4.33 | | | |
|   | 8 | 9.14 | 1.33 | 1.13 | " | | | |
|   | 9 | 9.27 | 1.35 | .50 | 5.58 | | | |
|   | 10 | 9.25 | 1.34 | 0.47 | " | | | |

# DETERMINATION OF CONSTANTS FOR PREDICTION EQUATION

FIGURE 3 shows the correlation between T/To and Qw, which is plotted for each of the groups A, B and C, i.e. for each case that Q and H are same by using those figures given in TABLE 1. From the gradient of the line, m = -2 is obtained.

Values for dM, h, Cpm $\gamma$o, To, $\gamma$w and g in this experiment are as follows.

dM = 1mm = 1 x $10^{-3}$m
h  = 3,780 kcal/kg (cedar wood with moisture content of 9 $\backsim$ 12%)
Cp = 0.24 kcal/kg °C
$\gamma$o = 1.161 kg/m³
to = 20°C
$\gamma$w = 998.2 kg/m³
g  = 9.807m/s² = 35.3 x $10^{3}$m/min²

Consequently

A = 1.172 x $10^{-4}$m⁷ min/kg$^{5/2}$

Value for $\gamma$ was obtained from the following equations (8), [6], and (9).

$$\Delta t = 43.9 \times \frac{(Q\ h/60)^{2/3}}{H^{5/3}} \qquad (8)$$

Where
$\Delta t$: difference between gas temperature in plume and compartment temperature °C
Q : kg/min
h : kcal/kg
H : m

$$\gamma (kg/m^3) = 1.2931 \times \frac{273}{293 + \Delta t} \qquad (9)$$

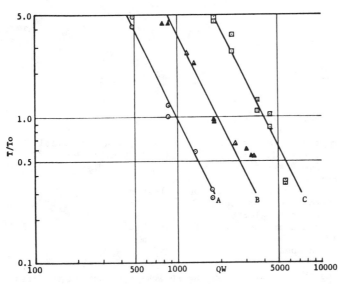

FIGURE 3  Correlation between T/To and QW

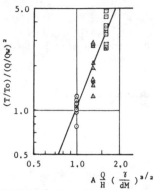

FIGURE 4 Correlation between $(T/To)/(Q/Qw)^2$ and $[A \cdot \frac{Q}{H} \cdot (\frac{\gamma}{dM})^{3/2}]$

FIGURE 4 shows the correlation between $(T/To)/(Q/Qw)^2$ and $[A \cdot \frac{Q}{H} \cdot (\frac{\gamma}{dM})^{3/2}]$ which is plotted by using the results and values obtained in the abovementioned experiment. From this figure and by regression analysis, k = 1.05, n = 2.5, and a correlation factor between the above two quantities is calculated to be 0.91. Therefore, the equation (7) becomes:

$$\frac{T}{To} = 1.05 \ (\frac{Q}{Qw})^2 \ [A \cdot \frac{Q}{H} \cdot (\frac{\gamma}{dM})^{3/2}]^{2.5} \qquad (10)$$

FIGURE 5 compares the equation (10) with the results of the experiment. With the weight loss of the fuel by fire until the sprinkler actuation being M = Q To, the equation (10) becomes:

$$T = 1.05 \ (\frac{Q}{Qw})^2 \ [A \cdot \frac{Q}{H} \cdot (\frac{\gamma}{dM})^{3/2}]^{2.5} \ (\frac{M}{Q})$$

$$= 1.05 \ A^{2.5} \ \frac{Q^{3.5} \ M}{Qw^2 \ H^{2.5}} \ (\frac{\gamma}{dM})^{3.75} \qquad (11)$$

where
T : time from sprinkler actuation to fire extinguishment      min

$$A = \frac{h}{Cp \ \gamma o \ to \ \gamma w^{3/2} \ g^{1/2}} \qquad\qquad m^7 \ min/kg^{5/2}$$

h : heat release rate per unit weight of fuel      kcal/kg
Cp: isopiestic specific heat of air      0.24 kcal/kg °C
$\gamma$o: density of air under standard ambient condition (20°C, 1 atm.)  1.161 kg/m³
to: temperature of air under standard ambient condition (20°C, 1 atm.)  20°C
$\gamma$w: density of water      998.2 kg/m³
g : gravitational acceleration      35.3 x 10³ m/min²
Q : mass burning rate before sprinkler actuation      kg/min
Qw: water delivery rate onto upper surface of fuel  kg/min
M : weight loss of fuel until sprinkler actuation   kg
H : distance from upper surface of fuel to ceiling surface      m
$\gamma$ : density of gas in plume      kg/m³
dM: mean diameter of spray drops      m

It is now possible to calculate the time from the sprinkler actuation to fire extinguishment by the equation(11).

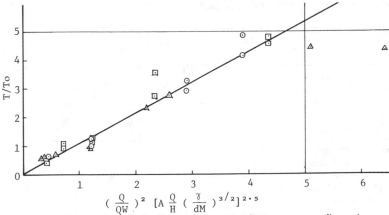

FIGURE 5  Correlation between T/To and $(\frac{\gamma}{QW})^2 [A\frac{Q}{H}(\frac{\gamma}{dM})^{3/2}]^{2 \cdot 5}$

NUMERICAL CALCULATION

Here, a crib fire burning at a constant rate is used as fire model and sprinkler actuation time and fire extinguishing time are calculated. Four sprinklers are mounted on the ceiling of a very large room and on a spacing of 3.25m in a cross pattern. The crib is located in the center of the four sprinklers and with a distance of 2.5m from the ceiling to the upper surface of the crib. Therefore, the horizontal distance from the center of the crib to sprinklers is 2.3m.

The temperature rise of gas at the ceiling, $\Delta Tg$, and velocity, $u$, are obtained from the following equation [6].

$$\Delta Tg = \frac{14 \dot{Q}^{2/3}}{H r^{2/3}} \ [°C] \qquad (12)$$

$$u = \frac{0.32 \dot{Q}^{1/3} H^{1/2}}{r^{5/6}} \ [m/s] \qquad (13)$$

where
r : horizontal distance from center of crib                    m
H : distance between upper surface of crib and ceiling      m
$\dot{Q}$ : heat release rate                                              kcal/s

Using these equations, gas temperature and velocity at the sprinklers are calculated for each of three cribs with six, eight and ten layers of wood sticks as shown in FIGURE 2. TABLE 2 shows results of the calculations. In this case the heat release rate was calculated with the heat release rate per unit weight of wood as being 3,870 kcal/kg. The sprinkler model discussed in this paper is to be the one we used for this experiment (i.e. model MHS-12B manufactured by Nohmi Bosai Kogyo Co., Ltd.). The sprinkler has a rated operating temperature of 72°C, and the relation between the time constant $\tau$, and air velocity, $u$, is expressed by the following equation.

$$\tau = 2.32 \ u^{-0 \cdot 66} \ [\tau:min, \ u: \ m/s] \qquad (14)$$

TABLE 2.  Sprinkler actuation time in fire models

| Number of layers of crib | 6 | 8 | 10 |
|---|---|---|---|
| Mass burning rate (kg/min.) | 0.95 | 1.40 | 2.06 |
| Heat release rate [Q̇] (kcal/s) | 61.3 | 90.3 | 132.9 |
| Distance from upper surface of crib to ceiling [H] (m) | 2.5 | | |
| Horizontal distance between crib and sprinkler [r] (m) | 2.3 | | |
| Rise in actuation temperature above ambient [ΔTa] (°C) | 52 (72-20) | | |
| Rise in gas temperature at sprinkler location above ambient [ΔTg] (°C) | 50 | 65 | 84 |
| Gas temperature at sprinkler location (°C) | 70 | 85 | 104 |
| Flow velocity at sprinkler location on ceiling (m/s) | 1.0 | 1.1 | 1.3 |
| Time constant of sprinkler (min.) | 2.32 | 2.18 | 1.95 |
| Actuation time of sprinkler (min.) | —— | 3.51 | 1.88 |

Generally the actuation time, tr, of the sprinkler exposed to air current with a constant temperature is expressed by the following equation.

$$tr = -\tau \ln (1 - \Delta Ta/\Delta Tg) \qquad (15)$$

where
$\tau$ : time constant of sprinkler
$\Delta Ta$: difference between initial temperature (20°C) of link and actuation temperature
$\Delta Tg$: difference between initial temperature (20°C) of link and gas temperature

TABLE 2 shows sprinkler actuation time for each of the fire models obtained from the equations (14) and (15). From this table it can be seen that with the 6-layer crib fire the gas temperature at the sprinkler does not reach the operating temperature, and consequently the sprinkler is not actuated.

TABLE 3.  Extinguishing time in fire models

| | | 8 | 10 |
|---|---|---|---|
| Number of layers of crib | | 8 | 10 |
| Mass burning rate (kg/min.) | | 1.40 | 2.06 |
| Water discharge rate [Q] (kg/min.) | 1 sprinkler | 0.76 | |
| | 2 " | 1.52 | |
| | 3 " | 2.28 | |
| | 4 " | 3.04 | |
| Distance from upper surface of crib to ceiling [H] (m) | | 2.5 | |
| Mean diameter of spray drops [dM] (m) | | $1 \times 10^{-3}$ | |
| Density of air in plume [γ] (kg/m³) | | 0.47 | 0.65 |
| Fuel weight loss at sprinkler actuation [M] (kg) | | 4.91 | 3.87 |
| Extinguishing time [T] (min.) | 1 sprinkler | —— | —— |
| | 2 " | 6.27 | —— |
| | 3 " | 2.79 | 5.22 |
| | 4 " | 1.57 | 2.93 |

Note:  Calculated time of extinguishment more than (5 x actuation time) is not listed because there is no such time in our experiment.

With a discharge pressure of 1kg/cm² the sprinkler has a water density of 2.1 ℓ/m²/min. (0.76 kg/min on the crib of 0.6m×0.6m) measured at the position of a vertical distance of 2.5m and a horizontal distance of 2.3m from the sprinkler, and its mean diameter of water drops is 1mm.  Using these data and values given in TABLE 2, calculation of extinguishing time for each of the fire models was made by the equation (11) with respect to the both cases that only one sprinkler is actuated and two to four sprinklers are actuated simultaneously, results of which are shown in TABLE 3.  In this case the weight loss of the fuel until actuation of the sprinkler, M(kg), was determined to be a product of the sprinkler actuation time and the mass burning rate (constant).

CONCLUSION

When sprinkler are used for the purpose of reducing physical damage, whether they will be capable of successfully extinguishing fires, and to what extent they will be able to suppress the burning loss until fires are extinguished are matters of our great concern. Nevertheless, it is of utmost importance that sprinklers used in houses, hotels, wards in hospitals etc., are capable of quickly responding to and extinguish fires from the viewpoint of assuring the safety of evacuation. Therefore, we consider that the extinguishing time for sprinklers will become one of the important factors as index for extinguishing effectiveness of sprinklers in the future.

ACKNOWLEDGMENTS

Dr. K. Akita, former professor of Tokyo University, has given advices on this study. The members of Laboratory of Nohmi Bosai Kogyo Co., LTD. have conducted the experiments. The auther would like to thank them for their kindness.

REFERENCES

1. Heskestad, G., Delichatsios M. A., Environment of Fire Detectors-Phase 1 : Effect of Fire Size, Ceiling Height and Material, Volume 2. Analysis, NBS-GCR-77-95, National Bureau of Standards, July 1977

2. Evans, David D., Calculating Sprinkler Actuation Time in Compartments, Society of Fire Protection Engineers Symposium, Computer Application in Fire Protection : Analysis, Modeling and Design, March 1984

3. Heskestad, G., Sprinkler Performance as related to Size and Design, Volume I -Laboratory Investigation, FMRC No.22437, Factory Mutual Research, February 1979

4. Kung, H. C., Hill, J. P., Extinction of Wood Crib and Pallet Fires, Combustion and Flame, Vol.24, June 1975

5. Akita, K., Study of Evaporation of Water Drops Falling down through the Air of High Temperature, Report of Fire Research Institute, Volume 3, No.4 March 1953 (in Japanese)

6. Alpert, R. L., Calculation of Response Time of Ceiling-Mounted Fire Detectors, Fire Technology, August 1972

# Experiments and Theory in the Extinction of a Wood Crib

**SATOSHI TAKAHASHI**
Fire Research Institute, Japan

ABSTRACT

Small scale extinction experiments and a theoretical analysis on the critical water application rate were performed using a wood crib fire and a water
stream. Water was manually applied from the top layer to the lower layer or in
reverse, and also with top layer variations of the initial degree of combustion
to eluciate the effects of heat input to the wet surface. On the supposition
that the extinction of the whole body follows fluctuation between local extinction and reignition, a dynamic equation was composed. The experimental results
seemed to be in reasonable agreement.

INTRODUCTION

In order to extinguish a wood crib fire, it is necessary for water to be
applied above the critical water application rate. Many precise investigations
have been carried out on this subject, including those by Kung [1], Tamanini
[2], Tyner [3], et al. [4,5]. A review of this past work is presented by
Heskestad [6]. In his paper critical rates ranging from 1.3 to 3.0 $g/m^2 \cdot s$ are
presented depending on the wood species, geometry and percent preburn. Magee
and Reitz's work [7] on the extinction of plastics fires under the influence of
radiation seems to suggest a similar effect as on wood fires. Fuchs and
Seeger's theoretical work [8] seems to indicate that the heat falling onto the
fuel surface is constant throughout the whole extinction period. Also in
reviewing the previous research and Usui's [9,10] experimental data on the relationships between reignition time and the water content in the char layer of the
wood ember, the author obtained a simple theoretical formula [11] which
following further experiments was revised to fit a wide range of heating temperatures [12,13].

The primary aim of this paper is to try to give a more fundamental and universal principle for extinction based on the dynamism between reignition and
local extinction. As may be seen in big scale fires or in small water application rates relative to fire objects, extinction seems to take place under a
dynamic balance between local extinction in front of the nozzle and reignition
after drying when the nozzle has moved on. No doubt the heat input onto the
local extinguished area decreases as the remaining fire area decreases, and the
longer the time it takes to suppress fire of the whole body, the more heat it
receives. The time necessary for igniting the wet ember is dependent on the
amount of the soaked water, and it is this time tolerance which makes the whole
object capable of being extinguished.

The reignition time of the wet ember, $\tau$, is correlated as a function of the heating temperature, T, the amount of water soaked, x, and the unique ignition time, $\tau_0$, of the non wet charred wood as follows [12,13],

$$\tau \simeq \frac{L}{G\sigma T^4} X + \tau_0 \qquad (1)$$

Where, L is the heat required for vaporizing water from ambient temperature (620 Kcal/Kg), G is the overall heat absorption coefficient, $\sigma$ the Stefan-Boltzmann constant.

Then the above said condition for total extinction can be expressed as,

Reignition time $\geq$ Time required for sweeping the whole object. $\qquad (2)$

Because $\tau_0$ is a complex function of the degree of carbonization and the heating condition etc., and since it is relatively short if the heating temperature is high [13], we eliminate it as a first step and approximate as,

Drying time $(Lx/G\sigma T^4) \geq$ Time required for sweeping the whole object. $\qquad (3)$

The author has already applied this dynamism for the extinction of a room fire and compared it successfully with the past data [14]. They will be presented elsewhere with other mechanisms of extinction.

EXPERIMENTAL APPARATUS AND PROCEDURE

Wood cribs were made of Japanese cedar (density, $0.4 \sim 0.5$ g/cm$^3$) with a moisture content of $9 \sim 13$ percent. The stick size was 3 cm. square $\times$ 21 cm long and the cribs were arranged in four sticks per layer, and there were eight tiers. The crib was placed on a netted steel frame with four legs and put on a platform type load cell. The whole experimental assembly is shown in Fig. 1. The weight change during the burning and extinction process was also recorded.

The water was manually applied from a glass-capillary. The stream was a thin cylindrical jet of moderate speed. It was directed towards the inside surfaces from grid-window openings. It could easily reach deep into inside surfaces. This jet was moved as smoothly as possible from one window from the same level to another, and then to the upper or lower level, depending on the method used. The water flow rate was adjusted by keeping the height from the water surface in a tank to the capillary tip nearly constant. The fluctuation of the flow rate by changing the tip height in the range of the crib height was measured, and was very small. The application of the water was started when the crib weight diminished to a pre-determined weight.

Extinction was judged to have been reached when all the glowing of the charcoal disappeared. And it was considered inextinguishable once the enlargement of the wet area seemed to stop. At this stage, the balance between the extinguished and the burning area was kept for a short time, although water was continually applied. After a time, the balance was destroyed and glowing material was scattered into the extinguished area and the recession of the extinguished area advanced and fell uncontrollably.

Reignition takes place after the drying by radiation from flames and from glowing hot char surfaces. Because flames move only upward, the type of

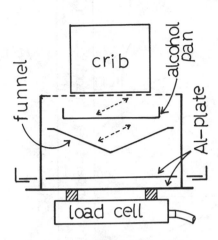

Fig. 1 Experimental assembly.

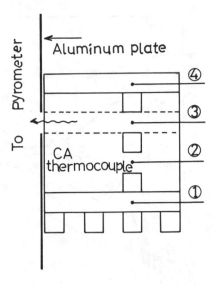

Fig. 2 Set up for measuring
heat flux from window and
temperature in the crib.

radiation must differ depending on whether the water is applied from the bottom
towards the top or from the top towards the bottom. To make this mechanism
clear is another objective.

We have to apply much water to the surfaces extinguished at an earlier
stage. At present, this is done empirically, and thus reignition occurred sometimes. In that case, the jet was quickly directed to extinguish and to
cover the shortage of the water. The jet was then restored to its former position.

For the purpose of providing a basis for the theoretical analysis, the
radiation intensity from the horizontal grid-window and the temperature in it
were also measured. An aluminum plate of the size 0.1 × 50 × 50 cm with a 3 cm
square window was pressed closely onto one side of a crib. An outline of the
set up is shown in Fig. 2. The plate was arranged so as to shield excessive
radiation to the pyrometer from flame and crib surface, but not from the window.
A pyrometer was placed on the axis of the window at a suitable distance. The
crib was placed on a load cell, as shown in Fig. 1. The weight and temperature
change in the grid were correlated with heat flux.

RESULTS AND DISCUSSION

Thermal Radiation from the Crib

In this theory, the author assumes black body radiation from the wall and
from the flame, irrespective of thickness. In the case of a wood fire where the
hot flame heats up the char surface and makes it glow, these will not separated.

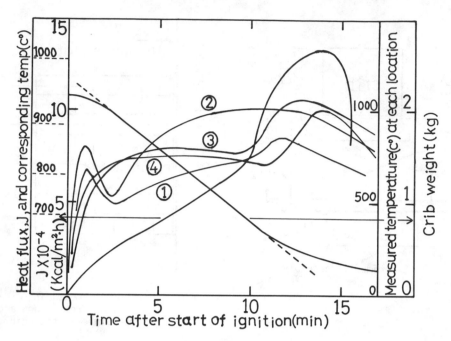

Fig. 3  Heat flux, temperature and weight change versus ignition time.

While it is not the aim of this paper to discuss this, it does try to define the overall intensity of radiation from the horizontal shaft window in order to contribute to the analysis of the following Case-1 and Case-2. As seen in Fig. 3, the temperature and the radiated energy from the third window coincide well if we assume black body radiation in this window space. In fact, the intensity is a function of shaft depth, although the maximum depth must be very short as seen in the experiment in Fig. 3. The threshold depth for giving black body radiation must be somewhat shorter than 21 cm for this geometry. As seen in Fig. 3, the radiation intensity increases as the burning time continues, and shortly before the end of pyrolysis, the intensity and temperature increase sharply. This indicates the shifting from flaming gas combustion to glowing surface combustion. It is not difficult to imagine that the critical water application rate should also vary in accordance with the burning degree.

Table-1  Observed critical water application rate, W, (g/min)

| Application method | Initial mass loss ratio $\phi = (M_0-M)/M_0$ | | |
|---|---|---|---|
| | 0.2 | 0.4 | 0.6 |
| Bottom to Top | 35 | 40 | 72 |
| Top to Bottom | 47 | 65 | 114 |

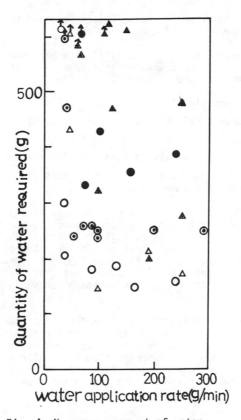

Fig. 4 Necessary amount of water
versus application rate for extinction

Arrow mark indicate inextinguishable.

| $\phi$ | 0.2 | 0.4 | 0.6 |
|---|---|---|---|
| Bottom to top | ○ | ◉ | ● |
| Top to bottom | △ | ◬ | ▲ |

## Experimental Results and Theory on the Critical Water Application Rate

The quantity of water (or the time) necessary for extinction versus the application rate as a function of the application method and the initial combustion rate, $\phi$, is illustrated in Fig. 4. The burning degree, $\phi$, is defined as $\phi = (M_0 - M)/M_0$, where $M_0$ is the initial crib weight and M is the weight at the beginning of water application. The critical application rate, W, was extrapolated from Fig. 4 and listed in Table-1. The critical rate is quite distinct. Above this point, as seen in Fig. 4, the necessary quantity for extinction is relatively constant depending on the method, but it sharply increases as it nears the critical rate almost vertically, so as to enable its definition within a very narrow range of accuracy. The constancy of the necessary quantity of water above the critical rate depends on, $\phi$, and the application method. This type of constancy is described in another paper by the author [15] and can also be seen in Tyner's experiment [3].

As seen in Fig. 4 and Table-1, more difficulty was observed (1) when the water was applied from the top downwards than from the bottom up, and (2) when the combustion degree increased.

The inner surface area (neglecting the vertical side wall) of this crib is about 0.5 $m^2$. So the critical rate may be converted, if like Heskestadt, we adopt the conventional expression by dividing the value in Table-1 by 0.5 $m^2$, to obtain 1.2 ~ 3.8 $g/m^2 \cdot s$. This value agrees with the previous works cited by him (1.3 ~ 3.0 $g/m^2 \cdot s$). Here again, the noteworthy facts are that this value depends on the application method and burning degree.

Theoretically, this critical rate can be derived as follows. Figure 5-(a) is a sectional plan of a crib with an enlarged vertical shaft. For the ease of computing heat transfer, let us assume that the vertical shaft is close to the cylinder, as shown in Fig. 5-(b). First we consider step by step the case when the water stream is applied horizontally from the bottom upwards. Because the flame moves from the burning area upwards, the wet shaft wall receives heat from the upward flame and from the glowing surfaces, as shown in Fig. 5-(b), through radiation only.

Deciding upon the proportion of heat from the flame and from the wall is very difficult, but at an early stage, the proportion from the flame must be bigger than at the final stage where the charcoal begins to glow. Fortunately according to computations, these discrepancies seem relatively small. In this paper two simplified cases, one where a flame does not exist and two where a flame exists and is dominant, are assumed and calculation of heat transfer was made as follows.


Case-1; Radiation from Hot Wall Only

The heat, H, input to the wet wall-1 was assumed irradiated from the hot wall-2, then,

$$H = JA_2F_{21} \tag{4}$$

where J is the heat flux from wall-2, $A_2$ is the surface area of the hot glowing wall-2, $F_{21}$ is the configuration factor from wall-2 to wall-1. And because,

$$A_2F_{21} = \frac{A_0}{2r^2}\left[s\sqrt{s^2 + 4r^2} + h\sqrt{h^2 + 4r^2} + 2Sh - \ell\sqrt{\ell^2 + 4r^2}\right] \tag{5}$$

Then the maximum quantity of heat input to wall-1 is when the wet area occupied one half of the total.

$$(A_2F_{21})_{max} = \frac{\ell}{4r^2}A_0\left[\ell + \sqrt{\ell^2 + 16r^2} - 2\sqrt{\ell^2 + 4r^2}\right] \tag{6}$$

Then,

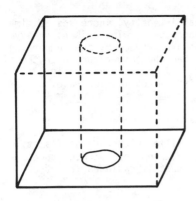

Fig. 5-(a)  Crib with vertical shafts.

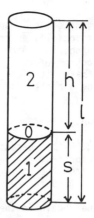

Fig. 5-(b)  Cylindrical shaft model.

$$H_{max} = J(A_2 F_{21})_{max} \qquad (7)$$

Here from Eq. (3), we have obtained the suggestion that the primary condition for extinguishing the whole object is to sweep the whole body before drying. This idea may be expressed as an energy relationship. Then the critical water application rate, w, is obtained as the rate which can remove the maximum heat from the wetted surface by evaporating. When $(r/1)^2$ is small enough,

$$w = \frac{H_{max}}{L} \approx \frac{J}{L} A_0 \qquad (8)$$

Where $A_0$ is the horizontal sectional area of the shaft and 1, s, h, are the crib, the wet-wall, the burning-wall height respectively.

Let's suppose for the heat flux, J, a value at 750°C ($\phi = 0.2 \sim 0.6$), which corresponds 891 Kcal/m$^2$·min, and putting the value, $A_0 = 9 \times 10^{-4}$ m$^2$, the critical application rate, w, per vertical shaft is then calculated from Eq. (8) as,

w = 1.29 (g/min per shaft)

In reality, the vertical shaft is not solely composed of wall, but there are also horizontal shaft spaces. All these inner wall surface areas were converted into that of a vertical shaft, and the equivalent number of vertical shafts was obtained. The total required critical rate, W, was obtained by multiplying $A_0$ by the equivalent shaft number (19.7) as,

W = 1.29 × 19.7 = 25.4 (g/min)

Referring this value to Table-1, the theoretical W is somewhat smaller than observed. Actually, some water must penetrate into the chink between layers and remain on the surface without contributing to extinction; it was applied in vain. Besides, the combustion degree, or the heat intensity, increases as time goes on, especially as critical conditions are approached. There is no steady state burning, thus elevating the experimental value. Moreover, the temperature near the bottom of the crib is generally higher than the temperature near the top which makes it difficult to obtain an average. The heat flux in Fig. 3 seems to be a better index for predicting the critical application rate.

Case-2; Radiation from the Hot Wall and the Flame

In Fig. 5-(b), because the upward cylinder is filled with burning flames, the radiation from the vertical shaft space must be taken into account together with that from the wall. Then, we can assume the heat is radiated from the round ceiling surface-0.

Then we obtain in the same manner,

$$H = JA_0 F_{01}$$

$$A_0 F_{01} = A_0 \left[ 1 - \frac{1}{2} \left\{ 2 + \left(\frac{\ell}{r}\right)^2 - \sqrt{\left\{ 2 + \left(\frac{\ell}{r}\right)^2 \right\}^2 - 4} \right\} \right]$$

$$- \frac{1}{2} A_0 \left(\frac{\ell}{r}\right)^2 \left[ \sqrt{1 + \left(\frac{2r}{\ell}\right)^2} - 1 \right] + \frac{1}{2} A_0 \left(\frac{S}{r}\right)^2 \left[ \sqrt{1 + \left(\frac{2r}{S}\right)^2} - 1 \right]$$

And,

$$(A_0 F_{01})_{max} = \frac{1}{2} A_0 \left[ \sqrt{\left\{ 2 + \left(\frac{\ell}{r}\right)^2 \right\}^2 - 4} - \left(\frac{\ell}{r}\right)^2 \right]$$

So,

$$w = \frac{H_{max}}{L} \approx \frac{J}{L} A_0 \quad \text{(Per shaft)} \tag{9}$$

The required quantity, W, is the same as in Case-1. This $H_{max}$ is attained when the extinguished wall-1 advances to the top of the crib.

As a matter of fact, the difficulty of extinguishing appeared when about one half of the height was suppressed; if it could pass this point, the total extinction seemed relatively easy. If it did not pass, the glowing combustion spots began to scatter around the lower portions and to fall uncontrollably. So case-1, where the radiation from the hot wall was assumed, seems more practical.

In the reverse situation when the water is applied from the top layer downwards, the wetted area is exposed to adjacent flame radiation as well as to the lower glowing surface throughout the extinguishing period, and receives more heat. Thus W must become somewhat larger than in the aforesaid case. This is obvious in Table-1. Exact computation for this case seems rather difficult, but the principle for extinction by the dynamism seems applicable.

If we want to obtain the critical rate, e, in the conventional form "weight per area per time" we merely need to divide Eq. (8) by the shaft wall area $2\pi r\ell$.

1204

$$e = \frac{W}{2\pi r \ell} = \frac{Jr}{2L\ell} \tag{10}$$

When $r = \sqrt{9/\pi}$ (cm), $J = 891$ (750°C) - 1540 (900°C) (Kcal/m$^2$·min), we obtain from Eq. (10) as,

$$e = 0.85 - 1.5 \ (g/m^2 \cdot s)$$

Critical application rate per unit area is again a function of the inner heat flux, J. Referring to Fig. 4, Eq. (10) also seems to explain Table-1 well.

CONCLUSIONS

The critical water application rate was derived experimentally as well as theoretically for wood crib fire extinction. In order to make clear the mechanism of heat transfer from glowing wall and from flame, the water stream was applied horizontally from bottom layer to top layer and the reverse. The rate was precisely obtained empirically and it was found that it varied much depending on the burning degree and the method of applying the water. The rate increased as the burning degree increased and the bottom to top method required a smaller rate than the reverse case.

A theoretical analysis was also performed. When the application rate is sufficiently small and critical, the reignition speed competes with the suppression speed. Then the fundamemental condition for the total extinction can be defined as,

Reignition time $\geq$ Time required for sweeping the whole object.

According to this definition, a dynamic formula for heat transfer was established, and the results seemed to agree reasonably well with the experiment and with past data. But this critical rate is theoretically a complex function of heat flux in the crib and the geometry and it is expressed by the dimension [MASS/TIME]. If we want to transform it into the conventional expression [MASS/AREA/TIME], we can obtain this by merely dividing it by the surface area in the crib.

REFERENCES

1. Kung, H. C., Hill, J. P.: "Extinction of Wood Crib and Pallet Fires," Combustion and Flame, 24: 305, 1975.

2. Tamanini, F.: "The Application of Water Sprays to the Extinguishment of Crib Fires," Combustion Science and Technology, 14: 17, 1976.

3. Tyner, H. D.: "Fire-Extinguishing Effectiveness of Chemicals in Water Solution," Industrial and Engineering Chemistry, 33: 1, 60, 1941.

4. Bryan, J., Smith, D. N.: "The Effect of Chemicals in Water Solution," Engineering, 457, 1945.

5. Kida, T.: "Extinction of Fires of Small Wooden Crib with Sprays of Water and Some Solutions," Report of Fire Research Institute of Japan, 36: 1, 1973.

6. Heskestad, G.: "The Role of Water in Suppression of Fires," Journal of Fire and Flammability, 11: 254, 1980.

7. Magee, R. S., Reitz, R.D.: "Extinguishment of Radiation Augmented Plastics Fires by Water Sprays," Fifteenth Symposium on Combustion, 337, 1974.

8. Fuchs, P., Seeger, P. G.: "Ein Mathematisches Modell zur Bestimmung der Löschwassermenge und Vergleich mit Experimenten," VFDB, 1: 3, 1981.

9. Usui, K.: "Transactions of the Architectural Institute of Japan," 41: 100, 1950.

10. ibid., 43: 77, 1951.

11. Takahashi, S.: "On the Reignition," Bulletin of Japanese Association of Fire Science and Engineering, 32: 1, 41, 1982.

12. Takahashi, S.: "Heat Transfer and Ignition of the Extinguished Wood Residue and the Wet Filter Paper," Report of Fire Research Institute of Japan, 56: 1, 1983.

13. Takahashi, S.: "Reignition of the Carbonized Wood and the Water Soaked Wood Ember," ibid, 58: 33, 1984.

14. Takahashi, S.: "A Theoretical Study on the Requirements of Water Application for a Compartment Fire Extinction," Bulletin of Japanese Association of Fire Science and Engineering, 34: 1, 1, 1984.

15. Takahashi, S.: "Extinction Mechanism and Efficiency by Sprays of Water and Chemically Improved Water," Report of Fire Research Institute of Japan, 56: 7, 1983.

# Analysis of Fire Suppression Effectiveness Using a Physically Based Computer Simulation

**L. M. PIETRZAK and G. A. JOHANSON**
Mission Research Corporation
P.O. Drawer 719
Santa Barbara, California 93102, USA

ABSTRACT

The Swedish Fire Research Board and the U.S. Federal Emergency Management Agency are sponsoring a project to further the understanding of the basic mechanisms involved, as well as to support the development of standards for and to seek ways of improving the performance of portable fire suppression systems used by fire departments. This effort includes both experiments and computer model development work.

This paper describes a physically based computer model developed to simulate one aspect of the problem: the manual suppression of post-flashover fires. This includes: (1) a discussion of the physical basis behind the model; (2) a comparison of model predictions with available experimental data; and (3) an analysis of fire suppression effectiveness using the model.

The analysis concludes that, when direct assess and extinguishment of the burning fuel is not possible, improved fire control occurs with water sprays having a Rosin-Rammler distribution of droplet sizes and volume medium drop diameters in the 0.15 to 0.35mm range. This agrees with available experimental data. It is also shown that firefighting venting and standoff distance requirements may lead to more severe fires requiring more water for control. Finally, the analysis shows that venting and water spray induced air/gas flow effects also serve to channel hot steam and gases away from the fire-fighter adding to his safety. Additional experimental work is also recommended before all these conclusions can be considered definitive.

## INTRODUCTION

The focus of the effort described in this paper is on the suppression of post-flashover or fully developed compartment fires using manually applied water spray. The paper begins by describing a fire suppression computer simulation developed as part of the research effort. This includes inputs, outputs, and physical effects and interactions modeled, both with and without suppression effects. Example model results and sensitivities for a single compartment geometry are then described including comparisons with available experimental data.

## MODEL DESCRIPTION

The fire suppression model described here is called the Fire Demand (FD) Model. Refs. 1 through 6 describe this FD Model and its applications in detail. The following briefly summarizes the physical effects and interactions incorporated in the FD Model.

Previous investigators (Refs. 7,8,9) successfully modeled the post-flashover fire in its freely burning phase (without suppression). In the FD Model, the freely burning segment of the fire history follows their methods and calculates the fire development in time in terms of lumped parameters describing the energy and mass balance of the compartment as a whole. The FD Model adds the effect of water suppression application to this work.

The FD Model is capable of simulating fires involving both char forming and non-charring solid fuels in compartments with single or multiple vents of different sizes and in different locations, and venting changes with time due to fire-fighting activities, including water spray induced air inflow.

In the freely burning period (without suppression effects), the fire behavior is determined by the room itself (dimensions, size, and shape of the ventilation openings, thickness and composition of bounding walls, etc.) and by fuel features (heat of combustion, weight of fuel, total surface area of fuel, etc.). The fire behavior is described by the average temperature of the room gas, the average temperature of the walls and ceilings, floor temperature, the retained heat in the room, and the burning rate of the fuel. The fundamental basis of the freely burning post-flashover fire model is a mass balance and a heat balance of the gas contained in the compartment. One key operational feature of the FD Model is the causal relation whereby the buoyancy of the hot room gas drives combustion products out of the ventilation opening and draws fresh air in. In turn, the rate of fresh air entry determines (for this post-flashover case) the combustion rate, which is the major source of heat for the room gas.

The factor of water application modifies the fire behavior drastically. The cooling of interior gases and interior surfaces by water vaporization, the choking of ventilation by the exit of steam, and the direct extinguishment of burning surfaces reachable by water are simulated by the FD Model. The FD Model also accounts for any additional air from outside being forced into the fire by the induced effects of the water spray. The relative magnitude of these effects determines whether fire control is achieved (with sufficient water) or whether the fire only stabilizes at lower temperatures (with insufficient water). To estimate these effects the FD Model requires the specification of the time of water application, the distance of the hose-nozzle from the vent, and water spray characteristics such as: the flow rate, pressure, the distribution and volume medium diameter of water drops in the spray pattern, the cone angle of the stream, the sweep time required for the stream to cover the interior of the compartment, and the fraction of fuel area accessible to water impact. Apart from the fire conditions--as determined by the compartment and fuel--these are the factors which determine the FD Model estimate of suppression effectiveness.

The level of physical detail incorporated into the FD Model is determined by its practical objectives and the desire for simplicity and computability. The suppression effects are accounted for using relatively simple submodels consistant with the lumped parameter nature of the overall model. These submodels are based on the overall assumption that on introduction of the water spary into a fully involved fire, the resulting steam expands and mixes rapidly with the compartment gases so that one can continue to represent the processes in terms of lumped parameters characterized by average temperatures and heat fluxes within the compartment.

Central to the estimation of water effects is the apportionment of the water volume into three parts: (1) a part which is blown away thorough failure to penetrate the updrafts in the compartment, (2) a part which is vaporized in the compartment gas, and (3) the remainder which impacts the fuel and interior surfaces in liquid form. This simplified submodel assumes a water drop of given initial diameter falls and evaporates in a compartment characterized by a uniform temperature and a uniform updraft velocity. The temperature is the gas temperature of the compartment and the updraft velocity is estimated from the room geometry and the air circulation rate in the compartment. The water drop is assumed to fall vertically at terminal velocity (relative to the gas) determined by its instantaneous diameter which changes as the drop falls. For given compartment conditions there are two critical drop diameters: the diameter of a drop whose terminal velocity equals the updraft velocity, and the diameter of a drop which will just reach the floor before its diameter has decreased by evaporation to a size small enough to be swept away by the updraft. Results for single drops are averaged over an assumed drop size distribution to produce a water partitioning. The model assumes a Rosin-Rammler (Ref. 10) distribution of drop sizes which may be completely characterized by the volume median drop diameter, half of the water volume occurs in drops below this size and half above. No account is taken of any further breakup of the spray by impact on surfaces. The fraction of water which is vaporized cools the compartment gas. The fraction which reaches the floor is re-interpreted as the fraction which reaches interior surfaces and fuel, and is distributed to them in proportion to wall/ceiling area, floor area, and exposed fuel surface area.

Regarding the cooling of hot, non-burning interior surfaces, account is taken of the fact that only a fraction of these surfaces are instantaneously impacted at any one time by a sweeping water spray of limited cone angle. There is therefore a residence time during which cooling of the surface can occur. In the case of walls the fraction vaporized there and the fraction which runs off is estimated. In estimating surface cooling the impacting water spray is assumed to coalesce into a thin sheet over the impacted surface. The average rate of heat extraction is then calculated as the limiting value obtained by either the amount of water available or by conduction from the interior. If conduction limits, the surface temperature under hose impact is assumed to equal 100°C and the vaporized and runoff water fraction is calculated based on this. If the available water is limiting all the water is assumed to vaporize and the surface temperature is calculated accounting for the cooling effect of this water and the residence time of the hose stream. For the water reaching and standing on the floor, another limiting condition accounted for is the rate of heat transfer possible by boiling.

The fraction of the total burning fuel surface area accessible to water impact may also vaporize liquid water and thereby reduce the rate of heat generation. Extinguishment of this fuel area occurs by different criteria depending upon the type of fuel. For charring cellulosic or plastic fuels, extinguishment is assumed to occur when the rate of heat extraction by water vaporization exceeds the heat generation rate by charring combustion alone.

For non-charring fuels, the extinguishment submodel follows the conditions examined experimentally by Magee and Reitz (Ref. 11) wherein critical water application rates were measured as a function of incident radiation. The model assumes that for post-flashover conditions, radiation from the fire plumes, hot surfaces and gases in the compartment control the rate of fuel pyrolysis or vaporization even as a portion of the fuel is directly extinguished by the water. The critical water application rate is therefore taken to be that required to counter the heat received by radiation--i.e., extinguishment is assumed to occur when the rate of heat extraction by water vaporization exceeds the net heating rate to the exposed fuel by radiation.*

The rate of heat extraction is calulated for the following two cases: For non-charring fuel surfaces burning in a rigid or softened state, the impacting water spray is assumed to coalesce into a thin sheet and to act as a thermal radiation barrier from above and a coolant that cools the hot fuel from below. For surfaces burning in a molten or liquid state the impacting water is assumed to penetrate the surface and cool the fuel from within. Emperical data from Reference 11 is used to account for possible burning rate enhancement due to splashing of droplets on impact and/or bubbling of the vaporizing water from within.

COMPARISON WITH EXPERIMENTS

Results from the FD Model compare favorably with the limited experimental data available which addresses the suppression of post-flashover fires using manually applied water spray (Refs. 13 through 18).

For the fire conditions examined in Ref. 13, the experiments found that water application rates greater than 25 ℓ/min were required for fire control. The FD Model predicts application rates for fire control of 34 ℓ/min to 57 ℓ/min depending on the volume medium drop size. In Ref. 14, the amount of water vaporized was also measured. For the experimental conditions tested, approximately 94 to 110 ℓ were required. The FD Model predicts 64 to 130 ℓ again depending on the volume medium drop size. Although the above results compare favorably, the experimental data in Refs. 13 and 14 do not record the values of several parameters (e.g. drop sizes) to which model results are sensitive, so reasonable parametric values were chosen for the simulation (Ref. 1). Further experimental calibration and validation of the FD Model is required.

In addition to the above, a series of fully envolved fire suppression experiments using water sprays were recently completed in Osaka, Japan (Refs. 15 and 16). These tests were conducted in compartments characteristic of residential occupancies. The objective of the Japanese tests was to establish the sizes of water droplets which controlled the fire with minimum water runoff and damage. A major conclusion of this work was that average droplet sizes in the range of 0.2 to 0.3mm achieved the best results. These droplet sizes were obtained from nozzles operating under a pressure of 10kg/cm$^2$ with a discharge of 180ℓ/min. The nozzle produces fine droplet sizes without resorting to high discharge pressures. The nozzle design diverts part of the water through a whirler to produce a flowing vortex of water. The water in the vortex is then mixed with a high velocity stream of the remaining water and ejected from the nozzle outlet. The nozzle is also capable of straight stream application. Unfortunately the information regarding vent size and other compartment characteristics reported in Refs. 15 and 16 is insufficient to allow direct comparison with the FD Model. Furthermore, these papers do not define specifically what is meant by "average" droplet size or how it was measured. Although the optimal range of 0.2 to 0.3mm appears to agrees with FD Model predictions as presented in the next section, definitive comparisons are not possible until the above uncertanties are resolved.

---

*For the case of suppression of small fires consisting of a single fire plume or multiple but non-interacting plumes, convection or conduction from the local plume rather than radiation can dominate. In Ref. 12, Rasbash calculates the critical rate at which a water spray abstracts heat from liquid fuels at the surface to reduce the vaporization below fire sustaining levels.

In addition to the above, it has recently come to the author's attention that a series of post-flashover fire suppression experiments were carried out in England in the later 1950's (Refs. 17, 18 and 19). Further work is required to adjust the input parameters of the FD Model to allow comparisons to these tests.

EXAMPLE MODEL SENSITIVITIES

The results presented in what follows uses the Rosin-Rammler distribution of drop sizes measured for sprinkler heads (Ref. 10). In the future, a more thorough analysis is required using actual measured flow rates and spray characteristics from currently used manual hose-nozzle equipment as well as relevant data reported in Refs. 18, and 20 through 24.

FD Model results and principal sensitivities are presented in terms of graphs that relate water application rate per unit total interior area of the room to the volume median drop diameter of the water spray. For example, for the room and fuel conditions listed in Table 1, simulation predictions of total water requirements are given in Fig. 1.

Table 1. Input conditions used in the example for charring fuels in a compartment having a single vent.

| COMPARTMENT DESCRIPTION | INPUT VALUE |
|---|---|
| Room Height | 2.44 m |
| Floor Area | 11.11 m$^2$ |
| Area of Walls and Ceiling | 42.85 m$^2$ |
| Window Area | 1.69 m$^2$ |
| Window Height | 1.12 m |
| Wall/Ceiling/Floor Thickness | 0.15 m |
| Wall/Ceiling/Floor Conductivity | 0.00833 kcal/m/min/°C |
| Wall/Ceiling/Floor Specific Heat | 250 kcal/m$^3$/°C |
| | |
| FUEL DESCRIPTION | |
| | |
| Fuel Type | Wood |
| Fuel Load | 21.95 Kg/m$^2$ |
| Fuel Surface Area | 28.52 m$^2$ |
| Fuel Surface Area Exposed to Water | 0% |
| Effective Heat of Combustion | 2575 Kcal/Kg |
| | |
| WATER APPLICATION DESCRIPTION | |
| | |
| Distance of Nozzle from Vent | 0 m |
| Time of Water Application | |
| After Flashover | 5 minutes |
| Nozzle Pressure | 1 Kg/cm$^2$ |
| Cone Angle of Hose Stream | 60°C |
| Sweep Time to Cover Compartment | 5 seconds |
| Volume Median Water Drop Diameter* | Varies |
| Flow Rate of Water | Varies |

*Assuming a Rosin-Rammler drop size distribution (Ref. 10).

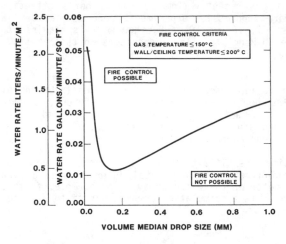

FIRE CONTROL CRITERIA
GAS TEMPERATURE ≤ 150° C
WALL/CEILING TEMPERATURE ≤ 200° C

FIRE CONTROL POSSIBLE

FIRE CONTROL NOT POSSIBLE

Fig. 1.  Example results for charring wood
fuels at 0% exposed fuel fraction.

The "control-failure" line appearing in Fig. 1 divides the graph into two regions characterized by combinations of water delivery rate and volume median drop diameters which are successful or unsuccessful in controlling the fire. The gas and surface temperatures which the model employs as fire control criteria are arbitrarily selected and results are sensitive to them (Refs. 1 and 2).

Note, in Fig. 1, that as the volume median drop diameters increase from about 0.05 mm to 1 mm, the bounding control failure line first slopes downward, achieving a minimum at about 0.15 mm, and then increases with drop size. This behavior reflects the sensitivity to drop size of fires controlled primarily by gaseous and non-burning surface cooling as well as vent choking mechanisms. For the example shown in Fig. 1, this means a significant fraction of the drops with sizes less than 0.15 mm are blown away by hot gases and are not effective in achieving fire control. On the other hand, droplets greater than 0.15 mm result in a significant fraction that fail to fully vaporize and, therefore, contribute to water runoff. In each case water is wasted and, because of this, the total water application rates required for fire control increases.

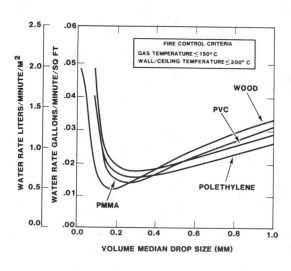

FIRE CONTROL CRITERIA
GAS TEMPERATURE ≤ 150° C
WALL/CEILING TEMPERATURE ≤ 200° C

WOOD

PVC

POLETHYLENE

PMMA

Fig. 2. Effect of fuel type (exposed fuel
fraction = 0%).

The previous results are for cellulosic fuels which form a char layer while burning. Fig. 2 presents results for three different non-charring plastic fuels compared to the wood case. Other than fuel properties, the compartment and water delivery characteristics are the same as for the wood case.

Unlike fuels with an oxidizing char layer, the mass loss rate due to the pyrolysis or vaporization of fuels that burn without developing an oxidizing char layer depends strongly on the degree of thermal heat transfer to the fuel bed from the hot or burning gases and surfaces in the compartment. For these materials, fuel-thermal coupling effects have also been modeled. In general, one can see from Fig. 2 that for the specific compartment examined the "failure-control" line tracts fairly closely for all the fuels examined.

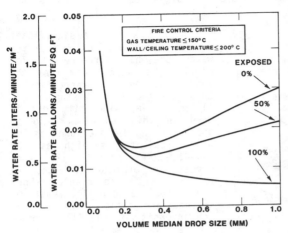

FIRE CONTROL CRITERIA
GAS TEMPERATURE ≤ 150°C
WALL/CEILING TEMPERATURE ≤ 200° C

EXPOSED
0%

50%

100%

Fig. 3.    Sensitivity to water exposed fuel
fraction--PMMA fuel.

Fig. 3 illustrates the effect of the "exposed fuel fraction". In general, the shape of the line defining the control-failure region is sensitive to the fraction of the burning fuel area accessible to direct water impact. If little or no fuel can be directly extinguished, control occurs by gaseous/surface cooling and vent choking. The gaseous cooling, which is a very sensitive function of drop size, produces the sharp minimum of the curve at the optimum drop size. If the exposed fuel fraction is high, direct extinguishment rather than gaseous/non-burning-surface cooling and vent choking dominate. For this case, the critical application rate required for fire control is similar at the lower drop sizes but after a certain point continues to decrease as drop size increases. Here fire control is effected by a decrease of heat production rate consequent to extinguishment of burning surfaces. The larger drops become more effective since they penetrate better, loose less of their volume by evaporation and carry more water to the burning surfaces. This means that if the seat of the fire is directly accessible, in order to minimize water usage, extinguishment is preferred to gaseous/non-burning surface cooling and vent choking. On the other hand, because of standoff requirements for safety, direct access is not always possible in post-flashover fire situations.

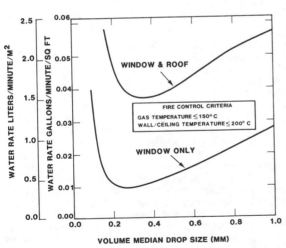

WINDOW & ROOF

FIRE CONTROL CRITERIA
GAS TEMPERATURE ≤ 150°C
WALL/CEILING TEMPERATURE ≤ 200° C

WINDOW ONLY

Fig. 4.    Effect of venting (PMMA
fuel--exposed fuel fraction = 0%).

The previous results are for a compartment having a single vent where the hose-nozzle is positioned within the open vent of the compartment. The FD Model can also handle more generalized venting conditions. This includes multiple wall and/or roof vents of different sizes and location and opened at different times. This also includes the opening of wall or roof vents occurring as part of the firefighting venting operations as well as water spray induced air inflow for hose-nozzles positioned away from the vent.

Fig. 4 shows the effect of a firefighter opening a 1 square meter hole in the roof of the compartment beginning one minute before water is applied and proceeding at a rate of 1m²/min.

Except for the added roof vent, the other compartment and water delivery characteristics are the same as given in Table 1. It is clear from Fig. 4 that venting has a significant effect on the flow-rates and drop sizes required to control the fire.

Fig. 5 shows the significant effect induced air inflow from the water spray has on the results for nozzles positioned away from the vent a distance of respectively, zero, 1m and 2m. A single vent (1.12mx1.509m) with PMMA fuel is used in this example. Fig. 5 applies to nozzle pressures as low as 15 psi to as high as 100 psi--i.e. essentially the same curves are obtained for each. An analysis of these results shows that after a certain level of induced air inflow, the burning and intensity of the fire becomes limited by the amount of available fuel rather than air. This limit is apparently exceeded by the air inflow generated by 100 psi as well as 15 psi for the water flow rates indicated in Fig. 5. In general, Fig. 5 shows the importance of positioning the nozzle as close as possible to the vent consistent with standoff distance requirements for safety. This

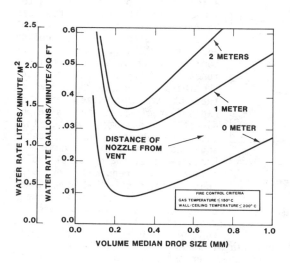

Fig. 5. Effect of water spray induced air inflow (PMMA--0% exposed fuel fraction).

minimizes water spray air induced effects reducing the water flow rates required by factors of three or more.

CONCLUSIONS

In general, the above results and those reported in Ref. 1 through 4 indicate that improved fire control is possible with water sprays having a Rosin-Rammler distribution of droplet sizes and volume median drop diameters in the 0.15 to 0.35 mm range. This optimal range applies only when direct assess and extinguishment of the burning fuel is not possible. This range also implies no further breakup of the water jet or spray by impact on solid surfaces. A more thorough analysis using actual measured manual hose-nozzle spray, a broader range of compartment and fuel characteristics and additional experimental verification work is required before these conclusions can be considered definitive.

The FD Model results also suggest that firefighting venting and standoff distance requirements can lead to more severe fires requiring more water for control. They also suggest that venting together with the enhanced gas/air velocities from water spray induced effects also serve to channel the hot steam and products of combustion away from the firefighter and therefore have important safety implications quite apart from fire control. More analysis and experimental work is required to understand the tradeoffs and identify if there is perhaps some better balance between, for example, fireground venting or similar activities and the resulting water spray requirements. This can help establish important previously unavailable quantitative rules of thumb to follow on the fireground as well as improvements in firefighting equipment performance.

REFERENCES

1. Pietrzak, L.M., Johanson, G.A., Ball, J.A. A Physically Based Fire Suppression Computer Simulation For Post-Flashover Compartment Fires -- Applications, Experimental Requirements, Software Documentation and User's Guide, Mission Research Corporation Report MRC-R-846, 1985.

2. Pietrzak, L.M., and Ball, J.A., "Investigation to Improve the Effectiveness of Water in the Suppression of Compartment Fires," Fire Research, 1, 291-300, 1978.

3. Pietrzak, L.M. and Ball, J.A., "Optimizing the Mobility and Fire Suppression Performance of Fire Engines Using Physically Based Computer Simulations," Fou-Brand, 1979.

4. Pietrzak, L.M., and Patterson, W.J., "Effect of Nozzles on Fires Studied in Terms of Flow Rate, Droplet Size," Fire Engineering, Vol. 132, No. 12, 1979.

5. Pietrzak, L.M., and Ball, J.A., "A Physically Based Fire Suppression Computer Simulation--Definition, Feasibility Assessment, Development Plan and Applications," MRC-R-732, Mission Research Corporation, 1983.

6. Pietrzak, L.M., et al., "Decision Related Research on Equipment Technology Utilized by Local Government: Fire Suppression--Phase II Research Report," National Science Foundation Report No. NSF/RA-770207, 1977.

7. Kawagoe, K., "Fire Behavior in Rooms," Building Research Institute Report No. 27, Tokyo, 1950.

8. Kawagoe, K., "Estimation of Fire Temperature Time Curve in Rooms," Building Research Paper No. 29, Building Research Institute, Tokyo, 1967.

9. Barbaruskas, V., and Williamson, R.B., "Post-Flashover Compartment Fires,"Report No. USBFRG 75-1, Fire Research Group, University of California, Berkeley, 1975.

10. Rosin, P. and Rammler, E.J., J. Institute Fuel, 7, 1933.

11. Magee, R.S., and Reitz, R.D., "Extinguishment of Radiation Augmented Plastic Fires by Water Sprays," Factory Mutual Research, FMRC Serial No. 22357-1, 1974.

12. Rasbash, D.J., "The Extinction of Fires by Water Spray," Fire Research Abstracts and Review, Vol. 4 (1 & 2), 28-53, 1962.

13. Salzberg, F., Vodvarka, F.J., and Maatman, G.C., "Minimum Water Requirements for Suppression of Room Fires," Fire Technology, 6 (1), 22 1970.

14. Fuchs, P., "Brand-und Loschversuche Mit Vershiedenen Loschmitteln in Einem Versuchsraum Naturlicher Gross," Fire Protection Seminar, Karsruhe, Germany, VFDB, 1976.

15. Osuga, Ichinosuke, "Development of a Fog Stream-Fire Extinguishing System for Medium and High Storied Buildings," OSAKA, No. 4, 1983.

16. "Osaka Fire Department Fights Water Damage," Urban Innovation Abroad, 1984.

17. Pickard, R. W., Hird, D., et al., "Use of High and Low Pressure Water Spray Against Fully Involved Room Fires," Fire Research Station Note 388, England, 1959.

18. Nichols, S. and Freeman, S., D"etermination of Drop Sizes of High and Low Pressure Water Sprays," Fire Research Station Note 373, England, 1959.

19. Thomas, P. H. and Smart, P.M., "Fire Extinction Tests in Rooms," Fire Research Station Note No. 121, England, 1954.

20. Rasbash, D. J., "The Properties of Sparys Produced by Batteries of Impinging Jets, "Fire Research Station Note No. 181, England, 1955.

21. Rasbash, D. J. and Stark, G. W. V., "Design of Sprays for Protective Installations, "Part III, Fire Research Station Note No. 303, 1958.

22. Fry, J.F. and Smart, P. M. T., "The Production of Water Sprays for Fire Extinction," Quart. Inst. Fire Engrs. Edinb., 13 (10) 82-104, 1953.

23. "Characteristics of water spray nozzles including those designated as fog nozzles for fire-fighting use," National Board of Fire Underwriters, Chicago, 1944.

24. Mobius, K., "Experience Obtained with the Testing of Spray Nozzles," V.F.D.B., 5 (2) 33-42, 1956.

# Author Index

# Subject Index

**Notes**